QUANTUM ERROR CORRECTION

Quantum computation and information are among the most exciting developments in science and technology of the last 20 years. To achieve large-scale quantum computers and communication networks it is essential to overcome noise not only in stored quantum information, but also in general faulty quantum operations. Scalable quantum computers require a far-reaching theory of fault-tolerant quantum computation.

This comprehensive text, written by leading experts in the field, focuses on quantum error correction, and thoroughly covers the theory as well as experimental and practical issues. The book is not limited to a single approach, but also reviews many different methods to control quantum errors, including topological codes, dynamical decoupling, and decoherence-free subspaces.

Basic subjects as well as advanced theory and a survey of topics from cutting-edge research make this book invaluable both as a pedagogical introduction at the graduate level and as a reference for experts in quantum information science.

DANIEL A. LIDAR is a Professor of Electrical Engineering, Chemistry, and Physics at the University of Southern California, and directs the USC Center for Quantum Information Science and Technology. He received his Ph.D. in Physics from the Hebrew University of Jerusalem, was a postdoctoral fellow at UC Berkeley, and a faculty member at the University of Toronto. He was elected a Fellow of the American Association for the Advancement of Science and of the American Physical Society for his contributions to the theory of decoherence control of open quantum systems for quantum information processing.

TODD A. BRUN is an Associate Professor of Electrical Engineering, Physics, and Computer Science at the University of Southern California. He received his Ph.D. from the California Institute of Technology, and held postdoctoral positions at the University of London, the Institute for Theoretical Physics in Santa Barbara, Carnegie Mellon University, and the Institute for Advanced Study in Princeton. He has worked broadly on decoherence and the quantum theory of open systems, quantum error correction, and related topics for quantum information processing.

QUANTUM ERROR CORRECTION

Edited by

DANIEL A. LIDAR
University of Southern California

TODD A. BRUN
University of Southern California

CAMBRIDGE
UNIVERSITY PRESS

University Printing House, Cambridge CB2 8BS, United Kingdom

Published in the United States of America by Cambridge University Press, New York

Cambridge University Press is part of the University of Cambridge.

It furthers the University's mission by disseminating knowledge in the pursuit of education, learning and research at the highest international levels of excellence.

www.cambridge.org
Information on this title: www.cambridge.org/9780521897877

© Cambridge University Press 2013

This publication is in copyright. Subject to statutory exception and to the provisions of relevant collective licensing agreements, no reproduction of any part may take place without the written permission of Cambridge University Press.

First published 2013

Printed in the United Kingdom by CPI Group Ltd, Croydon CR0 4YY

A catalog record for this publication is available from the British Library

ISBN 978-0-521-89787-7 Hardback

Cambridge University Press has no responsibility for the persistence or accuracy of URLs for external or third-party internet websites referred to in this publication, and does not guarantee that any content on such websites is, or will remain, accurate or appropriate.

Contents

List of contributors		*page* xi
Prologue		xv
Preface and guide to the reader		xix
Acknowledgements		xxi

	Part I Background	1
1	**Introduction to decoherence and noise in open quantum systems**	3
	Daniel A. Lidar and Todd A. Brun	
	1.1 Introduction	3
	1.2 Brief introduction to quantum mechanics and quantum computing	4
	1.3 Master equations	26
	1.4 Stochastic error models	32
	1.5 Conclusions	45
2	**Introduction to quantum error correction**	46
	Dave Bacon	
	2.1 Error correction	46
	2.2 From reversible classical error correction to simple quantum error correction	48
	2.3 The quantum error-correcting criterion	56
	2.4 The distance of a quantum error-correcting code	59
	2.5 Content of the quantum error-correcting criterion and the quantum Hamming bound	59
	2.6 Digitizing quantum noise	60
	2.7 Classical linear codes	61
	2.8 Calderbank, Shor, and Steane codes	64
	2.9 Stabilizer quantum error-correcting codes	65
	2.10 Conclusions	76
	2.11 History and further reading	76

3	**Introduction to decoherence-free subspaces and noiseless subsystems**	78
	Daniel A. Lidar	
	3.1 Introduction	78
	3.2 A "classical decoherence-free subspace"	78
	3.3 Collective dephasing decoherence-free subspace	79
	3.4 Decoherence-free subspace defined and characterized	81
	3.5 Initialization-free decoherence-free subspace	90
	3.6 Noiseless subsystems	92
	3.7 Initialization-free noiseless subsystems	98
	3.8 Protection against additional decoherence sources	101
	3.9 Conclusions	102
	3.10 History and further reading	102
4	**Introduction to quantum dynamical decoupling**	105
	Lorenza Viola	
	4.1 Motivation and overview	105
	4.2 Warm up: bang-bang decoupling of qubit dephasing	107
	4.3 Control-theoretic framework	110
	4.4 Bang-bang periodic decoupling	113
	4.5 The need for advanced decoupling design	119
	4.6 Bounded-strength Eulerian decoupling	120
5	**Introduction to quantum fault tolerance**	126
	Panos Aliferis	
	5.1 Quantum circuits and error discretization	127
	5.2 Noisy quantum computers	130
	5.3 Encoded quantum computation	142
	5.4 Coarse-grained noise and level reduction	152
	5.5 The quantum accuracy threshold	155
	5.6 Assessment	157
	5.7 History and further reading	158
	Part II Generalized approaches to quantum error correction	161
6	**Operator quantum error correction**	163
	David Kribs and David Poulin	
	6.1 Introduction	163
	6.2 Equivalent conditions for OQEC	165
	6.3 Stabilizer formalism for OQEC	169
	6.4 Examples	172
	6.5 Measuring gauge operators	175
	6.6 Bounds for subsystem codes	177
	6.7 Unitarily correctable codes	179
7	**Entanglement-assisted quantum error-correcting codes**	181
	Todd A. Brun and Min-Hsiu Hsieh	
	7.1 Introduction	181
	7.2 Constructing EAQECCs	184

	7.3	Constructing EAQECCs from classical linear codes	195
	7.4	Catalytic QECCs	197
	7.5	Conclusions	199
8	**Continuous-time quantum error correction**		**201**
	Ognyan Oreshkov		
	8.1	Introduction	201
	8.2	CTQEC in an encoded basis	204
	8.3	Quantum-jump CTQEC with weak measurements	207
	8.4	Schemes with indirect feedback	213
	8.5	Quantum jumps for Markovian and non-Markovian noise	218
	8.6	Outlook	226
	Part III Advanced quantum codes		**229**
9	**Quantum convolutional codes**		**231**
	Mark Wilde		
	9.1	Introduction	231
	9.2	Definition and operation of quantum convolutional codes	235
	9.3	Mathematical formalism of quantum convolutional codes	238
	9.4	Quantum shift-register circuits	244
	9.5	Examples of quantum convolutional codes	249
	9.6	Entanglement-assisted quantum convolutional codes	253
	9.7	Closing remarks	260
10	**Nonadditive quantum codes**		**261**
	Markus Grassl and Martin Rötteler		
	10.1	Introduction	261
	10.2	Stabilizer codes	262
	10.3	Characterization of nonadditive quantum codes	263
	10.4	Construction of nonadditive QECCs	268
	10.5	Quantum circuits	274
	10.6	Conclusions	277
11	**Iterative quantum coding systems**		**279**
	David Poulin		
	11.1	Introduction	279
	11.2	Decoding	284
	11.3	Turbo-codes	292
	11.4	Sparse codes	297
	11.5	Conclusion	305
12	**Algebraic quantum coding theory**		**307**
	Andreas Klappenecker		
	12.1	Quantum stabilizer codes	307
	12.2	Cyclic codes	317
	12.3	Quantum BCH codes	318
	12.4	Quantum MDS codes	325

13 Optimization-based quantum error correction 327
Andrew Fletcher

- 13.1 Limitation of the independent arbitrary errors model 327
- 13.2 Defining a QEC optimization problem 328
- 13.3 Maximizing average entanglement fidelity 331
- 13.4 Minimizing channel nonideality: the indirect method 336
- 13.5 Robustness to channel perturbations 338
- 13.6 Structured near-optimal optimization 340
- 13.7 Optimization for (approximate) decoherence-free subspaces 346
- 13.8 Conclusion 347

Part IV Advanced dynamical decoupling 349

14 High-order dynamical decoupling 351
Zhen-Yu Wang and Ren-Bao Liu

- 14.1 Introduction 351
- 14.2 Operator set preservation 351
- 14.3 Dynamical decoupling for multi-qubit systems 353
- 14.4 Concatenated dynamical decoupling 355
- 14.5 Uhrig dynamical decoupling 357
- 14.6 Concatenated Uhrig dynamical decoupling 365
- 14.7 Quadratic dynamical decoupling 366
- 14.8 Nested Uhrig dynamical decoupling 367
- 14.9 Pulses of finite amplitude 368
- 14.10 Time-dependent Hamiltonians 369
- 14.11 Randomized dynamical decoupling 372
- 14.12 Experimental progress 373
- 14.13 Discussion 374

15 Combinatorial approaches to dynamical decoupling 376
Martin Rötteler and Pawel Wocjan

- 15.1 Introduction 376
- 15.2 Combinatorial bang-bang decoupling 378
- 15.3 Combinatorial bounded strength decoupling 391
- 15.4 Conclusions and future directions 393

Part V Alternative quantum computation approaches 395

16 Holonomic quantum computation 397
Paolo Zanardi

- 16.1 Introduction 397
- 16.2 Holonomic quantum computation 398
- 16.3 HQC with quantum dots 400
- 16.4 Robustness 403
- 16.5 Hybridizing HQC and error-avoiding/correcting techniques 406
- 16.6 Conclusions 407
- Appendix: quantum holonomies 408

17	**Fault tolerance for holonomic quantum computation**	412
	Ognyan Oreshkov, Todd A. Brun, and Daniel A. Lidar	
	17.1 Holonomic quantum computation on subsystems	413
	17.2 FTHQC on stabilizer codes without additional qubits	415
	17.3 Conclusion and outlook	430
18	**Fault-tolerant measurement-based quantum computing**	432
	Debbie Leung	
	18.1 Introduction	432
	18.2 Common models for measurement-based quantum computation	433
	18.3 Apparent issues concerning fault tolerance in measurement-based quantum computation	442
	18.4 Simulation of an operation	442
	18.5 Fault tolerance in graph state model	444
	18.6 History and other approaches	451

	Part VI Topological methods	453
19	**Topological codes**	455
	Héctor Bombín	
	19.1 Introduction	455
	19.2 Local codes	455
	19.3 Surface homology	457
	19.4 Surface codes	461
	19.5 Color codes	471
	19.6 Conclusions	480
	19.7 History and further reading	480
20	**Fault-tolerant topological cluster state quantum computing**	482
	Austin Fowler and Kovid Goyal	
	20.1 Introduction	482
	20.2 Topological cluster states	482
	20.3 Logical initialization and measurement	486
	20.4 State injection	489
	20.5 Logical gates	491
	20.6 Topological cluster state error correction	492
	20.7 Threshold	496
	20.8 Overhead	499

	Part VII Applications and implementations	507
21	**Experimental quantum error correction**	509
	Dave Bacon	
	21.1 Experiments in liquid-state NMR	509
	21.2 Ion trap quantum error correction	514
	21.3 Experiments using linear optics quantum computation	517
	21.4 The future of experimental quantum error correction	518

22	**Experimental dynamical decoupling** *Lorenza Viola*		519
	22.1 Introduction and overview		519
	22.2 Single-axis decoupling		520
	22.3 Two-axis decoupling		532
	22.4 Recent experimental progress and outlook		534
23	**Architectures** *Jacob Taylor*		537
	23.1 The principles of fault tolerance		537
	23.2 Memory		539
	23.3 Building gates		543
	23.4 Entangling operations and transport		545
	23.5 Quantum networking		550
24	**Error correction in quantum communication** *Mark Wilde*		553
	24.1 Introduction		553
	24.2 Entanglement distillation		554
	24.3 Quantum key expansion		562
	24.4 Continuous-variable quantum error correction		570
	24.5 Implementations of quantum error correction for communication		578
	24.6 Closing remarks		579
	24.7 Historical notes		579
	Part VIII Critical evaluation of fault tolerance		583
25	**Hamiltonian methods in quantum error correction and fault tolerance** *Eduardo Novais, Eduardo R. Mucciolo, and Harold U. Baranger*		585
	25.1 Introduction		585
	25.2 Microscopic Hamiltonian models		588
	25.3 Time evolution with quantum error correction		590
	25.4 The threshold theorem in a critical environment		598
	25.5 The threshold theorem and quantum phase transitions		599
	25.6 An example: the simplified spin-boson model		601
	25.7 Conclusions		609
	Some useful results		609
26	**Critique of fault-tolerant quantum information processing** *Robert Alicki*		612
	26.1 Introduction		612
	26.2 Fault-tolerant quantum computation		613
	26.3 Fault tolerance and quantum memory		619
	26.4 Concluding remarks		624

References 625
Index 657

Contributors

Robert Alicki
Institute of Theoretical Physics and
 Astrophysics
University of Gdansk
ul. Wita Stwosza 57
80-952 Gdansk
Poland

Panos Aliferis
CFM S.A.
23 rue de l'Université
75007 Paris
France

Dave Bacon
Department of Computer science and
 Engineering
University of Washington
Box 352350
Seattle, WA 98195-2350
USA

Harold U. Baranger
Department of Physics
Duke University
Durham, NC 27708-0305
USA

Héctor Bombín
Perimeter Institute for Theoretical Physics
31 Caroline St. N.
Waterloo, Ontario
N2L 2Y5
Canada

Todd A. Brun
Departments of Electrical Engineering,
 Physics, and Computer Science
Center for Quantum Information Science nad
 Technology
University of Southern California
3740 McClintock Ave
Los Angeles, CA 90089
USA

Andrew Fletcher
MIT Lincoln Laboratory
244 Wood Street
Lexington, MA 02420-9108
USA

Austin Fowler
Centre for Quantum Computation and
 Communication Technology
School of Physics
The University of Melbourne
VIC 3010
Australia

Kovid Goyal
Previously at California Institute of
 Technology

Markus Grassl
Centre for Quantum Technologies (CQT)
National University of Singapore
S15 #06-09
3 Science Drive 2
Singapore 117543
Singapore

Min-Hsiu Hsieh
Centre for Quantum Computation and
　Intelligent Systems (QCIS)
Faculty of Engineering and Information
　Technology
University of Technology
Sydney
New South Wales 2007
Australia

Andreas Klappenecker
Department of Computer Science and
　Engineering
Texas A&M University
College Station, TX 77853-3112
USA

David Kribs
Department of Mathematics and Statistics
University of Guelph
Guelph, Ontario
N1G 2W1
Canada

Debbie Leung
Department of Combinatorics and
　Optimization
University of Waterloo
200 University Avenue West
Waterloo, Ontario
N2L 3G1
Canada

Daniel A. Lidar
Departments of Electrical Engineering,
　Chemistry, and Physics
Center for Quantum Information Science and
　Technology
University of Southern California
920 Bloom Walk
Los Angeles, CA 90089
USA

Ren-Bao Liu
Department of Physics
The Chinese University of Hong Kong
Shatin, N.T.
Hong Kong
China

Eduardo R. Mucciolo
Department of Physics
University of Central Florida
P.O. Box 162385
Orlando, FL 32816-2385
USA

Eduardo Novais
Centro de Ciências Naturais e Humanas
Universidade Federal do ABC
Rua Santa Adélia, 166
Santo André - SP
CEP 09210170
Brazil

Ognyan Oreshkov
Centre for Quantum Information and
　Communication (QuIC)
Université Libre de Bruxelles
50 av. F.D. Roosevelt – CP 165/59
B-1050 Brussels
Belgium

David Poulin
Département de Physique
Université de Sherbrooke
Sherbrooke, Québec
J1K 2R1
Canada

Martin Rötteler
NEC Laboratories America, Inc.
4 Independence Way, Suite 200
Princeton, NJ 08540
USA

Jacob Taylor
Joint Quantum Institute
National Institute of Standards and
  Technology
100 Bureau Drive MS 8410
Gaithersburg, MD 20899-8423
USA

Lorenza Viola
Department of Physics and
  Astronomy
Dartmouth College
6127 Wilder Laboratory, Room 247
Hanover, NH 03755-3528
USA

Zhen-Yu Wang
Department of Physics
The Chinese University of Hong Kong
Shatin, N.T.
Hong Kong
China

Mark Wilde
Department of Physics and Astronomy
Center for Computation and Technology
Louisiana State University
202 Nicholson Hall
Baton Rouge, LA 70803-4001
USA

Pawel Wocjan
Department of Electrical Engineering and
  Computer Science
University of Central Florida
Orlando, FL 32816
USA

Paolo Zanardi
Physics and Astronomy Department
Center for Quantum Information Science and
  Technology
University of Southern California
920 Bloom Walk
Los Angeles, CA 90089
USA

Prologue

For most of human history we maneuvered our way through the world based on an intuitive understanding of physics, an understanding wired into our brains by millions of years of evolution and constantly bolstered by our everyday experience. This intuition has served us very well, and functions perfectly at the typical scales of human life – so perfectly, in fact, that we rarely even think about it. It took many centuries before anyone even tried to formulate this understanding; centuries more before the slightest evidence suggested that these assumptions might not always hold. When the twin revolutions of relativity and quantum mechanics overturned twentieth-century physics, they also overturned this intuitive notion of the world.

In spite of this, the direct effect at the human scale has been small. Our cars do not run on Szilard engines. Very few freeways, even in Los Angeles, have signs saying "Speed Limit 300,000 km/s." And human intuition remains rooted in its evolutionary origins. It takes years of training for scientists to learn the habits of thought appropriate to quantum mechanics; and even then, surprises still come along in the areas we think we understand the best.

Technology has transformed how we live our lives. Computers and communications depend on the amazingly rapid developments of electronics, which in turn derive from our understanding of quantum mechanics: we use the microscopic movements of electrons in solids to do work and play games; pulses of coherent light in optical fibers tie the world together. But the theories underlying this technology – computer science and information theory – were built on a fundamentally classical view of the world. These two theories have been fantastically successful – so successful that for many years no one worried (and few even recognized) that they depended on physical assumptions that, at the smallest scales, did not hold. It was only in the past two decades that we realized these theories are only part of a vastly richer subject. This is the subject of quantum information science.

The precursors of quantum information science arrived in the 1980s, building on many years of earlier work on the foundations of quantum mechanics, and spurred by the development of new and powerful experimental methods, such as laser cooling, electromagnetic traps, and optical microcavities. But it was in the 1990s that the revolution began in earnest. Stimulated in large part by the stunning discovery of Peter Shor's quantum factoring algorithm, dozens and then hundreds of researchers began to examine the properties of quantum theory with a new eye. How could superposition, interference, and entanglement be combined to produce new and better forms of information processing – new algorithms, new communication protocols, and at the same time new ways of understanding the physical world? The result, nearly two decades later, has been an astonishing avalanche of new ideas. The century-old dog of quantum mechanics has learned a whole host of new tricks, and the output shows no signs of stopping.

In addition to its technological promise, quantum information science has dramatically changed our understanding of quantum physics. It had already been long recognized that there were deep connections between information theory and the laws of thermodynamics. Quantum measurement also seemed to connect information and physics in a fundamental way, where acquiring information about a quantum system was accompanied by unavoidable disturbance. These connections have now spread through the broad field of quantum mechanics. Long-studied systems in optical and condensed matter physics have yielded new insights to an analysis based on the presence and flow of quantum information. Just as quantum effects have been repurposed as resources for information processing, information concepts have provided a new window on physics.

The rapid early development of quantum information science, however, was accompanied by immediate skepticism. The same developments in the foundations of quantum theory that had led to the ideas of quantum information had brought a new appreciation of the power of decoherence – how unavoidable interactions between a quantum system and its external environment cause large-scale superpositions to decay and destroy interference effects. Quantum algorithms were based on idealized models, in which large quantum systems undergo perfect unitary evolution. The larger a quantum system, the more strongly it interacts with its environment, and the more rapidly decoherence will destroy the superpositions and interference effects on which quantum computing depends. Schrödinger's cat, far from being in a superposition of alive and dead, would decohere in an infinitesimal fraction of a second into a probabilistic mixture. The power of decoherence to screen quantum effects at macroscopic levels explains why quantum mechanics was never even suspected for the vast majority of human history.

From a computational point of view, we can think of decoherence as a source of computational errors. In the early days of classical computation, it was feared that large-scale computation might be impossible because of errors: however unlikely they might be, in a sufficiently large computation errors would eventually happen and corrupt the output. To address this concern, John von Neumann studied models of computation and concluded that in fact reliable computations of unlimited size were possible, provided that the rates of error were sufficiently low. The key was error correction: the information in a computer would be stored redundantly using error-correcting codes, and throughout the running of a program the computer would constantly check for errors and correct them when they occur.

Skeptics of quantum computation argued that a similar solution was impossible for a quantum computer, because of two constraints that do not apply to a classical digital computer. First, quantum information cannot be reliably copied – this is the famous no-cloning theorem. And second, errors are detected by measuring the system, but measurements destroy quantum superpositions and eliminate the interference effects on which quantum computers depend. Moreover, unlike digital computers, the allowed states of a quantum system form a continuum. Small imprecisions in the control of the computer could accumulate, eventually derailing the computation. This effect is what has prevented large-scale analog computation.

Fortunately, all these intuitive arguments proved to be incorrect. It is true that quantum information cannot be copied; but it can be spread redundantly over multiple subsystems in a way that protects it from local errors. It is possible to perform measurements that reveal the presence of errors without probing the actual state of the stored information. And while quantum states form a continuum, this rests on a discrete structure, just like a digital computer; the ability to correct a well-chosen set of discrete errors grants the ability to correct all of their linear combinations as well. Within a few years, the first quantum error-correcting codes were discovered; a theory of the broad class of stabilizer codes, analogous to classical linear codes, had been developed; and threshold theorems had been proven, showing that a quantum computer could be constructed fault-tolerantly in such a way that for sufficiently low rates of error (due to decoherence, imperfect control, or other sources) quantum computations of unlimited length were possible.

Since these early results the field of quantum error correction has dramatically expanded. Brand new codes, and new fault-tolerant techniques, have moved the stringent requirements on error rates closer and closer to reality, even as experiments have reduced the error rates to lower and lower levels. New techniques have also been found to prevent errors – by effectively eliminating the interaction with the environment, in dynamical decoupling; by rendering the system immune to noise with particular symmetries, in decoherence-free subspaces and noiseless subsystems; or by encoding quantum information into naturally robust degrees of freedom, in topological quantum computing.

The world of quantum error correction is already vastly rich, and growing richer all the time. This book will help you enter it, or act as a map and reference for those already knowledgeable. We invite you to come in.

Daniel A. Lidar and Todd A. Brun

Preface and guide to the reader

We were inspired to put this book together during the process of organizing the First International Conference on Quantum Error Correction at the University of Southern California (in December 2007, with a sequel in December 2011). With many of the world's foremost experts in the various branches of quantum error correction gathered together in Los Angeles, we solicited chapters on what we thought were the most important topics in the broad field. As editors, we then faced the difficult challenge of integrating material from many expert authors into one comprehensive and yet coherent volume. To achieve this feeling of coherence, we asked all our contributors to work with a single, common notation, and to work with the authors of other chapters in order to minimize overlap and maximize synchronicity. This proved hard to enforce, and while we made every effort to achieve consistency among the different chapters, this goal was surely only partly met. To the extent that the reader discovers inconsistencies, we as editors take full responsibility. The resulting book is not a textbook; for one, it doesn't include any exercises, and figures are not abundant. Moreover, it can only be a snapshot of such a rapidly evolving subject as quantum error correction. Nevertheless, we believe that the basic results in the field are now well enough established that this book, with its extensive index and list of references to the literature, will serve both as a reference for experts and as a guidebook for new researchers, for some years to come.

This book is organized into eight parts, containing a total of 26 chapters by 27 authors, some of whom wrote or co-wrote more than one chapter. Part I (with contributions by Panos Aliferis, Dave Bacon, Todd Brun, Daniel Lidar, and Lorenza Viola) is an introduction; it contains five chapters covering the basics of decoherence and quantum noise, quantum error-correcting codes, dynamical decoupling, decoherence-free subspaces and subsystems, and fault tolerance. Readers interested in a comprehensive introduction to the field, at the graduate course level, will do well to read this first part. It is roughly equivalent to two one-semester courses. Part II (with contributions by Todd Brun, Min-Hsiu Hsieh, David Kribs, Ognyan Oreshkov, and David Poulin) presents several important generalizations of quantum error correction (QEC), accounting for additional richness in the Hilbert space structure, and for the possibility of continuously correcting errors. It covers operator QEC, entanglement-assisted QEC, and continuous-time QEC. Part III (with contributions by Andrew Fletcher, Markus Grassl, Andreas Klappenecker, David Poulin, Martin Rötteler, and Mark Wilde) examines more advanced topics in quantum error-correcting codes, and presents several important classes of specialized quantum codes, in particular quantum convolutional codes, nonadditive quantum codes, iterative quantum coding systems (along with a discussion of optimal decoding strategies), algebraic quantum coding theory, and optimization-based QEC. Part IV (with contributions by Ren-Bao Liu, Martin Rötteler, Zhen-Yu Wang, and Pawel Wocjan) considers advanced topics in dynamical decoupling, focusing in particular on high-order methods that allow error cancellation to any desired order in perturbation theory, and on the efficient construction of dynamical decoupling sequences using combinatorial methods. Part V (with contributions by Todd Brun, Debbie Leung, Daniel Lidar, Ognyan Oreshkov, and Paolo Zanardi) considers alternatives to the standard "circuit model" of quantum computation, in particular holonomic and measurement-based quantum computation. Part VI (with contributions by Héctor Bombín, Austin Fowler, and Kovid Goyal) concerns topological methods in QEC, particularly topological codes and cluster state quantum computation. Part VII (with contributions by Dave Bacon, Jacob Taylor, Lorenza Viola, and Mark Wilde) deals with issues of practical implementation. It covers representative early experiments in QEC and dynamical decoupling, and discusses the implementation of error correction and fault-tolerance ideas in quantum computing architectures and quantum communication. Finally, Part VIII (with contributions by Robert Alicki, Harold Baranger, Eduardo Mucciolo, and Eduardo Novais) takes a critical look at the assumptions going into the fault-tolerant threshold theorems, and how they may not necessarily hold in realistic systems.

Acknowledgements

The impetus for this book was a visit in 2004 by Dr. Simon Capelin, Editorial Director at Cambridge University Press, to Daniel Lidar at the University of Toronto. Already at that time it seemed that writing a comprehensive book on quantum error correction would be a monumental task, requiring input from many authors. The opportunity came with the two QEC conferences at the University of Southern California in 2007 and 2011, which gathered a broad group of distinguished experts capable of collectively covering much of the field. The editors are profoundly grateful for Simon's patience, gentle guidance, and support throughout. Sincere and deep gratitude is due to all the authors who contributed to this volume, for their expertise, their hard work, and their tolerance of the long and sometimes erratic process of putting this book together. We would particularly like to thank our colleague Paolo Zanardi, who read many of the chapters and provided his insights, and Igor Devetak, who provided early input into the plans for this book. We would like to thank the staff members at USC who assisted us in carrying out the two QEC conferences and helped with the book in other ways: Lamia Dabuni, Anita Fung, Vicky de Los Reyes, Gerrielyn Ramos, and Mayumi Thrasher. Special thanks are due to Jennifer Ramos for tirelessly working on the extensive index, and to Sehar Tahir at Cambridge University Press for her professional and patient assistance with LaTeX typesetting questions. Many of our students, postdocs, and former students have made contributions, both to the writing and in reading over chapters and providing comments; these include José Gonzales, Min-Hsiu Hsieh, Kung-Chuan Hsu, Kaveh Khodjasteh, Wan-Jung Kuo, Ching-Yi Lai, Masoud Mohseni, Ognyan Oreshkov, Gregory Quiroz, Alireza Shabani, Gerardo Paz-Silva, Bilal Shaw, Soraya Taghavi, Mark Wilde, and Yicong Zheng.

Daniel Lidar would like to dedicate this book to his wife, Jennifer, daughters, Abigail and Nina, and parents, David and Rachel, without whose love and support a project of this magnitude would never have been possible. Abigail and Nina's infectious enthusiasm for quantum cats made the project fun even when the number of LaTeX errors exceeded the total number of shots of espresso required to bring it to completion.

Todd Brun would like to thank his parents, for their unflagging belief and support, and his wife Cara King, for all her love and understanding over many years.

Part I

Background

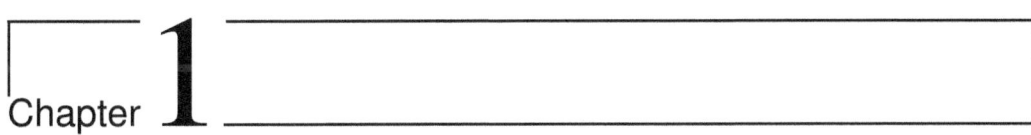

Chapter 1

Introduction to decoherence and noise in open quantum systems

Daniel A. Lidar and Todd A. Brun

1.1 Introduction

This chapter gives the physical and mathematical basis of decoherence in quantum systems, with particular emphasis on its importance in quantum information theory and quantum computation. We assume that the reader is already familiar with the basics of quantum mechanics, such as wave functions, the Schrödinger equation, and Hilbert spaces, which are covered in standard textbooks. Nevertheless, all basic concepts will be at least briefly defined and described. A working knowledge of graduate-level linear algebra and calculus is also assumed.

In quantum information processing, the term decoherence is often used loosely to describe *any* kind of noise that can affect a quantum system. This can include sources of noise that seem, qualitatively, to be quite different from each other. These noise sources include:

(i) Random driving forces from the environment (e.g., Brownian motion).
(ii) Interactions that produce entanglement between the system and the environment.
(iii) Statistical imprecision in the experimental controls on the system (e.g., timing errors, frequency fluctuations, etc.).

Remarkably, all of these sources of noise can be described by an almost identical mathematical framework, in which pure quantum states can evolve into mixed quantum states, and the quantum effects necessary for information processing – interference, entanglement, and reversible evolution – are disrupted.

The chapter begins by examining a highly schematic model of decoherence, involving systems of qubits, but uses this model to present the mathematical formalism used in describing more realistic systems as well. We will derive from this simple model the idea of the *completely positive trace-preserving map*, the most widely used description of quantum noise, and show

---

Quantum Error Correction, ed. Daniel A. Lidar and Todd A. Brun. Published by CAMBRIDGE UNIVERSITY PRESS.
© Cambridge University Press 2013.

how it is equivalent to a stochastic error model. From this basic framework, we will also describe a continuous noise model: the quantum *master equation*.

After describing basic models of decoherence, we will discuss experimental methods of characterizing the ill effects. In the simplest models, decoherence can often be characterized by a few simple time scales: commonly T_1 (the *thermalization* time scale) and T_2 (the *dephasing* time scale). On a less simplified basis, one can try to determine an exact mathematical description of the quantum evolution, including the effects of noise, by *quantum process tomography*.

In the remainder of the chapter we will briefly cover several additional topics. While the simplest case of decoherence is the case where we are simply trying to preserve quantum information, more generally we wish to process that information as well, generally by applying unitary *quantum gates*. We will show how even a single dominant source of noise can manifest itself as different noise processes for different gates. We will compare our simple model to more realistic environments, and briefly discuss new issues that may arise, such as non-Markovian evolution and Zeno-like effects. We will also touch on Hamiltonian environment modes, which are more suitable descriptions for some approaches to error correction or suppression (especially dynamical decoupling).

Various methods have been proposed to combat the effects of decoherence in quantum information processing, and these are the subject of the rest of this book. They include straightforward attempts to cool and isolate the quantum system from its environment; techniques based on error-correcting codes; the use of decoherence-free subspaces and subsystems when the decoherence has certain symmetry properties; dynamical decoupling methods of error suppression, using rapid unitary pulses to suppress the effects of the system–environment interaction; and the use of topological properties to encode protected quantum information so that it is robust against local perturbations.

1.2 Brief introduction to quantum mechanics and quantum computing

There are many outstanding general introductions to quantum mechanics, and our purpose here is not to try to add to this list, but rather to provide a quick and relatively technical introduction to many of the concepts used later in the book. For a general introduction we recommend Chapter 2 in the excellent book by Nielsen and Chuang [NC00]. Our introduction to some extent overlaps with that book, but also complements it by covering a number of topics not mentioned there, which are particularly relevant to recent developments in quantum decoherence and noise mitigation. In particular, we cover the Dyson and Magnus expansions of time-dependent perturbation theory, general maps on quantum states, and weak measurements. Another excellent introduction is the book by Breuer and Petruccione [BP02], which covers the theory of open quantum systems in great detail.

1.2.1 States and spaces

Quantum states reside in, or act on, Hilbert spaces. A *Hilbert space* $\mathcal{H}$ is a real or complex inner product space (always complex in quantum mechanics) that is also a complete metric space with respect to the distance function induced by the inner product. A metric space M is a set where a notion of distance (called a metric) between elements of the set is defined. M is called complete if every Cauchy sequence in M converges in M.

1.2.1.1 Quantum states Quantum mechanics makes predictions about the evolution of quantum states and the outcomes of measurements when a system is in a particular quantum state. A general quantum state is a positive operator ρ called a *density matrix* whose trace is equal to 1, and whose domain and range are Hilbert spaces. We denote the set of all states by

$$\mathscr{S}(\mathscr{H}) \equiv \{\rho \in \mathscr{B}(\mathscr{H}) : \rho \geq 0, \text{Tr}\rho = 1\}, \tag{1.1}$$

where $\mathscr{B}(\mathscr{H})$ is the set of all bounded linear operators on $\mathscr{H}$. "Bounded" means that the operator has a finite norm; we will define such norms later, in Section 1.2.6. A more general situation is when $\mathscr{B}$ maps between two different Hilbert spaces $\mathscr{H}_1$ and $\mathscr{H}_2$, in which case we write $\mathscr{B}(\mathscr{H}_1, \mathscr{H}_2)$. Of course, linear operators need not always be bounded (though in most of the applications in this book that will indeed be the case). We denote by $\mathscr{L}(\mathscr{H}_1, \mathscr{H}_2)$ the set of all linear operators from Hilbert space $\mathscr{H}_1$ to Hilbert space $\mathscr{H}_2$, and by $\mathscr{T}(\mathscr{H}_1, \mathscr{H}_2)$ the set of all *superoperators*: mappings from $\mathscr{L}(\mathscr{H}_1)$ to $\mathscr{L}(\mathscr{H}_2)$ [the notation $\mathscr{L}(\mathscr{H})$ is shorthand for $\mathscr{L}(\mathscr{H}, \mathscr{H})$]. The name "superoperator" reflects the fact that a map $\Phi \in \mathscr{T}(\mathscr{H}_1, \mathscr{H}_2)$ transforms operators acting on $\mathscr{H}_1$ into operators acting on $\mathscr{H}_2$. Note that the two Hilbert spaces $\mathscr{H}_1, \mathscr{H}_2$ may have different dimensions.

An operator A is *Hermitian* if all its eigenvalues are real, or equivalently if $A = A^\dagger$, where $A^\dagger$ denotes the operation of taking the transpose of A as well the complex conjugate of every matrix element, whichever basis A is represented in. The notation $\rho \geq 0$ means that ρ is *positive semidefinite*, i.e., it is Hermitian and all of its eigenvalues are nonnegative. The sum of all of ρ's diagonal elements is denoted $\text{Tr}\rho$.[1] Since $\text{Tr}\rho = 1$, clearly at least one of the eigenvalues must be strictly positive.

The properties $\rho \geq 0$ and $\text{Tr}\rho = 1$ are a consequence of the probabilistic interpretation of quantum mechanics, which states that a quantum state yields a valid probability distribution over possible measurement outcomes for any measurement. When ρ is written as a matrix in a basis in which it is diagonal, its diagonal elements are probabilities of mutually exclusive events. Therefore these diagonal elements must be positive and sum to one. When ρ has *rank* 1 (the rank is the number of linearly independent eigenvectors with nonzero eigenvalues) it is said to be *pure*. Otherwise it is *mixed*. A simple test for this, which one can easily verify, is a computation of $\text{Tr}(\rho^2)$. When $\text{Tr}(\rho^2) = 1$ the state is pure, and when $\text{Tr}(\rho^2) < 1$ it is mixed.

When ρ is pure it can be written as an outer product of two vectors, which reads $\rho = |\psi\rangle\langle\psi|$ in standard Dirac bra-ket notation. When it is written in an explicit basis representation the normalized vector (ket) $|\psi\rangle \in \mathscr{H}$ is called the wave function in many textbooks. It is convenient to think operationally of a bra $\langle\psi|$ as the row vector that is the complex conjugate and transpose of the column vector ket $|\psi\rangle$, i.e., $\langle\psi| = |\psi\rangle^\dagger$.

Let us pick an orthonormal basis $\{|i\rangle\}_{i=0}^{d-1}$ for the Hilbert space $\mathscr{H}$, where $d \equiv \dim(\mathscr{H})$ is the dimension of the Hilbert space. We write the inner product in Dirac notation as $\langle\cdot|\cdot\rangle$. This is in general a complex number. Orthonormality of the basis is then expressed as $\langle i|j\rangle = \delta_{ij}$, where δ_{ij} is the Kronecker symbol ($\delta_{ij} = 1$ if $i = j$, and 0 otherwise). Since $\{|i\rangle\}$ is a basis we can write $|\psi\rangle = \sum_{i=0}^{d-1} \alpha_i |i\rangle$ for any ket $|\psi\rangle$, while the corresponding bra becomes

---

[1] More formally, in this book we deal almost exclusively with linear trace-class bounded operators that map between separable Hilbert spaces. A Hilbert space $\mathscr{H}$ is separable if and only if it admits a countable orthonormal basis. A bounded linear operator $A : \mathscr{H} \mapsto \mathscr{H}$, where $\mathscr{H}$ is separable, is said to be in the trace class if for some (and hence all) orthonormal bases $\{|k\rangle\}_k$ of $\mathscr{H}$ the sum of positive terms $\sum_k \langle k|\sqrt{A^\dagger A}|k\rangle$ is finite. In this case, the sum $\sum_k \langle k|A|k\rangle$ is absolutely convergent and is independent of the choice of the orthonormal basis. This value is called the trace of A, denoted by $\text{Tr}(A)$.

$\langle\psi| = \sum_{i=0}^{d-1} \alpha_i^* \langle i|$. The numbers $\alpha_i = \langle i|\psi\rangle \in \mathbb{C}$ are called *amplitudes*. Since $|\psi\rangle$ is normalized we have $1 = \langle\psi|\psi\rangle = \sum_{i=0}^{d-1} |\alpha_i|^2$. This implies $\mathrm{Tr}(\rho) = 1$, since for a pure state $\mathrm{Tr}(\rho) = \langle\psi|\psi\rangle$. The positive numbers $|\alpha_i|^2 \leq 1$ play the role of a probability distribution, a fact we will explain in more detail when we discuss measurements, below.

The *outer product* of any two vectors $|\psi\rangle = \sum_{i=0}^{d-1} \alpha_i |i\rangle$ and $|\phi\rangle = \sum_{i=0}^{d-1} \beta_i |i\rangle$ can be written in the basis $\{|i\rangle\}$ as $|\psi\rangle\langle\phi| = \sum_{i,i'=0}^{d-1} \alpha_i \beta_{i'}^* |i\rangle\langle i'|$, where the operator $|i\rangle\langle i'| \in \mathcal{B}(\mathcal{H})$ acts as $(|i\rangle\langle i'|)|i''\rangle = \langle i'|i''\rangle|i\rangle$ on any basis element $|i''\rangle$.

1.2.1.2 Qubits The case $d=2$ of a two-level system is special: it defines a quantum bit, or *qubit*, for which a pure state can always be written as

$$|\psi\rangle = \alpha|0\rangle + \beta|1\rangle, \quad |\alpha|^2 + |\beta|^2 = 1. \tag{1.2}$$

The basis states $|0\rangle$ and $|1\rangle$ are defined to be the eigenstates of the Pauli spin matrix σ_z with eigenvalues 1 and -1, respectively. Here are all three Pauli spin matrices:

$$\sigma_x = \begin{pmatrix} 0 & 1 \\ 1 & 0 \end{pmatrix} = X, \quad \sigma_y = \begin{pmatrix} 0 & -i \\ i & 0 \end{pmatrix} = Y, \quad \sigma_z = \begin{pmatrix} 1 & 0 \\ 0 & -1 \end{pmatrix} = Z. \tag{1.3}$$

In addition, often the 2×2 identity matrix I is denoted by σ_0. A qubit is the simplest possible quantum mechanical system. There are many physical embodiments of such a system including the spin of a spin-1/2 particle (e.g., an electron), the polarization states of a photon, two hyperfine states of a trapped atom or ion, two neighboring levels of a Rydberg atom, or the presence or absence of a photon in a microcavity. All of these have been proposed in various schemes for quantum information and quantum computation, and used in actual experiments. The review [LJL+10] provides a comprehensive survey.

A global phase may be assigned arbitrarily to the state $|\psi\rangle = \alpha|0\rangle + \beta|1\rangle$ in Eq. (1.2), so that all physically distinct pure states of a single qubit form a two-parameter space. A useful parametrization is in terms of two angular variables θ and ϕ:

$$|\psi\rangle = \cos(\theta/2)e^{-i\phi/2}|0\rangle + \sin(\theta/2)e^{i\phi/2}|1\rangle, \tag{1.4}$$

where $0 \leq \theta \leq \pi$ and $0 \leq \phi \leq 2\pi$. These two parameters define a point on the *Bloch sphere*. The north and south poles of the sphere represent the eigenstates of σ_z, and the eigenstates of σ_x and σ_y lie on the equator. Orthogonal states always lie opposite each other on the sphere.

If we allow states to be mixed, we represent a qubit by a density matrix ρ; the most general qubit density matrix can be written as

$$\rho = p|\psi\rangle\langle\psi| + (1-p)|\bar\psi\rangle\langle\bar\psi|, \tag{1.5}$$

where $|\psi\rangle$ and $|\bar\psi\rangle$ are two orthogonal pure states, $\langle\psi|\bar\psi\rangle = 0$. The mixed states of a qubit form a three parameter family:

$$\rho = \left[\frac{1+r}{2}\cos^2(\theta/2) + \frac{1-r}{2}\sin^2(\theta/2)\right]|0\rangle\langle 0|$$
$$+ \left[\frac{1+r}{2}\sin^2(\theta/2) + \frac{1-r}{2}\cos^2(\theta/2)\right]|1\rangle\langle 1|$$
$$+ r\cos(\theta/2)\sin(\theta/2)\left[e^{i\phi}|0\rangle\langle 1| + e^{-i\phi}|1\rangle\langle 0|\right], \tag{1.6}$$

where θ and ϕ are the same angular parameters as before and $0 \leq r \leq 1$. The limit $r = 1$ is the set of pure states, parametrized as in Eq. (1.4), while $r = 0$ is the *completely mixed state* $\rho = I/2$, where $I = \sigma_0$ is the identity matrix. This state is the same as the state of a classical unbiased coin. Thus we can think of the Bloch sphere as having pure states on its surface and mixed states in its interior; and the distance r from the center is a measure of the state's purity. It is simply related to the parameter p in Eq. (1.5): $p = (1 + r)/2$.

For two qubits, the Hilbert space $\mathcal{H}_2 \otimes \mathcal{H}_2$ has a tensor-product basis:

$$|0\rangle_A \otimes |0\rangle_B \equiv |00\rangle, \tag{1.7a}$$
$$|0\rangle_A \otimes |1\rangle_B \equiv |01\rangle, \tag{1.7b}$$
$$|1\rangle_A \otimes |0\rangle_B \equiv |10\rangle, \tag{1.7c}$$
$$|1\rangle_A \otimes |1\rangle_B \equiv |11\rangle. \tag{1.7d}$$

We define tensor products formally in Section 1.2.4. Similarly, for N qubits we can define a basis $\{|j_{N-1}j_{N-2}\cdots j_0\rangle\}$, $j_k = 0, 1$. A useful labeling of these 2^N basis vectors is by the integers $0 \leq j < 2^N$ whose binary expressions are $j_{N-1}\cdots j_0$:

$$|j\rangle \equiv |j_{N-1}\cdots j_0\rangle, \quad j = \sum_{k=0}^{N-1} j_k 2^k. \tag{1.8}$$

1.2.1.3 Mixed states and the Bloch space Since a mixed state has rank at least 2, it can be written as

$$\rho = \sum_{i=1}^{d} r_i |\phi_i\rangle\langle\phi_i|, \tag{1.9}$$

where at least two of the r_is are nonzero and the set $\{|\phi_i\rangle\}$ is again an orthonormal basis for $\mathcal{H}$. This is called the *spectral decomposition* of ρ, since the r_i are its eigenvalues and the $|\phi_i\rangle$ are its eigenvectors. Since $\mathrm{Tr}(\rho) = 1$ we have $\sum_{i=1}^{d} r_i = 1$ and since $\rho \geq 0$ we have $r_i \geq 0$ for all i. If we pick an arbitrary orthonormal basis $\{|i\rangle\}$ for $\mathcal{H}$ then ρ need not be diagonal and can be written in the form $\rho = \sum_{i=1}^{d} a_{ij}|i\rangle\langle j|$. The matrix $A \equiv \{a_{ij}\}$ then satisfies $\mathrm{Tr}(A) = 1$ and $A \geq 0$. Every positive matrix can be diagonalized with a unitary transformation. When this is done we recover the spectral decomposition (1.9).

It is possible to give a geometric characterization of density matrices by using the *Bloch vector representation* for an arbitrary d-dimensional Hilbert space $\mathcal{H}$ [KK05]. This works as follows: Let $F_0 \equiv I$ and let $\{F_\mu : \mathrm{Tr}(F_\mu) = 0\}_{\mu=1}^{d^2-1}$ be a basis for the set of traceless Hermitian matrices in $\mathcal{B}(\mathcal{H})$. Assume further that $\mathrm{Tr}(F_\mu F_\nu) = d\delta_{\mu\nu}$ $\forall \mu, \nu$, i.e., the Fs are mutually orthogonal with respect to the *Hilbert–Schmidt inner product* $\langle A|B\rangle \equiv \mathrm{Tr}(A^\dagger B)$, which is the natural inner product on $\mathcal{B}(\mathcal{H})$, when this space is a Hilbert space. Hence any state ρ can be expanded as

$$\rho = \frac{1}{d}\left(I + \sum_{\mu=1}^{d^2-1} b_\mu F_\mu\right); \quad b_\mu = \mathrm{Tr}[\rho F_\mu] \equiv \langle F_\mu\rangle_\rho, \tag{1.10}$$

The vector $\mathbf{b} = (b_1, ..., b_{N^2-1}) \in \mathbb{R}^{N^2-1}$ of expectation values (more on this notion in the measurements, Section 1.2.5) is called the *Bloch vector*, and knowing its components is equivalent to complete knowledge of the corresponding state ρ, via the mapping $\mathbf{b} \mapsto \rho = \frac{1}{d}(I + \sum_{\mu=1}^{d^2-1} b_\mu F_\mu)$.

Let $\mathbf{n}$ denote a unit vector, i.e., $\mathbf{n} \in \mathbb{R}^{d^2-1}$ and $\sum_{i=1}^{d^2-1} n_i^2 = 1$, and define $F_{\mathbf{n}} \equiv \sum_{\mu=1}^{d^2-1} n_\mu F_\mu$. Let the minimum eigenvalue of each $F_{\mathbf{n}}$ be denoted $m(F_{\mathbf{n}})$. The "Bloch space" $\boldsymbol{B}(\mathbb{R}^{d^2-1})$ is the set of all Bloch vectors and is a closed convex set, since the set of states $\mathscr{S}(\mathscr{H})$ is closed and convex, and the map $\mathbf{b} \mapsto \rho$ is linear homeomorphic. The Bloch space is characterized in the "spherical coordinates" determined by $\{F_{\mathbf{n}}\}$ as

$$\boldsymbol{B}(\mathbb{R}^{d^2-1}) = \left\{ \mathbf{b} = r\mathbf{n} \in \mathbb{R}^{d^2-1} : r \leq \frac{1}{|m(F_{\mathbf{n}})|} \right\}. \tag{1.11}$$

This result is useful for visualization of quantum states. For example, for two qubits the Bloch space is given by Eq. (1.11) with $d = 4$, which corresponds to a certain 15-dimensional convex set. The Bloch space of a qubit is defined with the $\{F_\mu\}$ being the Pauli matrices; it is a simple sphere, since it so happens that for a qubit the minimum eigenvalues $m(F_{\mathbf{n}})$ are 1 for all $F_{\mathbf{n}}$.

1.2.2 Quantum gates and the dynamics of isolated (closed) systems

1.2.2.1 Schrödinger equation The evolution of a quantum state is governed by the Schrödinger equation:

$$\frac{\partial \rho}{\partial t} = -\frac{i}{\hbar}[H, \rho]. \tag{1.12}$$

The Hermitian operator $H \in \mathscr{B}(\mathscr{H})$ is called the Hamiltonian and is the "generator" of the evolution. It is the sum of the (quantized) kinetic and potential energies of the system. The symbol $\hbar \approx 10^{-34}$ joule second is the Planck constant and has units of energy multiplied by time (or, equivalently, units of action). From now on we will use units where $\hbar = 1$, which means that H has units of inverse time (frequency). The symbol $[\cdot, \cdot]$ is the commutator: $[H, \rho] \equiv H\rho - \rho H$. Since H is Hermitian the solution to the Schrödinger equation is

$$\rho(t) = U(t)\rho(0)U^\dagger(t), \tag{1.13}$$

where U is a *unitary* operator ($U^\dagger = U^{-1}$) sometimes called the propagator, and is related to H via

$$\frac{\partial U}{\partial t} = -iHU. \tag{1.14}$$

When H is time independent we have the simple solution $U(t) = \exp(-iHt)$, assuming the boundary condition $U(0) = I$, where I is the identity operator.

Note that given $U(t)$ for all times t we can compute the Hamiltonian as the logarithm of $U(t)$. However, rather than taking the actual log of $U(t)$, which would require proper handling of branch cuts in the complex plane, we can obtain $H(t)$ by computing the "logarithmic derivative":

$$H(t) = -i\frac{\partial U(t)}{\partial t} U^{-1}(t). \tag{1.15}$$

1.2.2.2 Dyson series When H is time dependent it need not commute with itself at different times and Eq. (1.14) has the formal solution

$$U(t) = T_+ \exp\left[-i\int_0^t H(t')dt'\right], \tag{1.16}$$

where T_+ is the *time-ordering operator*. This means that the solution can be expressed as an infinite sum, called the *Dyson series*:

$$U(t) = I + \sum_{n=1}^{\infty} S_n(t), \qquad (1.17)$$

where each term is a time-ordered multiple integral

$$S_n(t) \equiv (-i)^n \int_0^t dt_1 H(t_1) \int_0^{t_1} dt_2 H(t_2) \cdots \int_0^{t_{n-1}} dt_n H(t_n), \qquad (1.18)$$

where "time-ordered" means with early to late times being composed from right to left.

When H is time independent each of the multiple integrals becomes $S_n(t) = (-iHt)^n/n!$, so that $U(t) = \exp(-iHt)$, as it should.

The Dyson series (1.17) has the advantage of providing a simple recipe for generating terms to arbitrarily high order. As we will see in Section 1.2.6.1, it also converges for all t provided H is a bounded operator for all t, but it has the disadvantage that the partial sum $U_k(t) \equiv I + \sum_{n=1}^{k} S_n(t)$ is not unitary. This is sometimes a problem in quantum computation applications, where we are often interested in unitary transformations.

1.2.2.3 Magnus expansion An alternative to the Dyson series is the *Magnus expansion*, which has the advantage that it is unitary to all orders. This makes it particularly useful for applications in quantum computation, such as in dynamical decoupling, as we will see in Chapter 4. Like the Dyson expansion, the Magnus expansion at time t is an operator series

$$\Omega(t) \equiv \sum_{n=1}^{\infty} \Omega_n(t) \qquad (1.19)$$

such that $\Omega_n(t)$ is nth order in the Hamiltonian $H(t)$, but now the series appears in the exponential:

$$U(t) = \exp[\Omega(t)]. \qquad (1.20)$$

Thus, for a fixed time T, the time evolution generated by the time-dependent Hamiltonian $H(t)$ is equivalent to the time evolution generated by the time-independent effective Hamiltonian $H_{\text{eff}} \equiv \frac{i}{T}\Omega(T)$.

The first few terms in the Magnus expansion are

$$\Omega_1(t) = -i \int_0^t dt_1\, H(t_1), \qquad (1.21\text{a})$$

$$\Omega_2(t) = -\frac{1}{2} \int_0^t dt_1 \int_0^{t_1} dt_2\, [H(t_1), H(t_2)], \qquad (1.21\text{b})$$

$$\Omega_3(t) = \frac{i}{6} \int_0^t dt_1 \int_0^{t_1} dt_2 \int_0^{t_2} dt_3 ([H(t_1),[H(t_2),H(t_3)]] + [H(t_3),[H(t_2),H(t_1)]]). \qquad (1.21\text{c})$$

Higher-order terms can be computed using a recursive formula, but unfortunately, in contrast to the Dyson series, they do not have a simple structure. In general, $\Omega_n(t)$ is the time integral of a sum of $(n-1)$-nested commutators with coefficients related to the Bernoulli numbers, each with

n factors of $H(t)$. The Magnus expansion is thus an infinite series in Ht; a sufficient condition for convergence is

$$\int_0^t dt' \, \|H(t')\| < \pi, \tag{1.22}$$

but the Magnus expansion need not otherwise converge, which is a disadvantage it has compared to the Dyson expansion. The review [BCO+09] provides a comprehensive discussion of the Magnus and Dyson expansions.

1.2.2.4 Quantum gates You can verify that when $\rho = |\psi\rangle\langle\psi|$, i.e., a pure state, the Schrödinger equation (1.12) becomes the familiar

$$\frac{\partial|\psi\rangle}{\partial t} = -iH|\psi\rangle, \tag{1.23}$$

and has the solution $|\psi(t)\rangle = U(t)|\psi(0)\rangle$. The propagator $U(t)$ is the same unitary operator as the one we just discussed.

It is often convenient to treat time evolutions at the level of unitary transformations rather than explicitly solving the Schrödinger equation. In such cases time can be treated as a discrete variable:

$$|\psi_n\rangle = U_n U_{n-1} \cdots U_1 |\psi_0\rangle. \tag{1.24}$$

If the unitary operator U_n is *weak*, that is, close to the identity, one can always find a Hamiltonian operator H_n such that

$$U_n = e^{-iH_n \delta t} \approx I - iH_n \delta t, \tag{1.25}$$

where δt is an appropriately short time interval. Thus, one can easily recover the Schrödinger equation from a description in terms of unitary operators:

$$\delta|\psi_n\rangle = |\psi_n\rangle - |\psi_{n-1}\rangle = (U_n - I)|\psi_{n-1}\rangle \approx -iH_n|\psi_{n-1}\rangle \delta t \,. \tag{1.26}$$

Linear combinations of the Pauli operators σ_x, σ_y, and σ_z, together with the identity I, are sufficient to produce any operator on a single qubit. To specify any unitary transformation it suffices to give its effect on a complete set of basis vectors. We will consider only a fairly limited set of two-qubit transformations, and no transformations involving more than two qubits, but the simple formalism we derive readily generalizes to higher-dimensional systems.

Let us examine a couple of examples of two-qubit transformations. The controlled-NOT gate (or CNOT) is widely used in quantum computation; applied to the tensor-product basis vectors [Eqs. (1.7a)–(1.7d)] it gives

$$U_{\text{CNOT}}|00\rangle = |00\rangle, \tag{1.27a}$$
$$U_{\text{CNOT}}|01\rangle = |01\rangle, \tag{1.27b}$$
$$U_{\text{CNOT}}|10\rangle = |11\rangle, \tag{1.27c}$$
$$U_{\text{CNOT}}|11\rangle = |10\rangle \,. \tag{1.27d}$$

If the first qubit is in state $|0\rangle$, this gate leaves the second qubit unchanged; if the first qubit is in state $|1\rangle$, the second qubit is flipped $|0\rangle \leftrightarrow |1\rangle$. Hence the name: whether a NOT gate is performed on the second qubit is *controlled* by the first qubit. In terms of single-qubit operators, $U_{\text{CNOT}} = |0\rangle\langle 0| \otimes I + |1\rangle\langle 1| \otimes \sigma_x$.

Another important gate in quantum computation is the SWAP; applied to the tensor-product basis vectors it gives

$$U_{\text{SWAP}}|00\rangle = |00\rangle, \qquad (1.28a)$$
$$U_{\text{SWAP}}|01\rangle = |10\rangle, \qquad (1.28b)$$
$$U_{\text{SWAP}}|10\rangle = |01\rangle, \qquad (1.28c)$$
$$U_{\text{SWAP}}|11\rangle = |11\rangle. \qquad (1.28d)$$

As the name suggests, the SWAP gate just exchanges the states of the two bits: $U_{\text{SWAP}}(|\psi\rangle \otimes |\phi\rangle) = |\phi\rangle \otimes |\psi\rangle$.

CNOT and SWAP are examples of two-qubit quantum gates. Such gates are of tremendous importance in the theory of quantum computation. These two gates in particular have an additional useful property. Note that the operator $U_{\text{CNOT}} = U_{\text{CNOT}}^\dagger$, i.e., it is both unitary and Hermitian, as is also true of U_{SWAP}. The eigenvalues of unitary operators are roots of unity, and those of Hermitian operators are real. Therefore this is only possible if all of the operator's eigenvalues are ± 1. This also means that these operators are their own inverses, i.e., they square to identity: $U_{\text{CNOT}}^2 = U_{\text{SWAP}}^2 = I$. Among single-qubit operators, the Pauli matrices $\sigma_{x,y,z}$ also have this property, as does any operator $\mathbf{n} \cdot \sigma$, where $\sigma = (\sigma_x, \sigma_y, \sigma_z)$ and $\mathbf{n}$ is a unit vector. By Taylor expansion, it is simple to verify that any operator A that squares to identity satisfies the useful formula

$$e^{-i\theta A} = I\cos\theta - iA\sin\theta. \qquad (1.29)$$

Therefore

$$U_{\text{CNOT}}(\theta) = e^{-i\theta U_{\text{CNOT}}} = I\cos\theta - iU_{\text{CNOT}}\sin\theta \qquad (1.30)$$

and

$$U_{\text{SWAP}}(\theta) = e^{-i\theta U_{\text{SWAP}}} = I\cos\theta - iU_{\text{SWAP}}\sin\theta. \qquad (1.31)$$

These families range from the identity I for $\theta = 0$ to the full CNOT (or SWAP) gate for $\theta = \pi/2$, up to a global phase. For small values $\theta \ll 1$, this is a weak interaction, which leaves the state only slightly altered. Similarly, any single-qubit gate can be written as $e^{-i\theta \mathbf{n}\cdot\sigma} = I\cos\theta - i\mathbf{n}\cdot\sigma\sin\theta$.

1.2.3 Quantum correlations

1.2.3.1 Tensor products With the exception of our brief discussion of multi-qubit gates, so far we only allowed for the possibility of a single quantum system. However, in reality systems are never isolated and so any proper description of an experiment must account for the existence of an *environment* (or "bath," or "reservoir"). This means that we must embed our system of interest in a larger Hilbert space, which includes the environment. In principle, this larger Hilbert space should be that of the entire universe, but in practice one often settles for what are believed to be the relevant degrees of freedom that couple dominantly to the system of interest. To see how to embed our system S into the larger Hilbert space $\mathcal{H}_{SB}$, which includes both the system and the bath that forms the relevant environmental degrees of freedom, consider the quantum mechanical rule for forming the joint Hilbert space of any two systems S and B with respective

Hilbert spaces $\mathcal{H}_S$ and $\mathcal{H}_B$. As we have seen informally in our discussion of two-qubit gates above, the rule is to take the tensor product:

$$\mathcal{H}_{SB} = \mathcal{H}_S \otimes \mathcal{H}_B. \tag{1.32}$$

This rule is extremely important and gives rise to much of what is often deemed counterintuitive about quantum mechanics, including entanglement. Note that the classical rule for combining systems is to take the Cartesian product (the set of ordered pairs), which is formally why entanglement is absent classically. The tensor product rule extends to linear operators acting on $\mathcal{H}_S$ and $\mathcal{H}_B$, meaning that if $P \in \mathcal{L}(\mathcal{H}_S)$ and $Q \in \mathcal{L}(\mathcal{H}_S)$ then $P \otimes Q \in \mathcal{L}(\mathcal{H}_{SB})$, where $\mathcal{L}$ denotes the set of linear operators. Rather than stating an abstract definition of the tensor product, let us explain it in terms of a matrix representation known as the Kronecker product:

$$P \otimes Q = \begin{bmatrix} P_{11}Q & \cdots & P_{1n}Q \\ \vdots & P_{ij}Q & \vdots \\ P_{m1}Q & \cdots & P_{mn}Q \end{bmatrix}, \tag{1.33}$$

where P_{ij} are the matrix elements of P, and entries such as $P_{ij}Q$ are submatrices with matrix elements $P_{ij}Q_{kl}$, running over the matrix elements of Q. If we assume P to be an $m \times n$ matrix and Q is a $p \times q$ matrix then the tensor product matrix $P \otimes Q$ is $mp \times nq$. In particular, if P and Q are pure states, i.e., $m \times 1$ and $p \times 1$ matrices (column vectors), then their tensor product is a pure state whose dimension has grown to $mp \times 1$. Thus, if $|0\rangle = (1,0)^t$ (the superscript t denotes the transpose, so that $|0\rangle$ is a column vector) and if $|1\rangle = (0,1)^t$, then $|00\rangle = (1,0,0,0)^t$, and $|11\rangle = (0,0,0,1)^t$ [these are the two-qubit states of Eqs. (1.7a)–(1.7d)].

1.2.3.2 Entanglement and separability Pure state entanglement arises when we consider linear combinations of tensor product pure states, such as the normalized linear combination $\frac{|00\rangle + |11\rangle}{\sqrt{2}}$. Namely, if such a linear combination cannot itself be written as a tensor product state in the same basis then the state is said to be *entangled*.

Similarly, a bipartite mixed state is entangled if it cannot be written as a convex combination of pure product states, i.e., if it cannot be written in the form

$$\rho = \sum_i p_i |\psi_i\rangle\langle\psi_i| \otimes |\phi_i\rangle\langle\phi_i|, \tag{1.34}$$

where $\{p_i\}$ is a probability distribution, while if it can the state is called *separable* [W89]. Curiously, even if a state is separable this does not mean it is classical, as we will see in our discussion of quantum discord below.

Note that the qualifier "in the same basis" is important in the definition of pure state entanglement. To see this, let $|x\rangle \otimes |y\rangle \equiv |x,y\rangle$, ($x, y \in \{0, 1\}$) be the standard product basis for a two-qubit system. Each qubit forms a subsystem. With respect to this bipartition of the four-dimensional Hilbert space of two qubits the *Bell states*

$$|\Phi^\pm\rangle = (|00\rangle \pm |11\rangle)/\sqrt{2}, \quad |\Psi^\pm\rangle = (|01\rangle \pm |10\rangle)/\sqrt{2}, \tag{1.35}$$

are (maximally) entangled. That is, no matter how hard you try you will not be able to find a way to write any of these states as a tensor product in the basis $\{|00\rangle, |01\rangle, |10\rangle, |11\rangle\}$. Now note that these can be rewritten as $|\chi^\lambda\rangle := |\chi\rangle \otimes |\lambda\rangle$, where $\chi = \Phi, \Psi$ and $\lambda = +, -$. With respect to this new bi-partition the Bell states are product states, and the subsystems are the χ and λ degrees of freedom. On the other hand the states $|x,y\rangle$ are now entangled. The way out of this apparent conundrum is to realize that the tensor product structure is induced by the observables we can measure, and hence entanglement is a property that is defined relative to a given choice of observables [ZLL04].

1.2.3.3 Partial trace Suppose that our system is in the pure state $|\psi_S\rangle$ and the bath is in the pure state $|\psi_B\rangle$, so that their joint state is $|\Psi\rangle = |\psi_S\rangle \otimes |\psi_B\rangle \in \mathscr{H}_{SB}$. Clearly, to extract $|\psi_S\rangle$ from $|\Psi\rangle$ we should somehow remove $|\psi_B\rangle$. The general rule for doing this is called the *partial trace*. Note first that the outer product and tensor product commute, i.e., $(|i_S\rangle \otimes |k_B\rangle)(\langle j_S| \otimes \langle l_B|) = |i_S\rangle\langle j_S| \otimes |k_B\rangle\langle l_B|$. The partial trace over B, which is a map from operators acting on $\mathscr{H}_{SB}$ to operators acting on $\mathscr{H}_S$ alone, is defined as[2]

$$\mathrm{Tr}_B[|i_S\rangle\langle j_S| \otimes |k_B\rangle\langle l_B|] \equiv \mathrm{Tr}[|k_B\rangle\langle l_B|]|i_S\rangle\langle j_S| = \langle l_B|k_B\rangle |i_S\rangle\langle j_S|. \quad (1.36)$$

This definition extends naturally to linear combinations. That is,

$$\mathrm{Tr}_B\big[a|i_S\rangle\langle j_S| \otimes |k_B\rangle\langle l_B| + a'|i'_S\rangle\langle j'_S| \otimes |k'_B\rangle\langle l'_B|\big] = a\langle l_B|k_B\rangle|i_S\rangle\langle j_S| + a'\langle l'_B|k'_B\rangle|i'_S\rangle\langle j'_S|. \quad (1.37)$$

In essence, the partial trace is just the quantum version of creating a marginal probability distribution $p(x) = \sum_y p(x,y)$ from a joint classical probability distribution $p(x,y)$ over two variables.

1.2.3.4 Entropy and entanglement We are now in a position to quantify the entanglement of a state. Consider a state $\rho \in \mathscr{B}(\mathscr{H})$. Its *von Neumann entropy* is defined as

$$S(\rho) \equiv -\mathrm{Tr}[\rho \log_2(\rho)]. \quad (1.38)$$

The notation $f(\rho)$, with f any function, is shorthand for $\sum_i f(p_i)|\psi_i\rangle\langle\psi_i|$, where $|\psi_i\rangle$ and p_i are the eigenvectors and eigenvalues of ρ. The von Neumann entropy is the quantum analog of the classical *Shannon entropy*:

$$H(X) \equiv -\sum_x p(x) \log_2 p(x), \quad (1.39)$$

where $p(x) \equiv p(X = x)$ is a probability distribution for a random variable X, and where we define $0 \log 0 \equiv 0$.

Let

$$\rho_A \equiv \mathrm{Tr}_B[\rho_{AB}] \quad (1.40)$$

and similarly let $\rho_B \equiv \mathrm{Tr}_A[\rho_{AB}]$. These *reduced density matrices* are the quantum versions of classical marginals. The reduced von Neumann entropy of ρ_{AB} with respect to system A is $S(\rho_A)$, which, in explicit form is

$$S(\rho_A) = -\mathrm{Tr}\{\mathrm{Tr}_B[\rho_{AB}] \log_2 \mathrm{Tr}_B[\rho_{AB}]\}. \quad (1.41)$$

This is also called the *entropy of entanglement*.

---

[2] More formally, consider two separable Hilbert spaces $\mathscr{H}_1$ and $\mathscr{H}_2$ and let $A : \mathscr{H} \mapsto \mathscr{H}$ denote a linear trace-class bounded operator acting on the tensor product Hilbert space $\mathscr{H} := \mathscr{H}_1 \otimes \mathscr{H}_2$. Let $\{|k_i\rangle\}_k$ denote an orthonormal basis for $\mathscr{H}_i$, where $i \in 1, 2$. The partial trace operation over the first (second) Hilbert space is a map from $\mathscr{H}$ to the second (first) Hilbert space, and has the operational definition $\mathrm{Tr}_i(A) := \sum_{k_i} \langle k_i|A|k_i\rangle$. When A is decomposed in terms of the two orthonormal bases as $A = \sum_{k_1,k'_1,l_2,l'_2} \langle k_1 l_2|A|k'_1 l'_2\rangle|k_1 l_2\rangle\langle k'_1 l'_2|$, where $|k_1 l_2\rangle := |k_1\rangle \otimes |l_2\rangle$ etc., the partial trace over $\mathscr{H}_2$ can written as $\mathrm{Tr}_2(A) = \sum_{k_1,k'_1,l_2} \langle k_1 l_2|A|k'_1 l_2\rangle|k_1\rangle\langle k'_1|$. This makes it clear that $\mathrm{Tr}_2(A)$ is an operator that acts on $\mathscr{H}_1$.

For bipartite pure states, the entropy of entanglement is the unique measure of entanglement in the sense that it is the only function on the family of states that satisfies certain axioms required of an entanglement measure [DHR02]. For bipartite pure states $|\Psi\rangle \in \mathcal{H}_{AB}$ there is a very useful representation called the *Schmidt decomposition*:

$$|\Psi\rangle = \sum_{j=1}^{s} c_j |\psi_j\rangle_A \otimes |\psi_j\rangle_B, \qquad (1.42)$$

where $|\psi_j\rangle_A$ and $|\psi_j\rangle_B$ are orthonormal vectors in $\mathcal{H}_A$ and $\mathcal{H}_B$, respectively, and $c_j > 0$. Normalization requires that $\sum_j c_j^2 = 1$. The number of nonzero Schmidt coefficients s is a simple measure of entanglement.

The Schmidt decomposition is a consequence of the *singular value decomposition* (SVD), which states that every matrix M can be decomposed as $M = VDW^\dagger$, where D is the diagonal matrix of singular values of M, i.e., the positive square roots of the eigenvalues of the positive matrix $M^\dagger M$, where the columns of the unitary matrix W are the eigenvectors of $M^\dagger M$, and where the columns of the unitary matrix V are the eigenvectors of $MM^\dagger$. Note that M need not be square. To go from the SVD to the Schmidt decomposition, start by expanding $|\Psi\rangle$ in a basis $\{|k\rangle_S \otimes |p\rangle_B\}$ for $\mathcal{H}_{SB}$, i.e., let $|\Psi\rangle = \sum_{k,p} m_{kp} |k\rangle_S \otimes |p\rangle_B$, and consider the SVD of the coefficient matrix $M = \{m_{kp}\}$. It allows us to write $M = \sum_j d_j |v_j\rangle\langle w_j|$, where $d_j = (D)_{jj}$, $|v_j\rangle$ are the columns of V, and $|w_j\rangle$ are the columns of W. The Schmidt decomposition then follows by writing out the matrix elements m_{kp} in terms of this SVD, and forming appropriate linear combinations of the $|k\rangle_S$ and $|p\rangle_B$ (try it!).

With the Schmidt decomposition in hand it follows right away that for a bipartite pure state

$$S(\rho_A) = S(\rho_B) = -\sum_j c_j^2 \log_2 c_j^2 \equiv S, \qquad (1.43)$$

i.e., it does not matter which subsystem we trace over first.

For mixed bipartite states ρ, the entropy of entanglement is not the unique entanglement measure, and in general $S(\rho_A) \neq S(\rho_B)$. Indeed, suppose ρ has the simple bi-diagonal form $\rho = \sum_{i,p} \lambda_{ip} |i\rangle_S\langle i| \otimes |p\rangle_B\langle p|$. A straightforward calculation then reveals that while $S(\rho_A) = \sum_{pq} \sum_i \lambda_{ip} \log \lambda_{iq}$, the other reduced density matrix is $S(\rho_B) = \sum_{pq} \sum_i \lambda_{pi} \log \lambda_{qi}$. Since λ need not be symmetric, $S(\rho_A)$ need not equal $S(\rho_B)$. Many other entanglement measures exist for mixed states, but reviewing these would take us too far afield. See, for example, the comprehensive review [HHH+09].

A *maximally entangled* bipartite pure state $|\Psi\rangle \in \mathcal{H} \otimes \mathcal{H}$ [e.g., a Bell state (1.35)] is defined as a state whose reduced density matrices are maximally mixed: $\rho_A = \rho_B = I/d$, where $d = \text{Tr}[I] = \dim \mathcal{H}$. The Shannon entropy achieves its maximum at, and only at, the uniform probability distribution. Therefore, in this case the entropy of entanglement $S(\rho_A) = S(\rho_B) = \log_2 d$ is also maximal, which justifies the terminology "maximally entangled." For n qubits $d = 2^n$, so that $S(\rho_A) = S(\rho_B) = n$. Any smaller value of $S(\rho_A) = S(\rho_B)$ therefore signals that this state is not maximally entangled (provided we know that ρ_{AB} is pure), and this shows how the entropy of entanglement serves as a measure of entanglement. Indeed, you can easily check this explicitly for a state such as $a|00\rangle + b|11\rangle$, where $|a|^2 + |b|^2 = 1$ but $a \neq 1/\sqrt{2}$.

Some other useful properties of the von Neumann entropy and entropy of entanglement are worth noting:

- $S(\rho)$ is invariant under a basis transformation: $S(\rho) = S(U\rho U^\dagger)$ for any unitary U.
- $S(\rho)$ is concave: if $\rho = \sum_i a_i \rho_i$, where $\{a_i\}$ is a probability distribution and $\{\rho_i\}$ are themselves density matrices, then $S(\rho) \geq \sum_i a_i S(\rho_i)$.
- $S(\rho)$ is additive for independent subsystems: $S(\rho_A \otimes \rho_B) = S(\rho_A) + S(\rho_B)$. However, it is subadditive in general:

$$|S(\rho_A) - S(\rho_B)| \leq S(\rho_{AB}) \leq S(\rho_A) + S(\rho_B), \tag{1.44}$$

where $S(\rho_{A,B})$ are the entropies of entanglement. This "triangle inequality" is called the Araki–Lieb inequality [AL70]. In the case of three subsystems the entropy is "strongly subadditive:"

$$S(\rho_{ABC}) + S(\rho_B) \leq S(\rho_{AB}) + S(\rho_{BC}). \tag{1.45}$$

This is the Lieb–Ruskai inequality [LR73].

1.2.3.5 Quantum discord   As another natural generalization from the classical case, the *quantum mutual information* is defined as

$$\mathscr{I}(\rho_{AB}) \equiv S(\rho_A) + S(\rho_B) - S(\rho_{AB}). \tag{1.46}$$

The *classical mutual information* expression that has been generalized here is

$$\mathscr{I}(X:Y) = H(X) + H(Y) - H(X,Y), \tag{1.47}$$

where $H(X,Y)$ is the Shannon entropy of the joint probability distribution $p(x,y)$, Eq. (1.39). But, using Bayes' rule $p(x|y) = p(x,y)/p(y)$ for the conditional probability of $X = x$ given $Y = y$, it is not hard to show that $H(X|Y) = H(X,Y) - H(Y)$, where $H(X|Y) \equiv \sum_y p(y) H(X|Y=y)$ is the conditional entropy of X given Y. It follows that an equivalent expression for the classical mutual information is $\mathscr{I}(X:Y) = H(X) - H(X|Y)$. This expression shows that the mutual information measures the decrease of X's entropy upon learning Y. It turns out that the quantum version of this last expression for $\mathscr{I}(X:Y)$ differs from $\mathscr{I}(\rho_{AB})$ in general. The reason is that the conditional entropy $H(X|Y)$ requires knowledge of Y, which in the quantum case means that the corresponding system must have been measured. In fact, this quantity depends in general on exactly which measurement is done on system Y. This is not the case in the symmetric expression $H(X) + H(Y) - H(X,Y)$ for the mutual information. Thus, the quantum version of the asymmetric expression $H(X) - H(X|Y)$ is in general different from Eq. (1.46). The minimum difference between the two expressions is called the "quantum discord" [OZ02], and plays an important role in characterizing open system dynamics, as we explain below.

It turns out that separable states [Eq. (1.34)] can have nonvanishing quantum discord. In this sense we see that "separable" is not the same as "classical." The quantumness of such states is because any measurement on one of the parties will perturb the bipartite state if it is separable, yet has nonvanishing discord. In contrast, a bipartite state with vanishing discord is truly classical in the sense that information about one of the two parties can always be extracted from the state without disturbing it. A useful example is the Werner state $\rho = \frac{1-z}{4} I + z |\psi\rangle\langle\psi|$, where $|\psi\rangle = (|00\rangle + |11\rangle)/\sqrt{2}$ (a Bell state) and $z \in [0,1]$. Werner states are separable for $z < 1/3$, but the quantum discord vanishes only when $z = 0$.

In general, a state ρ_{AB} has vanishing quantum discord if and only if there exists a complete set of rank-one projectors $\{\Pi_j\}$ on either the A or the B subsystem such that $\rho_{AB} = \sum_j \Pi_j \rho_{AB} \Pi_j$. An operator A is a *projector* if $A^2 = A$. *Completeness* means that $\sum_j \Pi_j = I$. We will see much more of such projection operators when we discuss measurements in Section 1.2.5. Now consider again the class of separable states (1.34), and note that if the $|\psi_i\rangle$ are all mutually orthogonal and span $\mathcal{H}_A$, then they are a complete set of rank-one projectors. Therefore a state is separable but nonclassical (has discord) if, e.g., not all the $|\psi_i\rangle$ are mutually orthogonal. A further discussion of conditions for vanishing quantum discord can be found in [DVB10].

1.2.4 Open systems

1.2.4.1 Kraus operator sum representation Using the partial trace rule we can deduce what the dynamics of an open system is in general, given that we know the Hamiltonian H_{SB} for the system and its environment. Such a Hamiltonian can always be decomposed into a sum of three terms:

$$H_{SB} = H_S \otimes I_B + I_S \otimes H_B + H_I, \qquad (1.48)$$

where H_S, H_B and H_I are, respectively, the system, bath and system–bath interaction Hamiltonians. Without loss of generality the interaction Hamiltonian can be written as

$$H_I = \sum_\alpha S_\alpha \otimes B_\alpha, \qquad (1.49)$$

where S_α and B_α are, respectively, system and bath operators. As we will see later, it is this coupling between system and bath that causes decoherence, through nonclassical correlations with the bath. The time-evolved system–bath state is given according to Eq. (1.13) by

$$\rho_{SB}(t) = U(t)\rho_{SB}(0)U^\dagger(t), \qquad (1.50)$$

where U is the propagator generated by H_{SB}. The evolution of the system alone is given by

$$\rho_S(t) = \mathrm{Tr}_B[U(t)\rho_{SB}(0)U^\dagger(t)]. \qquad (1.51)$$

We would like an explicit representation of the map from $\rho_S(0)$ to $\rho_S(t)$, assuming that the system and bath are initially decoupled:

$$\rho_{SB}(0) = \rho_S(0) \otimes \rho_B(0). \qquad (1.52)$$

By writing the spectral decomposition for the initial bath density matrix, $\rho_B(0) = \sum_\nu \lambda_\nu |\nu_B\rangle\langle\nu_B|$, and evaluating the partial trace in the same basis, we obtain

$$\rho_S(t) = \sum_\mu \langle \mu_B | U(t) \left(\rho_S(0) \otimes \sum_\nu \lambda_\nu |\nu_B\rangle\langle\nu_B| \right) U^\dagger(t) | \mu_B \rangle$$

$$= \sum_a A_a \rho_S(0) A_a^\dagger, \qquad (1.53)$$

where the *operation elements*, or *Kraus operators*, are given by

$$A_a = \sqrt{\lambda_\nu} \langle \mu_B | U | \nu_B \rangle \; ; \qquad a = (\mu, \nu). \qquad (1.54)$$

Note that the matrix element $\langle\mu_B|U|\nu_B\rangle$ is a *partial* matrix element, which means that if U is written out explicitly in terms of a tensor product basis for $\mathcal{H}_{SB}$, then the matrix elements are to be evaluated only over the bath part of the expression, leaving us with an operator acting on $\mathcal{H}_S$.

Since $\text{Tr}\rho(t) = \sum_a \text{Tr}(A_a \rho_S(0) A_a^\dagger) = \sum_a \text{Tr}(\rho_S(0) A_a^\dagger A_a)$, a sufficient condition for normalization $\text{Tr}\rho(t) = 1$ is that the operation elements satisfy the constraint

$$\sum_a A_a^\dagger A_a = I_S. \tag{1.55}$$

If this constraint is satisfied then, when the sum in Eq. (1.53) includes only one term, the dynamics are unitary. In this case we recover the evolution of a closed system. Thus a simple criterion for a genuinely open system is the presence of multiple independent terms in the sum in Eq. (1.53).

If we demand that $\text{Tr}\rho(t) = 1$ for any possible initial state $\rho_S(0)$ then Eq. (1.55) is also necessary for normalization. To see this, let $\sum_a A_a^\dagger A_a = Q$, where Q is a system operator. Consider an arbitrary initial state, $\rho_S(0) = \sum_{ij} s_{ij}|i\rangle\langle j|$. The normalization condition becomes $1 = \sum_a \text{Tr}(A_a \rho_S(0) A_a^\dagger) = \sum_i s_{ii} \langle i|Q|i\rangle + \sum_{i\neq j} s_{ij}\langle j|Q|i\rangle$. Using $\text{Tr}\rho_S(0) = \sum_i s_{ii} = 1$ we then obtain $\sum_i s_{ii}(\langle i|Q|i\rangle - 1) + \sum_{i\neq j} s_{ij}\langle j|Q|i\rangle = 0$, which can only be true for arbitrary s_{ij} if $\langle i|Q|i\rangle = 1$ for all i and $\langle j|Q|i\rangle = 0$ for all $i \neq j$, i.e., $Q = I_S$.

The expression (1.53) is known as the *Kraus operator sum representation* and can also be derived from an axiomatic approach to quantum mechanics, without reference to Hamiltonians [K83]. Instead, the notions of positivity and linearity play a key role. Namely, we can ask: what is the most general linear transformation that maps a quantum state to a quantum state? Since a quantum state is a trace 1 positive operator, such a transformation must preserve both the trace and positivity. (Recall that a positive operator is a linear operator whose eigenvalues are all nonnegative.) Every such "positive map" has a positivity domain, meaning that in general some, but not all quantum states will be mapped to quantum states. States outside of the positivity domain will be mapped to nonpositive operators, i.e., operators with at least one negative eigenvalue. Maps whose positivity domain includes all quantum states are therefore special and are called *completely positive maps*. They play a central role in quantum information theory: it turns out that every completely positive (CP) map can be written in the form of the Kraus operator sum representation, and that the Kraus operator sum representation always yields a CP map. For this reason we will from now on refer to any map of the type

$$\rho \mapsto \sum_a A_a \rho A_a^\dagger \equiv \rho' \tag{1.56}$$

as a CP map. When $\text{Tr}\rho' = 1$ (or, more generally, when $\text{Tr}\rho' = \text{Tr}\rho$) the map is often called completely positive trace-preserving (CPTP). It will turn out to be convenient to sometimes drop the constraint that $\text{Tr}\rho' = 1$, so that we will be able to account for "probability leakage," or certain types of incomplete measurements (generally involving *postselection*). For this reason we will not insist that $\sum_a A_a^\dagger A_a = I$.

Let us quickly verify that positivity of ρ implies positivity of ρ'. Let $|v\rangle$ be any vector in the Hilbert space acted on by ρ. Then, using the spectral decomposition $\rho = \sum_i s_i |i\rangle\langle i|$, we have $\langle v|\rho'|v\rangle = \sum_a \langle v|A_a \rho A_a^\dagger|v\rangle = \sum_{a,i} s_i \langle v|A_a|i\rangle\langle i|A_a^\dagger|v\rangle = \sum_{a,i} s_i |\langle v|A_a|i\rangle|^2 \geq 0$, as required.

1.2.4.2 Maps How special is the form (1.56)? A map is an association of elements in a range with elements in a domain. A map $\Phi : \mathscr{B}(\mathscr{H}) \mapsto \mathscr{B}(\mathscr{H})$ is called linear if $\Phi[a\rho_1 + b\rho_2] = a\Phi[\rho_1] + b\Phi[\rho_2]$ for any pair of states $\rho_1, \rho_2 \in \mathscr{S}(\mathscr{H})$, and constants $a, b \in \mathbb{C}$. It turns out that a map $\Phi : \mathfrak{M}_n \mapsto \mathfrak{M}_m$ (where $\mathfrak{M}_n$ is the space of $n \times n$ matrices) is linear if and only if it can be represented as

$$\Phi(\rho) = \sum_a E_a \rho E_a'^{\dagger}, \tag{1.57}$$

where the "left and right operation elements" $\{E_a\}$ and $\{E_a'\}$ are, respectively, $m \times n$ and $n \times m$ matrices. That this condition is sufficient is straightforward to verify. For a proof of necessity see, e.g., [SL09a]. Thus CP maps have the special property that they have equal left and right operation elements.

A subclass of linear maps are Hermitian maps: a linear map is called Hermitian if it maps all Hermitian operators in its domain to Hermitian operators. It turns out that Φ is a Hermitian map if and only if [H73]

$$\Phi(\rho) = \sum_a c_a E_a \rho E_a^{\dagger}, \quad c_a \in \mathbb{R}. \tag{1.58}$$

Now we see that CP maps are Hermitian maps with $c_a \geq 0\ \forall a$, for then we can identify $A_a = \sqrt{c_a} E_a$. It turns out that there is a tight connection between CP and Hermitian maps: a map is Hermitian if and only if it can be written as the difference of two CP maps [JSS04].

Hermitian maps are important in quantum mechanics. Recall our prescription (1.51) for the "reduced" dynamics of the system alone. What if we had the system and bath start out in some general initial state $\rho_{SB}(0)$ rather than the special factorized initial state $\rho_S(0) \otimes \rho_B(0)$? It can be shown that then the partial trace prescription $\rho_S(t) = \text{Tr}_B[U(t)\rho_{SB}(0)U^{\dagger}(t)]$ always yields a *Hermitian* map from $\rho_S(0) = \text{Tr}_B[\rho_{SB}(0)]$ to $\rho_S(t)$, but this map need not always be CP [SL09a]. There appears to be something troubling about this statement, since a Hermitian map need not preserve the positivity of the initial system state, i.e., we can end up with an unphysical state with negative eigenvalues. The culprit is the initial system–bath state. It is not that there is something wrong with our quantum mechanical prescription of taking the partial trace after unitary evolution. It is that some (in fact most!) initial system–bath states $\rho_{SB}(0)$ are simply too correlated to allow for a meaningful separation between "system" and "bath." When such a separation is not possible our formalism signals that something is wrong by yielding unphysical maps, which can take an initially positive system state to a negative system state, reflected in the appearance of negative "probabilities."

At this point you may wonder under which conditions Hamiltonian dynamics followed by partial trace [Eq. (1.51)] gives rise to a CP map. It turns out that a sufficient condition for this is the vanishing of all quantum correlations between system and bath in the initial state $\rho_{SB}(0)$. Indeed, while the product state form $\rho_{SB}(0) = \rho_S(0) \otimes \rho_B(0)$ is sufficient for obtaining a CP map, it is not necessary. Instead, for Eq. (1.51) to yield a CP map it is sufficient that $\rho_{SB}(0)$ has vanishing quantum discord [R.-RMK+08]. Recall that the quantum discord is a measure of the difference between the quantum versions of two classically equal expressions for the quantum mutual information, and when it vanishes a state contains only classical correlations. When this is the case we can be sure that a physically sensible separation between system and bath exists, allowing us to always make sense of a map from the initial system state $\rho_S(0) = \text{Tr}_B[\rho_{SB}(0)]$ to the final system state $\rho_S(t)$.

1.2.5 Strong and weak measurements

1.2.5.1 Projective measurements In the standard description of quantum mechanics, *observables* are identified with Hermitian operators $O = O^\dagger$. A measurement of O for a system in the pure state $|\psi\rangle$ returns an eigenvalue o_n of O and leaves the system in the corresponding eigenstate $|\phi_n\rangle$, $O|\phi_n\rangle = o_n|\phi_n\rangle$, with probability $p_n = |\langle\phi_n|\psi\rangle|^2$. The *expectation value* of O is the weighted average of its eigenvalues, i.e., $\langle O \rangle = \sum_n p_n o_n = \sum_n \langle\psi|\phi_n\rangle o_n \langle\phi_n|\psi\rangle = \langle\psi|O|\psi\rangle$.

A more general formulation, which applies also if a particular eigenvalue of O is degenerate, is to instead use the projector P_n onto the eigenspace with eigenvalue o_n. Since O is Hermitian it has a spectral decomposition in the form

$$O = \sum_n o_n P_n. \tag{1.59}$$

The probability of the measurement outcome is then $p_n = \langle\psi|P_n|\psi\rangle$, and right after the measurement the system is left in the state $P_n|\psi\rangle/\sqrt{p_n}$.

Still more generally, for a mixed state ρ the state after the measurement is

$$\rho \mapsto \rho_n = \frac{P_n \rho P_n}{p_n}, \tag{1.60}$$

where the probability of outcome n is

$$p_n = \text{Tr}[P_n \rho]. \tag{1.61}$$

The expectation value of an observable O, for a system in the state ρ, is

$$\langle O \rangle = \sum_n p_n o_n = \text{Tr}\left[\sum_n o_n P_n \rho\right] = \text{Tr}[O\rho]. \tag{1.62}$$

Because two observables with the same eigenspaces are completely equivalent to each other (as far as measurement probabilities and outcomes are concerned), we will not worry about the exact choice of Hermitian operator O; instead, we will choose a complete set of orthogonal projections $\{P_n\}$ that represent the possible measurement outcomes. These satisfy

$$P_n P_{n'} = P_n \delta_{nn'}, \quad \sum_n P_n = I. \tag{1.63}$$

A set of projection operators that obey Eq. (1.63) is often referred to as an *orthogonal decomposition of the identity*.

For a single qubit, the only nontrivial measurements have exactly two outcomes, which we label $+$ and $-$, with probabilities p_+ and p_- and associated projectors of the form

$$P_\pm = \frac{I \pm \mathbf{n} \cdot \sigma}{2} = |\psi_\pm\rangle\langle\psi_\pm|. \tag{1.64}$$

Equation (1.64) is equivalent to choosing an axis $\mathbf{n}$ on the Bloch sphere and projecting the state onto one of the two opposite points. All such operators are projections onto pure states. The two projectors sum to the identity operator, $P_+ + P_- = I$. The average information obtained from a projective measurement on a qubit is just the Shannon entropy for the two measurement outcomes:

$$H_{\text{meas}} = -p_+ \log_2 p_+ - p_- \log_2 p_- \,. \tag{1.65}$$

The maximum information gain is precisely one bit, when $p_+ = p_- = 1/2$, and the minimum is zero bits when either p_+ or p_- is 0. After the measurement, the state is left in an eigenstate of $P_\pm$, so repeating the measurement will result in the same outcome. This repeatability is one of the most important features of projective measurements.

1.2.5.2 Weak measurements and POVMs
While projective measurements are the most familiar from introductory quantum mechanics, there is a more general notion of measurement, the *positive operator valued measure* (POVM) [P98a]. Instead of giving a set of projectors that sum to the identity, we give a set of *positive operators* E_n that sum to the identity

$$\sum_n E_n = I \,. \tag{1.66}$$

The probability of outcome n is $p_n = \langle \psi | E_n | \psi \rangle$, or for a mixed state

$$p_n = \text{Tr}[E_n \rho]. \tag{1.67}$$

Unlike a projective measurement, knowing the operators E_n is *not* sufficient to determine the state of the system after measurement. One must further know a set of operators A_{nk} such that

$$E_n = \sum_k A_{nk}^\dagger A_{nk} \,. \tag{1.68}$$

After measurement outcome n the state is

$$\rho \mapsto \rho' = \frac{1}{p_n} \sum_k A_{nk} \rho A_{nk}^\dagger \,. \tag{1.69}$$

This measurement will not preserve the purity of states, in general, unless there is only a single A_{nk} for each E_n (that is, $E_n = A_n^\dagger A_n$), in which case $|\psi'\rangle = A_n |\psi\rangle / \sqrt{p_n}$.

Because the positive operators E_n need not be projectors, one is not limited to only two possible outcomes; indeed, there can be an unlimited number of possible outcomes. However, if a POVM is repeated, the same result will not necessarily be obtained the second time. Projective measurements are clearly a special case of POVMs in which the results *are* repeatable. Most actual experiments do not correspond to projective measurements, but are described by some more general POVM.

Conversely, it is easy to show that any POVM can be performed in principle by allowing the system that is to be measured to interact with an additional system, or *ancilla*, and then doing a projective measurement on the ancilla. In this viewpoint, only some of the information obtained comes from the system; part can also come from the ancilla, which injects extra randomness into the result.

A particularly interesting kind of POVM is the *weak measurement* [AAV88]. This is a measurement that gives very little information about the system on average, but also disturbs the state very little. Loosely speaking, there are two ways a measurement can be considered weak. Suppose we have a qubit in a pure state of the form (1.2), and we perform a POVM with the following two operators:

$$E_0 \equiv |0\rangle\langle 0| + (1-\epsilon)|1\rangle\langle 1| = A_0^2, \tag{1.70a}$$
$$E_1 \equiv \epsilon|1\rangle\langle 1| = A_1^2, \tag{1.70b}$$
$$A_0 \equiv |0\rangle\langle 0| + \sqrt{1-\epsilon}|1\rangle\langle 1|, \tag{1.70c}$$
$$A_1 \equiv \sqrt{\epsilon}|1\rangle\langle 1|, \tag{1.70d}$$

where $\epsilon \ll 1$. Clearly E_0 and E_1 are positive and $E_0 + E_1 = I$, so this is a POVM. The probability $p_0 = \langle\psi|E_0|\psi\rangle = 1 - \epsilon|\beta|^2$ of outcome 0 is close to 1, while $p_1 = \langle\psi|E_1|\psi\rangle = \epsilon|\beta|^2$ is very unlikely. Thus, most such measurements will give outcome 0, and very little information is obtained about the system. We can bound the attainable information by the Shannon entropy (1.65), which for this measurement is $H_{\text{meas}} \leq h(\epsilon) = -\epsilon \log_2 \epsilon - (1-\epsilon)\log_2(1-\epsilon)$, where $h(\epsilon) \to 0$ as $\epsilon \to 0$.

The state changes very slightly after a measurement outcome 0, with $|0\rangle$ becoming slightly more likely relative to $|1\rangle$. The new state is

$$|\psi_0\rangle = A_0|\psi\rangle/\sqrt{p_0} = (\alpha|0\rangle + \beta\sqrt{1-\epsilon}|1\rangle)/\sqrt{p_0}. \tag{1.71}$$

However, after an outcome of 1, the state can change dramatically: a measurement outcome of 1 leaves the qubit in the state $|1\rangle$. We see that this type of measurement is weak in that it *usually* gives little information and has little effect on the state; but on rare occasions it can give a great deal of information and have a large effect.

By contrast, consider the following positive operators:

$$E_0 \equiv \left(\frac{1+\epsilon}{2}\right)|0\rangle\langle 0| + \left(\frac{1-\epsilon}{2}\right)|1\rangle\langle 1| = A_0^2, \tag{1.72a}$$
$$E_1 \equiv \left(\frac{1-\epsilon}{2}\right)|0\rangle\langle 0| + \left(\frac{1+\epsilon}{2}\right)|1\rangle\langle 1| = A_1^2, \tag{1.72b}$$
$$A_0 \equiv \sqrt{\frac{1+\epsilon}{2}}|0\rangle\langle 0| + \sqrt{\frac{1-\epsilon}{2}}|1\rangle\langle 1|, \tag{1.72c}$$
$$A_1 \equiv \sqrt{\frac{1-\epsilon}{2}}|0\rangle\langle 0| + \sqrt{\frac{1+\epsilon}{2}}|1\rangle\langle 1|. \tag{1.72d}$$

These operators also constitute a POVM. Both E_0 and E_1 are close to $I/2$, and so are almost equally likely for all states, $p_0 \approx p_1 \approx 1/2$; the information acquired, as measured by the Shannon entropy (1.65), is approximately one bit: $1 - \epsilon^2/2\ln 2 < H_{\text{meas}} \leq 1$. But the state of the system is almost unchanged for *both* outcomes, with the new state being $|\psi_j\rangle = A_j|\psi\rangle/\sqrt{p_j}$,

$$|\psi_0\rangle = \frac{1}{\sqrt{1+\epsilon(|\alpha|^2-|\beta|^2)}}(\alpha\sqrt{1+\epsilon}|0\rangle + \beta\sqrt{1-\epsilon}|1\rangle)$$
$$\approx \alpha(1+\epsilon|\beta|^2)|0\rangle + \beta(1-\epsilon|\alpha|^2)|1\rangle, \tag{1.73a}$$
$$|\psi_1\rangle = \frac{1}{\sqrt{1-\epsilon(|\alpha|^2-|\beta|^2)}}(\alpha\sqrt{1-\epsilon}|0\rangle + \beta\sqrt{1+\epsilon}|1\rangle)$$
$$\approx \alpha(1-\epsilon|\beta|^2)|0\rangle + \beta(1+\epsilon|\alpha|^2)|1\rangle, \tag{1.73b}$$

for outcomes 0 and 1, respectively. For this type of weak measurement, the measurement outcome is almost random, but does include a tiny amount of information about the state. In this case, we cannot quantify the state information using the Shannon entropy, which is always close to 1. Instead we look at how much the von Neumann entropy can be decreased: if we start with a maximally mixed qubit state $\rho = I/2$, which has von Neumann entropy 1, and perform this weak measurement, the von Neumann entropy decreases by $\epsilon^2/2\ln 2$.

By performing repeated weak measurements and examining the statistics of the results, one can in effect perform a strong measurement; for the particular case considered here, the state will tend to drift towards either $|0\rangle$ or $|1\rangle$ with overall probabilities $|\alpha|^2$ and $|\beta|^2$.

1.2.6 Norms, fidelity and distance measures

1.2.6.1 Unitarily invariant norms Unitarily invariant norms are norms that satisfy, for all unitary U, V [B97],

$$\|UAV\|_{\mathrm{ui}} = \|A\|_{\mathrm{ui}}. \tag{1.74}$$

We list some important examples.

(i) The trace norm:

$$\|A\|_1 \equiv \mathrm{Tr}|A| = \sum_i s_i(A), \tag{1.75}$$

where

$$|A| \equiv \sqrt{A^\dagger A}, \tag{1.76}$$

and $s_i(A)$ are the singular values (eigenvalues of $|A|$).

(ii) The operator norm: Let $\mathscr{V}$ be an inner product space equipped with the Euclidean norm $\|x\| \equiv \sqrt{\sum_i |x_i|^2 \langle e_i, e_i \rangle}$, where $x = \sum_i x_i e_i \in \mathscr{V}$ and $\mathscr{V} = \mathrm{Span}\{e_i\}$. Let $\Lambda : \mathscr{V} \mapsto \mathscr{V}$. The operator norm is

$$\|\Lambda\|_\infty \equiv \sup_{x \in \mathscr{V}} \frac{\|\Lambda x\|}{\|x\|} = \max_i s_i(\Lambda). \tag{1.77}$$

Therefore $\|\Lambda x\| \leq \|\Lambda\|_\infty \|x\|$. In this book $\mathscr{V} = \mathscr{L}(\mathscr{H})$ is the space of all linear operators on the Hilbert space $\mathscr{H}$, Λ is a superoperator, and $x = \rho$ is a normalized quantum state: $\|\rho\|_1 = \mathrm{Tr}\rho = 1$.

(iii) The Frobenius, or Hilbert–Schmidt norm:

$$\|A\|_2 \equiv \sqrt{\mathrm{Tr} A^\dagger A} = \sqrt{\sum_i s_i^2(A)}. \tag{1.78}$$

All unitarily invariant norms satisfy the important property of submultiplicativity:

$$\|AB\|_{\mathrm{ui}} \leq \|A\|_{\mathrm{ui}} \|B\|_{\mathrm{ui}}. \tag{1.79}$$

The norms of interest to us are also multiplicative over tensor products and obey an ordering:

$$\|A \otimes B\|_i = \|A\|_i \|B\|_i \quad i = 1, 2, \infty, \tag{1.80a}$$

$$\|A\|_\infty \leq \|A\|_2 \leq \|A\|_1, \tag{1.80b}$$

$$\|AB\|_{\text{ui}} \leq \|A\|_\infty \|B\|_{\text{ui}}, \|B\|_\infty \|A\|_{\text{ui}}. \tag{1.80c}$$

There is an interesting duality between the trace and operator norm:

$$\|A\|_1 = \max\{|\text{Tr}(B^\dagger A)| : \|B\|_\infty \leq 1\}, \tag{1.81a}$$

$$\|A\|_\infty = \max\{|\text{Tr}(B^\dagger A)| : \|B\|_1 \leq 1\}, \tag{1.81b}$$

$$|\text{Tr}(BA)| \leq \|A\|_\infty \|B^\dagger\|_1, \|B^\dagger\|_\infty \|A\|_1. \tag{1.81c}$$

In the last three inequalities A and B can map between spaces of different dimensions. If they map between spaces of the same dimension then

$$\|A\|_1 = \max_{B^\dagger B = I} |\text{Tr}(B^\dagger A)|. \tag{1.82}$$

As a simple application of unitarily invariant norms, let us revisit the convergence of the Dyson series, which we discussion in Section 1.2.2.2. Consider any unitarily invariant norm $\| \|_{\text{ui}}$. Since such norms are submultiplicative,

$$\|S_n(t)\| \leq \int_0^t dt_n \cdots \int_0^{t_2} dt_1 \|H(t_n)\|_{\text{ui}} \cdots \|H(t_1)\|_{\text{ui}} \leq \frac{(\lambda t)^n}{n!}, \tag{1.83}$$

where $\lambda = \max_t \|H(t_n)\|_{\text{ui}}$. Thus the Dyson series is bounded term by term by the power series of the exponential function with argument λt, which converges for all finite values of λt.

1.2.6.2 Uhlman fidelity and trace-norm distance The Uhlman fidelity between two density matrices ρ and σ is defined as

$$F(\rho, \sigma) \equiv \|\sqrt{\rho}\sqrt{\sigma}\|_1 = \text{Tr}\left[\sqrt{\rho^{\frac{1}{2}} \sigma \rho^{\frac{1}{2}}}\right]. \tag{1.84}$$

Sometimes one is concerned only with the fidelity between a density matrix ρ and the case where σ is a pure state $\sigma = |\psi\rangle\langle\psi|$:

$$F(|\psi\rangle, \rho) = \text{Tr}\left[\sqrt{\rho^{\frac{1}{2}} |\psi\rangle\langle\psi| \rho^{\frac{1}{2}}}\right] = \text{Tr}\left[\sqrt{|\psi\rangle\langle\psi|\rho|\psi\rangle\langle\psi|}\right] = \sqrt{\langle\psi|\rho|\psi\rangle}. \tag{1.85}$$

The fidelity is a measure of how close two states are. It is equal to 1 if and only if $\rho = \sigma$. Moreover, the fidelity can serve as an upper and lower bound on the trace distance, $D(\rho, \sigma)$,

$$1 - F(\rho, \sigma) \leq D(\rho, \sigma) \leq \sqrt{1 - F(\rho, \sigma)}. \tag{1.86}$$

The trace-norm distance between two density matrices is

$$\|\rho - \sigma\|_1 \equiv D(\rho, \sigma) = \frac{1}{2}\text{Tr}\left[|\rho - \sigma|\right]. \tag{1.87}$$

The trace-norm distance is a beautiful metric on density matrices, because it is related directly to how different the outcomes of a POVM on these states can be. In particular, assume that we perform a POVM with corresponding measurement operators E_i. Then on the two states ρ and σ, this results in measurement probabilities $p_i = Pr(i|\rho) = \text{Tr}[\rho E_i]$ and $q_i = Pr(i|\sigma) = \text{Tr}[\sigma E_i]$. The Kolmogorov distance between the two probability distributions produced by these measurements is

$$D(p,q) = \frac{1}{2}\sum_i |p_i - q_i|. \tag{1.88}$$

Let us show that $D(\rho,\sigma)$ is equal to the maximum over all possible generalized measurements of the Kolmogorov distance between the resulting probability distributions. Note that

$$D(p,q) = \frac{1}{2}\sum_i |\text{Tr}[E_i(\rho - \sigma)]|. \tag{1.89}$$

Decompose $\rho - \sigma$, which is Hermitian, into a difference of two positive Hermitian operators P and N with orthogonal support, i.e., $\rho - \sigma = P - N$ with $PN = NP = 0$, where P has support over the eigenvectors of $\rho - \sigma$ with non-negative eigenvalues, $-N$ over the eigenvectors with negative eigenvalues. Note that $|\rho - \sigma| = \sqrt{(P-N)^2} = \sqrt{P^2 + N^2} = \sqrt{(P+N)^2} = P+N$. Then

$$\begin{aligned} D(p,q) &= \frac{1}{2}\sum_i |\text{Tr}[E_i(P-N)]| \leq \frac{1}{2}\sum_i |\text{Tr}[E_i P]| + |\text{Tr}[E_i N]| \\ &= \frac{1}{2}\sum_i \text{Tr}[E_i(P+N)] = \frac{1}{2}\sum_i \text{Tr}[E_i|\rho-\sigma|] \\ &= \frac{1}{2}\text{Tr}\left[\left(\sum_i E_i\right)|\rho-\sigma|\right] = D(\rho,\sigma), \end{aligned} \tag{1.90}$$

where we used the completeness relation $\sum_i E_i = I$. Thus $D(p,q) \leq D(\rho,\sigma)$. But now choose a POVM with elements that project along the eigenvectors of $\rho - \sigma$. In this case $E_i(\rho - \sigma) = \lambda_i |i\rangle\langle i|$, where $|i\rangle$ is the ith eigenvector and λ_i is the corresponding eigenvalue.

$$\begin{aligned} D(p,q) &= \frac{1}{2}\sum_i |\text{Tr}[E_i(\rho-\sigma)]| \\ &= \frac{1}{2}\sum_i |\text{Tr}\lambda_i|i\rangle\langle i|| = \frac{1}{2}\sum_i |\lambda_i| = \frac{1}{2}\text{Tr}|\rho-\sigma| = D(\rho,\sigma). \end{aligned} \tag{1.91}$$

Thus $D(p,q)$ is never greater than $D(\rho,\sigma)$. It follows that $D(\rho,\sigma)$ is equal to the maximum over all POVMs of the Kolmogorov distance of the probabilities resulting from this measurement.

1.2.6.3 Distinguishing superoperators The Uhlman fidelity and the trace-norm distance are useful for comparing quantum states, but how about quantum transformations, or superoperators? Consider two CP maps Φ and Ψ. The *maximum output fidelity* of Φ and Ψ is defined as

$$F_{\max}(\Phi,\Psi) = \max\{F[\Phi(\rho),\Psi(\sigma)] : \rho,\sigma \in \mathscr{B}(\mathscr{H}), \rho,\sigma \geq 0\}. \tag{1.92}$$

Note that in this definition we do not need to assume that the states ρ and σ are normalized, just that they are positive semidefinite and bounded.

We can also define a distance between superoperators. First we need a norm: for a superoperator $\Phi \in \mathscr{T}(\mathscr{H}_1, \mathscr{H}_2)$, its *trace norm* is defined as

$$\|\Phi\|_1 = \max\{\|\Phi(\rho)\|_1 : \rho \in \mathscr{L}(\mathscr{H}_1), \|\rho\|_1 = 1\}. \tag{1.93}$$

However, it turns out that in order to distinguish superoperators we should consider a more general situation, accounting for the possibility of entanglement with an auxiliary Hilbert space. Consider a state ρ acting on the doubled Hilbert space $\mathcal{H}_1 \otimes \mathcal{H}_1$. Suppose we extend Φ in a trivial manner, by tensoring it with the identity superoperator on the auxiliary copy of $\mathcal{H}_1$, i.e., we consider $\Phi \otimes I_{\mathscr{L}(\mathcal{H}_1)}$. With this in mind we now define the *diamond norm* of a superoperator $\Phi \in \mathscr{T}(\mathcal{H}_1, \mathcal{H}_2)$ as

$$\|\Phi\|_\diamond = \|\Phi \otimes I_{\mathscr{L}(\mathcal{H}_1)}\|_1 \tag{1.94a}$$
$$= \max\{\|\Phi \otimes I_{\mathscr{L}(\mathcal{H}_1)}(\rho)\|_1 : \rho \in \mathscr{L}(\mathcal{H}_1 \otimes \mathcal{H}_1), \|\rho\|_1 = 1\}. \tag{1.94b}$$

The diamond norm satisfies several nice properties. One of these is that it is multiplicative with respect to tensor products. Namely, let $\Phi \in \mathscr{T}(\mathcal{H}_1, \mathcal{H}_2)$ and $\Psi \in \mathscr{T}(\mathcal{H}_3, \mathcal{H}_4)$. Then

$$\|\Phi \otimes \Psi\|_\diamond = \|\Phi\|_\diamond \|\Psi\|_\diamond. \tag{1.95}$$

The diamond norm also serves to bound extensions of superoperators. Namely, if $\Phi \in \mathscr{T}(\mathcal{H}_1, \mathcal{H}_2)$ and $\mathcal{H}_3$ is a Hilbert space of arbitrary dimension, then

$$\|\Phi \otimes I_{\mathscr{L}(\mathcal{H}_3)}\|_1 \leq \|\Phi\|_\diamond, \tag{1.96}$$

with equality if (but not necessarily only if) $\dim(\mathcal{H}_3) \geq \dim(\mathcal{H}_1)$.

Moreover, the diamond norm can be used to quantify the distance between CP maps. The possibility of entanglement with an auxiliary space means that we should define the distance between CP maps as

$$\operatorname{dist}(\Phi, \Psi)$$
$$\equiv \sup\left\{\|(\Phi \otimes I_{\mathscr{L}(\mathcal{H}_2)})\rho - (\Psi \otimes I_{\mathscr{L}(\mathcal{H}_2)})\rho\|_1 : \rho \in \mathscr{S}(\mathcal{H}_1 \otimes \mathcal{H}_2), \mathcal{H}_2 = \mathbb{C}^n, n \in \mathbb{N}\right\}. \tag{1.97}$$

If this definition of distance seems cumbersome, you will be glad to know that it be replaced with the much more compact expression

$$\operatorname{dist}(\Phi, \Psi) = \|\Phi - \Psi\|_\diamond. \tag{1.98}$$

It is also possible to relate the diamond norm to the maximum output fidelity $F_{\max}$, as follows. Let $\rho \in \mathscr{L}(\mathcal{H}_1)$ and consider two linear operators whose range is a Hilbert space with a tensor product structure: $A, B \in \mathscr{L}[\mathcal{H}_1, \mathcal{H}_2 \otimes \mathcal{H}_3]$. Now, any superoperator $\Phi \in \mathscr{T}(\mathcal{H}_1, \mathcal{H}_2)$ can be written in a *Stinespring representation*:

$$\Phi(\rho) = \operatorname{Tr}_{\mathcal{H}_3}\left[A\rho B^\dagger\right] \in \mathscr{L}(\mathcal{H}_2). \tag{1.99}$$

Next, we associate superoperators $\Phi_A, \Phi_B \in \mathscr{T}(\mathcal{H}_1, \mathcal{H}_3)$ with A and B:

$$\Phi_A(\rho) = \operatorname{Tr}_{\mathcal{H}_2}\left[A\rho A^\dagger\right] \in \mathscr{L}(\mathcal{H}_3), \tag{1.100a}$$
$$\Phi_B(\rho) = \operatorname{Tr}_{\mathcal{H}_2}\left[B\rho B^\dagger\right] \in \mathscr{L}(\mathcal{H}_3). \tag{1.100b}$$

We then have the remarkable alternative characterization of the diamond norm as

$$\|\Phi\|_\diamond = F_{\max}(\Phi_A, \Phi_B). \tag{1.101}$$

This immediately implies, via the multiplicativity (1.95) of the diamond norm, that if $\Phi_i, \Psi_i \in \mathscr{T}(\mathscr{H}_i, \mathscr{H}_i')$, where $i = 1, 2$, are CP maps, then

$$F_{\max}(\Phi_1 \otimes \Phi_2, \Psi_1 \otimes \Psi_2) = F_{\max}(\Phi_1, \Psi_1) F_{\max}(\Phi_2, \Psi_2). \tag{1.102}$$

For proofs and additional details about the various norms and fidelities we discussed in this section, see the excellent lecture notes by Watrous [W04].

1.3 Master equations

The Schrödinger equation (1.23) is a first-order linear differential equation whose solution maps pure states to pure states. As we saw in Sections 1.2.4.1 and 1.2.4.2, a system interacting with an environment usually cannot be described in terms of a pure state. Similarly, it is not usually possible to give a simple description of the time evolution of the system alone, without reference to the state of the environment. However, as we have seen, provided the initial system–bath state is perfectly classical (has vanishing quantum discord), one can find an effective evolution equation for the system alone. Can such an evolution equation be the result of a first-order linear differential equation, like the Schrödinger equation? This is the subject of quantum master equations.

1.3.1 Time-local and non-Markovian master equations

Master equations propagate density matrices to density matrices, but not necessarily pure states to pure states. Therefore, master equations are quite different from the Schrödinger equation (1.23). To be numerically and analytically tractable, a master equation should also generally be time-local: it should give the evolution of the density operator solely in terms of its state at the present time, without retarded terms, i.e., it should be of the form

$$\frac{\partial \rho_S}{\partial t} = \mathscr{L}(t) \rho_S(t), \tag{1.103}$$

where $\mathscr{L}(t)$ is a time-local generator, which need not in general be Hermitian or even diagonalizable. In contrast, a time-nonlocal equation is of the form

$$\frac{\partial \rho_S}{\partial t} = \int_{t_0}^{t} \mathscr{K}(t, s) \rho_S(s) ds, \tag{1.104}$$

which includes a memory kernel $\mathscr{K}(t, s)$, and expresses a history dependence, in the sense that the rate of change of ρ at the present time t depends on its history all the way to the initial time t_0. In fact, the nonlocal integro-differential equation (1.104) can be shown to arise from Eq. (1.51) using the so-called Nakajima–Zwanzig projection operator technique, provided the initial system–bath state is a tensor product. The projection is simply the operation $\rho_{SB} \mapsto \mathrm{Tr}_B[\rho_{SB}] \otimes \rho_B \equiv \rho_S \otimes \rho_B$, where ρ_B is some fixed state of the bath (e.g., a thermal equilibrium state). However, Eq. (1.104) is difficult to solve due to its nonlocal nature. In contrast, the time-local equation (1.103) is an ordinary first-order operator valued differential equation and has the formal solution

$$\rho_S(t) = \Lambda(t, t_0) \rho_S(t_0), \tag{1.105}$$

where the "dynamical map" $\Lambda(t, t_0)$ satisfies the first-order differential equation

$$\frac{\partial}{\partial t}\Lambda(t, t_0) = \mathscr{L}(t)\Lambda(t, t_0), \quad \Lambda(t_0, t_0) = I \qquad (1.106)$$

and can therefore be expressed as a time-ordered integral, similar to the Hamiltonian case of Eq. (1.16):

$$\Lambda(t, t_0) = T_+ \exp\left(\int_{t_0}^{t} \mathscr{L}(s)ds\right), \qquad (1.107)$$

Note that $\Lambda(t, t_0)$ satisfies the *composition law*

$$\Lambda(t, t_0) = \Lambda(t, t')\Lambda(t', t_0), \quad t \geq t' \geq t_0. \qquad (1.108)$$

In contrast, while the solution to the time nonlocal Eq. (1.104) can also be written in the form (1.105), the corresponding dynamical map $\Lambda'(t, t_0)$, which obeys

$$\frac{\partial}{\partial t}\Lambda'(t, t_0) = \int_{t_0}^{t} \mathscr{K}(t, s)\Lambda'(s, t_0)ds, \quad \Lambda'(t_0, t_0) = I, \qquad (1.109)$$

does not satisfy the composition law (1.108). This fact is the very signature of *non-Markovianity*, and is due to the memory effect embedded in the kernel $\mathscr{K}(t, s)$. Conversely, it is the absence of memory that allows the dynamical map $\Lambda(t, t_0)$ to satisfy the composition law (1.108). When the generator $\mathscr{L}$ is time independent we have *Markovian dynamics*.

Let us now assume that the kernel is time-homogeneous, i.e., $\mathscr{K}(t, s) = \mathscr{K}(t - s)$. It is interesting to note that then, just as we saw in Section 1.2.5.2 that any POVM can be performed by allowing the system that is to be measured to interact with an ancilla and then doing a projective measurement on the ancilla, any non-Markovian evolution can be viewed as Markovian evolution on an extended Hilbert space, followed by a partial trace over the ancilla. Formally, if ρ_a denotes a fixed state of the ancilla and $\mathscr{L}$ denotes a time-independent generator on the joint system–ancilla Hilbert space, then it is always possible to obtain a given non-Markovian dynamical map via the prescription

$$\Lambda'(t, t_0)\rho_S(t_0) = \text{Tr}_a\left[e^{(t-t_0)\mathscr{L}}(\rho_S(t_0) \otimes \rho_a)\right]. \qquad (1.110)$$

Actually, another interesting observation follows from this prescription: since the right-hand side depends on time only via the *difference* $t - t_0$, then so must Λ', i.e., non-Markovian dynamics is time-homogeneous (but not time-local). It can be shown [CK10] that it follows that in fact Λ' also satisfies a first-order differential equation:

$$\frac{\partial}{\partial t}\Lambda'(t, t_0) = \mathscr{L}'(t - t_0)\Lambda'(t, t_0), \quad \Lambda'(t_0, t_0) = I. \qquad (1.111)$$

Note the similarity, and the difference, when compared to the time-local Eq. (1.106). The time-dependent generator $\mathscr{L}'(s)$ is defined, similarly to the Hamiltonian case (1.15), by the following logarithmic derivative of the dynamical map:

$$\mathscr{L}'(s) = \frac{\partial \Lambda'(s)}{\partial s}[\Lambda'(s)]^{-1}. \qquad (1.112)$$

Clearly, this prescription requires the existence of the inverse of $\Lambda'(s)$. Under a different invertibility assumption it is possible to convert the general Eq. (1.104) into a *time-local* equation using the so-called time-convolutionless projection operator technique. This is described in detail in Chapter 9 of [BP02], and leads to a systematic time-dependent perturbation theory for Λ', in the (presumed small) system–bath coupling constant.

When one is given the generator $\mathscr{L}'$ the solution to Eq. (1.111) is again the time-ordered integral

$$\Lambda'(t, t_0) = T_+ \exp\left(\int_0^{t-t_0} \mathscr{L}'(s) ds\right). \quad (1.113)$$

As you can check, this dynamical map does not satisfy the composition law (1.108), so while it is time-homogeneous, it is non-Markovian.

From now on we will focus on the time-local case. A particularly simple case arises when $\mathscr{L}(t)$ is time independent. In this case the most general Markovian master equation is a master equation in Lindblad form [L76],

$$\frac{\partial \rho_S}{\partial t} = -i[H_S, \rho_S] + \sum_k \left(L_k \rho_S L_k^\dagger - \frac{1}{2} \{L_k^\dagger L_k, \rho_S\} \right) \equiv \mathscr{L} \rho_S, \quad (1.114)$$

where $\{A, B\} \equiv AB + BA$ (anticommutator), H_S is the Hamiltonian for the system alone, and the "Lindblad operators" $\{L_k\}$ describe the effects of the interaction with the environment when the latter is in some given initial state.

1.3.2 Derivation of the Lindblad equation

Let us derive the Lindblad equation (1.114). To this end let us first revisit the derivation of the Kraus operator sum representation (1.53). Suppose that a system and bath in an initial pure product state $|\psi_S\rangle \otimes |\varphi_B\rangle$ undergo some joint unitary transformation U:

$$|\Psi\rangle = |\psi_S\rangle \otimes |\varphi_B\rangle \mapsto |\Psi'\rangle = U(|\psi_S\rangle \otimes |\varphi_B\rangle). \quad (1.115)$$

How does the state of the system alone change? We can write any operator on $\mathscr{H}_S \otimes \mathscr{H}_B$ as a sum of product operators:

$$U = \sum_j S_j \otimes B_j. \quad (1.116)$$

The state of the system after the unitary transformation is

$$\begin{aligned}
\rho_S^{(1)} &= \mathrm{Tr}_B[U|\Psi\rangle\langle\Psi|U^\dagger] \\
&= \sum_{j,j'} \mathrm{Tr}_B[(S_j|\psi_S\rangle\langle\psi_S|S_{j'}^\dagger) \otimes (B_j|\varphi_B\rangle\langle\varphi_B|B_{j'}^\dagger)] \\
&= \sum_{j,j'} \langle\varphi_B|B_{j'}^\dagger B_j|\varphi_B\rangle S_j|\psi_S\rangle\langle\psi_S|S_{j'}^\dagger,
\end{aligned} \quad (1.117)$$

where in the last line we made use of Eq. (1.36). The Hermitian matrix M with matrix elements

$$M_{jj'} \equiv \langle\varphi_B|B_{j'}^\dagger B_j|\varphi_B\rangle \quad (1.118)$$

has a set of orthonormal eigenvectors $\vec{\mu}_k = \{\mu_{kj}\}$ with real eigenvalues λ_k such that

$$\sum_{j'} M_{jj'} \mu_{kj'} = \lambda_k \mu_{kj} \quad \Rightarrow \quad M_{jj'} = \sum_k \lambda_k \mu_{kj} \mu_{kj'}^*. \quad (1.119)$$

We define new (Kraus) operators A_k by

$$A_k \equiv \sqrt{\lambda_k} \sum_j \mu_{kj} S_j. \tag{1.120}$$

The expression (1.117) then simplifies to

$$\rho_S^{(1)} = \sum_k A_k |\psi\rangle\langle\psi| A_k^\dagger, \tag{1.121}$$

which is nothing but the Kraus operator sum representation we have already seen. Note, however, that this expression is quite like the outcome of a POVM [Eq. (1.69)], but without determining a particular measurement result; the system is left in a mixture of all possible outcomes. Also note that in the derivation above we can replace $M_{jj'} \equiv \langle\varphi_B|B_{j'}^\dagger B_j|\varphi_B\rangle$ by $M_{jj'} \equiv \text{Tr}[\rho_B B_{j'}^\dagger B_j]$ if the bath starts out in the mixed initial state ρ_B.

Suppose now that the system interacts with the bath in the same way, with the bath again in the *same* initial state $|\varphi_B\rangle$. This assumption means that the bath resets at each discrete time step, i.e., it retains no memory of the interaction with the system between successive time steps. This is the key signature of Markovian dynamics, in agreement with the idea that the generator of the open system dynamics, which must depend on the state of the bath, is time independent. The system state will become

$$\rho_S^{(2)} = \sum_k A_k \rho_S^{(1)} A_k^\dagger = \sum_{k_1,k_2} A_{k_2} A_{k_1} |\psi\rangle\langle\psi| A_{k_1}^\dagger A_{k_2}^\dagger, \tag{1.122}$$

i.e, the two Kraus maps are composed. After interacting successively with the bath n times in this manner, the system state is the result of n map compositions:

$$\rho_S^{(n)} = \sum_{k_1,\ldots,k_n} A_{k_n} \cdots A_{k_1} |\psi\rangle\langle\psi| A_{k_1}^\dagger \cdots A_{k_n}^\dagger. \tag{1.123}$$

This evolution is a type of discrete master equation. We see clearly that an initially pure state will in general evolve into a mixed state. Depending on the A_k, it may or may not converge to a unique final state. Notice the similarity to the discrete evolution equation (1.24).

Let us consider the special case when the unitary operator U is close to the identity,

$$U = \exp(-i\epsilon H_I), \quad H_I = \sum_j S_j \otimes B_j, \tag{1.124}$$

where $\epsilon \ll 1$ and the operators S_j and B_j are Hermitian and bounded. [Note that these operators S_j and B_j are different from the S and B operators in Eq. (1.116).] Thus, we are assuming a *weak-coupling limit* between the system and the bath. By expanding to second order in ϵ, we see that the new density matrix for the system is

$$\begin{aligned}\rho_S' &= \text{Tr}_B[U|\Psi_S\rangle\langle\Psi|U^\dagger] \\ &\approx |\psi_S\rangle\langle\psi_S| - i\epsilon[H_L, |\psi_S\rangle\langle\psi_S|] \\ &\quad + \frac{\epsilon^2}{2} \sum_{jj'} \langle\varphi_B|B_{j'}B_j|\varphi_B\rangle \left(2S_j|\psi_S\rangle\langle\psi_S|S_{j'} - \{S_{j'}S_j, |\psi_S\rangle\langle\psi_S|\}\right),\end{aligned} \tag{1.125}$$

where

$$H_L \equiv \sum_j \langle\varphi_B|B_j|\varphi_B\rangle S_j \tag{1.126}$$

is called the Lamb shift Hamiltonian. It is a system–bath interaction-induced system-only Hamiltonian which gets added to the original system Hamiltonian and "renormalizes" it. Let us make the simplifying assumption that the first-order term vanishes (it is always possible to do this by applying a unitary transformation on the system, generated by H_L). Let the duration of the interaction be δt. We can again define a matrix M with matrix elements as in Eq. (1.118), with orthonormal eigenvectors $\vec{\mu}_k = \{\mu_{kj}\}$ and real eigenvalues λ_k as in Eq. (1.119), and define Lindblad operators

$$L_k = \frac{\epsilon}{\sqrt{\delta t}} A_k, \tag{1.127}$$

with the Kraus operators A_k given in Eq. (1.120) above. In terms of these Lindblad operators, Eq. (1.125) can be rewritten as

$$\frac{\rho' - \rho}{\delta t} = \sum_k \left[L_k \rho L_k^\dagger - (1/2) L_k^\dagger L_k \rho - (1/2) \rho L_k^\dagger L_k \right], \tag{1.128}$$

which has exactly the same form as Eq. (1.114) with a vanishing system Hamiltonian.

1.3.3 Interpolating between Markovian and non-Markovian dynamics

We have just seen how, starting from a joint unitary evolution of system and bath, we can derive the exact, generally non-Markovian operator sum representation, or by performing a weak-coupling approximation we can derive the Markovian Lindblad equation. Can we somehow interpolate between these two cases? In this subsection we will show how this can be done by focusing on the two cases as representing different numbers of measurements of the bath state.

1.3.3.1 Kraus operator sum representation from measuring the bath once Imagine the bath acting as a probe coupled to the system at $t = 0$, via some interaction Hamiltonian H_I, which induces the joint unitary evolution $\rho_{SB}(t) = U(t)\rho_{SB}(0)U^\dagger(t)$. To study the state of the system a *single* projective measurement is performed on the *bath* at time t, with a complete set of projection operators $|k\rangle\langle k|$, where $\mathcal{H}_B = \text{Span}\{|k\rangle\}$. The measurement yields the result k and collapses the state of the bath to the corresponding eigenstate $|k\rangle$. This happens with probability $p_k = \text{Tr}_S[\langle k|\rho_{SB}(t)|k\rangle]$, and the system density matrix reduces to

$$\rho_S^{(k)}(t) = \langle k|\rho_{SB}(t)|k\rangle / p_k =: A_k^\dagger \rho_S(0) A_k / p_k, \tag{1.129}$$

where A_k are the Kraus operators. If we repeat this process for an identical ensemble initially prepared in the state $\rho_{SB}(0)$ the system density matrix becomes

$$\rho_S(t) = \sum_k p_k \rho_S^{(k)}(t) = \text{Tr}_B[U(t)\rho_{SB}(0)U^\dagger(t)], \tag{1.130}$$

which is just the usual trace-out-the-bath prescription, thus affirming the validity of this bath-measurement interpretation of open system dynamics. This procedure is illustrated in Fig. 1.1a.

1.3.3.2 Lindblad equation from measuring the bath quasi-continuously In deriving Eq. (1.128) we went from the Kraus representation to the Lindblad equation. Let us now go in the opposite direction and see how also the Lindblad equation can be given a measurement interpretation

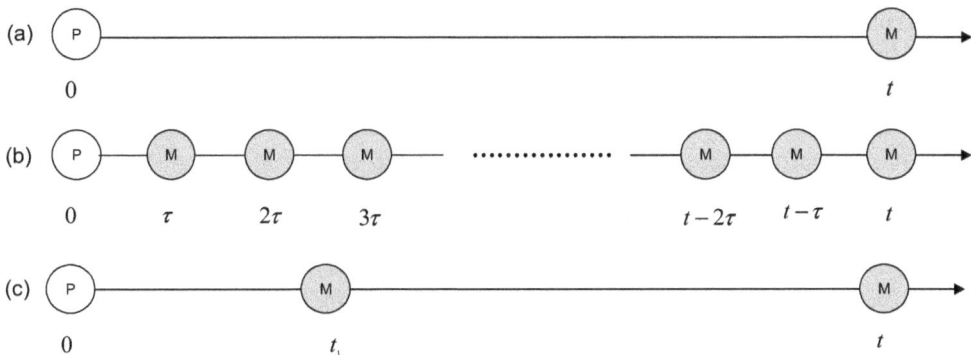

Fig. 1.1. Measurement approach to open system dynamics. P is preparation, M is measurement, time proceeds from left to right. (a) Exact Kraus operator sum representation, (b) Markovian approximation, (c) single-shot measurement. Reprinted with permission from [SL05]. Copyright 2005, American Physical Society.

using the Kraus representation. Setting $H_S = 0$ for simplicity and expanding Eq. (1.114) to first order in the short time interval δt yields

$$\rho_S(t+\delta t) = L_0 \rho_S(t) L_0^\dagger + \sum_{k>0} (\sqrt{\delta t} L_k) \rho_S(t) (\sqrt{\delta t} L_k)^\dagger, \quad (1.131)$$

where

$$L_0 \equiv I - \frac{\delta t}{2} \sum_{k>0} L_k^\dagger L_k. \quad (1.132)$$

To the same order we also have the normalization condition

$$L_0^\dagger L_0 + \sum_{k>0} (\sqrt{\delta t} L_k)^\dagger (\sqrt{\delta t} L_k) = I. \quad (1.133)$$

Thus, provided we identify L_0 and $\sqrt{\delta t} L_k$ as Kraus operators – in agreement with Eq. (1.127), with ϵ absorbed into the L_k – the Lindblad equation has been recast as an operator sum representation, but only to first order in δt, which we can think of as an appropriate "coarse-graining" time scale for which the Markovian approximation is valid [LBW01]. Recall that we just recognized in Section 1.3.3.1 that the Kraus operator index k can be interpreted as labeling the different measurement outcomes of the bath state. Thus the Kraus operators L_0 and $\sqrt{\delta t} L_k$ in the Lindblad equation correspond similarly to different measurement outcomes of the bath. This measurement process is repeated quasi-continuously as time advances in infinitesimal δt steps – see Fig. 1.1b. In order for the same set of Kraus operators to arise after each measurement we need the bath state to "reset" to the same state every time, just as we saw when we derived Eq. (1.122). In other words, the bath has no memory of the dynamics, which is precisely what we would expect from Markovian evolution. Alternatively, we can allow the bath to reset to a new random state each time, which would result in a time-local master equation as in Eq. (1.103).

It is also clear from the above description that the given measurement is a *weak* measurement for small δt, as described above, and the time evolution is a sequence of such weak measurements. What we have just described is the so-called quantum trajectories process, wherein

L_0 represents the "conditional evolution" and $\sqrt{\delta t}L_k$ represent the "jumps" [DCM92]. We will use this idea in the next section to justify the description of quantum noise (or "errors" in a quantum computer) as a sequence of stochastic events. The Lindblad equation can be recovered as the continuous limit of a sequence of weak measurements, where we average over all possible measurement outcomes. Even if we think of the environment as really "measuring" the system, in general the measurement outcomes are inaccessible, so this averaging is quite natural.

We have thus seen how a measurement picture leads to the two limits of (i) exact dynamics (via an evolution of the coupled system–bath followed by a single measurement at time t), and (ii) Markovian dynamics (via a series of measurements interrupting the joint evolution after each time interval δt). With this in mind it is now easy to see that by relaxing the many-measurements process one is led to a less restricted approximation than the Markovian one.

1.3.3.3 Post-Markovian master equation Let us return to the setting of the exact dynamics [Eq. (1.130)], and include one extra measurement in the time interval $[0, t]$. Thus we consider the following process: a probe (the bath) is coupled to the system at $t = 0$, they evolve jointly for a time t' ($0 \leq t' < t$) such that at t' the system state is $\Lambda(t')\rho(0)$, where $\Lambda(t')$ is a one-parameter dynamical map, at which moment the extra measurement is performed on the bath. Λ does not depend on t since the bath resets upon measurement. The system and the bath continue their coupled evolution between t' and t, upon which the final measurement is applied. This is illustrated in Fig. 1.1c. Since this intermediate measurement determines the system state $|\psi\rangle$ at t', after time $t - t'$ the system state will be $\rho(t) = \Lambda(t - t')\rho(t')$. To make further progress we may assume a Markovian form for the dynamical map: $\Lambda(t) = \exp(\mathscr{L}t)$. Here $\mathscr{L}$ can be interpreted as the Lindblad generator. These observations can be used to derive a "post-Markovian" master equation based on a probabilistic single-shot measurement process [SL05]:

$$\frac{\partial \rho}{\partial t} = \mathscr{L} \int_0^t dt'\, \mathscr{K}(t') \exp(\mathscr{L}t')\rho(t - t') = \mathscr{L}\mathscr{K}(t)\exp(\mathscr{L}t) * \rho(t), \qquad (1.134)$$

where $*$ denotes convolution and $\mathscr{K}$ plays the role of a memory kernel. Because of its relatively simple form, the post-Markovian master equation can be solved analytically using a Laplace transform, in terms of the left and right eigenvectors of $\mathscr{L}$, the so-called damping basis [BE93]. The post-Markovian master equation has some other pleasing properties, such as an explicit test for complete positivity. See [SL05] for a complete derivation and analysis.

1.4 Stochastic error models

A simple model of quantum noise, widely used in the design of quantum error-correcting codes, is the *stochastic error model*. In the simplest version of this model, there is a set of basic errors that can occur. Such an error is represented by multiplying the state of the system by one of a set of error operators $\{E_i\}$: $|\psi\rangle \to E_i|\psi\rangle$. (If the operators are not unitary, the state must also be renormalized.) Typically, these errors are assumed to occur as a Poisson process with some fixed (generally low) rate r_i, so that the probability of error E_i in a short interval Δt is $r_i \Delta t$. For example, in depolarizing noise, each qubit can be multiplied by X, Y, or Z, in the simplest case with the same rate $r/3$.

Introduction to decoherence and noise in open quantum systems 33

In this section we first show how such a stochastic error model can be equivalent, in discrete time, to some CPTP map. We then consider continuous time, where the noise can be described by Markovian master equations, and present an intuitive way to understand how the effects of this noise can be treated as a combination of stochastic "jumps," possibly interspersed with nonunitary "conditional" dynamics. We also consider the potential effects that are often neglected in simple stochastic error models with fixed error rates.

1.4.1 Completely positive map and generalized measurements

As we saw in Eq. (1.123), a quantum system undergoing decoherence can be described as evolving by a sequence of CP maps:

$$\rho \mapsto \sum_k A_k \rho A_k^\dagger \mapsto \cdots \mapsto \sum_{k_1,\ldots,k_n} A_{k_n} \cdots A_{k_1} \rho A_{k_1}^\dagger \cdots A_{k_n}^\dagger, \tag{1.135}$$

where the set of operators $\{A_k\}$ satisfy $\sum_k A_k^\dagger A_k = I$. This condition is just like the condition for a generalized measurement. What would happen if this actually *were* a measurement?

Suppose the state is initially pure, $\rho = |\psi\rangle\langle\psi|$. It follows from the general formalism for projective measurements given in Section 1.2.5.1 that after a measurement the state would become

$$|\psi\rangle \mapsto |\psi_k\rangle = A_k|\psi\rangle/\sqrt{p_k} \tag{1.136}$$

with probability

$$p_k = \text{Tr}\{A_k|\psi\rangle\langle\psi|A_k^\dagger\}. \tag{1.137}$$

If we consider the set of all outcome states with their probabilities $\{p_k, |\psi_k\rangle\}$ as an ensemble, it corresponds to the density matrix

$$\rho_1 = \sum_k p_k |\psi_k\rangle\langle\psi_k| = \sum_k A_k|\psi\rangle\langle\psi|A_k^\dagger, \tag{1.138}$$

which will in general be a mixed state.

Suppose the outcome of the first measurement is k_1, and we then perform a second measurement. Outcome k_2 has probability

$$p_{k_2|k_1} = \text{Tr}\{A_{k_2}|\psi_{k_1}\rangle\langle\psi_{k_1}|A_{k_2}^\dagger\} = \frac{1}{p_{k_1}}\text{Tr}\{A_{k_2}A_{k_1}|\psi\rangle\langle\psi|A_{k_1}^\dagger A_{k_2}^\dagger\}. \tag{1.139}$$

We can define the joint probability to be

$$p_{k_2 k_1} = p_{k_1} p_{k_2|k_1} = \text{Tr}\{A_{k_2} A_{k_1}|\psi\rangle\langle\psi|A_{k_1}^\dagger A_{k_2}^\dagger\}. \tag{1.140}$$

Then after two measurements with outcomes k_1 and k_2, the state is

$$|\psi_{k_2 k_1}\rangle = A_{k_2} A_{k_1}|\psi\rangle/\sqrt{p_{k_2 k_1}}. \tag{1.141}$$

Treating all these states as forming an ensemble, the corresponding density matrix is

$$\rho_2 = \sum_{k_1, k_2} A_{k_2} A_{k_1}|\psi\rangle\langle\psi|A_{k_1}^\dagger A_{k_2}^\dagger. \tag{1.142}$$

Similarly, after n steps of this process, we have n outcomes $k_1, \ldots, k_n$; the probability of the outcomes is

$$p_{k_n \ldots k_1} = \text{Tr}\{A_{k_n} \cdots A_{k_1} |\psi\rangle \langle \psi| A_{k_1}^\dagger \cdots A_{k_n}^\dagger\}, \tag{1.143}$$

and the resulting state is

$$|\psi_{k_n \ldots k_1}\rangle = A_{k_n} \cdots A_{k_1} |\psi\rangle / \sqrt{p_{k_n \ldots k_1}}. \tag{1.144}$$

This ensemble of states corresponds to a density matrix

$$\rho_n = \sum_{k_1, \ldots, k_n} A_{k_n} \cdots A_{k_1} |\psi\rangle \langle \psi| A_{k_1}^\dagger \cdots A_{k_2}^\dagger. \tag{1.145}$$

This means we can replace our density matrix evolution with a *pure state*, undergoing a random evolution resulting from repeated generalized measurements. By averaging over all possible outcomes with appropriate probabilities, the density matrix is reconstructed.

Such a random pure state evolution is called a *quantum trajectory*. If we rewrite the CP evolution as an average over trajectories, we say that we have *unraveled* the evolution.

1.4.2 Environmental measurements and conditional evolution

Suppose that the system qubit is in the initial state $|\psi\rangle = \alpha|0\rangle + \beta|1\rangle$ and it interacts with an environment qubit in state $|0\rangle$, so that a CNOT is performed from the system onto the environment qubit. The new joint state of the two qubits is

$$|\Psi'\rangle = \alpha|0\rangle_S \otimes |0\rangle_B + \beta|1\rangle_S \otimes |1\rangle_B. \tag{1.146}$$

If we trace out the environment qubit, the system is in the mixed state

$$\rho_S = |\alpha|^2 |0\rangle\langle 0| + |\beta|^2 |1\rangle\langle 1|. \tag{1.147}$$

This is a CPTP map. Like all such maps, it can be represented using different sets of Kraus operators. Two such sets are $\{|0\rangle\langle 0|, |1\rangle\langle 1|\}$, and $\{I/\sqrt{2}, Z/\sqrt{2}\}$.

Now suppose that we make a projective measurement on the *environment* qubit in the usual z basis. With probability $p_0 = |\alpha|^2$ ($p_1 = |\beta|^2$), we find this qubit in state $|0\rangle$ ($|1\rangle$), in which case the system qubit will *also* be projected into state $|0\rangle$ ($|1\rangle$). If the system interacts in the same way with more environment bits, which are also subsequently measured in the z basis, the same result will be obtained each time with probability 1. This scheme acts just like a projective measurement on the system – in other words, the measurement with operators $\{|0\rangle\langle 0|, |1\rangle\langle 1|\}$ that give one representation of the CPTP map.

This need not be the case in general. Suppose that instead of a CNOT, a SWAP gate is performed on the two qubits. In this case the new joint state after the interaction is

$$|\Psi'\rangle = |0\rangle_S \otimes (\alpha|0\rangle_B + \beta|1\rangle_B). \tag{1.148}$$

A subsequent z measurement on the environment yields 0 or 1 with the same probabilities as before, but now in both cases the system is left in state $|0\rangle$. Further interactions and measurements will always produce the result $|0\rangle$. Clearly, the choice of measurement on the environment qubit makes no difference to the state of the system – it will be $|0\rangle$ for any measurement result, or for no measurement at all. Note that U_{CNOT} can produce entanglement between two initially unentangled qubits, while U_{SWAP} cannot.

Returning to the case of the CNOT interaction, what happens if we vary our choice of measurement? Suppose that instead of measuring the environment qubit in the z basis, we measure it in the x basis $|x_\pm\rangle = (|0\rangle \pm |1\rangle)/\sqrt{2}$. In terms of this basis, we can rewrite the joint state after the interaction as

$$|\Psi'\rangle = \frac{1}{\sqrt{2}}(\alpha|0\rangle + \beta|1\rangle) \otimes |x_+\rangle + \frac{1}{\sqrt{2}}(\alpha|0\rangle - \beta|1\rangle) \otimes |x_-\rangle . \qquad (1.149)$$

The $+$ and $-$ results are equally likely; the $+$ result leaves the system state unchanged, while the $-$ flips the relative sign between the $|0\rangle$ and $|1\rangle$ terms. Aside from knowing whether a flip has occurred, this measurement result yields no information about the system state. In other words, we have done the generalized measurement with measurement operators $\{I/\sqrt{2}, Z/\sqrt{2}\}$, which also corresponds to the CPTP map described above.

If this state $|\psi\rangle$ is initially unknown, it will remain unknown after the interaction and measurement. By changing the choice of measurement, one can go from learning the exact state of the system to learning nothing about it whatsoever.

One important thing to note is that, whatever measurement is performed and whatever outcome is obtained, the system and environment qubit are afterwards left in a product state. No entanglement remains. Thus, any subsequent evolution of the environment qubit will have no further effect on the state of the system. Because of this, it is unnecessary to keep track of the environment's state to describe the system. This is an approximation, arising from the assumption of projective measurements; in real experiments, some residual entanglement with the environment almost certainly exists which the experimenter is unable to measure or control. It does, however, enormously simplify the description of the system's evolution. We will later relax this assumption, but for now we make use of it.

1.4.3 Stochastic Schrödinger equations with jumps

Far more complicated dynamics result if we replace the strong two-qubit gates of Section 1.4.2 with weak interactions, such as $U_{\text{CNOT}}(\theta)$ or $U_{\text{SWAP}}(\theta)$ from Eqs. (1.30) and (1.31), for $\theta \ll 1$. In this case the measurements on the environment qubits correspond to weak measurements on the system, and the system state evolves unpredictably according to the outcome of the measurements. Weak interaction with the environment is the norm for most laboratory systems; indeed, it is only the weakness of this interaction that justifies the division into system and environment in the first place, and enables one to treat the system as isolated to a first approximation, with the effects of the environment as a perturbation.

This problem is simple enough that we can analyze it for a general weak interaction. Suppose that the system and environment qubits are initially in the state $|\Psi\rangle = |\psi\rangle_S \otimes |0\rangle_B$. Let the interaction be a two-qubit unitary transformation of the form

$$U = \exp\left(-i\theta \sum_j A_j \otimes B_j\right), \qquad (1.150)$$

where $\theta \ll 1$, and $\text{Tr}[A_j^\dagger A_j] \sim \text{Tr}[B_j^\dagger B_j] \sim O(1)$, so that we can expand the exponential in powers of θ. The new state of the system and environment is

$$U|\Psi\rangle = |\psi\rangle_S \otimes |0\rangle_B - i\theta \sum_j A_j |\psi\rangle_S \otimes B_j |0\rangle_B$$

$$- \frac{\theta^2}{2} \sum_{jj'} A_{j'} A_j |\psi\rangle_S \otimes B_{j'} B_j |0\rangle_B + O(\theta^3). \tag{1.151}$$

Just as in Section 1.3.2, we define the matrix M with matrix elements $M_{jj'} = \langle 0|B_{j'} B_j|0\rangle$ with eigenvectors $\vec{\mu}_k$ and real eigenvalues λ_k, and use it to find a new set of operators

$$L_k = \sqrt{\frac{\theta^2 \lambda_k}{\delta t}} \sum_j \mu_{kj} A_j, \tag{1.152}$$

where δt is the time between interactions with successive environment qubits. Because we can decompose the matrix M in terms of two vectors u and v,

$$M_{jj'} = \langle 0|B_{j'}|0\rangle \langle 0|B_j|0\rangle + \langle 0|B_{j'}|1\rangle \langle 1|B_j|0\rangle = u_{j'} u_j + v_{j'}^* v_j, \tag{1.153}$$

it has at most two nonvanishing eigenvalues. In general, one of these will give an extra effective term in the system Hamiltonian. For simplicity in the present derivation, we can make the additional assumption, as in Section 1.3.2, that

$$\sum_j A_j \langle 0|B_j|0\rangle \equiv \sum_j A_j u_j = 0. \tag{1.154}$$

This leaves only a single nonvanishing Lindblad operator,

$$L = \sqrt{\frac{\theta^2}{\delta t}} \sum_j A_j \langle 1|B_j|0\rangle, \tag{1.155}$$

and no effective Hamiltonian term. In terms of L, the joint state of the system and environment is

$$U|\Psi\rangle = (I - (1/2)L^\dagger L \delta t)|\psi\rangle_S \otimes |0\rangle_B - i\sqrt{\delta t} L |\psi\rangle_S \otimes |1\rangle_B + O(\theta^3). \tag{1.156}$$

The time dependence of $L \sim \theta/\sqrt{\delta t}$ may seem a bit strange at first. It is chosen so that the master equation (1.128) properly approximates a time derivative. This does, however, imply a somewhat odd scaling. The strength of the interaction between the system and environment qubits can be parametrized by $\theta/\delta t$. This in turn implies that if the two bits interacted for only half as long, the Lindblad operators would have to be redefined, $L \to L/\sqrt{2}$. In the limit $\delta t \to 0$, the right-hand side of Eq. (1.128) would vanish, and the system would not evolve. This is actually true, and is an example of the so-called quantum Zeno effect [MS97]. This is a characteristic non-Markovian effect; similar effects are exploited in dynamical decoupling, as we will see later in the book.

Once the interaction has taken place, we measure the environment qubit in the $\{|0\rangle, |1\rangle\}$ basis. Using the assumption (1.154), the probabilities $p_{0,1}$ of the outcomes are

$$\begin{aligned} p_0 &= 1 - \theta^2 \sum_{jj'} \langle\psi|A_{j'}A_j|\psi\rangle\langle 0|B_{j'}|1\rangle\langle 1|B_j|0\rangle \\ &= 1 - \langle\psi|L^\dagger L|\psi\rangle\delta t \\ &= 1 - p_1 \, . \end{aligned} \qquad (1.157)$$

The change in the *system* state after a measurement outcome 0 on the *environment* is

$$\delta|\psi\rangle = |\psi'\rangle - |\psi\rangle = -\frac{1}{2}\left(L^\dagger L - \langle L^\dagger L\rangle\right)|\psi\rangle\delta t \, . \qquad (1.158)$$

Because the result 0 is highly probable, as the system interacts with successive environment bits in initial state $|0\rangle_B$, most of the time the system state will evolve according to the nonlinear (and nonunitary) continuous Eq. (1.158).

Occasionally, however, a measurement result of 1 will be obtained. In this case, the state of the system can change dramatically:

$$|\psi\rangle \mapsto |\psi'\rangle = \frac{L|\psi\rangle}{\sqrt{\langle L^\dagger L\rangle}} \, . \qquad (1.159)$$

Because these changes are large but rare, they are usually referred to as *quantum jumps*.

These two different evolutions – continuous and deterministic (after a measurement result 0) or discontinuous and random (after a measurement result 1) – can be combined into a single *stochastic Schrödinger equation*:

$$\begin{aligned} \delta|\psi\rangle = |\psi'\rangle - |\psi\rangle &= -\frac{1}{2}\left(L^\dagger L - \langle L^\dagger L\rangle\right)|\psi\rangle\delta t \\ &+ \left(\frac{L}{\sqrt{\langle L^\dagger L\rangle}} - I\right)|\psi\rangle\delta N \, , \end{aligned} \qquad (1.160)$$

where δN is a *stochastic variable* that is usually 0, but has a probability $p_1 = \langle\psi|L^\dagger L|\psi\rangle\delta t$ of being 1 during a given time step δt. The values of δN obviously represent measurement outcomes; when the environment is measured to be in state $|0\rangle$, the variable is $\delta N = 0$; when the measurement outcome is 1, $\delta N = 1$. Equation (1.160) combines the two kinds of system evolution into a single equation. Most of the time the δN term vanishes, and the system state evolves according to the deterministic nonlinear equation (1.158); however, when $\delta N = 1$, the δN terms completely dominate, and the system state changes abruptly according to Eq. (1.159). Generically, Eq. (1.160) also includes a Hamiltonian term, but this was eliminated by the assumption (1.154).

We can summarize the behavior of the stochastic variable by giving equations for its ensemble mean:

$$(\delta N)^2 = \delta N \, , \quad M[\delta N] = \langle L^\dagger L\rangle\delta t \, , \qquad (1.161)$$

where $M[\cdot]$ represents the ensemble mean over all possible measurement outcomes, and $\langle\cdot\rangle$ is the quantum expectation value in the state $|\psi\rangle$.

This quantum jump evolution is similar to the simple stochastic error model described at the beginning of this section, but differs in a couple of important respects. First, the rate of jumps (or "errors") is not a constant, but depends on the state of the system through the factor $\langle L^\dagger L \rangle$. So the jump process is generally not exactly a Poisson process. Second, the continuous (but nonunitary) term $(L^\dagger L - \langle L^\dagger L \rangle)|\psi\rangle$ is missing in the simple stochastic error model.

We can, however, recover the simple model for one important special case: if $L^\dagger L \propto I$, then the nonunitary term cancels out, and the rate of jumps becomes a constant. An obvious example would be where L is proportional to a Pauli operator. This is true for several common channel models, such as the bit-flip, phase-flip, and depolarizing channels. However, it is not true for some other important models, such as the amplitude damping channel.

The preceding analysis is rather abstract. Let us see how this works with a particular choice of two-qubit interaction. Suppose that the system interacts with a succession of environment qubits, via the transformation $Z_S(\theta)U_{\text{CNOT}}(\theta)$, where $Z_S(\theta) = \exp(-i\theta\sigma_z/2)_S \otimes I_B$ and $\theta \ll 1$; after each interaction the environment bits are measured in the z basis. If the environment bits all begin in state $|0\rangle$, and the system is in state (1.2) after interacting with the first environment qubit, the two bits will be in the joint state

$$|\Psi'\rangle = \alpha|00\rangle + \beta\cos\theta|10\rangle - i\beta\sin\theta|11\rangle , \tag{1.162}$$

modulo an overall phase. This state is entangled, with entropy of entanglement $S_B \approx -x\log_2 x$, where $x = |\alpha\beta|^2\theta^2$. If the environment is then measured in the z basis, it will be found in state $|0\rangle$ with probability $p_0 = |\alpha|^2 + |\beta|^2\cos^2\theta \approx 1 - |\beta|^2\theta^2$ and in state $|1\rangle$ only with the very small probability $p_1 = |\beta|^2\theta^2$. After a result of 0, the system will then be in the new state

$$\begin{aligned}|\psi_0\rangle &= (\alpha|0\rangle + \beta\cos\theta|1\rangle)/\sqrt{p_0} \\ &\approx \alpha(1 + |\beta|^2\theta^2/2)|0\rangle + \beta(1 - |\alpha|^2\theta^2/2)|1\rangle ,\end{aligned} \tag{1.163}$$

where we have kept terms up to second order in θ. The amplitude of $|0\rangle$ increases relative to $|1\rangle$. On the other hand, if the outcome 1 occurs, the system will be left in state $|1\rangle$. Further interactions and measurements will not alter this. We see that this combination of interactions and measurements is equivalent to a weak measurement on the system, of the first type discussed in Section 1.2.5.2. This bears out the fact that all POVMs are equivalent to an interaction with an ancillary system followed by a projective measurement, a fact known as Neumark's theorem.

Suppose that the system interacts successively with n environment qubits, each of which is afterwards measured to be in state $|0\rangle$. The state of the system is $|\psi_n\rangle = \alpha_n|0\rangle + \beta_n|1\rangle$, where

$$\frac{\beta_n}{\alpha_n} = \left(1 - \frac{\theta^2}{2}\right)^n \frac{\beta}{\alpha} \approx \exp(-n\theta^2/2)\frac{\beta}{\alpha} , \tag{1.164}$$

which, together with the normalization condition $|\alpha_n|^2 + |\beta_n|^2 = 1$, implies that

$$|\alpha_n|^2 = \frac{|\alpha|^2}{|\alpha|^2 + |\beta|^2\exp(-n\theta^2)}, \tag{1.165a}$$

$$|\beta_n|^2 = \frac{|\beta|^2\exp(-n\theta^2)}{|\alpha|^2 + |\beta|^2\exp(-n\theta^2)} . \tag{1.165b}$$

We see that $|\alpha_n|^2 \to 1$, which is what one would expect, because after many measurement outcomes without a single 1 result, one would estimate that the $|1\rangle$ component of the state must be very small. The probability of measuring 0 at the nth step, conditioned on having observed 0s at all previous steps, is $p_0(n) \approx 1 - |\beta_n|^2\theta^2 \to 1$.

The probability of at some point getting a result 1 and leaving the system in state $|1\rangle$ is therefore $1 - |\alpha|^2 = |\beta|^2$. The effect, after many weak measurements, is exactly the same as a single strong measurement in the z basis, with the same probabilities. (Note that, as discussed above, we do not need to consider *actual* measurements. This stochastic model is equivalent to the master equation when we average over all possible trajectories, or noise realizations, as we will show explicitly below.)

This evolution is described exactly by Eq. (1.160) with the Lindblad operator

$$L = \sqrt{\theta^2/\delta t}|1\rangle\langle 1| \,. \tag{1.166}$$

In this case, the evolution equation simplifies to

$$|\psi'\rangle - |\psi\rangle = -\frac{\theta^2}{2}\left(|1\rangle\langle 1| - |\langle 1|\psi\rangle|^2\right)|\psi\rangle + \left(\frac{|1\rangle\langle 1|}{|\langle 1|\psi\rangle|} - I\right)|\psi\rangle\delta N \,, \tag{1.167}$$

where $M[\delta N] = \theta^2|\langle 1|\psi\rangle|^2 = \theta^2|\beta|^2 = p_1$. Equation (1.160) may seem unnecessarily complicated. After all, L is Hermitian, so the distinction between L and $L^\dagger$ is unimportant; also, L is proportional to the projector $|1\rangle\langle 1|$, which simplifies Eq. (1.160) considerably. However, many other systems exist in which L is not so simple, and which still obey Eq. (1.160).

We can readily find such an alternative system. Suppose that $U_{\text{SWAP}}(\theta)$ is used instead of $U_{\text{CNOT}}(\theta)$ in the two-qubit interaction. The system and environment bits after the interaction are in the state

$$|\Psi'\rangle = \alpha|00\rangle + \beta\cos\theta|10\rangle - i\beta\sin\theta|01\rangle \,. \tag{1.168}$$

(Note that unlike the *strong* SWAP gate U_{SWAP}, the weak $U_{\text{SWAP}}(\theta)$ *can* produce entanglement.) From Eq. (1.168) we see that, just as in the CNOT case, a measurement of 0 on the environment leaves the system in a slightly altered state – in fact, exactly the same state that is produced by a 0 result in the CNOT case. The probabilities p_0 and p_1 for the two outcomes are identical to those for the CNOT case, $p_0 = 1 - p_1 \approx 1 - |\beta|^2\theta^2$. However, unlike the CNOT case, a measurement outcome 1 will put the system into the state $|0\rangle$ rather than $|1\rangle$.

This process is essentially a simplified model of spontaneous decay. Suppose that $|0\rangle$ is the ground state and $|1\rangle$ is the excited state. Initially, the system is in a superposition of these two energy levels. At each time step, there is a chance for the system to emit a quantum of energy into the environment. If it does, the system drops immediately into the ground state $|0\rangle$, while the environment goes into the excited state $|1\rangle$; if not, we revise our estimate of the system state, making it more probable that the system is unexcited. This kind of evolution is similar to quantum jumps in quantum optics, in which a photodetector outside a high-Q cavity clicks if a photon escapes, and gives no output if it sees no photon [C93, GZ04]. A measurement result of 1 on the environment corresponds to the photodetector click.

This model obeys exactly the same stochastic Schrödinger equation (1.160), but with the Lindblad operator $L = \sqrt{\theta^2/\delta t}|0\rangle\langle 1|$. Because this operator is *not* Hermitian, the distinction between $L^\dagger$ and L is important in this case. We compare these two stochastic evolutions in Figs. 1.2a and 1.2b. In both cases, the coefficient $|\beta|^2$ decays steadily away; however, in some trajectories the state suddenly jumps either to $|1\rangle$ or $|0\rangle$, while others continue to decay smoothly towards $|0\rangle$.

What happens (as is usually the case in practice) if we don't actually measure the environment bits? In this case the system qubit evolves into a mixed state that is the average over all

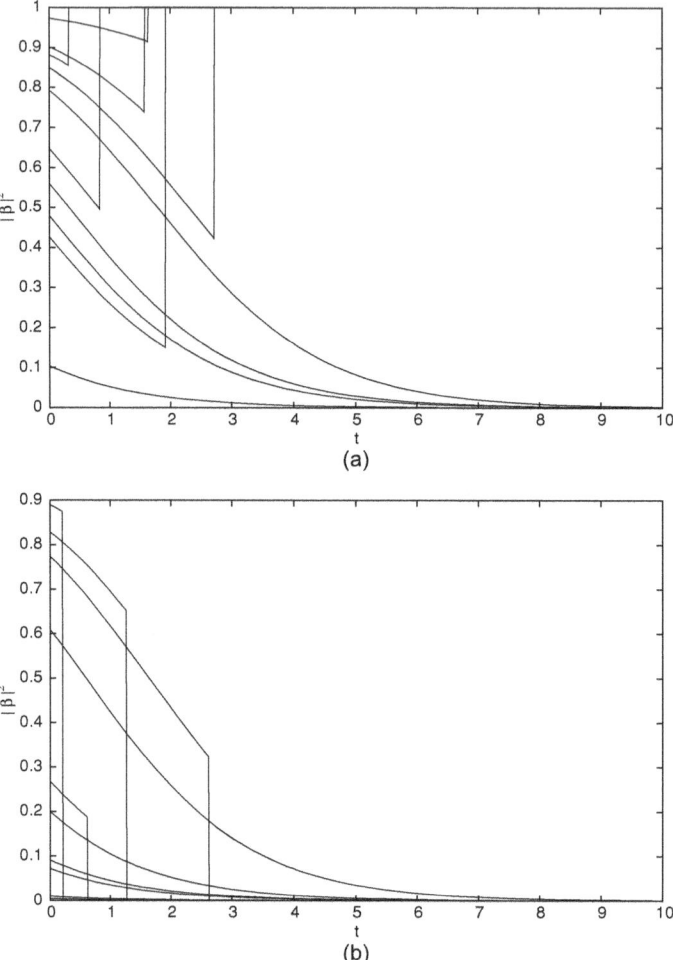

Fig. 1.2. Plot of the squared amplitude $|\beta|^2$ for the system to be in state $|1\rangle$ versus the time for a variety of quantum jump trajectories for different initial states. Part (a) assumes an interaction $\hat{U}_{\mathrm{CNOT}}$ with the environment, while (b) assumes $\hat{U}_{\mathrm{SWAP}}$. In both cases, $|\beta|^2$ decays towards zero. Some trajectories just decay smoothly to 0, while others undergo a jump. In (a) this jump is to 1, while in (b) they jump to 0, as in spontaneous emission. Note that the decays of $|\beta|^2$ are not simple exponentials due to the nonlinearity of the trajectory equations. As $|\beta|^2 \to 0$, however, the decay approaches an exponential. Reprinted with permission from [B02b]. Copyright 2002, American Association of Physics Teachers.

possible measurement outcomes. If we denote by ξ the n outcomes after n steps of the evolution, $p(\xi)$ the probability of those outcomes, and $|\psi_\xi\rangle$ the state conditioned on those outcomes (that is, the solution of the appropriate stochastic Schrödinger equation), then the system state will be a density operator

$$\rho^{(n)} = \sum_\xi |\psi_\xi\rangle p(\xi) \langle\psi_\xi| \,. \tag{1.169}$$

If we start with a state ρ, allow it to interact with exactly one environment qubit, and average over the possible outcomes 0 and 1 with their correct probabilities, we will find that the new state

ρ' is

$$\rho' = \sum_{k=0}^{1} A_k \rho A_k^\dagger , \qquad (1.170)$$

where for U_{CNOT}

$$A_0 = |0\rangle\langle 0| + \cos\theta |1\rangle\langle 1|, \qquad (1.171\text{a})$$
$$A_1 = \sin\theta |1\rangle\langle 1| . \qquad (1.171\text{b})$$

In the limit of small $\theta \ll 1$, this evolution gives an approximate Lindblad master equation:

$$\frac{\rho' - \rho}{\delta t} = L\rho L^\dagger - \frac{1}{2} L^\dagger L \rho - \frac{1}{2} \rho L^\dagger L . \qquad (1.172)$$

The Lindblad operator L is the same as in the jump equation for the CNOT interaction, Eq. (1.166). Similarly, for the SWAP interaction, the evolution equations are identical, but with $A_1 = \sin\theta |0\rangle\langle 1|$ and $L = \sqrt{\theta^2/\delta t}|0\rangle\langle 1|$.

We should note that, unlike the stochastic trajectory evolution, the evolution of the density matrix is perfectly deterministic, with no sign of jumps or other discontinuities. This is generally true; the jumps appear only if measurements are performed. However, this averaged master equation description and the stochastic description in terms of jumps are completely equivalent to each other. So error correction or suppression techniques that work against the errors arising in the jump equation will also work in the master equation picture. Because the stochastic error model is similar in spirit to the type of stochastic errors commonly assumed in classical error correction, it is helpful in developing intuition for designing schemes for quantum error correction.

This is a highly simplified and schematic picture of how stochastic Schrödinger equations can arise from system/environment interactions, conditioned on a particular choice of continuous environmental measurement. The actual measurement itself need not be done – the average behavior of the stochastic equation is the same as the master equation – but can be a useful guide to understanding what kinds of "errors" can arise in a quantum system. We will now look at the other main source of error in quantum computers, which is imprecision in control.

Exactly this kind of evolution can also be derived for realistic systems, at least in the limit of a Markovian approximation (see the previous section). Typically, such equations include Hamiltonian as well as nonunitary evolution terms, and can describe systems and environments in mixed as well as pure states, i.e., stochastic master equations.

1.4.4 Noisy gates

Environmental decoherence can lead to errors in quantum information processing; but it is not the *only* source of error. Another problem is imprecision in carrying out quantum gates.

Here is a simple example. Suppose we wish to carry out the one-qubit gate $\hat{R}_z(\theta)$. We could do this (for example) by turning on a magnetic field in the Z direction with a particular strength for a particular length of time.

Neither the field strength nor the timing will be absolutely precise. This means that the *actual* angle of rotation θ' will in general be a random variable, (hopefully) centered on the desired value

of θ:
$$\theta' \equiv \theta + \delta\theta, \tag{1.173}$$
where we assume $\delta\theta$ is a Gaussian random variable:
$$M[\delta\theta] = 0, \quad M[\delta\theta^2] = \epsilon \ll 1. \tag{1.174}$$

In place of the desired gate $\hat{R}_z(\theta)$ we have used $\hat{R}_z(\theta + \delta\theta) = \hat{R}_z(\delta\theta)\hat{R}_z(\theta)$. We do not know the value of $\delta\theta$; so the best we can do is describe the state by a density operator
$$\int d(\delta\theta)\, p(\delta\theta) \hat{R}_z(\delta\theta) \hat{R}_z(\theta) |\psi\rangle\langle\psi| \hat{R}_z^\dagger(\theta) \hat{R}_z^\dagger(\delta\theta). \tag{1.175}$$

This may look like an infinite ensemble; but, in fact, this integral can be replaced by the finite quantum operation
$$|\psi\rangle\langle\psi| \mapsto (1 - \epsilon) \hat{R}_z(\theta) |\psi\rangle\langle\psi| \hat{R}_z^\dagger(\theta) + \epsilon Z \hat{R}_z(\theta) |\psi\rangle\langle\psi| \hat{R}_z^\dagger(\theta) Z. \tag{1.176}$$

We can describe the imperfect gate as a *perfect* gate followed by a *weak* (i.e., near the identity) CP map.

What if there are environmental interactions that go on during the gate, as well as imprecision? We can use the same trick in that case as well; instead of the desired transformation
$$|\psi\rangle\langle\psi| \mapsto U|\psi\rangle\langle\psi|U^\dagger, \tag{1.177}$$
we instead get some transformation
$$|\psi\rangle\langle\psi| \mapsto \sum_k A_k U |\psi\rangle\langle\psi| U^\dagger A_k^\dagger, \tag{1.178}$$
where the particular CP transformation will (in general) depend on which unitary U we were trying to perform. If the decoherence and imprecision are both small, then this CP transformation will be weak.

If we were attempting a series of gates
$$|\psi\rangle\langle\psi| \mapsto U_n \cdots U_1 |\psi\rangle\langle\psi| U_1^\dagger \cdots U_n^\dagger, \tag{1.179}$$
we would instead get the state
$$\sum_{k_1,\ldots,k_n} A_{n,k_n} U_n \cdots A_{1,k_1} U_1 |\psi\rangle\langle\psi| U_1^\dagger A_{1,k_1}^\dagger \cdots U_n^\dagger A_{n,k_n}^\dagger. \tag{1.180}$$

Each unitary transformation is followed by a CP map; the particular map depends on which unitary was performed.

This equivalence between a continuous unitary noise process (due to imprecision in implementing gates) and a discrete error process is called *discretization of errors*, and is one of the key insights that makes quantum error correction possible. We do not have to separately correct a continuum of possible errors, but only a discrete, finite set of errors. This automatically gives us the ability to correct all linear combinations of those errors as well. This situation is in sharp contrast to *analog* computation, where the use of continuous physical quantities (not just continuous states) makes scalable error correction impossible.

Moreover, because both environmental interactions and imprecision of operations can be described by CPTP maps, we do not have to make distinctions between them in designing quantum error correction. A quantum error correction scheme that can correct the effects of a given CPTP map, or its equivalent stochastic evolution, can do so regardless of the exact cause of the errors.

1.4.5 Common error models

For practical purposes, we are usually interested in noise processes that are *weak*, i.e., that change the state little on average. We define a weak noise process as one that is close to the identity superoperator:

$$\rho' = \sum_k A_k \rho A_k^\dagger, \tag{1.181}$$

$$\|\rho - \rho'\|_1 \ll 1 \,\forall\, \rho, \tag{1.182}$$

where $\|\cdot\|_1$ denotes the trace norm (Eq. (1.75)). In looking at the stochastic evolution, a weak noise process does *not* imply that $A_k|\psi\rangle/\sqrt{p_k}$ is close to $|\psi\rangle$ for all k. In fact, the change can be very large. It does, however, imply that an operator A_k that causes a large change in the state must have a low probability; and operators that have high probability must be close to the identity.

1.4.5.1 Examples   *Dephasing errors* are one of the most common types of errors considered for quantum computers. They can arise due to environmental interactions, imprecise Z rotation gates, or fluctuations in the Hamiltonian. The CP map can be written

$$\rho \mapsto (1-\epsilon)\rho + \epsilon Z \rho Z. \tag{1.183}$$

This gives us two *error operators*:

$$A_0 = \sqrt{1-\epsilon}I, \quad A_1 = \sqrt{\epsilon}Z. \tag{1.184}$$

In this case, the error probability is fixed: a probability ϵ after each gate of a phase-flip error. Obviously, for rotations about different axes, the type of error will be different.

Another very common type of noise is the qubit-flip error:

$$\rho \mapsto (1-p)\rho + pX\rho X. \tag{1.185}$$

The form is exactly the same as that for dephasing, but with X instead of Z. These errors are of particular interest because they are essentially the same as the classical idea of a bit error: they can change a $|0\rangle$ to a $|1\rangle$ and vice versa.

A wide class of quantum error-correcting codes (QECCs) are designed specifically to correct X and Z errors – for instance, the CSS (Calderbank–Shor–Steane) codes described in Section 2.8. If a code can correct both X and Z errors, it can correct general qubit errors: any operator acting on a qubit can be written as a linear combination of I, X, Z, and $Y = iXZ$. Codes that can correct all such error operators are called *general* codes.

In principle, it makes sense to tailor an error-correcting code to the particular error model of the system in question. In practice, this is a complicated question. One might think that errors from environmental decoherence would be independent of the gate that is performed; but this is not true, in general. If an error occurs *during* the performance of a gate, the action of the gate can transform it into a different error.

Let us look at a simple example. Suppose we perform a gate U by turning on a Hamiltonian H for a time τ. (For the moment we ignore imprecision in the gate.)

$$U = \exp(-iH\tau/\hbar). \tag{1.186}$$

Suppose an error A_k occurs in the middle of this gate. Then the state becomes

$$|\psi\rangle \mapsto e^{-iHt/\hbar} A_k e^{-iH(\tau-t)/\hbar}|\psi\rangle/\sqrt{p_k} = \left(e^{-iHt/\hbar} A_k e^{iHt/\hbar}\right) U|\psi\rangle/\sqrt{p_k}. \tag{1.187}$$

Unless A_k commutes with H, the error A_k has become a new error $e^{-iHt/\hbar} A_k e^{iHt/\hbar}$.

1.4.5.2 Independent errors It is often assumed in quantum information processing that errors are *independent*. We assume the random deviations in different gates are *not* correlated with each other; and that each qubit interacts with a *separate* environment.

These are the usual assumptions:

(i) Every gate has an error process associated with it, which is essentially the same no matter to which qubits the gate is applied (e.g., a CNOT on bits i and j has the same error process for every value of i and j).
(ii) Errors in gates occur only to the bits to which the gate is applied (e.g., a gate on qubits 1 and 3 will not cause an error in qubit 2).
(iii) A qubit not undergoing a gate has some intrinsic error process (storage errors) that is the same and independent for all bits.

A canonical kind of independent noise for a qubit is *depolarizing* noise:

$$\rho \mapsto (1 - p_x - p_y - p_z)\rho + p_x X\rho X + p_y Y\rho Y + p_z Z\rho Z, \tag{1.188}$$

with error operators

$$A_0 = \sqrt{1 - p_x - p_y - p_z}\,I, \tag{1.189a}$$
$$A_1 = \sqrt{p_x}\,X, \tag{1.189b}$$
$$A_2 = \sqrt{p_y}\,Y, \tag{1.189c}$$
$$A_3 = \sqrt{p_z}\,Z. \tag{1.189d}$$

At long times, all initial states will tend towards the maximally mixed state $\rho = I/2$.

This type of noise is often used as a model for a *noisy quantum channel*. This is not generally because this is a particularly realistic model for most quantum channels – it is not. But it is a model that allows one to quantify the likelihood of errors, without making assumptions about their form. In general, it is assumed that if error correction can do well against depolarizing noise, it will do well against other independent noise models with similar error probability. This is similar to the practice in classical error correction of using performance against the binary symmetric channel (i.e., the bit-flip channel) as a benchmark for the quality of a general code.

The independent errors assumption can be violated. For instance, if different qubits represent different kinds of physical systems, they might be expected to have different error processes (e.g., electron spins versus nuclear spins). Some kinds of systems may have correlated errors (e.g., Pauli exchange errors for spin-1/2). Figuring out a good error model for a particular system may require very detailed experimental measurements, and may have important implications for which type of error correction strategy is best suited for that system.

1.5 Conclusions

In this chapter we have taken a quick tour of the fundamental tools of quantum mechanics of open quantum systems that will be needed for the rest of this book. We have seen that noise and decoherence can arise in a variety of forms. Whatever the noise process may be, however, the practical question is: what do we do about it? Errors will accumulate with time, which implies that quantum computations past a certain size will be impossible unless errors are somehow avoided, suppressed, or corrected. Fortunately, techniques have been devised to cope with this, even when the exact error model is unknown, so long as the error process is not too dramatic. Just how much drama we can cope with is the subject of the rest of this book.

Chapter 2

Introduction to quantum error correction

Dave Bacon

In Chapter 1 we saw that open quantum systems could interact with an environment and that this coupling could turn pure states into mixed states. This process will detrimentally impact any quantum computation, because it can lessen or destroy the interference effects that are vital to distinguishing a quantum computer from a classical computer. The problem of overcoming this effect is called the *decoherence problem*. Historically, the problem of overcoming decoherence was thought to be a major obstacle towards building a quantum computer. However, it was discovered that, under suitable conditions, the decoherence problem could be overcome. The main idea behind how this can be achieved is through the theory of quantum error correction (QEC). In this chapter we give an introduction into the way in which the decoherence problem can be overcome via the method of QEC. It is important to note that the scope of the introduction is not comprehensive, and focuses only on the basics of QEC without reference to the notion of fault-tolerant quantum computation, which is covered in Chapter 5. Quantum error correction should be thought of as a (major) tool in this larger theory of fault-tolerant quantum computing.

2.1 Error correction

When we are stumped about what to do in the quantum world, it is often useful to look to the classical world to see if there is an equivalent problem, and if so, how that problem is dealt with. Thinking this way, one realizes that the decoherence problem has an analogy in the classical world: classical noise. While this statement is not at all obvious, we will see that actually the analogy holds by thinking about decoherence and classical noise from a particular perspective.

Consider the following situation. We have a bit that we send from Seattle to New York over a phone line. This phone line, however, is noisy. With probability $1 - p$ nothing happens to our bit, but with probability p the bit is flipped: 0 is turned into 1 and 1 is turned into 0. This setup is an example of classical communication over a communication *channel*. This particular channel is called a *binary symmetric channel*. We call the operation of flipping the bit an *error*. Later the

---

Quantum Error Correction, ed. Daniel A. Lidar and Todd A. Brun. Published by Cambridge University Press.
© Cambridge University Press 2013.

term *error* will take on a more specific meaning in the theory of fault tolerance. If we use this channel once, then the probability that we will receive a wrong bit is p. If p is sufficiently close to $\frac{1}{2}$ this poses quite a problem as we want the information encoded in our bit to arrive safely at its destination. Thus we are naturally led to the question: is there a way to use this channel in such a way that we can decrease this probability? The answer to this question is yes, and the way to do this is actually rather simple.

The solution is to just use the channel multiple times. In other words, we use *redundancy*. Thus, if we want to send a 0, we use the *encoding* $0 \to 000$ and $1 \to 111$ and send each of these bits through the channel. Of course there will still be errors on the information. Assuming that the channel's noise acts independently (note this assumption!), with probability $(1-p)^3$ no errors occur on the bits, with probability $3(1-p)^2 p$ one error occurs on the bits, with probability $3(1-p)p^2$ two errors occur on the bits, and with probability p^3 three errors occur on the bits. Now assume that p is small for intuition's sake (we will calculate what small means in a second). Notice that the three probabilities we have listed above will then be in decreasing order. In particular, the probability of no or one error will be greater than there being two or three errors. But if a single error occurs on our bit, we can detect this and correct it. In particular if, on the other end of the channel, we *decode* the bit strings by $\{000, 001, 010, 100\} \to 0$ and $\{111, 110, 101, 011\} \to 1$, then in the case of no error or one bit-flip we will have correctly transmitted the bit. We can thus calculate the probability that this procedure – encoding, sending the bits individually through the channel, and decoding – fails. It is given by $3(1-p)p^2 + p^3 = 3p^2 - 2p^3$. Now if this is less than p, the failure probability with no encoding, we have decreased the probability of failing to transmit our bit. Indeed, this occurs when $3p^2 - 2p^3 \leq p$ and hence when $p < \frac{1}{2}$. Thus, if the probability of flipping our bit is less than $\frac{1}{2}$, then we will have decreased our failing (from p to $3p^2 - 2p^3$). This method of encoding information is known as a redundancy error-correcting code. The classical theory of error-correcting codes is devoted to expanding on this basic observation, that redundancy can be used to protect classical information. We will delve into some of the details of classical error-correcting codes in Section 2.7. The main take-home point at this juncture is that it is possible to use classical error correction to lessen the chance of errors destroying classical information.

2.1.1 Obstacles to quantum error correction

If classical error correction can be used to protect against noise in a noisy classical channel, a natural question to ask is whether a similar tactic can be used to protect quantum information. When we first encounter classical error correction and think about porting it over to the quantum world, there are some interesting reasons to believe that it will be impossible to make this transition. We list these here, since they are rather interesting (although claims that these were major blockades to discovering quantum error correcting are probably a bit exaggerated).

No cloning The no-cloning theorem [WZ82] states that there is no machine that can perform the operation $|\psi\rangle \to |\psi\rangle \otimes |\psi\rangle$ for all $|\psi\rangle$. Thus, a naive attempt to simply clone quantum information in the same way that we copy information in a redundancy code fails.

Measurement When we measure a quantum system, our description of the quantum system changes. Another way this is stated is that measurement disturbs the state of a quantum system. In error correction, we read out classical information in order to correctly recover our classical information. How do we perform measurements on quantum systems that do not destroy the quantum information we are trying to protect?

Quantum noise Quantum noise has a continuous set of parameters to describe it. So we might think that this will cause a problem, since in classical theory we could interpret the noise as probabilities of deterministic evolutions occurring, but in quantum theory we do not have such an interpretation (at least not yet.) Of course this feels like a little bit of a red herring, since classical noise also has continuous parameters (say the probabilities of the erring procedures) to describe it.

For these reasons we might expect that an equivalent to classical error correction in the quantum world does not exist. One of the surprising discoveries of the mid-nineties was that this is not true: protection of quantum information using QEC is possible.

2.2 From reversible classical error correction to simple quantum error correction

So where to begin in figuring out how to protect quantum information? Well, one place to begin is to try to understand how to perform the classical error correction we described above using classical reversible circuits, since reversible classical computation is the closest classical theory to quantum theory.

To this end let us work through using the classical three-bit redundancy code to protect classical information. The first part of our procedure is to encode our classical bit. Suppose that we represent our three bits by three wires. Then it is easy to check that an encoding procedure for taking a bit in the first wire to the encoded 000 and 111 configurations is

$$
\begin{array}{c}
b \longrightarrow\bullet\longrightarrow\bullet\longrightarrow b \\
0 \longrightarrow\oplus\longrightarrow\;\;\;\;\longrightarrow b \\
0 \longrightarrow\;\;\;\;\longrightarrow\oplus\longrightarrow b
\end{array}
\tag{2.1}
$$

where the gates diagrammed are CNOT gates, which act in the computational basis as $CX|x\rangle|y\rangle = |x\rangle|y+x\rangle$, where the addition is done modulo 2, and where time flows from left to right. Next we send each of these bits through the bit-flip channel. Each of these bits is sent independently. We denote this by the gate M on these bits:

(2.2)

Now we need to describe the procedure for diagnosing an error and fixing this error. Consider what the two CNOTs do in our encoding circuit. They take $000 \to 000$, $001 \to 001$, $010 \to 010$, $011 \to 011$, $100 \to 111$, $101 \to 110$, $110 \to 101$, and $111 \to 100$. Notice that, except in the case where the last two bits are 11, after this procedure the first bit has been restored to its proper value, given that only zero or one bit-flip has occurred on our three bits. And when the last two bits are 11, then we need to flip the first bit to perform the proper correction. This implies that the decoding and fixing procedure can be done by the two CNOTs, followed by a Toffoli gate (which acts as $C^2(X)|x\rangle|y\rangle|z\rangle = |x\rangle|y\rangle|z+xy\rangle$). In other words,

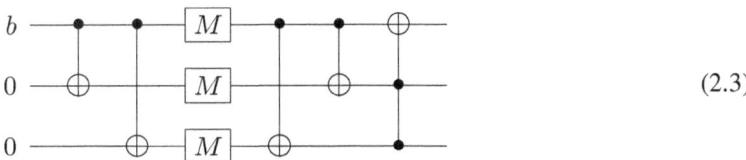
(2.3)

It is easy to check that if only one or no error occurs where the M gates are, then the output of the first bit will be b (and in the other cases the bit will be flipped). This is exactly the procedure we described in the previous section and we have done it using completely reversible circuit elements (except for the Ms, of course.)

2.2.1 When in Rome do as the classical coders would do

Now that we have a classical reversible circuit for our simple error-correcting procedure, we can see what happens when we use this circuit on quantum information instead of classical information. One thing we need to do is to choose the appropriate noise channel for our code (we will come back to more general noise eventually). A natural choice is the bit-flip channel that had the Kraus operator

$$E_0 = \sqrt{1-p}I, \quad E_1 = \sqrt{p}X. \tag{2.4}$$

The circuit we want to evaluate is now

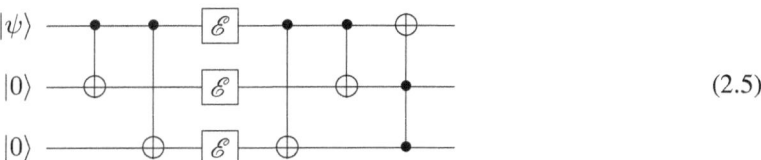
(2.5)

where $|\psi\rangle$ is now an arbitrary quantum state $\alpha|0\rangle + \beta|1\rangle$, which we wish to protect. So what does this circuit do to our quantum data? Well, after the first two CNOTs, we can see that the state is given by

$$\alpha|000\rangle + \beta|111\rangle. \tag{2.6}$$

Notice that we have done something like redundancy here: but we have not copied the state, we have just "copied" it in the computational basis.

Next what happens? Recall that we can interpret $\mathcal{E}$ as a bit-flip error X happening with probability p and nothing happening with probability $1-p$. It is useful to use this interpretation to say that with probability $(1-p)^3$ no error occurred on all three qubits, with probability $(1-p)^2 p$ a single error occurred on the first qubit, etc. So what happens if no error occurs on our system (the Kraus operator $I \otimes I \otimes I$ happens on our system)? Then we just need to run those two CNOTs on our state

$$(CX)_{13}(CX)_{12}(\alpha|000\rangle + \beta|111\rangle) = \alpha|000\rangle + \beta|100\rangle = (\alpha|0\rangle + \beta|1\rangle)|00\rangle. \tag{2.7}$$

The Toffoli then does nothing to this state and we see that our quantum information has survived. But this is not too surprising since no error occurred. What about when a single bit-flip error occurs? Let us say an error occurs on the second qubit. Then our state is $\alpha|010\rangle + \beta|101\rangle$. The effect of the CNOTs is then

$$(CX)_{13}(CX)_{12}(\alpha|010\rangle + \beta|101\rangle) = \alpha|010\rangle + \beta|110\rangle = (\alpha|0\rangle + \beta|1\rangle)|10\rangle. \quad (2.8)$$

Again, the Toffoli will do nothing to this state. And we see that our quantum information has survived its encounter with the bit-flip error! One can go through the other cases of a single bit-flip error. In the case where the bit-flip error is on the first qubit, the Toffoli is essential in correcting the error, but in the case where it is on the third qubit, then the Toffoli does nothing. But in all three cases the quantum information is restored!

One can go through and check what happens for the cases where two or three bit-flip errors occur. What one finds out is that in these cases the resulting state in the first qubit is $\beta|0\rangle + \alpha|1\rangle$. Thus, if we look at the effect of this full circuit, it will perform the evolution

$$\begin{aligned}
\rho \otimes |00\rangle\langle 00| \to\ & (1-p)^3 \rho \otimes |00\rangle\langle 00| + (1-p)^2 p\rho \otimes |01\rangle\langle 01| \\
& + (1-p)^2 p\rho \otimes |10\rangle\langle 10| + (1-p)^2 p\rho \otimes |11\rangle\langle 11| \\
& + (1-p)p^2 X\rho X \otimes |01\rangle\langle 01| + (1-p)p^2 X\rho X \otimes |10\rangle\langle 10| \\
& + (1-p)p^2 X\rho X \otimes |11\rangle\langle 11| + p^3 X\rho X \otimes |00\rangle\langle 00|.
\end{aligned} \quad (2.9)$$

Tracing over the second and third qubits, this amounts to the evolution

$$\rho \to [(1-p)^3 + 3p(1-p)^2]\rho + [3p^2(1-p) + p^3]X\rho X. \quad (2.10)$$

If we compare this with the evolution that would have occurred given no encoding,

$$\rho \to (1-p)\rho + pX\rho X, \quad (2.11)$$

we see that if $p < \frac{1}{2}$, then our encoding acts to preserve the state better than if we had not encoded the state.

Actually, how do we know that we have preserved the state better? What measure should we use to deduce this and why would this be a good measure? In particular, we might note that quantum errors will affect different states in different ways. The answer is given in terms of the fidelity discussed in Section 1.2.6.

2.2.2 Fidelity of classical error correction in the quantum model

What happens to the fidelity of the two cases we had above, one in which no error correction was performed and one in which error correction was performed? In the first case the fidelity, assuming we start in a pure state, is

$$F_1 = \left[\langle\psi|\left[(1-p)|\psi\rangle\langle\psi| + pX|\psi\rangle\langle\psi|X\right]|\psi\rangle\right]^{\frac{1}{2}} = \left[(1-p) + p|\langle\psi|X|\psi\rangle|^2\right]^{\frac{1}{2}}. \quad (2.12)$$

This is minimized (remember we want high fidelity) when $\langle\psi|X|\psi\rangle = 0$ and so

$$F_1 \geq \sqrt{1-p}. \quad (2.13)$$

Similarly, if we perform error correction using the three qubit scheme we obtain

$$\begin{aligned}
F_3 &= \left[\langle\psi|\left[((1-p)^3 + 3p(1-p)^2)|\psi\rangle\langle\psi| + (3p^2(1-p) + p^3)X|\psi\rangle\langle\psi|X\right]|\psi\rangle\right]^{\frac{1}{2}} \\
&= \left[(1-p)^3 + 3p(1-p)^2 + (3p^2(1-p) + p^3)|\langle\psi|X|\psi\rangle|^2\right]^{\frac{1}{2}},
\end{aligned} \quad (2.14)$$

which is again bounded by

$$F_3 \geq \sqrt{(1-p)^3 + 3p(1-p)^2}\,. \tag{2.15}$$

The fidelity is greater using error correction when $p < \frac{1}{2}$. So our naive analysis was not so much the dunce after all.

2.2.3 Morals

What lessons should we draw from our first success in QEC? One observation is that instead of *copying* the quantum information we *encoded* the quantum information into a *subspace*. In particular, we have encoded into the subspace spanned by $\{|000\rangle, |111\rangle\}$, i.e., we have encoded our quantum information as $\alpha|000\rangle + \beta|111\rangle$. This is our way of getting around the no-cloning theorem.

The second problem we brought up was measurement. Somehow we have made a measurement such that we could fix our quantum data (if needed) using the Toffoli gate. Let us examine what happens to our subspace basis elements, $|000\rangle$ and $|111\rangle$, under the errors that we could correct. Notice that they enact the evolution

$$\begin{array}{ccc}
|000\rangle & \xrightarrow{I \otimes I \otimes I} & |000\rangle \\
|111\rangle & & |111\rangle \\
|000\rangle & \xrightarrow{X \otimes I \otimes I} & |100\rangle \\
|111\rangle & & |011\rangle \\
|000\rangle & \xrightarrow{I \otimes X \otimes I} & |010\rangle \\
|111\rangle & & |101\rangle \\
|000\rangle & \xrightarrow{I \otimes I \otimes X} & |001\rangle \\
|111\rangle & & |110\rangle
\end{array}. \tag{2.16}$$

Now think about what this is doing. These error processes are mapping the subspace where we encoded the information into different *orthogonal* subspaces for each of the different errors. Further, when this map is performed, the orthogonality between the basis elements is not changed (i.e., $|000\rangle$ and $|111\rangle$ are orthogonal before and after the error occurs). Now this second fact is nice, because it means that the quantum information has not been *distorted* in an irreversible fashion. And the first fact is nice because, if we can measure which subspace our error has taken us to, then we will be able to fix the error by applying the appropriate operation to reverse the error operation. In particular, consider the operators $S_1 = Z \otimes Z \otimes I$ and $S_2 = Z \otimes I \otimes Z$. These operators square to identity and so have eigenvalues $+1$ and -1. In fact, we can see that these eigenvalues do not distinguish between states within a subspace, but do distinguish which of the four subspaces our state is in. That is to say, for example, that $|000\rangle$ and $|111\rangle$ have eigenvalues $+1$ for both S_1 and S_2. Further, the subspace that occurs if a single bit-flip occurs on our first qubit, $|100\rangle$ and $|011\rangle$, has eigenvalues -1 for S_1 and -1 for S_2. We can similarly calculate the other cases:

Basis states of subspace	S_1	S_2	Error
$\{\|000\rangle, \|111\rangle\}$	$+1$	$+1$	$I \otimes I \otimes I$
$\{\|100\rangle, \|011\rangle\}$	-1	-1	$X \otimes I \otimes I$
$\{\|010\rangle, \|101\rangle\}$	-1	$+1$	$I \otimes X \otimes I$
$\{\|001\rangle, \|110\rangle\}$	$+1$	-1	$I \otimes I \otimes X$

Thus we see that, if we could perform a measurement that projects onto the $+1$ and -1 eigenstates of S_1 and S_2, then we could use the results of this measurement to diagnose which subspace the error has taken us to and apply the appropriate X operator to recover the original subspace. So is it possible to measure S_1 and S_2? Well, we have already done it, but in a destructive way, in our circuit.

Consider the following circuit:

$$\text{(2.17)}$$

What does this circuit do? Well, if the input to this circuit is $\alpha|00\rangle + \beta|11\rangle$, then the measurement outcome will be $|0\rangle$ and if the circuit is $\alpha|01\rangle + \beta|10\rangle$, then the measurement outcome is $|1\rangle$. Associating $|0\rangle$ with $+1$ and $|1\rangle$ with -1, we thus see that this is equivalent to measuring the eigenvalue of the operator $Z \otimes Z$. Notice, however, that this is a destructive measurement, i.e., it does not leave the subspace intact after the measurement. In the circuit we have constructed above, then, the CNOTs after the errors have occurred would have measured the operators S_1 and S_2. This is enough to diagnose which error has occurred. Since this also does decoding of our encoded quantum information, only in the case where the error occurred on the first qubit do we need to do anything; and this is the case where the measurement outcomes are both $|1\rangle$ and so we use the Toffoli to correct this error. This suggests that a different way to implement this error-correcting circuit is to measure the second and third qubits. Since measurements commute through control gates turning them into classical control operations, we could thus have performed the following circuit:

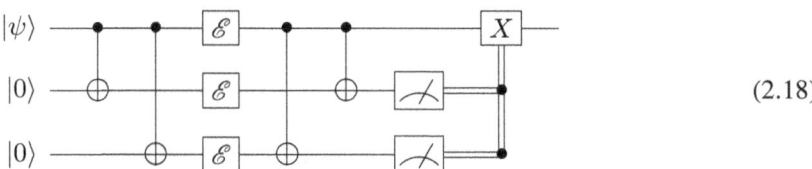

$$\text{(2.18)}$$

What do we learn from the above analysis? We learn that QEC avoids the fact that measuring disturbs a quantum system by performing measurements that project onto subspaces. These measurements do not disturb the information encoded into the subspaces since the measurements yield degenerate values outcomes for any state in this subspace. This technique, of performing measurements that do not fully project onto a basis, is essential to being able to perform quantum error correction.

2.2.4 Dealing with phase-flips

The third problem we brought up was the fact that quantum errors form a continuous set. For now, however, let us just move on to a different error model. In particular, let us consider, instead of a bit-flip model, a phase-flip model. In this model, the Kraus operators are given by

$$A_0 = \sqrt{1-p}I, \quad A_1 = \sqrt{p}Z. \quad (2.19)$$

The effect of Z on a quantum state is to change the phase between $|0\rangle$ and $|1\rangle$. So how are we going to correct this error? It changes a phase, not an amplitude! Well, we use the fact that phase changes are amplitude changes in a different basis. The basis where this occurs is the so-called plus minus basis: $|\pm\rangle = \frac{1}{\sqrt{2}}(|0\rangle \pm |1\rangle)$. In particular, the gate that transforms this basis into the computational basis is the Hadamard gate, W; and note that $WZW^\dagger = X$. This suggests that just prior to sending our information through the quantum channel that causes phase errors and just after receiving the quantum information we should apply Hadamard gates. Hence we are led to the circuit:

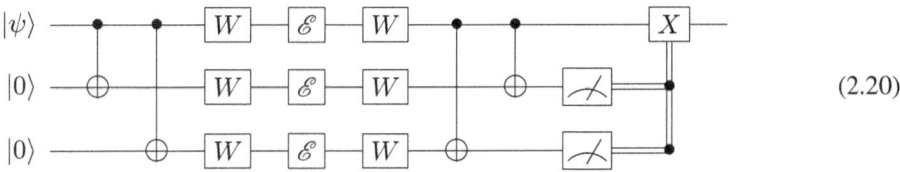
(2.20)

Now, will this work? Certainly it will work, this is just a basis change and the fidelities will be identical to the bit-flip analysis. What does this circuit do? Instead of encoding into the subspace spanned by $|000\rangle$ and $|111\rangle$, this code encodes into a subspace spanned by $|+++\rangle$ and $|---\rangle$, where $|\pm\rangle = \frac{1}{\sqrt{2}}(|0\rangle \pm |1\rangle)$. Thus we see that, by expanding our notion of encoding beyond encoding into something simple like the repeated computational basis states, we can deal with a totally different type of error, one that does not really have a classical analogy in the computational basis.

We are just putting off the question of what happens when we have arbitrary errors, of course; but notice something here. The phase-flip error model is equivalent, via a change in Kraus operators, to the phase damping model that has Kraus operators:

$$B_0 = \begin{bmatrix} \sqrt{1-q} & 0 \\ 0 & \sqrt{1-q} \end{bmatrix}, \quad B_1 = \begin{bmatrix} \sqrt{q} & 0 \\ 0 & 0 \end{bmatrix}, \quad B_2 = \begin{bmatrix} 0 & 0 \\ 0 & \sqrt{q} \end{bmatrix}. \quad (2.21)$$

When $\frac{p}{2} = q$, these were the same superoperator. We can express the above Kraus operators as

$$B_0 = \sqrt{1-q}I, \quad B_1 = \frac{\sqrt{q}}{2}(I+Z), \quad B_2 = \frac{\sqrt{q}}{2}(I-Z). \quad (2.22)$$

Surely, since this is the same superoperator, our analysis of the error-correcting procedure will be identical and we will be able to increase the probability of successfully correcting this model given $q \leq \frac{1}{4}$. But the B_i operators are sums of I and Z errors. Thus, a code designed to correct single Z errors seems to be working on superoperators that have Kraus operators which are sums of identity and Z errors. Why is this so? Well, we have some encoded information $|\psi\rangle$. Then a Kraus operator that is the sum of terms which we can correct occurs (plus terms we cannot correct). Then we perform the measurement to distinguish which subspace our error has taken us to. But, at this point, the Kraus operators that are sums of errors get projected onto one of the error subspaces. Thus, in effect, the fact that the error is a sum of errors gets erased when we do this projection. We will return to this fact later, but this is an important point. While quantum errors may be a continuous set, the error-correcting procedure can, in effect, digitize the errors and deal with them as if they formed a discrete set.

2.2.5 The Shor code

So far we have dealt with two models of errors, bit-flip errors and phase-flip errors. We have also seen that if a code corrects an error then it will be able to correct Kraus operator errors that have a sum of these errors. Thus, we expect that if we can design a quantum error-correcting code (QECC) that can correct a single X, Y, or Z error, then this can correct an arbitrary error on a single qubit. Indeed, Peter Shor designed [S95] just such a code (Steane independently arrived at the idea of QECC in [S96c]). How does this code work? Well we have already seen that if we encode into the subspace spanned by $|000\rangle$ and $|111\rangle$, then we can correct a single bit-flip error. What do single phase-flip errors do to this code? Well, notice that $ZII|000\rangle = IZI|000\rangle = IIZ|000\rangle = |000\rangle$ but $ZII|111\rangle = IZI|111\rangle = IIZ|111\rangle = -|111\rangle$. Thus we see that, unlike bit-flip errors, single phase-flip errors on this code act to *distort* the information encoded into the subspace. But also notice that these single phase-flip errors act like phase-flip errors on the encoded basis $|000\rangle$, $|111\rangle$. But we have seen how to deal with phase-flip errors. Suppose we define the states

$$|p\rangle = \frac{1}{\sqrt{2}}(|000\rangle + |111\rangle), \tag{2.23a}$$

$$|m\rangle = \frac{1}{\sqrt{2}}(|000\rangle - |111\rangle). \tag{2.23b}$$

Then a single phase-flip error on these two states sends $|p\rangle$ to $|m\rangle$ and vice versa, i.e., it looks like a bit-flip on these qubits. We can thus deal with these single phase-flip errors by using a bit-flip code. In particular, we define the two nine-qubit states:

$$|0_L\rangle = |ppp\rangle = \frac{1}{2\sqrt{2}}(|000\rangle + |111\rangle) \otimes (|000\rangle + |111\rangle) \otimes (|000\rangle + |111\rangle), \tag{2.24a}$$

$$|1_L\rangle = |mmm\rangle = \frac{1}{2\sqrt{2}}(|000\rangle - |111\rangle) \otimes (|000\rangle - |111\rangle) \otimes (|000\rangle - |111\rangle). \tag{2.24b}$$

Suppose we encode a qubit of quantum information into the subspace spanned by these two states. Now single phase-flip errors can be fixed by diagnosing what subspace the $|ppp\rangle$ and $|mmm\rangle$ subspace has been sent to. Further single bit-flip errors can be dealt with from within each $|p\rangle$ and $|m\rangle$ state: these states are encoded states in our original bit-flip code. Putting this together, we see that we should be able to use this error-correcting code to correct bit-flips and phase-flips. Actually it can do more, and indeed it can handle a single Y error as well.

To see this let us construct the circuit for encoding into this code and then performing the decoding and correction.

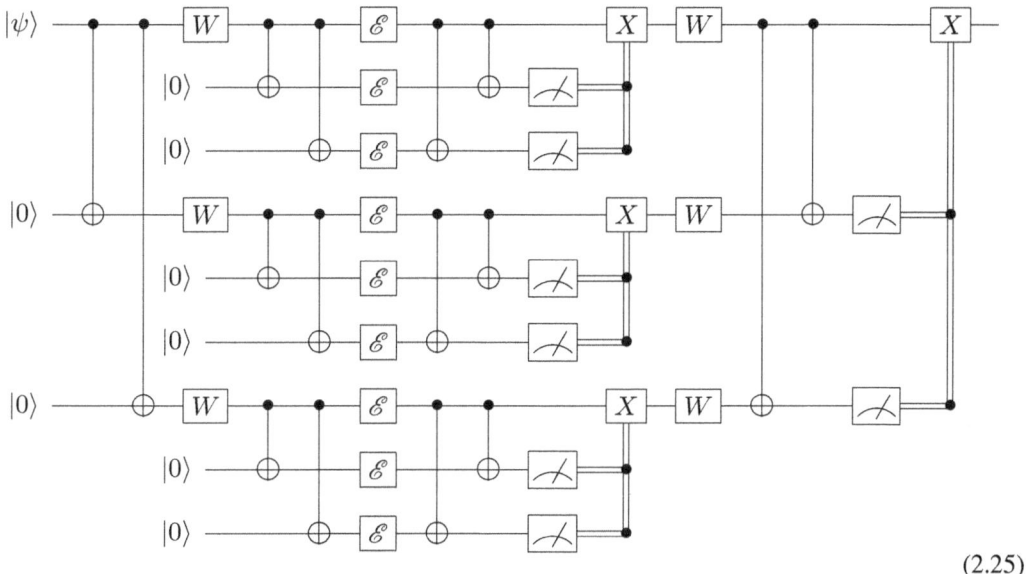

(2.25)

Now, first of all, isn't this beautiful! Notice the three blocks of three in this code. Also notice how these blocks, when there is either no error or a single X error in the superoperators,[1] produce a channel on the outer three wires (just after the first triple of Ws and before the second triple of Ws) which is the identity (because these are designed to correct just that error). But what about if a single Z error occurs. As we noted above, this means that on the encoded quantum information this acts like a Z error on the $|p\rangle$, $|m\rangle$ states. Since only a Z error occurs, we are not taken out of the $|p\rangle$ and $|m\rangle$ subspace by this error, and so the effect of the error correction in an inner block will be to produce a state that has a single-qubit Z error on the outer wire. Now we see how the code corrects this single Z error: this is just the phase-flip error-correcting code!

But what happens if, say, a single $Y = iXZ$ error occurs? First notice that the global phase i doesn't really matter. So we can assume that the error is XZ. Now the Z error acting on the encoded state acts within the encoded space as a Z error. Thus, imagine performing this error, and then running the bit-flip code. The bit-flip code will correct the X error, but the end result will be that a Z error has occurred on the encoded information. But then the Z error will be corrected by the outer code. Thus we see that a single Y error will be corrected by this code.

So what have we in Shor's code? We have a code that can correct any single-qubit error from the set $\{X, Y, Z\}$. But, as we have argued above, this means that any single-qubit error that is a sum of these errors will also be corrected (we will make this argument rigorous in Section 2.6). So if a single arbitrary error occurs on our qubit, Shor's code will correct it. We call this a QECC that can correct a single-qubit error.

What have we learned? We have learned that it is possible to circumvent the objections we raised early: cloning, measurement, and continuous sets of errors, and to enact error correction

---

[1] Of course, in general our superoperators will each contain error terms, but we will simply talk about the case where there is one such error since this is, to first order, the most important error, assuming that the errors are "weak enough." Yes, this is all rather vague right now, but we need to make progress without getting bogged down in the details.

on quantum information. What does this mean for practical implementations of a quantum computer? Well, it means that more sophisticated versions of QEC might be able to make a very robust computer out of many noisy components. There are many issues that we need to discuss in this theory, and the full story of how to build such a quantum computer is the subject of fault-tolerant quantum computation. A main aim in the next section will be to obtain an understanding of the theory of QEC.

2.3 The quantum error-correcting criterion

Above we have seen the simplest example of a QECC that is truly quantum in nature, Shor's code. The basic idea there was that we *encode* quantum information into a subspace of the Hilbert space of many quantum systems. This is the principle idea of a *quantum error correcting code*. In particular, this is an example of a subspace QECC where the information is encoded into a subspace of the larger Hilbert space of the system in question. Later we will encounter a variation on this idea known as subsystem QECCs. Give that a QECC is nothing less than a subspace, an important question to ask is under what conditions does a subspace act as a QECC.

Suppose that we have a quantum system that evolves according to some error process that we represent by the superoperator $\mathscr{E}$. We assume that this superoperator is given by some Krauss operator sum representation

$$\mathscr{E}[\cdot] = \sum_k E_k[\cdot]E_k^\dagger. \tag{2.26}$$

In general, our codes will not be able to reverse the effect of all errors on our system: the goal of QEC is to make the probability of error so small that it is effectively zero, not to eliminate the possibility of error completely (although philosophers will argue about the difference between the two). It is therefore useful to assume that the Kraus operators in the expansion for $\mathscr{E}$ are made up of some errors E_i, $i \in S$, which we wish to correct. This will be a good assumption because the real error process will contain these terms, which we will then be certain we have fixed, plus the errors that we might not fix. We will return to this point later. Thus, with this assumption, we may think about $\mathscr{E}$ as having Kraus operators, some of which are errors E_i that we wish to correct and some of which are not. Note that this assumption is not necessary, but is simply a convenience for the short term. Define $\mathscr{F}$ as the operator,

$$\mathscr{F}[\cdot] = \sum_{i \in S} E_i[\cdot]E_i^\dagger. \tag{2.27}$$

Notice that $\mathscr{F}$ will not necessarily preserve the trace of a density matrix. This will not stop us from considering reversing its operation.

So given $\mathscr{F}$ with some Kraus operators E_i we can ask, under what conditions is it possible to design a quantum code and recovery operations $\mathscr{R}$ such that

$$\mathscr{R} \circ \mathscr{F}[\rho_C] \propto \rho_C, \tag{2.28}$$

for ρ_C with support over the code subspace, $\mathscr{H}_C \subseteq \mathscr{H}$? Why do we use $\propto$ here instead of $=$? Well, because $\mathscr{E}$ is not trace-preserving. This means that there may be processes occurring in the full $\mathscr{D}$ that occur with some probability and we do not need to preserve ρ on these errors.

Let us call a basis for the code subspace $|\phi_i\rangle$. $\mathscr{H}_C = \mathrm{span}\{|\phi_i\rangle\}$. We will show that a necessary and sufficient condition for the recovery operations to preserve the subspace is given by the Knill–Laflamme *quantum error-correcting code criterion* [KL97]:

$$\langle \phi_i | E_k^\dagger E_l | \phi_j \rangle = C_{kl} \delta_{ij}, \tag{2.29}$$

where C_{kl} is a Hermitian matrix, sometimes called the code matrix. It tells us when our encoding into a subspace can protect us from quantum errors E_k. As such, it is a very important criterion for the theory of QEC. Lets show that this is a necessary and sufficient condition.

2.3.1 Sufficiency

Let us begin by showing that if this criterion is satisfied, we can construct a recovery operation $\mathscr{R}$ with the desired properties.

The first thing to do is to change the error operators. Instead of discussing the error operators E_k, define a new set of error operators $F_m = \sum_k u_{mk} E_k$, where u_{lk} are the elements of a unitary matrix u. It is a simple exercise to show that this means that F_l represents the same superoperator. Now we see that, since the E_i satisfy the error-correcting criterion,

$$\langle \phi_i | F_m^\dagger F_n | \phi_j \rangle = \sum_{k,l} \langle \phi_i | u_{mk}^* E_k^\dagger u_{nl} E_l | \phi_j \rangle = \sum_{k,l} u_{mk}^* C_{kl} u_{nl} \delta_{ij}. \tag{2.30}$$

Since C_{kl} is Hermitian, it is always possible to choose $u = \{u_{ij}\}$ such that it diagonalizes this matrix,

$$\langle \phi_i | F_m^\dagger F_n | \phi_j \rangle = d_m \delta_{m,n} \delta_{i,j}, \tag{2.31}$$

with $d_m \in R$. Now define the following operators for $d_k \neq 0$:

$$R_k = \frac{1}{\sqrt{d_k}} \sum_i |\phi_i\rangle \langle \phi_i | F_k^\dagger, \tag{2.32}$$

and if $d_k = 0$ then let $R_k = 0$. Here $\sum_i |\phi_i\rangle\langle\phi_i|$ is the projector onto the code subspace. We want to show that a recovery superoperator with R_k as its Kraus operators will correctly recover our erred quantum information:

$$\sum_k R_k \sum_l \left(F_l \rho_C F_l^\dagger \right) R_k^\dagger = \sum_{k | d_k \neq 0} \frac{1}{\sqrt{d_k}} \sum_i |\phi_i\rangle\langle\phi_i| F_k^\dagger \sum_l \left(F_l \rho_C F_l^\dagger \right) \frac{1}{\sqrt{d_k}} \sum_j F_k |\phi_j\rangle\langle\phi_j|. \tag{2.33}$$

If we can show that for $\rho_C = |\phi_m\rangle\langle\phi_n|$ this produces something proportional to ρ_C, then we will have shown that the recovery correctly restores information in the subspace (since by linearity this will work for the entire density matrix). Substituting this ρ_C in, we obtain

$$\sum_{k|d_k \neq 0} \frac{1}{d_k} \sum_i |\phi_i\rangle\langle\phi_i| F_k^\dagger \sum_l \left(F_l |\phi_m\rangle\langle\phi_n| F_l^\dagger \right) \sum_j F_k |\phi_j\rangle\langle\phi_j|. \tag{2.34}$$

Using the quantum error-correcting criterion, this becomes

$$\sum_{k|d_k\neq 0} \frac{1}{d_k} \sum_{ilj} |\phi_i\rangle d_k \delta_{lk} \delta_{im} d_k \delta_{lk} \delta_{jn} \langle\phi_j| = \sum_k d_k |\phi_m\rangle\langle\phi_n| = \left(\sum_k d_k \right) \rho_C. \tag{2.35}$$

Thus we see that, indeed, the recovery produces a state proportional to ρ_C. Notice that if $\mathscr{E}$ is trace-preserving, then $\sum_k d_k = 1$ and then we recover exactly ρ_C, as desired.

We need to check that R_k forms a valid superoperator. Check,

$$R = \sum_k R_k^\dagger R_k$$

$$= \sum_{k|d_k \neq 0} \frac{1}{d_k} \sum_{i,j} F_k |\phi_i\rangle\langle\phi_i||\phi_j\rangle\langle\phi_j| F_k^\dagger = \sum_{k|d_k \neq 0} \frac{1}{d_k} \sum_i F_k |\phi_i\rangle\langle\phi_i| F_k^\dagger. \quad (2.36)$$

Notice, using the quantum error-correcting criterion, that this operator is a projector:

$$R^2 = \sum_{k|d_k \neq 0} \frac{1}{d_k} \sum_i F_k |\phi_i\rangle\langle\phi_i| F_k^\dagger \sum_{k'|d_{k'} \neq 0} \frac{1}{d_{k'}} \sum_{i'} F_{k'} |\phi_{i'}\rangle\langle\phi_{i'}| F_{k'}^\dagger$$

$$= \sum_{k|d_k \neq 0} \frac{1}{d_k} \sum_i F_k |\phi_i\rangle \sum_{k'|d_{k'} \neq 0} \frac{1}{d_{k'}} \sum_{i'} d_k \delta_{k,k'} \delta_{i,i'} \langle\phi_{i'}| F_{k'}^\dagger$$

$$= \sum_{k|d_k \neq 0} \frac{1}{d_k} \sum_i F_k |\phi_i\rangle\langle\phi_i| F_k^\dagger = R. \quad (2.37)$$

Thus, if we add one extra (if necessary) projector to the R_ks that has support on the space orthogonal to this projector, $I - \sum_k R_k^\dagger R_k$, then we will obtain a complete set of Kraus operators that satisfy the proper normalization condition for the Kraus operators. Thus, we have seen that we have a valid recovery operator that does the proper recovery and that this valid recovery operator, with the addition of possibly one extra Kraus operator, is indeed a valid superoperator.

2.3.2 Necessity

Let us show the necessity of the quantum error-correcting criterion. Errors followed by recovery produces the following evolution on an encoded state:

$$\sum_k R_k \left(\sum_i E_i \rho_C E_i^\dagger \right) R_k^\dagger = c \rho_C. \quad (2.38)$$

We want to show that this implies the error-correcting criterion. Note that ρ_C by itself is equivalent to a superoperator in which no evolution has taken place. Express the above as

$$\sum_{k,l} (R_k E_i) \rho_C (E_i^\dagger R_k^\dagger) = c I \rho_C I. \quad (2.39)$$

Now let P_C be a projector onto the code subspace, $P_C = \sum_i |\phi_i\rangle\langle\phi_i|$. Then the above criterion is that, for all ρ,

$$\sum_{k,l} (R_k E_i P_C) \rho (P_C E_i^\dagger R_k^\dagger) = c P_C \rho P_C. \quad (2.40)$$

Invoke now the unitary degree of freedom of the Kraus operator sum representation. This says that there must exist a unitary transform on the superoperators on the left-hand side of the equation such that one obtains only the single superoperator on the right-hand side of the equation. This means that there must exist an orthogonal set of vectors with coefficients u_{ki} such that

$$R_k E_i P_C = u_{ki} c P_C. \quad (2.41)$$

Taking the conjugate transpose of this equation and setting $i = j$ yields

$$P_C E_j^\dagger R_k^\dagger = u_{kj}^* c^* P_C. \tag{2.42}$$

Multiplying this equation on the left of the original equation yields

$$P_C E_j^\dagger R_k^\dagger R_k^\dagger E_i P_C = u_{kj}^* u_{ki} |c|^2 P_C. \tag{2.43}$$

Summing this equation and using the fact that $\mathscr{R}$ must be a trace-preserving operator,

$$P_C E_i^\dagger E_j P_C = \sum_k u_{kj}^* u_{ki} |c|^2 P_C. \tag{2.44}$$

Defining $C_{ij} = \sum_k u_{kj}^* u_{ki} |c|^2$, this is just

$$P_C E_i^\dagger E_j P_C = C_{ij} P_C, \tag{2.45}$$

where we see that C_{ij} is Hermitian. Taking matrix elements of this equation and relabeling i and j as k and l then yields the quantum error-correcting criterion, Eq. (2.29). Thus we have established the necessity and sufficiency of the quantum error-correcting criterion.

2.4 The distance of a quantum error-correcting code

The general theory of QECCs is designed to deal with an arbitrary set of errors $\{E_i\}$ and an arbitrary size of the subspace into which we encode. However, in many cases we will specialize to the case where the information is encoded across qubits, and further where we consider $\{E_i\}$ to be made up of tensor products of Pauli operators of weight less than some fixed length t. Consider an operator P on n qubits that is made up of a tensor product of Pauli operators $\{I, X, Y, Z\}$. The number of nonidentity (i.e., not I) terms in this operator is called the weight of P. Thus, the situation we are concerned with is where the set of correctable errors is $\mathscr{E}_t = \{P|\text{ weight of } P \text{ is less than } t\}$. In the case where we have a QECC on n qubits, where we encode k qubits into the code, and the maximum number of errors the code can correct is t, then we call this a $[[n, k, d]]$ code, where $d = 2t + 1$ is the distance of the code. The distance of a code is the smallest weight operator that can be used to enact a transformation on the information encoded into the code.

2.5 Content of the quantum error-correcting criterion and the quantum Hamming bound

What is the content of the quantum error-correcting criterion, $\langle \phi_i | E_k^\dagger E_l | \phi_j \rangle = C_{kl} \delta_{ij}$, Eq. (2.29)? First look at the δ_{ij}. This implies that codewords after being changed by the error E_l are orthogonal to the codewords after being changed by the error E_k. If $l = k$ this implies that information in codewords is not distorted by the effect of error E_k: they may be rotated, but the inner product between all codewords will be the same before as after (up to a full normalization factor). In our example of QECCs for the bit-flip code, we saw that each possible error could act to take the error to an orthogonal subspace. If every such error acts this way for a code, then the code is said to be nondegenerate. In this case, C_{kl} will be diagonal. Some codes, however, do not possess this property: there are multiple errors that can produce the same syndrome, but the recovery procedure works in spite of this.

For nondegenerate codes there is a nice bound on the size of the codes. Suppose that we wish to encode k qubits into n bare qubits in a QECC that corrects errors on t or fewer qubits, (i.e., a $[[n, k, 2t+1]]$ code). In order for a nondegenerate QECC to correct all of these errors, for each error there must be an orthogonal subspace. There are $\binom{n}{j}$ places where j errors can occur. And in each of these places there are three different nontrivial Pauli errors. Thus, the total number of errors for such a code we have described is

$$\sum_{j=0}^{t} \binom{n}{j} 3^j. \tag{2.46}$$

For each of these errors, there must be a subspace as big as the size of the encoded space, 2^k, and these subspaces must be orthogonal. Thus, each subspace must fit into the full space of n qubits. We then obtain the bound

$$\sum_{j=0}^{t} \binom{n}{j} 3^j 2^k \leq 2^n. \tag{2.47}$$

This is called the *quantum Hamming bound*. Suppose that we want a code that corrects $t = 1$ error and encodes $k = 1$ qubit. Then we obtain the inequality $(1 + 3n)2 \leq 2^n$. This inequality cannot be satisfied for $n \leq 4$. Thus, for nondegenerate codes, the smallest code that can correct a single error and encodes a single qubit has $n = 5$. Indeed, we will find that just such a code exists (such codes that saturate this bound are called perfect codes).

2.6 Digitizing quantum noise

Suppose that we have an error-correcting code that corrects a set of errors $\{E_k\}$. What other errors will this code correct? It turns out that this code will correct any linear combination of these errors. To see this, work with the errors that satisfy the diagonal error-correcting criterion, as in the sufficiency construction above (the F_ls). Now suppose that the actual F_ls are written as a sum over the F_ls we can correct: $G_l = \sum_p f_{lp} F_p$. Then, using the recovery operation we defined in the sufficiency proof, we obtain that the action of recovery after the error is

$$\sum_k R_k \sum_l \left(G_l \rho_C G_l^\dagger \right) R_k^\dagger$$
$$= \sum_{k | d_k \neq 0} \frac{1}{\sqrt{d_k}} \sum_i |\phi_i\rangle\langle\phi_i| F_k^\dagger \sum_l \left(G_l \rho_C G_l^\dagger \right) \frac{1}{\sqrt{d_k}} \sum_j F_k |\phi_j\rangle\langle\phi_j|. \tag{2.48}$$

We wish to show that if we operate on $\rho_C = |\phi_m\rangle\langle\phi_n|$ we will again obtain something proportional to ρ_C. Thus we obtain

$$\sum_{k | d_k \neq 0} \frac{1}{d_k} \sum_i |\phi_i\rangle\langle\phi_i| F_k^\dagger \sum_l \left(G_l |\phi_m\rangle\langle\phi_n| G_l^\dagger \right) \sum_j F_k |\phi_j\rangle\langle\phi_j|. \tag{2.49}$$

Substituting our expression for G_l as a sum over F_k yields

$$\sum_{k|d_k \neq 0} \frac{1}{d_k} \sum_i |\phi_i\rangle\langle\phi_i|F_k^\dagger \sum_l \left(\sum_p f_{lp} F_p |\phi_m\rangle\langle\phi_n| \sum_q f_{lq}^* F_q^\dagger \right) \sum_j F_k |\phi_j\rangle\langle\phi_j|. \quad (2.50)$$

Using the quantum error-correcting criterion, we see that this becomes

$$\sum_{k|d_k \neq 0} \frac{1}{d_k} \sum_{iljpq} |\phi_i\rangle d_k \delta_{pk} \delta_{im} d_k \delta_{qk} f_{lp} f_{lq}^* \delta_{jn} \langle\phi_j| = \sum_{kl} d_k f_{lk} f_{lk}^* |\phi_m\rangle\langle\phi_n|$$

$$= \left(\sum_{kl} d_k f_{lk} f_{lk}^* \right) \rho_C. \quad (2.51)$$

So, even for this linear sum of errors, we correctly restore the coded subspace.

What have we done? We have shown that even though we have designed a code to correct E_k operators, it can in fact correct any linear sum of these operators. This is great! Why? Because, for example, if we want to correct a superoperator that has one qubit that has been arbitrarily erred (and only one qubit), then we need only consider a code that corrects X, Y, and Z errors, since every single-qubit error operator can be written as a sum of these errors (plus identity, which we, by default almost always include in our possible errors). This is what is known as making the errors discrete or digital. This discovery, that a code that was designed to correct a discrete set of errors can also correct a continuous set of errors, is one of the most important discoveries in all of quantum computing. The reason for this property is that quantum theory is linear. This linearity has a lot to do with why we can treat amplitudes like fancy probabilities and indeed, when we view quantum theory this way, we are not quite as surprised as if we thought about the components of a wave function as being some parameters with a reality all their own.

2.7 Classical linear codes

Above we have seen one of the most basic QECCs, the Shor code. The Shor code is a [[9,1,3]] QECC. In order to discuss further QECCs, it is useful to understand a class of traditional classical codes known as classical linear codes (of course they were known just as linear codes until quantum computing came along). In this section we will discuss classical binary linear codes, a subset of classical linear codes.

A *classical binary code* is a method for encoding k bits into n bits. In other words, for every one of the 2^k combinations of the k bits, a different n-bit vector is used to encode these k bits (so, obviously $k \leq n$). A classical binary code is called *linear* if the set of binary strings used in the code forms a closed linear subspace C of F_2^n. Recall that F_2 is the finite field with two elements, which we can call 0 and 1. Since F_2 is a field we can add and multiply these numbers using the normal rules of arithmetic with the only modification that addition is done modulo 2. We can then create a linear vector space of dimension n, F_2^n, by considering ordered sets of n elements of F_2: these are just binary vectors. In the linear vector space we can now add two elements, which corresponds to the process of bit-wise addition modulo 2 of the elements of the two elements. To be clear, an example of the addition of two vectors over F_2^5 is $01100 + 01010 = 00110$. Using these simple definitions, a closed linear subspace C of F_2^n is a set of vectors such that if the codeword v_1 is in C and v_2 is in C, then $v_1 + v_2$ is in C.

Since a classical linear binary code forms a closed linear subspace, we may express any codeword in a complete basis, $\{w_i\}_{i=1}^k$, for this subspace:

$$v = \sum_{i=1}^{k} a_i w_i, \qquad (2.52)$$

for $v \in C$, where $a_i \in \{0, 1\}$. From this expression we can explicitly see how to encode into such a code: if we let the a_i denote the k-bit binary string we wish to encode, then we see that, given a basis $\{w_i\}_{i=1}^k$, we can encode by simply taking the appropriate linear combination of the basis vectors. Of course there are different choices of basis vectors, and while these lead to different explicit schemes for encoding, the codes in different bases are essentially the same.

Given the basis expansion in Eq. (2.52) it is natural to define the *generator matrix*. This is a matrix with a set of basis vectors for the rows of the matrix:

$$G = \begin{pmatrix} w_1 \\ w_2 \\ \vdots \\ w_k \end{pmatrix}. \qquad (2.53)$$

Thus, the generator matrix for a k-bit linear code encoding into n bits is a k by n dimensional matrix. With the generator matrix one can immediately see how to encode a binary k-bit vector by simply taking the appropriate left multiplication. If a is the k-bit vector (a in F_2^k), then the encoding procedure is simply the matrix multiplication $a^T G$.

An example of a generator matrix is the matrix for the seven-bit Hamming code. In this case the generator matrix is

$$G = \begin{bmatrix} 1 & 0 & 1 & 0 & 1 & 0 & 1 \\ 0 & 1 & 1 & 0 & 0 & 1 & 1 \\ 0 & 0 & 0 & 1 & 1 & 1 & 1 \\ 1 & 1 & 1 & 0 & 0 & 0 & 0 \end{bmatrix}. \qquad (2.54)$$

To encode 0110 using this matrix we left multiply by the row vector $(0, 1, 1, 0)$ and obtain the result $(0, 1, 1, 1, 1, 0, 0)$. Thus 011 is encoded as 0111100 using this generator matrix.

The generator matrix is a nice way to specify a linear subspace, but what about the error-correcting properties of such codes? In order to discuss error correction on classical binary linear codes, it is useful to define the *parity check* matrix. The parity check matrix, H, is defined such that its null space is the code C. In other words, the check matrix is a $n - k$ set of constraints such that

$$Hv = 0, \qquad (2.55)$$

for all $v \in C$. H is an $n - k$ by n matrix. Here 0 is a k-dimensional column vector made of zeros. Note that these dimensions arise because H has rows made up of the maximal number of linearly independent vectors that are orthogonal to the subspace C.

What use is the parity check matrix? Well, as one can guess from its name, it is used to check whether an error has occurred on information encoded into the binary code and to (hopefully) correct this error. Classical errors are simply bit-flips, so any error on a codeword v can be represented by addition over F_2^n. Thus, if e is an n-bit binary vector with 1s where the bit-flips occur, then the result of this error on the codeword is $v + e$. If we apply H to this resulting erred codeword, we obtain $H(v+e) = He$, since v is in C. He is an $n-k$-bit binary string which we call the *syndrome*.

Suppose that we wish to correct a set of errors $\mathscr{S} = \{e_i\}$. If for every error e_i there is a unique syndrome He_i, then we can correct this set of errors. Suppose that $w = v + e_i$ is the error e_i acting on codeword v. Error correction is then simply performed by calculating the syndrome $Hw = He_i$. Since He_i uniquely identifies the error e_i, we can apply to w the bit-flips corresponding to e_i. In other words, we can perform the error-correction procedure $w + e_i$, which since $x + x = 0$ over F_2^n, is equal to $w + e_1 = v + e_1 + e_1 = v$. Thus, we see that Hw denotes the syndrome of what error has occurred. Contrariwise, if there are two errors, $e_1 \neq e_2$, that have the same syndrome, then if the error was e_1 we might apply e_2 to fix the error, resulting in $w + e_1 + e_2$. Since $e_1 + e_2$ does equal 0, this will result in an error on the encoded information.

For the seven-bit code described in Eq. (2.54), the parity check matrix is

$$H = \begin{bmatrix} 1 & 0 & 0 & 1 & 0 & 1 & 1 \\ 0 & 1 & 0 & 1 & 1 & 0 & 1 \\ 0 & 0 & 1 & 1 & 1 & 1 & 0 \end{bmatrix}. \tag{2.56}$$

One can check that all single-bit errors give different syndromes using this parity check matrix. To see this, note that single-bit-flip errors correspond to errors e with one 1 and six 0s. When applying H to these vectors one obtains as an output a three-bit syndrome made up of one of the columns of H. Since all of these columns are different, we see that each such error yields a distinct syndrome.

Finally we introduce the equivalent notion of the parameters of classical linear binary code as we did for a QECC. In particular, we say a code is an $[n, k, d]$ code if it encodes k bits into n bits and the distance of the code is d. The distance of a code is the weight (number of 1s in the vector describing the error) of the smallest error that transforms one codeword into another. A code with a distance of d can correct $t = \lfloor \frac{d-1}{2} \rfloor$ errors. (The picture you should have in mind is that the codewords are separated by a distance d and thus in balls of radius t there must be unique codewords that allow one to uniquely diagnose the error.)

One further concept we will find useful in QECCs is the notion of the *dual* of a classical linear code. Recall that the generator matrix, G, for an $[n, k, t]$ code C is a $k \times n$ matrix and the parity check matrix, H, is an $(n-k) \times n$ matrix. One can then define the dual of C, denoted $C^\perp$, by exchanging the role of the parity check and generator matrices. In particular, if we define $G^\perp = H$ and $H^\perp = G$, then this defines an $[n, n-k, t']$ classical linear binary code. The requirement that the parity check matrix acting on a codeword yields 0 can be seen to be satisfied by noting that this condition for the original code C is $HG^T = 0$, and taking the transpose of this equation yields the similar condition for the dual code, $GH^T = 0$. The dual code is made up of all the codewords that are orthogonal to the original codewords. Note that codewords in F_2^n can be orthogonal to itself (if it is made up of an even number of 1s it will be orthogonal to itself). Thus it is possible that a code C may contain as a subspace its dual code $C^\perp$. Further, it may even be possible, if $n = 2k$, that $C = C^\perp$. Such a code is called a *self-dual* code and will be used in the constructions described in the next section.

2.8 Calderbank, Shor, and Steane codes

We previously saw how Shor's nine-qubit QECC could be used to protect against bit-flip errors as well as against phase-flip errors. Remarkably, this code also protected against a combined bit-flip and phase-flip error and hence, by linearity, any combination of these errors. Here we will describe a similar set of codes that independently treat bit-flip and phase-flip errors. These codes are constructed from classical linear binary codes and are known as Calderbank, Shor, and Steane (CSS for short) codes [CS96, S96d].

Suppose we have an $[n, k, d]$ classical binary linear code C with generator matrix G and parity check matrix H. One could naturally define a QECC that can correct for bit-flips on this code, but it is unclear how this could be used to construct a code that deals with phase-flips. Consider, however, a state that is an equal superposition over the codewords in C:

$$|\psi\rangle = \frac{1}{\sqrt{2^k}} \sum_{v \in C} |v\rangle. \tag{2.57}$$

To understand how phase-flips act on this state, we apply a Hadamard transform to every qubit (such that the role of phase-flips and bit-flips is exchanged):

$$W^{\otimes n}|\psi\rangle = \frac{1}{\sqrt{2^{k+n}}} \sum_{v \in C} \sum_{w \in F_2^n} (-1)^{vw} |w\rangle, \tag{2.58}$$

where $vw = v_1 w_1 + v_2 w_2 + \cdots + v_n w_n \mod 2$. It is easy to verify that

$$\sum_{v \in C} (-1)^{vw} = \begin{cases} 2^k & \text{if } w \in C^\perp \\ 0 & \text{otherwise} \end{cases}, \tag{2.59}$$

and thus

$$W^{\otimes n}|\psi\rangle = \frac{1}{\sqrt{2^{n-k}}} \sum_{w \in C^\perp} |w\rangle. \tag{2.60}$$

This should give you an idea: since the Hadamard exchanged bit-flips and phase-flips, if the code $C^\perp$ corrects bit-flip errors, then we could protect $|\psi\rangle$ from phase-flip errors using this protection. Of course, at this point $|\psi\rangle$ is just a single state, not a QECC of any nontrivial dimension, but this is the basic idea of how CSS codes work.

Suppose that C_1 is an $[n, k_1, d_1]$ classical binary linear error-correcting code and C_2 is a subcode of C_1 that is a $[n, k_2, d_2]$ code (k_2 is less than k_1). Because C_2 is a subcode of C_1 we can define the different *cosets* of C_2 in C_1. In particular, consider the set $\mathscr{C}_w = \{v + w | v \in C_2\}$. These sets form a partition of C_2 into different cosets. In other words, it is possible to pick a set of $k_1 - k_2$ vectors such that each $\mathscr{C}_w$ is a unique set of vectors in C_1. We call these vectors *coset representatives*. Given this, we can now define the CSS codeword states. In particular, we define

$$|v\rangle = \frac{1}{\sqrt{2^{k_2}}} \sum_{w \in C_2} |w + v\rangle, \tag{2.61}$$

where v is a coset representative. Note that each of these states contains a superposition over codewords contained in C_1 and further, because the cosets do not overlap, these codewords are orthonormal.

The first idea here is that we will use the code C_1 to correct for bit-flip errors. Note that, as we have seen with the Shor code, the fact that we have a superposition of codewords is not an obstacle to protecting such states from bit-flip errors. The second idea is that we can use $C_2^\perp$ to protect for phase-flip errors. To see how this works, apply the n-qubit Hadamard transform to our codewords. A slight modification of our above calculation shows that

$$W^{\otimes n}|v\rangle = \frac{1}{\sqrt{2^{n-k_2}}} \sum_{w \in C_2^\perp} (-1)^{vw}|w\rangle. \tag{2.62}$$

Notice now that we have codewords that are superpositions over $C_2^\perp$. Thus, if $C_2^\perp$ can correct t_2 bit-flip errors, then we can use this correction (in the appropriately changed basis) to protect the quantum information in the above code space against t_2 phase-flip errors.

To summarize, if we have an $[n, k_1, d_1]$ code and an $[n, k_2, d_2]$ subcode whose dual is an $[n, n-k_2, d_3]$ code, the code described above is a QECC with parameters $[[n, k_1 - k_2, d]]$, where d is the minimum of d_1 and d_3. These codes are the *CSS codes*.

The most famous example of a CSS code is the $[[7, 1, 3]]$ Steane code. This code arises from the Hamming code we have described in the previous section. In particular, let C_1 be the $[7, 4, 3]$ Hamming code described in Section 2.7. Examining the parity check matrix for this code one can easily see that the dual of C_1 is contained in C_1 (this is the code whose generator matrix is the parity check matrix of C_1). For the Steane code we choose C_1 as the Hamming code and $C_2 = C_1^\perp$. The dual of C_2 is therefore C_1 again. Thus, since C_1 has a distance of 3, the Steane code has a distance of 3. Further, since C_2 encodes one less bit than C_1, the Steane code can encode 1 qubit. Thus we see, as claimed, that the Steane code is a $[[7, 1, 3]]$ QECC.

2.9 Stabilizer quantum error-correcting codes

CSS codes were among the first QECCs discovered. Following their discovery, a class of codes that are more general than the CSS codes was discovered, which had the property of also unifying other known QECCs [G96a, CRS+98]. These codes are known as stabilizer QECCs. The theory of stabilizer QECCs is a tool of great use in quantum computing today, both within the theory of QEC and elsewhere. It is, of course, important to remind oneself that this is not the only type of QECC out there. In this section we describe the basis of stabilizer QECCs.

2.9.1 Anticommuting

Suppose that we have a set of states $|\psi_i\rangle$ that are $+1$ eigenstates of a Hermitian operator S, $S|\psi_i\rangle = |\psi_i\rangle$. Further suppose that T is an operator that anticommutes with S, $ST = -TS$ (T is not zero). Then it is easy to see that $S(T|\psi_i\rangle) = -TS|\psi_i\rangle = -(T|\psi_i\rangle)$. Thus, the states $T|\psi\rangle$ are -1 eigenstates of S. Since the main idea of QEC is to detect when an error has occurred on a code space, such pairs of operators S and T can be used in such a manner: if we are in the $+1$ eigenvalue subspace of S then an error of T on these subspace vectors will move to a -1 eigenvalue subspace of S: we can detect that this error has occurred.

In fact, we have already seen an example of this in the bit-flip code. Recall that we noted that the code subspace for this code is spanned by $|000\rangle$ and $|111\rangle$ and that these two operators are $+1$ eigenstates of both $S_1 = Z \otimes Z \otimes I$ and $S_2 = Z \otimes I \otimes Z$. Further note that $(X \otimes I \otimes I)S_1 = -S_1(X \otimes I \otimes I)$ and $(X \otimes I \otimes I)S_2 = -S_2(X \otimes I \otimes I)$. Thus, if we start out in the $+1$ eigenvalue subspace of both S_1 and S_2 (like the bit-flip code), then if a single bit-flip occurs on the first qubit, we will now have a state that is in the -1 eigenvalue subspace of both S_1 and S_2. This at least fulfills our requirement that our errors should take us to orthogonal subspaces.

More generally, consider the following situation. Suppose that we have a set of operators S_i such that our code space is defined by $S_i|\psi\rangle = |\psi\rangle$ for $|\psi\rangle$ in the code subspace. Now suppose that we have errors E_i such that the products $E_k^\dagger E_l$ always anticommute with at least one S_i. Recall that the quantum error-correcting criterion was

$$\langle\phi_i|E_k^\dagger E_l|\phi_j\rangle = C_{kl}\delta_{ij}.$$

Since the codewords are $+1$ eigenvalue eigenstates of S_i, we find that

$$\langle\phi_i|E_k^\dagger E_l|\phi_j\rangle = \langle\phi_i|E_k^\dagger E_l S_i|\phi_j\rangle. \tag{2.63}$$

Suppose that S_i is one of the particular S_is that anticommute with $E_k^\dagger E_l$. Then this is equal to

$$\langle\phi_i|E_k^\dagger E_l|\phi_j\rangle = \langle\phi_i|E_k^\dagger E_l S_i|\phi_j\rangle = -\langle\phi_i|S_i E_k^\dagger E_l|\phi_j\rangle. \tag{2.64}$$

But, since S_i acts as $+1$ on the code space, this is just

$$\langle\phi_i|E_k^\dagger E_l|\phi_j\rangle = \langle\phi_i|E_k^\dagger E_l S_i|\phi_j\rangle = -\langle\phi_i|S_i E_k^\dagger E_l|\phi_j\rangle = -\langle\phi_i|E_k^\dagger E_l|\phi_j\rangle. \tag{2.65}$$

This implies that

$$\langle\phi_i|E_k^\dagger E_l|\phi_j\rangle = 0. \tag{2.66}$$

Thus, we have shown that, given the S_i and $E_k^\dagger E_l$ that properly anticommute with these S_i, the set of errors $\{E_k\}$ satisfies the quantum error-correcting criterion and therefore the code space is a valid QECC for these errors.

This trick, of defining the states as being the $+1$ eigenstates of some operators and then noting that if the product of error terms anticommutes then this is a valid quantum error-correcting code, is at the heart of the reason we use the stabilizer formalism. But what should we use for the S_is? Well, you might already be guessing what to use, because you recall that the Pauli group has nice commuting, anticommuting properties and eigenvalues that are ± 1 (or $\pm i$). Indeed, this is what we will use.

2.9.2 The Pauli and stabilizer groups

2.9.2.1 Pauli group First recall the definition of a group. A group is a set of objects $\mathscr{G}$ along with a binary operation of multiplication that satisfies (0) [closure] $g_1 g_2 \in \mathscr{G}$ for all $g_1, g_2 \in \mathscr{G}$, (1) [associativity] $g_1(g_2 g_3) = (g_1 g_2)g_3$ for all $g_1, g_2, g_3 \in \mathscr{G}$, (2) [identity] there exists an element $e \in \mathscr{G}$ such that for all $g_1 \in \mathscr{G}$, $g_1 e = g_1$, (3) [inverse] for every $g_1 \in \mathscr{G}$ there exists an element $g_2 \in \mathscr{G}$ such that $g_1 g_2 = e$, which we call the inverse of g_1, written $g_2 = g_1^{-1}$. The *Pauli group* is a particular group that satisfies these group axioms. Actually, people in quantum computing are very sloppy, and when they refer to the Pauli group they are usually refering to a particular representation of the Pauli group by unitary matrices. We will slip into this nomenclature soon enough; having learned a bad habit it is hard to go back.

What is this representation of the Pauli group, $\mathscr{P}_n$? Recall that the Pauli operators on a single qubit are $\{I, X, Y, Z\}$. The representation of the Pauli group we will deal with is the group formed by elements of the form $i^k P_1 \otimes P_2 \otimes \cdots \otimes P_n$, where each P_i is an element of $\{I, X, Y, Z\}$ and $k \in \{0, 1, 2, 3\}$. From this representation, we see that the Pauli group is a non-Abelian group, i.e., the elements of the group do not all commute with each other. Some important properties of elements of the Pauli group:

(i) Elements of the Pauli group square to $\pm I$: $P^2 = \pm I$.
(ii) Elements of a the Pauli group either commute $PQ = QP$ or anticommute $PQ = -QP$.
(iii) Elements of the Pauli group are unitary $PP^\dagger = I$.

2.9.2.2 Stabilizer group Define a *stabilizer group* $\mathscr{S}$ as a subgroup of $\mathscr{P}_n$ which has elements that all commute with each other and which does not contain the element $-I$. An example of a stabilizer group on three qubits is the group with elements $\mathscr{S} = \{III, ZZI, ZIZ, IZZ\}$. Notice that here we have dropped the tensor product between the elements, i.e., $ZZI = Z \otimes Z \otimes I$. We usually do not specify all of the elements of the stabilizer group. Instead we specify a minimal set of generators. A set of generators of a group is a set of elements of the group such that multiplication of these generators leads to the full group. A minimal set of such generators is a set of generators of minimal size that has this property. In the example of $\mathscr{S} = \{III, ZZI, ZIZ, IZZ\}$, this group is generated by ZZI and ZIZ: $(ZZI)(ZIZ) = IZZ$ and $(ZZI)^2 = III$. We write this fact as $\mathscr{S} = \langle ZZI, ZIZ \rangle$. For a stabilizer $\mathscr{S}$ we write a set of minimal generators as $S_1, S_2, \ldots, S_r$.

Now, since $-I$ is not in the stabilizer, all elements of our stabilizer must square to $+I$. Operators that square to $+I$ must have eigenvalues $+1$ or -1. Since $\mathscr{S}$ is Abelian, we can write a generic element of the stabilizer as $S_1^{a_1} S_2^{a_2} \cdots S_k^{a_r}$. Since $S_i^2 = I$, $a_i \in \{0, 1\}$. Further, for each $a \in Z_2^r$ the element of the stabilizer so specified is unique. Suppose that this was not true, that $S_1^{a_1} S_2^{a_2} \cdots S_k^{a_r} = S_1^{b_1} S_2^{b_2} \cdots S_k^{b_r}$ for $a \neq b$. Then $S_1^{a_1+b_1} S_2^{a_2+b_2} \cdots S_k^{a_r+b_r} = I$. But this is only true if $a_i = b_i$.

2.9.3 Stabilizer subspace and error correction

Now, given a stabilizer group $\mathscr{S}$, we can define a subspace on our n qubits. In particular we define this subspace as all states $|\psi\rangle$ that satisfy $S|\psi\rangle = |\psi\rangle$ for all stabilizer elements $S \in \mathscr{S}$. Actually, we do not need all of these equations to define the code space. All we need are the equations for the generators of the stabilizer: $S_i|\psi\rangle = |\psi\rangle$. Let us call the subspace defined by these equations $\mathscr{H}_S$. Such stabilizer subspaces are very nice. One reason that they are nice is that instead of specifying the states of the subspace we can just specify the generators of the stabilizer group. This is oftentimes much easier. Further, as we will see, it is easy to figure out the error-correcting properties of stabilizer subspaces.

In particular, suppose we have a stabilizer subspace for a code generated by $S_1, S_2, \ldots, S_r$. Then, suppose the Pauli operator P anticommutes with one of these generators S_i. Then, as above, for $\{|\psi_i\rangle\}$ a basis for $\mathscr{H}_S$,

$$\langle \psi_i | P | \psi_j \rangle = \langle \psi_i | P S_i | \psi_j \rangle = -\langle \psi_i | S_i P | \psi_j \rangle, \tag{2.67}$$

so

$$\langle \psi_i | P | \psi_j \rangle = 0. \tag{2.68}$$

Thus, as above, if $\{E_a\}$ is a set of Pauli group errors, if we consider the products $E_a^\dagger E_b$ and these anticommute with at least one of the generators of $\mathscr{S}$, then we have satisfied the error-correcting criterion for these errors:

$$\langle \psi_i | E_b^\dagger E_a | \psi_j \rangle = 0. \tag{2.69}$$

If these elements are themselves elements of the stabilizer, $E_a^\dagger E_b \in \mathscr{S}$, then

$$\langle \psi_i | E_b^\dagger E_a | \psi_j \rangle = \langle \psi_i | S | \psi_j \rangle = \delta_{ij}. \tag{2.70}$$

For an error set $\{E_a\}$, if all of the products $E_a^\dagger E_b$ either anticommute with generators of the stabilizer $S_1, S_2, \ldots, S_r$ or are elements of the stabilizer, then we see that the E_a satisfy the quantum error-correcting criterion.

In our example where $\mathscr{S} = \langle ZZI, ZIZ \rangle$, we can consider the set of errors comparising the identity operator and all single-qubit bit-flip errors, $\{III, XII, IXI, IIX\}$. Then the set of products of these errors is $\{III, XII, IXI, IIX, XXI, XIX, IXX\}$. Of these, the first III is in the stabilizer. All of the others, however, anticommute with either ZZI or ZIZ. For example, $(XXI)(ZIZ) = -(ZIZ)(XXI)$. How do we check whether two Pauli group elements commute or anticommute? Suppose these group elements are $P_1 \otimes P_2 \otimes \cdots \otimes P_n$ and $Q_1 \otimes Q_2 \otimes \cdots \otimes Q_n$. Then we count the locations where P_i and Q_i differ and neither P_i nor Q_i is I. If this number is even, then these two Pauli group elements commute, and if it is odd, then they anticommute.

If a stabilizer group has a minimal number of generators, which are $S_1, S_2, \ldots, S_r$, what is the dimension of the stabilizer subspace? Take the first stabilizer generator. This stabilizer generator squares to identity, so has ± 1 eigenvalues. Further, this stabilizer generator has trace zero. Why? All of the Pauli operators are trace 0 except I which has trace 2. Since the stabilizer generator cannot be identity (unless the stabilizer consists solely of the identity, a case we will disregard) it is a tensor product of terms, at least one of which must be a Pauli element not equal to I. Then, since $\text{Tr}[A \otimes B] = \text{Tr}[A]\text{Tr}[B]$, this implies that $\text{Tr}[P] = 0$, for all Pauli group elements except $\pm I$. Thus, if we take S_1 it must have 2^{n-1} eigenvalues $+1$ and 2^{n-1} eigenvalues -1. So $S_1|\psi\rangle = |\psi\rangle$ splits the Hilbert space of our n qubits in half. What happens when we impose $S_2|\psi\rangle = |\psi\rangle$? Note that $\frac{1}{2}(I + S_1)$ is the projector onto the $+1$ eigenvalue eigenspace of S_1. We can thus use $\frac{1}{2}(I + S_1)S_2$ to understand how much of this subspace has eigenvalues of S_2 that are $+1$. Note that $\text{Tr}[\frac{1}{2}(I + S_1)S_2] = 0$. Thus, we see that for the 2^n-dimensional subspace that satisfies $S_1|\psi\rangle = |\psi\rangle$, a subspace of dimension 2^{n-2} satisfies $S_2|\psi\rangle = |\psi\rangle$. Continuing inductively, we see that each $S_i|\psi\rangle = |\psi\rangle$ cuts the space of the previous $S_1|\psi\rangle = |\psi\rangle, \ldots, S_{i-1}|\psi\rangle = |\psi\rangle$ in half.

What does this mean? This means that the dimension of a stabilizer subspace for a stabilizer with r generators is 2^{n-r}. It is also useful to notice that the dimension of the subspace with a fixed set of ± 1 eigenvalues of the S_i operators is also 2^{n-r}. For our example of $S_1 = ZZI$ and $S_2 = ZIZ$, this implies that the stabilizer subspace is two-dimensional, which is correct.

2.9.4 Logical operators for stabilizer codes

So, given a stabilizer group $\mathscr{S}$ generated by r elements, we now know that this specifies a subspace of dimension 2^{n-r}. This subspace can be used to encode $k = n - r$ qubits. Is there a nice way to talk about this subspace as k qubits? Indeed there is.

The centralizer of the stabilizer group $\mathscr{S}$ in $\mathscr{P}_n$ is the set of operators $P \in \mathscr{P}_n$ that satisfy $PS = SP$ for all $S \in \mathscr{S}$. Since our stabilizer group does not contain $-I$, it turns out that the centralizer is equal to the normalizer. The normalizer $\mathscr{N}$ of $\mathscr{S}$ in $\mathscr{P}_n$ is the set of operators $P \in \mathscr{P}_n$ such that $PSP^\dagger \in S$ for all $S \in \mathscr{S}$. Notice that the stabilizer is automatically in the normalizer, since all of the elements of the stabilizer commute with each other and are all unitary. An important set of operators are those that are in the normalizer $\mathscr{N}$ but not in the stabilizer, $\mathscr{N} - \mathscr{S}$. Why are these elements important? Because they represent *logical Pauli* operators on the k encoded qubits of our subsystem code.

Let us see this for an example and then move on to understanding the logical operators in general. Our example is the stabilizer group generated by $S_1 = ZZI$ and $S_2 = ZIZ$. Then, elements of $\mathscr{N} - \mathscr{S}$ are $i^k \times \{XXX, YYX, YXY, XYY, ZII, IZI, IIZ, ZZZ, YYY, XXY, XYX, YXX\}$. Notice that if we take the group generated by the two operators XXX and ZII, each of these elements is equal to such a group member times an element of the stabilizer group. What does XXX do to our code space? Recall that the stabilizer subspace in this example is spanned by $|0_L\rangle = |000\rangle$ and $|1\rangle_L = |111\rangle$. Thus we see that $XXX|0_L\rangle = |1_L\rangle$, $XXX|1_L\rangle = |0_L\rangle$ and $ZII|0_L\rangle = |0_L\rangle$, $ZII|1_L\rangle = -|1_L\rangle$. In other words, XXX acts like an encoded X operation on the subspace and ZII acts like an encoded Z operation on the subspace. Similarly, one can calculate that YXX acts as Y on the code subspace. Also notice that these operators preserve the code subspace. Now, since all the other elements of the normalizer are either stabilizer elements or products of stabilizer elements, they will also preserve the code subspace.

Following this example, we are motivated to define the group $\mathscr{N}/\mathscr{S}$. This is the normalizer group quotient the stabilizer. We can write every element of $\mathscr{N}$ as RS, where S is a stabilizer element and R is not. Then, multiplication of these elements is like $R_1 S_1 R_2 S_2 = (R_1 R_2)(S_1 S_2)$. This defines a multiplication rule for the R_is. This group is the group $\mathscr{N}/\mathscr{S}$. It is possible to show that this group is equal to the Pauli group of size $k = n - r$. This means that it is possible to generate this group by a set of Pauli operators $\bar{X}_1, \bar{Z}_1, \ldots, \bar{X}_k, \bar{Z}_k$ (along with a phase $i^k I$). These operators are the encoded Pauli operators on the subspace. (One thing to note is that this choice of division into k different qubits is not unique. However, we will rarely deal with the case where there is more than one encoded qubit, so this will not matter much for us.)

Elements of $\mathscr{N}$ represent nontrivial operations on the encoded subspace. They are, then, exactly the type of operators whose action we cannot detect on our QECC. Using this fact, it is possible to give a very nice characterization of the quantum error-correcting criterion. Suppose that we have a stabilizer $\mathscr{S}$ with normalizer $\mathscr{N}$. Let E_a denote a set of Pauli errors on the qubits. If $E_a^\dagger E_b \notin \mathscr{N} - \mathscr{S}$ for all possible error pairs, then E_a is a set of correctable errors. Why is this so? Well, there are two cases. One is that $E_a^\dagger E_b$ is in $\mathscr{S}$. Then $\langle \phi_i | E_a^\dagger E_b | \phi_j \rangle = \langle \phi_i | \phi_j \rangle = \delta_{ij}$ since $E_a^\dagger E_b$ acts trivially on the stabilizer subspace. The other case is that $E_a^\dagger E_b$ is not in the stabilizer and it is also not in the normalizer. This implies that there must exist an element of the stabilizer S such that $E_a^\dagger E_b S \neq S E_a^\dagger E_b$ (recall the centralizer and normalizer are the same for stabilizers). But all elements of the Pauli group either commute or anticommute. This implies that $E_a^\dagger E_b S = -S E_a^\dagger E_b$. By our previous argument this implies that $\langle \phi_i | E_a^\dagger E_b | \phi_j \rangle = 0$.

2.9.5 Examples of stabilizer codes

Here we present some examples of stabilizer codes.

2.9.5.1 Three-qubit bit-flip and phase-flip codes The example we have been dealing with, which is generated by ZZI and ZIZ, is able to correct a single bit-flip. This code is called the three-qubit bit-flip code. Its stabilizer and logical operators are

Element	Operator
S_1	ZZI
S_2	ZIZ
$\bar{X}$	XXX
$\bar{Z}$	ZII

Similarly, we can construct the code that is designed to correct single phase flip errors. Recall that this code was related to the bit-flip code by a Hadamard change of basis. This implies that its stabilizer and logical operators are

Element	Operator
S_1	XXI
S_2	XIX
$\bar{X}$	XII
$\bar{Z}$	ZZZ

Let us get on to codes that can correct single-qubit errors. We have already seen an example of such a single-qubit error-correcting code, the Shor code. This is an example of a Shor [[9,1,3]] quantum code. It turns out, since this code is really a concatenation of the bit-flip and phase-flip codes, that the Shor code is also a stabilizer code. In fact, we see that in the Shor code we use three bit-flip codes plus a single phase-flip code on the encoded information. From this it is easy to deduce what the stabilizer of the Shor code is.

Element	Operator
S_1	$ZZIIIIIII$
S_2	$ZIZIIIIII$
S_3	$IIIZZIIII$
S_4	$IIIZIZIII$
S_5	$IIIIIIZZI$
S_6	$IIIIIIZIZ$
S_7	$XXXXXXIII$
S_8	$XXXIIIXXX$
$\bar{X}$	$XXXXXXXXX$
$\bar{Z}$	$ZZZZZZZZZ$

Notice that this is a degenerate code for single-qubit errors: operators like $ZZIIIIII$ act as identity on the code space even though they are products of two single-qubit errors.

The nine-qubit code that Shor came up with is a rather large code (although it has a certain beauty to it in its simplicity). What is the smallest code that can correct a single qubit error? This code is the five-qubit QECC [BDS+96, LMP+96]. It is specified by

Element	Operator
S_1	$XZZXI$
S_2	$IXZZX$
S_3	$XIXZZ$
S_4	$ZXIXZ$
$\bar{X}$	XXXXX
$\bar{Z}$	$ZZZZZ$

Notice that this code has stabilizer generators that are related to each other by a cyclic shift of the qubits. This makes it rather easy to remember its stabilizer. The five-qubit code is notable in that it is not an example of a CSS code. CSS codes have stabilizers that can be written as operators that consist of all X or all Z (and I) operators in an appropriate chosen computational basis. This complicates things for this code when we look at the issue of fault tolerance; so while this is a nice code, it is not as nice as we would like. The five-qubit code is a [[5,1,3]] code. It is useful to test your ability to spot when elements of the Pauli group commute or anticommute on the five-qubit code, since it has nontrivial stabilizer elements.

The [[7,1,3]] Steane code that we discussed above can be written in stabilizer form as follows:

Element	Operator
S_1	IIIXXXX
S_2	IXXIIXX
S_3	XIXIXIX
S_4	$IIIZZZZ$
S_5	$IZZIIZZ$
S_6	$ZIZIZIZ$
$\bar{X}$	XXXXXXX
$\bar{Z}$	$ZZZZZZZ$

The Steane code is nice since it is a CSS code. We will use it in a lot of our examples since it has some very useful properties and is easy to discuss, but not so big as to prevent us writing down statements about it in explicit form.

2.9.6 Measurement of Pauli group projectors

Now that we have defined stabilizer codes, it is useful to discuss how to use them in practice. One essential task we will need to perform is to make a projective measurement onto the $+1$

and -1 eigenvalue eigenspaces of a general Pauli operator P that squares to identity, $P^2 = I$. How do we do this? Well, we have already basically seen a trick for doing this. Consider the following circuit:

$$\begin{array}{c}\text{—}\boxed{P}\text{—}\\ |+\rangle \text{——}\bullet\text{——}\boxed{W}\text{—}\boxed{\measuredangle}\end{array} \qquad (2.71)$$

What is the effect of this circuit? We can calculate that if our measurement obtains the result $|0\rangle$, then the measurement operator we are measuring on the first qubit is

$$M_0 = \langle 0|_B U|+\rangle_B = \frac{1}{2}(I + P), \qquad (2.72)$$

and if the measurement outcome is $|1\rangle$, then the measurement operator is

$$M_1 = \langle 1|_B U|+\rangle_B = \frac{1}{2}(I - P). \qquad (2.73)$$

Thus, we see that this circuit can be used to make a projective measurement onto the $+1$ and -1 eigenvalue eigenspaces of P. This is a very useful primitive for stabilizer codes. Notice that, unlike in our introduction to error correction codes, here in order to make a measurement we need to attach an ancilla qubit and will end up with the result of the measurement in this ancilla qubit. Further note that the measurement is not destructive on the original qubit, in the sense that while the measurement projects onto the subspaces, it leaves the resulting state in the appropriate subspace.

2.9.7 Error recovery routine for stabilizer codes

Given a stabilizer code, how do we perform a procedure for QEC?

Suppose that we have a set of Pauli errors E_a that are correctible via the error-correcting routine. Now we know that there must be a recovery routine for these errors. What is this recovery routine?

The first possibility for an error E_a is that it is in the stabilizer. In this case there is no need to perform QEC, since the effect of a stabilizer operator on our encoded quantum information is identity $E_a|\psi\rangle = |\psi\rangle$. So that case was easy! A second case is that E_a is not in the stabilizer. Now we know that E_a, since it satisfies the error-correcting criterion, must take us to an orthogonal subspace. What is this orthogonal subspace? Suppose that the stabilizer generators are $S_1, \ldots, S_r$. Then if E_a is a correctable error not in the stabilizer, it will anticommute with some of these generators S_i. For these generators, the state $E_a|\psi\rangle$ will no longer be in the $+1$ eigenvalue subspace, but will be in the -1 eigenvalue eigenspace. To see this, simply note that $S_i E_a|\psi\rangle = -E_a S_i|\psi\rangle = -E_a|\psi\rangle$ for $|\psi\rangle \in \mathcal{H}_S$. Further, if E_a commutes with S_i, then it does not change the eigenvalue of S_i. Thus, we see that the subspace the error sends us to is labeled by the ± 1 eigenvalues of the stabilizer operators. Thus, to perform QEC on a stabilizer code, it is enough to make measurements that project onto the ± 1 eigenvalue eigenspaces of the stabilizers. From this information, one can deduce what the error was and then apply an appropriate recovery operation. We call the values of the measured stabilizer generators the syndrome of the error. Now, having diagnosed what the error is by measuring the syndrome, one then applies the appropriate Pauli operator to reverse the error. One interesting fact is that, for degenerate codes, there are often multiple errors corresponding to a syndrome. In this case, however, one just reverses one of these errors and this guarantees that the net effect of this procedure is to either apply I or a stabilizer element, which is the same as applying I to the code space.

Now, how do we perform a measurement onto the ± 1 eigenvalue eigenspace of the S_i operators? We can measure them using the circuit described above for measuring Pauli operators. Another important fact to note is that, since the stabilizer generators all commute with each other, measurements that project onto their eigenstates can be simultaneously performed. Here, for example, is the syndrome measurement circuit for the [[7,1,3]] code:

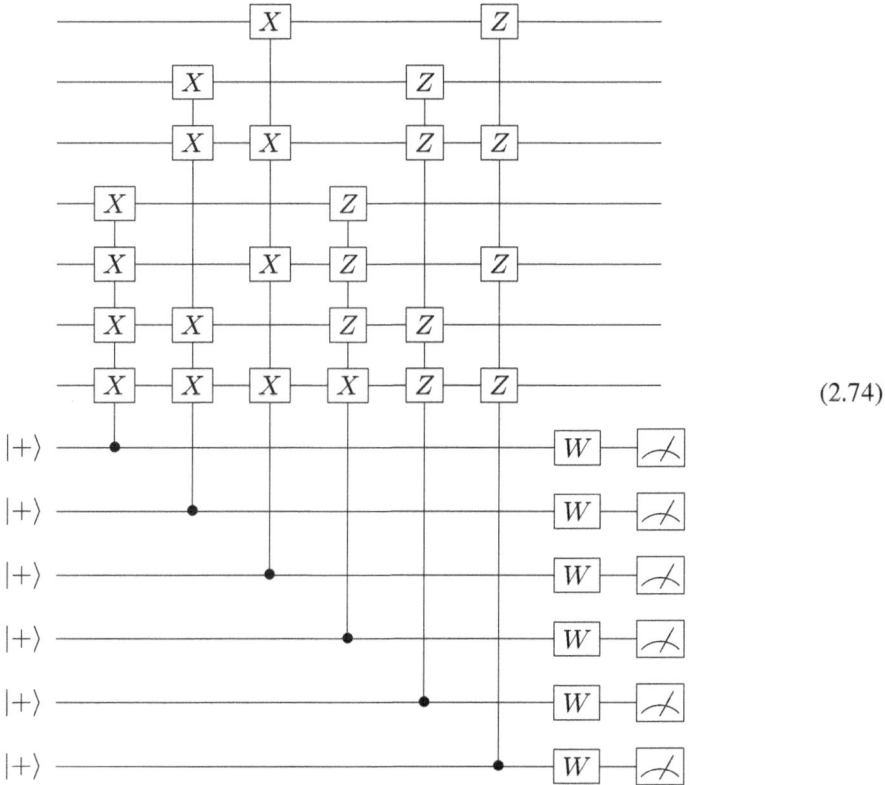

(2.74)

The top seven qubits are the encoded qubits. The bottom six qubits are the ancilla registers that will hold the syndrome operators.

2.9.8 Preparation and measurement of stabilizer states

How do we prepare a stabilizer state? Well, this is rather easy. For example, suppose we just prepare the $|0^n\rangle$ state for our bare, unencoded qubits. Now, if we measure the syndrome and apply the error correcting recovery procedure, we are guaranteed to be in the code subspace. Now if, say, we have one encoded qubit, we can measure in the encoded $|0\rangle$, $|1\rangle$ basis by measuring the eigenvalues of the $\bar{Z}$ operator (using the tricks we learned above). If this outcome is $+1$ we have prepared the encoded $|0\rangle$ state. If this outcome is -1 then we have prepared the encoded $|1\rangle$ state and application of the $\bar{X}$ operator will then turn this into the $|0\rangle$ state. For multiple encoded qubits a similar procedure can be enacted. Thus, we have seen how to prepare encoded stabilizer states starting in, say, the encoded $+1$ eigenstates of the $\bar{Z}$ operator. Similar procedures apply for preparation of eigenstates of the other encoded Pauli operators. Thus, we have seen how to prepare these simple states.

What about more general states? Well, those will be harder. We will discuss some examples of these preparation procedures in Chapter 5, when we talk about fault-tolerant quantum computation.

And a further note. We could also ask how to take a generic unencoded qubit and then design a circuit for encoding into a stabilizer code. But this primitive – encoding into a code – is not a very useful one for our purposes (building a quantum computer). Why? Well, because we really never want to work with unencoded quantum computation. We always want to work with encoded quantum information. Every once in a while you will come across an argument against quantum computing that discusses a decoherence mechanism that always leads to errors on a single qubit. Such arguments miss the point, simply because we never work with quantum information in a single qubit: we always work with encoded quantum information. (The same is true for computing with classical faulty elements: that this is true was first pointed out by John von Neumann.)

Finally, we can ask how do we perform measurements on our stabilizer states. In particular, suppose we have k qubits and we want to measure in the encoded computational basis. Then we can do this by performing our Pauli measuring circuit for the encoded operators $\bar{Z}_i$. It is thus simple to see how to measure encoded Pauli operators on our qubits.

So we have seen how to prepare computational basis states, and measure computational basis states for our encoded information. The next obvious question is: how do we perform quantum computation, i.e., unitary gates, on our encoded quantum information? We will take this up in the next section.

2.9.9 Gates on stabilizer codes

How do we perform unitary transforms on quantum information encoded into a stabilizer code? In some sense this question is trivial: just define unitary operations that carry out the desired unitary on the encoded subspace. On the other hand, this observation is not very useful for reasons that will become apparent when we talk about fault-tolerant quantum computation. Thus, here we will discuss a particularly beautiful and useful set of gates that we can implement on our stabilizer code.

Actually we have already seen some encoded operations on our code: the Pauli operators on our code. These operations were enacted on our encoded quantum information by the application of some Pauli group element. But in this section we will go beyond these simple gates.

2.9.9.1 The Clifford group Consider our Pauli group on n qubits, $\mathscr{P}_n$. This is a subgroup of the unitary group on n qubits, $U(2^n)$. An important group, and we will see why in a second, is the Clifford group (in an abuse of mathematical nomenclature we will not even begin to try to undo). The Clifford group is the normalizer of the Pauli group in the unitary group on n qubits. In other words, it is the group of operators N that satisfy $NPN^\dagger \in \mathscr{P}_n$ for all $P \in \mathscr{P}_n$. What is an example of such an operation? Our good friend the Hadamard operator. In fact, we can easily show that

$$WIW^\dagger = I, \quad WXW^\dagger = Z, \quad WYW^\dagger = -Y, \quad WZW^\dagger = X. \tag{2.75}$$

Thus, a Hadamard on a single qubit is an operator in the normalizer of the Pauli group in $U(2^n)$. Indeed, multiple Hadamards are also in this normalizer. What are other examples? Well, elements of $\mathscr{P}_n$ themselves are obviously in the normalizer. Perhaps the first really interesting example beyond the Hadamard operator is the CNOT operation. The effect of the CNOT on Pauli operators is so useful we will write the action down here:

$$\text{(circuit diagrams)} \tag{2.76}$$

$$\text{(circuit diagrams)} \tag{2.77}$$

Note that we can deduce the action of conjugating a CNOT about Y operations by simply applying the X and Z conjugations in turn. Thus we see that the CNOT operation is indeed in the Clifford group.

A great question to ask is, what is needed from the Clifford group on n qubits to generate the entire group? In fact, the Clifford group on n qubits can be generated by W and the CNOT, C_X, and one extra gate,

$$S = \begin{bmatrix} 1 & 0 \\ 0 & i \end{bmatrix}. \tag{2.78}$$

Note that this gate acts as $SXS^\dagger = Y$, $SYS^\dagger = -X$, and $SZS^\dagger = Z$. A proof of this result can be found in [G96a].

2.9.9.2 Clifford group gates on stabilizer codes Suppose that we have a stabilizer code generated by the stabilizer operators $S_1, \ldots, S_r$. We saw that elements of the normalizer of the stabilizer group in the Pauli group were logical operators on our stabilizer code. What if we now examine the normalizer of the stabilizer group, not just in the Pauli group, but now in the full unitary group $U(2^n)$. These gates will certainly be in the Clifford group, since the stabilizer group is made up of Pauli group operators. But now these operations might perform a nontrivial operation on our encoded qubits.

Let us look at an example. Consider the seven qubit Steane code with generators specified above. Next consider the operator $W^{\otimes 7}$. This operator when conjugated about stabilizer elements will produce another stabilizer element for the Steane code. How do we see this? Recall that $WXW = Z$ and vice versa. Now notice that if we conjugate $W^{\otimes 7}$ about one of the generators that is made up completely of X operations, then we obtain a new operator that is made up completely of Z operations. But now notice that there is a symmetry of the Steane code that guarantees that, for the generators of the stabilizer made up completely of X operations, there is an equivalent generator formed by replacing the X operations with Z operations. Thus we see that, on the generators of the stabilizer code, conjugating by $W^{\otimes 7}$ produces a stabilizer generator. Since all stabilizer elements are made as products of stabilizer generators, this implies that the stabilizer group is preserved under conjugation by $W^{\otimes 7}$.

So now the question arises as to what $W^{\otimes 7}$ does on our logical encoded qubit? Recall that the logical operators on the Steane code are $\bar{X} = X^{\otimes 7}$ and $\bar{Z} = Z^{\otimes 7}$. We can then see that

$W^{\otimes 7}\bar{X}W^{\otimes 7} = \bar{Z}$ and $W^{\otimes 7}\bar{Z}W^{\otimes 7} = \bar{X}$. Thus, we see that $W^{\otimes 7}$ acts in the same way on the encoded quantum gates as a single W does on a single Pauli. In fact, this is enough to imply that $W^{\otimes 7}$ is exactly the Hadamard gate on our encoded quantum information.

More generally, then, we can consider elements of the Clifford group that, when conjugated about the stabilizer generators, produce elements of the stabilizer. Such elements act to preserve the stabilizer subspace, since $NS_iN^\dagger = S_j$, and therefore $N|\psi_C\rangle = NS_i|\psi_C\rangle = NS_iN^\dagger N|\psi_C\rangle = S_jN|\psi_C\rangle$. Thus, $S_j(N|\psi_C\rangle) = N|\psi_C\rangle$, so again $N|\psi_C\rangle$ is still stabilized (further note that $NS_iN^\dagger$ preserves the group multiplication rule of the stabilizer and thus $NS_iN^\dagger$ if N is in the normalizer are again generators of the stabilizer group). Further, these gates also act on the encoded qubits as elements of the Clifford group on these encoded qubits.

It is important to note that, when it comes to Clifford group elements, not all stabilizer codes were created equal. In particular, implementing different Clifford group elements is often much more difficult on some codes than on other codes. This fact is one reason for choosing different stabilizer codes.

We have seen in this section how to perform some Clifford group elements on stabilizer codes. Even if a code supports all Clifford group operations on our code this is not enough to perform universal quantum computation on the code. Universal quantum computation is discussed much further in Chapter 5 on fault-tolerant quantum computation.

2.10 Conclusions

In this chapter we have defined the basics of QECCs. We have seen how it is possible to protect quantum information by suitably encoding the quantum information across multiple independently erred quantum systems. Since the origin of the destruction of quantum information by decoherence is often the effect of a system becoming entangled with its environment, the QEC procedure has been dubbed by John Preskill as "fighting entanglement with entanglement." We have shown that QECCs exist and derived the necessary and sufficient condition for such codes to succeed in correcting a set of errors. A class of QECCs derived from self-dual classical codes, the CSS codes, was described. A more general set of codes was then derived using the stabilizer formalism. Stabilizer codes are the bread and butter of many ideas for QEC in the real world, and are distinguished by the relatively easy formalism with which one can discuss these codes. However, one should note that the theory of QEC contains many more important codes than just stabilizer codes, and the development of "good" codes is an important ongoing endeavor (where good depends on the color of the crystal through which you look, i.e., the problem you wish to solve). Many of these codes will described in later chapters in this book.

2.11 History and further reading

Classical error correction has a long and important history, with a seminal piece of work being Shannon's paper founding the study of information theory [S48]. A good introduction to information theory and error-correction is the book of Cover and Thomas [CT91]. Quantum error-correcting codes were first discovered by Shor [S95] and Steane [S96c] in 1995 and 1996. Knill and Laflamme were the first to derive the necessary and sufficient quantum error-correcting criterion [KL97]. The five-qubit error-correcting code was discovered by Bennett et al. [BDS+96] and Laflamme et al. [LMP+96] in 1996. Stabilizer codes and a closely related approach using codes over $GF(4)$ were introduced by Gottesman [G96a] and by Calderbank *et al.* [CRS+98] . Gottesman's Ph.D. thesis on stabilizer codes is very readable and a good introduction to these codes [G96a]. The quantum Hamming bound was introduced by Ekert and Macchiavello [EM96] and shown to be achievable by Gottesman [G96a]. Gentle introductions to quantum computing, which include the basics of QEC, include the classic textbook of Nielsen and Chuang [NC00] and the more introductory books by Mermin [M07] and Loepp and Wootters [LW06].

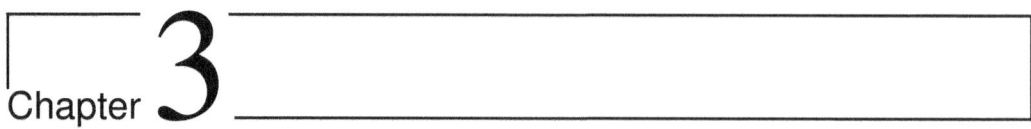

Introduction to decoherence-free subspaces and noiseless subsystems

Daniel A. Lidar

3.1 Introduction

In the previous chapter we saw how quantum information can be protected by *active* means. A combination of encoding and syndrome measurements, followed by corrective operations conditioned on the syndrome measurement outcomes, led to a method by which quantum information could be error corrected. In this chapter we will see how quantum information can be protected *passively*. Rather than attempting to fix errors, we will *hide* the quantum information from the environment, so that it never becomes corrupted. This "quiet corner" of the system's Hilbert space is called a decoherence-free subspace (DFS) or noiseless subsystem (NS). The main mechanism that will allow us to play this trick is *symmetry*. Every student of physics knows that symmetries lead to conservation laws. In our case the conserved quantity will turn out to be quantum information. Of course such symmetries must be exceedingly rare, or else protected quantum information would be all around us. Thus we will be forced to make some very strong assumptions, and ask the environment to "cooperate." While we will definitely allow the environment to couple to our system, we will only allow it to do so in a certain symmetric manner. We will later discuss how relaxing this "cooperation requirement" still allows quantum information to be passively protected for a short time. In Chapter 4 we will see how, by using dynamical decoupling techniques, it becomes possible to force the environment to cooperate, i.e., to *dynamically generate* symmetries leading to a DFS or NS. But perhaps the most important lesson of this chapter is the idea that pure symmetry can lead to absolute protection of quantum information. This idea is so powerful that it can be seen as the unifying theme of *all* methods for overcoming quantum errors, whether active or passive, dynamical or topological.

3.2 A "classical decoherence-free subspace"

As a warmup, consider the following problem. You have two coins and want to use them to store one bit of classical information. "Easy," you say, "since two coins represent two bits of

---

Quantum Error Correction, ed. Daniel A. Lidar and Todd A. Brun. Published by Cambridge University Press.
© Cambridge University Press 2013.

information." But now imagine that some nasty demon keeps flipping the coins at random, so that your efforts to store a bit are frustrated. Fortunately the demon can only flip both coins simultaneously. Is it still possible to reliably store a classical bit? A moment's reflection reveals that the answer is yes: define the two subspaces "equal" (even parity) and "opposite" (odd parity). The first is the subspace comprising the states {heads,heads} and {tails,tails}. The second subspace is {heads,tails} and {tails,heads}. Now call "equal" 0 and call "opposite" 1. Since the demon can only flip both coins together, these 0 and 1 are protected. Indeed, under the demon's action {heads,heads}↔{tails,tails} and {heads,tails}↔{tails,heads}, so that the two subspaces never get mixed. Thus, instead of encoding a bit into each coin, we should encode a bit into the parity of the two coins. What is special about parity? It is the fact that it respects the symmetry induced by the demon's inability to distinguish between the two coins, a permutation symmetry.

Let us contrast this situation with that of classical error correction against bit-flip errors. This situation was analyzed in detail in Chapter 2, where we saw that, provided the error probability p of a bit-flip is small enough ($p < 1/2$), we can actively recover from bit-flip errors by flipping bits back based on a majority vote decision. There are two essential ways in which this strategy differs from the one we have just analyzed for permutationally invariant noise. The first is that we do not need to assume that the error probability is small: our protection works regardless of the rate at which the bits flip, as long as our symmetry assumption is satisfied. The second is that we don't need to perform any recovery operations: once encoded, we don't need to measure which of the bits flipped, nor do we need to flip bits back. The encoding we have chosen guarantees complete protection without any intervention. This is a very appealing scenario, which of course depends entirely on the presence and preservation of the permutation symmetry we have assumed.

There is another noteworthy feature of our classical example. Suppose we label the four coin states as $|\lambda\rangle|\mu\rangle$, where λ represents the parity (0 for equal or 1 for opposite) and μ represents the identity of the first coin, i.e., $\mu =$ heads or tails. The environment can then alter the μ label (we don't mind), but not the λ label. This reveals the nature of our encoding: it is into a "virtual subsystem" (the λ label), not an actual physical state. We will see later how this type of encoding appears as the general method of quantum encoding into NSs.

Is it possible to generalize our warmup problem to more than two coins? Consider N coins, all subject to the same collective flipping noise. Thus, at any point in time all coins can be flipped simultaneously. Let us denote by x_i the value of the ith coin, where x is either heads or tails. Arrange the space of 2^N different coin states into complementary pairs, i.e., pairs with the property: $x_i =${heads or tails} in one member $\Longrightarrow x_i =${tails or heads} in the other member, respectively. Such pairs are interchanged under the action of the noise, but the noise cannot interchange members belonging to different pairs. We thus have 2^{N-1} protected pairs, for a total of $\log_2 2^{N-1} = N - 1$ protected logical bits.

Let us now move on to the quantum case.

3.3 Collective dephasing decoherence-free subspace

We begin by analyzing in detail the operation of the simplest DFS. Suppose that a system of N qubits is coupled to a bath in a symmetric way, and undergoes a dephasing process. Namely, qubit j undergoes the transformation

$$|0\rangle_j \to |0\rangle_j, \qquad |1\rangle_j \to e^{i\phi}|1\rangle_j, \qquad (3.1)$$

which puts a random phase ϕ between the basis states $|0\rangle$ and $|1\rangle$. This can also be described by the matrix $R_z(\phi) = \text{diag}\left(1, e^{i\phi}\right)$ acting on the $\{|0\rangle, |1\rangle\}$ basis. Notice how the phase ϕ – by assumption – has no space (j) dependence, i.e., the dephasing process is invariant under qubit permutations. Without encoding, a qubit initially in an arbitrary pure state $|\psi\rangle_j = a|0\rangle_j + b|1\rangle_j$ will decohere. One way to see this is by calculating its density matrix as an average over all possible values of ϕ,

$$\rho_j = \int_{-\infty}^{\infty} R_z(\phi)|\psi\rangle_j\langle\psi|R_z^\dagger(\phi)\, p(\phi) d\phi, \qquad (3.2)$$

where $p(\phi)$ is a probability density, and we assume the initial state of all qubits to be a product state. For a Gaussian distribution, $p(\phi) = (4\pi\alpha)^{-1/2} \exp(-\phi^2/4\alpha)$, it is simple to check that

$$\rho_j = \begin{pmatrix} |a|^2 & ab^* e^{-\alpha} \\ a^* b e^{-\alpha} & |b|^2 \end{pmatrix}. \qquad (3.3)$$

The decay of the off-diagonal elements in the computational basis is a signature of decoherence.

Let us now consider what happens in the two-qubit Hilbert space. The four basis states undergo the transformation

$$\begin{aligned}
|0\rangle_1 \otimes |0\rangle_2 &\to |0\rangle_1 \otimes |0\rangle_2, \\
|0\rangle_1 \otimes |1\rangle_2 &\to e^{i\phi}|0\rangle_1 \otimes |1\rangle_2, \\
|1\rangle_1 \otimes |0\rangle_2 &\to e^{i\phi}|1\rangle_1 \otimes |0\rangle_2, \\
|1\rangle_1 \otimes |1\rangle_2 &\to e^{2i\phi}|1\rangle_1 \otimes |1\rangle_2\rangle.
\end{aligned} \qquad (3.4)$$

Observe that the basis states $|0\rangle_1 \otimes |1\rangle_2$ and $|1\rangle_1 \otimes |0\rangle_2$ acquire the same phase, hence experience the same error. Let us define encoded states by $|0_L\rangle = |0\rangle_1 \otimes |1\rangle_2 \equiv |01\rangle$ and $|1_L\rangle = |10\rangle$. Then the state $|\psi_L\rangle = a|0_L\rangle + b|1_L\rangle$ evolves under the dephasing process as

$$|\psi_L\rangle \to a|0\rangle_1 \otimes e^{i\phi}|1\rangle_2 + be^{i\phi}|1\rangle_1 \otimes |0\rangle_2 = e^{i\phi}|\psi_L\rangle, \qquad (3.5)$$

and the overall phase thus acquired is clearly unimportant. This means that the two-dimensional subspace $\text{DFS}_2(1) = \text{Span}\{|01\rangle, |10\rangle\}$ of the four-dimensional Hilbert space of two qubits is *decoherence-free* (DF). The subspaces $\text{DFS}_2(2) = \text{Span}\{|00\rangle\}$ and $\text{DFS}_2(0) = \text{Span}\{|11\rangle\}$ are also (trivially) DF, since they each acquire a global phase as well, 1 and $e^{2i\phi}$ respectively. Since the phases acquired by the different subspaces differ, there is no coherence *between* the subspaces. Notice the similarities and differences between the quantum case and the case of two classical coins.

For $N = 3$ qubits a similar calculation reveals that the subspaces

$$\begin{aligned}
\text{DFS}_3(2) &= \text{Span}\{|001\rangle, |010\rangle, |100\rangle\}, \\
\text{DFS}_3(1) &= \text{Span}\{|011\rangle, |101\rangle, |110\rangle\}
\end{aligned}$$

are DF, as well as the (trivial) subspaces $\text{DFS}_3(3) = \text{Span}\{|000\rangle\}$ and $\text{DFS}_3(0) = \text{Span}\{|111\rangle\}$.

By now it should be clear how this generalizes. Let λ_N denote the number of 0s in a computational basis state (i.e., a bit string) over N qubits. Then it is easy to check that any subspace spanned by states with constant λ_N is DF against collective dephasing, and can be denoted $\text{DFS}_N(\lambda_N)$ in accordance with the notation above. The dimensions of these subspaces are given by the binomial coefficients: $d_N \equiv \dim[\text{DFS}_N(\lambda_N)] = \binom{N}{\lambda_N}$ and they each encode $\log_2 d_N$ qubits. It might seem that we lost a lot of bits in the encoding process. However, consider the encoding rate $r \equiv \frac{\log_2 d_N}{N}$, defined as the number of output bits to the number of input bits. Using Stirling's formula, $\ln x! \approx x \ln x - x + \ln \sqrt{2\pi}$, we find, for the case of the highest-dimensional DFS ($\lambda_N = N/2$, for even N):

$$r \stackrel{N \gg 1}{\approx} 1 - \frac{1}{2} \frac{\log_2 N}{N}, \tag{3.6}$$

where we neglected the constant $\log(\pi/2)$ compared to $\log N$. Thus, the rate approaches 1 with only a logarithmically small correction. In spite of its simplicity, the collective dephasing DFS is important. It has played a major role in early experimental demonstrations [KBA+00]. It has also found important applications in quantum cryptography and communication [WAS+03].

3.4 Decoherence-free subspace defined and characterized

It is time to give a formal definition of a DFS.

Definition 3.1. *Consider a system with Hilbert space $\mathcal{H}_S$. A subspace $\tilde{\mathcal{H}}_S \subset \mathcal{H}_S$ is called decoherence free if any state $\rho_S(0)$ of the system initially prepared in this subspace is* unitarily *related to the final state $\rho_S(t)$ of the system, i.e.,*

$$\rho_S(0) = \tilde{P}\rho_S(0)\tilde{P} \implies \rho_S(t) = U_S \rho_S(0) U_S^\dagger, \tag{3.7}$$

where $U_S : \tilde{\mathcal{H}}_S \mapsto \tilde{\mathcal{H}}_S$ is unitary and $\tilde{P}$ is a projection operator onto $\tilde{\mathcal{H}}_S \subset \mathcal{H}_S$.

More specifically, $\tilde{P}^2 = \tilde{P} = \sum_i |\tilde{\imath}\rangle\langle\tilde{\imath}|$, where the orthonormal set $\{|\tilde{\imath}\rangle\}$ forms a basis for $\tilde{\mathcal{H}}_S$. This definition captures the two key properties we saw in our examples: passivity and exactness. Namely, any state in $\tilde{\mathcal{H}}_S$ is automatically protected against decoherence without the need for active intervention, and this protection is nonperturbative: the evolution is unitary no matter what the strength of the interaction with the bath is.

While this definition of a DFS is intuitively appealing, it does lead to a couple of difficulties. First, the requirement that $\rho_S(0) = \tilde{P}\rho_S(0)\tilde{P}$ means that the state must be perfectly initialized in the subspace $\tilde{\mathcal{H}}_S$. This can be experimentally challenging. Second, merely demanding that the evolution inside $\tilde{\mathcal{H}}_S$ is unitary does not guarantee that it is error free. For example, the Lamb shift discussed in Chapter 1 [Eq. (1.126)] contributes to the unitary evolution, and can cause computational errors. We will see later how to modify the definition of a DFS in order to deal with these issues. However, we will not be able to do anything about the problem of unitary errors that may be the result of inaccurate implementation of quantum logic gates. This problem belongs squarely in the realm of fault tolerance, and we do not address it in this chapter.

With a definition in hand, let us now derive a sufficient condition for the existence of a DFS. We do so in the most general setting, of Hamiltonian dynamics. Later on we will present a formulation appropriate for quantum channels (CP maps), and exhibit the connection to quantum error-correcting codes (QECC).

3.4.1 Hamiltonian formulation

In the Hamiltonian formulation, which is the most general dynamical one, decoherence is the result of the entanglement between system and bath caused by the system–bath interaction term $H_{SB} = \sum_\alpha S_\alpha \otimes B_\alpha$. In other words, if $H_{SB} = 0$ then system and bath are decoupled and evolve independently and unitarily under their respective Hamiltonians H_S and H_B. Clearly, then, a sufficient condition for *decoherence-free* (DF) dynamics is that $H_{SB} = 0$. However, since one cannot simply switch off the system–bath interaction, we need to look for a special subspace $\tilde{\mathcal{H}}_S$ of the full system Hilbert space $\mathcal{H}_S$ over which $H_{SB} = 0$, a condition we can write as $H_{SB}|_{\tilde{\mathcal{H}}_S} = 0$ (read as: "H_{SB} restricted to $\tilde{\mathcal{H}}_S$"). In fact, this condition is a bit too strong. Instead, it is sufficient for H_{SB} to merely act as the identity operator on $\tilde{\mathcal{H}}_S$, i.e., $H_{SB}|_{\tilde{\mathcal{H}}_S} \propto I$. We will see that this *DFS condition* can be expressed by assuming that there exists a set $\{|\tilde{k}\rangle\}$ of eigenvectors of the S_αs with the property that

$$S_\alpha|\tilde{k}\rangle = c_\alpha|\tilde{k}\rangle \qquad \forall \alpha, |\tilde{k}\rangle. \tag{3.8}$$

Note that these eigenvectors are *degenerate*, i.e., the eigenvalue c_α depends only on the index α of the system operators in H_{SB}, but not on the state index k. This degeneracy is a strong indication that a DFS is associated with symmetry, as we saw in our examples above. But there is more to the DFS condition (3.8) than intuition. In fact, it is very useful for checking whether a given interaction supports a DFS: the operators S_α often generate a Lie algebra, and the condition (3.8) then translates into the problem of finding the *one-dimensional* irreducible representations (irreps) of this Lie algebra. Why one-dimensional? Clearly, because the S_α operators all act as scalars on the states $\{|\tilde{k}\rangle\}$.

So far we have focused on the interaction Hamiltonian H_{SB}. However, this is not enough, since damage can be done by H_S as well. We should make sure that a state that starts out in the protected subspace $\tilde{\mathcal{H}}_S$ does not leave it via the action of H_S, for if it does, there is no guarantee that the action of H_{SB} will not corrupt it. That is, we require that H_S leaves the subspace $\tilde{\mathcal{H}}_S = \mathrm{Span}[\{|\tilde{k}\rangle\}]$ invariant:

$$H_S|\tilde{\psi}\rangle \in \tilde{\mathcal{H}}_S, \quad \forall |\tilde{\psi}\rangle \in \tilde{\mathcal{H}}_S. \tag{3.9}$$

Finally, let us assume that we start the evolution from within $\tilde{\mathcal{H}}$, i.e., $\rho_S(0) = \sum_{ij} s_{ij}|\tilde{i}\rangle\langle\tilde{j}| = \tilde{P}\rho_S(0)\tilde{P}$, and that the initial system–bath state is decoupled: $\rho_{SB}(0) = \rho_S(0) \otimes \rho_B(0)$. With all these ingredients we can show that the evolution of the system will be DF. To do so, let us use Eq. (3.8) to write the combined operation of the bath and interaction Hamiltonians over $\tilde{\mathcal{H}}_S$ as

$$I_S \otimes H_B + H_{SB}|_{\tilde{\mathcal{H}}_S} = I_S \otimes \left[H_B + \sum_\alpha c_\alpha B_\alpha\right] = I_S \otimes H_e. \tag{3.10}$$

The second equality serves to define the effective Hamiltonian $H_e \equiv H_B + \sum_\alpha c_\alpha B_\alpha$. Clearly, $I_S \otimes H_e$ commutes with H_S over $\tilde{\mathcal{H}}_S$. Now let U be the joint unitary evolution generated by $H = H_S + H_{SB} + H_B$. Then, since neither H_S (by assumption) nor H_e takes states out of the subspace, we have for any pair of states $|\tilde{i}\rangle \in \tilde{\mathcal{H}}_S$ and $|\mu\rangle \in \mathcal{H}_B$ (the bath Hilbert space)

$$U[|\tilde{i}\rangle \otimes |\mu\rangle] = U_S|\tilde{i}\rangle \otimes U_e|\mu\rangle, \tag{3.11}$$

where $U_S = \exp(-iH_S t)$ and $U_e = \exp(-iH_e t)$. Let us expand the initial bath state in a basis for $\mathscr{H}_B$: $\rho_B(0) = \sum_{\mu\nu} b_{\mu\nu} |\mu\rangle\langle\nu|$. Then, using the assumption of a decoupled initial system–bath state,

$$\rho_{SB}(t) = \sum_{ij} s_{ij} U_S |\tilde{i}\rangle\langle\tilde{j}| U_S^\dagger \otimes \sum_{\mu\nu} b_{\mu\nu} U_e |\mu\rangle\langle\nu| U_e^\dagger. \tag{3.12}$$

Tracing over the bath then yields

$$\rho_S(t) = \mathrm{Tr}_B[\rho_{SB}(t)] = \sum_{ij} s_{ij} U_S |\tilde{i}\rangle\langle\tilde{j}| U_S^\dagger \mathrm{Tr}\left[\sum_{\mu\nu} b_{\mu\nu} U_e |\mu\rangle\langle\nu| U_e^\dagger\right]$$

$$= U_S \left(\sum_{ij} s_{ij} |\tilde{i}\rangle\langle\tilde{j}|\right) U_S^\dagger \mathrm{Tr}[\rho_B(0)] = U_S \rho_S(0) U_S^\dagger, \tag{3.13}$$

i.e., the DFS definition (3.7) is satisfied. We have thus shown that condition (3.8) is sufficient for a DFS:

Theorem 3.1. *Consider a system S coupled to a bath B, subject to the Hamiltonian* $H = H_S + H_{SB} + H_B$, *where* $H_{SB} = \sum_\alpha S_\alpha \otimes B_\alpha$. *If* $\rho_{SB}(0) = \rho_S(0) \otimes \rho_B(0)$ *and conditions (3.8) and (3.9) are satisfied, the system has a DFS according to Definition 3.1, with* U_S *generated by* H_S.

We now turn to examples.

3.4.2 Spin-boson model with collective decoherence

Let us illustrate the DFS condition (3.8) with an important example: the spin-boson model. Consider N spins interacting with a bosonic field via the Hamiltonian

$$H_{SB} = \sum_{i=1}^N \sum_k \left(g_{i,k}^+ \sigma_i^+ \otimes b_k + g_{i,k}^- \sigma_i^- \otimes b_k^\dagger + g_{i,k}^z \sigma_i^z \otimes (b_k + b_k^\dagger)\right). \tag{3.14}$$

Here $\{\sigma_i^+, \sigma_i^-, \sigma_i^z\}$ are Pauli operators acting on the ith spin, b_k ($b_k^\dagger$) is an annihilation (creation) operator for the kth bosonic mode, and $g_{i,k}^\alpha$ are coupling constants. The Hamiltonian H_{SB} describes a rather general interaction between a system of qubits (the spins) and a bath of bosons, exchanging energy through the $\sigma_i^+ \otimes b_k$ and $\sigma_i^- \otimes b_k^\dagger$ terms, and changing phase through the $\sigma_i^z \otimes (b_k + b_k^\dagger)$ term. As it stands, H_{SB} does not support a DFS: there are $3N$ system operators S_α in H_{SB}, i.e., the N triples of *local* $sl(2)$ algebras $\{\sigma_i^+, \sigma_i^-, \sigma_i^z\}$; each such algebra acts on a single qubit, and therefore has a two-dimensional irrep. The overall action of the total Lie algebra $\bigoplus_{i=1}^N sl_i(2)$ is represented by the irreducible N-fold tensor product of all local two-dimensional irreps. This implies that there are no one-dimensional irreps as required by Eq. (3.8), and hence no DFS.

The situation changes dramatically when a *permutation symmetry* is imposed on the system–bath interaction, by requiring that the coupling constants do not depend on the qubit index i:

$$g_{i,k}^\alpha \equiv g_k^\alpha. \tag{3.15}$$

This "collective decoherence" situation is relevant, e.g., in solid-state quantum computing proposals where decoherence due to interaction with a cold phonon bath is dominant. At low temperatures only long-wavelength phonons survive (since there is an energy gap for excitation of high-energy/short-wavelength phonons), and the model of collective decoherence becomes relevant provided the qubit spacing is small compared to the phonon wavelength.

Given the collective decoherence assumption, we can define three collective spin operators $S_\alpha = \sum_{i=1}^N \sigma_i^\alpha$ so that the interaction Hamiltonian becomes

$$H_{SB} = \sum_{\alpha=+,-,z} S_\alpha \otimes B_\alpha, \tag{3.16}$$

where $B_+ = \sum_k g_k^+ \sigma_i^+ \otimes b_k$, $B_- = B_+^\dagger$ and $B_z = \sum_k g_{i,k}^z \sigma_i^z \otimes (b_k + b_k^\dagger)$. In fact, the specific form of the bath operators B_α is irrelevant. The important point is that the system operators S_α now form a *global* $sl(2)$ angular momentum algebra, i.e., $[S_+, S_-] = S_z$ and $[S_z, S_\pm] = 2S_\pm$, with a highly *reducible* $2^N \times 2^N$ representation formed by its action on all N qubits at once. Note that the case of collective dephasing discussed in Section 3.3 corresponds to having just a single global angular momentum operator (S_z). The big difference is that now we no longer have a scalar interaction between system and bath, and the resulting Lie algebra is richer. In fact, it is the Lie algebra familiar from the theory of angular momentum. The S_z and $S_\pm$ operators are simply the total angular momentum operators of N spin-$1/2$ particles, and Eq. (3.8) is telling us that we need to identify the states that transform as scalars under the action of these total angular momentum operators. It is not hard to see that there are no such states for N odd (since such states always have half-integer total angular momentum), but for N even these are just the states with zero total angular momentum, also known as singlets. Thus, we are looking for all states $|\tilde{\psi}\rangle$ satisfying

$$S_\alpha |\tilde{\psi}\rangle = 0, \qquad \alpha \in \{z, \pm\}. \tag{3.17}$$

Their explicit form is well known for the case $N = 2$, in which case there is only one singlet:

$$|\tilde{0}\rangle_{N=2} \equiv |s\rangle_{12} = \frac{1}{\sqrt{2}}(|01\rangle - |10\rangle) = |J=0, m_J=0\rangle, \tag{3.18}$$

where the notation $|s\rangle_{12}$ should be read as "singlet state of qubits 1 and 2." In writing $\frac{1}{\sqrt{2}}(|01\rangle - |10\rangle)$ we have used standard computational basis notation, and in writing $|J = 0, m_J = 0\rangle$ we have used angular momentum notation, where J is the total angular momentum and m_J is its projection on the z-axis (in general $-J \le m_J \le J$). It is easy to check that the state $\bigotimes_{m=1}^{N/2} |s\rangle_{2m-1,2m}$ is in the N-qubit DFS, since adding pairwise $|J = 0, m_J = 0\rangle$ states again produces a state with total $J = 0$. Let us use angular momentum addition rules to find the singlet states for $N = 4$. One of the states is $|\tilde{0}\rangle \equiv |s\rangle_{12} \otimes |s\rangle_{34}$. The second, orthogonal, state must be a combination of the triplet states $|t_-\rangle \equiv |00\rangle = |J = 1, m_J = -1\rangle$, $|t_+\rangle \equiv |11\rangle = |J = 1, m_J = 1\rangle$, $|t_0\rangle \equiv \frac{1}{\sqrt{2}}(|01\rangle + |10\rangle) = |J = 1, m_J = 0\rangle$. Indeed, the combination $|\tilde{1}\rangle = \frac{1}{\sqrt{3}}(|t_-\rangle_{12} \otimes |t_+\rangle_{34} - |t_0\rangle_{12} \otimes |t_0\rangle_{34} + |t_+\rangle_{12} \otimes |t_-\rangle_{34})$ clearly has total $J = 0$, and has the correct Clebsch–Gordan coefficients. Having found two orthonormal states (note that $\langle \tilde{1}|\tilde{0}\rangle = 0$), we see that this DFS *encodes one logical qubit*. In fact, we will later show that

$$d_N \equiv \dim[\text{DFS}(N)] = \frac{N!}{(N/2+1)!(N/2)!} \tag{3.19}$$

is the number of singlet states for N qubits, and hence $\dim[DFS(4)] = 2$, meaning that our four-qubit DFS does not include any more independent DF states than the two we have already found. Using this dimensionality formula along with Stirling's formula we can find the code rate $r \equiv \frac{\log_2 d_N}{N}$, just as we did for the collective dephasing case:

$$r \stackrel{N \gg 1}{\approx} 1 - \frac{3}{2}\frac{\log_2 N}{N}. \tag{3.20}$$

Thus, again, the rate approaches 1 with only a logarithmically small correction, which is three times larger than the correction in the collective dephasing case [Eq. (3.6)]. This is not surprising, given that there are three times as many independent error operators in the collective decoherence model than in the collective dephasing model.

How can we identify all states in a DFS of N qubits? We postpone the answer to this question to when we return to the collective decoherence example from the more general perspective of noiseless subsystems.

3.4.3 Decoherence-free subspace for atoms in a cavity

Let us now consider another example, where the condition (3.9) on the system Hamiltonian plays a role. We motivated the collective decoherence model in terms of a solid-state setting, but it arises also in a very different, quantum optical context. This is known as the Jaynes–Cummings Hamiltonian, wherein N identical two-level atoms are trapped in a cavity, and couple to a single mode of the radiation field in the cavity, with bosonic annihilation or creation operator b or $b^\dagger$. The Hamiltonian is very similar to that of Eq. (3.14), namely

$$H_{SB} = \sum_{i=1}^{N} \left(g_i^+ \sigma_i^+ \otimes b + g_i^- \sigma_i^- \otimes b^\dagger \right). \tag{3.21}$$

When the atoms are closer together than the wavelength of the radiation field, the coupling parameter g_i becomes independent of the atom index i, and one can write Eq. (3.21) in terms of the collective spin operators $S_+ = \sum_{i=1}^{N} \sigma_i^+$ and $S_- = \sum_{i=1}^{N} \sigma_i^-$. The DFS we find in this case is in fact identical to the one we found for the collective decoherence model above, since $S_\pm |\tilde{\psi}\rangle = 0$ already implies $S_z |\tilde{\psi}\rangle = 0$. However, it is not very natural to consider the cavity mode as part of the bath. Let us therefore instead incorporate the single field mode into the system, and distinguish it from the real environment, consisting of the external electromagnetic field modes with bosonic annihilation and creation operators and $\{a_\lambda^\dagger\}$, respectively. The Hamiltonian is now

$$H = \sum_{i=1}^{N} \left(g_i^+ \sigma_i^+ \otimes b + g_i^- \sigma_i^- \otimes b^\dagger \right) + \sum_\lambda \left(s_\lambda^+ b \otimes a_\lambda^\dagger + s_\lambda^- b^\dagger \otimes a_\lambda \right), \tag{3.22}$$

where $s_\lambda^\pm$ are coupling constants. H_{SB} consists only of the second sum, since the atom–cavity mode interaction is now within the expanded system of atom plus cavity mode, i.e., the first sum is part of the system Hamiltonian H_S (note that for simplicity we have neglected the bath Hamiltonian $\sum_\lambda \omega_\lambda a_\lambda^\dagger a_\lambda$ and some other terms in H_S and H_B). In this case one of the DFS conditions is $b^\dagger |\psi\rangle |n\rangle \propto |\psi\rangle |n\rangle$, where $|\psi\rangle$ is the collective atomic state and $|n\rangle$ is the cavity number state. This condition cannot be satisfied, since $b^\dagger |n\rangle \propto |n+1\rangle$ for all values of n. This looks like bad news, but let us carry on for now despite that. The second DFS condition is $b|\psi\rangle |n\rangle \propto |\psi\rangle |n\rangle$, which implies that $b|n\rangle = 0$, i.e., the cavity mode must be empty, but does not otherwise place any restrictions on the atomic component of the system state. It is only when we impose the further requirement that the system Hamiltonian does not cause system states to evolve outside the DFS that we arrive at specific atomic states. In particular, this requirement tells us that both S_+b and $S_-b^\dagger$ must act on $|\psi\rangle|0\rangle$ in such as a way as to conserve the emptiness of the cavity, i.e., so as to yield another state $|\psi'\rangle|0\rangle$. Since $b|0\rangle = 0$, the first condition is trivially satisfied, while the second condition reduces to $S_-|\psi\rangle = 0$ because $b^\dagger|0\rangle = |1\rangle$, which would take the joint state outside the DFS. Thus, the requirement for a DFS of the coupled N atom/cavity system is simply $S_-|\psi\rangle = 0$, with a zero cavity occupation number. We see that this also resolves our conundrum regarding the condition $b^\dagger|\psi\rangle|n\rangle \propto |\psi\rangle|n\rangle$ ($b^\dagger$ appears together with S_-). However, there is still one more dangerous term in the Hamiltonian (3.22): the term $s_\lambda^- b^\dagger \otimes a_\lambda$ can re-occupy the cavity by annihilating an environment photon. We therefore must impose the additional condition that the environment modes are also unoccupied ($|n_\lambda = 0\rangle$ for all λ), which means that the environment must be at zero temperature. For $N = 2$ atoms we then find a two-dimensional DFS given by the atomic states $|J = 1, m = -1\rangle = |00\rangle$ and $|J = 0, m = 0\rangle = (1/\sqrt{2})[|10\rangle - |01\rangle]$ (combined, of course, with the zero cavity mode state $|n = 0\rangle$). The corresponding DFS for higher values of N may be readily generated using angular momentum algebra. The dimensionality of these atom–cavity DFSs differs from that of collective decoherence given above [Eq. (3.19)], being given by the expression

$$\dim[\mathrm{DFS}(N, J)] = \frac{(2J+1)N!}{(N/2+J+1)!(N/2-J)!}, \qquad (3.23)$$

which reduces to Eq. (3.19) when $J = 0$. We shall prove this important dimensionality formula in Section 3.6.3.

3.4.4 CP map formulation

In many cases it is convenient to analyze decoherence in terms of CP maps. Let us now see what the DFS conditions look like in this formulation. Recall from Chapter 1 that under a CP map the initial system state $\rho_S(0)$ undergoes the transformation

$$\rho(t) = \sum_i E_i \, \rho(0) \, E_i^\dagger, \qquad (3.24)$$

where the Kraus operators, or operation elements, are given by

$$E_i = \sqrt{\lambda_\nu} \langle \mu | U | \nu \rangle, \qquad i = (\mu, \nu), \qquad (3.25)$$

with U the joint system–bath unitary evolution operator, and $\{\lambda_\nu\}$ being the eigenvalues of the initial bath state $\rho_B(0) = \sum_\nu \lambda_\nu |\nu\rangle\langle\nu|$. Also recall that the Kraus operators obey the normalization rule $\sum_i E_i^\dagger E_i = I_S$. Let us now assume that the Kraus operators all have the following block-diagonal representation, where the upper block acts on a subspace $\tilde{\mathcal{H}}_S$ of $\mathcal{H}_S$, and the lower block acts on the orthogonal complement:

$$E_i = \begin{pmatrix} g_i \tilde{U} & 0 \\ 0 & \bar{E}_i \end{pmatrix}, \qquad g_i = \sqrt{\nu} \langle \mu | U_e | \nu \rangle. \qquad (3.26)$$

Here $\tilde{U} : \tilde{\mathcal{H}}_S \mapsto \tilde{\mathcal{H}}_S$ is unitary, U_e is the same unitary we encountered in the Hamiltonian formulation, and $\bar{E}_i$ is an arbitrary matrix. It is easy to see why this form of the Kraus operators guarantees that $\tilde{\mathcal{H}}_S$ is a DFS. Consider the set of system states $\{|\tilde{k}\rangle\}$ satisfying

$$E_i|\tilde{k}\rangle = g_i \tilde{U}|\tilde{k}\rangle \qquad \forall i, \tag{3.27}$$

where $\tilde{U}$ is an arbitrary, i-independent but possibly time-dependent unitary transformation, and g_i is a complex constant. Under this condition, an initially pure state belonging to $\tilde{\mathcal{H}}_S = \text{Span}[\{|\tilde{k}\rangle\}]$, i.e., $|\tilde{\psi}(0)\rangle = \sum_k \gamma_k |\tilde{k}\rangle$, will be DF, since

$$|\phi_a\rangle = E_i|\tilde{\psi}(0)\rangle = \sum_k \gamma_k g_i \tilde{U}|\tilde{k}\rangle = g_i \tilde{U}|\tilde{\psi}(0)\rangle \tag{3.28}$$

so

$$\rho(t) = \sum_i E_i \rho(0) E_i^\dagger = \sum_i g_i \tilde{U}|\tilde{\psi}(0)\rangle\langle\tilde{\psi}(0)|\tilde{U}^\dagger g_i^* = \tilde{U}|\tilde{\psi}(0)\rangle\langle\tilde{\psi}(0)|\tilde{U}^\dagger, \tag{3.29}$$

where we used the normalization rule to set $\sum_i |g_i|^2 = 1$. This means that the output state $\rho(t)$ is pure, and its evolution is governed by $\tilde{U}$. This argument is easily generalized to a mixed initial state $\rho(0) = \sum_{kj} \rho_{kj} |\tilde{k}\rangle\langle\tilde{j}|$, in which case $\rho(t) = \tilde{U}\rho(0)\tilde{U}^\dagger$. We can summarize this as follows:

Theorem 3.2. *A subspace $\tilde{\mathcal{H}}_S$ is a DFS if all Kraus operators have an identical unitary representation upon restriction to $\tilde{\mathcal{H}}_S$, up to a multiplicative constant.*

3.4.5 Decoherence-free subspace from a group algebra

A useful alternative formulation of the DFS condition using Kraus operators can be given, which is slightly less general. Consider a group $\mathscr{G} = \{G_n\}$ and expand the Kraus operators as a linear combination over the group elements: $E_i = \sum_{n=1}^N b_{i,n} G_n$ (i.e., the Kraus operators belong to the group algebra of $\mathscr{G}$). Then the following theorem gives an alternative sufficient condition for a DFS:

Theorem 3.3. *If a set of states $\{|\tilde{k}\rangle\}$ belong to a given* one-dimensional *irrep Γ^k of $\mathscr{G}$, then $E_i|\tilde{k}\rangle \propto |\tilde{k}\rangle$.*

Clearly, this implies condition (3.26), hence is sufficient for a DFS. It is not always possible to expand the Kraus operators over a given group. However, when this is possible, the last theorem can be used to find a class of DFSs under relaxed symmetry assumptions. Let us see an example.

Collective decoherence involves the assumption that all qubits couple to the bath in a symmetric manner, which is the consequence of full permutational symmetry. However, the symmetry that gives rise to a DFS can be less strict. Consider the case of $N = 4$ qubits that can undergo bit-flips σ_j^x: $|0\rangle_j \longleftrightarrow |1\rangle_j$. Suppose that the qubits are arranged in pairs, such that the bit-flip error affects at least two at a time. Specifically, let the allowed errors be the group algebra of the Abelian group $\mathscr{G} = Q_X(4) \equiv \{I, XXII, IIXX, XXXX\}$. Note that the error IXXI is assumed not to happen. Since $Q_X(4)$ is Abelian, all its irreps are one-dimensional:

	I^4	XXII	IIXX	XXXX
Γ^1	1	1	1	1
Γ^2	1	1	-1	-1
Γ^3	1	-1	1	-1
Γ^4	1	-1	-1	1

It is then simple to verify that the encoding

$$\begin{aligned}
|\widetilde{00}\rangle &= \frac{1}{2}\left(|0000\rangle + |1100\rangle + |0011\rangle + |1111\rangle\right), \\
|\widetilde{01}\rangle &= \frac{1}{2}\left(|0001\rangle + |1101\rangle + |0010\rangle + |1110\rangle\right), \\
|\widetilde{10}\rangle &= \frac{1}{2}\left(|0100\rangle + |1000\rangle + |0111\rangle + |1011\rangle\right), \\
|\widetilde{11}\rangle &= \frac{1}{2}\left(|1001\rangle + |0101\rangle + |1010\rangle + |0110\rangle\right)
\end{aligned} \quad (3.30)$$

is a four-dimensional DFS (denoted $\text{DFS}_{Q_X(4)}$), thus encoding two logical qubits. To see this, check that all states in $\text{DFS}_{Q_X(4)}$ are eigenstates with the same eigenvalue $+1$ of any linear combination of operators over $Q_X(4)$. This means that the states in $\text{DFS}_{Q_X(4)}$ all belong to the irrep Γ^1 and hence, in accordance with Theorem 3.3, they belong to a DFS. The other irreps also have DFSs associated with them, and it is not hard to show that they are also four-dimensional.

In order to generalize this result, note that the group $Q_X(4)$ is generated (under multiplication) by $\sigma_1^x \sigma_2^x$ and $\sigma_3^x \sigma_4^x$; clearly the group $Q_X(N)$ generated by $\{\sigma_{2j-1}^x \sigma_{2j}^x\}_{j=1}^{N/2}$ also supports a DFS with the same defining property, namely that the DF states are simultaneous eigenstates of all $Q_X(N)$ generators.

Even more generally, the type of DFS we have just considered here is an example of a *DFS for stabilizer errors*. The error model can be stated abstractly as follows: Starting from the Pauli group on N qubits (the group of N-fold tensor products of the Pauli matrices $\{\sigma_x, \sigma_y, \sigma_z\}$), look for *Abelian subgroups*, i.e., stabilizer subgroups of the type we encountered in the theory of quantum error correction in Chapter 2. Every such subgroup defines an error model, in the sense that its group algebra forms the Kraus operators. We are looking for Abelian subgroups in order to have one-dimensional irreps. This, in turn, is a sufficient condition for a DFS by Theorem 3.3. Thus, to every such stabilizer error model there corresponds a DFS, which is found by studying the one-dimensional irreps of the subgroup over N qubits. The DFS basis states are states that transform according to a one-dimensional irrep with fixed eigenvalue. The dimension of such a DFS therefore turns out to be the multiplicity of the one-dimensional irreps of the subgroup (this result will become clearer when we discuss the general theory of NS. The result seen above for $Q_X(4)$ is general: the entire Hilbert space splits up into equi-dimensional, independent DFSs.

3.4.6 Quantum error correction formulation

Information encoded in a DFS is immune to errors. In quantum error correction (QEC), information is encoded into states that can be perturbed by the environment, but can be recovered. Thus

DFSs can be viewed as a special type of "degenerate" quantum error-correcting code (QECC), where the perturbation acts trivially and the recovery is trivial (it is the identity). The errors are represented by the Kraus operators $\{E_i\}$. To decode the quantum information after the action of the bath, we need "recovery" operators $\{R_k\}$. Now recall the quantum error-correcting criterion from Chapter 2, which we will rewrite in matrix form: It is possible to correct the errors induced by a given set of Kraus operators $\{E_i\}$, (i) if and only if

$$R_k E_i = \begin{pmatrix} \lambda_{ki} I_{\mathscr{C}} & 0 \\ 0 & B_{ki} \end{pmatrix} \quad \forall k, i, \tag{3.31}$$

or equivalently, (ii) if and only if

$$E_i^\dagger E_j = \begin{pmatrix} \gamma_{ij} I_{\mathscr{C}} & 0 \\ 0 & \bar{E}_i^\dagger \bar{E}_j \end{pmatrix} \quad \forall i, j. \tag{3.32}$$

In both conditions the first block acts on the code subspace $\mathscr{C}$; B_{ki} and $\bar{E}_i$ are arbitrary matrices acting on $\mathscr{C}^\perp$ ($\mathscr{H}_S = \mathscr{C} \oplus \mathscr{C}^\perp$). Let us now explore the relationship between DFSs and QECCs. First of all, it is immediate that DFSs are indeed a valid QECC. For, given the representation of the Kraus operators E_i as in Eq. (3.26), it follows that Eq. (3.32) is satisfied with $\gamma_{ij} = g_i^* g_j$. Note, however, that unlike the general QECC case where there is no restriction on the rank of the matrix γ_{ab}, in the DFS case this matrix has rank 1 (since the ith row equals row 1 upon multiplication by g_1^*/g_i^*). A QECC is nondegenerate if it has full rank. A DFS, therefore, is a completely degenerate QECC. This, of course, is due to the assumption of symmetry. Thus, the complete characterization of DFSs as a QECC is given by the following theorem:

Theorem 3.4. *Let $\mathscr{C}$ be a QECC for error operators $\{E_i\}$, with recovery operators $\{R_k\}$. Then $\mathscr{C}$ is a DFS if and only if, upon restriction to $\mathscr{C}$, $R_k \propto \tilde{U}_S^\dagger$ for all k, where $\tilde{U}_S$ is a unitary acting on $\mathscr{C}$.*

Proof. First suppose $\mathscr{C}$ is a DFS. Then by Eqs. (3.26) and (3.31),

$$R_k \begin{pmatrix} g_i \tilde{U}_S & 0 \\ 0 & \bar{E}_i \end{pmatrix} = \begin{pmatrix} \lambda_{ki} I_{\mathscr{C}} & 0 \\ 0 & B_{ki} \end{pmatrix}.$$

To satisfy this equation, it must be true that

$$R_k = \begin{pmatrix} \frac{\lambda_{ki}}{g_i} \tilde{U}_S^\dagger & C_k \\ D_k & F_k \end{pmatrix},$$

with the same block structure. Multiplying, the condition $g_i D_k \tilde{U}_S = 0$ implies $D_k = 0$ by unitarity of $\tilde{U}_S$. Also, since $\bar{E}_i$ is arbitrary, generically the condition $C_k \bar{E}_i = 0$ implies $C_k = 0$. Thus, upon restriction to $\mathscr{C}$, indeed $R_k \propto \tilde{U}_S^\dagger$ (by unitarity of $\tilde{U}_S$, $|\lambda_{ki}/g_i| = 1$). Now suppose $R_k \propto \tilde{U}_S^\dagger$. The very same argument applied to E_i in Eq. (3.31) yields $E_i \propto \tilde{U}_S$ upon restriction to $\mathscr{C}$. Since this is exactly the condition defining a DFS in Eq. (3.26), the theorem is proved. □

Thus, in the sense of reversal of quantum operations on a subspace, DFSs are a particularly simple instance of general QECCs, where, upon restriction to the code subspace, all recovery operators are proportional to the inverse of the system evolution operator.

3.4.7 Stabilizer formulation

Let us briefly discuss an alternative way to make the connection between DFSs and error-correction theory. Recall the stabilizer formulation of QEC we encountered in Chapter 2. By analogy with QEC, we define the *DFS stabilizer* $\mathscr{S}$ as a set of operators D_β that act as the identity on the DFS states:

$$D_\beta |\psi\rangle = |\psi\rangle \quad \forall D_\beta \in \mathscr{S} \quad \text{iff} \quad |\psi\rangle \in \text{DFS}. \tag{3.33}$$

Here β can be a discrete or continuous index; $\mathscr{S}$ can be a group or some other set. This $\mathscr{S}$ is therefore a generalization of the stabilizers of QEC, which are Abelian subgroups of the Pauli group. While some DFSs can also be specified by a stabilizer in the Pauli group (e.g., the $Q_X(4)$ example we saw above), many DFSs are specified by non-Abelian groups, and hence are *nonadditive* codes.

Consider now the following continuous-index stabilizer, where the S_α are again the system operators of the system–bath interaction Hamiltonian:

$$D(v_0, v_1, ..., v_A) = D(\vec{v}) = \exp\left[\sum_{\alpha=1}^{A} (c_\alpha I - S_\alpha) v_\alpha\right]. \tag{3.34}$$

Clearly, the DFS condition (3.8) implies that $D(\vec{v})|\psi\rangle = |\psi\rangle$. Conversely, if $D(\vec{v})|\psi\rangle = |\psi\rangle$ for all $\vec{v}$, then in particular it must hold that for each α, $\exp\left[(c_\alpha I - S_\alpha) v_\alpha\right] |\psi\rangle = |\psi\rangle$. Since $\exp(A)$ is a one-to-one continuous mapping of a small neighborhood of the zero matrix 0 onto a small neighborhood of the identity matrix I, it follows that there must be a sufficiently small v_α such that $(c_\alpha I - S_\alpha)|\psi\rangle = 0$. Therefore the DFS condition (3.8) holds if and only if $D(\vec{v})|\psi\rangle = |\psi\rangle$ for all $\vec{v}$.

3.4.8 Section summary

We have defined a DFS as a subspace that is fully protected from noise, i.e., any system state that is initially prepared in this subspace undergoes purely unitary dynamics. With this definition we have seen a number of different formulations of the conditions for a DFS. These formulations make contact with different approaches to the theory of open quantum systems, from the traditional Hamiltonian, via the CP map formulation, to the theory of error-correcting codes. We have also seen a number of interesting examples of models permitting DFSs, even of a size that asymptotically fills the entire system Hilbert space. It is now time to relax some of the constraints imposed by our strict "perfect initialization" requirement, and by the requirement that protected information should reside in a subspace. As we will see next, these are not the most general ways in which we can construct a "quiet corner" in Hilbert space. Moreover, adopting this more general viewpoint will also allow us to prove necessary conditions for noiseless quantum information.

3.5 Initialization-free decoherence-free subspace

As we already remarked, definition (3.7) poses a practical problem: perfect initialization of a quantum system inside a DFS might be challenging in many cases. In other words, what if we cannot prepare an initial state with the property $\rho(0) = \tilde{P}\rho(0)\tilde{P}$? It pays to introduce a generalized definition to relax this constraint.

The system density matrix ρ is an operator on the entire system Hilbert space $\mathcal{H}_S$, which we assume to be decomposable into a direct sum as $\mathcal{H}_S = \tilde{\mathcal{H}}_S \oplus \tilde{\mathcal{H}}_S^\perp$. It is convenient to represent the system state in a matrix form whose block structure corresponds to this decomposition of the Hilbert space. Thus the system density matrix takes the form

$$\rho = \begin{pmatrix} \tilde{\rho} & \rho_2 \\ \rho_2^\dagger & \rho_3 \end{pmatrix}. \tag{3.35}$$

The corresponding projection operators onto $\tilde{\mathcal{H}}_S$ and $\tilde{\mathcal{H}}_S^\perp$ are, respectively,

$$\tilde{P} = \begin{pmatrix} \tilde{I} & 0 \\ 0 & 0 \end{pmatrix}, \quad \tilde{P}^\perp = \begin{pmatrix} 0 & 0 \\ 0 & \tilde{I}^\perp \end{pmatrix}, \tag{3.36}$$

where I denotes the identity operator, as usual. We are now ready to give the general definition of a DFS:

Definition 3.2. *Let the system Hilbert space $\mathcal{H}_S$ decompose into a direct sum as $\mathcal{H}_S = \tilde{\mathcal{H}}_S \oplus \tilde{\mathcal{H}}_S^\perp$, and partition the system state ρ accordingly into blocks, as in Eq. (3.35). Assume $\tilde{\rho}(0) \neq 0$. Then $\mathcal{H}_{\text{DFS}}$ is called decoherence free if the initial and final DFS blocks of ρ are unitarily related:*

$$\tilde{\rho}(t) = \tilde{U}\tilde{\rho}(0)\tilde{U}^\dagger, \tag{3.37}$$

where $\tilde{U} : \tilde{\mathcal{H}}_S \mapsto \tilde{\mathcal{H}}_S$ is unitary. We call a DFS "perfectly initialized" if $\rho_2 = 0$ and $\rho_3 = 0$. We call a DFS "imperfectly initialized" if ρ_2 and/or ρ_3 are nonvanishing.

Thus, our discussion of sufficient conditions for a DFS so far has centered entirely on the case of perfect initialization. Having introduced the possibility of imperfect initialization, we are now in a position to state both necessary and sufficient conditions. Interestingly, some "leakage" is allowed into the DFS from the orthogonal subspace. The following theorem and its corollary make this clear.

Theorem 3.5. *Assume imperfect initialization. Let $\tilde{U}$ be unitary, $\{g_i\}$ arbitrary scalars, and $\{\bar{E}_i\}$ arbitrary operators on $\tilde{\mathcal{H}}_S^\perp$. A necessary and sufficient condition for the existence of a DFS with respect to CP maps is that the Kraus operators have a matrix representation of the form*

$$E_i = \begin{pmatrix} g_i \tilde{U} & 0 \\ 0 & \bar{E}_i \end{pmatrix}. \tag{3.38}$$

Note that due to the sum rule $\sum_i E_i^\dagger E_i = I$ the otherwise arbitrary operators $\{\bar{E}_i\}$ satisfy the constraint $\sum_i \bar{E}_i^\dagger \bar{E}_i = \tilde{I}^\perp$ and the scalars $\{g_i\}$ satisfy $\sum_i |g_i|^2 = 1$.

The form (3.38) is identical to the previous result (3.26), with the important distinction that, due to the new definition of a DFS, the theorem holds not just for states initialized perfectly into $\mathcal{H}_{\text{DFS}}$, but for *arbitrary initial states*. This is a very substantial relaxation of the constraints on preparation of a DFS. In fact, there is no constraint at all, surely a delight to any experimentalist!

In light of this we can revisit the original definition, wherein the system *is* initialized inside the DFS. As we will see, this situation admits more general Kraus operators. Specifically:

Theorem 3.6. *Assume perfect initialization. Let $\tilde{U}$ be unitary, $\{g_i\}$ arbitrary scalars, $\{\bar{E}_i\}$ arbitrary operators on $\tilde{\mathcal{H}}_S^\perp$, and $\{A_i\}$ arbitrary operators from $\tilde{\mathcal{H}}_S^\perp$ to $\tilde{\mathcal{H}}_S$. Then the necessary and sufficient DFS condition with respect to CP maps is*

$$E_i = \begin{pmatrix} g_i \tilde{U} & A_i \\ 0 & \bar{E}_i \end{pmatrix}. \tag{3.39}$$

Note that due to the sum rule $\sum_i E_i^\dagger E_i = I$ the otherwise arbitrary operators A_i and $\bar{E}_i$ satisfy the constraints (i) $\sum_i A_i^\dagger A_i + \bar{E}_i^\dagger \bar{E}_i = \tilde{I}^\perp$ and (ii) $\sum_i g_i^* A_i = 0$, and where additionally the scalars g_i satisfy (iii) $\sum_i |g_i|^2 = 1$.

In contrast to the diagonal form in the imperfect initialization scenario, Eq. (3.39) allows for the existence of the off-diagonal term A_i, which represent a *leakage error* from $\tilde{\mathcal{H}}_S^\perp$ into $\tilde{\mathcal{H}}_S$. This more general form of the Kraus operators implies that a larger class of noise processes allow for the existence of DFSs, as compared to the imperfect initialization scenario. Thus, there is a trade-off: work hard to initialize perfectly and thou shalt be protected against leakage; be lazy and initialize imperfectly, and thou shalt not tolerate leakage. We will prove Theorems 3.5 and 3.6 below, as special cases of the NS setting.

3.6 Noiseless subsystems

The notion of a *subspace* that remains protected throughout the evolution of a system is not the most general method for providing DF encoding of information in a quantum system. Instead of encoding into a subspace, it is possible to encode protected information into a *subsystem*. A subsystem is a tensor factor in a tensor product, and this does not have to be a subspace (e.g., in general it is not closed under addition). A simple example will serve to illustrate this. Suppose that we can prepare a pure product state of two qubits, $|\psi\rangle_1 \otimes |\varphi\rangle_2$, and that the bath acts only on the second qubit, transforming the joint state to $|\psi\rangle_1\langle\psi| \otimes \rho_2$. Thus, after the action of the bath the state of the second qubit can be anything, even the completely mixed state, but as long as the bath doesn't affect the first qubit, its state remains pure and continues to store a qubit of information. The first qubit is then a *noiseless subsystem*. A more general situation is this: The system Hilbert space decomposes into two subspaces, one (or more) of which has a tensor product structure

$$\mathcal{H}_S = (\mathcal{H}_{\text{NS}} \otimes \mathcal{H}_{\text{in}}) \oplus \mathcal{H}_{\text{out}}. \tag{3.40}$$

As long as the bath acts only on the factor $\mathcal{H}_{\text{in}}$ of the first subspace, and has arbitrary action on the second subspace $\mathcal{H}_{\text{out}}$, the information in the factor $\mathcal{H}_{\text{NS}}$ is protected. These examples may seem a bit fantastic – after all, why would we ever expect a bath to interact only with one qubit or factor in a tensor product? But as we see next, this situation arises whenever there is a symmetry in the system–bath interaction.

3.6.1 Noiseless subsystems from symmetry

To see how NSs arise from symmetry we need to delve a little deeper into abstract representation theory. This is most easily done in the Hamiltonian formulation of decoherence. Let $\mathcal{A}$ denote the associative algebra (i.e., the algebra of all polynomials) generated by the system Hamiltonian H_S and the system components of the interaction Hamiltonian, the S_αs. We assume that H_S is expressed in terms of the S_αs as well (if needed we can always enlarge this set), and that the identity operator is included as $S_0 = I_S$ and $B_0 = I_B$. This will have no observable consequence but allows for the use of an important representation theorem. Because the Hamiltonian is Hermitian the S_αs must be closed under Hermitian conjugation: $\mathcal{A}$ is a "†-closed" operator algebra. A basic theorem of such operator algebras, which include the identity operator, states that, in general, $\mathcal{A}$ will be a reducible subalgebra of the full algebra of operators on the system Hilbert space $\mathcal{H}_S$. This means that the algebra is isomorphic to a direct sum of $d_J \times d_J$ complex matrix algebras $\mathcal{M}(d_J)$, each with multiplicity n_J:

$$\mathcal{A} \cong \bigoplus_{J \in \mathcal{J}} I_{n_J} \otimes \mathcal{M}(d_J). \tag{3.41}$$

Here $\mathcal{J}$ is a finite set labeling the irreducible components of $\mathcal{A}$.

The structure implied by Eq. (3.41) is illustrated schematically as follows, for some system operator S_α:

$$S_\alpha = \begin{bmatrix} \boxed{J=1} & & & \\ & \boxed{J=2} & & \\ & & \ddots & \\ & & & \boxed{J=|\mathcal{J}|} \end{bmatrix} \tag{3.42}$$

(note that J need not always be an integer; e.g., for half-integer spin systems J may range over half-integer values as well). In this block-diagonal matrix representation, a typical block with given J may have a further block diagonal structure:

$$\begin{bmatrix} \boxed{M_\alpha} & & & & \\ & \boxed{M_\alpha} & & \mu \begin{smallmatrix} 0 \\ \vdots \\ d_J - 1 \end{smallmatrix} & \\ & \mu' : 0 \cdots d_J - 1 & \ddots & \\ & & & \boxed{M_\alpha} \end{bmatrix} \begin{matrix} \lambda = 0 \\ \\ \lambda = 1 \\ \\ \\ \end{matrix} \tag{3.43}$$

Here λ labels the different degenerate sub-blocks, and μ labels the states inside each sub-block. Associated with this decomposition of the algebra $\mathcal{A}$ is the decomposition over the system Hilbert space:

$$\mathcal{H}_S = \sum_{J \in \mathcal{J}} \mathbb{C}^{n_J} \otimes \mathbb{C}^{d_J}. \tag{3.44}$$

We can now encode quantum information into one of the component $\mathbb{C}^{n_J}$, and call this a NS with dimension n_J. The decomposition in Eq. (3.41) reveals that information encoded in such a subsystem will always be affected as the identity on the component $\mathbb{C}^{n_J}$, and thus this information will not decohere. Indeed, since $S_\alpha \in \mathcal{A}$ for all system operators, Eq. (3.41) implies that we can write the system–bath Hamiltonian in the form

$$H_{SB} = \sum_\alpha \left[\bigoplus_{J \in \mathcal{J}} I_{n_J} \otimes \mathcal{M}(d_J) \right]_\alpha \otimes B_\alpha \qquad (3.45)$$

for some unspecified matrices $\mathcal{M}(d_J)$. Because of the presence of the identity matrices I_{n_J}, this shows that the leftmost factors $\mathbb{C}^{n_J}$ in the system Hilbert space decomposition will not be affected by the action of H_{SB}. This deep fact is the basis for the theory of NSs.

Before giving formal proofs and connecting noiselessness to initialization issues, let us state a few general observations:

(i) The tensor product structure that gives rise to the name subsystem in Eq. (3.41) will not in general correspond to the natural tensor product of (spatially distinguishable) qubits. Indeed, often the noiseless factors will correspond to *virtual* degrees of freedom.

(ii) The subsystems in the different irreps can be used simultaneously only in a classical sense: (phase) decoherence will occur between the different irreducible components of the Hilbert space labeled by $J \in \mathcal{J}$.

(iii) DFSs are now easily connected to NSs. DFSs correspond to those NSs possessing *one-dimensional* irreducible matrix algebras: $\mathcal{M}(1)$. The multiplicity of these one-dimensional irreducible algebras is the dimension of the DFSs. In other words, a DFS arises when for a given irrep label J the dimension $d_J = 1$. Such an n_J-dimensional subspace of $\mathcal{H}_S$ is of the form $\mathbb{C}^{n_J} \otimes \mathbb{C}^1$, and $\mathbb{C}^1$ is of course just a scalar.

(iv) In fact, it is easy to see how NSs arise out of a noncommuting generalization of the DFS condition (3.8). Let $\{|\lambda_\mu\rangle\}$, $1 \leq \lambda \leq n_J$ and $1 \leq \mu \leq d_J$, denote a subsystem of $\mathcal{H}_S$ with given J. Then the condition for the existence of an irreducible decomposition as in Eq. (3.41) is

$$S_\alpha |\lambda_\mu\rangle = \sum_{\mu'=1}^{d_J} M_{\mu\mu',\alpha} |\lambda_{\mu'}\rangle, \qquad (3.46)$$

for all S_α, λ and μ. Note that $M_{\mu\mu',\alpha}$ is *not* dependent on λ, in the same way that c_α in Eq. (3.8) is the same for all states $|\tilde{k}\rangle$ (there, $\mu = 1$ and fixed). Thus, for a fixed λ, the subsystem spanned by $|\lambda_\mu\rangle$ is acted upon in some nontrivial way. However, because $M_{\mu\mu',\alpha}$ is not dependent on λ, each subsystem defined by a fixed μ and running over λ is acted upon in an *identical manner* by the decoherence process.

(v) The subsystem concept is crucial in establishing a general theory of universal quantum computation over DFSs, which is beyond the scope of this book (see the "History and further reading" section at the end of this chapter). Very briefly, for computation, it is necessary to introduce the *commutant* $\mathcal{A}'$ of $\mathcal{A}$. This is the set of operators that commutes with the algebra $\mathcal{A}$, $\mathcal{A}' = \{X : [X, A] = 0, \forall A \in \mathcal{A}\}$. They also form a †-closed algebra, which is reducible to

$$\mathcal{A}' = \bigoplus_{J \in \mathcal{J}} \mathcal{M}(n_J) \otimes I_{d_J} \qquad (3.47)$$

over the same basis as $\mathcal{A}$ in Eq. (3.41). Comparing to Eq. (3.41) it is clear that nontrivial operators in $\mathcal{A}'$ are exactly the transformations that preserve the DFS. Thus, these are the operations that can be used to perform computations over a NS, or a DFS.

(vi) Finally, what determines the nature of the decomposition into subsystems? This is, how are the irrep labels J and the corresponding dimensions d_J and multiplicities n_J determined? The answer is symmetry. More specifically, it is the symmetry group of the algebra $\mathscr{A}$, namely the closure as a group of the set of operators that commute with $\mathscr{A}$, which can be used to figure out the subsystem decomposition. We will see this more directly through examples.

3.6.2 Examples

To illustrate the DF states that emerge from the subsystem formulation it is convenient to use once more the collective decoherence model. Recall that collective decoherence on N qubits is characterized by the three system operators S_+, S_-, and S_z. These operators form a $2^N \times 2^N$ representation of the semisimple Lie algebra $sl(2)$, the traceless (special) Lie algebra of all 2×2 matrices. The algebra $\mathscr{A}$ generated by these operators can be decomposed as

$$\mathscr{A} \cong \bigoplus_{J=0(1/2)}^{N/2} I_{n_J} \otimes gl(2J+1), \qquad (3.48)$$

where J labels the total angular momentum of the corresponding Hilbert space decomposition (and hence the 0 or 1/2 depending on whether N is even or odd respectively) and $gl(2J+1)$ is the general linear algebra of all $(2J+1) \times (2J+1)$ matrices. The resulting decomposition of the system Hilbert space is

$$\mathscr{H}_S \cong \bigoplus_{J=0(1/2)}^{N/2} \mathbb{C}^{n_J} \otimes \mathbb{C}^{2J+1}. \qquad (3.49)$$

The degeneracy n_J for each J is given by Eq. (3.23). The right-hand factor

$$\mathbb{C}^{2J+1} \cong \text{span}\{|J, M_J\rangle, | -J \leq M_J \leq J\}$$

represents all the states with a fixed value of J and a given value of the degeneracy label λ, of which there are $d_J = 2J + 1$ due to the different eigenvalues M_J of the angular momentum component J_z. Eqution (3.48) shows that, given J, a state $|J, \lambda, \mu\rangle$ ($1 \leq \lambda \leq n_J$ and $1 \leq \mu \leq d_J$,) is acted upon as identity on its λ component. Thus, an NS is defined by fixing J and μ, where μ can now be naturally interpreted as being in direct correspondence to M_J, i.e., $\mu = J + M_J + 1$. As we will show shortly, λ corresponds to the degeneracy of paths leading to a given vertex (N, J) on the "Bratteli diagram" of Fig. 3.1. We denote the N-qubit NS labeled by a particular angular momentum J, by $\text{NS}_N(J)$, and retain the subspace designation DFS_N when $J = 0$.

The NSs corresponding to the different J values for a given N can be computed using standard methods for the addition of angular momentum. We use the convention that $|1\rangle$ represents a $|j = 1/2, m_j = 1/2\rangle$ particle and $|0\rangle$ represents a $|j = 1/2, m_j = -1/2\rangle$ particle in this decomposition although, of course, one should be careful to treat this labeling as strictly symbolic and not related to the physical angular momentum of particles.

The smallest N that supports an NS and encodes at least a qubit of information is $N = 3$. In this case there are two possible values of the total angular momentum: $J = 3/2$ or $J = 1/2$. The four $J = 3/2$ states $|J, \lambda, \mu\rangle = |3/2, 0, \mu\rangle$ $((\mu - 1 - J)/2 = m_J = \pm 3/2, \pm 1/2)$ are singly degenerate; the $J = 1/2$ states have degeneracy 2. They can be constructed by either adding a $J_{12} = 1$ (triplet) or a $J_{12} = 0$ (singlet) state to a $J_3 = 1/2$ state. These two possible methods of adding the angular momentum to obtain a $J = 1/2$ state are exactly the degeneracy of the representation. The four $J = 1/2$ states are

$$\begin{aligned}
|\tfrac{1}{2}, 0, \tfrac{1}{2}\rangle &= \tfrac{1}{\sqrt{2}} (|010\rangle - |100\rangle), \\
|\tfrac{1}{2}, 0, -\tfrac{1}{2}\rangle &= \tfrac{1}{\sqrt{2}} (|011\rangle - |101\rangle), \\
|\tfrac{1}{2}, 1, 0\rangle &= \tfrac{1}{\sqrt{6}} (-2|001\rangle + |010\rangle + |100\rangle), \\
|\tfrac{1}{2}, 1, 1\rangle &= \tfrac{1}{\sqrt{6}} (2|110\rangle - |101\rangle - |011\rangle),
\end{aligned} \quad (3.50)$$

where on the left we used the $|J, \lambda, M_J\rangle$ notation and on the right the states are expanded in terms of the single-particle $|j = 1/2, m_j = \pm 1/2\rangle$ basis using Clebsch–Gordan coefficients. The NS qubit can now be written compactly as

$$\begin{aligned}
|0_L\rangle &= \alpha |\tfrac{1}{2}, 0, \tfrac{1}{2}\rangle + \beta |\tfrac{1}{2}, 0, -\tfrac{1}{2}\rangle, \\
|1_L\rangle &= \alpha |\tfrac{1}{2}, 1, \tfrac{1}{2}\rangle + \beta |\tfrac{1}{2}, 1, -\tfrac{1}{2}\rangle,
\end{aligned} \quad (3.51)$$

where $|\alpha|^2 + |\beta|^2 = 1$, i.e., the encoding is into the degeneracy of the two $J = 1/2$ subsystems. By the decomposition of Eqs. (3.48) and (3.49) it is clear that collective errors can change the α, β coefficients, but have the same effect on the $|0_L\rangle, |1_L\rangle$ states, which is why this encoding is an NS.

The smallest DFS (as opposed to NS) supporting a full encoded qubit comes about for $N = 4$. Subspaces for collective decoherence correspond to the degeneracy of the zero total angular momentum eigenstates (for $N = 4$ there are also two NS with degeneracy 1 and 3). As we already noted in Section 3.4.2, this subspace is spanned by the states:

$$\begin{aligned}
|0_L\rangle &= |0, 0, 0\rangle = |0, 0\rangle \otimes |0, 0\rangle = \tfrac{1}{2}(|01\rangle - |10\rangle)(|01\rangle - |10\rangle), \\
|1_L\rangle &= |0, 1, 0\rangle = \tfrac{1}{\sqrt{3}}(|1, 1\rangle \otimes |1, -1\rangle - |1, 0\rangle \otimes |1, 0\rangle + |1, -1\rangle \otimes |1, 1\rangle) \\
&= \tfrac{1}{\sqrt{12}}(2|0011\rangle + 2|1100\rangle - |0101\rangle - |1010\rangle - |0110\rangle - |1001\rangle).
\end{aligned} \quad (3.52)$$

The notation is the same as in Eq. (3.51), except that we have also used the notation $|J_{12}, m_{J_{12}}\rangle \otimes |J_{34}, m_{J_{34}}\rangle$, which makes it easy to see how the angular momentum is added.

There is a simple graphical interpretation of these results, shown in Fig. 3.1. Starting from the origin (no spins), begin adding spin-1/2 particles. The first particle ($N = 1$) has total angular momentum $J = 1/2$. This single particle has the trivial $NS_1(1/2)$. When the second particle is added, form either a singlet ($J = 0$) by subtracting 1/2 or form a triplet by adding 1/2. In this manner the subspace DFS_2 and subsystem $NS_2(1)$ are formed, containing a single state ($m_J = 0$) and three states ($m_J = -1, 0, 1$), respectively. The first interesting NS is formed when $N = 3$, since now there are two paths, $\lambda = 0, 1$, in Fig. 3.1 leading to the same total $J = 1/2$. These two paths correspond exactly to the $|0_L\rangle$ and $|1_L\rangle$ states of Eq. (3.51), and define the subsystem $NS_3(1/2)$, which encodes a single noiseless qubit. For $N = 4$ spins there is the subspace DFS_4 and also a three-dimensional subsystem $NS_4(1)$. It should be clear that the DFSs are simply all the vertices that lie on the N-axis in Fig. 3.1, while the NSs are all the vertices above this axis. Interpreted in this manner, Fig. 3.1 is a partitioning of the entire system Hilbert space into disjoint (i.e., a direct sum of) DFSs and NS.

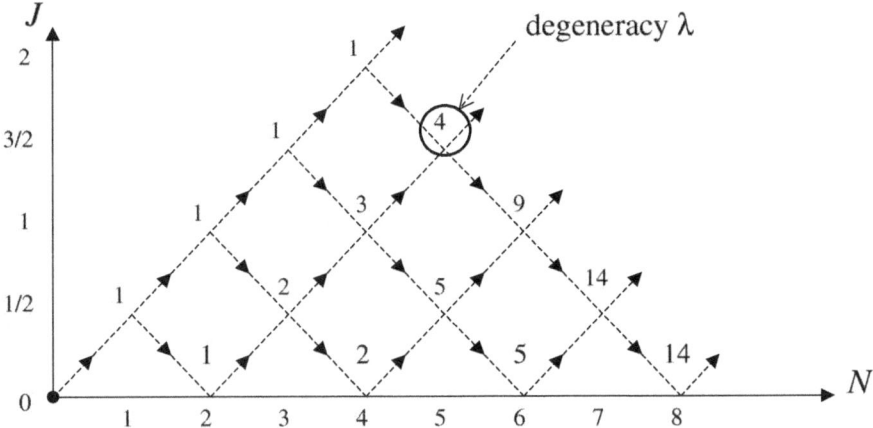

Fig. 3.1. Partitioning of the Hilbert space of N spin-$1/2$ particles into DFSs (nodes on the N-axis) and NS (nodes above the N-axis). The integer above a node represents the number of paths leading from the origin to that node, i.e., the degeneracy quantum number λ. See text for full details. Reprinted with kind permission of Springer Science and Business Media from Lecture Notes in Physics Volume 622, *Irreversible Quantum Dynamics*, edited by F. Benatti and R. Floreanini (Springer, Berlin, 2003).

3.6.3 Dimension of noiseless subsystems and decoherence-free subspaces for collective decoherence

Let us first derive Eq. (3.23) for the DFS case, i.e., $J = 0$. One way to do this is to use combinatorial arguments relating to the permutation group of N objects. The result follows straightforwardly from the Young diagram technique. The singlet states of $sl(2)$ belong to the rectangular Young tableaux of $N/2$ columns and 2 rows. The multiplicity λ of such states is the number of "standard tableaux" (tableaux containing an arrangement of numbers that increase from left to right in a row and from top to bottom in a column), which is also the dimension of the irreducible representation of the permutation group corresponding to the Young diagram $\eta_{N/2,2}$ (an empty tableau) of $N/2$ columns and 2 rows. This number is found using the "hook recipe," where one writes the "hook length" g_i (the sum of the number of positions to the right of box i, plus the number of positions below it, plus one) of each box i in the Young diagram:

$$\lambda(\eta) = \frac{K!}{\prod_{i=1}^{K} g_i}. \tag{3.53}$$

For example, for $\eta_{c,2}$ the hook lengths are

$c+1$	c	$c-1$	$\cdots$	3	2
c	$c-1$	$c-2$	$\cdots$	2	1

(3.54)

and one finds, with $c = N/2$:

$$\lambda(\eta_{N/2,2}) = \frac{N!}{(N/2+1)!(N/2)!}, \tag{3.55}$$

which is indeed the $J = 0$ case of the general degeneracy formula, Eq. (3.23).

Next let us derive Eq. (3.23) for general J, by using angular momentum addition rules. Suppose we have N spin-1/2 particles and we are interested in all states with fixed J_z eigenvalue M. We are allowing J to vary. For example, for $N = 3$ and $M = -1/2$, we have the state $|J, \lambda, M_J\rangle = |3/2, 0, -1/2\rangle$ and the two degenerate states $|J, \lambda, M_J\rangle \in \{|1/2, 0, -1/2\rangle, |1/2, 1, -1/2\rangle\}$. To get $M = -1/2$, each of these states must have two down-spins and one up-spin, which is another way to see why there should be three states in total. More generally, let there be x down-spins and $N - x$ up-spins. Each down-spin contributes $-1/2$ to M, while each up-spin contributes $+1/2$. Thus $M = -\frac{1}{2}x + \frac{1}{2}(N - x)$. The total number of ways to arrange x down-spins among N spins is $\binom{N}{x} = \frac{N!}{(N/2-M)!(N/2+M)!}$, so this is the dimension d_M of the fixed-M subspace. On the other hand, as we just saw in the $N = 3$ example, for each permissible value of J there are n_J degenerate states (having the same value of M_J) contributing to this subspace. We say permissible, since for example when $M = 3/2$ there can be no contribution from $J = 1/2$ (recall that $-J \leq M_J \leq J$). The rule is simply that the lowest J that can contribute to a given fixed-M subspace is $J = |M|$. The highest value J can attain is always $N/2$. Therefore, we conclude that $d_M = \sum_{J'=|M|}^{N/2} n_{J'}$ is another expression for the dimension of a fixed-M subspace. Now let $M = J > 0$ and note that $n_J = d_J - d_{J+1}$. Therefore

$$n_J = \frac{N!}{(N/2 - J)!(N/2 + J)!} - \frac{N!}{(N/2 - J - 1)!(N/2 + J + 1)!}$$
$$= \frac{(2J + 1)N!}{(N/2 - J)!(N/2 + J + 1)!}, \qquad (3.56)$$

which is Eq. (3.23).

3.7 Initialization-free noiseless subsystems

Just as we were able to generalize the notion of a DFS by relaxing the initialization constraint, we will gain by introducing a similar relaxation into NS theory. In accordance with the splitting of the system Hilbert space, as in Eq. (3.40), let us define projection operators:

$$\tilde{P} = \begin{pmatrix} I_{\text{NS}} \otimes I_{\text{in}} & 0 \\ 0 & 0 \end{pmatrix}, \quad \tilde{P}^\perp = \begin{pmatrix} 0 & 0 \\ 0 & I_{\text{NS}} \otimes I_{\text{in}} \end{pmatrix}. \qquad (3.57)$$

These generalize the DFS projection operators defined in Eq. (3.36), to which the current $\tilde{P}$ and $\tilde{P}^\perp$ reduce when $I_{\text{in}} = 1$ (hence there is no risk of confusion). The system density matrix takes the corresponding block form

$$\rho = \begin{pmatrix} \rho_{\text{NS-in}} & \rho' \\ \rho'^\dagger & \rho_{\text{out}} \end{pmatrix}. \qquad (3.58)$$

Definition 3.3. *Let the system Hilbert space $\mathcal{H}_S$ decompose as $\mathcal{H}_S = \mathcal{H}_{\text{NS}} \otimes \mathcal{H}_{\text{in}} \oplus \mathcal{H}_{\text{out}}$, and partition the system state ρ_S accordingly into blocks, as in Eq. (3.58). Assume $\rho_{\text{NS-in}}(0) \neq 0$. Then the factor $\mathcal{H}_{\text{NS}}$ is called a noiseless subsystem if the following condition holds:*

$$\text{Tr}_{\text{in}}\{\rho_{\text{NS-in}}(t)\} = U_{\text{NS}} \text{Tr}_{\text{in}}\{\rho_{\text{NS-in}}(0)\} U_{\text{NS}}^\dagger, \qquad (3.59)$$

where U_{NS} is a unitary matrix acting on $\mathcal{H}_{\text{NS}}$.

Definition 3.4. *Perfect initialization (noiseless subsystems): $\rho' = 0$ and $\rho_{\text{out}} = 0$ in Eq. (3.58).*

Definition 3.5. *Imperfect initialization (noiseless subsystems):* ρ' *and/or* ρ_{out} *in Eq. (3.58) are non-vanishing.*

According to Definition 3.3, a quantum state encoded into the $\mathcal{H}_{\text{NS}}$ factor at some time t is unitarily related to the $t = 0$ state. The factor $\mathcal{H}_{\text{in}}$ is unimportant, and hence is traced over. Clearly, an NS reduces to a DFS when $\mathcal{H}_{\text{in}}$ is one-dimensional, i.e., when $\mathcal{H}_{\text{in}} = \mathbb{C}$.

We now present the necessary and sufficient conditions for NSs. In stating constraints on the form of the Kraus operators, below, it is understood that in addition they must satisfy the normalization rule $\sum_i E_i^\dagger E_i = I$, which we do not specify explicitly.

Theorem 3.7. *Assume imperfect initialization. Then a subsystem* $\mathcal{H}_{\text{NS}}$ *in the decomposition* $\mathcal{H}_S = \mathcal{H}_{\text{NS}} \otimes \mathcal{H}_{\text{in}} \oplus \mathcal{H}_{\text{out}}$ *is noiseless with respect to CP maps if and only if the Kraus operators have the matrix representation*

$$E_i = \begin{pmatrix} U \otimes C_i & 0 \\ 0 & B_i \end{pmatrix}. \tag{3.60}$$

Theorem 3.8. *Assume perfect initialization. Then a subsystem* $\mathcal{H}_{\text{NS}}$ *in the decomposition* $\mathcal{H}_S = \mathcal{H}_{\text{NS}} \otimes \mathcal{H}_{\text{in}} \oplus \mathcal{H}_{\text{out}}$ *is noiseless with respect to CP maps if and only if the Kraus operators have the matrix representation*

$$E_i = \begin{pmatrix} U \otimes C_i & A_i \\ 0 & B_i \end{pmatrix}. \tag{3.61}$$

Note that, as in the DFS case, there is a trade-off between the quality of preparation and the amount of leakage that can be tolerated.

3.7.1 Proof of noiseless subsystems condition for CP maps

Let us now prove Theorem 3.7 and then Theorem 3.8, for NSs. The DFS results will follow as special cases.

3.7.1.1 Proof for an arbitrary initial state Assume the system evolution due to its interaction with a bath is described by a CP map with Kraus operators $\{E_i\}$: $\rho(t) = \sum_i E_i \rho(0) E_i^\dagger$. Note that here ρ is an operator on the entire system Hilbert space $\mathcal{H}_S$, which we assume to be decomposable as $\mathcal{H}_{\text{NS}} \otimes \mathcal{H}_{\text{in}} \oplus \mathcal{H}_{\text{out}}$. From the NS definition, Eq. (3.59), we have

$$\text{Tr}_{\text{in}}\{U \otimes I(\tilde{P}\rho_S(0)\tilde{P})U^\dagger \otimes I\} = \text{Tr}_{\text{in}}\left\{\sum_i \left(\tilde{P}E_i\right)\rho_S(0)(E_i^\dagger \tilde{P})\right\}. \tag{3.62}$$

Let us represent the Kraus operators as matrices with the same block structure as that of the system state, i.e., corresponding to the decomposition $\mathcal{H}_S = \mathcal{H}_{\text{NS}} \otimes \mathcal{H}_{\text{in}} \oplus \mathcal{H}_{\text{out}}$, where the blocks correspond to the subspaces $\mathcal{H}_{\text{NS}} \otimes \mathcal{H}_{\text{in}}$ (upper-left block) and $\mathcal{H}_{\text{out}}$ (lower-right block). Then

$$\rho_S = \begin{pmatrix} \rho_1 & \rho_2 \\ \rho_2^\dagger & \rho_3 \end{pmatrix}, \tag{3.63}$$

$$E_i = \begin{pmatrix} P_i & A_i \\ D_i & B_i \end{pmatrix}, \tag{3.64}$$

with appropriate normalization constraints, considered below. Equation (3.62) simplifies in this matrix form as

$$\mathrm{Tr}_{\mathrm{in}}\{U \otimes I \rho_1 U^\dagger \otimes I\} = \mathrm{Tr}_{\mathrm{in}}\left\{\sum_i P_i \rho_1 P_i^\dagger + P_i \rho_2 A_i^\dagger + A_i \rho_2^\dagger P_i^\dagger + A_i \rho_3 A_i^\dagger\right\}, \quad (3.65)$$

which must hold for arbitrary $\rho_S(0)$. To derive constraints on the various terms we therefore consider special cases, which yield necessary conditions. First, consider an initial state $\rho_S(0)$ such that $\rho_2 = 0$. Then, as the left-hand side of Eq. (3.65) is independent from ρ_3, the last term must vanish:

$$\sum_i A_i \rho_3 A_i^\dagger = 0 \implies A_i = 0. \quad (3.66)$$

Further assume $\rho_1 = |i\rangle\langle i| \otimes |i'\rangle\langle i'|$. Note that the partial matrix element $\langle j'|P_\alpha|i'\rangle$ is an operator on the $\mathcal{H}_{\mathrm{NS}}$ factor, $|i\rangle\langle i|$. Then Eq. (3.65) reduces to

$$|i\rangle\langle i| = \sum_{\alpha,j'} [U^\dagger \langle j'|P_\alpha|i'\rangle] |i\rangle\langle i| [\langle i'|P_\alpha^\dagger|j'\rangle U]. \quad (3.67)$$

Taking matrix elements with respect to $|i^\perp\rangle$, a state orthogonal to $|i\rangle$, yields

$$\begin{aligned} 0 &= \sum_{\alpha,j'} |\langle i^\perp| [U^\dagger \langle j'|P_\alpha|i'\rangle] |i\rangle|^2 \\ &\implies \langle i^\perp| [U^\dagger \langle j'|P_\alpha|i'\rangle] |i\rangle = 0, \end{aligned} \quad (3.68)$$

which in turn implies that $[U^\dagger \langle j'|P_\alpha|i'\rangle] |i\rangle$ is proportional to $|i\rangle$, i.e.,

$$[\langle j'|P_\alpha|i'\rangle] |i\rangle \propto U|i\rangle. \quad (3.69)$$

Since $|i'\rangle, |j'\rangle$ are arbitrary, this condition implies that the submatrix P_α must be of the form $P_\alpha = U \otimes C_\alpha$. Substituting $P_\alpha = U \otimes C_\alpha$ into Eq. (3.65) we have $\mathrm{Tr}_{\mathrm{in}}\{(U \otimes I)\rho_1(U^\dagger \otimes I)\} = \mathrm{Tr}_{\mathrm{in}}\{\sum_\alpha (U \otimes C_\alpha)\rho_1(U^\dagger \otimes C_\alpha^\dagger)\}$, so that

$$\mathrm{Tr}_{\mathrm{in}}\{\rho_1\} = \mathrm{Tr}_{\mathrm{in}}\left\{\sum_\alpha I_{\mathrm{NS}} \otimes C_\alpha \rho_1 I_{\mathrm{NS}} \otimes C_\alpha^\dagger\right\}. \quad (3.70)$$

Now suppose $\rho_1 = \sum_{iji'j'} \lambda_{iji'j'} |i\rangle\langle j| \otimes |i'\rangle\langle j'|$; then from Eq. (3.70) we find

$$\sum_{iji'} \lambda_{iji'i'} |i\rangle\langle j| = \sum_{iji'j'k'\alpha} \lambda_{iji'j'} |i\rangle\langle j| \langle k'|C_\alpha|i'\rangle\langle j'|C_\alpha^\dagger|k'\rangle. \quad (3.71)$$

Using $\sum_{k'} |k'\rangle\langle k'| = I_{\mathrm{in}}$, Eq. (3.71) becomes

$$\sum_{iji'} \lambda_{iji'i'} |i\rangle\langle j| = \sum_{iji'j'} \lambda_{iji'j'} |i\rangle\langle j|\langle j'| \sum_\alpha C_\alpha^\dagger C_\alpha |i'\rangle. \quad (3.72)$$

It follows that

$$\sum_\alpha C_\alpha^\dagger C_\alpha = I_{\mathrm{in}}. \quad (3.73)$$

Next consider the normalization constraint $\sum_\alpha E_\alpha^\dagger E_\alpha = I$ for the Kraus operators, together with the additional constraints we have derived ($A_\alpha = 0$, $P_\alpha = U \otimes C_\alpha$):

$$\sum_\alpha P_\alpha^\dagger P_\alpha + D_\alpha^\dagger D_\alpha = I_{\text{NS}} \otimes I_{\text{in}}$$

$$\implies I_{\text{NS}} \otimes \sum_\alpha C_\alpha^\dagger C_\alpha + \sum_\alpha D_\alpha^\dagger D_\alpha = I_{\text{NS}} \otimes I_{\text{in}}. \tag{3.74}$$

But, from Eq. (3.73) we have $\sum_\alpha P_\alpha^\dagger P_\alpha = I_{\text{NS}} \otimes I_{\text{in}}$. Therefore $D_\alpha = 0$.

Taking all these conditions together finalizes the matrix representation of the Kraus operators as

$$E_\alpha = \begin{pmatrix} U \otimes C_\alpha & 0 \\ 0 & B_\alpha \end{pmatrix}. \tag{3.75}$$

For a scalar C_α we recover the DFS condition (3.38). These considerations establish the necessity of the representation (3.75); it is simple to show that this representation is also sufficient, by substitution and checking that the NS and DFS conditions are satisfied. Therefore we have proved Theorem 3.7 and, as an immediate corollary, Theorem 3.5. □

3.7.1.2 Perfect initialization Theorems 3.8 and 3.6 for NS/DF-initialized states of the form $\rho_S(0) = \tilde{P}\rho_S(0)\tilde{P}$ now follow as easy corollaries. We simply need to prove that $D_i = 0$ in Eq. (3.64). Indeed, when $\rho_S(0) = \tilde{P}\rho_S(0)\tilde{P}$ we have that $\rho_2 = 0$ and $\rho_3 = 0$ and Eq. (3.65) reduces to

$$\text{Tr}_{\text{in}}\{U \otimes I \rho_1 U^\dagger \otimes I\} = \text{Tr}_{\text{in}}\left\{\sum_i P_i \rho_1 P_i^\dagger\right\}. \tag{3.76}$$

The argument leading to the vanishing of the A_i [Eq. (3.66)] then does not apply, and indeed the A_i need not vanish. However, the arguments leading to $P_i = U \otimes C_i$ and $\sum_i P_i^\dagger P_i = I_{\text{NS}} \otimes I_{\text{in}}$ do apply. Hence $D_i = 0$. □

3.8 Protection against additional decoherence sources

Noiseless subsystems and DFSs will provide complete protection when the conditions for which they are designed are satisfied. However, these conditions require that certain symmetries are perfectly respected, and in general nature is not so generous. One is therefore motivated to consider how to add protection against symmetry-breaking perturbations, which appear as sources of decoherence. A detailed discussion of this subject is beyond the scope of this chapter (see the "History and further reading" section). We are content merely to mention here that several approaches have been taken to address this issue, and that one can prove a certain degree of stability of DFSs to symmetry-breaking perturbations (no such results have been obtained for NSs, probably for lack of trying). In particular, it can be shown that DFSs are stable to symmetry-breaking perturbations to second order in the strength of the perturbation. To be specific, let us define a symmetry-breaking perturbation as any term added to the system–bath Hamiltonian whose system operators do not satisfy Eq. (3.8). Consider the memory fidelity $f(t) \equiv \text{Tr}[\rho_S(0)\rho_S(t)]$, a measure of the overlap between the initial system state $\rho_S(0)$ and its time-evolved self. Expanding this fidelity in a power series in t as $f(t) = 1 - \sum_{n=1}^\infty \frac{1}{n!}(t/\tau_n)^n$, with the "decoherence rates" $1/\tau_n = \left\{ \text{Tr} \left[\rho_S(0) \left. \frac{d\rho_S}{dt} \right|_{t=0} \right] \right\}^{1/n}$, it is possible to show that $\frac{1}{\tau_1} = 0$ for any symmetry-breaking perturbation, and that the leading order correction is $O(\epsilon^2 t^2)$, where ϵ is the strength of the perturbation.

Of course, this level of robustness is only a start, as we would like to ensure that the fidelity remains high for arbitrarily long times. This is not possible using NS or DFS protection alone, in the presence of symmetry-breaking perturbations. Additional *active* protection must be included, e.g., in the form of QECCs. For example, in the collective decoherence DFS, independent single-qubit errors can be shown to cause either independent errors acting on the encoded qubit states, or leakage of the DFS encoded qubit states to states lying outside the DFS. Both of these types of errors may be corrected by concatenating the DFS encoding with a standard QECC that corrects independent single-qubit errors. For example, the four-qubit DFS for collective decoherence can be concatenated with the five-qubit QECC to produce a 20-qubit encoding that provides protection against both collective and independent errors on the physical qubits. Concatenation thus provides one way to correct leakage errors that take the encoded quantum information out of the DFS.

A completely different approach to dealing with errors not specifically protected against by the DFS encoding is to "engineer" the system Hamiltonian. The idea here is to supplement this Hamiltonian by additional interactions that impose an energy spectrum on the DFS states. The additional Hamiltonian terms are specifically engineered to result in an energy spectrum that eliminates or thermally suppresses additional errors. For example, this can done using the four-qubit DFS against collective decoherence by addition of a specifically designed set of exchange interactions between the physical qubits, which creates an energy gap from the degenerate ground state that provides the two singlets for the DFS encoding. The effect of this additional Hamiltonian is to transform all independent single-qubit errors into energy nonconserving errors that cause excitation out of the ground-state encoding. The energy-nonconserving errors can be suppressed by going to low temperatures, or alternatively used as an error-detecting code.

3.9 Conclusions

In this chapter we have seen that quantum information can be perfectly protected in a passive manner against arbitrarily strong coupling to the environment. The central idea behind this DFS or NS protection is to exploit dynamical symmetries. This chapter focused on the problem of preservation of quantum information, as opposed to its manipulation (see Sections 16.5 and 22.2.2 for discussions of computation on a DFS). We illustrated information preservation with examples from quantum optics and general models of interacting quantum systems. Ultimately, we observed that passive protection alone cannot suffice, since symmetry breaking will cause leakage out of the protected subspace or subsystem. However, the concept of an NS has turned out to be very powerful and central to a unification of a variety of quantum information protection methods, as we will see later throughout this book.

3.10 History and further reading

Perhaps the earliest study employing symmetries in order to reduce the coupling of subsets of system states to the environment is Alicki's work on limited thermalization [A88]. The search for systematic ways to bypass decoherence in the context of quantum information processing, based on identification of states that might be immune to certain decohering interactions, started with observations of Palma, Suominen, and Ekert [PSE96] in a study of the effects of pure dephasing, that two qubits possessing identical interactions with the environment do not decohere. Palma *et al.* used the term "subdecoherence" to describe this phenomenon, suggested using the corresponding states to form a "noiseless" encoding into logical qubits, and noted that the set of states robust against dephasing will depend on the specific form of qubit–environment coupling. This model was subsequently studied using a different method by Duan and Guo [DG98b], with similar conclusions and a change of terminology to "coherence preserving states." The idea of pairing qubits as a means for preserving coherence was further generalized by Duan and Guo in [DG97], where it was shown that both collective dephasing and dissipation could be prevented. However, this assumed knowledge of the system–environment coupling strength. These early studies were subsequently cast into a general mathematical framework for DFSs of more general system/environment interactions, first for the spin-boson model [LCD+87] by Zanardi and Rasetti in [ZR97b], where the important collective decoherence model was introduced, then for general Hamiltonians [ZR97a]. Their algebraic analysis established the importance of identifying the dynamical symmetries in the system–environment interaction, and provided the first general formal condition for decoherence-free (DF) states that did not require knowledge of the system–environment coupling strength. In the work of Zanardi and Rasetti these are referred to as "error-avoiding codes." Several papers focusing on collective dissipation appeared subsequently [Z97, DG97, DG98a], as well as applications to encoding information in quantum dots [ZR98, ZR99b]. The Jaynes–Cummings model (3.21) was treated extensively in [DG97, DG98a, Z98, Z97], building on the Dicke model of quantum optics [D54], where Eq. (3.23) was first derived. Zanardi [Z98] and independently Lidar, Chuang, and Whaley [LCW98] showed that DF states could also be derived from very general considerations of Markovian master equations. Lidar *et al.* introduced the term "decoherence-free subspace," analyzed their robustness against perturbations, and pointed out that the absence of decoherence for DF states can be spoiled by evolution under the system Hamiltonian, identifying a second major requirement for viable use of the DF states for either quantum memory or quantum computation [LCW98]. A completely general condition for the existence of DF states was subsequently provided in terms of CP maps by Lidar, Bacon, and Whaley [LBW99]. All these studies share essentially the same canonical example of system–environment symmetry: a qubit-permutation symmetry in the system–environment coupling, which gives rise to collective dephasing, dissipation, or decoherence. The other main example of a symmetry giving rise to large DFSs was provided by Lidar *et al.* in [LBK+01a] (which includes a proof of Theorem 3.3), using the group algebra construction.

Several papers reported various generalizations of DFSs, e.g., by extending DFSs to quantum groups [DMO01], using the rigged Hilbert space formalism [QRZ02], and deriving DFSs from a scattering S-matrix approach [SB02]. However, the next major step forward in generalizing the DFS concept was taken by Knill, Laflamme, and Viola [KLV00], who introduced the notion of a noiseless subsystem. Whereas previous DFS work had characterized DF states as singlets (one-dimensional irreducible representations) of the algebra generating the dynamical symmetry in the system–bath interaction, the work by Knill *et al.* showed that higher-dimensional irreducible representations can support DF states as well. An important consequence was the reduction of the number of qubits needed to construct a DFS under collective decoherence from four to three. This was noted independently by De Filippo [F00] and by Yang and Gea-Banacloche [YG.-B01]. The generalization from subspaces to subsystems has provided a basis for unifying essentially all known decoherence-suppression/avoidance strategies [Z00, ZL03]. It also paved the way towards a comprehensive theory of universal quantum computation on DFSs and NSs [BKL+00, KBL+01, LBK+01b].

Following the initial studies establishing the conditions for DFSs, Bacon, Lidar, and Whaley made a thorough investigation of the robustness of DF states to symmetry-breaking perturbations [BL+99]. These authors also showed that the passive error-avoidance properties of a DFS can be combined with the active error-correction protocols provided by quantum error correction by concatenation of a DFS inside a QECC, resulting in an encoding capable of protecting against both collective and independent errors [LBW99]. The DFS for collective decoherence offers a natural energy barrier against other decoherence processes, a phenomenon termed "supercoherence" by Bacon, Brown, and Whaley [BBW01]. Another approach complementing naturally occurring DFSs is to *actively generate* the symmetries required for NS or DFS protection by use of dynamical decoupling, a strategy explored in detail by Wu, Byrd, and Lidar [WL02, BL02a].

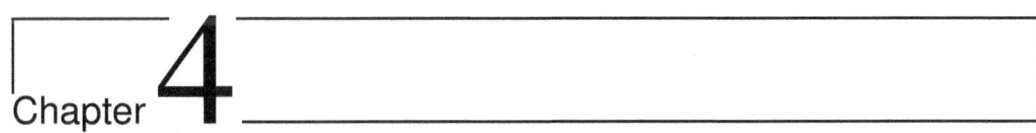

Chapter 4

Introduction to quantum dynamical decoupling

Lorenza Viola

4.1 Motivation and overview

The aim of this chapter is to illustrate the basic physical principles and mathematical framework of *dynamical decoupling* (DD) techniques for open quantum systems, as relevant to quantum information processing (QIP) applications. Historically, the physical origins of DD date back to the idea of *coherent averaging* of interactions, as pioneered in high-resolution solid-state nuclear magnetic resonance (NMR) by Haeberlen and Waugh using elegantly designed multiple-pulse sequences [HW68, WHH68, H76]. It was in the same landmark work [HW68] that *average Hamiltonian theory* was developed as a formalism on which the design and analysis of DD sequences has largely relied since then. In the original context of NMR spectroscopy, decoupling serves the purpose of enhancing resolution by simplifying complex spectra. This is achieved by realizing that an otherwise static spin Hamiltonian "can be made to appear time-dependent in a controlled way," so that "as the characteristic repetition period of the pulses becomes [sufficiently] small, the spin system comes to behave over long times as though under the influence of a time-independent average Hamiltonian" [HW68]. Some basic insight may be gained by revisiting the paradigmatic example offered by the so-called Hahn echo [H50] and Carr–Purcell (CP) sequences [CP54] in the simplest setting of a two spin-1/2 system. Consider a Hamiltonian of the form

$$H = \omega_1 \sigma_z^{(1)} + \omega_2 \sigma_z^{(2)} = \omega_1 Z \otimes I + \omega_2 I \otimes Z, \quad (4.1)$$

where, physically, the parameters ω_ℓ account for the Zeeman splitting in a static applied magnetic field $\vec{B}$ along the z direction. If $\omega_1 \neq \omega_2$ as a consequence of magnet inhomogeneity, spin precession under H results in relative dephasing, causing the magnetization free induction signal to decay in time. In an actual experiment, the intrinsic width of the resonance line is obscured by unwanted "inhomogeneous broadening" [S90, EBW94]. Suppose, however, that a pair of instantaneous π-rotations is "nonselectively" applied to both spins at instants $t_1 = \Delta t > 0$ and

---

Quantum Error Correction, ed. Daniel A. Lidar and Todd A. Brun. Published by Cambridge University Press.
© Cambridge University Press 2013.

$t_2 = 2\Delta t$, the action of each pulse corresponding to ideal spin inversion according to $\sigma_z^{(\ell)} \mapsto -\sigma_z^{(\ell)}$. Then, intuitively (and formally, as we discuss in Section 4.2), evolution in the second time interval may be viewed as occurring under a *time-reversed* Hamiltonian $H' = -H$, in such a way that the undesired phase evolution is effectively *refocused* and an "echo" forms at time $T = 2\Delta t$. If, as in a CP sequence, a train of repeated spin echoes is implemented, the net spin evolution occurs as though over the corresponding time scale, $\overline{H} = 0$. Similarly, if only spin 1 is flipped at time t_1, spin 1 is effectively "traced out" – as though $\overline{H} = \omega_2 \sigma_z^2$ over time T. Interestingly, if a coupling with strength J is present between the two spins, so that Hamiltonian (4.1) is modified to

$$H = \omega_1 \sigma_z^{(1)} + \omega_2 \sigma_z^{(2)} + J\sigma_z^{(1)}\sigma_z^{(2)} = \omega_1 Z \otimes I + \omega_2 I \otimes Z + J Z \otimes Z, \qquad (4.2)$$

then such a "selective" pulse sequence still yields $\overline{H} = \omega_2 \sigma_z^2$. This shows how refocusing of couplings that involve a specific subset of spins effectively results in *decoupling*. Building on similar intuition, multiple-pulse sequences have now evolved into a rich and sophisticated toolbox – including super-cycle techniques, noise decoupling, composite pulses, broadband decoupling schemes, and more, see e.g., [FKL80, L83, SK87, EBW94] – whose exploitation largely underpins the exquisite degree of coherent control achieved in manipulating nuclear-spin Hamiltonians in liquids and solids.

Within QIP, the need to achieve accurate dynamical control in the presence of both operational and decoherence errors has prompted us to re-interpret coherent averaging ideas in the general framework of *open* quantum systems, and to seek a sound control-theoretical foundation for DD (Section 4.3). As a result, in the last decade DD techniques have both substantially enlarged their meaning and serve a broader range of purposes as compared to the original NMR setting. A DD scheme may be generally understood as an *open-loop* control protocol involving a sequence of *unitary* operations drawn from an available *fixed* (typically finite) repertoire – according to various possible design criteria. Compared to QEC, DD methods have the important advantage of avoiding the need for quantum measurements and, in their basic form, encoding overhead – which makes them considerably less resource-intensive and practically attractive for a large class of devices. Applications of DD methods span a wide range of Hamiltonian engineering problems in QIP. From a quantum error control perspective, a task of chief relevance is decoherence control, as sought through active error-suppression schemes that approximately remove or symmetrize the coupling to the environment, paving the way to the dynamical generation of DFSs and NSs. For closed systems, DD schemes are likewise employed to effectively switch off undesired qubit dynamics and/or qubit couplings, with objectives ranging from Hamiltonian "time-suspension" to Hamiltonian inversion, and universal quantum simulation.

Throughout this chapter, the focus is on discussing the basic DD approaches for robust quantum information *storage* in non-Markovian open quantum systems, building as much as possible on illustrative QIP-motivated examples. In particular, Section 4.4 is devoted to analyzing the DD problem to its lowest order of accuracy within the idealized setting of unbounded-strength "bang-bang" (BB) control. Important limitations of "low-level" DD design and DD schemes capable of operating beyond the BB assumption are highlighted in Section 4.5 and Section 4.6, respectively. While experimental implementations of some basic DD protocols are surveyed in Chapter 22, the present analysis serves as the starting point for the advanced DD approaches presented in Chapters 14 and 15.

4.2 Warm up: bang-bang decoupling of qubit dephasing

In preparation for more complicated systems and control settings, and more formal treatments, it is useful to develop physical intuition about DD by revisiting the simplest case of decoherence suppression in a single qubit coupled to a purely dephasing quantum environment (or bath). The joint open-system Hamiltonian may be taken to read

$$H = H_S \otimes I_B + I_S \otimes H_B + H_{SB} = \omega_0 Z \otimes I_B + H_B + Z \otimes B_Z, \quad (4.3)$$

with $2\omega_0 > 0$ giving the energy difference between computational basis states $|0\rangle, |1\rangle$ and H_B, B_Z being suitable bath operators. For the purpose of realizing a single-qubit memory, both H_S and H_{SB} induce unwanted phase evolution, whose effective refocusing is the DD task. While formal analogy with Eq. (4.2) (by treating spin 2 as an environment for spin 1) *suggests* that the desired averaging effect should still be attainable through a suitable sequence of spin-flips, it is important to appreciate similarities and differences between the two situations. In both cases, $[H_S, H_{SB}] = 0$: while this implies the existence of a preferred (z) basis for decoherence and underlies the exact solvability of the resulting dynamics for arbitrary (possibly unknown) H_B, B_Z, from a control standpoint this ensures that *single-axis* DD will indeed suffice for refocusing – as opposed to *two-axis* DD that will be required for an arbitrary *unknown* direction; see Section 4.4.3. In the case of (4.3), however, $[H_B, H_{SB}] \neq 0$, reflecting the presence of a genuinely *dynamical* quantum environment. From a control standpoint, this interplay will be responsible for introducing *time-scale requirements* that are not present in Eq. (4.2).

A paradigmatic example for quantitative DD analysis is provided by the diagonal linear spin-boson Hamiltonian examined in [VS98], whereby

$$H_B = \sum_k \omega_k b_k^\dagger b_k, \quad B_Z = B_+ + B_- = \sum_k (g_k b_k^\dagger + g_k^* b_k), \, g_k \in \mathbb{C}, \quad (4.4)$$

with $b_k, b_k^\dagger$ denoting canonical bosonic operators for mode k. Under the assumption that the bath B is initially uncorrelated from S and in thermal equilibrium at temperature T, the evolution of the qubit coherence $\rho_{01}(t) = \langle 0|\rho(t)|1\rangle$ in the absence of control is described by [PSE96]

$$\rho_{01}(t) = \rho_{01}(0) e^{i2\omega_0 t - \Gamma_0(t)}, \quad \Gamma_0(t) = \int_0^\infty d\omega \, J(\omega) \frac{1 - \cos \omega t}{\omega^2}, \quad (4.5)$$

where the power spectrum $J(\omega) = I(\omega)[2\bar{n}(\omega, T) + 1]$ encapsulates the influence of B in terms of the oscillator spectral density $I(\omega)$ and the average number $\bar{n}$ of thermal and vacuum excitations [BP02] (both defined in Chapter 1). Physically, $J(\omega)$ is proportional to the Fourier transform of the environment equilibrium correlation function $\langle B_Z(t) B_Z(0) \rangle = \text{Tr}_B\{e^{iH_B t} B_Z e^{-iH_B t} B_Z\}$. By analogy with the NMR CP sequence described above, let DD be implemented by a train of resonant, instantaneous and infinitely strong BB π-pulses about, say, the x-axis, with constant separation Δt. We will refer to such a control scheme as *CP DD* henceforth. If $P = \exp(-i\pi X/2)$ and $U_0(\Delta t)$ denote, respectively, the unitary propagator describing each pulse and a free evolution period of duration Δt under H, the joint propagator after an elementary spin-flip cycle of duration $T_c = 2\Delta t$ is given by:

$$\begin{aligned} U(T_c) &= P U_0(\Delta t) P U_0(\Delta t) = (PP)[P^\dagger U_0(\Delta t) P] U_0(\Delta t) \\ &= I_S \, e^{-iH'\Delta t} e^{-iH\Delta t} \equiv e^{-iH_{\text{eff}} T_c}, \end{aligned} \quad (4.6)$$

where, explicitly,
$$H' = P^\dagger H P = -H_S \otimes I_B + I_S \otimes H_B - H_{SB},$$
and, by invoking the Campbell–Baker–Hausdorff formula, H_{eff} may be read off the following expansion:
$$e^{-iH'\Delta t}e^{-iH\Delta t} = e^{-i(H+H')\Delta t - \frac{1}{2}[H',H]\Delta t^2 + \frac{i}{12}([H',[H',H]]+[H,[H,H']])\Delta t^3 \ldots}. \quad (4.7)$$

Two main observations emerge: First, evolution during the second control interval can still be viewed as occurring under a transformed Hamiltonian H'; however, $H' \neq -H$ if $H_B \neq 0$ – implying that an *approximate* time-reversal can be effected, provided that Δt is sufficiently short to ensure that B has not appreciably evolved, and retaining first-order terms in the expansion (4.7) is justified. Second, to the extent that *commutators* between H and H' can be neglected in (4.7), $H_{\text{eff}} \approx I_S \otimes H_B$ – implying that effective decoupling of S is achieved at T_c upon partial trace over B. Under the BB assumption, the separation between the evolution under the natural Hamiltonian and the evolution under the applied control makes it possible to obtain exact analytical expressions for H_{eff} and the qubit coherence after an arbitrary number of DD cycles, $t = NT_c$, $N \in \mathbb{N}$. Remarkably, it turns out that, similarly to Eq. (4.5), the result can be cast in terms of a controlled decoherence function $\Gamma_c(t)$:
$$\rho_{01}(t) = \rho_{01}(0)e^{-\Gamma_c(t)}, \quad \Gamma_c(t) = \int_0^\infty d\omega \, J_c(\omega) \frac{1-\cos\omega t}{\omega^2}, \quad (4.8)$$
where now stroboscopic sampling is understood and the relevant spectral density function $J_c(\omega)$ is *renormalized* by a dynamical interference factor due to the control pulses [VS98, G.-S02, SL04]:
$$J(\omega) \to J_c(\omega) = J(\omega)\tan^2\left(\frac{\omega T_c}{4}\right). \quad (4.9)$$
In the limit of *arbitrarily fast control*, where evolution over a finite time t is implemented as an infinite sequence of elementary time slices each of infinitesimal duration $T_c = t/N$, Eqs. (4.8) and (4.9) lead to the exact result
$$\lim_{T_c \to 0, N \to \infty} \rho_{01}(t = NT_c) = \rho_{01}(0), \quad (4.10)$$
which implies, in principle, the possibility to *stroboscopically* preserve arbitrary qubit states indefinitely in time. Representative results for the important class of an Ohmic bath, $I(\omega) = \alpha\omega e^{-\omega/\omega_c}$, $\alpha \in \mathbb{R}$, are shown in Fig. 4.1.

While the above setting is nongeneric from both an open-system and a control perspective, it is instrumental to highlight some of the basic DD physics as also relevant to more complicated scenarios. In particular:

(i) In the BB limit, exact evaluation of the qubit dynamics under CP multi-pulse control is possible for an *arbitrary* pure-dephasing model of the form (4.3) [UA02]. In fact, for a spin-boson dephasing Hamiltonian as in (4.4), exact results can still be derived for single-axis DD with *arbitrary* pulse separations [U07, U08]. A modification of the way in which the time-dependent bath correlation functions enter the reduced qubit dynamics in the presence of control remains the basic mechanism by which decoherence can be actively manipulated and completely inhibited in principle. For more general decoherence processes and control protocols, thinking of DD as inducing an effective *renormalization of open-system properties* has proved useful both to gain

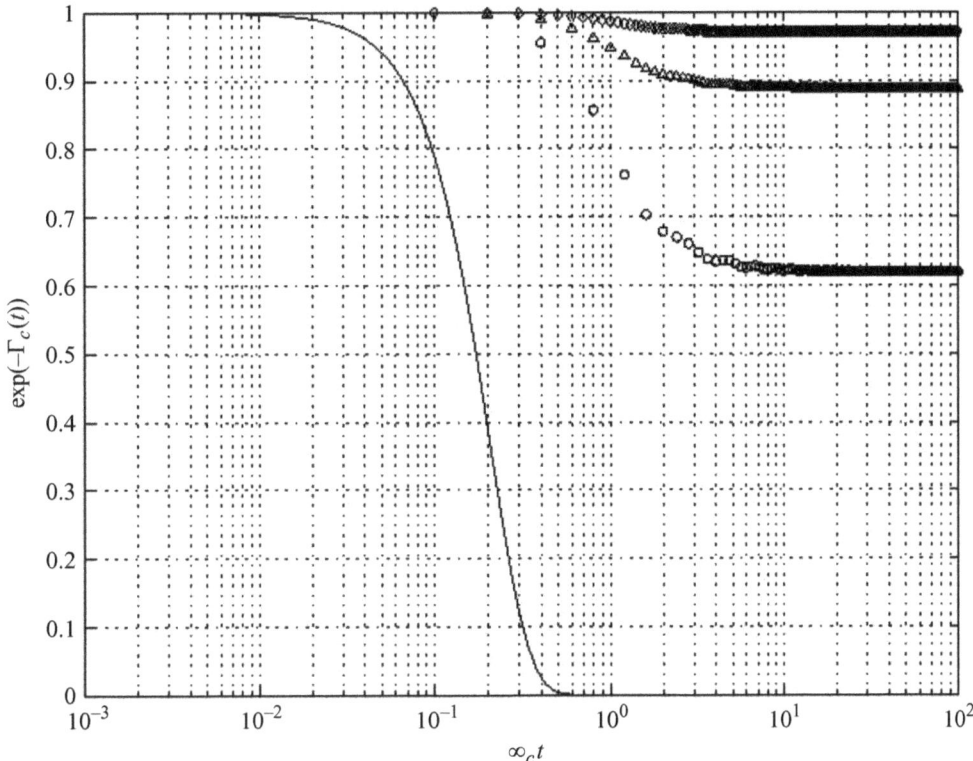

Fig. 4.1. BB-controlled qubit coherence as a function of time for an Ohmic bath in the high-temperature limit, $J(\omega) = \alpha\omega e^{-\omega/\omega_c}\coth[\omega/(2k_B T)]$, with $\alpha = 1$, $\omega_c/T = 10^{-2}$, $\omega_c = 100$ in units where $\hbar = k_B = 1$. Solid line: free decoherence dynamics, Eq. (4.5). Bottom to top: stroboscopic controlled decoherence dynamics, Eq. (4.8), for $\omega_c t_N = 2N\omega_c \Delta t$, $N = 1, 2, \ldots$, and decreasing pulse separation, $\omega_c \Delta t = 0.2$ (circles), $\omega_c \Delta t = 0.1$ (triangles), $\omega_c \Delta t = 0.05$ (diamonds). In each case, coherence decays freely up to $\omega_c t = \omega_c \Delta t$ when the first pulse is applied, with the first "coherence echo" forming at $\omega_c t_1 = 2\omega_c \Delta t$. Note that the spectral density $J_c(\omega)$ of the driven system is "upshifted" in frequency and develops a "resonance" at $\omega_{\text{res}} = \pi/\Delta t$, which is much higher than ω_c in the limit of good DD. A stroboscopic plateau of the controlled coherence is observed in the long-time regime, reflecting the fact that $J_c(\omega)$ becomes supra-Ohmic [PSE96], and $\Gamma_c(t)$ is independent of the total elapsed time t in this limit, with $\Gamma_c(t) = \mathcal{O}(\alpha T \omega_c \Delta t^2)$, see e.g. Section V in [SV05] and discussion in [HVD10].

physical insight and to derive DD performance bounds in various regimes, see e.g. [KL07, C07, ZKD+08].

(ii) The limit of arbitrary fast control is a mathematical idealization that is clearly impossible to attain in practice. For real systems, such a limiting situation can be approached, however, if the control time scale T_c is *sufficiently short with respect to the shortest time scale* of the physical degrees of freedom we wish to average out. Since, on physical grounds, an upper bound on this time scale is set by the existence of an upper energy cutoff – say, $\omega_c = \max_k\{\omega_k\} \gg \omega_0$ in Eq. (4.4), the condition $T_c \lesssim \tau_c \sim \omega_c^{-1}$ always *suffices* for regarding the relevant environment degrees of freedom as *approximately frozen* – allowing for H_{SB} to be *coherently* manipulated and suppressed by repeated partial time-reversal. While this criterion can be very stringent for realistic environments, it offers an alternative picture of DD in the frequency domain, where the control may be viewed as effectively shifting the support of the power spectrum beyond the highest frequency at which appreciable spectral components are found. This is in line with point (i) above as well as with general results for dynamically modified decay rates under arbitrary time-dependent modulation [A00a, KK01, GKL08]. In practice, the identification of realistic criteria that are also *necessary* to ensure good DD performance can depend sensitively on system and control features – including the competition between multiple time scales characterizing the internal system and/or bath individual dynamics as related to the system–bath coupling strength, and details of the noise power spectrum. While suppression of so-called $1/f$ dephasing provides an instructive instance in this respect [SL04, FV04], a more concrete feeling on factors determining the DD efficacy will be gained upon inspection of representative experimental settings in Chapter 22. In all cases, however, a hallmark of DD methods is their intrinsic suitability to handle *slowly correlated error processes* – thereby *non-Markovian* decoherence settings.

(iii) From a mathematical standpoint, the continuous limit invoked in Eq. (4.10) points to a suggestive analogy between dynamical freezing as achieved by unitary control via BB pulses and by frequent measurement via the *quantum Zeno effect* [VS98]. Formalizing this analogy, which is physically rooted in the ability of both mechanisms to disrupt the coherent evolution in a controlled fashion, has allowed unified insight on different control methodologies to be gained [FLP04]. In particular, this has shed light on the crossover from DD-induced decoherence suppression to *decoherence acceleration*, which Eq. (4.8) supports in the low-temperature regime (see Fig. 3 in [VS98]).

As a final remark, the application of CP DD to a qubit oscillator is worth mentioning, as implemented through suitable "parity kicks" [VT99]. In this case, $H_S = \omega_0 a^\dagger a$ is the oscillator Hamiltonian, and the spin-boson interaction in Eq. (4.4) is modified to $H_{SB} = a^\dagger \otimes B_- + a \otimes B_+$. Although $[H_S, H_{SB}] \neq 0$ and both dephasing and genuine dissipation effects emerge in this case, single-axis DD is still effective for DD since a simultaneous (approximate) reversal of $a, a^\dagger$ may be achieved by a single control operation: mapping to a pseudo-spin language, a sequence of π-pulses about the z-axis still causes both σ_+, σ_- to be inverted, and $H_{\text{eff}} \approx I_S \otimes H_B$ over each control cycle.

4.3 Control-theoretic framework

Extending DD techniques beyond the specific setting analyzed thus far requires formulating the problem within the general control-theoretic framework of *open-loop model-based Hamiltonian engineering*. While we can refer, for instance, to [D07] for an up-to-date introduction to quantum control theory, the essential background may be summarized as follows.

4.3.1 System and control assumptions

Let the target system S be a finite-dimensional open quantum system, defined on a state space $\mathcal{H}_S$ of dimension d, with $d = 2^n$ for an n-qubit system, and let $\mathcal{B}(\mathcal{H}_S)$ denote the space of all bounded operators on S. In general, S will be coupled to an *uncontrollable* quantum environment B, resulting in a joint evolution governed by a time-independent "drift" Hamiltonian of the form

(see Chapter 1)

$$H = H_S \otimes I_B + I_S \otimes H_B + H_{SB}, \quad H_{SB} = \sum_a S_a \otimes E_a. \qquad (4.11)$$

Without loss of generality, both the "bare" system Hamiltonian H_S and the *error generators* S_a may be taken to be traceless. Since, in practice, it is often the case that no complete knowledge is available of the environment operators H_B and E_a, DD strategies specifically aim at error correcting the unwanted decoherence effects induced by the error set $\{S_a\}$ under *minimal knowledge*, by largely relying on the exploitation of symmetry principles. In order for error estimates to be established, however, it is necessary for H_B and S_a to be bounded. While care must be taken if B is infinite dimensional, restricting to an effectively finite-dimensional subspace is possible by imposing a physical upper energy cutoff (see e.g. Lemma 8 in [DKS+07] for a formal justification).

In open-loop schemes, control over the system dynamics is achieved by adjoining to S a *semi-classical controller*, specified by a time-dependent control Hamiltonian that acts nontrivially on S alone, that is,

$$H \to H + H_c(t) \otimes I_B, \quad H_c(t) = \sum_m u_m(t) H_m, \qquad (4.12)$$

in terms of a (finite) set of admissible control inputs $\{u_m(t)\}$ and primitive control Hamiltonians $\{H_m\}$ whose specification is dictated by different assumptions on the overall resources at hand (Section 4.3.2). In constructing DD schemes, primary emphasis is placed on the *control propagator* as the basic object for control design:

$$U_c(t) = \mathcal{T}_+ \exp\left\{ -i \int_0^t dt' H_c(t') \right\}, \qquad (4.13)$$

where $\mathcal{T}_+$ denotes time ordering. Formally, a DD problem may be viewed as a steering problem for the unitary propagator $U(t)$ describing the joint dynamics under the combined influence of H and H_c, with the goal of synthesizing a target effective dynamics of interest, over a predetermined (finite) control time scale. In particular, for the prototypical task of arbitrary state preservation we are focusing on, the objective is to effectively remove arbitrary evolution induced by H, thereby implementing a "no-op" gate on S. Since, in a situation where this objective is achieved perfectly, S evolves independently of B, exclusively under the action of the controller, it is natural to seek a representation where the intended control action is isolated from the rest. Following NMR standard practice [H76, EBW94], this is accomplished by effecting a transformation to a *toggling* (or logical) frame that explicitly removes the applied control field. If $\rho_{SB}(t) = U(t)[\rho_S(0) \otimes \rho_B(0)]U^\dagger(t) \equiv U_c(t)\tilde{\rho}_{SB}(t)U_c^\dagger(t)$, evolution of the toggling-frame state $\tilde{\rho}_{SB}(t)$ is governed by a *control-modulated* Hamiltonian $\tilde{H}(t) = U_c^\dagger(t) H U_c(t)$, in such a way that the net evolution in the Schrödinger picture may be exactly separated into the product

$$U(t) = U_c(t)\tilde{U}(t) = U_c(t)\mathcal{T}_+ \exp\left\{ -i \int_0^t dt' \tilde{H}(t') \right\}. \qquad (4.14)$$

It is useful to realize that, formally, one may always write

$$\tilde{U}(t) = e^{-itH_{\text{eff}}(t)} \equiv e^{-i\Phi(t)}, \qquad (4.15)$$

in terms of a suitable (possibly time-dependent) effective Hamiltonian $H_{\text{eff}}(t)$ or an Hermitian "error action operator" Φ, so that upper bounds on the distance between the intended and actual dynamics can be directly established in terms of operator norms [LZK08] (recall Chapter 1): in particular, a DD storage protocol over time T aims at *minimizing* the net error $\Phi(T)$, up to pure bath contributions that have no direct effect on the reduced dynamics of S. Depending on specific applications, different metrics of control performance may be relevant and/or computationally advantageous, common choices being worst-case pure-state fidelity or gate entanglement fidelity; see e.g. [V06, LZK08, ZKD+08, SV08] for representative discussions in open- and closed-system settings.

4.3.2 A plethora of decoupling schemes

Different assumptions on the target Hamiltonian and available control resources entering Eq. (4.12) influence DD design in several ways. While covering the large number of existing variants is beyond our current scope, a broad classification may be useful for later reference (Chapters 14 and 15) as well as to identify suitable points of entry into the literature. From the point of view of the controller specification, three main features are relevant:

(i) *Mode of applying control operations*: A first important distinction arises depending on whether the available control operations are effected according to a *deterministic* or *randomized* sequence. The simplest representatives of the first category are *cyclic* DD schemes [VKL99, VLK99], whereby the controller is periodic with a cycle time T_c, whilst the most prominent example of deterministic aperiodic DD is provided by *concatenated* DD [KL05, KL07]. In randomized DD [VK05, KA05], the future control path is not fully determined by the past history, thus access to a classical register is necessary, in order to track the applied control trajectory. In passing, it is worth noting that if access to auxiliary *quantum* resources is available (for instance, a single controllable ancilla qubit), qualitatively different DD strategies may be envisioned, for instance augmented by discrete-time feedback capabilities [TV06].

(ii) *Type of control operations*: DD schemes can be further categorized on the basis of different constraints specifying how the available control Hamiltonians H_m are used to effect primitive DD operations. A basic distinction in this sense is between *unbounded-* and *bounded-strength* protocols. Unbounded control amplitudes (or, equivalently, power) are necessarily involved in achieving instantaneous "δ-pulses" – as demanded in the BB setting, along with arbitrarily fast switching rates. While this provides a convenient starting point (Section 4.4), analytical procedures for achieving DD using finite power are provided by *Eulerian* design [VK03], discussed in Section 4.6.

(iii) *Accuracy of control operations*: In practice, even for a closed quantum system, different types of operational *classical* imperfections limit the achievable degree of control, causing the actual control Hamiltonian to deviate from the intended one due to both systematic and random control errors. Formally, $H'_c(t) = H_c(t) + \delta H_c(t)$ with respect to a suitable parametrization that isolates the ideal control Hamiltonian $H_c(t)$, and seeking DD schemes that minimize sensitivity to relevant classes of nonidealities $\delta H_c(t)$ is essential to ensure *robust* performance. For an open quantum system, additional limitations in the achievable control accuracy stem from *quantum* errors induced by the system–bath coupling *while* control is applied. Beside Section 4.6, this will be central to the development of dynamically corrected quantum gates and quantum computation, see [KV09a, KV09b, KLV10].

In respect to assumptions on the underlying target Hamiltonian H, two factors are worth mentioning in the present context: first, limited knowledge about the internal Hamiltonian and/or error generators (e.g., due to *time-dependent* parameter drifts) may require that robustness against *model uncertainty* is also incorporated in DD design. Randomized methods have so far proved superior in that respect [SV06]. Second, for QIP applications, a key consideration is whether DD may be achieved *efficiently*, that is, with a minimum number of control operations that scales polynomially with the number of qubits n. Provided that controls with a sufficient degree of qubit selectivity are available, this is achieved though *combinatorial* methods that make explicit reference to the multipartite structure of $\mathscr{H}_S$ [SM01, RW06], as discussed in Chapter 15.

4.4 Bang-bang periodic decoupling
4.4.1 Tools for analysis: average Hamiltonian theory

Periodic DD (PDD) provides the simplest yet most important approach to "low level" DD schemes, making its detailed understanding both important on its own and useful as a preparation for more advanced approaches. As mentioned above, the defining feature of PDD is cyclicity of the controller, in the sense that

$$U_c(t + NT_c) = U_c(t), \quad N \in \mathbb{N}. \tag{4.16}$$

If, as explicitly assumed in writing Eq. (4.11), the open-system Hamiltonian H is time independent, such periodicity is transferred to the toggling-frame Hamiltonian $\tilde{H}(t)$ in Eq. (4.14). The basic observation of average Hamiltonian theory [EBW94] is the fact that, under these conditions and recalling Eq. (4.15), a *time-independent* effective Hamiltonian $\overline{H} \equiv H_{\text{eff}}(T_c)$ also exactly describes the evolution over arbitrary time periods, provided that observation is synchronized with T_c. That is, taking advantage of the fact that $U_c(NT_c) = I_S$, the stroboscopic evolution becomes

$$U(NT_c) = \tilde{U}(NT_c) = [\tilde{U}(T_c)]^N = e^{-iNT_c\overline{H}}. \tag{4.17}$$

In most cases, only an approximate calculation of the time-ordered product that defines $\tilde{U}(T_c)$ (hence $\overline{H}$) is feasible. While different methods have been employed (see in particular [MK06]), the standard approach within average Hamiltonian theory is to invoke a Magnus expansion (Section 1.2.2.3) in terms of a suitable "small" parameter proportional to the control period T_c. Consistent with the fact that all operators entering H are bounded, we may assume that $\|H\| < \kappa$, where $\kappa > 0$ has, say, units of frequency, $\kappa \equiv \tau_{\min}^{-1}$, and $\|\cdot\|$ is an appropriate norm.[1] Then the Magnus series reads:

$$\overline{H} = \sum_{\ell=0}^{\infty} \overline{H}^{(\ell)}, \quad \|\overline{H}^{(\ell)}\| = \mathcal{O}[\kappa(\kappa T_c)^\ell)], \tag{4.18}$$

where the ℓth term contains ℓ nested commutators of toggling-frame Hamiltonians at different times and may be generated recursively based on a rooted-tree expansion [I02]. In particular, the

---

[1] While in practice the success of average Hamiltonian theory may be quite sensitive to the choice of a correct norm [M82], any unitarily invariant norm is suitable at this stage, a good choice being the operator norm, $\|H\|_\infty$.

two lowest-order corrections are given in explicit form by[2]

$$\overline{H}^{(0)} = \frac{1}{T_c} \int_0^{T_c} dt' \, \tilde{H}(t'), \tag{4.19}$$

$$\overline{H}^{(1)} = -\frac{i}{2T_c} \int_0^{T_c} dt'' \int_0^{t''} dt' \, [\tilde{H}(t''), \tilde{H}(t')]. \tag{4.20}$$

Convergence conditions for the Magnus series have been extensively studied in the mathematical-physics literature. Recall that a *sufficient* condition for absolute convergence over t is [Eq. (1.22)] $\int_0^t dt' \|\tilde{H}(t')\| < \pi$, which may be interpreted as requiring that the error action operator per cycle is norm-bounded by $\|\Phi(T_c)\| < \pi$. Alternatively, this translates into $\kappa T_c = T_c/\tau_{\min} \lesssim 1$, which directly (although loosely) expresses the need for the external control to impose a modulation that is faster than that afforded by the natural dynamics to be removed.

It is important to appreciate that the structure of the first-order contribution to the effective Hamiltonian, Eq. (4.19), is essentially *classical* in the sense that it neglects arbitrary commutators of transformed Hamiltonians at different times where, in our case,

$$\tilde{H}(t) = [U_c^\dagger(t) H_S U_c(t)] \otimes I_B + I_S \otimes H_B + \sum_a [U_c^\dagger(t) S_a U_c(t)] \otimes E_a.$$

In particular, since commutators between H_B and the transformed error generators are neglected, the quantum structure of B does not directly appear in the DD problem to the lowest order. Likewise, Eq. (4.20) also makes it clear that whenever different toggling-frame Hamiltonians commute – such is the case if $[H_S, H_{SB}] = 0$ *and* $[H_B, H_{SB}] = 0$, recall the CP example in Eq. (4.2) – then $\overline{H} = \overline{H}^{(0)}$, $\overline{H}^{(\ell)} = 0$ for all $\ell \geq 1$: the series truncates, and no time-scale requirements for control emerge. Basically, DD attempts to enforce conditions under which the Magnus series can be *approximately* truncated to some order, so that an accurate implementation of a desired effective Hamiltonian may be achieved by engineering a finite number of terms. Specifically, a DD scheme is said to achieve *mth order decoupling* if the unwanted error evolution is effectively removed up to the $(m-1)$th order in the Magnus expansion for $\overline{H}$ (or, equivalently, up to the mth order in the corresponding expansion for $\Phi(T_c)$). To formalize this notion in the open-system context, it is useful to introduce a definition for *operators to be equal modulo arbitrary pure bath terms* [KV09a]: Given A_1, A_2 acting on $\mathcal{H}_S \otimes \mathcal{H}_B$, let

$$A_1 = A_2 \bmod B \quad \text{if and only if} \quad A_1 - A_2 = I_S \otimes E_B, \tag{4.21}$$

where E_B acts only on $\mathcal{H}_B$. Then mth order DD satisfies that

$$\overline{H}^{(\ell)} = 0 \bmod B, \quad \ell = 0, \ldots, m-1.$$

In particular, in PDD schemes DD is achieved to the lowest order, $m = 1$. Under convergence conditions, this first-order approximation becomes exact in the fast control limit introduced in Section 4.2, thus the DD problem reduces to constructing a suitable $\overline{H}^{(0)}$.

---

[2] Different conventions are found in the QIP literature. In particular, a Magnus expansion may be directly applied to the error action $\Phi(T_c) = T_c \overline{H}$, in which case the analog of Eq. (4.18) ranges over $\ell' = 1, \ldots, \infty$, the lowest-order term being of order $\mathcal{O}(\kappa T_c)$.

The above average Hamiltonian theory formalism may be easily specialized to the BB setting, whereby each PDD cycle consists of a sequence $\mathscr{P} = \{P_k, \Delta t_k\}$ of M BB pulses separated by time delays $\Delta t_k = t_k - t_{k-1}$, $k = 1, \ldots, M$, in such a way that $\sum_k \Delta t_k = T_c$ and the cyclicity condition (4.16) is ensured by the requirement $\prod_k P_k = I_S$. The cycle propagator through which $\overline{H}$ is defined via Eq. (4.17) may be written as

$$\tilde{U}(T_c) = \prod_{k=1}^{M} U_k(\Delta t_k) = \prod_{k=1}^{M} e^{-i H_k \tau_k T_c}, \quad \tilde{H}_k = Q_k^\dagger H Q_k, \qquad (4.22)$$

where the relative separations $\tau_k = \Delta t_k / T_c$, $\sum_k \tau_k = 1$, and the unitary operator Q_k determining the kth stepwise toggling Hamiltonian $\tilde{H}_k$ is the integrated value of the control propagator after $(k-1)$ pulses, that is, $Q_k = P_{k-1} \ldots P_0$, where $P_0 = U_c(0) = I_S$. By recalling Eqs. (4.19) and (4.20), the lowest-order contributions to $\overline{H}$ now become

$$\overline{H}^{(0)} = \sum_{k=1}^{M} \tau_k \tilde{H}_k = \sum_{k=1}^{M} \tau_k Q_k^\dagger H Q_k \equiv (\mathcal{Q}_S \otimes I_B)(H), \qquad (4.23)$$

$$\overline{H}^{(1)} = -\frac{iT_c}{2} \sum_{k=1}^{M} \sum_{j=1}^{k} \tau_k \tau_j [\tilde{H}_k, \tilde{H}_j], \qquad (4.24)$$

where, in Eq. (4.23), we have emphasized how the most general transformation that BB control induces to the first order may be viewed as a trace-preserving, unital superoperator acting non-trivially only on S [V02]. DD schemes that use *uniform* time intervals, $\Delta t_k \equiv \Delta t$, $\tau_k = 1/M$, are called *regular*. It is useful to note that, while different BB pulse placements within a cycle have no effect on $\overline{H}^{(0)}$ in Eq. (4.23), higher-order corrections depend sensitively on that. The simplest example emerges for CP two-pulse sequences. As implemented in Eq. (4.6), CP DD corresponds to a regular scheme with $\tau_1 = \tau_2 = 1/2$, which decouples to the first order. By rearranging the two π-pulses in such a way that $\tau_1 = \tau_3 = 1/4$ and $\tau_2 = 1/2$, the control propagator becomes symmetric with respect to the cycle middle point:

$$U_c(T_c - t) = U_c(t), \quad t \in [0, T_c]. \qquad (4.25)$$

It is then possible to show [H76, EBW94] that all *odd-order* corrections vanish, $\overline{H}^{(2\ell+1)} = 0 \mod B$, immediately yielding a time-symmetrized second-order DD protocol with the same T_c.

4.4.2 Group-based decoupling design

Pictorially, in a BB protocol the control propagator $U_c(t)$ follows a piecewise constant path in the space of unitary transformations on S, with jumps between consecutive values, $Q_k \to Q_{k+1} = (Q_{k+1} Q_k^\dagger) Q_k$, corresponding to the application of a BB pulse $P_k = Q_{k+1}^\dagger Q_k$, $k = 1, \ldots, M$. An important class of PDD schemes widely used in QIP is obtained by constraining the available set of propagators $\{Q_k\}$ (thus the BB pulses $\{P_0, P_k\}$) to be a discrete *group* of unitary operations. While the restriction to the group setting is *not* a fundamental requirement for DD design (notable exceptions arising, for instance, for DD schemes not based on inversion pulses such as in WAHUHA sequences [WHH68]; see also Section 22.3.1), it has the advantage of allowing powerful connections to symmetry properties to be made.

Let $\mathcal{G} = \{g_j\}$ be a discrete group of order $|\mathcal{G}|$ (so-called *DD group* henceforth), which acts on $\mathcal{H}_S$ via a faithful, unitary, projective representation $\{U_{g_j}\}$, $j = 1, \ldots, |\mathcal{G}|$, where we take $U_{g_1} = I_S$. That is, loosely speaking, the set $\{U_{g_j}\}$ is itself a "group up to phase factors,"

$$U_{g_j g_k} = e^{i\alpha(g_j, g_k)} U_{g_j} U_{g_k}, \quad \alpha(g_j, g_k) \in \mathbb{R}, \; \forall j, k.$$

Then PDD based on $\mathcal{G}$ is implemented by iterating a regular DD cycle of duration $T_c = |\mathcal{G}| \Delta t$, $\Delta t > 0$, in which $U_c(t)$ is sequentially changed over each element of $\mathcal{G}$ according to a predetermined order. Formally, we may write [VKL99, Z99]

$$U_c[(k-1)\Delta t + s] = U_{g_k}, \quad s \in [0, \Delta t), \; k = 1, \ldots, |\mathcal{G}|. \tag{4.26}$$

Comparing with Eq. (4.22), this prescription corresponds to identifying $Q_k \equiv U_{g_k}$, $M \equiv |\mathcal{G}|$, and applying a sequence of BB pulses $P_k = U_{g_{k+1}}^\dagger U_{g_k}$ whose order depends on the specific "group path" that is being chosen over $\mathcal{G}$, beginning from the group identity g_1.

As a main consequence of the group assumption, the time average that defines the lowest-order effective Hamiltonian in Eq. (4.23) is mapped to an average over $\mathcal{G}$, the net DD action $\mathcal{Q}_S$ effectively corresponding to the projection superoperator that extracts the $\mathcal{G}$-invariant component of H; that is,

$$\overline{H}^{(0)} = \frac{1}{|\mathcal{G}|} \sum_{j=1}^{|\mathcal{G}|} U_{g_j}^\dagger H U_{g_j} \equiv (\Pi_\mathcal{G} \otimes I_B)(H). \tag{4.27}$$

Let $\mathbb{C}\mathcal{G}' = \{A \,|\, [A, U_{g_j}] = 0, \forall j\} \subset \mathcal{B}(\mathcal{H}_S)$ denote the (complex) subspace of all operators on S commuting with arbitrary elements in the representation, the so-called *commutant* (sometimes also called *centralizer*) of the "group algebra" $\mathbb{C}\mathcal{G}$ [F01]. Then, formally, $\Pi_\mathcal{G}(X) \in \mathbb{C}\mathcal{G}'$ for every operator X on S. If, in particular, $\mathcal{G}$ acts *irreducibly* on $\mathcal{H}_S$, then by Schur's lemma the commutant is trivial, $\mathbb{C}\mathcal{G}' = \mathbb{C} I_S$, and we have

$$\Pi_\mathcal{G}^{\text{irred}}(X) = \frac{\text{Tr}(X)}{d} I_S, \quad \forall X \in \mathcal{B}(\mathcal{H}_S). \tag{4.28}$$

The symmetrization of the lowest-order controlled dynamics implied by Eq. (4.27) is the basis for understanding how, in group-based PDD, symmetry can be exploited to approximately "filter out" unwanted evolution. In general, if Ω_{err} is the *error subspace* that contains all (traceless) system operators we wish to remove, a good DD group for Ω_{err} is defined by the *decoupling condition*:

$$\Pi_\mathcal{G}(E_{\text{err}}) = 0, \quad \forall E_{\text{err}} \in \Omega_{\text{err}}. \tag{4.29}$$

Since, in our case,

$$\overline{H}^{(0)} = \Pi_\mathcal{G}(H_S) \otimes I_B + I_S \otimes H_B + \sum_a \Pi_\mathcal{G}(S_a) \otimes E_a,$$

arbitrary state preservation is ensured, to the first order, provided that both H_S and each of the error generators S_a obey Eq. (4.29), yielding $\overline{H}^{(0)} = 0 \mod B$, as desired. Two relevant scenarios are worth highlighting:

(i) Given Eq. (4.28) and the tracelessness assumption on H_S, S_a, *nonselective DD* (or complete decoupling, or "annihilation") is always achievable for $d < \infty$ by letting $\{U_{g_j}\}$ be a *unitary error basis* on $\mathcal{H}_S$ [VKL99, WRJ+02b]. Conceptually, this is remarkable as it implies that decoherence from an *arbitrary* (in principle completely unknown) quantum environment may be suppressed in an appropriate limit. A potential problem, however, arises from the fact that the resulting DD schemes are inefficient, as the number of required BB operations is $|\mathcal{G}| = d^2 = 4^n$, which scales exponentially with n. Achieving good DD with smaller groups requires additional knowledge about H, so that the relevant error set Ω_{err} can be restricted to operators that, for instance, have known symmetry properties and/or locality structure (such as in combinatorial DD, see Chapter 15).

(ii) *Selective* DD is achieved by groups that act *reducibly* on $\mathcal{H}_S$, in such a way that the DD condition (4.29) is still obeyed, but $\mathbb{C}\mathcal{G}' \supset \mathbb{C}I_S$. Aside from reducing the complexity of DD, selective DD offers important advantages in regard to dynamical generation of DFS/NS [VKL00] and the combination of BB DD with computation [VLK99, KL08].

Rigorous lower bounds on the worst-case pure-state fidelity (or other fidelity metrics) may be obtained starting from bounds on the trace distance between the intended and the actual evolution under DD, as established in [LZK08] and Chapter 1. More simply for the current purposes, a direct extension of the PDD bound formally derived in [VK05] in the closed-system limit yields

$$F_{\min}(T) \geq 1 - \mathcal{O}[(\kappa^2 T T_c)^2], \quad \text{for } T T_c \kappa^2 \ll 1 - \kappa T_c. \tag{4.30}$$

Tighter estimates are possible in specific settings, notably single-qubit decoherence scenarios such as investigated in detail in [KL07, ZKD+08]. The analysis of [KL07], in particular, points to the need for more accurately describing the interplay between the direct system–environment interaction, whose strength is proportional to $\|H_{SB}\| \equiv J$, and the internal environment evolution, with norm $\|H_B\| \equiv \beta$. By identifying $\kappa = \max(J, \beta)$, Eq. (4.30) is then replaced by

$$F_{\min}(T) \geq 1 - \mathcal{O}[(\kappa J T T_c)^2], \quad \text{for } \kappa T \ll 1, \tag{4.31}$$

which gives a better bound in the limit of fast-environment dynamics, $\beta \gg J$. Intuitively, notice that the quantity $\kappa^2 T T_c$ in Eq. (4.30) [or $\kappa^2 J T_c$ in Eq. (4.31)] may be thought of as bounding the total *error amplitude* that arises from T/T_c PDD cycles, each contributing with an error amplitude of order $|\mathcal{G}|^2 (\kappa \Delta t)^2$ due to a dominant $\overline{H}^{(1)} \neq 0$ correction. In this sense, the generic *quadratic* fidelity decay with respect to the total elapsed time T may be taken as a signature of coherent error buildup from deterministically repeated DD cycles.

We conclude our general discussion of BB DD methods by referring to [BL02b] for a suggestive geometric interpretation based on coherence-vector representation, and proceed to examine illustrative situations of special significance to QIP.

4.4.3 Bang-bang decoupling by example

4.4.3.1 Single qubit with pure dephasing Without loss of generality, a single-qubit pure dephasing scenario is described by a one-dimensional error space, $\Omega_{\text{err}} = \text{span}\{Z\}$. A minimal DD group is provided by $\mathcal{G} = \mathbb{Z}_2$, reducibly represented in $\mathcal{H}_S = \mathbb{C}^2$ as $\{U_g\} = \{I, X\}$. Note that another possibility is to base DD on the representation $\{U'_g\} = \{I, Y\}$, corresponding to a different choice of a DD axis orthogonal to the preferred error basis.[3] Using Eq. (4.26), each PDD cycle based, for instance, on $\{U_g\}$ consists of a pair of BB π_x-pulses separated by an interval of free evolution (simply denoted by f henceforth) of length Δt, with $T_c = 2\Delta t$. That is,

---

[3] While different group representations are, at this level, entirely equivalent, their existence may prove useful in both designing randomized DD schemes [SV08] and controlling decoupled evolutions [VLK99] in the presence of reducible groups.

$\mathscr{P} = \text{f}X\text{f}X$, which clearly recovers the CP example we started with (Section 4.2). Explicitly, the action of the projection superoperator $\Pi_{\mathscr{G}}$ reads

$$\Pi_{\mathscr{G}}(Z) = \frac{1}{2}\Big(IZI + XZX\Big) = \frac{1}{2}(Z - Z) = 0, \qquad (4.32)$$

consistent with the interpretation in terms of an instantaneous time-reversal action (exact to lowest order) underlying Eq. (4.6). As mentioned in Section 4.4.1, rearranging the pulses so that the control path is time-symmetrized, $\mathscr{P} = \text{f}'X\text{f}X\text{f}'$, with f' of duration $\Delta t/2$, has the effect of changing the order of PDD from first to second. Either way, group reducibility is reflected by a nontrivial space of $\mathscr{G}$-invariant, preserved operators, which belong to $\mathbb{C}\mathscr{G}' = \text{span}\{I, X\}$.

4.4.3.2 Single qubit with arbitrary decoherence In this case, both the qubit Hamiltonian H_S and each error generator S_a in Eq. (4.11) may be taken as suitable linear combinations of Pauli matrices. In particular, arbitrary single-qubit decoherence may always be derived from a system–environment coupling Hamiltonian of the form

$$H_{SB} = X \otimes B_X + Y \otimes B_Y + Z \otimes B_Z, \qquad (4.33)$$

for appropriate environment operators. Thus, $\Omega_{\text{err}} = \text{span}\{X, Y, Z\}$. The smallest DD group ensuring that Eq. (4.29) is obeyed for generic elements in Ω_{err} is $\mathscr{G} = \mathbb{Z}_2 \times \mathbb{Z}_2$ [VKL99], acting on $\mathscr{H}_S = \mathbb{C}^2$ via the irreducible projective representation $\{U_{g_j}\} = \{I, X, Y, Z\}$, leading to what we call *Pauli DD*. Irreducibility is manifest in the fact that $\mathbb{C}\mathscr{G}' = \text{span}\{I\}$, thus Pauli DD is "universal," in the sense that arbitrary single-qubit evolution is suppressed. Using again Eq. (4.26), each PDD cycle is now realized by a sequence of four BB π-pulses, alternated along two orthogonal control axes whose choice is dictated by the specific path chosen to traverse $\mathscr{G}$. Recalling that group elements specify the values of the control propagator, and BB pulses $P_k = U_{k+1}^\dagger U_k$ instantaneously update such values at the end of each control slot, we have universal DD sequences with $T_c = 4\Delta t$ of the form

$$\mathscr{P} = I\text{f}IX\text{f}XY\text{f}YZ\text{f}Z = \text{f}X\text{f}Z\text{f}X\text{f}Z, \quad \text{corresponding to } (I, X, Y, Z),$$
$$\mathscr{P}' = I\text{f}IX\text{f}XZ\text{f}ZY\text{f}Y = \text{f}X\text{f}Y\text{f}X\text{f}Y, \quad \text{corresponding to } (I, X, Z, Y),$$

where time is understood to flow from left to right. In NMR terminology, a sequence such as, say, $\mathscr{P}'$ is simply referred to as XY-4 [GBC90], and so on.

By construction, all of the above $|\mathscr{G}|! = 24$ possibilities are equivalent to the lowest order, resulting in a DD projection map of the form

$$\Pi_{\mathscr{G}}(\sigma_a) = \frac{1}{4}\Big(I\sigma_a I + X\sigma_a X + Y\sigma_a Y + Z\sigma_a Z\Big) = 0, \quad \sigma_a = X, Y, Z.$$

As already remarked, however, higher-order contributions to $\overline{H}$ are, in general, fairly sensitive to the implemented control path. In particular, as one may verify by direct computation using Eq. (4.24), the leading second-order correction $\overline{H}^{(1)}$ for a universal PDD cycle coincides with the *half-cycle* direction of the sequence, that is, $\overline{H}^{(1)} \propto Y$ for path $\mathscr{P}$, whereas $\overline{H}^{(1)} \propto Z$ for $\mathscr{P}'$. On the one hand, such corrections may be eliminated by time-reversing the control path in such a way that Eq. (4.25) is obeyed, and second-order "symmetric DD" (SDD) schemes (and corresponding XY-8 or XZ-8 sequences) are obtained. On the other hand, the dependence upon the control path may also be exploited in different ways. If, for instance, the control objective is the stabilization of a *known* qubit state, storage fidelity may be *optimized* by ensuring that $\overline{H}$ has the desired state as an approximate eigenstate [ZDS+07, ZKD+08, KDV11]. For arbitrary state preservation, sensitivity to the applied control path underlies the efficiency of path-randomized DD protocols [SV06].

4.4.3.3 Quantum register with arbitrary linear decoherence A generic linear decoherence model on n qubits may be specified by a coupling Hamiltonian of the form

$$H_{SB} = \sum_{i=1}^{n} \sum_{a=x,y,z} \sigma_a^{(i)} \otimes E_a^{(i)}, \qquad (4.34)$$

for appropriate environment operators that may or may not explicitly depend on the qubit index. The latter case corresponds to the permutation-invariant limit of *collective decoherence*, whereas $n = 1$ clearly recovers Eq. (4.33). The error set corresponding to (4.34) is spanned by all $3n$ single-qubit operators, $\Omega_{\text{err}} = \text{span}\{\sigma_a^{(i)}\}$, and may still be decoupled by the DD group $\mathscr{G} = \mathbb{Z}_2 \times \mathbb{Z}_2$, provided that the n-fold tensor-power representation in terms of collective Pauli rotations is chosen [VKL99]: $\{U_{g_j}\} = \{I^{\text{coll}}, X^{\text{coll}}, Y^{\text{coll}}, Z^{\text{coll}}\}$, where $X^{\text{coll}} = \bigotimes_{i=1}^{n} \sigma_x^{(i)} = X \otimes \cdots \otimes X$, and so on. This representation is projective for n odd, and regular for n even. Control-wise, this translates into DD sequences and error cancellation very similar to the above-discussed single-qubit setting, except that the required π-pulses are now realized as "hard" (nonselective) pulses, which simultaneously rotate all qubits along the intended axis. Unlike the case $n = 1$, however, the representation of $\mathscr{G}$ in $\mathscr{H}_S = (\mathbb{C}^2)^{\otimes n}$ is highly reducible, with commutant $\mathbb{C}\mathscr{G}' \supset \mathbb{C}I$. For n even, for instance, $\mathbb{C}\mathscr{G}'$ may itself be viewed as a group under operator multiplication, whose $2n$ generators include two generators for $\mathscr{G}$, plus arbitrary *bilinear* generators of the form $\sigma_x^{(i)} \otimes \sigma_x^{(j)}$, $\sigma_z^{(i)} \otimes \sigma_z^{(j)}$, $i, j = 1, \ldots, n$ [VKL00]. This has the important consequence that Heisenberg Hamiltonians $\vec{\sigma}_i \cdot \vec{\sigma}_j = \sum_a \sigma_a^{(i)} \otimes \sigma_a^{(j)}$ are preserved under this DD protocol, with implications for universal decoupled control and decoherence-protected quantum computation [VKL00].

Clearly, the above selective DD scheme for linear decoherence is successful for quantum storage as long as the internal register Hamiltonian H_S does itself only involve single-qubit terms. In situations where H_S or H_{SB} are generic, arbitrary n-body qubit operators might be present in principle. In such a "worst-case" (or "adversarial") scenario, DD can only be implemented *inefficiently*, based on the group $\mathscr{G} = \mathbb{Z}_d \times \mathbb{Z}_d$, irreducibly represented as a tensor product $\{U_{g_j}\} = \bigotimes_{i=1}^{n} \{U_{g_j}^{(i)}\}$, consisting of all possible Pauli strings of length n, with $|\mathscr{G}| = 4^n$. If, however, as often happens in realistic systems, H is known to include only linear and bilinear terms (possibly with a known coupling topology, for instance nearest neighbor), efficient PDD schemes for removing both linear and quadratic errors may be constructed based on combinatorial design, see Chapter 15.

4.5 The need for advanced decoupling design

While PDD is conceptually very attractive owing to its design simplicity, it is inadequate as a tool for high-fidelity quantum storage in realistic control scenarios, for a number of reasons.

On the one hand, some major shortcomings are inherent to PDD as a low-level deterministic DD scheme, in particular: (i) Even if the control cycle T_c is sufficiently short to ensure that the Magnus series converges (as discussed in Section 4.4.1), unwanted error terms are suppressed only to the *first order*, causing the loss of fidelity to be directly dominated by the second-order term; (ii) no matter how small, residual error amplitudes arising from imperfect averaging at each cycle tend to add up *coherently* in deterministic DD (as mentioned in Section 4.4.2), severely limiting the fidelity achievable for realistically long storage times T.

On the other hand, a different set of limitations arises from the fact that the required control resources are too idealized from both a practical and a control-theory standpoint. Leaving aside that neither systematic nor random control imperfections have been incorporated within the analysis thus far, the BB assumption is itself highly questionable to begin with, since practically accessible operations always entail a *bounded* control amplitude, thereby a *finite* duration. Even in situations where BB δ-pulses can be approximated reasonably well, it is very desirable for theoretical consistency to recover the BB setting as a limiting case of a formulation based on physical (nonsingular) admissible controls.

Can we devise advanced DD schemes that achieve higher fidelity, yet operate under more realistic control resources? While no single strategy can counteract all the above difficulties as effectively as desired, important advances are possible. Within the BB setting, *high-level DD* protocols have been constructed, aiming specifically at boosting performance by reducing the sensitivity to high-order corrections and/or residual error accumulations. For generic open quantum systems, this is accomplished by invoking different strategies, which rely on concatenated and randomized design [KL05, VK05], as well as on the optimization of the control timings [U07, WFL10], and which will be addressed in Chapter 14. By remaining with *low-level DD* for the time being, we devote the rest of this introductory survey to describing how the BB requirement can always be lifted, at the expense of slightly complicating the control design [VK03].

4.6 Bounded-strength Eulerian decoupling

The basic idea underlying bounded-strength DD is to avoid discontinuous changes of the control propagator along a cycle [as implied by the BB prescription (4.26)], by steering $U_c(t)$ along a smooth path that still ensures first-order DD. Formally, this may be accomplished by following suitable closed paths on the so-called "Cayley graph" of the DD group $\mathscr{G}$, which prompts a brief excursion into combinatorial group theory [MKS04].

4.6.1 Tools for design: Eulerian cycles on graphs

Let $\mathscr{G} = \{g_j\}$, $j = 1, \ldots, |\mathscr{G}|$ be a discrete group of order $|\mathscr{G}|$, as before, and let $\Gamma = \{\gamma_\ell\}$, $\ell = 1, \ldots, |\Gamma|$, be a set of generators for $\mathscr{G}$, with the property that every element in $\mathscr{G}$ may be uniquely expressed as a product of generators in Γ. The *Cayley graph* $G(\mathscr{G}, \Gamma)$ of $\mathscr{G}$ with respect to Γ is the directed (multi)graph constructed by assigning a vertex to each group element, and by joining vertex g_{j-1} to vertex g_j with an edge labeled ("colored") by the generator γ_ℓ if and only if $g_j = \gamma_\ell g_{j-1}$. Basically, at least for finite groups, the Cayley graph $G(\mathscr{G}, \Gamma)$ may be regarded as a compact way for visually conveying the information contained in the multiplication table of $\mathscr{G}$. For instance, two elements $g_1, g_2 \in \mathscr{G}$ are equal if and only if the corresponding walks beginning at the identity element on G end at the same vertex.

Given a graph G, various types of *paths* in G may be specified as suitable sequence of edges [B98a]. An *Eulerian cycle* (or circuit), in particular, is a closed sequence of connected edges that uses each edge exactly *once*.[4] Graphs that allow the construction of Eulerian circuits are called Eulerian graphs. If G is connected, one may prove that a necessary and sufficient condition for it to be Eulerian is that every vertex has an equal number of incident and outgoing edges. A *regular* graph satisfies the stronger properties that all degrees are the same. Because a Cayley graph is regular, it always possesses Eulerian cycles, which have length $L = |\mathscr{G}| \times |\Gamma|$.

4.6.2 Eulerian decoupling design

In *Euler DD* (EDD), $\mathscr{G}$ is identified with the DD group under consideration, group generators represent the control operations achievable with *bounded* control, and an Euler cycle on $G(\mathscr{G}, \Gamma)$ dictates the sequence according to which control Hamiltonians are turned on to effect these operations within a cycle. Specifically, let us assume that in the absence of H_{SB}, we have the ability to implement each generator U_{γ_ℓ} (up to an irrelevant phase factor) via the application of a control Hamiltonian $H_\ell(t)$ over $\Delta t > 0$,

$$u_\ell(s) = \mathscr{T}_+ \exp\left\{-i \int_0^s dt' H_\ell(t')\right\}, \quad U_{\gamma_\ell} \equiv u_\ell(\Delta t), \quad \ell = 1, \ldots, |\Gamma|, \quad (4.35)$$

where the "partial" propagator $u_\ell(s)$, $s \in [0, \Delta t]$, is introduced for later convenience. If $H_{SB} \neq 0$, each generator will deviate from the intended one due to error evolution induced during the finite duration Δt, resulting in a nonzero error phase Φ_ℓ. Thus, in general, the challenge of first-order EDD is to ensure that unwanted error terms are compensated for both *in between* and *during* the applied control operations.

Let $\mathscr{C} = (p_1, \ldots, p_L)$ be an Eulerian cycle on $G(\mathscr{G}, \Gamma)$ beginning and ending at the identity, with $p_k = \gamma_{\ell_k}$ for each k and $\gamma_{\ell_k} \in \Gamma$. Then EDD according to $\mathscr{G}$ is implemented by iterating a regular DD cycle of duration $T_c = L\Delta t$, during which $U_c(t)$ is assigned by following the chosen path $\mathscr{C}$. That is, we may write

$$U_c[(k-1)\Delta t + s] = u_{\ell_k}(s) U_c[(k-1)\Delta t], \quad k = 1, \ldots, L, \quad (4.36)$$

meaning that during the kth time slot we apply the control Hamiltonian H_{ℓ_k} that implements the generator appearing in the kth element of the Euler path. The lowest-order effective Hamiltonian under EDD is obtained by evaluating the time-average defining $\overline{H}^{(0)}$ in Eq. (4.19) with the control propagator given by Eq. (4.36). While we refer to [VK03] for a more formal derivation (see also [KV09a]), the basic idea is to partition the L terms into $|\Gamma|$ families, each corresponding to a fixed generator γ_ℓ. Because for each such generator the cycle $\mathscr{C}$ contains exactly one γ_ℓ-colored edge ending at any given vertex g_j, each family leads to a sum over all group elements as in Eq. (4.27). The net EDD action may still be described in terms of a superoperator on S, of the form

$$\mathscr{Q}_{\mathscr{C}}(X) = \Pi_{\mathscr{G}}(F_\Gamma(X)), \quad \forall X \in \mathscr{B}(\mathscr{H}_S), \quad (4.37)$$

where the map F_Γ implements an average over both the group generators and the control interval, according to

$$F_\Gamma(X) = \frac{1}{|\Gamma|} \sum_{\ell=1}^{|\Gamma|} \frac{1}{\Delta t} \int_0^{\Delta t} ds\, u_\ell^\dagger(s) X u_\ell(s). \quad (4.38)$$

---

[4] Historically, the origin of Euler cycles is rooted in a famous problem in discrete mathematics, the so-called "Seven Bridges of Königsberg" problem, whose negative resolution by Euler in 1736 laid the foundations of modern graph theory.

Notice that, by definition, the integral over the control time slot of the toggling-frame Hamiltonian $u_\ell^\dagger(s)Hu_\ell(s)$ quantifies, to the first order, the error action Φ_ℓ arising from implementing the generator γ_ℓ in the presence of B. Altogether, Eqs. (4.37) and (4.38) may then be interpreted as requiring that a first-order DD condition of the form (4.29) be still obeyed by each error action along the cycle, that is,

$$(\Pi_\mathscr{G} \otimes I_B)(\Phi_\ell) = 0 \mod B, \quad \ell = 1, \ldots, |\Gamma|. \tag{4.39}$$

Formally, the BB limit described by (4.26) is recovered by letting F_Γ be the identity map, meaning that no error has time to develop during instantaneous control operations. Provided that the appropriate DD condition (4.39) is obeyed, EDD allows $\mathscr{G}$-symmetrization of the lowest-order effective dynamics to be attained by avoiding unbounded controls. While this comes at the expense of lengthening the cycle time, the overhead $|\Gamma|$ is always polynomial in the group size $|\mathscr{G}|$.

As an added advantage of the Euler approach, the fact that control operations are spread along finite intervals allows for a degree of robustness with respect to control imperfections that is not present in the BB setting. As mentioned in Section 4.3.2, let the applied control Hamiltonian be partitioned into the sum of ideal and faulty components:

$$H_c'(t) = H_c(t) + \delta H_c(t).$$

By working, as before, in the toggling frame that explicitly removes $H_c(t)$, the evolution under H with faulty control $H_c'(t)$ corresponds to evolution under $[H + \delta H_c(t)]$ with ideal control $H_c(t)$. Provided that the error component $\delta H_c(t)$ does itself respect cyclicity [H76], this allows the stroboscopic evolution in the Schrödinger picture, Eq. (4.17), to be evaluated by replacing H with $[H + \delta H_c(t)]$ in the time-average of (4.19). Assume, in particular, that $\delta H_c(t)$ is *systematic*, in the sense that every time a given generator U_{γ_ℓ} is implemented, the same control imperfection occurs, irrespective of the location of γ_ℓ along $\mathscr{C}$. Then effectively $\delta H_c[(k-1)\Delta t + s] \equiv \delta H_\ell(s)$, and a calculation similar to the ideal case leads to a modified Eq. (4.38) as follows:

$$\mathscr{Q}_\mathscr{C}'(X) = \mathscr{Q}_\mathscr{C}(X) + \mathscr{Q}_\mathscr{C}(\delta H_\ell), \quad \forall X \in \mathscr{B}(\mathscr{H}_S), \ \ell = 1, \ldots, |\Gamma|. \tag{4.40}$$

An immediate consequence of Eq. (4.40) is the fact that, without extra assumptions, systematic control errors are automatically symmetrized, since $\mathscr{Q}_\mathscr{C}(\delta H_\ell) \in \mathbb{C}\mathscr{G}'$. Thus, full fault-tolerance is ensured (to the lowest order) if $\mathscr{G}$ acts irreducibly. Even if this is not the case, error symmetrization is generally helpful in facilitating further error protection by means of suitable encodings [VK03].

While Eulerian design will be reconsidered in Chapter 15, we conclude by seeking a more concrete understanding in the illustrative DD scenarios previously addressed in Section 4.4.3 in the BB setting.

4.6.3 Eulerian decoupling by example

4.6.3.1 Single qubit with pure dephasing revisited   The DD group $\mathscr{G} = \mathbb{Z}_2 = \{0, 1\}$ for $\Omega_{\text{err}} = \{Z\}$ has a single generator $\Gamma = \{1\}$, represented in $\mathscr{H}_S = \mathbb{C}^2$ as $\{U_\gamma\} = \{X\}$. The relevant Cayley graph is depicted in Fig. 4.2, along with the Euler cycle $\mathscr{C} = (X, X)$.

Since $L = |\mathscr{G}| = 2$, EDD can be implemented in this case with no time overhead with respect to the BB limit. Let the generator X be realized through the application of a time-dependent

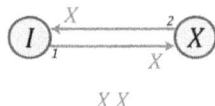

XX

Fig. 4.2. Cayley graph of $\mathscr{G} = \mathbb{Z}_2$, represented as $\{I, X\}$ in $\mathbb{C}^2$, with respect to the single group generator $\{U_\gamma\} = \{X\}$. Note that according to Eq. (4.41), the points reached after an elapsed time s in each of the two intervals are correlated from a DD point of view, see Eq. (4.42).

Hamiltonian $H_x(t) = h_x(t)X$, that is [cf. Eq. (4.35)],

$$u_x(s) = \exp\left\{-i\int_0^s dt'\, h_x(t')X\right\}, \quad U_x = X = u_x(\Delta t).$$

Note that aside from obeying the end-point constraint of achieving (up to a phase) the generator X over the interval Δt, EDD allows for full flexibility in specifying the "pulse shape" $h_x(t)$. For instance, a simple choice is provided by rectangular pulses of length τ, $h_x(t) = \pi/(2\tau)$, in which case the "fully stretched" (windowless) profile $\tau = \Delta t$ corresponds to minimum control amplitude and $\tau \to 0^+$ recovers a BB π_x-pulse. The EDD control prescription of Eq. (4.36) explicitly reads as follows:

$$U_c(t) = \begin{cases} u_x(t)I, & t \in [0, \Delta t) \equiv \Delta t_1, \\ u_x(s)X, & t \in [\Delta t, \Delta t + s) \equiv \Delta t_2,\ s \in [0, \Delta t). \end{cases} \quad (4.41)$$

Calculation of the EDD superoperator $\mathscr{Q}_\mathscr{C}$ using the fact that $t = \Delta t + s$ and $[u_x(t), X] = 0$ in the second integral yields [V04]

$$\mathscr{Q}_\mathscr{C}(Z) = \frac{1}{2\Delta t}\left(\int_0^{\Delta t} ds\, u_x^\dagger(s) Z u_x(s) + \int_0^{\Delta t} ds\, u_x^\dagger(s)[XZX] u_x(s)\right)$$

$$= \frac{1}{2\Delta t}\int_0^{\Delta t} ds\, [u_x^\dagger(s) Z u_x(s) - u_x^\dagger(s) Z u_x(s)] = 0. \quad (4.42)$$

By comparing with the BB counterpart, Eq. (4.32), one may intuitively understand EDD in this case as a "continuous-time spin echo," which is achieved by suitably pairing points along the Euler cycle. Upon explicit calculation of the functional $\mathscr{Q}'_\mathscr{C}$, one sees that systematic control errors proportional to Y, Z are also canceled to the lowest order.

Note that the idea of exploiting time-symmetrization of the control propagator as in Eq. (4.25) to achieve second-order averaging is generally applicable to EDD, by paying proper attention to implementing each generator. If, for instance, the above Eulerian cycle is implemented using piece-wise constant, fully stretched control inputs, $u_x(t) = e^{-i(\pi/2\Delta t)Xt}$, then a time-symmetric "Euler supercycle" (so-called SEDD) of length $T_s = 2T_c = 4\Delta t$ is obtained by cascading the forward path in (4.41) with a mirror path where $u_{\bar{x}}(t) = e^{i(\pi/2\Delta t)Xt} = u_x(t)^\dagger$, that is, the phase of *both* pulses is reversed. Needless to say, this has been noticed by the NMR community, where "reversing the phase of alternating pairs of pulses" is mentioned in [H76] as a way to reduce the impact of finite-width errors in the CP sequence.

4.6.3.2 Arbitrary linear decoherence revisited The DD group $\mathscr{G} = \mathbb{Z}_2 \times \mathbb{Z}_2$, which is appropriate for generic linear decoherence as in (4.34), has two independent generators, which may be chosen as $\gamma_1 = (0, 1)$, $\gamma_2 = (1, 0)$. Under the tensor power representation used in the BB

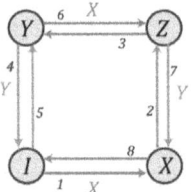

$$XYXYYXYX$$

Fig. 4.3. Cayley graph of $\mathscr{G} = \mathbb{Z}_2 \times \mathbb{Z}_2$, represented in terms of collective Pauli operators $\{I, X, Y, Z\}$ in $(\mathbb{C}^2)^{\otimes n}$, with respect to the generating set $\{U_{\gamma_\ell}\} = \{X, Y\}$. A possible Eulerian cycle is specified at the bottom and marked by numbers along the graph edges.

case, let these generators be realized as $U_{\gamma_1} = X$, $U_{\gamma_2} = Y$, respectively, where for notational convenience we identify $X \equiv X^{\text{coll}}$, and so on. The corresponding Cayley graph is shown in Fig. 4.3, along with a possible Eulerian cycle,

$$\mathscr{C} = (\gamma_1, \gamma_2, \gamma_1, \gamma_2, \gamma_2, \gamma_1, \gamma_2, \gamma_1) \to XYXYYXYX,$$

which in this case has length $L = 8$. If u_x, u_y denote the intended control propagators for the generators as in Eq. (4.35), EDD with respect to the above Euler cycle corresponds to the following control sequence:

$$U_c(t) = \begin{cases} u_x(s)I, & t \in \Delta t_1, \\ u_y(s)X, & t \in \Delta t_2, \\ u_x(s)Z, & t \in \Delta t_3, \\ u_y(s)Y, & t \in \Delta t_4, \\ u_y(s)I, & t \in \Delta t_5, \\ u_x(s)Y, & t \in \Delta t_6, \\ u_y(s)Z, & t \in \Delta t_7, \\ u_x(s)X, & t \in \Delta t_8, \end{cases} \qquad (4.43)$$

where $s \in [0, \Delta t)$ and $\Delta t_k = [(k-1)\Delta t, (k-1)\Delta t + s)$, $k = 1, \ldots, 8$. The above equation makes it clear how in EDD the desired cancellation is ensured with respect to *each control generator*.

Explicit calculations of the EDD superoperator are most transparent in the single-qubit case, whereby $\mathscr{G}$ acts irreducibly. By using Eq. (4.43), and by grouping together terms labeled by the same generator, one is left with two contributions, which separately correspond to a group projection. That is, we find

$$\mathscr{D}_\mathscr{C}(\sigma_a) = \frac{1}{8\Delta t} \left[\sum_{U_{g_j} \in \{I,X,Y,Z\}} U_{g_j}^\dagger \Phi_x(\Delta t) U_{g_j} + \sum_{U_{g_j} \in \{I,X,Y,Z\}} U_{g_j}^\dagger \Phi_y(\Delta t) U_{g_j} \right]$$

$$= \frac{1}{\Delta t} \left[\Pi_\mathscr{G}(\Phi_x(\Delta t)) + \Pi_\mathscr{G}(\Phi_y(\Delta t)) \right] = 0 + 0 = 0, \qquad (4.44)$$

where $a = x, y, z$, and the error action per generator is given by

$$\Phi_\ell(\Delta t) = \int_0^{\Delta t} ds\, u_\ell^\dagger(s) \sigma_a u_\ell(s), \quad \ell = x, y.$$

Note that, by construction, the resulting EDD scheme is insensitive to arbitrary systematic errors as long as the description given by the lowest-order $\overline{H}^{(0)}$ is accurate. Similarly to the pure-dephasing case, second-order EDD may be achieved by a symmetrized protocol of length $T_s = 16\Delta t$, whereby the control path in Eq. (4.43) is first effected forward and then backward, with time-reversed implementations $u_{\bar{x}}(t), u_{\bar{y}}(t)$ of the generators in the second half.

For n qubits, $\mathscr{G}$ acts reducibly, thus care must be taken in ensuring that the relevant DD condition of Eq. (4.39) is obeyed, which may depend upon the details of the generators implementation. A general *sufficient* condition, established in [VK03], is to restrict control Hamiltonians to be in the group algebra of the representation, that is, the linear span$_\mathbb{R}\{U_{g_j}\}$. In our case, this would require X (or, respectively, Y) to be implemented by turning on an n-body Hamiltonian proportional to $X = \bigotimes_i \sigma_x^{(i)}$ (or, respectively, Y), which is physically unrealistic for $n > 2$. Fortunately, this condition is not necessary for Φ_ℓ to be correctable by $\mathscr{G}$. In particular, for linear decoherence as examined here, one may show [KV09a] that an adequate choice for implementing collective generators is provided by collective *single-qubit* Hamiltonians, that is, we may let

$$H_\ell(t) = u_\ell(t) \sum_{i=1}^n \sigma_\ell^{(i)}, \quad \ell = x, y.$$

Since, in the presence of faulty control implementation, residual control errors are also projected onto $\mathbb{C}\mathscr{G}'$ under $\mathscr{Q}'_\mathscr{C}$ [Eq. (4.40)], compensation of arbitrary systematic single-qubit errors is still guaranteed to the lowest order, as in the single-qubit case.

Chapter 5

Introduction to quantum fault tolerance

Panos Aliferis

One of the applications of quantum error correction is protecting quantum computers from noise. Certainly, there is nothing particularly quantum mechanical in the idea of protecting information by encoding it. Even ordinary digital computers use various *fault tolerance* methods at the software level to correct errors during the storage or the transmission of information; e.g., the integrity of the bits stored in hard disks is verified by using parity checks (checksums). In addition, for critical computing systems such as those inside airplanes or nuclear reactors, software fault tolerance methods are also applied during the *processing* of information; e.g., airplane control computers compare the results from multiple parallel processors to detect faults. In general, however, the hardware of modern computers is remarkably robust to noise, so that for most applications the processing of information can be executed with high reliability without using software error correction.

In contrast to the ease and robustness with which classical information can be processed,[1] the processing of quantum information appears at present to be much more challenging. Although constructing reliable quantum computing hardware is certainly a daunting task, we have nevertheless strong hopes that large-scale quantum computers, able to implement useful long computations, can in fact be realized. This optimism is founded on methods of *quantum fault tolerance*, which show that scalable quantum computation is, in principle, possible against a variety of noise processes. Demonstrating that these methods work effectively in practice is a major challenge for contemporary science, a challenge whose outcome will depend on our progress in understanding the physical noise processes in experiments, and on our ability to design and optimize fault-tolerance methods according to the limitations and the noise characteristics of experimental devices.

---

[1] Of course, this was not always the case; photographs of ENIAC, the first universal electrical computer, speak volumes about how difficult the first steps of classical computing were.

Quantum Error Correction, ed. Daniel A. Lidar and Todd A. Brun. Published by Cambridge University Press.
© Cambridge University Press 2013.

This chapter is an introduction to software methods of quantum fault tolerance. Broadly speaking, these methods describe strategies for using the noisy hardware components of a quantum computer to perform computations while continually monitoring and actively correcting the hardware faults. The methods we will discuss are general and apply independently of how the hardware components are physically realized in the laboratory. Nevertheless, one should not lose sight of the fact that what we describe in this chapter as elementary hardware components are *not* elementary from an experimental point of view. Already, at the level of the realization of qubits in the laboratory, the experimenter strives to choose implementations with high inherent robustness to noise, such as qubits encoded in decoherence-free subspaces or noiseless subsystems, or qubits that are topologically protected. In addition, as discussed in the preceding chapters, noise in the elementary hardware operations can be suppressed by using various open-loop techniques such as refocusing or dynamical decoupling. Even though these various qubit encodings and noise-suppression techniques can be highly effective, some residual noise will always remain; it is this residual effective noise that needs to be treated by the error correction and fault tolerance methods we will discuss in this chapter.

The basic conceptual ideas of fault tolerance for quantum computation are very similar to the case of classical computation: First, a code is chosen and each logical step of the computation is implemented by a fault-tolerant *gadget* which acts on the encoded information; these gadgets comprise many elementary hardware operations, and they are designed to implement the desired logical transformation on the encoded information while at the same time detecting and correcting errors. And second, the protection from noise is increased by designing a hierarchy of encoding layers such that errors become progressively weaker as we pass from one layer to the next.

Despite these similarities, there are two major differences between quantum and classical fault tolerance. The first difference is that in the quantum case error correction needs to be implemented *coherently*, i.e., in a way that preserves the quantum superpositions in the encoded information that is processed by the quantum computer – this requirement has no analog in classical fault tolerance, since quantum superpositions and quantum interference play no role in classical computation. The second difference relates to the types of noise that are of concern in the two cases. For systems that store classical information digitized in bits, noise can simply be viewed as causing abrupt changes in the value of each bit (bit-flips). For systems that store quantum information, on the other hand, the relevant noise processes are more general: The state of such a system is in general a superposition $a_1 e^{i\phi_1}|\psi_1\rangle + a_2 e^{i\phi_2}|\psi_2\rangle + \cdots$ of various states $|\psi_i\rangle$ with real coefficients a_i and ϕ_i; then not only can noise cause changes in the amplitudes a_i (which is analogous to the bit-flip errors for classical information, as a_i^2 is the probability of occupation of the state $|\psi_i\rangle$ if the superposed states are all mutually orthogonal) but noise can also cause changes to the phases ϕ_i (phases are irrelevant when storing classical information but they are important quantum mechanically as they determine the ability of the superposed states to interfere).

5.1 Quantum circuits and error discretization

The precise character of noise in future quantum computers will depend on the particular hardware implementation. There is a wide variety of prospective implementation schemes that are being experimentally investigated at present, but in this chapter we restrict the discussion to

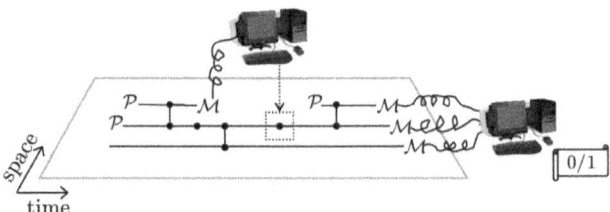

Fig. 5.1. A quantum circuit is a sequence of qubit preparations ($\mathscr{P}$), qubit measurements ($\mathscr{M}$), and quantum gates (a • denotes a gate applied on a single qubit, and two •s connected vertically denote a gate applied on a pair of qubits). The measurement outcomes are processed by classical computers alongside the quantum computer; intermediate measurement outcomes condition the application of future quantum gates, while the final measurement outcomes encode the answer of the quantum computation (0 or 1).

those schemes that fall under the quantum circuit model for which methods of fault tolerance are better understood.[2]

Quantum circuits are a generalization of classical circuits: A classical circuit computing a Boolean function f on n bits is a prescription for expressing f as a composition of functions or *gates* that act on a fixed, independent of n, number of bits at a time; gates are chosen from a finite set that is *universal*, allowing any function to be computed; e.g., the NOT gate (which flips the value of a bit) together with the AND gate (which computes the conjunction of the value of two bits) form a universal gate set. Similarly, a quantum circuit is a prescription for implementing a physical operation on the Hilbert space $\mathscr{H}_n = (\mathbb{C}^2)^{\otimes n}$ of n qubits as a composition of elementary physical operations that are applied on a fixed number of qubits at a time. Although $\mathscr{H}_n$ is continuous, there exist *finite* sets of physical operations acting on at most two qubits that are *quantum universal*, allowing the approximation of any physical operation in $\mathscr{H}_n$ to any desired accuracy; these universal sets comprise preparations of single qubits in certain pure states, certain unitary transformations or *quantum gates* on single qubits or between pairs of qubits, and measurements of single-qubit observables. Our diagrammatic representation of a quantum circuit is shown in Fig. 5.1.

We will not discuss examples of quantum universal sets here; the interested reader should consult the references in Section 5.7. Conceptually, what is important is that a finite number of elementary operations suffices for implementing any quantum computation. Therefore, a quantum computer is a discrete machine just like a classical computer; in the classical case, the elementary hardware components are gates on one or a few bits, while in the quantum case they are physical operations on one or a few qubits. Discreteness is essential both classically and quantumly because it implies that fault tolerance can be achieved by constructing fault-tolerant gadgets for each one of the operations in a universal set and composing these gadgets together.

The second essential ingredient for fault tolerance is the ability to discretize errors so that error correction becomes possible. In classical computation, the basis of error discretization is the digital encoding of information. Once a bit of information is represented in a physical quantity taking the value v_0 to encode 0 and v_1 to encode 1, noise in gates can be described in terms

---

[2] In particular, we will not discuss quantum computations realized purely by *adiabatic* evolution, for which a general theory of fault tolerance is still lacking.

of discrete errors taking v_0 to v_1 or vice versa, and small fluctuations around the values v_0 and v_1 can in practice be ignored; e.g., a NOT gate can be implemented by a CMOS inverter in saturation, the input and output bit values 0 and 1 are encoded as different voltages v_0 and v_1, and the output voltage is essentially insensitive to small variations $\delta v \ll v_0, v_1$ in the input voltage.

In the case of quantum computation, it is important to recognize that there is no unambiguous way to say which qubits of an entangled multi-qubit quantum state processed by the quantum computer are erroneous and which are not; e.g., if noise acts on the maximally entangled two-qubit state

$$|\Phi_0\rangle = \frac{1}{\sqrt{2}}(|0\rangle|0\rangle + |1\rangle|1\rangle) , \quad (5.1)$$

there is no good way to say which of the two qubits is erroneous because for any single-qubit operator E, $(E \otimes I)|\Phi_0\rangle = (I \otimes E^T)|\Phi_0\rangle$, and whether the state has suffered an error cannot be determined by just observing the properties of any one of the two qubits in isolation. It is therefore helpful to avoid using a *semantic* language where the notion of an error depends on the quantum state on which errors act, and instead to adopt a *syntactic* language where the notion of an error is defined operationally independent of the actual quantum state. More concretely, instead of associating errors with individual qubits, we can associate errors with mutually orthogonal *subspaces* of the entire Hilbert space of a collection or a *block* of several encoded qubits; for a block comprising n encoded qubits

$$\mathcal{H}_n = \bigoplus_s \mathcal{H}_n^s , \quad (5.2)$$

where the superscript s, called the *syndrome*, is a label for the different subspaces $\mathcal{H}_n^s$ for the n-qubit Hilbert space $\mathcal{H}_n$. One of the subspaces, the *code space* $\mathcal{H}_n^0$, is the preferred one in the sense that quantum information is encoded in a quantum state that is supported in $\mathcal{H}_n^0$. Because of noise, the encoded quantum state will tend to escape from the code space towards other subspaces. But because all subspaces are mutually orthogonal, there is a generalized measurement, called a *syndrome measurement*, which allows different subspaces to be distinguished unambiguously. By performing a syndrome measurement, we can then use the measurement outcome μ to deduce s and therefore learn about the subspace on which the noisy encoded quantum state is supported. Since there is only a finite number of orthogonal subspaces, we can execute the computation that yields s given μ digitally by processing the measurement outcome in a classical computer. If the result of this computation is a nontrivial syndrome value which indicates that the noisy encoded quantum state is supported in a subspace different than the code space, we can apply a *recovery* operation on the encoded qubits, which is conditioned on the result of the classical computation and which returns the encoded quantum state to the code space. Because classical digital computers are in practice extremely robust to noise, we usually assume that the classical processing of μ to obtain s and the classical control of the quantum computer conditioned on s can be implemented perfectly without faults. Of course, no matter how improbable, faults in the classical processing of the measurement outcomes can lead to a failure to apply the appropriate recovery operation, so that the accuracy of the quantum computer is ultimately limited by the accuracy of the on-the-side classical computer.

5.2 Noisy quantum computers

We have seen that discretizing the entire Hilbert space of the qubits processed by the quantum computer allows us to encode quantum information in a quantum state supported in the code space and to protect against noise that takes this encoded quantum state to other orthogonal subspaces. But for what types of noise processes is this method of encoding quantum information effective? Certainly, we have very little hope of protecting information against noise that acts collectively on many hardware components of the quantum computer and whose strength is not moderated as a function of the number of qubits it affects; e.g., we would be helpless if a power outage or an earthquake were to hit our quantum computer affecting many qubits all at once.

If we exclude such malicious types of collective noise against which no error-correction method can be effective either for quantum or for classical computation, we are left with several other contributions to noise that need to be considered: First, there is noise due to imprecisions in the implementation of each elementary hardware operation; e.g., noise in the control parameters during the implementation of a unitary gate U might result in realizing another unitary $U + \delta U$ instead, where δU may be systematic or it may vary stochastically. Second, there are unwanted interactions among the qubits in the quantum computer, e.g., an electromagnetic coupling of nearby quantum-dot qubits, which decays as a power of their relative distance. Third, there are interactions between the qubits of the quantum computer and an *environment* representing external degrees of freedom that are not under our control, e.g., a coupling of integrated superconducting qubits to nuclear spins in the substrate. And finally, in settings where qubits are realized by selecting a two-dimensional subspace inside a multi-dimensional system, there is noise that couples the two-level qubit subspace to other levels of the same system; e.g., if a qubit is realized by using the ground-state hyperfine splittings of a trapped ion, noise can induce transitions between these hyperfine levels and other higher-energy levels of the ion.

A useful classification of these noise processes concerns their spatial and temporal locality. Intuitively, we say that noise is spatially local or simply *local* if, during any time interval, it acts collectively only on qubits that are interacting in the ideal quantum circuit at the same time interval; i.e., if two-qubit gates are applied in parallel to several pairs of qubits during a specific time interval, local noise can act collectively on qubits that belong to the same pair but not on those that belong to different pairs. Locality is a desirable property because it implies that noise cannot inflict global damage by causing the simultaneous failure of many hardware components. Although in general error correction fails for nonlocal noise, there are in fact some nonlocal noise processes for which effective error correction is possible: First, certain types of collective noise whose nature is known in advance can be suppressed effectively by using the techniques of decoherence-free subspaces and noiseless subsystems, as explained in Chapter 3. Second, there are types of nonlocal noise that can be treated as if the noise were local and for which fault-tolerance methods designed to protect against local noise are effective; we will discuss two such examples in Section 5.2.1.2.

With regard to temporal locality, the question is: how correlated is the noise that acts on different hardware components that are executed at different time intervals? We say that noise is temporally local or *Markovian* if the noisy evolution can be described by using a sequence of *superoperators*, each taking the density matrix at the end of one time interval to the density matrix at the end of the following time interval.

When the quantum computer implements a quantum circuit of depth D,[3] we can discretize the total computation time T in D intervals $T_1, T_2, \ldots, T_D$, each of duration t_0 equal to the time it takes to execute an elementary operation. The Markovian property then translates to the requirement that noise has a typical correlation time comparable to t_0. Since the interaction between the quantum computer and the environment is incoherent across different time intervals, we can trace over the state of the environmental degrees of freedom after each interval to obtain a reduced density matrix describing the state of the quantum computer. In this case, the noisy evolution is described as a mapping between the reduced density matrices at different intervals:

$$\rho_j = \mathscr{S}_j(\rho_{j-1}) \,, \tag{5.3}$$

where ρ_j is the reduced density matrix at the end of interval T_j, and $\mathscr{S}_j$ is a superoperator describing the evolution from the end of interval T_{j-1} to the end of interval T_j.

On the other hand, if there are noise processes with typical correlation times longer than t_0, we cannot obtain an accurate description of the noisy evolution by tracing out the external degrees of freedom after every time interval. In this case, the information that the environment exchanges with the quantum computer could in principle be retained for long times, so that we cannot simply describe the entire noisy evolution as a composition of superoperators. For this reason, the analysis of the effects of non-Markovian noise is more demanding than for simple Markovian noise and, as we will see in the next section, our conclusions for the effectiveness of fault-tolerance methods against non-Markovian noise are generally weaker than for Markovian noise.

5.2.1 Noise models

In this section, we will discuss several concrete examples of noise models that have been analyzed in the context of fault-tolerant quantum computation.

In all these examples, we consider the noisy implementation of an ideal quantum circuit comprising L elementary operations followed by the final qubit measurements whose outcome encodes the result of the computation. The ideal quantum circuit produces the quantum state

$$\rho^{\text{ideal}} = \mathscr{O}_L \circ \cdots \circ \mathscr{O}_2 \circ \mathscr{O}_1 \,, \tag{5.4}$$

where $\circ$ denotes composition and the superoperators $\mathscr{O}_j$ correspond to either a qubit preparation, a unitary gate, or an intermediate qubit measurement that conditions subsequent operations; the operation for the preparation of a qubit in the pure state $|\psi\rangle$ is

$$\mathscr{P}_{|\psi\rangle} = |\psi\rangle\langle\psi| \,, \tag{5.5}$$

the operation for a quantum gate U applied on input X is

$$\mathscr{U}(X) = UXU^\dagger \,, \tag{5.6}$$

---

[3] The *depth* of a quantum circuit is the maximum number of elementary operations applied on any qubit of the quantum computer (including the identity operation that is implicitly applied when a qubit is stored while operations are applied on other qubits).

and the operation for a projective measurement of an observable $\hat{a}$ applied on input X with projector M_a corresponding to measurement outcome a,

$$\mathcal{M}_{\{a\}}(X) = \sum_a M_a \, X M_a \, . \tag{5.7}$$

For simplicity and since noise on the final measurements can be modeled by noise acting in the immediately preceding operations, we may assume that the final measurements are implemented ideally without faults. Finally, as we have noted, we will consider performing the processing of the outcomes of both the intermediate and the final measurements in a classical computer operating alongside the quantum computer, and we will assume there are practically no faults in this classical hardware.

5.2.1.1 Local Markovian noise Our first example is noise that is both local and Markovian. The Markovian property implies that we can describe the noisy evolution as a composition of superoperators, and the locality property implies that the superoperator for each time interval can be expressed as a tensor product of superoperators, each superoperator corresponding to one of the different elementary operations that are implemented in parallel during that interval.

We may express the superoperator describing the noisy implementation of each elementary operation as $\mathcal{N}_j \circ \mathcal{O}_j$, where $\mathcal{O}_j$ is the ideal superoperator and $\mathcal{N}_j$ is a superoperator describing deviations from the ideal due to noise. By definition, the support of $\mathcal{N}_j$ is contained in the support of $\mathcal{O}_j$ when noise is local.[4] Therefore, because of noise, instead of the ideal quantum state in Eq. (5.4), the quantum computer really prepares the state

$$\rho^{\text{noisy}} = \mathcal{N}_L \circ \mathcal{O}_L \circ \cdots \circ \mathcal{N}_2 \circ \mathcal{O}_2 \circ \mathcal{N}_1 \circ \mathcal{O}_1 \, . \tag{5.8}$$

Since all $\mathcal{N}_j$ would be trivial were there no noise, a natural measure for the noise is the distance between $\mathcal{N}_j$ and the identity superoperator $\mathcal{I}$, and we can define the noise *strength*

$$\varepsilon = \max_j ||\mathcal{N}_j - \mathcal{I}||_\diamond \, , \tag{5.9}$$

where $|| \cdot ||_\diamond$ is a suitable superoperator norm, the *diamond* norm, defined and discussed in Chapter 1, Eq. (1.94).[5] If we now write $\mathcal{N}_j = \mathcal{I} + \mathcal{F}_j$ for some fault operator, or simply *fault*, $\mathcal{F}_j$ and substitute in Eq. (5.8), we find

$$\rho^{\text{noisy}} = (\mathcal{I} + \mathcal{F}_L) \circ \mathcal{O}_L \circ \cdots \circ (\mathcal{I} + \mathcal{F}_2) \circ \mathcal{O}_2 \circ (\mathcal{I} + \mathcal{F}_1) \circ \mathcal{O}_1 \, . \tag{5.10}$$

By opening all parentheses, we obtain a sum of terms corresponding to different *fault paths*; in each fault path, faults $\mathcal{F}_j$ have occurred in a specific subset of the L elementary operations, while the identity superoperators are applied on all remaining operations. In particular, we can write

$$\rho^{\text{noisy}} = \rho^{\text{ideal}} + \zeta^{\text{faulty}} \, , \tag{5.11}$$

where ρ^{ideal} corresponds to the unique fault path where identity superoperators are applied everywhere and ζ^{faulty} contains all other fault paths for which there is at least one insertion

---

[4] The support of a superoperator $\mathcal{O}$ (or operator O) acting on density matrices (or quantum states respectively) defined on various subsystems is the tensor product of the Hilbert spaces of all the subsystems on which $\mathcal{O}$ (or O respectively) acts nontrivially; in our case, the subsystems are the qubits of the quantum computer and any subsystems in the environment.

[5] To optimize our estimate for the noise strength, we can take $\mathcal{I}$ to be proportional to the identity superoperator with a proportionality constant of magnitude between 0 and 1; but here we will not discuss this generalization.

of a fault; we use ζ instead of ρ in ζ^{faulty} to emphasize that it is not a density matrix but rather the difference of two density matrices.

The fault path expansion is helpful for understanding how accurate is the noisy circuit. More precisely, we would like to know what would be the distance δ between the probability distribution $\{q_\mu^{\text{noisy}}\}$ for the outcomes $\{\mu\}$ of the final measurements on the noisy circuit and the distribution $\{q_\mu^{\text{ideal}}\}$ if these measurements were applied on the ideal circuit instead. We can express δ in terms of the 1-norm between the two probability distributions:

$$\delta = \sum_\mu |q_\mu^{\text{noisy}} - q_\mu^{\text{ideal}}| = \sum_\mu \left| \text{Tr}\left(M_\mu \left(\rho^{\text{noisy}} - \rho^{\text{ideal}}\right)\right)\right| ; \tag{5.12}$$

then

$$\delta = \sum_\mu \left| \sum_\nu v_\nu \langle v_\nu | M_\mu | v_\nu \rangle \right| \leq \sum_\mu \sum_\nu |v_\nu| |\langle v_\nu | M_\nu | v_\nu \rangle| \leq \sum_\nu |v_\nu| = ||\zeta^{\text{faulty}}||_1 , \tag{5.13}$$

where the projectors M_μ define the final measurements,[6] and $|v_\nu\rangle$ are the eigenvectors of $\zeta^{\text{faulty}} = \rho^{\text{noisy}} - \rho^{\text{ideal}}$ with corresponding eigenvalues v_ν. Now, how do we upper bound the norm of ζ^{faulty}?

If we let C denote the set of all L elementary operations in the quantum circuit,

$$\zeta^{\text{faulty}} = \sum_{r=1}^L (-1)^{r-1} \sum_{C_r \subseteq C} \zeta(C_r) , \tag{5.14}$$

where the second sum is over all subsets C_r of C with cardinality r, and $\zeta(C_r)$ denotes a *sum* of all the fault paths with faults applied on all operations in the set C_r. Equation (5.14) can be derived from the *inclusion-exclusion* trick of combinatorics: Since ζ^{faulty} is the sum of all the fault paths with at least one fault, the sum of all $\zeta(C_1)$ counts correctly all the fault paths with exactly one fault but overcounts the fault paths with at least two faults; to amend the overcounting, we subtract the sum of all $\zeta(C_2)$ that corrects the overcounting of all the fault paths with exactly two faults but introduces an undercounting of the fault paths with at least three faults; and so on.

For each specific set C_r, $\zeta(C_r)$ is nothing but the composition of the ideal superoperators $\mathcal{O}_j$ interspersed with faults $\mathcal{F}_j$ applied on all operations in C_r and the full noise superoperators $\mathcal{N}_j$ applied on all the remaining operations. Of course, since superoperators have unity norm, $||\mathcal{O}_j||_\diamond = ||\mathcal{N}_j||_\diamond = 1$ and thus

$$||\zeta(C_r)||_1 \leq \varepsilon^r . \tag{5.15}$$

By using the triangle inequality, and since there are $\binom{L}{r}$ distinct subsets C_r of C, Eqs. (5.13) and (5.14) now imply that

$$\delta \leq \sum_{r=1}^L \binom{L}{r} \varepsilon^r \leq L\varepsilon \left(1 + \frac{1}{L^2}\binom{L}{2} + \frac{1}{L^3}\binom{L}{3} + \cdots + \frac{1}{L^L}\binom{L}{L}\right) \leq (e-1)L\varepsilon , \tag{5.16}$$

where in the last two steps we assumed that $\varepsilon \leq 1/L$.

Our derivation of Eq. (5.16) via Eq. (5.14) was made in order to introduce a simple application of the inclusion-exclusion trick that is also being used later in this chapter. An improved

---

[6] The projectors M_μ are nonnegative (i.e., $\langle u|M_\mu|u\rangle \geq 0$ for any $|u\rangle$), and they are normalized so that $\sum_\mu M_\mu = I$.

upper bound on δ can in fact be derived without the assumption $\varepsilon \leq 1/L$ by simply noting that we can group fault paths depending on their *earliest* faulty operation:

$$\zeta^{\text{faulty}} = \sum_{r=1}^{L} \zeta(\mathscr{O}_r) \,, \qquad (5.17)$$

where $\zeta(\mathscr{O}_r)$ is the composition of the ideal superoperators $\mathscr{O}_j$ interspersed with the identity superoperators applied on the operations 1 to $r-1$, a fault $\mathscr{F}_r$ applied on the rth operation, and the full noise superoperators $\mathscr{N}_j$ applied on the operations $r+1$ to L. Since superoperators have unity norm, $\|\zeta(\mathscr{O}_r)\|_\diamond \leq \varepsilon$ and thus Eq. (5.17) implies

$$\delta \leq L\varepsilon \,. \qquad (5.18)$$

We conclude that, for a constant error strength ε, the accuracy $1 - \delta$ of the noisy quantum circuit decreases at most *linearly* with the circuit size L, in accordance with what is expected for a discrete model of computation.[7] Of course, as we will discuss in the following sections, the goal of implementing the quantum computation by using fault-tolerance methods is to replace ε in Eq. (5.16) by a smaller – in fact, an arbitrarily small – *effective* noise strength, thus making the accuracy of the noisy circuit approach as close to unity as desired.

Assessment and examples. At this point, we can step back to note the two essential assumptions that allowed us to derive Eq. (5.18): First, we assumed that the superoperators describing the noisy evolution can be expanded perturbatively as a sum over fault paths. Second, we assumed that fault paths with many faults are exponentially suppressed in the sense of Eq. (5.15). In fact, we may view Eq. (5.15) as the *defining* property of local Markovian noise, even if the superoperators that describe the noisy evolution are not strictly local. Thus, we generally say:

Definition (Local Markovian noise). *Noise is local and Markovian if the noisy evolution can be expanded as a sum over fault paths, where faults are described as (differences of) superoperators and the norm of the sum of all the fault paths with faults in any r specific elementary operations is upper bounded by ε^r for some constant noise strength ε.*

This relaxed definition has the advantage that it can describe correlated noise both in space and in time: Subject to the constraint that fault paths must satisfy Eq. (5.15), the fault operators comprising each fault path are otherwise unconstrained; in particular, the various fault operators can be controlled by an adversary who may chose to act collectively on all the faulty operations any way she pleases.

The local Markovian noise model captures several noise processes of interest; below, we discuss three simple but important examples.

---

[7] For discrete models of computation, to achieve a constant accuracy $1 - \delta$, the number of bits of precision required to specify the physical parameters associated with each elementary operation to within ε – e.g., the amplitude and timing of a voltage pulse used to control a CMOS gate – grows logarithmically with the size L of the computation. (With the bits of precision growing logarithmically with L, ε decreases polynomially with L.) In contrast, for analog models, the number of bits of precision grows polynomially or even exponentially with L.

Control noise. A common source of noise is due to imprecision in the control parameters during the implementation of each elementary operation, e.g., noise in the timing or the intensity of external magnetic fields used to manipulate the state of a superconducting qubit.

In the simplest case, consider the implementation of a single-qubit gate corresponding to a rotation by an angle 2θ around the z direction; this operation is described by the superoperator

$$\mathscr{R}_\theta^z(X) = e^{i\theta\sigma_z} X e^{-i\theta\sigma_z} . \tag{5.19}$$

Because of imprecisions in the control parameters, a rotation by a different angle $2\theta' = 2(\theta + \delta\theta)$ for some small fixed deviation $\delta\theta$ may be implemented instead. We can express the noisy superoperator as $\mathscr{R}_{\theta'}^z = \mathscr{N}_{\text{ctrl}} \circ \mathscr{R}_\theta^z$, where $\mathscr{N}_{\text{ctrl}} = \mathscr{I} + \mathscr{F}_{\text{ctrl}}$ and

$$\mathscr{F}_{\text{ctrl}}(X) = i\,\delta\theta\,(\sigma_z X - X\sigma_z) + O(\delta\theta^2) , \tag{5.20}$$

so that control noise satisfies Eq. (5.15) with $\varepsilon = O(\delta\theta)$.

A similar conclusion also holds if the deviation angle is not fixed but varies stochastically, and also for control errors in multi-qubit gates, preparations, or measurements.

Relaxation. Another common source of noise is due to thermal relaxation; e.g., in systems where $|0\rangle$ and $|1\rangle$ are encoded in different energy eigenlevels, the state $|1\rangle$ may spontaneously relax to the lower-energy state $|0\rangle$. To first approximation, relaxation can be expected to act independently on each qubit during the execution of a quantum computation; then, for each qubit and during each time interval, relaxation with a characteristic time scale T_1 can be modeled by the *amplitude damping* superoperator $\mathscr{N}_{\text{relax}}(X) = M_0 X M_0^\dagger + M_1 X M_1^\dagger$, where

$$M_0 = \frac{1+\sqrt{1-\gamma}}{2} I + \frac{1-\sqrt{1-\gamma}}{2} \sigma_z ,\quad M_1 = \frac{\sqrt{\gamma}}{2}\sigma_x(1-\sigma_z) , \tag{5.21}$$

and $\gamma = 1 - e^{-t_0/T_1}$. We can write $\mathscr{N}_{\text{relax}} = \mathscr{I} + \mathscr{F}_{\text{relax}}$, where

$$\mathscr{F}_{\text{relax}}(X) = \frac{\gamma}{4}\sigma_z X M_0^\dagger + \frac{\gamma}{4} M_0 X \sigma_z + M_1 X M_1^\dagger + O(\gamma^2) , \tag{5.22}$$

so that relaxation noise satisfies Eq. (5.15) with $\varepsilon = O(\gamma)$.

Probabilistic noise. In many cases noise can be modeled as a random processes, e.g., shot noise in the laser fields used to control trapped ionic qubits. Ignoring the details of the underlying random process, and letting p denote the probability of a fault during the implementation of each elementary operation (if r is the fault rate, $p = rt_0$), noise can be modeled by the superoperator

$$\mathscr{N}_{\text{rand}}(X) = (1-p)X + p E X E^\dagger , \tag{5.23}$$

where E is an arbitrary operator acting on the support of the ideal operation (subject to the constraint $E^\dagger E = I$ required for $\mathscr{N}_{\text{rand}}$ to be trace-preserving).

We can write $\mathscr{N}_{\text{rand}} = \mathscr{I} + \mathscr{F}_{\text{rand}}$, where

$$\mathscr{F}_{\text{rand}}(X) = -pX + p E X E^\dagger , \tag{5.24}$$

so that probabilistic noise satisfies Eq. (5.15) with $\varepsilon = O(p)$.

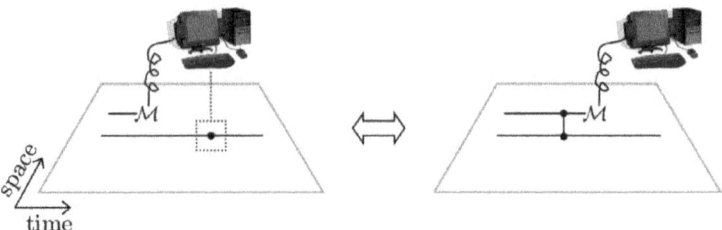

Fig. 5.2. On the left, a unitary gate U in a quantum circuit is conditioned on the outcome of a preceding single-qubit measurement along the orthonormal basis $\{|\tilde{0}\rangle, |\tilde{1}\rangle\}$. On the right, a mathematically equivalent circuit where the two qubits interact via a unitary gate $\tilde{U}$ followed by a single-qubit measurement along the basis $\{|0\rangle, |1\rangle\}$; here, $\tilde{U} = \Lambda(U)(V \otimes I)$, where the single-qubit unitary V acts on the measured qubit as $V|\tilde{0}\rangle = |0\rangle$, $V|\tilde{1}\rangle = |1\rangle$, and the two-qubit $\Lambda(U)$ applies U on the second qubit conditioned on the state of the measured qubit being $|1\rangle$.

5.2.1.2 Local non-Markovian noise Our local Markovian noise model is powerful enough to capture several important noise processes such as systematic control errors or thermal relaxation. However, requiring that fault paths are associated with superoperators is rather limiting because it constrains the possible noise correlations between different fault paths; while the fault operators in any specific fault path may be arbitrarily correlated, there can be only classical but no quantum correlations – i.e., no quantum interference – between the fault operators in different fault paths.

To go beyond Markovian noise, we can no longer trace over the environmental degrees of freedom to obtain a superoperator description of the noisy evolution. Now, our description will need to include explicitly the quantum state of the environment and its joint unitary evolution with the qubits of the quantum computer during the course of the entire quantum computation; of course, while we assume that we have control over all the qubits of our quantum computer, the environmental degrees of freedom are inaccessible and in many cases their precise nature is unknown.

We assume that the qubits of the quantum computer can be initialized in a pure state $|\psi_0\rangle_{QC}$ – e.g., we can prepare all qubits in their lowest-energy eigenstate (at least, come very close to it) by cooling – so that the state at the beginning of the quantum computation, including the environment, is $|\psi_0\rangle_{QC} \otimes |\phi_0\rangle_E$ for some unspecified pure state $|\phi_0\rangle_E$.[8] If there were no noise, implementing the ideal quantum circuit would then correspond to implementing a sequence of unitary operators U_j producing the final state

$$|\psi\rangle_{QC}^{\text{ideal}} = (U_L \cdots U_2 \cdot U_1) |\psi_0\rangle_{QC} , \qquad (5.25)$$

on which state we finally apply measurements that give the result of the computation. (If there are unitary gates that are conditioned on the outcome of a preceding measurement (see Fig. 5.1), we can mathematically replace them in our analysis by different unitary gates, which are followed by measurements as in Fig. 5.2.)

We may describe the noisy implementation of each elementary unitary operator U_j as $N_j \cdot U_j$, where N_j is a unitary operator on the quantum computer *and* the environment and it describes deviations from the ideal due to noise; by definition, the support of N_j is contained in the union

---

[8] We can always obtain a representation of the environment in terms of a pure state since any mixed state can be purified by introducing an auxiliary Hilbert space.

of the support of U_j and the environment when noise is local. Since all N_j would be trivial were there no noise, we can now define the noise strength as

$$\varepsilon = \max_j \min_{I_{QC}} ||N_j - I_{QC}||_\infty \,, \qquad (5.26)$$

where we vary over all unitary I_{QC} that act trivially on the quantum computer and in an arbitrary way on the environment.

By expanding each N_j as the sum of the I_{QC} that minimizes the norm in Eq. (5.26) and a fault, $N_j = I_{QC} + F_j$, we can substitute in Eq. (5.25) to find that the joint state of the quantum computer and the environment prior to the final measurements is

$$|\psi\rangle_{QCE}^{\text{noisy}} = (I_{QC} + F_L) \cdot U_L \cdots (I_{QC} + F_2) \cdot U_2 \cdot (I_{QC} + F_1) \cdot U_1(|\psi_0\rangle_{QC} \otimes |\phi_0\rangle_E), \qquad (5.27)$$

where it is understood that the U_j act trivially on the environment. We may now open all the parentheses to obtain a fault-path expansion just like we did in the previous section; here, each fault path identifies a specific subset of the L elementary operations where faults F_j have occurred, where the F_j are differences of unitary operators instead of differences of superoperators as in the case of Markovian noise. We can then write

$$|\psi\rangle^{\text{noisy}} = |\psi\rangle^{\text{ideal}} + |\vartheta\rangle^{\text{faulty}} \,, \qquad (5.28)$$

where for succinctness we have dropped the QCE subscripts but it is understood that all states are supported in the Hilbert space of the qubits of the quantum computer and also the environment. Here, $|\psi\rangle^{\text{ideal}}$ corresponds to the unique fault path where operators I_{QC} are applied everywhere and $|\vartheta\rangle^{\text{faulty}}$ contains all other fault paths for which there is at least one insertion of a fault; we use ϑ instead of ψ in $|\vartheta\rangle^{\text{faulty}}$ to emphasize that it is not normalized but it is rather the difference of two normalized pure states.

How can we estimate the accuracy of the noisy quantum circuit in the presence of local non-Markovian noise? From Eq. (5.12), it suffices to evaluate the trace norm (Eq. (1.75)) of the difference between the noisy and the ideal density matrices; since the final states are pure,

$$\delta = |||\psi\rangle\langle\psi|^{\text{noisy}} - |\psi\rangle\langle\psi|^{\text{ideal}}||_1 \leq 2|||\vartheta\rangle^{\text{faulty}}|| \,, \qquad (5.29)$$

where we used that

$$|||n\rangle\langle n| - |i\rangle\langle i|||_1 = 2\sqrt{1 - |\langle n|i\rangle|^2} \leq 2|||n\rangle - |i\rangle|| \qquad (5.30)$$

for any normalized pure states $|n\rangle$ and $|i\rangle$. It remains to obtain an upper bound on the norm of $|\vartheta\rangle^{\text{faulty}}$.

In analogy to Eq. (5.14),

$$|\vartheta\rangle^{\text{faulty}} = \sum_{r=1}^{L} (-1)^{r-1} \sum_{C_r \subseteq C} |\vartheta(C_r)\rangle \,, \qquad (5.31)$$

where $|\vartheta(C_r)\rangle$ denotes a *sum* of all the fault paths with faults applied on all r operations in the set C_r; since $|\vartheta(C_r)\rangle$ is obtained by applying unitary operators everywhere except at the faulty operations,

$$|||\vartheta(C_r)\rangle|| \leq \varepsilon^r \,. \qquad (5.32)$$

Alternatively, in analogy to Eq. (5.17), we may group the fault paths depending on the earliest faulty operation:

$$|\vartheta\rangle^{\text{faulty}} = \sum_{r=1}^{L} |\vartheta(U_r)\rangle \,, \qquad (5.33)$$

where $|||\vartheta(U_r)\rangle|| \leq \varepsilon$. We conclude that

$$\delta \leq 2L\varepsilon \,, \qquad (5.34)$$

so that we obtain for local non-Markovian noise a similar result as for local Markovian noise: in both cases, the accuracy $1 - \delta$ of the noisy quantum circuit decreases at most linearly with the circuit size L. This illustrates that fully coherent noise, where the environment can store and process quantum information allowing different fault paths to interfere quantum mechanically, does not alter our view of quantum computation as a discrete model of computation similar to the model of modern digital computers (or, more abstractly, classical Turing machines). As we will discuss in the following sections, methods of quantum fault tolerance can replace ε in Eq. (5.34) by an arbitrarily small effective noise strength, showing that, just like Markovian noise, non-Markovian noise is not in principle an obstacle to large-scale quantum computation.

Assessment and examples. We note that Eq. (5.34) was derived based on two essential assumptions: First, we assumed that the final noisy quantum state can be expanded perturbatively as a sum over fault paths. Second, we assumed that fault paths with many faults are suppressed in the sense of Eq. (5.32). We can in fact *define* local non-Markovian noise in terms of these two assumptions, even if the noisy unitary evolution is not strictly local. We then generally say:

Definition 5.1 (Local non-Markovian noise). *Noise is local and non-Markovian if the noisy evolution can be expanded as a sum over fault paths, where faults are described as (differences of)* unitaries *acting between the quantum computer and the environment and the norm of the sum of all the fault paths with faults in any r specific elementary operations is upper bounded by ε^r for some constant noise strength ε.*

This definition is very similar to our definition of local Markovian noise in the previous section: In both cases, the noisy evolution is expanded as a sum over fault paths and also noise is *weak* in the sense that, as the total number of faults in a fault path increases, the fault path norm is suppressed exponentially. In addition, in both cases noise can be correlated both in space and in time since we place no restrictions on the form of the fault operators that appear in each fault path, which are allowed to be arbitrarily and even adversarially correlated.

The important distinction between the two cases is that while, for local Markovian noise, different fault paths do not interfere, for local non-Markovian noise, we make the worst-case assumption that all the fault paths *do* interfere coherently and the environment is *not* traced over until the end of the quantum computation. Therefore, while for local Markovian noise the strength ε can be viewed as a probability (the probability for the occurrence of a single fault), for local non-Markovian noise the strength ε corresponds in essence to a quantum *amplitude* (the amplitude for a term with a single fault in the final quantum state). If we were to use distinct symbols for the two cases, ε and ε' respectively for Markovian and non-Markovian noise, then, since probabilities are squares of amplitudes, we expect $\varepsilon \sim (\varepsilon')^2$; thus, requiring that the fault amplitude ε' is small (say, less than 1.0×10^{-3}) implies that the fault probability ε is even smaller (in this case of order 1.0×10^{-6}).

The local non-Markovian noise model describes several noise processes for which the environment interacts *coherently* with the quantum computer over long time scales; below, we discuss four examples of noise processes that give rise to local non-Markovian noise.

Local Hamiltonian noise. The ideal unitary evolution in Eq. (5.25) is generated by a time-dependent Hamiltonian

$$H_{QC} = \sum_j H_{QC}^j \text{ , such that } U_j = \exp\left(-it_0 H_{QC}^j\right). \tag{5.35}$$

The Hamiltonian that describes the noisy evolution of the quantum computer and the environment then has the general form

$$H = H_{QC} + H_E + H_{QCE}, \tag{5.36}$$

where H_E generates the evolution of the environmental degrees of freedom, and H_{QCE} describes the interaction of the quantum computer and the environment that introduces noise. If noise is local, H_{QCE} has the same locality as H_{QC}, i.e.,

$$H_{QCE} = \sum_j H_{QCE}^j, \tag{5.37}$$

where the support of H_{QCE}^j is contained in the union of the support of H_{QC}^j *and* the environment.

We can study the noisy evolution generated by H during a time interval of duration t_0 perturbatively; if we divide this interval into N micro-intervals each of duration $\Delta t_0 = t_0/N$, then

$$U_{QCE}(t_0, 0) = \lim_{N \to \infty} \prod_{n=1}^{N} U_{QC}^n U_{QCE}^n U_E^n, \tag{5.38}$$

where U_{QC}^n, U_{QCE}^n, and U_E^n denote the evolution during the nth micro-interval according to H_{QC}, H_{QCE}, and H_E respectively. After expanding[9]

$$U_{QCE}^n \approx \prod_j \left(I - i\Delta t_0 H_{QCE}^j\right), \tag{5.39}$$

and substituting in Eq. (5.38), we can open the parentheses to obtain a perturbative fault-path expansion. The noisy implementation of the ideal unitary U_j then takes the form

$$U_j^{\text{noisy}} = (I_{QC} + F_j) U_j; \tag{5.40}$$

here, I_{QC} denotes a sum of all the fault paths where, in every micro-interval, we insert either the identity or a *micro-fault* $f_{j'}$ acting on an operation with label $j' \neq j$, where

$$f_{j'} = -i\Delta t_0 H_{QCE}^{j'}. \tag{5.41}$$

It follows that F_j includes all remaining fault paths where a micro-fault f_j acting on the operation with label j is inserted in at least one micro-interval.

---

[9] Since we take the limit $N \to \infty$, we have $\Delta t_0 \to 0$ and we may keep only the linear term in the expansion.

We can express F_j as a sum of terms labeled by the micro-interval where the *earliest* micro-fault f_j is applied on the operation with label j; if we denote by $U_{QCE}(\Delta t_0^j, \Delta t_0^i)$ the entire evolution generated by H_{QCE} between the ith and jth micro-intervals with $i < j$, then

$$F_j = \lim_{N \to \infty} \sum_{r=1}^{N} \left(U_{QCE}(\Delta t_0^N, \Delta t_0^{r+1}) \prod_{q=1}^{r} \left(U_{QC}^q f_j^{\delta_{r,q}} \prod_{j' \neq j} (I + f_{j'}) U_E^q \right) \right) U_j^\dagger . \quad (5.42)$$

Now, since the operator norm is unitarily invariant, each term in the sum in Eq. (5.42) has norm $\|f_j\|_\infty = \Delta t_0 \|H_{QCE}^j\|_\infty$, and so $\|F_j\|_\infty \leq t_0 \|H_{QCE}^j\|_\infty$.

In fact, we can perform a similar perturbative expansion to analyze faults in any specific subset of the L elementary operations in the quantum circuit. It follows that local Hamiltonian noise satisfies Eq. (5.32) with

$$\varepsilon = t_0 \cdot \max_j \|H_{QCE}^j\|_\infty . \quad (5.43)$$

As expected, the noise strength ε depends on the strength of the interaction term H_{QCE}^j between the quantum computer and the environment and also the time during which the interaction is acting. On the contrary, we observe that ε does *not* depend on the strength of the term H_E, which describes the internal evolution of the environment. Moreover, we note that while H_{QCE}^j is assumed to act locally on the quantum computer in the sense of Eq. (5.37), the derivation of Eq. (5.43) did not rely on making any assumptions about H_E, which is completely arbitrary.

Long-range static noise. In certain systems, noise can arise due to static, i.e., time-independent, interactions among pairs of qubits of the quantum computer, where these interactions do not depend on the ideal circuit that is being implemented. Such nonlocal noise can be modeled by the Hamiltonian in Eq. (5.36) with

$$H_{QCE} = \sum_{(j,k)} H_{(j,k)} , \quad (5.44)$$

where $H_{(j,k)}$ is supported on qubits j, k and the environment and we sum all unordered pairs (j, k) of qubits.

We can perform a similar perturbative expansion as in the case of local Hamiltonian noise, except that now the two qubits in the support of any micro-fault may not be directly interacting via a unitary gate. Despite this difference, which necessitates a more complicated combinatorial analysis (see the references in Section 5.7), it can be shown that long-range static noise satisfies Eq. (5.32) with

$$\varepsilon = \sqrt{c t_0 \max_j \sum_k \|H_{(j,k)}\|_\infty} , \quad (5.45)$$

where $c = 2e$ provided $\varepsilon^2 \leq e$ and it is understood that, if H_{QCE} is time dependent, the maximum is also taken over all times.

Gaussian noise. In a variety of physical settings where the qubits of the quantum computer are coupled to a large number of environmental degrees of freedom, the environment can be well approximated as a collection of uncoupled harmonic oscillators obeying Gaussian statistics; the

Hamiltonian of the environment is

$$H_E = \sum_k \omega_k a_k^\dagger a_k , \qquad (5.46)$$

where a_k are bosonic annihilation operators satisfying $[a_k, a_{k'}^\dagger] = \delta_{kk'}$. In this *spin-boson model* of the noise, the interaction between the quantum computer and the environment is described by a coupling of each qubit to a linear combination of oscillator amplitudes:

$$H_{QCE} = \sum_{x,m} \sigma_m(x) \otimes \tilde{\phi}_m(x,t) \qquad (5.47)$$

with

$$\phi_m(x,t) = e^{itH_E} \tilde{\phi}_m(x,t) e^{-itH_E} = \sum_k \left(g_{k,m}(x,t) a_k e^{-it\omega_k} + g_{k,m}^*(x,t) a_k^\dagger e^{it\omega_k} \right) , \qquad (5.48)$$

where x labels a qubit's position and $\sigma_m(x)$ with $m \in \{x, y, z\}$ are the three Pauli operators on the qubit with label x.

The statistics of the environment amplitudes $\phi_m(x,t)$ are Gaussian in the sense that the n-point correlation functions with respect to the environment state $|\phi_0\rangle_E$ vanish for n odd, while for n even they obey Wick's theorem:

$$\langle \phi_{m_1}(x_1, t_1) \cdots \phi_{m_n}(x_n, t_n) \rangle = \sum_{(i_1,i_2),\ldots,(i_{n-1},i_n)} \Delta(i_1, i_2) \cdots \Delta(i_{n-1}, i_n) , \qquad (5.49)$$

where $\Delta(p,q) = \langle \phi_{m_p}(x_p, t_p) \phi_{m_q}(x_q, t_q) \rangle$, and we sum all ways of dividing the label 1 to n into $n/2$ unordered pairs. By performing a perturbative analysis similar to the case of long-range static noise (again, see the references in Section 5.7), it can be shown that Gaussian non-Markovian noise satisfies Eq. (5.32) with

$$\varepsilon = \sqrt{c \max_j \int_{(x_1,t_1) \in U_j} \int_{(x_2,t_2) \in \cup_l U_l} \sum_{m_1, m_2} |\Delta(1,2)|} , \qquad (5.50)$$

where the first integral denotes an integration over the qubits in the support of the unitary U_j and the time interval during which this gate is implemented, and the second integral denotes an integration over all the qubits of the quantum computer and the total duration of the quantum computation.

Local leakage noise. The qubits of the quantum computer are in practice always realized as two-dimensional subspaces inside a multi-dimensional system; the Hilbert space $\mathscr{H}_{QC}$ of the quantum computer then has a natural extension to

$$\mathscr{H}_{QC}^{\text{ext}} = \mathscr{H}_{QC} \oplus \mathscr{H}_{QC}^\perp , \qquad (5.51)$$

where the *leakage* space $\mathscr{H}_{QC}^\perp$ includes all states outside the two-dimensional qubit subspaces; in most settings, $\mathscr{H}_{QC}^\perp$ is a tensor product over leakage spaces corresponding to each qubit.

Now, the Hamiltonian that describes the noisy evolution of the quantum computer and the environment has the same general form as in Eq. (5.36),

$$H = H_{QC} + H_{QC}^\perp + H_E + H_{QCE}^{\text{ext}} , \qquad (5.52)$$

where $H_{QC}^\perp$ generates the evolution in the leakage space, and H_{QCE}^{ext} describes the interaction between the extended space of the quantum computer (the qubits and their leakage spaces) and the environment. If noise is local, H_{QCE}^{ext} has the same locality as H_{QC}, i.e.,

$$H_{QCE}^{\text{ext}} = \sum_j {}^j H_{QCE}^{\text{ext}}, \qquad (5.53)$$

where the support of ${}^j H_{QCE}^{\text{ext}}$ is contained in the union of the support of H_{QC}^j, the leakage space, and the environment. By repeating the same analysis as for local Hamiltonian noise, we find that local leakage noise satisfies Eq. (5.32) with

$$\varepsilon = t_0 \cdot \max_j ||{}^j H_{QCE}^{\text{ext}}||_\infty. \qquad (5.54)$$

5.3 Encoded quantum computation

When we desire to implement long computations – i.e., when the size L of the quantum circuit is large – an accuracy that decreases linearly with L as in Eqs. (5.16) and (5.34) is not satisfactory. Of course, we could achieve an accuracy independent of L if ε were a decreasing function of L, thus making $L\varepsilon$ a constant, but this is certainly not a physically reasonable assumption; we cannot hope that the hardware of the quantum computer will get less and less noisy the longer we keep quantum computing!

In order to obtain the results of a quantum computation with constant accuracy, some method for detecting and correcting the errors that are introduced by the noisy hardware is necessary; we say that such a method of computation is *fault tolerant*. The basic idea of fault-tolerant computation, whether classical or quantum, is the use of *redundancy*: Every hardware operation in the circuit to be implemented is replaced by a collection of several hardware operations that are designed to be more robust to local noise than a single hardware operation alone.

A formal method for introducing redundancy is via the use of error-correcting codes; see Chapter 2 for an introduction. For classical computation, the simplest example of a redundant encoding of information is based on the repetition code: "To protect information traveling from gate to gate, we replace each wire of the noiseless circuit by a *cable* of n wires (where n is chosen appropriately); each wire within the cable is supposed to carry the same bit of information, and we hope that a majority will carry this bit even if some of the wires fail."[10] To protect information during the execution of each gate, we also replace each gate in the noiseless circuit by an *organ* comprising several gates. The organ operates on the information carried by the wires inside the cables in the same way that the initial unencoded gate operated on the information carried by single wires; e.g., a NOT gate must be replaced by n NOT gates acting in parallel on every wire in a cable, as in Fig. 5.3, and similarly an AND gate must be replaced by n AND gates acting in parallel on corresponding pairs of wires in two cables. Organs also include a procedure for detecting and correcting faults in the noisy hardware; as discussed by von Neumann (see the references in Section 5.7), this may be implemented for each wire by copying the value of every bit to a larger number of k bits, randomly permuting all the resulting $n \cdot k$ bits, computing the majority function n times independently on these bits, and having the n outputs form the output wire.

---

[10] Quote from Gacs's paper "Reliable computation"; see [G05] and the references in Section 5.7.

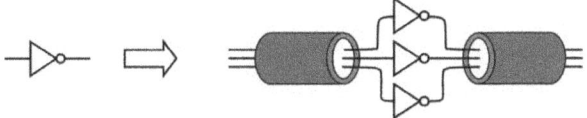

Fig. 5.3. On the left, a NOT gate is applied on the bit value carried in a single wire; e.g., if the NOT operation is implemented by a CMOS inverter, the input wire controls the gate voltage and the output is taken as the drain voltage. The input and output wires on the left are replaced on the right by cables each comprising n wires, and also the NOT gate on the left is replaced on the right by an organ comprising n parallel NOT gates (here, $n = 3$).

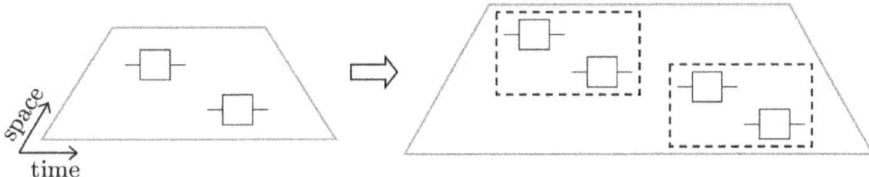

Fig. 5.4. Two elementary operations (single-qubit preparations, unitary gates, or measurements) in the noiseless quantum circuit on the left are replaced in the actual noisy quantum circuit on the right by two gadgets. For illustration, the elementary operations comprising each gadget are enclosed by dashed-line rectangles and only two elementary operations per gadget are shown.

Similar redundant encodings are also possible for quantum information where, for historical reasons, cables are now called *blocks* and organs are called *gadgets*.[11] Each qubit in the noiseless quantum circuit is replaced by a *block* of n encoded qubits; the joint state of the encoded qubits is supposed to carry the same quantum information as the state of the initial qubit, and we hope that this quantum information can be recovered even if faults occur on some of the encoded qubits. In addition, each elementary operation in the noiseless quantum circuit is replaced by a *gadget* that comprises several elementary operations acting on the encoded qubits in a block or across multiple blocks. A gadget is designed to operate on the quantum information carried by the encoded qubits in the same way that the initial unencoded operation acted on the quantum information carried by single qubits, and it also includes a procedure for detecting and correcting faults in the noisy hardware. The encoding of quantum circuits is shown schematically in Fig. 5.4.

But what do we mean when we say that the joint state of the encoded qubits carries the same quantum information as the state of the initial unencoded qubit? And how do we hope to recover the quantum information that is carried in a block if local noise acts on the encoded qubits? To answer the first question, consider a pure state $|\psi\rangle_{B_1 R}$ supported on $\mathcal{H}_{B_1} \otimes \mathcal{H}_R$, where $\mathcal{H}_{B_1}$ is the Hilbert space of a qubit B_1 and $\mathcal{H}_R$ is the Hilbert space of a *reference* system R. The encoded version of $|\psi\rangle_{B_1 R}$ is a pure state $|\psi\rangle_{BR}$ supported on $\mathcal{H}_n^0 \otimes \mathcal{H}_R$, where $\mathcal{H}_n^0$ is the code space of a block B of n encoded qubits $B_1, \ldots, B_n$. We say that the states $|\psi\rangle_{B_1 R}$ and $|\psi\rangle_{BR}$ carry the same quantum information because $|\psi\rangle_{BR}$ can be obtained from $|\psi\rangle_{B_1 R}$ by applying an isometry that maps $\mathcal{H}_{B_1}$ to $\mathcal{H}_n^0$; or in other words, there exists a unitary *decoding* unitary operator U_{dec} acting on the block such that

$$\mathrm{Tr}_{(RB_1)^\perp}(U_{\text{dec}} \otimes I_R)|\psi\rangle_{BR} = |\psi\rangle_{B_1 R}, \tag{5.55}$$

---

[11] The literature also uses the term *rectangle* instead of the term *organ* or *gadget*.

where I_R is the identity operator on the reference system and $\mathrm{Tr}_{(RB_1)^\perp}$ denotes a trace over everything else except for the reference system and the qubit B_1.

To answer the second question, we recall that the basic idea of quantum error correction is error discretization; see Section 5.1 and Chapter 2. To monitor the effects of noise, we partition the entire Hilbert space of the encoded qubits into mutually orthogonal subspaces and, if there is no noise, we demand that the support of $|\psi\rangle_{BR}$ coincides at all times with the code space $\mathcal{H}_n^0$ (and the reference system). In the presence of noise, our strategy is to detect periodically whether $|\psi\rangle_{BR}$ develops a nonzero overlap with any other subspace $\mathcal{H}_n^s$ labeled by a nontrivial syndrome s, in which case we apply a recovery operation that returns the support of $|\psi\rangle_{BR}$ to $\mathcal{H}_n^0$. Physically, distinguishing on which subspace the state $|\psi\rangle_{BR}$ is supported can be implemented by performing a generalized measurement $\mathcal{M}_{\{\mu\}}$ jointly on the encoded qubits, processing the measurement outcome μ to determine the syndrome s and hence the subspace $\mathcal{H}_n^s$ on which $|\psi\rangle_{BR}$ has been projected by the measurement,[12] and applying an operation $\mathcal{R}_s$ on the encoded qubits conditioned on the value of s that maps $\mathcal{H}_n^s$ to the code space $\mathcal{H}_n^0$.

We may revise Eq. (5.55) to include cases when the quantum information has been afflicted by noise. We now say that $|\psi\rangle_{B_1 R}$ and the noisy $|\tilde{\psi}\rangle_{BR}$ carry the same quantum information if

$$\mathrm{Tr}_{(RB_1)^\perp}\left(\mathcal{D}(|\tilde{\psi}\rangle_{BR})\right) = |\psi\rangle_{B_1 R} \tag{5.56}$$

with

$$\mathcal{D} = \mathcal{U}_{\mathrm{dec}} \circ \mathcal{R}_s \circ \mathcal{M}_{\{\mu\}} \otimes \mathcal{I}_R , \tag{5.57}$$

where $\mathcal{U}_{\mathrm{dec}}, \mathcal{I}_R$ are the physical operations corresponding to applying the unitaries U_{dec}, I_R, respectively, and we have suppressed the classical on-the-side computation that determines s given μ. It is noteworthy that the classical bits carrying the outcome μ of the syndrome measurement, which are traced over in Eq. (5.56), carry information about the subspace on which the quantum information was encoded prior to the decoding; for this reason, they are often referred to as *syndrome bits*.

The combined operation $\mathcal{D}$ is called a *decoder*. Because decoders output qubits that are unencoded and therefore unprotected from noise, we will *never* use a decoder in our actual noisy quantum circuits – encoded quantum information will never be decoded. A noiseless *ideal* decoder is, however, very useful as a tool for formalizing the requirement that a gadget operates on the encoded qubits in the same way that the operation that was replaced by the gadget acted on the initial unencoded qubits.

It is convenient to denote noiseless ideal operations using wavy boxes and noisy operations using square boxes; e.g., an *ideal* decoder is shown in Fig. 5.5. With this notation, (a) an operation $\mathcal{P}_{|\psi\rangle}$ that prepares the single-qubit pure state $|\psi\rangle$ in the noiseless quantum circuit is replaced in the noisy quantum circuit by a gadget implementing the operation $\mathcal{P}_{|\psi\rangle}^L$ on the encoded qubits where

$$\boxed{\mathcal{P}_{|\psi\rangle}^L} \; \vdots \; \mathcal{D} \;\rangle\!\!\!\sim \;\; = \;\; \mathcal{P}_{|\psi\rangle} \;\rangle\!\!\!\sim \;\; , \tag{5.58}$$

---

[12] This processing may be performed in a classical on-the-side computer.

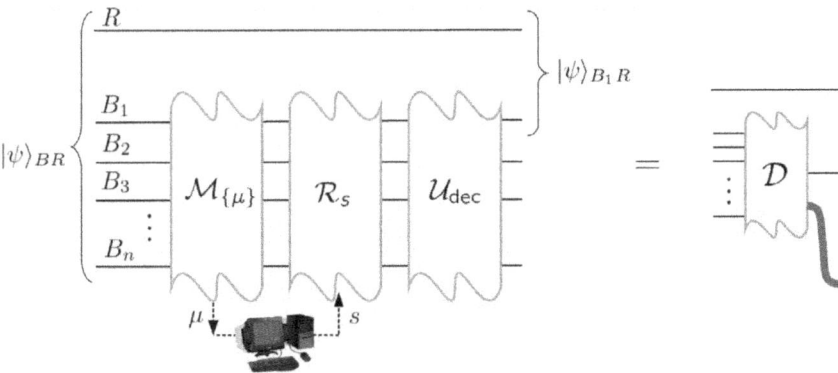

Fig. 5.5. A decoder $\mathscr{D}$ comprises a syndrome measurement $\mathscr{M}_{\{\mu\}}$, followed by a classical computation that outputs the syndrome s with input the measurement outcome μ, followed by a recovery operation $\mathscr{R}_s$ conditioned on the syndrome, followed by a decoding operation $\mathscr{U}_{\text{dec}}$ mapping the code space of the block of encoded qubits $B_1, B_2, \ldots, B_n$ to the Hilbert space of the qubit B_1. Our notation for $\mathscr{D}$ shows that the input consists of n encoded qubits and the output of one decoded qubit, qubit B_1. As denoted by the bold gray line, the decoder output also includes the qubits $B_2, \ldots, B_n$ and the *syndrome bits* that carry the outcome of the syndrome measurement; the state of these systems ends up being always a tensor product with the joint state of qubit B_1 and the reference R. When we speak of an *ideal* decoder, which is an imaginary noiseless operation, we draw $\mathscr{D}$ and the operations comprising it inside wavy boxes.

(b) an operation $\mathscr{U}$ that applies the single-qubit unitary U is replaced by a gadget implementing the operation $\mathscr{U}_L$ on the encoded qubits where

$$\mathscr{U}_L \; \mathscr{D} \;=\; \mathscr{D} \; \mathscr{U} \;, \tag{5.59}$$

(and similarly for multi-qubit unitary operations), and (c) an operation $\mathscr{M}_{\{a\}}$ that measures a single-qubit observable $\hat{a}$ with eigenvalues $\{a\}$ is replaced by a gadget implementing the operation $\mathscr{M}^L_{\{a\}}$ on the encoded qubits where

$$\mathscr{M}^L_{\{a\}} \;=\; \mathscr{D} \; \mathscr{M}_{\{a\}} \;. \tag{5.60}$$

Equations (5.58), (5.59), and (5.60) capture what we mean when we say that a gadget implementing the operation $\mathscr{O}_L$ *simulates* an unencoded operation $\mathscr{O}$.

But apart from simulating the desired operation on the encoded qubits, gadgets need also to contain a mechanism for detecting and correcting faults that may occur in the elementary operations comprising them. We can introduce such a mechanism by inserting an *error recovery* operation inside every gadget. Specifically, the gadget that implements $\mathscr{U}_L$ now becomes

$$\boxed{\mathcal{U}_L} \implies \boxed{\mathcal{U}_L}\boxed{\mathcal{ER}} \quad , \tag{5.61}$$

where $\mathscr{ER}$ comprises a syndrome measurement followed by a recovery operation,

$$\boxed{\mathcal{ER}} = \boxed{\mathcal{M}_{\{\mu\}}} \boxed{\mathcal{R}_s} \quad . \tag{5.62}$$

For the gadget that implements $\mathscr{P}^L_{|\psi\rangle}$, we insert similarly an error recovery operation following $\mathscr{P}^L_{|\psi\rangle}$, while the gadget that implements $\mathscr{M}^L_{\{a\}}$ is *not* modified since its output is a classical number, the measured eigenvalue a, which is protected from noise provided the classical computer that processes the measurement outcome is robust.

We do not have enough time and space to discuss explicit gadget constructions here, but the interested reader can find useful information in the references in Section 5.7. What is important at this abstract level is that each gadget simulates an operation in the noiseless quantum circuit, and that we hope that gadgets can be more robust to noise than unencoded operations, because faults inside each gadget can be detected and corrected by the error recovery operation.

5.3.1 Properties of noisy gadgets

To make progress, we need a method for formalizing the degree to which gadgets are protected from noise. We recall from Section 5.2.1 that if the noisy quantum circuit is afflicted by local noise, we can expand the noisy evolution perturbatively as a sum of fault paths; each fault path identifies a specific subset of all the elementary operations where faults have been inserted, while there are no fault insertions in all remaining elementary operations. The idea of fault-tolerant constructions is to ensure that gadgets operate reliably for all the fault paths with no more than a certain number $t > 1$ of insertions of faults inside them; the intuition is that each gadget will then be more robust to noise than any single elementary operation alone because the first contribution to a gadget's failure comes at order $t + 1$ of our perturbative fault-path expansion.

5.3.1.1 Good gadgets But what do we mean when we say that gadgets *operate reliably* for fault paths with at most t faults inside them? To formalize this requirement we need to consider each gadget *together* with the error recovery operations of the immediately preceding gadgets; we will refer to a gadget together with its preceding error recovery operations as an *extended gadget*. The idea is to construct gadgets such that (a) for each fault path with at most t insertions of faults inside a measurement extended gadget, the noisy gadget is equivalent to an ideal measurement of the gadget's ideally decoded input:

$$\mathcal{ER} \cdot \mathcal{M}^L_{\{a\}} \big|_{\leq t \text{ insertions}} = \mathcal{ER} \cdot \mathcal{D} - \mathcal{M}_{\{a\}} \quad ; \tag{5.63}$$

(b) for each fault path with at most t insertions of faults inside an extended gadget simulating a single-qubit unitary gate, the gadget is equivalent to applying the ideal unitary gate to the gadget's ideally decoded input:[13]

$$\mathcal{ER} \cdot \mathcal{U}_L \cdot \mathcal{ER} \cdot \mathcal{D} \big|_{\leq t \text{ insertions}} = \mathcal{ER} \cdot \mathcal{D} - \mathcal{U} - \quad ; \tag{5.64}$$

and (c) for each fault path with at most t insertions of faults inside a preparation gadget,[14] the gadget is equivalent to the ideal preparation:

$$\mathcal{P}^L_{|\psi\rangle} \cdot \mathcal{ER} \cdot \mathcal{D} \big|_{\leq t \text{ insertions}} = \mathcal{P}_{|\psi\rangle} \quad . \tag{5.65}$$

Here, we have illustrated insertions of faults as couplings between the noisy elementary operations inside the gadgets and the environment denoted by a thick dark gray line.

Figuratively, Eq. (5.63) allows us to *create* ideal decoders out of measurement extended gadgets that contain at most t faults, Eq. (5.64) allows us to *propagate* ideal decoders to the left through unitary-gate extended gadgets that contain at most t faults, and Eq. (5.65) allows us to *annihilate* ideal decoders in preparation gadgets that contain at most t faults. As ideal decoders are created in measurement gadgets, propagated through unitary-gate gadgets, and annihilated in preparation gadgets, the noisy encoded quantum circuit is transformed to a noiseless *unencoded* quantum circuit.

We conclude that for all the fault paths with at most t faults in each and every gadget, a noisy quantum computer is equivalent to a noiseless ideal quantum computer, in the sense that they both produce the same probability distribution for the final measurements that determine the computation result. However, for fault paths for which there are more than t faults inside one of the extended gadgets, the output probability distributions from the noisy and the noiseless quantum computers are not guaranteed to agree; thus it appears that the accuracy $1-\delta$ of a noisy *encoded* quantum circuit composed of fault-tolerant gadgets scales with

$$\delta \sim \varepsilon^{t+1} \, , \tag{5.66}$$

[13] And similarly for multi-qubit unitary gates where we include all preceding error recovery operations.
[14] The extended preparation gadget coincides with the preparation gadget since preparation gadgets have no input and, therefore, they are not preceded by any error recovery operations.

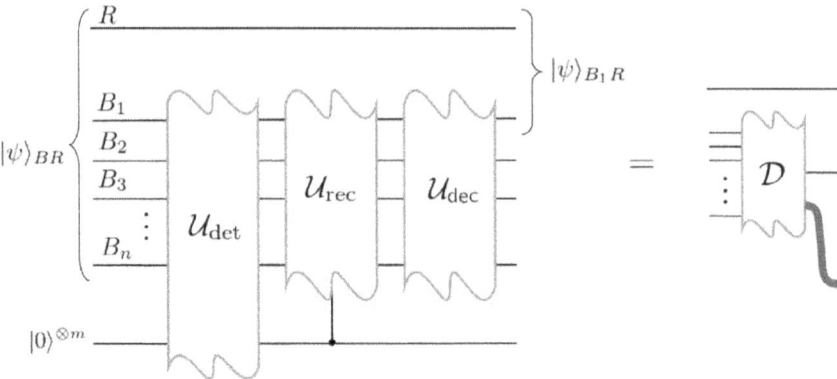

Fig. 5.6. A reversible ideal decoder, to be contrasted with the irreversible ideal decoder in Fig. 5.5. First, the syndrome measurement $\mathcal{M}_{\{\mu\}}$ is replaced by a unitary operation $\mathcal{U}_{\text{det}}$ that acts on a larger Hilbert space which includes ancillary *syndrome qubits*; this unitary maps coherently the state of the syndrome qubits to a state corresponding to the subspace s on which the input block is supported. Second, the recovery operation $\mathcal{R}_s$ is replaced by a controlled unitary operation $\mathcal{U}_{\text{rec}}$ that applies the appropriate recovery unitary on the block depending on the syndrome s carried by the syndrome qubits. In the figure, we assume there are m syndrome qubits and that each one is prepared in the state $|0\rangle$.

which should be compared with the scaling $\delta \sim \varepsilon$ for a noisy *unencoded* quantum circuit; cf., Eq. (5.16) for local Markovian noise and Eq. (5.34) for local non-Markovian noise. To prove that this scaling indeed holds and to determine the proportionality constant in Eq. (5.66) requires that we analyze what happens when a gadget is afflicted by more than t faults.

5.3.1.2 Bad gadgets Equations (5.63), (5.64), and (5.65) show that extended gadgets that contain at most t faults can be viewed as implementing ideal operations acting on unencoded qubits. But if more than t faults occur inside a gadget, there are no guarantees what may happen. Although we cannot in general say much about what *actually* happens unless we know details about the noise and how gadgets are constructed, we will be satisfied if we can show that extended gadgets that contain more than t faults can be viewed as implementing some *noisy* operations acting on unencoded qubits.

The basic tool we will need is a decomposition of the identity, i.e. do nothing, operation in terms of an ideal *decoder–encoder pair*:

$$\vdots \quad = \quad \mathcal{D} \quad \mathcal{D}^{-1} \vdots \quad , \tag{5.67}$$

where the operation $\mathcal{D}^{-1}$ is an *ideal encoder*. Of course, for ideal decoders to have an inverse, they need to be implemented by a reversible circuit. In fact, we can easily modify our definition of ideal decoders in Section 5.3 to achieve reversibility. Figure 5.6 shows such a reversible ideal decoder; an ideal encoder can then be implemented by simply executing the circuit in this figure backward in time.

Decoder–encoder pairs are trivial (they multiply to the identity operation) but they can be useful for understanding the properties of our noisy circuits if we insert them strategically at appropriate places. As a start, consider how to modify Eq. (5.64) in the case of more than t faults. The following property now holds: For each fault path with more than t insertions of faults inside an extended gadget simulating a single-qubit unitary gate, the gadget is equivalent to applying some noisy gate to the extended gadget's ideally decoded input:

$$\underbrace{\mathcal{ER}\;\mathcal{U}_L\;\mathcal{ER}\;\mathcal{D}}_{>t\text{ insertions}} \;=\; \mathcal{D}\;\mathcal{U} \;, \qquad (5.68)$$

where the noisy gate is

$$\mathcal{U} \;=\; \mathcal{D}^{-1}\;\mathcal{ER}\;\mathcal{U}_L\;\mathcal{ER}\;\mathcal{D} \;. \qquad (5.69)$$

There are two points about Eq. (5.68) deserving emphasis. First, this property replaces the entire extended gadget by a (noisy) gate rather than replacing the gadget alone; cf. Eq. (5.64) and see Section 5.3.1.3 below. Second, the noisy gate that replaces the extended gadget depends on both the faults inside the extended gadget and also on the syndrome bits (found inside the bold gray lines) that are input to the ideal encoder. As our notation in Eq. (5.69) is intended to illustrate, the syndrome bits can be viewed as a fictitious environment that operates together with the actual environment associated with the noise.

Similar properties to Eq. (5.68) can be derived for measurement extended gadgets and preparation gadgets; schematically,

$$\underbrace{\mathcal{ER}\;\mathcal{M}^L_{\{a\}}}_{>t\text{ insertions}} \;=\; \mathcal{D}\;\mathcal{M}_{\{a\}} \;, \qquad (5.70)$$

where the noisy measurement on the right-hand side can be obtained by inserting a decoder–encoder pair on the left-hand side, and

$$\underbrace{\mathcal{P}^L_{|\psi\rangle}\;\mathcal{ER}\;\mathcal{D}}_{>t\text{ insertions}} \;=\; \mathcal{P}_{|\psi\rangle} \;. \qquad (5.71)$$

5.3.1.3 Truncation

Figuratively, Eqs. (5.68), (5.70), and (5.71) allow us to create, to propagate to earlier times, and to annihilate ideal decoders in the case when extended gadgets are *bad*, containing more than t faults. These properties are therefore complementary to Eqs. (5.63), (5.64), and (5.65) that apply to *good* extended gadgets containing at most t faults.

Of course, one difference between the two cases is that in one case noisy unencoded operations appear on the right-hand side, while in the other case the unencoded operations are noiseless and ideal. But here we would like to discuss a second difference that was mentioned in the previous section: namely, while Eqs. (5.68), (5.70), and (5.71) replace the *entire* bad extended gadgets by some (noisy) unencoded operations, Eqs. (5.63), (5.64), and (5.65) only replace the *gadgets* contained inside the good extended gadgets by the ideal unencoded operations. This modification is important in order to prevent overcounting faults in successive, and therefore overlapping, bad extended gadgets.

To be concrete, imagine that we are to encode a quantum circuit comprising just a single-qubit preparation $\mathscr{P}_{|\psi\rangle}$ followed by a single-qubit measurement $\mathscr{M}_{\{a\}}$. The encoded quantum circuit comprises two overlapping extended gadgets:

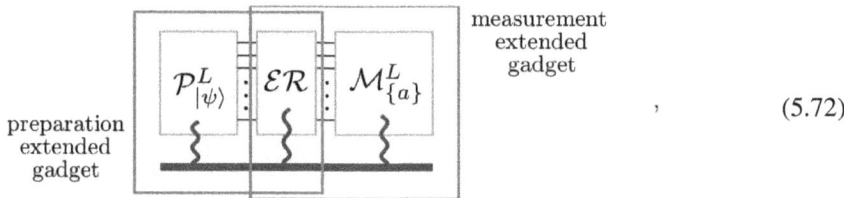

(5.72)

where the error recovery step is contained in both extended gadgets enclosed in the large square gray boxes. Now, consider a fault path with more than t insertions of faults in each extended gadget. Considering the two extended gadgets in isolation, we may think that the noisy encoded circuit equals some noisy preparation followed by some noisy measurement:

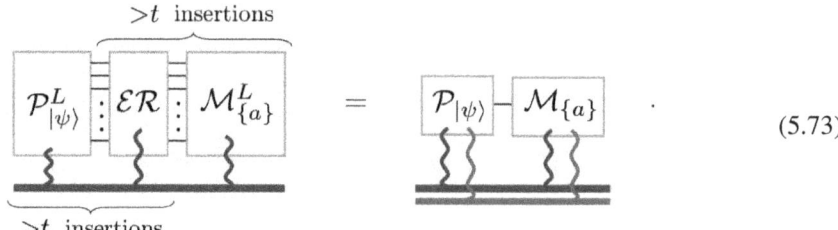

(5.73)

However, we soon realize that this may not always be a satisfying answer. Because the two extended gadgets overlap, it is possible that the *total* number N_f of faults in the given fault path is less than $2(t+1)$; thus the two noisy unencoded operations on the right-hand side appear at order ε^{N_f} in our perturbative fault path expansion, which is less than the order $\left(\varepsilon^{t+1}\right)^2$ we would expect based on the fact that each extended gadget by itself fails at order ε^{t+1}. Clearly, the problem is that our naive estimate $\left(\varepsilon^{t+1}\right)^2$ double counts each fault inside the error recovery step shared by the two overlapping extended gadgets.

Nevertheless, this complication is a red herring, and our naive estimate can actually be justified. To formally obtain Eq. (5.73), we first need to use Eq. (5.70), thereby replacing the entire measurement extended rectangle by a noisy unencoded measurement:

$$\mathcal{P}^L_{|\psi\rangle} \; \mathcal{ER} \; \mathcal{M}^L_{\{a\}} \underbrace{}_{>t \text{ insertions}} = \mathcal{P}^L_{|\psi\rangle} \; \mathcal{D} \text{---} \mathcal{M}_{\{a\}} \quad , \tag{5.74}$$

where we observe that the error recovery step has been removed from the preparation extended gadget – we say that the gadget has been *truncated*. To annihilate the ideal decoder at the next step, we need to consider how many faults are contained inside the *truncated* preparation gadget. If there are at most t faults then we can use Eq. (5.65) to obtain

$$\mathcal{P}^L_{|\psi\rangle} \; \mathcal{ER} \; \mathcal{M}^L_{\{a\}} \underbrace{}_{\leq t} \underbrace{}_{>t \text{ insertions}} = \mathcal{P}_{|\psi\rangle} \text{---} \mathcal{M}_{\{a\}} \quad , \tag{5.75}$$

whereas, if there are more than t faults, we can use Eq. (5.71) to obtain Eq. (5.73). Since in this second step what matters is the number of faults in the truncated preparation gadget, there is no double counting of faults; thus the two noisy unencoded operations on the right-hand side of Eq. (5.73) do appear at order $\left(\varepsilon^{t+1}\right)^2$ in our perturbative fault path expansion, as desired.

We need to note that Eqs. (5.63), (5.64), and (5.65) for good extended gadgets and Eqs. (5.68), (5.70), and (5.71) for bad extended gadgets, which have been formulated for the full extended gadgets, apply in the same way to truncated extended gadgets. Indeed, the ideal decoders contain a noiseless ideal error recovery step – cf. Fig. 5.5 – which can be used to replace the truncated noisy error recovery steps, thereby reassembling the full gadgets for which the properties apply.

Although we have illustrated the concept of truncation with a simple example, a similar truncation procedure can be used for any fault path that leads to more than two successive bad extended gadgets, including gadgets that simulate unencoded operations on more than one qubit: Starting from the latest bad extended gadgets that are not succeeded by any other bad extended gadgets, we progressively move to all earlier bad extended gadgets, one gadget at a time. At each step, we truncate the bad extended gadget under consideration from the error recovery steps it shares with all its succeeding bad extended gadgets (which may themselves be truncated or not), and we label the truncated gadget as good or bad depending on the number of faults it contains after the truncation (if it contains at most t faults, it is declared good, otherwise it is declared bad). Eventually we reach the earliest bad extended gadgets, and we truncate the good extended gadgets that preceded them.[15] In the end, the successive bad extended gadgets are divided into *nonoverlapping* truncated extended gadgets, which have been declared good or bad depending on the number of faults they contain and which can be replaced by either noisy or ideal unencoded operations respectively by using the corresponding gadget properties.

---

[15] These extended gadgets remain good after the truncation as they were already good prior to it.

5.4 Coarse-grained noise and level reduction

We have now assembled all the properties we need to characterize noisy gadgets. If we combine all the pieces together, we can arrive at a helpful description of the noise acting on the encoded quantum computer.

We recall from Section 5.2.1 that, without encoding, the only fault path leading to the noiseless ideal evolution is the trivial fault path that contains absolutely no faults; we are then forced to decompose the noisy evolution into an ideal and a faulty part as in Eq. (5.11) for local Markovian noise and Eq. (5.28) for local non-Markovian noise. In contrast, in an encoded quantum computation, many more fault paths lead to the noiseless ideal evolution; now, for local Markovian noise, we can decompose the noisy evolution as

$$\rho^{\text{noisy}} = \rho^{\text{good}} + \zeta^{\text{bad}} , \tag{5.76}$$

and similarly for local non-Markovian noise,

$$|\psi\rangle^{\text{noisy}} = |\psi\rangle^{\text{good}} + |\vartheta\rangle^{\text{bad}} , \tag{5.77}$$

where the unnormalized density matrix ρ^{good} and the unnormalized pure state $|\psi\rangle^{\text{good}}$ are sums of all the fault paths with at most t faults in each and every extended gadget, while ζ^{bad} and $|\vartheta\rangle^{\text{bad}}$ are sum of all the remaining fault paths.

In what sense are the fault paths included in ζ^{good} and $|\vartheta\rangle^{\text{good}}$ good? For each (*good*) fault path with at most t faults in each and every extended gadget, Eqs. (5.63), (5.64), and (5.65) apply. We now consider the entire encoded quantum circuit and, by using these properties, we first create ideal decoders in the measurement gadgets. Then, we propagate the ideal decoders to earlier times through unitary-gate gadgets. Finally, we annihilate the ideal decoders in preparation gadgets. As the ideal decoders appear, move to earlier times, and finally disappear, the entire encoded quantum circuit afflicted by the given good fault path is shown to be formally equal to the ideal quantum circuit that the gadgets simulate. Since this is true for every good fault path separately, by linearity it is also true for the sum of all of them; thus ζ^{good} and $|\vartheta\rangle^{\text{good}}$ lead (after normalization) to the same probability distribution for the final computation result as a noiseless ideal quantum computer would.

We can now estimate the accuracy $1 - \delta$ of the encoded quantum computation. For local Markovian noise, δ can be bounded as in Eq. (5.13) by the norm of the difference of the final noisy superoperator minus its (normalized) good part:

$$\delta \leq \left\| \rho^{\text{noisy}} - \frac{\rho^{\text{good}}}{\text{Tr}(\rho^{\text{good}})} \right\|_1 \leq \left(1 + \frac{1}{1 - \|\zeta^{\text{bad}}\|_1}\right) \|\zeta^{\text{bad}}\|_1 , \tag{5.78}$$

and we have used the triangle inequality multiple times.[16] Similarly, for local non-Markovian noise, δ can be bounded as in Eq. (5.29) by the norm of the difference of the final noisy pure quantum state minus its (normalized) good part:

$$\delta \leq 2 \left\| |\psi\rangle^{\text{noisy}} - \frac{|\psi\rangle^{\text{good}}}{\||\psi\rangle^{\text{good}}\|} \right\| \leq 2 \left(1 + \frac{1}{1 - \||\vartheta\rangle^{\text{bad}}\|}\right) \||\vartheta\rangle^{\text{bad}}\| . \tag{5.79}$$

It remains to obtain upper bounds on ζ^{bad} and $|\vartheta\rangle^{\text{bad}}$ that are sums of all (*bad*) fault paths with more than t faults in at least one gadget. For each bad fault path, the noisy encoded quantum

---

[16] We have $\delta \leq \|\alpha \rho^{\text{good}} + \zeta^{\text{bad}}\|_1 \leq |\alpha| + \|\zeta^{\text{bad}}\|_1$, where $|\alpha| = \text{Tr}^{-1}(\rho^{\text{good}}) - 1 \leq \|\zeta^{\text{bad}}\|_1 / (1 - \|\zeta^{\text{bad}}\|_1)$.

circuit can be analyzed by using the gadget properties. We first consider whether each extended gadget contains at most t faults or more than t faults, declaring the former gadgets good and the latter bad. If multiple successive extended gadgets are declared bad, we use the truncation procedure described in Section (5.3.1.3) to divide them into non-overlapping truncated extended gadgets that are good or bad depending on the number of faults they contain. Eventually, we use Eqs. (5.68), (5.70), and (5.71) for the good extended gadgets (truncated or not) and Eqs. (5.63), (5.64), and (5.65) for the bad extended gadgets (also, truncated or not). Every good extended gadget is thereby replaced by the noiseless ideal operation the gadget simulates, every bad extended gadget is replaced by some noisy operation, and thus the noisy encoded quantum circuit as a whole is replaced by a noisy unencoded quantum circuit.

If we now let $C^{(1)}$ denote the set of all L extended gadgets in the encoded quantum circuit, then by analogy with Eq. (5.14) we may write

$$\zeta^{\text{bad}} = \sum_{r=1}^{L} (-1)^{r-1} \sum_{C_r^{(1)} \subseteq C^{(1)}} \zeta(C_r^{(1)}) , \qquad (5.80)$$

where the second sum is over all subsets $C_r^{(1)}$ of $C^{(1)}$ of cardinality r, and $\zeta(C_r^{(1)})$ denotes a sum of all the fault paths for which all the extended gadgets in $C_r^{(1)}$ are declared bad.[17]

To gain intuition about how to proceed, consider the simplest case $r=1$ when $\zeta(C_1^{(1)})$ is a sum of all the fault paths for which the single extended gadget in a *specific* set $C_1^{(1)}$ is bad. Since for an extended gadget (truncated or not) to be bad it needs to contain more than t faults, we can generalize Eq. (5.14) to obtain

$$\zeta(C_1^{(1)}) = \sum_{s=t+1}^{L_0} (-1)^{s-t-1} \binom{s-1}{t} \sum_{C_s} \zeta(C_s) , \qquad (5.81)$$

where L_0 is the number of elementary operations in the extended gadget in $C_1^{(1)}$, the second sum is over all subsets C_s of s of these operations, and $\zeta(C_s)$ is a sum of all the fault paths with faults applied on all operations in C_s.[18]

For the general case of r bad extended gadgets, it suffices to perform a similar inclusion-exclusion analysis independently in each gadget:

$$\zeta(C_r^{(1)}) = \prod_{j=1}^{r} \left(\sum_{s_j=t+1}^{L_0} (-1)^{s_j-t-1} \binom{s_j-1}{t} \sum_{C_{s_j} \subseteq C_r^{(1)}(j)} \zeta(C_{s_j}) \right) , \qquad (5.82)$$

---

[17] Whether each extended gadget in $C_r^{(1)}$ is truncated depends on the fault path; however, for each specific fault path, we can first use the truncation procedure to decide which extended gadgets need to be truncated, and then we can unambiguously declare every extended gadget (truncated or not) as being either good or bad.

[18] We first sum all $\zeta(C_{t+1})$ accounting correctly for all the fault paths with exactly $t+1$ faults; however, all the fault paths with exactly $t+2$ faults are overcounted $\binom{t+2}{t+1} - 1 = \binom{t+1}{t}$ times. So, next, we subtract the sum of all $\zeta(C_{t+2})$ multiplied by $\binom{t+1}{t}$, but in doing so we undercount all the fault paths with exactly $t+3$ faults $\binom{t+3}{t+1} - \binom{t+1}{t}\binom{t+3}{t+2} - 1 = -\binom{t+3-1}{t}$ times. And so on.

where L_0 now denotes the number of elementary operations in the *largest* extended gadget,[19] $C_r^{(1)}(j)$ denotes the set of elementary operations in the jth bad extended gadget in $C_r^{(1)}$, and C_{s_j} denotes a subset of s_j of the elementary operations in $C_r^{(1)}(j)$.

But $\|\zeta(C_{s_j})\|_1 \leq \varepsilon^{s_j}$ by the definition of local Markovian noise, and we find

$$\|\zeta(C_r^{(1)})\|_1 \leq \prod_{j=1}^{r} \sum_{s_j=t+1}^{L_0} \binom{s_j-1}{t}\binom{L_0}{s_j}\varepsilon^{s_j}$$

$$\leq \left(\binom{L_0}{t+1}\varepsilon^{t+1} \sum_{\omega=0}^{\infty} \frac{(L_0-t-1)^\omega \varepsilon^\omega}{\omega!}\right)^r \leq \left(\varepsilon^{(1)}\right)^r , \quad (5.83)$$

with

$$\varepsilon^{(1)} = \xi \binom{L_0}{t+1}\varepsilon^{t+1} \quad (5.84)$$

for some constant $\xi \geq e^{(L_0-t-1)\varepsilon}$ (as typically we are interested in small values $\varepsilon \leq 1/(L_0 - t - 1)$, we may take ξ to be e). By replacing ζ with $|\vartheta\rangle$, we may repeat a similar calculation for local non-Markovian noise to obtain

$$\||\vartheta(C_r^{(1)})\rangle\| \leq \left(\varepsilon^{(1)}\right)^r , \quad (5.85)$$

with $\varepsilon^{(1)}$ again as in Eq. (5.84).

Equations (5.83) and (5.85) tell us that if we choose any r extended gadgets, then the sum of all the fault paths for which all of them are bad has norm that is exponentially suppressed with r. This is exactly the condition we imposed for noise to be local, except that in this case we think of noise as afflicting the gadgets themselves instead of the elementary operations. While the strength of the noise acting on the elementary operations is ε, the strength of the coarse-grained noise acting on the gadgets is $\varepsilon^{(1)}$, which scales as ε^{t+1} because each gadget can tolerate up to t faults. We often refer to the encoded computation executed by the gadgets as a *level-1* simulation of an unencoded *level-0* quantum circuit; in this language, what we have shown is that a noisy level-1 simulation afflicted by local noise with strength ε can be viewed as a *level reduced* noisy level-0 simulation afflicted by a coarse-grained local noise with renormalized strength $\varepsilon^{(1)}$. Figure 5.7 illustrates this coarse-graining level reduction procedure.

The accuracy of an encoded computation afflicted by local Markovian noise can now be determined from Eq. (5.78) by combining Eqs. (5.80) and (5.83); we find

$$\delta \approx \|\zeta^{\text{bad}}\|_1 \leq \sum_{r=1}^{L} \binom{L}{r}\left(\varepsilon^{(1)}\right)^r \leq (e-1)L\varepsilon^{(1)} , \quad (5.86)$$

where, in the first step, we have kept only the leading order contribution. By comparing with Eq. (5.16), which corresponds to the case no encoding is used, we conclude that if $\varepsilon^{(1)} < \varepsilon$ then the encoding is in fact a good idea since it improves the accuracy $1 - \delta$ of the final computation result. The same conclusion also holds for local non-Markovian noise.

---

[19] The number of elementary operations may vary among gadgets. In addition, some gadgets may be truncated depending on the fault path. By taking L_0 to correspond to the largest extended gadget, we thus unavoidably include in the sum some extra fault paths that should not be counted. However, since eventually we will take the norm of both sides and use the triangle inequality, including these additional fault paths merely weakens our bounds.

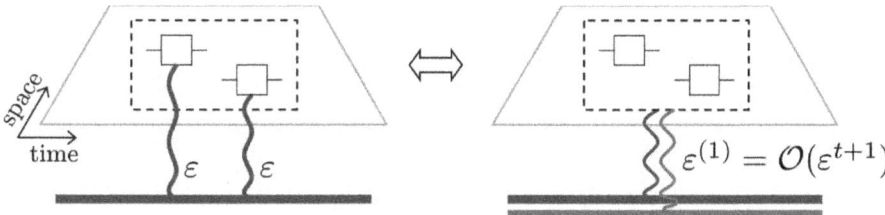

Fig. 5.7. The encoded operations in an encoded quantum circuit are executed by using gadgets (here, one gadget is shown, along with two of the elementary operations it contains); the elementary operations inside each gadget are afflicted by local noise (either Markovian or non-Markovian) of strength ε. The physical noise can be coarse-grained into a local effective noise with renormalized strength $\varepsilon^{(1)}$ of order ε^{t+1} acting on the gadgets themselves. The noise coarse-graining allows us to concentrate on the gadgets alone and forget about the elementary operations inside them; we say that the encoded quantum circuit is *level reduced* to an equivalent unencoded quantum circuit, where the effect of coding is to map the physical noise strength ε to the effective noise strength $\varepsilon^{(1)}$.

5.5 The quantum accuracy threshold

We are one breath away from the central result in the theory of quantum fault tolerance: the threshold theorem. You must have guessed the next step... If an encoded quantum circuit is more accurate than a quantum circuit that is not encoded, then why not apply the encoding to the encoded circuit itself, taking every elementary operation inside it and replacing it by a gadget; this doubly encoded quantum circuit should be even more accurate. In fact, why stop here? If we continue recursively re-encoding our encoded circuits, we expect their accuracy to steadily increase, reaching any limit we please.

To formalize this idea, let us consider the recursive construction of these multiply encoded quantum circuits we imagined above. At the base of our construction is the unencoded quantum circuit corresponding to our quantum algorithm; we say that this is our level-0 circuit in the sense that it does not use any coding. The next step is to replace every elementary operation in the level-0 circuit by the corresponding gadget; we say that this is our level-1 circuit, performing a level-1 simulation of the level-0 circuit. Instead of physically implementing the level-1 circuit as is, we may next replace every elementary operation in the level-1 circuit by the corresponding gadget to obtain our level-2 circuit, and so on. Figure 5.8 illustrates this replacement procedure repeated k times; the final encoded quantum circuit, which is the one we *do* physically implement, performs a level-k simulation of the level-0 circuit.

The question is: what is the accuracy of the level-k circuit as a function of k? Estimating this accuracy is actually especially easy if we use the noise coarse-graining concept from Section 5.4. The noise afflicting the elementary operations in the level-k circuit can be coarse-grained to give an effective noise that acts on the gadgets; if the physical noise is local and has strength ε, the coarse-grained noise is also local and has renormalized strength $\varepsilon^{(1)}$ as in Eq. (5.84). The noise coarse-graining level reduces the level-k circuit to an equivalent level-$(k-1)$ circuit; this level-$(k-1)$ circuit produces the same probability distribution for the computation outcome as the initial level-k circuit, but it is afflicted by an effective local noise of strength $\varepsilon^{(1)}$. Thus we have reduced the problem of estimating the accuracy of the level-k circuit (the circuit we actually

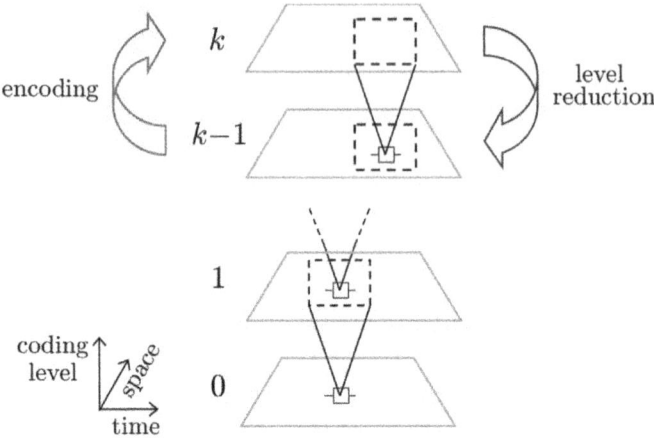

Fig. 5.8. A recursive construction of a multiply encoded quantum circuit. At the base, the unencoded level-0 circuit corresponds to our quantum algorithm. One level higher, the encoded level-1 circuit is obtained by simulating every elementary operation in the level-0 circuit by using a gadget. By repeating this replacement rule, we eventually obtain an encoded level-k circuit which is the circuit we physically implement. To understand the effect of local noise on the level-k circuit, we can level reduce it to a level-$(k-1)$ circuit which is acted by noise of renormalized strength. If level reduction is repeated k times, the level-k circuit can be level reduced to an unencoded level-0 circuit; we can view the strength of noise acting on this level-0 circuit as the *effective* noise strength acting on the multiply encoded operations in the level-k circuit.

physically implement) to estimating the accuracy of the level-reduced level-$(k-1)$ circuit (which is, of course, imaginary as it represents the result of our noise coarse-graining procedure).

We can next coarse-grain the noise in the level-$(k-1)$ circuit, thereby level reducing the initial level-k circuit to a level-$(k-2)$ circuit afflicted by local noise of strength $\varepsilon^{(2)}$; because of the self-similarity of our recursive circuit construction, the map from $\varepsilon^{(1)}$ to $\varepsilon^{(2)}$ is the same as from ε to $\varepsilon^{(1)}$. The noise in the level-$(k-2)$ circuit can in turn be coarse-grained, thereby level reducing the initial level-k circuit to a level-$(k-3)$ circuit. And so on, where at the l-th coarse-graining step the input noise strength $\varepsilon^{(l-1)}$ is renormalized to an output strength

$$\varepsilon^{(l)} \leq \xi \binom{L_0}{t+1} \left(\varepsilon^{(l-1)}\right)^{t+1} . \qquad (5.87)$$

After k coarse-graining steps, the initial level-k circuit is eventually level reduced to an *unencoded* level-0 circuit afflicted by local noise of strength $\varepsilon^{(k)}$; this level-0 circuit corresponds to the quantum algorithm that is simulated by the initial encoded level-k circuit, since every level reduction takes us down one level in the ladder in Fig. (5.8). If we use the recursion Eqs. (5.87), where $\varepsilon^{(0)} = \varepsilon$ is the physical noise strength, we find

$$\varepsilon^{(k)} \leq \varepsilon_0 \left(\frac{\varepsilon}{\varepsilon_0}\right)^{(t+1)^k} , \text{ for a constant } \varepsilon_0 = \left(\xi \binom{L_0}{t+1}\right)^{-1/t} . \qquad (5.88)$$

The constant ε_0 is the critical noise strength below which the recursive encoding scheme we have described is effective; if $\varepsilon < \varepsilon_0$ then $\varepsilon^{(k)}$ decreases double exponentially with the coding level k. The critical noise strength ε_0 is often referred to as the *threshold* for fault-tolerant quantum computation.

The k successive level reduction steps tell us that we can view the encoded operations in the level-k circuit as being afflicted by an effective local noise of strength $\varepsilon^{(k)}$. Thus, the accuracy $1 - \delta$ of the level-k circuit can be estimated as in Section 5.4 where we estimated the accuracy of a level-1 circuit afflicted by local noise of strength $\varepsilon^{(1)}$. For local Markovian noise, we now have

$$\delta \lesssim (e-1)L\varepsilon^{(k)} , \qquad (5.89)$$

where we used Eq. (5.86) with $\varepsilon^{(1)}$ replaced by $\varepsilon^{(k)}$. If the physical noise has a strength below the threshold, $\varepsilon < \varepsilon_0$, then δ can become as small as desired by recursively re-encoding the level-0 circuit of our quantum algorithm sufficiently many times k. The same conclusion also holds for local non-Markovian noise. This is the accuracy threshold theorem.

In particular, imagine that we desire to obtain the computation output with an accuracy $1 - \delta \geq 1 - \delta_0$ for some constant (error) δ_0, independent of the size L of the quantum algorithm. We can arrange to have $\delta \leq \delta_0$ by choosing k so that

$$(t+1)^k \leq \frac{\log\left((e-1)L\varepsilon_0/\delta_0\right)}{\log\left(\varepsilon_0/\varepsilon\right)} . \qquad (5.90)$$

Because of the recursiveness of our encoding construction, each of the L encoded operations in the level-k circuit can be implemented by using at most $(L_0)^k$ elementary operations, where L_0 is the number of elementary operations in the largest gadget. The ratio then of the number L^* of elementary operations in the entire level-k circuit over the number L of elementary operations in the quantum algorithm scales as

$$\frac{L^*}{L} \leq (L_0)^k = O\left(\log L\right)^a , \text{ with } a = \frac{\log L_0}{\log(t+1)} . \qquad (5.91)$$

Thus, not only does the level-k circuit achieve the desired accuracy $1 - \delta_0$, but it does so very efficiently; the level-k circuit is only larger than the unencoded level-0 circuit by a polynomial in the *logarithm* of the size L of the quantum algorithm.

5.6 Assessment

The idea in the previous section was to establish the existence of a critical noise strength by considering a specific fault-tolerance scheme; in particular, we chose to study the recursive scheme illustrated in Fig. 5.8 because its self-similar nature greatly simplified our analysis. Although recursive schemes are easier to analyze, it is clearly possible that they are not optimal from a practical point of view, and other more complex schemes may have higher thresholds and/or more favorable overhead costs. Proposing and analyzing improved schemes for fault-tolerant quantum computation is a major focus of current research; see the references in Section 5.7.

The existence of a critical noise strength is significant because it implies that the quest to build a reliable quantum computer is not a mere fantasy, but it is based on firm foundations. We have learned that if we find a physical setting allowing us to experimentally implement quantum circuits with local noise of strength ε_0 or less, then the noisy operations can be assembled efficiently to perform an encoded quantum computation and obtain the computation result to as high an accuracy as desired.

Although this knowledge gives us confidence and encouragement to research further how quantum computers can be constructed, it is possible that the outcome of this endeavor may ultimately be failure. We can contemplate several possibilities under which such a failure might occur. First, it is possible that the entire concept of what it means to quantum compute, a concept that is based on the laws of conventional quantum mechanics, is flawed when applied to quantum computers with either a very large number of qubits or very long running times; clearly, in all our considerations we have assumed (as the majority of physicists currently believe) that the framework of quantum mechanics can be extrapolated without change to the (long) time and (large) length scales relevant for quantum computers implementing useful computations. Perhaps there are fundamental, as yet unknown, principles that prevent the realization of the highly entangled multi-particle quantum states required to implement useful quantum algorithms. In this sense, the project of quantum computing can be seen in a different light, the light of testing quantum mechanics in new regions of the parameter space; even if nothing useful as regards computation comes out, we may uncover puzzles forcing us to revise our approach to quantum mechanics and physics in general.

A second possibility for failure relates to the conditions we imposed on the noise as we formulated our theoretical analysis. It is possible that, as we design and test quantum computing devices of increasing complexity, we will eventually find that the physical noise is not captured by our local noise models or that, even if noise is local, its strength cannot be upper bounded by a small constant number; see Chapter 26 for an assessment of these issues. Ultimately, the question of whether methods of quantum fault tolerance can in practice be as effective as our theoretical analysis indicates will be decided by the progress of the future experiments. In the meantime, theoretical research still has ample room for further progress. Fruitful new research can attempt to relax the requirements under which reliable quantum computation can be provably shown to be possible; e.g., one may consider more precise models for the noise during qubit preparation and measurement, one may specialize to noise models that more closely describe the particular characteristics of observed decoherence in modern prototype experimental devices, etc.

At present, we have no evidence either that quantum mechanics is violated at the length and time scales relevant for long useful quantum computations, or that the physical noise in prospective implementations of quantum computation has features that prevent quantum fault tolerance from working. Certainly, there are formidable technical difficulties for building a large-scale quantum computer with present technology, and it is possible that the engineering requirements may prove too challenging to overcome for a long time in the future. Nevertheless, experimental efforts during the past decade have shown great progress, and there is a great sense of optimism among experimentalists that this progress will continue even more rapidly as they gain more insight and intuition about their systems.

5.7 History and further reading

Computer engineering is the art and science of translating user requirements we do not fully understand; into hardware and software we cannot precisely analyze; to operate in environments we cannot accurately predict; all in such a way that the society at large is given no reason to suspect the extent of our ignorance.

Adapted from Ralph Caplan's *By Design: Why There Are No Locks on the Bathroom Doors in the Hotel Louis XIV and Other Object Lessons*, New York: Fairchild Books, 2004, p. 229.

In the hope of making the flow of thought in this chapter as smooth as possible, we have avoided interruptions to discuss the history of the subject of quantum fault tolerance and we have also omitted discussing a number of technical but important details. This final section provides some of this historical context and references to published works – most of which are available freely on the internet – where further information can be obtained. Of course, knowing that our historical account and our list of references cannot be perfectly complete, our aspiration is not to provide an exhaustive list of all relevant publications but rather to guide the interested reader in his/her first steps in the large bibliography.

The question whether logical operations can be implemented fault tolerantly despite noise was central from the early days of the development of classical computing. Shannon's master's thesis [S40] laid the foundations of digital circuit design, and von Neumann's analysis of noisy cellular automata [N55, N66] showed how unreliable components can be assembled to implement reliable computations. A more recent exposition of methods for reliable classical computation can be found in Gacs' work [G05]; see also Gacs' works [G83, G01a] on noisy cellular automata and Gray's guide [G01b] on [G83].

The corresponding study for quantum computing was pioneered by Shor [S96b], who described the first gadget constructions for universal quantum computation. Soon after, the existence of a critical noise strength based on a recursive scheme as in Fig. 5.8 was discussed by Aharonov and Ben-Or [AB.-O08], by Kitaev [K97a, K97b], and by Knill, Laflamme, and Zurek [KLZ98b]. All these works considered local Markovian noise and made a series of additional assumptions about the experimental quantum computing devices. Most notably, one assumes that one can supply fresh ancillary qubits or refresh existing qubits at any point in time during the noisy quantum computation (this is a necessary assumption; if it is dropped, there is no critical noise strength [AB.-OI+96]). In addition, one assumes that there is maximum parallelism, i.e., it is possible to apply gates in parallel on disjoint sets of qubits (also a necessary assumption [P96]). Finally, one assumes that multi-qubit gates can be applied between any set of qubits irrespective of their geometric distance (this is not a necessary assumption; a critical noise strength exists even when geometric constraints are taken into account [G00]).

A new proof for the existence of a critical noise strength was described by Aliferis, Gottesman, and Preskill [AGP06]; see also Aliferis' doctoral thesis [A07, G09]. This proof is significantly simpler than earlier proofs, and it applies to both Markovian and non-Markovian local noise (the analysis for non-Markovian local noise extends prior results by Terhal and Burkard [TB05]). Building on this new proof, Aharonov, Kitaev, and Preskill [AKP06] later analyzed long-range static noise, Aliferis and Terhal [AT07] analyzed leakage noise, Aliferis and Preskill [AP08] analyzed biased noise with dephasing being much more dominant than relaxation or leakage, and Ng and Preskill [NP09] analyzed Gaussian noise.

The value of the critical noise strength has been estimated both analytically by means of combinatorial analyses and also, for simple probabilistic local noise models, by performing numerical simulations in a classical computer. The highest numerical estimates to date (of order 1.0×10^{-2}) have been obtained for Knill's postselection and Fibonacci schemes [K05b] and for Raussendorf, Harrington, and Goyal's scheme based on surface codes [RHG07]. The highest analytical estimates to date (of order 1.0×10^{-3}) have been obtained by Reichardt [R06], by Aliferis, Gottesman, and Preskill [AGP08], and by Aliferis and Preskill [AP09], all by analyzing Knill's schemes and modifications of them. It is interesting to note that the schemes with the highest known critical noise strengths share two features: First, they make use of quantum teleportation [BBC+93] for implementing quantum error correction [K05b] and for simulating certain gates [GC99]. Second, they use a method by Bravyi and Kitaev [BK05] for *distilling* high-accuracy copies of certain ancillary quantum states out of noisier copies of the same states.

Part II

Generalized approaches to quantum error correction

Chapter 6

Operator quantum error correction

David Kribs and David Poulin

6.1 Introduction

In this chapter we review the basic theory of operator quantum error correction (OQEC) and present a selection of applications and examples.

OQEC [KLP05, KLP+06] began with an attempt to bring passive techniques for quantum error correction (QEC), such as DFS and NS, together with standard active QEC (the subjects of Chapters 3 and 2, respectively). Broadly put, this meshes two distinct disciplinary approaches to QEC rooted in physics (DFS/NS) and computer science. Technically, the approach involves a generalization of encoding quantum information into *subsystems*; that is, a system B (or A) that arises as a tensor factor $\mathscr{C} = A \otimes B$ of a subspace inside a (potentially) larger system Hilbert space $\mathscr{H} = \mathscr{C} \oplus \mathscr{C}^\perp$. Hence, in one sense OQEC may be regarded as a formalization of the *subsystem principle* for encoding quantum information, first elucidated by Knill, Laflamme, and Viola [KLV00]. In recognition of this point, often such codes B are referred to as *subsystem codes*.

The resulting theory has led to significant advances in QEC. Conceptually, the reasons for this include a simplification of recovery procedures, as subsystem codes can be more degenerate than their subspace counterpart, and a simplification of the syndrome measurement that can result. These features of subsystem codes have, for instance, led to threshold improvements in fault-tolerant quantum computing. Additionally, OQEC gives a clean mathematical setting that more readily allows for the application of mathematical tools from operator theory and operator algebras to problems in QEC.

6.1.1 Error avoidance versus error correction

As the terminology suggests, passive QEC, as manifested by DFS and NS, involves finding and then encoding quantum information into sectors of the system Hilbert space that remain

---

Quantum Error Correction, ed. Daniel A. Lidar and Todd A. Brun. Published by Cambridge University Press.
© Cambridge University Press 2013.

immune to the overall noise of a quantum map. As we have seen in Chapter 3, it is easy to find simple mathematical examples of DFS and NS; some Pauli noise models yield such codes for instance. Typically passive codes arise physically when the noise is constrained by a symmetry. An important example of this phenomenon arises in the noise model that describes photons of light traveling through an optical fiber, wherein the photon polarizations are exposed to correlated rotation errors. In such "collective noise" channels, the errors are the same on each photon. This permutation symmetry protects certain sectors of the Hilbert space from decoherence.

Thus, NS and DFS are a form of error avoidance, in contrast to standard active error correction (QEC). OQEC formalizes the distinction between the two settings, while at the same time allowing us to see them as special cases of a general framework. Let us begin with the definition of passive codes in OQEC. Our standing assumption is that we have a subspace $\mathscr{C}$ of a larger Hilbert space $\mathscr{H}$, which has a factorization $\mathscr{C} = A \otimes B$.

Definition 6.1. *A subsystem code B is a* passive code *for a quantum channel $\mathscr{E}$ if there is a channel $\mathscr{F}_A$ on A such that*

$$\mathscr{E} \circ \mathscr{P}_\mathscr{C} = (\mathscr{F}_A \otimes \mathrm{id}_B) \circ \mathscr{P}_\mathscr{C}, \tag{6.1}$$

where we define the map $\mathscr{P}_\mathscr{C}(\sigma) = P_\mathscr{C} \sigma P_\mathscr{C}$.

The OQEC notion of NS is captured by this definition in the case that $\dim A > 1$, and DFS when $\dim A = 1$.

We can easily establish the following testable conditions for passive codes strictly in terms of the Kraus operators for the channel. The result readily follows from the unitary freedom of operator-sum decompositions for completely positive (CP) maps.

Proposition 6.1. *Let $\mathscr{E}$ be a channel on $\mathscr{H}$ with Kraus operators $\{E_i\}$ and let B be a subsystem of $\mathscr{H}$. Then B is a passive code for $\mathscr{E}$ if and only if there exist operators F_i on A such that $E_i P_\mathscr{C} = (F_i \otimes I_B) P_\mathscr{C}$ for all i.*

This provides a test for passivity when a candidate code has been identified. However, actually *finding* passive codes for a given channel (or class of channels) is a more delicate problem. Fortunately, recent work has clarified the issue.

First consider the case of a *unital* channel ($\mathscr{E}(I) = I$) with Kraus operators $\{E_i\}$. Observe that any operator ρ that belongs to the *noise commutant* $\mathscr{A}' := \{\sigma : [\sigma, E_i] = 0 \,\forall i\}$ for $\mathscr{E}$ satisfies $\mathscr{E}(\sigma) = \sigma \mathscr{E}(I) = \sigma$. Thus, $\mathscr{A}'$ is an algebra contained inside the operator space given by the fixed point set $\mathrm{Fix}(\mathscr{E}) = \{\sigma : \mathscr{E}(\sigma) = \sigma\}$. In fact, a basic result for unital channels shows that the converse is true, $\mathrm{Fix}(\mathscr{E}) = \mathscr{A}'$ [K03b]. Therefore, as a †-closed algebra (by which we mean a set of operators that is closed under addition, multiplication, and the † operation; in other words a finite-dimensional C*-algebra [D96]), this set induces an orthogonal decomposition of the Hilbert space $\mathscr{H} = \oplus_k (A_k \otimes B_k) \oplus \mathscr{K}$, such that the algebra consists of all operators of the form $\oplus (I_{A_k} \otimes X_k) \oplus 0_\mathscr{K}$, where X_k is an arbitrary operator on B_k and $0_\mathscr{K}$ is the zero operator on $\mathscr{K}$. It is not hard to see that each subsystem B_k (of dimension at least 2) defines a passive code for $\mathscr{E}$. In fact, it has been shown that *all* passive codes for a unital channel $\mathscr{E}$ are encoded into the algebra $\mathscr{A}'$ in this way [CK06].

For the general (not necessarily unital) case, the fixed point set can still be seen to encode all the passive codes for a channel. This set may not be an algebra, nevertheless in [B.-KNP+08] it was shown that a more general operator space structure for the set can be obtained. As the fixed point set for a channel is simply its eigenvalue-1 eigenspace when the channel is viewed as a superoperator, this set can be readily computed. See [B.-KNP+08] for further details and [K06a, CK06] for other related results on passive codes.

6.1.2 OQEC: Basic definition

As a straightforward generalization of the definition for passive subsystem codes above, we arrive at the definition of a general OQEC code.

Definition 6.2. *A subsystem code B is* correctable *for a quantum channel $\mathcal{E}$ on $\mathcal{H}$ if there are channels $\mathcal{R}$ on $\mathcal{H}$ and $\mathcal{F}_A$ on A such that*

$$\mathcal{R} \circ \mathcal{E} \circ \mathcal{P}_\mathcal{C} = (\mathcal{F}_A \otimes \mathrm{id}_B) \circ \mathcal{P}_\mathcal{C}. \tag{6.2}$$

In other words, B is correctable for $\mathcal{E}$ if there is a channel $\mathcal{R}$ such that B is a passive code for $\mathcal{R} \circ \mathcal{E}$. In terms of states, this is equivalent to the following quantification:

$$\forall \rho_A \ \forall \rho_B \ \exists \tau_A : (\mathcal{R} \circ \mathcal{E})(\rho_A \otimes \rho_B) = \tau_A \otimes \rho_B, \tag{6.3}$$

where we have written ρ_A for states on A, etc., and $\rho_A \otimes \rho_B$ for the operator on $\mathcal{H}$ given by $(\rho_A \otimes \rho_B) \oplus 0_{\mathcal{C}^\perp}$.

It is evident that passive DFS and NS are captured by the special case in this definition with $\mathcal{R} = \mathrm{id}$ the identity channel. Standard (subspace) QEC is obtained when one focuses on the subsystem codes that can be regarded as subspace codes; that is, when the ancilla is trivial, with $\dim A = 1$.

6.2 Equivalent conditions for OQEC

In this section we present a number of characterizations of OQEC codes, ranging from testable conditions in terms of Kraus operators for a given channel, to an entirely information theoretic characterization, and a description based in operator algebra and representation theory. The conditions are all equivalent; however, they come from very different perspectives. We will sketch the proofs and point the reader to the original articles for full details.

6.2.1 Kraus operator testable conditions

As in QEC, the need for testable conditions in OQEC is obvious. Indeed, the condition Eq. (6.2) for a subsystem code cannot be tested explicitly as it a priori involves an uncountable number of conditions; $\forall \rho_A$, etc. Fortunately, as in QEC and the Knill–Laflamme condition [Eq. (2.29)], testable conditions for subsystem codes may be derived in terms of Kraus operators.

Theorem 6.1. *Let $\mathcal{E}$ be a quantum channel on $\mathcal{H}$ with Kraus operators $\{E_i\}$ and let B be a subsystem of $\mathcal{H}$. Then $\mathcal{E}$ is correctable on B if and only if there exist operators F_{ij} on A such that*

$$P_\mathcal{C} E_i^\dagger E_j P_\mathcal{C} = (F_{ij} \otimes I_B) P_\mathcal{C} \quad \forall i, j. \tag{6.4}$$

In the QEC case (dim $A = 1$) the operators F_{ij} are simply complex scalars that determine a nonnegative matrix C. One direction of the general subsystem proof is straightforward and essentially the same as for QEC: If $\mathscr{R}$ is a correction operation for $\mathscr{E}$ on B with Kraus operators $\{R_k\}$, then by Proposition 6.1 there are operators G_{ik} on A such that $R_k E_i P_{\mathscr{C}} = (G_{ik} \otimes I_B) P_{\mathscr{C}}$, and we have

$$P_{\mathscr{C}} E_j^\dagger E_i P_{\mathscr{C}} = \sum_k P_{\mathscr{C}} E_j^\dagger R_k^\dagger R_k E_i P_{\mathscr{C}} = \left(\left(\sum_k G_{jk}^\dagger G_{ik} \right) \otimes I_B \right) P_{\mathscr{C}}.$$

There are multiple ways to establish the other direction. The argument from [NP07] proceeds as follows. Fix a state $\sigma = |\psi\rangle\langle\psi|$ of A, and define a quantum operation $\mathscr{E}_\sigma(\rho) := \mathscr{E}(\sigma \otimes \rho)$ on B. Then Eqs. (6.4) imply the existence of a single universal recovery operation $\mathscr{R}$ that acts as a recovery operation for all $\mathscr{E}_\sigma$. Linearity then implies that B is correctable for $\mathscr{E}$. To prove this, note that a set of Kraus operators for $\mathscr{E}_\sigma$ is the set $E_{j,\sigma} : B \to \mathscr{H}$ defined by $E_{j,\sigma} := E_j P|\psi\rangle$. That is, $\mathscr{E}_\sigma(\rho) = \sum_j E_{j,\sigma} \rho E_{j,\sigma}^\dagger$. Then one can check by direct calculation that the set of errors $E_{j,\sigma}$, where j and σ are both allowed to vary over all possible values, is a correctable set of errors mapping B to $\mathscr{H}$, in the sense of standard error correction. This suffices to establish the existence of a single universal recovery operation $\mathscr{R}$ that acts as a recovery operation for all $\mathscr{E}_\sigma$.

Another approach to proving this result from [KS06, CJK09] links both of these conditions via a technical result that we will use below, hence we state it for completeness.

Lemma 6.1. *Let $\mathscr{E}$ be a channel on $\mathscr{H}$ and let B be a subsystem of $\mathscr{H}$. Then B is correctable for $\mathscr{E}$ if and only if there are unitary operators U_i on $\mathscr{H}$ and mutually commuting positive operators D_i on A such that $\mathscr{E} \circ \mathscr{P}_{\mathscr{C}} = \mathscr{F} \circ \mathscr{P}_{\mathscr{C}}$ where $\mathscr{F}$ is the channel with Kraus operators $\{U_i(D_i \otimes I_B)\}$ and the subspaces $\mathrm{range}(D_i) \otimes B$ are mutually orthogonal.*

In the case of QEC, this argument arises in the proof of the Knill–Laflamme condition, Eq. (2.29), when the code matrix C is diagonalized (the positive operators D_i are square roots of probabilities in that case). In the easy direction of this proof, to see how the technical condition implies that B is correctable, for each i, let $P_{\mathscr{C}_i}$ be the projection of $\mathscr{H}$ onto the subspace $U_i(\mathrm{Ran}(D_i) \otimes B)$. Then a correction operation $\mathscr{R}$ for $\mathscr{E}$ on B is defined by using the Kraus operators $\{U_i^\dagger P_{\mathscr{C}_i}\}$. Thus, conceptually one can see that the correction operation is obtained in a manner similar to QEC; measurement followed by a conditioned unitary.

6.2.2 Information-theoretic conditions

In this section, we will derive a set of information theoretic conditions for OQEC [NP07]. The correlations in a system subject to an open evolution can be drastically affected: existing correlations between the system and another system can be transferred to the environment, new correlations can be created with the environment, etc. The conditions we will demonstrate in this section set restrictions on these effects that are necessary and sufficient for the existence of a recovery map that will correct a subsystem of the noisy quantum system.

To state these conditions, we introduce three new systems: an environment E and reference systems R_A and R_B for the subsystems A and B respectively. The environment E will be used to purify the quantum map; i.e., the joint evolution of E and S is given by a unitary matrix U_{SE}

such that

$$U_{SE}|\psi\rangle_S \otimes |e_0\rangle_E = \sum_a E_a|\psi\rangle_S \otimes |e_a\rangle \quad (6.5)$$

for all $|\psi\rangle_S$. It can be verified as a special case of the Stinespring theorem [S55] that such a unitary matrix always exists and that tracing out E reproduces the effect of $\mathcal{E}$ on the state $|\psi\rangle_S$ [this is a special case of the Stinespring representation, Eq (1.99)]. Thus, the dimension of E is equal to the number of Kraus operators (at most the square of the dimension of S), and it is assumed to be initially in the pure state $|e_0\rangle$.

The reference systems are used to model correlations that the quantum system could have with other systems. To account for all possible such correlations, we assume that A and R_A are initially in a maximally entangled state $|\Phi_A\rangle = \sqrt{\frac{1}{d_A}} \sum_i |i\rangle_A \otimes |i\rangle_{R_A}$, and similarly for B and R_B. Thus, the dimensions of R_A and R_B are the same as the dimensions of A and B respectively.

The initial state of all these systems is

$$|\Psi\rangle_{ABR_AR_BE} = |\Phi_A\rangle_{AR_A} \otimes |\Phi_B\rangle_{BR_B} \otimes |e_0\rangle_E.$$

The purified open evolution transforms this state into

$$|\Psi\rangle'_{ABR_AR_BE} = (U_{SE} \otimes I_{R_AR_B})|\Psi\rangle_{ABR_AR_BE}.$$

We denote the corresponding density matrices $\rho_{ABR_AR_BE}$ and $\rho'_{ABR_AR_BE}$, and use the appropriate subscripts to denote its marginals obtained by partial trace; e.g., $\rho'_{R_BE} = \text{Tr}_{ABR_A}\{\rho_{ABR_AR_BE}\}$. Similarly, we denote the entropy of a system before or after the application of a map using the associated unprimed or primed letter respectively; e.g., $S(R'_BE') = \text{Tr}\{\rho'_{R_BE} \log \rho'_{R_BE}\}$.

With these definitions in hand, we can state a necessary and sufficient condition for OQEC:

Theorem 6.2. *Let $\mathcal{E}$ be a quantum channel on $\mathcal{H}$ and let B be a subsystem of $\mathcal{H}$. Then $\mathcal{E}$ is correctable on B if and only if*

$$I(R_B : S) = I(R'_B : S'). \quad (6.6)$$

The mutual information is defined $I(A : B) = S(A) + S(B) - S(AB)$. The mutual information $I(A : B)$ can only decrease under local operations on the two systems. This reflects the fact that we cannot increase the correlations between the two systems without having them interact either directly or indirectly. Thus, the above condition simply states that the system must not lose its correlations with the reference system R_B, and that provided these correlations are kept, the information content of B can be restored.

To prove the theorem, let us expand the mutual information and re-express the condition as

$$S(R_B) + S(S) - S(R_BS) = S(R'_B) + S(S') - S(R'_BS') \quad (6.7)$$
$$\Leftrightarrow S(S) - S(R_BS) = S(S') - S(R'_BS') \quad (6.8)$$
$$\Leftrightarrow S(R_B) = S(S') - S(R'_BS') \quad (6.9)$$
$$\Leftrightarrow S(R'_B) = S(S') - S(R'_BS') \quad (6.10)$$
$$\Leftrightarrow S(R'_B) = S(R'_AR'_BE') - S(R'_BS') \quad (6.11)$$
$$\Leftrightarrow S(R'_B) = S(R'_AR'_BE') - S(R'_AE') \quad (6.12)$$
$$\Leftrightarrow I(R'_B : R'_AE') = 0. \quad (6.13)$$

Here we have used the fact that the reference systems do not evolve, the fact that the global state is pure – which implies that the entropy of one part is equal to the entropy of its complement – and the fact that $R_B R_A$ are initially in a maximally entangled state with S. The conclusion is that after the evolution the reference system R_B is not correlated with the reference system R_A and the environment, i.e.,

$$\rho'_{R_A R_B E} = \rho'_{R_B} \otimes \rho'_{R_A E} = \frac{1}{d_B} I_{R_B} \otimes \rho'_{R_A E}. \tag{6.14}$$

It is now straightforward to demonstrate that this condition is necessary and sufficient for B to be correctable. Using the fact that $(M_S \otimes I_{R_A R_B E})|\Psi\rangle_{ABR_A R_B E} = (M^T_{R_A R_B} \otimes I_{ABE})|\Psi\rangle_{ABR_A R_B E}$ for any operator M, we obtain

$$\rho'_{R_A R_B E} = \sum_{ab} (P E_a^T E_b^* P)_{R_A R_B} \otimes |a\rangle\langle b|_E. \tag{6.15}$$

Given this equality, it is easily seen that the condition formulated in Eq. (6.14) is equivalent to the condition Eq. (6.4) that is necessary and sufficient for B to be correctable. We have thus demonstrated that in order for the system B to be correctable, the mutual information between the system and B's reference must not be affected by the quantum map.

6.2.3 Operator algebraic conditions

One reason the result of the previous section is powerful is that it characterizes the OQEC condition strictly in terms of the code and the map itself, without reference to Kraus operators. We next focus on a set of equivalent conditions for OQEC that includes conditions that have this important feature. In addition, it is shown that correction of subsystems is equivalent to the precise correction of certain operator algebras. These conditions are somewhat abstract, and perhaps because of this they have proved to be particularly useful as tools to establish further results on special classes of OQEC codes. We state the conditions as a single result, and then briefly sketch the proofs. Two notational points first. We use notation such as $\mathscr{F}_{A'|A}$ to denote a channel mapping from A to A'. Also, by a representation π of an algebra $\mathfrak{A}$ we mean a linear map that is multiplicative and positive (and hence automatically completely positive) on $\mathfrak{A}$.

Theorem 6.3. *Let $\mathscr{E}$ be a quantum channel on $\mathscr{H}$ and let B be a subsystem of $\mathscr{H}$. Then the following are equivalent:*

(i) *$\mathscr{E}$ is correctable on B.*
(ii) *The algebra $\mathfrak{A}$ of operators of the form $I_A \otimes \rho_B$ is precisely correctable for $\mathscr{E}$; that is, there is a channel $\mathscr{R}$ such that $\mathscr{R} \circ \mathscr{E}(\sigma) = \sigma$ for all $\sigma \in \mathfrak{A}$.*
(iii) *There is a representation π of $\mathfrak{A}$ on $\mathscr{H}$ such that*

$$\mathscr{E}(\rho) = \pi(\rho)\mathscr{E}(P_\mathscr{C}) = \mathscr{E}(P_\mathscr{C})\pi(\rho) \quad \forall \rho \in \mathfrak{A}. \tag{6.16}$$

(iv) *There is a subspace $\mathscr{C}' = A' \otimes B'$ of $\mathscr{H}$ with subsystems A' and $B' \cong B$, a channel $\mathscr{F}_{A'|A}$ and a unitary channel $\mathscr{V}_{B'|B}$ such that*

$$\mathscr{E} \circ \mathscr{P}_\mathscr{C} = (\mathscr{F}_{A'|A} \otimes \mathscr{V}_{B'|B}) \circ \mathscr{P}_\mathscr{C}. \tag{6.17}$$

It is clear that correctability implies condition (ii) since we can always randomize the subsystem A after the correction to produce I_A. However, at first sight condition (ii) may appear weaker. How can correctability in the special case when subsystem A is in the maximally mixed state imply correctability for arbitrary states ρ_A? One way to see this is to think of the maximally mixed state that arises from the ensemble where either ρ_A or $\rho'_A \propto (I/N - \rho_A)$ are prepared with appropriate probabilities. If the subsystem B can be corrected for the mixture of both terms, then certainly knowing whether A is prepared in ρ_A or ρ'_A can only help.

With regard to condition (iii), given the unitaries U_i and projections $P_{\mathscr{C}_i}$ from Lemma 6.1 when B is correctable, we can define a family of partial isometries $V_i = U_i P_{\mathscr{C}}$. It can be verified that the map $\pi(\sigma) = \sum_i V_i \sigma V_i^\dagger$ acts as a representation, and the identity Eq. (6.16) can be directly verified. A similar plan of attack can be used to prove the converse implication. Note that a subtlety imbedded in the proof is that the operator $\mathscr{E}(P_{\mathscr{C}})$ must commute with the range operators of the representation.

It is clear that if condition (iv) is satisfied, then B is correctable for $\mathscr{E}$ with any channel $\mathscr{R}$ such that $\mathscr{R} \circ \mathscr{P}_{\mathscr{C}'} = (\mathscr{G}_{A|A'} \otimes \mathscr{V}^\dagger_{B'|B}) \circ \mathscr{P}_{\mathscr{C}'}$ for some channel $\mathscr{G}_{A|A'}$ as a viable correction channel. Conversely, if B is correctable, Lemma 6.1 can be used to "splice" together a subsystem pair $\mathscr{C}' = A' \otimes B'$ and isometric map $\mathscr{V}_{B'|B}$ so that condition (iv) is satisfied.

Finally, we note that each of these conditions gives a characterization of correctable subsystem codes in the Schrödinger picture for quantum dynamics (evolution of states). Recent work investigates QEC in the Heisenberg picture (evolution of observables) [BKK07a, BKK07b, B.-KNP+08] and makes use of this perspective to extend the theory to infinite-dimensional Hilbert spaces [BKK09].

6.3 Stabilizer formalism for OQEC

The stabilizer formalism has proven extremely useful for the design of quantum codes. In particular, it establishes a connection between quantum and classical codes, and is particularly well suited for fault-tolerant protocols. In this section, we present a stabilizer formalism for OQEC codes, or subsystem codes [P05].

Let us briefly review the ordinary stabilizer formalism. Although it can be done at a more abstract level, we choose to work with bases because they greatly simplify the presentation. In this language, an ordinary $[[n, k, d]]$ stabilizer code is entirely specified by a canonical basis of the n-qubit Pauli group. We say that a basis of the Pauli group is canonical if its elements can be paired up in such a way that any element of the basis commutes with every other element except with its partner, with which it anticommutes. For instance, the $2n$ single-qubit operators X_i, Z_i $i = 1, \ldots, n$ form a canonical basis. However, this is not the canonical basis we use to describe a QECC. Instead, we use the following $2n$ basis elements that fall into three categories:

(i) The $2k$ logical operators $\overline{X}_i$ and $\overline{Z}_i$ that obey canonical commutation relations: $\{\overline{X}_i, \overline{Z}_i\} = 0$ and all other pairs commute.
(ii) The $n - k$ stabilizer generators S_j that mutually commute, and commute with all logical operators.
(iii) The $n - k$ pure errors T_j that mutually commute, commute with all logical operators, and are canonically conjugated to the stabilizer generators: $\{S_j, T_j\} = 0$ and $[S_j, T_{j'}] = 0$ for $j \neq j'$.

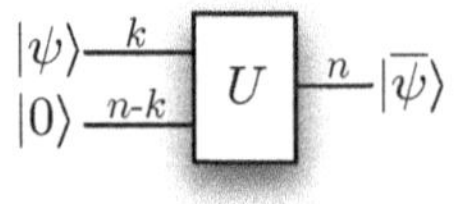

$$|\psi\rangle \xrightarrow{k} \boxed{U} \xrightarrow{n} |\overline{\psi}\rangle$$
$$|0\rangle \xrightarrow{n-k}$$

Fig. 6.1. Encoding circuit for QECC.

Because they both form canonical bases, there exists a one-to-one unitary Clifford transformation U that takes the elements of one basis to the other:

$$U : \begin{cases} X_i \to \overline{X}_i & \text{for } i = 1, \ldots k \\ Z_i \to \overline{Z}_i & \text{for } i = 1, \ldots k \\ X_{i+k} \to T_i & \text{for } i = 1, \ldots n-k \\ Z_{i+k} \to S_i & \text{for } i = 1, \ldots n-k \end{cases}. \tag{6.18}$$

This transformation is the encoder for the QECC, i.e., it takes the unencoded k-qubit state $|\psi\rangle$ and additional $n - k$ qubits in the state $|0\rangle$, and produces an encoded state $|\overline{\psi}\rangle$, see Fig. 6.1. Thus, the code space is a 2^k-dimensional subspace of the n-qubit Hilbert that is the common $+1$ eigenspace of the $n - k$ stabilizer generators. Hence the stabilizer generators act trivially on the code space. The logical operators map the code space to itself in a nontrivial way. Finally, the pure errors map the code space to an orthogonal subspace. More precisely, all $2^{n-k} - 1$ distinct nontrivial combinations of the pure errors map the code space to distinct mutually orthogonal subspaces that span the entire Hilbert space.

This choice of basis is motivated by the fact that it simplifies the error analysis. We can decompose any Pauli errors E in this basis and write $E = \mathscr{L}(E)\mathscr{S}(E)\mathscr{T}(E)$, where $\mathscr{L}(E)$ contains all the logical operators entering in the decomposition of E, similarly $\mathscr{S}(E)$ contains all the stabilizer generators, and $\mathscr{T}(E)$ contains the pure errors. The error syndrome $s(E)$ is an $n-k$-bit string that encodes the commutation relations of the errors with the stabilizer generators: $ES_j + (-1)^{s_j(E)} S_j E = 0$. Because of the canonical commutation relations, the error syndrome is in one-to-one correspondence with the pure error component: $\mathscr{T}(E) = \prod_j T_j^{s_j(E)}$. Thus, the syndrome measurement completely reveals the pure error component. On the other hand, the stabilizer component $\mathscr{S}(E)$ acts trivially on the code space. Thus, syndrome-based decoding consists in identifying the logical component $\mathscr{L}(E)$ of the error given its pure error component $\mathscr{T}(E)$. Errors that only differ in their syndrome component need not be distinguished because they have the same effect on the code space. Such errors are said to be degenerate.

6.3.1 Gauge operator

We now transpose this formalism to the case of OQEC. Similarly to an ordinary QECC, an $[[n, k, r, d]]$ subsystem code can be specified by a canonical basis of the n-qubit Pauli group. The $2n$ basis elements are now divided into four groups:

(i) The $2k$ logical operators $\overline{X}_i$ and $\overline{Z}_i$ that obey canonical commutation relations: $\{\overline{X}_i, \overline{Z}_i\} = 0$ and all other pairs commute.
(ii) The $n - k - r$ stabilizer generators S_j that mutually commute, and commute with all logical operators.

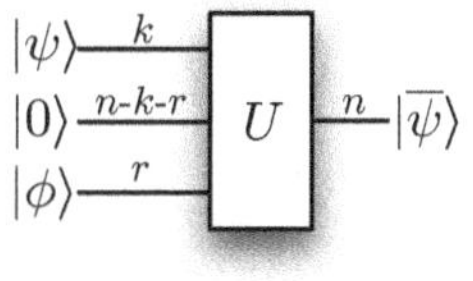

Fig. 6.2. Encoding circuit for OQECC.

(iii) The $n - k - r$ pure errors T_j that mutually commute, commute with all logical operators, and are canonically conjugated to the stabilizer generators: $\{S_j, T_j\} = 0$ and $[S_j, T_{j'}] = 0$ for $j \neq j'$.

(iv) The $2r$ gauge operators G_j^x and G_j^z that commute with all logical operators, stabilizer generators, and pure errors, and form canonically conjugate pairs: $\{G_j^x, G_j^z\} = 0$ and all other pairs of gauge operators commute.

Once again, there exists a Clifford unitary transformation taking the elements of the canonical basis formed of single-qubit Pauli operators to this basis:

$$U : \begin{cases} X_i \to \overline{X}_i & \text{for } i = 1, \ldots k \\ Z_i \to \overline{Z}_i & \text{for } i = 1, \ldots k \\ X_{i+k} \to T_i & \text{for } i = 1, \ldots n - k - r \\ Z_{i+k} \to S_i & \text{for } i = 1, \ldots n - k - r \\ X_{i+k+r} \to G_i^x & \text{for } i = 1, \ldots r \\ Z_{i+k+r} \to G_i^z & \text{for } i = 1, \ldots r \end{cases}. \quad (6.19)$$

As before, this transformation can serve as an encoder for the OQECC where the first k qubits contain the unencoded information $|\psi\rangle$ and the $n - k - r$ stabilizer qubits are in the all-zero state, see Fig. 6.2. The novelty is that the following r "gauge" qubits can be in any state $|\phi\rangle$. The coding scheme does not specify the state of these qubits and they can be in any mixed or pure state. They correspond to the A factor of the Hilbert space in the decomposition $\mathscr{C} = A \otimes B$.

We can decompose any Pauli error E in this basis and write

$$E = \mathscr{L}(E)\mathscr{S}(E)\mathscr{T}(E)\mathscr{G}(E)$$

with the same meaning as above: $\mathscr{L}$, $\mathscr{S}$, and $\mathscr{T}$ contain the logical, stabilizer, and pure error components of the error respectively and $\mathscr{G}$ contains the gauge component. The syndrome measurement reveals the pure error component $\mathscr{T}(E)$ and decoding consists in identifying the logical component $\mathscr{L}(E)$ given this information. Errors that only differ by their stabilizer and/or *gauge* components have the same syndrome and have the same effect on the encoded information, so they are regarded as equivalent. Thus, in a sense, the gauge operators play a role analogous to the stabilizer generators: they generate equivalence classes among the errors because they act trivially on the encoded information. But contrarily to the stabilizer generators, the gauge operators are not (a priori) used to extract information about the error that has corrupted the information. In fact, since the gauge operators generate a non-Abelian group, they could not possibly all be measured simultaneously without disturbing each other's outcome.

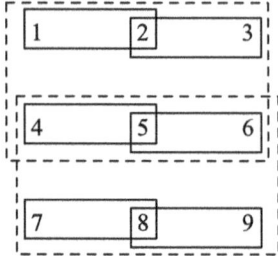

Fig. 6.3. Shor's nine-qubit QECC. The full lines represent Z stabilizer generators and the dash lines X stabilizer generators.

6.4 Examples

6.4.1 Bacon–Shor code

An important example of a stabilizer OQECC, often referred to as the Bacon–Shor code [B06, P05], is obtained from a modification of Shor's original QECC [S95]. Recall that the stabilizer generators for this nine-qubit code are given by $Z_i Z_{i+1}$ for $i = 1, 2, 4, 5, 6, 7$ and $X_i X_{i+1} X_{i+2} X_{i+3} X_{i+4} X_{i+5}$ for $i = 1, 4$. These can be visualized if we place the qubits on a 3×3 lattice as shown on Fig. 6.3. There is an obvious asymmetry between the X and the Z stabilizer generators. This is because the code is constructed from the concatenation of a three-qubit repetition code in the Z basis with a three-qubit repetition code in the X basis. Bit-flip errors are corrected independently in each row of three qubits and are not degenerate. Phase errors are corrected collectively and are highly degenerate: it is only necessary to identify the row in which a phase error occurred, since phase errors on distinct qubits in a given row have the same effect on the encoded information.

On the other hand, the logical operators of the code are completely symmetric, $\overline{X} = X^{\otimes 9}$ and $\overline{Z} = Z^{\otimes 9}$. This is in contrast with the stabilizer generators and the recovery procedure. The Bacon–Shor code restores this symmetry. Its stabilizer generators are shown in Fig. 6.4. This code has fewer stabilizer generators than Shor's code. As a consequence, it cannot distinguish between as many errors. Contrarily to Shor's code, these stabilizers cannot identify the precise qubit on which a bit-flip occurs, they can only identify the column in which it occurs. This loss is compensated by new gauge operators that make bit-flip errors in a column all equivalent to one another. In more detail, the complete canonical basis associated with the Bacon–Shor code is

$$\begin{aligned}
&\overline{X} = X^{\otimes 9} & &\overline{Z} = Z^{\otimes 9} \\
&S_1 = Z_1 Z_2 Z_4 Z_5 Z_7 Z_8 & &T_1 = X_5 X_6 \\
&S_2 = Z_2 Z_3 Z_5 Z_6 Z_8 Z_9 & &T_2 = X_4 X_5 \\
&S_3 = X_1 X_2 X_3 X_4 X_5 X_6 & &T_3 = Z_2 Z_5 \\
&S_4 = X_4 X_5 X_6 X_7 X_8 X_9 & &T_4 = Z_5 Z_8 \\
&G_1^z = Z_1 Z_2 & &G_1^x = X_1 X_4 \\
&G_1^z = Z_2 Z_3 & &G_1^x = X_3 X_6 \\
&G_1^z = Z_7 Z_8 & &G_1^x = X_4 X_7 \\
&G_1^z = Z_8 Z_9 & &G_1^x = X_6 X_9
\end{aligned} \qquad (6.20)$$

By combining the stabilizer generators and the gauge operators, we can obtain any pair of Z operators in the same row and any pair of X operators from the same column. Thus, all single

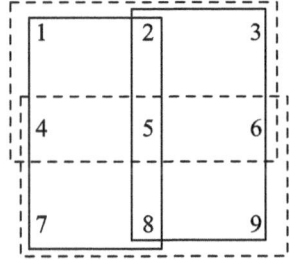

Fig. 6.4. The Bacon–Shor code. The full lines represent Z stabilizer generators and the dash lines X stabilizer generators.

phase errors in a given row are equivalent, and all single qubit-flip errors in a given column are equivalent.

6.4.2 Generalization: Klappenecker–Sarvepalli–Bacon–Casaccino

To understand how to generalize the Bacon–Shor code, we first need to generalize the concatenation scheme that led to Shor's code. This code is obtained from the concatenation of two classical codes, one used to correct bit-flip errors and one used to correct phase errors. The bit-flip code, which we refer to as "code A," has stabilizer generators $S_1^A = ZZI$ and $S_2^A = IZZ$ and logical operators $\overline{X}^A = XXX$ and $\overline{Z}^A = ZZZ$. The second code, "code B," is also a repetition code, but it is used as a phase code. Its stabilizers are $S_1^B = XXI$ and $S_2^B = IXX$ and its logical operators are $\overline{X}^B = XXX$ and $\overline{Z}^B = ZZZ$. Shor's code, which we denote $A \triangleright B$, is obtained from the concatenation of these two codes. It is defined on a 3×3 array of qubits and has two types of stabilizer generators:

(i) The stabilizer generators of A acting independently on each row of the array, represented by the full rectangles in Fig. 6.3. There are six of those.
(ii) The operators $\overline{X}_i^A \triangleright S_j^B$ obtained by forming a column with the operator S_j^B, and expanding each X operator in the horizontal direction as a logical operator X_i^A of the code A. These two stabilizers are represented by the dotted rectangles in Fig. 6.3.

The logical operators of $A \triangleright B$ are $L_i^A \triangleright L_j^B$, obtained by forming a column with the operator L_j^B, and expanding each single qubit operator by a logical operator of the code A, i.e., $X \to \overline{X}_i^A$ and $Z \to \overline{Z}_i^A$.

This construction can be generalized using any two classical codes. Let A and B be two classical codes with parameters $[n^A, k^A, d^A]$ and $[n^B, k^B, d^B]$ respectively. Code A is used as a bit-flip code and code B as a phase code. In other words, the stabilizer generators of A are obtained from the columns of the parity check matrix of the associated classical code by substituting each 1 by a Z. The logical operators $\overline{X}_i^A$ are obtained from the lines of the generating matrix of the associated classical code by substituting each 1 by an X. The commutation properties of these operators follow directly from the orthogonality of the parity check matrix and generator matrix. The logical operators $\overline{Z}_i^A$ are composed uniquely of Zs and Is and can be found by Gaussian elimination as the operators that are canonically conjugated to the $\overline{X}$s (they will hence automatically be independent from the stabilizer generators). The stabilizer generators and logical operators of code B are obtained in a similar way, but with the roles of Z and X reversed.

Given these two codes, we can follow the same steps that led to Shor's code. The code $A \triangleright B$ is defined on an $n^A \times n^B$ qubit array. It has two kinds of stabilizer generators:

(i) The stabilizer generators of A acting independently on each row of the array. We denote them $S^A_{j,r}$, where j labels a stabilizer generator of A and r a row for the qubit array. There are $(n^A - k^A)n^B$ of those.

(ii) The operators $\overline{X}^A_i \triangleright S^B_j$ obtained by forming a column with the operator S^B_j, and expanding each X operator in the horizontal direction as a logical operator X^A_i of the code A. There are $k^A(n^B - k^B)$ of those.

Thus, there are a total of $n^A n^B - k^A k^B$ stabilizer generators for $A \triangleright B$. The logical operators of the concatenated code are $L^A_i \triangleright L^B_j$, of which there are $k^A k^B$.

That these stabilizers generate an Abelian group can be verified straightforwardly. On each row of the array, the stabilizer generators of $A \triangleright B$ are formed by either stabilizer generators of A or logical operators $\overline{X}^A$. These operators all commute with one another, and this commutativity is inherited by the generators of $A \triangleright B$. Similarly, the canonical commutation of the logical operators of the concatenated code is inherited from the canonical commutation relations of the logical operators of the constituent codes.

The code $A \triangleright B$ has minimal distance $d^{A \triangleright B} = \min\{d^A_c, d^B_c\}$. Indeed, it is obvious that the code has a bit-flip distance equal to d^A_c since code A is independently used on each row of the qubit array. On the other hand, any phase errors on a given row decompose into S^A_js and $\overline{Z}^A_i$s on that row. Since B has a phase-flip distance d^B_c, it takes that many rows with phase errors to produce an undetectable phase error. Thus, this concatenation scheme can be used to produce a quantum code from any two classical codes, without any orthogonality condition. The code obtained from this scheme has parameters $[[n^A n^B, k^A k^B, \min\{d^A, d^B\}]]$.

Again we observe an asymmetry in the two codes. Another concatenated code $A \triangleleft B$ can be constructed with stabilizers formed of the stabilizers of B acting independently on each column of the qubit array, denoted $S^B_{j,c}$, and the operators $S^A_j \triangleleft \overline{Z}^B_i$ obtained by forming a row with the operator S^A_j, and expanding each of its Z operators in the vertical direction as a logical operator $\overline{Z}_i$ of the code B. The logical operators of $A \triangleleft B$ are $L^A_i \triangleleft L^B_j$, which is the same as $L^A_i \triangleright L^B_j$ used for the code $A \triangleright B$. We will use the notation $L^A_i \triangle L^B_j$ to emphasize this symmetry. Thus, the asymmetry is only in the code, not in the logical operators.

The concatenated OQECC, denoted $A \triangle B$, restores this symmetry. It has the following parameters: $[[n^A n^B, k^B k^B, (n^A - k^A)(n^B - k^B), \geq \min\{d^A, d^B\}]]$. Its logical operators $L^A_i \triangle L^B_j$ are the same as those obtained from the two distinct asymmetric constructions $A \triangleright B$ and $A \triangleleft B$. It also uses the second set of stabilizer generators $\overline{X}^A_i \triangleright S^B_j$ of $A \triangleright B$. However, the first set of stabilizers of $A \triangleright B$ is modified. Instead of using the stabilizers of A independently on each row $S^A_{j,c}$, only the combinations $S^A_j \triangleleft \overline{Z}^B_i$ are used. These correspond to the first set of stabilizers of the code $A \triangleleft B$. The other combinations are recycled as gauge operators. What do these operators look like? In general, they are operators of the form $S^A_i \triangle S^B_j$ obtained by substituting each Z operator of S^B_j by S^A_i, or equivalently substituting each X operator of S^A_i by S^B_j.

The minimal distance of this code is also $\min\{d_A, d_B\}$. This can easily be seen from the fact that both $A \triangleright B$ and $A \triangleleft B$ have that minimal distance and correct bit-flip and phase errors independently, and the fact that $A \triangle B$ has the same bit-flip stabilizers as the code $A \triangleleft B$, and the same phase stabilizers as the code $A \triangleright B$.

6.5 Measuring gauge operators

The previous section presented a family of OQECCs that are a modified version of generalized Shor codes. The modification consisted in removing a subset of stabilizer generators and promoting them to gauge operators. Removing some stabilizers implies that less information about the error is available to the decoder. However, this loss is compensated by an increase in the code's degeneracy and the minimal distance is unaffected. The main advantage of this construction is that fewer stabilizer generators need to be measured to diagnose the error. The drawback, however, is that the weight of the stabilizer generators increases. Indeed, if w^A and w^B denote the minimal weight of the stabilizers of code A and B respectively, then the minimal weight of a stabilizers of $A \triangle B$ is $\min\{w^A d^B, w^B d^A\}$.

Having low-weight stabilizers is a serious advantage from an implementation and fault-tolerance perspective. Indeed, performing collective measurements on a very large number of qubits is a daunting experimental task, and requires the purification of complex ancillary states to be executed fault tolerantly. A code with low-weight stabilizer generators is thus highly favorable.

On the other hand, the bit-flip stabilizers $S_{j,r}^A$ of the code $A \triangleright B$ have minimum weight w^A, and its phase stabilizers $\overline{X}_i^A \triangleright S_j^B$ have minimum weight $d^A w^B$. Thus, it has some low-weight and some high-weight stabilizers. Similarly, the phase stabilizers $S_{j,c}^B$ of code $A \triangleleft B$ have minimum weight w^B, and its bit-flip stabilizers $S_j^A \triangleleft \overline{Z}_i^B$ have minimum weight $d^B w^A$. The code $A \triangle B$ is obtained from the phase stabilizers of the code $A \triangleright B$ and the bit-flip stabilizers of the code $A \triangleleft B$, so it keeps only the high-weight stabilizers. Can we combine instead the low-weight stabilizers of both codes?

The short answer is no. These operators do not all mutually commute, so they cannot be used to define a code. However, it is possible to use these operators in a clever way to extract the syndrome information of the code $A \triangle B$. We start by measuring the operators $S_{j,rc}^B$. These operators are not stabilizers of the code $A \triangle B$, so the measurement outcomes are random even in the absence of errors. However, the correlations between these random measurement outcomes reveal the syndrome information.

Indeed, these are products of various subsets of the commuting operators $S_{j,r}^A$, so their values can be obtained by measuring the operators $S_{j,r}^A$ directly and taking the corresponding product of the outcomes. This is because the operators $S_{j,c}^B$ mutually commute and because each stabilizer $S_j^A \triangleleft \overline{Z}_i^B$ decomposes as a product of a subset of the $S_{j,c}^B$. More precisely

$$S_j^A \triangleleft \overline{Z}_i^B = \prod_{c:(\overline{Z}_i^B)_c = Z} S_{j,c}^B, \qquad (6.21)$$

where $(Q)_c$ denotes the cth tensor factor of the Pauli operator Q. Because they commute, measuring each operator $S_{j,c}^B$ and taking the product of their measurement outcomes gives the same answer as measuring the operator equal to their product. Thus, we can extract the syndrome associated with the high-weight stabilizers $S_j^A \triangleleft \overline{Z}_i^B$ through a measurement of the low-weight operators $S_{j,c}^B$. Similarly, the syndrome associated with the stabilizers $\overline{X}_i^A \triangleleft S_j^B$ can be extracted through the measurement of $S_{j,r}^A$.

Note that this procedure is somewhat wasteful. The low-weight stabilizers $S_{j,r}^A$ and $S_{j,c}^B$ cannot be used jointly to define a code, but it is not necessary to replace both sets by high-weight operators; we can choose to replace only one of them. This restores the asymmetry in the code. Such an asymmetric code can be relevant when one type of noise dominates. It is indeed often the case that phase errors are much more prominent than bit-flip errors. Thus, we can choose the code $A \triangleleft B$. In that case, we must begin by measuring the stabilizers $S_{j,c}^B$. These directly reveal the phase error syndrome; there is no need to coarse-grain the measurement outcomes as was done in the previous scheme, so there is more information about phase errors. The measurement of the bit-flip stabilizers $S_j^A \triangleleft \overline{Z}_i^B$ proceeds indirectly as above through a measurement of the $S_{j,r}^A$. These measurements will take the state outside the code space defined by the operators $S_{j,c}^B$, but without affecting its logical content because the measurements consist of gauge operators. Thus, the state can simply be returned to the code space before the next error-correction round.

6.5.1 Sparse quantum codes?

What we have learned so far is that it is possible to construct OQECC from any two linear classical codes, and that, moreover, the syndrome can be extracted by measuring operators of weight equal to the weight of the columns of the associated parity check matrix. This is exciting because among the best classical codes are sparse (or low-density parity check, LDPC) codes that, as their names suggest, have sparse parity check matrices. Thus, they can be used to produce OQECCs with very simple syndrome extraction schemes. These classical codes are nearly capacity achieving on a variety of channels and have efficient decoding algorithms (see Chapter 11).

Unfortunately, these nice properties do not all transpose to the quantum setting. First, near-capacity achieving codes must have a minimal distance that grows proportionally to the code length n. (Atypical low-weight codewords can be tolerated.) Assume that codes A and B are good in this sense, so $d^A \sim n^A$ and $d^B \sim n^B$. The resulting OQECC has $n = n^A n^B$ and $d = \min\{d^A, d^B\} \sim \sqrt{n}$. Thus, as n grows, the failure probability on the depolarizing channel will approach unity.

The second difficulty has to do with the decoding algorithm. This algorithm is used to identify the most likely error on the encoded information given the syndrome. The decoding algorithms used for sparse classical codes are described in Chapter 11. Unfortunately, they are not suitable for the OQECC presented in the previous section. In fact, this problem is not limited to sparse codes. The decoding algorithm used for most classical coding schemes will not be suitable for the derived OQECC.

The problem stems from the fact that the error model changes under concatenation. Suppose that we use the coding scheme $A \triangleleft B$ and that the noise depolarizes each qubit independently. The phase errors can be handled by the decoding algorithm. Indeed, this consists in using the code B independently on each column of the qubit array, i.e., with stabilizers $S_{j,c}^B$. Thus, as far as decoding is concerned, this is just like decoding the corresponding classical code with a bit error rate $2p/3$ independently on each column.

For the bit-flip errors, we measure the operators S_j^A and extract from them the syndrome associated with the stabilizers $S_j^A \triangleleft \overline{Z}_i^B$. The outcome of each S_j^A individually is meaningless, they are random because they do not commute with the phase stabilizers. Thus, the decoder uses the error syndrome associated with $S_j^A \triangleleft \overline{Z}_i^B$. From this perspective, the problem is just like decoding a single copy of the code A, but with an error model given by the probability that an $\overline{X}_i^B$ operation was applied. In other words, this is the probability that the error that has affected the qubits anticommutes with $\overline{Z}_i^B$. These logical errors are highly correlated, something that decoding algorithms are usually not tailored to. Even if we ignore these correlations, we see that the logical error rate is greater than $w_i 2p/3$, where w_i is the weight of $\overline{X}_i^B$. Since $w_i \sim n^B$ for good codes, the code's error threshold is decrease by a factor proportional to $n^B \sim \sqrt{n}$.

Despite this negative analysis, the coding scheme can find useful applications. The solution is to keep the size relatively small, such that the error rate increase $w_i 2p/3$ does not become problematic. This rules out for the time being the design of capacity-achieving OQECC based on this scheme. However, fault-tolerant protocols using this coding scheme have a significantly larger error threshold than protocols based on conventional QEC because they can be realized by measuring only low-weight operators.

6.6 Bounds for subsystem codes

We next discuss the quantum Hamming and Singleton bounds for subsystem codes considered in [KS07]. For this section we consider the more general setting of n-qudit Hilbert space $\mathscr{H} = (\mathbb{C}^q)^{\otimes n} = \mathbb{C}^{q^n}$, where q is a power of a prime number p. Let $\mathbb{F}_q$ be a finite field with q elements and characteristic p. (For instance, if $q = p$, then the integers $\{1, 2, \ldots, p\}$ with addition and multiplication modulo p form such a field.) Let $\mathscr{C} \subseteq \mathbb{F}_q^n$ be an $\mathbb{F}_q$-linear classical code denoted by $[n, k, d]_q$, where k is the dimension of $\mathscr{C}$ over $\mathbb{F}_q$ and d is the minimum distance of $\mathscr{C}$, where $\text{wt}(\mathscr{C}) = \min\{\text{wt}(c) | 0 \neq c \in \mathscr{C}\} = d$ and $\text{wt}(c)$ is the natural extension of the Hamming weight, defined as the sum of nonzero letters in c. If $\mathscr{C}$ is a linear subspace over $\mathbb{F}_q$, then it is called an *additive* code.

The noise models considered here are defined as follows. Let $\mathscr{B} = \{|x\rangle : x \in \mathbb{F}_q\}$ be an orthonormal basis for $\mathbb{C}^q$. Let $X(a)$ and $Z(b)$ be the unitary operators on $\mathbb{C}^q$ defined by

$$X(a)|x\rangle = |x + a\rangle \quad \text{and} \quad Z(b)|x\rangle = \omega^{(bx)_p}|x\rangle,$$

where $\omega = e^{j2\pi/p}$ is a primitive pth root of unity and $(bx)_p$ is multiplication modulo p. Let $\mathfrak{E}$ be the error group on $\mathscr{H}$, the natural generalization of the Pauli group, defined as the tensor product of n such operators;

$$\mathfrak{E} = \{\omega^c E_1 \otimes \cdots \otimes E_n \mid E_i = X(a_i)Z(b_i); a_i, b_i \in \mathbb{F}_q; c \in \mathbb{F}_p\}.$$

As before, the *weight* $\text{wt}(E)$ of an error $E \in \mathfrak{E}$ is equal to the number of E_i not equal to the identity operator.

Every nontrivial normal subgroup N of $\mathfrak{E}$ defines a subsystem code $\mathscr{C}$ as follows. Let $C_\mathfrak{E}(N)$ be the centralizer of N inside $\mathfrak{E}$ and $Z(N)$ the center of N. The subsystem code $\mathscr{C}$ defined by N is precisely the same as the stabilizer code defined by $Z(N)$. It follows that $\mathscr{C}$ can be decomposed as $A \otimes B$ where $\dim A = |N : Z(N)|^{1/2}$ (here $|N : Z(N)| = |N|/|Z(N)|$ is the index of the subgroup $Z(N)$ inside the group N) and

$$\dim B = |Z(\mathfrak{E}) : N||\mathfrak{E} : Z(\mathfrak{E})|^{1/2}|N : Z(N)|^{1/2}/|N|.$$

An error E in $\mathfrak{E}$ is detectable by B if and only if E is contained in the set $\mathfrak{E} - (N\,C_{\mathfrak{E}}(N) - N)$. The distance of the code is defined as

$$d = \min\{\text{wt}(E) | I \neq E \in NC_{\mathfrak{E}}(N) - N\} = \text{wt}(NC_{\mathfrak{E}}(N) - N).$$

If $NC_{\mathfrak{E}}(N) = N$, then the code distance is defined as $\text{wt}(N)$. The group N is the *gauge group* of $\mathscr{C}$ and $Z(N)$ is its *stabilizer*. Observe that the gauge group acts trivially on B. We can construct a special class of such subsystem codes, a class that includes the Bacon–Shor codes, such that the subsystem B can detect all errors in $\mathfrak{E}$ of weight less than d, and can correct all errors in $\mathfrak{E}$ of weight at most $\lfloor (d-1)/2 \rfloor$. This construction is contained in [KS08], where these codes were called *Clifford subsystem codes*.

There is a natural notion of purity for Clifford subsystem codes. Let N be the gauge group of a subsystem code $\mathscr{C}$ with distance d. Then $\mathscr{C}$ is *pure to d'* if there is no error of weight less than d' in N. The code is said to be *exactly pure to d'* if $\text{wt}(N) = \min_{g \in N} \text{wt}(g) = d'$ and it is *pure* if $d' \geq d$. The code is said to be *degenerate* (or *impure*) if it is exactly pure to $d' < d$.

The following upper bound for Clifford subsystem codes is proved in [KS07]. As before, r refers to the number of gauge bits, and we use the generalized notation $[[n, k, r, d]]_q$ to describe the code parameters. This result generalizes the quantum Singleton bound obtained for the subspace version of the stabilizer formalism (which is captured in the case that $r = 0$; i.e., N is Abelian).

Theorem 6.4. *An $\mathbb{F}_q$-linear $[[n, k, r, d \geq 2]]_q$ Clifford subsystem code satisfies*

$$k + r \leq n - 2d + 2. \tag{6.22}$$

Observe that one implication of this result is that the number $n - k - r$ of syndrome measurements is bounded by $2d - 2$, which indicates for a fixed distance d a tradeoff between the code dimension k and the difference $n - r$ between length and number of gauge qubits.

The following quantum Hamming bound for pure Clifford subsystem codes has also been established.

Theorem 6.5. *A pure Clifford subsystem code $\mathscr{C} = A \otimes B$ of distance d, with $R = \dim A$ and $K = \dim B$, satisfies*

$$\sum_{j=0}^{\lfloor (d-1)/2 \rfloor} \binom{n}{j} (q^2 - 1)^j \leq q^n / KR. \tag{6.23}$$

On the one hand, this result (along with its subspace version) suggests it might be unlikely that highly degenerate subsystem codes could lead to much more efficiency in terms of packing subspaces more compactly in Hilbert space. On the other hand, and quite surprisingly, there exist impure subsystem codes that do not satisfy this Hamming bound. The $[[9,1,4,3]]_2$ Bacon–Shor code provides a nice example of this fact. Notice first that, as predicted by Theorem 6.4, it satisfies (in fact optimally) the Singleton bound for Clifford subsystem codes since

$$k + r = 1 + 4 = n - 2d + 2 = 9 - 6 + 2.$$

However, substituting these parameters into Eq. (6.23) yields

$$\sum_{j=0}^{1} \binom{9}{j} 3^j = 27 > 2^{9-5} = 16,$$

and hence the Bacon–Shor code beats the quantum Hamming bound for pure subsystem codes.

The reasons why impure codes can pack more efficiently than pure codes are somewhat subtle. One important reason is that impure subsystem codes can have more degeneracy. For instance, in a pure single error-correcting code all single errors must take the code subspace to mutually orthogonal subspaces. In an impure code, this is not required as two or more distinct errors can take the code subspace to the same subspace. In the case of the Bacon–Shor code, a phase-flip error on any of the first three qubits takes the code to the same subspace, and so the errors cannot be distinguished on the full code subspace. But this is not an issue, since the A qubit subsystem can still be recovered.

6.7 Unitarily correctable codes

Recall that the typical recovery operation in QEC consists of a measurement followed by a unitary reversal conditioned on the result of the measurement. A situation of experimental importance in certain scenarios is one where we have a system with good unitary control but poor measurement capabilities. Thus, we find motivation for consideration of error-correcting codes that do not require a measurement as part of the recovery process. In other words, codes that have evolved unitarily, and hence can be recovered by a unitary operation. Note that this does not mean the noise model itself is unitary, only that it acts unitarily when restricted to the code.

Thus, in OQEC we say that a subsystem B is a *unitarily correctable code* (UCC) for a channel $\mathscr{E}$ if B can be corrected with a unitary recovery operation $\mathscr{R} = \mathscr{U}$, where $\mathscr{U}(\rho) = U\rho U^\dagger$. That is, there is a channel $\mathscr{F}_A$ such that

$$\mathscr{U} \circ \mathscr{E} \circ \mathscr{P}_C = (\mathscr{F}_A \otimes \mathrm{id}_B) \circ \mathscr{P}_C. \tag{6.24}$$

Observe that DFS and NS are precisely the UCC with trivial unitary correction $\mathscr{U} = \mathrm{id}$. Interestingly, the analysis of UCC connects with certain aspects of the theory of CP maps. We now briefly describe this connection in the case of unital channels, and show how it leads to a technique to compute UCC for arbitrary unital channels.

Notice that if $\mathscr{E}$ is both unital and trace-preserving, then so is $\mathscr{E}^\dagger$ and consequently so is $\mathscr{E}^\dagger \circ \mathscr{E}$. It is also easy to see that if B is a noiseless subsystem for $\mathscr{E}^\dagger \circ \mathscr{E}$, and hence $\mathscr{E}^\dagger \circ \mathscr{E} \circ \mathscr{P}_C = (\mathscr{G}_A \otimes \mathrm{id}_B) \circ \mathscr{P}_C$, then B is a correctable subsystem for $\mathscr{E}$ because $\mathscr{E}^\dagger$ is trace-preserving and thus constitutes a recovery operation that satisfies the OQEC definition. But more than this is true, it turns out that B is a UCC for $\mathscr{E}$. Together with its converse, this fact gives us the following result.

Theorem 6.6. *Let $\mathscr{E}$ be a unital quantum channel. Then B is a UCC for $\mathscr{E}$ if and only if B is a NS for $\mathscr{E}^\dagger \circ \mathscr{E}$.*

The proof relies on a number of ancillary results for unital trace-preserving maps, and previously stated characterizations of OQEC codes. We briefly sketch the proof here.

First consider the case that B is a UCC for $\mathscr{E}$. Given that B is a correctable subsystem, it follows from the testable conditions of Theorem 6.1 or condition (iv) of Theorem 6.3 that $\mathscr{P}_\mathscr{C} \circ \mathscr{E}^\dagger \circ \mathscr{E} \circ \mathscr{P}_\mathscr{C} = (\mathscr{G}_A \otimes \mathrm{id}_B) \circ \mathscr{P}_\mathscr{C}$ for some channel $\mathscr{G}_A$. Now, it is clear that if not for the leading $\mathscr{P}_\mathscr{C}$ in this expression, B would satisfy the definition of a NS for $\mathscr{E}^\dagger \circ \mathscr{E}$. The rest of the proof for this direction of the theorem is focused on establishing that the leading $\mathscr{P}_\mathscr{C}$ can be dropped from this expression in the case of UCC and unital channels. A surprising amount of technical effort is required to do so, and we point the reader to [KS06] for details.

For the other direction, note that if B is a passive code for $\mathscr{E}^\dagger \circ \mathscr{E}$, then B is correctable for $\mathscr{E}$ since the dual map $\mathscr{E}^\dagger$ is a valid recovery operation. Thus, by condition (iv) of Theorem 6.3, there exist subsystems $\mathscr{C}' = A' \otimes B'$, a channel $\mathscr{F}_{A'|A}$, and a unitary channel $\mathscr{V}_{B'|B}$ such that $\mathscr{E} \circ \mathscr{P}_\mathscr{C} = (\mathscr{F}_{A'|A} \otimes \mathscr{V}_{B'|B}) \circ \mathscr{P}_\mathscr{C}$. The technical component in this direction of the proof again uses properties of unital channels to show that $\text{rank}(\mathscr{F}_{A'|A}(I_A)) = \text{rank}(I_A)$. This implies the existence of a unitary $\mathscr{U}$ such that $\mathscr{U} \circ \mathscr{E} \circ \mathscr{P}_\mathscr{C} = \mathscr{U} \circ (\mathscr{F}_{A'|A} \otimes \mathscr{V}_{B'|B}) \circ \mathscr{P}_\mathscr{C} = (\mathscr{F}_A \otimes \text{id}_B) \circ \mathscr{P}_\mathscr{C}$ for some channel $\mathscr{F}_A$ on A, and hence B is a UCC for $\mathscr{E}$.

There is an important object from the theory of CP maps that connects with the UCC theory for channels [CJK09]. The *multiplicative domain* of a CP map $\mathscr{E}$ was first considered in operator theory in the 1970s, and is defined as $MD(\mathscr{E}) := \{\sigma : \mathscr{E}(\sigma\gamma) = \mathscr{E}(\sigma)\mathscr{E}(\gamma)$ and $\mathscr{E}(\gamma\sigma) = \mathscr{E}(\gamma)\mathscr{E}(\sigma) \ \forall \gamma\}$. It is evident that this set forms an algebra, and in the case that $\mathscr{E}$ is a unital CP map, it is defined through its internal structure as follows [C74]:

$$MD(\mathscr{E}) := \{\sigma : \mathscr{E}(\sigma\sigma^\dagger) = \mathscr{E}(\sigma)\mathscr{E}(\sigma)^\dagger, \ \mathscr{E}(\sigma^\dagger\sigma) = \mathscr{E}(\sigma)^\dagger\mathscr{E}(\sigma)\}. \tag{6.25}$$

Interestingly, the subsystem codes defined via the algebra structure of $MD(\mathscr{E})$ are *precisely* the UCC for $\mathscr{E}$. In the arbitrary (not necessarily unital) case, one can still consider the multiplicative domain, though the characterization of Eq. 6.25 no longer holds. In that case, the algebra $MD(\mathscr{E})$ encodes a proper subclass of UCC that can be computed from properties of the map.

By combining Theorem 6.6 with the fixed point theorem for unital channels [K03b], we obtain a method for finding the UCC for *any* unital channel $\mathscr{E}$. Specifically, the DFS and NS of $\mathscr{E}^\dagger \circ \mathscr{E}$ are obtained from the fixed point set $\text{Fix}(\mathscr{E}^\dagger \circ \mathscr{E})$ via its algebra structure, as discussed above in the case of passive codes for unital channels. We point the reader to [SMK+08] for a detailed analysis of UCC for an experimentally relevant class of examples. Here we present an example.

Example 6.1. *First consider the swap operation $|\psi\rangle \otimes |\phi\rangle \mapsto |\phi\rangle \otimes |\psi\rangle$ on a composite quantum system $\mathscr{H} = \mathscr{A} \otimes \mathscr{R}_A$ made up of a subsystem $\mathscr{A}$ and a reference system $\mathscr{R}_A = \mathscr{A}$. It is clear that both the subsystem $\mathscr{A}$ and its copy can be returned to their initial locations by simply applying the swap operation again (which is equal to its dual). Now, one could note that the swap operation itself has an NS of the same size; namely the symmetric space $|\psi\rangle \otimes |\psi\rangle$. But it is easy to find examples of channels with no passive codes, for which the composition of the map with its dual has a nontrivial passive code. To this end, consider a two-qubit system exposed to decoupled phase-flips. The associated error model satisfies $\mathscr{E}(\rho) = pZ_1\rho Z_1 + (1-p)Z_2\rho Z_2$ for some fixed probability $0 < p < 1$ and $Z_1 = Z \otimes I_2$, $Z_2 = I_2 \otimes Z$. In this case $\mathscr{E}$ has no passive codes. This follows from the fact that the noise commutant $\{Z_1, Z_2\}'$ is isomorphic to the algebra $\mathbb{C} \oplus \mathbb{C} \oplus \mathbb{C} \oplus \mathbb{C}$. Thus, only classical information can be safely sent through the channel. However, the operators supported on the subspace spanned by $|0_L\rangle = |00\rangle$ and $|1_L\rangle = |11\rangle$ form a DFS for $\mathscr{E}^\dagger \circ \mathscr{E}$. Indeed, the set of operators $\sigma = a|00\rangle\langle 00| + b|00\rangle\langle 11| + c|11\rangle\langle 00| + d|11\rangle\langle 11|$ form a subalgebra of the commutant $\text{Fix}(\mathscr{E}^\dagger \circ \mathscr{E}) = \{Z_1^\dagger Z_2, Z_1^\dagger Z_1, Z_2^\dagger Z_1 Z_2^\dagger Z_2\}' = \{Z_1 Z_2\}'$. A unitary correction operation guaranteed by Theorem 6.6 in this case happens to be the controlled phase-flip operation $U = |00\rangle\langle 00| + |01\rangle\langle 01| + |10\rangle\langle 10| - |11\rangle\langle 11|$.*

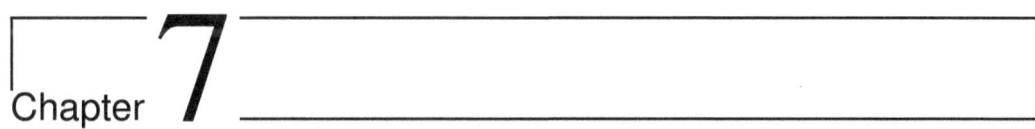

Chapter 7

Entanglement-assisted quantum error-correcting codes

Todd A. Brun and Min-Hsiu Hsieh

7.1 Introduction

Quantum error-correcting codes (QECCs) have turned out to have many applications; but their primary purpose, of course, is to protect quantum information from noise. This need manifests itself in two rather different situations. The first is in quantum computing. The qubits in a quantum computer are subject to noise due both to imprecision in the operations (or gates) and to interactions with the external environment. They undergo errors when they are acted upon, and when they are just being stored (though hopefully not at the same rate). All the qubits in the computer must be kept error-free long enough for the computation to be completed. This is the principle of fault tolerance.

A rather different situation occurs in quantum *communication*. Here, the sender and receiver (Alice and Bob) are assumed to be physically separated, and the qubits travel from the sender to the receiver over a (presumably noisy) channel. This channel is assumed to be the dominant source of errors. While the qubits may undergo processing before and after transmission, errors during this processing are considered negligible compared to the channel errors. This picture of transmission through a channel is similar in spirit to the paradigm underlying much of classical information theory.

The types of codes that are of interest in quantum computation and quantum communication may be quite different. In computation, reliability is key. Therefore, codes with a high distance (such as concatenated codes) are used. Also, the choice of code may be highly constrained in other ways: for instance, it is very important that the codes allow efficient circuits for encoded logic gates (such as the transversal gates allowed by the Steane code). The more efficient these circuits, the better the fault-tolerant threshold. (See Chapter 5 for a fuller discussion of this.) By contrast, in communication one often imagines sending a very large number of qubits, perhaps

---

Quantum Error Correction, ed. Daniel A. Lidar and Todd A. Brun. Published by Cambridge University Press.
© Cambridge University Press 2013.

even a continuous stream of them (as in the convolutional codes of Chapter 9). One is generally trying to maximize the rate of transmission, often by encoding many qubits into a large block, within the constraint of a low error probability for the block. In the asymptotic limit, one would like to achieve the actual capacity of the channel – the maximum rate of communication possible.

Because the sender and receiver are separated, joint unitary transformations are impossible. However, they might be able to draw on other resources, such as extra classical communication, pre-shared randomness, or pre-shared entanglement. In this chapter we study the use of pre-shared entanglement in quantum communication, and how we can design QECCs that use entanglement to boost either the rate of communication or the number of errors that can be corrected. Codes that use pre-shared entanglement are called *entanglement-assisted quantum error-correcting codes* (EAQECCs). We will study a large class of these codes that generalizes the usual stabilizer formalism from Chapter 2. We will also see how these codes can be readily constructed from classical linear codes, and how they can be useful tools in building standard QECCs.

7.1.1 Entanglement-assisted codes

In a standard QECC, the encoding operation proceeds in two steps. First, to the quantum state $|\psi\rangle$ of k qubits that one wishes to encode, one appends some number of ancilla qubits in a standard state (usually $|0\rangle$), and then applies an encoding unitary U_{enc}:

$$|\psi\rangle \to |\psi\rangle \otimes |0\rangle^{\otimes n-k} \to |\Psi_L\rangle = U_{\text{enc}}|\psi\rangle \otimes |0\rangle^{\otimes n-k}, \qquad (7.1)$$

where $|\Psi_L\rangle$ is the encoded or logical state. The systems in states $|\psi\rangle$ and $|0\rangle$ are initially in the possession of Alice, the sender. The encoding unitary acts on the space of the input qubits and ancillas together.

In an EAQECC, one can append not only ancillas, but also ebits,[1] before doing the encoding unitary:

$$|\psi\rangle \to |\psi\rangle \otimes |0\rangle^{\otimes n-k-c} \otimes |\Phi_+\rangle_{AB}^{\otimes c} \to (U_{\text{enc}} \otimes \hat{I}_B)|\psi\rangle \otimes |0\rangle^{\otimes n-k-c} \otimes |\Phi_+\rangle_{AB}^{\otimes c}. \qquad (7.2)$$

The states $|\Phi_+\rangle_{AB}$ are EPR pairs shared between the sender (Alice) and the receiver (Bob). The encoding operation U_{enc} acts on the qubits in Alice's possession; we write $(U_{\text{enc}} \otimes \hat{I}_B)$ above to indicate that the encoding acts as the identity on Bob's halves of the ebits. Obviously, in order to append c ebits to the information qubits, Alice and Bob must have c ebits of pre-shared entanglement. After Alice does the encoding, all n of her qubits are sent through the channel. So this procedure consumes c ebits of preshared entanglement. However, note that Bob's halves of the c ebits do not pass through the channel. These ebits were prepared ahead of time, by entanglement distillation or a similar procedure (see [BBP+96b, BDS+96] and Chapter 22). Because Bob's qubits do not pass through the channel, they are assumed to be error-free. A schematic picture of this protocol is shown in Fig. 7.1.

Why should we wish to use shared entanglement in our encoding? There are two ways to see how this could enhance the power of the code. First, we can compare EAQECCs to *teleportation*.

---

[1] An ebit is often taken to mean one unit of bipartite entanglement, the amount of entanglement that is contained in a maximally entangled two-qubit state (Bell state). Here we use ebit to denote a maximally entangled pair.

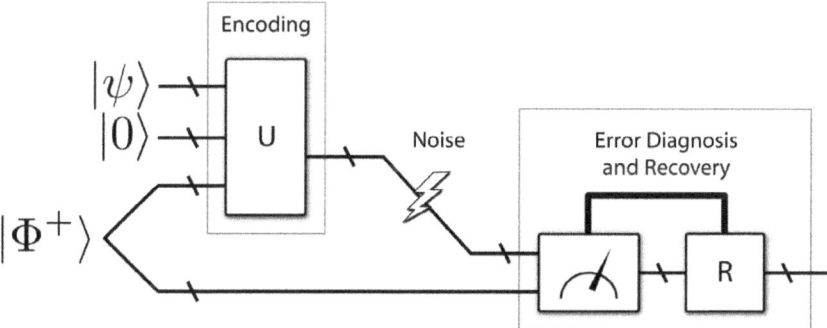

Fig. 7.1. Schematic structure of an entanglement-assisted quantum error-correcting code. Alice and Bob share ebits that Alice uses in encoding her quantum information; the qubits that are initially on Alice's side all pass through the noisy channel, but Bob's halves of the initial ebits do not. Reprinted with permission from [WKB07]. Copyright 2007, American Physical Society.

In teleportation, an ebit can be used (in combination with classical communication) to transmit one qubit perfectly from the sender to the receiver. We might therefore expect that making use of ebits in error correction could boost the rate of transmission.

Second, we can compare EAQECCs to *superdense coding*. In superdense coding, by using an ebit the sender can send two classical bits of information to the receiver by means of a single qubit. Each ancilla bit in a standard QECC can be thought of as holding one classical bit of information about any errors that have occurred. Replacing an ancilla with one half of an ebit could, in principle, allow the receiver to extract *two* bits of classical information about the errors; with more information, more errors can be corrected. We will see, in the constructions that follow, that we can think of the enhanced power of EAQECCs in both of these ways, but the superdense coding interpretation is more often the best fit. EAQECCs can achieve some communication tasks with fewer resources (or less chance of error) than would be needed by a standard QECC plus teleportation [HW10].

The possibility of constructing entanglement-assisted codes was suggested in [BDS+96], though no practical codes were constructed. From asymptotic results in quantum information theory, we know that having pre-shared entanglement allows an enhanced rate of quantum communication between a sender and receiver [DHW04, DHW08]. EAQECCs are the finite-length realization of this idea. By constructing larger and larger code blocks, in principle one can approach the entanglement-assisted quantum capacity of the channel. An example of such a code was constructed by Bowen in [B02a], starting from a standard stabilizer QECC (see Chapter 2). In [BDH06a, BDH06b] an essentially complete theory of stabilizer EAQECCs was constructed. That theory is presented in this chapter.

Note, however, that in general shared entanglement does not come for free: entanglement must be established between the sender and receiver, and this in general requires the use of quantum channels, and perhaps additional entanglement purification. Therefore, EAQECCs do not outperform standard quantum codes under all circumstances. The enhanced communication rates that come from the use of entanglement must be paid for in establishing the entanglement in the first place. For many purposes, what is important is the *net rate* of the code: the number of qubits transmitted minus the number of ebits consumed. With this measure, EAQECCs tend to perform about the same as standard codes. Even in this case, though, EAQECCs sometimes have advantages: as we will see, there are many fewer restrictions on the construction of EAQECCs than standard codes.

7.2 Constructing EAQECCs

7.2.1 Noncommuting "stabilizers"

First, let us recall the stabilizer formalism for conventional QECCs, as presented in Chapter 2. Let G_n be the n-qubit Pauli group [NC00]. Every operator in G_n has eigenvalues either ± 1 or $\pm i$. Let $S \subset G_n$ be an Abelian subgroup that does not contain $-I$. Then this subgroup has a common eigenspace $C(S)$ of $+1$ eigenvectors, which we call the *code space* determined by the stabilizer S. Later on, we will just use C to denote the code space. Typically, the stabilizer is represented by a minimal generating set $\{g_1, \ldots, g_m\}$, which makes this a compact way to specify a code (analogous to specifying a classical linear code by its parity check matrix). We write $S = \langle g_1, \ldots, g_m \rangle$ to denote that S is generated by $\{g_1, \ldots, g_m\}$.

Let $E \subset G_n$ be a set of possible errors. If a particular error $E_1 \in E$ anticommutes with any of the generators of S, then the action of that error can be detected by measuring the generators; if the measurement returns -1 instead of 1, we know an error has occurred. On the other hand, if the error is actually *in* the stabilizer S, then it leaves all the states in C unchanged. The code C can correct any error in E if either $E_2^\dagger E_1 \notin Z(S)$ or $E_2^\dagger E_1 \in S$ for all pairs of errors E_1 and E_2 in E, where $Z(S)$ is the *centralizer* of S.

We can now generalize this description to the entanglement-assisted case. Given a *non-Abelian* subgroup $S \subset G_n$ of size 2^m, there exists a set of generators $\{\bar{Z}_1, \ldots, \bar{Z}_{s+c}, \bar{X}_1, \ldots, \bar{X}_c\}$ for S with the following commutation relations:

$$\begin{aligned}
[\bar{Z}_i, \bar{Z}_j] &= 0 & \forall i, j, \\
[\bar{X}_i, \bar{X}_j] &= 0 & \forall i, j, \\
[\bar{X}_i, \bar{Z}_j] &= 0 & \forall i \neq j, \\
\{\bar{X}_i, \bar{Z}_i\} &= 0 & \forall i.
\end{aligned} \quad (7.3)$$

(We will see shortly how to find this set of generators for any subgroup.) Here, $[A, B]$ is the commutator and $\{A, B\}$ the anticommutator of A with B. The parameters s and c satisfy $s + 2c = m$. Let S_I be the *isotropic* subgroup generated by $\{\bar{Z}_{c+1}, \ldots, \bar{Z}_{c+s}\}$ and S_E be the *entanglement* subgroup generated by $\{\bar{Z}_1, \ldots, \bar{Z}_c, \bar{X}_1, \ldots, \bar{X}_c\}$. The numbers of generators of S_I and pairs of generators of S_E describe the number of ancillas and the number of ebits, respectively, needed by the corresponding EAQECC. The pair of subgroups (S_I, S_E) defines an $[[n, k; c]]$ EAQECC $C(S)$. We use the notation $[[n, k; c]]$ to denote an EAQECC that encodes $k = n - s - c$ logical qubits into n physical qubits, with the help of s ancillas and c shared ebits. These n qubits are transmitted from Alice (the sender) to Bob (the receiver), who measures them together with his halves of the c ebits to correct any errors and decode the k logical qubits. We define k/n as the *rate* and $(k-c)/n$ as the *net rate* of the code. (Sometimes we will write $[[n, k, d; c]]$ to indicate that the distance of the code is d, meaning it can correct at least $\lfloor \frac{d-1}{2} \rfloor$ single-qubit errors.)

7.2.2 The Gram–Schmidt procedure

We now illustrate the idea of the entanglement-assisted stabilizer formalism by an example. Let S be the group generated by the following noncommuting set of operators:

$$\begin{aligned} M_1 &= Z\ X\ Z\ I, \\ M_2 &= Z\ Z\ I\ Z, \\ M_3 &= X\ Y\ X\ I, \\ M_4 &= X\ X\ I\ X. \end{aligned} \quad (7.4)$$

It is easy to check the commutation relations of this set of generators: M_1 anticommutes with the other three generators, M_2 commutes with M_3 and anticommutes with M_4, and M_3 and M_4 anticommute. We begin by finding a different set of generators for S with a particular class of commutation relations. We then relate S to a group B with a particularly simple form, and discuss the error-correcting conditions using B. Finally, we relate these results back to the group S.

To see how this works, we need two lemmas. (We only sketch the proofs – these results are well known.) The first lemma shows that there exists a new set of generators for S such that S can be decomposed into an "isotropic" subgroup S_I, generated by a set of commuting generators, and a "symplectic" subgroup S_S, generated by a set of anticommuting generator pairs [FCY+04].

Lemma 7.1. Given any arbitrary subgroup V in G_n that has 2^m distinct elements up to overall phase, there exists a set S of m independent generators for V of the forms $\{\bar{Z}_1, \bar{Z}_2, \ldots, \bar{Z}_\ell, \bar{X}_1, \ldots, \bar{X}_{m-\ell}\}$ where $m/2 \leq \ell \leq m$, such that $[\bar{Z}_i, \bar{Z}_j] = [\bar{X}_i, \bar{X}_j] = 0$, for all i, j; $[\bar{Z}_i, \bar{X}_j] = 0$, for all $i \neq j$; and $\{\bar{Z}_i, \bar{X}_i\} = 0$, for all i. Let $V_I = \langle \bar{Z}_{m-\ell+1}, \ldots, \bar{Z}_\ell \rangle$ denote the isotropic subgroup generated by the set s_I of commuting generators, and let $V_S = \langle \bar{Z}_1, \ldots, \bar{Z}_{m-\ell}, \bar{X}_1, \ldots, \bar{X}_{m-\ell} \rangle$ denote the symplectic subgroup generated by the set s_S of anticommuting generator pairs. Then, with a slight abuse of notation, $V = \langle V_I, V_S \rangle$ denotes that V is generated by subgroups V_I and V_S.

Proof. We sketch a constructive procedure for finding the sets of generators, which is analogous to the Gram–Schmidt procedure for finding an orthonormal basis for a vector space. Suppose one has any set of generators $\{g_1, \ldots, g_m\}$ for the subgroup V. We will successively find new sets of generators for V while assigning the generators we find to either the set s_I of isotropic generators or the set s_S of symplectic generators. The generators of s_S come in anticommuting pairs.

We start by assigning the anticommuting pairs. Suppose that one has so far assigned k pairs of generators to s_S. Go through the list of as-yet-unassigned generators, and find any two generators g and h that anticommute. These will be our next symplectic pair. First, however, we must make them commute with all the remaining generators. Go through the whole list of unassigned generators (except g and h). If a generator anticommutes with g, then multiply that generator by h and replace it. The new generator will commute with g, and the new set generates the same group. Similarly, if a generator anticommutes with h, multiply it by g and replace it. (Note that some generators may be multiplied by both g and h, and some by neither.) Once we have gone through the entire list, we will have found a new set of generators such that g and h anticommute and all other generators commute with them both. Relabel g as $\bar{Z}_{k+1}$ and h as $\bar{X}_{k+1}$, assign the pair to s_S, and let $k \to k+1$.

Continue this procedure until either all the generators have been assigned to s_S, or all of the remaining unassigned generators commute. If there are a total of $m - \ell$ pairs of generators in s_S, we relabel the remaining generators $\bar{Z}_{m-\ell+1}, \ldots, \bar{Z}_\ell$ and assign them all to s_I. This procedure is optimal, in the sense that it produces the minimum number of symplectic pairs (and hence the minimum number of ebits for the code). □

For the group S that we are considering, generated by (7.4), we can follow the steps of the procedure outlined in the lemma to construct such a set of independent generators. We start by taking the first generator M_1 and labeling it $\bar{Z}_1$. We then find the first anticommuting generator – which happens to be M_2 – and label it $\bar{X}_1$. We must then eliminate any anticommutation with the remaining generators. $\bar{Z}_1$ anticommutes with M_3 and M_4, so we multiply each of them by $\bar{X}_1$. M_4 anticommutes with $\bar{X}_1$, so we multiply it by $\bar{Z}_1$. The two new generators that result commute with $\bar{Z}_1$ and $\bar{X}_1$, and also commute with each other; so we label them $\bar{Z}_2$ and $\bar{Z}_3$. The resulting set of generators is

$$\begin{aligned} \bar{Z}_1 &= Z\ X\ Z\ I, \\ \bar{X}_1 &= Z\ Z\ I\ Z, \\ \bar{Z}_2 &= Y\ X\ X\ Z, \\ \bar{Z}_3 &= Z\ Y\ Y\ X. \end{aligned} \quad (7.5)$$

so that $S_S = \langle \bar{Z}_1, \bar{X}_1 \rangle$, $S_I = \langle \bar{Z}_2, \bar{Z}_3 \rangle$, and $S = \langle S_I, S_S \rangle$.

The choice of the notation $\bar{Z}_i$ and $\bar{X}_i$ is not accidental: these generators have exactly the same commutation relations as Pauli operators Z_i and X_i on a set of qubits labeled by i. We now see that the subgroup they generate matches one-to-one with this simpler subgroup. Let B be the group generated by the following set of Pauli operators:

$$\begin{aligned} Z_1 &= Z\ I\ I\ I, \\ X_1 &= X\ I\ I\ I, \\ Z_2 &= I\ Z\ I\ I, \\ Z_3 &= I\ I\ Z\ I. \end{aligned} \quad (7.6)$$

From the previous lemma, $B = \langle B_I, B_S \rangle$, where $B_S = \langle Z_1, X_1 \rangle$ and $B_I = \langle Z_2, Z_3 \rangle$. Therefore, groups B and S are *isomorphic*, which is denoted $B \cong S$. We can relate S to the group B by the following lemma [BFG06].

Lemma 7.2. If B and S are both subgroups of G_n, and $B \cong S$, then there exists a unitary U such that for all $g \in B$ there exists an $h \in S$ such that $g = UhU^{-1}$ up to an overall phase.

Proof. We only sketch the proof here. First, apply Lemma 7.1 to both B and S to find a set of generators for each group in the standard form. We call the generators of B $\{\bar{Z}_1, \ldots, \bar{Z}_r, \bar{X}_1, \ldots, \bar{X}_c\}$, and the generators of S $\{\hat{Z}_1, \ldots, \hat{Z}_r, \hat{X}_1, \ldots, \hat{X}_c\}$, where $r = s + c$. (Because $B \cong S$ the parameters r and c must be the same for both groups.) The symplectic subgroup of B is generated by $\bar{Z}_1, \ldots, \bar{Z}_c, \bar{X}_1, \ldots, \bar{X}_c$, the isotropic subgroup of B is generated by $\bar{Z}_{c+1}, \ldots, \bar{Z}_r$, and similarly for S.

Starting from the set of generators for B, we add generators to get a complete set of generators for G_n: first we add symplectic partners $\bar{X}_{c+1}, \ldots, \bar{X}_r$ for the isotropic generators $\bar{Z}_{c+1}, \ldots, \bar{Z}_r$, and then we add $n - r$ additional symplectic pairs $(\bar{Z}_{r+1}, \bar{X}_{r+1}), \ldots, (\bar{Z}_n, \bar{X}_n)$ to get a complete set of generators. In exactly the same way, we add new generators $\hat{X}_{c+1}, \ldots,$

$\hat{X}_r$ and $(\hat{Z}_{r+1}, \hat{X}_{r+1}), \ldots, (\hat{Z}_n, \hat{X}_n)$ to the generators for S to get a different complete set of generators for G_n.

If any of the generators have eigenvalues $\pm i$, multiply them by i, so that all generators have eigenvalues ± 1. (This is allowed because we are ignoring the overall phases.) Now define the following two states: $|\bar{0}\rangle$ is the simultaneous $+1$ eigenstate of the generators $\bar{Z}_1, \ldots, \bar{Z}_n$, and $|\hat{0}\rangle$ is the simultaneous $+1$ eigenstate of the generators $\hat{Z}_1, \ldots, \hat{Z}_n$. Starting from these two states, we construct two orthonormal bases. Let $\mathbf{b} = (b_1, \ldots, b_n)$ be a string of n bits. Define the basis states $|\bar{\mathbf{b}}\rangle = \bar{X}_1^{b_1} \cdots \bar{X}_n^{b_n} |\bar{0}\rangle$ and $|\hat{\mathbf{b}}\rangle = \hat{X}_1^{b_1} \cdots \hat{X}_n^{b_n} |\hat{0}\rangle$. It is easy to show that these form two orthonormal bases. Therefore we can define a unitary operator

$$U = \sum_{\mathbf{b}} |\bar{\mathbf{b}}\rangle \langle \hat{\mathbf{b}}|.$$

One can now show that $\bar{Z}_j = U \hat{Z}_j U^\dagger$ and $\bar{X}_j = U \hat{X}_j U^\dagger$ for all j. This implies that any operator in S can be mapped to a corresponding operator in B, by writing the two operators as products of corresponding generators. This completes our sketch of the proof. □

As a consequence of this lemma, the error-correcting codes $C(B)$ and $C(S)$ are also related by a unitary transformation. In what follows, we will use the straightforward group B to discuss the error-correcting conditions for an EAQECC, and then translate the results back to the code $C(S)$. As we will see, the unitary U constructed in the lemma can be thought of as the *encoding* operator for the code $C(S)$. Note that while this unitary has been expressed in abstract terms, there are efficient techniques to directly find a quantum circuit for U in terms of CNOTs, Hadamards, and phase gates [CG97, GRB03, WB10a, WB10b].

7.2.3 Anticommuting pairs and entanglement

What is the code space $C(B)$ described by B in (7.6)? Because B is not a commuting group, the usual definition of a QECC $C(B)$ does not apply, since the generators do not have a common $+1$ eigenspace. However, by *extending* the generators, we can find a new group that *is* Abelian, and for which the usual definition of code space *does* apply. The qubits of the codewords will be embedded in a larger space. Notice that we can append a Z operator at the end of Z_1, an X operator at the end of X_1, and an identity at the end of Z_2 and Z_3, to make B into a new Abelian group B_e:

$$\begin{array}{rccccc|c}
Z_1' = & Z & I & I & I & Z \\
X_1' = & X & I & I & I & X \\
Z_2' = & I & Z & I & I & I \\
Z_3' = & I & I & Z & I & I
\end{array}. \quad (7.7)$$

The four original qubits are possessed by Alice (the sender), but the additional qubit is possessed by Bob (the receiver) and is not subject to errors. Let B_e be the extended group generated by $\{Z_1', X_1', Z_2', Z_3'\}$. We define the code space $C(B)$ to be the simultaneous $+1$ eigenspace of all elements of B_e, and we can write it down explicitly in this case:

$$C(B) = \{|\Phi\rangle^{AB} |0\rangle |0\rangle |\psi\rangle\}, \quad (7.8)$$

where $|\Phi\rangle^{AB}$ is the maximally entangled state $(|00\rangle + |11\rangle)/\sqrt{2}$ shared between Alice and Bob, and $|\psi\rangle$ is an arbitrary single-qubit pure state. (Bob's qubit corresponds to the fifth column in

(7.7).) Because entanglement is used, this is an EAQECC. The number of ebits c needed for the encoding is equal to the number of anticommuting pairs of generators in B_S. The number of ancilla bits s equals the number of independent generators in B_I. The number k of encoded qubits is equal to $n - c - s$. Therefore, $C(B)$ is a [[4,1;1]] EAQECC with zero net rate: $n = 4$, $c = 1$, $s = 2$, and $k = 1$. Note that zero net rate does not mean that no qubits are transmitted by this code! Rather, it implies that the number of ebits needed is equal to the number of qubits transmitted. In general, $k - c$ can be positive, negative, or zero.

Now we will see how the error-correcting conditions are related to the generators of B. If an error $E_a \otimes I^B$ anticommutes with one or more of the operators in $\{Z'_1, X'_1, Z'_2, Z'_3\}$, it can be detected by measuring these operators. This will only happen if the error E_a on Alice's qubits anticommutes with one of the operators in the original set of generators $\{Z_1, X_1, Z_2, Z_3\}$, since the entangled bit held by Bob is assumed to be error-free. Alternatively, if $E_a \otimes I^B \in B_e$, or equivalently $E_a \in B_I$, then E_a does not corrupt the encoded state. (In this case we call the code *degenerate*.) Altogether, $C(B)$ can correct a set of errors E_0 if and only if $E_a^\dagger E_b \in B_I \cup (G_4 - Z(B))$ for all $E_a, E_b \in E_0$.

With this analysis of B, we can go back to determine the error-correcting properties of our original stabilizer S. We can construct a QECC from a non-Abelian group S if entanglement is available, just as we did for the group B. We add extra operators Z and X on Bob's side to make S Abelian as follows:

$$\begin{array}{rcccccc|c}
\bar{Z}'_1 = & Z & X & Z & I & & & Z \\
\bar{X}'_1 = & Z & Z & I & Z & & & X \\
\bar{Z}'_2 = & Y & X & X & Z & & & I \\
\bar{Z}'_3 = & Z & Y & Y & X & & & I
\end{array} \qquad (7.9)$$

where the extra qubit is once again assumed to be possessed by Bob and to be error-free. Let S_e be the group generated by the above operators. Since $B \cong S$, let U^{-1} be the unitary from Lemma 7.2. Define the code space $C(S)$ by $C(S) = (U^{-1} \otimes I^B)C(B)$, where the unitary U^{-1} acts on the qubits on Alice's side. This unitary U^{-1} can be interpreted as the encoding operation of the EAQECC defined by S. Observe that the code space $C(S)$ is a simultaneous eigenspace of all elements of S_e. As in the case of $C(B)$, the code $C(S)$ can correct a set of errors E if and only if

$$E_a^\dagger E_b \in S_I \cup (G_4 - Z(S)), \qquad (7.10)$$

for all $E_a, E_b \in E$.

The algebraic description is somewhat abstract, so let us translate this into a physical picture. Alice wishes to encode a single ($k = 1$) qubit state $|\psi\rangle$ into four ($n = 4$) qubits, and transmit them through a noisy channel to Bob. Initially, Alice and Bob share a single ($c = 1$) maximally entangled pair of qubits – one ebit. Alice performs the encoding operation U^{-1} on her bit $|\psi\rangle$, her half of the entangled pair, and two ($s = 2$) ancilla bits. She then sends the four qubits through the channel to Bob. Bob measures the extended generators $\bar{Z}'_1, \bar{X}'_1, \bar{Z}'_2$, and $\bar{Z}'_3$ on the four received qubits plus his half of the entangled pair. The outcome of these four measurements gives the error syndrome; as long as the error set satisfies the error-correcting condition (7.10), Bob can correct the error and decode the transmitted qubit $|\psi\rangle$. We can see schematically how this procedure works in Fig. 7.1.

In fact, this particular example is a [[4,1,3;1]] EAQECC: it can correct any single-qubit error. No standard QECC with $n < 5$ can correct an arbitrary single-qubit error. Clearly, the use of an ebit enhances the error-correcting power of the code. However, the need for a shared ebit is also a limitation, since the ebit must presumably have been shared originally by extra channel uses – this is reflected in the fact that the net rate of the code is zero.

We have worked out the procedure for a particular example, but any entanglement-assisted quantum error correction code will function in the same way. If we have c anticommuting pairs of generators $\bar{Z}_j$ and $\bar{X}_j$ in S_E, we resolve their anticommutation by adding a Z and X operator, respectively, to the two operators. This additional operator acts on an extra qubit on Bob's side. We will need one such extra qubit for each pair, or c in all; these correspond to Bob's halves of the c ebits initially shared between Alice and Bob. The particular parameters n, k, c, s will vary depending on the code. It should be noted that the first example of entanglement-assisted quantum error correction produced a [[3, 1, 3; 2]] EAQECC based on the [[5, 1, 3]] standard QECC [B02a]. We have now produced a completely general description, which also avoids the need for a commuting stabilizer group.

7.2.4 The symplectic representation of EAQECCs

There is another very useful representation for stabilizer QECCs, in which tensor products of Pauli operators are represented by pairs of bit strings: $e^{i\phi}X^{\mathbf{a}}Z^{\mathbf{b}} \to (\mathbf{a}|\mathbf{b})$, where $\mathbf{a}$ and $\mathbf{b}$ are strings of bits $\mathbf{a} \equiv a_1 a_2 \cdots a_n$ and $\mathbf{b} \equiv b_1 b_2 \cdots b_n$, and we use the power notation:

$$X^{\mathbf{a}} \equiv X^{a_1} \otimes X^{a_2} \otimes \cdots \otimes X^{a_n}, \quad Z^{\mathbf{b}} \equiv Z^{b_1} \otimes Z^{b_2} \otimes \cdots \otimes Z^{b_n}. \tag{7.11}$$

The overall phase is lost in this representation, but for our present purposes this is not important. We also define a map in the other direction, from a pair of bit strings to a Pauli operator: $\mathbf{w} \to N_{\mathbf{w}} \equiv X^{\mathbf{a}} Z^{\mathbf{b}}$, where $\mathbf{w} = (\mathbf{a}|\mathbf{b})$. This *symplectic* representation is also used in Chapters 9, 12, and 22.

The symplectic representation has many advantages. Up to a phase, multiplication of Pauli operators is given by binary vector addition. A set of generators $g_1, \ldots, g_{n-k}$ for the stabilizer is given by a set of $n-k$ binary strings of length $2n$, which we write as the rows of a matrix. We can thus represent a quantum stabilizer code by a quantum check matrix, quite analogous to a classical linear code:

$$g_1, \ldots, g_{n-k} \longrightarrow \hat{H} = (H_X | H_Z), \tag{7.12}$$

where H_X and H_Z are $(n-k) \times n$ binary matrices. The rows of $\hat{H}$ represent the stabilizer generators, and the row space of $\hat{H}$ represents the full set of stabilizer operators.

While the symplectic representation does not include the overall phase of Pauli group elements, it does capture the commutation relations between different elements of the Pauli group. If two group elements g and g' are represented by the strings $(\mathbf{a}|\mathbf{b})$ and $(\mathbf{a}'|\mathbf{b}')$, respectively, then their commutation relation is given by the *symplectic inner product* of the two strings:

$$(\mathbf{a}|\mathbf{b}) \odot (\mathbf{a}'|\mathbf{b}') = \mathbf{a} \cdot \mathbf{b}' + \mathbf{a}' \cdot \mathbf{b}, \tag{7.13}$$

where $\mathbf{a} \cdot \mathbf{b}'$ denotes the usual binary inner product. The symplectic inner product between two strings is either 0 or 1; if it is 0, then the operators they represent commute, otherwise they anticommute:

$$N_{\mathbf{w}} N_{\mathbf{w'}} = (-1)^{\mathbf{w} \odot \mathbf{w'}} N_{\mathbf{w'}} N_{\mathbf{w}}.$$

As we will see, this enables us to derive a compact formula for the amount of entanglement needed by a given code. For a standard stabilizer QECC, the symplectic inner product between any two rows of the quantum check matrix $\hat{H}$ must be 0.

For EAQECCs, we typically use the symplectic representation in two slightly different but related ways. First, we can represent the (in general noncommuting) generators on Alice's qubits by an $(n - k) \times 2n$ check matrix, whose rows can have nonzero symplectic inner product. So, for example, the generators in Eq. (7.5) are represented by the following quantum check matrix:

$$\begin{array}{c} \bar{Z}_1 = Z\ X\ Z\ I \\ \bar{X}_1 = Z\ Z\ I\ Z \\ \bar{Z}_2 = Y\ X\ X\ Z \\ \bar{Z}_3 = Z\ Y\ Y\ X \end{array} \longrightarrow \begin{pmatrix} 0 & 1 & 0 & 0 & | & 1 & 0 & 1 & 0 \\ 0 & 0 & 0 & 0 & | & 1 & 1 & 0 & 1 \\ 1 & 1 & 1 & 0 & | & 1 & 0 & 0 & 1 \\ 0 & 1 & 1 & 1 & | & 1 & 1 & 1 & 0 \end{pmatrix}. \tag{7.14}$$

It is also useful to represent the augmented operators, in which c extra qubits have been added on Bob's side in order to resolve the anticommutativity of the stabilizer group. In this case, we include the operators on Bob's side as well, with the convention that the bits representing operators on Bob's side will be listed in the c rightmost columns of H_X and H_Z. Taking the augmented generators from (7.9) we get the symplectic representation

$$\begin{array}{c} \bar{Z}'_1 = Z\ X\ Z\ I\ |\ Z \\ \bar{X}'_1 = Z\ Z\ I\ Z\ |\ X \\ \bar{Z}'_2 = Y\ X\ X\ Z\ |\ I \\ \bar{Z}'_3 = Z\ Y\ Y\ X\ |\ I \end{array} \longrightarrow \begin{pmatrix} 0 & 1 & 0 & 0 & 0 & | & 1 & 0 & 1 & 0 & 1 \\ 0 & 0 & 0 & 0 & 1 & | & 1 & 1 & 0 & 1 & 0 \\ 1 & 1 & 1 & 0 & 0 & | & 1 & 0 & 0 & 1 & 0 \\ 0 & 1 & 1 & 1 & 0 & | & 1 & 1 & 1 & 0 & 0 \end{pmatrix}, \tag{7.15}$$

where the fifth and tenth columns represent the X and Z parts, respectively, of the operators on Bob's qubit.

7.2.5 The canonical code

Consider the trivial encoding operation defined by

$$|\varphi\rangle \mapsto |\Phi\rangle^{\otimes c}|0\rangle|\varphi\rangle. \tag{7.16}$$

In other words, registers containing $|0\rangle$ (of size $s = n - k - c$ qubits) and $|\Phi\rangle^{\otimes c}$ (c ebits shared beween Alice and Bob) are appended to the register containing the "encoded" information $|\varphi\rangle$ (of size k qubits). The group B_e analyzed above is the stabilizer for a code of this type, with $n = 4$, $k = 1$, and $c = 1$. To make the comparison to other codes, we can think of Alice appending the ancillas and her halves of the entangled pairs, and then applying an encoding unitary that is simply the identity I. It might seem at first glance that such a simple-minded encoding could not correct any set of errors, but in fact it can, as we will see now.

Proposition 7.1. *The code given by (7.16) and a suitably defined decoding map D_0 can correct an error set (represented in symplectic form)*

$$E_0 = \{(\mathbf{a}_1, \mathbf{b}_1, \alpha(\mathbf{a}_1, \mathbf{a}_2, \mathbf{b}_1) | \mathbf{a}_2, \mathbf{b}_2, \beta(\mathbf{a}_1, \mathbf{a}_2, \mathbf{b}_1)) : \mathbf{a}_1, \mathbf{a}_2 \in (\mathbb{Z}_2)^c, \mathbf{b}_1, \mathbf{b}_2 \in (\mathbb{Z}_2)^s\},$$

$$\tag{7.17}$$

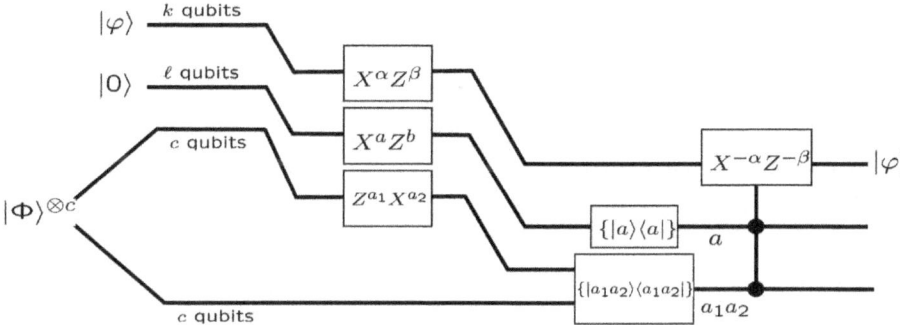

Fig. 7.2. The canonical code. Note that this code differs from a general EAQECC only in that the encoding unitary used is the identity, and we assume a particular form for the errors. Reprinted with kind permission of Springer Science and Business Media from *New Trends in Mathematical Physics*, edited by V. Disoravicus (Springer, Berlin, 2009).

for any pair of known functions $\alpha, \beta : (\mathbb{Z}_2)^c \times (\mathbb{Z}_2)^c \times (\mathbb{Z}_2)^s \to (\mathbb{Z}_2)^k$. *To put this more intuitively: an error set* E_0 *is correctable by the canonical code if every error in* E_0 *leaves a record of itself in the appended registers that determines exactly what has been done to the information qubits, and how to undo it. (This is why the functions* α *and* β *must be known functions.)*

Proof. The protocol is shown in Fig. 7.2. After applying an error $N_\mathbf{u}$ with

$$\mathbf{u} = (\mathbf{a}_1, \mathbf{b}_1, \alpha(\mathbf{a}_1, \mathbf{a}_2, \mathbf{b}_1)|\mathbf{a}_2, \mathbf{b}_2, \beta(\mathbf{a}_1, \mathbf{a}_2, \mathbf{b}_1)) \tag{7.18}$$

the encoded state $|\Phi\rangle^{\otimes c}|0\rangle|\varphi\rangle$ becomes (up to a phase factor)

$$(X^{\mathbf{a}_1} Z^{\mathbf{a}_2} \otimes I)|\Phi\rangle^{\otimes c} \otimes X^{\mathbf{b}_1} Z^{\mathbf{b}_2}|0\rangle \otimes X^{\alpha(\mathbf{a}_1, \mathbf{a}_2, \mathbf{b}_1)} Z^{\beta(\mathbf{a}_1, \mathbf{a}_2, \mathbf{b}_1)}|\varphi\rangle$$
$$= |\mathbf{a}_1, \mathbf{a}_2\rangle \otimes |\mathbf{b}_1\rangle \otimes X^{\alpha(\mathbf{a}_1, \mathbf{a}_2, \mathbf{b}_1)} Z^{\beta(\mathbf{a}_1, \mathbf{a}_2, \mathbf{b}_1)}|\varphi\rangle, \tag{7.19}$$

where $|\mathbf{b}_1\rangle = X^{\mathbf{b}_1}|0\rangle$ and $|\mathbf{a}_1, \mathbf{a}_2\rangle = (X^{\mathbf{a}_1} Z^{\mathbf{a}_2} \otimes I)|\Phi\rangle^{\otimes c}$. As the vector $(\mathbf{a}_1, \mathbf{a}_2, \mathbf{b}_1, \mathbf{b}_2)^T$ completely specifies the error $\mathbf{u}$, it is called the *error syndrome*. The state (7.19) only depends on the *reduced syndrome* $\mathbf{r} = (\mathbf{a}_1, \mathbf{a}_2, \mathbf{b}_1)^T$. In effect, $\mathbf{b}_1$ and $(\mathbf{a}_1, \mathbf{a}_2)$ have been recorded in the ancillas and ebits using plain and superdense coding, respectively. Bob, who holds the entire state (7.19) at the end, may identify the reduced syndrome by a simple projective measurement. Bob simultaneously measures the $Z^{\mathbf{e}_1} \otimes Z^{\mathbf{e}_1}, \ldots, Z^{\mathbf{e}_c} \otimes Z^{\mathbf{e}_c}$ observables to decode $\mathbf{a}_1$ (where the operators to the right of the $\otimes$ symbol act on Bob's halves of the ebits), the $X^{\mathbf{e}_1} \otimes X^{\mathbf{e}_1}, \ldots, X^{\mathbf{e}_c} \otimes X^{\mathbf{e}_c}$ observables to decode $\mathbf{a}_2$, and the $Z^{\mathbf{e}_{c+1}}, \ldots, Z^{\mathbf{e}_{c+s}}$ observables to decode $\mathbf{b}_1$, where $\{\mathbf{e}_j\}$ are the standard basis vectors. He then performs a correction $Z^{(\mathbf{a}_1, \mathbf{a}_2, \mathbf{b}_1)} X^{(\mathbf{a}_1, \mathbf{a}_2, \mathbf{b}_1)}$ on the remaining k qubit system, leaving it in the original state $|\varphi\rangle$.

Since the goal is the transmission of quantum information, no actual measurement is necessary. Instead, Bob can perform the CPTP map D_0 consisting of the controlled unitary

$$U_{0\,\mathrm{dec}} = \sum_{\mathbf{a}_1, \mathbf{a}_2, \mathbf{b}_1} P_{\mathbf{a}_1, \mathbf{a}_2} \otimes P_{\mathbf{b}_1} \otimes Z^{\beta(\mathbf{a}_1, \mathbf{a}_2, \mathbf{b}_1)} X^{\alpha(\mathbf{a}_1, \mathbf{a}_2, \mathbf{b}_1)}, \tag{7.20}$$

followed by discarding the first two subsystems. Here we use projection operators

$$P_{\mathbf{a}_1, \mathbf{a}_2} = |\mathbf{a}_1, \mathbf{a}_2\rangle\langle\mathbf{a}_1, \mathbf{a}_2|, \quad P_{\mathbf{b}_1} = |\mathbf{b}_1\rangle\langle\mathbf{b}_1|.$$

[Note that equations very similar to (7.18)–(7.20) arise in the theory of continuous-variable error-correcting codes, as described in Chapter 22.] □

The above code is *degenerate* with respect to the error set E_0, which means that the error can be corrected without knowing the full error syndrome. This is because the Z^{b_2} operator acting on the ancillas has no effect beyond a trivial global phase. It is our assumption that the errors acting on the information qubits did not depend on b_2 that makes this error set correctable. This kind of restriction will always be necessary – no error-correcting code can correct every possible error. Part of the art of error correction is to choose a code whose correctable error set matches, as closely as possible, the actual physical errors that are likely to happen.

We can characterize our code in terms of the check matrix $\hat{H}$ given by

$$\hat{H} = \begin{pmatrix} \hat{H}_I \\ \hat{H}_S \end{pmatrix}, \qquad (7.21)$$

$$\hat{H}_I = \begin{pmatrix} \mathbf{0}_{s \times c} & \mathbf{0}_{s \times s} & \mathbf{0}_{s \times k} & | & \mathbf{0}_{s \times c} & \mathbf{I}_{s \times s} & \mathbf{0}_{s \times k} \end{pmatrix}, \qquad (7.22)$$

$$\hat{H}_S = \begin{pmatrix} \mathbf{I}_{c \times c} & \mathbf{0}_{c \times s} & \mathbf{0}_{c \times k} & | & \mathbf{0}_{c \times c} & \mathbf{0}_{c \times s} & \mathbf{0}_{c \times k} \\ \mathbf{0}_{c \times c} & \mathbf{0}_{c \times s} & \mathbf{0}_{c \times k} & | & \mathbf{I}_{c \times c} & \mathbf{0}_{c \times s} & \mathbf{0}_{c \times k} \end{pmatrix}, \qquad (7.23)$$

with $s = n - k - c$.

The row space of $\hat{H}$ decomposes into a direct sum of the *isotropic* subspace, given by the row space of $\hat{H}_I$ and the *symplectic* subspace, given by the row space of $\hat{H}_S$. Define the *symplectic code* corresponding to $\hat{H}$ by

$$C_0 = \text{rowspace}(\hat{H})^\perp \qquad (7.24)$$

where

$$V^\perp = \{\mathbf{w} : \mathbf{w} \odot \mathbf{u}^T = 0, \ \forall \mathbf{u} \in V\}. \qquad (7.25)$$

Note that $(V^\perp)^\perp = V$. Then $C_0^\perp = \text{rowspace}(\hat{H})$, $\text{iso}(C_0^\perp) = \text{rowspace}(\hat{H}_I)$ and $\text{symp}(C_0^\perp) = \text{rowspace}(\hat{H}_S)$, where iso and symp denote the isotropic and symplectic subspaces, respectively.

Just as with standard stabilizer codes (as described earlier in this chapter), each row of the check matrix $\hat{H}$ maps to a stabilizer generator, and the complete set of vectors in the row space of $\hat{H}$ maps onto the complete set of stabilizer operators (up to phases). The rows of $\hat{H}_S$ come in nonorthogonal pairs under the symplectic inner product; these nonorthogonal pairs of rows map into anticommuting pairs of generators in the stabilizer. Each such pair requires one extra qubit on Bob's side to resolve the noncommutativity of the stabilizer, and hence requires one shared ebit between Alice and Bob at the encoding step. The number of ebits used in the code is therefore

$$c = \frac{1}{2} \dim \text{rowspace}(\hat{H}_S), \qquad (7.26)$$

and the number of encoded qubits is

$$k = n - \dim \text{rowspace}(\hat{H}_I) - \frac{1}{2} \dim \text{rowspace}(\hat{H}_S)$$
$$= n - \dim \text{rowspace}(\hat{H}) + c. \qquad (7.27)$$

The code parameter $\hat{k} := k - c$, which is the number of encoded qubits minus the number of ebits used, is independent of the symplectic structure of $\hat{H}$:

$$\hat{k} = n - \dim \text{rowspace}(\hat{H}). \tag{7.28}$$

We see that the quantity k/n is the rate of the code, and $\hat{k}/n$ is the net rate.

The correctable error set E_0 of this symplectic code can be described in terms of $\hat{H}$:

Proposition 7.2. *The set E_0 of errors correctable by the canonical code (in symplectic form) is such that, if $\mathbf{u}, \mathbf{u}' \in E_0$ and $\mathbf{u} \neq \mathbf{u}'$, then either*

(i) $\mathbf{u} - \mathbf{u}' \notin C_0$ *(equivalently: $\hat{H} \odot (\mathbf{u} - \mathbf{u}')^T \neq \mathbf{0}^T$), or*
(ii) $\mathbf{u} - \mathbf{u}' \in \text{iso}(C_0^\perp)$ *(equivalently: $\mathbf{u} - \mathbf{u}' \in \text{rowspace}(\hat{H}_I)$).*

Proof. If $\mathbf{u}$ is given by (7.18) then $\hat{H} \odot \mathbf{u}^T = \mathbf{r} = (\mathbf{a}_1, \mathbf{a}_2, \mathbf{b}_1)^T$, the reduced error syndrome. By definition (7.17), two distinct elements of E_0 either have different reduced syndromes $(\mathbf{a}_1, \mathbf{a}_2, \mathbf{b}_1)$ (condition 1) or they differ by a vector of the form $(\mathbf{0}, \mathbf{0}, \mathbf{0}|\mathbf{0}, \mathbf{b}_2, \mathbf{0})$ (condition 2). In the first case, the two errors can be distinguished, and appropriate different corrections applied. In the second case, the two errors have the same reduced syndrome and hence cannot be distinguished, but they are both correctable by the same operation. Observe that condition 1 is analogous to the usual error-correcting condition for classical codes [MS77], while condition 2 is an example of the quantum phenomenon of degeneracy. □

The parity check matrix $\hat{H}$ also specifies the encoding and decoding operations. The space $\mathcal{H}_2^{\otimes k}$ is encoded into the *code space* defined by

$$C_0 = \{|\Phi\rangle^{\otimes c}|\mathbf{0}\rangle|\varphi\rangle : |\varphi\rangle \in \mathcal{H}_2^{\otimes k}\}.$$

It is not hard to see that the code space is the simultaneous $+1$ eigenspace of the commuting operators:

(i) $I \otimes Z^{\mathbf{e}_i} \otimes I \otimes I$, $i = 1, \ldots, s$;
(ii) $Z^{\mathbf{e}_j} \otimes I \otimes I \otimes Z^{\mathbf{e}_j}$, $j = 1, \ldots, c$;
(iii) $X^{\mathbf{e}_j} \otimes I \otimes I \otimes X^{\mathbf{e}_j}$, $j = 1, \ldots, c$.

For operators written above in the form $A_1 \otimes A_2 \otimes A_3 \otimes B$, the first three operators A_1, A_2, A_3 act on Alice's qubits, and the fourth operator B on Bob's. Define the check matrix

$$H_B = \begin{pmatrix} \mathbf{0}_{s \times c} & \mathbf{0}_{s \times c} \\ \mathbf{I}_{c \times c} & \mathbf{0}_{c \times c} \\ \mathbf{0}_{c \times c} & \mathbf{I}_{c \times c} \end{pmatrix}. \tag{7.29}$$

Define the *augmented* parity check matrix $\hat{H}_{\text{aug}}$ by adding the columns of H_B to $\hat{H}$:

$$\hat{H}_{\text{aug}} = \begin{pmatrix} \mathbf{0}_{s \times c} & \mathbf{0}_{s \times s} & \mathbf{0}_{s \times k} & \mathbf{0}_{s \times c} & \mathbf{0}_{s \times c} & \mathbf{I}_{s \times s} & \mathbf{0}_{s \times k} & \mathbf{0}_{s \times c} \\ \mathbf{I}_{c \times c} & \mathbf{0}_{c \times s} & \mathbf{0}_{c \times k} & \mathbf{I}_{c \times c} & \mathbf{0}_{c \times c} & \mathbf{0}_{c \times s} & \mathbf{0}_{c \times k} & \mathbf{0}_{c \times c} \\ \mathbf{0}_{c \times c} & \mathbf{0}_{c \times s} & \mathbf{0}_{c \times k} & \mathbf{0}_{c \times c} & \mathbf{I}_{c \times c} & \mathbf{0}_{c \times s} & \mathbf{0}_{c \times k} & \mathbf{I}_{c \times c} \end{pmatrix}. \tag{7.30}$$

Observe that rowspace($\hat{H}_{\text{aug}}$) is purely isotropic. The code space is now described as the simultaneous $+1$ eigenspace of

$$\{N_{\mathbf{w}} : \mathbf{w} \in \text{rowspace}(\hat{H}_{\text{aug}})\},$$

or, equivalently that of

$$S_0 = \langle N_{\mathbf{w}} : \mathbf{w} \text{ is a row of } \hat{H}_{\text{aug}} \rangle.$$

The decoding operation D_0 is also described in terms of $\hat{H}$. The reduced syndrome $\mathbf{r} = \hat{H} \odot \mathbf{u}^{\mathsf{T}}$ is obtained by simultaneously measuring the observables in S_0. The reduced error syndrome corresponds to an equivalence class of possible errors $\mathbf{u} \in E_0$ that all have an identical effect on the code space, and can all be corrected by the same operation. Bob performs $\hat{N}_{\mathbf{u}} \equiv \hat{N}_{-\mathbf{u}}$ to undo the error.

7.2.6 General codes and code parameters

The canonical code described above is not very useful for practical error correction. It corrects a set of highly nonlocal errors, of a type that are not likely to occur in reality. However, it is very helpful in clarifying how the process of error correction works, and the role of the resources used in encoding. Each added ancilla can contain, in principle, one bit of information about any errors that occur. Each shared ebit between the sender and receiver can contain two bits of information about errors. This information about the errors is retrieved by measuring the stabilizer generators. After Alice's qubits have passed through the channel, Bob has all the qubits and can perform these measurements.

A general EAQECC has exactly the same algebraic structure as the canonical code. The encoded state, before going through the channel, has c ebits of entanglement with the c qubits on Bob's side. Corresponding to this entanglement are c pairs of stabilizer generators. Each pair of generators has the following structure: a pair of anticommuting operators on Alice's qubits, tensored with a Z and an X, respectively, on one of the qubits on Bob's side to "resolve" the anticommutativity. These pairs of generators are the symplectic pairs of the EAQECC. In addition to these pairs, there can be single generators that act as the identity on all of Bob's qubits. These are the isotropic generators. For an $[[n, k, d; c]]$ EAQECC, each of the stabilizer elements acts on $n + c$ qubits, n on Alice's side and c on Bob's side; there is a set of $n - k$ generators for the stabilizer chosen in a standard form, with $2c$ symplectic pairs of generators and $s = n - k - c$ isotropic generators.

By Lemma 7.2, if two sets of Pauli operators can be put in one-to-one correspondence so that they have the same commutation relations, there is a unitary transformation that turns one set into the other (up to a phase). The $n - k$ generators of an $[[n, k; c]]$ EAQECC can be mapped by a unitary operator $U_E^{\dagger}$ to the generators of a canonical code.

It is important to realize that the canonical code describes the *input* of the EAQECC *before encoding*. It describes a set of k information qubits, c ebits, and s ancillas. The unitary U_E transforms the stabilizer for this input into the stabilizer for the EAQECC. Therefore U_E is a unitary encoding operator for this code. It transforms the localized states of the input qubits into nonlocal, entangled states spread over all the qubits. By this transformation, the correctable error set is transformed from the strange, nonlocal errors of the canonical code to a more physically realistic set of errors (most commonly, localized errors on the individual qubits).

7.3 Constructing EAQECCs from classical linear codes
7.3.1 Mapping $GF(4)$ to Pauli operators

We will now examine the [[4,1;1]] EAQECC used as an example in Section 7.2 above, and show that it can be derived from a classical non-dual-containing quaternary $[4,2]$ code. This is a generalization of the well-known CRSS construction for standard QECCs [CRS+98].

First, note that this [[4,1;1]] code is nondegenerate, and can correct an arbitrary one-qubit error. (The distance d of the code $C(S)$ is 3.) This is because the 12 errors X_i, Y_i, and Z_i, $i = 1, \ldots, 4$, have distinct nonzero error syndromes. X_i denotes the bit-flip error on ith qubit, Z_i denotes the phase error on ith qubit, and Y_i means that both a bit-flip and phase-flip error occur on the ith qubit. It suffices to consider only these three standard one-qubit errors, since any other one-qubit error can be written as a linear combination of these three errors and the identity.

Next, we define the following map between the Pauli operators, symplectic strings, and elements of $GF(4)$, the field with four elements:

G	I	X	Y	Z
Symplectic	00	10	11	01
$GF(4)$	0	$\bar{\omega}$	1	ω

Addition of symplectic strings is by bit-wise XOR. The elements of $GF(4)$ obey a simple addition rule: $x + 0 = x$ and $x + x = 0$ for all $x \in GF(4)$, $\bar{\omega} + 1 = \omega$, $1 + \omega = \bar{\omega}$, and $\omega + \bar{\omega} = 1$, where addition is commutative. Multiplication is also commutative: $0 \cdot x = 0$ and $1 \cdot x = x$ for all $x \in GF(4)$, $\omega \cdot \omega = \bar{\omega}$, $\bar{\omega} \cdot \bar{\omega} = \omega$, and $\omega \cdot \bar{\omega} = 1$. Note that under this map, addition in $GF(4)$ (or addition of symplectic strings) corresponds to multiplication of the Pauli operators, up to an overall phase. So multiplication of two elements of G_n corresponds to addition of two n-vectors over $GF(4)$, up to an overall phase. The same correspondence applies between multiplication of the Pauli operators and bit-wise addition (or XOR) of the symplectic strings.

We can define a linear code over $GF(4)$ in exactly the same way we do for a binary or symplectic code, by writing down a check matrix. Consider the matrix

$$H_4 = \begin{pmatrix} 1 & \omega & 1 & 0 \\ 1 & 1 & 0 & 1 \end{pmatrix}. \tag{7.31}$$

The code is the null space of H_4 over $GF(4)$ – the set of vectors orthogonal to all the rows of the check matrix. H_4 is the check matrix of a classical $[4, 2, 3]$ quaternary code whose rows are not orthogonal, and 3 is the minimum distance between codewords.

From this starting point we can define a new matrix:

$$\tilde{H}_4 = \begin{pmatrix} \omega H_4 \\ \bar{\omega} H_4 \end{pmatrix} = \begin{pmatrix} \omega & \bar{\omega} & \omega & 0 \\ \omega & \omega & 0 & \omega \\ \bar{\omega} & 1 & \bar{\omega} & 0 \\ \bar{\omega} & \bar{\omega} & 0 & \bar{\omega} \end{pmatrix}. \tag{7.32}$$

As a quaternary code, this defines exactly the same code space as the original H_4, since the last two rows are multiples of the first two rows. However, we now apply our map from $GF(4)$ to G_4, where each row of $\tilde{H}_4$ is mapped to a Pauli operator in G_4. The resulting set of operators is exactly the set of generators $\{M_i\}$ given in (7.4). If we map the matrix $\tilde{H}_4$ to a symplectic matrix, using the correspondence given above, we see that we get exactly the same symplectic representation for the code as in Section 7.2.4. We can carry out this procedure with any linear quaternary code.

The generators corresponding to the rows of the matrix $\tilde{H}_4$ will not commute, in general. In that case, we must apply Lemma 7.1 to find a new set of generators comprising isotropic generators (that commute with all other generators) and symplectic pairs of generators (that anticommute with each other but commute with all other generators). Extra operators can be added on Bob's side to resolve the anticommutativity, as described above. This produces an EAQECC using c ebits, where c is the number of symplectic pairs.

In the example above, we get a $[[4, 1, 3; 1]]$ EAQECC from a classical $[4, 2, 3]$ quaternary code. This outperforms the best four-bit self-dual QECC currently known, which is $[[4, 0, 2]]$, in both rate and distance [CRS+98]. This connection between EAQECCs and quaternary classical codes is quite general [BDH06a]. Given an arbitrary classical $[n, k, d]$ quaternary code, we can use (7.32) to construct a nondegenerate $[[n, 2k - n + c, d; c]]$ EAQECC. The rate becomes $(2k - n)/n$ because the $n-k$ classical parity checks give rise to $2(n-k)$ quantum stabilizer generators. This gives us a tremendous advantage in constructing quantum codes, because the parameters of the quantum code depend exactly on those for the classical code: the minimum distance is the same, and high-rate classical codes produce high-rate (though not as high) quantum codes.

The one parameter that is not simply determined by the parameters of the classical code is c, the number of ebits used. For a standard QECC, which uses no entanglement and hence has $c = 0$, the rows of the classical check matrix are all orthogonal. Intuitively, we might expect that as the matrix moves away from being self-orthogonal, the resulting code will require more entanglement. In fact, for a CRSS code of the type described above, the amount of entanglement needed is

$$c = \operatorname{rank}\left(H_4 H_4^\dagger\right), \tag{7.33}$$

where the conjugate $H_4^\dagger$ is defined by taking the transpose of H_4 and interchanging $\omega \leftrightarrow \bar{\omega}$ [WB08a, W09a]. For a standard QECC, in which the rows are all orthogonal, $H_4 H_4^\dagger$ will be the zero matrix. Since this matrix is $(n - k) \times (n - k)$, the maximum rank (and hence the maximum possible entanglement used by the code) is $n - k$. Such a quantum code will transmit $2k - n + c = k$ qubits – the same rate as the original classical code.

A special case of this construction is where the original classical code is *binary*; that is, when the check matrix H uses only 0s and 1s, and has no elements ω or $\bar{\omega}$. In that case, the procedure for constructing a quantum code is to create a symplectic check matrix of the form

$$\left(\begin{array}{c|c} H & 0 \\ 0 & H \end{array}\right).$$

The stabilizer generators corresponding to the rows of this matrix either contain only Zs and Is or only Xs and Is, and the two sets of generators are identical in form. In a sense, such a code can be thought of as the intersection of two classical codes, one to correct bit-flips (X errors) and the other to correct phase-flips (Z errors). This is the CSS construction, analogous to the way it is defined in Chapter 2 for standard QECCs.

Somewhat more generally, we could define a QECC using two different classical binary codes with check matrices H_1 and H_2:

$$\left(\begin{array}{c|c} H_1 & 0 \\ 0 & H_2 \end{array}\right).$$

Table 7.1. *Table of codes up to length 10*

$n \backslash \hat{k}$	0	1	2	3	4	5	6	7	8	9	10
3	2	2*	1	1							
4	3*	2	2	1	1						
5	3	3	2	2*	1	1					
6	4	3	2	2	2	1	1				
7	3	3	2	2	2	2*	1	1			
8	4	3	3	3	2	2	2	1	1		
9	4	4*	3	3	2	2	2	2*	1	1	
10	5*	4	4	3	3	2	2	2	2	1	1

The best minimum distance is given for a code with length n and net transmission $\hat{k} = k - c$, where k is the number of information qubits and c the number of ebits used. The entries marked with an asterisk are better than the best known standard stabilizer code of the same size and net rate. See [BDH06a].

Once again the generators would use either Xs or Zs, and not both; but the generators will no longer share the same structure. This approach may be useful when the noise is *asymmetric*; for instance, dominated by phase errors with occasional rare bit-flips. The simpler CSS code is obviously the special case where $H_1 = H_2 = H$. It is not necessary that the check matrices H_1 and H_2 be the same size; H_1 can be $(n - k_1) \times n$ and H_2 can be $(n - k_2) \times n$. Together they produce a quantum code that encodes $k_1 + k_2 - n + c$ information qubits in n physical qubits with the use of c ebits. For a CSS code of this type the amount of entanglement needed [WB08a, W09a] is

$$c = \operatorname{rank}\left(H_1 H_2^T\right). \tag{7.34}$$

7.3.2 Performance

In [CRS+98] a table of the best known QECCs was given. In Table 7.1 we show an updated table that includes EAQEC codes. In order to make sensible comparisons between codes that use varying amounts of entanglement we use the net number of encoded qubits $\hat{k} = k - c$. The table entries with an asterisk mark improvements over the table of standard codes in [CRS+98]. All the codes in this table were constructed from classical quaternary codes by the construction in Section 7.3.1, and were found by numerical search. The corresponding classical quaternary codes are available online at http://www.codetables.de.

7.4 Catalytic QECCs

EAQECCs make use of an additional resource – prior shared entanglement – to boost the power of an error-correcting code. This paradigm raises both theoretical and practical questions as to how these codes might be useful in practice, and how they fit into families of codes that draw on other resources.

The first practical question is: how was this entanglement shared in the first place? If the sender and receiver communicate by a noisy quantum channel, then the entanglement would have to be shared through this channel. Alice locally prepares EPR pairs, and sends half of each pair through the channel, either protected by a standard QECC, or making use of entanglement purification techniques [BBP+96b, BDS+96] to produce a smaller number of perfect EPR pairs. So we see that in this context the enhanced rate of an EAQECC must be paid for by additional channel communication to establish the entanglement in the first place. This means that for many purposes it is the *net rate* $(k-c)/n$ of the EAQECC that is the appropriate figure of merit. This net rate need not be better than that achieved by a standard QECC; indeed, it is quite possible for the net rate to be zero or negative.

Even a code with a negative net rate may actually be useful in practice. One great advantage of shared entanglement as a resource is that it is independent of the message being sent, and can in principle be prepared well ahead of time. One natural application of EAQECCs would therefore be to a quantum network where usage varies at different times. During periods of low usage, shared entanglement could be built up between the sender and receiver. During periods of high usage, this shared entanglement could be drawn on to increase the rate of transmission without sacrificing error-correcting power.

EAQECCs with positive net rates can be used in other ways to improve the power and flexibility of quantum communications. To illustrate this, let us consider the idea of a *catalytic code*. Here, in addition to n uses of a noisy channel, the sender may send c qubits noise-free. One obvious approach is to send c information qubits using the noise-free bits, and then send k more qubits through the channel by means of an $[[n, k]]$ standard QECC. However, another way of proceeding would be to use the c noise-free bits to establish c ebits between Alice and Bob, and then use an $[[n, k'; c]]$ EAQECC. So long as $k' \geq k + c$ (and both codes are assumed to correct the errors in the channel), one does as well or better with the second approach as with the straightforward one. In this case, the noise-free qubits are used to boost the rate of the noisy qubits. We call a procedure of this type a *catalytic code*; the c noise-free qubits enhance transmission by an amount equal to at least c.

In practice, noise-free transmission is not really possible. However, it is possible to simulate it by using a standard QECC to protect the c noise-free qubits. Suppose that we start with an $[[n, k; c]]$ EAQECC. To establish c ebits with Bob, Alice would need to send c qubits through the channel. Suppose that she protects them with an $[[n', c]]$ standard QECC. Then we can think of the full block of $n + n'$ qubits that go through the channel as a single codeword; by putting these two codes together we have constructed an $[[n + n', k]]$ standard QECC.

Of course, the second code could also be an EAQECC. If we have an $[[n, k; c]]$ EAQECC and an $[[n', k'; c']]$ EAQECC, we can send Bob's halves of the c ebits in the second code block, so long as $k' \geq c$. By putting these two code blocks together, we have constructed an $[[n + n', k + k' - c; c']]$ EAQECC.

Nor need we combine only two codes in this way. One rather simple construction is as follows: take an $[[n, k; c]]$ EAQECC with positive net rate, and encode Bob's halves of the c ebits in another copy of the same code. Repeat this M times, and one quickly builds up a much larger block code: an $[[Mn, M(k - c) + c; c]]$ EAQECC. This procedure is called *bootstrapping*. Two observations of this new code are quickly evident: First, as M becomes large, the rate of the code approaches the net rate $(k - c)/n$ of the original code. Second, as M becomes large, the rate of entanglement usage c/Mn becomes small. By bootstrapping we have built a large code whose rate is the net rate of the original code, and which uses relatively little entanglement.

Practically speaking there is a limit to how big M can be; if the block size becomes too large, the probability that one of the sub-blocks of n qubits will have an uncorrectable error becomes high, and the code will no longer be useful. At this point, little is known about the practical performance of catalytic codes; more research is needed.

By this approach we see another use of EAQECCs; rather than being used themselves directly, they can be used as building blocks in the construction of standard codes. Because EAQECCs can be constructed from classical codes without the need to satisfy the dual-containing constraint, in some cases they will outperform standard codes with otherwise similar parameters.

One particularly interesting example is the case of LDPC codes. Classically, these codes can achieve capacity, and are decoded using iterative belief-propagation algorithms (see Chapter 11). These algorithms are suboptimal, but very effective in most practical cases. Certain codes, however – especially codes with small loops in the Tanner graph – tend not to work as well with iterative decoding, and it is difficult to construct quantum versions of LDPC codes that do not have such loops. By relaxing the dual-containing constraint, we can find entanglement-assisted codes without such small loops, and some families have been found that use very little entanglement [HBD09, HYH11].

Combining EAQECCs with each other, or with standard codes, may allow the relatively easy construction of standard codes with particular desired properties. Moreover, this may lead to applications of EAQECCs beyond the quantum communication paradigm, using them to construct standard codes that are useful in other applications (such as quantum computing). However, at present this is pure speculation: such applications have yet to be studied.

7.5 Conclusions

EAQECCs expand the usual paradigm of quantum error correction by allowing the sender and receiver to make use of pre-shared entanglement. This entanglement can increase either the rate of communication or the number of correctable errors. While entanglement can be used to improve quantum communication in other ways – for example, by sending classical information through the channel and teleporting extra qubits – in many cases EAQECCs have better performance [HW10].

In this chapter we have presented an extension of the usual stabilizer formalism for QECCs to include entanglement assistance. This extension includes the usual theory of stabilizer codes as a special case, and has a number of other advantages; in particular, it allows one to construct quantum codes from classical linear codes without having to impose the dual-containing constraint, while retaining the usual relation between the net rate and distance of the classical and quantum codes. Moreover, if entanglement assistance is allowed, one can in some cases outdo the performance of the best standard QECC by optimizing without the dual-containing constraint. Some examples are illustrated in Table 7.1.

In quantum communication protocols involving entanglement, the applications of EAQECCs are obvious. However, the construction may also be useful in other cases, by using EAQECCs as *catalytic* codes. In this case, we are essentially using EAQECCs as building blocks for standard codes, or for codes that use only a small amount of pre-existing entanglement. In these constructions, the net rate of the EAQECCs is the most important parameter; this net rate can be directly calculated from the classical linear code, for codes constructed from classical precursors.

Because these constructions do not require dual-containing codes, it is possible to directly derive quantum versions of highly efficient modern codes, such as LDPC codes [HBD09, HYH11] and turbo-codes [WH10]. Some examples have been studied, comparing specific EAQECCs to standard QECCs, where the EAQECCs show improved decoding behavior over the standard codes; however, much more work remains to be done on this problem.

Additional research has been done on combining the idea of entanglement assistance with more general types of quantum codes. In particular, it is possible to construct an even more general class of stabilizer codes that include both entanglement-assisted and operator (or subsystem) codes as well as standard stabilizer codes. (See Chapter 6 for more details about operator codes.) Generic codes in this class are entanglement-assisted operator quantum error-correcting codes [HDB07]. It is also possible to define codes that carry both quantum and classical information, with or without entanglement assistance [KHB08]. The possibility of adding entanglement to standard quantum codes in order to increase the number of errors that can be corrected has also been studied [LB10], as well as further explicating the properties of this class of codes [LBW10].

This chapter has considered only block codes. Some work has already been done on entanglement-assisted quantum convolutional codes as well [WB10a, WB10b]; standard convolutional QECCs are described in Chapter 9. Many questions remain open; in particular, the minimum entanglement needed to produce a quantum version of a classical convolutional code is unknown, though some answers have been conjectured [WB08a]. The field of entanglement-assisted codes is relatively new, and active research is ongoing to find more applications of these codes.

Chapter 8

Continuous-time quantum error correction

Ognyan Oreshkov

8.1 Introduction

In the standard theory of quantum error correction, both the noise and the error-correcting operations are represented by discrete transformations. If $\mathscr{B}(\mathscr{H})$ denotes the space of bounded operators over a Hilbert space $\mathscr{H}$, and $\mathscr{H}^S$ is the (finite-dimensional) Hilbert space of the controlled system, we say that the code subsystem $\mathscr{H}^A$ in the decomposition

$$\mathscr{H}^S = \mathscr{H}^A \otimes \mathscr{H}^B \oplus \mathscr{K} \tag{8.1}$$

is correctable under the completely positive trace-preserving (CPTP) noise map $\mathscr{E} : \mathscr{B}(\mathscr{H}^S) \to \mathscr{B}(\mathscr{H}^S)$, if there exists a CPTP error-correcting map $\mathscr{R} : \mathscr{B}(\mathscr{H}^S) \to \mathscr{B}(\mathscr{H}^S)$, such that

$$\mathrm{Tr}_B\{(\mathscr{P}^{AB} \circ \mathscr{R} \circ \mathscr{E})(\sigma)\} = \mathrm{Tr}_B\{\sigma\}, \tag{8.2}$$
$$\text{for all } \sigma \in \mathscr{B}(\mathscr{H}^S), \sigma = \mathscr{P}^{AB}(\sigma) ,$$

where $\mathscr{P}^{AB}(\cdot)$ denotes the projection superoperator on $\mathscr{B}(\mathscr{H}^A \otimes \mathscr{H}^B)$.

This formalism is fundamental for the understanding of preserved information under CPTP dynamics, but it depicts an idealized version of the error-correction process. It represents both the noise and the error-correcting operations as discrete CPTP maps, and assumes that error correction is applied *after* the noise. Such a picture is a good approximation for the case when we are concerned with error correction at a single instant via an operation that is fast on the time scale of the noise, or in the case of repeated error correction with fast operations in a regime where the accumulation of uncorrectable errors can be ignored. In general, however, a full error-correcting operation takes a finite time interval during which the noise process is on. Furthermore, even if we assume that error-correcting operations are instantaneous, deviations from perfect correctability between repeated corrections are unavoidable in any real situation. Thus, in the case of non-Markovian dynamics, the system may develop correlations with the environment and the effective error maps between successive corrections need not be completely positive. Therefore,

---

Quantum Error Correction, ed. Daniel A. Lidar and Todd A. Brun. Published by Cambridge University Press.
© Cambridge University Press 2013.

a complete description must take into account the continuous nature of both the decoherence and the error-correction processes. Situations in which both these processes are regarded as continuous in time are the subject of continuous-time quantum error correction (CTQEC).

The first CTQEC model was proposed by Paz and Zurek (PZ) [PZ98] as a method of studying the performance of repeated error correction with fast operations in the presence of Markovian decoherence. Rather than describing the overall evolution as a continuous decoherence process interrupted by instantaneous error-correcting operations at discrete intervals, the authors proposed to model the error-correcting procedure as a continuous quantum-jump process, which allows a description of the evolution of the system in terms of a continuous master equation in Lindblad form [L76]. In this model, the infinitesimal error-correcting transformation that the density matrix of the controlled system undergoes during a time step dt is

$$\rho \to (1 - \kappa\, dt)\rho + \kappa\, dt \mathscr{R}(\rho), \tag{8.3}$$

where $\mathscr{R}(\rho)$ is the CPTP map describing a full error-correcting operation, and κ is the error-correction rate. The full error-correcting operation $\mathscr{R}(\rho)$ can be thought of as consisting of a syndrome detection, followed (if necessary) by a unitary operation conditioned on the syndrome. The master equation describing the evolution of a system subject to Markovian decoherence plus error correction is then

$$\frac{d\rho}{dt} = \mathscr{L}(\rho) + \kappa \mathscr{J}(\rho), \tag{8.4}$$

where $\mathscr{L}(\rho)$ is the Lindblad generator describing the noise process, and

$$\mathscr{J}(\rho) = \mathscr{R}(\rho) - \rho \tag{8.5}$$

is the quantum-jump error-correction generator. The Lindblad generator has the form

$$\mathscr{L}(\rho) = -i[H, \rho] + \frac{1}{2}\sum_j \lambda_j (2L_j \rho L_j^\dagger - L_j^\dagger L_j \rho - \rho L_j^\dagger L_j), \tag{8.6}$$

where H is a system Hamiltonian and the $\{L_j\}$ are suitably normalized Lindblad operators describing different error channels with decoherence rates λ_j. For example, the Lindbladian

$$\mathscr{L}(\rho) = \sum_j \lambda_j (X_j \rho X_j - \rho), \tag{8.7}$$

where X_j denotes a local bit-flip operator acting on the jth qubit, describes independent Markovian bit-flip errors.

The quantum-jump model can be viewed as a smoothed version of the discrete scenario of repeated error correction, in which instantaneous full error-correcting operations are applied at random times with rate κ. It can also be looked upon as arising from a continuous sequence of infinitesimal CPTP maps of the type (8.3). In practice, such a weak map is never truly infinitesimal, but rather has the form

$$\rho \to (1 - \epsilon^2)\rho + \epsilon^2 \mathscr{R}(\rho), \tag{8.8}$$

where $\epsilon \ll 1$ is a small but finite parameter, and the weak operation takes a small but finite time τ_c. For times t much greater than τ_c, the weak error-correcting map (8.8) is well approximated by the infinitesimal form (8.3), where the rate of error correction is

$$\kappa = \epsilon^2/\tau_c. \tag{8.9}$$

A weak map of the form (8.8) could be implemented, for example, by a weak coupling between the system and an ancilla via an appropriate Hamiltonian, followed by discarding of the ancilla. The continuous process in such a case corresponds to coupling the system to a stream of fresh ancillas, which continuously pump out the entropy accumulated due to correctable errors. A closely related scenario, where the ancilla is continuously cooled in order to reset it to its initial state, was studied by Sarovar and Milburn in [SM05]. Another possible implementation of the above scheme is via weak measurements and weak unitary operations, as we will see in this chapter.

If the set of errors $\{L_j\}$ are correctable by the code, the effect of the described CTQEC procedure is to slow down the rate at which information is lost, and in the limit of infinite error-correction rate (strong error-correcting operations applied continuously often) the state of the system freezes and is protected from errors at all times [PZ98]. The effect of freezing can be understood by noticing that the transformation arising from decoherence during a short time step Δt, is

$$\rho \to \rho + \mathscr{L}(\rho)\Delta t + O(\Delta t^2), \tag{8.10}$$

i.e., the weight of correctable errors emerging during this time interval is proportional to Δt, whereas uncorrectable errors (higher-order terms) are of order $O(\Delta t^2)$. Thus, if errors are constantly corrected, in the limit $\Delta t \to 0$ uncorrectable errors cannot accumulate and the evolution stops.

The idea of using continuous weak operations for error correction was developed further by Ahn, Doherty, and Landahl (ADL) who proposed a scheme for CTQEC based on continuous measurements of the error syndromes and feedback operations conditioned on the measurement record [ADL02]. A continuous measurement is one resulting from the continuous application of weak measurements, i.e., measurements whose outcomes change the state by a small amount [AAV88, L89, P89, AV89, AV90, B02b, OB05, O08]. As shown in [OB05, O08], weak measurements can be used to generate any quantum operation and therefore provide a natural tool for approaching the problem of error correction in continuous time. In the ADL scheme, the evolution of the density matrix of the system subject to Markovian noise with Lindbladian $\mathscr{L}$ and CTQEC is described by the stochastic differential equation

$$d\rho(t) = \mathscr{L}(\rho(t))dt + \frac{\kappa}{4}\sum_l \mathscr{D}[M_l](\rho(t))dt + \frac{\sqrt{\kappa}}{2}\sum_l \mathscr{F}[M_l](\rho(t))dW_l(t)$$
$$- i\sum_r \lambda_r(\rho(t))[H_r, \rho(t)]dt, \tag{8.11}$$

where $\mathscr{D}[A](\rho) = A\rho A^\dagger - \frac{1}{2}(A^\dagger A\rho + \rho A^\dagger A)$, $\mathscr{F}[A](\rho) = A\rho + \rho A^\dagger - \rho\text{Tr}[A\rho + \rho A^\dagger]$, M_l are the stabilizer generators of the code, W_l are Wiener processes (see Section (8.4.1)), and H_r are correcting Hamiltonians that are turned on with strength $\lambda_r(\rho)$ dependent on the state of the system. Note that the encoded information is in principle unknown, but the feedback is not conditioned on properties of the state related to the encoded information. Thus, in order to estimate the state of the system at the present moment for the purpose of applying feedback, one can assume that the encoded state was initially the maximally mixed state. The parameters $\lambda_r(\rho)$ are chosen to maximize the instantaneous increase of the code-space fidelity, and are given by $\lambda_r(\rho) = \lambda \,\text{sgn}\,\text{Tr}([\Pi_c, H_r]\rho)$, where λ is the maximum strength of the control Hamiltonians and

Π_c is the projector on the code subspace. (Here the code is assumed to be a standard stabilizer code.)

Following the ADL scheme, a number of variations of this approach were proposed (see, e.g., [AWM03, SAJ+04, CLG08]). All these schemes are to a large extent heuristic, and their workings are not thoroughly understood. The difficulty in rigorously motivating the construction of error-correction protocols based on weak measurements and feedback is that stochastic evolutions are generally too complicated to study analytically. This is further complicated by the large dimension of the Hilbert space of all qubits participating in the code (note that even the problem of controlling a single qubit generally requires numerical treatment [J04]). However, numerical simulations have shown that these schemes often lead to a better performance in the presence of continuous noise than the application of strong operations at finite time intervals. Therefore, the use of continuous measurements and feedback seems to offer a promising tool for decoherence control.

In this chapter, we will try to understand CTQEC and how to approach the problem of constructing CTQEC protocols by looking at the evolution of the state of the system in an encoded basis in which the subsystem containing the protected information is explicit. We will see that this point of view reduces the problem to that of protecting a known state, and allows for designing CTQEC procedures from protocols for the protection of a single qubit. We will show how the PZ quantum-jump model and the ADL and similar schemes with indirect feedback can be obtained from strategies for the protection of a single qubit based on weak measurements and weak unitary operations. We will also study the performance of CTQEC of the quantum-jump type in the case of Markovian and non-Markovian decoherence. We will show that due to the existence of a Zeno regime in non-Markovian dynamics, the performance of CTQEC can exhibit a quadratic improvement if the time resolution of the weak error-correcting operations is sufficiently high to reveal the non-Markovian character of the noise process.

8.2 CTQEC in an encoded basis

As discussed in Chapter 6, correctable information is always contained in subsystems of the system's Hilbert space [K06a, B.-KNP+08]. This means, in particular, that if the information initially encoded in the subsystem $\mathscr{H}^A$ in Eq. (8.1) is correctable after the noise map $\mathscr{E}$, it is *unitarily recoverable* [KS06], i.e., there exists a unitary map $\mathscr{U}(\cdot) = U(\cdot)U^\dagger$, $U \in \mathscr{B}(\mathscr{H}^S)$, such that

$$\mathscr{U} \circ \mathscr{E}(\rho \otimes \tau) = \rho \otimes \tau', \quad \tau' \in \mathscr{B}(\mathscr{H}^{B'}), \tag{8.12}$$
$$\text{for all } \rho \in \mathscr{B}(\mathscr{H}^A), \tau \in \mathscr{B}(\mathscr{H}^B),$$

where the subsystem $\mathscr{H}^{B'}$ can be different from $\mathscr{H}^B$. Complete correction generally requires an additional CPTP map that transforms the operators on $\mathscr{H}^{B'}$ into operators on $\mathscr{H}^B$. As shown in [OLB08], Eq. (8.12) is equivalent to the condition that the Kraus operators M_α of $\mathscr{E}$ satisfy

$$M_\alpha P^{AB} = U^\dagger I^A \otimes C_\alpha^{B \to B'}, \quad C_\alpha^{B \to B'} : \mathscr{H}^B \to \mathscr{H}^{B'}, \quad \forall \alpha. \tag{8.13}$$

Observe that if a particular set of error operators $\{M_i\}$ is correctable by the code, that is, if any CPTP map whose Kraus operators are linear combinations of $\{M_i\}$ is correctable, then there is a common recovery unitary U for all such CPTP maps. Note also that if the identity is among the correctable errors for which the code is designed (this is the case, in particular, for all stabilizer codes), from condition (8.13) it follows that the unitary U must leave the subsystem $\mathcal{H}^A$ in $\mathcal{H}^A \otimes \mathcal{H}^B$ invariant up to a transformation of the co-subsystem, $\mathcal{H}^B \to \mathcal{H}^{\tilde{B}}$ (dim$\mathcal{H}^{\tilde{B}}$ = dim$\mathcal{H}^B$). This means that if we change the basis by the unitary map U, the effect of the error operators $M'_\alpha = U M_\alpha U^\dagger$ in the new basis is

$$M'_\alpha P^{A\tilde{B}} = U M_\alpha P^{AB} U^\dagger = I^A \otimes C_\alpha^{\tilde{B} \to B'}, \quad C_\alpha^{\tilde{B} \to B'} : \mathcal{H}^{\tilde{B}} \to \mathcal{H}^{B'}, \quad \forall \alpha, \qquad (8.14)$$

i.e., the errors leave the code subsystem invariant up to a transformation of the co-subsystem. A method of obtaining U can be found in [KS06].

In what follows, we will imagine for concreteness the case of an $[[n, 1, r, d]]$ operator stabilizer code. This is a code that encodes 1 qubit into n, has r gauge qubits, and has distance d. In the encoded basis defined above, the Hilbert space of all n qubits can be written as

$$\mathcal{H}^S = \mathcal{H}^A \otimes \bigotimes_{i=1}^{n-r-1} \mathcal{H}_i^s \otimes \bigotimes_{j=1}^{r} \mathcal{H}_j^g, \qquad (8.15)$$

where $\mathcal{H}^A$ is a subsystem that corresponds to the logical qubit, $\mathcal{H}_i^s$ are the subsystems of the *syndrome* qubits, and $\mathcal{H}_j^g$ are the subsystems of the *gauge* qubits. Up to a redefinition of the basis of the syndrome qubits, we can assume that the subspace $\mathcal{H}^A \otimes \mathcal{H}^B$ in Eq. (8.1) corresponds to

$$\mathcal{H}^A \otimes \mathcal{H}^B = \mathcal{H}^A \otimes \bigotimes_{i=1}^{n-r-1} |0\rangle_i^s \otimes \bigotimes_{j=1}^{r} \mathcal{H}_j^g. \qquad (8.16)$$

We will refer to this subspace loosely as the *code space*, since this is where the state of the system is initialized, but we must keep in mind that the information of interest is contained in the tensor factor $\mathcal{H}^A$ in Eq. (8.15). If each of the syndrome qubits is initialized in the state $|0\rangle$, any correctable error will leave the subsystem $\mathcal{H}^A$ invariant and will only affect the co-subsystem, most generally transforming density operators on $\bigotimes_{i=1}^{n-r-1} |0\rangle_i^s \otimes \bigotimes_{j=1}^{r} \mathcal{H}_j^g$ into density operators on $\bigotimes_{i=1}^{n-r-1} \mathcal{H}_i^s \otimes \bigotimes_{j=1}^{r} \mathcal{H}_j^g$. In this basis, an error-correcting operation is simply a map on the syndrome qubits, which returns them to the state $|00...0\rangle$. In the language of stabilizer codes, a measurement of the syndrome is a measurement of the state of all syndrome qubits in the $\{|0\rangle, |1\rangle\}$ basis, and a correcting operation is any operation that effectively realizes a bit flip to those qubits that are in the state $|1\rangle$.

If the syndrome qubits are not properly initialized (as, for example, after the occurrence of an error), a subsequent error generally would not leave the code subsystem invariant. Most generally, after a system subject to decoherence and error correction evolves for a given time t, the state of the system becomes

$$\rho^S(t) = \alpha(t) \rho^{Ag}(t) \otimes \bigotimes_{i=1}^{n-r-1} |0\rangle\langle 0|_i^s + (1-\alpha(t))\widetilde{\rho}^{Ags}(t) + \text{cross terms}. \qquad (8.17)$$

Here $\rho^{Ag}(t)$ is a density matrix on the Hilbert space $\mathcal{H}^A \otimes \bigotimes_{j=1}^{r} \mathcal{H}_j^g$, $\widetilde{\rho}^{Ags}(t)$ is a density matrix with support on the orthogonal complement $\widetilde{\mathcal{H}}$ of the code space ($\mathcal{H}^A \otimes \bigotimes_{i=1}^{n-r-1} |0\rangle_i^s \bigotimes_{j=1}^{r} \mathcal{H}_j^g \oplus \widetilde{\mathcal{H}} = \mathcal{H}^S$), $\alpha(t) \in [0,1]$ is the code-space fidelity, and "cross terms" refers to linear combinations of terms of the form $|\psi_i\rangle\langle\widetilde{\phi}_j|$ and $|\widetilde{\phi}_j\rangle\langle\psi_i|$, where $\{|\psi_i\rangle\}$

is an orthonormal basis of $\mathscr{H}^A \otimes \bigotimes_{i=1}^{n-r-1} |0\rangle_i^s \otimes \bigotimes_{j=1}^{r} \mathscr{H}_j^g$ and $\{|\widetilde{\phi}_i\rangle\}$ is an orthonormal basis of $\widetilde{\mathscr{H}}$. The density matrix of the logical subsystem is $\rho^A(t) = \alpha(t)\text{Tr}_g(\rho^{Ag}(t)) + (1 - \alpha(t))\text{Tr}_{gs}(\widetilde{\rho}^{Ags}(t))$, where Tr_g denotes partial tracing over the gauge qubits and Tr_{gs} denotes partial tracing over the gauge qubits and the syndrome qubits. This density matrix is a transformed version of the state initially encoded in the code subsystem, where the transformation is the result of accumulation of uncorrectable errors. (Note that any transformation inside the subsystem $\mathscr{H}^A$ in Eq. (8.15) is by definition uncorrectable.)

Let us see how the density matrix ρ^A changes as a result of the action of the generator of noise during a time step Δt. Since by assumption the action of the noise generator leaves the code subsystem invariant up to a transformation of the co-subsystem, its effect on the term $\alpha \rho^{Ag}(t) \otimes \bigotimes_{i=1}^{n-r-1} |0\rangle\langle 0|_i^s$ in Eq.(8.17) during a time step Δt does not give rise to a nontrivial change in $\rho^A(t)$, but only to a decrease in the code-space fidelity,

$$\alpha(t) \to \alpha(t) - \gamma(t)\alpha(t)\Delta t + O(\Delta t^2), \qquad (8.18)$$

where $\gamma(t) \geq 0$ is a parameter that depends on the characteristics of the noise process, such as the rates of different errors, and possibly on the current density matrix of the gauge qubits inside the code space, $\text{Tr}_A(\rho^{Ag}(t))$. Note that if the noise is non-Markovian, the leading-order correction to $\alpha(t)$ due to the action of the noise on $\alpha \rho^{Ag}(t) \otimes \bigotimes_{i=1}^{n-r-1} |0\rangle\langle 0|_i^s$ is $O(\Delta t^2)$, i.e., $\gamma(t) = 0$ (see Section 8.5.2). The only way errors can arise inside the subsystem $\mathscr{H}^A$ is by the action of the noise mechanism on the other terms in Eq. (8.17). The weight of the second term is $(1-\alpha(t))$, and during a single time step the noise generator can give rise to a change in $\rho^A(t)$:

$$\rho^A(t) \to \rho^A(t) + \delta\rho^A(t), \qquad (8.19)$$

where

$$\| \delta\rho^A(t) \| \leq B(1-\alpha(t))\Delta t + O(\Delta t^2), \quad B \geq 0. \qquad (8.20)$$

The constant B depends on the rate of the noise process, its characteristics, and the characteristics of the code. From the positivity of the density matrix ρ^S one can show that the coefficients in front of the cross terms $|\psi_i\rangle\langle\widetilde{\phi}_j|$ and $|\widetilde{\phi}_j\rangle\langle\psi_i|$ are at most $\sqrt{\alpha(1-\alpha)}$ in magnitude, and therefore the change that can result in ρ^A due to the action of the noise generator on the third term in Eq. (8.17) is limited by

$$\| \delta\rho^A(t) \| \leq C\sqrt{\alpha(1-\alpha(t))}\Delta t + O(\Delta t^2), \qquad (8.21)$$

where $C \geq 0$ is another constant dependent on the characteristics of the noise and the code. Thus we see that the rate of change of the density matrix ρ^A is upper bounded as follows:

$$\left\| \frac{d\rho^A}{dt} \right\| \leq B(1-\alpha(t)) + C\sqrt{\alpha(1-\alpha(t))}. \qquad (8.22)$$

In other words, if we manage to keep $(1-\alpha(t))$ small, we will suppress the rate of accumulation of uncorrectable errors. The goal of CTQEC can thus be understood as that of keeping the state of every syndrome qubit close to the state $|0\rangle$.

Notice that a strong error-correcting operation in this basis can be realized by bringing each of the syndrome qubits to the state $|0\rangle$ independently. Therefore, the problem of implementing a strong error-correcting operation in terms of weak operations can be reduced to the problem of implementing the corresponding single-qubit operations via weak single-qubit operations. Of course, this is not the most general way of realizing collective initialization of the syndrome qubits, but it is appealing because it reduces the task to that of addressing several independent qubits individually. We will see, however, that the performance can be enhanced if, instead of addressing each of the syndrome qubits individually, we address each syndrome that can be associated with a qubit subspace in the space of the syndrome qubits. This will be discussed in the next section. Here we note that the operations in the original basis can be obtained by applying the inverse of the basis transformation to the operations in the encoded basis.

To get an idea of what the transformation between bases looks like, let us consider as an example the three-qubit bit-flip code with stabilizer generated by $\{IZZ, ZZI\}$. This code has logical codewords $|0_L\rangle = |000\rangle$ and $|1_L\rangle = |111\rangle$ and even though it only corrects bit-flip errors and does not have gauge qubits, it captures all the characteristics of nontrivial codes that are pertinent to our discussion. It can be verified that a correcting unitary for this code is $U = U_c CX_{1,2} CX_{1,3}$, where

$$U_c = X_1 \otimes |11\rangle\langle 11|_{23} + I_1 \otimes (I_2 \otimes I_3 - |11\rangle\langle 11|_{23}), \tag{8.23}$$

and $CX_{i,j}$ denotes the "controlled not" with qubit i being the control and qubit j the target. This unitary transforms the single-qubit bit-flip error operators as

$$\begin{aligned} XII &\to I \otimes (|00\rangle\langle 11| + |11\rangle\langle 00|) + X \otimes (|01\rangle\langle 10| + |10\rangle\langle 01|), \\ IXI &\to I \otimes X \otimes |0\rangle\langle 0| + X \otimes X \otimes |1\rangle\langle 1|, \\ IIX &\to I \otimes |0\rangle\langle 0| \otimes X + X \otimes |1\rangle\langle 1| \otimes X. \end{aligned} \tag{8.24}$$

In this basis, when the second and third qubits are in the state $|0\rangle$, the error operators leave the state of the first qubit invariant. Going back to the original basis is achieved by applying the basis transformation backwards, i.e., by applying the unitary $CX_{1,3} CX_{1,2} U_c$.

8.3 Quantum-jump CTQEC with weak measurements
8.3.1 The single-qubit problem

In this section we will show how to implement the PZ quantum-jump error-correction scheme [Eq. (8.3)] using weak measurements in the encoded basis. We start with the problem of protecting a single qubit in the state $|0\rangle$ from noise using weak measurements. The state $|0\rangle$ can be thought of as a trivial stabilizer code with stabilizer generated by Z. We will first consider the case of Markovian bit-flip decoherence, since this model is simple and provides a good intuition. Later, we will extend the result to general noise models.

A Markovian bit-flip process is described by the master equation

$$\frac{d\rho(t)}{dt} = \gamma(X\rho X - \rho), \tag{8.25}$$

where γ is the bit-flip rate. The general solution to this equation is

$$\rho(t) = \frac{1 + e^{-2\gamma t}}{2} \rho(0) + \frac{1 - e^{-2\gamma t}}{2} X\rho(0)X. \tag{8.26}$$

If the system starts in the state $|0\rangle\langle 0|$, without error correction it will decay down the Z-axis towards the maximally mixed state.

In the language of stabilizer codes, an error-correcting operation for this code consists of a measurement of the stabilizer generator Z followed by a unitary correction. If the result is $|1\rangle$, we apply a bit-flip operation X, and if the result is $|0\rangle$, we do nothing. The CP map corresponding to this strong error-correcting operation is

$$\mathscr{R}(\rho) = X|1\rangle\langle 1|\rho|1\rangle\langle 1|X + |0\rangle\langle 0|\rho|0\rangle\langle 0| = |0\rangle\langle 1|\rho|1\rangle\langle 0| + |0\rangle\langle 0|\rho|0\rangle\langle 0|. \quad (8.27)$$

One heuristic approach to making the above procedure continuous is to consider weak measurements of the stabilizer generator Z and weak rotations around the X-axis of the Bloch sphere conditioned on the measurement record. This is exactly the approach considered in the feedback procedures of the ADL type, and we will discuss it in Section 8.4.1.

Observe that the transformation (8.27) can also be written as

$$\mathscr{R}(\rho) = |0\rangle\langle +|\rho|+\rangle\langle 0| + |0\rangle\langle -|\rho|-\rangle\langle 0|$$
$$= ZW|+\rangle\langle +|\rho|+\rangle\langle +|WZ + XW|-\rangle\langle -|\rho|-\rangle\langle -|WX, \quad (8.28)$$

where $|\pm\rangle = (|0\rangle \pm |1\rangle)/\sqrt{2}$ and W is the Hadamard gate. Therefore, the same error-correcting operation can be implemented as a measurement in the $\{|+\rangle, |-\rangle\}$ basis (measurement of the operator X), followed by a unitary conditioned on the outcome: if the outcome is $|+\rangle$, we apply ZW; if the outcome is $|-\rangle$, we apply XW. This choice of unitaries is not unique; for example, we could apply just W instead of ZW after outcome $|+\rangle$. But this particular choice has a convenient geometric interpretation – the unitary ZW corresponds to a rotation around the Y-axis by an angle $\pi/2$, $ZW = e^{i\frac{\pi}{2}\frac{Y}{2}}$, and XW corresponds to a rotation around the same axis by an angle $-\pi/2$, $ZW = e^{-i\frac{\pi}{2}\frac{Y}{2}}$.

A weak version of the above error-correcting operation can be constructed by taking the corresponding weak measurement of the operator X, followed by a weak rotation around the Y-axis, whose direction is conditioned on the outcome:

$$\rho \to \frac{I + i\epsilon'Y}{\sqrt{1+\epsilon'^2}}\sqrt{\frac{I+\epsilon X}{2}}\rho\sqrt{\frac{I+\epsilon X}{2}}\frac{I - i\epsilon'Y}{\sqrt{1+\epsilon'^2}}$$
$$+ \frac{I - i\epsilon'Y}{\sqrt{1+\epsilon'^2}}\sqrt{\frac{I-\epsilon X}{2}}\rho\sqrt{\frac{I-\epsilon X}{2}}\frac{I + i\epsilon'Y}{\sqrt{1+\epsilon'^2}}. \quad (8.29)$$

Here ϵ and ϵ' are small parameters. Note that the fact that we describe the net result of the transformation by a CPTP map means that after we apply feedback we discard information about the outcome of the measurement, or rather, we do not condition any future operations on that information and therefore the transformation of the average density matrix during a single time step is given by Eq. (8.29). Such a scheme is said to be based on *direct* feedback, i.e., the feedback Hamiltonian depends only on the outcome of the most recent measurement, which does not require information processing of the measurement record. Generally, discarding information leads to suboptimal protocols, and we will discuss the possibility of improving that scheme in Section 8.4.1.

From the symmetry of the map (8.29) it can be seen that if the map is applied to a state that lies on the Z-axis, it will keep the state on the Z-axis. Whether the state will move towards $|0\rangle\langle 0|$ or towards $|1\rangle\langle 1|$ depends on the relation between ϵ and ϵ'. Since our goal is to protect the state from drifting away from $|0\rangle\langle 0|$ due to bit-flip decoherence, for now we will assume that the state lies on the Z-axis in the northern hemisphere. We would like, if possible, to choose the relation between the parameters ϵ and ϵ' in such a way that the effect of this map on any state on the Z-axis to be to move that state towards $|0\rangle\langle 0|$.

In order to calculate the effect of this map on a given state, it is convenient to write the state in the $\{|+\rangle, |-\rangle\}$ basis. For a state on the Z-axis, $\rho = \alpha|0\rangle\langle 0| + (1-\alpha)|1\rangle\langle 1|$, we have

$$\rho = \frac{1}{2}|+\rangle\langle +| + \frac{1}{2}|-\rangle\langle -| + (2\alpha - 1)\left(\frac{1}{2}|+\rangle\langle -| + \frac{1}{2}|-\rangle\langle +|\right). \tag{8.30}$$

For the action of our map on the state (8.30) we obtain:

$$\rho \to \frac{1}{2}|+\rangle\langle +| + \frac{1}{2}|-\rangle\langle -| \\ + \frac{(1-\epsilon'^2)\sqrt{1-\epsilon^2}(2\alpha-1) + 2\epsilon\epsilon'}{1+\epsilon'^2}\left(\frac{1}{2}|+\rangle\langle -| + \frac{1}{2}|-\rangle\langle +|\right). \tag{8.31}$$

Thus, we can think that upon this transformation the parameter α transforms to α', where

$$2\alpha' - 1 = \frac{(1-\epsilon'^2)\sqrt{1-\epsilon^2}(2\alpha-1) + 2\epsilon\epsilon'}{1+\epsilon'^2}. \tag{8.32}$$

If it is possible to choose the relation between ϵ and ϵ' in such a way that $\alpha' \geq \alpha$ for every $0 \leq \alpha \leq 1$, then clearly the state must remain invariant when $\alpha = 1$. Imposing this requirement, we obtain

$$\epsilon = \frac{2\epsilon'}{1+\epsilon'^2}, \tag{8.33}$$

or equivalently

$$\epsilon' = \frac{1-\sqrt{1-\epsilon^2}}{\epsilon}. \tag{8.34}$$

Substituting back in (8.32), we can express

$$\alpha' - \alpha = \frac{4\epsilon'^2}{(1+\epsilon'^2)^2}(1-\alpha) \geq 0. \tag{8.35}$$

We see that the coefficient α (which is the fidelity of the state with $|0\rangle\langle 0|$) indeed increases after every application of our weak CP map. The amount by which it increases for fixed ϵ' depends on α and becomes smaller as α approaches 1.

Since we will be taking the limit $\epsilon \to 0$, we can write Eq. (8.34) as

$$\epsilon' = \frac{\epsilon}{2} + O(\epsilon^3). \tag{8.36}$$

If we define the relation between the time step τ_c and ϵ as in Eq. (8.9), for the effect of the CPTP map (8.29) on an arbitrary state of the form $\rho = \alpha|0\rangle\langle 0| + \beta|0\rangle\langle 1| + \beta^*|1\rangle\langle 0| + (1-\alpha)|1\rangle\langle 1|$, $\alpha \in \mathbb{R}, \beta \in \mathbb{C}$, we obtain

$$\alpha \to \alpha + (1-\alpha)\kappa\tau_c, \tag{8.37}$$

$$\beta \to \sqrt{1-\kappa\tau_c}\beta = \beta - \frac{1}{2}\kappa\beta\tau_c + O(\tau_c^2). \tag{8.38}$$

This is exactly the map (8.8) for $\mathcal{R}(\rho)$ given by Eq. (8.27).

We see that for an infinitesimal time step dt, the effect of the noise is to decrease $\alpha(t)$ by the amount $\lambda(2\alpha(t)-1)dt$ and that of the correcting operation is to increase it by $\kappa(1-\alpha(t))dt$.

Combining both effects, we obtain the net master equation that describes the evolution of the qubit subject to Markovian bit-flip errors and the quantum-jump error-correction scheme:

$$\frac{d\alpha(t)}{dt} = -(\kappa + 2\lambda)\alpha(t) + (\kappa + \lambda). \tag{8.39}$$

The solution is

$$\alpha(t) = (1 - \alpha_*)e^{-(\kappa+2\lambda)t} + \alpha_*, \tag{8.40}$$

where

$$\alpha_* = 1 - \frac{1}{2+r}, \tag{8.41}$$

and $r = \kappa/\lambda$ is the ratio between the rate of error correction and the rate of decoherence. We see that the fidelity decays, but it is confined above its asymptotic value α_*, which can be made arbitrarily close to 1 for sufficiently large r.

Finally, let us show that this procedure works for any kind of decoherence where the state need not remain on the Z-axis at all times. From Eq. (8.38) we see that the effect of a single application of the map to a general state is to transfer a small portion of the $|1\rangle\langle 1|$-component to $|0\rangle\langle 0|$, and to decrease the magnitude of the off-diagonal components by multiplying them by $\sqrt{1 - \kappa\tau_c}$. If there is noise, the most general negative effect of a single step of the noise process is to increase the magnitude of β and decrease α. For a realistic physical map, the amounts by which these components change during a time step Δt should tend to zero when $\Delta t \to 0$. Since ultimately any noise process is driven by a Hamiltonian acting on the system and its environment, this means that for small Δt each of these amounts can be upper-bounded by $\gamma_{max}\Delta t$, where γ_{max} is some finite positive number. Therefore, if the system is simultaneously subject to decoherence and error correction, $|\beta|$ and $(1 - \alpha)$ will not increase above certain values for which the single-step effects of decoherence and error correction exactly cancel each other. We can upper-bound these quantities by

$$(1 - \alpha)_{max} = \frac{\gamma_{max}}{\kappa}, \tag{8.42}$$

$$|\beta|_{max} = \frac{2\gamma_{max}}{\kappa}. \tag{8.43}$$

This means that the state can be kept arbitrarily close to $|0\rangle\langle 0|$ for sufficiently high rates of error correction κ. In Section 8.5 we will see that if the noise is non-Markovian $(1 - \alpha)_{max}$ scales as $\frac{1}{\kappa^2}$ for large κ!

We remark that one way of implementing the weak measurement of the X operator used in this scheme is by coupling the system qubit to an ancilla qubit prepared in the state $|+\rangle\langle +|$ for a short time, via the Hamiltonian $H_X = -X \otimes Y$, where X acts on the system qubit and Y acts on the ancilla, followed by a measurement of the ancilla in the $\{|0\rangle, |1\rangle\}$ basis (the latter can be destructive). It can be verified that if we first apply the unitary transformation $U_X(\epsilon) = \exp i\frac{\epsilon}{2}X \otimes Y$ followed by a measurement of the ancilla, up to second order in ϵ the resulting measurement on the system is

$$\rho \to \frac{\sqrt{\frac{I \pm \epsilon X}{2}}\rho\sqrt{\frac{I \pm \epsilon X}{2}}}{p_\pm}, \tag{8.44}$$

with probabilities $p_\pm = \frac{1}{2}(1 \pm \epsilon \text{Tr}(X\rho))$. Since we are interested in the limit where $\epsilon \to 0$, only the lowest-order nontrivial contributions to the error-correcting CPTP map are important, and they are of order ϵ^2.

8.3.2 General codes

How do we extend this approach to general codes? As we mentioned earlier, one way is to simply apply the described operation to each of the syndrome qubits in the encoded basis. According to the argument in the previous subsection, no matter what the exact form of the noise process on the syndrome qubits is, this scheme will keep each of them close to the state $|0\rangle\langle 0|$ within some distance that can be made arbitrarily small for sufficiently large error-correction rates. This in turn would ensure that the code-space fidelity is close to 1, which would suppress the rate of accumulation of uncorrectable errors as argued in Section 8.2. This approach is particularly attractive because of its conceptual simplicity and the fact that it involves operations only on each of the syndrome qubits whose number $n-r-1$ is smaller than the number of different nontrivial correctable errors, which can be up to $2^{n-r-1} - 1$. Furthermore, it is obvious that the operations on the different qubits commute and therefore can be applied simultaneously. However, it is not difficult to see that even though the equivalent infinitesimal map has the form (8.3), the effective $\mathscr{R}(\rho)$ is not equal to the error-correcting map for this code, where the latter acts as

$$\mathscr{R}(\rho^s) = \bigotimes_{i=1}^{n-r-1} |0\rangle\langle 0|_i^s \tag{8.45}$$

for any state ρ^s of all syndrome qubits. This is because, if we apply error correction separately on the different qubits, up to first order in dt only those terms in which there is one qubit in the state $|1\rangle$ and all the rest are in the state $|0\rangle$ (e.g., $|10...0\rangle\langle 10...0|$) will get mapped to $|00...0\rangle\langle 00...0|$. The full error-correcting map, however, maps all states to the state $|00...0\rangle\langle 00...0|$ and therefore it is more powerful. Is there a way to construct the full map based on the single-qubit operations described in the previous subsection?

It turns out that the answer is yes. The idea is to associate an abstract qubit to each nontrivial error syndrome in the code as follows. As was mentioned earlier, each syndrome corresponds to a state of the syndrome qubits of the form $|\nu_1\nu_2\cdots\nu_{n-r-1}\rangle$, where ν_i can be either 0 or 1. Let us label these different syndrome states by $|i_s\rangle$, $i_s = 0, ..., 2^{n-r-1} - 1$, with $|0_s\rangle = |00...0\rangle$ being the trivial syndrome corresponding to "no error." The density matrix of the entire system can then be written

$$\rho^S = \alpha(t)\rho^{Ag}(t) \otimes |0_s\rangle\langle 0_s| + \sum_{i_s \geq 1} \beta_{i_s} \widetilde{\rho}_{i_s}^{Ag}(t) \otimes |i_s\rangle\langle i_s|$$
$$+ \sum_{i_s \neq j_s} \sigma_{i_s j_s}^{Ag}(t) \otimes |i_s\rangle\langle j_s|, \tag{8.46}$$

where $\widetilde{\rho}_{i_s}^{Ag}(t)$ are density matrices on $\mathscr{H}^A \otimes \bigotimes_{j=1}^r \mathscr{H}_j^g$, $\beta_{i_s} \geq 0$ are the weights of the state inside the different error subspaces, and $\sigma_{i_s j_s}^{Ag}(t)$ are operators on $\mathscr{H}^A \otimes \bigotimes_{j=1}^r \mathscr{H}_j^g$.

To each nontrivial syndrome we can associate a qubit subspace of the space of all syndrome qubits, which is spanned by the state $|0_s\rangle$ and the state $|i_s\rangle$ corresponding to that syndrome. Let us take for concreteness one of these qubits – the subspace spanned by $|0_s\rangle$ and $|1_s\rangle$. If we apply the single-qubit operations described in the previous subsection to this subspace while acting trivially on its orthogonal complement, the effect of the resulting operation on the terms $\beta_{1_s}\tilde{\rho}_{1_s}^{Ag}(t) \otimes |1_s\rangle\langle 1_s| + \sigma_{0_s 1_s}^{Ag}(t) \otimes |0_s\rangle\langle 1_s| + \sigma_{1_s 0_s}^{Ag}(t) \otimes |1_s\rangle\langle 0_s|$ in Eq. (8.46) will be the same as that of the quantum-jump error-correcting map (8.3) with $\mathscr{R}(\rho)$ given by Eq. (8.45). At the same time, the effect on the rest of the terms will be trivial. Therefore, if we apply the analogous operation to each of the qubit subspaces spanned by $|i_s\rangle$ and $|0_s\rangle$, we will effectively realize the desired quantum-jump error-correcting map.

Observe that all these single-qubit maps commute and so do the generators they give rise to in the corresponding continuous quantum-jump equation. If we think of the resulting processes as being driven by the action of the quantum-jump generators, then it is obvious that all of them can be implemented simultaneously. However, if we think of each of these maps as resulting from weak measurements and weak unitary operations as described in the previous subsection, the measurements and unitaries do *not* commute. For example, the X operator for the qubit j_s has the form $X_{j_s} = |j_s\rangle\langle 0_s| + |0_s\rangle\langle j_s|$, and therefore $[X_{i_s}, X_{j_s}] = |i_s\rangle\langle j_s| - |j_s\rangle\langle i_s|$. This means that the measurements of the X operators cannot be implemented simultaneously on all qubits. The same holds for the rotations around the Y-axes. Does this mean that we have to apply the different operations in series? This would require the ability to precisely turn on and off, on a very short time scale, the couplings to the external fields needed for the different measurements, which does not correspond to a continuous measurement.

It turns out that alternating the different couplings is not needed; the same couplings that one would use for implementing the weak measurements on the individual qubits can be turned on simultaneously, and so can the feedback Hamiltonians that one would use depending on the outcomes of the different measurements. This is because all extra terms that arise from the fact that the operations on the different qubits do not commute cancel out when we average over the outcomes. We outline how this can be verified using the implementation of the weak measurement via a qubit ancilla described at the end of Section 8.3.1. For each of the qubits corresponding to different syndromes, we will need to turn on a different Hamiltonian that couples that qubit to a separate ancilla initially prepared in the state $|+\rangle$. Let us label the ancilla corresponding to the qubit j_s also by j_s. If we turn on all of these Hamiltonians simultaneously, the overall Hamiltonian is

$$H_{meas} = -\sum_{j_s} X_{j_s} \otimes Y_{j_s}^a, \qquad (8.47)$$

where the $Y_{j_s}^a$ act on the different ancilla systems but the X_{j_s} do not act on different systems and do not commute. Imagine that this Hamiltonian acts for time $\frac{\epsilon}{2}$, i.e., it gives rise to the unitary $U = \exp(\frac{i\epsilon}{2} \sum_{j_s} X_{j_s} \otimes Y_{j_s}^a)$. At this point we can measure projectively each of the ancillas in the $\{|0\rangle, |1\rangle\}$ basis and turn on the corresponding single-qubit correction Hamiltonians $\xi_{j_s} Y_{j_s}$, where $\xi_{j_s} = \pm 1$ is the sign of the Hamiltonian, which depends on the outcome of the measurement, and $Y_{j_s} = i|j_s\rangle\langle 0_s| - i|0_s\rangle\langle j_s|$. The overall feedback Hamiltonian is

$$H_{fb} = \sum_{j_s} \xi_{j_s} Y_{j_s}. \qquad (8.48)$$

One can verify that, up to second order in ϵ, the resulting operation after averaging over the outcomes is exactly equal to the quantum-jump operation (8.3) with $\mathscr{R}(\rho)$ given by Eq. (8.45). The easiest way to see this is to observe that all unwanted terms in the resulting density matrix are proportional to $\xi_{i_s}\xi_{j_s}$, $i_s \neq j_s$, and therefore when we sum over all different outcomes, these terms disappear.

To get an idea of what the weak measurements and feedback unitaries mean in the original basis, let us look again at the three-qubit bit-flip code. Observe that the syndrome states $|i\rangle^s$ in the encoded basis are $|1_s\rangle = |10\rangle$, $|2_s\rangle = |01\rangle$, $|3_s\rangle = |11\rangle$, i.e., the three abstract qubits corresponding to these syndromes have X and Y operators:

$$X_{1_s} = I \otimes (|10\rangle\langle 00| + |00\rangle\langle 10| + |01\rangle\langle 01| + |11\rangle\langle 11|),$$
$$X_{2_s} = I \otimes (|01\rangle\langle 00| + |00\rangle\langle 01| + |10\rangle\langle 10| + |11\rangle\langle 11|), \quad (8.49)$$
$$X_{3_s} = I \otimes (|11\rangle\langle 00| + |00\rangle\langle 11| + |01\rangle\langle 01| + |10\rangle\langle 10|);$$

$$Y_{1_s} = I \otimes (i|10\rangle\langle 00| - i|00\rangle\langle 10| + |01\rangle\langle 01| + |11\rangle\langle 11|),$$
$$Y_{2_s} = I \otimes (i|01\rangle\langle 00| - i|00\rangle\langle 01| + |10\rangle\langle 10| + |11\rangle\langle 11|), \quad (8.50)$$
$$Y_{3_s} = I \otimes (i|11\rangle\langle 00| - i|00\rangle\langle 11| + |01\rangle\langle 01| + |10\rangle\langle 10|).$$

By applying the inverse basis transformation $CX_{1,3}CX_{1,2}U_c$ with U_c given by Eq. (8.23), we obtain these operators in the original basis:

$$X'_{1_s} = \frac{1}{2}ZXZ + \frac{1}{2}IXI + \frac{1}{2}III - \frac{1}{2}ZIZ,$$
$$X'_{2_s} = \frac{1}{2}ZZX + \frac{1}{2}IIX + \frac{1}{2}III - \frac{1}{2}ZZI, \quad (8.51)$$
$$X'_{3_s} = \frac{1}{2}XZZ + \frac{1}{2}XII + \frac{1}{2}III - \frac{1}{2}IZZ;$$

$$Y'_{1_s} = \frac{1}{2}ZYI + \frac{1}{2}IYZ + \frac{1}{2}III - \frac{1}{2}ZIZ,$$
$$Y'_{2_s} = \frac{1}{2}IZY + \frac{1}{2}ZIY + \frac{1}{2}III - \frac{1}{2}ZZI, \quad (8.52)$$
$$Y'_{3_s} = \frac{1}{2}YZI + \frac{1}{2}YIZ + \frac{1}{2}III - \frac{1}{2}IZZ.$$

We see that implementing the PZ scheme using weak measurements and unitary operations requires the ability to apply Hamiltonians that are complicated sums of different elements of the Pauli group. We will postpone the analysis of the performance of that scheme in the presence of decoherence until Section 8.5. We now turn to look at alternative methods for protecting a single qubit from noise using weak measurements, and their corresponding generalizations to multi-qubit codes.

8.4 Schemes with indirect feedback

8.4.1 The single-qubit problem

We already mentioned that another way of "continuization" of the discrete single-qubit error-correcting map (8.27) is to apply continuous measurements of the stabilizer generator Z and rotations around the X-axis conditioned on the measurement record. A continuous measurement of the operator Z can be achieved by an infinite repetition of a weak measurement with measurement operators

$$M_{\pm}^Z(\epsilon) = \sqrt{\frac{I \pm \tanh(\epsilon)Z}{2}}. \tag{8.53}$$

The evolution of the state of the system under such observation can be described by a random walk along a curve parametrized by $x \in R$. The state at any moment during the procedure can be written in the form

$$\rho(x) = \frac{M^Z(x)\rho(0)M^Z(x)}{\text{Tr}(M^Z(x)\rho(0)M^Z(x))} \tag{8.54}$$

for some value of x, where $M^Z(x) = \sqrt{(I + \tanh(x)Z)/2}$ and $\rho(0)$ is the initial state. After every application of the weak measurement $M_{\pm}^Z(\epsilon)$, the parameter x changes to $x \pm \epsilon$ depending on the outcome. The two projective measurement outcomes of the strong measurement of Z correspond to $x = \pm\infty$. The procedure is continued until $|x| \geq X$ for some X which is sufficiently large that $M^Z(X) \approx |0\rangle\langle 0|$ and $M^Z(-X) \approx |1\rangle\langle 1|$ to any desired precision [O08].

In the limit when $\epsilon \to 0$, the evolution of the state of the system can be described by a continuous stochastic differential equation. We can introduce a time step δt and a rate

$$\kappa = \epsilon^2/\delta t. \tag{8.55}$$

Then we can define a mean-zero increment δW as follows:

$$\delta W = (\delta x - M[\delta x])/\sqrt{\kappa}, \tag{8.56}$$

where $\delta x = \pm \epsilon$ and $M[\delta x]$ is the mean of δx,

$$M[\delta x] = \epsilon(p_+(x) - p_-(x)). \tag{8.57}$$

Here $p_\pm(x)$ are the probabilities for the two outcomes of the weak measurement $M_{\pm}^Z(\epsilon)$ at the point x,

$$p_\pm(x) = \frac{1}{2}(1 \pm \epsilon \langle Z \rangle_x), \tag{8.58}$$

with $\langle Z \rangle_x = \text{Tr}(Z\rho(x))$. Note that $M[(\delta W)^2] = \delta t + O(\delta t^2)$.

Expanding the change of the state under the measurement $M_{\pm}^Z(\epsilon)$ up to second order in δW, and taking the limit $\delta W \to 0$ while keeping the rate κ fixed, it can be shown that the evolution of the state of the system subject to such a continuous observation is described by the following stochastic differential equation:

$$d\rho(t) = \frac{\kappa}{4}\mathscr{D}[Z](\rho(t))dt + \frac{\sqrt{\kappa}}{2}\mathscr{F}[Z](\rho(t))dW(t). \tag{8.59}$$

Here $dW(t)$ is a Wiener increment, i.e., a mean-zero normally distributed random variable with variance dt. The evolution of the parameter x is given by

$$dx(t) = \kappa \langle Z \rangle_t dt + \sqrt{\kappa}dW(t), \tag{8.60}$$

where $\langle Z \rangle_t = \text{Tr}(Z\rho(t))$. From $x(t)$ one can define the average measurement current as the mean of $dx(t)/dt$,

$$I_x^{ave}(t) = \kappa \langle Z \rangle_t. \tag{8.61}$$

If we apply no error correction to our qubit (initially in the state $|0\rangle\langle 0|$), under bit-flip decoherence its state will drift down the Z-axis of the Bloch sphere towards the center of the sphere (the maximally mixed state). According to the scheme proposed in [ADL02], at a given moment we apply a weak measurement of the stabilizer Z and a weak rotation around the X-axis, which depends on the state of the system at that moment. In the simplified version of that scheme in [SAJ+04], the feedback is conditioned only on an estimate of the average measurement current. If at a given moment the state is somewhere along the Z-axis, i.e., $\rho = \alpha|0\rangle\langle 0| + (1-\alpha)|1\rangle\langle 1|$, $0 \leq \alpha \leq 1$, the effect of a weak measurement would be to move the state slightly up or down along the axis depending on the outcome. It is easy to see that the result of such a measurement does not change the value of α on average, because $M_+^Z(\epsilon)\rho M_+^Z(\epsilon) + M_-^Z(\epsilon)\rho M_-^Z(\epsilon) = \rho$. One is then led to ask whether including feedback could improve the average fidelity. The answer depends on whether the state lies in the northern or the southern hemisphere of the Bloch sphere. If the state lies on the Z-axis in the northern hemisphere, it is not possible to improve its fidelity by feedback. Assuming that the measurement is sufficiently weak, so that the negative outcome $M_-^Z(\epsilon)\rho M_-^Z(\epsilon)/\text{Tr}(M_-^Z(\epsilon)\rho M_-^Z(\epsilon))$ is still in the northern hemisphere, no unitary operation can bring any of the two outcomes closer to the north pole since unitary operations preserve the distance from the center. On the contrary, a unitary rotation around the X-axis would move both outcomes away from the Z-axis and therefore away from the target.

In the ADL scheme there is no risk of the feedback decreasing the fidelity with the target state because the feedback is conditioned on the current state and always tends to increase the fidelity with the code space; if the state lies on the Z-axis in the northern hemisphere, no rotation would be applied. However, during initial times that scheme would not be helpful for increasing the average value of α either, because a weak measurement keeps the state on the Z-axis in the northern hemisphere. If we go to the continuous limit, $\epsilon \to 0$, the Wiener parameter is normally distributed and during an infinitesimal time step the state may enter the southern hemisphere, but with a negligible probability. Thus, during initial times, the scheme would not be helpful with respect to the average fidelity, and only after the probability for the state to enter the southern hemisphere becomes significant will it start to have an effect. This intuition is confirmed by the numerical simulations of a generalization of this protocol to multi-qubit codes presented in [ADL02].

In the scheme in [SAJ+04], the feedback is not conditioned on the state but on an estimate of the average measurement current (8.61). The idea is that, by filtering the noisy measurement data obtained during some short time interval before a given moment t, we can try to obtain an estimate of the average change of $x(t)$ with time at that moment, i.e., an estimate of $\langle Z \rangle_t$. But clearly such an estimate cannot be precise, because it would mean that we could measure the expectation value of an observable almost without disturbing the state. Therefore, any such estimate inevitably carries imprecision. For example, it could be that the state of the system is $|0\rangle\langle 0|$ but we obtain a sequence of negative outcomes, which give rise to the effective measurement operator $M^Z(x) = \sqrt{I + \tanh(x)Z}$ with $x < 0$. This can occur with finite probability and it would suggest that the state lies in the southern hemisphere, while the state will remain $|0\rangle\langle 0|$ under this measurement. In such a case, this scheme would apply a rotation that would take the state away from the target state, i.e., during short initial times this scheme could have a negative effect. Nevertheless, as time progresses, more and more trajectories enter the southern hemisphere and the scheme may lead to an improvement of the average fidelity with the target state at later times. Indeed, numerical simulations have confirmed the efficiency of this scheme and its generalization to multi-qubit codes in certain parameter regimes [SAJ+04].

We point out that the two general strategies for the protection of a qubit that we considered – the one involving continuous measurement of the X operator and direct feedback (the quantum-jump scheme), and the one involving continuous measurements of the Z operator and indirect feedback (the ADL and similar schemes) – strongly resemble two optimal protocols for the purification of a qubit discussed in [J04] and [WR06]. In [J04] it was shown that the fastest increase on average of the purity of a single qubit using weak measurements is achieved if the qubit is measured in a basis perpendicular to the axis in the Bloch sphere that connects the current state with the center of the sphere. If we assume that we can apply fast unitary rotations on the time scale of the measurements, the fastest preparation of a qubit in the state $|0\rangle\langle 0|$ can be achieved by measuring the state in the eigenbasis of X, and after every weak measurement apply a rotation around the Y-axis that brings this state to the Z-axis. This is almost the same as the quantum-jump scheme, except that we did not assume that we can apply an arbitrarily strong and precise rotation that brings each outcome on the Z-axis, but only a rotation that would bring the state to the north pole if it was there before the measurement.

In [J04], on the other hand, it was shown that, if we are interested in the *average time* that it would take to purify the qubit to a certain degree, we have to measure it along the axis that connects it with the center of the Bloch sphere. Again, if we assume that we can apply arbitrarily fast rotations, the optimal average time for preparing a qubit in the state $|0\rangle\langle 0|$ with some precision can be achieved if we measure the qubit in the eigenbasis of Z and, whenever the qubit enters the southern hemisphere, apply rotations around the X-axis that bring it to the northern half of the Z-axis. The difference of the ADL scheme from this approach is again that the ADL scheme does not assume infinitely fast and precise rotations. Thus we see that the two competitive error-correction schemes we discussed can be regarded as originating from two optimal protocols for the preparation of a qubit in a known state – one that optimizes the average fidelity with the target state, and another that optimizes the average time to reach the target state.

Of course, this does not mean that the two schemes we described are optimal for the resources they use. In the quantum-jump scheme, for example, we discard information about the outcome of the measurement after every feedback operation. If we keep this information and estimate the current state, we can in principle improve the performance of the scheme. Let us say that the state is somewhere far from the Z-axis. Since each of the outcomes of the weak measurement change the state by a small amount, after either outcome we will have to apply rotations in the same direction in order to bring the state closer to the Z-axis. If we did not keep track of the actual state, however, we would apply rotations in opposite directions after the two different outcomes. But it turns out that the improvement we can gain by keeping track of the actual states is small. It can be verified that even if we assume that we are able to apply infinitely fast and precise rotations, i.e., that we can bring the state on the Z-axis after every weak measurement outcome, if the measurement strength is fixed, the correction to the quantity $(1-\alpha_*^M)$ [Eq. (8.41)] we can obtain is of order $O((1-\alpha_*^M)^2)$. But as we argued in Section 8.2 and will discuss further in Section 8.5, this is the quantity that is responsible for the effective decrease of the error rate in a general code. In that sense, the performance of the quantum-jump scheme is very close to optimal when $(1-\alpha_*^M)$ is small, even though the scheme requires no side information processing. Note, however, that we assumed that at the level of a single weak operation we can ensure a particular relation between the measurement strength and the strength of the correcting rotation – Eq. (8.36). If we cannot apply a sufficiently strong rotation to keep the state $|0\rangle\langle 0|$ invariant, the equilibrium fidelity with the target state α_* would be lower.

8.4.2 Generalizations to multi-qubit codes

A natural extension of the single-qubit schemes with indirect feedback to nontrivial codes can be obtained simply by applying these schemes to the syndrome qubits in the encoded basis with the purpose of keeping each of them close to the state $|0\rangle\langle 0|$. It is not hard to see that the operators Z_i^s on the syndrome qubits in the encoded basis are actually the stabilizer generators for the code. For example, by applying the inverse of the basis transformation for the bit-flip code described in Section 8.2, one can see that the operators $I^A \otimes Z_1^s \otimes I_2^s$ and $I^A \otimes I_1^s \otimes Z_2^s$ correspond to the generators ZZI and ZIZ, respectively.

The Hamiltonians X_i^s needed for the feedback, however, do not have simple forms in the original basis. In particular, for the bit-flip code, the operators $I^A \otimes X_1^s \otimes I_2^s$ and $I^A \otimes I_1^s \otimes X_2^s$ correspond to $\frac{1}{2}XIX + \frac{1}{2}YIY + \frac{1}{2}ZXZ + \frac{1}{2}IXI$ and $\frac{1}{2}XXI + \frac{1}{2}YYI + \frac{1}{2}ZZX + \frac{1}{2}IIX$, respectively. The models considered in [ADL02, SAJ+04, CLG08] also measure continuously the stabilizer generators of the code, but the feedback Hamiltonians are assumed to be single-qubit operators in the original basis. However, note that in the general formulation of the ADL scheme – Eq. (8.11) – the correcting Hamiltonians H_r are not specified, and in that sense the possibility we discuss here can be regarded as a special case of the ADL scheme.

In the case of the bit-flip code, the authors in [ADL02, SAJ+04] take the correcting Hamiltonians to be XII, IXI, and IIX. This choice is motivated on one hand by its analogy with the strong version of the error-correcting operation for this code, and on the other by its simplicity. In the encoded basis, however, these operators are correlated and act on subsystem $\mathcal{H}^A$ as well. More precisely, XII, IXI, and IIX are equal to $\frac{1}{2}IXX - \frac{1}{2}IYY + \frac{1}{2}XXX + \frac{1}{2}XYY$, $\frac{1}{2}IXZ + \frac{1}{2}IXI + \frac{1}{2}XXI - \frac{1}{2}XXZ$, and $\frac{1}{2}IZX + \frac{1}{2}IIX + \frac{1}{2}XIX - \frac{1}{2}XZX$, respectively. Naturally, since the code is designed to correct single-qubit bit-flips, these operators leave the factor $\mathcal{H}^A$ in the code space $\mathcal{H}^A \otimes |0\rangle_1^s \otimes |0\rangle_2^s$ invariant by definition. A similar property holds for codes that can correct arbitrary single-qubit errors. But these operators can introduce errors to the code subsystem through their nontrivial action on the orthogonal complement of the code space. In particular, imagine that the system undergoes just a single perfectly correctable error, say, a single bit-flip. Then a strong error-correcting operation must be able to correct it. But if we apply a continuous scheme in which the correcting Hamiltonians act nontrivially on the complement of the code space, this scheme would generally apply nontrivial transformations to the subsystem $\mathcal{H}^A$ in the error subspace, which are by definition uncorrectable. (Note that this cannot occur with a scheme that uses operations acting locally on the syndrome qubits.) Nevertheless, in the case of continuous decoherence where uncorrectable errors inevitably arise, this property is not of crucial significance. As we argued earlier, the way CTQEC works is by keeping the weight outside of the code space small, which suppresses the effective accumulation of uncorrectable errors. As long as the scheme is able to keep that weight small, it will still have an effect according to our earlier arguments. Indeed, numerical simulations show that with the use of single-qubit feedback Hamiltonians one can achieve a significant improvement of the codeword fidelity with respect to that of an unprotected qubit, and outperform the approach of single-shot error correction in various regimes. For details about the numerical results, we refer the reader to [ADL02, SAJ+04, CLG08].

8.5 Quantum jumps for Markovian and non-Markovian noise

In this section we will look at the performance of the quantum-jump scheme in the cases of Markovian and non-Markovian decoherence. We will consider the bit-flip code in the case of simple noise models for which the evolution is exactly solvable. The conclusions we obtain, however, hold for general codes and noise models.

8.5.1 Markovian decoherence

The model described by Eq. (8.4) represents the noise as driven by a Lindblad generator, which is valid under the Markovian assumption of bath correlation times that are much shorter than any characteristic time scale of the system [BP02]. In the case of protecting a single qubit from Markovian bit-flip decoherence, we already found the solution for this model – Eq. (8.40). We saw that the equilibrium fidelity to which the qubit decays scales as $1/\kappa$ for large error-correction rates κ.

For the bit-flip code, we will assume that all qubits decohere through identical independent bit-flip channels, i.e., $\mathscr{L}(\rho)$ is of the form (8.7) with $\lambda_1 = \lambda_2 = \lambda_3 = \lambda$. Then one can verify that the density matrix at any moment can be written

$$\rho(t) = a(t)\rho(0) + b(t)\rho_1 + c(t)\rho_2 + d(t)\rho_3, \tag{8.62}$$

where

$$\begin{aligned}\rho_1 &= \frac{1}{3}(X_1\rho(0)X_1 + X_2\rho(0)X_2 + X_3\rho(0)X_3),\\ \rho_2 &= \frac{1}{3}(X_1X_2\rho(0)X_1X_2 + X_2X_3\rho(0)X_2X_3 + X_1X_3\rho(0)X_1X_3),\\ \rho_3 &= X_1X_2X_3\rho(0)X_1X_2X_3,\end{aligned} \tag{8.63}$$

are equally weighted mixtures of single-qubit, two-qubit, and three-qubit errors on the original state.

The evolution of the system subject to decoherence plus error correction is described by the following system of first-order linear differential equations:

$$\begin{aligned}\frac{da(t)}{dt} &= -3\lambda a(t) + (\lambda + \kappa)b(t),\\ \frac{db(t)}{dt} &= 3\lambda a(t) - (3\lambda + \kappa)b(t) + 2\lambda c(t),\\ \frac{dc(t)}{dt} &= 2\lambda b(t) - (3\lambda + \kappa)c(t) + 3\lambda d(t),\\ \frac{dd(t)}{dt} &= (\lambda + \kappa)c(t) - 3\lambda d(t).\end{aligned} \tag{8.64}$$

The exact solution was found in [PZ98] and we will not present it here. We only note that for the initial conditions $a(0) = 1, b(0) = c(0) = d(0) = 0$, the exact solution for the weight outside the code space is

$$b(t) + c(t) = \frac{3}{4+r}(1 - e^{-(4+r)\gamma t}), \tag{8.65}$$

where $r = \kappa/\lambda$. We see that similarly to what we obtained for the single-qubit code, the weight outside the code space quickly decays to its asymptotic value $\frac{3}{4+r}$, which scales as $1/r$. But note that this value is roughly three times greater than that for the single-qubit model. This corresponds to the fact that there are three single-qubit channels. More precisely, it can be verified that, if for a given κ the uncorrected weight by the single-qubit scheme is small, then the uncorrected weight by a multi-qubit code using the same κ and the same kind of decoherence for each qubit scales approximately linearly with the number of qubits. Similarly, the ratio r required to preserve a given overlap with the code space scales linearly with the number of qubits in the code.

The most important difference from the single-qubit model is that in this model there are non-correctable errors that cause a decay of the state inside the code space. Due to the finiteness of the resources employed by our scheme, there always remains a finite portion of the state outside the code space, which gives rise to noncorrectable three-qubit errors. To understand how the state decays inside the code space, one can ignore terms of the order of the weight outside the code space in the exact solution. The result is

$$a(t) \approx \frac{1 + e^{-\frac{6}{r}2\gamma t}}{2}, \quad b(t) \approx 0, \quad c(t) \approx 0, \quad d(t) \approx \frac{1 - e^{-\frac{6}{r}2\gamma t}}{2}. \tag{8.66}$$

Comparing with the expression for the fidelity of a single decaying qubit without error correction, which can be seen from (8.40) for $\kappa = 0$, we see that the encoded qubit decays roughly as if subject to bit-flip decoherence with rate $\frac{6}{r}\gamma$. Therefore, for large r this error-correction scheme can reduce the rate of decoherence approximately $\frac{r}{6}$ times. In the limit $r \to \infty$, it leads to perfect protection of the state for all times.

8.5.2 Non-Markovian decoherence

8.5.2.1 The Zeno effect: error correction versus error prevention The effect of freezing of the evolution in the limit of infinite error-correction rate bears a strong similarity to the quantum Zeno effect [MS97], where frequent measurements slow down the evolution of a system and freeze the state in the limit where they are applied continuously. The Zeno effect arises when the system and its environment are initially decoupled and they undergo a Hamiltonian-driven evolution, which leads to a quadratic change with time of the state during the initial moments [NNP96] (the so-called Zeno regime). Let the initial state of the system plus the bath be $\rho^{SB}(0) = |0\rangle\langle 0|^S \otimes \rho^B(0)$. For small times, the fidelity $\alpha = \text{Tr}\{(|0\rangle\langle 0|^S \otimes I^B)\rho^{SB}(t)\}$ of the system's density matrix with the initial state can be approximated as

$$\alpha(t) = 1 - Ct^2 + \mathcal{O}(t^3). \tag{8.67}$$

In terms of the Hamiltonian H^{SB} acting on the entire system, the coefficient C is

$$\begin{aligned} C = {} & \text{Tr}\{(H^{SB})^2|0\rangle\langle 0|^S \otimes \rho^B(0)\} \\ & - \text{Tr}\{H^{SB}|0\rangle\langle 0|^S \otimes I^B H^{SB}|0\rangle\langle 0|^S \otimes \rho^B(0)\}. \end{aligned} \tag{8.68}$$

According to (8.67), if after a time step Δt the state is measured in an orthogonal basis that involves the initial state, the probability for not projecting it on the initial state is of order $\mathcal{O}(\Delta t^2)$. Thus, if the state is continuously measured ($\Delta t \to 0$), this prevents the system from evolving.

It has been proposed to utilize the quantum Zeno effect in schemes for error prevention [Z84, BBD+97, VGW96], in which an unknown encoded state is protected from errors simply by frequent measurements that keep it inside the code space. From the point of view of the encoded basis, this approach can be understood as measuring the operators Z_i^s, which prevents the syndrome qubits from leaving the state $|00...0\rangle$. The approach is similar to error correction, in that the errors for which the code is designed send a codeword to a space orthogonal to the code space. The difference is that the subsystem containing the protected information generally does not remain invariant under the errors, since the procedure does not involve correction of errors but only their prevention. In [VGW96] it was shown that with this approach it is possible to use codes of smaller redundancy than those needed for error correction and a four-qubit encoding of a qubit was proposed, which is capable of preventing arbitrary independent errors arising from Hamiltonian interactions. The workings of this approach are based on the existence of a Zeno regime, and fail if we assume Markovian decoherence for all times. This is because the errors emerging during a time step dt in a Markovian model are proportional to dt and they accumulate with time if not corrected.

By the above observations, error correction can achieve results in noise regimes where error prevention fails. Of course, this advantage is at the expense of a more complicated procedure; in addition to the measurements that constitute error prevention, error correction involves correcting unitaries, and in general is based on codes with higher redundancy. At the same time, we see that in the Zeno regime it is possible to reduce decoherence using weaker resources than those needed for Markovian noise. This suggests that in this regime error correction may exhibit higher performance than it does for Markovian decoherence. In many situations of practical significance, the memory of the environment cannot be neglected, and the evolution is highly non-Markovian [BP02, QWJ+97, BBP04, KOR+07]. Furthermore, no evolution is strictly Markovian and for a system initially decoupled from its environment a Zeno regime is always present, short though it may be [NNP96]. Therefore, if the time resolution of error-correcting operations is high enough so that they "see" the Zeno regime, this could give rise to different behavior.

One important difference between Markovian and non-Markovian noise is that, in the latter case, the error correction and the effective noise on the reduced density matrix of the system cannot be treated as independent processes. One could derive an equation for the effective evolution of the system alone subject to interaction with the environment, such as the Nakajima–Zwanzig [N58, Z60] or the time-convolutionless (TCL) [STH77, SA80] master equations, but the generator of transformations at a given moment in general will depend (implicitly or explicitly) on the entire history up to that moment. Therefore, adding error correction can affect the effective error model nontrivially. This means that in order to describe the evolution of a system subject to non-Markovian decoherence and error correction, either one has to derive an equation for the effective evolution taking into account error correction from the very beginning, or one has to look at the evolution of the entire system including the bath, where the error generator and the generator of error correction can be considered independent. In the latter case, for sufficiently small τ_c, the evolution of the entire system including the bath can be described by

$$\frac{d\rho}{dt} = -i[H, \rho(t)] + \kappa \mathscr{J}(\rho), \qquad (8.69)$$

where ρ is the density matrix of the system plus the bath, H is the total Hamiltonian, and the error-correction generator $\mathscr{J}$ acts locally on the encoded system. We will consider a description in terms of Eq. (8.69) for a sufficiently simple bath model that allows us to find a solution for the evolution of the entire system. To gain understanding of how the scheme works, we will again look at the single-qubit model first.

8.5.2.2 The single-qubit code We choose the simple scenario of a system coupled to a single bath qubit via the Hamiltonian

$$H = \gamma X \otimes X, \tag{8.70}$$

where γ is the coupling strength. This can be a good approximation for situations in which the coupling to a single spin from the bath dominates over other interactions [KOR+07].

We assume that the bath qubit is initially in the maximally mixed state, which can be thought of as an equilibrium state at high temperature. From (8.69) one can verify that, if the system is initially in the state $|0\rangle$, the state of the system plus the bath at any moment will have the form

$$\rho(t) = (\alpha(t)|0\rangle\langle 0| + (1-\alpha(t))|1\rangle\langle 1|) \otimes \frac{I}{2} - \beta(t) Y \otimes \frac{X}{2}. \tag{8.71}$$

In the tensor product, the first operator belongs to the Hilbert space of the system and the second to the Hilbert space of the bath. We have $\alpha(t) \in [0,1]$, and $|\beta(t)| \leq \sqrt{\alpha(t)(1-\alpha(t))}, \beta(t) \in R$. The reduced density matrix of the system has the same form as the one for the Markovian case. The part proportional to $\beta(t)$ can be thought of as a "hidden" part, which nevertheless plays an important role in the error-creation process, since errors can be thought of as being transferred to the "visible" part from the "hidden" part (and vice versa). This can be seen from the fact that, during an infinitesimal time step dt, the Hamiltonian changes the parameters α and β as follows:

$$\begin{aligned} \alpha &\to \alpha - 2\beta\gamma\, dt, \\ \beta &\to \beta + (2\alpha - 1)\gamma\, dt. \end{aligned} \tag{8.72}$$

The effect of an infinitesimal error-correcting operation is

$$\begin{aligned} \alpha &\to \alpha + (1-\alpha)\kappa\, dt, \\ \beta &\to \beta - \beta\kappa\, dt. \end{aligned} \tag{8.73}$$

Note that the "hidden" part is also being acted upon. Putting it all together, we get the system of equations

$$\begin{aligned} \frac{d\alpha(t)}{dt} &= \kappa(1-\alpha(t)) - 2\gamma\beta(t), \\ \frac{d\beta(t)}{dt} &= \gamma(2\alpha - 1) - \kappa\beta(t). \end{aligned} \tag{8.74}$$

The solution for the fidelity $\alpha(t)$ is

$$\alpha(t) = \frac{2\gamma^2 + \kappa^2}{4\gamma^2 + \kappa^2} + e^{-\kappa t}\left(\frac{\kappa\gamma}{4\gamma^2 + \kappa^2}\sin 2\gamma t + \frac{2\gamma^2}{4\gamma^2 + \kappa^2}\cos 2\gamma t\right). \tag{8.75}$$

We see that, as time increases, the fidelity stabilizes at the value

$$\alpha_*^{NM} = \frac{2 + R^2}{4 + R^2} = 1 - \frac{2}{4 + R^2}, \tag{8.76}$$

where $R = \kappa/\gamma$ is the ratio between the error-correction rate and the coupling strength. Figure 8.1 shows the fidelity as a function of the dimensionless parameter γt for three different

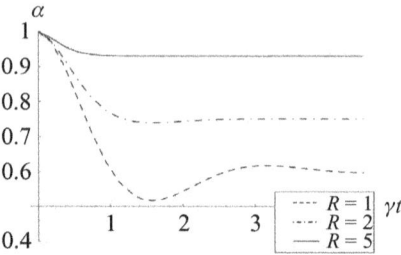

Fig. 8.1. Fidelity of the single-qubit code with continuous bit-flip errors and correction, as a function of dimensionless time γt, for three different values of the ratio $R = \kappa/\gamma$. Reprinted with permission from [OB07]. Copyright 2007, American Physical Society.

values of R. For error-correction rates comparable to the coupling strength ($R = 1$), the fidelity undergoes a few partial recurrences before it stabilizes close to α_*^{NM}. For larger $R = 2$, however, the oscillations are already heavily damped, and for $R = 5$ the fidelity seems confined above α_*^{NM}. As R increases, the evolution becomes closer to a decay like the one in the Markovian case.

A remarkable difference, however, is that the asymptotic weight outside the code space $(1 - \alpha_*^{NM})$ decreases with κ as $1/\kappa^2$, whereas in the Markovian case the same quantity decreases as $1/\kappa$. The asymptotic value can be obtained as an equilibrium point at which the infinitesimal weight flowing out of the code space during a time step dt is equal to the weight flowing into it. The latter corresponds to vanishing right-hand sides in Eqs. (8.39) and (8.75). In Section 8.5.3 we will see that the difference in that quantity for the two different types of decoherence arises from the difference in the corresponding evolutions during initial times.

8.5.2.3 The three-qubit bit-flip code We will consider a model where each qubit independently undergoes the same kind of non-Markovian decoherence as the one we studied for the single-qubit code. Here the system we look at consists of six qubits – three for the codeword and three for the environment. We assume that all system qubits are coupled to their corresponding environment qubits with the same coupling strength, i.e., the Hamiltonian is

$$H = \gamma \sum_{i=1}^{3} X_i^S \otimes X_i^B, \tag{8.77}$$

where the operators X^S act on the system qubits and X^B act on the corresponding bath qubits, which are initially in the maximally mixed state. The subscripts label on which particular qubit they act. Obviously, the types of effective single-qubit errors on the density matrix of the system that can result from this Hamiltonian at any time, CP or not, will have operator elements that are linear combinations of the identity and X^S. According to the error-correction conditions for non-CP maps obtained in [SL09a], these errors are correctable by the code. Considering the form of the Hamiltonian (8.77) and the error-correcting map, one can see that the density matrix of the entire system at any moment is a linear combination of terms of the type

$$\varrho_{lmn,pqr} \equiv X_1^l X_2^m X_3^n \rho(0) X_1^p X_2^q X_3^r \otimes \frac{X_1^{l+p}}{2} \otimes \frac{X_2^{m+q}}{2} \otimes \frac{X_3^{n+r}}{2}. \tag{8.78}$$

Here the first term in the tensor product refers to the Hilbert space of the system, and the following three refer to the Hilbert spaces of the bath qubits that couple to the first, the second, and the third qubits from the code respectively. The power indices l, m, n, p, q, r take values 0 and 1 in all possible combinations, and $X^1 = X$, $X^0 = X^2 = I$. (Note that $\varrho_{lmn,pqr}$ should not be mistaken for the components of the density matrix in the computational basis.) More precisely, we can write the density matrix in the form

$$\rho(t) = \sum_{l,m,n,p,q,r} (-i)^{l+m+n}(i)^{p+q+r} C_{lmn,pqr}(t) \times \varrho_{lmn,pqr}, \tag{8.79}$$

where the coefficients $C_{lmn,pqr}(t)$ are real. The coefficient $C_{000,000}$ is less than or equal to the codeword fidelity (with equality when $\rho(0) = |\bar{0}\rangle\langle\bar{0}|$ or $\rho(0) = |\bar{1}\rangle\langle\bar{1}|$). Since the scheme aims at protecting an unknown codeword, we will be interested in its performance in the worst case and we will assume that the codeword fidelity is $C_{000,000}$.

The exact equations for the coefficients $C_{lmn,pqr}(t)$ and their solutions were obtained in [OB07]. Here we present an approximation that can be obtained by perturbation theory for $\gamma\delta t \ll 1 \ll \kappa\delta t$ [OB07]. The approximate system of equations reads

$$\begin{aligned}\frac{dC_{000,000}}{dt} &= \frac{24}{R^2}\gamma C_{111,000}, \\ \frac{dC_{111,000}}{dt} &= -\frac{12}{R^2}\gamma(2C_{000,000} - 1).\end{aligned} \tag{8.80}$$

Comparing with (8.73), we see that the encoded qubit undergoes approximately the same type of evolution as that of a single qubit without error correction, but the coupling constant is effectively decreased $R^2/12$ times. The solution of (8.81) yields for the codeword fidelity

$$C_{000,000}(t) = \frac{1 + \cos(\frac{24}{R^2}\gamma t)}{2}. \tag{8.81}$$

This solution is valid only with precision $\mathcal{O}(\frac{1}{R})$ for times $\gamma t \ll R^3$. If one carries out the perturbation to fourth order in γ, one obtains the approximate equations

$$\begin{aligned}\frac{dC_{000,000}}{dt} &= \frac{24}{R^2}\gamma C_{111,000} - \frac{72}{R^3}\gamma(2C_{000,000} - 1), \\ \frac{dC_{111,000}}{dt} &= -\frac{12}{R^2}\gamma(2C_{000,000} - 1) - \frac{144}{R^3}\gamma C_{111,000},\end{aligned} \tag{8.82}$$

which yield for the fidelity

$$C_{000,000}(t) = \frac{1 + e^{-\frac{144}{R^3}\gamma t}\cos(\frac{24}{R^2}\gamma t)}{2}. \tag{8.83}$$

We see that in addition to the effective error process, which is of the same type as that of a single qubit, there is an extra Markovian bit-flip process with rate $\frac{72}{R^3}\gamma$. This Markovian behavior is due to the Markovian character of our error-correcting procedure which, at this level of approximation, is responsible for the direct transfer of weight between $\varrho_{000,000}$ and $\varrho_{111,111}$, and between $\varrho_{111,000}$ and $\varrho_{000,111}$. The exponential factor explicitly reveals the range of applicability of solution (8.81); with precision $\mathcal{O}(\frac{1}{R})$, it is valid only for times γt of up to order R^2. For times of the order of R^3, the decay becomes significant and cannot be neglected. The exponential factor may also play an important role for short times of up to order R, where its contribution is bigger than that of the cosine. But in the latter regime the difference between the cosine and the exponent is of order $\mathcal{O}(\frac{1}{R^2})$, which is negligible for the precision that we consider.

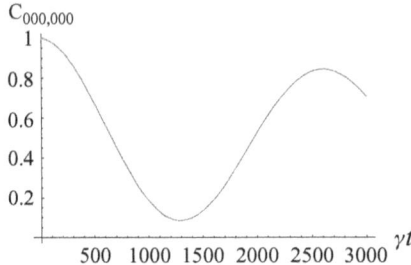

Fig. 8.2. Long-time behavior of three-qubit system with bit-flip noise and continuous error correction. The ratio of correction rate to decoherence rate is $R = \kappa/\gamma = 100$. Reprinted with permission from [OB07]. Copyright 2007, American Physical Society.

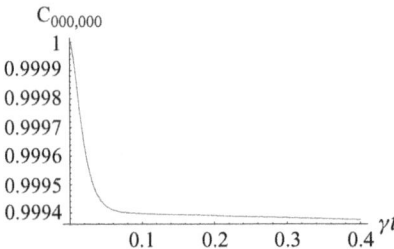

Fig. 8.3. Short-time behavior of three-qubit system with bit-flip noise and continuous error correction. The ratio of correction rate to decoherence rate is $R = \kappa/\gamma = 100$. Reprinted with permission from [OB07]. Copyright 2007, American Physical Society.

Figure 8.2 presents the exact solution for the codeword fidelity $C_{000,000}(t)$ as a function of the dimensionless parameter γt for $R = 100$. For very short times after the beginning ($\gamma t \sim 0.1$), one can see a fast but small in magnitude decay (Fig. 8.3). The maximum magnitude of this quickly decaying term obviously decreases with R, since in the limit of $R \to \infty$ the fidelity should remain constantly equal to 1.

We see that in the limit $R \to \infty$, the evolution approaches an oscillation with an angular frequency $\frac{24}{R^2}\gamma$. This is the same type of evolution as that of a single qubit interacting with its environment, but the coupling constant is effectively reduced $R^2/12$ times. While the coupling constant can serve to characterize the decoherence process in this particular case, such a description is not valid in general. As a general measure of the effect of noise one can use the instantaneous rate of decrease of the codeword fidelity F_{cw} (in our case $F_{cw} = C_{000,000}$):

$$\Lambda(F_{cw}(t)) = -\frac{dF_{cw}(t)}{dt}. \tag{8.84}$$

This quantity does not coincide with the decoherence rate in the Markovian case (which can be defined naturally from the Lindblad equation), but it is a good estimate of the rate of loss of fidelity and can be used for any decoherence model. We will refer to it simply as an error rate. Since the goal of error correction is to preserve the codeword fidelity, the quantity (8.84) is a useful indicator of the performance of a given scheme. Note that $\Lambda(F_{cw})$ is a function of the codeword fidelity and therefore it makes sense to use it for a comparison between different cases only for identical values of F_{cw}. For our example, the fact that the coupling constant is effectively reduced approximately $R^2/12$ times implies that the error rate for a given value of F_{cw} is also reduced $R^2/12$ times. Similarly, the reduction of λ by the factor $r/6$ in the Markovian case implies a reduction of Λ by the same factor. We see that the effective reduction of the error rate increases quadratically as κ^2 in the non-Markovian case, whereas it increases only linearly as κ in the Markovian case.

8.5.3 The role of the Zeno regime

The effective continuous evolution (8.81) is derived under the assumption $\gamma \delta t \ll 1 \ll \kappa \delta t$. The first inequality implies that δt can be considered within the Zeno time scale of the system's evolution without error correction. On the other hand, from the relation between κ and τ_c in (8.9) we see that $\tau_c \ll \delta t$. Therefore, the time for implementing a weak error-correcting operation has to be sufficiently small so that on the Zeno time scale the error-correction procedure can be described approximately as a continuous Markovian process. This suggests a way of understanding the quadratic enhancement in the non-Markovian case based on the properties of the Zeno regime.

Let us consider again the single-qubit code from Section 8.5.2.2, but this time let the error model be any Hamiltonian-driven process. We assume that the qubit is initially in the state $|0\rangle\langle 0|$, i.e., the state of the system including the bath has the form $\rho(0) = |0\rangle\langle 0| \otimes \rho_B(0)$. For times smaller than the Zeno time δt_Z, the evolution of the fidelity without error correction can be described by (8.67). Equation (8.67) naturally defines the Zeno regime in terms of α itself:

$$\alpha \geq \alpha_Z \equiv 1 - C\delta t_Z^2. \tag{8.85}$$

For a single time step $\Delta t \ll \delta t_Z$, the change in the fidelity is

$$\alpha \to \alpha - 2\sqrt{C}\sqrt{1-\alpha}\Delta t + \mathcal{O}(\Delta t^2). \tag{8.86}$$

On the other hand, the effect of error correction during time Δt is

$$\alpha \to \alpha + \kappa(1-\alpha)\Delta t + \mathcal{O}(\Delta t^2), \tag{8.87}$$

i.e., it tends to oppose the effect of decoherence. If both processes happen simultaneously, the effect of decoherence will still be of the form (8.86), but the coefficient C may vary with time. This is because the presence of error correction opposes the decrease of the fidelity, and consequently can lead to an increase in the time for which the fidelity remains within the Zeno regime. If this time is sufficiently long, the state of the environment could change significantly under the action of the Hamiltonian, thus giving rise to a different value for C in (8.86) according to (8.68). Note that the strength of the Hamiltonian puts a limit on C, and therefore this constant can vary only within a certain range. The equilibrium fidelity α_*^{NM} that we obtained for the error model in Section 8.5.2.2 can be thought of as the point at which the effects of error and error correction cancel out. For a general model, where the coefficient C may vary with time, this leads to a quasi-stationary equilibrium. From (8.86) and (8.87), one obtains the equilibrium fidelity

$$\alpha_*^{NM} \approx 1 - \frac{4C}{\kappa^2}. \tag{8.88}$$

In agreement with the result in Section 8.5.2.2, the equilibrium fidelity differs from 1 by a quantity proportional to $1/\kappa^2$. If one assumes a Markovian error model, for short times the fidelity changes linearly with time, which leads to $1 - \alpha_*^M \propto 1/\kappa$. Thus, the difference can be attributed to the existence of a Zeno regime in the non-Markovian case.

This argument readily generalizes to the case of nontrivial codes if we look at the picture in the encoded basis. There, each syndrome qubit undergoes a Zeno-type evolution and so do the abstract qubits associated with each error syndrome. Then, using only the properties of the Zeno behavior as we did above, we can conclude that the weight outside the code space will be kept at a quasi-stationary value of order $1/\kappa^2$. As we argued in Section 8.2, this in turn would lead to an effective decrease of the uncorrectable error rate at least by a factor proportional to $1/\kappa^2$.

Finally, let us make a remark about the resources needed to achieve the effect of quadratic reduction of the error rate. As it was pointed out, there are two conditions involved – one concerns the magnitude of the error-correction rate, the other concerns the time resolution of the weak error-correcting operations. Both of these quantities should be sufficiently large. There is, however, an interplay between the two, which involves the strength of the interaction required to implement the weak error-correcting map (8.8). Let us imagine that the weak map is implemented by making the system interact weakly with an ancilla in a given state, after which the ancilla is discarded. The error-correction procedure consists of a sequence of such interactions and can be thought of as a cooling process. If the time for which a single ancilla interacts with the system is τ_c, one can verify that the parameter ε in (8.8) would be proportional to $g^2 \tau_c^2$, where g is the coupling strength between the system and the ancilla. From (8.9) we then obtain that

$$\kappa \propto g^2 \tau_c. \tag{8.89}$$

The two parameters that can be controlled are the interaction time and the interaction strength, and they determine the error-correction rate. Thus, if g is kept constant, a decrease in the interaction time τ_c leads to a proportional decrease in κ, which may be undesirable. Therefore, in order to achieve a good working regime, one generally may need to adjust both τ_c and g. But in some situations decreasing τ_c alone can prove advantageous, since this may lead to a time resolution that reveals the non-Markovian character of an error model that was previously treated as Markovian. Then the quadratic enhancement of the performance as a function of κ may compensate the decrease in κ, thus leading to a seemingly paradoxical result – better performance with a lower error-correction rate.

8.6 Outlook

In this chapter we saw that the subsystem principle can be useful for understanding various aspects of the workings of CTQEC and its performance under different noise models, as well as for the design of CTQEC protocols using protocols for the protection of a known state. However, further research is needed to understand how to construct optimal CTQEC protocols. In the case of the quantum-jump model, the code-space fidelity reaches a quasi-equilibrium value, which can be used to estimate the performance of the scheme. It would be interesting to see whether an analog of the equilibrium fidelity exists for schemes with indirect feedback. This could prove useful since stochastic evolutions are generally too complicated for analytical treatment. The equilibrium code-space fidelity can be useful also in assessing the performance of CTQEC under non-Markovian decoherence, where the description of the evolution of a system subject to CTQEC can be difficult due to the large number of environment degrees of freedom.

We discussed two main methods for obtaining CTQEC protocols from protocols for the protection of a single qubit: one based on the application of single-qubit protocols to the separate syndrome qubits, and another based on the application of single-qubit protocols to qubit subspaces associated with the different syndromes. An interesting question is whether the performance of CTQEC protocols obtained by these methods can be related to the performance of the underlying single-qubit protocols. A difficulty in the case of indirect feedback is that the noise in the encoded basis is correlated, and the effective noise on a given qubit can depend on the outcomes of the measurements on the rest of the qubits.

Another interesting direction for future investigation is to explore CTQEC for specific physical models and limitations of the control parameters (for recent experimental proposals and connections to classical control theory, see [KBS+09, M09]). We saw that applying single-qubit schemes to the syndrome qubits in the encoded basis generally requires multi-qubit operations in the original basis, but numerical simulations show that single-qubit feedback Hamiltonians in the original basis are also efficient. It would be interesting to see whether it is possible to construct efficient CTQEC protocols for nontrivial codes assuming only one- and two-qubit Hamiltonians. The ability to apply CTQEC with Hamiltonians of limited locality would be important for the scalability of this approach.

So far, CTQEC has been considered only as a method of protecting quantum memory. A natural next step is to combine this approach with universal quantum computation. An important question in this respect is whether CTQEC can be made fault tolerant. In the theory of quantum fault tolerance, logical operations and error correction are implemented mainly in terms of transversal operations between physical qubits from different blocks, where the basic operations are assumed to be discrete. Is something similar possible for CTQEC? One way of approaching this problem could be to look for fault-tolerant implementations of a universal set of weak operations using only weak transversal unitary operations and projective ancilla measurements.

Undoubtedly, the area of CTQEC offers a variety of interesting problems for future investigation. As quantum operations with limited strength or limited rate are likely to be the tools available in many quantum computing architectures in the near term, developing further the approach to protecting quantum information from noise via continuous-time feedback seems a promising direction for research.

Part III

Advanced quantum codes

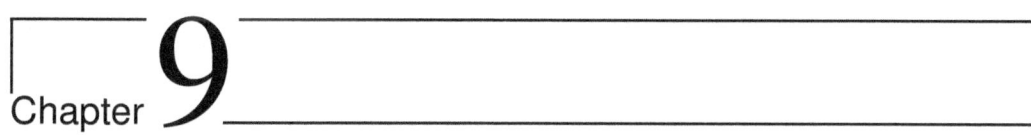

Chapter 9

Quantum convolutional codes

Mark Wilde

9.1 Introduction

Many of the quantum codes discussed in this book are quantum block codes. Quantum block codes are useful both in quantum computing and in quantum communication, but one of the drawbacks for quantum communication is that, in general, the sender must have her block of qubits ready before encoding takes place. This preparation may be a heavy demand on the sender when the size of the block code is large.

Quantum convolutional coding theory [OT03, OT04, FGG07] offers a paradigm different from quantum block coding and has numerous benefits for quantum communication. The convolutional structure is useful when a sender possesses a stream of qubits to transmit to a receiver. It is possible to encode and decode quantum convolutional codes with quantum shift-register circuits [W09b], the natural "quantization" of a classical shift register circuit. These quantum circuits ensure a low complexity for encoding and decoding while also providing higher performance than a block code with equivalent encoding complexity [FGG07]. Quantum shift-register circuits have the property that the sender Alice and the receiver Bob can respectively send and receive qubits in an "online" fashion. Alice can encode an arbitrary number of information qubits without worrying beforehand how many she may want to send over the quantum communication channel.

The actual operation and design of quantum convolutional stabilizer codes differs significantly from that of classical convolutional codes, but in some aspects, quantum convolutional codes borrow heavily from the structure of their classical counterparts. They are similar to classical convolutional codes because they have a memory structure, as is explicit in a quantum shift-register circuit that encodes or decodes a quantum convolutional code [W09b]. The mathematical description of a quantum convolutional code is similar to that of a classical convolutional code in the sense that it admits a parity check matrix of binary polynomials or binary rational functions.

Quantum Error Correction, ed. Daniel A. Lidar and Todd A. Brun. Published by Cambridge University Press.
© Cambridge University Press 2013.

Quantum convolutional coding theory may be one step along the way to finding quantum error-correcting codes that approach the capacity of a noisy quantum communication channel for sending quantum information [L97, S02b, D05, HHW+08, HSW08, HLW08, K08]. Some authors have incorporated some of this well-developed theory of quantum convolutional coding into a theory of quantum serial-turbo coding [PTO09, WH10] with the goal of designing quantum codes that come close to achieving capacity. Yet, certainly more investigation in this direction is necessary.

This chapter discusses the mathematical description of quantum convolutional codes and how to construct encoding and decoding circuits for them. We do not discuss error estimation algorithms or maximum-likelihood error correction, but instead refer the reader to Chapter 11 for an overview.

We structure this chapter as follows. We first briefly review notions from classical convolutional coding theory that are important for its formal "quantization." In Section 9.2, we introduce the definition of a quantum convolutional code and discuss how it operates. Section 9.3 presents much of the mathematics that is helpful in manipulating the matrices corresponding to quantum convolutional codes. In Section 9.4, we discuss quantum shift-register circuits that can implement the encodings and decodings for quantum convolutional codes. Section 9.5 gives two examples of methods for constructing quantum convolutional codes from classical codes that satisfy a particular constraint. The final section introduces the entanglement-assisted quantum convolutional coding theory as a natural extension of quantum convolutional coding theory and gives some examples to highlight the main aspects of the theory. The upshot of this latter theory is that we can produce entanglement-assisted quantum convolutional codes from arbitrary classical convolutional codes.

9.1.1 Brief review of classical convolutional codes

We first briefly review the theory of classical convolutional coding before moving on to the quantum case. Classical convolutional codes have found many applications in communication, especially in deep space missions such as Voyager and Pioneer [ADD+07]. Some of the major concepts from classical convolutional coding theory play an important role in quantum convolutional coding theory.

A rate-k/n classical convolutional code converts k streams of input information bits to n streams of output encoded bits that a sender transmits over a noisy classical channel. For simplicity, let us consider the case where the sender encodes one input bit stream into two output bit streams. Let $x[i]$ denote the input bit stream, where i is a discrete time index for the bits in the input stream. Let $y_1[i]$ and $y_2[i]$ similarly denote the respective two output bit streams. A classical convolutional code obeys three constraints: linearity, time-invariance, and causality. These constraints dramatically simplify the theory of classical convolutional coding because they allow us to write the two output bit streams $y_1[i]$ and $y_2[i]$ as a convolution of the input stream $x[i]$ with two respective impulse response bit streams $g_1[i]$ and $g_2[i]$:

$$y_1[i] = \sum_{j=0}^{\infty} g_1[j] x[i-j], \qquad y_2[i] = \sum_{j=0}^{\infty} g_2[j] x[i-j]. \tag{9.1}$$

Each of the above sums begins at time index $j = 0$ because we assume that the linear, time-invariant systems are causal. In a causal system, only past and present values of the input or output bit streams affect a present value of the output bit stream.

As is common in any theory of linear systems, transforming a problem into a frequency domain simplifies analysis, and the case of classical convolutional coding is no exception. The appropriate transform for bit streams is the delay or D-transform, defined as follows:

$$x(D) \equiv \sum_{i=0}^{\infty} D^i x[i].$$

In the above and throughout this overview of classical convolutional codes, we assume that the bit streams vanish for any index below zero (they are *zero on the past*). The above sum is a *formal power series* if the bit stream $x[i]$ is not finite or does not possess any periodic structure. If the bit stream $x[i]$ is finite, then $x(D)$ is a *polynomial*, and if $x[i]$ is periodic, then $x(D)$ is a *rational function*. Similar definitions apply for $y_1(D)$ and $y_2(D)$. We can think of the variable D in $x(D)$ simply as a placeholder that tells us if the ith bit is "on" or "off." For example, the following bit stream on the left translates to the polynomial on the right:

$$010110000\ldots \quad \leftrightarrow \quad D + D^3 + D^4.$$

The input–output relations in (9.1) translate to the relations $y_1(D) = g_1(D)x(D)$ and $y_2(D) = g_2(D)x(D)$ in the D-domain because a convolution in the time domain leads to a multiplication in the frequency domain in any linear system theory. Both $g_1(D)$ and $g_2(D)$ can be either rational functions or finite polynomials for a linear, time-invariant system. We can rewrite the above relation so that the output encoded streams arise from the multiplication of a generator matrix $G(D)$ with $x(D)$: $Y(D) = G(D)x(D)$, where

$$Y(D) \equiv \begin{bmatrix} y_1(D) \\ y_2(D) \end{bmatrix}, \qquad G(D) \equiv \begin{bmatrix} g_1(D) \\ g_2(D) \end{bmatrix}.$$

The utility of the generator matrix $G(D)$ is that it allows us to determine any output codeword of the code by multiplying it with the D-transform of the input stream. More generally, let us now suppose that we have k input streams $x_1[i], \ldots, x_k[i]$ that we would like to encode into n output streams $y_1[i], \ldots, y_n[i]$. It is straightforward to show that the constraints of linearity and time invariance imply that the encoder for the classical convolutional code corresponds to an $n \times k$ generator matrix:

$$\begin{bmatrix} y_1(D) \\ \vdots \\ y_n(D) \end{bmatrix} = \begin{bmatrix} g_{1,1}(D) & \cdots & g_{1,k}(D) \\ \vdots & \ddots & \vdots \\ g_{n,1}(D) & \cdots & g_{n,k}(D) \end{bmatrix} \begin{bmatrix} x_1(D) \\ \vdots \\ x_k(D) \end{bmatrix}.$$

We can now formally define a classical convolutional code. It is the set of all output streams that one can realize by multiplying a generator matrix with the D-transforms (formal power series) of k arbitrary input bit streams $x_1[i], \ldots, x_k[i]$. In particular, this definition implies that a given generator matrix does not uniquely identify a classical convolutional code. Two generator matrices $G_1(D)$ and $G_2(D)$ correspond to the same code if and only if a rational polynomial matrix $A(D)$ relates them by $G_1(D) = A(D)G_2(D)$ [JZ99].

An important concept in both the classical and quantum convolutional worlds is the notion of catastrophicity. A classical generator matrix is catastrophic if it maps any infinite duration input bit stream to a finite duration coded bit stream. The reason that catastrophicity is such a problem (and named so strikingly) is that a finite duration error stream acting on the coded bit stream could map to an infinite error sequence when the receiver decodes the corrupted stream. Such a scenario would in fact be catastrophic for any parties trying to communicate.

A parity check matrix $H(D)$ of a classical convolutional code is a rational matrix such that the product of the time-reversed matrix $H(D^{-1})$ and $G(D)$ vanishes: $H(D^{-1})G(D) = 0$. Thus, the product of any codeword with the time-reversed check matrix vanishes, and this gives a way to identify when errors occur. Note that the above "time-reversed" convention for the parity check matrix is not standard, but it is important when we move to the case of quantum convolutional codes. We define a classical convolutional code to be self-dual if $H(D^{-1})H^T(D) = 0$, i.e., if the matrix $H^T(D)$ can act as a generator matrix for a "dual" code and this dual code is the same as the original code. Two classical convolutional codes with respective check matrices $H_1(D)$ and $H_2(D)$ are dual to each other if $H_1(D)H_2^T(D^{-1}) = 0$. These concepts become important when constructing quantum convolutional codes from classical ones.

There is no reason for us to restrict ourselves to convolutional codes over the binary field. We can also generate classical convolutional codes whose symbols are from the Galois field $GF(4)$ over four elements. The four elements of this field are 0, 1, ω, and $\bar{\omega}$, where $\bar{\omega}$ is the conjugate of ω. The algebraic rules for manipulating elements of this field are as follows:

$$0 + 0 = 1 + 1 = \omega + \omega = \bar{\omega} + \bar{\omega} = 0,$$
$$1 + \omega = \bar{\omega}, \quad \omega + \bar{\omega} = 1, \quad 1 + \bar{\omega} = \omega,$$
$$\omega \cdot \bar{\omega} = 1, \quad 1 \cdot \omega = \omega, \quad 1 \cdot \bar{\omega} = \bar{\omega}, \quad 0 \cdot x = 0,$$

where x is any element over the field. The "trace" of any element x over $GF(4)$ is $\text{Tr}\{x\} \equiv x + \bar{x}$, and is always equal to zero or one. The trace induces a natural trace product between any two elements x and y of $GF(4)$, defined to be $\text{Tr}\{x\bar{y}\}$. Classical convolutional codes over $GF(4)$ are useful for constructing quantum convolutional codes.

We do not delve much further into the details of classical convolutional codes, but comment briefly on how one might decode them. A receiver can employ one of many algorithms for decoding a classical convolutional code, but the Viterbi algorithm has been by for the most popular algorithm to decode classical convolutional codes when the complexity of the code is not too high [V67]. This algorithm is an online, maximum-likelihood decoding algorithm that finds the most likely code sequence corresponding to the observations that the receiver obtains at the output of a noisy classical channel. There are ways to adapt this algorithm so that it is a syndrome-based algorithm, making decisions with the observed syndromes rather than with the observed sequences, but we do not delve into the details here. Though, note that such a syndrome-based Viterbi algorithm is important for error correction in quantum convolutional codes.

We now close our brief review with an example of a classical convolutional code. Consider the device depicted in Fig. 9.1. It is a classical shift register circuit that encodes a rate-1/2 classical convolutional code. By inspection, the two output streams $y_1[i]$ and $y_2[i]$ are the following linear, time-invariant functions of the input bit stream $x[i]$:

$$y_1[i] = x[i] + x[i - 2], \quad y_2[i] = x[i] + x[i - 3].$$

The impulse responses giving the input–output relations are then as follows:

$$g_1[i] = \delta[i] + \delta[i - 2], \quad g_2[i] = \delta[i] + \delta[i - 3],$$

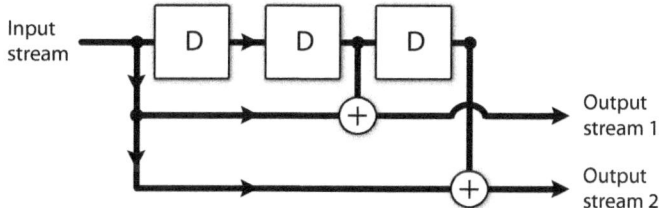

Fig. 9.1. A classical shift register circuit that encodes a rate-1/2 classical convolutional code. The boxes labeled with a "D" are memory elements that each store a single bit. At the initial time, all memory elements are set to zero and the two output bits are thus zero as well. On the first clock cycle, the circuit feeds the first bit of the input stream into the first memory element, and it feeds the bit in the first memory element into the second memory element, and so on. In this way, the device acts as a shift register. The two output bits are a function of the memory and the present value of the input. Successive clock cycles operate similarly to the first clock cycle to encode two output bit streams that the sender feeds into the noisy classical channel.

where $\delta[i]$ is the discrete-time delta function equal to unity when $i = 0$ and vanishing for all other i. By taking the D-transform of these two functions, we arrive at a generator matrix for the code:

$$G(D) = \begin{bmatrix} 1 + D^2 \\ 1 + D^3 \end{bmatrix},$$

and, evidently, a parity check matrix for this code is

$$H(D) = \begin{bmatrix} 1 + D^{-3} & 1 + D^{-2} \end{bmatrix}$$

because $H(D^{-1})G(D) = 0$. The sender transmits the encoded bit streams over the channel, and the receiver performs syndrome processing, error correction, and decoding to recover the transmitted information bits.

9.2 Definition and operation of quantum convolutional codes

We now begin our development of quantum convolutional coding theory. We introduce the theory of convolutional stabilizer codes by considering a set of Pauli matrices that stabilize a stream of encoded qubits. We first give the mathematical definition of a quantum convolutional code and follow by discussing the various steps involved in its operation.

9.2.1 Definition

A quantum convolutional stabilizer code acts on a Hilbert space $\mathcal{H}$ that is a countably infinite tensor product of two-dimensional qubit Hilbert spaces $\{\mathcal{H}_i\}_{i \in \mathbb{Z}^+}$ where $\mathcal{H} \equiv \bigotimes_{i=0}^{\infty} \mathcal{H}_i$, and $\mathbb{Z}^+ \equiv \{0, 1, \ldots\}$. A sequence $\mathbf{A}$ of Pauli matrices $\{A_i\}_{i \in \mathbb{Z}^+}$, where $\mathbf{A} \equiv \bigotimes_{i=0}^{\infty} A_i$ can act on states in $\mathcal{H}$. Let $\Pi^{\mathbb{Z}^+}$ denote the set of all Pauli sequences. The support $\mathrm{supp}(\mathbf{A})$ of a Pauli sequence $\mathbf{A}$ is the set of indices of the entries in $\mathbf{A}$ that are not equal to the identity. The weight of a sequence

A is the size $|\text{supp}(\mathbf{A})|$ of its support. The delay $\text{del}(\mathbf{A})$ of a sequence **A** is the smallest index for an entry not equal to the identity. The degree $\deg(\mathbf{A})$ of a sequence **A** is the largest index for an entry not equal to the identity. For example, the Pauli sequence $I\ X\ I\ Y\ Z\ I\ I\ \ldots$ has support $\{1, 3, 4\}$, weight three, delay one, and degree four. A sequence has finite support if its weight is finite. Let $F(\Pi^{\mathbb{Z}^+})$ denote the set of Pauli sequences with finite support. The following definition for a quantum convolutional code utilizes the set $F(\Pi^{\mathbb{Z}^+})$ in its description.

Definition 9.1. *A rate k/n-convolutional stabilizer code with $0 \leq k \leq n$ is specified by a commuting set $\mathscr{G}$ of all n-qubit shifts of a basic generator set $\mathscr{G}_0$. The commutativity requirement is necessary for the same reason that standard stabilizer codes require it. The basic generator set $\mathscr{G}_0$ has $n - k$ Pauli sequences of finite support:*

$$\mathscr{G}_0 \equiv \{\mathbf{G}_i \in F(\Pi^{\mathbb{Z}^+}) : 1 \leq i \leq n - k\}. \tag{9.2}$$

A frame of the code consists of n qubits. The definition of a quantum convolutional code as n-qubit shifts of the basic set $\mathscr{G}_0$ is what gives the code its periodic structure.

In the above, $\mathscr{G}$ is analogous to a classical parity check matrix (we would have used the symbol $\mathscr{H}$, but this symbol usually denotes "Hilbert space.")

A quantum convolutional code admits an equivalent definition in terms of the delay operator or D-operator. This operator is analogous to the role of D in classical convolutional codes from the previous section. The D-operator captures shifts of the basic generator set $\mathscr{G}_0$. Let us define the n-qubit delay operator D acting on any Pauli sequence $\mathbf{A} \in \Pi^{\mathbb{Z}^+}$ as follows:

$$D(\mathbf{A}) = I^{\otimes n} \otimes \mathbf{A}. \tag{9.3}$$

We can write j repeated applications of D as a power of D: $D^j(\mathbf{A}) = I^{\otimes jn} \otimes \mathbf{A}$. Let $D^j(\mathscr{G}_0)$ be the set of shifts of elements of $\mathscr{G}_0$ by j. Then the full stabilizer $\mathscr{G}$ for the convolutional stabilizer code is

$$\mathscr{G} \equiv \bigcup_{j \in \mathbb{Z}^+} D^j(\mathscr{G}_0). \tag{9.4}$$

Example 9.1. *We provide a simple example of a rate-1/3 quantum convolutional code called the Forney–Grassl–Guha (FGG) code [FG05, FGG07]. The basic stabilizer and its first shift are as follows:*

$$\begin{array}{|c|c|c|c|} XXX & XZY & III & III \\ ZZZ & ZYX & III & III \\ III & XXX & XZY & III \\ III & ZZZ & ZYX & III \end{array} \cdots \tag{9.5}$$

The code consists of all three-qubit shifts of the above generators. The vertical bars are a visual aid to illustrate the three-qubit shifts of the basic generators. The code can correct for an arbitrary single-qubit error in every other frame by a simple syndrome-based, table-lookup algorithm.

9.2.2 Operation

Figure 9.2 illustrates the basic operation of a quantum convolutional code. The operation of a rate-k/n quantum convolutional code consists of several steps:

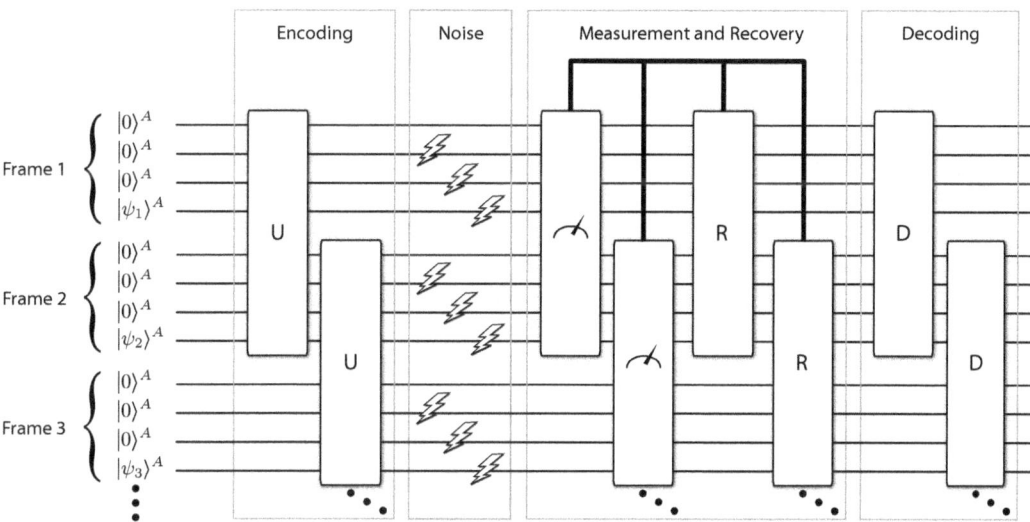

Fig. 9.2. The operation of a quantum convolutional code. The sender applies the same unitary successively to a stream of information qubits and ancilla qubits. The convolutional structure implies that the unitary overlaps some of the same qubits. The sender transmits her qubits as soon as the unitary finishes processing them. The noisy quantum channel corrupts the transmitted qubits. The receiver performs overlapping multi-qubit measurements to diagnose channel errors and corrects for them. The receiver performs an online decoding circuit to recover the sender's original stream of information qubits.

Encoding. The protocol begins with the sender encoding a stream of information qubits with an online encoding circuit. The sender encodes $n - k$ ancilla qubits and k information qubits per frame [GR06a, GR07]. The encoding circuit is "online" if each unitary U in it (see Fig. 9.2) acts on a few frames at a time.

Transmission over a noisy channel. The sender transmits a set of initial qubits as soon as the first unitary finishes processing them and continues to transmit later qubits after the next part of the online encoding circuit finishes processing them.

Syndrome measurements. The above basic set $\mathscr{G}_0$ and all of its n-qubit shifts act like a parity check matrix for the quantum convolutional code. The receiver measures the generators in the stabilizer to determine an error syndrome. It is important that the generators in $\mathscr{G}_0$ have finite weight so that the receiver can perform the measurements and produce an error syndrome. It is also important that the generators have a block-band form so that the receiver can perform the measurements online as the noisy encoded qubits arrive.

Syndrome processing. The receiver processes the error syndromes with an online classical error estimation algorithm such as the Viterbi algorithm [V67] or any other decoding algorithm [JZ99] to determine the most likely error for each frame of quantum data. Syndrome-based versions of the Viterbi algorithm that are appropriate for quantum coding are available in [OT03, OT04, OT06].

Error correction. The receiver performs unitary recovery operations that reverse the estimated errors.

Decoding. The receiver finally processes the encoded qubits with a decoding circuit to recover the original stream of information qubits. The qubits decoded from this convolutional procedure should be error-free and ready for quantum computation at the receiving end. It is again possible to implement the decoding circuit as a quantum shift-register circuit.

In the above description, each step is online. Processing of one step can begin once the previous step has finished some amount of processing. All steps above are always active during processing once the initial rounds have been completed.

Figure 9.2 exploits a "quantum circuit notation" to depict the encoding circuit. This circuit is "unwrapped" in the sense that the figure shows how it acts on the entire qubit stream. It is possible to implement this encoding with a quantum shift-register circuit (discussed in Section 9.4), and the graphical depiction of quantum shift-register circuits is different, but equivalent, to the "quantum circuit" depiction in Fig. 9.2.

9.3 Mathematical formalism of quantum convolutional codes

9.3.1 The Pauli-to-binary isomorphism

This section discusses the isomorphism from the set of Pauli sequences to the module over the ring of binary polynomials [OT04, GR06a, FGG07]. We also name it the Pauli-to-binary (P2B) isomorphism. The P2B isomorphism is important because it is easier to perform manipulations with vectors of binary rational functions than with Pauli sequences.

We first define the phase-free Pauli group $[\Pi^{\mathbb{Z}}]$ on a sequence of qubits. As defined in (9.3), the delay operator D shifts a Pauli sequence to the right by n qubits. Let $\Pi^{\mathbb{Z}}$ denote the set of all countably infinite Pauli sequences. The set $\Pi^{\mathbb{Z}}$ is equivalent to the set of all n-qubit shifts of arbitrary Pauli operators:

$$\Pi^{\mathbb{Z}} = \left\{ \prod_{i \in \mathbb{Z}} D^i(A_i) : A_i \in \Pi^n \right\}, \tag{9.6}$$

where Π^n is the Pauli group over n qubits. We remark that $D^i(A_i) = D^i(A_i \otimes I^{\otimes \infty})$. We make this same abuse of notation in what follows. We can define the equivalence class $[\Pi^{\mathbb{Z}}]$ of phase-free Pauli sequences:

$$[\Pi^{\mathbb{Z}}] = \{\beta \mathbf{A} \mid \mathbf{A} \in \Pi^{\mathbb{Z}}, \beta \in \mathbb{C}, |\beta| = 1\}. \tag{9.7}$$

We develop a relation between binary rational functions and Pauli sequences that is useful for representing the shifting nature of quantum convolutional codes. As specified in Definition 9.1, a quantum convolutional code in general consists of generators with n qubits per frame. One can obtain the full set of generators of the quantum convolutional code by shifting the basic generator set by integer multiples of the frame size n. Let the delay operator D shift a Pauli sequence to the right by an arbitrary integer n. Consider a $2n$-dimensional vector $\mathbf{u}(D)$ of binary polynomials where $\mathbf{u}(D) \in (\mathbb{Z}_2(D))^{2n}$. Let us write $\mathbf{u}(D)$ as follows:

$$\mathbf{u}(D) = (\mathbf{z}(D)|\mathbf{x}(D)),$$
$$= \left(z_1(D) \cdots z_n(D) \mid x_1(D) \cdots x_n(D) \right),$$

where $\mathbf{z}(D), \mathbf{x}(D) \in (\mathbb{Z}_2(D))^n$. Suppose

$$z_i(D) = \sum_j z_{i,j} D^j, \quad x_i(D) = \sum_j x_{i,j} D^j.$$

Define a map $\mathbf{N} : (\mathbb{Z}_2(D))^{2n} \to \Pi^{\mathbb{Z}}$:

$$\mathbf{N}(\mathbf{u}(D)) = \prod_j D^j(Z^{z_{1,j}} X^{x_{1,j}}) D^j(I \otimes Z^{z_{2,j}} X^{x_{2,j}}) \cdots D^j(I^{\otimes n-1} \otimes Z^{z_{n,j}} X^{x_{n,j}}).$$

Suppose $\mathbf{v}(D) = (\mathbf{z}'(D)|\mathbf{x}'(D))$, where $\mathbf{v}(D) \in (\mathbb{Z}_2(D))^{2n}$. The map $\mathbf{N}$ induces an isomorphism

$$[\mathbf{N}] : (\mathbb{Z}_2(D))^{2n} \to [\Pi^{\mathbb{Z}}]$$

because addition of binary polynomials is equivalent to multiplication of Pauli elements up to a global phase:

$$[\mathbf{N}(\mathbf{u}(D) + \mathbf{v}(D))] = [\mathbf{N}(\mathbf{u}(D))] [\mathbf{N}(\mathbf{v}(D))].$$

Example 9.2 in the next section provides an example of the P2B isomorphism.

9.3.2 The shifted symplectic product

We consider the commutative properties of quantum convolutional codes in this section and develop some mathematics for the important "shifted symplectic product." The shifted symplectic product reveals the commutation relations of an arbitrary number of shifts of a set of Pauli sequences.

Recall that a commuting set comprising a basic set of finite-weight generators and all of their shifts specifies a quantum convolutional code. How can we capture the commutation relations of a Pauli sequence and all of its shifts? The shifted symplectic product $\odot$, where

$$\odot : (\mathbb{Z}_2(D))^{2n} \times (\mathbb{Z}_2(D))^{2n} \to \mathbb{Z}_2(D), \tag{9.8}$$

does so in an elegant way. It maps vectors of binary polynomials to a finite-degree and finite-delay binary polynomial:

$$(\mathbf{u} \odot \mathbf{v})(D) = \mathbf{z}(D^{-1}) \cdot \mathbf{x}'(D) - \mathbf{x}(D^{-1}) \cdot \mathbf{z}'(D), \tag{9.9}$$

where $\cdot$ represents the standard inner product. The shifted symplectic product is not a proper symplectic product because it fails to be "alternating." The alternating property requires that $(\mathbf{u} \odot \mathbf{v})(D) = -(\mathbf{v} \odot \mathbf{u})(D)$, but we find instead that the following holds:

$$(\mathbf{u} \odot \mathbf{v})(D) = -(\mathbf{v} \odot \mathbf{u})(D^{-1}).$$

Every vector $\mathbf{u}(D) \in \mathbb{Z}_2(D)^{2n}$ is auto-time-reversal symmetric with respect to $\odot$ because addition and subtraction are the same over $\mathbb{Z}_2$:

$$(\mathbf{u} \odot \mathbf{u})(D) = (\mathbf{u} \odot \mathbf{u})(D^{-1}) \quad \forall \mathbf{u}(D) \in \mathbb{Z}_2(D)^{2n}. \tag{9.10}$$

We employ the addition convention from now on and drop the minus signs. The shifted symplectic product for vectors of binary polynomials is a binary polynomial in D. We write its coefficients as follows:

$$(\mathbf{u} \odot \mathbf{v})(D) = \sum_{i \in \mathbb{Z}} (\mathbf{u} \odot \mathbf{v})_i D^i. \tag{9.11}$$

The coefficient $(\mathbf{u} \odot \mathbf{v})_i$ captures the commutation relations of two Pauli sequences for i n-qubit shifts of one of the sequences:

$$\mathbf{N}(\mathbf{u}(D))D^i(\mathbf{N}(\mathbf{v}(D))) = (-1)^{(\mathbf{u} \odot \mathbf{v})_i} D^i(\mathbf{N}(\mathbf{v}(D)))\mathbf{N}(\mathbf{u}(D)).$$

Example 9.2. *We consider the case where the frame size $n = 4$. Consider the following vectors of polynomials:*

$$\begin{bmatrix} \mathbf{u}(D) \\ \mathbf{v}(D) \end{bmatrix} = \begin{bmatrix} 1+D & D & 1 & D & 0 & 1 & 0 & 0 \\ 0 & 1 & 0 & 0 & 1+D & 1+D & 1 & D \end{bmatrix}. \quad (9.12)$$

The P2B isomorphism maps $\mathbf{u}(D)$ and $\mathbf{v}(D)$ to the following Pauli sequences:

$$\mathbf{N}(\mathbf{u}(D)) = (\cdots |IIII|ZXZI|ZZIZ|IIII|\cdots),$$
$$\mathbf{N}(\mathbf{v}(D)) = (\cdots |IIII|XYXI|XXIX|IIII|\cdots). \quad (9.13)$$

We can determine the commutation relations by inspection of the above Pauli sequences. $\mathbf{N}(\mathbf{u}(D))$ anticommutes with itself shifted by one to the left or right, $\mathbf{N}(\mathbf{v}(D))$ anticommutes with itself shifted by one to the left or right, and $\mathbf{N}(\mathbf{u}(D))$ anticommutes with $\mathbf{N}(\mathbf{v}(D))$ shifted by one to the left. The following shifted symplectic products confirm the above commutation relations:

$$(\mathbf{u} \odot \mathbf{u})(D) = D^{-1} + D, \quad (\mathbf{v} \odot \mathbf{v})(D) = D^{-1} + D, \quad (\mathbf{u} \odot \mathbf{v})(D) = D^{-1}.$$

9.3.3 Parity check matrix representation

In general, we represent a rate-k/n quantum convolutional code with an $(n-k) \times 2n$-dimensional quantum check matrix $H(D)$ according to the P2B isomorphism. The entries of $H(D)$ are binary polynomials where

$$H(D) = [\, Z(D) \,|\, X(D) \,], \quad (9.14)$$

and $Z(D)$ and $X(D)$ are both $(n-k) \times n$-dimensional binary polynomial matrices. It is this representation that allows us to think of a quantum convolutional code in a classical sense, but a quantum convolutional code has more constraints than a classical one does (namely, the commutativity constraint).

The following matrix $\Omega(D)$ captures the commutation relations of the generators in $H(D)$:

$$\Omega(D) \equiv Z(D)X^T(D^{-1}) + X(D)Z^T(D^{-1}).$$

The reader can verify that the matrix elements $[\Omega(D)]_{ij}$ of $\Omega(D)$ are the shifted symplectic products between the ith and jth respective rows $h_i(D)$ and $h_j(D)$ of $H(D)$:

$$[\Omega(D)]_{ij} = (h_i \odot h_j)(D).$$

We call the matrix $\Omega(D)$ the *shifted symplectic product matrix* because it encodes all of the shifted symplectic products or the commutation relations of the code. This matrix vanishes when its corresponding quantum convolutional code is valid because all generators commute with themselves and with all n-qubit shifts of themselves and each other. For a general set of generators, $\Omega(D)$ does not vanish and obeys the symmetry: $\Omega(D) = \Omega^T(D^{-1})$. This latter case arises when we construct entanglement-assisted quantum convolutional codes in Section 9.6

(these codes are an extension of standard quantum convolutional codes that allow us to construct quantum codes even when the generators do not commute).

9.3.4 Row operations

We can perform row operations on the matrix in (9.14) that corresponds to a quantum convolutional code. A row operation is merely a "mental" operation that has no effect on the states in the code space or on the error-correcting properties of the code. It just changes the rows of the check matrix for a code. We have two types of row operations:

(i) An elementary row operation multiplies a row times an arbitrary binary polynomial and adds the result to another row. This additive invariance holds for any code that admits a description within the stabilizer formalism. Additive codes are invariant under multiplication of the stabilizer generators in the "Pauli picture" or under row addition in the "binary-rational picture."
(ii) We also employ row operations that multiply a row by an arbitrary rational function. This type of row operation occurs when we have generators with infinite weight that we would like to reduce to finite weight so that the receiver can perform measurements in an online fashion as qubits arrive from the noisy channel.

We can encode all of the above types of row operations into a full-rank matrix $R(D)$ with rational polynomial entries. Let $H'(D)$ denote the resulting check matrix after performing a set of row operations in the matrix $R(D)$ where $H'(D) = R(D)H(D)$. The resulting effect on the shifted symplectic product matrix $\Omega(D)$ is to change it to another shifted symplectic product matrix $\Omega'(D)$ related to $\Omega(D)$ by

$$\Omega'(D) = R(D)\Omega(D)R^T(D^{-1}). \tag{9.15}$$

Row operations do not change the commutation relations of a valid quantum convolutional code because its shifted symplectic product matrix vanishes. But row operations do change the commutation relations of a set of generators whose corresponding shifted symplectic product matrix does not vanish. This ability to change the commutation relations through row operations is crucial for constructing entanglement-assisted quantum convolutional codes from an arbitrary set of generators (discussed in Section 9.6).

9.3.5 Column operations

We can also perform column operations on the matrix in (9.14) that corresponds to a quantum convolutional code. Column operations do change the error-correcting properties of the code and are important in the theoretical development of an encoding circuit for the code. We have three types of column operations:

(i) An elementary column operation multiplies one column by an arbitrary binary polynomial and adds the result to another column. We implement elementary column operations with gates from the shift-invariant Clifford group [GR07, GR06a].
(ii) Another column operation is to multiply column i in both the "X" and "Z" matrix by D^l where $l \in \mathbb{Z}$. We perform this operation by delaying or advancing the processing of qubit i by l frames relative to the original frame.

(iii) An infinite-depth column operation multiplies one column in the "X" matrix by a rational function whose numerator is one and multiplies the corresponding column in the "Z" matrix by a corresponding finite polynomial.

A column operation implemented on the "X" side of the binary polynomial matrix has a corresponding effect on the "Z" side of the binary polynomial matrix. This corresponding effect is a manifestation of the Heisenberg uncertainty principle because commutation relations remain invariant with respect to the action of unitary quantum gates. The shifted symplectic product is therefore invariant with respect to column operations from the shift-invariant Clifford group. The next two sections describe possible column operations for implementing encoding circuits.

9.3.6 Generalized Clifford operations

One of the main advantages of a quantum convolutional code is that we can implement its encoding circuit with a quantum shift-register circuit. We can encode a stream of quantum information with the same physical routines or devices and therefore gain an improvement in the performance/complexity trade-off for a code.

In this section, we describe several specific column operations and the effect that they have on the parity check matrix of a quantum convolutional code. We show how each column operation corresponds to a binary polynomial or binary rational matrix that postmultiplies the parity check matrix. Our discussion in this section focuses on the linear algebra, and we later show in Section 9.4 how we can actually implement these column operations with quantum shift-register circuits.

Column operations fall into two classes: finite-depth and infinite-depth. We first discuss finite-depth operations.

Definition 9.2. *A finite-depth operation transforms every finite-weight stabilizer generator to one with finite weight.*

This property is important for a decoding circuit because we do not want it to propagate uncorrected errors into the information qubit stream. It is also important for an encoding circuit that may suffer from noise.

The finite-depth operations we give below form the shift-invariant Clifford group, an extension of the Clifford group operations mentioned several times throughout this book. Below we describe how five finite-depth operations in the shift-invariant Clifford group affect the binary polynomial matrix for a code.

Finite-depth CNOT gate. The sender performs a CNOT from qubit i to qubit j (where $i \neq j$) in every frame where qubit j is in a frame delayed by k. The effect on the binary polynomial matrix is to multiply column i by D^k and add the result to column j in the "X" matrix and to multiply column j by D^{-k} and add the result to column i in the "Z" matrix. It corresponds to the following matrix transformation:

$$\text{CNOT}(i,j)(D^k) \equiv \begin{bmatrix} 1 & 0 & 0 & 0 \\ D^{-k} & 1 & 0 & 0 \\ 0 & 0 & 1 & D^k \\ 0 & 0 & 0 & 1 \end{bmatrix}. \tag{9.16}$$

Hadamard gate. A Hadamard on qubit i in every frame swaps column i in the "X" matrix with column i in the "Z" matrix. It corresponds to the following matrix transformation:

$$\text{H}(i) \equiv \left[\begin{array}{c|c} 0 & 1 \\ 1 & 0 \end{array} \right]. \tag{9.17}$$

Phase gate. A phase gate on qubit i in every frame adds column i from the "X" matrix to column i in the "Z" matrix. It corresponds to the following matrix transformation:

$$\text{P}(i) \equiv \left[\begin{array}{c|c} 1 & 1 \\ 1 & 0 \end{array} \right]. \tag{9.18}$$

Controlled-phase gate I. A controlled-phase gate from qubit i to qubit j (where $i \neq j$) in a frame delayed by k multiplies column i in the "X" matrix by D^k and adds the result to column j in the "Z" matrix. It also multiplies column j in the "X" matrix by D^{-k} and adds the result to column i in the "Z" matrix. It corresponds to the following matrix transformation:

$$\text{C-PHASE}(i,j)(D^k) \equiv \left[\begin{array}{cc|cc} 1 & 0 & 0 & 0 \\ 0 & 1 & 0 & 0 \\ 0 & D^k & 1 & 0 \\ D^{-k} & 0 & 0 & 1 \end{array} \right]. \tag{9.19}$$

Controlled-phase gate II. A controlled-phase gate from qubit i to qubit i in a frame delayed by k multiplies column i in the "X" matrix by $D^k + D^{-k}$ and adds the result to column i in the "Z" matrix. It corresponds to the following matrix transformation:

$$\text{C-PHASE}(i)(D^k) \equiv \left[\begin{array}{c|c} D^k + D^{-k} & 1 \\ 1 & 0 \end{array} \right]. \tag{9.20}$$

We now introduce infinite-depth operations. We must be delicate when using them because they can propagate errors to an infinite number of neighboring qubits in the qubit stream. Though an infinite-depth operation gives more flexibility when designing encoding circuits – similar to the way in which an infinite-impulse response filter gives more flexibility in the design of classical convolutional circuits. A decoding circuit with infinite-depth operations on qubits sent over the noisy channel is undesirable because it spreads uncorrected errors infinitely into the decoded information qubit stream. But an encoding circuit with infinite-depth operations is acceptable if we assume a communication paradigm in which the only noisy process is the noisy quantum channel.

Definition 9.3. *An infinite-depth operation can transform a finite-weight stabilizer generator to one with infinite weight (but does not necessarily do so to every finite-weight generator).*

Below we give two variations of an infinite-depth operation.

Infinite-depth CNOT gate I. A CNOT gate from qubit i to qubit i in a frame delayed by k (and repeated for all frames) multiplies column i in the "X" matrix by $1/\left(1 + D^k\right)$ and it multiplies column i in the "Z" matrix by $1 + D^{-k}$. It corresponds to the following transformation:

$$\text{CNOT}(i)\left(1/(1+D^k)\right) \equiv \left[\begin{array}{c|c} 1 + D^{-k} & 0 \\ 0 & 1/(1+D^k) \end{array} \right].$$

Infinite-depth CNOT gate II. A CNOT gate from qubit i in a frame delayed by k to qubit i in the original frame (and repeated for all frames) multiplies column i in the "Z" matrix by

$1/(1 + D^k)$ and it multiplies column i in the "X" matrix by $1 + D^{-k}$. Note that this variation merely flips the direction of the CNOT gates in the previous construction. It corresponds to the following transformation:

$$\text{CNOT}(i)(1 + D^{-k}) \equiv \begin{bmatrix} 1/(1+D^k) & 0 \\ 0 & 1+D^{-k} \end{bmatrix}.$$

9.4 Quantum shift-register circuits

Section 9.1.1 demonstrated that the encoding circuit for a classical convolutional code has a particularly simple form. We showed the translation from a generator matrix to a shift-register implementation for an example only, but in general, one can easily write down a shift-register implementation for the encoding circuit given a mathematical description of a classical convolutional code [JZ99].

A natural question is whether there exists such a simple mapping from the mathematical description of a quantum convolutional code to a quantum shift-register implementation. In this section, we give an example that illustrates the operation of a quantum shift-register circuit, and refer the reader to [W09b] to see how to implement an arbitrary generalized Clifford operation with a quantum shift-register circuit.

9.4.1 Example

Let us begin with a simple example to show how we can build up a primitive quantum shift-register circuit. Consider the following set of Pauli operators on two qubits: ZI, IZ, XI, IX. We can form a symplectic representation of these Pauli operators with the following matrix:

$$\begin{bmatrix} 1 & 0 & 0 & 0 \\ 0 & 1 & 0 & 0 \\ 0 & 0 & 1 & 0 \\ 0 & 0 & 0 & 1 \end{bmatrix}, \qquad (9.21)$$

where the entries to the left of the vertical bar correspond to the Z operators and the entries to the right of the vertical bar correspond to the X operators, so that the first row corresponds to ZI and so on. Suppose that we perform a CNOT gate from the first qubit to the second qubit conditional on a classical control bit f_0. We perform the gate if $f_0 = 1$ and do not perform it otherwise. The above Pauli operators transform as follows:

$$\begin{bmatrix} 1 & 0 & 0 & 0 \\ f_0 & 1 & 0 & 0 \\ 0 & 0 & 1 & f_0 \\ 0 & 0 & 0 & 1 \end{bmatrix}.$$

Figure 9.3 depicts the "quantum shift-register circuit" that implements this transformation (this device is not really a quantum shift-register circuit because it does not exploit a set of memory qubits).

Let us now incorporate one frame of memory qubits so that the circuit really becomes a quantum shift-register circuit. Consider the circuit in Fig. 9.4. The first two qubits are fed into the

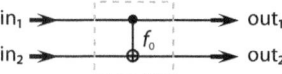

Fig. 9.3. The above figure depicts a simple CNOT transformation conditional on the bit f_0. The circuit does not apply the gate if $f_0 = 0$ and applies it if $f_0 = 1$. Reprinted with permission from [W09b]. Copyright 2009, American Physical Society.

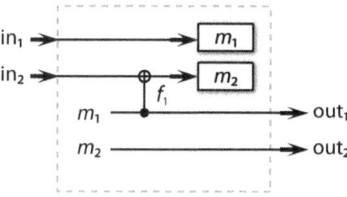

Fig. 9.4. A quantum shift-register device that incorporates one frame of memory qubits. Reprinted with permission from [W09b]. Copyright 2009, American Physical Society.

device and the second one is the target of a CNOT gate from a future frame of qubits (conditional on the bit f_1). The two qubits are then stored as two memory qubits (swapped out with what was previously there). On the next cycle, the two qubits are fed out and the first qubit that was previously in memory acts on the second qubit in a frame that is in the past with respect to itself. We would expect the X variable of the first outgoing qubit to propagate one frame into the past with respect to itself and the Z variable of the second incoming qubit to propagate one frame into the future with respect to itself. We make this idea more clear in the analysis below.

We can analyze this situation with a set of recursive equations. Let $x_1[n]$ denote the bit representation of the X Pauli operator for the first incoming qubit at time n and let $z_1[n]$ denote the bit representation of the Z Pauli operator for the first incoming qubit at time n. Let $x_2[n]$ and $z_2[n]$ denote similar quantities for the second incoming qubit at time n. Let $m_1^x[n]$ denote the bit representation of the X Pauli operator acting on the first memory qubit at time n and let $m_1^z[n]$ denote the bit representation of the Z Pauli operator acting on the first memory qubit at time n. Let $m_2^x[n]$ and $m_2^z[n]$ denote similar quantities for the second memory qubit. In the symplectic bit vector notation, we denote the "Z" part of the Pauli operators acting on these four qubits at time n as

$$\mathbf{z}[n] \equiv [\, z_1[n] \ \ z_2[n] \ \ m_1^z[n-1] \ \ m_2^z[n-1] \,],$$

and the "X" part by

$$\mathbf{x}[n] \equiv [\, x_1[n] \ \ x_2[n] \ \ m_1^x[n-1] \ \ m_2^x[n-1] \,].$$

The symplectic vector for the inputs is then

$$[\, \mathbf{z}[n] \mid \mathbf{x}[n] \,]. \tag{9.22}$$

This notation is flexible for quantum shift-register circuits and allows us to capture the evolution of an arbitrary tensor product of Pauli operators acting on these four qubits at time n.

At time n, the two incoming qubits and the *previous* memory qubits from time $n-1$ are fed into the quantum shift-register device and the CNOT gate acts on them. The notation in Fig. 9.4 indicates that there is an implicit swap at the end of the operation. The incoming qubits get fed into the memory, and the previous memory qubits get fed out as output. Let $x'_1[n]$, $z'_1[n]$, $x'_2[n]$, and $z'_2[n]$ denote the respective output variables. The symplectic transformation for the CNOT gate is

$$\begin{bmatrix} 1 & 0 & 0 & 0 & 0 & 0 & 0 & 0 \\ 0 & 1 & f_1 & 0 & 0 & 0 & 0 & 0 \\ 0 & 0 & 1 & 0 & 0 & 0 & 0 & 0 \\ 0 & 0 & 0 & 1 & 0 & 0 & 0 & 0 \\ 0 & 0 & 0 & 0 & 1 & 0 & 0 & 0 \\ 0 & 0 & 0 & 0 & 0 & 1 & 0 & 0 \\ 0 & 0 & 0 & 0 & 0 & f_1 & 1 & 0 \\ 0 & 0 & 0 & 0 & 0 & 0 & 0 & 1 \end{bmatrix}. \quad (9.23)$$

The above matrix postmultiplies the vector in (9.22) to give an output vector. We denote the "Z" part of the output Pauli operators acting on these four qubits at time n as

$$\mathbf{z}'[n] \equiv [\, m_1^z[n]\; m_2^z[n]\; z'_1[n]\; z'_2[n] \,],$$

and the "X" part by

$$\mathbf{x}'[n] \equiv [\, m_1^x[n]\; m_2^x[n]\; x'_1[n]\; x'_2[n] \,],$$

with the change of locations corresponding to the implicit swap. The symplectic vector for the outputs is then $[\mathbf{z}'[n]|\mathbf{x}'[n]]$. It is simpler to describe the above transformation as a set of recursive "update" equations, by exploiting the transformation in (9.23):

$$x'_1[n] = m_1^x[n-1],$$
$$z'_1[n] = m_1^z[n-1] + f_1 z_2[n],$$
$$x'_2[n] = m_2^x[n-1],$$
$$z'_2[n] = m_2^z[n-1],$$
$$m_1^x[n] = x_1[n],$$
$$m_1^z[n] = z_1[n],$$
$$m_2^x[n] = x_2[n] + f_1 m_1^x[n-1],$$
$$m_2^z[n] = z_2[n].$$

Some substitutions simplify this set of recursive equations so that it becomes the following set:

$$x'_1[n] = x_1[n-1],$$
$$z'_1[n] = z_1[n-1] + f_1 z_2[n],$$
$$x'_2[n] = x_2[n-1] + f_1 x_1[n-2],$$
$$z'_2[n] = z_2[n-1].$$

We can transform this set of equations into the D-domain with the D-transform from Section 9.1.1. The set transforms as follows:

$$x'_1(D) = D x_1(D),$$
$$z'_1(D) = D(z_1(D) + f_1 D^{-1} z_2(D)),$$
$$x'_2(D) = D(x_2(D) + f_1 D x_1(D)),$$
$$z'_2(D) = D z_2(D).$$

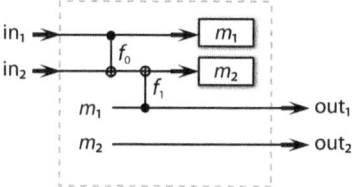

Fig. 9.5. The circuit in the above figure combines the circuits in Figs. 9.3 and 9.4. Reprinted with permission from [W09b]. Copyright 2009, American Physical Society.

This set of transformations is linear, and we can write them as the following matrix equation:

$$D \begin{bmatrix} 1 & 0 & 0 & 0 \\ f_1 D^{-1} & 1 & 0 & 0 \\ 0 & 0 & 1 & f_1 D \\ 0 & 0 & 0 & 1 \end{bmatrix}.$$

The factor of D accounts for the unit delay necessary to implement this device, but it is not particularly relevant for the purposes of the transformation (we might as well say that this quantum shift-register device implements the transformation without the factor of D). Postmultiplying the vector

$$[\, z_1(D)\, z_2(D) \,|\, x_1(D)\, x_2(D)\,]$$

by the above matrix gives the output vector

$$[\, z_1'(D)\, z_2'(D) \,|\, x_1'(D)\, x_2'(D)\,].$$

The above transformation confirms our intuition concerning the propagation of X and Z variables. The D term on the right side of the transformation matrix indicates that the X variable of the first qubit propagates one frame into the past with respect to itself, and the D^{-1} term on the left side of the matrix indicates that the Z variable of the second qubit propagates one frame into the future with respect to itself.

We now consider combining the different quantum shift-register circuits together. Suppose that we connect the outputs of the device in Fig. 9.3 to the inputs of the device in Fig. 9.4. Figure 9.5 depicts the resulting quantum shift-register circuit, and it follows that the resulting transformation in the D-domain is

$$D \begin{bmatrix} 1 & 0 & 0 & 0 \\ f_0 + f_1 D^{-1} & 1 & 0 & 0 \\ 0 & 0 & 1 & f_0 + f_1 D \\ 0 & 0 & 0 & 1 \end{bmatrix}. \tag{9.24}$$

Now consider the "two-delay transformation" in Fig. 9.6. The circuit is similar to the one in Fig. 9.4, with the exception that the first outgoing qubit acts on the second incoming qubit and the second incoming qubit is delayed two frames with respect to the first outgoing qubit. We now expect that the X variable propagates two frames into the past, while the Z variable propagates

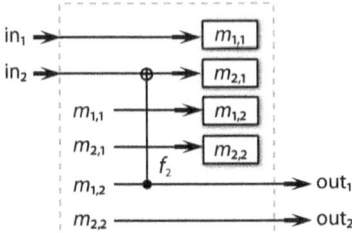

Fig. 9.6. The circuit in the above figure implements a two-delay CNOT transformation. Reprinted with permission from [W09b]. Copyright 2009, American Physical Society.

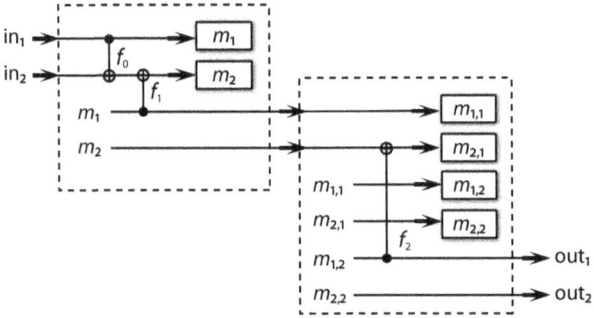

Fig. 9.7. The above circuit connects the outputs of the circuit in Fig. 9.5 to the inputs of the circuit in Fig. 9.6. Reprinted with permission from [W09b]. Copyright 2009, American Physical Society.

two frames into the future. The transformation should be as follows:

$$D^2 \begin{bmatrix} 1 & 0 & 0 & 0 \\ f_2 D^{-1} & 1 & 0 & 0 \\ 0 & 0 & 1 & f_2 D \\ 0 & 0 & 0 & 1 \end{bmatrix}. \qquad (9.25)$$

An analysis similar to the one for the "one-delay" CNOT transformation shows that the circuit indeed implements the above transformation.

Let us connect the outputs of the device in Fig. 9.5 to the inputs of the device in Fig. 9.6. The resulting D-domain transformation should be the multiplication of the transformation in (9.24) with that in (9.25), and an analysis with recursive equations confirms that the transformation is the following one:

$$D^3 \begin{bmatrix} 1 & 0 & 0 & 0 \\ f_0 + f_1 D^{-1} + f_2 D^{-2} & 1 & 0 & 0 \\ 0 & 0 & 1 & f_0 + f_1 D + f_2 D^2 \\ 0 & 0 & 0 & 1 \end{bmatrix}. \qquad (9.26)$$

The resulting device uses three frames of memory qubits to implement the transformation. This amount of memory seems like it may be too much, considering that the output data only depends on the input from two frames into the past. Is there any way to save on memory consumption?

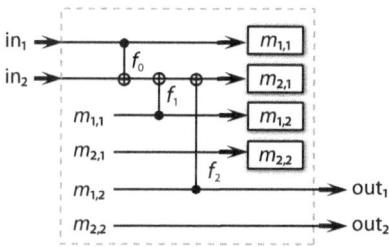

Fig. 9.8. The above circuit reduces the amount of memory required to implement the transformation in (9.26). Reprinted with permission from [W09b]. Copyright 2009, American Physical Society.

First, let us connect the outputs of the circuit in Fig. 9.5 to the inputs of the circuit in Fig. 9.6. Figure 9.7 depicts the resulting device. In this "combo" device, the target of the CNOT gate conditional on f_2 does not act on the source of the CNOT gate conditional on f_1. Therefore, we can commute the "f_2-gate" with the "f_1-gate." Now, we can actually then "commute this gate through the memory" because it does not matter whether this CNOT gate acts on the qubits before they pass through the memory or after they come out. It then follows that the last frame of memory qubits are not necessary because there is no gate that acts on these last qubits. Figure 9.8 depicts the simplified transformation. It is also straightforward to check that the resulting transformation is as follows:

$$D^2 \begin{bmatrix} 1 & 0 & 0 & 0 \\ f_0 + f_1 D^{-1} + f_2 D^{-2} & 1 & 0 & 0 \\ 0 & 0 & 1 & f_0 + f_1 D + f_2 D^2 \\ 0 & 0 & 0 & 1 \end{bmatrix}, \quad (9.27)$$

where the premultiplying delay factor in (9.27) is now D^2 instead of D^3 as in (9.26).

The above procedure allows us to simplify the circuit by eliminating the last frame of memory qubits. This procedure of determining whether we can "commute gates through the memory" is a general one that we can employ for reducing memory in quantum shift-register circuits. In general, it is possible to break any encoding or decoding circuit for a quantum convolutional code into a sequence of elementary operations. This procedure is known as the Grassl–Rötteler algorithm [GR06a], and we highlight a simple example of it in Section 9.5.1. The general procedure implements each elementary operation with a quantum shift-register circuit, connects the outputs of one quantum shift-register circuit to the inputs of the next, and determines if it is possible to "commute gates through memory" as shown in the above example. This procedure produces a quantum shift-register encoding circuit that uses the minimal amount of memory to realize a particular transformation. Reference [HH.-KW10] elaborates the development in this section by giving a general algorithm to determine the minimal memory implementation for a particular sequence of finite-depth CNOT operations.

9.5 Examples of quantum convolutional codes

This section presents a simple example of a quantum shift-register circuit that encodes a Calderbank–Shor–Steane (CSS) quantum convolutional code. We then follow with another example that encodes the FGG quantum convolutional code from Example 9.1. The first example demonstrates the main ideas for constructing quantum shift-register encoding circuits. First, build a quantum shift-register circuit for each elementary encoding operation in the Grassl–Rötteler encoding algorithm. Then, connect the outputs of the first quantum shift-register circuit to the inputs of the next one and so on for all of the elementary quantum shift-register circuits. Finally, simplify the device by determining how to "commute gates through the memory" of the larger quantum shift-register circuit (discussed in more detail later). This last step allows us to reduce the amount of memory that the quantum shift-register circuit requires.

9.5.1 A CSS code

The general construction for a CSS quantum convolutional code is similar to that for the block case, except that we import classical convolutional codes rather than classical block codes. Suppose that we have an $[n, k_1]$ classical convolutional code and an $[n, k_2]$ code with respective check matrices $H_1(D)$ and $H_2(D)$. Suppose further that the codes are orthogonal to each other: $H_1(D)H_2^T(D^{-1}) = 0$. Then we can build a CSS quantum convolutional code according to the following construction:

$$\left[\begin{array}{c|c} H_1(D) & 0 \\ 0 & H_2(D) \end{array} \right].$$

The above matrix is a valid check matrix for an $[[n, k_1 + k_2 - n]]$ quantum convolutional code. The quantum convolutional code also inherits the error-correcting properties of the two classical convolutional codes.

Let us consider a particular example of a CSS quantum convolutional code. Suppose that one classical convolutional code has the following check matrix: $[\,D\ 1\ 1+D\,]$, and that another has the following check matrix: $[\,1\ D\ 1+D\,]$, where a quick check reveals that these two codes are orthogonal. For the purposes of presentation here, we do not care much about the error-correcting properties of these codes but use them merely to illustrate the main concepts. The stabilizer matrix for the resulting quantum convolutional code is then as follows:

$$\left[\begin{array}{ccc|ccc} 0 & 0 & 0 & 1 & D & 1+D \\ D & 1 & 1+D & 0 & 0 & 0 \end{array} \right]. \tag{9.28}$$

We now show how to encode the above quantum convolutional code with the Grassl–Rötteler encoding algorithm for CSS codes [GR07]. One begins with the stabilizer matrix for two ancilla qubits per frame:

$$\left[\begin{array}{ccc|ccc} 0 & 0 & 0 & 1 & 0 & 0 \\ 0 & 1 & 0 & 0 & 0 & 0 \end{array} \right]. \tag{9.29}$$

The first ancilla qubit of every frame is in the state $|+\rangle$ and the second ancilla qubit of every frame is in the state $|0\rangle$. First send the three qubits through a quantum shift-register device that implements a CNOT$(3, 2)\,(1 + D^{-1})$ as defined in (9.16). The stabilizer becomes

$$\left[\begin{array}{ccc|ccc} 0 & 0 & 0 & 1 & 0 & 0 \\ 0 & 1 & 1+D & 0 & 0 & 0 \end{array} \right], \tag{9.30}$$

after acting on the matrix in (9.29) with the binary polynomial matrix corresponding to CNOT$(3, 2)\,(1 + D^{-1})$. Then send the three qubits through a quantum shift-register device

Quantum convolutional codes

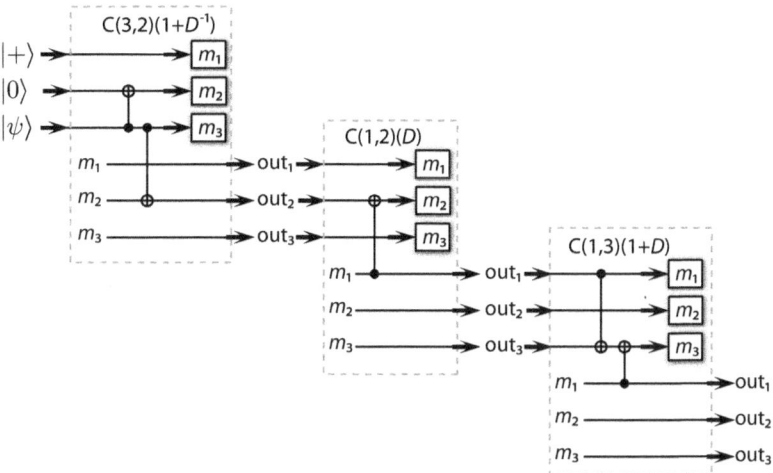

Fig. 9.9. The above circuit implements the set of transformations outlined in (9.30) and (9.31). Reprinted with permission from [W09b]. Copyright 2009, American Physical Society.

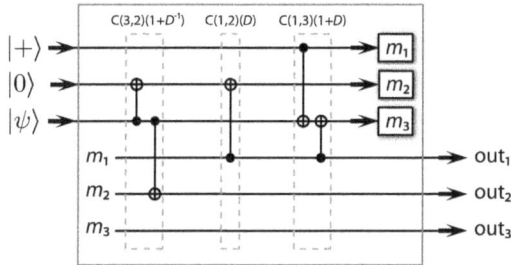

Fig. 9.10. The circuit in the above figure is a quantum shift-register encoding circuit for the CSS quantum convolutional code in (9.28). Reprinted with permission from [W09b]. Copyright 2009, American Physical Society.

that performs a $\text{CNOT}(1,2)(D)$ and another quantum shift-register device that performs a $\text{CNOT}(1,3)(1+D)$. The stabilizer becomes

$$\left[\begin{array}{ccc|ccc} 0 & 0 & 0 & 1 & D & 1+D \\ D & 1 & 1+D & 0 & 0 & 0 \end{array}\right], \qquad (9.31)$$

and is now encoded. Fig. 9.9 depicts the quantum shift-register circuit corresponding to the above operations.

The circuit in Fig. 9.9 seems to be wasteful in memory consumption. Is there anything we can do to simplify it? First notice that the target qubit of the CNOT gate in the second quantum shift register is the same as the target qubit of the second CNOT gate in the first quantum shift register. It follows that these two gates commute so that we can act with the CNOT gate in the second quantum shift register before acting with the second CNOT gate of the first quantum shift register. But we can do even better. Acting first with the CNOT gate in the second quantum shift register is equivalent to having it act before the first frame of memory qubits gets delayed. After a further simplification, we can apply a series of similar steps to remove another frame of memory qubits that is not necessary. Figure 9.10 depicts the final form of the encoding circuit for this code.

The overall encoding matrix for this code is

$$\text{CNOT}(3,2)(1+D^{-1}) \text{ CNOT}(1,2)(D) \text{ CNOT}(1,3)(1+D).$$

The Grassl–Rötteler algorithm applies to the more general case of any CSS quantum convolutional code [GR07]. Reference [FGG07] provides many constructions of CSS quantum convolutional codes from classical binary convolutional codes.

9.5.2 A CRSS code: the Forney–Grassl–Guha code

A Calderbank–Rains–Shor–Sloane (CRSS) quantum convolutional code is one that we produce from a classical convolutional code over $GF(4)$. Before constructing such codes, let us first consider the following mapping between Pauli matrices, symplectic binary vectors, and elements of $GF(4)$:

Π	$(\mathbb{Z}_2)^2$	$GF(4)$
I	00	0
X	01	ω
Y	11	1
Z	10	$\bar{\omega}$

(9.32)

Let γ denote the mapping from binary vectors to elements of $GF(4)$ (second column to the third column). It is straightforward to check that the symplectic product between binary vectors is equivalent to the trace product of their $GF(4)$ representations:

$$h_i \odot h_j = \text{Tr}\left\{\gamma^{-1}(h_i) \cdot \overline{\gamma^{-1}(h_j)}\right\}, \qquad (9.33)$$

where h_i and h_j are any two symplectic binary vectors, $\cdot$ denotes the inner product, the overbar denotes the conjugate operation, and "Tr" denotes the $GF(4)$ trace operation defined before.

Suppose now that $H(D)$ is a check matrix for an $[n, k]$ classical convolutional code over $GF(4)$. Suppose further that $H(D)$ is orthogonal with respect to the $GF(4)$ trace operation, i.e.,

$$\text{Tr}\left\{H(D)H^{\dagger}(D^{-1})\right\} = 0,$$

where † represents a conjugate transpose of a matrix over $GF(4)$ and "Tr" above indicates an element-wise trace operation on its matrix argument. Then the construction for a CRSS quantum convolutional code is to form the following check matrix:

$$\begin{bmatrix} \omega H \\ \bar{\omega} H \end{bmatrix}, \qquad (9.34)$$

and substitute the $GF(4)$ entries with Pauli matrices according to the map in (9.32). The resulting quantum convolutional code has parameters $[[n, 2k - n]]$ and inherits the error-correcting properties of the $GF(4)$ classical convolutional code.

The Forney–Grassl–Guha (FGG) code from Example 9.1 is a CRSS quantum convolutional code, with its parent classical code as follows:

$$(\cdots \mid 000 \mid 111 \mid 1\omega\bar{\omega} \mid 000 \mid \cdots).$$

One can check that constructing a quantum convolutional code according to the CRSS prescription in (9.34) leads to the following two Pauli generators for the FGG quantum convolutional code:

$$\cdots \left| \begin{array}{c} III \\ III \end{array} \right| \left. \begin{array}{c} XXX \\ ZZZ \end{array} \right| \left. \begin{array}{c} XZY \\ ZYX \end{array} \right| \left. \begin{array}{c} III \\ III \end{array} \right| \cdots \quad (9.35)$$

We now present the quantum shift-register encoding circuit for the FGG code. The code has three qubits per frame, and its stabilizer matrix is

$$\begin{bmatrix} 1+D & 1 & 1+D & 0 & D & D \\ 0 & D & D & 1+D & 1+D & 1 \end{bmatrix}, \quad (9.36)$$

by translating the generators in (9.35) to the binary polynomial picture. The Grassl–Rötteler algorithm [GR06b] determines a sequence of transformations using the elementary operations in (9.16)–(9.20) to map the polynomial matrix in (9.36) to the following one:

$$\begin{bmatrix} 1 & 0 & 0 & 0 & 0 & 0 \\ 0 & 1 & 0 & 0 & 0 & 0 \end{bmatrix}. \quad (9.37)$$

Note that the sequence of elementary transformations does not have to be unique. The matrix in (9.37) stabilizes a state of the form $|0\rangle |0\rangle |\psi\rangle$ where we see that the first two qubits of every frame are ancilla qubits and the last qubit of every frame is an information qubit. Then *reversing the order* of the elementary operations found from the Grassl–Rötteler algorithm gives an encoding circuit. The sequence of encoding operations for the FGG code is

$$\text{H}(1) \ \text{H}(2) \ P(1) \ \text{C-PHASE}\,(1,3)\, (D^{-1} + 1 + D)$$
$$\text{C-PHASE}(1,2)(D^{-1}) \ \text{C-PHASE}\,(2,3)\, (1 + D + D^2)$$
$$\text{CNOT}\,(2,3)\,(1) \ \text{CNOT}\,(3,2)\,(D) \ \text{CNOT}\,(2,3)\,(D)$$
$$\text{CNOT}(1,2)(1) \ \text{CNOT}\,(1,3)\,(1+D) \ \text{CNOT}\,(2,1)\,(D),$$

where the ordering of gates is from left to right then top to bottom. The reader should verify that the above set of transformations indeed takes the matrix in (9.37) to the one in (9.36). Then, using the technique of cascading several quantum shift-register circuits and commuting gates through memory (discussed in Sections 9.4.1 and 9.5.1) gives a quantum shift-register circuit encoding for the code. Figure 9.11 depicts the quantum shift-register circuit that encodes the FGG code using five frames of memory qubits.

9.6 Entanglement-assisted quantum convolutional codes

In this section, we introduce the theory of entanglement-assisted quantum convolutional coding [WB10a, WB09, WB10b, WB08c]. This theory naturally extends the theory of quantum convolutional coding, just as entanglement-assisted stabilizer codes extend quantum stabilizer codes. The major result of this theory is that we can produce an entanglement-assisted quantum convolutional code from two *arbitrary* classical binary convolutional codes or from an *arbitrary* classical convolutional code over $GF(4)$. The resulting quantum convolutional codes respectively admit a CSS or CRSS structure. The rates and error-correcting properties of the classical convolutional

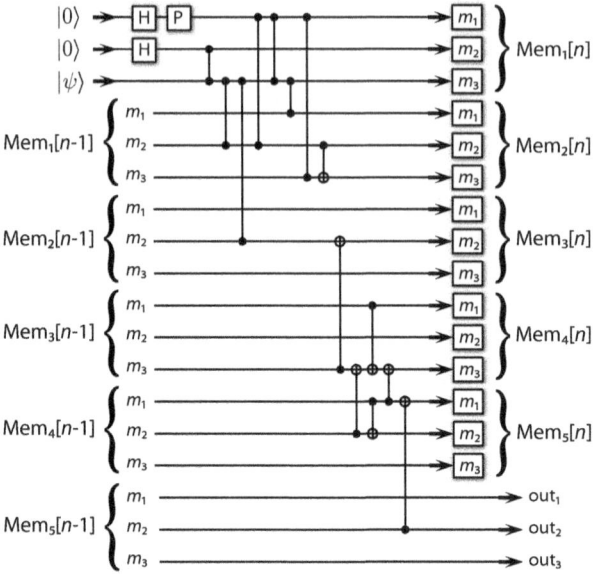

Fig. 9.11. The above circuit encodes the Forney–Grassl–Guha code. Reprinted with permission from [W09b]. Copyright 2009, American Physical Society.

codes directly determine the corresponding properties of the entanglement-assisted quantum convolutional codes. This theory gives a way of producing high-rate quantum codes from high-rate classical convolutional codes.

We do not delve much into the details of the theory of entanglement-assisted quantum convolutional coding, but only present the major concepts and three examples to highlight the major ideas. The construction of entanglement-assisted quantum convolutional codes is a subtle business, mostly because the commutation relations of an arbitrary set of generators become much more complicated for the convolutional case than they are for the block case.

9.6.1 Operation

Figure 9.12 highlights the main features of the operation of an entanglement-assisted quantum convolutional code. Such a code operates similarly to a standard quantum convolutional code, but the main difference is that the sender and receiver share entanglement in the form of ebits before quantum communication begins. An $[[n, k; c]]$ entanglement-assisted quantum convolutional code encodes k information qubits per frame with the help of c ebits and $n - k - c$ ancilla qubits per frame.

9.6.2 The entanglement-assisted CSS construction

Recall that the construction in Section 9.5.1 allows us to produce CSS quantum convolutional codes from two classical binary convolutional codes that satisfy an orthogonality constraint – the polynomial parity check matrices $H_1(D)$ and $H_2(D)$ of the two classical codes are orthogonal with respect to the shifted symplectic product: $H_1(D)H_2^T(D^{-1}) = 0$. The resulting

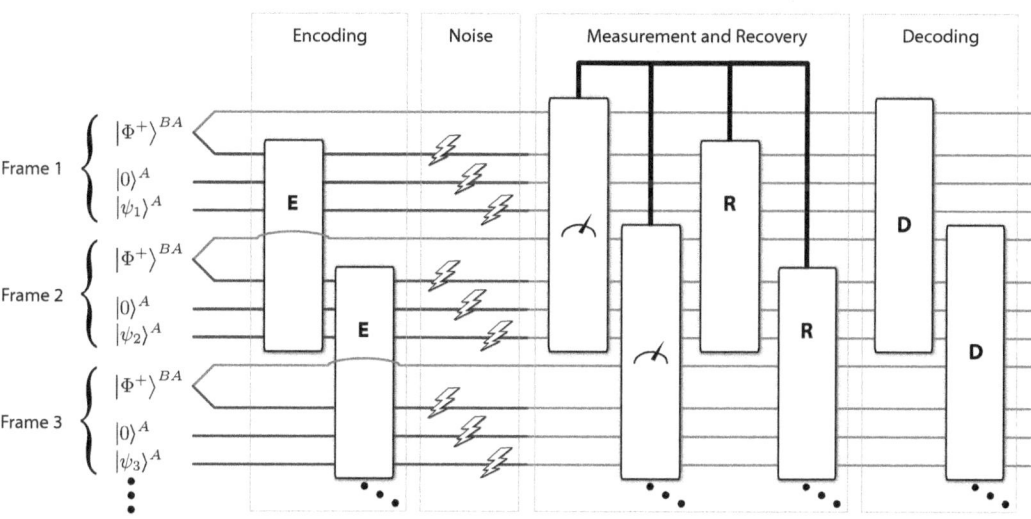

Fig. 9.12. An entanglement-assisted quantum convolutional code operates on a stream of qubits partitioned into a countable number of frames. The sender encodes the frames of information qubits, ancilla qubits, and half of shared ebits with a repeated, overlapping encoding circuit E. The noisy channel affects the sender's encoded qubits but does not affect the receiver's half of the shared ebits. The receiver performs overlapping measurements on both the encoded qubits and his half of the shared ebits. These measurements produce an error syndrome which the receiver can process to determine the most likely error. The receiver reverses the errors on the noisy qubits from the sender. The final decoding circuit operates on all qubits in a frame and recovers the original stream of information qubits. Reprinted with permission from [WB10a]. Copyright 2010, American Physical Society.

symplectic code has a self-orthogonal parity check matrix when we join them together using the CSS construction.

The CSS construction for entanglement-assisted quantum convolutional codes allows us to import two *arbitrary* binary classical convolutional codes for use as a quantum code – the codes do not necessarily have to obey the self-orthogonality constraint. CSS entanglement-assisted quantum convolutional codes divide into two classes based on certain properties of the classical codes from which we produce them. These properties of the classical codes determine the structure of the encoding and decoding circuit for the code, and the structure of the encoding and decoding circuit in turn determines the class of the entanglement-assisted quantum convolutional code.

(i) Codes in the first class admit both finite-depth encoding and decoding circuits.
(ii) Codes in the second class have an encoding circuit that employs both finite-depth and infinite-depth operations. Their decoding circuits have finite-depth operations only.

We overview examples of these two classes in the forthcoming subsections, but first state the main theorem for CSS entanglement-assisted quantum convolutional codes. The theorem gives a direct way to compute the amount of entanglement that the code requires. The number of ebits required is equal to the rank of a particular matrix derived from the check matrices of the two classical codes. It generalizes theorems discussed in Chapter 7 that determine the amount of entanglement required for an entanglement-assisted quantum block code.

Theorem 9.1 also provides a formula to compute the performance parameters of the entanglement-assisted quantum convolutional code from the performance parameters of the two classical codes. This formula ensures that high-rate classical convolutional codes produce high-rate entanglement-assisted quantum convolutional codes.

Theorem 9.1. *Let $H_1(D)$ and $H_2(D)$ be the respective check matrices corresponding to classical binary convolutional codes C_1 and C_2. Suppose that classical code C_i encodes k_i information bits with n bits per frame where $i = 1, 2$. The respective dimensions of $H_1(D)$ and $H_2(D)$ are thus $(n - k_1) \times n$ and $(n - k_2) \times n$. Then the resulting entanglement-assisted quantum convolutional code encodes $k_1 + k_2 - n + c$ information qubits per frame and is an $[[n, k_1 + k_2 - n + c; c]]$ entanglement-assisted quantum convolutional code. The code requires c ebits per frame where c is equal to the rank of $H_1(D) H_2^T(D^{-1})$.*

The proof of the above theorem is somewhat involved so we instead refer the reader to [WB10a] for the full proof. We now proceed to providing an example code for each class above.

9.6.2.1 Example of a Type-I CSS EAQCC Consider a classical convolutional code with the following check matrix:
$$H(D) = \begin{bmatrix} 1 + D^2 & 1 + D + D^2 \end{bmatrix}.$$

We can use $H(D)$ in an entanglement-assisted quantum convolutional code to correct for both bit-flip errors and phase-flip errors. We form the following quantum check matrix according to the CSS construction:
$$\begin{bmatrix} 1 + D^2 & 1 + D + D^2 & 0 & 0 \\ 0 & 0 & 1 + D^2 & 1 + D + D^2 \end{bmatrix}. \tag{9.38}$$

This code falls into the first class of entanglement-assisted quantum convolutional codes because $H(D) H^T(D^{-1}) = 1$ (see [WB10a] for more details).

We show how to encode the above code starting from a stream of information qubits and ebits. Each frame has one ebit and one information qubit. Let us begin with a polynomial matrix that stabilizes the unencoded state:
$$\begin{bmatrix} 1 & 1 & 0 & 0 & 0 & 0 \\ 0 & 0 & 0 & 1 & 1 & 0 \end{bmatrix}.$$

Alice possesses the two qubits on the "right" and Bob possesses the qubit on the "left." Each unencoded frame then has the structure $|\Phi^+\rangle^{BA} |\psi\rangle^A$, where $|\Phi^+\rangle^{BA}$ is an ebit shared between Bob and Alice and $|\psi\rangle^A$ is the information qubit that Alice wants to transmit (though, note that information qubits can be entangled across frames). We label the middle qubit as "qubit one" and the rightmost qubit as "qubit two." The sequence of transformations that Alice performs on her qubits are as follows:
$$\text{CNOT}(1, 2)(D + D^2) \; \text{H}(1) \; \text{H}(2) \; \text{CNOT}(1, 2)(D) \tag{9.39}$$
$$\text{CNOT}(2, 1)(D) \; \text{CNOT}(1, 2)(1),$$

where the order of operations is again left to right and top to bottom. The stabilizer after all these operations is
$$\begin{bmatrix} 1 & 0 & 0 & 0 & 1 + D^2 & 1 + D + D^2 \\ 0 & 1 + D^2 & 1 + D + D^2 & 1 & 0 & 0 \end{bmatrix},$$

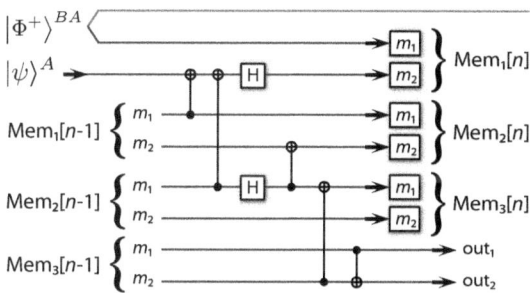

Fig. 9.13. The finite-depth encoding circuit for the entanglement-assisted quantum convolutional code in Section 9.6.2.1. Each clock cycle requires a fresh ebit and information qubit to be input into the device. Bob must be careful to synchronize the processing of his half of the ebits with what Alice transmits over the channel.

and a simple row operation that switches the first row with the second row gives the following stabilizer:

$$\begin{bmatrix} 0 & 1+D^2 & 1+D+D^2 & 1 & 0 & 0 \\ 1 & 0 & 0 & 0 & 1+D^2 & 1+D+D^2 \end{bmatrix}.$$

The entries on Alice's side of the above stabilizer have equivalent error-correcting properties to the quantum check matrix in (9.38). One can then devise a quantum shift-register circuit to encode this entanglement-assisted code. Figure 9.13 illustrates how the above operations encode a stream of ebits and information qubits for our example.

We can devise a decoding circuit for Bob simply by considering the operations in (9.39) in reverse order. This class of entanglement-assisted convolutional codes does not require Bob's decoding circuit to incorporate his halves of the ebits in the decoding. He only requires his halves of the ebits to perform measurements that diagnose channel errors.

Codes in the first class are more useful in practice than those in the second because their encoding and decoding circuits are both finite depth. An uncorrected error propagates only to a finite number of information qubits in the decoded qubit stream. Codes in the first class therefore do not require any assumptions about noiseless encoding or decoding.

9.6.2.2 Example of an infinite-depth CSS EAQCC The other class of CSS entanglement-assisted quantum convolutional codes requires both infinite-depth and finite-depth operations in the encoding circuit but requires only finite-depth operations in the decoding circuit. Here, we present an example of this class.

Consider a classical convolutional code with the following check matrix:

$$H(D) = \begin{bmatrix} 1 & 1+D \end{bmatrix}.$$

We can use the above check matrix in an entanglement-assisted quantum convolutional code to correct for both bit-flips and phase-flips. We form the following quantum check matrix:

$$\begin{bmatrix} 1 & 1+D & 0 & 0 \\ 0 & 0 & 1 & 1+D \end{bmatrix}. \quad (9.40)$$

We first perform the following finite-depth operations:

$$\text{CNOT}(1,2)\,(1+D)\ \text{H}(1)\ \text{H}(2)\ \text{CNOT}(1,2)(1)\ \text{H}(1)\ \text{H}(2). \quad (9.41)$$

These operations put the above quantum check matrix into a particular form:
$$\begin{bmatrix} 1+D+D^2 & 1 & 0 & 0 \\ 0 & 0 & 1 & 0 \end{bmatrix}. \tag{9.42}$$

Note that we have implicitly performed a row operation that delays the first row by D.

We show how to realize the above stabilizer with both infinite-depth and finite-depth operations. We begin encoding with one ebit and one information qubit per frame. The stabilizer matrix for the unencoded stream is
$$\begin{bmatrix} 1 & 1 & 0 & 0 & 0 & 0 \\ 0 & 0 & 0 & 1 & 1 & 0 \end{bmatrix},$$
and the "information-qubit matrix" is
$$\begin{bmatrix} 0 & 0 & 0 & 0 & 0 & 1 \\ 0 & 0 & 1 & 0 & 0 & 0 \end{bmatrix},$$
where Bob possesses the first qubit of each frame and Alice possesses the other two. The "information-qubit matrix" corresponds to the logical operators of the information qubit. The following sequence of operations can realize the stabilizer matrix in (9.42):

$$H(1)\ \text{CNOT}(1,2)\ \text{CNOT}(2)(1/(1+D+D^2))\ H(1)\ H(2). \tag{9.43}$$

Note that the rational polynomial $1/\left(1+D+D^2\right)$ corresponds to an infinite-depth operation. The stabilizer matrix then becomes
$$\begin{bmatrix} 1 & 1 & 1/(1+D+D^2) & 0 & 0 & 0 \\ 0 & 0 & 0 & 1 & 1 & 0 \end{bmatrix}, \tag{9.44}$$
and the information-qubit matrix becomes
$$\begin{bmatrix} 0 & 0 & 1/(1+D+D^2) & 0 & 0 & 0 \\ 0 & 0 & 0 & 0 & 1 & 1+D^{-1}+D^{-2} \end{bmatrix}. \tag{9.45}$$

At this point, the stabilizer matrix in (9.44) is actually equivalent to that in (9.42) by a row operation on the first row (multiplication by $1+D+D^2$). Performing the finite-depth operations in (9.41) in reverse order transforms the stabilizer to
$$\begin{bmatrix} D^{-1} & \frac{1}{1+D+D^2} & \frac{1+D}{1+D+D^2} & 0 & 0 & 0 \\ 0 & 0 & 0 & 1 & 1 & 1+D \end{bmatrix},$$
and the information-qubit matrix to
$$\begin{bmatrix} 0 & \frac{D^{-1}+D^{-2}}{1+D+D^2} & \frac{1}{1+D+D^2} & 0 & 0 & 0 \\ 0 & 0 & 0 & 0 & D^{-1}+D^{-2} & D^{-1} \end{bmatrix}.$$

The above stabilizer is equivalent to the desired quantum check matrix in (9.40) by a row operation that multiplies its first row by $1+D+D^2$. Thus, to recap, the sequence of operations that encodes the desired check matrix is

$$H(1)\ \text{CNOT}(1,2)\ \text{CNOT}(2)(1/(1+D+D^2))$$
$$\text{CNOT}(1,2)(1)\ H(1)\ H(2)\ \text{CNOT}(1,2)(1+D),$$

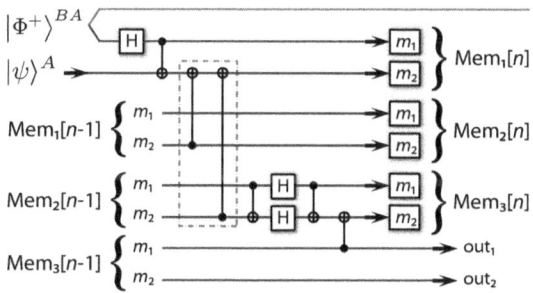

Fig. 9.14. The encoding circuit for the entanglement-assisted quantum convolutional code in Example 9.6.2.2. The gates in the dotted box form an infinite-depth operation. The other operations in the encoding circuit are finite-depth.

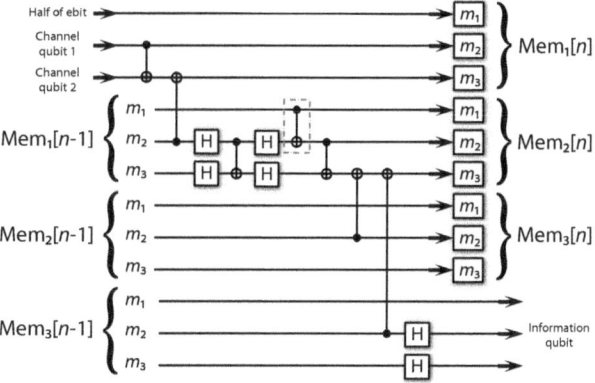

Fig. 9.15. Bob feeds his half of each ebit and his channel qubits into the above quantum shift-register circuit. The decoding involves finite-depth operations only but the dotted box indicates that there is a non-trivial operation between his half of the ebit and the other channel qubits. This type of decoding is typical for the second class of CSS entanglement-assisted quantum convolutional codes.

where we are able to cancel some Hadamard gates. Figure 9.14 depicts a quantum shift-register implementation of these encoding operations.

The receiver decodes by first performing the finite-depth operations in (9.41) and gets the stabilizer in (9.44) and the information-qubit matrix in (9.45). The receiver performs a CNOT from his half of the ebit to the first channel qubit and follows with a CNOT from the first channel qubit to the second channel qubit in the same frame, in an advanced frame, and in a twice-advanced frame. Finally perform a Hadamard gate on the two channel qubits. The stabilizer becomes

$$\begin{bmatrix} 0 & 0 & 0 & | & 0 & 0 & 1/(1+D+D^2) \\ 0 & 0 & 0 & | & 0 & 0 & 0 \end{bmatrix},$$

and the information-qubit matrix becomes

$$\begin{bmatrix} 0 & 0 & 0 & | & 0 & 1 & 1/(1+D+D^2) \\ 0 & 1 & 0 & | & 0 & 0 & 0 \end{bmatrix}.$$

The receiver decodes the information qubits successfully because a row operation from the first row of the stabilizer to the first row of the information-qubit matrix gives the proper logical operators for the information qubits. Figure 9.15 displays a quantum shift-register circuit for decoding the information qubits.

9.7 Closing remarks

For the moment, the role of quantum convolutional codes in fault-tolerant quantum computation is not entirely clear, mainly because it is not clear that the circuits used for encoding and decoding have desirable fault-tolerant properties. Further investigation in this direction might suggest that quantum convolutional codes could be useful for fault tolerance. Quantum convolutional codes should be an important part of future quantum communication devices, either on their own or as constituent codes in a more elaborate scheme such as a quantum turbo code. They might one day be useful when we have the ability to control large streams of qubits for transmission over noisy channels.

Since the original writing of this chapter, Houshmand et al. have made good progress on encoders of quantum convolutional codes [WHH.-K11, HH.-KW11]. In particular, they have addressed how to minimize the memory required for an encoder of a quantum convolutional code, a question not considered in depth in this chapter and in prior work on encoding algorithms [GR06a, GR07]. Their technique gives a dramatic reduction in the memory required for encoding; for example, Houshmand et al. have found an encoder for the Forney–Grassl–Guha code that requires only one memory qubit, as opposed to the one in Fig. 9.11 that requires 15 memory qubits.

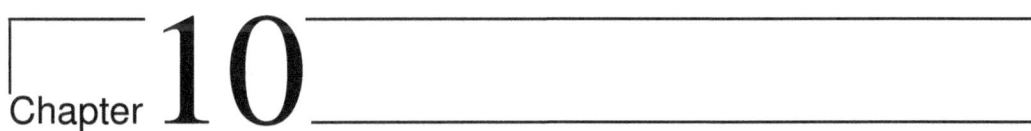

Chapter 10

Nonadditive quantum codes

Markus Grassl and Martin Rötteler

10.1 Introduction

Most quantum error-correcting codes (QECCs) that are designed to correct local errors are stabilizer codes that correspond to additive classical codes. Because of this correspondence, stabilizer codes are also referred to as additive codes, while nonstabilizer codes are called nonadditive codes. Shortly after the invention of stabilizer codes by Gottesman [G96a] and independently Calderbank *et al.* [CRS+98], the nonadditive $((5, 6, 2))$ code was found [RHS+97], which constitutes the first example of a nonadditive quantum code with a higher dimension than any stabilizer code of equal length and minimum distance.[1] About ten years later, the first one-error-correcting $((9, 12, 3))$ code and the optimal $((10, 24, 3))$ code were found [YCL+09, YCO07], both better than any stabilizer code of the same length.

Here we describe a framework that not only allows a joint description of these codes, but also enables the construction of new nonadditive codes with better parameters than the best known stabilizer codes. These codes are based on the classical nonlinear Goethals and Preparata codes, which themselves are better than any classical linear code.

In Section 10.3, the framework will be presented from two different points of view, namely union stabilizer codes and codeword stabilized codes, each highlighting different aspects of the construction. To illustrate the relationship of these nonadditive codes to stabilizer codes we first recall the main aspects of the latter. Section 10.4 is devoted to methods for obtaining new codes, including the aforementioned quantum Goethals–Preparata codes. Finally, quantum circuits for encoding and syndrome computation are discussed in Section 10.5.

---

[1] Note that in this chapter we use the notation $((n, K, d))$ for a general QECC of length n, i.e., using n qubits, dimension K, and minimum distance d; stabilizer codes will continue to be denoted using the standard $[[n, k, d]]$ notation.

Quantum Error Correction, ed. Daniel A. Lidar and Todd A. Brun. Published by Cambridge University Press.
© Cambridge University Press 2013.

10.2 Stabilizer codes

We recall some of the details of the stabilizer formalism for QECC (see Chapter 2) and briefly discuss the connection to additive codes over $GF(4)$ (see Chapter 12 and, e.g., [CRS+98, G96a]). Here we restrict ourselves to codes for qubit systems, although the constructions can easily be generalized to quantum codes for higher-dimensional systems (see, e.g., [AK01]). A stabilizer code encoding k qubits into n qubits having minimum distance d, denoted by $\mathscr{C} = [[n, k, d]]$, is a subspace of dimension 2^k of the complex Hilbert space $(\mathbb{C}^2)^{\otimes n}$ of dimension 2^n. The code is the joint eigenspace of a set of $n - k$ commuting operators $g_1, \ldots, g_{n-k}$ that are tensor products of the Pauli matrices or identity. The operators g_i generate an Abelian group $\mathscr{S}$ with 2^{n-k} elements, called the *stabilizer* of the code. It is a subgroup of the n-qubit Pauli group $\mathscr{P}_n$, which itself is generated by the tensor product of n Pauli matrices and identity. We further require that $\mathscr{S}$ does not contain any nontrivial multiple of identity. The *normalizer* of $\mathscr{S}$ in $\mathscr{P}_n$, denoted by $\mathscr{N}$, acts on the code $\mathscr{C} = [[n, k, d]]$. It is possible to identify $2k$ logical operators $\overline{X}_1, \ldots, \overline{X}_k$ and $\overline{Z}_1, \ldots, \overline{Z}_k$ such that these operators commute with any element in the stabilizer $\mathscr{S}$, and such that together with $\mathscr{S}$ they generate the normalizer $\mathscr{N}$ of the code. The operators $\overline{X}_i$ mutually commute, and so do the operators $\overline{Z}_j$. The operator $\overline{X}_i$ anticommutes with the operator $\overline{Z}_j$ if $i = j$ and otherwise commutes with it.

It has been shown that the n-qubit Pauli group corresponds to a symplectic geometry, and that one can reduce the problem of constructing stabilizer codes to finding additive codes over $GF(4)$ that are self-orthogonal with respect to a symplectic inner product [CRS+97, CRS+98]. Up to a scalar multiple, the elements of $\mathscr{P}_1$ can be expressed as $X^a Z^b$ where $(a, b) \in GF(2)^2$ is a binary vector. Choosing the basis $\{1, \omega\}$ of $GF(4)$, where ω is a primitive element of $GF(4)$ with $\omega^2 + \omega + 1 = 0$, we get the following correspondence between the Pauli matrices, elements of $GF(4)$, and binary vectors of length two:

Operator	$GF(4)$	$GF(2)^2$
I	0	(00)
$X = \sigma_x$	1	(10)
$Y = \sigma_y$	ω^2	(11)
$Z = \sigma_z$	ω	(01)

This defines a mapping from Pauli operators to binary vectors that extends naturally to tensor products of n Pauli matrices being mapped to vectors of length n over $GF(4)$ or binary vectors of length $2n$. We rearrange the latter in such a way that the first n coordinates correspond to the exponents of the operators X and write the vector as $(a|b)$, i.e.,

$$g = X^{a_1} Z^{b_1} \otimes \cdots \otimes X^{a_n} Z^{b_n} \stackrel{\wedge}{=} (a|b) = (g^X | g^Z). \quad (10.1)$$

Two operators corresponding to the binary vectors $(a|b)$ and $(c|d)$ commute if and only if the symplectic inner product $a \cdot d - b \cdot c = 0$. In terms of the binary representation, the stabilizer corresponds to a binary code C that is self-orthogonal with respect to this symplectic inner product, and the normalizer corresponds to the symplectic dual code C^*. In terms of the correspondence to vectors over $GF(4)$, the stabilizer and normalizer correspond to an additive code over $GF(4)$ and its dual with respect to a symplectic inner product, respectively, which we will also denote by C and C^*. The term *additive quantum code* refers to this correspondence. The minimum distance d of the quantum code is given as the minimum weight in the set $C^* \setminus C \subset GF(4)^n$ which is lower bounded by the minimum distance d^* of the additive code C^*. If $d = d^*$, the code is said to be *pure* or *non-degenerate*, and for $d \geq d^*$, the code is said to be *pure up to* d^*. Note that the minimum distance of the quantum code primarily depends on the dual code C^*, while the code C defines the stabilizer, and hence the quantum code itself.

Fixing the logical operators $\overline{X}_i$ and $\overline{Z}_j$, there is a canonical basis for the additive quantum code $\mathscr{C}$. The stabilizer group $\mathscr{S}$ of the quantum code together with the logical operators $\overline{Z}_j$ generate an Abelian group of order 2^n which corresponds to a self-dual additive code. The joint $+1$-eigenspace is one-dimensional, hence there is a unique quantum state $|(00\ldots0)_L\rangle \in \mathscr{C}$ stabilized by all elements of $\mathscr{S}$ and all $\overline{Z}_j$. An orthonormal basis of the code $\mathscr{C}$ is given by the states

$$|(i_1 i_2 \ldots i_k)_L\rangle = \overline{X}_1^{i_1} \overline{X}_2^{i_2} \cdots \overline{X}_k^{i_k} |(00\ldots0)_L\rangle, \qquad (10.2)$$

where $(i_1 i_2 \ldots i_k) \in GF(2)^k$.

10.3 Characterization of nonadditive quantum codes

In the following we will describe two equivalent constructions for nonadditive quantum codes. The first construction presented in [GR08a, GR08b] is based on stabilizer codes $[[n, k, d]]$. The second construction [CSS+09, CZC08] starts with so-called graph states or stabilizer states, which are stabilizer codes $[[n, 0, d]]$.

Both approaches provide a framework for the construction of nonadditive quantum codes that constitutes a generalization of the stabilizer formalism. This allows the reduction of the problem of finding good quantum codes to the search for certain good nonadditive classical codes. Furthermore, it allows us to generalize certain constructions such as CSS codes [CS96, S96d] and the enlargement of CSS codes [S99b].

10.3.1 Union stabilizer codes

The stabilizer group $\mathscr{S}$ gives rise to an orthogonal decomposition of the space $(\mathbb{C}^2)^{\otimes n}$ into common eigenspaces of equal dimension. The stabilizer code $\mathscr{C}$ is the joint $+1$-eigenspace of dimension 2^k. In general, the joint eigenspaces of $\mathscr{S}$ can be labeled by the eigenvalues of a set of $n - k$ generators of $\mathscr{S}$. Moreover, the n-qubit Pauli group $\mathscr{P}_n$ operates transitively on the eigenspaces. Hence one can identify a set $\mathscr{T} \subset \mathscr{P}_n$ of 2^{n-k} operators such that

$$(\mathbb{C}^2)^{\otimes n} = \bigoplus_{t \in \mathscr{T}} t\mathscr{C}. \qquad (10.3)$$

Note that each of the spaces $t\mathscr{C}$ is a QECC with the same parameters as the code $\mathscr{C}$ and stabilizer group $t\mathscr{S}t^{-1}$. Two spaces $t_1\mathscr{C}$ and $t_2\mathscr{C}$ are either identical if $t_2^{-1}t_1\mathscr{C} = \mathscr{C}$, i.e., the operator $t_2^{-1}t_1$ is an element of the normalizer $\mathscr{N}$, or for $t_2^{-1}t_1 \notin \mathscr{N}$, the spaces $t_1\mathscr{C}$ and $t_2\mathscr{C}$ are orthogonal. Hence the decomposition (10.3) corresponds to the decomposition of the n-qubit Pauli group $\mathscr{P}_n$ into cosets with respect to the normalizer $\mathscr{N}$ of the code $\mathscr{C}$, and likewise to the decomposition of the full vector space $GF(4)^n$ into cosets of the additive code C^*.

The main idea of union stabilizer codes is to find a subset $\mathscr{T}_0$ of the *translations* $\mathscr{T}$ such that the space $\bigoplus_{t \in \mathscr{T}_0} t\mathscr{C}$ is a good quantum code (see [GB97, GR08a, GR08b]).

$$\left(\begin{array}{c|c} S_1^X & S_1^Z \\ \vdots & \vdots \\ S_{n-k}^X & S_{n-k}^Z \\ \hline \overline{Z}_1^X & \overline{Z}_1^Z \\ \vdots & \vdots \\ \overline{Z}_k^X & \overline{Z}_k^Z \\ \hdashline \overline{X}_1^X & \overline{X}_1^Z \\ \vdots & \vdots \\ \overline{X}_k^X & \overline{X}_k^Z \end{array}\right)$$

$$\left\{\begin{array}{c|c} t_1^X & t_1^Z \\ \vdots & \vdots \\ t_K^X & t_K^Z \end{array}\right\}$$

Fig. 10.1. Arrangements of the vectors associated with a union stabilizer code.

Definition 10.1 (Union stabilizer code). *Let $\mathscr{C}_0 = [[n,k]]$ be a stabilizer code and let furthermore $\mathscr{T}_0 = \{t_1, \ldots, t_K\}$ be a subset of the coset representatives of the normalizer $\mathscr{N}_0$ of the code $\mathscr{C}_0$ in $\mathscr{P}_n$. Then the* union stabilizer code *is defined as*

$$\mathscr{C} = \bigoplus_{t \in \mathscr{T}_0} t\mathscr{C}_0.$$

Without loss of generality we assume that $\mathscr{T}_0$ contains identity. The dimension of $\mathscr{C}$ is $K \times 2^k$, and we will use the notation $\mathscr{C} = ((n, K \times 2^k, d))$.

Similarly to (10.2), a canonical basis of the union stabilizer code $\mathscr{C}$ is given by

$$|(j; i_1 i_2 \ldots i_k)_L\rangle = t_j \overline{X}_1^{i_1} \overline{X}_2^{i_2} \cdots \overline{X}_k^{i_k} |(00\ldots 0)_L\rangle, \qquad (10.4)$$

where $j = 1, \ldots, K$, $(i_1 i_2 \ldots i_k) \in GF(2)^k$, and $\overline{X}_l$ are logical operators of the stabilizer code $\mathscr{C}_0$.

The binary vectors corresponding to the various operators are illustrated in Fig. 10.1. The first $n-k$ rows correspond to the binary vectors [cf. (10.1)] associated with the generators S_i of the stabilizer $\mathscr{S}$ of the code $\mathscr{C}_0$. They generate the classical code C_0. The next k rows correspond to the logical operators $\overline{Z}_m$, which together with the stabilizer $\mathscr{S}$ generate a maximal set of 2^n commuting operators whose joint $+1$-eigenspace is the logical basis state $|(00\ldots 0)_L\rangle$. The next k rows between the dashed horizontal line and the second horizontal line correspond to the k logical operators $\overline{X}_i$ giving rise to 2^k logical basis states as in (10.2). The last K rows correspond to the K translations t_j defining the cosets of the classical code C_0^* and the unitary images of the stabilizer code $\mathscr{C}_0$, respectively. We use curly brackets to stress the fact that the set of operators $\mathscr{T}_0$ need not be closed under group operation. In general, the quantum code is not invariant under these *generalized logical X-operators*. On the other hand, if $\mathcal{T}_0$ is closed under group operation, the resulting code will be a stabilizer code where a basis of $\mathcal{T}_0$ defines an additional set of logical X-operators.

In order to compute the minimum distance of this code, we first consider the distance between two spaces $t_1 \mathcal{C}_0$ and $t_2 \mathcal{C}_0$. Since for a fixed stabilizer code $\mathcal{C}_0$ two spaces $t_1 \mathcal{C}_0$ and $t_2 \mathcal{C}_0$ are either identical or orthogonal, we can define the distance of them as follows:

$$\operatorname{dist}(t_1 \mathcal{C}_0, t_2 \mathcal{C}_0) := \min\{\operatorname{wgt}(p) \colon p \in \mathcal{P}_n \mid p t_1 \mathcal{C}_0 = t_2 \mathcal{C}_0\}. \tag{10.5}$$

Here $\operatorname{wgt}(p)$ is the number of tensor factors in the n-qubit Pauli operator p that are different from identity. Clearly, $\operatorname{dist}(t_1 \mathcal{C}_0, t_2 \mathcal{C}_0) = \operatorname{dist}(t_2^{-1} t_1 \mathcal{C}_0, \mathcal{C}_0)$. The two spaces are identical if and only if $t_2^{-1} t_1$ is an element of the normalizer group $\mathcal{N}_0$, or equivalently, if the cosets $C_0^* + t_1$ and $C_0^* + t_2$ of the additive normalizer code C_0^* are identical. (Note that we denote both an n-qubit Pauli operator and the corresponding vector over $GF(4)$ by t_j.) Hence the distance (10.5) can also be expressed in terms of the associated vectors over $GF(4)$.

Lemma 10.1. *The distance of the spaces $t_1 \mathcal{C}_0$ and $t_2 \mathcal{C}_0$ equals the minimum weight in the coset $C_0^* + t_1 - t_2$.*

Proof. Direct computation shows

$$\begin{aligned}
\operatorname{dist}(t_1 \mathcal{C}_0, t_2 \mathcal{C}_0) &= \operatorname{dist}(C_0^* + t_1, C_0^* + t_2) \\
&= \operatorname{dist}(C_0^* + (t_1 - t_2), C_0^*) \\
&= \min\{\operatorname{wgt}(c + t_1 - t_2) \colon c \in C_0^*\} \\
&= \min\{\operatorname{wgt}(v) \colon v \in C_0^* + t_1 - t_2\}. \qquad \square
\end{aligned}$$

While the distance between the cosets $C_0^* + t_j$ is an upper bound on the minimum distance of the union code $\mathcal{C}$, the true minimum distance can be derived from the following code over $GF(4)$.

Definition 10.2 (Union normalizer code). *With the union stabilizer code $\mathcal{C}$ we associate the (in general nonadditive) classical* union normalizer code *given by*

$$C^* = \bigcup_{t \in \mathcal{T}_0} C_0^* + t = \{c + t_j \colon c \in C_0^*, \, j = 1, \ldots, K\},$$

where C_0^* denotes the additive code associated with the normalizer $\mathcal{N}_0$ of the stabilizer code $\mathcal{C}_0$. We will refer to both the vectors t_j and the corresponding unitary operators as translations.

Theorem 10.1. *The minimum distance of a union stabilizer code with union normalizer code C^* is given by*

$$\begin{aligned}
d &= \min\{\operatorname{wgt}(v) \colon v \in (C^* - C^*) \setminus \widetilde{C}_0\} \tag{10.6} \\
&\geq d_{\min}(C^*) \\
&= \min\{\operatorname{dist}(c + t_j, c' + t_{j'}) \colon t_j, t_{j'} \in \mathcal{T}_0, \, c, c' \in C_0^*, c + t_j \neq c' + t_{j'}\},
\end{aligned}$$

where $C^* - C^* := \{a - b \colon a, b \in C^*\}$ denotes the set of all differences of vectors in C^*, and $\widetilde{C}_0 \leq C_0$ is the additive code that corresponds to all elements of the stabilizer group $\mathcal{S}$ that commute with all $t_j \in \mathcal{T}_0$.

Proof. Let $E \in \mathscr{P}_n$ be an n-qubit Pauli error of weight $0 < \text{wgt}(E) < d$. For two canonical basis states $|\psi_a\rangle$ and $|\psi_b\rangle$ as given in (10.4) we consider the inner product

$$\begin{aligned}\langle\psi_a|E|\psi_b\rangle &= \langle(j;i_1i_2\ldots i_k)_L|E|(j';i'_1i'_2\ldots i'_k)_L\rangle \\ &= \langle(00\ldots 0)_L|\overline{X}_1^{i_1}\cdots\overline{X}_k^{i_k}t_jEt_{j'}\overline{X}_1^{i'_1}\cdots\overline{X}_k^{i'_k}|(00\ldots 0)_L\rangle \\ &= \pm\langle(00\ldots 0)_L|\overline{X}_1^{i_1+i'_1}\cdots\overline{X}_k^{i_k+i'_k}t_jt_{j'}E|(00\ldots 0)_L\rangle.\end{aligned}$$

If $E \in \mathscr{S}$ commutes with all $t_j \in \mathscr{T}_0$, then $\langle\psi_a|E|\psi_b\rangle = \delta_{ab}$. Otherwise, $E \notin C^* - C^*$ since $0 < \text{wgt}(E) < d$, and hence the inner product vanishes. □

Note that the code $\widetilde{C}_0 \subseteq C_0$ in (10.6) corresponds to all elements of the stabilizer of $\mathscr{C}_0$ that commute with all translations t_j, and hence also with the product of translations. Therefore the code $\widetilde{C}_0$ is the dual code of the additive closure of the in general nonadditive union normalizer code C^*, i.e., $\widetilde{C}_0 = \langle C^*\rangle^*$. In many cases, $\widetilde{C}_0$ will have small dimension or is even the trivial code containing only the all-zero code. We are not aware of any degenerate truly nonadditive code, i.e., a code that is not equivalent to a stabilizer code and whose minimum distance is strictly larger than $d_{\min}(C^*)$.

For the construction of nonadditive quantum codes we can start with a classical additive code C_0^* that contains its symplectic dual, i.e., $C_0 \subseteq C_0^*$. Then we combine some of its cosets in order to obtain a larger code, which should of course still have a good minimum distance. The resulting code will be the classical union normalizer code as given in Definition 10.2.

10.3.2 Codeword stabilized codes

Next we describe the framework of codeword stabilized codes (CWS) [CSS+09], which is an alternative way to derive the union stabilizer codes of the previous section. Here we will only consider CWS codes in standard form, since any CWS code is locally equivalent to a code in standard form. The starting point of the construction is a so-called graph state, which is a special case of a graph code (see [SW02a]). Since any graph code is equivalent to a stabilizer code and vice versa [GKR02, S02a], the graph state corresponds to a stabilizer code with parameters $[[n,0,d]]$. The stabilizer of the code is given by an undirected simple graph $\mathscr{G} = (V, E)$ with n vertices V and edges E. For any vertex $v \in V$ we define a generator g_v of the stabilizer by

$$g_v = X_v \prod_{(v,w)\in E} Z_w,$$

where X_v and Z_w denote operators X and Z acting on position v and w, respectively. An example is given in Fig. 10.2a. The stabilizer is generated by

$$g_1 = X_1Z_2Z_3Z_4, \quad g_2 = X_2Z_1Z_5, \quad g_3 = X_3Z_1Z_6, \quad g_4 = X_4Z_1Z_5Z_6,$$
$$g_5 = X_5Z_2Z_4Z_7, \quad g_6 = X_6Z_3Z_4Z_7, \quad g_7 = X_7Z_5Z_6.$$

The representation of the stabilizer as a binary matrix is shown in Fig. 10.3. Note that the Z-part of this matrix is given by the adjacency matrix of the underlying graph. Furthermore, as the X-part of the generator matrix is the identity matrix, it follows that the set of translation operators $\mathscr{T}$ in the decomposition (10.3) is given by the 2^n operators that are tensor products of Z and identity (Z-only operators).

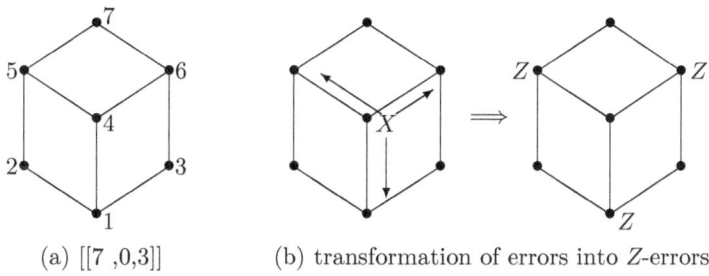

(a) [[7,0,3]]   (b) transformation of errors into Z-errors

Fig. 10.2. Graph corresponding to a stabilizer state $[[7,0,3]]$ (a) and transformation of an X-error at position 4 into Z-errors at the neighbors (b).

$$\left(\begin{array}{ccccccc|ccccccc} 1 & 0 & 0 & 0 & 0 & 0 & 0 & 0 & 1 & 1 & 1 & 0 & 0 & 0 \\ 0 & 1 & 0 & 0 & 0 & 0 & 0 & 1 & 0 & 0 & 0 & 1 & 0 & 0 \\ 0 & 0 & 1 & 0 & 0 & 0 & 0 & 1 & 0 & 0 & 0 & 0 & 1 & 0 \\ 0 & 0 & 0 & 1 & 0 & 0 & 0 & 1 & 0 & 0 & 0 & 1 & 1 & 0 \\ 0 & 0 & 0 & 0 & 1 & 0 & 0 & 0 & 1 & 0 & 1 & 0 & 0 & 1 \\ 0 & 0 & 0 & 0 & 0 & 1 & 0 & 0 & 0 & 1 & 1 & 0 & 0 & 1 \\ 0 & 0 & 0 & 0 & 0 & 0 & 1 & 0 & 0 & 0 & 0 & 1 & 1 & 0 \end{array}\right).$$

Fig. 10.3. Generator matrix of the stabilizer of a graph state $[[7,0,3]]$ in binary form.

Using the graphical description of the stabilizer state, we can transform any error given by an element of the Pauli group into Z-only form. This is illustrated in Fig. 10.2b for an X-error acting on position 4. The single X-error is replaced by Z-errors acting on all neighbors of vertex 4 in the graph. Hence we can describe both the translation operators and the errors in terms of Z-only operators, which in turn can be represented by binary vectors of length n. For the $[[7,0,3]]$ code given by the graph in Fig. 10.2, the single qubit errors are translated as shown in Table 10.1.

For a general graph state, first note that the generators of the stabilizer are given by

$$g_i = X_i \prod_{j=1}^{n} Z_j^{\Gamma_{ij}},$$

where Γ is the adjacency matrix of the underlying graph, i.e., $\Gamma_{ij} = 1$ if and only if $(i,j) \in E$. All errors in the same coset of the stabilizer have the same effect. Hence, a single error X_i has the same effect as

$$X_i g_i = \prod_{j=1}^{n} Z_j^{\Gamma_{ij}},$$

which is a Z-only operator.

It remains to find a classical code $\mathscr{T}_Z$ that can correct all the classical error patterns given in Table 10.1. It is not hard to verify that $\mathscr{T}_Z = \{0000000, 0110001\}$ is such a code. The resulting CWS code $\mathscr{C}$ is given by $|0_L\rangle = |\psi_\mathscr{G}\rangle$ and $|1_L\rangle = Z_2 Z_3 Z_7 |\psi_\mathscr{G}\rangle$, where $|\psi_\mathscr{G}\rangle$ denotes the graph state given by the graph in Fig. 10.2. It turns out that the code $\mathscr{C}$ is actually a stabilizer code since the classical code $\mathscr{T}_Z$ is a linear code. Moreover, $\mathscr{C}$ is equivalent to Steane's code $[[7,1,3]]$.

Table 10.1. *Translation of the one-qubit errors at the different positions into Z-only errors using the propagation rule illustrated in Fig. 10.2b*

pos.	Z	X	Y
1	1000000	0111000	1111000
2	0100000	1000100	1100100
3	0010000	1000010	1010010
4	0001000	1000110	1001110
5	0000100	0101001	0101101
6	0000010	0011001	0011011
7	0000001	0000110	0000111

10.4 Construction of nonadditive QECCs

10.4.1 Exhaustive search for good codes

Many of the new nonadditive quantum codes have been found by randomized or even exhaustive search using the framework of CWS codes or an equivalent description (see, e.g., [CCS+09, CSS+09, HTZ+08, LYG+08, YCO07, YCL+09]). For this, the problem of finding a nondegenerate code with prescribed minimum distance d is translated into the problem of finding a large or even maximum clique in the following graph associated to a stabilizer code.

Definition 10.3 (search graph). *Let $\mathscr{C}_0 = [[n, k, d_0]]$ with $d_0 \geq d$ be a stabilizer code and let $\mathscr{T}$ be a set of translation operators such that $\bigoplus_{t \in \mathscr{T}} t\mathscr{C}_0$ is a decomposition of the full space into mutually orthogonal translates of the code $\mathscr{C}_0$. The search graph $\mathscr{G}_{\mathscr{C}_0} = (\mathscr{T}, E)$ is an undirected graph with the translation operators $\mathscr{T}$ as its vertex set. There is an edge (t_t, t_2) between the vertices t_1 and t_2 if and only if the distance between the spaces $t_1\mathscr{C}_0$ and $t_2\mathscr{C}_0$ is not smaller than the prescribed minimum distance d.*

Clearly, if we find a clique $\mathscr{T}_0 \subset \mathscr{T}$ of size K in the search graph, i.e., a subset of translation operators such that there is an edge between any pair of them, then we obtain a union stabilizer code $((n, K \times 2^k, d))$ of distance d and dimension $K \times 2^k$. Unfortunately, the problem of finding a maximum clique is NP-hard [GJ79], so this approach is only feasible for codes of small length. In order to reduce the complexity, we can without loss of generality fix one of the translation operators to be identity.

For example, starting with a particular graph state $[[10, 0, 4]]$, the simplified search graph for distance $d = 3$ has 668 vertices and 142 233 edges. A maximum clique with 20 vertices (including identity) can be found using the program Cliquer [NÖ03] in about 1000 seconds (on a computer with 2.5 GHz clock speed). Starting with a stabilizer code $[[10, 1, 3]]$, the simplified search graph obtained by fixing one of the translations has only 214 vertices and 8706 edges. On the same computer, a maximum clique of size 10 is found in less than one second. Clearly, imposing more structure on the code reduces the search complexity. However, the optimal code $((10, 24, 3))$ can only be found when starting with a graph state $[[10, 0, d \geq 3]]$. It cannot be obtained when starting with a stabilizer code $[[10, 1, 3]]$.

In the following we describe some constructions of nonadditive quantum codes that directly use good classical nonlinear codes.

$$\begin{pmatrix} H_2 & 0 \\ 0 & H_1 \\ \hline G_{12} & 0 \\ \hdashline 0 & G_{21} \end{pmatrix}$$

$$\left\{ \begin{array}{c|c} t_1^{(1)} & t_1^{(2)} \\ \vdots & \vdots \\ t_1^{(1)} & t_{K_2}^{(2)} \\ \vdots & \vdots \\ t_{K_1}^{(1)} & t_1^{(2)} \\ \vdots & \vdots \\ t_{K_1}^{(1)} & t_{K_2}^{(2)} \end{array} \right\}$$

Fig. 10.4. Arrangements of the vectors associated with a CSS-like union stabilizer code.

10.4.2 CSS-like codes

The so-called CSS construction (named after Calderbank and Shor [CS96] and Steane [S99b], see also Chapter 2) is based on two linear binary codes $C_1 = [n, k_1, d_1]$ and $C_2 = [n, k_2, d_2]$ with $C_2^\perp \subset C_1$. The resulting QECC $\mathscr{C} = [[n, k_1+k_2-n, d]]$ has minimum distance $d \geq \min(d_1, d_2)$. We can generalize this construction by replacing the linear codes C_1 and C_2 by unions of their cosets, i.e.,

$$\widetilde{C}_i = \bigcup_{t^{(i)} \in \mathscr{T}_i} C_i + t^{(i)}$$

such that the minimum distance of the codes $\widetilde{C}_i$ is at least $\widetilde{d} \leq d$. Combining the coset representatives $t^{(1)} \in \mathscr{T}_1$ and $t^{(2)} \in \mathscr{T}_2$ we obtain a set of $|\mathscr{T}_1| \times |\mathscr{T}_2|$ translations $\{(t^{(1)}|t^{(2)}): t^{(1)} \in \mathscr{T}_1, t^{(2)} \in \mathscr{T}_2\}$. The resulting quantum code is a CSS-like union stabilizer code of dimension $|\mathscr{T}_1| \times |\mathscr{T}_2| \times 2^{k_1+k_2-n}$ whose minimum distance is at least $\widetilde{d}$. Similarly to Fig. 10.1, the description of the CSS-like union stabilizer code in terms of the corresponding binary vectors is given in Fig. 10.4. The generator matrices $G_1 = \left(\frac{H_2}{G_{12}}\right)$ and $G_2 = \left(\frac{H_1}{G_{21}}\right)$ of the codes C_i are chosen such that H_i is a generator matrix of the dual code $C_i^\perp$.

Both the Goethals and the Preparata codes (see below) are classical nonlinear binary codes that can be described as cosets of binary Reed–Muller codes $\mathrm{RM}(m-3, m)$, which for $m \geq 5$ contain their dual code. Hence we can derive CSS-like union stabilizer codes from them [GR08b]. The starting point is a CSS code $\mathscr{C}_0 = [[2^m, 2^m - 2\binom{m}{2} - 2m - 2, 8]]$ derived from the Reed–Muller code $\mathrm{RM}(m-3, m)$. For even $m \geq 6$, the Goethals code $\mathscr{G}(m)$ is the union of $K_\mathscr{G} = 2^{\binom{m}{2} - 2m + 2}$ cosets of $\mathrm{RM}(m-3, m)$. The minimum distance of the resulting nonadditive code of length 2^m is 8 and its dimension is $K_\mathscr{G}^2 \dim(\mathscr{C}_0) = 2^{2^m - 6m + 2}$.

Replacing the Goethals code by the Preparata code $\mathscr{P}(m)$, we have $K_\mathscr{P} = 2^{\binom{m}{2} - m + 1}$ cosets of $\mathrm{RM}(m-3, m)$. This results in a CSS-like union stabilizer code with minimum distance 6 and dimension $K_\mathscr{P}^2 \dim(\mathscr{C}_0) = 2^{2^m - 4m}$. The parameters of the codes for small values of m

are as follows:

Enlarged RM	Goethals	Preparata
$[[64, 35, 6]]$	$((64, 2^{30}, 8))$	$((64, 2^{40}, 6))$
$[[256, 210, 6]]$	$((256, 2^{210}, 8))$	$((256, 2^{224}, 6))$
$[[1024, 957, 6]]$	$((1024, 2^{966}, 8))$	$((1024, 2^{984}, 6))$
$[[2^m, 2^m - \binom{m}{2} - 2m - 2, 6]]$	$((2^m, 2^{2^m-6m+2}, 8))$	$((2^m, 2^{2^m-4m}, 6))$

Nonadditive quantum codes with similar parameters have been obtained by Ling and Solé using a CSS-like construction for $\mathbb{Z}_4$-linear codes [LS08]. Both the codes derived from Goethals codes and those derived from Preparata codes have better parameters than the codes obtained by Steane's enlargement construction from Reed–Muller codes [S99c] given in the first column. However, applying the enlargement construction to extended primitive BCH codes results in stabilizer codes with better parameters $[[2^m, 2^m-5m-2, 8]]$ and $[[2^m, 2^m-3m-2, 6]]$ (see [S99b]). In the next section we will present a family of nonadditive codes with even better parameters.

10.4.3 Enlargement construction

Steane has presented a construction that allows us to increase the dimension of a CSS code [S99b]. For this, he starts with the CSS construction applied to a binary code $C = [n, k, d]$ that contains its dual, yielding a CSS code $\mathscr{C}_0 = [[n, 2k-n, d]]$. Using a classical code $C' = [n, k' > k+1, d']$, which contains C, he obtains a quantum code $[[n, k+k'-n, \min(d, \lceil 3d'/2 \rceil)]]$.

Theorem 10.2 ([S99b]). *Let $C = [n, k, d]$ and $C' = [n, k' > k+1, d']$ be linear binary codes with $C^\perp \leq C < C'$. Then there exists an additive quantum code $\mathscr{C} = [[n, k+k'-n, \geq \min(d, 3d'/2)]]$. Given a generator matrix G of the code C and a generator matrix D of the complement of C in C', the normalizer of the code $\mathscr{C}$ is generated by*

$$\begin{pmatrix} G & 0 \\ 0 & G \\ D & AD \end{pmatrix},$$

where A is an invertible linear transformation without trivial fixed points.

The resulting code can also be considered as a union stabilizer code derived from the stabilizer code $\mathscr{C}_0$, which is a CSS code derived from the classical binary code C. With the same notation as above, a set of translations can be defined as

$$\mathscr{T}_0 = \{(iD|iAD) : i \in GF(2)^{k'-k}\}.$$

Since the set $\mathscr{T}_0$ is closed under addition, the resulting union stabilizer code is an additive stabilizer code.

The key observation for proving the lower bound on the minimum distance is that the weight of an operator $g = (g^X|g^Z)$ can be expressed in terms of the Hamming weight of the binary vectors and their sum:

$$\text{wgt}((g^X|g^Z)) = \frac{1}{2}\left(\text{wgt}(g^X) + \text{wgt}(g^Z) + \text{wgt}(g^X + g^Z)\right). \tag{10.7}$$

As $\mathcal{T}_0$ is closed under addition and the properties of A ensure that $0 \neq t^X \neq t^Z \neq 0$ for any nonzero element $(t^X|t^Z) \in \mathcal{T}_0$, the weight of all three binary vectors in (10.7) is lower bounded by d'.

10.4.4 Quantum Goethals–Preparata codes

Before describing the quantum Goethals–Preparata codes, a family of nonadditive QECCs presented in [GR08b], we discuss some properties of the nonlinear binary Goethals codes [G74] and the Preparata codes [P68]. It has been shown that variations of these codes have a simple description as $\mathbb{Z}_4$-linear codes [HKC+94], but in our context the description in terms of cosets of linear binary codes is used. For classical error-correcting codes in general, see, e.g., [MS77].

In the following, m is an even integer ($m \geq 4$) and $n = 2^{m-1} - 1$. Let α be a primitive element of the finite field $GF(2^{m-1})$. By $\mu_i(z)$ we denote the minimal polynomial of α^i over $GF(2)$, i.e., the polynomial with roots α^j for $j = i2^k$. The idempotent $\theta_i(z)$ is the unique polynomial over $GF(2)$ satisfying

$$\theta_i(\alpha^i) = 1 \quad \text{and} \quad \theta_i(\alpha^j) = 0 \text{ for } j \neq i2^k.$$

Codewords of a cyclic code can be represented by polynomials $f(z)$, and we use $(f(z); f(1))$ to denote the codeword of the extended cyclic code obtained by adding an overall parity check. Similar, we use $(f(z); f(1); g(z); g(1))$ to denote the juxtaposition of codewords of two extended cyclic codes.

Proposition 10.1 ($|u|u+v|$ construction [MS77]). *Let $C_1 = [n, k_1, d_1]$ and $C_2 = [n, k_2, d_2]$ be linear codes of equal length. The Plotkin sum of C_1 and C_2 is*

$$C = \{(u; u+v): u \in C_1, v \in C_2\}.$$

The code C obtained by this $|u|u+v|$ construction has parameters $C = [2n, k_1 + k_2, \min(2d_1, d_2)]$.

With this preparation, we are ready to define the Goethals code.

Definition 10.4 (Goethals code [G74]). *The Goethals code $\mathcal{G}(m)$ of length 2^m is the union of 2^{m-1} cosets of the linear binary code $C_\mathcal{G} = [2^m, 2^m - 4m + 2, 8]$. The code $C_\mathcal{G}$ is obtained via the $|u|u+v|$ construction applied to the extended cyclic codes $\overline{C_1}$ and $\overline{C_2}$. The cyclic code C_1 is a single-error-correcting code with generator polynomial $\mu_1(z)$, and C_2 is generated by $\mu_1(z)\mu_r(z)\mu_s(z)$ where $r = 1 + 2^{m/2-2}$ and $s = 1 + 2^{m/2-1}$. The nonzero coset representatives are given by $(z^i; 1; z^i\theta_1(z); 0)$ for $i = 1, \ldots, n-1$.*

Using the framework of [BLW83], it can be shown that the Goethals code is contained in the Preparata code [GR08b]. Similarly to Definition 10.4, we can describe the Preparata code as the union of cosets of a linear binary code $C_\mathcal{P}$ which contains the linear binary code $C_\mathcal{G}$.

Definition 10.5 (Preparata code). *The extended Preparata code $\mathcal{P}(m)$ of length 2^m is the union of 2^{m-1} cosets of the linear binary code $C_\mathcal{P} = [2^m, 2^m - 3m + 1, 6]$. The code $C_\mathcal{P}$ is obtained via the $|u|u+v|$ construction applied to the extended cyclic codes $\overline{C_1}$ and $\overline{C_3}$. The cyclic code C_1 is a single-error-correcting code with generator polynomial $\mu_1(z)$, and C_3 is generated by $\mu_1(z)\mu_s(z)$ where $s = 1 + 2^{m/2-1}$. The nonzero coset representatives are given by $(z^i; 1; z^i\theta_1(z); 0)$.*

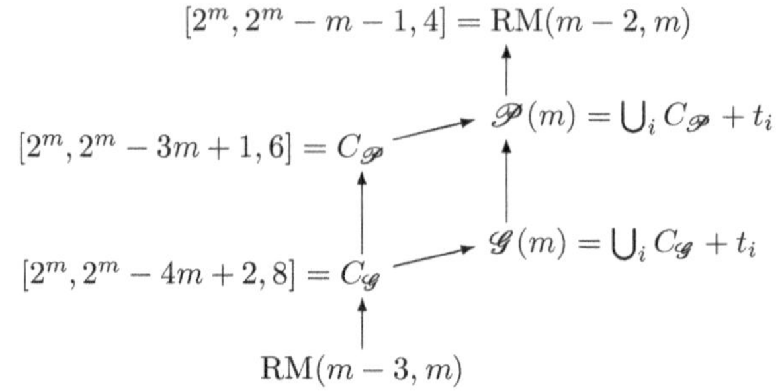

$$[2^m, 2^m - m - 1, 4] = \mathrm{RM}(m-2, m)$$

$$[2^m, 2^m - 3m + 1, 6] = C_{\mathscr{P}} \longrightarrow \mathscr{P}(m) = \bigcup_i C_{\mathscr{P}} + t_i$$

$$[2^m, 2^m - 4m + 2, 8] = C_{\mathscr{G}} \longrightarrow \mathscr{G}(m) = \bigcup_i C_{\mathscr{G}} + t_i$$

$$\mathrm{RM}(m-3, m)$$

Fig. 10.5. Nested structure of the Goethals and Preparata codes as well as the associated linear binary codes. Arrows point from the smaller to the larger of the codes.

Comparing Definitions 10.4 and 10.5 we see that we can use the very same coset representatives to construct the Goethals and the Preparata codes as unions of cosets of the linear binary codes $C_{\mathscr{G}}$ and $C_{\mathscr{P}}$, respectively. Moreover, all codes lie between codes that are equivalent to the Reed–Muller codes $\mathrm{RM}(m-3, m)$ and $\mathrm{RM}(m-2, m) = [2^m, 2^m - m - 1, 4]$ (see [H90]). This is illustrated by the diagram in Fig. 10.5. The components of the codes are summarized as follows:

C_1: cyclic code generated by $\mu_1(z)$,
C_3: cyclic code generated by $\mu_1(z)\mu_s(z)$,
C_2: cyclic code generated by $\mu_1(z)\mu_r(z)\mu_s(z)$, where $r = 1 + 2^{m/2-2}$, $s = 1 + 2^{m/2-1}$,
$C_{\mathscr{G}}$: $|u|u+v|$ construction applied to the extended cyclic codes $\overline{C_1}$ and $\overline{C_2}$,
$C_{\mathscr{P}}$: $|u|u+v|$ construction applied to the extended cyclic codes $\overline{C_1}$ and $\overline{C_3}$,
t_i: $n + 1 = 2^{m-1}$ coset representatives with

$$t_i = \begin{cases} (z^i; 1; z^i \theta_1(z); 0) & \text{for } i = 0, \ldots, n-1, \\ (0, \ldots, 0) & \text{for } i = n. \end{cases}$$

Since the code $C_{\mathscr{G}}$ contains a code that is isomorphic to the Reed–Muller code $\mathrm{RM}(m-3, m)$ it follows that $C_{\mathscr{G}}^\perp \leq C_{\mathscr{G}}$. Hence, we can apply Steane's construction [S99b] to the chain $C_{\mathscr{G}}^\perp \leq C_{\mathscr{G}} < C_{\mathscr{P}}$ of linear binary codes and obtain an additive quantum code with parameters $\mathscr{C}_0 = [[2^m, 2^m - 7m + 3, 8]]$. In a second step we use the $K = 2^{m-1}$ coset representatives t_i of the decomposition of both the Goethals and the Preparata codes. This yields a nonadditive code with dimension $K^2 2^{2^m - 7m + 3} = 2^\ell$, where $\ell = 2^m - 5m + 1$.

Theorem 10.3 (Quantum Goethals–Preparata codes). *For even $m \geq 6$, let $\mathscr{C}_0 = [[2^m, 2^m - 7m + 3, 8]]$ be the additive quantum code obtained from the chain of linear binary codes $C_{\mathscr{G}}^\perp \leq C_{\mathscr{G}} < C_{\mathscr{P}}$ using Steane's enlargement construction. Furthermore, let $\mathscr{T}_0 = \{(t_i|t_j) \colon i, j = 0, \ldots, 2^{m-1} - 1\}$, where t_i are the coset representatives used to obtain the Goethals and Preparata codes. Then the quantum Goethals–Preparata code is a union stabilizer code given by $\mathscr{C}_0$ and $\mathscr{T}_0$. The minimum distance of the quantum Goethals–Preparata code is eight.*

Proof. Let G denote a generator matrix of the code $C_{\mathscr{G}}$ and let D be such that $\binom{G}{D}$ generates $C_{\mathscr{P}}$. The structure of the nonadditive union-normalizer code of the quantum Goethals–Preparata codes is illustrated in Fig. 10.6. A generator matrix of the normalizer of the additive quantum code $\mathscr{C}_0$

$$\begin{pmatrix} G & 0 \\ 0 & G \\ D & AD \end{pmatrix}$$

$$\left.\begin{matrix} t_1 & t_1 \\ \vdots & \vdots \\ t_1 & t_K \\ \vdots & \vdots \\ t_K & t_1 \\ \vdots & \vdots \\ t_K & t_K \end{matrix}\right\}$$

Fig. 10.6. Structure of the nonadditive union normalizer code of the quantum Goethals–Preparata codes.

is given above the horizontal line, while the set of translations is listed below the horizontal line. Every codeword of the nonadditive union normalizer code is of the form

$$g = (g^X|g^Z) = (c_1 + v + t_i|c_2 + w + t_j),$$

where $c_1, c_2 \in C_{\mathcal{G}} = [2^m, 2^m - 4m + 2, 8]$ and $v, w \in C_{\mathcal{P}}/C_{\mathcal{G}}$. For $g, g' \in C^*, g \neq g'$ we compute

$$\begin{aligned}\text{dist}(g, g') &= \text{dist}((c_1 + v + t_i|c_2 + w + t_j), (c_1' + v' + t_i'|c_2' + w' + t_j')) \\ &= \text{wgt}((c_1'' + v'' + t_i - t_i'|c_2'' + w'' + t_j - t_j')),\end{aligned}$$

where $c_1'' = c_1 - c_1'$ and $c_1'' = c_1 - c_1'$ are codewords of $C_{\mathcal{G}}$, and $v'' = v - v'$, $w'' = w - w'$ are codewords of $C_{\mathcal{P}}/C_{\mathcal{G}}$. In general, the weight of $g = (g^X|g^Z)$ is given by

$$\text{wgt}((g^X|g^Z)) = \frac{1}{2}(\text{wgt}(g^X) + \text{wgt}(g^Z) + \text{wgt}(g^X + g^Z)).$$

Hence we get

$$\text{dist}(g, g') = \frac{1}{2} \text{wgt}(c_1'' + v'' + t_i - t_i') \quad (10.8\text{a})$$

$$+ \frac{1}{2} \text{wgt}(c_2'' + w'' + t_j - t_j') \quad (10.8\text{b})$$

$$+ \frac{1}{2} \text{wgt}(c_1'' + c_2'' + v'' + w'' + t_i - t_i' + t_j - t_j'). \quad (10.8\text{c})$$

By Steane's construction the vectors v'' and w'' are either both zero, or both nonzero and different. For $v'' = w'' = 0$, we can assume without loss of generality that the vectors in (10.8a) and (10.8b) are both nonzero. The weight of these vectors equals the distance between two codewords of the Goethals code, so it is at least 8. For $v'' \neq 0 \neq w''$ the terms (10.8a) and (10.8b) equal the distance of two codewords of the Preparata code, so they are lower bounded by 6. We will show that for $v'' \neq w''$, the vector in (10.8c) is a nonzero codeword of the linear code isomorphic to the Reed–Muller code $\text{RM}(m - 2, m)$, hence its weight is at least 4. For this, consider the vectors $a = (a_1; a_2) = c_1'' + c_2'' + v'' + w'' \neq 0$ and $b = (b_1; b_2) = t_i - t_i' + t_j - t_j'$. The coset representatives are of the form $t_i = (z^i; 1; z^i \theta_1(z); 0)$, so the second half b_2 of b is a codeword of the extended cyclic code generated by $\theta_1(z)$, while a_2 is a codeword of the extended cyclic code generated by $\mu_1(z)$. The intersection of the two codes is trivial, so $a_2 = b_2$ only if $a_2 = b_2 = 0$. Then $\mathrm{wgt}(b) \leq 4$ while $\mathrm{wgt}(a) \geq 6$ since $0 \neq a \in C_{\mathscr{P}}$. Hence $a \neq b$. □

To the best of our knowledge, the best additive quantum code with the same length and minimum distance as the quantum Goethals–Preparata code has dimension $2^m - 5m - 2$, i.e., it encodes three fewer qubits. Codes with these parameters can, e.g., be obtained by applying Steane's construction to extended primitive $[2^m, 2^m - 2m - 1, 8]$ and $[2^m, 2^m - 3m - 1, 6]$ BCH codes (see [S99b]). In the following table we give the parameters of the first codes in these families. Additionally, we give the parameters of the CSS-like nonadditive quantum codes derived from Goethals codes in the previous section.

Goethals	Enlarged BCH	Goethals–Preparata
$((64, 2^{30}, 8))$	$[[64, 32, 8]]$	$((64, 2^{35}, 8))$
$((256, 2^{210}, 8))$	$[[256, 214, 8]]$	$((256, 2^{217}, 8))$
$((1024, 2^{966}, 8))$	$[[1024, 972, 8]]$	$((1024, 2^{975}, 8))$
$((2^m, 2^{2^m - 6m + 2}, 8))$	$[[2^m, 2^m - 5m - 2, 8]]$	$((2^m, 2^{2^m - 5m + 1}, 8))$

10.5 Quantum circuits

10.5.1 Encoding circuits

Quantum circuits for encoding a union stabilizer code can be decomposed into two parts, a purely classical quantum circuit Q_c related to the translation operators, and an encoding circuit Q_S for the underlying stabilizer code $\mathscr{C}_0$.

In [GRB03] it was shown how to compute a polynomial-size quantum circuit consisting of Clifford gates only that transforms any stabilizer $\mathscr{S}$ given by the binary $(n - k) \times 2n$ matrix $(X|Z)$ into the stabilizer of a trivial code given by $(0|I0)$, where I is an identity matrix of size $n - k$. The corresponding trivial code corresponds to the mapping $|\varphi\rangle \mapsto |0 \ldots 0\rangle|\varphi\rangle$. We denote the resulting quantum circuit that corresponds to the inverse encoding circuit of $(X|Z)$ by Q_S. Note that we can apply Q_S to all the operators defining the code as illustrated in Fig. 10.7. Further note that, for the trivial stabilizer code, the logical X- and Z-operators are weight-one Pauli operators X and Z, respectively, acting on the last k qubits. Since the transformed translations $\mathscr{T}_0^{Q_S}$ define cosets of the normalizer, we may choose them such that they are tensor products of operators X and identity acting on the first $n - k$ qubits only. Then we have the trivial union code spanned by a set of $K2^k$ basis states of the form $|c_j\rangle|i\rangle$, where $\{|i\rangle : i = 0, \ldots, 2^k - 1\}$ is the computational basis of k qubits and $\{c_j : j = 0, \ldots, K - 1\}$ is a set of bit strings of length $n - k$. In order to obtain a standard basis for our input space of dimension $K2^k$, we need a quantum circuit Q_c mapping $|c_j\rangle \mapsto |j\rangle$ for $j = 0, \ldots, K - 1$. Note that this is a purely classical circuit that can be realized, e.g., using X, CNOT gates, and Toffoli gates. The encoding circuits presented in [CSS+09] follow the same principle, i.e., they use an encoding circuit for a graph state and a classical circuit.

$$
\left(\begin{array}{c|c} X & Z \\ \hline \multicolumn{2}{c}{\overline{Z}} \\ \hdashline \multicolumn{2}{c}{\overline{X}} \end{array}\right)
\left\{ \mathscr{T}_0 \right\}
\xrightarrow{Q_S}
\left(\begin{array}{c|c} 00 & I0 \\ \hline 00 & 0I \\ \hdashline 0I & 00 \end{array}\right)
\left\{ \mathscr{T}_0^{Q_S} \right\}
\xrightarrow{Q_c}
\left(\begin{array}{c|c} 00 & I0 \\ \hline 00 & 0I \\ \hdashline 0I & 00 \end{array}\right)
\left\{\begin{array}{c|c} c_1 0 & 00 \\ \vdots & \vdots \\ c_K 0 & 00 \end{array}\right\}
$$

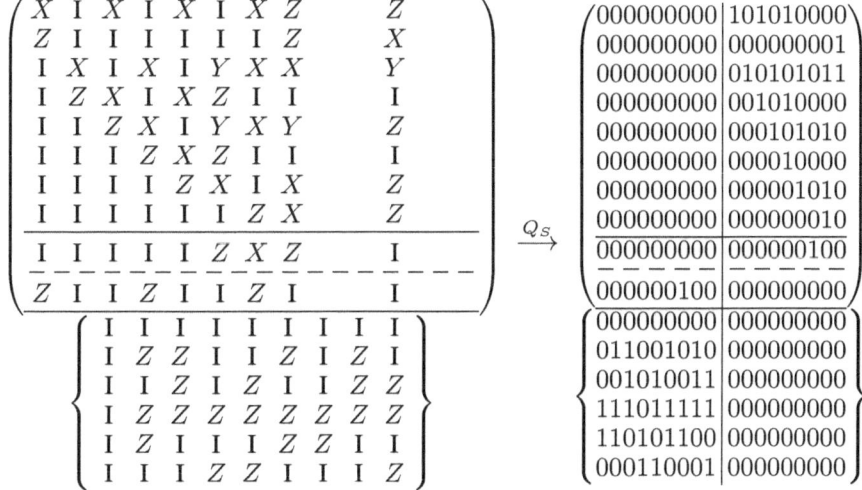

Fig. 10.7. Transformation of the union stabilizer code given by the inverse encoding circuits Q_S and Q_c.

$$
\left(\begin{array}{ccccccccc}
X & \mathrm{I} & X & \mathrm{I} & X & \mathrm{I} & X & Z & Z \\
Z & \mathrm{I} & \mathrm{I} & \mathrm{I} & \mathrm{I} & \mathrm{I} & \mathrm{I} & Z & X \\
\mathrm{I} & X & \mathrm{I} & X & \mathrm{I} & Y & X & X & Y \\
\mathrm{I} & Z & X & \mathrm{I} & X & Z & \mathrm{I} & \mathrm{I} & \mathrm{I} \\
\mathrm{I} & \mathrm{I} & Z & X & \mathrm{I} & Y & X & Y & Z \\
\mathrm{I} & \mathrm{I} & \mathrm{I} & Z & X & Z & \mathrm{I} & \mathrm{I} & \mathrm{I} \\
\mathrm{I} & \mathrm{I} & \mathrm{I} & \mathrm{I} & Z & X & \mathrm{I} & X & Z \\
\mathrm{I} & \mathrm{I} & \mathrm{I} & \mathrm{I} & \mathrm{I} & \mathrm{I} & Z & X & Z \\
\hline
\mathrm{I} & \mathrm{I} & \mathrm{I} & \mathrm{I} & \mathrm{I} & Z & X & Z & \mathrm{I} \\
\hdashline
Z & \mathrm{I} & \mathrm{I} & Z & \mathrm{I} & \mathrm{I} & Z & \mathrm{I} & \mathrm{I}
\end{array}\right)
\left\{\begin{array}{ccccccccc}
\mathrm{I} & \mathrm{I} & \mathrm{I} & \mathrm{I} & \mathrm{I} & \mathrm{I} & \mathrm{I} & \mathrm{I} & \mathrm{I} \\
\mathrm{I} & Z & Z & \mathrm{I} & \mathrm{I} & Z & \mathrm{I} & Z & \mathrm{I} \\
\mathrm{I} & \mathrm{I} & Z & \mathrm{I} & Z & \mathrm{I} & \mathrm{I} & Z & Z \\
\mathrm{I} & Z & Z & Z & Z & Z & Z & Z & Z \\
\mathrm{I} & Z & \mathrm{I} & \mathrm{I} & \mathrm{I} & Z & Z & \mathrm{I} & \mathrm{I} \\
\mathrm{I} & \mathrm{I} & \mathrm{I} & Z & Z & \mathrm{I} & \mathrm{I} & \mathrm{I} & Z
\end{array}\right\}
\xrightarrow{Q_S}
\left(\begin{array}{c|c}
000000000 & 101010000 \\
000000000 & 000000001 \\
000000000 & 010101011 \\
000000000 & 001010000 \\
000000000 & 000101010 \\
000000000 & 000010000 \\
000000000 & 000001010 \\
000000000 & 000000010 \\
\hline
000000000 & 000000100 \\
\hdashline
000000100 & 000000000
\end{array}\right)
\left\{\begin{array}{c|c}
000000000 & 000000000 \\
011001010 & 000000000 \\
001010011 & 000000000 \\
111011111 & 000000000 \\
110101100 & 000000000 \\
000110001 & 000000000
\end{array}\right\}
$$

Fig. 10.8. A nonadditive code $((9, 6\times 2^1, 3))$ as union stabilizer code (left) and the corresponding trivial code obtained by the inverse encoding circuit Q_S (right).

We illustrate the encoding circuit for the nonadditive $((9, 12, 3))$ code. In [YCL+09] such a code is derived as a CWS code using a graph state $\mathscr{C}_0' = [[9,0,3]]$ corresponding to a circle graph with nine vertices. Furthermore, an encoding circuit for the code is presented that was obtained using a different technique than what follows.

It turns out that we can construct a nonadditive $((9, 6 \times 2^1, 3))$ code starting with a [[9,1,3]] stabilizer code. Generators for the stabilizer, the logical Z- and X-operators, and the six translation operators are shown in the left part of Fig. 10.8.

The essentially quantum part of the encoding circuit can be taken from the encoding circuit Q_S for this stabilizer code. All that remains is to find a (classical) quantum circuit Q_c that maps the six binary strings

$$
\begin{array}{ccc}
000000000 & 011001010 & 001010011 \\
111011011 & 110101000 & 000110001
\end{array}
\tag{10.9}
$$

obtained from the transformed translation operators (see Fig. 10.8) to the binary representations of $j = 0, \ldots, 5$. Note that we have set position 7 to zero, as this position is the input for the

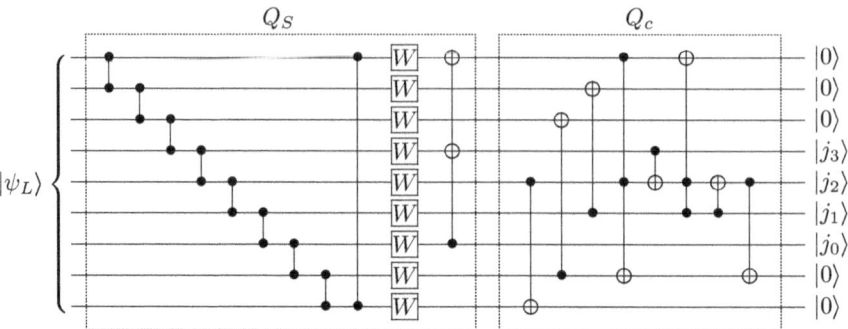

Fig. 10.9. Inverse encoding circuit for the union stabilizer code $((9, 6 \times 2^1, 3))$. The first block shows the inverse encoding circuit Q_S for the stabilizer code $\mathscr{C}_0 = [[9, 1, 3]]$. The second block is a purely classical circuit mapping the binary strings of (10.9) to $000j_3j_2j_1j_000$ for $j = 0, \ldots, 5$, where $i = j_3 2^2 + j_2 2^1 + j_1 2^0$. The least significant bit j_0 serves as an input to the stabilizer code.

encoding circuit of the stabilizer code. So effectively we search for a quantum circuit acting on eight qubits. There are several degrees of freedom that we can use when searching for such a circuit. First, we can map the number $j = 0, \ldots, 5$ in any order to the strings in (10.9). Furthermore, the binary representation of j can be at any of the eight positions $1, \ldots, 6, 8, 9$. Finally, the mapping is only defined for six of the $2^8 = 256$ strings – the remaining strings can be transformed arbitrarily. Using a breadth-first search among all circuits composed of CNOT and Toffoli gates, we found the realization shown together with the circuit Q_S in Fig. 10.9.

10.5.2 Syndrome computation

The computation of an error syndrome for a union stabilizer code is somewhat more involved than the construction of an encoding circuit. Nonetheless, we can again split the quantum circuit into one part related to the stabilizer code and a purely classical part.

The first step is to compute the error syndrome of the underlying stabilizer code $\mathscr{C}_0 = [[n, k, d]]$ using $n - k$ ancillae, but without measuring them. Since the union stabilizer code is the span of several translations of the code $\mathscr{C}_0$, this error syndrome is not constant. For the eight generators of the stabilizer of the code $\mathscr{C}_0 = [[9, 1, 3]]$ given in Fig. 10.8, we get the following six different error syndromes when using the encoding circuit shown in Fig. 10.9:

$$S_0 = \{11010100, 10111001, 11001101, 10101010, 01000111, 00001000\}.$$

Note that this set does not contain the all-zero syndrome 00000000. This is because we have ignored phase factors when deriving the encoding circuit. Similarly, we get the syndrome sets S_i, $i = 1, \ldots, 27$, for the 27 different one-qubit errors. Each set contains six different syndrome values, and all 28×6 values are different. The second step is to compute a function $f : \{0,1\}^8 \to \{0, \ldots, 27\}$ such that $f(x) = i$ for all $x \in S_i$, where the numbering i of the sets is arbitrary. This is a purely classical task, yet the function is already quite complicated for our example. Note that this function f is related to the Boolean function used in [AC08] to construct a projection

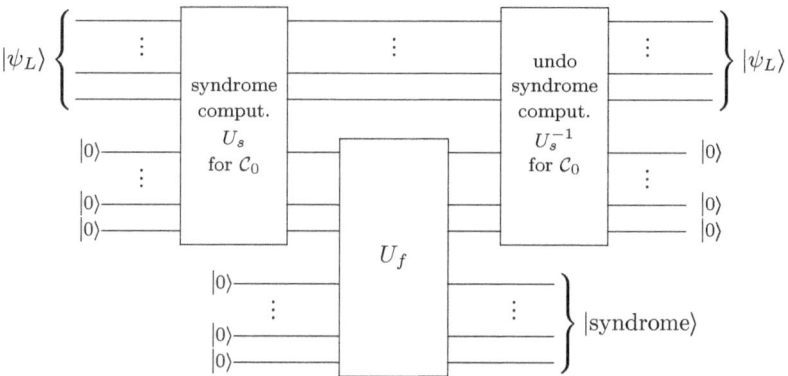

Fig. 10.10. Outline of a quantum circuit computing an error syndrome for a union stabilizer code. The syndrome of the underlying stabilizer code $\mathcal{C}_0 = [[n, k, d_0]]$ is computed using U_S. The transformation U_f maps $|x\rangle|0\rangle$ to $|x\rangle|f(x)\rangle$, where f is a classical function that separates the different syndrome sets.

operator onto the code space. Here we need a function that additionally describes projections onto the error spaces. The final step is to uncompute the syndrome of the code $\mathcal{C}_0$. The overall process is illustrated in Fig. 10.10.

Finally, we sketch a two-step approach to error correction for CWS codes given in [LDP10, LDG+10]. The main idea is based on the observation that the errors acting on a fixed set $\mathcal{I}$ of t positions form a group. Hence the original CWS code $\mathcal{C}$ together with the translated spaces resulting from the errors acting on the positions in $\mathcal{I}$ correspond to a union stabilizer code $\mathcal{C}_\mathcal{I}$ with higher dimension. Then a measurement that either projects onto $\mathcal{C}_\mathcal{I}$ or its orthogonal complement can be used to decide whether the actual error acts only on the positions in $\mathcal{I}$, provided that $2t < d$. After the correct set $\mathcal{I}$ has been found, the error can be determined using $2t$ additional measurements. Correctable errors of weight $t > d/2$ have to be treated separately.

10.6 Conclusions

The concept of union stabilizer codes and codeword stabilized (CWS) codes provides a framework that allows the construction of nonadditive quantum codes based on classical nonlinear codes. For small length, these codes can be found using the search graph, but there are also families of nonadditive quantum codes that can be derived from known families of classical nonlinear codes. While it is an open question whether the nonadditive quantum Goethals–Preparata codes presented here have better parameters than any additive quantum codes, some nonadditive codes that have a higher dimension than any stabilizer code of the same length and minimum distance have been constructed in [GSS+09].

Both encoding circuits and quantum circuits for the computation of an error syndrome for union stabilizer codes can be decomposed into a part derived from the underlying stabilizer code and a purely classical part. While the first part has an efficient polynomial-size realization, little can be said about the complexity of the classical functions. It can be expected that the classical functions become more simple the more structure the code has. The extremal case is when the functions are all linear, which is for example the case for stabilizer codes.

Aspects of fault tolerance, such as fault-tolerant realizations of encoding and decoding networks for these codes as well as implementations of logical operations on the codes, have to be addressed in future research.

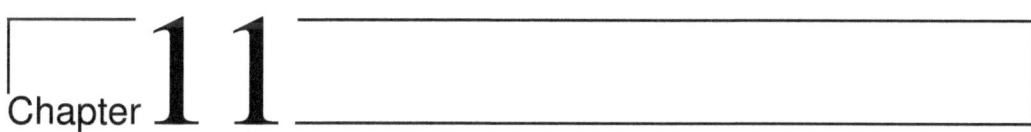

Chapter 11

Iterative quantum coding systems

David Poulin

11.1 Introduction

This chapter concerns quantum coding schemes based on iterative decoding. So far, most of the focus of this book has been on quantum error-correcting codes (QECCs) that use a relatively small number of qubits, and concatenation thereof. The decoding strategy adopted for these codes consists of identifying the lowest-weight error that is consistent with the syndrome and reversing its effect. The lowest-weight error associated with each syndrome can, in the worst case, be found by exhaustive search and stored in a look-up table for future reference. Each time a syndrome is measured, the appropriate recovery can be rapidly extracted from the table to complete the error-correction cycle.

In contrast, the codes described in this chapter use a very large number of qubits; the larger the better! For a fixed rate $r = k/n$, the performance of the codes improves as n increases. For such large codes however, more clever decoding schemes are required. For a code with parameters $[[n, k, d]]$, the size of the look-up table – i.e. the number of distinct syndromes – is $2^{n-k} = 2^{(1-r)n}$. Thus, look-up tables are only suitable for codes of a few dozen qubits. For larger codes, the decoding process involves a highly nontrivial computation that needs to be executed "on the fly," so fast and reliable algorithms are needed. Iterative decoding is one such algorithm and it has produced some of the best coding schemes – in particular turbo-codes and sparse codes – widely used in classical wireless communication.

In this chapter, we will generalize iterative coding schemes to the quantum setting, discuss the additional challenges presented in this setting, and finally present numerical results of benchmark performance tests.

11.1.1 Shannon's random coding

There is a fundamental limit to the amount of information that can be transmitted in a noisy channel. A memoryless channel $\mathcal{E}$ acts independently on each input, so it is modeled by a stochastic

---

Quantum Error Correction, ed. Daniel A. Lidar and Todd A. Brun. Published by Cambridge University Press.
© Cambridge University Press 2013.

matrix $\Pr(B|A)$, where A and B represent the channel's input and output, respectively. Given a probability distribution of the input $\Pr(A)$, the joint probability is $\Pr(A, B) = \Pr(B|A)\Pr(A)$. The maximum amount of information that can be transmitted in the channel, called the Shannon capacity, is

$$C = \max_{\Pr(A)} I(A, \mathscr{E}), \tag{11.1}$$

where, as we saw in Eq. (1.47), the mutual information can be expressed in terms of Shannon's entropy $I(A, \mathscr{E}) = I(A : B) = H(A) + H(B) - H(A, B)$. For the binary symmetric channel – the channel that maps the bit b to itself with probability $1 - p$ and to $\bar{b}$ with probability p – the input distribution that achieves this maximum is the uniform one and we obtain $C = 1 - h(p)$.

A random linear classical error-correcting code (CECC) encoding k bits into n bits can be constructed by choosing k independent codewords at random from the set of all 2^n bit strings of length n and setting the code to be their linear span. For large values of n, such a random linear code will typically have a minimum distance d that saturates the Gilbert–Varshamov bound:

$$\frac{k}{n} \geq 1 - h\left(\frac{d}{n}\right), \tag{11.2}$$

where $h(x) = -x \log x - (1 - x) \log(1 - x)$ is Shannon's binary entropy, a special case of Eq. (1.39). If we use such a random code with a bounded distance decoder – i.e., a decoder that finds the unique codeword that is within a distance $d/2$ of the received message and declares forfeit otherwise – then we will successfully correct all errors that affect fewer than $d/2$ bits. For the binary symmetric channel, the typical weight of errors is np. Thus, a random code used with a bounded distance decoder can transmit information with arbitrarily small failure probability on the binary symmetric channel with any rate $\frac{k}{n} < 1 - h(2p)$. This is much less than the channel's capacity. Indeed, the factor of 2 means that this coding scheme can tolerate only half as much noise as what is predicted by Shannon's upper bound.

To obtain a better performance, we must abandon the idea of a bounded distance decoder and focus only on correcting typical errors. On the binary symmetric channel, there are $2^{nh(p)}$ typical errors. A random $[n, k]$ code assigns a random $n - k$ bit syndrome to each of these errors. The probability that two typical errors be assigned the same syndrome (in which case they could not be corrected) is thus upper bounded by $2^{nh(p)} 2^{k-n}$. This failure probability vanishes in the limit of large n provided that $\frac{k}{n} \leq 1 - h(p)$, which is known as the Hamming bound and coincides with the channel's capacity. Thus, random linear codes asymptotically achieve the Shannon limit, so they are optimal in that sense [S48]. However, this cannot be achieved with a bounded distance decoder.

11.1.2 Quantum random coding

The situation in quantum mechanics is more complicated. A natural generalization of the mutual information $I(A, \mathscr{E})$ is the coherent information defined $I^c(A, \mathscr{E}) = I^c((\mathscr{E} \otimes \mathscr{I})(|\phi_{AB}\rangle\langle\phi_{AB}|))$ where $|\phi_{AB}\rangle$ is a purification of the input state ρ_A, the channel $\mathscr{E}$ acts on the subsystem A, the trivial channel $\mathscr{I}$ acts on subsystem B, and $I^c(\rho_{AB}) = S(\rho_A) - S(\rho_{AB})$. Thus, the quantum analog of Eq. (11.1) for the capacity of a memoryless channel is obtained by maximizing the coherent information over all input states:

$$Q_1 = \max_{\rho_A} I^c(A, \mathscr{E}). \tag{11.3}$$

The depolarizing channel – which acts by leaving the state of the qubit untouched with probability $1 - p$ and otherwise applies one of the three Pauli matrices at random – is a natural generalization of the binary symmetric channel. Analogously to the classical case, the coherent information of this channel is maximized with a totally mixed input state, where it takes the value $1 - h(p) - p \log 3$.

The analog of an $[n, k]$ random linear code is an $[[n, k]]$ random stabilizer code: a code whose $n - k$ stabilizer generators are selected at random from the n-qubit Pauli group, with the constraint that they form an Abelian group not containing -1. In terms of minimum distance d, random codes saturate in the limit of large n the quantum Gilbert–Varshamov bound:

$$\frac{k}{n} \geq 1 - \frac{d}{n}\log 3 - h\left(\frac{d}{n}\right). \tag{11.4}$$

As in the classical case, a bounded distance decoder will fail to achieve the capacity Q_1. Instead, we must focus only on correcting typical errors. The same reasoning as above leads to the quantum Hamming bound:

$$\frac{k}{n} \leq 1 - h(p) - p\log 3. \tag{11.5}$$

Thus, a random code of rate slightly below Q_1 will correct all typical errors, and so asymptotically attain the capacity Q_1. It is important to note that this bound is derived assuming that each typical error is assigned a distinct syndrome, i.e., that the code is nondegenerate. This assumption is justified because the typical weight of a random element of the Pauli group is $3n/4$, which is well above the weight of typical errors that can be corrected by any code. Hence, random stabilizer codes are nondegenerate.

However, Q_1 is *not* the true capacity of a quantum channel. This is due to the possibility that a code can correct more errors that it can identify: errors can be degenerate. The true capacity is instead obtained by regularizing Eq. (11.3) by taking several copies of the channel

$$Q = \lim_{N \to \infty} \frac{1}{N} \max_{\rho_{A_N}} I^c(A_N, \mathscr{E}^{\otimes N}), \tag{11.6}$$

where ρ_{A_N} can be an entangled state of the N input systems [L97, S02b, D05]. Only in a few special cases can this quantity be computed. The capacity of the depolarizing channel for instance is currently unknown.

We learn that random quantum codes are interesting because they achieve the Hamming rate Q_1 in the limit of large n. But contrary to the classical case, they do not achieve the true channel capacity, which requires the use of degenerate codes.

11.1.3 Decoding problem

Why are not random codes, used in practice? The answer has to do with the decoding problem. Although random codes are nearly optimal from the code design perspective, there exists no efficient algorithm to decode them. Moreover, it is very unlikely that an efficient algorithm will ever be found because the classical decoding problem, i.e., finding the most likely error given the syndrome, is NP-complete [BMT78]. The complexity of the decoding problem for quantum codes is not known, but it is at least as hard as in the classical case (CECCs being a subset of QECCs).[1]

Thus, a coding scheme must be judged not only by the quality of its constituent code (e.g., minimum distance, weight enumerator) but also by the quality of the companion decoding algorithm. Of course, the decoding algorithm must be sufficiently fast for the targeted application. In a communication setting, the decoder introduces communication delays that can be problematic if they are too long. When the code is used in an information processor such as a quantum computer, the decoder needs to be sufficiently fast to prevent errors from accumulating.

Speed is not the only important quality of a decoder. In some cases, the decoder can be a heuristic algorithm, providing an approximate solution to the statistical inference problem. In other words, its output may be different from the actual most likely error. This is not necessarily a big problem as long as it doesn't happen too often. In the limit of large code lengths, all that really matters is that the syndromes associated with typical errors be correctly decoded. A few exceptions will not affect the average performance of the coding scheme.

Thus, the performance of a coding scheme on a particular channel should be judged by the speed of its decoder and the probability that it fails to correct the error, both of which depend on the particular channel. Note that there are two types of failures. On the one hand, the channel can generate errors that are beyond the error-correcting capabilities of the code. These are errors in an information-theoretical sense since the information is simply not present at the output of the channel. On the other hand, there can be errors that could *in principle* be corrected by the code, but that are misdiagnosed by the decoder. These are computational errors. Ultimately, it is the limited time available to the decoder that is responsible for those failures.

We will say that a coding scheme is "good" for a certain channel when it has a decoder that runs in a time polynomial in n and a failure probability that goes to 0 in the limit of large n. In this chapter, when the channel is not specified, it is implicitly assumed that it is the binary symmetric channel in the classical case or the depolarizing channel in the quantum case. A good coding scheme can then be characterized by a critical channel error probability p^* below which the failure probability goes to zero as n increases.

11.1.4 Two schools: Algebraic versus probabilistic codes

There are essentially two very broad approaches that have been pursued to design good coding schemes: algebraic and probabilistic coding (the latter terminology is used by some authors, e.g., [FC07], but it is not standard).

Algebraic codes make use of regular structures from finite fields or groups to design codes that have large minimum distance and can be efficiently decoded. These include Hamming, Golay, Reed–Muller, BCH, and Reed–Solomon codes, to name a few. The regularity in the algebraic structure plays a double role. On the one hand it guarantees a minimum distance between codewords. On the other hand, it greatly simplifies the decoding problem that can be solved in polynomial time in contrast to the exponential time required by random codes. Being completely regular, algebraic codes are in some sense at the opposite end of the code spectrum from Shannon's optimal random code construction. They come in a wide variety, with different lengths, minimum distances and rates, and offer unequaled performance on certain types of channels.

---

[1] For nondegenerate codes, such as random codes on the depolarizing channel, the quantum and classical decoding problems are equivalent.

A large number of algebraic codes have been transposed to the quantum setting; this topic is covered in Chapter 12.

The codes we will consider in this chapter, sometimes referred to as "probabilistic codes," are of a very different nature. Much in the spirit of Shannon, there is an element of randomness in their construction. The main focus is not the minimum distance of the code but rather the average performance of a code ensemble on a given channel. The codes are not completely random; there is a minimal amount of structure imposed on a code ensemble. Nonetheless, these regularities are insufficient to enable an efficient minimum distance decoder, so the decoding problem remains NP-hard.

Although this may seem like a major drawback at first sight, the situation is not as bad as it appears. The reason is that there exist good heuristic algorithms to tackle the decoding problem. These are iterative procedures that exploit the little structure available in the code ensemble. The iterative decoders would not succeed on truly random codes, but when the ensembles are carefully chosen their performance is remarkable. In fact, probabilistic coding schemes used with an iterative decoding algorithm often come near to achieving capacity on a wide variety of channels.

Although they appear different at first sight, the iterative algorithms used to decode these codes all have a common foundation. In fact, they belong to a very broad family of algorithms known as belief propagation, with application in all fields of science (often under different names) ranging from coding theory to image recognition and from statistical physics to medical diagnostics. Belief propagation is a general heuristic method to approximate the solution of complex statistical inference problems, see e.g., [AM00, Y01, N04a]. Belief propagation has been generalized to the quantum setting [LP08], leading to decoding algorithms for quantum probabilistic codes.

To close this introduction, let us use a (somewhat imperfect) analogy to explain how probabilistic codes can sometime outperform algebraic codes. Suppose we are faced with the problem of packing as many billiard balls as possible into a rectangular box of dimension $l_1 \times l_2 \times l_3$. A mathematically inclined person might attempt to arrange the balls in a regular structure. If she chooses for instance a cubic lattice, she will manage to insert $\lfloor l_1/d \rfloor \times \lfloor l_2/d \rfloor \times \lfloor l_3/d \rfloor$ balls, where d is the diameter of a ball. This is clearly not the optimal configuration, and other regular structures can be examined.

Having exhausted all the regular structures that she could think of, our mathematician wants to test her solution. She finds a shoe box and some tennis balls of the appropriate dimension and places the balls according to her solution. It fits!

Later that evening, the custodian is cleaning the mathematician's office and finds a bunch of tennis balls along with the shoe box containing the balls placed according to the optimal regular arrangement. He shakes the shoe box and manages to squeeze in the remaining tennis balls.

What are the lessons learned from this analogy? Of course, the problem it describes is equivalent to optimizing the rate of a code (the number of codewords that can fit in the space $\{0,1\}^n$) for a fixed minimum distance d (the codewords have a radius $d/2$ and cannot overlap). First, we learn that randomness can help to attain (or at least get closer to) the optimal solution. In this sense, random codes sometimes offer superior rates compared with algebraic codes of the same minimum distance. Second, tennis balls, as opposed to billiard balls, are compressible. Thus, by cheating a little bit, i.e., by slightly squeezing some of the tennis balls, the custodian managed to insert many more balls into the box. Similarly, random codes do not necessarily obey a strict minimum distance and there may be a few atypical low-weight codewords. But as long as there are not too many of them, these codewords will not directly affect the average performance of the code in the limit of large block size.

11.2 Decoding

Before describing the families of codes that we will be studying in this chapter, we will first rigorously state the decoding (or error diagnosis) problem. To simplify the presentation, we will start with the decoding of classical codes where everything is quite simple and intuitive. Then, we will consider the decoding problem for quantum codes on Pauli channels. This decoding problem is quite similar to the classical one, but the possibility of degenerate errors adds a level of complication. Finally, we will state the decoding problem for general quantum channels.

11.2.1 Decoding a linear classical code

Let G be the $k \times n$ generating matrix of a binary linear classical code. The k-bit information word L is encoded as $A = LG$. The channel $\mathcal{E}$ is modeled by a stochastic matrix $P(B|A)$ representing the probability that the input A produces the output B. Then, given the channel output B, maximum-likelihood decoding consists in identifying the most likely input A that could have produced this output. To compute this value, we need to know the distribution $P(A)$ of inputs. Then, using Bayes' rule, we can compute the input probability given the output

$$\Pr(A|B) = \frac{\Pr(A,B)}{\Pr(B)} = \frac{\Pr(B|A)\Pr(A)}{\Pr(B)}, \tag{11.7}$$

where the marginal probability on the output is defined the usual way: $\Pr(B) = \sum_A \Pr(A,B) = \sum_A \Pr(B|A)\Pr(A)$. Thus, the maximum-likelihood decoder is

$$A^*(B) = \underset{A}{\operatorname{argmax}} \Pr(A|B). \tag{11.8}$$

It is often assumed that the input is uniformly distributed (that the information has been compressed), in which case optimizing $\Pr(A|B)$ over A is the same as optimizing $\Pr(B|A)$ over A.

When the errors that occur in the channel are independent of the transmitted symbol, we say that the channel is symmetric. In that case, the stochastic matrix has the form $\Pr(B|A) = \Pr(E = A - B)$ where the error probability $\Pr(E)$ completely specifies the channel. In other words, the channel acts by adding the error E to the input word, and E is distributed according to $\Pr(E)$. For instance, the binary symmetric channel has $\Pr(0) = 1 - p$ and $\Pr(1) = p$.

In this case, maximum-likelihood decoding can be done via the error syndrome. Remember that the parity check matrix is a $(n-k) \times n$ full-rank matrix H that is orthogonal to the generator matrix $HG^T = 0$. The syndrome associated to the output $B = A + E$ is $T = HB = HE$ since $A = LG$ is a codeword and hence annihilated by H. In that case, $A^*(B) = A + E^*(T)$ where

$$E^*(T) = \underset{E}{\operatorname{argmax}} \Pr(E|T) \tag{11.9}$$

is the most likely error given the syndrome T. Thus, all that is needed to restore the most likely codeword A^* is to apply the inverse of the most likely error $-E^*$ to the received message B.

(Note that this can be done without knowing the value of A^* or B; only the syndrome and its associated most likely errors are required. The possibility of such an oblivious error correction is crucial for quantum error correction, where a direct measurement of the received message would collapse its information.)

The conditional error probability can be computed directly from the channel's error probability $\Pr(E)$ and the syndrome using Bayes' rule:

$$\Pr(E|T) = \frac{\Pr(T|E)\Pr(T)}{\Pr(E)} = \frac{\delta(T, HE)\Pr(E)}{\sum_{E'} \delta(T, HE')\Pr(E')}, \tag{11.10}$$

where δ denotes the Kronecker delta. In other words, $\Pr(E|T)$ is 0 when the syndrome HE associated with E differs from T, and otherwise it is proportional to $\Pr(E)$. On the binary symmetric channel, the probability of an error E is a decreasing function of its weight (for $p < \frac{1}{2}$):

$$\Pr(E) = p^{\text{weight}(E)}(1-p)^{n-\text{weight}(E)}. \tag{11.11}$$

Thus, maximum-likelihood decoding consists in that case of identifying the lowest-weight error that is consistent with the syndrome T.

The above procedure involves a maximization over the set of all possible errors, a set that grows exponentially with the input size. Unless the code is very carefully chosen, this optimization cannot be performed. A simpler decoding method consists of optimizing each output symbol individually. Instead of finding the n-symbol error E that optimizes $\Pr(E|T)$, we compute n marginal probabilities:

$$\Pr(E_i|T) = \sum_{E_1,\ldots \hat{E}_i,\ldots E_n} \Pr(E_1 E_2 \cdots E_n | T) \tag{11.12}$$

(where the hat means that the variable E_i is not summed over) and maximize each of these probabilities individually. This procedure is called physical symbol-wise (or bit-wise) maximum-likelihood decoding, and can produce results that are different from maximum-likelihood decoding. Indeed, the global maximum does not necessarily maximize the marginal probabilities. The term "physical" is used to specify that the marginal is taken over the physical symbols constituting the message. Another possibility would be to consider the marginal distribution on the symbols of the information word L. The conditional probability on the physical errors $\Pr(E|T)$ can be transformed into a conditional probability on the "logical errors" using the encoding matrix

$$\Pr(L|T) = \sum_E \Pr(E|T)\delta(LG, E). \tag{11.13}$$

The logical symbol-wise decoder is obtained as above:

$$\Pr(L_i|T) = \sum_{L_1,\ldots \hat{L}_i,\ldots L_k} \Pr(L_1 L_2 \cdots L_k | T). \tag{11.14}$$

While optimizing $\Pr(E|T)$ and $\Pr(L|T)$ are equivalent, the corresponding marginal optimizations may differ from the global optimization and from each other. In the quantum setting, all four optimizations give different results.

Whether one should optimize the probability of correcting each bit individually or the probability of correcting the entire bit string depends on the application. One advantage of bit-wise decoders is the possibility of soft decoding. Instead of outputting the symbols that maximize the conditional probability $\Pr(E_i|T)$, a soft decoder simply outputs the marginal probabilities themselves. This would not be possible with the joint probability $\Pr(E|L)$ because it has a number of parameters growing exponentially with the input size, in contrast to the linear growth of the marginals. Soft decoders play a crucial role in iterative decoding systems.

11.2.2 Connection to spin models

We have mentioned that the decoding problem is generally intractable, i.e., that it is NP-hard. In this section, we will demonstrate this by establishing various connections between the decoding problem and the problem of computing ground-state and thermal properties of an Ising spin model, and in particular spin-glasses. This is a notoriously hard problem in condensed-matter physics that is well known to be NP-complete [B82]. The connection between error correction and statistical physics goes well beyond the introduction we present here, see for instance [N01, MM08].

The main motivation behind this section is not to establish the NP-hardness of the problem, but rather to show how numerical methods developed in the context of condensed-matter physics can serve as decoding algorithms. This connection has indeed led to important developments in both of the fields that it relates.

For this section only, we use a notation closer to physics and assume that bits take values ± 1 rather that 0 and 1. Denote the k-bit information word, previously denoted L, by σ. The information word σ is encoded in a message A taking the value

$$A_i = \prod_{j:G_{ji}=1} \sigma_j \tag{11.15}$$

(this is equivalent to LG in the previous notation). This encoded message is transmitted over a binary symmetric channel that acts by "flipping" each bit with probability p and leaving bits intact otherwise. The output B is therefore distributed according to

$$\Pr(B_i|\sigma) = \begin{cases} 1-p & \text{if } B_i = A_i = \prod_{j:G_{ji}=1} \sigma_j, \\ p & \text{if } B_i = -A_i = -\prod_{j:G_{ji}=1} \sigma_j. \end{cases} \tag{11.16}$$

Defining the Nishimori inverse temperature β_p by $e^{-2\beta_p} = \frac{1-p}{p}$, we can rewrite the output probability conditioned on the input by a simple algebraic manipulation:

$$P(B_i|\sigma) = \frac{\exp\{-\beta_p B_i \prod_{j:G_{ji}=1} \sigma_j\}}{2\cosh\beta_p}. \tag{11.17}$$

For the decoding problem, we are interested instead in the probability of the input σ given the output B, which can be obtained using Bayes' rule:

$$\begin{aligned}
\Pr(\sigma|B) &= \frac{\Pr(B|\sigma)\Pr(\sigma)}{\Pr(B)} \\
&= \frac{\exp\{-\beta_p \sum_i B_i \prod_{j:G_{ji}=1} \sigma_j\}}{\sum_\sigma \exp\{-\beta_p \sum_i B_i \prod_{j:G_{ji}=1} \sigma_j\}} \\
&= \frac{e^{-\beta_p H\{\sigma\}}}{Z(\beta_p, B, G)}.
\end{aligned} \tag{11.18}$$

This expression can be recognized as the Gibbs distribution of a collection of interacting Ising spins with Hamiltonian $H\{\sigma\} = \sum_i B_i \prod_{j:G_{ji}=1} \sigma_j$, and $Z = \sum_\sigma e^{-\beta H\{\sigma\}}$ denotes the partition function. The connections between the spin model and the error-correcting code are:

- The information word $\sigma \in \{-1, +1\}^k$ labels the different spin configurations.
- Each column i of the encoding matrix G determines a coupling in the spin model. The Hamiltonian is the sum of terms $H = \sum_i h_i$ and h_i couples a number of spins equal to the weight of the ith column of G.
- The received bits B_i determine the sign, either ferromagnetic or anti-ferromagnetic, of the coupling h_i.
- The inverse Nishimori temperature β_p is fixed by the channel's bit-flip rate p.

From this connection, it is clear that the spin configuration – or information word – that is assigned the largest probability is the ground state of the system. In other words, determining the most likely information word given the received message is equivalent to the problem of finding the ground state of the associated Ising model. Letting $\langle \sigma_j \rangle_\beta = \frac{1}{Z} \sum_\sigma \sigma_j e^{-\beta H\{\sigma\}}$ denote the magnetization of spin j at inverse temperature β, we see that the maximum-likelihood decoding sets satisfy

$$\sigma_j^*(B) = \langle \sigma_j \rangle_{\beta \to \infty}. \qquad (11.19)$$

On the other hand, the magnetization at the Nishimori temperature is seen to be related to marginal probability Eq. (11.14) used by the logical bit-wise maximum-likelihood decoder:

$$\Pr(\sigma_j | B) = \frac{\langle \sigma_j \rangle_{\beta_p} + 1}{2}. \qquad (11.20)$$

All these calculations were done assuming a fixed channel output B. This output is in fact a random variable and the average performance of the code is obtained by computing each spin's magnetization averaged over the different noise patterns. In physical terms, the sign B_i of each coupling h_i is a random variable distributed according to the distribution $\{p, 1-p\}$. The introduction of this randomness in the physical system can drive a phase transition. At low error rates p and low temperatures, the system is in an ordered magnetic phase (See Fig. 11.1). As the noise rate or temperature increases, the system disorders and enters a glassy phase. This implies that the decoder is incapable of choosing the most likely input and rather gets trapped in local optima. At even greater noise rate and temperature, the system becomes paramagnetic and the information is completely lost.

What we learn from these examples is that there is a direct relation between decoding and Ising spin models. Methods to compute the magnetization at different temperatures are in fact decoding algorithms. What temperature should be used for the decoding? Ideally, the magnetization should be computed at zero temperature since this corresponds to the most likely information word. However, physics practitioners know that ground state properties are often difficult to compute, particularly in the vicinity of a phase transition, i.e., when the code is used near its error threshold. Thus, for the sake of time constraints, in practice we must often resort to working at higher temperatures.

In short, any method developed in statistical physics to estimate the magnetization of Ising spin models can serve as decoders. Algorithms that can handle lower temperatures have lower failure probabilities, but typically use a longer running time so compromises are made.

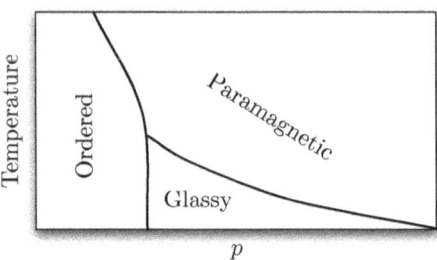

Fig. 11.1. Sketch of the phase diagram of an Ising spin model as a function of temperature (representing different choices of decoders) and the disorder (representing the channel error probability).

11.2.3 Decoding a stabilizer code on Pauli channels

With these classical preliminaries behind, let us now address the decoding problem in the quantum setting restricted to Pauli channels, i.e., when the channel's Kraus operators are elements of the Pauli group. In that case, the channel can be entirely specified by a probability $\Pr(E)$, the error probability, over the Pauli group. We will also assume that the code is a stabilizer code based on the Pauli group.

It is then convenient to choose a set of generators of the Pauli group that are tailored to the specific code. Usually, we choose the X_i and Z_i as generators of the Pauli group. This choice is particularly nice because the generators come in canonically conjugate pairs (X_i, Z_i): each generator commutes with all the other generators, with the exception of its conjugate partner with which it anticommutes. We call a set of generators with this property a canonical basis. A stabilizer code is specified by its stabilizer generators $\{S_j\}_{j=1,\ldots,n-k}$. Given these, we can choose logical operators $\{\overline{X}_i, \overline{Z}_i\}_{i=1,\ldots k}$ that form conjugate pairs and commute with all stabilizer generators. These operators do not generate the entire n-qubit Pauli group. To complete the set, we must throw in additional generators $\{T_j\}_{j=1,\ldots,n-k}$ that can be chosen to commute with the logical operators and form conjugate pairs with the stabilizer generators, i.e., $\{S_j, T_j\} = 0$ and $[S_i, T_j] = 0$ when $i \neq j$. We call these the "pure errors." Thus, $\{\overline{X}_i, \overline{Z}_i, S_j, T_j\}$ forms a canonical basis of the Pauli group and will be quite convenient in the analysis of the decoding problem.

Any error E can thus be decomposed in this canonical basis and we write

$$E = \mathscr{L}(E)\mathscr{S}(E)\mathscr{T}(E), \qquad (11.21)$$

where $\mathscr{L}(E)$ contains all logical generators $\overline{X}_i, \overline{Z}_i$ entering in the decomposition of E, $\mathscr{S}(E)$ contains all stabilizer generators S_j, and finally $\mathscr{T}(E)$ contains the pure errors T_j.

The error syndrome T is extracted by measuring the stabilizer generators. The syndrome completely reveals the pure error component $\mathscr{T}(E)$ of the error. Indeed, the pure errors are the only elements from the canonical basis that anticommute with the stabilizers, so only they can cause a nontrivial syndrome. Moreover, the jth syndrome bit will be 1 if and only if T_j enters the decomposition of the error that has affected the qubits. This follows from the canonical construction. Thus, the error probability can be updated conditioned on the syndrome, exactly as it was done in the classical setting in Eq. (11.10):

$$\Pr(E|T) = \frac{\Pr(T|E)\Pr(E)}{\Pr(T)} = \frac{\delta(\mathscr{T}(E'),T)\Pr(E)}{\sum_{E'} \delta(\mathscr{T}(E'),T)\Pr(E')}. \quad (11.22)$$

In other words, among all the errors E that could have corrupted the qubits, only those in agreement with the measured syndrome are kept as potential suspects, and the "truncated" probability is re-normalized. At this stage, we can identify the error with the largest probability, and output it as the result of the decoding algorithm. Similarly to the classical case, on the depolarizing channel this corresponds to choosing the lowest-weight error that is consistent with the syndrome.

In contrast to the classical case however, this decoding method is not optimal. After the syndrome measurement, the pure error component is known and what remains unknown are the values of the logical component $\mathscr{L}$ and the stabilizer component $\mathscr{S}$. But, by definition, the code space is invariant under the action of the stabilizer group. Thus, errors that differ only by their stabilizer components $\mathscr{S}$ have the same effect on code, and they can be regarded as equivalent. Such equivalent errors are called "degenerate." To be successful, the decoder needs only to identify the correct logical component $\mathscr{L}$ of the error. Hence, optimal decoding consists in identifying the equivalence class of errors – formed by errors with the same logical $\mathscr{L}$ and pure error $\mathscr{T}$ component but arbitrary stabilizer component $\mathscr{S}$ – with the largest probability. Formally, optimal decoding consists in optimizing the probability

$$\Pr(L|T) = \sum_E \Pr(E|T)\delta(\mathscr{L}(E),L)$$
$$= \frac{\sum_E \Pr(E)\delta(\mathscr{T}(E),T)\delta(\mathscr{L}(E),L)}{\sum_E \Pr(E)\delta(\mathscr{T}(E),T)}. \quad (11.23)$$

Qubit-wise maximum decoders can be defined similarly to the classical case by computing marginals. We can define a physical qubit-wise decoder by taking the marginal of $\Pr(E|T)$ or a logical qubit-wise decoder by taking the marginal of $\Pr(L|T)$. The former corresponds to the marginal error probability on each of the n physical bits. By definition, it cannot take degeneracy into account. The latter is a marginal on each k logical qubits – viewing each logical operator L as a tensor product of logical single qubit operators – that makes use of the equivalence between errors. As in the classical setting, a soft decoder outputs the marginal probability directly instead of the optimal argument. Soft physical and logical qubit-wise decoders will be used in the iterative coding systems presented in this chapter.

Finally, we note that the difference between the (suboptimal) decoder based on $\Pr(E|T)$ (essentially equivalent to the decoder for classical codes) and the optimal decoder based on $\Pr(L|T)$ (which takes into account the equivalence between errors) is mostly relevant when some elements of the stabilizer group have low weight. Assume the contrary, i.e., that all stabilizers have large weight $\geq w$. Then, for each equivalence class of errors, there will typically be one single error of small weight (equal or slightly larger than the code's minimum distance) and the weight of the other errors will be greater by roughly w (there can be cancellations that make this difference smaller). Thus, the probability will typically be dominated by a single error, and adding the probability of all errors in the equivalence class has hardly any effect. So, in that case, the two decoders will typically give the same answer. When some elements of the stabilizer group have low weight, however, there can be many errors that contribute to the equivalence class's probability. It then becomes possible that the error with the largest probability $\Pr(E|T)$ does not belong to the most likely equivalence class of errors. This is the situation where the optimal decoder makes a difference.

11.2.4 Decoding a quantum code: general statement

Although the rest of this chapter will be devoted to codes on Pauli channels, we take a moment to state the decoding problem in the most general case. In fact, the decoding algorithms that will be presented below can all be adapted to this more general situation. As was done in the previous section, the general case can be decomposed in two steps. (1) An update of the channel based on the observed syndrome [cf. Eq. (11.22)] and (2) a projection of the channel on the logical subalgebra [cf. Eq. (11.23)].

When a syndrome bit is extracted, we learn whether the error that occurred commuted or anticommuted with the associated stabilizer generator. For Pauli channels, this meant that some Kraus operators were discarded and others were kept [as indicated by the $\delta(\mathscr{T}(E), T)$ in Eq. (11.22)]. But for general channels, the different Kraus operators do not fall into either of these categories. As a consequence, each Kraus operator is modified following a syndrome measurement. For instance, if the jth syndrome bit is ± 1, then each Kraus operator should be updated according to

$$K_a \to \begin{cases} K'_a = P_j^{(+)} K_a P_j^{(+)} + P_j^{(-)} K_a P_j^{(-)} & \text{if } s_j = +1 \\ K'_a = P_j^{(+)} K_a P_j^{(-)} + P_j^{(-)} K_a P_j^{(+)} & \text{if } s_j = -1 \end{cases} \quad (11.24)$$

where $P_j^{(\pm)} = \frac{1}{2}(I \pm S_j)$. In other words, when the syndrome is positive, we only keep the error components that preserve the eigenspaces of S_j, while with a negative syndrome, we keep the error components that invert the two eigenspaces of S_j. After these updates, the channel will in general not be trace-preserving. This is expected since measurements (projectors) are trace-decreasing.

After the channel has been updated conditioned on the syndromes, it must be projected onto the logical subalgebra. This reflects the fact that we are only concerned with the effect of the noise on the encoded information, so we need only to know how the conditional channel acts on the algebra generated by the logical operators. In the previous section, this step was performed by identifying the Kraus operators that differed only by their stabilizer components into a unique operator, see Eq. (11.23). In the more general setting, it is also convenient to work with an operator basis, so let $\{L_i\}$ be a basis for the logical algebra (typically given as a canonical basis $\overline{X}_i$ and $\overline{Z}_i$, as above), and let $\{Q_j\}$ be additional operators required to complete the operator basis. We choose these generators to be mutually orthogonal with respect to the trace inner product. For an ordinary stabilizer code, this last set would be formed of the stabilizer generators and the pure errors. For an operator QECC (or subsystem code), this set would also contain the gauge generators. Then, the projected channel is given by,

$$\mathscr{E}'(\rho) = \sum_{i,i'} \chi_{i,i'} L_i \rho L_{i'}^\dagger, \quad (11.25)$$

where

$$\chi_{i,i'} = \sum_{j,j',a} \text{Tr}\{L_i^\dagger Q_j^\dagger K'_a\} \text{Tr}\{L_{i'} Q_{j'} K'^\dagger_a\}. \quad (11.26)$$

It can be verified that the combination of Eqs. (11.24)–(11.26) is equivalent to Eq. (11.23) for a Pauli channel.

Once the conditional channel on the logical algebra has been identified, its effect can be optimally reversed. This means finding a quantum operation $\mathscr{R}$ composed of the logical operators L_j that brings $\mathscr{R} \circ \mathscr{E}'$ as close as possible to the identity channel. If the closeness is measured in terms of the L_1 norm, the optimization problem is convex and can be solved exactly [FSW07]. The transpose channel $\mathscr{E}'^T$ defined in [BKN00] can also be used directly and yields near-optimal recovery.

11.2.5 Heuristic methods

The decoding problem, as formulated in the two previous sections, requires resources that grow exponentially with the system size. We have seen in Chapter 10 that it is possible to design codes for which this problem, or at least some version of it, becomes easy. In fact, it is often a minimum distance decoder that can be efficiently implemented in those cases which, as we have seen, is not optimal.

The method we will adopt in this chapter instead is to use heuristic decoding algorithms. These algorithms will not exactly solve the decoding problem, nor will they even solve well-defined approximations of the problem. Instead, they use gross approximations for certain quantities, hoping to capture the essential features of the problem. This is a broad strategy often used in physics, e.g., in perturbative approximations. These methods are formulated within a rigorous framework where they can be shown to give a controlled approximation, but often they are heuristically used outside such a framework. The two broad types of gross approximations that are used in the heuristic decoding algorithms are to (1) ignore certain correlations and (2) use uncontrolled loopy inference methods.

The reason why the simple probability update Eq. (11.23) requires exponential resources is because error correction creates correlations. The error probability on the physical qubits $\Pr(E)$ is typically assumed to be uncorrelated, meaning that errors occur independently on each qubit. Such an uncorrelated distribution requires a number of parameters proportional to the number of qubits, so it is not problematic. However, the measurement of the syndromes – associated with stabilizer operators acting on many qubits – induces correlations in the error distribution. Thus, while $\Pr(E)$ is uncorrelated, $\Pr(E|T)$ is typically highly correlated and this is where the exponential resources become necessary. Thus, only the marginal qubit-wise distributions $\Pr(E_i|T)$ are kept and their correlations are ignored for the sake of the efficiency of the algorithm.

But even the seemingly simpler task of computing marginal distributions can be complicated and sometimes requires additional approximation methods. For instance, one qubit can become correlated with many other qubits. When the correlations among the qubits have a clear causal structure – for instance qubit A depends on qubit B, which depends on qubit C, etc. – the marginals can easily be computing by moving up along the causal structure. This is equivalent to the transfer matrix method used for one-dimensional systems in many-body physics, and it is the main ingredient of the trellis-based decoding for convolutional codes, see Chapter 9. But when there exists no such causal structure – for instance qubit A depends on qubit B, ..., which depends on qubit Z, which depends on A – then the calculation of the marginal becomes a loopy inference problem and there is in general no efficient method to solve it exactly. One heuristic method that we will employ consists of blindly using the transfer matrix, even though it is not

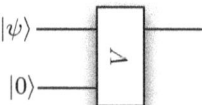

Fig. 11.2. Circuit representation of a convolutional code encoder. The circuit takes Nk to-be-encoded input qubits in state $|\psi\rangle$, $N(n-k) + Rn + m$ input qubits in the state $|0\rangle$, and outputs $(N+R)n + m$ qubits in a code state.

prescribed for this situation. While this is not justified rigorously and may appear sloppy, it has led to some of the best classical coding schemes. The rest of this chapter describes two of these schemes.

11.3 Turbo-codes

Turbo-codes were introduced in the classical setting by Berrou, Glavieux, and Thitimajshima in 1993 as an intricate form of concatenation of two convolutional codes [BGT93]. Roughly speaking, the novelty of the approach is that a byproduct of one code's decoder is used to enhance the performance of the other code's decoder. This is reminiscent of the workings of a turbo engine that uses some of its own exhaust gases to power a turbine that forces a greater air mass into the engine, hence generating more power. This turbo-procedure is repeated in an iterative fashion, and the overall code's performance improves with the number of iterations.

Although the construction of Berrou *et al.* was based on a parallel concatenation scheme, other concatenation schemes were later considered and yield equal or sometimes superior performance. In the quantum setting, only the serial concatenation of convolutional codes has been considered for technical reasons [OT05, PTO09]. Thus, we henceforth use turbo-code to mean serial turbo-codes. Quantum convolutional codes were covered in detail in Chapter 9, and we will not reproduce this theory here. In the construction of turbo-codes, we use convolutional codes of finite length. There are several ways to truncate a convolutional code. The simplest perhaps is to use the convolutional code normally for a duration of N frames and then use R additional termination frames in which the encoded qubits are all in the state $|0\rangle$. The precise number of these termination frames depends on the details of the code, but what matters is that it is independent of N. Thus, in the large N limit, the additional qubits used for the termination have a negligible effect on the code's rate, which converges to k/n asymptotically.

Thus, for our purpose, a quantum convolutional code is an ordinary error-correcting code of adjustable length. We will represent it by a quantum circuit, an encoder, taking Nk to-be-encoded input qubits, $N(n-k) + Rn + m$ input qubits in the state $|0\rangle$, and outputs $(N+R)n + m$ qubits in a code state, see Fig. 11.2.

The other important ingredient of a turbo-code is an interleaver. This is simply a transformation that shuffles the qubits. More precisely, an n-qubit interleaver is a random permutation of n qubits combined with a random permutation of the three Pauli matrices X_i, Y_i, and Z_i for each of the n qubits. The correlations between qubits are not taken into account by the decoders, and the essential goal of the interleaver is to scramble existing correlations to minimize their effect.

With these ingredients in hand, a quantum turbo-code can be defined as the interleaved serial concatenation of two convolutional codes. The qubits are encoded a first time using a convolutional code, the outer code. The encoded information is then passed through the interleaver that

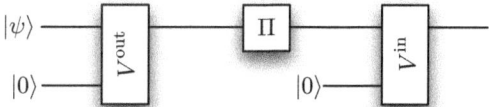

Fig. 11.3. Circuit representation of a quantum turbo-code encoder obtained from the interleaved serial concatenation of two convolutional codes. The encoding circuit of the outer code is used to encode the information a first time. The encoded message is then passed through an interleaver that shuffles its qubits. The resulting message is then encoded a second time with the inner code's encoding circuit.

shuffles its qubits. Finally, the resulting message is encoded a second time with an inner encoder, see Fig. 11.3.

The properties of the constituent convolutional codes have important effects on the quality of the resulting turbo-code. The properties of the outer code are not so important, provided that it has a large minimum distance. In fact, it is possible to vary the construction and use other types of outer code, including a sequence of small-size block codes, and obtain very good results. On the other hand, the inner code must be chosen very carefully. In the classical theory, it can be shown that a recursive inner code yields a turbo-code whose minimum distance grows almost linearly with its length [KU98]. More precisely, a turbo-code using a random interleaver typically has a minimum distance equal to $L^{\frac{d}{d+1}}$, where L is the total number of qubits in the code and d is the outer code's minimum distance (or free distance for convolutional codes). It is also necessary to use a noncatastrophic inner code to obtain good iterative decoding performance.

Unfortunately, it can be shown that all recursive quantum convolutional codes have catastrophic error propagation [PTO09]. Thus, it is not possible to build quantum turbo-codes whose performance relies on the same principles as classical turbo-codes. If we abandon the recursiveness, then the associated turbo-codes will have a finite minimum distance. On the other hand, if we use catastrophic inner codes, then the iterative decoding scheme will fail. It is conceivable that a different decoding scheme could cope with this problem, but at present there exist no quantum turbo-codes with an efficient decoder and a provably large minimum distance. In what follows, we opt for the first option and utilize inner codes that are noncatastrophic but also nonrecursive. As we will see, the performance achieved with this construction is quite remarkable.

11.3.1 Turbo-decoding

The decoding procedure for a convolutional code was explained in Chapter 9. It is a trellis-based decoding that has many variants. The classical Viterbi algorithm [V67] can be used almost straightforwardly to obtain the most likely physical error given the syndrome, i.e., the error that optimizes $\Pr(E|T)$. This algorithm can be modified to take into account the code's degeneracy in order to obtain the most likely logical error instead, i.e., the logical operator that optimizes $\Pr(L|T)$. This is achieved by summing over the stabilizer operators in each frame as we move along the trellis. It is therefore an algorithm that combines max-sum and sum-product. Note that turbo-codes have low-weight stabilizers – the stabilizers of the outer convolutional code – so degeneracy is significant.

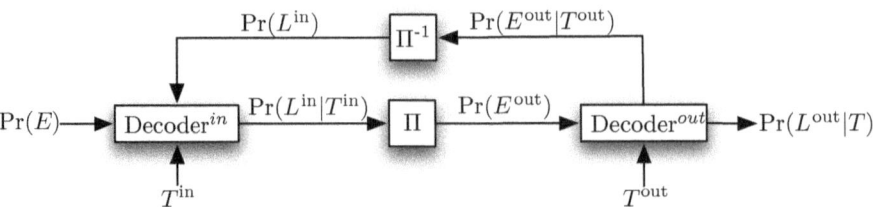

Fig. 11.4. Diagrammatic representation of the information flow in the turbo-decoder.

The decoding algorithms we will make use of for turbo-decoding are soft-output decoders based on qubit-wise maximum-likelihood decoding. Again, there are two type of qubit-wise decoders: a physical qubit-wise decoder obtained from the marginal of $\Pr(E|T)$, and a logical qubit-wise decoder obtained from the marginal of $\Pr(L|T)$. We will make use of both of them.

The first step of the decoding procedure consists of decoding the inner code using only the syndromes associated directly with that code. This is a soft logical qubit-wise decoding algorithm that returns the conditional error probability of each logical qubit of the inner code $\Pr(L_i^{\text{in}}|T^{\text{in}})$. These probabilities are then sent through the interleaver resulting in $\Pr(\Pi(L_i^{\text{in}})|T^{\text{in}})$, which permutes the labels i and also changes each qubit's Pauli frame. The logical qubits of the inner code correspond to the physical qubits of the outer code, so we set $\Pr(E_i^{\text{out}}) = \Pr(\Pi(L_i^{\text{in}})|T^{\text{in}})$. The outer code can thus be decoded using this probability induced by the inner code for an error model. At this point we can use the modified Viterbi algorithm to output the most likely logical error conditioned on the outer code's syndrome, based on the error model inherited from the inner code. This could produce the final output of the decoder. The only approximation in this scheme comes from the correlations that are ignored among the logical qubits of the inner code. The goal of the interleaver is precisely to scramble these correlations in order to minimize their effect.

Instead of following the above proposal, the turbo-decoder decodes the outer code with a soft physical qubit-wise decoder, yielding $\Pr(E_i^{\text{out}}|T^{\text{out}})$. This distribution is sent through the interleaver resulting in $\Pr(\Pi^{-1}(E_i^{\text{out}})|T^{\text{out}})$. This represents information on the inner code's logical qubits, which is fed to the inner code's decoder. Using this "new" information, the inner code produces a soft output that is sent back to the outer code. The outer code will use this information as an a-priori distribution on its logical operators as if it were statistically independent from its syndromes. This iterative procedure, illustrated in Fig. 11.4, is repeated a certain number of times. The final step can make use of a modified Viterbi algorithm as described in the previous paragraph or more simply use a physical qubit-wise decoder as we have chosen to do. A good halting condition is that the error estimate remains stable between consecutive iterations.

We see that two sorts of approximations are used in this scheme. First, the correlations among the logical qubits of the inner code and among the physical qubits of the outer code are ignored each time marginal probabilities are used. Second, the information passed between the two codes is treated as statistically independent from the observed syndrome, while in reality it is not. This is most easily visualized by the loop in Fig. 11.4. The information on the logical operators of the inner code truly has its origin in the inner code's decoder. The interleaver is

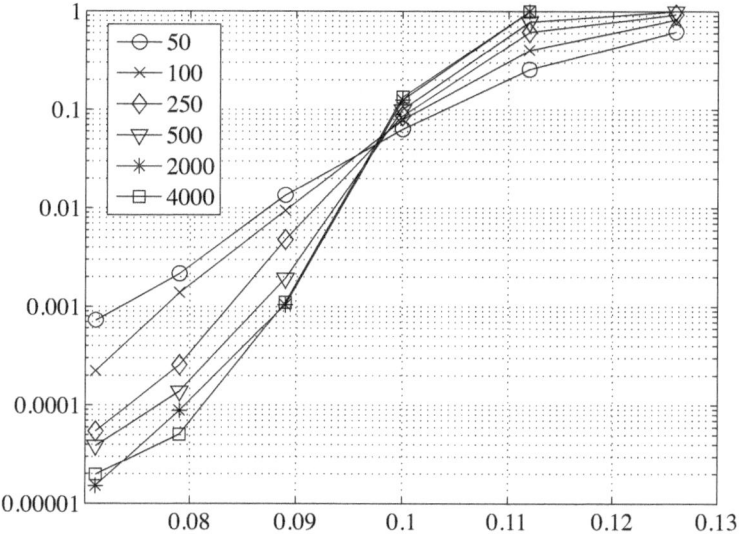

Fig. 11.5. Failure probability of a quantum turbo-code on a depolarizing channel as a function of the channel error probability p. The different lines are obtained from different code lengths, parametrized by the number of encoded qubits ranging from 50 to 4000. The quantum turbo-code is constructed by concatenating a convolutional code with itself. The convolutional code used has rate 1/3, meaning that the turbo-code has rate 1/9. The convolutional code has three memory qubits. Reprinted with permission from [PTO09]. © 2009, IEEE.

treated as a causal shield and information is regarded as statistically independent despite the presence of this causal loop.

The turbo-decoder can be described in terms of statistical physics. A convolutional code can be seen as a ribbon of interacting spins. The ribbon's width is equal to the code's memory. In that case, trellis-based decoding is equivalent to a transfer matrix, which is suitable for a one-dimensional system. In the case of a ribbon, the transfer matrix runs in a time exponential in the ribbon's width, hence the need to restrict to relatively small memories. A turbo-code comprises two convolutional codes, which corresponds to interactions between the associated ribbons. Because of the interleaver, the interactions between the ribbons are not local, in fact they obey no geometric constraint at all. In the decoding algorithm, these interactions are treated using a mean self-consistent field approximation.

11.3.2 Example

The performance of quantum turbo-codes can be assessed by numerical Monte Carlo experiments. Errors are pseudo-randomly generated according to a specified noise model $\Pr(E)$ and the error-correction cycle is performed. This is repeated a large number of times and the fraction of incorrect decodings provides an estimate of the code's failure probability for that specific noise model.

The results of Monte Carlo experiments are shown in Fig. 11.5. The turbo-code used for this numerical experiment is obtained from the interleaved concatenation of a convolutional code

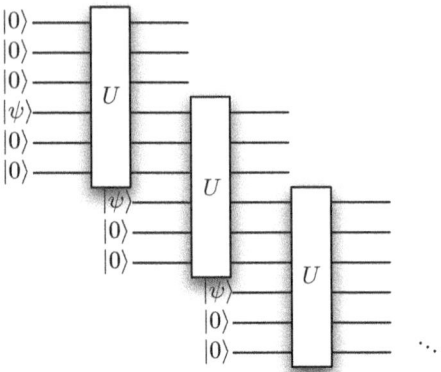

Fig. 11.6. General encoding circuit for a convolutional code with parameters $n = 3$, $k = 1$, and $m = 3$. The wires with $|\psi\rangle$ labels are where the quantum information is to be input.

with itself. The interleaver is chosen at random for each Monte Carlo trial. The convolutional code's parameters were set to $n = 3$, $k = 1$, and $m = 3$ (see Chapter 9) and the other details of the construction were chosen at random. More precisely, the encoding circuit of a convolutional code with those parameters has the form shown in Fig. 11.6. The seed transformation U used for the code in the example is chosen at random from all six-qubit Clifford transformations. The only criterion that was systematically imposed is that the resulting code be noncatastrophic, which can be efficiently verified given the matrix U. The specific Clifford transformation U used in the example corresponds to the symplectic matrix

$$U = \left(\begin{array}{cccccc|cccccc} 1 & 0 & 0 & 1 & 0 & 0 & 0 & 0 & 0 & 0 & 1 & 1 \\ 1 & 0 & 0 & 0 & 0 & 0 & 0 & 0 & 0 & 0 & 1 & 1 \\ 0 & 1 & 1 & 0 & 1 & 1 & 0 & 1 & 0 & 0 & 1 & 0 \\ 0 & 0 & 1 & 0 & 1 & 0 & 1 & 1 & 1 & 0 & 1 & 0 \\ 1 & 0 & 1 & 0 & 1 & 0 & 1 & 0 & 0 & 0 & 1 & 0 \\ 0 & 0 & 0 & 0 & 0 & 0 & 0 & 0 & 1 & 0 & 1 & 0 \\ \hline 0 & 1 & 1 & 0 & 1 & 1 & 0 & 1 & 0 & 1 & 1 & 0 \\ 0 & 0 & 1 & 0 & 1 & 1 & 1 & 1 & 0 & 1 & 0 & 0 \\ 1 & 1 & 0 & 1 & 0 & 1 & 1 & 1 & 1 & 0 & 1 & 1 \\ 1 & 1 & 0 & 0 & 0 & 0 & 0 & 1 & 1 & 0 & 0 & 1 \\ 1 & 0 & 1 & 0 & 1 & 1 & 0 & 0 & 1 & 0 & 0 & 0 \\ 0 & 0 & 0 & 0 & 1 & 1 & 0 & 1 & 1 & 0 & 0 & 1 \end{array}\right). \quad (11.27)$$

The most important feature of the results shown in Fig. 11.5 is the existence of a crossover probability $p^* \approx 0.098$ below which the code's failure probability decreases as a function of the block size. We emphasize that this is achieved at a fixed code rate $k/n = 1/9$. We can interpret this crossover probability as the finite-size precursor of a phase transition: in the limit of infinite code length, the information would be perfectly transmitted with probability 1 whenever the channel error probability is below the critical value p^*. We note, however, that the codes described in this section cannot have this asymptotic behavior because they have finite minimum distances. Thus, there should exist a second (size-dependent) crossover probability at smaller p where the size effect becomes detrimental. The Monte Carlo data give weak indication of such a second crossover.

Thus, although the codes are not good in a technical sense, they offer remarkable error suppression for a wide range of channel error probabilities and code lengths. For the sake of comparison, a random code of rate $1/9$ has a critical error probability $p^* \approx 0.16024$, which is not that far from the pseudo-threshold observed in Fig. 11.5. This pseudo-threshold also compares favorably with that obtained from other code constructions (see e.g., Fig. 10 in [MMM04]). The turbo-code construction appears to be robust in the sense that the details of the constituent convolutional codes and interleaver have little effect on the overall performance, provided that the inner code is noncatastrophic. Varying the convolutional code's memory size m has hardly any effect on the crossover probability, but a larger memory generally leads to a sharper transition.

11.4 Sparse codes

Sparse codes, also known as low-density parity check (LDPC) codes, were introduced in the classical setting in Gallager's 1963 Ph.D. thesis [G63]. Because of the limited computational resources available at the time, the iterative decoding scheme was impractical, and sparse codes were nearly forgotten for about 30 years. With the great advance of modern information processors, sparse codes have become widely used for a broad range of applications.

Sparse quantum codes can be defined as stabilizer codes with the following properties:

(i) They have a set of low-weight stabilizer generators.
(ii) The number of these low-weight stabilizer generator acting nontrivially on any given qubit is small.

Here, "low-weight" and "small" should be taken as meaning bounded by a constant. What matters is that they do not scale with the code length n. In other words, if we represent the stabilizer generators as an $(n-k) \times n$ matrix over $\mathbb{F}_4$ in the usual way, we obtain a sparse matrix with a bounded number of nonzero elements in each line and column. The weight of row c is equal to the weight of a stabilizer generator S_c, while the weight of column q is the number of stabilizer generators acting nontrivially on qubit q.

11.4.1 Tanner graph

Sparse quantum codes can be represented graphically by a decorated Tanner graph. This is a bi-partite graph, with the nodes of one subset representing the qubits, and the nodes of the other subset representing the sparse stabilizer generators. We refer to these two type of nodes as qubit and check nodes respectively. In fact, it is sometimes useful to have an over-complete set of stabilizer generators, so the number of check nodes s may be larger than $n - k$. The graph has an edge (c, q) if the qth tensor factor of stabilized S_c differs from the identity. The edge (c, q) is then decorated by the value of the qth tensor factor of stabilizer S_c.

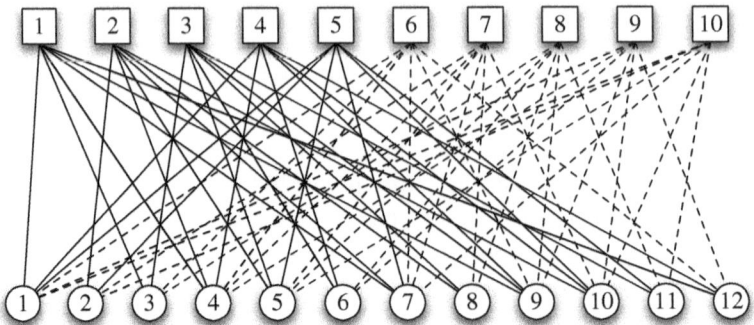

Fig. 11.7. Tanner graph of a sparse quantum code. Circles are qubit nodes and squares are check nodes. The solid lines represent X labels, while the dashed lines represent Z labels.

For example, consider the [[12,2,3]] stabilizer code with the following set of generators.

$$\begin{array}{cccccccccccc} X & I & X & X & I & I & X & I & X & I & I & X \\ I & X & I & X & X & I & X & X & I & X & I & I \\ I & I & X & I & X & X & I & X & X & I & X & I \\ X & I & I & X & I & X & I & I & X & X & I & X \\ X & X & I & I & X & I & X & I & I & X & X & I \\ Z & I & Z & Z & I & I & Z & I & Z & I & I & Z \\ I & Z & I & Z & Z & I & Z & Z & I & Z & I & I \\ I & I & Z & I & Z & Z & I & Z & Z & I & Z & I \\ Z & I & I & Z & I & Z & I & I & Z & Z & I & Z \\ Z & Z & I & I & Z & I & Z & I & I & Z & Z & I \end{array}$$

The decorated Tanner graph associated with this code is shown in Fig. 11.7. The stabilizer generators all have weight equal to 6. On the other hand, qubits 1, 4, 5, 7, 9, and 10 are each acted upon by six generators, while qubits 2, 3, 6, 8, 11, and 12 qubits are each acted upon by four generators. The stabilizers are independent, so $s = n - k = 10$.

A sparse code is characterized by its degree distribution. Given a Tanner graph with n qubit nodes and s stabilizer vertices, the qubit degree distribution λ_i is the fraction of qubit nodes of degree i, so that $\sum_i \lambda_i = 1$. Similarly, the check degree distribution p_i is the fraction of check nodes of degree i so that $\sum_i p_i = 1$. The number of edges in the graph is then $n \sum_i i\lambda_i = s \sum_i i p_i$. (These are the degree distributions from a node perspective. Some authors also define a degree distribution from an edge perspective.) The x, y, and z degree distributions $\lambda^{x,y,z}$ and $p^{x,y,z}$ are defined similarly, but only the edges with the specified decoration are counted.

In the example of Fig. 11.7, we have $p_6 = 1$, with all other $p_i = 0$, and $\lambda_4 = \lambda_6 = \frac{1}{2}$, with all other $\lambda_i = 0$. In terms of decoration, we have the check node distribution $p_6^x = \frac{1}{2}$, $p_6^z = \frac{1}{2}$ with all other $p_i^{x,y,z} = 0$ and the qubit degree distribution $\lambda_2^x = \frac{1}{4}$, $\lambda_3^x = \frac{1}{4}$, $\lambda_2^z = \frac{1}{4}$, $\lambda_3^z = \frac{1}{4}$ with all other $\lambda_i^{x,y,z} = 0$ for the qubits nodes.

In terms of their Tanner graphs, the defining properties of a family of sparse quantum codes with varying parameter n are

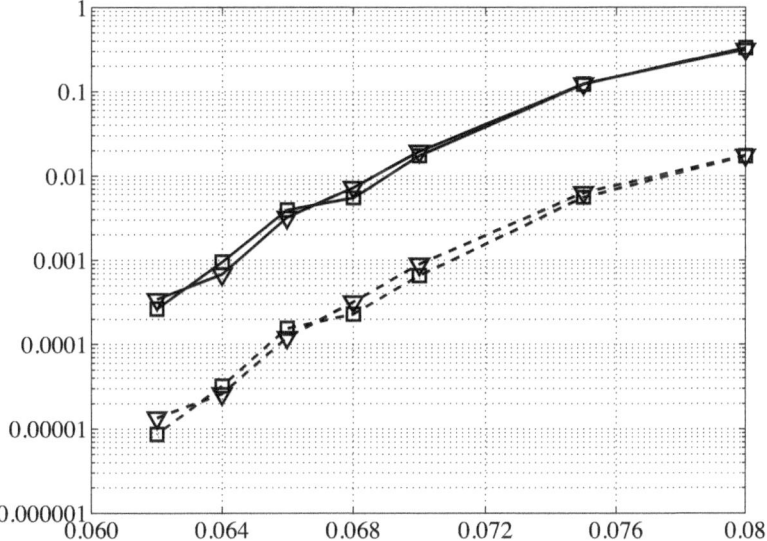

Fig. 11.8. Performance of classical sparse codes, constructed according to the random procedure explained in the main text, on a binary symmetric channel. Block error rate (full line) and bit error rate (dash line) are shown as a function of the channel error probability p for two randomly generated codes. The bit degree is 3 and the check degree is 6. Thus, the rate of the code is $1/2$. The length of the code is $n = 3000$, so $k = 1500$ bits are encoded.

(i) the check degree distribution p_i vanishes for $i > c_1$,
(ii) the qubit degree distribution λ_i vanishes for $i > c_2$,

where c_1 and c_2 are constants independent of the code parameters n and k.

11.4.2 Code construction

Classically, sparse codes with any desired degree distributions (λ, p) can be constructed at random. Here is perhaps the simplest way to do this. Imagine that each node is equipped with a certain number of ports that can receive an edge. To construct a Tanner graph with N_e edges, randomly assign N_e ports to the n qubit nodes according to the distribution λ and randomly assign N_e ports to the s check nodes according to the distribution p. Then, randomly pair up the ports of the qubit nodes with those of the check nodes. To obtain a Tanner graph, we delete some edges between the pairs of nodes that are linked by more than one edge. This will slightly modify the weight distribution of the graph, but the effect is suppressed by a factor $1/n$ (the likelihood of a collision).

This simple construction will typically yield reasonably good error-suppressing codes for n of a few thousands. For example, Fig. 11.8 shows the block and bit error rates of two randomly constructed rate-$1/2$ sparse classical codes of size $n = 3000$.

In contrast, it is much more difficult to construct sparse quantum codes. The main obstacle is that the stabilizers must generate an Abelian group. Suppose, for instance, that we construct the Tanner graph according to the above prescription and then decorate the edges randomly. Since each check overlaps with at least one other check (since the qubit degree must be strictly greater than 1) and the average weight of a check is say d_c, then there are roughly n/d_c pairs of distinct overlapping checks in the code. The probability that two random overlapping checks commute is less than $1/2$ (it is $1/3$ when there is only one overlap qubit and tends to $1/2$ for larger overlaps). Thus, the probability that this random procedure generates an Abelian group is less than $(1/2)^{\frac{n}{d_c}}$ (this is a very loose upper bound). Clearly this technique will fail for large n.

Of course, the above negative conclusion could be circumvented by carefully assigning the decorations on the graph. For instance, if the decoration on edge (c, q) was determined uniquely by the value of q, then we would obtain an Abelian group. However, such a code could only correct one type of error – i.e., either X, Y, or Z – on each qubit. Indeed, suppose that all the edges attached to qubit q have decoration X. Then, an X error on that qubit would not be detected. This would result in a code of minimum distance 1, unless the operator X_q belongs to the stabilizer group. But in that case, qubit q would not participate in the encoding at all because it would need to be in the state $|0\rangle + |1\rangle$, unentangled with the other qubits forming the code. Thus, we arrive at the conclusion that

(iii) the edges connected to any given qubit node must carry at least two different type of decorations.[2]

Together with conditions (i) and (ii) above, this forms the complete list of basic properties required for sparse quantum codes.

The CSS construction (see Section 2.8) has been very successful for generating quantum codes from classical codes. However, it has serious limitations for the construction of sparse quantum codes. Remember that for the CSS construction we need a pair of codes C_1 and C_2 such that $C_2^\perp \subseteq C_1$. To obtain a sparse code, it is moreover necessary that both $C_1^\perp$ and $C_2^\perp$ be generated by a set of low-weight words, i.e., that C_1 and C_2 be classical sparse codes. But classical sparse codes, being good codes, generally have large minimum distances. So if C_2 has a large minimum distance, it cannot contain $C_2^\perp$, since this last subspace is generated by low-weight words.

Thus, what is needed are two sparse classical codes C_1 and C_2 such that $C_2^\perp \subseteq C_1$ and the minimum weight of elements of $C_1 - C_2^\perp$ is high. In other words, the code C_1 can have low-weight codewords but only if they are in $C_2^\perp$. These would correspond to degenerate errors of the corresponding quantum code. An equivalent condition must hold with the roles of C_1 and C_2 inverted.

It has proven very difficult to construct quantum codes, CSS or other, satisfying all the desired properties. In fact, there are no constructions to date that meet all the basic requirements and have (typical) minimum distances proportional to the code length. Postol [P01] has constructed quantum sparse codes using finite geometry. This idea was pursued by Aly [A08], but the resulting codes have a small (constant) provable minimum distance and no numerical assessment of the code has been realized. Constructions based on the CSS form using cyclic classical codes have been proposed [MMM04, HI07]. The next section describes the construction of bicycle codes, which are the simplest example of this class. Computer searches have also been used to

---

[2] To obtain good iterative decoding performance, it is necessary to have multiple distinct decorations at each qubit node. More precisely, at each qubit node, there must be at least two edge decorations that anticommute with every Pauli operator. For instance, two Xs and two Zs would suffice, or an X, a Y, and a Z.

find good sparse quantum codes [MMM04]. Finally, the sparse quantum codes that offer the best performance to date on the depolarizing channel were obtained by Camera, Ollivier, and Tillich [COT07] and are based on group theoretic methods. For block lengths ranging from ten to a hundred thousand, their iterative decoding offers interesting error suppression on the depolarizing channel.

It is possible to construct sparse operator QECCs [KLP05, P05, NP07] using the technique [BC06, KS08] described in Chapter 6. These codes are constructed from any pair of classical codes, with no dual-containment constraint. Combining a $[n_1, k_1, d_1]$ and a $[n_2, k_2, d_2]$ CEC code produces a $[[n_1 n_2, k_1, k_2, \min(d_1, d_2)]]$ QEC code. When the two constituent codes have minimum distances proportional to their length, the resulting quantum code has a minimum distance growing only as the square-root of its length, so it is not good in a technical sense. Indeed, for a fixed error probability, the block error rate will tend to 1 as the code length increases. But for finite block size, this construction could offer interesting error suppression. Unfortunately, the standard iterative decoding scheme does not apply to these codes because the dual-containment constraint is not observed. There is to date no known efficient technique to decode such codes.

Another setting where sparse quantum codes can be constructed [HBD09] is entanglement-assisted QEC, see Chapter 7. This setting also relaxes the dual-containment constraint by making use of error-free entangled qubits [BDH06b]. Thus, when this additional resource is available, good sparse quantum codes can be constructed.

11.4.2.1 Bicycle codes Bicycle codes [MMM04] are CSS codes that make use of cyclic matrices. A simple method to create a $d \times d$ cyclic binary matrix C consists of generating a binary vector A of length d with random entries and defining $(C)_{i,j} = (A)_{j-i}$. A sparse cyclic matrix is obtained when the vector A is sparse.

For a bicycle code with row weight w, block length n, and number of checks s, first generate a random $n/2 \times n/2$ cyclic matrix C with row weight $w/2$. (As in [MMM04], this matrix can be constructed from a difference set, but the construction outlined above seems to work just as well.) Then, construct a matrix H_0 by merging the matrices C and $C^\dagger$, i.e., $H_0 = (C|C^\dagger)$. This matrix is self-dual:

$$(H_0 H_0^\dagger)_{i,j} = (CC^\dagger + C^\dagger C)_{i,j}$$
$$= \sum_k A_{i+k} A_{j+k} + A_{k+i} A_{k+j} = 0. \quad (11.28)$$

Deleting $n/2 - s/2$ of the rows from H_0 yields a self-dual matrix H with $s/2$ rows and n columns. The code C associated to the parity check matrix H is dual-containing, i.e., $C^\perp \subseteq C$, so it is suitable for the CSS construction. We obtain a sparse quantum code of rate $\frac{n-s}{n}$ with stabilizer generators

$$S_c = \begin{cases} \bigotimes_{j=1}^n Z_j^{(H)_{c,j}} & \text{for } 0 < c \leq s/2, \\ \bigotimes_{j=1}^n X_j^{(H)_{c,j}} & \text{for } s/2 < c \leq s, \end{cases} \quad (11.29)$$

where $Z^0 = X^0 = I$. The commutativity of the checks follows from the self-duality of H.

The construction of bicycle codes gives complete freedom in controlling the size and weight of the code. However, all the deleted rows of H_0 are low-weight codewords that do not lie in the dual (unless the stabilizer generators were not linearly independent). More precisely, if x is a deleted row, then

$$\bigotimes_{j=1}^{n} Z_j^{x_j} \quad \text{and} \quad \bigotimes_{j=1}^{n} X_j^{x_j} \tag{11.30}$$

are low-weight errors that commute with all stabilizers. Thus, bicycle codes have a small minimum distance (less than or equal to the check weight w). Nonetheless, they offer good performance under iterative decoding on the depolarizing channel.

The code shown in Fig. 11.7 is a bicycle code constructed from the vector $A = (011001)$. This produces the matrix

$$C = \begin{pmatrix} 0 & 1 & 1 & 0 & 0 & 1 \\ 1 & 0 & 1 & 1 & 0 & 0 \\ 0 & 1 & 0 & 1 & 1 & 0 \\ 0 & 0 & 1 & 0 & 1 & 1 \\ 1 & 0 & 0 & 1 & 0 & 1 \\ 1 & 1 & 0 & 0 & 1 & 0 \end{pmatrix}. \tag{11.31}$$

Then $H_0 = (C|C^\dagger)$ is a self-dual matrix. Deleting the first row from this matrix and using the CSS construction yields the code in Fig. 11.7.

11.4.3 Message passing decoding

The iterative decoding algorithm used for sparse quantum codes is a message passing algorithm, also known as belief propagation, operated on the Tanner graph. The messages are probability distributions on the four Pauli matrices (including the identity), i.e., they are arrays of four positive numbers adding to 1.[3] Messages are exchanges along the edges of the Tanner graph. Roughly speaking, the message sent from a qubit node to a check node summarizes the information received at that qubit node from all other check nodes. Similarly, the message sent from a check node to a qubit node summarizes the information received from all the other qubit nodes conditioned on the observed syndrome at that check node. The algorithm is exact when the underlying graph is a tree, as is the case for instance for concatenated block codes [P06]. The presence of loops in the graph creates self-induction loops, and in that case the method gives uncontrolled approximations.

We assume a memoryless Pauli channel, such that errors are elements of the Pauli group and are distributed independently on each qubit. The error distribution on qubit q is denoted $\Pr(E_q)$. A depolarizing channel for instance would have $\Pr(I) = 1 - p$ and $\Pr(\sigma) = p/3$ for $\sigma = X, Y, Z$. To initialize the algorithm, each qubit sends out a message to its neighbors equal to its prior error probability $m_{q \to c}(E_q) = \Pr(E_q)$. Upon reception of these messages, each check sends out a message to its neighboring qubits given by

$$m_{c \to q}(E_q) \propto \sum_{\substack{E_{q'} \\ q' \in n(c) \setminus q}} \left(\delta(T_c, \mathscr{T}(E_c)) \prod_{q' \in n(c) \setminus q} m_{q' \to c}(E_{q'}) \right), \tag{11.32}$$

---

[3] This is for Pauli channels. For more general channels, the messages are 4×4 unit-trace positive operators.

where $n(c)\backslash q$ denotes all neighbors of c except q. The sum is over all Pauli operators on the neighbors of c but with E_q held fixed. The proportionality factor can be fixed by normalization $\sum_{E_c} m_{c\rightarrow q} = 1$. Note that the function $\delta(T_c, \mathcal{T}(E_c))$ can be evaluated locally, by verifying whether the error E_c commutes with S_c or not, and comparing the result with T_c. Thus, $m_{c\rightarrow q}$ is a function of the syndrome bit T_c associated with check c, and the messages $m_{q'\rightarrow c}$ received from all neighbors of c, except q.

Upon reception of these messages, each qubit sends out a message to its neighboring checks given by

$$m_{q\rightarrow c}(E_q) \propto \Pr(E_q) \prod_{c'\in n(q)\backslash c} m_{c'\rightarrow q}(E_q), \tag{11.33}$$

where $n(q)\backslash c$ denotes all neighbors of q except c. Again, the proportionality factor can be fixed by normalization. Thus, $m_{q\rightarrow c}$ is a function of the qubit prior error probability $\Pr(E_q)$, and the messages $m_{c'\rightarrow q}$ received from all neighbors of q, except c.

Equations (11.32) and (11.33) define an iterative procedure that is at the core of the decoding algorithm. The beliefs $b_q(E_q)$ – which are meant to represent an approximation to the marginal physical qubit conditional probabilities, Eq. (11.12) – are computed from the messages as follows:

$$b_q(E_q) \propto \Pr(E_q) \prod_{c\in n(q)} m_{c\rightarrow q}(E_q). \tag{11.34}$$

The recovery can be chosen as the product of qubit-wise maximum-belief Pauli matrices $E^{MB} = \bigotimes_{q=1}^{n} E_q^{MB}$ where

$$E_q^{MB} = \underset{E_q}{\mathrm{argmax}}\{b_q(E_q)\}. \tag{11.35}$$

This message passing algorithm is not a maximum-likelihood decoder. It has important limitations.

(1) The algorithm is fundamentally a physical qubit decoder, based on the probability $\Pr(E|T)$ rather than $\Pr(L|T)$. In other words, it makes no attempt at grouping equivalent errors. This is a serious limitation because sparse codes are by definition highly degenerate and so would benefit from a logical error decoder.

(2) The algorithm is based on marginal probabilities rather than joint probabilities. In other words, it decodes at the Nishimori temperature instead of $\beta = \infty$. Thus, it is designed to optimize the probability of correctly decoding each physical qubit independently rather than the probability of correctly decoding the entire qubit string.

(3) These marginal error probabilities are not computed exactly but instead make use of a heuristic method. The message passing algorithm yields exact marginals only when the Tanner graph is a tree. However, the defining properties of a sparse quantum code prohibit this from being the case. Condition (iii) combined with the fact that the stabilizers must commute imply the existence of length-4 loops in the graph. Indeed, the only way that two stabilizers can intersect on a given qubit with different decorations and commute is if they intersect on at least one other qubit again with different labels, generating a 4-loop.

Problems 2 and 3 are present in the classical setting and, despite those, sparse classical codes are nearly capacity achieving on a wide range of channels. Problem 3 is somewhat weaker in the classical setting however. Although their Tanner graphs contain loops, sparse code designs try to minimize the number of small loops. In many contexts, message passing has been observed to provide reliable estimates, particularly when the typical loop-size is large. Intuitively, one expects that such a local algorithm should be relatively insensitive to the large-scale topology of the graph. In this sense, sparse quantum codes represent the worst-case scenario for message passing algorithms.

Problem 1 has no classical analog, and is indeed a rather serious problem. Sparse quantum codes are highly degenerate: there are always many different operators of roughly equal weight that can correct an error. Instead of regarding all these corrections as equivalent and grouping them as a united decoding option, the decoder will distribute the probability roughly equally among all these options. Thus, the probability of each of these individual corrections is lowered. On the other hand, when the noise level is low, the decoder is naturally biased towards the trivial recovery: the identity matrix is the error that is a-priori assigned the largest probability. As a consequence of the division of the probability among all equivalent errors, it can happen that the single error probability that prevails is the one associated with the identity. This would result in a failed decoding.

The failure scenario described above is indeed observed in practice. Its one major positive aspect is that it typically leads to a detectable error, i.e., the error correction does not reset the syndrome to 0. While the probability $\Pr(E|T)$ assigns nonzero probability only to the E that have the right syndrome, it is not the case for its marginals that typically assign positive probabilities to all errors. Thus, when the decoder's output does not return a trivial syndrome, we can assume we are in the above scenario and act accordingly. Ideally, we would want to transfer all the belief assigned to the various equivalent errors to a single one of them. This is in general difficult, but it is possible to create a bias that will partially accomplish this task. We can first identify potential qubits that are responsible for the failed decoding. If the belief on one of these qubits is dominated by the identity – i.e., $b_q(I) > b_q(\sigma)$ for $\sigma = X, Y, Z$ – and two other beliefs are nonnegligible and roughly equal – say $b_q(X) \approx b_q(Y)$ – then we can artificially introduce a random bias in the prior probability $\Pr(E_q)$ that will favor either one of these two errors. This bias will propagate to the other qubits under message passing and can produce a new decoding output. Heuristic methods based on this approach have been implemented and produce substantial improvements.

11.4.4 Example

Figure 11.9 shows the performance of a bicycle quantum LDPC code on the depolarizing channel obtained by Monte Carlo numerical experiments. Three different decoding algorithms have been applied to the same code. The first one, denoted "plain," is the basic message passing algorithm defined by Eqs. (11.33) and (11.34). The other two methods use different forms of perturbations to create a bias that will hopefully lead to a concentration of the beliefs onto a single error, as described above.

The code offers an interesting noise suppression; the failure probability after decoding is suppressed by several orders of magnitude. Moreover, the two heuristic methods used to concentrate the belief lead to significant improvements of the block error rates. This corroborates our diagnostic of the failure of the decoding algorithm. It is important to note that all failures found in

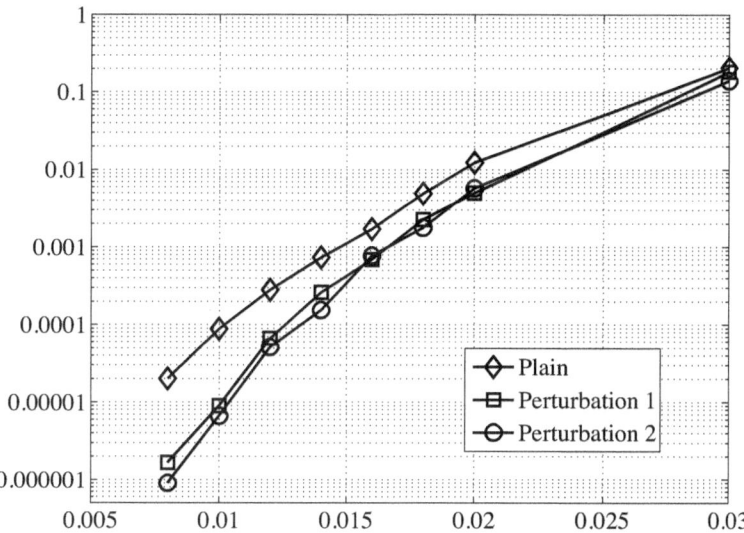

Fig. 11.9. Performance of quantum sparse codes, constructed using the bicycle model. The codes encode 400 qubits in 800 qubits for a rate of $1/2$. The check nodes all have degree 30 and the average degree of the qubit nodes is 15. Block error rates are shown as a function of the channel error probability p for three different decoding algorithms. The plain algorithm is the message passing algorithm described in the main text. The two other algorithms use different forms of perturbations to create a bias that will lead to a concentration of the belief on one of many equivalent errors as described in the main text.

our Monte Carlo experiments returned nontrivial syndromes. This means that 100% of the code's failures are attributed to the decoder and that it is known to the user that the decoding algorithm has failed. This can be an important feature in some settings. In a communication scenario, for instance, the receiver can ask the sender to send another copy of the information when he knows that the decoder has failed.

11.5 Conclusion

Coding schemes based on iterative decoding offer some of the best performance on a wide range of channels in the classical setting. Transposing these designs and techniques to the quantum setting is difficult for numerous reasons. First, the design of quantum codes is highly constrained because the quantum checks must generate an Abelian group. Consequently, it is not possible to design quantum turbo-codes whose quality relies on the same principles as classical turbo-codes, namely a recursive noncatastrophic inner code. This constraint also seriously limits the design of sparse quantum codes, and implies the presence of 4-loops in their Tanner graphs, which seriously impair iterative decoding.

The other important novelty is the existence of degenerate errors. Both turbo-codes and sparse codes have low-weight stabilizers, so they are highly degenerate. In the case of turbo-codes, it is possible to design a decoding algorithm that explicitly treats degeneracy. This is not the case for sparse codes, and as a consequence their decoding performance is significantly impaired. Heuristic methods can partially overcome this problem and lead to significant improvements, but the decoding algorithm remains the bottleneck.

Despite these important limitations, from a pragmatic point of view we have seen that quantum iterative coding schemes offer impressive error reduction at relatively high encoding rate. In fact, quantum turbo-codes compare favorably to other coding schemes over a reasonable range of block length and error rates.

Chapter 12

Algebraic quantum coding theory

Andreas Klappenecker

In algebraic quantum coding theory, the quantum codes are derived from algebraic and combinatorial structures such as vector spaces, ideals, group rings, algebraic curves, and finite geometries. This chapter gives some examples of quantum stabilizer codes that are derived from algebraic structures. After reviewing the main constructions of stabilizer codes and their relations to classical codes, it is shown how one can obtain quantum codes from classical Reed–Muller, Bose–Chaudhuri–Hocquenghem and Reed–Solomon codes.

12.1 Quantum stabilizer codes

Notation. Let $\mathbf{C}$ denote the field of complex numbers. For a positive integer n, we denote by $\mathscr{B}(\mathbf{C}^n)$ the set of bounded operators on $\mathbf{C}^n$. Let $\mathbb{F}_q$ denote a finite field with q elements. We use Dirac notation and denote by $\{|x\rangle \mid x \in \mathbb{F}_q\}$ an orthonormal basis of $\mathbf{C}^q$, called the computational basis. We denote by $\{\langle y| \mid y \in \mathbb{F}_q\}$ the dual basis of the computational basis, that is, the elements $|y\rangle$ are linear functionals on $\mathbf{C}^q$ such that $\langle y||x\rangle = 0$ if $x \neq y$, and $\langle x||x\rangle = 1$. As is customary, we abbreviate $\langle y||x\rangle$ to $\langle y \mid x\rangle$.

Error bases. Let $\mathbb{F}_q$ be a finite field with q elements. Let p be the characteristic of this field. Thus, $q = p^m$ for some positive integer m. Let a and b be elements of the finite field $\mathbb{F}_q$. We define unitary operators $X(a)$ and $Z(b)$ on $\mathbf{C}^q$ by

$$X(a)|x\rangle = |x + a\rangle, \quad Z(b)|x\rangle = \omega^{\operatorname{Tr}(bx)}|x\rangle,$$

where $\omega = \exp(2\pi i/p)$ and $\operatorname{Tr}\colon \mathbb{F}_q \to \mathbb{F}_p$ is the absolute trace given by

$$\operatorname{Tr}(x) = x + x^p + \cdots + x^{p^{m-1}}.$$

Any unitary operator on $\mathbf{C}^q$ is in the linear span of the operators $X(a)Z(b)$ with $a, b \in \mathbb{F}_q$. In fact, these operators even form an orthonormal basis.

Quantum Error Correction, ed. Daniel A. Lidar and Todd A. Brun. Published by Cambridge University Press.
© Cambridge University Press 2013.

Proposition 12.1. *The set* $\mathscr{E} = \{X(a)Z(b) \mid a, b \in \mathbb{F}_q\}$ *forms an orthonormal basis of* $\mathscr{B}(\mathbf{C}^q)$ *with respect to the inner product*

$$\langle A \mid B \rangle = \frac{1}{q} \mathrm{Tr}(A^\dagger B),$$

where $\mathrm{Tr}(A^\dagger B) = \sum_{x \in \mathbb{F}_q} \langle x | A^\dagger B | x \rangle$ *denotes the trace of the operator* $A^\dagger B$.

Proof. Let $A = X(a)Z(b)$ and $B = X(a')Z(b')$ be two arbitrary elements of the error basis $\mathscr{E}$. If these elements differ in the first parameter, $a \neq a'$, then

$$\langle A \mid B \rangle = \frac{1}{q} \mathrm{Tr}(Z(-b)X(a'-a)Z(b')) = 0,$$

since $\langle x | Z(-b)X(a'-a)Z(b') | x \rangle = 0$ for all x in $\mathbb{F}_q$ when $a \neq a'$. If A and B coincide in the first parameter but differ in the second, i.e., when $a = a'$ but $b \neq b'$, then

$$\langle A \mid B \rangle = \frac{1}{q} \mathrm{Tr}(Z(b'-b)) = \frac{1}{q} \sum_{x \in \mathbb{F}_q} \omega^{\mathrm{Tr}((b'-b)x)} = 0,$$

since the latter sum averages over all values of the nontrivial additive character $x \mapsto \omega^{\mathrm{Tr}((b'-b)x)}$ of the finite field $\mathbb{F}_q$. Finally, if $A = B$, then $\langle A \mid A \rangle = q^{-1}\mathrm{Tr}(A^\dagger A) = q^{-1}\mathrm{Tr}(I) = 1$ since A is unitary. $\square$

We will call the set $\mathscr{E}$ an *error basis* of $\mathscr{B}(\mathbf{C}^q)$, and will refer to the elements in $\mathscr{E}$ as *error operators*. We note that in dimension $q = 2$, the set $\{X(a)Z(b) \mid a, b \in \mathbb{F}_2\}$ is equal to the Pauli basis $\{I, X, Z, XZ\}$. Therefore, the error basis $\mathscr{E}$ of $\mathscr{B}(\mathbf{C}^q)$ indeed generalizes the Pauli basis.

The Pauli basis has the following convenient property: All elements commute up to a scalar factor. The error operators in $\mathscr{E}$ enjoy the same property, as the next proposition shows.

Proposition 12.2. *Let* a, b *be elements of a finite field* $\mathbb{F}_q$. *Then the operators* $X(a)$ *and* $Z(b)$ *satisfy the commutation relation*

$$\omega^{\mathrm{Tr}(ba)} X(a)Z(b) = Z(b)X(a),$$

where $\omega = \exp(2\pi i/p)$ *and* p *denotes the characteristic of the field* $\mathbb{F}_q$.

Proof. For all elements x in $\mathbb{F}_q$, we get $Z(b)X(a)|x\rangle = Z(b)|x+a\rangle = \omega^{\mathrm{Tr}(b(x+a))}|x+a\rangle$ as well as $\omega^{\mathrm{Tr}(ba)} X(a)Z(b)|x\rangle = \omega^{\mathrm{Tr}(ba)} X(a) \omega^{\mathrm{Tr}(bx)}|x\rangle = \omega^{\mathrm{Tr}(b(a+x))}|x+a\rangle$. Thus, the operators coincide on a basis, so they must be the same. $\square$

These commutation relations simplify many calculations. We illustrate this fact by determining the power of an element $X(a)Z(b)$.

Corollary 12.1. *Let* a, b *be elements of a finite field* $\mathbb{F}_q$. *Then the exponent formula*

$$(X(a)Z(b))^n = \omega^{\mathrm{Tr}\left(\frac{(n-1)n}{2}ba\right)} X(na)Z(nb)$$

holds for all positive integers n.

Proof. If $n = 1$, then the assertion obviously holds. Suppose that the statement holds for $n - 1$. Then

$$\begin{aligned}(X(a)Z(b))^n &= (X(a)Z(b))^{n-1}X(a)Z(b)\\ &= \omega^{\operatorname{Tr}(\frac{(n-2)(n-1)}{2}ba)}X((n-1)a)Z((n-1)b)X(a)Z(b)\\ &= \omega^{\operatorname{Tr}(\frac{(n-2)(n-1)}{2}ba)}\omega^{\operatorname{Tr}((n-1)ba)}X(na)Z(nb)\\ &= \omega^{\operatorname{Tr}(\frac{n(n-1)}{2}ba)}X(na)Z(nb).\end{aligned}$$

Therefore, the claim follows by induction on n. □

We can view the set $\mathscr{E}$ as a way to discretize the errors affecting a single q-dimensional quantum system. We can obtain error operators acting on the state space of n quantum systems of dimension q by tensoring the error operators in $\mathscr{E}$.

Let us first introduce some convenient notation. For $a = (a_1, \ldots, a_n)$ and $b = (b_1, \ldots, b_n)$ in $\mathbb{F}_q^n$, we define unitary operators $X(a)$ and $Z(b)$ acting on the vector space $\mathbf{C}^q \otimes \cdots \otimes \mathbf{C}^q \cong \mathbf{C}^{q^n}$ by

$$X(a) = X(a_1) \otimes \cdots \otimes X(a_n) \quad \text{and} \quad Z(b) = Z(b_1) \otimes \cdots \otimes Z(b_n).$$

Thus, the error operators $X(a)$ and $Z(b)$ are now parametrized by vectors of length n over the finite field $\mathbb{F}_q$.

Corollary 12.2. *The set $\mathscr{E}_n = \{X(a)Z(b) \mid a, b \in \mathbb{F}_q^n\}$ is an orthonormal basis of $\mathscr{B}(\mathbf{C}^{q^n})$ with respect to the inner product $\langle A \mid B \rangle = q^{-n}\operatorname{Tr}(A^\dagger B)$.*

Proof. This follows from Proposition 12.1 and the fact that $\operatorname{Tr}(M_1 \otimes M_2) = \operatorname{Tr}(M_1)\operatorname{Tr}(M_2)$ holds all for operators M_1 in $\mathscr{B}(\mathbf{C}^k)$ and M_2 in $\mathscr{B}(\mathbf{C}^\ell)$. This concludes the proof. □

Stabilizer codes. Let us consider n q-dimensional quantum systems. The state space of these quantum systems is given by

$$H = \mathbb{C}^q \otimes \cdots \otimes \mathbb{C}^q \cong \mathbb{C}^{q^n}.$$

In general, a quantum error-correcting code is a nontrivial subspace Q of H. The theory of stabilizer codes provides a geometric approach to the construction of such quantum error-correcting codes.

Corollary 12.2 shows that $\mathscr{E}_n$ is an orthonormal basis of $\mathscr{B}(\mathbb{C}^{q^n})$. This error basis has the interesting property that the product of two elements in $\mathscr{E}_n$ is a scalar multiple of an element in $\mathscr{E}_n$. However, $\mathscr{E}_n$ fails to be a group, since it is not closed under multiplication. It will be convenient to embed the error basis $\mathscr{E}_n$ into a suitably chosen group $G_{n,q}$.

Let $\mathbb{F}_q$ denote a finite field of characteristic p, and let n be a positive integer. We denote by $G_{n,q}$ the finite group

$$G_{n,q} = \begin{cases} \langle X(a), Z(b) \mid a, b \in \mathbb{F}_q^n \rangle & \text{if the characteristic } p \text{ is odd,} \\ \langle i\mathbf{1}, X(a), Z(b) \mid a, b \in \mathbb{F}_q^n \rangle & \text{if the characteristic } p \text{ is even.} \end{cases}$$

Recall that the *exponent* of a finite group G is the smallest positive integer n such that $x^n = 1$ holds for all x in G. The *center* $Z(G)$ of a group G consists of all elements z in G that commute with all group elements in G. We denote by $[x, y] = x^{-1}y^{-1}xy$ the *commutator* of two elements

x and y in G. The commutator bears its name because $yx[x,y] = xy$ holds for x,y in G. The *derived subgroup* G' of a group G is the subgroup generated by commutators of G, that is, $G' = \langle [x,y] \mid x,y \in G \rangle$.

Let r denote the exponent of the group generated by $\mathscr{E}_n$. The group $G_{n,q}$ is the smallest group containing the error basis $\mathscr{E}_n$ and having a center of order divisible by r. The motivation for imposing the order condition on the center will become clear once we investigate the relation between stabilizer codes and classical codes. Let us first record some key properties of the group $G_{n,q}$.

Proposition 12.3. *Let $\mathbb{F}_q$ be a finite field of characteristic p. Let $r = 4$ if p is even, and $r = p$ if p is odd. Let $\omega = \exp(2\pi i/p)$ and $\eta = \exp(2\pi i/r)$ be complex constants. Then:*
  (i) *An element of $G_{n,q}$ can be written in the form*

$$\eta^c X(a) Z(b)$$

  for some a,b in $\mathbb{F}_q^n$ and some c in $\{x \in \mathbb{Z} \mid 0 \leq x < r\}$.
  (ii) *The group $G_{n,q}$ has rq^{2n} elements.*
  (iii) *The exponent of the group $G_{n,q}$ is equal to r.*
  (iv) *The center of $G_{n,q}$ is equal to the cyclic group of order r generated by the element $\eta \mathbf{1}$.*
  (v) *The derived subgroup $G'_{n,q}$ is equal to the cyclic group of order p generated by the element $\omega \mathbf{1}$.*
  (vi) *The quotient group $G_{n,q}/Z(G_{n,q})$ is isomorphic to $(\mathbb{F}_q^{2n}, +)$.*

Proof. The six assertions can be proved as follows:
  (i) The form of the group elements is a consequence of Proposition 12.2 and the definition of the group $G_{n,q}$.
  (ii) The elements given in (i) are all pairwise distinct, whence the formula for the order of the group.
  (iii) It follows from Corollary 12.1 that $(\eta^c X(a) Z(b))^r = X(ra) Z(rb) = \mathbf{1}$, so each element of $G_{n,q}$ has order dividing r. The element $\eta \mathbf{1}$ has order r, hence the exponent of $G_{n,q}$ is equal to r.
  (iv) It follows from Proposition 12.2 that an element $\eta^c X(a) Z(b)$ of $G_{n,q}$ commutes with all elements in $G_{n,q}$ if and only if $a = 0$ and $b = 0$ holds.
  (v) Let $x = \eta^c X(a) Z(b)$ and $y = \eta^{c'} X(a') Z(b')$ be two arbitrary elements of $G_{n,q}$. Then $x^{-1} = Z(-b) X(-a) \eta^{-c}$ and $y^{-1} = Z(-b') X(-a') \eta^{-c'}$; hence, it follows from Proposition 12.2 that the commutator of x and y is of the form $[x,y] = x^{-1} y^{-1} xy = \omega^d \mathbf{1}$ for some integer d in the range $0 \leq d < p$. Since $G_{n,q}$ is non-Abelian, it follows that $G'_{n,q} = \langle \omega \mathbf{1} \rangle$.
  (vi) For all a,b,a',b' in $\mathbb{F}_q$, we have

$$X(a) Z(b) X(a') Z(b') \equiv X(a+a') Z(b+b') \bmod Z(G_{n,q})$$

by Proposition 12.2. Therefore, $G_{n,q}/Z(G_{n,q}) \cong \mathbb{F}_q^n \times \mathbb{F}_q^n \cong \mathbb{F}_q^{2n}$. □

Remark 12.1. In general, a p-group G with cyclic center $Z(G)$ and derived subgroup G' of order p is called a *generalized extraspecial p-group*. Thus, the preceding proposition shows that the group $G_{n,q}$ is a generalized extraspecial p-group.

We will now define the key notion. A *stabilizer code* Q is a nonzero subspace of $\mathbb{C}^{q^n}$ that satisfies
$$Q = \bigcap_{E \in S} \{v \in \mathbb{C}^{q^n} \mid Ev = v\} \tag{12.1}$$
for some subgroup S of the group $G_{n,q}$. We call S the *stabilizer group* of the stabilizer code Q.

The next proposition shows that a stabilizer code determines its stabilizer group and vice versa.

Proposition 12.4. *Let S be a subgroup of the error group $G_{n,q}$.*

(i) If S is the stabilizer group of a stabilizer code Q, then S is an Abelian group satisfying $S \cap Z(G_{n,q}) = 1$.

(ii) If S is an Abelian group satisfying $S \cap Z(G_{n,q}) = 1$, then
$$P = \frac{1}{|S|} \sum_{E \in S} E \tag{12.2}$$

is an orthogonal projector onto a stabilizer code with stabilizer group S.

(iii) The stabilizer code in $\mathbb{C}^{q^n}$ has dimension K if and only if its stabilizer group has q^n/K elements.

Proof. (i) Seeking a contradiction, we assume that E and F are noncommuting elements in S. By Proposition 12.2 there exists some complex constant λ such that $\lambda \neq 1$ and $EF = \lambda FE$. For any nonzero vector v in Q, we get $v = EFv = \lambda FEv = \lambda v \neq v$, contradicting our assumption that there exist noncommuting elements in S. Therefore, we can conclude that S is an Abelian group. A nonidentity element in $Z(G_{n,q})$ cannot satisfy Eq. (12.1), hence $S \cap Z(G_{n,q}) = 1$ holds.

(ii) For all F in S, we have $\sum_{E \in S} FE = \sum_{E \in S} F$, since a group acts on itself by permutation. It follows that $P^2 = P$. Since S is mapped to itself by the map $E \mapsto E^\dagger$, we have $P = P^\dagger$. Therefore, P is an orthogonal projector onto some subspace Q of $\mathbb{C}^{q^n}$. For all v in Q, we have $Ev = EPv = Pv = v$. Conversely, if $Ev = v$ holds for all E in S, then $Pv = v$, so v is contained in Q. Therefore, Q is a stabilizer code satisfying
$$\operatorname{img} P = Q = \bigcap_{E \in S} \{v \in \mathbb{C}^{q^n} \mid Ev = v\}.$$

(iii) The last assertion follows from $K = \dim Q = \operatorname{Tr} P = q^n/|S|$. □

Detectable errors. A quantum code Q is able to *detect* an error E in $\mathscr{B}(\mathbb{C}^{q^n})$ if and only if there exists a constant λ_E depending only on E such that for all u, v in Q the condition $\langle u|E|v\rangle = \lambda_E \langle u|v\rangle$ is satisfied, see [KL97] or in Chapter 2 of this volume.

Recall that for a subset T of a group G, the *centralizer* of T in G is defined as $C_G(T) = \{g \in G \mid gt = tg \text{ for all } t \in T\}$.

The next proposition characterizes the errors in the group $G_{n,q}$ that are detectable by a stabilizer code that has stabilizer group S.

Proposition 12.5. *For brevity, set $G = G_{n,q}$. Let the subgroup S of G be the stabilizer group of a stabilizer code Q. An error E in G is detectable by the quantum code Q if and only if E is an element of $SZ(G) \cup (G \setminus C_G(S))$.*

Proof. "$\Longleftarrow$". An element E in $SZ(G)$ is a scalar multiple of a stabilizer; thus, it acts by multiplication with a scalar λ_E on Q. It follows that E is a detectable error.

Suppose now that E is an error in G that does not commute with some element F of the stabilizer S; it follows that $EF = \lambda FE$ for some complex number $\lambda \neq 1$ by Proposition 12.2. All vectors u and v in Q satisfy the condition

$$\langle u|E|v\rangle = \langle u|EF|v\rangle = \lambda \langle u|FE|v\rangle = \lambda \langle u|E|v\rangle; \tag{12.3}$$

hence, $\langle u|E|v\rangle = 0$. It follows that the error E is detectable.

"$\Longrightarrow$". Conversely, suppose that E is an element of $C_G(S) \setminus SZ(G)$. Seeking a contradiction, we assume that E is detectable; this implies that there exists a complex scalar λ_E such that $Ev = \lambda_E v$ for all v in Q. The scalar λ_E cannot be zero because E commutes with the elements of S, so $EP = PEP = \lambda_E P$ and clearly $EP \neq 0$. Let S^* denote the Abelian group generated by $\lambda_E^{-1} E$ and by the elements of S. The joint eigenspace of S^* with eigenvalue 1 has dimension $q^n/|S^*| < \dim Q = q^n/|S|$. This implies that not all vectors in Q remain invariant under $\lambda_E^{-1} E$, in contradiction to the detectability of E. This concludes the proof. $\square$

Remark 12.2. *The assertion of the previous proposition can be rephrased as follows: An element in $G_{n,q}$ is detectable by a stabilizer code with stabilizer group S if and only if (a) it does not commute with some element in S or (b) it is scalar multiple of an element in S.*

We define the *weight* of an element in the error group $G_{n,q}$ by

$$\text{wt}(\eta^c X(a)Z(b)) = |\{k \,|\, (a_k, b_k) \neq (0,0) \text{ and } 1 \leq k \leq n\}|.$$

The *minimum distance* d of a stabilizer code of dimension K with stabilizer group S is defined by

$$d = \begin{cases} \min\{\text{wt}(E) \,|\, E \in C_{G_{n,q}}(S) \setminus SZ(G_{n,q})\} & \text{if } K > 1, \\ \min\{\text{wt}(E) \,|\, E \in C_{G_{n,q}}(S) \setminus Z(G_{n,q})\} & \text{if } K = 1. \end{cases} \tag{12.4}$$

Thus, a stabilizer code Q of dimension $K > 1$ has minimum distance d if and only if it can detect all errors in $G_{n,q}$ of weight less than d, but cannot detect some error of weight d.

A quantum code Q is called *pure to* t if and only if its stabilizer group S does not contain non-scalar matrices of weight less than t. A quantum code is called *pure* if and only if it is pure to its minimum distance.

We say that Q is an $((n, K, d))_q$ code if and only if Q is a K-dimensional subspace of $\mathbb{C}^{q^n}$ that has minimum distance d. An $((n, q^k, d))_q$ code is also called an $[[n, k, d]]_q$ code.

Remark 12.3. *The reason for the dichotomy in the definition of the minimum distance is that a stabilizer code of dimension $K = 1$ with stabilizer group S satisfies $C_{G_{n,q}}(S) = SZ(G_{n,q})$. In this case, the definition of minimum distance used in the case of dimension $K > 1$ would be vacuous. A more meaningful definition is obtained by requiring that quantum codes of dimension 1 have to be pure. The definition given above allows one to devise a measurement that can detect all errors in $G_{n,q}$ with weight below the minimum distance; this would not be possible if the definition for dimension $K > 1$ were used for the case of dimension $K = 1$.*

Symplectic geometry. The detectability of errors in $G_{n,q}$ by a stabilizer code with stabilizer group S is completely determined by commutation properties. In this paragraph, we will explain a fundamental connection between the error group $G_{n,q}$ and a symplectic geometry on $\mathbb{F}_q^{2n}$. This connection will allow us to relate stabilizer codes to classical codes and vice versa.

Let V be a vector space over $\mathbb{F}_p$. Recall that $f\colon V \times V \to \mathbb{F}_p$ is called a *symplectic form* if and only if f is bilinear, and $f(x,y) = -f(y,x)$ and $f(x,x) = 0$ hold for all x,y in V. A symplectic form is called *nondegenerate* if and only if $f(x,y) = 0$ for all y in V implies that $x = 0$. The pair (V, f) is called a *symplectic space*.

Recall that a finite p-group is called *elementary Abelian* if and only if it is a finite Abelian group in which every element has order p. We can regard an elementary Abelian p-group as a vector space over $\mathbb{F}_p$.

The following well-known fact from group theory shows how symplectic spaces arise from finite groups.

Lemma 12.1. *Let G be a finite p-group with cyclic center $Z(G)$ and elementary Abelian factor group $G/Z(G)$. Then G' is a cyclic group of order p generated by an element z in $Z(G)$. If we set*

$$f(\overline{x},\overline{y}) = a \text{ for } \overline{x} = xZ(G),\ \overline{y} = yZ(G),\ \text{and } [x,y] = z^a$$

and regard a as an element in $\mathbb{F}_p$, then $V = G/Z(G)$ can be viewed as a vector space over $\mathbb{F}_p$ that is equipped with a nondegenerate symplectic form $f\colon V \times V \to \mathbb{F}_p$.

Proof. Since $G/Z(G)$ is Abelian, we have $G' \subseteq Z(G)$. Since commutators are central elements in G, we have

$$[xx',y] = [x,y][x',y] \quad \text{and} \quad [x,yy'] = [x,y][x,y'] \tag{12.5}$$

for all x, x', y, y' in G. By definition of G, we have $x^p \in Z(G)$ for all x in G. Therefore,

$$1 = [x^p, y] = [x,y]^p$$

holds for all x, y in G. Since $G' \subseteq Z(G)$ is a cyclic group, we can conclude that the derived subgroup G' is of order p.

The function f is well-defined, since central elements do not affect a commutator. It follows from the equations (12.5) that f is bilinear. Since $[x,y] = [y,x]^{-1}$ and $[x,x] = 1$, we have $f(\overline{x},\overline{y}) = -f(\overline{y},\overline{x})$ and $f(\overline{x},\overline{x}) = 0$. Therefore, f is a symplectic form on V. For x, y in G, let $\overline{x} = xZ(G)$ and $\overline{y} = yZ(G)$. If $f(\overline{x},\overline{y}) = 0$ for all $\overline{y}$ in $G/Z(G)$, then $[x,y] = 1$ for all y in G, so x must be an element of $Z(G)$. Therefore, the form $f\colon G/Z(G) \times G/Z(G) \to \mathbb{F}_p$ is nondegenerate. □

Remark 12.4. *The key feature of a group G satisfying the hypothesis of the preceding lemma is that two elements x and y in G commute if and only if their images $\overline{x}$ and $\overline{y}$ in $G/Z(G)$ are orthogonal with respect to the form f.*

By Proposition 12.3, the error group $G_{n,q}$ satisfies the hypothesis of the previous lemma. The next lemma shows that in this case the symplectic form $f\colon \mathbb{F}_q^{2n} \times \mathbb{F}_q^{2n} \to \mathbb{F}_p$ is given by

$$f((a|b),(a'|b')) = \operatorname{Tr}(b \cdot a' - b' \cdot a). \tag{12.6}$$

Since in this case f is obtained by taking the trace of the standard $\mathbb{F}_q$-linear symplectic form $b \cdot a' - b' \cdot a$ on $\mathbb{F}_q^{2n}$, we will refer to (12.6) as the *trace symplectic form* on $\mathbb{F}_q^{2n}$.

Lemma 12.2. *Let ω and η denote the constants defined in Proposition 12.3. The commutator of two elements $E = \eta^c X(a)Z(b)$ and $E' = \eta^{c'} X(a')Z(b')$ of the error group $G_{n,q}$ is given by*

$$[E, E'] = \omega^{\mathrm{Tr}(b \cdot a' - b' \cdot a)} \mathbf{1}.$$

In particular, the elements E and E' in $G_{n,q}$ commute if and only if the trace symplectic form $\mathrm{Tr}(b \cdot a' - b' \cdot a)$ vanishes.

Proof. By Proposition 12.2, we have

$$EE' = \eta^{c+c'} \omega^{\mathrm{Tr}(b \cdot a')} X(a+a') Z(b+b')$$

and

$$E'E = \eta^{c+c'} \omega^{\mathrm{Tr}(b' \cdot a)} X(a+a') Z(b+b').$$

Therefore,

$$[E, E'] = E^{-1} E'^{-1} EE' = \omega^{\mathrm{Tr}(b \cdot a' - b' \cdot a)} \mathbf{1}. \qquad \square$$

We will write $(a|b) \perp_s (a'|b')$ if and only if $\mathrm{Tr}(b \cdot a' - b' \cdot a) = 0$. If X is a subset of $\mathbb{F}_q^{2n}$, we will denote by $X^{\perp_s}$ the set

$$X^{\perp_s} = \{(a|b) \in \mathbb{F}_q^{2n} \mid (a|b) \perp_s (a'|b') \text{ for all } (a'|b') \in X\}.$$

We define the *symplectic weight* swt of a vector $(a|b)$ in $\mathbb{F}_q^{2n}$ as

$$\mathrm{swt}((a|b)) = |\{\, k \in \{1, 2, \ldots, n\} \mid (a_k, b_k) \neq (0, 0)\,\}|,$$

that is, the symplectic weight equals the number of coordinates k such that a_k or b_k are nonzero. It follows from the definitions that

$$\mathrm{wt}(\eta^c X(a)Z(b)) = \mathrm{swt}((a|b)).$$

If X is a subset of $G_{n,q}$, then we define $\mathrm{wt}(X) = \min\{\mathrm{wt}(x) \mid x \in X\}$. Similarly, if S is a subset of $\mathbb{F}_q^{2n}$, then we define $\mathrm{swt}(S) = \min\{\mathrm{wt}(s) \mid s \in S\}$.

Theorem 12.1. *An $((n, K, d))_q$ stabilizer code exists if and only if there exists an additive code C that is a subset of $\mathbb{F}_q^{2n}$ of cardinality $|C| = q^n/K$ satisfying the self-orthogonality relation $C \subseteq C^{\perp_s}$ and the weight condition*

$$d = \begin{cases} \mathrm{swt}(C^{\perp_s} \setminus C) & \text{if } K > 1, \\ \mathrm{swt}(C^{\perp_s} \setminus \{0\}) & \text{if } K = 1. \end{cases} \tag{12.7}$$

Proof. "$\Longrightarrow$". Suppose that an $((n, K, d))_q$ stabilizer code Q exists. By Proposition 12.4, the stabilizer group S of Q is an Abelian subgroup of $G_{n,q}$ of order $|S| = q^n/K$ satisfying $S \cap Z(G_{n,q}) = 1$. It follows that the quotient group $C \cong SZ(G_{n,q})/Z(G_{n,q})$ is an additive subgroup of $\mathbb{F}_q^{2n}$ such that $|C| = |S| = q^n/K$. We have $C^{\perp_s} = C_{G_{n,q}}(S)/Z(G_{n,q})$ by Lemma 12.2. Since S is an Abelian group, $SZ(G_{n,q}) \leq C_{G_{n,q}}(S)$, hence $C \leq C^{\perp_s}$. If $K = 1$, then Q is a pure quantum code, thus $d = \mathrm{wt}(C_{G_{n,q}}(S) \setminus Z(G_{n,q})) = \mathrm{swt}(C^{\perp_s} \setminus \{0\})$. If $K > 1$, then $d = \mathrm{wt}(C_{G_{n,q}}(S) \setminus SZ(G_{n,q})) = \mathrm{swt}(C^{\perp_s} \setminus C)$ by Proposition 12.5.

"$\Longleftarrow$". Conversely, suppose that C is an additive subcode of $\mathbb{F}_q^{2n}$ of order $|C| = q^n/K$ satisfying $C \subseteq C^{\perp_s}$ and the weight condition (12.7). If we define N by

$$N = \{\eta^c X(a)Z(b) \,|\, c \in \{0, \ldots, r-1\} \text{ and } (a|b) \in C\},$$

then N is an Abelian normal subgroup of $G_{n,q}$, because it is the pre-image of $C = N/Z(G_{n,q})$. Choose a character χ of N such that $\chi(\eta^c I) = \eta^c$. Then

$$P_N = \frac{1}{|N|} \sum_{E \in N} \chi(E^{-1})E$$

is an orthogonal projector onto a vector space Q, because P_N is an idempotent in the group ring $\mathbb{C}[G_{n,q}]$. We have

$$\dim Q = \operatorname{Tr} P_N = |Z(G_{n,q})|q^n/|N| = q^n/|C| = K.$$

Each coset of N modulo $Z(G_{n,q})$ contains exactly one matrix E such that $Ev = v$ for all v in Q. Set $S = \{E \in N \,|\, Ev = v$ for all $v \in Q\}$. Then S is an Abelian subgroup of $G_{n,q}$ of order $|S| = |C| = q^n/K$. Consider the vector space

$$Q^* = \bigcap_{E \in S} \{v \,|\, Ev = v \text{ for all } E \in S\}.$$

By definition, Q is a subspace of Q^*. Since $\dim Q^* = q^n/|S| = K = \dim Q$, we have $Q = Q^*$. Therefore, $Q = Q^*$ is a stabilizer code with stabilizer group S. If (12.4) does not hold, then (12.7) does not hold either, contradicting our assumption. Therefore, Q is an $((n, K, d))_q$ stabilizer code. □

Remark 12.5. *The previous theorem fixes an error with respect to characteristic 2 that I made in my prior work. In [KKK+06, Theorem 13], I incorrectly used an extraspecial 2-group instead of a generalized extraspecial 2-group with a center of order 4 as an error group.*

Recall that the *Hamming weight* $\operatorname{wt}(v)$ of a vector v in $\mathbb{F}_q^n$ is equal to the number of nonzero components of the vector v. We use the same symbol for the Hamming weight as for the weight of stabilizer elements, since there is no risk of confusion.

For vectors u and v in $\mathbb{F}_q^n$, the dot inner product $u \cdot v$ is defined as $u \cdot v = u_1 v_1 + \cdots + u_n v_n$. If $u \cdot v = 0$, then we write $u \perp v$. For a subset V of $\mathbb{F}_q^n$, we denote by $V^\perp$ the vector space $V^\perp = \{u \in \mathbb{F}_q^n \,|\, u \cdot v = 0 \text{ for all } v \in V\}$. We call $V^\perp$ the *Euclidean dual* of V.

A simple consequence of the previous theorem is the CSS construction. This elegant construction of quantum codes was introduced by Calderbank and Shor [CS96] and Steane [S96e].

Corollary 12.3 (CSS construction). *(a) Let C_1 and C_2 denote two classical linear codes with parameters $[n, k_1, d_1]_q$ and $[n, k_2, d_2]_q$ such that $C_2^\perp \leq C_1$. Then there exists a $[[n, k_1 + k_2 - n, d]]_q$ stabilizer code with minimum distance $d = \min\{\operatorname{wt}(c) \,|\, c \in (C_1 \setminus C_2^\perp) \cup (C_2 \setminus C_1^\perp)\}$ that is pure to $\min\{d_1, d_2\}$.*
(b) If C is a classical linear $[n, k, d]_q$ code containing its dual, $C^\perp \leq C$, then there exists an $[[n, 2k - n, \geq d]]_q$ stabilizer code that is pure to d. If the minimum distance of $C^\perp$ exceeds d, then the stabilizer code is pure and has minimum distance d.

Proof. For the proof of assertion (a), let $C = C_1^\perp \times C_2^\perp \le \mathbb{F}_q^{2n}$. If $(c_1 \mid c_2)$ and $(c_1' \mid c_2')$ are two elements of C, then we observe that

$$\mathrm{Tr}(c_2 \cdot c_1' - c_2' \cdot c_1) = \mathrm{Tr}(0 - 0) = 0.$$

Therefore, $C \le C^{\perp_s}$. Furthermore, the trace-symplectic dual of C contains $C_2 \times C_1$, and a dimensionality argument shows that $C^{\perp_s} = C_2 \times C_1$. Since the Cartesian product $C_1^\perp \times C_2^\perp$ has $q^{2n-(k_1+k_2)}$ elements, the stabilizer code has dimension $q^{k_1+k_2-n}$ by Theorem 12.1. The claim about the minimum distance and purity of the code is obvious from the construction.

Assertion (b) is a direct consequence of assertion (a). $\square$

Classical coding theory deals almost exclusively with the Hamming metric. Unfortunately, the CSS construction is only applicable to a limited class of quantum stabilizer codes. Our next goal is to find an isometric isomorphism from $\mathbb{F}_q^{2n}$ with the metric induced by the symplectic weight to $\mathbb{F}_{q^2}^n$ with the metric induced by the Hamming weight.

Let β be an element of $\mathbb{F}_{q^2}$ such that (β, β^q) is a basis of $\mathbb{F}_{q^2}$ over $\mathbb{F}_q$. For example, a primitive element β of $\mathbb{F}_{q^2}$ generates such a basis. We define a bijective map $\phi \colon \mathbb{F}_q^{2n} \to \mathbb{F}_{q^2}^n$ by $\phi((a|b)) = \beta a + \beta^q b$. Then

$$\mathrm{swt}\big((a|b)\big) = \mathrm{wt}\big(\phi((a|b))\big)$$

holds for all $(a|b)$ in $\mathbb{F}_q^{2n}$, so ϕ is the desired isometry.

We define a symplectic form $a \colon \mathbb{F}_{q^2}^n \times \mathbb{F}_{q^2}^n \to \mathbb{F}_p$ by

$$a(v, w) = \mathrm{Tr}_{q/p}\left(\frac{v \cdot w^q - v^q \cdot w}{\beta^{2q} - \beta^2}\right). \tag{12.8}$$

Notice that the argument of the trace is invariant under the Galois automorphism $x \mapsto x^q$, so it is indeed an element of $\mathbb{F}_q$, which shows that (12.8) is well-defined. Regarding $\mathbb{F}_{q^2}^n$ as an $\mathbb{F}_{q^2}$ vector space, we can interpret the form a as a trace form. We will refer to a as the *trace-alternating form*. We write $u \perp_a w$ if and only if $a(u, w) = 0$ holds.

Theorem 12.2. *An $((n, K, d))_q$ stabilizer code exists if and only if there exists an additive subcode D of $\mathbb{F}_{q^2}^n$ of cardinality $|D| = q^n/K$ satisfying the self-orthogonality constraint $D \subseteq D^{\perp_a}$ and the weight condition*

$$d = \begin{cases} \mathrm{wt}(D^{\perp_a} \setminus D) & \text{if } K > 1, \\ \mathrm{wt}(D^{\perp_a} \setminus \{0\}) & \text{if } K = 1. \end{cases}$$

Proof. We begin by showing that the isometry ϕ preserves the value of the trace symplectic form. If $c = (a|b)$ and $d = (a'|b')$ are elements of $\mathbb{F}_q^{2n}$, then

$$\phi(c) \cdot \phi(d)^q = \beta^{q+1} a \cdot a' + \beta^2 a \cdot b' + \beta^{2q} b \cdot a' + \beta^{q+1} b \cdot b',$$
$$\phi(c)^q \cdot \phi(d) = \beta^{q+1} a \cdot a' + \beta^{2q} a \cdot b' + \beta^2 b \cdot a' + \beta^{q+1} b \cdot b'.$$

Therefore, the trace-alternating form of $\phi(c)$ and $\phi(d)$ is given by

$$a(\phi(c), \phi(d)) = \mathrm{Tr}_{q/p}\left(\frac{\phi(c) \cdot \phi(d)^q - \phi(c)^q \cdot \phi(d)}{\beta^{2q} - \beta^2}\right),$$
$$= \mathrm{Tr}_{q/p}(b \cdot a' - a \cdot b'),$$

which is precisely the trace-symplectic form. In particular, this shows that $c \perp_s d$ holds if and only if $\phi(c) \perp_a \phi(d)$ holds.

By Theorem 12.1, an $((n, K, d))_q$ stabilizer code exists if and only if there exists an additive code $C \subseteq \mathbb{F}_q^{2n}$ of cardinality $|C| = q^n/K$ satisfying $C \leq C^{\perp_s}$, $\operatorname{swt}(C^{\perp_s} \setminus C) = d$ if $K > 1$, and $\operatorname{swt}(C^{\perp_s}) = d$ if $K = 1$. By the first part of the proof, we obtain the statement of the theorem by applying the isometry ϕ. □

We obtain the following convenient condition for the existence of a stabilizer code as a direct consequence of the previous theorem.

Corollary 12.4. *If there exists a classical $[n, k]_{q^2}$ additive code $D \subseteq \mathbb{F}_{q^2}^n$ such that $D \subseteq D^{\perp_a}$ and $d^{\perp_a} = \operatorname{wt}(D^{\perp_a} \setminus \{0\})$, then there exists a stabilizer code with parameters $[[n, n - 2k, \geq d^{\perp_a}]]_q$ that is pure to $d^{\perp_a}$.*

12.2 Cyclic codes

This section gives some background on classical cyclic codes. We will merely introduce the terminology and the result needed in the subsequent section. More details on cyclic codes can be found in any standard textbook on coding theory, see e.g., [HP03] or [MS77].

A classical linear $[n, k]_q$ code C is a k-dimensional subspace of the vectors space $\mathbb{F}_q^n$. A classical linear code C is called *cyclic* if and only if it contains $(a_{n-1}, a_1, a_2, \ldots, a_{n-2})$ for each codeword $(a_0, a_1, \ldots, a_{n-2}, a_{n-1})$ in C.

A more convenient algebraic characterization of classical cyclic codes of length n over $\mathbb{F}_q$ can be obtained as follows. Let

$$\langle x^n - 1 \rangle = \{q(x)(x^n - 1) \mid q(x) \in \mathbb{F}_q[x]\}$$

denote the principal ideal generated by the polynomial $x^n - 1$ in the univariate polynomial ring $\mathbb{F}_q[x]$. We denote by R the quotient ring $R = \mathbb{F}_q[x]/\langle x^n - 1 \rangle$. Let us define a bijective map $p \colon \mathbb{F}_q^n \to R$ by

$$p((a_0, \ldots, a_{n-1})) = \sum_{j=0}^{n-1} a_j \, x^j.$$

One can show that a nonzero subset C of $\mathbb{F}_q$ is a cyclic code if and only if $p(C)$ is an ideal in R.

Since $\mathbb{F}_q[x]$ is a Euclidean domain, every nonzero ideal J in $\mathbb{F}_q[x]$ is generated by the unique monic polynomial of least degree contained in J. It follows that every nonzero ideal I in R is a principal ideal generated by the quotient class $g(x) + \langle x^n - 1 \rangle$ of the unique monic polynomial $g(x)$ of least degree contained in the ideal $J = I + \langle x^n - 1 \rangle$ in $\mathbb{F}_q[x]$.

If C is a cyclic code of length n over the alphabet $\mathbb{F}_q$, then the unique monic polynomial $g(x)$ of least degree in the ideal $p(C) + \langle x^n - 1 \rangle$ of $\mathbb{F}_q[x]$ is called the *generator polynomial* of the code C.

One can show that the generator polynomial divides $x^n - 1$. The polynomial $h(x) = (x^n - 1)/g(x)$ is called the *check polynomial* of the cyclic code C. A cyclic $[n, k]_q$ code has a generator polynomial of degree $n - k$ and a check polynomial of degree k. The generator polynomial of the Euclidean dual code $C^\perp$ is given by $h^\dagger(x) = x^k h(1/x)$, the reciprocal polynomial of the check polynomial of C.

The generator polynomial $g(x)$ provides a succinct description of a cyclic code. If the length n and the size q of the finite field are coprime, then an alternate description of the code can be obtained by so-called defining sets.

Assume that $\gcd(n,q) = 1$ so that the polynomial $x^n - 1$ has simple roots. If m is the smallest positive integer such that n divides $q^m - 1$, then $\mathbb{F}_{q^m}$ is the splitting field of $x^n - 1$ over $\mathbb{F}_q$. The field $\mathbb{F}_{q^m}$ contains a primitive nth root of unity β. One can describe a cyclic code with generator polynomial $g(x)$ in terms of its *defining set*

$$Z = \{k \mid g(\beta^k) = 0 \text{ for } 0 \leq k < n\}.$$

The defining set depends on the choice of the root of unity β. If α is another primitive nth root of unity in $\mathbb{F}_{q^m}$, then $\alpha = \beta^a$ for some integer a in the range $1 \leq a < n$ such that $\gcd(a,n) = 1$ and the defining set Z_α with respect to α is given by $Z_\alpha = \{a^{-1}k \pmod{n} \mid k \in Z\}$.

Cyclic codes that contain their Euclidean duals can be nicely characterized in terms of their generator polynomials and defining sets.

Lemma 12.3. *Suppose that q is a power of a prime and n is a positive integer such that $\gcd(n,q) = 1$. Let C be an $[n,k,d]_q$ cyclic code with defining set Z and generator polynomial $g(x)$. Then the following conditions are equivalent:*

(i) $C^\perp \subseteq C$;
(ii) $x^n - 1 \equiv 0 \pmod{g(x)g^\dagger(x)}$, where $g^\dagger(x) = x^{n-k}g(1/x)$;
(iii) $Z \subseteq \{-z \mid z \in N \setminus Z\}$, where $N = \{0, 1, \ldots, n-1\}$;
(iv) $Z \cap Z^{-1} = \emptyset$, where $Z^{-1} = \{-z \pmod{n} \mid z \in Z\}$.

If any of the previous conditions are satisfied, then there exists an $[[n, 2k - n, \geq d]]_q$ stabilizer code that is pure to d.

Proof. The code C has the check polynomial $h(x) = (x^n - 1)/g(x)$. Recall that the generator polynomial $h^\dagger(x)$ of $C^\perp$ is the reciprocal of the check polynomial of C. Thus, $C^\perp \subseteq C$ if and only if $g(x)$ divides $h^\dagger(x)$. We can rewrite the generator polynomial $h^\dagger(x)$ of $C^\perp$ in the form

$$h^\dagger(x) = x^k h(x^{-1}) = (1 - x^n)/(x^{n-k}g(x^{-1})) = -(x^n - 1)/g^\dagger(x).$$

Therefore, $C^\perp \subseteq C$ holds if and only if $g(x)$ divides $(x^n - 1)/g^\dagger(x)$. Since the latter condition is equivalent to $x^n - 1 \equiv 0 \mod g(x)g^\dagger(x)$, we have established the equivalence of conditions (i) and (ii).

The defining set of $C^\perp$ is given by $\{-z \bmod n \mid z \in N \setminus Z\}$, where $N = \{0, 1, \ldots, n-1\}$. Thus, $C^\perp \subseteq C$ if and only if $Z \subseteq \{-z \bmod n \mid N \setminus Z\}$. This proves the equivalence of conditions (i) and (iii).

Since $Z \subseteq \{-z \bmod n \mid N \setminus Z\}$ means that $Z^{-1} \subseteq N \setminus Z$, we can express this condition also in the form $Z \cap Z^{-1} = \emptyset$. Hence, conditions (iii) and (iv) are equivalent.

The existence of an $[[n, 2k-n, \geq d]]_q$ stabilizer code follows from Corollary 12.3. $\square$

Notice that conditions (iii) and (iv) of Lemma 12.3 do not depend on the choice of the primitive nth root.

12.3 Quantum BCH codes

In this section, we illustrate the construction of cyclic stabilizer codes. We will use classical Bose–Chaudhuri–Hocquenghem (BCH) codes. This class of codes is rich enough to construct many interesting quantum codes. Binary quantum BCH codes were studied in [CRS+98, CEL99,

GB99, S99b]. We will discuss quantum BCH codes whose alphabet size is a prime power, following [A08].

Let q be a power of a prime and n a positive integer that is coprime to q. If an integer x is in the range $0 \leq x < n$, then the *cyclotomic coset* C_x of x modulo n is defined as

$$C_x = \{xq^r \bmod n \mid r \in \mathbb{Z}, r \geq 0\}.$$

A *BCH code* C of length n and designed distance δ over $\mathbb{F}_q$ is a cyclic code whose defining set Z is given by a union of $\delta - 1$ subsequent cyclotomic cosets,

$$Z = \bigcup_{x=b}^{b+\delta-2} C_x.$$

The generator polynomial of the code is of the form

$$g(x) = \prod_{z \in Z} (x - \beta^z),$$

where β is a primitive nth root of unity of some extension field of $\mathbb{F}_q$.

The definition of a BCH code ensures that $g(x)$ generates a cyclic $[n, k, d]_q$ code of dimension $k = n - |Z|$ and minimum distance $d \geq \delta$. The parameter δ is called the *designed distance* of the BCH code, and the parameter b is called the *offset*.

If the offset $b = 1$, then the code C is called a *narrow-sense BCH code*, and if $n = q^m - 1$ for some $m \geq 1$, then the code is called *primitive*. We will denote by $\mathrm{BCH}(n, q; \delta)$ a narrow-sense BCH code of length n over $\mathbb{F}_q$ with designed distance δ.

A BCH code over $\mathbb{F}_q$ of length $q - 1$ is called a *Reed–Solomon code*. A Reed–Solomon code that is a narrow-sense BCH code is called a *narrow-sense Reed–Solomon code*. Arguably, Reed–Solomon codes are the most important subclass of BCH codes.

One of the main reasons why BCH codes have been so widely studied in classical coding theory is that they have some highly desirable features:
- One can specify a BCH code of length n by merely giving its offset b and designed distance δ.
- The dimension k of a BCH code of length n of designed distance δ is bounded by $k \geq n - m(\delta - 1)$, where m is the multiplicative order of q modulo n.
- Since the minimum distance d is lower-bounded by the designed distance δ, one can easily control the error-correction capabilities of the code.

As far as classical codes are concerned, it is often sufficient to know that a BCH code of length n with design distance δ yields an

$$[n, k \geq n - m(\delta - 1), d \geq \delta]_q$$

code. However, for the construction of quantum error-correcting codes using Lemma 12.3 or Corollary 12.3, it would be desirable to obtain the following information:
◦ For which choice of the length n, offset b, and designed distance δ are we guaranteed that the resulting BCH code contains its dual code?
◦ Can we determine the dimension of the BCH code precisely without leaving it to the reader to calculate the defining set?
◦ Can we determine the true dimensions of the resulting code (or at least get some fairly sharp bounds)?

The next paragraphs address these questions for some classes of BCH codes.

Reed–Solomon codes and more. We begin by considering narrow-sense BCH codes of length n such that the multiplicative order of q modulo n equals 1. Among others, Reed–Solomon codes belong to this class of codes. It turns out that this is a particularly simple case, but it nicely illustrates the proof technique that will be used throughout this section.

Proposition 12.6. *Suppose that q is a power of a prime and n is a positive integer such that $q \equiv 1 \pmod{n}$. We have $\mathrm{BCH}(n,q;\delta)^\perp \subseteq \mathrm{BCH}(n,q;\delta)$ if and only if the designed distance δ is in the range $2 \leq \delta \leq \delta_{\max} = \lfloor (n+1)/2 \rfloor$.*

Proof. Since q has multiplicative order 1 modulo n, it follows that all cyclotomic cosets C_j are Singleton sets, $C_j = \{j\}$. Therefore, the defining set Z of $\mathrm{BCH}(n,q;\delta)$ is given by $Z = \{1,\ldots,\delta-1\}$. By Lemma 12.3, the inclusion $\mathrm{BCH}(n,q;\delta)^\perp \subseteq \mathrm{BCH}(n,q;\delta)$ holds if and only if the defining set Z satisfies $Z \cap Z^{-1} = \emptyset$.

Suppose that $Z \cap Z^{-1} = \emptyset$. If $x \in Z$, then $n - x \notin Z$, hence we have in particular $n - x > x$. Therefore, $\delta_{\max} \leq \lfloor (n+1)/2 \rfloor$.

Conversely, if $\delta \leq \lfloor (n+1)/2 \rfloor$, then $\min Z^{-1} = \min\{n-1,\ldots,n-\delta+1\} = n-\delta+1 \geq n - \lfloor (n+1)/2 \rfloor + 1 = \lceil (n+1)/2 \rceil \geq \delta_{\max}$; hence, $Z \cap Z^{-1} = \emptyset$ and Lemma 12.3 implies that $\mathrm{BCH}(n,q;\delta)^\perp \subseteq \mathrm{BCH}(n,q;\delta)$. □

Theorem 12.3. *Suppose that q is a power of a prime and n is a positive integer such that $q \equiv 1 \pmod{n}$. If the designed distance δ is in the range $2 \leq \delta \leq \lfloor (n+1)/2 \rfloor$, then there exists a pure $[[n, n-2\delta+2, \delta]]_q$ stabilizer code.*

Proof. Since q has multiplicative order 1 modulo n, a narrow-sense BCH code over $\mathbb{F}_q$ of designed distance δ has defining set $Z = \{1,2,\ldots,\delta-1\}$, that is, the defining set does not contain any excess exponents. Therefore, the code $C = \mathrm{BCH}(n,q;\delta)$ has the parameters $[n,k,d \geq \delta]$ with $k = n - |Z| = n - \delta + 1$ and $d = \delta$ by the Singleton bound for classical codes. By Proposition 12.6, the code C satisfies $C^\perp \subseteq C$. Therefore, by Lemma 12.3, there exists an $[[n, n-2\delta+2, d' \geq \delta]]_q$ stabilizer code that is pure to δ. By the quantum Singleton bound, the minimum distance d' must be equal to δ. □

Dual-containing BCH codes. If the multiplicative order m of q modulo n is larger than 1, then the defining set of the code has a more intricate structure, so proofs become more involved. The next theorem gives a sufficient condition on the designed distances for which the dual code of a narrow-sense BCH code is self-orthogonal.

Theorem 12.4. *Suppose that $m = \mathrm{ord}_n(q)$. If the designed distance δ is in the range $2 \leq \delta \leq \delta_{\max} = \lfloor \kappa \rfloor$, where*

$$\kappa = \frac{n}{q^m - 1}(q^{\lceil m/2 \rceil} - 1 - (q-2)[m \text{ odd}]), \qquad (12.9)$$

then $\mathrm{BCH}(n,q;\delta)^\perp \subseteq \mathrm{BCH}(n,q;\delta)$.

Proof. It suffices to show that $\mathrm{BCH}(n,q;\delta_{\max})^\perp \subseteq \mathrm{BCH}(n,q;\delta_{\max})$ holds, since $\mathrm{BCH}(n,q;\delta)$ contains $\mathrm{BCH}(n,q;\delta_{\max})$, and the claim follows from these two facts.

Seeking a contradiction, we assume that $\mathrm{BCH}(n,q;\delta_{\max})$ does not contain its dual. Let $Z = C_1 \cup \cdots \cup C_{\delta_{\max}-1}$ be the defining set of $\mathrm{BCH}(n,q;\delta_{\max})$. By Lemma 12.3, $Z \cap Z^{-1} \neq \emptyset$, which means that there exist two elements $x,y \in \{1,\ldots,\delta_{\max}-1\}$ such that $y \equiv -xq^j$

(mod n) for some $j \in \{0, 1, \ldots, m-1\}$, where m is the multiplicative order of q modulo n. Since $\gcd(q, n) = 1$ and $q^m \equiv 1 \pmod{n}$, we also have $x \equiv -yq^{m-j} \pmod{n}$. Thus, exchanging x and y if necessary, we can even assume that j is in the range $0 \leq j \leq \lfloor m/2 \rfloor$. It follows from (12.9) that

$$1 \leq xq^j \leq (\delta_{\max} - 1)q^j$$
$$\leq \frac{n}{q^m - 1}(q^m - q^j - q^j(q-2)[m \text{ odd}]) - q^j$$
$$< n,$$

for all j in the range $0 \leq j \leq \lfloor m/2 \rfloor$. Since $1 \leq xq^j < n$ and $1 \leq y < n$, it follows from $y \equiv -xq^j \pmod{n}$ that $y = n - xq^j$. But this implies that

$$y \geq n - xq^{\lfloor m/2 \rfloor}$$
$$\geq n - \frac{n}{q^m - 1}(q^m - q^{\lfloor m/2 \rfloor} - q^{\lfloor m/2 \rfloor}(q-2)[m \text{ odd}]) + q^{\lfloor m/2 \rfloor}$$
$$= \frac{n}{q^m - 1}(q^{\lfloor m/2 \rfloor} - 1 + q^{\lfloor m/2 \rfloor}(q-2)[m \text{ odd}]) + q^{\lfloor m/2 \rfloor}$$
$$\geq \delta_{\max},$$

contradicting the fact that $y < \delta_{\max}$. □

Narrow-sense BCH codes containing their Euclidean dual code cannot have a designed distance signficantly larger than the range of the previous theorem, as the following fact shows.

Theorem 12.5. *Suppose that $m = \text{ord}_n(q)$. If the designed distance δ exceeds $\delta_{\max} = \lfloor qn^{1/2} \rfloor$, then* $\text{BCH}(n, q; \delta)^\perp \not\subseteq \text{BCH}(n, q; \delta)$.

For the proof see [A08, Theorem 4].

For primitive narrow-sense BCH codes, one can derive necessary and sufficient conditions on the designed distance to guarantee that the code contains its Euclidean dual code.

Theorem 12.6. *A primitive narrow-sense BCH code of length $n = q^m - 1$, $m \geq 2$, over the finite field $\mathbb{F}_q$ contains its Euclidean dual code if and only if its designed distance δ satisfies*

$$2 \leq \delta \leq \delta_{\max} = q^{\lceil m/2 \rceil} - 1 - (q-2)[m \text{ odd}].$$

For the proof see [A08, Corollary 6].

Dimension and minimum distance. While the results in the previous section are sufficient to tell us when we can construct quantum BCH codes, they are still unsatisfactory because we do not know the dimension of these codes. To this end, we determine the dimension of narrow-sense BCH codes of length n with minimum distance $d = O(n^{1/2})$. It turns out that these results on dimension also allow us to sharpen the estimates of the true distance of some BCH codes.

First, we make some simple observations about cyclotomic cosets that are essential in our proof.

Lemma 12.4. *Let n be a positive integer and q be a power of a prime such that $\gcd(n, q) = 1$ and $q^{\lfloor m/2 \rfloor} < n \leq q^m - 1$, where $m = \text{ord}_n(q)$. The cyclotomic coset $C_x = \{xq^j \bmod n \mid 0 \leq j < m\}$ has cardinality m for all x in the range $1 \leq x \leq nq^{\lceil m/2 \rceil}/(q^m - 1)$.*

Proof. If $m = 1$, then $|C_x| = 1$ for all x and the statement is trivially true. Therefore, we can assume that $m > 1$. Seeking a contradiction, we suppose that $|C_x| < m$, meaning that there exists a divisor j of m such that $xq^j \equiv x \bmod n$, or, equivalently, that $x(q^j - 1) \equiv 0 \bmod n$ holds.

Suppose that m is even. The divisor j of m must be in the range $1 \le j \le m/2$. However, $x(q^j - 1) \le nq^{m/2}(q^{m/2} - 1)/(q^m - 1) < n$; hence $x(q^j - 1) \not\equiv 0 \bmod n$, contradicting the assumption $|C_x| < m$.

Suppose that m is odd. The divisor j of m must be in the range $1 \le j \le m/3$. Since $q^{(m+1)/2} \le q^{2m/3}$ for $m \ge 3$, we have $x(q^j - 1) \le nq^{(m+1)/2}(q^{m/3} - 1)/(q^m - 1) \le nq^{2m/3}(q^{m/3}-1)/(q^m-1) < n$. Therefore, $x(q^j-1) \not\equiv 0 \bmod n$, contradicting the assumption $|C_x| < m$. $\square$

The following observation tells us when some cyclotomic cosets are disjoint.

Lemma 12.5. *Let $n \ge 1$ be an integer and q be a power of a prime such that $\gcd(n, q) = 1$ and $q^{\lfloor m/2 \rfloor} < n \le q^m - 1$, where $m = \mathrm{ord}_n(q)$. If x and y are distinct integers in the range $1 \le x, y \le \min\{\lfloor nq^{\lceil m/2 \rceil}/(q^m - 1) - 1 \rfloor, n - 1\}$ such that $x, y \not\equiv 0 \bmod q$, then the q-ary cyclotomic cosets of x and y modulo n are distinct.*

Proof. If $m = 1$, then clearly $C_x = \{x\}$, $C_y = \{y\}$ and distinct x, y implies that C_x and C_y are disjoint. If $m > 1$, then $x, y \le \lfloor nq^{\lceil m/2 \rceil}/(q^m - 1) - 1 \rfloor < n - 1$. The set $S = \{xq^j \bmod n, yq^j \bmod n \,|\, 0 \le j \le \lfloor m/2 \rfloor\}$ contains $2(\lfloor m/2 \rfloor + 1) \ge m + 1$ elements, since $q^{\lfloor m/2 \rfloor} \times \lfloor nq^{\lceil m/2 \rceil}/(q^m - 1) - 1 \rfloor < n$ and, thus, no two elements are identified modulo n. If we assume that $C_x = C_y$, then the preceding observation would imply that $|C_x| = |C_y| \ge |S| \ge m + 1$, which is impossible since the maximal size of a cyclotomic coset is m. Hence, the cyclotomic cosets C_x and C_y must be disjoint. $\square$

With these results in hand, we can now derive the dimension of narrow-sense BCH codes.

Theorem 12.7. *Let q be a prime power and $\gcd(n, q) = 1$ with $\mathrm{ord}_n(q) = m$. Then a narrow-sense BCH code of length $q^{\lfloor m/2 \rfloor} < n \le q^m - 1$ over $\mathbb{F}_q$ with designed distance δ in the range $2 \le \delta \le \min\{\lfloor nq^{\lceil m/2 \rceil}/(q^m - 1) \rfloor, n\}$ has dimension*

$$k = n - m\lceil (\delta - 1)(1 - 1/q) \rceil. \tag{12.10}$$

Proof. Let the defining set of $\mathrm{BCH}(n, q; \delta)$ be $Z = C_1 \cup C_2 \cdots \cup C_{\delta-1}$; a union of at most $\delta - 1$ consecutive cyclotomic cosets. However, when $1 \le x \le \delta - 1$ is a multiple of q, then $C_{x/q} = C_x$. Therefore, the number of cosets is reduced by $\lfloor (\delta - 1)/q \rfloor$. By Lemma 12.5, if $x, y \not\equiv 0 \bmod q$ and $x \ne y$, then the cosets C_x and C_y are disjoint. Thus, Z is the union of $(\delta-1) - \lfloor (\delta-1)/q \rfloor = \lceil (\delta-1)(1-1/q) \rceil$ distinct cyclotomic cosets. By Lemma 12.4, all these cosets have cardinality m. Therefore, the degree of the generator polynomial is $m\lceil (\delta-1)(1-1/q) \rceil$, which proves our claim about the dimension of the code. $\square$

As a consequence of the dimension result, we can tighten the bounds on the minimum distance of narrow-sense BCH codes, generalizing a result due to Farr, see [MS77, p. 259].

Corollary 12.5. *A $\mathrm{BCH}(n, q; \delta)$ code*
  (i) *with length in the range $q^{\lfloor m/2 \rfloor} < n \le q^m - 1$, $m = \mathrm{ord}_n(q)$,*
  (ii) *and designed distance in the range $2 \le \delta \le \min\{\lfloor nq^{\lceil m/2 \rceil}/(q^m - 1) \rfloor, n\}$*

(iii) such that

$$\sum_{i=0}^{\lfloor(\delta+1)/2\rfloor} \binom{n}{i}(q-1)^i > q^{m\lceil(\delta-1)(1-1/q)\rceil}, \qquad (12.11)$$

has minimum distance $d = \delta$ or $\delta + 1$; if $\delta \equiv 0 \bmod q$, then $d = \delta + 1$.

Proof. Seeking a contradiction, we assume that the minimum distance d of the code satisfies $d \geq \delta + 2$. We know from Theorem 12.7 that the dimension of the code is $k = n - m\lceil(\delta-1)(1-1/q)\rceil$. If we substitute this value of k into the sphere-packing bound $q^k \sum_{i=0}^{\lfloor(d-1)/2\rfloor} \binom{n}{i}(q-1)^i \leq q^n$, then we obtain

$$\sum_{i=0}^{\lfloor(\delta+1)/2\rfloor} \binom{n}{i}(q-1)^i \leq \sum_{i=0}^{\lfloor(d-1)/2\rfloor} \binom{n}{i}(q-1)^i$$
$$\leq q^{m\lceil(\delta-1)(1-1/q)\rceil},$$

but this contradicts condition (12.11); hence, $\delta \leq d \leq \delta + 1$.

If $\delta \equiv 0 \bmod q$, then the cyclotomic coset C_δ is contained in the defining set Z of the code because $C_\delta = C_{\delta/q}$. Thus, the BCH bound implies that the minimum distance must be at least $\delta + 1$. □

We conclude this section with a minor result on the dual distance of BCH codes that will be needed later for determining the purity of quantum codes.

Lemma 12.6. *Suppose that C is a narrow-sense BCH code of length n over $\mathbb{F}_q$ with designed distance $2 \leq \delta \leq \delta_{\max} = \lfloor n(q^{\lceil m/2 \rceil} - 1 - (q-2)[m \text{ odd}])/(q^m - 1)\rfloor$, then the dual distance $d^\perp \geq \delta_{\max} + 1$.*

Proof. Let $N = \{0, 1, \ldots, n-1\}$ and Z_δ be the defining set of C. We know that $Z_{\delta_{\max}} \supseteq Z_\delta \supset \{1, \ldots, \delta-1\}$. Therefore $N \setminus Z_{\delta_{\max}} \subseteq N \setminus Z_\delta$. Further, we know that $Z \cap Z^{-1} = \emptyset$ if $2 \leq \delta \leq \delta_{\max}$ from Lemma 12.3 and Theorem 12.4. Therefore, $Z_{\delta_{\max}}^{-1} \subseteq N \setminus Z_{\delta_{\max}} \subseteq N \setminus Z_\delta$.

Let T_δ be the defining set of the dual code. Then $T_\delta = (N \setminus Z_\delta)^{-1} \supseteq Z_{\delta_{\max}}$. Moreover $\{0\} \in N \setminus Z_\delta$ and therefore T_δ. Thus there are at least $\delta_{\max}$ consecutive roots in T_δ. Thus the dual distance $d^\perp \geq \delta_{\max} + 1$. □

Families of quantum BCH codes. In this paragraph, we study the construction of (nonbinary) quantum BCH codes. Calderbank, Shor, Rains, and Sloane outlined the construction of binary quantum BCH codes in [CRS+98]. Grassl, Beth, and Pellizari developed the theory further by formulating a nice condition for determining which BCH codes can be used for constructing quantum codes [GBP97, GB99]. The dimension and the purity of the quantum codes constructed were determined by numerical computations. Steane simplified it further for the special case of binary narrow-sense primitive BCH codes [S99b] and gave a very simple criterion based on the design distance alone. Very little was done with respect to the nonprimitive and nonbinary quantum BCH codes.

In this paragraph, we show how the results we have developed in the previous paragraphs help us to generalize the previous work on quantum codes and give very simple conditions based on design distance alone. Further, we give precisely the dimension and tighten results on the purity of the quantum codes. But, first we review the methods of constructing quantum codes from classical codes.

Theorem 12.8. *Let* $m = \text{ord}_n(q) \geq 2$, *where q is a power of a prime and δ_1, δ_2 are integers such that $2 \leq \delta_1 < \delta_2 \leq \delta_{\max}$ where*

$$\delta_{\max} = \frac{n}{q^m - 1}(q^{\lceil m/2 \rceil} - 1 - (q-2)[m \text{ odd}]),$$

then there exists a quantum code with parameters

$$[[n, m(\delta_2 - \delta_1 - \lfloor (\delta_2 - 1)/q \rfloor + \lfloor (\delta_1 - 1)/q \rfloor), \geq \delta_1]]_q$$

pure to δ_2.

Proof. By Theorem 12.7, there exist BCH codes $\text{BCH}(n, q; \delta_i)$ with the parameters $[n, n - m(\delta_i - 1) + m\lfloor (\delta_i - 1)/q \rfloor, \geq \delta_i]_q$ for $i \in \{1, 2\}$. Further, $\text{BCH}(n, q; \delta_2) \subset \text{BCH}(n, q; \delta_1)$. Therefore, by the CSS construction (see Corollary 12.3), there exists a quantum code with the parameters

$$[[n, m(\delta_2 - \delta_1 - \lfloor (\delta_2 - 1)/q \rfloor + \lfloor (\delta_1 - 1)/q \rfloor), \geq \delta_1]]_q.$$

The purity follows due to the fact that $\delta_2 > \delta_1$ and Lemma 12.6 by which the dual distance of either BCH code is $\geq \delta_{\max} + 1 > \delta_2$. □

When the BCH codes contain their duals, then we can derive the following codes. Note that these cannot be obtained as a consequence of Theorem 12.8.

Theorem 12.9. *Let* $m = \text{ord}_n(q)$ *where q is a power of a prime and $2 \leq \delta \leq \delta_{\max}$, with*

$$\delta_{\max} = \frac{n}{q^m - 1}(q^{\lceil m/2 \rceil} - 1 - (q-2)[m \text{ odd}]),$$

then there exists a quantum code with parameters

$$[[n, n - 2m\lceil (\delta - 1)(1 - 1/q) \rceil, \geq \delta]]_q$$

pure to $\delta_{\max} + 1$

Proof. Theorems 12.4 and 12.7 imply that there exists a classical BCH code with parameters $[n, n - m\lceil (\delta - 1)(1 - 1/q) \rceil, \geq \delta]_q$ which contains its dual code. By Corollary 12.3 an $[n, k, d]_q$ code that contains its Euclidean dual code implies the existence of the quantum code with parameters $[[n, 2k - n, \geq d]]_q$. The purity follows from Lemma 12.6 by which the dual distance $\geq \delta_{\max} + 1 > \delta$. □

These are not the only possible families of quantum codes that can be derived from BCH codes. As pointed out in [GB99], we can expand BCH codes over $\mathbb{F}_{q^l}$ to get codes over $\mathbb{F}_q$. Once again the dimension and duality results of BCH codes makes it very easy to specify such codes.

Corollary 12.6. *Let* $m = \text{ord}_n(q^l)$ *where q is a power of a prime and $2 \leq \delta \leq \delta_{\max}$, with*

$$\delta_{\max} = \frac{n}{q^{lm} - 1}(q^{l\lceil m/2 \rceil} - 1 - (q^l - 2)[m \text{ odd}]),$$

then there exists a quantum code with parameters

$$[[ln, ln - 2lm\lceil (\delta - 1)(1 - 1/q^l) \rceil, \geq \delta]]_q$$

that is pure up to δ.

Proof. By Theorem 12.9 there exists a quantum BCH code with parameters $[[n, n - 2m\lceil(\delta - 1)(1 - 1/q^l)\rceil, \geq \delta]]_{q^l}$. An $[[n, k, d]]_{q^l}$ quantum code implies the existence of the quantum code with parameters $[[ln, lk, \geq d]]_q$ by [KKK+06, Lemma 76] and the code follows. □

12.4 Quantum MDS codes

An $[[n, k, d]]_q$ quantum code satisfies the *quantum Singleton bound*

$$n \geq k + 2d - 2.$$

An $[[n, k, d]]_q$ quantum code satisfying $n = k + 2d - 2$ is called a *quantum maximum distance separable code* or *quantum MDS code* for short.

Since a quantum MDS code attains the quantum Singleton bound with equality, one can expect that a quantum code with such extremal parameters will have interesting features. The following result by Rains shows that this is indeed the case.

Proposition 12.7 (Rains). *All quantum MDS stabilizer codes are pure.*

Proof. An $[[n, k, d]]_q$ quantum MDS code with $k = 0$ is pure by definition; if $k \geq 1$ then Rains showed that this quantum code is pure up to $n - d + 2$, see [R99, Theorem 2]. By the quantum Singleton bound $n - 2d + 2 = k \geq 0$, hence $n - d + 2 \geq d$. Therefore, a quantum MDS code is pure up to the minimum distance, which proves the claim. □

This is an interesting property, since pure quantum codes are easier to study than impure ones.

Lemma 12.7. *For any $[[n, n - 2d + 2, d]]_q$ quantum MDS stabilizer code with $n - 2d + 2 > 0$, the corresponding classical codes $C \subseteq C^{\perp_a}$ are also MDS.*

Proof. If an $[[n, n-2d+2, d]]_q$ stabilizer code exists, then Theorem 12.2 implies the existence of an additive $[n, d-1]_{q^2}$ code C such that $C \subseteq C^{\perp_a}$. Corollary 12.7 shows that $C^{\perp_a}$ has minimum distance d, so $C^{\perp_a}$ is an $[n, n - d + 1, d]_{q^2}$ MDS code. Since an $[[n, n - 2d + 2, d]]_q$ quantum MDS stabilizer code is pure to $\geq n - d + 2$ by [R99, Theorem 2], the minimum distance of C is $\geq n - d + 2$; hence, C is an $[n, d - 1, n - d + 2]_{q^2}$ MDS code. □

A classical $[n, k, d]_q$ MDS code is said to be trivial if $k \leq 1$ or $k \geq n - 1$. A trivial MDS code can have arbitrary length, but a nontrivial one cannot. The next lemma is a straightforward generalization from linear to additive MDS codes.

Lemma 12.8. *Assume that there exists a classical additive $(n, q^k, d)_q$ MDS code C.*
  (i) *If the code is trivial, then it can have arbitrary length.*
  (ii) *If the code is nontrivial, then its code parameters must be in the range $2 \leq k \leq \min\{n - 2, q - 1\}$ and $n \leq q + k - 1 \leq 2q - 2$.*

Proof. The first statement is obvious. For (ii), we note that the weight distribution of the code C and its dual are related by the MacWilliams relations. The proof given in [MS77, pp. 320–321] for linear codes applies without change, and one finds that the number of codewords of weight $n - k + 2$ in C is given by

$$A_{n-k+2} = \binom{n}{k-2}(q-1)(q-n+k-1).$$

Since A_{n-k+2} must be a nonnegative number, we obtain the claim. □

We say that a quantum $[[n,k,d]]_q$ MDS code is trivial if and only if its minimum distance $d \leq 2$. The length of trivial quantum MDS codes is not bounded, but the length of nontrivial ones is, as the next lemma shows.

Theorem 12.10 (Maximal length of MDS stabilizer codes). *A nontrivial $[[n,k,d]]_q$ MDS stabilizer code satisfies the following constraints:*
  (i) *its length n is in the range $4 \leq n \leq q^2 + d - 2 \leq 2q^2 - 2$;*
  (ii) *its minimum distance satisfies $\max\{3, n - q^2 + 2\} \leq d \leq \min\{n-1, q^2\}$.*

Proof. By definition, a quantum MDS code attains the Singleton bound, so $n - 2d + 2 = k \geq 0$; hence, $n \geq 2d - 2$. Therefore, a nontrivial quantum MDS code satisfies $n \geq 2d - 2 \geq 4$.

By Lemma 12.7, the existence of an $[[n, n-2d+2, d]]_q$ stabilizer code implies the existence of classical MDS codes C and $C^{\perp_a}$ with parameters $[n, d-1, n-d+2]_{q^2}$ and $[n, n-d+1, d]_{q^2}$, respectively. If the quantum code is a nontrivial MDS code, then the associated classical codes are nontrivial classical MDS codes. Indeed, for $n \geq 4$ the quantum Singleton bound implies $d \leq (n+2)/2 \leq (2n-2)/2 = n-1$, so C is a nontrivial classical MDS code.

By Lemma 12.8, the dimension of C satisfies the constraints $2 \leq d-1 \leq \min\{n-2, q^2-1\}$, or equivalently $3 \leq d \leq \min\{n-1, q^2\}$. Similarly, the length n of C satisfies $n \leq q^2 + (d-1) - 1 \leq 2q^2 - 2$. If we combine these inequalities then we get our claim. □

Example 12.1. *The length of a nontrivial binary MDS stabilizer code cannot exceed $2q^2 - 2 = 6$. In [CRS+98] the nontrivial MDS stabilizer codes for $q = 2$ were found to be $[[5,1,3]]_2$ and $[[6,0,4]]_2$, so there cannot exist further nontrivial MDS stabilizer codes.*

In [GBR04], the question of the maximal length of MDS codes was raised. All MDS stabilizer codes provided in that reference had a length of q^2 or less; this prompted us to look at the following famous conjecture for classical codes (see [HP03, Theorem 7.4.5] or [MS77, pp. 327–328]).

MDS conjecture. *If there is a nontrivial $[n,k]_q$ MDS code, then $n \leq q+1$ except when q is even and $k = 3$ or $k = q-1$ in which case $n \leq q+2$.*

If the MDS conjecture is true (and much supporting evidence is known), then we can improve upon the result of Theorem 12.10.

Corollary 12.7. *If the classical MDS conjecture holds, then there are no nontrivial MDS stabilizer codes of lengths exceeding $q^2 + 1$ except when q is even and $d = 4$ or $d = q^2$ in which case $n \leq q^2 + 2$.*

Chapter 13

Optimization-based quantum error correction

Andrew Fletcher

The purpose of quantum error correction (QEC) is to preserve a quantum state despite the presence of decoherence. As we see throughout this book, the desired robustness exacts a cost in resources, most commonly the inclusion of redundant qubits. The following are reasonable engineering queries: *How much will it cost to provide the desired robustness? For a fixed cost, what is the best performance I can achieve?* In this chapter, we present some numerical tools to illuminate these kinds of questions.

To understand the cost/performance trade-off quantitatively, we need a clear measure of performance and a model for permissible operations. With this in mind, we will revisit the concepts of fidelity and quantum operations. As it turns out, this quantitative approach can yield very well structured convex optimization problems. Using powerful numerical tools, we determine optimal encodings and recovery operations. Even if the optimal results are not directly implemented, the optimization tools can provide insight into practical solutions by providing the ultimate performance limits.

13.1 Limitation of the independent arbitrary errors model

As pointed out in Chapter 2, our rich history of classical error correction has provided a significant foundation for QEC methods. As such, the initial QEC breakthroughs involved importing classical coding concepts into a framework that made robust quantum codes (such as CSS codes or the stabilizer code formalism). We learned that we could create general purpose codes that made minimal assumptions about the structure of the decoherence process. Specifically, a QEC procedure that corrects for Pauli X, Y, and Z errors on a physical qubit is automatically robust to an arbitrary error on that qubit. This discretization of the decoherence showed an essential connection between classical, digital error correction and quantum error correction.

The generic nature of the arbitrary errors model has been instrumental in advancing both QEC and fault-tolerant quantum computing. However, the generic assumption pushes us away

---

Quantum Error Correction, ed. Daniel A. Lidar and Todd A. Brun. Published by Cambridge University Press.
© Cambridge University Press 2013.

from efficient code design. This is quite intuitive – if a code must be robust to a wide variety of decoherence, it is likely to be quite inefficient in protecting from a specific channel. In other words, the generic code will correct the errors, but with a high cost in terms of resources. Furthermore, a physical realization of quantum information processing will likely lead to a specific, rather than a generic, decoherence model.

This intuitive concept has been known since the first explorations of QEC. Consider the paper [LNC+97] by Leung *et al.* They noted that the amplitude damping channel was a physically relevant quantum channel, where the decoherence is described by the decay of the excited state ($|1\rangle$) to the ground state ($|0\rangle$) with probability γ. For this particular channel, they presented a simple [[4,1,2]] code and decoding operation. In contrast, the shortest block length generic code is the [[5,1,3]] stabilizer code in Chapter 2.

Leung *et al.* described this code as an example of *approximate quantum error correction*, since the error correction criteria were not perfectly met. They also noted that approximate QEC has no analogous construction in classical error correction; it is a quantum phenomenon that arises from the near-orthogonality of some quantum states. Such phenomena proved challenging to explore through analytic methods. Instead, approximate error correction has been explored through numerical optimization techniques, and is sometimes referred to as *channel-adapted quantum error correction*.

13.2 Defining a QEC optimization problem

In order to design an optimization problem for QEC, we need to establish two criteria. First, we need a numerical measure of performance to score the effectiveness of a QEC procedure. This is often called the objective function. Second, we need to define what constitutes a valid QEC procedure. This will be the constraint set. We will see that an application of the *fidelity* defined in Chapter 1 will provide a useful performance measure, while the completely positive trace-preserving (CPTP) nature of quantum operations (also defined in Chapter 1) makes an effective constraint set.

13.2.1 Fidelity

To define an optimization problem, we need to establish an appropriate measure of performance for QEC. In Chapter 1, the *fidelity* defined a measure of similarity between two quantum states. Restating here, quantum states ρ and σ have fidelity

$$F(\rho,\sigma) = \left(\text{Tr}\sqrt{\rho^{\frac{1}{2}}\sigma\rho^{\frac{1}{2}}}\right)^2. \tag{13.1}$$

If ρ and σ are identical quantum states, the fidelity is 1; if they are orthogonal, the fidelity is 0. Note that, for the purposes of this chapter, we have redefined the fidelity as the square of the corresponding expression in Eq. (1.84).

Obviously, a successful QEC system seeks to preserve a quantum state. A logical performance measure is therefore to maximize the fidelity of the input state with the recovered state. However, one needs to be careful: it is not sufficient to design a QEC procedure that protects only a given quantum state. The procedure must effectively protect all potential logical states. If

$\mathscr{A}$ represents the composite operation of the encoder, channel, and recovery, we want a measure of the channel fidelity of $\mathscr{A}$.

We define the *minimum fidelity* of $\mathscr{A}$ as the worst-case scenario over all input states $|\psi\rangle$:[1]

$$F_{\min}(\mathscr{A}) = \min_{|\psi\rangle} F(|\psi\rangle\langle\psi|, \mathscr{A}(|\psi\rangle\langle\psi|)). \tag{13.2}$$

By virtue of the minimization over $|\psi\rangle$, one need not assume anything about the input state. This was the metric of choice in [KL97] first establishing the theory of QEC, and translates nicely to the idea of perfectly correcting a set of errors. The disadvantage arises through the complexity of the metric; indeed, computation requires minimizing over all inputs. This drawback makes minimum fidelity a difficult choice for optimization.

Entanglement fidelity and *ensemble average fidelity* both provide more tractable metrics for $\mathscr{A}$. To use them, we must make some assumption about the ensemble of input states. We may define an ensemble E consisting of states ρ_i each with probability p_i. The ensemble average fidelity is naturally defined as

$$\bar{F}(E, \mathscr{A}) = \sum_i p_i F(\rho_i, \mathscr{A}(\rho_i)). \tag{13.3}$$

When ρ_i are pure states, $\bar{F}$ is linear in $\mathscr{A}$.

Entanglement fidelity [S96a] is defined for a mixed state ρ in terms of a purification to a reference system. Defining $\mathscr{L}(\mathscr{H})$ as the set of bounded linear operators on $\mathscr{H}$, $\rho \in \mathscr{L}(\mathscr{H})$ can be understood as a pure quantum state over $\mathscr{H}_R \otimes \mathscr{H}$, where $\mathscr{H}_R$ is a reference system; this is known as a *purification*. If $|\psi\rangle \in \mathscr{H}_R \otimes \mathscr{H}$ is a purification of ρ, then $\rho = \text{Tr}_{\mathscr{H}_R}|\psi\rangle\langle\psi|$. The purification captures all of the information in ρ. The entanglement fidelity is the measure of how well the channel $\mathscr{A}$ preserves the state $|\psi\rangle$, or in other words, how well $\mathscr{A}$ preserves the entanglement of the state with its reference system. We write the entanglement fidelity as

$$F_e(\rho, \mathscr{A}) = \langle\psi|\mathscr{I} \otimes \mathscr{A}(|\psi\rangle\langle\psi|)|\psi\rangle, \tag{13.4}$$

where $\mathscr{I}$ is the identity map on $\mathscr{L}(\mathscr{H}_R)$. We have used the fact that $|\psi\rangle$ is pure to express (13.4) in a more convenient equation for the fidelity than the generic mixed state form of (13.1). The entanglement fidelity is linear in $\mathscr{A}$ for any input ρ, and is a lower bound to the ensemble average fidelity for any ensemble E such that $\sum_i p_i \rho_i = \rho$.

The definition of entanglement fidelity given in (13.4) is intuitively useful, but awkward for calculations. An easier form arises when operator elements $\{A_i\}$ for $\mathscr{A}$ are given. The entanglement fidelity is then

$$F_e(\rho, \mathscr{A}) = \sum_i |\text{Tr}(\rho A_i)|^2. \tag{13.5}$$

The linearity of both ensemble average fidelity and entanglement fidelity in $\mathscr{A}$ is particularly useful for optimization. It enables the use of the convex optimization problems called semidefinite programs, which will be summarized in the next section. While all of the optimization problems in this chapter could be performed using either metric, we will follow the lead of [BK02] and derive based on the *average entanglement fidelity*, given by

$$\bar{F}_e(E, \mathscr{A}) = \sum_i p_i F_e(\rho_i, \mathscr{A}). \tag{13.6}$$

---

[1] One might suppose we should have to minimize over all mixed states ρ. In fact, it is sufficient to minimize over pure state inputs [NC00].

By so doing, all of the algorithms can be trivially converted to either entanglement fidelity or ensemble average fidelity with pure states, as both are special cases of average entanglement fidelity.

13.2.2 Structure of the completely positive trace-preserving constraint

A straightforward description of a QEC procedure consists of an encoding operation (usually unitary) U_C and a recovery operation $\mathscr{R}$. In designing a system, we may presume to have a model of the channel behavior $\mathscr{E}$, and we will assume (for now) to have complete freedom in the selection of U_C and $\mathscr{R}$. These operations are only constrained in that they are physically realizable; we saw in Chapter 1 that both U_C and $\mathscr{R}$ must be CPTP.

The most common representation of a CPTP map uses the operator sum, or Kraus, representation. An operation $\mathscr{A} : \mathscr{L}(\mathscr{H}) \mapsto \mathscr{L}(\mathscr{K})$ is defined by operator elements $\{A_i\}$, where the input–output relation is given by $\mathscr{A}(\rho) = \sum_i A_i \rho A_i^\dagger$. However, the operator elements are not a unique representation of the operation; a unitary recombination of $\{A_i\}$ forms an equivalent operator element representation of the channel. This many-to-one representation of the channel is often inconvenient for optimization.

We will find it more useful to use the Jamiolkowski isomorphism [C75, DL01, H03, C99, d67], which allows us to describe an operation $\mathscr{A} : \mathscr{L}(\mathscr{H}) \mapsto \mathscr{L}(\mathscr{K})$ in terms of a unique operator $X_\mathscr{A} \in \mathscr{L}(\mathscr{K} \otimes \mathscr{H}^*)$. $X_\mathscr{A}$ is often called the *Choi matrix*, a nomenclature we will follow. $\mathscr{H}^*$ is defined below.

The Jamiolkowski isomorphism represents a linear operator as a vector denoted by the symbol $|\cdot\rangle\!\rangle$. For intuition's sake, if we think of an operator as a matrix, the isomorphism is equivalent to writing all of the matrix elements "stacked" on top of each other as a vector. In fact, for a specific choice of basis vectors, this is precisely the isomorphism. However, we will follow the conventions of [T03] (also [YHT05]), which results in an isomorphism that is independent of the choice of basis. For convenience, we restate the relevant results here.

Let $A = \sum_{ij} a_{ij} |i\rangle\langle j|$ be a bounded linear operator from $\mathscr{H}$ to $\mathscr{K}$ (i.e., $A \in \mathscr{L}(\mathscr{H}, \mathscr{K})$), where $\{|i\rangle\}$ and $\{|j\rangle\}$ are bases for $\mathscr{K}$ and $\mathscr{H}$, respectively. Let $\mathscr{H}^*$ be the dual of $\mathscr{H}$. This is also a Hilbert space, generally understood as the space of *bras* $\langle j|$. If we relabel the elements as $\overline{|j\rangle} = \langle j|$, then we represent A as a vector in the space $\mathscr{K} \otimes \mathscr{H}^*$ as

$$|A\rangle\!\rangle = \sum_{ij} a_{ij} |i\rangle \overline{|j\rangle}. \tag{13.7}$$

It is useful to note the following facts. The inner product $\langle\!\langle A|B\rangle\!\rangle$ is the Hilbert–Schmidt inner product $\mathrm{Tr} A^\dagger B$. Also, the partial trace over $\mathscr{H}^*$ yields the useful operator relation

$$\mathrm{Tr}_{\mathscr{H}^*} |A\rangle\!\rangle\langle\!\langle B| = AB^\dagger. \tag{13.8}$$

Finally, index manipulation yields the relation

$$A \otimes \overline{B} |C\rangle\!\rangle = |ACB^\dagger\rangle\!\rangle, \tag{13.9}$$

where $\overline{B}$ is the conjugate of B such that $\overline{B|\psi\rangle} = \overline{B}\,\overline{|\psi\rangle}$ for all $|\psi\rangle$. (For a specific basis, $\overline{B}$ is the conjugation of all of the matrix elements B^*, or $B^{\dagger T}$.)

The Choi matrix is calculated from the Kraus elements $\{A_k\}$ of $\mathscr{A}$ as

$$X_{\mathscr{A}} = \sum_k |A_k\rangle\!\rangle\langle\!\langle A_k|. \tag{13.10}$$

The operation output is given by $\mathscr{A}(\rho) = \text{Tr}_{\mathscr{H}^*}(I \otimes \overline{\rho})X_{\mathscr{A}}$ and the CPTP constraint requires that $X_{\mathscr{A}} \geq 0$ and $\text{Tr}_{\mathscr{H}^*} X_A = I$.

As a final note on the Choi matrix form, let us restate the entanglement fidelity calculation when the channel $\mathscr{A}$ is expressed via the Choi matrix. Recalling the definition of the Hilbert–Schmidt inner product, we see that $\text{Tr}\, A_i\rho = \langle\!\langle \rho | A_i \rangle\!\rangle$. Inserting this into (13.5), we obtain the entanglement fidelity in terms of $X_{\mathscr{A}}$:

$$\begin{aligned}F_e(\rho, \mathscr{A}) &= \sum_i \langle\!\langle \rho | A_i \rangle\!\rangle \langle\!\langle A_i | \rho \rangle\!\rangle \\ &= \langle\!\langle \rho | X_{\mathscr{A}} | \rho \rangle\!\rangle. \end{aligned} \tag{13.11}$$

This expression is a quadratic form in $|\rho\rangle\!\rangle$ and is thus linear in $X_{\mathscr{A}}$. It is trivial to extend to the average entanglement fidelity given an ensemble E:

$$\bar{F}_e(E, \mathscr{A}) = \sum_k p_k \langle\!\langle \rho_k | X_{\mathscr{A}} | \rho_k \rangle\!\rangle. \tag{13.12}$$

13.3 Maximizing average entanglement fidelity

Let us restate the intuitive description of our optimization problem. In order to preserve quantum information in $\mathscr{H}$, we design an encoding operation $\mathscr{U}_C : \mathscr{L}(\mathscr{H}) \mapsto \mathscr{L}(\mathscr{K})$.[2] Our encoded information then resides in $\mathscr{K}$; we model the physical decoherence as $\mathscr{E} : \mathscr{L}(\mathscr{K}) \mapsto \mathscr{L}(\mathscr{K})$. A recovery operation $\mathscr{R} : \mathscr{L}(\mathscr{K}) \mapsto \mathscr{L}(\mathscr{H})$ attempts to undo the effects of $\mathscr{E}$ and returns our information to $\mathscr{H}$. $\mathscr{U}_C$ and $\mathscr{R}$ are chosen to maximize the average entanglement fidelity of an ensemble of states E in $\mathscr{L}(\mathscr{H})$. We write this as

$$\max_{\mathscr{U}_C, \mathscr{R}} \bar{F}_e(E, \mathscr{R} \circ \mathscr{E} \circ \mathscr{U}_C), \tag{13.13}$$

where $\mathscr{U}_C$ and $\mathscr{R}$ are constrained to be CPTP and $\circ$ is defined to be the composite operation; thus $\mathscr{R} \circ \mathscr{E} \circ \mathscr{U}_C(\rho) = \mathscr{R}(\mathscr{E}(\mathscr{U}_C(\rho)))$. In the remainder of the chapter, we will explore methods to solve this and related optimization problems.

13.3.1 Bi-convex structure

As it turns out, the optimization problem in (13.13) has a particularly useful structure. We noted that the average entanglement fidelity is linear in the composite channel; in fact, we will show that it is linear in both $\mathscr{U}_C$ and $\mathscr{R}$. Furthermore, we saw that all CPTP maps can be written as a positive semidefinite operator with an additional equality constraint. The combination of these two attributes means that (13.13) has a bi-convex structure. That is, if $\mathscr{U}_C$ is fixed, then the optimization over $\mathscr{R}$ is convex. Similarly, if $\mathscr{R}$ is fixed, the optimization over $\mathscr{U}_C$ is convex.

---

[2] While the encoding is almost always a unitary operation, we will for now merely constrain it to be CPTP, hence the notation $\mathscr{U}_C$ instead of the unitary U_C.

We first derive the convexity in $\mathscr{R}$; the derivation in $\mathscr{U}_C$ follows the same steps. Let us consider the operator elements of $\mathscr{R} \circ \mathscr{E} \circ \mathscr{U}_C$, which can be written as $\{R_i E_j U_{Ck}\}$. We can write the Jamiolkowski isomorphism as

$$|R_i E_j U_{Ck}\rangle\!\rangle = I \otimes \overline{U}_{Ck}^\dagger \overline{E}_j^\dagger |R_i\rangle\!\rangle, \qquad (13.14)$$

where we have used the relation given in (13.9). To evaluate the entanglement fidelity, we must evaluate terms of the form $\langle\!\langle \rho | R_i E_j U_{Ck}\rangle\!\rangle$. Again using (13.9), we can write this as

$$\langle\!\langle \rho | R_i E_j U_{Ck}\rangle\!\rangle = \langle\!\langle \rho | I \otimes \overline{U}_{Ck}^\dagger \overline{E}_j^\dagger |R_i\rangle\!\rangle$$
$$= \langle\!\langle \rho U_{Ck}^\dagger E_j^\dagger |R_i\rangle\!\rangle. \qquad (13.15)$$

This allows us to write the average entanglement fidelity as

$$\bar{F}_e(E, \mathscr{R} \circ \mathscr{E} \circ \mathscr{U}_C) = \sum_{ijkm} p_m \langle\!\langle \rho_m | R_i E_j U_{Ck}\rangle\!\rangle \langle\!\langle R_i E_j U_{Ck} | \rho_m\rangle\!\rangle$$
$$= \sum_{ijkm} p_m \langle\!\langle \rho_m U_{Ck}^\dagger E_j^\dagger | R_i \rangle\!\rangle \langle\!\langle R_i | \rho_m U_{Ck}^\dagger E_j^\dagger \rangle\!\rangle. \qquad (13.16)$$

With a little more manipulation of terms, we can highlight the convex structure. First, we use the definition of the Choi matrix to write

$$\bar{F}_e(E, \mathscr{R} \circ \mathscr{E} \circ \mathscr{U}_C) = \sum_{jkm} p_m \langle\!\langle \rho_m U_{Ck}^\dagger E_j^\dagger | X_\mathscr{R} | \rho_m U_{Ck}^\dagger E_j^\dagger \rangle\!\rangle. \qquad (13.17)$$

We can then use the cyclic property of the operator trace to further condense the expression.

$$\bar{F}_e(E, \mathscr{R} \circ \mathscr{E} \circ \mathscr{U}_C) = \sum_{jkm} p_m \mathrm{Tr} X_\mathscr{R} |\rho_m U_{Ck}^\dagger E_j^\dagger\rangle\!\rangle \langle\!\langle \rho_m U_{Ck}^\dagger E_j^\dagger|$$
$$= \mathrm{Tr} X_\mathscr{R} C_{E, \mathscr{U}_C \mathscr{E}}, \qquad (13.18)$$

where we have defined

$$C_{E, \mathscr{U}_C \mathscr{E}} = \sum_{jkm} p_m |\rho_m U_{Ck}^\dagger E_j^\dagger\rangle\!\rangle \langle\!\langle \rho_m U_{Ck}^\dagger E_j^\dagger|. \qquad (13.19)$$

The last few steps may seem unnecessary algebraic manipulations, but they allow us to write a straightforward expression for the optimal recovery operation:

$$X_\mathscr{R}^\star = \arg\max_X \mathrm{Tr}(X C_{E, \mathscr{U}_C \mathscr{E}})$$
$$\text{such that } X \geq 0, \quad \mathrm{Tr}_\mathscr{H} X = I. \qquad (13.20)$$

The constraint $X \geq 0$ is known as a linear matrix inequality (LMI); it means that X is positive semidefinite. With the optimization problem written in this form, it is quite simple to see the convexity. For $0 \leq p \leq 1$ and X_1 and X_2 that satisfy the constraint, we can immediately see that $pX_1 + (1-p)X_2 \geq 0$ and that $\mathrm{Tr}_\mathscr{H}(pX_1 + (1-p)X_2) = I$. Furthermore, the objective function $\mathrm{Tr}(X C_{E, \mathscr{U}_C \mathscr{E}})$ is obviously linear in X. A convex optimization with this particular structure is known as a semidefinite program (SDP).

An example of the value of optimizing the recovery operation can be seen in Fig. 13.1. In this case, we have begun with the [[5,1,3]] code and the [[4,1,2]] "approximate" code discussed

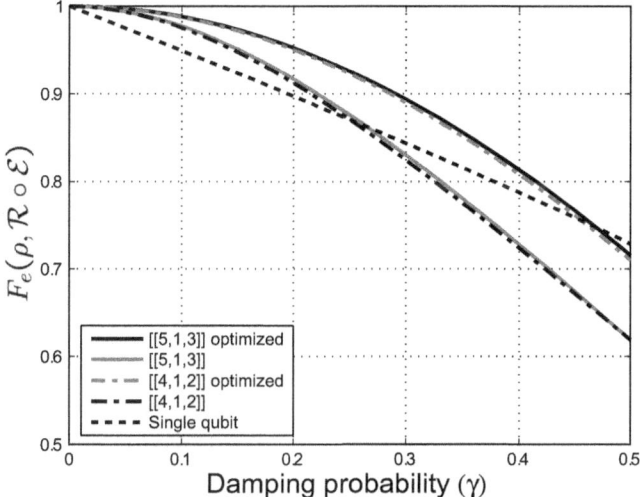

Fig. 13.1. Entanglement fidelity with the amplitude damping channel for the [[4,1,2]] "approximate" code of Leung et al. [LNC+97] and the [[5,1,3]] code. For both codes, we observe the performance when the recovery operation is an optimized recovery or "standard" recovery (meaning the stabilizer recovery for the [[5,1,3]] and the recovery proposed in [LNC+97] for the [[4,1,2]] code). Entanglement fidelity for a single qubit and no error correction is included as a performance baseline.

above. From both of these encodings, we have optimized a recovery operation based on the amplitude damping channel. We can note two features: first, the [[4,1,2]] code well approximates the [[5,1,3]] performance. Second, when considering the amplitude damping channel, the optimal recovery operation is appreciably superior to the recovery operation normally associated with these codes.

Before further discussion of semidefinite programming, let us state the optimization problem for the optimal encoding $\mathscr{U}_C^\star$ given a fixed recovery operation. Following the same steps as above, this can be written

$$X^\star_{\mathscr{U}_C} = \arg\max_X \text{Tr}(X C_{E,\mathscr{R}\mathscr{E}})$$
$$\text{such that } X \geq 0, \quad \text{Tr}_{\mathscr{K}} X = I,$$
(13.21)

where

$$C_{E,\mathscr{R}\mathscr{E}} = \sum_{ijm} p_m |E_j^\dagger R_i^\dagger \rho\rangle\!\rangle \langle\!\langle E_j^\dagger R_i^\dagger \rho|.$$
(13.22)

13.3.2 Semidefinite programming

Semidefinite programming is a useful construct for convex optimization problems; efficient routines have been developed to numerically evaluate SDPs. The theory of SDPs is sufficiently mature that the numerical solution can be considered a "black-box routine" for our purposes. That is, we do not need to explore the various computational algorithms for solving SDPs. Instead, we will concisely state the definition of an SDP and refer the interested reader to the review article [VB96] and the textbook [BV04a] for more extensive treatments.

A semidefinite program is defined as the minimization of a linear function of the variable $x \in \mathbf{R}^N$ subject to a matrix inequality constraint. In its standard form ([BV04a] Section 4.6.2) the optimization problem is

$$\min_X \mathrm{Tr} CX, \quad \text{such that} \quad \mathrm{Tr} A_i X = b_i, i = 1\ldots, p \quad \text{and} \quad X \geq 0. \qquad (13.23)$$

Just as we saw before, the inequality $\geq$ in (13.23) is an LMI that constrains X to be positive semidefinite. The SDP is convex as both the objective function and the constraint are convex: if $X \geq 0$ and $Y \geq 0$, we see that $\lambda X + (1-\lambda) Y \geq 0$ for all $\lambda \in [0,1]$. Convex optimization is particularly valuable, as any locally optimal point is also globally optimal. This is in contrast to the multiple nonglobal local minima that often arise in nonconvex optimization.

From this definition, it should be clear that the optimization problems in (13.20) and (13.21) are both semidefinite programs. SDPs have been applied to several quantum information topics including distillable entanglement [R01, DPS05, BV04b], quantum detection [ESH04, E03b, EMV03, E03a, JŘF02], optimizing CP maps (including channel-adapted QEC) [AD02, YHT05, FSW07, KL09, KSL08], and quantum algorithms for the ordered search problem [CLP07].

While we can think of an SDP algorithm as a black-box process, it is still important to note the computational burden. In the SDP of (13.20) and (13.21), the optimization variable is the matrix X. If $d_\mathcal{H}$ and $d_\mathcal{K}$ are the dimensions of the Hilbert spaces $\mathcal{H}$ and $\mathcal{K}$, then X has dimension $d_\mathcal{H} d_\mathcal{K} \times d_\mathcal{H} d_\mathcal{K}$. For an $[[n, k, d]]$ code, this means that there are 4^{n+k} variables. From [BV04a], an SDP with m variables and a $p \times p$ semidefinite matrix constraint requires $\mathcal{O}(\max\{mp^3, m^2 p^2, m^3\})$ flops per iteration (with typically 10–100 iterations necessary). For our case, this yields $\mathcal{O}(2^{5(n+k)})$ flops per iteration.

The computational burden of an SDP is a significant problem for optimization approaches to QEC. In particular, we find ourselves restricted to relatively short code blocks. However, even for the relatively small code blocks, optimization efforts can provide useful intuition and methods. Furthermore, we will explore ways to partially mitigate the challenging computational burden.

13.3.3 Iterative optimization of codes

From the bi-convex structure in both the encoding and recovery operations, a simple iterative algorithm is evident. For a fixed encoding, we may determine via the SDP the optimum recovery. Holding the recovery operation fixed, we may determine the optimum encoding. The procedure is iterated until convergence to a local maximum is achieved. We can only claim a local maximum as the overall optimization over both $\mathcal{U}_C$ and $\mathcal{R}$ is no longer convex.

Iterative optimization of error-correcting codes has been suggested and applied by several authors in recent years. The idea was suggested in the context of calculating channel capacities in [S03a], though without discussion of the form of the optimization problem. An iterative procedure based on eigen-analysis was laid out in [RW05]. Convex optimization of optimal recovery and encoding was derived independently by [FSW07], [RWA06], and [KL09].

Iteratively optimized codes are sometimes criticized as impractical. Much of the QEC literature has focused on encodings and recoveries that have a systematic or straightforward description. In many cases (such as stabilizer codes), the procedures can be described using only a small set of quantum gates (e.g., the Clifford group). These theoretical "implementations" provide both intuition and a path towards a physical implementation. In contrast, the codes and recoveries resulting from numerical optimization are not always operations that provide clear intuition. Similarly, the physical implementation as a quantum circuit may require complicated combinations of standard quantum gates.

While these criticisms are valid, they do not undermine the worth of numerically optimized codes. First of all, physical implementation of quantum operations is still in its infancy. It remains to be seen what operation descriptions are practical. Second, the optimized results can highlight areas of inefficiency and suggest more efficient solutions. In essence, we can think of optimized encodings and recoveries as ultimate performance bounds as well as guidance on how to reach those bounds.

13.3.4 The dual

Every optimization problem has an associated dual problem [BV04a]. Derived from the objective function and constraints of the original optimization problem (known as the *primal* problem), the dual problem optimizes over a set of dual variables often subject to a set of dual constraints. The dual problem has several useful properties. First of all, the dual problem is always convex. In many cases, calculation of the dual function is a useful method for constructing optimization algorithms. Most important for our purposes, the dual function provides a bound for the value of the primal function. We define a *dual feasible point* as any set of dual variables satisfying the dual constraint. The dual function value for any dual feasible point is less than or equal to the primal function at any primal feasible point. (We have implicitly assumed the primal function to be a minimization problem, which is the canonical form.)

The primal problem as given in (13.20) can be stated succinctly as

$$\min_X -\text{Tr} X C_{E, \mathcal{U}_C \mathcal{E}}, \text{ such that } X \geq 0 \text{ and } \text{Tr}_{\mathcal{H}^*} X = I. \qquad (13.24)$$

The negative sign on the $\text{Tr} X C_{E, \mathcal{U}_C \mathcal{E}}$ terms casts the primal problem as a minimization, which is the canonical form. The Lagrangian is given by

$$L(X, Y, Z) = -\text{Tr} X C_{E, \mathcal{U}_C \mathcal{E}} + \text{Tr} Y (\text{Tr}_{\mathcal{H}^*} X - I) - \text{Tr} Z X, \qquad (13.25)$$

where Y and $Z \geq 0$ are operators that serve as the Lagrange multipliers for the equality and generalized inequality constraints, respectively. The dual function is the (unconstrained) infimum over X of the Lagrangian:

$$\begin{aligned} g(Y, Z) &= \inf_X L(X, Y, Z) \\ &= \inf_X -\text{Tr} X (C_{E, \mathcal{U}_C \mathcal{E}} + Z - Y \otimes I) - \text{Tr} Y, \end{aligned} \qquad (13.26)$$

where we have used the fact that $\text{Tr}(Y \text{Tr}_{\mathcal{H}^*} X) = \text{Tr}(Y \otimes I) X$. Since X is unconstrained, note that $g(Y, Z) = -\infty$ unless $Z = Y \otimes I - C_{E, \mathcal{U}_C \mathcal{E}}$, in which case the dual function becomes $g(Y, Z) = -\text{Tr} Y$. Y and $Z \geq 0$ are the dual variables, but we see that the dual function depends only on Y. We can therefore remove Z from the function as long as we remember the constraint implied by $Z = Y \otimes I - C_{E,\mathcal{U}_C\mathcal{E}}$. Since Z is constrained to be positive semidefinite, this can be satisfied as long as $Y \otimes I - C_{E,\mathcal{U}_C\mathcal{E}} \geq 0$. (This expression is an LMI.)

We now have the bounding relation $-\text{Tr}(XC_{E,\mathcal{U}_C\mathcal{E}}) \geq \text{Tr}(-Y)$ for all X and Y that are primal and dual feasible points, respectively. If we now reverse the signs so that we have a more natural fidelity maximization, we write

$$\bar{F}_e(E, \mathcal{R} \circ \mathcal{E}) = \text{Tr}(X_\mathcal{R} C_{E,\mathcal{U}_C\mathcal{E}}) \leq \text{Tr}Y, \tag{13.27}$$

where $\mathcal{R}$ is CPTP and $I \otimes Y - C_{E,\mathcal{U}_C\mathcal{E}} \geq 0$. To find the best bounding point Y, we solve the dual optimization problem

$$\min_Y \text{Tr}Y, \text{ such that } Y \otimes I - C_{E,\mathcal{U}_C\mathcal{E}} \geq 0. \tag{13.28}$$

Notice that the constraint implies that $Y = Y^\dagger$. Note also that $Y \in \mathcal{L}(\mathcal{K})$. As a final point, since our problem is convex, we satisfy *strong duality*, where the dual and primal problems have the same value at their respective optimal points.

13.3.5 Minimum fidelity optimization

We have devoted the majority of our efforts towards the average entanglement fidelity. This is justified for two reasons: First, this performance measure leads to convex optimization. Second, as long as we select a representative ensemble, the average entanglement fidelity provides a relatively comprehensive evaluation of QEC.

It would perhaps have been preferred to derive all of the above methods using the minimum fidelity. In QEC we often wish to protect arbitrary quantum information; it is therefore satisfying to claim good performance even in the face of a worst-case input state. Unfortunately, the minimum fidelity performance measure breaks the useful convex structure, thus complicating optimization efforts.

The SDP has been shown to be somewhat useful in this context. In [YHT05], we learn that a suboptimal recovery operation is calculable. The ideal optimization is over all input states and all recovery operations. While this is not a convex problem, we can expand it into a convex problem.

If $\rho = |\phi\rangle\langle\phi|$ is a pure state in the code subspace $\mathcal{C} \subset \mathcal{K}$, then $|\rho\rangle\rangle = |\phi\rangle\overline{|\phi\rangle}$. A search over pure quantum states must satisfy this constraint. In order to achieve a convex optimization, Yamamoto *et al.* relax this constraint. They require ρ to be an element of $\mathcal{C} \otimes \mathcal{C}^*$. This set contains all pure quantum states, but also includes elements that are not quantum states at all. With this constraint choice, the SDP solution becomes a suboptimal recovery procedure for the minimum fidelity problem. While an interesting result, we will not make further mention of it in this chapter.

13.4 Minimizing channel nonideality: the indirect method

In the preceding section, we derived an optimization problem in terms of fidelity maximization. The QEC operations $\mathcal{R}$ and $\mathcal{U}_C$, together with the channel $\mathcal{E}$, form a composite operation $\mathcal{R} \circ \mathcal{E} \circ \mathcal{U}_C$. Since we want to preserve the fidelity of quantum information perturbed by this operation, we wrote the optimization problem as a direct maximization of the fidelity.

There is an alternate formulation. In the ideal case, we would like to constrain $\mathscr{R} \circ \mathscr{E} \circ \mathscr{U}_C$ to be the identity operation.[3] Failing that, we would like to minimize the "distance" between $\mathscr{R} \circ \mathscr{E} \circ \mathscr{U}_C$ and I. In this construction, we seek to minimize the nonideality of the operation. This method was presented in [KSL08] and expanded in [TKL10], and was labeled the *indirect* method of maximizing fidelity.

To motivate the form of the indirect method, recall the conditions for perfect quantum error correction. In this case we have

$$R_i E_j U_C = \alpha_{ij} I, \tag{13.29}$$

where $\sum_{ij} |\alpha_{ij}|^2 = 1$. Notice that we have returned to the assumption that the encoding is a unitary operator U_C. Since we do not anticipate perfect error correction, we define an objective function to minimize an operator norm, in this case the Hilbert–Schmidt (or Frobenius) norm:

$$d_{HS} = \sum_{ij} \|R_i E_j U_C - \alpha_{ij} I\|_F^2. \tag{13.30}$$

We have introduced the constrained parameters α_{ij}. We will find, in fact, that these are variables we wish to optimize.

It is convenient to define matrices $\mathbf{R}$ and $\mathbf{E}$ to describe the recovery and the channel. Defined on some basis for $\mathscr{H}$ and $\mathscr{K}$, $\mathbf{E}$ is a $d_{\mathscr{K}} \times m_E$ matrix

$$\mathbf{E} = \begin{bmatrix} E_1 & E_2 & \cdots & E_{m_E} \end{bmatrix}, \tag{13.31}$$

where m_E is the number of operator elements required to represent $\mathscr{E}$. Similarly, we can define $\mathbf{R}$ as the stacking of the operator elements of $\mathscr{R}$:

$$\mathbf{R} = \begin{bmatrix} R_1^\dagger & R_2^\dagger & \cdots & R_{m_R}^\dagger \end{bmatrix}^\dagger. \tag{13.32}$$

If we define I_{m_E} as the $m_E \times m_E$ identity matrix and α as the matrix with elements α_{ij}, we can write the Hilbert–Schmidt distance as

$$d_{HS} = \|\mathbf{RE}(I_{m_E} \otimes U_C - \alpha \otimes I_{\mathscr{H}})\|_F^2. \tag{13.33}$$

The indirect optimization problem minimizes d_{HS} subject to the constraints $\mathbf{R}^\dagger \mathbf{R} = I_{\mathscr{K}}$, $U_C^\dagger U_C = I_{\mathscr{H}}$ and $\|\alpha\|_F^2 = 1$.

We approach the indirect problem much as we did the direct: we will show that for a fixed encoding, the optimization is convex. Recall that $\|X\|_F^2 \equiv \mathrm{Tr} X^\dagger X$; we thus write out d_{HS} as

$$\begin{aligned} d_{HS} &= \mathrm{Tr}(I_{m_E} \otimes U_C^\dagger \mathbf{E}^\dagger \mathbf{R}^\dagger \mathbf{R} \mathbf{E} I_{m_E} \otimes U_C + \alpha^\dagger \alpha \otimes I_{\mathscr{H}}) \\ &\quad - 2\mathrm{Re}\{\mathrm{Tr} \mathbf{RE}(\alpha^\dagger \otimes U_C)\}. \end{aligned} \tag{13.34}$$

Noting the constraints on $\mathbf{R}^\dagger \mathbf{R}$ and $\|\alpha\|_F^2$, this simplifies to

$$d_{HS} = \mathrm{Tr}(I_{m_E} \otimes U_C U_C^\dagger \mathbf{E}^\dagger \mathbf{E}) + d_{\mathscr{H}} - 2\mathrm{Re}\{\mathrm{Tr} \mathbf{RE}(\alpha^\dagger \otimes U_C)\}. \tag{13.35}$$

As only the last term depends on our free parameters α and $\mathbf{R}$, minimizing d_{HS} is accomplished by maximizing $\mathrm{Re}\{\mathrm{Tr} \mathbf{RE}(\alpha^\dagger \otimes U_C)\}$.

---

[3] Throughout the chapter, we have assumed a communications model where the encoding and recovery operations minimize distortion in stored or transmitted information. All of these methods apply where the encoding and recovery preserve a quantum operation. In this case, we would constrain $\mathscr{R} \circ \mathscr{E} \circ \mathscr{U}_C$ to be the desired operator.

At this stage, we can significantly reduce the dimensionality of our optimization problem by optimizing only over α. By applying the singular value decomposition (SVD), we have

$$\mathbf{E}(\alpha^\dagger \otimes U_C) = \mathbf{U\Sigma V}^\dagger, \tag{13.36}$$

where $\mathbf{U}^\dagger \mathbf{U} = \mathbf{V}^\dagger \mathbf{V} = I$ and $\mathbf{\Sigma}$ is diagonal. We can therefore write the equivalence

$$\text{Re}\{\text{Tr}\mathbf{R}\mathbf{E}(\alpha^\dagger \otimes U_C)\} = \text{Re}\{\text{Tr}\mathbf{\Sigma X}\}, \tag{13.37}$$

where $\mathbf{X} = \mathbf{V}^\dagger \mathbf{R}\mathbf{U}$ is a square matrix. With a little thought about the constraint $\mathbf{R}^\dagger \mathbf{R} = I_{\mathscr{K}}$, we see that all elements of $\mathbf{R}$ (and hence all elements of $\mathbf{X}$) are ≤ 1. With that in mind, we can see that $\text{Re}\{\text{Tr}\mathbf{\Sigma X}\} \leq \text{Tr}\mathbf{\Sigma}$ with equality if and only if $\mathbf{X} = I_{\mathscr{K}}$, which is achieved if and only if $\mathbf{R} = \mathbf{V}\mathbf{U}^\dagger$.

Minimizing d_{HS} is therefore equivalent to maximizing $\text{Tr}\mathbf{\Sigma}$, which is the sum of the singular values of $\mathbf{E}(\alpha^\dagger \otimes U_C))$. This is calculated as $\text{Tr}\sqrt{\mathbf{E}(\alpha^\dagger \alpha \otimes U_C U_C^\dagger)\mathbf{E}^\dagger}$. We have succeeded in reducing our problem to an optimization over $\gamma = \alpha^\dagger \alpha$:

$$\max_\gamma \text{Tr}\sqrt{\mathbf{E}(\gamma \otimes U_C U_C^\dagger)\mathbf{E}^\dagger} \tag{13.38}$$

such that $\gamma \geq 0$, $\text{Tr}\gamma = 1$.

Having solved for γ, we can compute α up to a unitary rotation, and the resulting $\mathbf{R}$ via the SVD above. (The unitary freedom of α is the same as we see in choosing a Kraus operator sum representation for $\mathscr{R}$.)

The constraints in (13.38) are clearly convex; we have the same type of semidefinite cone that we saw in the direct optimization problem. In this case, the objective function is not linear in γ, but we do still satisfy a second-order convexity constraint. The determination of γ is, therefore, a convex optimization problem, which we can approach with sophisticated tools. For discussion of these methods, see [BV04a]. In fact, it was shown in [TKL10] that this can be reduced to an equivalent SDP.

As a final note on the indirect method, the authors of [TKL10] observed an interesting phenomenon. Instead of computing the optimal recovery via an SDP, they solved a constrained least squares problem whose form was suggested by the above derivation. This optimization problem is less resource-intensive than the SDP. Furthermore, in all of the computed cases, the resulting fidelity was identical to that achieved with the SDP method. The mathematical equivalence has not been proven, but this result is a further example of powerful computational methods for QEC.

13.5 Robustness to channel perturbations

Optimized QEC requires an accurate description of the noise model. The gains in performance are due to channel adaptation. This raises a significant issue: how robust are the encoding and recovery to uncertainty in the channel? Fortunately, the optimization routine can be derived to incorporate this uncertainty, generating a robust solution.

The formulation of the SDP can be adjusted to account for uncertainty in the channel. Consider a channel $\mathscr{E}_\Lambda$ that can be parametrized by a random variable Λ with density $f_\Lambda(\lambda)$. We can write the output state (to be corrected by $\mathscr{R}$) as

$$\mathscr{E}_\Lambda(\rho) = \int d\lambda f_\Lambda(\lambda) \mathscr{E}_\lambda(\rho) \tag{13.39}$$

due to the linearity of quantum operations. The linearity carries through the entire problem treatment and we can write the same optimization problem of (13.20) as

$$X^\star_{\mathscr{R}} = \arg \max_X \mathrm{Tr}(X C_{E,\mathscr{U}_C,\mathscr{E},\Lambda}) \tag{13.40}$$
$$\text{such that } X \geq 0, \quad \mathrm{Tr}_{\mathscr{H}} X = I,$$

where

$$C_{E,\mathscr{U}_C,\mathscr{E},\Lambda} = \int d\lambda f_\Lambda(\lambda) \sum_{jkm} p_m |\rho_m U_{Ck}^\dagger E_j^{\lambda\dagger}\rangle\!\rangle\!\langle\!\langle \rho_m U_{Ck}^\dagger E_j^{\lambda\dagger}|. \tag{13.41}$$

This is the direct method, as the variable to be optimized is the average entanglement fidelity of the channel. Robustness is also achievable through the indirect method; see [KSL08] and [TKL10] for details.

13.5.1 Channels represented by non-CP maps

Most of QEC has been developed under the assumption that the decoherence $\mathscr{E}$ can be described by a CP map. As discussed in Section 1.2.4.2, this corresponds to initial system–bath states with vanishing quantum discord. While this has been a useful model for the development of much QEC and fault-tolerant computing theory, it is quite possible that the CP assumption is overly optimistic and not representative of the behavior of quantum computers and quantum channels.

We can define a general linear channel $\mathscr{E}'$ by the operators $\{E_j, E'_j\}$, which has the output relation

$$\mathscr{E}'(\rho) = \sum_j E_j \rho E'_j. \tag{13.42}$$

We can represent this relation with an adjusted Choi matrix

$$\mathscr{E}'(\rho) = \sum_j \mathrm{Tr}_{\mathscr{H}^*} I \otimes \bar{\rho} |E_j\rangle\!\rangle\!\langle\!\langle E'_j| \tag{13.43}$$
$$= \mathrm{Tr}_{\mathscr{H}^*} I \otimes \bar{\rho} X_{\mathscr{E}'},$$

noting that $X_{\mathscr{E}'}$ has no further constraints. The entanglement fidelity can still be written $F_e(\rho, \mathscr{E}') = \langle\!\langle \rho | X_{\mathscr{E}'} | \rho \rangle\!\rangle$. From this point, we can follow through the same derivation as above. We state the optimization problem for a fixed encoding U_C:

$$\max_X \mathrm{Tr} X C_{\rho,\mathscr{E}',U_C}, \quad \text{such that } X \geq 0, \quad \mathrm{Tr}_{\mathscr{K}} X = I. \tag{13.44}$$

In this case, we define

$$C_{\rho,\mathscr{E}',U_C} = \sum_j |\rho U_C^\dagger E_j^\dagger\rangle\!\rangle\!\langle\!\langle \rho U_C^\dagger E_j'^\dagger| \tag{13.45}$$

using both sets of operators for the general linear map. We still have the same form that allows use of a semidefinite program: the objective function is linear in the optimization variable X, and X is restricted to a semidefinite cone.

There is an intuitive explanation for why the SDP still applies with a general linear channel. We have relaxed the CP constraint on the channel operation, but not on the behavior of the recovery operation. It is the CPTP constraint of $\mathscr{R}$ that constrains our optimization; having established that entanglement fidelity is linear in $\mathscr{E}'$, the rest follows naturally.

13.6 Structured near-optimal optimization

During our treatment of optimal recoveries and encodings, we touched on two significant drawbacks to optimized QEC using an SDP. First, the computational burden of the SDP grows very quickly with the code dimensions. Some of this growth is inevitable, as the Hilbert space dimension grows exponentially with the number of qubits. Nevertheless, the SDP is a resource intensive operation.

A second critique is in the unstructured nature of the codes and recovery operations. We constrained both to be CPTP maps, but further constraints were not enforced. While the encodings naturally tend to be isometries (even without being so constrained), they lack the convenient structure of stabilizer codes or other classically derived quantum codes. Additionally, the recovery operations have no particular constraints. Unlike most quantum code recoveries, optimized recovery operations begin with a generalized syndrome measurement, not a projection measurement. Such an operation may be resource intensive to implement; it is often less intuitive.

In order to address these issues, we will explore optimization methods that are less computationally intensive than the SDP and which do impose useful structure on the resulting operations. Specifically, we replace the SDP with an iterative algorithm based on eigen-analysis. We also constrain recovery operations to begin with a projective syndrome measurement. In imposing these constraints, we wish to remain near to the optimal solution, and present performance bounds to illustrate how well that goal is met.

In this section, we will focus on recovery operations – this is a choice for clarity of presentation. The algorithms are equally applicable for deriving encoding operations, as the projective syndrome measurement constraint is the equivalent of constraining the encoding operation to be unitary.

13.6.1 EigQER

We wish to constrain our recovery operation to begin with a projective measurement defined by a set of projection operators $\{P_i\}$. We have seen in more traditional approaches to QEC that a proper syndrome measurement is key to faithfully preserving the quantum information. We are seeking a computational method to derive such a measurement.

To do so, let us rewrite the objective function in (13.20) in terms of the operator elements of the recovery:

$$\max_{\{R_i\}} \sum_i \langle\langle R_i | C_{E,\mathscr{U}_C \mathscr{E}} | R_i \rangle\rangle. \tag{13.46}$$

If we require $\mathscr{R}$ to begin with a projective syndrome measurement, how do we write the constraint on $\{R_i\}$? Such an operation would have a set of operator elements that could be written as $\{R_i = U_i P_i\}$ where U_i are unitary and $\sum_i P_i = I$. Furthermore, we know that $P_i P_j = \delta_{ij} P_i$, which implies that $\langle\langle R_i | R_j \rangle\rangle = 0$ for $i \neq j$.

If the orthogonality constraint on $\{|R_i\rangle\rangle\}$ were sufficient, the optimization in (13.46) would have a well-known solution: $\{|R_i\rangle\rangle\}$ would be the eigenvectors associated with the largest eigenvalues of $C_{E,\mathcal{U}_C\mathcal{E}}$. While we know that the orthogonality constraint is not sufficient to generate the desired structure (we require that $R_i = U_i P_i$), it turns out, however, that the eigenvectors of $C_{E,\mathcal{U}_C\mathcal{E}}$ help identify a good syndrome measurement.

Consider the following algorithm, dubbed 'EigQER' for eigen quantum error recovery. The algorithm proceeds as follows:

(i) Initialize $C_1 = C_{E,\mathcal{E}}$.
   For the kth iteration:
(ii) Determine $|X_k\rangle\rangle$, the eigenvector associated with the largest eigenvalue of C_k.
(iii) As $X_k \in \mathcal{L}(\mathcal{K},\mathcal{H})$, determine the subspace of $\mathcal{K}$ that is the support of X_k. This can be identified by the projector P_k.
(iv) For the syndrome P_k, calculate the appropriate recovery operator R_k. Call R_k an operator element of $\mathcal{R}$.
(v) Determine C_{k+1} by projecting the subspace defined by P_k out of C_k.
(vi) Return to step (ii) until the recovery operation is complete.

The EigQER algorithm is guaranteed to generate a CPTP recovery operation, and will satisfy the criterion that it can be implemented by a projective syndrome measurement followed by a syndrome-dependent unitary operation.

Steps (iii) and (iv) in the above algorithm require further explanation. Given an operator $X \in \mathcal{L}(\mathcal{K},\mathcal{H})$, how should we calculate P_k and R_k? A straightforward answer uses the SVD. Let $X_k = U\Sigma V^\dagger$ be the SVD, where $U, \Sigma \in \mathcal{L}(\mathcal{H})$ and $V \in \mathcal{L}(\mathcal{H},\mathcal{K})$. They are constrained so that $U^\dagger U = V^\dagger V = I$ and Σ is the diagonal matrix of singular values (arranged in descending order). We first need to identify the subspace in $\mathcal{K}$ for this recovery operation; this is identified by the support of X_k, which is the range of V. If Σ is rank $d_\mathcal{H}$ (i.e., X_k is full rank), then the projector P_k should be defined by $VV^\dagger$. If X_k has rank $d < d_\mathcal{H}$,[4] we define P_k as $VI_d V^\dagger$, where I_d is a diagonal matrix with the 1 for the first d diagonal entries and 0 for the remainder.

Having defined P_k, we need now to define R_k. X_k was calculated as the operator that provides maximum contribution to the entanglement fidelity. We want R_k to best approximate this contribution, while having our desired structure. If we choose $R_k = UV^\dagger$ ($UI_d V^\dagger$ for the reduced rank case), this choice minimizes the Hilbert–Schmidt norm $\|X_k - R_k\|$.

To understand the update to C_k in Step (v), we remember that we are building up a projective measurement, step-by-step. Once we have determined a subspace defined by the projector P_i, all subsequent subspaces must be orthogonal. To achieve such orthogonality, we require that

$$X_k P_i = 0 \Leftrightarrow |X_k P_i\rangle\rangle = I \otimes \overline{P_i}|X_k\rangle\rangle = 0 \qquad (13.47)$$

for $i < k$. All $|X\rangle\rangle$ for which this is not satisfied should be in the nullspace of C_k. Thus, after each iteration we update the data matrix as

$$C_{k+1} = (I - I \otimes \overline{P_k})C_k(I - I \otimes \overline{P_k}). \qquad (13.48)$$

---

[4] Inclusion of reduced rank subspaces may seem unnecessary or even undesirable – after all, such a projection would collapse superpositions within the encoded information. We allow the possibility since such operator elements are observed in the optimal recovery operations. Furthermore, it is often useful to define rank liberally; singular values that are moderately close to 0 should be treated as if they were 0.

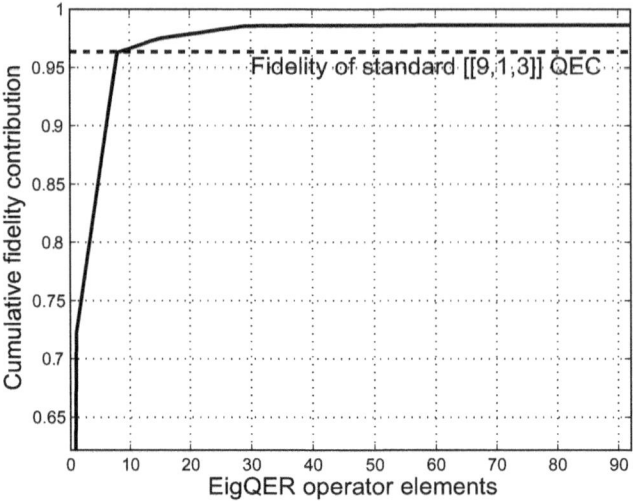

Fig. 13.2. Fidelity contribution of EigQER recovery operators for the amplitude damping channel ($\gamma = 0.09$) and the Shor code. Notice that the standard QEC performance is equaled with only nine optimized operator elements, and the relative benefit of additional operators goes nearly to zero after 30. Reprinted with permission from [FSW08]. Copyright 2008, American Physical Society.

The algorithm terminates when the recovery operation is complete, i.e., $\sum_k R_k^\dagger R_k = \sum_k P_k = I$. Given the structure of the recovery operations, this can be determined with a simple counter that is increased by the rank d of X_k at each step k. When the counter reaches $d_\mathcal{H}$, the recovery is complete.

In fact, we do not have to continue computing recovery operators until we reach $d_\mathcal{H}$. Each R_k contributes $\langle\!\langle R_k | C_{E,\mathcal{U}_C\mathcal{E}} | R_k \rangle\!\rangle$ to the average entanglement fidelity. EigQER is a "greedy" algorithm, which seeks to maximize its gain at each step; consequently, the performance return of each R_k diminishes as k grows. This is illustrated in Fig. 13.2, where we show the cumulative contribution for each recovery operator element with the Shor code and the amplitude damping channel. The greedy construction results in simplifications in both computation and implementation. When the contribution $\langle\!\langle R_k | C_{E,\mathcal{U}_C\mathcal{E}} | R_k \rangle\!\rangle$ passes below some selected threshold, the algorithm may terminate and thus reduce the computational burden. This results in an under-complete recovery operation where $\sum_k R_k^\dagger R_k \leq I$. An under-complete specification for the recovery operation may significantly reduce the difficulty in physical implementation. In essence, an under-complete recovery operation will have syndrome subspaces whose occurrence is sufficiently rare that the recovery operation may be left as a "don't care."

Let us count the computational burden of the EigQER algorithm. The limiting step is the calculation of $|X_k\rangle\!\rangle$, the eigenvector associated with the largest eigenvalue of C_k. C_k is a $2^{n+k} \times 2^{n+k}$ dimensional matrix, but the eigenvector has only 2^{n+k} dimensions. Using the *power method* [MS00] for calculating the dominant eigenvector requires $\mathcal{O}(2^{2(n+k)})$ flops for each iteration of the power method. For contrast, recall that the SDP requires $\mathcal{O}(2^{5(n+k)})$ flops per iteration. While both problems grow exponentially with n, the eigenvector problem is significantly simpler than the SDP.

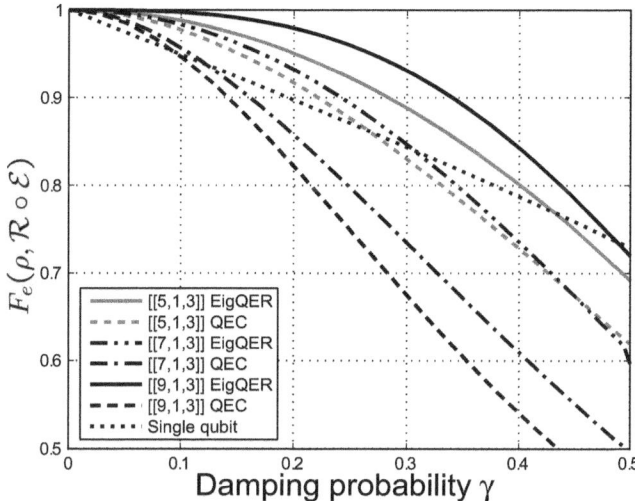

Fig. 13.3. EigQER and standard QEC recovery performance for the [[5,1,3]], [[7,1,3]], and [[9,1,3]] codes and the amplitude damping channel. Note that generic QEC performance decreases for longer codes, as multiple-qubit errors become more likely. While the EigQER performance for the nine-qubit Shor code is excellent, the seven-qubit Steane code shows only modest improvement, with performance similar to the generic five-qubit QEC recovery. Reprinted with permission from [FSW08]. Copyright 2008, American Physical Society.

We should note that the eigenvector computation must be repeated for each operator element of $\mathcal{R}$. If we were to compute all of them, not truncating early due to the diminishing returns of the greedy algorithm, this would require iterating the algorithm approximately $d_{\mathcal{K}}/d_{\mathcal{H}} = 2^{n-k}$ times. In fact, we have a further reduction as the algorithm iterates. At the jth iteration we are calculating the dominant eigenvector of C_j which lives on a $(d_{\mathcal{K}} - jd_{\mathcal{H}})d_{\mathcal{H}} = 2^k(2^n - j2^k)$ dimensional subspace. We can therefore reduce the size of the eigenvector problem at each iteration of EigQER.

Figure 13.3 demonstrates the performance of several codes and the amplitude damping channel. We compare the EigQER performance for the [[5,1,3]] code, the [[7,1,3]] Steane code, and the [[9,1,3]] Shor code, contrasting each with the generic QEC performance. Notice first the pattern with the standard QEC recovery: the entanglement fidelity decreases with the length of the code. Each of these codes is designed to correct a single error on an arbitrary qubit, and fails only if multiple qubits are corrupted. For a fixed γ, the probability of a multiple qubit error rises as the number of physical qubits n increases.

The QEC performance degradation with code length is a further illustration of the value of channel adaptivity. All three codes in Fig. 13.3 contain one qubit of information, so longer codes include more redundant qubits. Intuitively, this should better protect the source from error. When we channel-adapt, this intuition is confirmed for the Shor code, but not for the Steane code. In fact, the EigQER entanglement fidelity for the Steane code is only slightly higher than the generic QEC recovery for the [[5,1,3]] code. From this example, it appears that the Steane code is not particularly well suited for adapting to amplitude damping errors. We see that the choice of encoding significantly impacts channel-adapted recovery. This is further justification for the iterative optimization of both recovery and encoding.

13.6.2 Performance bounds

While we have seen some examples of good performance from the EigQER algorithm, we would like more than anecdotal evidence for the technique's efficacy. For this reason, we present here a technique for computing an upper bound on the entanglement fidelity.

We accomplish this by using the Lagrange dual function derived in Section 13.3.4. Specifically, we will use the bounding properties of the dual function. Recall that $Y \in \mathscr{L}(\mathscr{K})$ is a *dual feasible point* if and only if $Y \otimes I - C_{E,\mathscr{U}_C \mathscr{E}} \geq 0$. For a dual feasible Y, the bounding property implies that $\bar{F}_e(E, \mathscr{R} \circ \mathscr{E} \circ \mathscr{U}_C) \leq \text{Tr} Y$; Y provides quantitative evidence of the near-optimality of a particular recovery operation.

To provide a good performance bound, it is desirable to find a dual feasible point with a small dual function value. Indeed, the best such bound is the solution to (13.28), that is to find the dual feasible point with the smallest trace. However, finding the optimal Y is the equivalent of solving for the optimal recovery due to the strong duality of the SDP. This suffers the same computational burden as computing the optimal recovery. Instead, we want a less computationally demanding procedure to generate a dual feasible point.

We can iteratively compute a reasonable dual feasible point. We begin with an initial dual point $Y^{(0)}$ (we will discuss initialization options later) that is presumably not dual feasible. At the kth iteration, we update the dual point to produce $Y^{(k)}$ until we achieve feasibility. For convenience we define

$$Z^{(k)} \equiv Y^{(k)} \otimes I - C_{E,\mathscr{U}_C \mathscr{E}}. \qquad (13.49)$$

Let x and $|x\rangle\rangle$ be the smallest eigenvalue and associated eigenvector of $Z^{(k)}$. We stop if $x \geq 0$, as $Y^{(k)}$ is already dual feasible. If $x \leq 0$, we wish to update $Y^{(k)}$ a small amount to ensure that $\langle\langle x|Z^{(k+1)}|x\rangle\rangle \geq 0$. Essentially, we are adjusting the operator to eliminate any subspaces associated with negative eigenvalues. Given no constraints on the update, we could accomplish this as $Z^{(k+1)} = Z^{(k)} + x|x\rangle\rangle\langle\langle x|$ but we must instead update $Y^{(k)}$ with the tensor product structure implicit.

We determine the properly constrained update by means of the Schmidt decomposition of the eigenvector:

$$|x\rangle\rangle = \sum_i \lambda_i |\tilde{x}_i\rangle_\mathscr{K} |\hat{x}_i\rangle_{\mathscr{H}^*}. \qquad (13.50)$$

As we can only perturb $Z^{(k)}$ in the $\mathscr{K}$ slot, we choose the smallest perturbation guaranteed to achieve $\langle\langle x|Z^{(k+1)}|x\rangle\rangle \geq 0$. Let

$$Y^{(k+1)} = Y^{(k)} + \frac{|x|}{|\lambda_1|^2} |\tilde{x}_1\rangle\langle\tilde{x}_1|. \qquad (13.51)$$

Then

$$\begin{aligned}\langle\langle x|Z^{(k+1)}|x\rangle\rangle &= x + \frac{|x|}{|\lambda_1|^2} \langle\langle x|(|\tilde{x}_1\rangle\langle\tilde{x}_1| \otimes I)|x\rangle\rangle \\ &= x + \frac{|x|}{|\lambda_1|^2} |\lambda_1|^2 \\ &= 0, \end{aligned} \qquad (13.52)$$

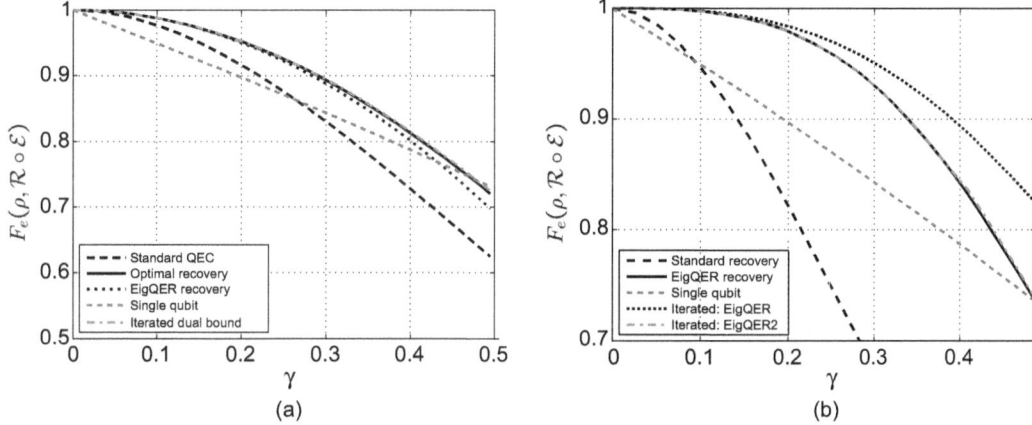

Fig. 13.4. Dual bound comparison for the amplitude damping channel and (a) the [[5,1,3]] and (b) the [[9,1,3]] Shor code. In both cases, we can see that an iterated dual bound produces a bound that is tight to the optimal fidelity. For the [[9,1,3]] case, this further demonstrates that the EigQER recovery operation is essentially optimal in this case. Notice that the iterated bound initialized with the EigQER recovery operation does not generate a tight bound. Instead, the bound is initialized with a modified EigQER recovery; for details, see [FSW08].

since $x < 0$. While we have not yet guaranteed that $Z^{(k+1)} \geq 0$, $|x\rangle\!\rangle$ is no longer associated with a negative eigenvalue. By repeatedly perturbing $Y^{(k)}$ in this manner, we iteratively approach a dual feasible point while adding as little as possible to the dual function value $\mathrm{Tr} Y^{(k)}$.

As a final point, we demonstrate that the iterative procedure will converge to a dual feasible point. Let us consider the effect of the kth iteration on the space orthogonal to $|x\rangle\!\rangle$. Let $|y\rangle\!\rangle \in \mathcal{K} \otimes \mathcal{H}^*$ be orthogonal to $|x\rangle\!\rangle$. Then, for $Z^{(k+1)}$ we see that

$$\langle\!\langle y|Z^{(k+1)}|y\rangle\!\rangle = \langle\!\langle y|Z^{(k)}|y\rangle\!\rangle + \frac{|x|}{|\lambda_1|^2}\langle\!\langle y|(I \otimes |\tilde{x}_1\rangle\langle\tilde{x}_1|)|y\rangle\!\rangle. \tag{13.53}$$

But since $|\tilde{x}_1\rangle\langle\tilde{x}_1| \otimes I \geq 0$ we see that

$$\langle\!\langle y|Z^{(k+1)}|y\rangle\!\rangle \geq \langle\!\langle y|Z^{(k)}|y\rangle\!\rangle \tag{13.54}$$

for all $|y\rangle\!\rangle \in \mathcal{K} \otimes \mathcal{H}^*$. We have eliminated a negative eigenspace, and the complementary subspace has only increased; no additional negative spaces are generated as a result.

Some example bounds can be seen in Fig. 13.4. The dual feasible points generated in this manner are often quite good bounds, though they can be dependent on the initialization choice $Y^{(0)}$. We can begin with $Y^{(0)} = 0$, but this will often require a significant computational cost. If computations must be reduced, we can derive an initial point based on the syndrome projection $\{P_i\}$ obtained via EigQER. As it turns out, we can calculate a reduced dimension SDP within each subspace, determining optimal $Y_k^\star$ dual feasible points. We can define $Y^{(0)} = \sum_k Y_k^\star$; this is not guaranteed to be dual feasible, but is known to be optimal within these subspaces.

The interested reader is referred to [FSW08] for further initialization options and alternative near-optimal channel-adapted methods.

13.7 Optimization for (approximate) decoherence-free subspaces

Throughout this chapter, we have focused on the most straightforward construction of QEC: an encoding operation (usually unitary) preceding the decoherence and a recovery operation following. We see throughout this book that this formulation of QEC, while powerful, is not comprehensive. It does not capture entanglement-assisted QEC (Chapter 7), decoherence-free subspaces (DFSs), or noiseless subsystems (NSs) (Chapter 3). We saw that optimization methods for the vanilla QEC formulation suggest more efficient approaches beyond obvious extensions of classical error correction. In a similar way, we are motivated to explore optimization in the more comprehensive QEC scenarios.

We will sketch here an approach taken by Yamamoto and Fazel [YF07] to determine an approximate DFS for a channel $\mathscr{E} \in \mathscr{L}(\mathscr{H})$. The *purity* of a state ρ can be defined as $p(\rho) = \mathrm{Tr}\rho^2$. The purity is 1 if and only if ρ is a pure state; if $p(\mathscr{E}(|\psi\rangle\langle\psi|)) \approx 1$, this indicates that $\mathscr{E}$ has introduced very little decoherence to $|\psi\rangle$. A subspace $\mathscr{C} \subset \mathscr{H}$ for which all states have a purity nearly 1 is a good candidate for a DFS. If $U_C \in \mathscr{L}(\mathscr{H}, \mathscr{K})$ is a unitary encoding operation, we can define the purity maximization problem as

$$\max_{U_C} \min_{|\psi\rangle \in \mathscr{H}} \mathrm{Tr}\mathscr{E}(U_C|\psi\rangle\langle\psi|U_C^\dagger)^2, \qquad (13.55)$$

where we are maximizing the purity for the worst-case encoded input state. Not surprisingly, this max-min problem is difficult to solve.

From our earlier derivations, we can already state some of the constraints. The encoding operation is completely positive; we therefore must have $X_{\mathscr{U}_C} \geq 0$ and $\mathrm{Tr}_{\mathscr{K}} X_{\mathscr{U}_C} = I_{\mathscr{H}^*}$. Furthermore we want the encoding to be unitary, so $X_{\mathscr{U}_C} = |U_C\rangle\!\rangle\langle\!\langle U_C|$ must be a rank 1 operator.

To continue, we need a short diversion to state a consequence of the Jamiolkowski isomorphism. We already saw one representation of a channel $\mathscr{A}$ with a positive operator $X_{\mathscr{A}}$. We now define an alternate operator $\tilde{X}_{\mathscr{A}}$ that is particularly convenient for input–output relations. Noting that $|\mathscr{A}(\rho)\rangle\!\rangle = \sum_k |A_k \rho A_k^\dagger\rangle\!\rangle$, we can use (13.9) to write

$$|\mathscr{A}(\rho)\rangle\!\rangle = \sum_k A_k \otimes \overline{A_k}|\rho\rangle\!\rangle \equiv \tilde{X}_{\mathscr{A}}|\rho\rangle\!\rangle. \qquad (13.56)$$

We also note that $\langle\!\langle \rho|\rho\rangle\!\rangle = \mathrm{Tr}\rho^2$, giving us an easy expression for the purity. Finally, for $\rho = |\psi\rangle\langle\psi|$, $|\rho\rangle\!\rangle = |\psi\rangle\overline{|\psi\rangle}$.

These results allow us to rewrite our desired optimization problem as

$$\max_{\mathscr{U}_C} \min_{|\psi\rangle} \langle\psi|\overline{\langle\psi|}\tilde{X}_{\mathscr{E}}^\dagger \tilde{X}_{\mathscr{U}_C}^\dagger \tilde{X}_{\mathscr{U}_C} \tilde{X}_{\mathscr{E}}|\psi\rangle\overline{|\psi\rangle}. \qquad (13.57)$$

In this case, it is useful to rewrite the objective function as a constraint. The maximum purity is 1 by definition; we can therefore minimize our departure from unity as

$$\min_{\mathscr{U}_C, \epsilon} \epsilon \text{ such that}$$
$$\langle\psi|\overline{\langle\psi|}\tilde{X}_{\mathscr{E}}^\dagger \tilde{X}_{\mathscr{U}_C}^\dagger \tilde{X}_{\mathscr{U}_C} \tilde{X}_{\mathscr{E}}|\psi\rangle\overline{|\psi\rangle} \geq 1 - \epsilon \; \forall |\psi\rangle \in \mathscr{H}. \qquad (13.58)$$

For the next step, we will describe the result but refer the interested reader to [YF07] for the details. Having now written the objective function as an inequality constraint, we would like the constraint to be an LMI, just as we have seen in our previous problems. By introducing a scalar variable τ, we can write the constraint in (13.58) as $T(\tilde{X}_{\mathcal{U}_C}, \epsilon, \tau) \geq 0$, where T is a matrix that is a linear function of $\tilde{X}_{\mathcal{U}_C}, \epsilon$, and τ. The constraints are exactly equivalent when $\mathcal{H}$ is two-dimensional (i.e., a qubit); the LMI constraint is a lower bound for the exact expression for larger subspaces.

Let us summarize the results so far. We are selecting $\tilde{X}_{\mathcal{U}_C}$, ϵ, and τ in an effort to minimize ϵ. We have the CPTP constraints on $\mathcal{U}_C$ ($X_{\mathcal{U}_C} \geq 0$ and $\mathrm{Tr}_{\mathcal{K}} X_{\mathcal{U}_C} = I_{\mathcal{H}^*}$), we have the constraint $T(\tilde{X}_{\mathcal{U}_C}, \epsilon, \tau) \geq 0$ which captures the purity calculation, and we have $\mathrm{rank}(X_{\mathcal{U}_C}) = 1$ which constrains us to a unitary encoding. The first three constraints are all convex, but the rank 1 constraint is more challenging.

To find a computational method, Yamamoto and Fazel propose yet another constraint relaxation. Instead of requiring a rank 1 $X_{\mathcal{U}_C}$, they jointly minimize $\mathrm{rank}(X_{\mathcal{U}_C})$ and ϵ, where ϵ is now constrained between 0 and 1. In essence, they are allowing both a departure from purity ($\epsilon \geq 0$) and a departure from unitary encoding ($\mathrm{rank}(X_{\mathcal{U}_C}) \geq 1$) while simultaneously minimizing both. The new objective function is $\mathrm{rank}(X_{\mathcal{U}_C}) + \gamma \epsilon$, where $\gamma > 0$ is a tuning parameter to relatively weight the two objectives.

This final reformulation is done because rank minimization subject to convex constraints is a well-researched problem. While computing the global solution is NP-hard, there are well-defined heuristics. The one used here is called the log det heuristic [FHB01, FHB03, FHB04]. In this case, the function $\log\det(X_{\mathcal{U}_C} + \delta I)$ (where det refers to the determinant) is a *smooth surrogate* for $\mathrm{rank}(X_{\mathcal{U}_C})$. The constant δ is selected to indicate the order of eigenvalues that can be considered effectively 0.

With the log det formulation, we can iteratively minimize our objective function. If we take the linear expansion of $\log\det(X_{\mathcal{U}_C} + \delta I)$ about a point X_i, we get

$$\log\det(X_{\mathcal{U}_C} + \delta I) \approx \log\det(X_i + \delta I) + \mathrm{Tr}[(X_i + \delta I)^{-1}(X - X_i)]. \tag{13.59}$$

We can minimize the function by iteratively minimizing the linear approximation, subject to our constraints:

$$X_{i+1} = \arg\min_{X, \epsilon, \tau} \mathrm{Tr}(X_i + \delta I)^{-1} X + \gamma \epsilon. \tag{13.60}$$

This step (finally!) can be accomplished via a semidefinite program.

Before concluding this discussion, we must emphasize that the iterative solution derived here is a local solution, not the global minimum. As with most nonconvex, iterative optimizations, the solution is significantly dependent on the initialization point X_0. A reasonably comprehensive solution should test several initialization points, including (if possible) those suggested by intuition regarding $\mathcal{E}$.

13.8 Conclusion

In this chapter, we have provided a few optimization tools to further QEC. In truth, researchers have only begun to scratch the surface of optimization methods for QEC. Nevertheless, the tools we have are powerful. The semidefinite program is often relevant, given its connection to the CPTP constraint on quantum operations. Armed with the SDP, researchers have a strong means towards highly efficient, channel-adapted QEC procedures.

As a final note, we should emphasize that optimized QEC does not purport to be the best practical QEC implementation. As we noted through the chapter, there are currently computational complexity issues in computing an optimal error correction as well as hardware constraints that need to be included. Alleviating these issues is an on-going area of research. At the same time, it is not clear how much strain this causes as the actual building blocks of a quantum computer are still embryonic. At present, however, it is very useful to lean on optimization as a process of informing our intuition about QEC, pointing us in directions that might otherwise have remained hidden.

Part IV

Advanced dynamical decoupling

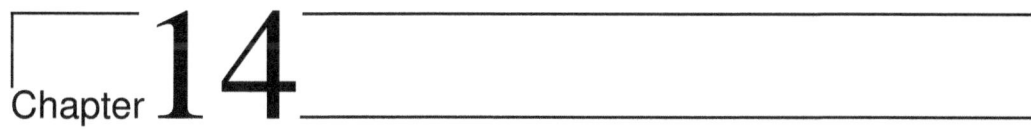

High-order dynamical decoupling

Zhen-Yu Wang and Ren-Bao Liu

14.1 Introduction

After the introduction to dynamical decoupling (DD) in Chapter 4, this chapter focuses on advanced DD techniques for the protection of quantum systems to higher orders of accuracy in the short-time limit, either by concatenation or by optimization of pulse timing. In Section 14.11, we briefly discuss randomized decoupling schemes, applicable beyond the short-time limit.

Our setting is that of a quantum system coupled to a quantum bath bounded by the (sup-operator) norm, so that the controlled system evolution can be faster than the error evolution driven by the uncontrolled Hamiltonian. Without time-dependent control, the time-independent system–bath Hamiltonian can always be written as

$$H = H_S + H_B + H_{SB}, \qquad (14.1)$$

where H_S and H_B are the system and bath Hamiltonians, respectively, and H_{SB} is the system–bath interaction Hamiltonian. To suppress errors due to the system–bath interactions, we apply a pulsed external control to the system, as discussed in Chapter 4. In this chapter (except Sections 14.9 and 14.10), we only consider bang-bang control, that is, the unitary control pulses are instantaneous and arbitrarily strong. DD based on bounded control (finite pulses) has been discussed in Chapter 4 and is also included in Chapter 15. Generalization of the results to time-dependent Hamiltonians will be given in Section 14.10.

14.2 Operator set preservation

In the short-time limit, we aim to find a sequence of instantaneous operations $P_1, P_2, \ldots, P_M$ applied at moments $T_1, T_2, \ldots, T_M$, such that after the controlled evolution from $T_0 = 0$ to $T_M = T$, a set of system operators $\{A_j\} \equiv \{A_1, A_2, \ldots\}$ is conserved up to errors of $O\left(T^{N+1}\right)$,

---

Quantum Error Correction, ed. Daniel A. Lidar and Todd A. Brun. Published by Cambridge University Press.
© Cambridge University Press 2013.

that is, for (throughout this chapter, $\hbar = 1$)

$$U(T) = P_M e^{-iH(T-T_{M-1})} P_{M-1} \cdots P_2 e^{-iH(T_2-T_1)} P_1 e^{-iHT_1}, \qquad (14.2)$$

we wish to have

$$U^\dagger A_j U = A_j + O\left(T^{N+1}\right). \qquad (14.3)$$

This type of control is an Nth-order DD scheme for preservation of $\{A_j\}$.

When a set of operators $\{A_j\}$ is preserved to a certain decoupling order, then all the operator products $A_{j_1} \ldots A_{j_m}$ from $\{A_j\}$ and their linear combinations are also protected to at least the same decoupling order. When a complete system operator basis is preserved, then the arbitrary system states are preserved. This amounts to realizing a quantum memory with storage errors $O\left(T^{N+1}\right)$.

Defining $Q_1 \equiv I_S$ (the identity system operator) and $Q_j \equiv P_{j-1} Q_{j-1}$ for $j \geq 2$, we rewrite Eq. (14.2) as

$$U(T) = \mathcal{T} \exp\left[-i \int_0^T H_p(t) dt\right], \qquad (14.4)$$

where $\mathcal{T}$ is the time-ordering operator, and $H_p(t) = Q_j^\dagger H Q_j$ for $t \in (T_{j-1}, T_j]$. Without loss of generality, here $P_M \cdots P_2 P_1 = I_S$ has been assumed, so that in the absence of the control-free Hamiltonian [Eq. (14.1)] the control protocol would realize the identity operation on the system. In the standard time-dependent perturbation theory formalism,

$$U(T) = 1 + \sum_{n=1}^{\infty} \sum_{M \geq j_1 > \cdots > j_n \geq 1} \sum_{p_1=1}^{\infty} \cdots \sum_{p_n=1}^{\infty} \left(\prod_{r=1}^n \frac{1}{p_r!}\right) h_{j_1}^{p_1} \cdots h_{j_n}^{p_n}, \qquad (14.5)$$

where

$$h_j \equiv -i(T_j - T_{j-1}) Q_j^\dagger H Q_j, \qquad (14.6)$$

and each expansion term is of $O(T^{\sum_{r=1}^n p_r})$. For example, the propagator expanded up to the second order reads

$$U(T) = 1 + \sum_{j=1}^{M} h_j + \sum_{M \geq j > k \geq 1} h_j h_k + \sum_{j=1}^{M} \frac{1}{2} h_j^2 + O(T^3). \qquad (14.7)$$

As shown in Chapter 4, if the set $\{Q_1, \ldots, Q_M\}$ forms a group (up to phase factors), a first-order DD protocol is realized by applying equally spaced operations P_j. The controlled propagator reads

$$U^{[1]}(T) = P_M U^{[0]}(\tfrac{T}{M}) P_{M-1} \cdots P_2 U^{[0]}(\tfrac{T}{M}) P_1 U^{[0]}(\tfrac{T}{M})$$
$$= \exp\left[-i \bar{H}^{(0)} T\right] + O(T^2) \equiv e^{-i H^{[1]}(T) T}, \qquad (14.8)$$

where $\bar{H}^{(0)} \equiv \frac{1}{M} \sum_{j=1}^{M} Q_j^\dagger H Q_j$ and $U^{[0]}(\tfrac{T}{M}) = e^{-iHT/M}$ is the free evolution propagator without control. The superscript "[1]" indicates that this is first-order decoupling. Since group-based symmetrization [VKL99, Z99] gives $[Q_j, \bar{H}^{(0)}] = 0$ for $j = 1, \ldots, M$, we realize first-order protection

$$U^\dagger(T) Q_j U(T) = Q_j + O\left(T^2\right). \qquad (14.9)$$

By choosing a suitable group $\{Q_j\}$, arbitrary quantum states can be preserved to first order in principle (see Chapter 4), and with higher efficiency when additional structure in the control-free Hamiltonian H is utilized (see Chapter 15). Repeated sequences are called periodic DD (PDD) (see Chapter 4).

If we have a sequence $e^{h_M} \cdots e^{h_2} e^{h_1}$ that preserves a set of system operators $\{A_j\}$ up to the first order (i.e., $[\sum_{i=1}^{M} h_i, A_j] = 0$), to realize second-order protection, we can use symmetric DD (SDD) [VKL99, SV08]. The sequence is

$$U_{\text{SDD}}^{[2]}(2T) = e^{h_1} e^{h_2} \cdots e^{h_{M-1}} e^{h_M} e^{h_M} e^{h_{M-1}} \cdots e^{h_2} e^{h_1}$$

$$= 1 + 2\sum_{i=1}^{M} h_i + 2\left(\sum_{i=1}^{M} h_i\right)^2 + O\left(T^3\right). \tag{14.10}$$

It commutes with $\{A_j\}$ up to second order, since $\sum_{i=1}^{M} h_i$ commutes with $\{A_j\}$.

To preserve a quantum system to a higher order N, in principle, we can obtain the corresponding DD sequences by solving Eq. (14.5) so that $U(T)$ commutes with a given set of system operators up to $O\left(T^{N+1}\right)$. However, finding solutions becomes formidable when the DD order is high and the Hilbert space of the quantum system is large. A general method to go beyond first-order protection to arbitrary Nth-order protection is desirable, and will be described later in this chapter.

14.3 Dynamical decoupling for multi-qubit systems

Before proceeding to high-order DD, we first discuss general DD of multi-qubit systems. We denote $\sigma_0^{(j)} = I_j$ the identity operator and $\{\sigma_x^{(j)}, \sigma_y^{(j)}, \sigma_z^{(j)}\}$ the Pauli operators of the jth qubit. The 4^L tensor products $\{\sigma_{\alpha_1}^{(1)} \otimes \cdots \otimes \sigma_{\alpha_L}^{(L)}\}$ with $\alpha_j \in \{0, x, y, z\}$ form a complete system operator basis of an L-qubit system. To realize an L-qubit quantum memory, we need to preserve this operator basis. Sometimes it suffices to preserve a subset, for example, the operator set $\{\sigma_x^{(1)}, \sigma_y^{(1)}, \ldots, \sigma_x^{(L)}, \sigma_y^{(L)}\}$. Note that all the tensor products $\sigma_{\alpha_1}^{(1)} \otimes \cdots \otimes \sigma_{\alpha_L}^{(L)}$ are unitary and Hermitian and they either commute or anticommute. As these operators are important in DD for multi-qubit systems, we give a generalized definition of this kind of operator set.

Definition 14.1. *(Mutually orthogonal operation set (MOOS).) A MOOS is a set of system operators that are unitary and Hermitian and have the property that each pair of elements either commutes or anticommutes.*

The elements Ω_l in a MOOS are parity kick operators [VT99] as they satisfy $\Omega_l^2 = \Omega_l \Omega_l^\dagger = I_S$. Obviously any subset of $\{\sigma_{\alpha_1}^{(1)} \otimes \cdots \otimes \sigma_{\alpha_L}^{(L)}\}$ is a MOOS. The commutation property of operators in a MOOS is important in constructing higher-order DD schemes, via the following theorem [WL11b].

Theorem 14.1. *For two operators Ω_1 and Ω_2 that either commute or anticommute, if a unitary evolution $U(T)$ commutes with Ω_1 up to an error of $O\left(T^{N+1}\right)$, then $\Omega_2 U(T) \Omega_2$ also commutes with Ω_1 up to an error of $O\left(T^{N+1}\right)$.*

Proof. We directly calculate the commutator

$$[\Omega_1, \Omega_2 U(T) \Omega_2] = \eta \Omega_2 [\Omega_1, U(T)] \Omega_2 = O\left(T^{N+1}\right), \tag{14.11}$$

where $\eta = +1 \, (-1)$ for Ω_1 commuting (anticommuting) with Ω_2. $\square$

With this theorem, certain DD sequences preserving a given element Ω_l in a MOOS can be used as construction units for an outer-level DD protection of another element in the MOOS, without affecting the DD order of the inner-level pulse sequences. For operators that do not form a MOOS, the outer-level control operations in general may interfere with those at the inner level [WL11b].

Here we describe an explicit scheme to protect system operators in a MOOS, which facilitates the construction of higher-order DD in the later sections of this chapter. Let us first consider protection of a single unitary Hermitian operator Ω_1. The Hamiltonian can be separated into two parts,

$$H = C_{\Omega_1} + A_{\Omega_1}, \qquad (14.12)$$

with $C_{\Omega_1} \equiv (H + \Omega_1 H \Omega_1)/2$ and $A_{\Omega_1} \equiv (H - \Omega_1 H \Omega_1)/2$. Ω_1 commutes with C_{Ω_1} and anticommutes with A_{Ω_1}, that is, $\Omega_1 C_{\Omega_1} \Omega_1 = C_{\Omega_1}$ and $\Omega_1 A_{\Omega_1} \Omega_1 = -A_{\Omega_1}$. With an instantaneous control pulse Ω_1 applied at the middle of the evolution time, the evolution operator becomes

$$U_1(T) = [\Omega_1] U^{[0]}(\tfrac{T}{2}) \Omega_1 U^{[0]}(\tfrac{T}{2}) = [\Omega_1] \Omega_1 e^{-i[C_{\Omega_1} + O(T)]T}, \qquad (14.13)$$

where $U^{[0]}(\tau) \equiv e^{-iH\tau}$ is the free evolution operator over time τ, and the brackets around the operation at the end of the sequence ($[\Omega_1]$) indicate that the operation is optional. $U_1(T)$ commutes with Ω_1 up to an error of $O(T^2)$. Therefore Ω_1 is protected to the first order. This parity kick method is equivalent to symmetrizing the time evolution of the system with respect to the group $\{I_S, \Omega_1\}$ [VT99]. Following the method given above, we can preserve more operators $\{\Omega_j\}$ in a MOOS by nesting. The second-level control is

$$U_2(T) = [\Omega_2] U_1(\tfrac{T}{2}) \Omega_2 U_1(\tfrac{T}{2}) = [\Omega_2] \Omega_2 e^{-i H_2 T}. \qquad (14.14)$$

The resultant effective Hamiltonian H_2 commutes with Ω_2 up to an error of $O(T)$. And according to Theorem 14.1, H_2 also commutes with Ω_1 up to $O(T)$. A general first-order scheme is achieved by using Eq. (14.13) iteratively,

$$U_L(T) = [\Omega_L] U_{L-1}(\tfrac{T}{2}) \Omega_L U_{L-1}(\tfrac{T}{2}) = [\Omega_L] \Omega_L e^{-i H_L T}, \qquad (14.15)$$

where the effective Hamiltonian H_L commutes with all operators in the MOOS $\{\Omega_1, \ldots, \Omega_L\}$ up to $O(T)$. Therefore we protect the MOOS by iteratively symmetrizing the system evolution. If we write Eq. (14.15) in the form of Eq. (14.4), we will find that $\{Q_j\}$ forms a group with Q_j being products of elements in the MOOS $\{\Omega_1, \ldots, \Omega_L\}$.

Note that, for multi-qubit systems, the pulses $\sigma_{\alpha_1}^{(1)} \otimes \cdots \otimes \sigma_{\alpha_L}^{(L)}$ (up to a trivial phase factor) are local control propagators. We denote the single-qubit gates for the lth qubit $X_l = \sigma_x^{(l)}$, $Y_l = \sigma_y^{(l)}$, and $Z_l = \sigma_z^{(l)}$ up to trivial phase factors. We may use single-qubit operations $\{X_1, Y_1, \ldots, X_L, Y_L\}$ to preserve the system states. The operations X_l and Y_l can be replaced with any two anticommuting Pauli operators of the lth qubit. For a single qubit coupled to a bath, we may choose $\{X, Y\}$ as the MOOS. Using Eqs. (14.13) and (14.14), we obtain the first-order sequence

$$Y \left(e^{-iHT/4} X e^{-iHT/4} \right) Y \left(e^{-iHT/4} X e^{-iHT/4} \right), \qquad (14.16)$$

where the parentheses indicate the inner-level pulse sequences. This reproduces the single-qubit universal DD sequence in Chapter 4.

If the Hamiltonian includes arbitrary interactions, we need to protect a MOOS of size $2L$, and to realize first-order DD, the pulse number is $O(4^L)$, which is indeed the minimum number of control pulses required for protecting a general L-qubit system [WRJ+02b]. When the Hamiltonian has some internal symmetry, we can reduce the size of the MOOS. For example, when the qubits undergo pure dephasing only (i.e., the Hamiltonian commutes with all $\sigma_z^{(l)}$ operators), DD of the MOOS $\{X_1, \ldots, X_L\}$ suffices to preserve arbitrary system states, and the pulse number is $O(2^L)$. If the Hamiltonian only involves certain kinds of interaction, such as pairwise interactions only, the number of pulses can be greatly reduced by combinatorial methods (see Chapter 15).

14.4 Concatenated dynamical decoupling

The possibility of realizing arbitrary order DD for specific systems was mentioned in [VKL99]. Khodjasteh and Lidar proposed the first explicit scheme to eliminate arbitrary qubit–bath couplings (including both diagonal and off-diagonal couplings) using a recursive construction called concatenated DD (CDD) [KL05, KL07]. The idea of CDD was further developed by incorporation of randomness into the sequence for improvement of long-time performance, so-called randomized DD (RDD) [KA05, SV06, VS06]. CDD schemes against pure dephasing were investigated for electron-spin qubits in realistic solid-state systems with nuclear spins as baths [YLS07, WD07, ZDS+07]. CDD for arbitrary quantum systems was discussed in [SV08].

14.4.1 General construction

A key observation is that after the group-based symmetrization [VKL99, Z99], the effective Hamiltonian $H^{[1]}(T)$ in Eq. (14.8) commutes with the operators $\{Q_j\}$ to a higher order, that is, the errors are reduced from $O(T^0)$ for the original Hamiltonian H to $O(T^1)$ for the resultant effective Hamiltonian $H^{[1]}(T)$. If we insert the first-order protected evolution $U^{[1]}(\frac{T}{M})$ [Eq. (14.8)] into the group-based symmetrization sequence, we have the concatenated sequence

$$U^{[2]}(T) = P_M U^{[1]}(\tfrac{T}{M}) P_{M-1} \cdots P_2 U^{[1]}(\tfrac{T}{M}) P_1 U^{[1]}(\tfrac{T}{M}). \tag{14.17}$$

Therefore, the new effective Hamiltonian in $U^{[2]}(T) = e^{-iH^{[2]}(T)T}$ commutes with $\{Q_j\}$ to the second order of T. Iteratively, the Nth-order CDD

$$U^{[N]}(T) = P_M U^{[N-1]}(\tfrac{T}{M}) P_{M-1} \cdots P_2 U^{[N-1]}(\tfrac{T}{M}) P_1 U^{[N-1]}(\tfrac{T}{M}), \tag{14.18}$$

preserves the operator group $\{Q_j\}$ to the Nth order, that is,

$$U^{[N]\dagger}(T) Q_j U^{[N]}(T) = Q_j + O\left(T^{N+1}\right). \tag{14.19}$$

A periodic version may be obtained by repeating the Nth-order CDD sequences [SV08].

14.4.2 Construction for qubits

We first consider an example of a Hamiltonian describing a qubit under pure dephasing,

$$H = H_B + \sigma_z \otimes B_z, \tag{14.20}$$

where H_B and B_z are bath operators. As the Hamiltonian commutes with σ_z, preservation of the operator X or Y will preserve arbitrary qubit states. The first-order sequence reads

$$U^{[1]}(T) = Xe^{-iHT/2}Xe^{-iHT/2}. \tag{14.21}$$

The second-order CDD sequence is

$$U^{[2]}(T) = XU^{[1]}(\tfrac{T}{2})XU^{[1]}(\tfrac{T}{2}) = e^{-iHT/4}Xe^{-iHT/2}Xe^{-iHT/4}, \tag{14.22}$$

which is just the repeating unit of Carr–Purcell (CP) sequences used in NMR spectroscopy [CP54]. The propagator of a CP sequence with $2n$ X pulses is

$$U^{\mathrm{CP}(2n)}(T) = \left[U^{[2]}\left(\frac{T}{n}\right)\right]^n. \tag{14.23}$$

The Nth-order ($N \geq 2$) CDD sequence is constructed iteratively:

$$U^{[N]}(T) = XU^{[N-1]}(\tfrac{T}{2})XU^{[N-1]}(\tfrac{T}{2}), \tag{14.24}$$

which preserves arbitrary quantum states up to an error of $O(T^{N+1})$.

For the most general qubit–bath Hamiltonian

$$H = \sum_{\alpha=0,x,y,z} \sigma_\alpha \otimes B_\alpha, \tag{14.25}$$

we need to protect two anticommuting Pauli operators, e.g., X and Z. Using Eqs. (14.13) and (14.14), we have the first-order sequence

$$U^{[1]}(T) = Ze^{-iHT/4}Xe^{-iHT/4}Ze^{-iHT/4}Xe^{-iHT/4}. \tag{14.26}$$

High-order CDD sequences can be constructed iteratively. We have

$$U^{[N]}(T) = ZU^{[N-1]}(\tfrac{T}{4})XU^{[N-1]}(\tfrac{T}{4})ZU^{[N-1]}(\tfrac{T}{4})XU^{[N-1]}(\tfrac{T}{4}), \tag{14.27}$$

which removes any qubit–bath interactions to the Nth order [KL05, KL07].

For multi-qubit systems, similar CDD sequences can be constructed iteratively. We first find an appropriate first-order DD sequence, for example by the nesting construction [Eqs. (14.13) to (14.15)] or the combinatorial methods (Chapter 15). Then high-order CDD sequences can again be constructed iteratively. We may also start the CDD construction from a DD sequence of any order $N > 1$.

The first-order DD given by Eq. (14.15) is 2^L free evolution operators $U^{[0]}(T/2^L)$ embedded in a sequence of control pulses $\{\Omega_l\}$. We denote this structure as

$$U_L^{[1]}(T) \equiv \mathscr{C}_\Omega\left\{U^{[0]}(T/2^L)\right\} = e^{-iH_L^{[1]}T}. \tag{14.28}$$

The resultant first-order effective Hamiltonian $H_L^{[1]}$ commutes with the operators $\{\Omega_1, \ldots, \Omega_L\}$ up to an error of $O(T)$. Here the sequence $\mathscr{C}_\Omega\{\cdots\}$ makes the effective Hamiltonian commute with $\{\Omega_1, \ldots, \Omega_L\}$ to a higher order. Iteratively, the Nth-order CDD reads

$$U_L^{[N]}(T) = \mathscr{C}_\Omega\left\{U_L^{[N-1]}(T/2^L)\right\}, \tag{14.29}$$

which preserves any operators in $\{\Omega_1, \ldots, \Omega_L\}$ up to $O(T^{N+1})$.

Based on Theorem 14.1, an alternative construction of CDD is given as follows [WL11b]. We first construct a CDD sequence to protect Ω_1 up to an error of $O\left(T^{N_1+1}\right)$ by the recursion

$$U_{\Omega_1}^{C[N_1]}(T) = \Omega_1 U_{\Omega_1}^{C[N_1-1]}(\tfrac{T}{2})\Omega_1 U_{\Omega_1}^{C[N_1-1]}(\tfrac{T}{2}), \tag{14.30}$$

with $U_{\Omega_1}^{C[0]}(T) \equiv U^{[0]}(T)$ and the superscript C denoting the nesting scheme of CDD. By defining $U_{\Omega_1,\Omega_2}^{C[N_1,0]}(T) \equiv U_{\Omega_1}^{C[N_1]}(T)$, we can construct a further level of CDD to protect Ω_2 up to $O\left(T^{N_2+1}\right)$, by the recursion

$$U_{\Omega_1,\Omega_2}^{C[N_1,N_2]}(T) = \Omega_2 U_{\Omega_1,\Omega_2}^{C[N_1,N_2-1]}(\tfrac{T}{2})\Omega_2 U_{\Omega_1,\Omega_2}^{C[N_1,N_2-1]}(\tfrac{T}{2}). \tag{14.31}$$

Similarly, we have the propagator by recursion

$$U_{\Omega_1,\Omega_2,\ldots,\Omega_l,\Omega_{l+1}}^{C[N_1,N_2,\ldots,N_l,0]}(T) \equiv U_{\Omega_1,\Omega_2,\ldots,\Omega_l}^{C[N_1,N_2,\ldots,N_l]}(T),$$

$$U_{\Omega_1,\ldots,\Omega_L}^{C[N_1,\ldots,N_L]}(T) = \Omega_L U_{\Omega_1,\ldots,\Omega_{L-1},\Omega_L}^{C[N_1,\ldots,N_{L-1},N_L-1]}(\tfrac{T}{2})\Omega_L U_{\Omega_1,\ldots,\Omega_{L-1},\Omega_L}^{C[N_1,\ldots,N_{L-1},N_L-1]}(\tfrac{T}{2}). \tag{14.32}$$

According to Theorem 14.1, the inner levels of DD are unaffected by the outer levels of control. Thus the evolution $U_{\Omega_1,\Omega_2,\ldots,\Omega_L}^{C[N_1,N_2,\ldots,N_L]}(T)$ commutes with $\Omega_1, \Omega_2, \ldots, \Omega_L$ to the orders $N_1, N_2, \ldots, N_L$, in turn. We may choose the single-qubit operations $\{X_1, Y_1, \ldots, Z_L\}$ as the pulses $\{\Omega_l\}$. An advantage of this construction is that the errors in $\{\Omega_l\}$ are eliminated independently and to different orders $\{N_l\}$, and it allows protecting operators with stronger error sources to higher orders. For example, usually for spin qubits under strong external magnetic field, the pure dephasing is much faster than the population relaxation, so it is favorable to protect the phase correlation on the inner level and to a higher CDD order.

The number of operations required in preserving the operators is $\sim 2^{NL}$ for the CDD scheme in Eq. (14.29) or $\sim 2^{\sum_{l=1}^{L} N_l}$ for that in Eq. (14.32). They increase exponentially with the DD order. Even though the exponentially increasing number of control pulses does yield significant improvement of precision through reduction of the coefficient in front of the power of time T^{N+1} [KL05, LYS07, NLP11] (see discussions in Section 14.5.4), accurate implementation of CDD to high orders may be challenging in experiments since errors are inevitably introduced by each control pulse. It is desirable to minimize the number of control pulses used to achieve a given order of decoupling.

14.5 Uhrig dynamical decoupling

Uhrig DD (UDD) [U07] is a remarkable advance in DD theory. UDD protects a unitary Hermitian operator Ω up to the Nth order of the evolution time using N instantaneous pulses of Ω applied at

$$T_j = T \sin^2 \frac{j\pi}{2(N+1)}, \quad (j = 1, 2, \ldots, N), \tag{14.33}$$

during the joint evolution of the system and bath from 0 to T. UDD is optimal in the short-time limit as it uses the minimum number of control pulses for a given DD order. Such pulse sequences for $N \leq 5$ were first noticed by Dhar et al. in designing control of the quantum Zeno effect [DGR06]. The general solution [Eq. (14.33)] to arbitrary decoupling order was first discovered by Uhrig for a pure dephasing spin-boson model [U07]. Then with computer algebra verification up to finite orders, it was conjectured that UDD may work for a general pure dephasing model [LWD08, U08]. Finally, UDD was rigorously proved to be universal for any order N [YL08] and was also extended to the case of population relaxation [YL08]. The performance bounds for UDD against pure dephasing were also established [UL10]. UDD has been generalized to protect a particular multi-qubit quantum state [MSS+10b] and more generally an arbitrary unitary Hermitian operator Ω [YWL11, WL11b].

By construction, UDD is optimal for a finite system (or a system with a hard spectral cutoff) in the "high-fidelity" regime where a short-time expansion of the qubit–bath propagator converges. For the "low-fidelity" regime, further theoretical works [CLN+08, PU10a] show that UDD is optimal when the noise spectrum has a hard cutoff, while CP performs better than CDD and UDD when the noise has a soft high-frequency cutoff or when the hard cutoff is not reached by the spectral filter functions of DD sequences (see Section 14.5.2).

14.5.1 Spin-boson model: the discovery of UDD

Equation (14.33) was first obtained by considering the spin-boson model under pure dephasing:

$$H = \sum_k \omega_k b_k^\dagger b_k + \frac{\sigma_z}{2} \sum_k \left(\lambda_k b_k^\dagger + \lambda_k^* b_k \right) = H_B + H_{SB}, \qquad (14.34)$$

where b_k is the bosonic operator for the kth mode, characterized by a coupling parameter λ_k. After M instantaneous X pulses applied at moments $\{T_j\}$, the propagator for the evolution from $T_0 = 0$ to $T_{M+1} = T$ is

$$U(T) = (X)^M e^{-iH(T-T_M)} X \cdots X e^{-iH(T_2-T_1)} X e^{-iHT_1}, \qquad (14.35)$$

where for quantum memory a final pulse is applied when M is odd. By introducing the modulation function

$$F(t) \equiv \begin{cases} (-1)^j, & \text{for } t \in [T_j, T_{j+1}) \text{ with } 0 \leq j \leq M; \\ 0, & \text{otherwise,} \end{cases} \qquad (14.36)$$

we formally write the propagator as

$$U(T) = e^{-iH_B T} \mathscr{T} \exp \left\{ -i \int_0^T F(t) \frac{\sigma_z}{2} B_z(t) dt \right\}, \qquad (14.37)$$

with $B_z(t) = \sum_k \left(\lambda_k b_k^\dagger e^{i\omega_k t} + \lambda_k^* b_k e^{-i\omega_k t} \right)$.

For an operator $A(t)$ with $[A(t), A(t')]$ being a c-number, the time-ordered exponential function

$$\mathscr{T} e^{\int_0^T A(t) dt} = \lim_{\frac{T}{n}=\epsilon \to 0} e^{A(n\epsilon)\epsilon} \cdots e^{A(2\epsilon)\epsilon} e^{A(\epsilon)\epsilon}, \qquad (14.38)$$

can be written as

$$\mathscr{T} e^{\int_0^T A(t) dt} = e^{\int_0^T A(t) dt} e^{\frac{1}{2} \int_0^T dt \int_0^t dt' [A(t), A(t')]}. \qquad (14.39)$$

This can be shown by using Baker–Hausdorff formula recursively, e.g., the first step

$$e^{A(2\epsilon)\epsilon} e^{A(\epsilon)\epsilon} = e^{[A(2\epsilon)+A(\epsilon)]\epsilon} e^{\frac{1}{2}[A(2\epsilon), A(\epsilon)]\epsilon^2}. \qquad (14.40)$$

Since the commutator of $F(t)\frac{\sigma_z}{2}B_z(t)$ at different times t is a c-number, dropping the time-ordering operator by using Eq. (14.39), we have

$$U(T) = e^{-iH_BT}e^{i\Phi}e^{-i\int_0^T F(t)\frac{\sigma_z}{2}B_z(t)dt}, \qquad (14.41)$$

where the global phase

$$\Phi = \frac{1}{4}\int_0^\infty d\omega \int_0^T dt_2 \int_0^{t_1} dt_1 F(t_2)F(t_1)I(\omega)\sin[\omega(t_2-t_1)], \qquad (14.42)$$

with the boson spectral density

$$I(\omega) \equiv \sum_k |\lambda_k|^2 \delta(\omega - \omega_k). \qquad (14.43)$$

The off-diagonal element of the reduced density matrix of the qubit $\rho_{+-}(T) = \langle +|\text{Tr}_B\rho(T)|-\rangle$ measures the qubit coherence, where $|\pm\rangle$ denote the two eigenstates of the qubit operator σ_z. For the initial system–bath state $\rho(0) = \rho_S \otimes \rho_B$, we obtain

$$\rho_{+-}(T) = \rho_{+-}(0)\text{Tr}_B\left[e^{-i\int_0^T F(t)B_z(t)dt}\rho_B\right], \qquad (14.44)$$

by using Eq. (14.41). For the thermal equilibrium bath state $\rho_B = \prod_k (1-e^{-\beta\omega_k})e^{-\beta\omega_k b_k^\dagger b_k}$, we get [U08, A10]

$$\rho_{+-}(T) = \rho_{+-}(0)e^{-\chi(T)}, \qquad (14.45)$$

where the coherence integral

$$\chi(T) = \frac{1}{2\pi}\int_0^\infty d\omega S(\omega)|f(\omega)|^2, \qquad (14.46)$$

with the noise spectrum

$$S(\omega) = 2\pi\left(\bar{n}(\omega) + \tfrac{1}{2}\right)I(\omega), \qquad (14.47)$$

the average number of the kth boson mode

$$\bar{n}(\omega_k) = \frac{1}{e^{\beta\omega_k}-1}, \qquad (14.48)$$

and the Fourier transform of $F(t)$

$$f(\omega) \equiv \int_0^T F(t)e^{i\omega t}dt = \frac{1}{i\omega}\sum_{j=0}^M (-1)^j \left(e^{i\omega T_{j+1}} - e^{i\omega T_j}\right). \qquad (14.49)$$

Note that the global phase Φ is canceled in the result [e.g., Eq. (14.44)] (see [U07, U08, A10]).

We expand $f(\omega)$ as

$$f(\omega) = \sum_{p=1}^\infty \frac{(i\omega)^{p-1}}{p!}\Lambda_p T^p, \qquad (14.50)$$

where

$$\Lambda_p = \sum_{j=0}^M (-1)^j \left[\left(\frac{T_{j+1}}{T}\right)^p - \left(\frac{T_j}{T}\right)^p\right]. \qquad (14.51)$$

Thus, assuming that the noise spectrum $S(\omega)$ has a high-frequency cutoff, the condition $\chi(T) = O\left(T^{N+1}\right)$ is equivalent to $M = N$ coupled algebra equations

$$\Lambda_p = 0, \quad p = 1, 2, \ldots, N, \tag{14.52}$$

whose unique physical solution is the UDD sequence in Eq. (14.33).

14.5.2 Filter functions

It is helpful to consider a qubit coupled to classical noise. The qubit–environment interaction is described by the pure dephasing Hamiltonian

$$H = \frac{1}{2}[\Delta E + \beta(t)]\sigma_z, \tag{14.53}$$

where ΔE is the qubit energy splitting and $\beta(t)$ describes random, zero-mean energy fluctuations with stationary statistics (i.e., the system has translational symmetry in time). The decoherence of the qubit is caused by the accumulation of a random phase $\phi(T) = \int_0^T \beta(t)dt$. Under the control of a sequence of instantaneous X pulses, the random phase is suppressed by the modulation function as

$$\phi(T) = \int_0^T \beta(t)F(t)dt. \tag{14.54}$$

If the noise is Gaussian, the decoherence function of the qubit is [A54, K54, CLN+08]

$$W(T) = \left\langle e^{-i\phi(T)} \right\rangle = e^{-\frac{1}{2}\langle \phi(T)\phi(T)\rangle} = e^{-\chi(T)}, \tag{14.55}$$

where the coherence integral $\chi(T)$ can be written in terms of the modulated noise correlation function as

$$\chi(T) = \frac{1}{2}\int_0^T dt_1 \int_0^T dt_2 \langle \beta(t_1)\beta(t_2)\rangle F(t_1)F(t_2). \tag{14.56}$$

The aim of DD is to make $\chi(T)$ as small as possible. Using the Fourier transform of the correlation function

$$S(\omega) = \int_{-\infty}^{\infty} dt \langle \beta(t)\beta(0)\rangle e^{i\omega t}, \tag{14.57}$$

the random phase correlation $\chi(T)$ is written as the overlap of the noise spectrum $S(\omega)$ and the filter function $|f(\omega)|^2$ in the frequency domain [cf. Eq. (14.46)],

$$\chi(T) = \frac{1}{2\pi}\int_0^{\infty} d\omega S(\omega)|f(\omega)|^2. \tag{14.58}$$

An efficient DD protocol minimizes the overlap as much as possible. If the noise spectrum is known, we can optimize the pulse sequences by minimizing the overlap [GKL08, BUV+09a, UBB09, CBK10, KEV11].

To give a feeling for the filter function, we plot the functions $|f(\omega)|^2$ with $T = 1$ for the tenth-order UDD, fourth-order CDD [Eq. (14.24)], and ten-pulse CP [Eq. (14.23)] sequences in Fig. 14.1. All these sequences use ten X pulses. For comparison, $|f(\omega)|^2$ for the case that there is no DD control is also shown. The expressions [CLN+08] of $|f(\omega)|^2$ for some DD sequences are listed in Table 14.1. The DD sequences act as high-pass filters for environmental

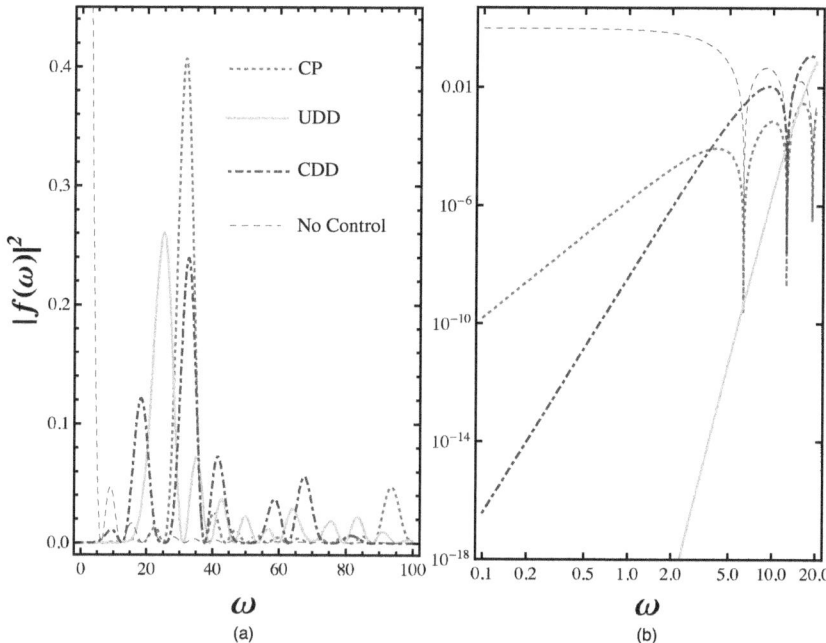

Fig. 14.1. Filter functions $|f(\omega)|^2$ for tenth-order UDD (gray line), fourth-order CDD (dot-dashed line), and ten-pulse CP (dotted line) sequences with $T = 1$. $|f(\omega)|^2$ for the case without DD control is also shown (dashed fine line). (b) shows $|f(\omega)|^2$ in the low-frequency regime on a logarithmic scale.

noise [SL04, CLN+08]. From another point of view, the decoupling pulses drive the quantum system so that it becomes off-resonant with the noise [KK04]. From Fig. 14.1, we see that $|f(\omega)|^2$ with DD control becomes much larger than that without control at frequencies near the driving frequencies of the pulse sequences (e.g., around $\omega_d = 2\pi N/(2T)$ for CP sequences). It means that the quantum system becomes sensitive to the noise at these frequencies, and the errors due to this noise accumulate coherently. Therefore, when there is noise near the driving frequencies of DD, DD is not very efficient. This coherent accumulation also has advantages, for example, in detecting NMR of single molecules by enhancing the weak noise with many-pulse DD [ZHH+11]. By introducing randomization in the pulse sequences, the noise will not accumulate coherently and DD schemes may become more efficient (see Section 14.11). In the short-time limit, it was noted that when the noise spectrum has a soft high-frequency cutoff or when the noise spectrum cutoff is not reached by DD filter functions, CP performs better than CDD and UDD [CLN+08, PU10a]. For telegraph-like noise, it was proved that CP is the most efficient scheme in protecting the qubit coherence in the short-time limit [CL10]. The Nth-order UDD sequence makes the first N derivatives of $f(\omega)$ with respect to ω vanish, thus UDD is optimized to filter out low-frequency noise efficiently [see Fig. 14.1b]. The slopes of the curves at $\omega \ll 1$ in Fig. 14.1b correspond to the decoupling orders. Therefore UDD is the most efficient scheme when the noise spectrum cutoff is reached. In the following section, we prove that UDD is universal for arbitrary bounded baths in the short-time limit.

Table 14.1. *The expressions of $|f(\omega)|^2$ for various pulse sequences*

Sequence	$|f(\omega)|^2$
No control	$\frac{4}{\omega^2}\sin^2(\frac{\omega T}{2})$
UDD	$\frac{1}{\omega^2}\left|\sum_{k=-M-1}^{M}(-1)^k \exp\left[\frac{i\omega T}{2}\cos\frac{\pi k}{M+1}\right]\right|^2$
CDD	$\frac{2^{2N+2}}{\omega^2}\sin^2\frac{\omega T}{2^{N+1}}\prod_{k=1}^{N}\sin^2\frac{\omega T}{2^{k+1}}$
CP (even M)	$\frac{16}{\omega^2}\sin^4\frac{\omega T}{4M}\sin^2\frac{\omega T}{2}/\cos^2\frac{\omega T}{2M}$

M is the number of pulses and N is the order of CDD.

14.5.3 Universality of Uhrig dynamical decoupling

Rather than reproducing the original proof for UDD [YL08], here we give a proof that applies in a more general setting [WL11b]. We aim to protect an element of a MOOS. The Hamiltonian may not be analytically time-dependent, but between two adjacent pulses of the UDD sequence the Hamiltonians are symmetric and have the same functional form of the relative time between the adjacent pulses. This general proof will yield a nesting scheme of UDD sequences to protect more system operators (see Section 14.8).

Theorem 14.2. *For a bounded time-dependent Hamiltonian $H(t)$ defined in $[0, T]$, an Nth order UDD sequence with N pulses realized as a unitary Hermitian operator Ω applied at $T_1, T_2, \ldots, T_N$ preserves Ω up to an error of $O\left(T^{N+1}\right)$, if*

$$H(T_n + s\tau_n) = H(T_{n+1} - s\tau_n) = H(sT_1), \tag{14.59}$$

for $s \in [0, 1]$ and $\tau_n = T_{n+1} - T_n$, that is, the Hamiltonian has the same form as a function of the relative time between adjacent operations and is symmetric within each interval.

Note: The case of time-independent Hamiltonians is just a special case of Theorem 14.2 with the condition $H(sT_1) = H$. For the pure dephasing Hamiltonian [Eq. (14.20)], if we use the control pulses $\Omega = X$ in the UDD sequence, X and hence all system operators are preserved up to an error of $O\left(T^{N+1}\right)$. For the general qubit–bath Hamiltonian, applying the UDD sequence of Z will cancel the qubit–bath coupling terms $\sigma_x \otimes B_x$ and $\sigma_y \otimes B_y$ which induce qubit relaxation to the Nth order. Applying a sequence of unitary operations $\Omega = 2|\psi\rangle\langle\psi| - I_S$ on a multi-level quantum system according to the timing of UDD, Ω, and hence the initial quantum state $|\psi\rangle$, is preserved to the order of $O\left(T^{N+1}\right)$ [MSS+10b]. Thus the proof of Theorem 14.2 will automatically give the proof of UDD in these cases.

Proof. The evolution under the application of Ω reads

$$U(T) = \Omega^N V_N \Omega V_{N-1} \cdots \Omega V_1 \Omega V_0, \tag{14.60}$$

with the evolution operator

$$V_n \equiv \mathscr{T} \exp\left[-i \int_{T_n}^{T_{n+1}} H(t)dt\right] \equiv \mathscr{T}_\theta \exp\left[-i\left(T_{n+1} - T_n\right) \int_{\frac{n}{N+1}\pi}^{\frac{n+1}{N+1}\pi} H^{\text{rel}}(\theta)d\theta\right], \quad (14.61)$$

where $\mathscr{T}_\theta$ stands for ordering in θ, and

$$H^{\text{rel}}(\theta) = \frac{N+1}{\pi} H(t), \quad (14.62)$$

where $\theta = \frac{n\pi}{N+1} + \frac{t - T_n}{T_{n+1} - T_n} \frac{\pi}{N+1}$ for $t \in (T_n, T_{n+1}]$. The symmetry requirements given in Eq. (14.59) are transformed to

$$H^{\text{rel}}\left(\frac{n\pi}{N+1} + \theta\right) = H^{\text{rel}}\left(\frac{n+1}{N+1}\pi - \theta\right) = H^{\text{rel}}(\theta). \quad (14.63)$$

The Hamiltonian $H^{\text{rel}}(\theta)$ can be separated into two parts,

$$H^{\text{rel}}(\theta) = C(\theta) + A(\theta), \quad (14.64)$$

with

$$C(\theta) = \left[H^{\text{rel}}(\theta) + \Omega H^{\text{rel}}(\theta)\Omega\right]/2, \quad (14.65)$$
$$A(\theta) = \left[H^{\text{rel}}(\theta) - \Omega H^{\text{rel}}(\theta)\Omega\right]/2. \quad (14.66)$$

$C(\theta)$ and $A(\theta)$ commute and anticommute with the operator Ω, respectively. Using Eq. (14.33), we rewrite the propagator as

$$U(T) = \mathscr{T}_\theta \exp\left[-iT \int_0^\pi G(\theta)\Big(C(\theta) + \tilde{F}(\theta)A(\theta)\Big)d\theta\right], \quad (14.67)$$

where the functions

$$G(\theta) = \frac{1}{2}\left[\cos\frac{n\pi}{N+1} - \cos\frac{(n+1)\pi}{N+1}\right] \quad \text{for } \theta \in \left(\frac{n\pi}{N+1}, \frac{(n+1)\pi}{N+1}\right], \quad (14.68)$$

$$\tilde{F}(\theta) = (-1)^n \quad \text{for } \theta \in \left(\frac{n\pi}{N+1}, \frac{(n+1)\pi}{N+1}\right]. \quad (14.69)$$

Thus, the part of Hamiltonian $C(\theta)$ that commutes with Ω is modulated by the step function $G(\theta)$, which has step heights given by the UDD intervals, and the part of Hamiltonian $A(\theta)$ that anticommutes with Ω is modulated by $G(\theta)$ and the periodic function $\tilde{F}(\theta)$. Furthermore, both $C(\theta)$ and $A(\theta)$ have the same symmetries as $H^{\text{rel}}(\theta)$ in Eq. (14.63). The symmetries of the time-dependent Hamiltonian and the modulation functions make them have particular Fourier expansions, which lead us to a proof of the theorem using a procedure similar to the proof of UDD in [YL08].

The Fourier expansions of $G(\theta)$, $\tilde{F}(\theta)$, and the time-dependent Hamiltonians are

$$G(\theta) = \sum_{k=0}^{\infty} g_k \sin[2k(N+1)\theta \pm \theta], \qquad (14.70)$$

$$\tilde{F}(\theta) = \sum_{k=0}^{\infty} f_k \sin[(2k+1)(N+1)\theta], \qquad (14.71)$$

$$C(\theta) = \sum_{k=0}^{\infty} c_k \cos[2k(N+1)\theta], \qquad (14.72)$$

$$A(\theta) = \sum_{k=0}^{\infty} a_k \cos[2k(N+1)\theta]. \qquad (14.73)$$

Here the operators c_k and a_k commute and anticommute with Ω, respectively. The features of these Fourier expansions to be used in the proof below are: (i) Both C and A contain only cosine harmonics of order of even multiple of $(N+1)$; (ii) $\tilde{F}(\theta)$ contains only sine harmonics of order of odd multiple of $(N+1)$; $G(\theta)$ contains only sine harmonics of an order differing from an even multiple of $(N+1)$ by $+1$ or -1.

With the product-to-sum trigonometric formulas, we have

$$U(T) = \mathscr{T}_\theta \exp\left[-iT \int_0^\pi \left(\tilde{C}(\theta) + \tilde{A}(\theta)\right) d\theta\right], \qquad (14.74)$$

where

$$\tilde{C}(\theta) \equiv G(\theta)C(\theta) = \sum_k \tilde{c}_k \sin[2k(N+1)\theta \pm \theta], \qquad (14.75)$$

$$\tilde{A}(\theta) \equiv G(\theta)\tilde{F}(\theta)A(\theta) = \sum_k \tilde{a}_k \cos[(2k+1)(N+1)\theta \pm \theta]. \qquad (14.76)$$

The operators $\tilde{c}_k$ and $\tilde{a}_k$ commute and anticommute with Ω, respectively.

Expanding $U(T)$ according to the standard time-dependent perturbation theory, the terms that do not commute with Ω must contain an odd number of $\{\tilde{a}_k\}$ (since Ω anticommutes with $\{\tilde{a}_k\}$). The expansion coefficients can be written as

$$(-iT)^n \int_0^\pi y_{k_1}^{\alpha_1,\eta_1}(\theta_1) \int_0^{\theta_1} y_{k_2}^{\alpha_2,\eta_2}(\theta_2) \cdots \int_0^{\theta_{n-1}} y_{k_n}^{\alpha_n,\eta_n}(\theta_n) d\theta_1 \cdots d\theta_n, \qquad (14.77)$$

with $y_k^{\text{s},\pm}(\theta) \equiv \sin[2k(N+1)\theta \pm \theta]$ associated with an operator $\tilde{c}_k$, and $y_k^{\text{c},\pm}(\theta) \equiv \cos[(2k+1)(N+1)\theta \pm \theta]$ associated with an operator $\tilde{a}_k$, for $\alpha_j \in \{\text{c},\text{s}\}$ and $\eta_j \in \{+,-\}$. By induction and repeatedly using the product-to-sum trigonometric formulas, one can straightforwardly verify that the coefficients in Eq. (14.77) vanish for $n \leq N$ and $y_k^{\text{c},\pm}$ appearing an odd number of times. Thus vanish any terms in the expansion that contain products of an odd number of operators in $\{\tilde{a}_k\}$ and have a power of T lower than $(N+1)$. $\square$

14.5.4 Comparison with concatenated dynamical decoupling

We consider a DD sequence of M pulses, with a total evolution time T and a minimum pulse interval τ. In CDD, the decoupling order is $N \sim \log_2 M$ and $\tau = T/M$. In UDD, the decoupling order is $N = M$ and $\tau \sim T/M^2$. To be specific, our discussion is based on the pure dephasing model. The situation for the general decoherence model is similar. We compare the efficiency of UDD and CDD in the two following scenarios.

Case I: The total evolution time T is fixed. The decoupling precision (defined as the effective coupling under the DD control relative to the original one) in UDD was derived as [UL10]

$$\epsilon_{\text{UDD}} \sim (\lambda T)^M / M!, \tag{14.78}$$

where λ is the sup-operator norm of H. In CDD, it scales with the time and the decoupling order as [KL07, LYS07, NLP11]

$$\epsilon_{\text{CDD}} \sim (\lambda T/\sqrt{M})^N \sim (\lambda T)^N / 2^{N^2/2}. \tag{14.79}$$

Thus with T fixed, increasing the decoupling order and hence the number of pulses always increases the decoupling precision. An arbitrarily high decoupling precision can be achieved simply by choosing a sufficiently high order of DD (and correspondingly a sufficiently small pulse interval τ). In the high-fidelity regime (λT is small), the decoupling precision of UDD scales with the number of pulses much faster than that of CDD. However, if we compare the efficiency of UDD and CDD for the same decoupling order N, i.e., the Nth-order UDD (containing N pulses) and the Nth-order CDD (containing 2^N pulses), CDD has a much higher decoupling precision than UDD does ($T^N/2^{N^2/2} \ll T^N/N!$ for large N), since the minimum pulse interval $\tau = T/2^N$ in CDD is much smaller than that in UDD ($\tau \sim T/N^2$). For the same reason (namely, reduction of τ), to achieve a given order of precision, CDD indeed requires far less than the seemingly exponential cost.

Case II: The minimum pulse interval τ is fixed, which is a frequently encountered restriction in realistic experiments. In this situation, increasing the order of DD leads to two competing effects [UL10, HVD10]. First, the qubit–bath coupling is eliminated to a higher order, which tends to increase the decoupling precision. Second, the total evolution time T increases and the bath has more time to inflict qubit decoherence. Competition between these two effects leads to the existence of an optimal decoupling order, beyond which further increasing the order of DD does not improve the decoupling precision any longer. For a given minimum pulse interval τ, the optimal order of UDD is [UL10]

$$N_{\text{opt,UDD}} \sim 1/(\lambda \tau), \tag{14.80}$$

and that of CDD is [LYS07, KL07, NLP11]

$$N_{\text{opt,CDD}} \sim -\log_2(\lambda \tau). \tag{14.81}$$

UDD has a much higher optimal level than CDD for a small minimum pulse interval, and therefore the maximum decoupling precision that can be achieved by UDD is much higher than that by CDD.

14.6 Concatenated Uhrig dynamical decoupling

For a qubit coupled to an arbitrary bath as given by Eq. (14.25), CDD can eliminate all the errors (including pure dephasing and population relaxation) up to an arbitrary order N at the cost of exponentially increasing number ($\sim 4^N$) of decoupling pulses. In contrast, the UDD sequence uses the smallest possible number (i.e., N) of decoupling pulses to eliminate either pure dephasing or population relaxation (but not both) to the desired order N. Based on a combination of CDD and UDD, concatenated UDD (CUDD) was proposed [U09] to suppress both pure dephasing and population relaxation to order N with a much smaller number ($\sim N2^N$) of pulses. The essential idea of CUDD is to use Nth-order UDD sequences (instead of free evolutions) as the building blocks of a new CDD sequence.

The propagator for the qubit–bath evolution driven by the general Hamiltonian Eq. (14.25) under the Nth-order UDD sequence of Z pulses is $U_z^{\text{UDD}[N]}(T) = e^{-iH_{\text{eff}}(T)T + O(T^{N+1})}$, where $H_{\text{eff}}(T) = H_B(T) + \sigma_z \otimes B_z(T)$ is a pure dephasing Hamiltonian. The pure dephasing can be eliminated by embedding $U_z^{\text{UDD}[N]}$ into the CDD sequences of X pulses. The propagator for the N'th-order concatenation of $U_z^{\text{UDD}[N]}$ is

$$U_{z,x}^{\text{CUDD}[N,N']}(T) = XU_{z,x}^{\text{CUDD}[N,N'-1]}(\tfrac{T}{2})XU_{z,x}^{\text{CUDD}[N,N'-1]}(\tfrac{T}{2}), \tag{14.82}$$

with $U_{z,x}^{\text{CUDD}[N,0]}(\tau) \equiv U_z^{\text{UDD}[N]}(\tau)$. In the CUDD scheme, $U_{z,x}^{\text{CUDD}[N,N]}$ eliminates both the pure dephasing and population relaxation up to the Nth order with $O(N2^N)$ pulses.

14.7 Quadratic dynamical decoupling

While CUDD is an improvement over CDD, it still requires an exponential number of pulses. To further reduce the number of pulses, West, Fong, and Lidar proposed a near-optimal DD protocol [WFL10] by nesting UDD sequences, dubbed quadratic DD (QDD), to protect a qubit against general noise. The inner N'th-order UDD eliminates population relaxation and the outer Nth-order UDD eliminates the pure dephasing, so that both pure dephasing and population relaxation are eliminated up to the $\min[N', N]$th order of the evolution time. Using $U_z^{\text{UDD}[N']}(\tau)$ to denote the qubit–bath propagator driven by the general Hamiltonian Eq. (14.25) under the N'th-order UDD sequence of Z pulses, the propagator of the (N', N)th-order QDD,

$$U_{z,x}^{\text{QDD}[N',N]}(T) = (X)^N U_z^{\text{UDD}[N']}(T - T_N) X U_z^{\text{UDD}[N']}(T_N - T_{N-1})$$
$$\cdots X U_z^{\text{UDD}[N']}(T_2 - T_1) X U_z^{\text{UDD}[N']}(T_1), \tag{14.83}$$

is obtained by setting the timing T_j given by Eq. (14.33). Thus $U_{z,x}^{\text{QDD}[N,N]}$ eliminates both the pure dephasing and population relaxation up to the Nth order using only $O(N^2)$ pulses, an exponential improvement over both CDD and CUDD. Computer algebra shows that for $N \leq 4$ QDD differs from the optimal solutions by no more than two pulses [WFL10]. A detailed numerical study of (N', N)th-order QDD was reported in [QL11].

Theorem 14.2 states that UDD still works if the Hamiltonians in different intervals have the same functional form of the relative time and are symmetric, regardless of the analytic properties of the Hamiltonians. This means that if we use even-order UDD sequences as the inner-level DD control, the decoupling order of the outer-level DD sequences will not change. By using Theorem 14.1, the outer-level DD pulse sequence will not affect the inner-level DD control if the DD pulses are elements of a MOOS. Therefore, Theorems 14.2 and 14.1 provide a proof of the validity of QDD with even-order UDD on the inner level. These theorems also enable a construction of higher-order DD by nesting UDD sequences of even orders to protect more system operators (see Section 14.8). A general proof of the universality of QDD with arbitrary inner and outer orders was given in [KL11].

14.8 Nested Uhrig dynamical decoupling

It is natural to generalize the idea of QDD to higher-order nesting. This scheme is called nested UDD (NUDD), and can be used to protect multiple qubits or even multiple multi-level systems [WL11b]. Theorems 14.2 and 14.1 can be used to prove that NUDD protects all the elements in a MOOS when all the inner nested UDD sequences are of even order. Numerical evidence of NUDD sequences in protecting two-qubit entangled states showed that the order of the inner sequences can be odd when all the nested UDD sequences have the same decoupling order [MSS+10a]. The validity of NUDD has been proven for the case that all the nested UDD sequences have the same order, whether odd or even [JI11].

Consider first a generalization of QDD to an arbitrary pair of MOOS operators. For a time-independent Hamiltonian H under DD control of a sequences of unitary Hermitian operators Ω applied at $t_1, t_2, \ldots, t_{N'}$, the evolution $U(\tau)$ from $t_0 = 0$ to $t_{N'+1} = \tau$ is equivalent, after a toggling frame transformation, to the evolution under a time-dependent Hamiltonian $H(t) = \Omega^n H \Omega^n$ for $t \in (t_n, t_{n+1}]$. Such a time dependence is not analytic. If N' is an even number and the sequence of time intervals is symmetric, the time-dependent Hamiltonian $H(t)$ is time symmetric in $[0, \tau]$. In a UDD sequence of an operator Ω applied at $T_1, T_2, \ldots, T_N$ between $T_0 = 0$ and $T_{N+1} = T$, we replace each interval of free evolution $e^{-iH(T_{n+1}-T_n)}$ with the evolution due to a sequence of another operation Ω' in a MOOS applied at $T_{n,1}, T_{n,2}, \ldots, T_{n,N'}$ between $T_{n,0} \equiv T_n$ and $T_{n,(N'+1)} \equiv T_{n+1}$, with the same symmetric structure in all intervals, that is,

$$T_{n+1} - T_{n,N'-k} = T_{n,k+1} - T_n, \tag{14.84a}$$

$$\frac{T_{n,k} - T_{n,k'}}{T_{n+1} - T_n} = \frac{T_{m,k} - T_{m,k'}}{T_{m+1} - T_m}. \tag{14.84b}$$

In particular, the inner-level control of Ω' can be chosen as even-order UDD. In this case, Theorems 14.2 and 14.1 validate that we achieve the desired protection orders N and N' of the operators Ω and Ω', respectively.

Next we describe the construction of NUDD for protecting a set of unitary Hermitian operators $\{\Omega_1, \ldots, \Omega_L\}$. First, the N_Lth-order UDD sequence of Ω_L is constructed with pulses applied at

$$T_{n_L} = T \sin^2 \frac{n_L \pi}{2N_L + 2}, \tag{14.85}$$

between $T_0 = 0$ and $T_{N_L+1} = T$ on the outermost level. Then the free evolution in each interval is substituted by the N_{L-1}th-order UDD sequence of Ω_{L-1} applied at

$$T_{n_L, n_{L-1}} = T_{n_L} + (T_{n_L+1} - T_{n_L}) \sin^2 \frac{n_{L-1} \pi}{2N_{L-1} + 2}, \tag{14.86}$$

in each interval between $T_{n_L,0} \equiv T_{n_L}$ and $T_{n_L, N_{L-1}+1} \equiv T_{n_L+1}$. So on and so forth, the lth level is constructed by N_l applications of Ω_l in each interval between $T_{n_L,\ldots,n_{l+1},0} \equiv T_{n_L,\ldots,n_{l+2},n_{l+1}}$ and $T_{n_L,\ldots,n_{l+1},N_l+1} \equiv T_{n_L,\ldots,n_{l+2},n_{l+1}+1}$ at

$$T_{n_L,\ldots,n_{l+1},n_l} = T_{n_L,\ldots,n_{l+1}} + \left[T_{n_L,\ldots,n_{l+2},n_{l+1}+1} - T_{n_L,\ldots,n_{l+1}}\right] \sin^2 \frac{n_l \pi}{2N_l + 2}. \tag{14.87}$$

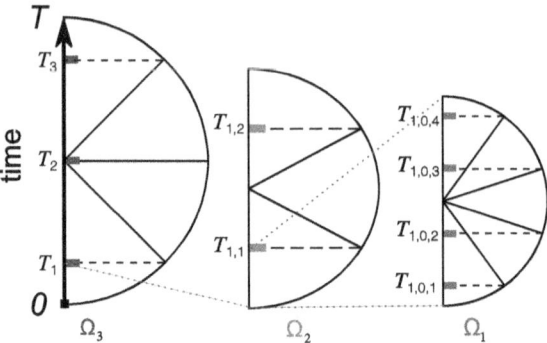

Fig. 14.2. Illustration of the timing structure of an NUDD sequence. The rectangles denote the control pulses. The outermost level is a third-order UDD sequence of Ω_3. The second and innermost levels consist of second-order UDD sequences of Ω_2 and fourth-order UDD sequences of Ω_1, respectively. The inner levels are magnified. The pulses for an Nth-order UDD sequence are applied at the times corresponding to the vertical coordinates of the N points equally dividing a semicircle.

We denote the evolution under such NUDD by $U_{\Omega_1,\Omega_2,\ldots,\Omega_L}^{\mathrm{U}[N_1,N_2,\ldots,N_L]}(T)$, where the superscript U denotes the nesting of UDD sequences. An NUDD sequence is illustrated in Fig. 14.2.

When the inner-level sequences are even-order UDD sequences, i.e., with N_l being an even number for $1 \leq l \leq L-1$, Theorems 14.2 and 14.1 validate that all $\{\Omega_l\}$ are preserved to the desired protection orders $\{N_l\}$. The number of control intervals is

$$M_{\mathrm{pulse}}^{\mathrm{U}[N_1,N_2,\ldots,N_L]} = (N_1+1)(N_2+1)\cdots(N_L+1), \qquad (14.88)$$

increasing polynomially with the decoupling order (e.g., as N^L when all orders $N_l = N$). As the errors in $\{\Omega_l\}$ are eliminated independently and to different orders $\{N_l\}$, we may protect operators with stronger error sources to higher orders and on the inner levels. If there are odd-order UDD sequences on the inner levels, the protection order is limited by the lowest protection order $\min[N_1,\ldots,N_L]$ [QL11, JI11, KL11]. A complete analytical proof of the universality of NUDD was given in [KQP+12].

14.9 Pulses of finite amplitude

In experiments, the pulses have finite durations and amplitudes, which introduce additional errors. Here we consider how pulse shaping for the DD schemes discussed above can improve their performance when the finiteness of the pulse width is accounted for. Let us consider a short-pulse operation described by the Hamiltonian

$$H_\Omega(t) = v(t)\Omega, \qquad (14.89)$$

where Ω is a unitary Hermitian operator. We aim to design the pulse shape of $v(t)$ such that the evolution during the pulse control approximates the ideal δ-pulse up to a certain order of the pulse duration τ_p, that is,

$$U(\tau_p, 0) = \mathcal{T} \exp\left[-i \int_0^{\tau_p} [H + H_\Omega(t)] \, dt\right]$$
$$= e^{-i(\tau_p - \tau_\Omega)H} P_\Omega e^{-i\tau_\Omega H} + O(\tau_p^{M_p}), \quad (14.90)$$

where $P_\Omega \equiv \exp\left[-i \int_0^{\tau_p} H_\Omega(t) dt\right]$ is the desired instantaneous pulse applied at the time τ_Ω. In particular, we need the pulse $P_\Omega = \Omega$ up to a trivial global phase factor. Note that the bath evolution is not perturbed up to an error $O(\tau_p^{M_p})$ in Eq. (14.90). Therefore this pulse shaping is applicable to all the DD schemes discussed in the previous sections. Unfortunately, a no-go theorem established in [PFK+08, PU08] restricts such that an instantaneous π pulse cannot be simulated by a finite-amplitude pulse of the form Eq. (14.90) with errors lower than $O(\tau_p^2)$. For first-order pulse shaping with $M_p = 2$, the leading-order error is eliminated when the following equations are satisfied [WL11b]:

$$\int_0^{\tau_p} (t - \tau_\Omega) v(t) \cos[\phi_0 - \psi(t)] \, dt = 0, \quad (14.91a)$$

$$\int_0^{\tau_p} (t - \tau_\Omega) v(t) \sin[\phi_0 - \psi(t)] \, dt = 0, \quad (14.91b)$$

with $\psi(t) \equiv 2 \int_{\tau_\Omega}^t v(s) ds$ and $\phi_0 = \int_{\tau_\Omega}^{\tau_p} v(s) ds - \int_0^{\tau_\Omega} v(s) ds$. For a solution of these equations see [PFK+08].

However, to realize some DD schemes, it is not necessary to exactly simulate the instantaneous π pulses but it suffices to satisfy some symmetry requirements. For example, as we have discussed in Section 14.5.3, the symmetry requirements of $\tilde{F}(\theta)$ automatically guarantee the performance of UDD. By appropriately designing the pulses, the qubit–bath Hamiltonian describing pure dephasing can be transformed into the form [UP10]

$$H = H_B + F'(t)\sigma_z \otimes B_z + O(\tau_p^M), \quad (14.92)$$

with the modulation function $F'(t)$ taking values from $\{-1, 0, 1\}$. The scaled modulation function $\tilde{F}'(\theta) = F'\left(T \sin^2(\theta/2)\right)$ is designed to have the symmetries required in the proof of Section 14.5.3 and therefore can be expanded in terms of odd harmonics of $\sin[(N+1)\theta]$. Thus, the dephasing is suppressed up to the order $O(T^{N+1}) + O(\tau_p^M)$. This sequence can also, alternatively, suppress longitudinal relaxation [YL08, UP10] or any element in a MOOS. An explicit construction for $M = 3$ can be found in [UP10]. A different approach is to construct arbitrarily accurate quantum gates by a recursive scheme based on concatenation [KLV10].

14.10 Time-dependent Hamiltonians

We call a quantum dynamical control protocol universal if it realizes a desired system evolution for arbitrary baths coupled to the system in a given form, up to errors in a given order of the short evolution time T. All the DD schemes we have discussed above are universal in this sense. However, they are designed for time-independent Hamiltonians. In this section we will consider universality for time-dependent Hamiltonians (see [DG99, KK04, VK05, SV05, SV06] for early work on specific time-dependent models and [PU10b] for UDD on time-dependent Hamiltonians). In particular, we show that if a dynamical control protocol has errors to the Nth order in the short evolution time T for a generic time-independent system, it will automatically achieve the same order of control precision for analytically time-dependent systems [WL11a].

14.10.1 Universal control of time-independent systems

Let us first consider a target system coupled to a bath through a time-independent Hamiltonian [Eq. (14.1)]:

$$H = H_S \otimes I_B + I_S \otimes H_B + H_{SB} \equiv \sum_{\alpha=0}^{D-1} S_\alpha \otimes B_\alpha, \qquad (14.93)$$

where S_α and B_α are operators of the system and bath, respectively. The coupling is generic, that is, the details of bath operators are unspecified. For convenience, we have defined $S_0 = I_S$ as the identity system operator and $B_0 = H_B$ the bath internal interaction. The identity bath operator I_B is absorbed into B_α, as details of B_α are not specified. S_α and B_α are bounded by the norm so that a perturbative expansion of the system–bath propagator driven by H converges for a short evolution time T. The set $\{S_0, \ldots, S_{D-1}\}$ does not have to be the basis of the full Lie algebra of the system. For example, in the pure dephasing Hamiltonian of a qubit coupled to a bath, $\{S_0, \ldots, S_{D-1}\} = \{I_S, \sigma_z\}$, which only generates a subalgebra of a qubit.

Control of the system is implemented by applying a Hamiltonian $V_{c,T}(t)$, such as instantaneous pulses, on the system. To realize a desired system evolution U_{gate} (e.g., a quantum gate or a quantum memory) over a given duration of time T, $V_{c,T}(t)$ scales with T so that

$$\begin{aligned} U_{c,T}(t) &\equiv \mathscr{T} \exp\left[-i \int_0^t V_{c,T}(\tau) d\tau\right] \\ &= \mathscr{T}_\theta \exp\left[-i \int_0^{t/T} V_c(\theta) d\theta\right] \equiv U_c(t/T), \end{aligned} \qquad (14.94)$$

where $V_c(\theta) \equiv T V_{c,T}(T\theta)$. We consider the case of perfect control, that is, $U_{c,T}(T) = U_c(1) = U_{\text{gate}}$ is the desired control of the system. Under the influence of the environment, the system–bath propagator reads

$$U(T) = \mathscr{T} \exp\left(-i \int_0^T [H + V_{c,T}(t)] \, dt\right). \qquad (14.95)$$

The errors induced by H can be extracted in the interaction picture by writing $U(T) = U_{\text{gate}} U_E(T)$, where the error propagator

$$\begin{aligned} U_E(T) &\equiv \mathscr{T} \exp\left[-i \int_0^T U_{c,T}^\dagger(t) H U_{c,T}(t) dt\right] \\ &= \mathscr{T}_\theta \exp\left[-iT \int_0^1 U_c^\dagger(\theta) H U_c(\theta) d\theta\right]. \end{aligned} \qquad (14.96)$$

We suppose that the control $V_{c,T}(t)$ has been designed to suppress the errors due to H up to the Nth order of the evolution time T, which is assumed short on the time scale set by the relevant Hamiltonian norms, that is,

$$U_E(T) = U_A \left[1 + O\left(T^{N+1}\right)\right], \qquad (14.97)$$

where U_A is an operator commuting with a certain set of system operators $\{A_j\}$ (A_j is not necessarily an element of a MOOS). The control $V_{c,T}(t)$ is assumed to be universal such that Eq. (14.97) holds for arbitrary time-independent B_α.

Reference [KLV10] shows that $V_{c,T}(t)$ can be designed to achieve Eq. (14.97) with arbitrary order of precision N and arbitrary U_{gate} for a general time-independent model [Eq. (14.93)]. For the special case of DD for quantum memory, $U_{\text{gate}} = I_S$.

14.10.2 Generalization to time-dependent systems

An arbitrary time-dependent version of Eq. (14.93) reads

$$H'(t) = \sum_{\alpha=0}^{D-1} S_\alpha \otimes B'_\alpha(t), \tag{14.98}$$

where the generic bath operators are assumed analytic in time. We also assume that the bath operators $B'_\alpha(t)$ are bounded by the norm for all $t \in [0, T]$. We have the following theorem.

Theorem 14.3. *If $V_{c,T}(t)$ realizes Eq. (14.97) for an arbitrary time-independent Hamiltonian in Eq. (14.93), it will realize the control with the same order of precision for an arbitrary time-dependent Hamiltonian in Eq. (14.98), that is, the system–bath propagator commutes with the system operator set $\{A_j\}$ up to an error $O(T^{N+1})$,*

$$\begin{aligned} U'(T) &\equiv \mathscr{T} \exp\left(-i \int_0^T [H'(t) + V_{c,T}(t)]\, dt\right) \\ &\equiv U_{\text{gate}} U'_E(T) = U_{\text{gate}} U'_A \left[1 + O\left(T^{N+1}\right)\right], \end{aligned} \tag{14.99}$$

where U'_A commutes with the operator set $\{A_j\}$.

Readers may refer to [WL11a] for the details of the proof. A brief sketch is the following. In the interaction picture of a pure bath Hamiltonian, $\tilde{U}_E(T) \equiv e^{iB_0T} U_E(T)$ and $\tilde{U}'_E(T) \equiv \left(\mathscr{T} e^{-i\int_0^T B'_0(t) dt}\right)^\dagger U'_E(T)$ have a similar expansion form. That is,

$$\tilde{U}_E(T) = 1 + \sum_{n=1}^{\infty} \sum_{\vec{\alpha},\vec{p}} T^{n+|\vec{p}|} S_n^{\vec{\alpha},\vec{p}} \otimes B_n^{\vec{\alpha},\vec{p}}, \tag{14.100a}$$

$$\tilde{U}'_E(T) = 1 + \sum_{n=1}^{\infty} \sum_{\vec{\alpha},\vec{p}} T^{n+|\vec{p}|} S_n^{\vec{\alpha},\vec{p}} \otimes B_n'^{\vec{\alpha},\vec{p}}, \tag{14.100b}$$

where the short-hand notations $\sum_{\vec{\alpha},\vec{p}} \equiv \sum_{\alpha_1=1}^{D-1} \cdots \sum_{\alpha_n=1}^{D-1} \sum_{p_1=0}^{\infty} \cdots \sum_{p_n=0}^{\infty}$ and $|\vec{p}| \equiv \sum_{j=1}^n p_j$. Here $B_n^{\vec{\alpha},\vec{p}}$ and $B_n'^{\vec{\alpha},\vec{p}}$ are bath operators, and the system operators

$$\begin{aligned} S_n^{\vec{\alpha},\vec{p}} \equiv \int_0^1 U_c^\dagger(\theta_1) S_{\alpha_1} U_c(\theta_1) \theta_1^{p_1} \int_0^{\theta_1} U_c^\dagger(\theta_2) S_{\alpha_2} U_c(\theta_2) \theta_2^{p_2} \\ \times \cdots \int_0^{\theta_{n-1}} U_c^\dagger(\theta_n) S_{\alpha_n} U_c(\theta_n) \theta_n^{p_n} d\theta_1 d\theta_2 \cdots d\theta_n. \end{aligned} \tag{14.101}$$

We can show by a specific construction of bath operators $\{B_\alpha\}$ that the universal control $V_{c,T}(t)$ realizes Eq. (14.97) if and only if $S_n^{\vec{\alpha},\vec{p}}$ commutes with the system operator set $\{A_j\}$ for $n+|\vec{p}| \leq N$. This completes the proof since both $\tilde{U}_E(T)$ and $\tilde{U}'_E(T)$ are an expansion of $T^{n+|\vec{p}|}S_n^{\vec{\alpha},\vec{p}}$.

Thus, all the DD schemes discussed above achieve the same decoupling order for analytically time-dependent Hamiltonians.

14.11 Randomized dynamical decoupling

So far we have discussed deterministic decoupling sequences, and we considered their performance in the short-time limit. When the sequence time T is long, the situation becomes different. As shown in Section 14.5.2, when the decoupling sequences are not fast enough, the coherently accumulated errors cannot be filtered out and may accelerate decoherence. To tackle the coherently accumulated errors, we can introduce randomized decoupling schemes [KAS05, VK05], which preserve quantum systems with a linear decay of fidelity in time (while the errors, for example, in a PDD scheme coherently accumulate quadratically in time). Randomized decoupling schemes for combining deterministic sequences (for efficient short-time protection) and randomized strategies (for avoiding errors coherently added up) were developed in [KA05, SV06, VS06].

The simplest randomized DD protocol is obtained by randomly picking the elements of a decoupling group. A possible random realization r of the propagator of duration time $T = n\Delta t$ reads

$$U_{\text{NRD}}(T) = e^{-iH_{r(n)}\Delta t} \cdots e^{-iH_{r(2)}\Delta t} e^{-iH_{r(1)}\Delta t}, \quad (14.102)$$

where $H_{r(j)} = Q^\dagger_{r(j)} H Q_{r(j)}$ with $Q_{r(j)}$ an element randomly drawn from the decoupling group. For this so-called naive random decoupling (NRD) [VK05], the average fidelity decay is $O\left(\lambda^2 \Delta t T\right)$, where λ is the sup-operator norm of H. At long evolution times, NRD is expected to outperform PDD, since the errors only add up probabilistically and accumulate linearly in time due to randomization. Note that the construction of NRD is independent of the size of the decoupling group. When the decoupling group size is too large to fully implement a first-order group-based decoupling sequence in a given time, NRD can be a choice.

To ensure good performance at both short and long times, we combine deterministic sequences and randomized strategies. For example, let $U^{[1]}(T_c)$ [Eq. (14.8)] be a first-order group-based decoupling sequence of duration T_c. Combining with randomization, an embedded decoupling (EMD) sequence is [KA05]

$$U_{\text{EMD}}(nT_c) = \prod_{k=1}^n Q^\dagger_{r(k)} U^{[1]}(T_c) Q_{r(k)}, \quad (14.103)$$

where $Q_{r(k)}$ may not be drawn from the decoupling group that realizes the first-order DD sequence $U^{[1]}(T_c)$. Another way to combine deterministic sequences is to use the random path DD (RPD) [VK05, SV06, VS06]. We construct possible first-order sequences of duration time $T_c = M\Delta t$:

$$U_a^{[1]}(T_c) = e^{-iH_{a(M)}\Delta t} \cdots e^{-iH_{a(2)}\Delta t} e^{-iH_{a(1)}\Delta t}, \quad (14.104)$$

where $H_{a(j)} = Q^\dagger_{a(j)} H Q_{a_k(j)}$ and $\{Q_{a(j)}\}$ is a random arrangement a of a first-order decoupling group of a size M. The RPD sequence reads

$$U_{\text{RPD}}(nT_c) = \prod_{a=1}^{n} U_a^{[1]}(T_c). \tag{14.105}$$

The average fidelity decay of EMD and RPD sequences is $O\left(\lambda^4 T_c^3 T\right)$, where $T = nT_c$ is the total evolution time, which may be long. The errors also accumulate linearly in time T. The sequences also have good performance in short time by eliminating first-order errors.

It is straightforward to combine higher-order DD sequences with randomization. Let $U_k^{[N]}(T_{c,k})$ be an Nth-order randomly constructed sequence labeled by k, a randomized DD sequence of duration time $T = \sum_{k=1}^{n} T_{c,k}$ is

$$U_{\text{Rand}}(T) = \prod_{k=1}^{n} U_k^{[N]}(T_{c,k}). \tag{14.106}$$

For example, when $U_k^{[2]}(T_c)$ is an SDD sequence, $\prod_{k=1}^{n} U_k^{[2]}(T_c)$ is the so-called symmetric random path (SRPD) sequence [KA06, SV06, VS06]. The average errors accumulate as $O\left(\lambda^6 T_c^5 T\right)$ for SRPD.

We may obtain other randomized DD by embedding other deterministic sequences. The key idea is that, by using randomly chosen high-order sequences of duration $T_{c,k}$ as construction units, errors for different intervals $T_{c,k}$ are not coherently accumulated and thus scale linearly with the whole decoupling time $T = \sum_{k=1}^{n} T_{c,k}$. Readers may find more details on randomized DD in [SV08] and [K09].

14.12 Experimental progress

Recently, there have been a lot of experimental investigations on the efficiency of various DD schemes (see Chapter 22). Some of the recent progress in DD experiments is remarkable.

UDD was first realized in experiments by Biercuk *et al.* in an array of $\sim$1000 Be$^+$ ions in a Penning ion trap [BUV+09a, UBB09] with noise mimicked by artificially introduced random modulation of the control fields. UDD was compared with the Carr–Purcell–Meiboom–Gill (CPMG) sequence [MG58], which is the CP sequence with the rotation axis the same as the orientation of the initial state of a qubit, in the "low-fidelity" regime (short-time limit not reached) for various classical noise spectra. The data showed that UDD dramatically outperforms CPMG for Ohmic noise (noise spectrum $S(\omega) \propto \omega$) with a sharp cutoff, while for the ambient magnetic field fluctuations whose noise spectrum $S(\omega) \propto 1/\omega^4$ has a soft high-frequency cutoff, UDD performs similarly to CPMG. The first experimental realization of UDD against realistic noise was achieved by Du *et al.* for radical electron spins in irradiated single-crystal malonic acid [DRZ+09], in which the electron spin coherence time was prolonged from 0.04 µs to about 30 µs by a seven-pulse UDD sequence. The experimental data from different samples under various conditions demonstrate that UDD performs better than CPMG in this solid-state system. For ^{13}C spin qubits in a ^{1}H spin bath of which the high-frequency cutoff was not reached by the DD sequences, CPMG outperforms UDD [AAS11]. These experimental results are consistent with the discussion in Section 14.5.2.

In a GaAs double quantum dot, using 16-pulse CPMG sequences, Bluhm *et al.* extended spin coherence times of individual electron pairs to more than 200 μs [BFN+11]. Barthel *et al.* also demonstrated long coherence times in similar systems using CP, UDD, and CDD sequences [BMM+10].

Coherence preservation of nitrogen-vacancy centers in diamond by DD was also investigated. Ryan *et al.* demonstrated over two orders of magnitude increase of the spin coherence time at room temperature [RHC10]. Using 136 DD pulses, de Lange *et al.* observed spin coherence time enhanced more than 25 times compared with that obtained with one-pulse spin echo [dWR+10]. By using DD, a spin coherence time of nitrogen-vacancy centers in diamond up to 2.44 ms was achieved by Naydenov *et al.* [NDH+11].

In experiments on nitrogen-vacancy centers [RHC10, dWR+10] as well as experiments on ^{13}C spin qubits in ^{1}H spin baths [AAP+10], the performance of XY pulse sequences [GBC90] was investigated (XY pulse sequences are the PDD sequences based on single-qubit universal DD sequences [Eq. (14.16)] and can suppress general errors). Reference [PSL11] gave the first experimental test of CDD against general noise on a qubit. In [AAP+10], CDD provides better overall performance than PDD, UDD, and CP when the initial states are not known. Reference [AAP+10] also confirms the theoretical predictions [KL07, LYS07, NLP11, HVD10] that there is an optimal CDD order due to pulse imperfections and minimum pulse intervals. Wang *et al.* experimentally increased the lifetime of bi-partite pseudo-entanglement in phosphorus donors in a silicon system from 0.4 μs to 30 μs by DD [WRF+11].

It should be pointed out that so far there is still a lack of experimental investigation in the short-time limit, i.e., the high-fidelity regime, which is particularly relevant to quantum computing. The main difficulties are the total evolution time limited by the finite durations of pulses, the fidelity of the pulse control, and the detection inefficiency.

14.13 Discussion

One of the primary aims in designing DD schemes is the reduction of pulse numbers. Numerical checks up to the fourth order indicate that QDD (two-level NUDD) is nearly optimal, differing from the optimal solutions by fewer than three control pulses [WFL10]. However, NUDD of a larger MOOS is far from optimal. For example, let us consider the case of second-order decoupling: For an L-qubit system suffering pure dephasing, an L-level NUDD sequence requires $O(3^L)$ control pulses, while SDD requires only $O(2^{L+1})$ pulses. Finding higher-order DD schemes more efficient than NUDD is desirable and is an interesting open problem. Furthermore, if the Hamiltonian only involves certain types of interaction or is invariant under certain symmetry groups, it is possible to significantly reduce the number of pulses in the higher-order schemes using the idea of combinatorial methods (Chapter 15). We may also analyze the structure of the Hamiltonian and find the smallest MOOS. But an explicit and efficient design improving over NUDD has not been developed yet.

In this chapter, we focused mostly on DD schemes with instantaneous pulses. Although arbitrary-order DD schemes with bounded control can be designed based on concatenation [KLV10], the price to be paid is an exponential growth in pulse complexity. Therefore, it is of interest to ask whether arbitrarily accurate universal control can be achieved with a more modest scale-up in resources, as in the transition from CDD to UDD and NUDD. Furthermore, in experiments the control cannot be perfect. It is important to design efficient DD schemes with some degree of inherent fault tolerance, such as CDD [KL05]. The design involves not only the timing of the DD pulses, but also the rotation axes. A notable early example was the generalization from CP sequences to CPMG sequences by a change of the rotation axes. However, in this example there is a dependence on the initial states, and for quantum information processing it is necessary to develop fault-tolerant techniques that work for arbitrary initial states, similarly to the case of fault-tolerant quantum error correction discussed in this book (e.g., Chapter 2 and Part II). We expect that hybrid approaches, combining ideas from dynamical decoupling and quantum error correction, will lead to additional significant advances and improvements in both fields.

Chapter 15

Combinatorial approaches to dynamical decoupling

Martin Rötteler and Pawel Wocjan

15.1 Introduction

In this chapter we continue the introduction to dynamical decoupling techniques for open quantum systems that was started in Chapter 4. Here we focus on the construction of efficient schemes for dynamical decoupling and we highlight some combinatorial constructions. Efficiency of decoupling schemes is measured in terms of the number of control operations to be applied to the system. This number should be small, ideally a polynomial in the number of qubits in the system.

If we assume that the quantum system is governed by a general Hamiltonian H_S, having interactions involving *any* subset of the qubits, then any scheme that achieves decoupling of H_S is necessarily inefficient in the above sense. The efficient schemes we consider here arise in physically realistic situations where there are much more stringent restrictions on the types of interactions. The most important example is the case of pair-interaction Hamiltonians. These Hamiltonians can be expressed as sums of interaction terms that involve at most two qubits.

While there is a large body of work on dynamical decoupling schemes for pair-interaction Hamiltonians [S90, EBW94, WHH68, H76], which historically has been used mainly in the context of nuclear magnetic resonance (NMR) theory, the past few years have seen a development of new techniques for decoupling that are based on group theory. We therefore call such techniques, or rather the pulse sequences that an experimenter can apply to a given Hamiltonian in order to selectively switch off unwanted couplings, combinatorial decoupling schemes. It turns out that the task of decoupling interactions is an important primitive for the task of simulation of Hamiltonians [LCY+00, DVC+01, BCL+02, WJB02, NBD+02, DNB+02, L02, CLV04, BDN+04, BBN05]. Therefore, it is of great interest to find decoupling schemes that are efficient in the sense that the number of pulses to be applied in the scheme scales polynomially in the number of qubits.

To describe the context, we begin by considering a quantum system S of n interacting qubits. The joint evolution of the system S and the environment (bath) B is described by a total drift

---

Quantum Error Correction, ed. Daniel A. Lidar and Todd A. Brun. Published by Cambridge University Press.
© Cambridge University Press 2013.

Hamiltonian H of the form

$$H = H_S \otimes I_B + I_S \otimes H_B + H_{SB}, \quad H_{SB} = \sum_a S_a \otimes B_a, \tag{15.1}$$

where H_S and H_B characterize the isolated dynamics of the system and the bath, respectively, and the interaction term H_{SB} is responsible for introducing unwanted decoherence effects and dissipation effects in the reduced dynamics of S alone. Without loss of generality we always choose the system Hamiltonian H_S and the error operators B_a to be traceless.

Chapter 4 describes ways of designing periodic dynamical decoupling schemes that use bang-bang and bounded strength controls and effectively change the drift Hamiltonian H to $\bar{H} = 0$ up to first-order and modulo environment terms. The goal of this chapter is to show that there are efficient decoupling schemes provided that we can selectively apply the control operations to the individual qubits. We say that a family of decoupling schemes is efficient if the number of control operations grows polynomially with the number of qubits. To obtain such efficient schemes, we rely on the fact that in most physically relevant situations the system Hamiltonian H_S and the error operators S_a are 2-local in the following sense:

$$H_S := \sum_{k<\ell} \sum_{\alpha\beta} J_{k\ell;\alpha\beta}\, \sigma_\alpha^k \sigma_\beta^\ell + \sum_k \sum_\alpha \omega_{k;\alpha}\, \sigma_\alpha^k, \tag{15.2}$$

$$S_a := \sum_{k<\ell} \sum_{a;\alpha\beta} J_{a;k\ell;\alpha\beta}\, \sigma_\alpha^k \sigma_\beta^\ell + \sum_k \sum_\alpha \omega_{a;k;\alpha}\, \sigma_\alpha^k, \tag{15.3}$$

where σ_α are Pauli matrices, i.e., $\alpha \in \{x, y, z\}$. In this case, we also say that H_S and S_a are pair-interaction operators.

A simple example of a pair-interaction Hamiltonian is given by

$$H_S = J\sigma_z^1 \sigma_z^2 + \omega_1 \sigma_z^1 + \omega_2 \sigma_z^2,$$

which was also considered in Chapter 4 to describe the Carr–Purcell sequence. Here ω_1 and ω_2 describe the strength of the local terms (Zeeman terms) and J describes the strength of the coupling term.

The difference between the scenarios considered in Chapter 4 and this chapter is summarized by the following table:

Control	Arbitrary interactions	Pair interactions
hard	$O(4^n)$; see Chapter 4	$O(n)$; this chapter
soft	$O(n4^n)$; see Chapter 4	$O(n^2)$; this chapter

In [W06] the theory of decoupling schemes from orthogonal arrays was combined with ideas for soft decoupling for systems with bounded strength controls in [VK03]. The resulting schemes have a number of $O(n^2)$ pulses.

As we will show in the following, the assumption about locality of the interactions in H_S as well as the terms S_a in H_{SB} leads to a dramatic improvement in terms of the efficiency of the schemes, which directly translates into the number of pulses that have to be applied to the system in order to achieve decoupling from the environment and decoupling of the system's qubits from each other.

The chapter is organized as follows. In Section 15.2 we provide some basic motivation for decoupling schemes and give necessary definitions. Also in this section we present a plethora of decoupling schemes that can be obtained from combinatorial objects such as Hadamard matrices, difference matrices, and more generally from orthogonal arrays. In Section 15.3 we then pick up the theme of implementing the control operations by means of bounded-strength controls that was motivated in Chapter 4. We show that this idea of bounded-strength Eulerian decoupling can be merged with the combinatorial schemes presented in the previous section. This gives rise to efficient bounded-strength Eulerian decoupling schemes that can be used for decoupling with realistic pulses.

15.2 Combinatorial bang-bang decoupling

15.2.1 Motivating examples

We consider examples of pair-interaction Hamiltonians to demonstrate the basic principles of decoupling. The examples are chosen to be Hamiltonians with $\sigma_z\sigma_z$ couplings only. These Hamiltonians can be visualized as (weighted) graphs. For instance the Hamiltonian

$$H = \sigma_z^1 \sigma_z^2 + \sigma_z^2 \sigma_z^3 + \sigma_z^1 \sigma_z^3 + \sigma_z^3 \sigma_z^4$$

is represented by the following graph:

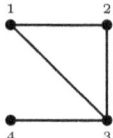

We introduce the notation ⊙ to denote the application of σ_x to a qubit. As an example, apply σ_x to one qubit of a pair-interaction $\sigma_z \otimes \sigma_z$. The result is given by

$$(I_2 \otimes \sigma_x)(\sigma_z \otimes \sigma_z)(I_2 \otimes \sigma_x) = \sigma_z \otimes (\sigma_x \sigma_z \sigma_x) = -\sigma_z \otimes \sigma_z.$$

Pictorially, this corresponds to •—⊙ $\sigma_z\sigma_z$ = •—• $-\sigma_z\sigma_z$ using the notation for σ_x introduced above. As a slightly more involved example, we consider the effect of applying a local operation $V = I \otimes \sigma_x \otimes I \otimes \sigma_x$ on the following pair-interaction Hamiltonian H of four qubits:

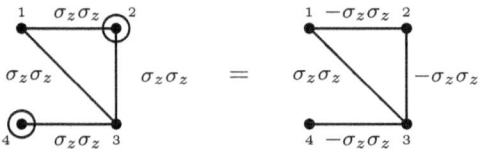

Hence the effect of the local σ_x operations is to invert the sign of some of the edges in the graph. This in turn can be used to selectively switch off some of the terms in the pair-interaction Hamiltonian. For instance, we can apply $\mathscr{P} = \mathsf{f} V \mathsf{f} V^\dagger$ (using the notation f for "free evolution" as in Chapter 4), which leads to the following result for the effective Hamiltonian $\bar{H}$:

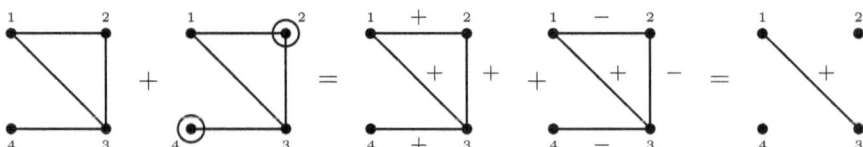

Algebraically, we obtain the following calculation for $\bar{H}$:

$$\frac{\frac{1}{2}\sigma_z^1\sigma_z^2 + \frac{1}{2}\sigma_z^2\sigma_z^3 + \frac{1}{2}\sigma_z^1\sigma_z^3 + \frac{1}{2}\sigma_z^3\sigma_z^4}{-\frac{1}{2}\sigma_z^1\sigma_z^2 - \frac{1}{2}\sigma_z^2\sigma_z^3 + \frac{1}{2}\sigma_z^1\sigma_z^3 - \frac{1}{2}\sigma_z^3\sigma_z^4}$$
$$+\ \sigma_z^1\sigma_z^3$$

Hence, we have simulated the Hamiltonian $\bar{H} = \sigma_z^1 \sigma_z^3$, which means that we have switched off all other coupling terms except for the term between qubits 1 and 3. In the subsequent sections we show how decoupling schemes can be designed in a systematic fashion.

15.2.2 Definition of combinatorial decoupling schemes

We formally define combinatorial bang-bang decoupling schemes, explaining what control operations we assume are available to the experimenter and also what approximations we use in analyzing the performance of decoupling schemes.

We consider a quantum system of n interacting qudits that are coupled to some uncontrollable environment. We assume that we can implement unitaries from some finite subset of the unitary group $U(d)$ on any of the n qudits (d-dimensional subsystems). The assumption that the control operations are given by unitaries is referred to as fast control limit or bang-bang control. This means that the pulses can be made so strong that during the short time it takes to implement the desired control unitaries the effects due to the system Hamiltonian and couplings to the environment are negligible. We assume that we can apply the control operations to individual qubits simultaneously. This is also referred to as the ability to perform selective pulses.

Following the approach of average Hamiltonian theory [EBW94] (see Section 4.4.1), we consider the first-order effect of applying a sequence of selective pulses as follows: Letting the system evolve for time t (in the following referred to as t f, which stands for "free evolution" for time t; see also Chapter 4) has the effect of applying the unitary $\exp(-iHt)$. Let $t_1, t_2, \ldots, t_N$ be a sequence of time durations and $V_1, V_2, \ldots, V_N \in U(d)^{\otimes n}$ a sequence of control operations. Then the control sequence

$$t_1 \text{ f}, V_1, t_2 \text{ f}, V_2, t_3 \text{ f}, \ldots, t_N \text{ f}, V_N$$

gives rise to the evolution

$$\prod_{j=1}^{N} \exp(-iU_j^\dagger H U_j t_j),$$

where $U_j = \prod_{i=1}^{j} V_i$ and we have assumed that the V_i are chosen such that $U_N = \prod_{i=1}^{N} V_i = I$. The latter condition means that we have a periodic control sequence. The unitaries are referred to as control propagators. We say that the sequence consists of N *time slots* and use the shorthand notation $(t_1, U_1; t_2, U_2; \ldots; t_N, U_N)$. Note that the pulses V_j can be determined from the propagators U_j. If the times t_j are small compared to the time scale of the natural evolution according to H, then the total time evolution of the sequence can be approximated by

$$\exp(-i\bar{H}t/\tau),$$

where

$$\bar{H} := \sum_{j=1}^{N} t_j U_j^\dagger H U_j \tag{15.4}$$

denotes the average Hamiltonian. The quantity $\tau := \sum_j t_j$ is called the slow-down factor. In other words, the slow-down factor corresponds to the ratio of speed of the time evolution according to $\bar{H}$ and that according to H (see, e.g., [BCL+02, WJB02]).

In this chapter, we do not consider the general problem of Hamiltonian simulation. We consider the task of dynamical decoupling, i.e., the task of finding sequences for which the simulated Hamiltonian $\bar{H}$ is equal to the zero Hamiltonian (modulo environment terms). Since this implies that the time evolution of the system is effectively stopped, these schemes are also used in dynamical suppression of decoherence in open quantum systems ("bang-bang" control), see Chapter 4 and [VS98, VKL99, VLK99].

Definition 15.1 (Bang-bang decoupling schemes). *A bang-bang decoupling scheme is a sequence $D = (t_1, U_1; \ldots; t_N, U_N)$ of control propagators $U_j \in U(d)^{\otimes n}$ and delays $t_j \in \mathbb{R}_+$ such that*

$$\bar{H} = \sum_{j=1}^{N} t_j U_j^\dagger H U_j = \mathbf{0}. \tag{15.5}$$

A decoupling scheme D is called regular *if $t_1 = \cdots = t_N$, i.e., if the lengths of the time slots are the same.*

Definition 15.1 includes decoupling schemes consisting of time-slots of different length. Indeed, practical decoupling schemes with unequal time intervals exist and are used, most notably the famous WAHUHA sequence [WHH68, HW68, EBW94]. Recently, an optimal decoupling scheme for a qubit using time slots of unequal length was devised [U07], which is universal for pure dephasing in the sense shown in [YL08]. This irregular scheme was extended to general decoherence of a qubit in [WFL10].

However, most decoupling schemes in the literature [JK99, LCY+00, SM01, L02, WRJ+02a] are equal length schemes ($t_j \equiv \Delta$). Throughout the rest of this chapter, we only consider regular decoupling schemes.

We now establish the connection to the notation of Chapter 4. A regular decoupling scheme $D = (\Delta, U_1; \ldots; \Delta, U_N)$ specifies the control propagators $U_c(t)$ over each of the N equally long subintervals. A control cycle is defined by

$$U_c\big((j-1)\Delta + \tau\big) = U_j, \tag{15.6}$$

where $\tau \in [0, \Delta)$, $T_c = N\Delta$ for some $\Delta > 0$, and $j = 1, \ldots, N$.

If the system consists of only one qudit, then an arbitrary unitary error basis in dimension d defines a regular decoupling scheme. Recall that a unitary error basis (or error basis for short) is a collection $\{U_1 := I, U_2, \ldots, U_{d^2}\}$ of unitary matrices that are orthogonal with respect to the trace inner product (also called the Hilbert–Schmidt inner product), i.e.,

$$\text{Tr}(U_i^\dagger U_j) = d \delta_{ij} \tag{15.7}$$

for all $1 \leq i, j \leq d^2$. It follows that

$$\frac{1}{d} \sum_{j=1}^{d^2} U_j X U_j^\dagger = \mathrm{Tr}(X) I_d \tag{15.8}$$

for any operator X acting on $\mathbb{C}^d$.

A nice error basis is the unitary error basis with an underlying group-theoretic structure. Let G be a group of order d^2. We say that $\{U_g : g \in G\}$ is a nice error basis with index group G if there is a function $\omega : G \times G \to \mathbb{C}^\times$ such that

$$U_g U_h = \omega(g, h) U_{gh} \tag{15.9}$$

for all $g, h \in G$, where $\mathbb{C}^\times$ denotes the set of complex numbers of modulus one. It is known that nice error bases are equivalent to irreducible, unitary, projective representations of G. An example of such a representation is given in the following. The discrete Fourier transform of length $d \in \mathbb{N}$ is the unitary transformation defined by

$$\mathrm{DFT}_d := \frac{1}{\sqrt{d}} \sum_{k,\ell=0}^{d-1} \omega^{k \cdot \ell} |k\rangle\langle \ell|, \tag{15.10}$$

where ω denotes the primitive dth root of unity $e^{2\pi i/d}$. Define operators

$$X := \sum_{k=0}^{d-1} |k\rangle\langle k+1|, \tag{15.11}$$

$$Z := \mathrm{DFT}_d^\dagger \cdot X \cdot \mathrm{DFT}_d = \sum_{k=0}^{d-1} \omega^k |k\rangle\langle k|, \tag{15.12}$$

where the indices are reduced modulo d. Then the map

$$\rho : \begin{cases} \mathbb{Z}_d \times \mathbb{Z}_d \to U(d) \\ (a, b) \mapsto X^a Z^b \end{cases} \tag{15.13}$$

is an irreducible, unitary, projective representation of $\mathbb{Z}_d$. Note that for $d = 2$ one obtains the identity I and the Pauli matrices $\sigma_x, \sigma_y, \sigma_z$.

The advantage of using nice error bases is that it suffices to be able to implement only pulses U_s to generate all propagators U_g for $g \in G$, where the elements s labeling the pulses are taken from a generating set S of the index group G.

Let us now consider a quantum system of n qudits. Then, the collection of all n-fold tensor products of elements of a unitary error basis of $\mathbb{C}^d$ clearly forms a unitary error basis of $(\mathbb{C}^d)^{\otimes n}$. The obvious disadvantage of a decoupling scheme derived from such an error basis is that the number of control operations grows exponentially with the number of qudits. In the following, we discuss how to obtain more efficient decoupling schemes, which make use of the fact that physically relevant system Hamiltonians and error terms are not arbitrary operators, but have some kind of local structure.

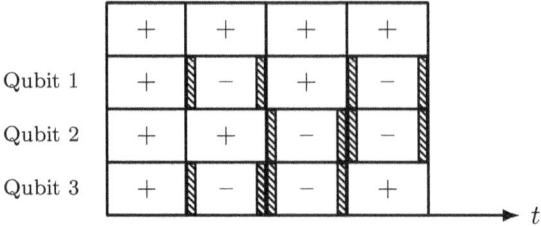

Fig. 15.1. Example for Hadamard scheme. The signs $\pm$ correspond to the propagators I and σ_x, respectively. The hatched bar corresponds to the control pulse σ_x.

15.2.3 Decoupling schemes based on Hadamard matrices

First, we describe efficient decoupling schemes based on Hadamard matrices [JK99]. These can be used to decouple arbitrary Hamiltonians of the form

$$H_S = \sum_{k<l} J_{k\ell} \sigma_z^k \sigma_z^\ell + \sum_k \omega_k \sigma_z^k \quad (15.14)$$

on a system of n qubits. Second, we describe how to generalize the construction based on Hadamard matrices to obtain efficient schemes that can be used to decouple arbitrary Hamiltonians of the form in Eq. (15.2). These efficient decoupling schemes are based on so-called *sign matrices* [L02]. While we develop here the qubit case, it should be noted that the case of higher-dimensional systems can be handled *mutatis mutandis*, see [RW06] for more details.

Recall that a Hadamard matrix of size N is a ± 1 matrix H_N of size $N \times N$ with the property that $H_N H_N^T = N I_N$. Hadamard matrices have been studied extensively in combinatorics, see [BJL99, S03c, CD96] for background information and constructions of Hadamard matrices. Simple examples of Hadamard matrices of small sizes are given by (here and in the following the entries ± 1 have been abbreviated to $+/-$):

$$\begin{pmatrix} + & + \\ + & - \end{pmatrix}, \quad \begin{pmatrix} + & + & + & + \\ + & - & + & - \\ + & + & - & - \\ + & - & - & + \end{pmatrix}, \quad \begin{pmatrix} - & + & + & + \\ + & - & + & + \\ + & + & - & + \\ + & + & + & - \end{pmatrix}. \quad (15.15)$$

It is known that a necessary condition for the existence of a Hadamard matrix is that either $N = 2$ or $N \equiv 0 \mod 4$. A long-standing conjecture is whether for any $N \equiv 0 \mod 4$ a Hadamard matrix of size N exists [BJL99]. Let H_2 be the leftmost Hadamard matrix in Eq. (15.15). To obtain a Hadamard matrix of size $N = 2^s$, construct $H_{2^s} := H_2 \otimes \cdots \otimes H_2$ (s tensor factors).

A decoupling scheme for $n < N$ qubits can be obtained from a Hadamard matrix H_N in a straightforward way as follows. We may assume, without loss of generality, that the first row of H_N contains only $+1$ as entries. To obtain this normal form, we simply multiply all elements in the jth column by the jth entry of the first row. The qubits correspond to the rows $2, \ldots, N$ of the matrix and each column corresponds to a composite pulse that is applied to the system. The scheme corresponding to the matrix shown in the middle of Eq. (15.15) is shown in Fig. 15.1.

15.2.4 Decoupling schemes based on sign matrices

We now describe how to obtain efficient decoupling schemes for general pair-interaction n-qubit Hamiltonians

$$H_S = \sum_{k<\ell} \sum_{\alpha\beta} J_{k\ell;\alpha\beta} \sigma_\alpha^k \sigma_\beta^\ell \qquad (15.16)$$

using Schur-closed orthogonal triples of *sign matrices* [L02]. Sign matrices can be generalized to higher-dimensional systems [RW06] by allowing more general complex phases. This leads to so-called phase matrices, which can be used to decouple interacting qudits.

A sign matrix $S_{n,N}$ is a matrix with entries ± 1. We say two sign matrices A and B of size $n \times N$ are orthogonal if $AB^T = NI_n$. If a set of $n \times N$ sign matrices has the property that any two distinct matrices are orthogonal, then we say that the set is *orthogonal*. Examples of orthogonal sign matrices are obtained by selecting n rows of a Hadamard matrix of size N.

Recall that the Schur product of two $n \times N$ matrices A and B, denoted by $C := A \circ B$, is defined as the entry-wise product: $C_{i,j} := A_{i,j} B_{i,j}$. If a set of $n \times N$ sign matrices has the property that all possible pairwise Schur products are already contained in the set, then we say that the set is closed under taking Schur products, or *Schur-closed* for short.

We obtain regular decoupling schemes that only use tensor products of the Pauli matrices I_2, σ_x, σ_y, and σ_z as control propagators. In this case, both the terms in the expansion of H in Eq. (15.16) and the control propagators are tensor products of Pauli matrices, so conjugation of each term by each control operation results in multiplying the term by either $+1$ or -1 in each time-slot. The entries in the following table are the resulting signs when conjugating the column's Pauli operators by the row's Pauli operators:

$$\begin{array}{c|cccc} & I_2 & \sigma_x & \sigma_y & \sigma_z \\ \hline I_2 & + & + & + & + \\ \sigma_x & + & + & - & - \\ \sigma_y & + & - & + & - \\ \sigma_z & + & - & - & + \end{array} \qquad (15.17)$$

To achieve decoupling we use an idea from spin-echo experiments on n qubits [JK99]: we design the scheme in such a way that any fixed term in Eq. (15.2) picks up either a $+$ or a $-$ sign in each time slot and such that the sum of these signs is zero, i.e., the term is canceled. Sufficient conditions for decoupling of general 2-local n-qubit Hamiltonian are given by the following theorem. These conditions have been given in [L02].

Theorem 15.1. *A set $\mathscr{S} = \{S_{id}, S_x, S_y, S_z\}$ of sign matrices of size $n \times N$ defines a decoupling scheme for an arbitrary pair-interaction n-qubit Hamiltonian if it is Schur-closed and orthogonal.*

Proof. We first show to how to read off the N local control propagators from the sign matrices and then prove that the decoupling property holds. Denote the (k,j)th entry of S_α for $\alpha \in \{x,y,z\}$ by $S_{\alpha;kj}$. Schur-closedness implies that the only possibilities for the tuple $(S_{\alpha;kj})_{\alpha \in \{id,x,y,z\}}$ are

$$(+++ +), \quad (++--), \quad (+-+-), \quad (+--+)$$

for each fixed (k, j). Each of these possibilities corresponds to one of the rows in Eq. (15.17). Therefore, each defines a unique Pauli matrix $\sigma_{(k,j)}$. The signs acquired when conjugating the Pauli matrices I_2, σ_x, σ_y, σ_z by $\sigma_{(k,j)}$ are given by the (k, j)th entries of S_{id}, S_x, S_y, S_z, respectively. Based on this correspondence we define the local control propagators by

$$U_j := \sigma_{(1,j)} \otimes \cdots \otimes \sigma_{(n,j)}$$

for $j = 1, \ldots, N$. The resulting scheme is $D := (\Delta, U_1; \ldots; \Delta, U_N)$ for some fixed $\Delta > 0$.

It remains to prove that D is a decoupling scheme, i.e., all terms in the expansion in Eq. (15.16) are switched off. Observe that the sign acquired by the local term σ_α^k in Eq. (15.16) when conjugated by U_j is equal to $S_{\alpha;kj}$ and that the sign acquired by the 2-local term $\sigma_\alpha^k \sigma_\beta^l$ in Eq. (15.16) when conjugated by U_j is equal to the product of the signs $S_{\alpha;kj}$ and $S_{\beta;\ell j}$. The effect of applying the whole sequence D to a local or 2-local term is to multiply it with the sum of the resulting signs. The property of the triple $\mathscr{S}$ to be orthogonal is equivalent to the two conditions

$$\sum_{j=1}^N S_{\alpha;kj} = 0 \quad \text{and} \quad \sum_{j=1}^N S_{\alpha;kj} S_{\beta;\ell j} = 0 \tag{15.18}$$

for all $\alpha, \beta \ne id$ and all $k < \ell$.

The first condition in (15.18) ensures that all local terms are removed and the second ensures that all 2-local terms are removed. $\square$

15.2.5 Decoupling schemes based on difference matrices

We now describe efficient decoupling schemes based on difference matrices [SM01]. These schemes can be used to decouple arbitrary diagonal Hamiltonians and error terms

$$H = \sum_{k<\ell} \sum_\alpha J_{k\ell;\alpha} \sigma_\alpha^k \sigma_\alpha^\ell + \sum_k \sum_\alpha \omega_{k;\alpha} \sigma_\alpha^k, \tag{15.19}$$

$$S_a = \sum_{k<\ell} \sum_\alpha J_{a;k\ell;\alpha} \sigma_\alpha^k \sigma_\alpha^\ell + \sum_k \sum_\alpha \omega_{a;k;\alpha} \sigma_\alpha^k. \tag{15.20}$$

The decoupling schemes based on difference matrices are slightly more efficient than those based on sign matrices because they take advantage of the fact that the couplings are diagonal.

This advantage is based on the following simple observation. Let A be an arbitrary diagonal operator acting on two qubits, i.e.,

$$A = \sum_\alpha J_\alpha \sigma_\alpha \otimes \sigma_\alpha. \tag{15.21}$$

Then, for any $\sigma \in \{I, \sigma_x, \sigma_y, \sigma_z\}$ we have

$$(\sigma \otimes I) A (\sigma \otimes I) = (I \otimes \sigma) A (I \otimes \sigma). \tag{15.22}$$

We give the general definition of a difference matrix over an arbitrary group G.

Definition 15.2 (Difference matrix). *Let G be a finite group and let $m, N \in \mathbb{N}$. An $n \times N$ array D with entries from G is a difference matrix with multiplicity λ if and only if for all $1 \le k < \ell \le m$ the list*

$$\left(g_{k,1} \cdot g_{\ell,1}^{-1}, \ldots, g_{k,N} \cdot g_{\ell,1}^{-1} \right) \tag{15.23}$$

contains each element of G exactly λ times. We use the notation $D(m, N, G)$ to denote such a difference matrix.

Similarly to the situation for Hadamard matrices, we may, without loss of generality, assume that the first row contains only the identity element of G. We simply multiply each entry $g_{k,j}$ of $D(n, N, G)$ by $g_{1,j}^{-1}$ to achieve this normal form. It is important that each row of a difference matrix in such normal form contains each element of G exactly λ times.

For the purposes of decoupling diagonal Hamiltonians it suffices to consider the special case $G = \mathbb{Z}_2 \times \mathbb{Z}_2$. We identify the elements of the index group $G = \mathbb{Z}_2 \times \mathbb{Z}_2$ with Pauli matrices as follows follows: $(0,0) \mapsto I$, $(1,0) \mapsto \sigma_x$, $(0,1) \mapsto \sigma_z$, and $(1,1) \mapsto \sigma_y$. Let U_g denote the Pauli matrix corresponding to $g \in G$. We can obtain a decoupling scheme for n qubits from a difference matrix $D(m, N)$ with $n = m + 1$ as follows. Bring the difference matrix in normal form and remove the first row and denote by $(g_{1,j}, \ldots, g_{n,j})^T$ the corresponding columns. These columns specify the control propagators

$$U_j := U_{g_{1j}} \otimes U_{g_{2j}} \otimes \cdots \otimes U_{g_{nj}} \tag{15.24}$$

for $j = 1, \ldots, N$. The resulting decoupling scheme is $D = (\Delta, U_1; \ldots, \Delta, U_N)$ for some fixed $\Delta > 0$.

The following theorem shows that the prescription in (15.24) allows to decouple diagonal Hamiltonians and couplings to the environment.

Theorem 15.2 (Decoupling with difference schemes). *Let $D(n+1, N)$ be a difference matrix over $\mathbb{Z}_2 \times \mathbb{Z}_2$. Denote by Π_D the corresponding control action (15.24). We have*

$$\Pi_D(H_S) = \mathbf{0} \tag{15.25}$$
$$\Pi_D(S_a) = \mathbf{0} \tag{15.26}$$

for all system Hamiltonians H_S of the form in Eq. (15.19) and all error terms S of the form in Eq. (15.20).

Proof. Due to linearity of Π_D it suffices to prove that $\Pi_D(\sigma_\alpha^k \sigma_\alpha^\ell) = 0$ for all $k \leq \ell$ and $\Pi_D(\sigma_\alpha^k) = 0$ for all k.

We have

$$\Pi_D(\sigma_\alpha^k \sigma_\alpha^\ell) = \frac{1}{N} \sum_{j=1}^N \left(U_{g_{1j}} \otimes U_{g_{2j}} \otimes \cdots \otimes U_{g_{nj}}\right)^\dagger \sigma_\alpha^k \sigma_\alpha^\ell \left(U_{g_{1j}} \otimes U_{g_{2j}} \otimes \cdots \otimes U_{g_{nj}}\right)$$

$$= \left[\frac{1}{N} \sum_{j=1}^N \left(U_{g_{k,j}} \otimes U_{g_{\ell,j}}\right)^\dagger (\sigma_\alpha \otimes \sigma_\alpha) \left(U_{g_{k,j}} \otimes U_{g_{\ell,j}}\right)\right]^{k,\ell}$$

$$= \left[\frac{1}{N} \sum_{j=1}^N \left(I \otimes U_{g_{k,j} g_{\ell,j}^{-1}}\right)^\dagger (\sigma_\alpha \otimes \sigma_\alpha) \left(I \otimes U_{g_{k,j} g_{\ell,j}^{-1}}\right)\right]^{k,\ell}$$

$$= \left[\frac{\lambda}{N} \sum_{g \in G} \left(I \otimes U_g\right)^\dagger (\sigma_\alpha \otimes \sigma_\alpha) \left(I \otimes U_g\right)\right]^{k,\ell}$$

$$= \left[\sigma_\alpha \otimes \Pi_G(\sigma_\alpha)\right]^{k,\ell}$$

$$= \mathbf{0}^{k,\ell} = \mathbf{0}. \tag{15.27}$$

The equality $\Pi_D(\sigma_\alpha^k) = 0$ for all k is proved in a similar way by using that each row of D contains each element of G exactly λ times. $\square$

15.2.6 Decoupling schemes based on orthogonal arrays

We now describe efficient decoupling schemes based on orthogonal arrays (OAs). These schemes can be used to decouple arbitrary t-local system Hamiltonians and error operators S_a acting on qudits.

Orthogonal arrays first appeared in statistics where they were used in the design of experiments for collecting statistical data systematically. We refer the reader to the books [BJL99, CD96, HSS99] for applications and constructions of OAs. Stollsteimer and Mahler first used OAs for the construction of decoupling schemes and selective coupling schemes [SM01] for qubit systems with pair-interactions. This method was generalized to qudit systems with t-local interactions in [WRJ+02a, RW06].

We need to introduce some notation to define what it means for an operator to be t-local. Choose a collection $\sigma_1, \sigma_2, \ldots, \sigma_{d^2-1}$ of traceless self-adjoint matrices such that together with the identity matrix $\sigma_0 = \mathbf{1}_d$ they form a basis of self-adjoint matrices acting on $\mathbb{C}^d$. Let $\mathscr{B} := \{\sigma_0, \sigma_1, \sigma_2, \ldots, \sigma_{d^2-1}\}$. Consider $\sigma := \sigma_{i_1} \otimes \sigma_{i_2} \otimes \cdots \otimes \sigma_{i_n}$, an arbitrary tensor product with $\sigma_{i_j} \in \mathscr{B}$ for $j = 1, \ldots, n$. We define the weight of σ to be the number of components for which σ_{i_j} is not equal to the identity matrix. We call an operator acting on $(\mathbb{C}^d)^{\otimes n}$ t-local if and only if it can be written as a linear combination of such tensor products of weight less than or equal to t.

We consider system Hamiltonians and error operators of the form

$$H_S := \sum_{w=1}^{t} \sum_{k_1 < \cdots < k_w} \sum_{\alpha_1, \ldots, \alpha_w} J_{\underline{k};\underline{\alpha}} \, \sigma_{\underline{\alpha}}^{\underline{k}}, \tag{15.28}$$

$$S_a = \sum_{w=1}^{t} \sum_{k_1 < \cdots < k_w} \sum_{\alpha_1, \ldots, \alpha_w} J_{a;\underline{k};\underline{\alpha}} \, \sigma_{\underline{\alpha}}^{\underline{k}}, \tag{15.29}$$

where $\underline{k} = (k_1, \ldots, k_w)$, $\underline{\alpha} = (\alpha_1, \ldots, \alpha_w)$, and $\sigma_{\underline{\alpha}}^{\underline{k}} = \sigma_{\alpha_1}^{k_1} \otimes \cdots \otimes \sigma_{\alpha_w}^{k_w}$.

Definition 15.3 (Orthogonal array of strength t). *Let $\mathscr{A}$ be a finite alphabet and let $n, N \in \mathbb{N}$. An $n \times N$ array M with entries from $\mathscr{A}$ is an orthogonal array with $s = |\mathscr{A}|$ different pulses, locality t, and multiplicity λ if and only if every $t \times N$ sub-array of M contains each possible t-tuple of elements in $\mathscr{A}^t$ precisely λ times as a column. We use the notation $OA_\lambda(N, n, s, t)$ to denote a corresponding orthogonal array. If λ, s, and t are understood we also use the shorthand notation $OA(N, n)$.*

The terminology used for parameters of OAs in decoupling differs from that used in statistics. We provide a list of the different languages in Table 15.1. Note that as a convention we write OAs as $n \times N$ matrices, whereas most authors in design theory prefer to write the same OAs as $N \times n$ matrices. Besides typographic reasons we find the presentation using $n \times N$ matrices more useful since it establishes a correspondence with pulse sequences in NMR, which are typically read from left to right like a musical score.

Table 15.1. *Terminology used in the theory of design of experiments to describe an orthogonal array $OA(N, n, s, t)$ over $\mathscr{A}$ and terminology used for decoupling schemes for t-local n-qudit Hamiltonians*

Parameter	Design of experiments	Decoupling schemes for qudit systems
n	factors	subsystem (qudits)
N	runs	intervals
$\mathscr{A}$	levels	different pulses
s	number of levels	(dimension of the subsystems)^2
t	strength	locality
λ	index	multiplicity

Example 15.1. *The orthogonal array with parameters $OA(16, 5, 4, 2)$ is an example of small size. This means that we have 16 intervals, five qubits, four different pulses, and locality 2. The array is given by the matrix*

$$\begin{pmatrix} 1 & 1 & 1 & 1 & 2 & 2 & 2 & 2 & 3 & 3 & 3 & 3 & 4 & 4 & 4 & 4 \\ 1 & 2 & 3 & 4 & 4 & 3 & 2 & 1 & 1 & 2 & 3 & 4 & 4 & 3 & 2 & 1 \\ 1 & 3 & 4 & 2 & 1 & 3 & 4 & 2 & 3 & 1 & 2 & 4 & 3 & 1 & 2 & 4 \\ 1 & 3 & 4 & 2 & 4 & 2 & 1 & 3 & 4 & 2 & 1 & 3 & 1 & 3 & 4 & 2 \\ 1 & 2 & 3 & 4 & 2 & 1 & 4 & 3 & 4 & 3 & 2 & 1 & 3 & 4 & 1 & 2 \end{pmatrix} \quad (15.30)$$

over the alphabet $\mathscr{A} = \{1, 2, 3, 4\}$. It is straightforward to check that indeed every pair of rows contains all the possible pairs of symbols precisely once. This array was obtained from a Hamming code over $\mathbb{F}_4$. We explore this construction in more detail in Theorem 15.4.

An important special case arises if the strength t is 2. This means that each pair of elements of $\mathscr{A}$ occurs λ times in the list $((a_{kj}, a_{lj}) \mid j = 1, \ldots, N)$ for $1 \leq k < l \leq n$. Most known constructions actually yield arrays of strength 2 [HSS99]. It turns out that for many physical systems it is sufficient to study arrays of small strength since the strength relates to the degree of the interactions, i.e., for pair-interaction Hamiltonians it is sufficient to consider arrays of strength $t = 2$.

We obtain a decoupling scheme from an orthogonal array M with parameters $OA(N, n, d^2, 2)$ over an alphabet $\mathscr{A}$ of size d^2 as follows. Recall that d denotes the dimension of the qudits. The elements of $\mathscr{A}$ are identified with the elements of the index group $G := \mathbb{Z}_d \times \mathbb{Z}_d$ that acts irreducibly on $\mathbb{C}^d$ via the generalized Pauli matrices, i.e., the representation defined on the generators of G via $(1, 0) \mapsto \sigma_x = \sum_i |i\rangle\langle i+1|$ and $(0, 1) \mapsto \sigma_z = \sum_i \omega_d^i |i\rangle\langle i|$. The columns $(g_{1j}, \ldots, g_{nj})^T$ of M specify the control propagators as follows

$$U_j := U_{g_{1j}} \otimes U_{g_{2j}} \otimes \cdots \otimes U_{g_{nj}}, \quad (15.31)$$

$j = 1, \ldots, N$. The resulting scheme is $D := (\Delta, U_1; \ldots; \Delta, U_N)$ for some fixed $\Delta > 0$.

The following theorem shows that the prescription in (15.31) makes it possible to decouple t-local Hamiltonians and couplings to the environment. We denote by $G^{\times w}$ the direct product group $G \times \cdots \times G$ with w copies of G.

Theorem 15.3 (Decoupling with OAs). *Let $M = (g_{kj})$ be an $OA(n, N)$ orthogonal array of strength t over the group G. Let Π_M denote the control action defined by Eq. (15.31). Then we have*

$$\Pi_M(H_S) = \mathbf{0}, \tag{15.32}$$

$$\Pi_M(S_a) = \mathbf{0}, \tag{15.33}$$

for all system Hamiltonians H_S of the form in Eq. (15.28) and all error terms S_a of the form in Eq. (15.29).

Proof. Due to linearity of Π_M it suffices to show $\Pi_M(\sigma_{\underline{\alpha}}^{\underline{k}}) = \mathbf{0}$ for all tuples $\underline{k} = (k_1, \ldots, k_w)$ with $k_1 < \cdots < k_w$ and all tuples $\underline{\alpha} = (\alpha_1, \ldots, \alpha_w)$ for all $w = 1, \ldots, t$. We have

$$\Pi_M(\sigma_{\underline{\alpha}}^{\underline{k}}) = \frac{1}{N} \sum_{j=1}^{N} \left(U_{g_{1j}} \otimes U_{g_{2j}} \otimes \cdots \otimes U_{g_{nj}} \right)^{\dagger} \sigma_{\underline{\alpha}}^{\underline{k}}$$

$$\times \left(U_{g_{1j}} \otimes U_{g_{2j}} \otimes \cdots \otimes U_{g_{nj}} \right)$$

$$= \left[\frac{1}{N} \sum_{j=1}^{N} \left(U_{g_{k_1,j}} \otimes U_{g_{k_2,j}} \otimes \cdots \otimes U_{g_{k_w,j}} \right)^{\dagger} \left(\sigma_{\alpha_1} \otimes \sigma_{\alpha_2} \otimes \cdots \otimes \sigma_{\alpha_w} \right) \right.$$

$$\left. \times \left(U_{g_{k_1,j}} \otimes U_{g_{k_2,j}} \otimes \cdots \otimes U_{g_{k_w,j}} \right) \right]^{\underline{k}}$$

$$= \left[\frac{\lambda}{N} \sum_{(h_1, \ldots, h_w) \in G^{\times w}} \left(U_{h_1} \otimes U_{h_2} \otimes \cdots \otimes U_{h_w} \right)^{\dagger} \left(\sigma_{\alpha_1} \otimes \sigma_{\alpha_2} \otimes \cdots \otimes \sigma_{\alpha_w} \right) \right.$$

$$\left. \times \left(U_{h_1} \otimes U_{h_2} \otimes \cdots \otimes U_{h_w} \right) \right]^{\underline{k}}$$

$$= \left[\Pi_{G^{\times w}} \left(\sigma_{\alpha_1} \otimes \sigma_{\alpha_2} \otimes \cdots \otimes \sigma_{\alpha_w} \right) \right]^{\underline{k}}$$

$$= \mathbf{0}^{\underline{k}} = \mathbf{0}. \tag{15.34}$$

For the last step in Eq. (15.34) we used that since M is an OA of strength w, the list $((g_{k_1,j}, \ldots, g_{k_w,j}))_{j=1}^{N}$ contains every element of $G^{\times w}$ exactly λ times. Hence, the average $\Pi_{G^{\times w}}(\sigma_{\alpha_1} \otimes \sigma_{\alpha_2} \otimes \cdots \otimes \sigma_{\alpha_w})$ is an average over the group $G^{\times w}$ acting via an irreducible representation and therefore equal to $\mathbf{0}$. $\square$

The following theorem [HSS99, Theorem 4.6] establishes a close relationship between OAs and classical error-correcting codes. We recall some notions before stating it. A linear code over the finite field $\mathbb{F}_q$ is a k-dimensional subspace of the vector space $\mathbb{F}_q^n$. We consider finite fields of $q = d^2$ only. In this case the additive group of the finite field $\mathbb{F}_q$ is isomorphic to $\mathbb{Z}_d \times \mathbb{Z}_d$; we again use the irreducible representation of the generalized Pauli group. The space $\mathbb{F}_q^n$ is endowed with a metric called the Hamming distance. It is defined as follows: for $x = (x_1, \ldots, x_n) \in \mathbb{F}_q^n$ we have that $\text{wt}(x) := |\{i \in \{1, \ldots, n\} : x_i \neq 0\}|$. The minimum distance of a linear code C is defined by $d = d_{\min} := \min\{\text{wt}(c) : c \in C, c \neq \mathbf{0}\}$, where $\mathbf{0}$ denotes the zero vector. In this situation we say that C is an $[n, k, d]_q$ code. We need the fact that a $[n, k, d]_q$ linear code can be described by a generator matrix G of size $n \times k$ with entries from $\mathbb{F}_q$. The matrix G defines the embedding from $\mathbb{F}_q^k$ to $\mathbb{F}_q^n$; the code words $c \in C$ are the images of the vectors $m \in \mathbb{F}_q^k$, i.e., $c = Gm$. We need one more definition, which is the dual code $C^\perp$ of C defined by $C^\perp := \{x \in \mathbb{F}_q^n : x \cdot y = 0 \text{ for all } y \in C\}$; the dot product $x \cdot y$ is given by $\sum_{i=1}^n x_i y_i$. In the following we refer to the minimum distance $d^\perp$ of the dual code as the dual distance.

Theorem 15.4 (OAs from linear codes). *Let C be a linear $[n, k, d]_q$ code over $\mathbb{F}_q$ with dual distance $d^\perp$. Arrange the codewords of C into the columns of a matrix $A \in \mathbb{F}_q^{n \times q^k}$. Then A is an $OA(q^k, n, q, d^\perp - 1)$.*

To obtain a decoupling scheme using a minimal number of pulses we have to find a code $[n, k, d]_q$ such that k is minimal and the dual distance $d^\perp$ is at least $t+1$. This may be formulated in terms of the dual code that has parameters $C^\perp = [n, n-k, d^\perp]_q$. The dual code $C^\perp$ should contain the maximal possible number of code words for given n and $d^\perp$. This question is one of the central optimization problems in the theory of error-correcting codes. To find such optimal or best known codes one could, e.g., use the computer algebra system Magma [BCP97] that contains a table of the best linear codes known (i.e., with the maximal number of codewords) for given length and minimal distance.

We now consider a quantum system consisting of n qubits that are governed by a pair-interaction Hamiltonian. For such a system we can construct decoupling schemes using N pulses from an array $OA(N, n, 4, 2)$. Hence, in order to apply Theorem 15.4 and Theorem 15.7 we have to find a linear code C over $\mathbb{F}_4$ for which the parameters are $[n, k, d]$ and for which the dual distance is at least 3. This can be done with the help of Hamming codes [R08, RW06]. For every $m \in \mathbb{N}$ there is an array $OA(4^m, (4^m - 1)/3, 4, 2)$. The columns of this OA are codewords of the dual code of a Hamming code.

To obtain a decoupling scheme for a quantum system consisting of n qubits, where n is an arbitrary natural number, i.e., not necessarily of the form $n = (4^m - 1)/3$, we proceed as follows: First let $m \in \mathbb{N}$ be the unique integer such that $n \leq \frac{4^m - 1}{3} \leq 4n$. Then construct the orthogonal array with parameters $OA(4^m, (4^m - 1)/3, 4, 2)$ for bang-bang controls. These results shows that the complexity of decoupling for general pair-interaction Hamiltonians acting on n qubits scales at most linearly in n for bang-bang controls.

15.2.7 Decoupling schemes with few different pulses

The construction of orthogonal arrays from Hamming codes over the alphabet $\mathbb{F}_{2^{2\alpha}}$, where $\alpha = \lceil \log d \rceil \in \mathbb{N}$, can be combined together with Gray codes for $\mathbb{F}_{2^{2m}}$ to obtain a specific sequence which has the advantage that only very few different pulses have to be actually applied. An example of a scheme that requires only two different pulses for decoupling of a Hamiltonian on one qudit (i.e., a Hamiltonian on $\mathbb{C}^d$) is shown in Fig. 15.2. This idea can be generalized to the case of several qudits, which is described in more detail in [R08]. The general construction yields schemes for systems consisting of $n = (2^{2m} - 1)/3$ qudits (each of dimension 2^α), which use only $4 \log n$ different pulses. For α constant, the total number of pulses required by these schemes is $O(n)$. To summarize, we have the following theorem:

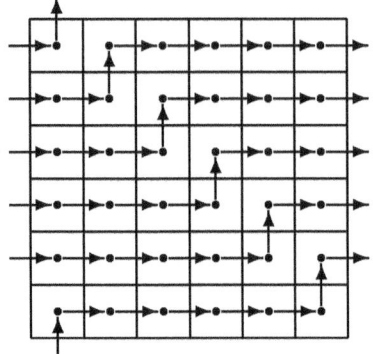
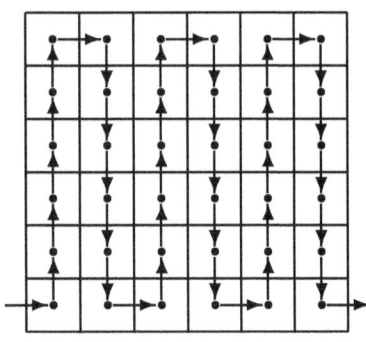

Fig. 15.2. Decoupling of a Hamiltonian on $\mathbb{C}^d$ can be achieved using two different pulses (the case $d = 6$ is shown). Pulses correspond to elements in $\mathbb{Z}_d^2$, which are generated (up to overall phases) by σ_x and σ_z. The left square shows the sequence in which the pulses have to be applied. In each square a horizontal move $\rightarrow$ corresponds to multiplication with $\sigma_x^{(i)}$ and a vertical move $\uparrow$ to multiplication with $\sigma_z^{(i)}$ for $i = 1, 2$. An alternative sequence is given in the right square. Decoupling of a bi-partite Hamiltonian on $\mathbb{C}^d \otimes \mathbb{C}^d$ can be achieved using four different pulses. Here pulses correspond to elements in $\mathbb{Z}_d^4 = \mathbb{Z}_d^2 \times \mathbb{Z}_d^2$ generated by $\sigma_x^{(i)}$, $\sigma_z^{(i)}$ for $i = 1, 2$. A cyclic sequence can be obtained by performing one step in the first copy of $\mathbb{Z}_d^2$ followed by one full cycle in the second copy and so on. Reprinted with permission from [R08]. Copyright 2008, American Institute of Physics.

Theorem 15.5. *Let H be a pair-interaction Hamiltonian on a system that consists of n interacting d-dimensional subsystems, where d is constant. Then there exists a decoupling scheme for H that uses a total number of $O(n)$ pulses. Moreover, the sequence of pulses in this decoupling scheme consists of only $O(\log n)$ different pulses which have to be applied in a certain order to achieve decoupling.*

Table of best known decoupling schemes for small systems. For small numbers of qubits we can determine the best known decoupling schemes that are based on orthogonal arrays. Tables of orthogonal arrays can for instance be found in [HSS99], in particular we make use of Table 12.3 from [HSS99]. Usually for a fixed number of qubits – which corresponds to the number of runs in the language of orthogonal arrays – there are several constructions available that may result in decoupling schemes having different numbers of pulses. Table 15.2 lists the best known decoupling schemes based on orthogonal arrays for a number of qubits between 2 and 85. Interestingly, it is possible that for a given number of qubits two different schemes might exist, where one scheme has a smaller total number of pulses that have to be applied to achieve decoupling, whereas the other has a smaller total number of different pulses out of which the decoupling scheme is formed. To have a small number of different pulses can be an advantage, especially when optimizing the pulse is a hard task.

A few comments are in order about the constructions mentioned in Table 15.2. The names of the constructions are taken from [HSS99]. Rao–Hamming refers to the construction of orthogonal arrays from Hamming codes. For $m = 1, 2, \ldots$, it yields schemes with 4^m pulses and up to $(4^m - 1)/3$ qubits. Addleman–Kempthorne refers to a construction described in [HSS99, Section 3.3] and for $m = 1, 2, \ldots$, it yields schemes with $2 \cdot 4^m$ pulses and up to $2(4^m - 1)/3 - 1$

Table 15.2. *Best known decoupling schemes for small numbers of qubits*

# qubits	# pulses	# different pulses	Construction
2–5	16	4	Rao–Hamming
6–9	32	32	Addleman–Kempthorne
6–21	64	6	Rao–Hamming
10–13	48	48	Difference matrix
22–41	128	128	Addleman–Kempthorne
22–85	256	8	Rao–Hamming

qubits. Finally, the difference matrix mentioned in the table is obtained by a construction due to de Launey [D86], see also [HSS99, Theorem 6.64]. Regarding the number of different pulses, given in the third column, we can show a significant reduction as compared to the total number of pulses only for the class of Rao–Hamming schemes. As shown in [R08] using Gray codes, the number of pulses given is computed as $\log N$, where N is the length of the pulse sequence. For the Addleman-Kempthorne and difference matrix constructions an analogous way of cycling through the set of columns in the orthogonal array is not known to exist, therefore the worst-case bound for the number of different pulses has been used.

15.3 Combinatorial bounded strength decoupling

15.3.1 Eulerian decoupling schemes

We now combine the ideas of Eulerian decoupling and orthogonal arrays. Let G be the index group of a nice error basis $\{U_g\}_{g \in G}$ and S a generating set of G, i.e., any element of G can be written as as product of elements of S. The Cayley graph $\Gamma(G, S)$ of G with respect to S is a directed graph whose vertices are labeled by the group elements and whose edges are labeled by the generators. More precisely, the vertex g is joined to the vertex h if and only if $gh^{-1} = s$ for some $s \in S$, i.e., $g = sh$.

We assume that we have the ability to physically implement the generators $s \in S$, i.e., to implement the unitaries U_s by the application of some suitably chosen control Hamiltonians $h_s(t)$ over Δ:

$$U_s = u_s(\Delta), \qquad (15.35)$$

where

$$u_s(\delta) = \mathcal{T}\left\{\exp\left(-i\int_0^\delta h_s(\tau)d\tau\right)\right\} \qquad (15.36)$$

for $\delta \in [0, \Delta]$. The choice of the control Hamiltonians $h_s(t)$ is not unique. This allows for additional flexibility for the concrete implementation. Once a choice of the control Hamiltonians is made, the control action is determined by assigning a cycle time and a rule for switching the Hamiltonians $h_s(t)$ during the cycle subintervals.

Chapter 4 shows that decoupling can be achieved by sequentially implementing generators so that they follow an Eulerian cycle in $\Gamma(G, S)$. An Eulerian cycle is defined as a cycle that uses each edge exactly once. Because a Cayley graph is regular, it always has an Eulerian cycle, whose length is necessarily $N = |G||S|$. For our purposes, we use a slightly more general definition: an Eulerian cycle with multiplicity λ is a cycle that uses each edge exactly λ times.

We are now ready to define Eulerian orthogonal arrays:

Definition 15.4 (Eulerian orthogonal array). *A $n \times N$-matrix $M = (g_{kj})$ with entries from the group G is said to be an Eulerian orthogonal array of strength t if and only if for all t-tuples $(k_1, k_2, \ldots, k_t)$ with $1 \leq k_1 < \cdots < k_t \leq n$ there is a generating set $S_{k_1,k_2,\ldots,k_t}$ of $G^{\times t}$ such that the list of group elements*

$$\left((g_{k_1,j}, g_{k_2,j}, \ldots, g_{k_t,j})\right)_{j=1}^{N} \tag{15.37}$$

defines an Eulerian cycle in the Cayley graph $\Gamma(G^{\times t}, S_{k_1,k_2,\ldots,k_t})$.

Note that these conditions imply that M is a (usual) orthogonal array of strength t.

We assume that we have the ability to implement the group elements $s \in S$, i.e., to implement the unitaries U_s on the individual qudits by the application of control Hamiltonians $h_s(t)$ over Δ. This means that we have the ability to switch on the control Hamiltonians $h_g(t)$ on any qudit, i.e., $I \otimes \cdots \otimes I \otimes h_g(t) \otimes I \otimes \cdots \otimes I$.

We define decoupling according to an Eulerian orthogonal array $M = (g_{kj})$ by setting the cycle time $T_c = N\Delta$ and by assigning the control propagators as follows:

$$U_c((j-1)\Delta + \delta) = \left(u_{s_{1j}}(\delta) \otimes \cdots \otimes u_{s_{nj}}(\delta)\right) U_c((j-1)\Delta) \tag{15.38}$$

where $\delta \in [0, \Delta)$ and $s_{kj} = g_{kj}^{-1} g_{k,j+1}$ for $j = 1, \ldots, N-1$ and $s_{kN} = g_{kN}^{-1} g_{k1}$. The tuples $(s_{k_1,j}, \ldots, s_{k_t,j})$ are the edges in the Eulerian cycle defined by the rows $k_1, \ldots, k_t$ of M.

Theorem 15.6 (Decoupling with Eulerian OAs). *Let $M = (g_{kj})$ be an Eulerian orthogonal array over G of size $n \times N$ and strength t. Let $\mathscr{Q}_M$ denote the control action that results from the control propagator defined in (15.38). Then we have*

$$\mathscr{Q}_M(H_S) = 0 \tag{15.39}$$
$$\mathscr{Q}_M(S_a) = 0 \tag{15.40}$$

for all system Hamiltonians H_S of the form in Eq. (15.28) and all error terms S_a of the form in Eq. (15.29).

Proof. Due to linearity of $\mathscr{Q}_M$ it suffices to show that $\mathscr{Q}_M(\sigma_{\underline{\alpha}}^{\underline{k}}) = 0$ for all tuples $\underline{k} = (k_1, \ldots, k_w)$ with $k_1 < \cdots < k_w$ and all tuples $\underline{\alpha} = (\alpha_1, \ldots, \alpha_w)$ for all $w = 1, \ldots, t$. Denote by $M_{\underline{k}}$ the submatrix of M with rows in $\underline{k}$.

Then we have

$$\mathscr{Q}_M(\sigma_{\underline{\alpha}}^{\underline{k}}) = \left[\mathscr{Q}_{M_{\underline{k}}}(\sigma_{\alpha_1} \otimes \cdots \otimes \sigma_{\alpha_w})\right]^{\underline{k}}$$
$$= \left[\Pi_{G^{\times w}}\left(F_{S_{\underline{k}}}(\sigma_{\alpha_1} \otimes \cdots \otimes \sigma_{\alpha_w})\right)\right]^{\underline{k}}$$
$$= 0^{\underline{k}} = 0. \tag{15.41}$$

Similarly to the situation in the bang-bang case, Eq. (15.41) is true because $\sigma_{\underline{k}}^{\alpha}$ acts on qudits $k_1, \ldots, k_w$ only. The decomposition $\mathscr{D}_{M_{\underline{k}}} = \Pi_{G^{\times w}} \circ F_{S_{\underline{k}}}$ directly corresponds to the decomposition in Chapter 4 into the operator defined by the generators and the symmetrization operator. $M_{\underline{k}}$ defines an Eulerian cycle in the Cayley graph $\Gamma(G^{\times w}, S_{\underline{k}})$ since M is an Eulerian orthogonal matrix. This implies that the term $\sigma_{\alpha_1} \otimes \cdots \otimes \sigma_{\alpha_w}$ is mapped onto $\mathbf{0}$ by the application of $\mathscr{D}_{M_{\underline{k}}}$. □

15.3.2 Schemes from linear error-correcting codes

We construct Eulerian orthogonal arrays using linear error-correcting codes.

Theorem 15.7 (Eulerian OAs from linear codes). *Let C be a $[n, k, d]_q$-code with dual distance $d^\perp$ and G be a generator matrix for C. Let $\mathscr{C} := (m_0, \ldots, m_{N-1})$ be an Eulerian cycle in the Cayley graph $\Gamma(V, S)$ with multiplicity 1, where the group is $V := \mathbb{F}_q^k$ and the generating set is the group itself, i.e., $S := \mathbb{F}_q^k$. The length of such an Eulerian cycle is necessarily $N = q^{2k}$. Set $t := d^\perp - 1$. Then the $n \times N$ matrix M whose columns are defined to be Gm_j for $j = 0, \ldots, N-1$ is an Eulerian orthogonal array over $\mathbb{F}_q$ of strength t. Furthermore, we have $S_{k_1, \ldots, k_t} = G^{\times t}$ for all t-tuples $(k_1, \ldots, k_t)$ with $1 \leq k_1 < \cdots < k_t \leq n$.*

Proof. Since $\mathscr{C} = (m_0, \ldots, m_{N-1})$ is an Eulerian cycle all elements of $V = \mathbb{F}_q^k$ appear exactly q^k (corresponding to the size $|S| = q^k$ of the generating set S) times in $\mathscr{C}$. Therefore, the column vector Gm appears exactly q^k times in M for all $m \in \mathbb{F}_q^k$. It now follows from Theorem 15.4 that M is an orthogonal array; its multiplicity is just q^k times the multiplicity of an orthogonal array constructed as in Theorem 15.4.

Let m be an arbitrary element of $V = \mathbb{F}_q^k$ and $I_m := \{j \mid m_j = m\}$. Then every element of $S = \mathbb{F}_q^k$ appears exactly once in the list $(m_{j+1} - m_j \mid j \in I_m)$ because $\mathscr{C}$ is an Eulerian cycle in $\Gamma(V, S)$ with multiplicity one (the addition $j + 1$ is done modulo N). Consequently, the list of transitions that occur in M from all columns of the form Gm, i.e., $(Gm_{j+1} - Gm_j \mid j \in I_m)$ is independent of m and is equal (up to reordering the columns) to the orthogonal array $M' := (Ge \mid e \in \mathbb{F}_q^k)$; it follows from Theorem 15.4 that M' is an orthogonal array. This proves that M is Eulerian and also that $S_{k_1, \ldots, k_t} = G^{\times t}$ for all t-tuples since M' is an orthogonal array of strength t. □

15.4 Conclusions and future directions

We have demonstrated how combinatorial structures such as Hadamard matrices, difference matrices, and orthogonal arrays can be used to construct efficient first-order decoupling schemes for composite quantum systems. An interesting question for future research is whether efficient higher-order decoupling schemes can be obtained with the help of the above or other combinatorial structures. Higher-order here refers to the higher-order terms in the Magnus expansion of a time-dependent Hamiltonian. It is known that it is possible to achieve second-order decoupling by picking a symmetric sequence of control pulses that can, e.g., be obtained by first running a control sequence forward and then backward (in the case of Eulerian decoupling we have to choose control Hamiltonians that are symmetric), but there might be more efficient ways to achieve second-order decoupling. The case of higher-order decoupling is an interesting field of research.

The focus in this chapter was on generality, often resulting in the schemes that can be used to decouple arbitrary (even unknown) Hamiltonians and couplings to the environment. For future research, it would be important to construct more efficient schemes for concrete cases by taking into account the *special* structure of the system Hamiltonian and the error operators. The majority of the schemes we have presented here both switch off the system Hamiltonian and remove the couplings. It would be critical for applications in quantum information processing to examine for concrete cases how to retain enough coupling between the subsystems to be able to implement gates, while at the same time completely removing the couplings to the environment to avoid decoherence.

Part V

Alternative quantum computation approaches

Chapter 16

Holonomic quantum computation

Paolo Zanardi

16.1 Introduction

In the standard picture of quantum information processing, quantum computation is physically realized by fast switching of external fields coupled to the quantum computer S, as well as by modulation of *local* interactions among the subsystems of S. In other words, the experimenter controls a basic set of time-dependent Hamiltonians which are activated at will in order to implement suitable sequences of quantum logic gates. As an alternative to this standard *dynamical* view of quantum information processing, *geometrical* and *topological* approaches have been developed in which quantum gates are realized in terms of operations of a purely geometrical/topological nature (see Chapters 19 and 20). These approaches are based, on the one hand, on the use of Abelian Berry phases [B84] or non-Abelian adiabatic quantum holonomies [WZ84], or, on the other hand, on the use of quantum statistics (anyons) [K03a] and the unusual properties of topological quantum field theory.

A striking feature of these geometric and topological proposals is this: a nontrivial quantum evolution can be obtained over the vector space $\mathscr{C}$ of quantum codewords even in the presence of a trivial Hamiltonian, e.g., $H|_{\mathscr{C}} = 0$. This peculiarity is due to the existence of an underlying *global* geometrical/topological structure. One way to understand how a theory with vanishing Hamiltonian can exhibit nontrivial unitary transformations lies in the Feynman path-integral approach to quantum field theory, where the theory is constructed from a Lagrangian that involves only first derivatives in time, but may nevertheless have nontrivial global features, as in the Aharonov–Bohm effect.

Aside from being conceptually fascinating, these computational strategies have several inherent *fault-tolerant* features. This latter attractive characteristic stems, basically, from the fact that some topological as well as geometric quantities are inherently stable against local perturbations. This in turn suggests that quantum information processing based on these ingredients might be inherently stable against special classes of computational errors.

Quantum Error Correction, ed. Daniel A. Lidar and Todd A. Brun. Published by Cambridge University Press.
© Cambridge University Press 2013.

The goal of this chapter is to give an account of one of these approaches: *holonomic quantum computation* (HQC) [ZR99a, PZ01]. We first provide an introduction to the main ideas (Section 16.2); these will be illustrated by means of the implementation proposal using semiconductor quantum dots [SZZ+03a, SZZ+03b] (Section 16.3). We then briefly address the issue of inherent fault tolerance using the same implementation scheme as a testbed (Section 16.4). In Section 16.5, we discuss the viability of hybrid strategies merging HQC with the other error-correcting and error-avoiding techniques. This issue is taken up in much greater detail in Chapter 17. The chapter concludes with a few final remarks. A concise summary of some basic notions concerning quantum geometric phases is provided in the Appendix at the end of this chapter.

In the spirit of this book, our goal is pedagogical, and no claim of completeness in covering this field is made. In particular, of the many proposals for physical implementation of HQC, only the semiconductor one mentioned above is considered, for the sake of giving concrete illustrations of the general, somewhat abstract ideas involved in HQC.

16.2 Holonomic quantum computation

Holonomic quantum computation is a fully geometric approach to quantum information processing wherein *the entire quantum logic network is built by means of quantum non-Abelian holonomies* (see Appendix).

The computational subspace $\mathscr{C}$ can be thought of as the lowest-energy manifold of a highly symmetric quantum system; from this point of view, HQC is a kind of *ground-state* computation. This fact points to the potential existence of a fault-tolerant feature of HQC due to energy gaps and even spontaneous relaxation mechanisms.

The general quantum information processing setup can be easily formulated in a geometric fashion. Any quantum evolution generated by a time-dependent Hamiltonian $H(t) := H(\lambda(t))$ can be formally written as

$$U(T) = \mathbf{T} \exp\left[-i \int_0^T dt H(t)\right]. \tag{16.1}$$

One can associate a path in the space $\mathscr{M}$ to this operator, whose points describe the configurations of suitable "control fields" λ, on which the Hamiltonian depends. The control parameter space $\mathscr{M}$ is hereafter referred to as the control manifold. If one is able to drive the control field configuration $\lambda \in \mathscr{M}$ through a (smooth) path $\gamma \colon [0, T] \to \mathscr{M}$, then a family $H(t) := H_{\gamma(t)}$ is defined along with the associated unitary U_γ.

In this framework, quantum information processing can be described as the experimenter's capability to generate a small set $\{\gamma_i\}_{i=1}^g$ of basic paths such that sequences of the corresponding U_{γ_i}s (basic quantum gates) approximate with arbitrarily high accuracy any unitary transformation on the quantum state space. *HQC is based on a series of restrictions: the paths in $\mathscr{M}$ are loops, i.e., $\gamma(T) = \gamma(0)$, they are adiabatically traversed, and the degeneracy structure of the family $\{H(\gamma(t))\}_{t=0}^T$ is constant, i.e., there are no level-crossings.*

We will see how, by encoding quantum information in one of the eigenspaces of a degenerate Hamiltonian H, one can in principle attain full quantum computational power by using holonomies only. These holonomic computations are realized by moving along loops in a suitable space $\mathscr{M}$ of control parameters labeling the family of Hamiltonians to which H belongs.

Attached to each point $\lambda \in \mathcal{M}$ there is a quantum code, and this bundle of codes is endowed with a nontrivial global topology described by a non-Abelian gauge field potential A. For generic A, the associated holonomies will allow for universal quantum computing.

Let us consider the case when $H_\lambda = H_{\gamma(t)}$ has R different eigenvalues $\{\varepsilon_i\}_{i=1}^R$ with degeneracies $\{n_i\}$. If $\Pi_i(\lambda)$ denotes the projector over the eigenspace $\mathcal{H}_i(\lambda) := \text{span}\{|\psi_i^\alpha(\lambda)\rangle\}_{\alpha=1}^{n_i}$ of H_λ, one has the spectral λ-dependent resolution $H_\lambda = \sum_{i=1}^R \varepsilon_i(\lambda)\Pi_i(\lambda)$. The state vector evolves according to the time-dependent Schrödinger equation $i\partial_t|\psi(t)\rangle = H_{\gamma(t)}|\psi(t)\rangle$. We restrict ourselves to the case in which the loops γ are *adiabatic*. Then it turns out [WZ84] that any initial preparation $|\psi_0\rangle \in \mathcal{H}$ will be mapped (neglecting corrections to the adiabatic limit associated with the finiteness of T), after the period T, onto: $|\psi(T)\rangle = U(T)|\psi_0\rangle$, $U(T) = \oplus_{l=1}^R e^{i\phi_l(T)}\Gamma_{A_l}(\gamma)$, where $\phi_l(T) := \int_0^T d\tau\, \varepsilon_l(\lambda_\tau)$ is the dynamical phase and

$$\Gamma_{A_l}(\gamma) := \mathbf{P}\exp\int_\gamma A_l \in U(n_l), \quad (l = 1, \ldots, R) \tag{16.2}$$

is called the *holonomy* associated with the loop γ (here $\mathbf{P}$ denotes path ordering). In particular, when $|\psi_0\rangle \in \mathcal{H}_l$, the final state belongs to the *same* eigenspace. In the following we will drop dynamical phases and focus on the geometric contribution (16.2). Notice that for $n = 1$ this term is just the Berry phase, while for $n_l > 1$ the holonomy $\Gamma_{A_l}(\gamma)$ is sometimes referred to as a *non-Abelian* geometric phase. The matrix-valued form A_l appearing in Eq. (16.2) is known as the *adiabatic connection*, and it is given by $A_l = \Pi_l(\lambda)\, d\,\Pi_l(\lambda) = \sum_\mu A_{l,\mu}\, d\lambda_\mu$, where

$$(A_{l,\mu})^{\alpha\beta} := \langle\psi_l^\alpha(\lambda)|\partial/\partial\lambda^\mu|\psi_l^\beta(\lambda)\rangle \tag{16.3}$$

$((\lambda_\mu)_{\mu=1}^d$ are local coordinates on $\mathcal{M}$). The A_ls are non-Abelian gauge potentials that allow for parallel transport of vectors over $\mathcal{M}$. Indeed, the linear mapping (16.2) of the fiber $\mathcal{H}_l$ onto itself is nothing but the parallel transport of the vector $|\psi_0\rangle$ associated with the connection form A_l.

We should stress that the necessity of a (large) degenerate eigenspace is very demanding from an experimental point of view. Due to the symmetry constraints that it involves, degeneracy is a singular case, while nondegeneracy is the generic one. If one slightly perturbs a nondegenerate Hamiltonian H, the resulting operator is, generically, still nondegenerate. Conversely, degeneracy is lifted by generic perturbations.

In the following we will take degeneracy for granted and will fix our attention on a given n-dimensional eigenspace $\mathcal{C}$ of H. The state-vectors in $\mathcal{C}$ will be our quantum codewords, and $\mathcal{C}$ will be referred to as the *code*.

Let us recall what computational universality means in this context. For all $U \in U(n)$ and given ϵ, one wants to find a (classically computable) finite sequence $\gamma_i(k)$, $k = 1, \ldots, N(U, \epsilon)$ of basic gates such that $\|U - \prod_{k=1}^N U_{\gamma_i(k)}\| \leq \epsilon$. A particular computation U is efficiently implementable if $N(U, \epsilon)$ has a mild dependence (poly-logarithmic) on ϵ and on the state-space dimension.

Here our aim is to carry out the largest number of computations over $\mathcal{C}$ by exploiting only non-Abelian holonomies (16.2) generated by adiabatic loops in $\mathcal{M}$. The first question is: *How many transformations can be obtained, by Eq. (16.2), as γ varies over the space of loops in $\mathcal{M}$?*

To address this point, let us begin by considering the properties of the holonomy map $\Gamma_A \colon L_{\lambda_0} \mapsto U(n)$, where $L_{\lambda_0} := \{\gamma \colon [0, 1] \mapsto \mathcal{M} / \gamma(0) = \gamma(1) = \lambda_0\}$ is the space of loops over $\mathcal{M}$ ($T = 1$). From Eq. (16.2) one finds:

(i) $\Gamma_A(\gamma_2 \cdot \gamma_1) = \Gamma_A(\gamma_2)\,\Gamma_A(\gamma_1)$; by composing loops in $\mathcal{M}$ one obtains a unitary evolution that is the product of the evolutions associated with the individual loops,
(ii) $\Gamma_A(\gamma_0) = I$; remaining at rest in the parameter space corresponds to no evolution at all,
(iii) $\Gamma_A(\gamma^{-1}) = \Gamma_A^{-1}(\gamma)$; traversing the path γ with reversed orientation yields the inverse holonomy,
(iv) $\Gamma_A(\gamma \circ \varphi) = \Gamma_A(\gamma)$, where φ is any diffeomorphism of $[0, 1]$; as long as adiabaticity holds, the holonomy does not depend on the speed at which the path is traveled but just on the geometry of the path.

From the properties listed above it is easy to show that the set $\mathrm{Hol}(A) := \Gamma_A(L_{\lambda_0})$ is a *subgroup* of $U(n)$. Such a subgroup is known as the *holonomy group* of the connection A. When the holonomy group coincides with $U(n)$, the connection A is called *irreducible*. The notion of irreducibility plays a crucial role in HQC in that it clearly corresponds to the computational notion of universality. In order to assess whether this condition is fulfilled by a given connection, it is useful to consider the *curvature* 2-form F associated with the 1-form connection A, whose components are

$$F_{\mu\nu} = \partial_\mu A_\nu - \partial_\nu A_\mu + [A_\mu, A_\nu]. \tag{16.4}$$

The relation between curvature and irreducibility is given by the following statement [N90]: *the linear span of the components $F_{\mu\nu}$ is the Lie algebra of the holonomy group*. It follows in particular that when the $F_{\mu\nu}$s span $u(n)$, the connection is irreducible.

Of course, this result is merely existential and it does not provide an explicit recipe for obtaining the desired transformations. Nevertheless, it is rather remarkable conceptually. Indeed, in view of restrictions involved in the HQC scheme, universal quantum information processing over the whole state space $\mathcal{H}$ is unattainable. The best one can hope for is some sort of *encoded universality*. In other words, one should be able to identify a subspace $\mathcal{C} \subset \mathcal{H}$ (the code) such that universality can be attained once one restricts to $\mathcal{C}$. This is exactly what we have just shown, in the irreducible case, for all the Hamiltonian eigenspaces $\mathcal{C}$.

Notice that so far no assumption about a possible underlying multi-partite (i.e., tensor-product) structure of $\mathcal{C}$ has been made. On the other hand, entanglement is believed to be one of the essential ingredients of quantum information processing, so one is naturally led to consider the viability of HQC involving different subsystems. Using standard results for quantum information processing on multi-partite systems, HQC is universal if, by using holonomies only, arbitrary single-qubit gates and an entangling gate between arbitrary pairs of qubits can be obtained.

In order to illustrate HQC universality more constructively and at the same time give an example of a candidate physical architecture proposed for implementation, we discuss in the next section the HQC scheme based on semiconductor quantum dots introduced in [SZZ+03a, SZZ+03b], and to which the reader is referred for further details.

16.3 HQC with quantum dots

In the implementation proposal for HQC that we consider in this section, a central role is played by the holonomic structure introduced in [DCZ01], as well as by the exciton–exciton interaction mechanism exploited in the all-optical semiconductor-based quantum information processing scheme proposed in [BIZ+00]. *The proposed quantum hardware is an array of semiconductor quantum dots (QDs) and the computational operations are interband optical excitations, also called excitonic transitions.*

An exciton is a Coulomb-correlated electron-hole pair produced by promoting an electron from the valence band with total angular momentum $J_{tot} = 3/2$ to the conduction band with $J_{tot} = 1/2$. For a GaAs-based quantum-dot structure, the confining potential along the growth (z) direction breaks the symmetry and lifts the degeneracy in the valence band; the states ($|J_{tot}, J_z\rangle$) of the quadruplet $J_{tot} = 3/2$ are then energetically separated into $J_z = \pm 3/2$ [heavy holes (HH)] and $J_z = \pm 1/2$ [light holes (LH)].

A properly tailored laser excitation may promote electrons from the valence to the conduction band in an energy-selective fashion. For the HH, the only allowed transitions are $|\frac{3}{2}, \frac{3}{2}\rangle \to |\frac{1}{2}, \frac{1}{2}\rangle$, $|\frac{3}{2}, -\frac{3}{2}\rangle \to |\frac{1}{2}, -\frac{1}{2}\rangle$. Here, the first transition is produced by light with left circular polarization (usually referred to as σ^-) while the second transition is produced by light with right circular polarization (σ^+). In contrast, due to the different structure of their wave functions for the LH we have more allowed transitions; in the HH, we have $|\frac{3}{2}, \frac{1}{2}\rangle \to |\frac{1}{2}, -\frac{1}{2}\rangle$, $|\frac{3}{2}, -\frac{1}{2}\rangle \to |\frac{1}{2}, \frac{1}{2}\rangle$. These transitions may be induced by light propagating along the z-direction with circular (left or right) polarization. Moreover, for light propagating in the xy plane with polarization along z (σ^0), the following transitions are also allowed (and experimentally observed): $|\frac{3}{2}, \frac{1}{2}\rangle \to |\frac{1}{2}, \frac{1}{2}\rangle$, $|\frac{3}{2}, -\frac{1}{2}\rangle \to |\frac{1}{2}, -\frac{1}{2}\rangle$. As a result, we see that by exciting LH electrons with three different kinds of light, namely left and right circular polarization as well as linear polarization along z, we can induce three different transitions with the same energy: $|G\rangle \mapsto |E^\alpha\rangle$ ($\alpha = \pm, 0$), where $|G\rangle$ denotes the ground state of the semiconductor crystal. The allowed optical transitions as well as the corresponding energy-level structure for HH and LH are schematically depicted in Fig. 16.1a.

For the case of a laser excitation resonant with the three degenerate LH transitions, the corresponding light–matter coupling Hamiltonian is (in the rotating frame) of the form:

$$H_{int} = -\hbar \sum_{\mu=0,\pm} (\Omega_{\mu, LH} |E^\mu\rangle\langle G| + \text{h.c.}). \tag{16.5}$$

This Hamiltonian has the same level structure as the one for trapped-ion internal levels analyzed in [DCZ01]: for each value of the Rabi couplings Ω_μ it admits a pair of *dark* states, i.e., two states $|D_\alpha(\Omega)\rangle$ ($\alpha = 0, 1$) corresponding to a zero eigenvalue. These dark states – at a distinguished point in the Ω space – will encode our qubit.

Quantum control operations are realized by the holonomies associated with the $u(2)$-valued connection A defined by $(A_\mu)_{\alpha\beta} = \langle D_\alpha|\partial/\partial\Omega^\mu|D_\beta\rangle$ ($\alpha, \beta = 0, 1; \mu = 0, \pm$). Our computational basis is given by $|1\rangle := |E^+\rangle$ and $|0\rangle := |E^-\rangle$. The state $|E^0\rangle$ plays the role of an *ancilla*, used as an auxiliary resource.

To achieve a complete set of single-qubit gates, it is sufficient to enact a pair of non-commuting single-qubit gates U_1 and U_2. For the first gate we choose $\Omega_- = 0$, $\Omega_+ = -\Omega \sin(\theta/2) e^{i\varphi}$ and $\Omega_0 = \Omega \cos(\theta/2)$. The dark states are given by $|E^-\rangle$ and $|\psi\rangle = \cos(\theta/2)|E^+\rangle + \sin(\theta/2) e^{i\varphi}|E^0\rangle$. By evaluating the connection associated with this two-dimensional degenerate eigenspace, it is not difficult to see that the unitary transformation $U_1 = e^{i\phi_1|E^+\rangle\langle E^+|}$ ($\phi_1 = \frac{1}{2} \oint \sin\theta \, d\theta \, d\psi$) can be realized as a holonomy. For the second gate we choose $\Omega_- = \Omega \sin\theta \cos\varphi$, $\Omega_+ = \Omega \sin\theta \sin\varphi$ and $\Omega_0 = \Omega \cos\theta$. The dark states are

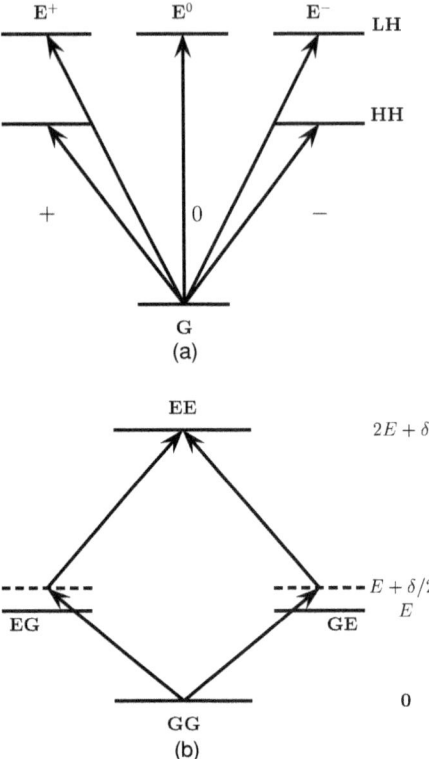

Fig. 16.1. Schematic diagram of the energy-level structure of LH and HH valence-band states (a) and of a typical two-photon process (b) in GaAs-based semiconductor macroatoms. Reprinted with permission from [SZZ+03b]. Copyright 2003, American Physical Society.

now given by $|\psi_1\rangle = \cos\theta\cos\varphi|E^-\rangle + \cos\theta\sin\varphi|E^+\rangle - \sin\theta|E^0\rangle$ and $|\psi_2\rangle = \cos\varphi|E^+\rangle - \sin\varphi|E^-\rangle$. In this case, the unitary transformation $U_2 = e^{i\phi_2\sigma_y}$ ($\phi_2 = \oint \sin\theta d\theta d\psi$) can be implemented.

The next key step is the implementation of the two-qubit gate. In order to achieve this goal, a physical coupling mechanism between excitons is needed. We resort to exciton–exciton dipole coupling in semiconductor quantum dots, exploited in the excitonic quantum information processing proposal [BIZ+00]. Coulomb dipole–dipole interaction between quantum dots implies that the presence of an exciton in one dot (e.g., dot b) produces a shift in the energy level of the other dot (e.g., dot a) from E to $E + \delta$; the total energy in the process is $2E + \delta$. The interaction energy δ is called the bi-excitonic shift; in realistic semiconductor nanostructures it is of the order of a few meV.

Let us consider the two dots in the ground state $|GG\rangle$; if we illuminate them with light having frequency $(E + \delta/2)/\hbar$, we should be able to produce two excitons $|EE\rangle$. This is a second-order, two-photon process, i.e., it involves a virtual transition to the intermediate states $|EG\rangle$ and $|GE\rangle$. Due to energy conservation this is the only possible transition (the first-order, single-photon absorption is at energy E). Using different polarizations ($\sigma_+, \sigma_-, \sigma_0$) all the degenerate second-order transitions $|GG\rangle \to |E^\alpha E^\beta\rangle$ ($\alpha, \beta = 0, +, -$) can be excited (see Fig. 16.1b).

This process may be described by the following (effective) two-photon Hamiltonian:

$$H_{int} = -\frac{2\hbar^2}{\delta} \sum_{\alpha,\beta=0,+} (\Omega_\alpha \Omega_\beta |E^\alpha, E^\beta\rangle\langle G, G| + \text{h.c.}) , \quad (16.6)$$

where $\Omega_{+,0}$ is the Rabi frequency for the single-photon process within second-order perturbation theory. Here we have a three-dimensional dark state manifold; by evaluating the associated $u(3)$-valued connection form, one can verify straightforwardly that universal control in this dark space can be achieved in a fully holonomic fashion.

In passing, we note that the adiabaticity requirement, along with the condition necessary for the validity of a second-order perturbative approximation, implies that the operation times for the two-qubit gates are necessarily longer than the ones for single-qubit gates. In view of the fast dephasing times in excitonic systems, the latter fact reduces fidelity; at the first level of defense, this drawback must be mitigated by a careful parameter optimization. Later in this chapter we will see how inherent fault tolerance of the HQC approach provides an additional defense layer. The issue will be addressed further in this chapter by resorting to passive protection, and in Chapter 17 by the construction of a hybrid HQC–QECC fault-tolerant scheme.

To test the viability of the proposed HQC implementation scheme in state-of-the-art semiconductor nanostructures, time-dependent simulations of gates U_1, U_2 and a two-qubit gate were reported in [SZZ+03a, SZZ+03b, SZZ04, PSS+06]. These simulations show that the adiabatic limit is attainable; the all-optical scheme described above allows for picosecond gating times. Since exciton dephasing takes place on a nanosecond time scale, this suggests that the HQC implementation scheme proposed above should allow for a few operations within the dephasing time. In this respect, let us stress that the aim so far was not to achieve the error rate threshold for fault-tolerant quantum computation, but rather to demonstrate how nontrivial non-Abelian quantum phases can be used to realize elementary quantum state manipulations in semiconductor-based nanostructures.

16.4 Robustness

Generally and qualitatively, geometry-based quantum information processing schemes can be argued to exhibit robustness against operational imperfections due to one or more of the following mechanisms:

(i) The matrix elements of the unitary operator describing the quantum gate depend only on some global *geometric* feature of the path in the space of control parameters $\mathcal{M}$. For example, in the holonomic case one can consider loops in two-dimensional submanifolds of $\mathcal{M}$. The resulting quantum evolutions depend just on the *area* enclosed by the loop itself. Any imperfections or perturbations changing all features of the control loop but its area would leave the gate unaffected.

(ii) The quantum control process is realized in an *adiabatic* fashion. The encoding is realized in a given degenerate energy eigenspace, e.g., the ground state, and geometric gating is enacted by slowly changing the control parameters. Excitation to other energy levels is then inhibited, resulting in strong suppression of leakage errors. The latter affect other quantum

computing schemes in view of the fact that the qubit is embedded in a higher-dimensional space.
(iii) The computation takes place in the ground state. This implies, for sufficiently low temperatures, an exponential suppression of leakage due to thermal fluctuations.

In the next two subsections we briefly summarize a couple of studies addressing the robustness question for the HQC proposal based on quantum dots, and described in Section 16.3. These results clearly show how the error resilience of HQC might depend on details of the particular implementation scheme and noise model.

16.4.1 Stochastic fluctuations of control parameters

The first problem one must address in order to asses the viability of a quantum information processing scheme is its robustness against control errors. This is a type of classical noise that is present even for systems that are totally decoupled from the environment, and thus suffer no decoherence. In [SZZ04] a numerical study of the robustness of non-Abelian holonomic quantum gates described above was presented, addressing stochastic errors in control parameters (laser intensities and phase). The outcome of this analysis was that the robustness of logical gates exhibits three regimes, depending upon the variations of the noise correlation time T_n. In describing these regimes we use T_{ad} to denote the adiabatic operation time needed to enact the quantum gate.

(i) Slowly varying random fluctuations ($T_{ad}/T_n \approx 1$): the loop basically maintains its shape and is simply shifted. This situation does not affect the gate much.
(ii) Intermediate regime ($50 \leq T_{ad}/T_n \leq 100$): the intense fluctuations severely modify the loop shape and alter the gate operator.
(iii) Fast-varying random fluctuations ($T_{ad}/T_n \gg 1$): the fluctuations effectively cancel out and do not change the operator.

A possible interpretation of these regimes can be given on the basis of the geometric dependence of the holonomic operator (i.e., the solid angle swept in the parameter space). For fast-varying random fluctuations, the holonomic gate tends to be robust since the fluctuations during the loop tend to cancel out. For random fluctuations in the intermediate regime, the holonomic gates are significantly corrupted because the fluctuations strongly deform the parameter loop. For slowly varying random fluctuations, the performance improves again. This fact is not as surprising as it may seem: indeed the loop in the parameter space turns out in this case to be simply shifted rather than deformed, so that similar solid angles are swept. This analysis suggests that the main noise source is due to fluctuations in the intensity of the control parameters, whereas phase fluctuations do not seem to sizeably affect the gate studied. The effect of noise decreases with the variance of the intensity of the fluctuations, and for $\delta\Omega/\Omega = 0.01$ we have an average fidelity that is close to 1. A first comparison shows that holonomic and dynamical gates have comparable performance in the $T_{ad}/T_n \gg 1$ region. Similar simulations were performed for two single-qubit gates and, in order to complete the set of universal quantum gates, for a two-qubit gate. For the single-qubit gates similar results were obtained. For the two-qubit gate the three regimes described above are present but are less evident.

16.4.2 HQC in a decohering environment

Besides unwanted fluctuations of control parameters, the other main source of errors in quantum information processing is associated with decoherence. Let us analyze this problem in the context of the excitonic HQC proposal described above [SZZ04]. The Hamiltonian (16.5) has four eigenstates: two eigenstates $|B_\pm\rangle$ with time-dependent eigenvalues (Schrödinger picture) $\epsilon_\pm$ (called *bright states*) and two eigenstates $|D_{1,2}\rangle$ with a constant and degenerate eigenvalue ϵ (called *dark states*). To construct a complete set of holonomic quantum gates, it is sufficient to restrict the Rabi frequencies $\Omega_j(t)$ such that the norm Ω of the vectors $\{\Omega_j(t)\}$ is time-independent. Under this condition, it can be easily shown that the two dark states have energy ϵ and the two bright states have time-independent energies $\epsilon_\pm = (\epsilon \pm \sqrt{\epsilon^2 + 4\Omega^2})/2$. The adiabatic condition is simply $\Omega T_{ad} \gg 1$. The environment is described as an ensemble of harmonic oscillators linearly coupled to the system. The operator A coupling the system to the environment can be taken to be of the form $A = \text{diag}(0, 1, 0, -1)$ in the basis $|G\rangle$, $|+\rangle$, $|0\rangle$, and $|-\rangle$. We now consider the time evolution of the reduced density matrix of the system. Using standard methods (Born–Markov approximation) one obtains the master equation $\dot{\rho}(t) = -i[H_0(t), \rho(t)] - \mathscr{L}(\rho)$ [SZZ04], where

$$\mathscr{L}(\rho) = \int_0^\infty d\tau \{g(\tau)[AA''\rho(t) - A''\rho(t)A] + g(-\tau)[\rho(t)A''A - A\rho(t)A'']\}, \quad (16.7)$$

where $A'' = U^\dagger(t-\tau, t)AU(t-\tau, t)$, $g(\tau) = \int_0^\infty J(\omega)[\coth(\frac{\omega}{2T})\cos(\omega\tau) - i\sin(\omega\tau)]d\omega$ and, in the adiabatic limit, $U(t-\tau, t) \approx \exp(i\tau H_0(t))$. The spectral density $J(\omega)$ is defined in a standard way in terms of the oscillator bath parameters. Its typical behavior, for physical environments in the low-frequency regimes, is proportional to ω^s, with $s \geq 1$; the asymptotic decay of the real part of $g(\tau)$ defines the characteristic memory time τ_E of the environment. For electronic states in quantum dots decoherence is principally due to coupling to phonons. Single-phonon processes are described by a *super-Ohmic* spectral density, with $J(\omega) = k_3 \omega^3 e^{-(\omega/\omega_c)^2}$. The high-frequency cutoff ω_c is due to planar confinement in the quantum dot. The dimensionless coupling constant k_3 allows for the description of different kinds of phonon–carrier interactions in semiconductor materials, including deformation potential, piezoelectric, and spin-orbit coupling.

Let $|\psi_{id}(T_{ad})\rangle$ be the final state one would obtain under the action of $H_0(t)$ only, i.e., under an ideal gate in the absence of decoherence, and let $\rho(T_{ad})$ denote the solution of the master equation, computed at the same time, and for the *same* initial (pure) state. The numerical results for the fidelity $\mathscr{F} = \sqrt{\langle \psi_{id}(T_{ad}) | \rho(T_{ad}) | \psi_{id}(T_{ad}) \rangle}$ (averaged over the initial states) obtained for the single-qubit operations U_1 and U_2 can be fit by $\mathscr{F} = 1 - T_{ad} \sum_{j=\pm} \eta_j \Gamma^j$ (see Fig. 16.2). For a constant adiabatic parameter $\Omega T_{ad} = \alpha = \text{const.}$, one has $\Gamma^\pm \propto 1/T_{ad}^6 \coth(\alpha^2/(\epsilon T_{ad}^2 T)) \exp(-(\alpha^2/(T_{ad}^2 \epsilon \omega_c))^2)$. Thus, it follows that *the fidelity should have a pronounced minimum as a function of T_{ad}. By varying Ω and T_{ad} (e.g., by laser control) the position of the fidelity minimum can be shifted and the effect of super-Ohmic environment can be suppressed.*

Since the possibility of suppressing the *super-Ohmic* effects is somewhat surprising, one may wonder whether a similar possibility arises for other environments, e.g., for Ohmic environments. In [PSS+06] it was found that the different Ohmic rates $\Gamma^\pm$ lead to completely different results. This is due to the presence, in the Ohmic case, of transitions between degenerate states

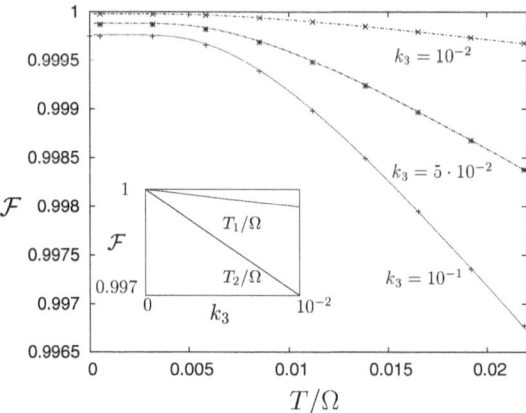

Fig. 16.2. Fidelity $\mathscr{F}$ for gate 1 as a function of temperature for different k_3 (expressed in $(\text{meV})^{-2}$) in the presence of a *super-Ohmic* environment. The data points are numerical simulation results and the curves have the form $\mathscr{F} = 1 - \sum_{j=\pm} \eta_j \Gamma^j t_{ad}$ ($\eta_- = 3 \cdot 10^{-2}$ and $\eta_+ = 0.7$). Inset: Fidelity $\mathscr{F}$ as a function of k_3 for two different temperatures ($T_1/\Omega = 1.6 \times 10^{-2}$ and $T_2/\Omega = 5 \cdot 10^{-3}$). Parameters: $\epsilon = 1$ eV, $\Omega = 25$ meV, $t_{ad} = 7.5$ ps and $\omega_c = 0.5$ meV. Reprinted with permission from [PSS+06]. Copyright 2006, American Physical Society.

which are absent in the case of a super-Ohmic environment. The fidelity behavior in the presence of an Ohmic environment changes dramatically: it is no longer possible to optimize the fidelity by changing the laser parameters.

16.5 Hybridizing HQC and error-avoiding/correcting techniques

Our discussion so far addressed natural robustness and simple control techniques. We now proceed to consider the viability of *hybrid* strategies, i.e., strategies where HQC is merged with the other error-avoidance techniques covered in this book. For example, in [WZL05] a hybrid decoherence-free subspace (DFS) and HQC technique was introduced. The appeal of such a strategy should be evident: *try to bring together the best of two worlds, namely the resilience of the DFS approach against environment-induced decoherence and the operational robustness of HQC*. The results in [WZL05] were formulated by using rather generic Hamiltonians, so the proposed scheme appears to be a suitable candidate for experimental demonstration in a variety of systems, including trapped ions and quantum dots. It was also shown in [WZL05] how to realize universal quantum computation over a scalable DFS against collective dephasing by using adiabatic holonomies only, and an extension was given to the general collective decoherence case, based only on the controllability of exchange Hamiltonians.

Let us now see how one may use the holonomic approach to enact information processing within a noiseless subsystem (Chapter 3), or generally within a *virtual* subsystem [Z01, ZLL04]. The argument we present here builds upon the one provided in [WZL05]. Let us assume that one has the following decomposition of the state space into noiseless subsystems (NSs) $\mathscr{H} = \sum_J \mathbf{C}^{n_J} \otimes \mathbf{C}^{d_J}$ (here $\mathbf{C}^{n_J}$ are NSs, and $\mathbf{C}^{d_J}$ are irreducible spaces of the noise algebra $\mathscr{A}$). Moreover, we assume that there exists at least one n_J equal to 4 (this is true, e.g., in the space of five qubits affected by collective decoherence). This NS might then encode two noiseless qubits, but since we wish to perform quantum information processing with holonomies, we will instead use this $\mathbf{C}^4$ space as a code for just one noiseless qubit $|\widetilde{\alpha}\rangle_L$ ($\alpha = 0, 1$) and two ancillary states $|\widetilde{a}_i\rangle$ ($i = 1, 2$). Suppose now that one is able to enact the controllable Hamiltonian $H_{\text{NS}} = J'|\widetilde{a}_1\rangle\langle\widetilde{a}_2| + J_0''e^{i\varphi}|\widetilde{a}_1\rangle\langle\widetilde{0}|_L + J_1''e^{i\varphi}|\widetilde{a}_1\rangle\langle\widetilde{1}|_L + \text{h.c.}$ By resorting to the same constructions as in the case of HQC with quantum dots described above, it is easy to see that complete one-qubit control by holonomies can be achieved in this case as well.

To realize the required multi-level controllable Hamiltonian, we observe that the $\mathbf{C}^4$ space under consideration is a four-dimensional irreducible representation of the commutant of the noise algebra, $\mathscr{A}'$. Universal control over this irreducible representation space amounts to the ability to switch on and off a pair of *generic* Hamiltonians in the group algebra of $\mathscr{A}'$. In the case of collective decoherence, the commutant of the noise algebra is given by permutation operations, and as shown in [KBL+01], universal control over the NSs can be achieved by switching on and off Heisenberg *exchange* Hamiltonians, i.e., $\sum_{l<m} J_{lm}\mathbf{S}_l \cdot \mathbf{S}_m$ [where $\mathbf{S}_l = (X_l, Y_l, Z_l)/2$]. This construction is *existential* in nature and realizes a fully geometric instance of *encoded universality in a protected subsystem*.

A second very important type of hybridization one can conceive involves HQC and quantum error-correcting codes (QECCs). For the sake of concreteness we will focus on the case of *stabilizer codes*. We recall that a stabilizer code for n qubits is a 2^k-dimensional subspace $\mathscr{C}$ of the whole $\mathscr{H}_n \cong \mathbf{C}^{\otimes n}$ state space ($k \leq n$), such that

$$\mathscr{C} := \{|\psi\rangle \mid X_i|\psi\rangle = |\psi\rangle\}, \ i = 1, \ldots, n - k\}, \tag{16.8}$$

where the stabilizer operators X_is are the generators of an Abelian subgroup $\mathscr{S}$ of the Pauli group of order 2^{n-k}. One has the following decomposition of the n-qubit space: $\mathscr{H}_n \cong \oplus_{i=1}^{2^{n-k}} (\mathbf{C}^2)^{\otimes k} \otimes \mathscr{H}_i$, where each $\mathscr{H}_i \cong \mathbf{C}$ corresponds to one of the 2^{n-k} distinct joint eigenvalues of the elements of $\mathscr{S}$. The code $\mathscr{C}$ is simply the $(\mathbf{C}^2)^{\otimes k}$ factor corresponding to all the X_is having eigenvalue one. All the degeneracy factors $(\mathbf{C}^2)^{\otimes k}$ can be regarded as virtual subsystems [Z01] composed of k qubits. Suppose now that *one has a family of controllable Hamiltonians $H(\lambda)$ commuting with all the $\mathscr{S}$ elements* $[H(\lambda), X_i] = 0, (i = 1, \ldots, k; \lambda \in \mathscr{M})$ (here $\mathscr{M}$ is the manifold of control parameters). In this case $\mathscr{C}$ is invariant under the action of the $H(\lambda)$s. Now it should be clear that a way to realize HQC within the code $\mathscr{C}$ is to mimic the scheme outlined above for NSs: design a family $H(\lambda)$ with a degenerate multi-level structure, e.g., admitting a dark-state manifold. For example, following the same four-state construction we had for NSs, if $n - k = 2m$, one would be able to encode m logical holonomic qubits. This argument is, as in the NS case, just an existential one. To provide constructive schemes suitable for experimental implementations is an interesting open problem. The techniques presented in Chapter 17 present a partial answer.

16.6 Conclusions

In this chapter we provided an account of the holonomic approach to quantum computation. The existence of such a strategy shows that quantum holonomies associated with non-Abelian gauge connections can be used for universal quantum information processing. It is quite intriguing that gauge fields, beyond the key role they play in the study of fundamental interactions, might also prove to be useful in the arena of quantum information processing. This is so regardless of the actual degree of fault tolerance provided by these geometric manipulations might have, inherently, or hybridized with the other error-correcting techniques described in this book. We might say that holonomic quantum computation shows that information, besides being "physical," can also be "geometric."

Appendix: quantum holonomies

This appendix contains basic mathematical facts about Abelian as well as non-Abelian geometric phases (holonomies). A very useful reference for the role of geometric phases in quantum physics is [SW89]. The reader can refer to the book [N90] for the required background on differential geometry, in particular fiber bundles.

Abelian phases

Pure quantum states are represented mathematically by *rays* in a separable Hilbert space $\mathcal{H}$ (defined in Section 1.2.1). There is therefore a one-to-one correspondence between physical states and the points of the *projective* space $\mathbf{P}(\mathcal{H})$. The latter is defined as the quotient space of $\mathcal{H}$ with respect to the equivalence relation $x \sim y \leftrightarrow \exists \lambda \in \mathbf{C} - \{0\} \,|\, y = \lambda x, \forall x, y \in \mathcal{H}$. One can also consider the unit sphere in $\mathcal{H}$, $S^\infty := \{|\psi\rangle \,/\, \||\psi\rangle\| = 1\}$, factored by the $U(1)$ action $(|\psi\rangle, e^{i\theta}) \mapsto e^{i\theta}|\psi\rangle$, i.e., $\mathbf{P}(\mathcal{H}) = S^\infty / U(1)$.

A $U(1)$ principal bundle over the base space $\mathbf{P}(\mathcal{H})$ is defined by the mapping

$$\pi \colon S^\infty \to \mathbf{P}(\mathcal{H}). \tag{16.9}$$

This fiber bundle is the relevant one for the appearance of Abelian holonomies, i.e., Berry phases. The fiber over the point $|\psi\rangle\langle\psi| \in \mathbf{P}(\mathcal{H})$ is given by $F_{|\psi\rangle} := \{e^{i\theta}|\psi\rangle \,/\, \theta \in [0, 2\pi)\}$.

This bundle is known as a *classifying* one. This means that any other principal $U(1)$ bundle can be obtained from (16.9). Let Φ be a smooth map from the parameter manifold $\mathcal{M}$ in $\mathbf{P}(\mathcal{H})$. Then one can construct the *pull-back* bundle $\Phi^* S^\infty$, with total space $\bigcup_{\lambda \in \mathcal{M}} F_{\Phi(\lambda)}$ and projection $\pi^\Phi \colon F_{\Phi(\lambda)} \to \lambda$. Sections of the bundle $\Phi^* S^\infty$ are then maps $\lambda \mapsto |\psi_\lambda\rangle \in F_{\Phi(\lambda)}$.

Let us now consider the quantum dynamics associated with the Schrödinger equation $i\,\partial/\partial t\, |\psi\rangle = H |\psi\rangle$. The temporal evolution defines the map $t \mapsto |\psi(t)\rangle$ in the total space S^∞ and, using π, in the base manifold $\mathbf{P}(\mathcal{H})$. Connections are then introduced in order to lift evolutions in $\mathbf{P}(\mathcal{H})$ to evolutions in S^∞. The $U(1)$ action over S^∞ defines the *vertical* direction at each point along the fiber. The connection $u(1)$-valued 1-form A enables one to introduce a *horizontal* direction as well. Once this field of directions is given, one can realize the horizontal *lift* of any curve in the base.

The natural way of introducing horizontal directions, i.e., a connection, in this $U(1)$-bundle derives from the observation that S^∞ inherits from $\mathcal{H}$ a natural scalar product structure. We can then define the horizontal directions at each point as those orthogonal to the fiber, i.e., to $|\psi\rangle$. Given the curve $t \mapsto |\psi(t)\rangle \in S^\infty$, we can split the tangent vector $|\dot\psi\rangle := d|\psi(t)\rangle/dt$ as follows: $|\dot\psi\rangle = \langle\psi|\dot\psi\rangle\,|\psi\rangle + |h_\psi\rangle$, where the horizontal component $|h_\psi\rangle$ satisfies, by definition, the relation $\langle\psi|h_\psi\rangle = 0$.

To obtain an explicit evaluation of the connection, we decompose the operator d/dt into its vertical and horizontal components: $d/dt = \alpha\,\partial/\partial\theta + \sum_\mu B^\mu D_\mu$, where the indexes μ label the local coordinates (λ_μ) of the base manifold $\mathcal{M}$. The horizontal operators D_μ are nothing but the *covariant* derivatives, given by $D_\mu := \partial/\partial\lambda_\mu + A_\mu\,\partial/\partial\theta$. The $u(1)$-valued 1-form $A := \sum_\mu A_\mu\,d\lambda_\mu$ is the *connection form*. By applying d/dt to $|\psi(t)\rangle$, the horizontal part gives

$\langle\psi|\,D_\mu\,|\psi\rangle = \langle\psi|\,\partial/\partial\lambda_\mu + A_\mu\,\partial/\partial\theta\,|\psi\rangle = 0$, from which $A_\mu\,\langle\psi|\partial/\partial\theta\,|\psi\rangle = -\langle\psi|\,\partial/\partial\lambda_\mu\,|\psi\rangle$. From $\partial/\partial\theta\,|\psi\rangle = i\,|\psi\rangle$ one obtains

$$A_\mu = i\,\langle\psi|\frac{\partial}{\partial\lambda_\mu}|\psi\rangle\,. \qquad (16.10)$$

Using the connection form (16.10) we can lift loops in the base manifold to the total space S^∞. Let us consider now a loop $\gamma\colon [0,\,T] \mapsto \mathbf{P}(\mathcal{H})$ in the base space and let $|\widetilde{\psi}(t)\rangle$ be its horizontal lift. One can write $|\widetilde{\psi}(t)\rangle = e^{i\,f(t)}\,|\psi(t)\rangle$, where $|\psi(t)\rangle = (s\circ\gamma)(t)$ is a *closed* curve in S^∞ obtained by composing γ with a bundle section s. One has $|\widetilde{\psi}(T)\rangle = e^{i\,\beta}\,|\widetilde{\psi}(0)\rangle$, where $\beta := f(T) - f(0)$. These relations, along with the horizontality condition $\langle\widetilde{\psi}(t)|\dot{\widetilde{\psi}}(t)\rangle = 0$, give $\beta = \int_0^T dt\langle\psi(t)|\dot{\psi}(t)\rangle$. From $|\dot{\psi}\rangle = (\dot{\theta}\partial/\partial\theta + \sum_\mu \dot{\lambda}_\mu\partial/\partial\lambda_\mu)\,|\psi\rangle$, and $\theta(T) = \theta(0)$ mod 2π, by using Eq. (16.10) one obtains

$$\beta = \sum_\mu \int_\gamma d\lambda_\mu A_\mu = \int_\gamma A. \qquad (16.11)$$

Note that under a local coordinate change the connection form changes according to $A_\mu \mapsto A_\mu - \partial\theta/\partial\lambda_\mu$, therefore according to the expression (16.11) the $U(1)$-holonomy is manifestly gauge-invariant; moreover, it does not depend on the Hamiltonian. The phase is purely *geometric* in nature.

It is worth noting that, in general, the state vector is *not* horizontal: $\langle\psi|\dot{\psi}\rangle = -i\,\langle\psi|H(t)|\psi\rangle \neq 0$. Therefore, besides the geometric term (16.11), one has a *dynamical*, Hamiltonian dependent, contribution to the phase, given by $\alpha := -\int_0^T \langle\psi(t)|H(t)|\psi(t)\rangle$. This dynamical contribution can be "gauged away" by the transformation $|\psi(t)\rangle \mapsto U(t)\,|\psi(t)\rangle$, where

$$U(t) := \exp\left[i\int_0^t d\tau\,\langle\psi(\tau)|H(\tau)|\psi(\tau)\rangle\right]. \qquad (16.12)$$

One can introduce also the *curvature* 2-form $F := dA = \sum_{\mu\nu} F_{\mu\nu} d\lambda_\mu \wedge d\lambda_\nu$, $F_{\mu\nu} = \partial A_\mu/\partial\lambda_\nu - \partial A_\nu/\partial\lambda_\mu$. By using Stokes' theorem, Eq. (16.11) can be cast in the form $\beta := \int_\Sigma F$, where Σ is any surface having $\gamma([0,\,T])$ as a boundary. A nonvanishing F accounts for a nontrivial geometry of the bundle (16.9).

Non-Abelian case

The above constructions can be generalized to the case in which the fiber is an n-dimensional complex space over which the group $U(n)$ acts [WZ84]. Let us consider a family of Hamiltonians $\{H(\lambda)\}_{\lambda\in\mathcal{M}}$ where $\mathcal{M}$ is the control manifold. We denote loops in $\mathcal{M}$ by γ, and we use the notation $H(t) := H_{\gamma(t)}$, $t \in [0,\,T]$. In general $H(\lambda) = H_{\gamma(t)}$ has R different eigenvalues $\{\varepsilon_l\}_{l=1}^R$ with degeneracies $\{n_l(\lambda)\}$. We restrict our attention to the case where *no* level crossing occurs, i.e., $n_l(\lambda) = n_l$. Let $\Pi_l(\lambda)$ denote the projector over the eigenspace $\mathcal{H}_l(\lambda) := \mathrm{span}\{|\psi_l^\alpha(\lambda)\rangle\}_{\alpha=1}^{n_l}$ of $H(\lambda)$. The spectral, λ-dependent, resolution of the Hamiltonians is then $H(\lambda) = \sum_{l=1}^R \varepsilon_l(\lambda)\,\Pi_l(\lambda)$. The mapping $\lambda \mapsto \{|\psi_l^\alpha(\lambda)\rangle\}_{\alpha=1}^{n_l}$ defines a section of the bundle

$$\frac{U(N)}{U(N - n_l)} \to \frac{U(N)}{U(n_l) \times U(N - n_l)}, \qquad (16.13)$$

where $N := \dim \mathcal{H}$. The total (base) space of the $U(n_l)$-principal bundle (16.13) is known as the Stiefel (or Grassmann) manifold and it is denoted by V_{N,n_l} (G_{N,n_l}). Intuitively speaking, the total (base) space of the bundle (16.13) is the set of n_l-dimensional orthonormal frames (subspaces). As for the Abelian case discussed above, the Hermitian structure over $\mathcal{H}$ provides a natural way to introduce horizontal directions: tangent vectors are horizontal if they are orthogonal to the fiber. Notice that one recovers the Abelian ($\mathcal{H} \cong \mathbf{C}^N$) case by setting $n_l = 1$; indeed $G_{N,1} = \mathbf{CP}^{N-1} = \mathbf{P}(\mathcal{H})$.

The state vector evolves according to the time-dependent Schrödinger equation $i\partial_t |\psi(t)\rangle = H_{\gamma(t)} |\psi(t)\rangle$. In the *adiabatic* limit any initial preparation $|\psi_0\rangle \in \mathcal{H}$ in the energy eigenspace $\mathcal{H}_l$ will evolve, after time T, into $|\psi(T)\rangle = U_\gamma |\psi_0\rangle \in \mathcal{H}_l$. For the sake of concreteness, let us focus on the eigenspace associated with the eigenvalue $\varepsilon_l = 0$. Also, let $\{|\psi_\alpha(t)\rangle\}$ be the corresponding orthonormal basis at the instant t. The analog of $|\tilde{\psi}\rangle$ is now a map $\eta_\alpha(t) = \sum_\beta U_{\alpha\beta}(t) \psi_\beta(t)$, i.e., the solution of the Schrödinger equation with initial condition $\eta_\alpha(0) = \psi_\alpha$.

The horizontality conditions are now obtained by imposing, at each instant, the orthogonality of the η_αs: $\langle \eta_\beta | \eta_\alpha \rangle = \delta_{\alpha\beta}$. By differentiation one obtains

$$0 = \langle \eta_\beta | \dot{\eta}_\alpha \rangle = \sum_\delta (\dot{U}_{\alpha\delta} \langle \eta_\beta | \psi_\delta \rangle + U_{\alpha\delta} \langle \eta_\beta | \dot{\psi}_\delta \rangle) \tag{16.14a}$$

$$= \dot{U}_{\alpha\delta} U^*_{\beta\delta} + \sum_{\delta\tau} U_{\alpha\delta} U^*_{\beta\tau} \langle \psi_\tau | \dot{\psi}_\delta \rangle, \tag{16.14b}$$

from which it follows that $(U^{-1} \dot{U})_{\beta\alpha} = A_{\alpha\beta}$, where we have defined $A_{\alpha\beta} := \langle \psi_\beta | \dot{\psi}_\alpha \rangle$. We have an equation for $U(t)$ whose solution is given by

$$U(t) = T_+ \exp \int_0^T d\tau A(\tau) = P_+ \exp \int_\gamma A. \tag{16.15}$$

The $u(n)$-valued 1-form $A = \sum_\mu A_\mu d\lambda_\mu$ is given by $(A_\mu)_{\alpha\beta} = \langle \psi_\alpha | \partial/\partial \lambda_\mu | \psi_\beta \rangle$. Note that the time ordering T_+ has been replaced by the path ordering symbol P_+. This is required by the non-Abelian (matrix) nature of the connection A.

In order to provide more physical insight, we now present an intuitive proof for deriving (16.15) [PZ01]. We assume the family $\mathscr{F}$ to be *iso-spectral*. This implies that for any $\lambda \in \mathcal{M}$ there exists a unitary transformation $\mathscr{U}(\lambda)$ such that $H(\lambda) = \mathscr{U}(\lambda) H_0 \mathscr{U}(\lambda)^\dagger$, where $H_0 := H(\lambda_0)$. Upon dividing the time interval $[0, T]$ into N equal segments Δt, for $\mathscr{U}_i = \mathscr{U}(\gamma(\lambda(t_i)))$ one obtains the evolution operator in the form

$$U_\gamma = T_+ e^{-i \int_0^T \mathscr{U}(\lambda) H_0 \mathscr{U}^\dagger(\lambda) dt} = T_+ \lim_{N \to \infty} e^{-i \sum_{i=1}^N \mathscr{U}_i H_0 \mathscr{U}_i^\dagger \Delta t}$$

$$= T_+ \lim_{N \to \infty} \prod_{i=1}^N \mathscr{U}_i e^{-i H_0 \Delta t} \mathscr{U}_i^\dagger. \tag{16.16}$$

The third equality holds due to the smallness of the interval Δt in the limit of large N. The product $\mathscr{U}_i^\dagger \mathscr{U}_{i+1}$ of two successive unitaries gives rise to an infinitesimal rotation of the form $\mathscr{U}_i^\dagger \mathscr{U}_{i+1} \approx I + \vec{A}_i \cdot \Delta \vec{\lambda}_i$, where $(A_i)_\mu \equiv \mathscr{U}_i^\dagger \frac{\Delta \mathscr{U}_i}{\Delta (\lambda_i)_\mu}$. The connection A has, at time t_i, the components $(A_i)_\mu$ with $\mu = 1, \ldots, d$. Hence the evolution operator (16.16) becomes

$$T_+ \lim_{N \to \infty} \mathscr{U}_N \left(I - i H_0 N \cdot \Delta t + \sum_{i=1}^{N-1} \vec{A}_i \cdot \Delta \vec{\lambda}_i \right) \mathscr{U}_1^\dagger. \tag{16.17}$$

Now we consider an initial state $|\psi_{in}\rangle$ belonging to an eigenspace $\mathcal{H}_0$ with associated eigenvalue, e.g., $\varepsilon_0 = 0$, and we demand adiabaticity. Then at each time t_i the state $|\psi(t_i)\rangle$ will remain in the $\varepsilon_0 = 0$ energy level. This allows us to factor out the action of H in Eq. (16.17), thus obtaining

$$U_\gamma = T_+ \lim_{N \to \infty} \left(1 + \sum_{i=1}^{N-1} \vec{A}_i \cdot \Delta \vec{\lambda}_i \right) = P_+ \exp \oint_\gamma A , \qquad (16.18)$$

where A is projected into the subspace $\mathcal{H}_0$. Note that we replaced the time-ordering operation with path ordering $\mathbf{P}$, as the integration parameter in the last expression is the position on the loop γ.

Chapter 17

Fault tolerance for holonomic quantum computation

Ognyan Oreshkov, Todd A. Brun, and Daniel A. Lidar

In Chapter 16 it was shown how holonomic quantum computation (HQC) can be combined with the method of decoherence-free subspaces (DFSs), leading to passive protection against certain types of correlated errors. However, this is not enough for fault tolerance since other types of errors can accumulate detrimentally unless corrected. Scalability of HQC therefore requires going beyond that scheme, e.g., by combining the holonomic approach with *active* error correction. One way of combining HQC with active quantum error-correcting codes, which is similar to the way HQC is combined with DFSs, was also mentioned in Chapter 16. This approach, however, is not scalable since it requires Hamiltonians that commute with the stabilizer elements, and when the code increases in size, this necessitates couplings that become increasingly nonlocal.

In this chapter, we will show how HQC can be made fault tolerant by combining it with the techniques for fault-tolerant quantum error correction (FTQEC) on stabilizer codes using Hamiltonians of finite locality. The fact that the holonomic method can be mapped directly to the circuit model allows us to construct procedures that resort almost entirely to these techniques. We will discuss an approach that makes use of the encoding already present in a stabilizer code and does not require additional qubits [O08, OBL08]. An alternative approach, which requires ancillary qubits for implementing transversal operations between qubits in the code, can be found in [O09].

Since protected information is contained in subsystems [K06a, B.-KNP+08] (see Chapter 6), the problem of implementing fault-tolerant HQC can be understood as that of manipulating fault-tolerantly the subsystem containing the protected information by holonomic means. We therefore begin by first introducing a generalization of the standard HQC method, which is applicable to encoding in subsystems.

17.1 Holonomic quantum computation on subsystems

As pointed out in Chapter 6, protected quantum information is most generally contained in the subsystems $\mathcal{H}_i^A$ in a decomposition of the Hilbert space of the system of the form

$$\mathcal{H}^S = \bigoplus_{i=1}^{m} \mathcal{H}_i^A \otimes \mathcal{H}_i^B \oplus \mathcal{K}. \tag{17.1}$$

Here the dimensions of the subsystems are related by $\dim \mathcal{H}^S = \sum_{i=1}^{m} \dim \mathcal{H}_i^A \times \dim \mathcal{H}_i^B + \dim \mathcal{K}$. In the formalism of operator quantum error correction [KLP05, KLP+06, P05, OLB08], the subsystems $\mathcal{H}_i^A$ contain the logical information, and $\mathcal{H}_i^B$ contain the syndrome and gauge degrees of freedom. In the most general case, all subsystems $\mathcal{H}_i^A$ can be used for encoding and computation [BKK07a]. By using adiabatic holonomies it is possible to apply arbitrary computations in the subsystems $\mathcal{H}_i^A$, a result that is summarized in the following theorem [O09].

Theorem 17.1. *Consider a nontrivial decomposition into subsystems of the form* (17.1). *Choose an initial Hamiltonian in the form*

$$H(0) = \bigoplus_{i=1}^{m} I_i^A \otimes H_i^B \oplus H_\mathcal{K}, \tag{17.2}$$

where H_i^B are operators on $\mathcal{H}_i^B$ such that all eigenvalues of H_i^B are different from the eigenvalues of $H_\mathcal{K}$ and the eigenvalues of H_j^B for $i \neq j$. In the case when $\dim \mathcal{K} = 0$ and $m = 1$, we impose the additional requirement that H_1^B has at least two different eigenvalues. By varying this Hamiltonian adiabatically along suitable loops in a sufficiently large control manifold, it is possible to generate a unitary of the form

$$U = \bigoplus_{i=1}^{m} W_i^A \otimes V_i^B \oplus V_\mathcal{K}, \tag{17.3}$$

where $\{W_i^A\}$ is any desired set of geometric unitary transformations on $\{\mathcal{H}_i^A\}$.

The proof of the theorem is based on the following lemma.

Lemma 17.1. *By varying a Hamiltonian adiabatically along suitable loops in a sufficiently large control manifold, it is possible to implement holonomically any combination of unitary transformations in its eigenspaces.*

Comment. As discussed in Chapter 16, if we have sufficient control over the parameters of a Hamiltonian, we can generate holonomically any unitary operation in a given eigenspace. This lemma concerns the question of whether it is possible to generate any *combination* of holonomies in the different eigenspaces. Although intuitively expected based on considerations concerning the generic irreducibility of the adiabatic connection, the property may not be obvious. For example, in the case of a two-level Hamiltonian, the evolution of one of the eigenspaces completely determines the evolution of the other one, which raises the question of whether it is possible to obtain independent holonomies in the two eigenspaces. We now show that this is possible. Note that even though the proof is constructive and can serve as a general prescription for simultaneous HQC in different eigenspaces, it is primarily meant as a proof of principle.

Proof of Lemma 17.1. It is sufficient to show that it is possible to generate an arbitrary operation in any given eigenspace while at the same time generating the identity operation in the rest of the eigenspaces. Without loss of generality, we will assume that there are only two eigenspaces; if there are more, we can operate within the subspace spanned by two of them at a time, by varying only the restriction of the Hamiltonian on that subspace. Then the initial Hamiltonian can be written as

$$H(0) = \varepsilon_1 \Pi_1 + \varepsilon_2 \Pi_2, \tag{17.4}$$

where $\Pi_{1,2}$ are the projectors on the ground and excited eigenspaces, and $\varepsilon_1 < \varepsilon_2$ are their corresponding eigenvalues. Notice that this Hamiltonian is invariant under unitary transformations of the form

$$V = V_1 \oplus V_2, \tag{17.5}$$

where $V_{1,2}$ are unitaries on the subspaces with projectors $\Pi_{1,2}$, respectively. Let the Hamiltonian vary along a loop $H(t)$, $H(0) = H(T)$, which satisfies the adiabatic requirement to some satisfactory precision. To this precision, the resulting unitary transformation can be written as

$$U(T) = \mathscr{T} \exp\left(-i \int_0^T dt H(t)\right) = e^{-i\omega_1} U_1 \oplus e^{-i\omega_2} U_2, \tag{17.6}$$

where U_1 and U_2 are the holonomies resulting in the two eigenspaces, and $\omega_{1,2} = \int_0^T dt \varepsilon_{1,2}(t)$ are dynamical phases. Observe that the Hamiltonian $VH(t)V^\dagger$, where $V = V_1 \oplus V_2$, is a valid loop based on $H(0)$ with the same spectrum as that of $H(t)$, which gives rise to the holonomies $V_1 U_1 V_1^\dagger$ and $V_2 U_2 V_2^\dagger$, respectively. This follows from the fact that the overall unitary transformation generated by $VH(t)V^\dagger$ is equal to $VU(t)V^\dagger$, where $U(t)$ is the unitary generated by $H(t)$.

Imagine that we want to generate holonomically the unitary transformation W_1 in the ground space of the Hamiltonian while at the same time obtaining the identity holonomy I_2 in the excited space. Choose any loop $H(t)$ that gives rise to the holonomy $W_1^{\frac{1}{d_2}}$ in the ground space, where d_2 is the dimension of the excited space (we know that such a loop can be found). Let this loop result in the holonomy W_2 in the excited space. The latter can be written as $W_2 = \sum_{j=1}^{d_2} e^{i\alpha_j} |j\rangle\langle j|$, where $\{|j\rangle\}$ is an eigenbasis of W_2 and $e^{i\alpha_j}, \alpha_j \in R$, are the corresponding eigenvalues. Consider the unitary transformation C_2 that cyclically permutes the eigenvectors $\{|j\rangle\}$: $C_2|j\rangle = |j+1\rangle$, where we define $|d_2 + 1\rangle \equiv |1\rangle$. We can implement the desired combination of holonomies in the two eigenspaces as follows. First apply $H(t)$. This results in the holonomies $W_1^{\frac{1}{d_2}}$ and W_2 in the ground and excited spaces, respectively. Next, apply $(I_1 \oplus C_2) H(t) (I_1 \oplus C_2)^\dagger$. This generates the holonomies $W_1^{\frac{1}{d_2}}$ and $C_2 W_2 C_2^\dagger = \sum_{j=1}^{d_2} e^{i\alpha_{j-1}} |j\rangle\langle j|$ (we have defined $\alpha_{1-1} \equiv \alpha_{d_2}$). The combined effect of these two operations is $W_1^{\frac{2}{d_2}}$ and $\sum_{j=1}^{d_2} e^{i(\alpha_j + \alpha_{j-1})} |j\rangle\langle j|$. We next apply $(I_1 \oplus C_2^2) H(t) (I_1 \oplus C_2^2)^\dagger$, which generates the holonomies $W_1^{\frac{1}{d_2}}$ and $C_2^2 W_2 C_2^{2\dagger} = \sum_{j=1}^{d_2} e^{i\alpha_{j-2}} |j\rangle\langle j|$. The net result becomes $W_1^{\frac{3}{d_2}}$

and $\sum_{j=1}^{d_2} e^{i(\alpha_j+\alpha_{j-1}+\alpha_{j-2})}|j\rangle\langle j|$. We continue this for a total of d_2 rounds, which results in the net holonomic transformations $W_1^{\frac{d_2}{d_2}} = W_1$ and $e^{i(\alpha_1+\alpha_2+\ldots+\alpha_{d_2})}\sum_{j=1}^{d_2}|j\rangle\langle j| \propto I_2$. $\square$

Proof of Theorem 17.1. For the purposes of this theorem, $\mathscr{K}$ can be regarded as a particular $\mathscr{H}_i^B$ whose co-factor $\mathscr{H}_i^A$ is one-dimensional, so we can assume that $\dim \mathscr{K} = 0$ for simplicity. Let us denote the eigenvalues of H_i^B in Eq. (17.2) by ω_{α_i}, where $\alpha_i = 1, 2, \ldots, d_i$, and the projectors on their corresponding eigenspaces $\mathscr{H}_{\alpha_i}^B$ by $\Pi_{\alpha_i}^B$. Then the spectral decomposition of the initial Hamiltonian reads $H(0) = \sum_{i=1}^{m}\sum_{\alpha_i=1}^{d_i}\omega_{\alpha_i}\Pi_i^A \otimes \Pi_{\alpha_i}^B$, where Π_i^A is the projector on $\mathscr{H}_i^A$. According to Lemma 17.1, we can implement holonomically any combination of unitary transformations in the different eigenspaces of $H(0)$ up to an overall phase. If we want to implement the set of unitary operations $\{W_i^A\}$ in the different subsystems $\mathscr{H}_i^A$, we can do this by implementing the holonomy $W_i^A \otimes W_{\alpha_i}^B$ in each of the eigenspaces $\mathscr{H}_i^A \otimes \mathscr{H}_{\alpha_i}^B$ for $\alpha_i = 1, 2, \ldots, d_i$ where $W_{\alpha_i}^B$ are arbitrary unitaries on $\mathscr{H}_{\alpha_i}^B$. This results in the net unitary

$$U = \bigoplus_i W_i^A \otimes (\bigoplus_{\alpha_i} e^{i\phi_{\alpha_i}} W_{\alpha_i}^B) \equiv \bigoplus_i W_i^A \otimes V_i^B, \quad (17.7)$$

where $e^{i\phi_{\alpha_i}}$ are dynamical phases resulting in the eigenspaces $\mathscr{H}_i^A \otimes \mathscr{H}_{\alpha_i}^B$. $\square$

To summarize, HQC on a subsystem can be realized by adiabatically varying a Hamiltonian that acts locally on the corresponding co-subsystem. During the evolution, the information initially encoded in the subsystem transforms to a different subsystem that is related to the initial one via a geometric unitary operation. The dynamical part of the unitary factors out as a transformation on the correspondingly transformed co-subsystem. The problem of FTHQC can be understood as that of finding a fault-tolerant realization of this approach.

17.2 FTHQC on stabilizer codes without additional qubits
17.2.1 The main idea

The developed techniques for FTQEC on stabilizer codes provide a prescription for how to encode information and how to transform the subsystem containing the information (hereafter referred to as the code subsystem) so that the class of errors for which the code is designed remains correctable. We will try to find realizations of the same transformations that the code subsystem follows in a standard fault-tolerant scheme using the generalized method of HQC on subsystems. There are two difficulties in this respect that have to be considered. First, not every unitary evolution of a subsystem can be realized by holonomic means. For example, a general evolution inside a fixed subsystem cannot be implemented holonomically because the HQC method requires that the encoded states leave the original code in order to undergo nontrivial geometric transformations. Second, the holonomic approach unavoidably gives rise to dynamical transformations on the co-subsystem (see below), and these could jeopardize the fault tolerance of the scheme. We will see that neither of these features is a fundamental obstacle to the realization of fault-tolerant HQC.

The standard fault-tolerant techniques are primarily based on the use of transversal operations. In addition, there are nontransversal operations for preparing and verifying a special ancillary state such as Shor's "cat" state $(|00...0\rangle + |11...1\rangle)/\sqrt{2}$. Since single-qubit unitaries together with the CNOT gate form a universal set of gates, fault-tolerant computation can be realized entirely in terms of single-qubit gates and transversal CNOT gates, assuming that the ancillary state can be prepared reliably. Thus we can aim at constructing holonomic gates on the code subsystem via transformations that in the original basis of the full Hilbert space are equivalent to transversal one- and two-qubit gates or to operations for the preparation of the "cat" state.

It turns out, however, that holonomic transformations on the code subsystem cannot be realized using purely transversal operations without the use of extra qubits, even if the encoded gate has a purely transversal implementation in the dynamical case. This is because the Hamiltonians that leave the code subsystem invariant are linear combinations of elements of the stabilizer or, more generally (in the case of operator codes), the gauge group of the code, and they unavoidably couple qubits in the same block. But transversal operations are not the most general class of operations that do not lead to propagation of errors. A transformation that at every moment is equal to a transversal operation followed by a syndrome-preserving transformation on the co-factor of the subsystem that contains the protected information is also fault tolerant. In fact, any transversal operation in a given fault-tolerant protocol can be safely replaced by an operation of the latter type. It is this latter type of transformations by means of which we can realize fault-tolerant HQC without the use of extra qubits.

We will show that by choosing as a starting Hamiltonian a suitable element of the stabilizer or the gauge group of the code and varying this Hamiltonian along appropriately chosen paths in parameter space, we can generate operations that – from the perspective of the full Hilbert space – transform both the ground and the excited spaces via the same transversal operation. This means, in particular, that the transformation of the code subsystem is the same as the one that would result from the application of the transversal operation, up to a transformation on the co-subsystem. The relative dynamical phase that accumulates between the ground and excited spaces is equal to a phase on the co-subsystem, which is irrelevant for the fault-tolerance of the scheme since it is either projected out when a measurement of the syndrome is performed, or is equivalent to a gauge transformation. Thus, by an appropriate sequence of such transformations we can take the code subsystem adiabatically along a loop such that the resultant holonomy at the end is by construction equal to a given encoded gate. This is the main idea behind our scheme. We note, however, that our scheme does not use an isodegenerate Hamiltonian along a given loop. This is because for simplicity we use Hamiltonians that are equal to a single element of the stabilizer or the gauge group at a given time, and we change the Hamiltonians along different portions of a loop. In this respect, the scheme we will present is slightly more general than the one described in Section 17.1, but the main idea behind its workings is the same – the full Hilbert space splits into different subspaces, each of which is adiabatically transported around a loop such that all subspaces undergo the same holonomy which factors out as an operation on the code subsystem. In the case of standard stabilizer codes, the subspaces in question are the syndrome subspaces. In the case of subsystem codes, each syndrome subspace further splits into subspaces containing redundant information. In the latter case, if along the loop we change between Hamiltonians that are noncommuting elements of the gauge group, these redundant subspaces may seem to undergo dynamical transformations in addition to the geometric ones. However, these dynamical effects are equivalent to gauge transformations and do not affect the holonomy taking place inside the code subsystem.

For concreteness, we will consider an $[[n, 1, r, 3]]$ stabilizer code. This is a code that encodes 1 qubit into n, has r gauge qubits, and can correct arbitrary single-qubit errors. As we saw in the previous section, in order to apply holonomic transformations on the subsystem that contains the logical information, we need a nontrivial starting Hamiltonian that leaves this subsystem invariant. It is easy to verify that the only Hamiltonians that satisfy this property are linear combinations of the elements of the stabilizer or the gauge group.

Note that the stabilizer and the gauge group transform during the course of the computation under the operations being applied. At any stage when we complete an encoded operation, they return to their initial forms modulo a gauge transformation. Our scheme will follow the same transversal operations as those used in a standard dynamical fault-tolerant scheme, but, as explained above, in addition we will have extra dynamical phases that multiply each syndrome subspace or are equivalent to more general gauge transformations. It is easy to see that these phases leave the stabilizer invariant and transform the gauge group via a gauge transformation, so without loss of generality we can omit them from our analysis of the transformation of these groups. During the implementation of a standard encoded gate, the Pauli group G_n on a given codeword may spread over other codewords, but it can be verified that this spreading can be limited to at most four other codewords including the "cat" state. This is because the encoded CNOT gate can be implemented fault-tolerantly on any stabilizer code by a transversal operation on four encoded qubits [G97a], and any encoded Clifford gate can be realized using only the encoded CNOT, provided that we are able to do fault-tolerant measurements (the encoded Clifford group is generated by the encoded Hadamard, phase and CNOT gates). Encoded gates outside of the Clifford group, such as the encoded $\pi/8$ or Toffoli gates, can be implemented fault-tolerantly using encoded CNOT gates conditioned on the qubits in a "cat" state, so they may require transversal operations on a total of five blocks. For CSS codes, however, the spreading of the Pauli group of one block during the implementation of a basic encoded operation can be limited to a total of three blocks, since the encoded CNOT gate has a transversal implementation [G97a].

It also should be noted that fault-tolerant encoded Clifford operations can be implemented using only Clifford gates on the physical qubits [G97a]. These operations transform the stabilizer and the gauge group into subgroups of the Pauli group, and their elements remain in the form of tensor products of Pauli matrices. The fault-tolerant implementation of encoded gates outside of the Clifford group, however, involves operations that take these groups outside of the Pauli group. We will, therefore, consider separately two cases: encoded operations in the Clifford group, and encoded operations outside of the Clifford group.

17.2.2 Encoded operations in the Clifford group

17.2.2.1 Single-qubit unitary operations For applying transformations on a given qubit, say, the first one, we will use as a starting Hamiltonian an element of the stabilizer (with a minus sign) or the gauge group of the code, which acts nontrivially on that qubit. Since we are considering codes that can correct arbitrary single-qubit errors, one can always find an element of the initial stabilizer or the initial gauge group that has a factor $\sigma_0 = I$, $\sigma_1 = X$, $\sigma_2 = Y$ or $\sigma_3 = Z$ acting on the first qubit, i.e.,

$$\widehat{G} = \sigma_i \otimes \widetilde{G}, \quad i = 0, 1, 2, 3, \tag{17.8}$$

where $\widetilde{G}$ is a tensor product of Pauli matrices and the identity on the remaining $n-1$ qubits. It can be verified that under Clifford gates the stabilizer and the gauge group transform in such a way that this is always the case except that the factor $\widetilde{G}$ may spread to qubits in other blocks. We can assume that the stabilizer spreads on at most five blocks including the "cat" state, since this is sufficient to implement any encoded operation. Henceforth, we will use "hat" to denote operators on all qubits on which the stabilizer spreads, and "tilde" to denote operators on all of these qubits except the first one.

Without loss of generality we will assume that the chosen stabilizer or gauge-group element for that qubit has the form

$$\widehat{G} = Z \otimes \widetilde{G}. \tag{17.9}$$

As initial Hamiltonian, we will take the operator

$$\widehat{H}(0) = -\widehat{G} = -Z \otimes \widetilde{G}. \tag{17.10}$$

Proposition 17.1. *If the initial Hamiltonian (17.10) is varied adiabatically so that only the factor acting on the first qubit changes,*

$$\widehat{H}(t) = -H(t) \otimes \widetilde{G}, \tag{17.11}$$

where

$$Tr\{H(t)\} = 0, \tag{17.12}$$

the transformation that each of the eigenspaces of this Hamiltonian undergoes will be equivalent to that driven by a local unitary on the first qubit, $\widehat{U}(t) \approx U(t) \otimes \widetilde{I}$.

Proof. Observe that (17.11) can be written as

$$\widehat{H}(t) = H(t) \otimes \widetilde{P}_0 - H(t) \otimes \widetilde{P}_1, \tag{17.13}$$

where

$$\widetilde{P}_0 = \frac{\widetilde{I} - \widetilde{G}}{2}, \quad \widetilde{P}_1 = \frac{\widetilde{I} + \widetilde{G}}{2}, \tag{17.14}$$

are orthogonal complementary projectors. The evolution driven by $\widehat{H}(t)$ is therefore

$$\widehat{U}(t) = U_0(t) \otimes \widetilde{P}_0 + U_1(t) \otimes \widetilde{P}_1, \tag{17.15}$$

where

$$U_{0,1}(t) = \mathscr{T}\exp\left(-i\int_0^t \pm H(\tau)d\tau\right). \tag{17.16}$$

Let $|\phi_0(t)\rangle$ and $|\phi_1(t)\rangle$ be the instantaneous ground and excited states of $H(t)$ with eigenvalues $E_{0,1}(t) = \mp E(t)$ ($E(t) > 0$). Then in the adiabatic limit we have

$$U_{0,1}(t) = e^{i\omega(t)}U_{A_{0,1}}(t) \oplus e^{-i\omega(t)}U_{A_{1,0}}(t), \tag{17.17}$$

where $\omega(t) = \int_0^t d\tau E(\tau)$ and

$$U_{A_{0,1}}(t) = e^{\int_0^t d\tau \langle \phi_{0,1}(\tau)| \frac{d}{d\tau} |\phi_{0,1}(\tau)\rangle} |\phi_{0,1}(t)\rangle\langle\phi_{0,1}(0)|. \tag{17.18}$$

The projectors on the ground and excited eigenspaces of $\widehat{H}(0)$ are

$$\widehat{P}_0 = |\phi_0(0)\rangle\langle\phi_0(0)| \otimes \widetilde{P}_0 + |\phi_1(0)\rangle\langle\phi_1(0)| \otimes \widetilde{P}_1 \tag{17.19}$$

and
$$\widehat{P}_1 = |\phi_1(0)\rangle\langle\phi_1(0)| \otimes \widetilde{P}_0 + |\phi_0(0)\rangle\langle\phi_0(0)| \otimes \widetilde{P}_1, \tag{17.20}$$

respectively. Using Eq. (17.17) and Eq. (17.18), one can see that the effect of the unitary (17.15) on each of these projectors is

$$\widehat{U}(t)\widehat{P}_0 = e^{i\omega(t)}(U_{A_0}(t) \oplus U_{A_1}(t)) \otimes \widetilde{I}\,\widehat{P}_0, \tag{17.21}$$
$$\widehat{U}(t)\widehat{P}_1 = e^{-i\omega(t)}(U_{A_0}(t) \oplus U_{A_1}(t)) \otimes \widetilde{I}\,\widehat{P}_1, \tag{17.22}$$

i.e., up to an overall dynamical phase its effect on each of the eigenspaces is the same as that of the unitary

$$\widehat{U}(t) = U(t) \otimes \widetilde{I}, \tag{17.23}$$

where

$$U(t) = U_{A_0}(t) \oplus U_{A_1}(t). \tag{17.24}$$

□

We next show how by suitably choosing $H(t)$ we can implement all necessary single-qubit gates. We will identify a set of points in parameter space, such that by interpolating between these points we can draw various paths resulting in the desired transformations.

Consider the single-qubit unitary operator

$$V^{\theta\pm} = \frac{1}{\sqrt{2}}\begin{pmatrix} 1 & \mp e^{-i\theta} \\ \pm e^{i\theta} & 1 \end{pmatrix}, \tag{17.25}$$

where θ is a real parameter. Note that $V^{\theta\mp} = (V^{\theta\pm})^\dagger$. Define the following single-qubit Hamiltonian:

$$H^{\theta\pm} \equiv V^{\theta\pm} Z V^{\theta\mp}. \tag{17.26}$$

Let $H(t)$ in Eq. (17.11) be a Hamiltonian that interpolates between $H(0) = Z$ and $H(T) = H^{\theta\pm}$ (up to a factor) as follows:

$$H(t) = f(t)Z + g(t)H^{\theta\pm} \equiv H^{\theta\pm}_{f,g}(t), \tag{17.27}$$

where $f(0), g(T) > 0$, $f(T) = g(0) = 0$. To simplify our notation, we will drop the indices f and g of the Hamiltonian, since the exact form of these functions is not important for our analysis as long as they are sufficiently smooth (see discussion in Section 17.2.2.2). This Hamiltonian has eigenvalues $\pm\sqrt{f(t)^2 + g(t)^2}$ and its energy gap is nonzero unless the entire Hamiltonian vanishes. It can be shown that in the adiabatic limit, the Hamiltonian (17.11) with $H(t) = H^{\theta\pm}(t)$ gives rise to the effective transformation

$$\widehat{U}^{\theta\pm}(T) = V^{\theta\pm} \otimes \widetilde{I} \tag{17.28}$$

on each eigenspace. Details of the proof can be found in [OBL09].

We will use this result to construct a set of standard gates by sequences of operations of the form $V^{\theta\pm}$, which can be generated by interpolations of the type (17.27) run forward or backward. For single-qubit gates in the Clifford group, we will only need three values of θ: 0, $\pi/2$ and $\pi/4$. For completeness, however, we will also demonstrate how to implement the $\pi/8$ gate, which together with the Hadamard gate is sufficient to generate any single-qubit unitary transformation [BMP+99]. For this we will need $\theta = \pi/8$. Note that

$$H^{\theta\pm} = \pm(\cos\theta X + \sin\theta Y), \tag{17.29}$$

so for these values of θ we have $H^{0\pm} = \pm X$, $H^{\pi/2\pm} = \pm Y$, $H^{\pi/4\pm} = \pm(\frac{1}{\sqrt{2}}X + \frac{1}{\sqrt{2}}Y)$, $H^{\pi/8\pm} = \pm(\cos\frac{\pi}{8}X + \sin\frac{\pi}{8}Y)$.

Consider the adiabatic interpolations between the following Hamiltonians:

$$-Z \otimes \widetilde{G} \to -Y \otimes \widetilde{G} \to Z \otimes \widetilde{G}. \tag{17.30}$$

According to the above result, the first interpolation yields the transformation $V^{\pi/2+}$. The second interpolation can be regarded as the inverse of $Z \otimes \widetilde{G} \to -Y \otimes \widetilde{G}$, which is equivalent to $-Z \otimes \widetilde{G} \to Y \otimes \widetilde{G}$ since $\widehat{H}(t)$ and $-\widehat{H}(t)$ yield the same geometric transformations. Thus the second interpolation results in $(V^{\pi/2-})^\dagger = V^{\pi/2+}$. The net result is therefore $V^{\pi/2+}V^{\pi/2+} = iX$. We see that up to a global phase the above sequence results in an implementation of the X gate in each eigenspace.

Similarly, one can verify that the Z gate can be realized via the loop

$$-Z \otimes \widetilde{G} \to -X \otimes \widetilde{G} \to Z \otimes \widetilde{G} \to Y \otimes \widetilde{G} \to -Z \otimes \widetilde{G}. \tag{17.31}$$

The phase gate P can be realized by applying

$$-Z \otimes \widetilde{G} \to -\left(\frac{1}{\sqrt{2}}X + \frac{1}{\sqrt{2}}Y\right) \otimes \widetilde{G} \to Z \otimes \widetilde{G}, \tag{17.32}$$

followed by the X gate.

The Hadamard gate W can be realized by first applying Z, followed by

$$-Z \otimes \widetilde{G} \to -X \otimes \widetilde{G}. \tag{17.33}$$

Finally, the $\pi/8$ gate T can be implemented by first applying $Y = iXZ$, followed by

$$Z \otimes \widetilde{G} \to -\left(\cos\frac{\pi}{8}X + \sin\frac{\pi}{8}Y\right) \otimes \widetilde{G} \to -Z \otimes \widetilde{G}. \tag{17.34}$$

17.2.2.2 A note on the adiabatic condition Before we show how to implement the CNOT gate, let us comment on the conditions under which the adiabatic approximation assumed in the above operations is satisfied. Because of the form (17.15) of the overall unitary, the adiabatic approximation depends on the extent to which each of the unitaries (17.16) approximate the expressions (17.17). The latter depends only on the adiabatic properties of the nondegenerate two-level Hamiltonian $H(t)$. For such a Hamiltonian, the simple version of the adiabatic condition [M65] reads

$$\frac{\varepsilon}{\Delta^2} \ll 1, \tag{17.35}$$

where

$$\varepsilon = \max_{0 \le t \le T} |\langle \phi_1(t)| \frac{dH(t)}{dt} |\phi_0(t)\rangle|, \tag{17.36}$$

and

$$\Delta = \min_{0 \le t \le T} (E_1(t) - E_0(t)) = \min_{0 \le t \le T} 2E(t) \tag{17.37}$$

is the minimum energy gap of $H(t)$.

Along the segments of the parameter paths we described, the Hamiltonian is of the form (17.27) and its derivative is

$$\frac{dH^{\theta\pm}(t)}{dt} = \frac{df(t)}{dt}Z + \frac{dg(t)}{dt}H^{\theta\pm}, \quad 0 < t < T. \tag{17.38}$$

This derivative is well defined as long as $\frac{df(t)}{dt}$ and $\frac{dg(t)}{dt}$ are well defined. The curves we described, however, may not be differentiable at the points connecting two segments. In order for the Hamiltonians (17.27) that interpolate between these points to be differentiable, the functions $f(t)$ and $g(t)$ have to satisfy $\frac{df(T)}{dt} = 0$ and $\frac{dg(0)}{dt} = 0$. This means that the change of the Hamiltonian slows down to zero at the end of each segment (except for a possible change in its strength), and increases again from zero along the next segment. We point out that when the Hamiltonian stops changing, we can turn it off completely by decreasing its strength. This can be done arbitrarily fast and it would not affect a state that belongs to an eigenspace of the Hamiltonian. Similarly, we can turn on another Hamiltonian for the implementation of a different operation.

The above condition guarantees that the adiabatic approximation is satisfied with precision $1 - O((\frac{\varepsilon}{\Delta^2})^2)$. It is known, however, that under certain conditions on the Hamiltonian, we can obtain better results [HJ02]. Let us write the Schrödinger equation as

$$i\frac{d}{dt}|\psi(t)\rangle = H(t)|\psi(t)\rangle \equiv \frac{1}{\epsilon}\bar{H}(t)|\psi(t)\rangle, \tag{17.39}$$

where $\epsilon > 0$ is small. Assume that $\bar{H}(t)$ is smooth and all its derivatives vanish at the end points $t = 0$ and $t = T$ (note that $\bar{H}(t)$ is nonanalytic at these points, unless it is constant). Then if we keep $\bar{H}(t)$ fixed and vary ϵ, the adiabatic error would scale super-polynomially with ϵ, i.e., the error will decrease with ϵ faster than $O(\epsilon^N)$ for any N. (Notice that $\frac{\varepsilon}{\Delta^2} \propto \epsilon$, i.e., the standard adiabatic condition guarantees error $O(\epsilon^2)$.)

In our case, the smoothness condition translates directly to the functions $f(t)$ and $g(t)$. For any smooth $f(t)$ and $g(t)$ we can further ensure that the condition at the end points is satisfied by the reparametrization $f(t) \to f(y(t))$, $g(t) \to g(y(t))$, where $y(t)$ is a smooth function of t that satisfies $y(0) = 0$, $y(T) = T$, and has vanishing derivatives at $t = 0$ and $t = T$. Then by slowing down the change of the Hamiltonian by a constant factor ϵ, which amounts to an increase of the total time T by a factor $1/\epsilon$, we can decrease the error super-polynomially in ϵ. We will use this result to obtain a low-error interpolation in Section 17.2.5, where we estimate the time needed to implement a holonomic gate with a given precision.

17.2.2.3 The CNOT gate The stabilizer (or gauge group) on multiple blocks of the code is a direct product of the stabilizers (or gauge groups) of the individual blocks. Therefore, from Eq. (17.8) it follows that one can always find an element of the initial stabilizer or gauge group on multiple blocks that has any desired combination of factors σ^i, $i = 0, 1, 2, 3$, on the first qubits in these blocks. It can be verified that applying transversal Clifford operations on the blocks does not change this property. Therefore, we can find an element of the stabilizer or the gauge group that has the form (17.9), where the factor Z acts on the target qubit and $\widetilde{G}$ acts trivially on the control qubit. We now explain how to implement the CNOT gate adiabatically starting from such a Hamiltonian.

Notice that a Hamiltonian of the form

$$\widehat{\bar{H}}(t) = |0\rangle\langle 0|^c \otimes H_0(t) \otimes \widetilde{G} + |1\rangle\langle 1|^c \otimes H_1(t) \otimes \widetilde{G}, \tag{17.40}$$

where the superscript c denotes the control qubit, gives rise to the unitary transformation

$$\widehat{\widehat{U}}(t) = |0\rangle\langle 0|^c \otimes \widehat{U}_0(t) + |1\rangle\langle 1|^c \otimes \widehat{U}_1(t), \qquad (17.41)$$

where

$$\widehat{U}_{0,1}(t) = \mathcal{T}\exp\left(-i\int_0^t d\tau H_{0,1}(\tau) \otimes \widetilde{G}\right). \qquad (17.42)$$

If $H_0(t)$ and $H_1(t)$ have the same nondegenerate instantaneous spectra, and $\text{Tr}\{H_{0,1}(t)\} = 0$, then it follows that in the adiabatic limit each of the eigenspaces of $\widehat{\widehat{H}}(t)$ will undergo the transformation

$$\widehat{\widehat{U}}_g(t) = |0\rangle\langle 0|^c \otimes V_0(t) \otimes \widetilde{I} + |1\rangle\langle 1|^c \otimes V_1(t) \otimes \widetilde{I}, \qquad (17.43)$$

where $V_{0,1}(t) \otimes \widetilde{I}$ are the effective transformations generated by $H_{0,1}(t) \otimes \widetilde{G}$ according to Proposition 17.1.

Our goal is to find $H_0(t)$ and $H_1(t)$, $H_0(0) = H_1(0) = Z$, such that at the end of the transformation the unitary (17.43) will be equal to the CNOT gate. In other words, we want $V_0(2T) = I$ and $V_1(2T) = X$ (we choose the total time of evolution to be $2T$ for convenience).

We already saw how to generate adiabatically the X gate up to a phase – Eq. (17.30). We can use the same Hamiltonian in place of $H_1(t)$:

$$H_1(t) = \begin{cases} H^{\pi/2+}(t), & 0 \leq t \leq T \\ H^{\pi/2-}(2T-t), & T \leq t \leq 2T. \end{cases} \qquad (17.44)$$

Now we want to find a Hamiltonian $H_0(t)$ with the same spectrum as $H_1(t)$, which gives rise to a trivial geometric transformation, $V_0(2T) = I$ (possibly up to a phase, which can be undone later). Since all Hamiltonians of the type $H^{\theta\pm}(t)$ have the same instantaneous spectrum (for fixed $f(t)$ and $g(t)$), we can simply choose

$$H_0(t) = \begin{cases} H^{\pi/2+}(t), & 0 \leq t \leq T \\ H^{\pi/2+}(2T-t), & T \leq t \leq 2T, \end{cases} \qquad (17.45)$$

which corresponds to applying a given transformation from $t = 0$ to $t = T$ and then undoing it (running it backwards) from $t = T$ to $t = 2T$. This results exactly in $V_0(2T) = I$.

Since, as we saw in Section 17.2.2.1, the Hamiltonian $H_1(t) \otimes \widetilde{G}$ gives rise to the transformation $iX \otimes \widetilde{I}$ in each of its eigenspaces, the above choice for the Hamiltonians (17.45) and (17.44) in Eq. (17.40) will result in the transformation

$$|0\rangle\langle 0|^c \otimes I \otimes \widetilde{I} + i|1\rangle\langle 1|^c \otimes X \otimes \widetilde{I}, \qquad (17.46)$$

which is the desired CNOT gate up to a phase gate on the control qubit. We can correct the phase by applying the inverse of the phase gate to the control qubit, either before or after the described transformation.

Notice that from $t = 0$ to $t = T$ the Hamiltonians (17.45) and (17.44) are identical, i.e., during this period the Hamiltonian (17.40) has the form

$$I^c \otimes H^{\pi/2+}(t) \otimes \widetilde{G}, \qquad (17.47)$$

so we are simply applying the single-qubit operation $V^{\pi/2+}$ to the target qubit according to the method for single-qubit gates described before. It is easy to verify that during the second period, from $t = T$ to $t = 2T$, the Hamiltonian (17.40) realizes the interpolation

$$-I^c \otimes Y \otimes \widetilde{G} \to -Z^c \otimes Z \otimes \widetilde{G}, \qquad (17.48)$$

which is understood as in Eq. (17.27).

To summarize, the CNOT gate can be implemented by first applying the inverse of the phase gate ($P^\dagger$) on the control qubit, as well as the transformation $V^{\pi/2+}$ on the target qubit, followed by the transformation (17.48). Due to the form (17.40) of $\widehat{\widetilde{H}}(t)$, the extent to which the adiabatic approximation is satisfied during this transformation depends only on the adiabatic properties of the single-qubit Hamiltonians $H^{\pi/2\pm}(t)$, which we discussed in the previous subsection.

17.2.3 Encoded operations outside of the Clifford group

For universal fault-tolerant computation we also need at least one encoded gate outside of the Clifford group. The fault-tolerant implementation of such gates is based on the preparation of a special encoded state [S96b, G97a, KLZ98b, BMP+99, ZLC00], which involves a measurement of an encoded operator in the Clifford group. For example, the $\pi/8$ gate requires the preparation of the encoded state $\frac{|0\rangle + \exp(i\pi/4)|1\rangle}{\sqrt{2}}$, which can be realized by measuring the encoded operator $e^{-i\pi/4}PX$ [BMP+99]. Equivalently, the state can be obtained by applying the encoded operation $WP^\dagger$ on the encoded state $\frac{\cos(\pi/8)|0\rangle + \sin(\pi/8)|1\rangle}{\sqrt{2}}$, which can be prepared by measuring the encoded Hadamard gate [KLZ98b]. The Toffoli gate requires the preparation of the three-qubit encoded state $\frac{|000\rangle + |010\rangle + |100\rangle + |111\rangle}{2}$ and involves a similar procedure [ZLC00]. In all these instances, the measurement of the encoded Clifford operator is realized by applying transversally the operator conditioned on the qubits in a "cat" state.

We now describe a general method that can be used to implement adiabatically any conditional transversal Clifford operation with conditioning on the "cat" state. Let O be a Clifford gate acting on the first qubits from some set of blocks. As we discussed in the previous section, under this unitary the stabilizer and the gauge group transform in such a way that we can always find an element with an arbitrary combination of Pauli matrices on the first qubits. If we write this element in the form

$$\widehat{G} = G_1 \otimes G_{2,\ldots,n}, \qquad (17.49)$$

where G_1 is a tensor product of Pauli matrices acting on the first qubits from the blocks, and $G_{2,\ldots,n}$ is an operator on the rest of the qubits, then applying O conditioned on the first qubit in a "cat" state transforms this stabilizer or gauge-group element as follows:

$$I^c \otimes G_1 \otimes G_{2,\ldots,n} = |0\rangle\langle 0|^c \otimes G_1 \otimes G_{2,\ldots,n} + |1\rangle\langle 1|^c \otimes G_1 \otimes G_{2,\ldots,n}$$
$$\to |0\rangle\langle 0|^c \otimes G_1 \otimes G_{2,\ldots,n} + |1\rangle\langle 1|^c \otimes OG_1O^\dagger \otimes G_{2,\ldots,n}, \qquad (17.50)$$

where the superscript c denotes the control qubit from the "cat" state. We can implement this operation by choosing the factor G_1 to be the same as the one we would use if we wanted to implement the operation O according to the previously described procedure. Then we can apply the following Hamiltonian:

$$\widehat{\widetilde{H}}_{C(O)}(t) = -|0\rangle\langle 0|^c \otimes G_1 \otimes G_{2,\ldots,n} - \alpha(t)|1\rangle\langle 1|^c \otimes H_O(t) \otimes G_{2,\ldots,n}, \qquad (17.51)$$

where $H_O(t) \otimes G_{2,...,n}$ is the Hamiltonian that we would use for the implementation of the operation O and $\alpha(t)$ is a real parameter chosen such that at every moment the operator $\alpha(t)|1\rangle\langle 1|^c \otimes H_O(t) \otimes G_{2,...,n}$ has the same instantaneous spectrum as the operator $|0\rangle\langle 0|^c \otimes G_1 \otimes G_{2,...,n}$. This guarantees that the overall Hamiltonian is degenerate and the transformation on each of its eigenspaces is

$$\widehat{\widetilde{U}}_g(t) = |0\rangle\langle 0|^c \otimes I_1 \otimes I_{2,...,n} + |1\rangle\langle 1|^c \otimes U_O(t) \otimes I_{2,...,n}, \tag{17.52}$$

where $U_O(t)$ is the transformation on the first qubits generated by $H_O(t) \otimes G_{2,...,n}$. Since we presented the constructions of the basic Clifford operations up to an overall phase, the operation $U_O(t)$ may differ from the desired operation by a phase. This phase can be corrected by applying a suitable gate on the control qubit from the "cat" state (we explain how this can be done in the next section). We remark that a Hamiltonian of the type (17.51) requires fine tuning of the parameter $\alpha(t)$ and generally can be complicated. In Section 17.2.6 we will show that depending on the code one can find more natural implementations of these operations.

If we want to apply a second conditional Clifford operation Q on the first qubits in the blocks, we can do this as follows. Imagine that if we had to apply the operation Q following the operation O, we would use the Hamiltonian $\widehat{H}_Q(t) = -H_Q(t) \otimes G'_{2,...,n}$, where $\widehat{H}_Q(0) = OG'_1 O^\dagger \otimes G'_{2,...,n}$ is a suitable element of the stabilizer or the gauge group after the application of O. Before the application of O, that element would have had the form $G'_1 \otimes G'_{2,...,n}$. Under the application of a conditional O, the element $G'_1 \otimes G'_{2,...,n}$ transforms to $|0\rangle\langle 0|^c \otimes G'_1 \otimes G'_{2,...,n} + |1\rangle\langle 1|^c \otimes OG'_1 O^\dagger \otimes G'_{2,...,n}$, which can be used (with a minus sign) as a starting Hamiltonian for a subsequent operation. In particular, we can implement the conditional Q following the conditional O using the Hamiltonian

$$\widehat{\widetilde{H}}_{C(Q)}(t) = -|0\rangle\langle 0|^c \otimes G'_1 \otimes G'_{2,...,n} - \beta(t)|1\rangle\langle 1|^c \otimes H_Q(t) \otimes G'_{2,...,n}, \tag{17.53}$$

where the factor $\beta(t)$ guarantees that there is no splitting of the energy levels. Subsequent operations can be applied analogously. Using this general method, we can implement a unitary whose geometric part is equal to any transversal Clifford operation conditioned on the "cat" state.

17.2.4 Preparing and using the "cat" state

In addition to transversal operations, a complete fault-tolerant scheme requires the ability to prepare, verify and use a special ancillary state such as the "cat" state $(|00...0\rangle + |11...1\rangle)/\sqrt{2}$. This can also be done in the spirit of the described holonomic scheme. Since the "cat" state is known and its construction is non-fault-tolerant, we can prepare it by simply treating each initially prepared qubit as a simple code (with $\widetilde{G}$ in Eq. (17.9) being trivial), and updating the stabilizer of the code via the applied transformation as the operation progresses. The stabilizer of the prepared "cat" state is generated by $Z_i Z_j$, $i < j$. Transversal unitary operations between the "cat" state and other codewords are applied as described in the previous sections.

We also have to be able to measure the parity of the state, which requires the ability to apply successive CNOT operations from two different qubits in the "cat" state to the same ancillary qubit initially prepared in the state $|0\rangle$. We can regard a qubit in state $|0\rangle$ as a simple code with stabilizer $\langle Z \rangle$, and we can apply the first CNOT as described before. Even though after this operation the state of the target qubit is unknown, the second CNOT gate can be applied via the same interaction, since the transformation in each eigenspace of the Hamiltonian is the same and at the end when we measure the qubit we project onto one of the eigenspaces.

17.2.5 Fault tolerance of the scheme

We saw how we can transform the code subsystem by any transversal operation adiabatically, which allows us, by a sequence of such operations, to generate a holonomy in the subsystem that is equivalent to any encoded gate. This was based on the assumption that the state has not undergone an error. But what if an error occurs on one of the qubits?

At any moment, we can distinguish two types of errors – those that result in transitions between the ground and the excited spaces of the current Hamiltonian, and those that result in transformations inside the eigenspaces. Due to the discretization of errors in quantum error correction, it suffices to prove correctability for each type separately. The key property of the construction we described is that each of the eigenspaces undergoes the same transversal transformation. Because of this, if we are applying a unitary on the first qubit, an error on that qubit will remain localized regardless of whether it causes an excitation or not. If the error occurs on one of the other qubits, at the end of the transformation the result would be the desired single-qubit unitary gate plus the error on the other qubit, which is correctable.

We see that even though the Hamiltonian couples qubits in the same block, single-qubit errors do not propagate. This is because the coupling between the qubits amounts to a change in the relative phase between the ground and excited spaces, but the latter is irrelevant since either it is equivalent to a gauge transformation, or when we apply a correcting operation we project on one of the eigenspaces. In the case of the CNOT gate, an error can propagate between the control and the target qubits, but it never results in two errors within the same codeword.

17.2.6 FTHQC with the Bacon–Shor code using 3-local Hamiltonians

The weight of the Hamiltonians needed for the scheme we described depends on the weight of the stabilizer or gauge-group elements. Remarkably, certain codes possess stabilizer or gauge-group elements of low weight covering all qubits in the code, which allows us to perform holonomic computation using low-weight Hamiltonians. Here we will consider as an example a subsystem generalization of the nine-qubit Shor code [S95] – the Bacon–Shor code [B06, BC06] – which has particularly favorable properties for fault-tolerant computation [AC07, A07]. In the nine-qubit Bacon–Shor code, the gauge group is generated by the weight-two operators $Z_{k,j}Z_{k,j+1}$ and $X_{j,k}X_{j+1,k}$, where the subscripts label the qubits by row and column when they are arranged in a 3×3 square lattice. Since the Bacon–Shor code is a CSS code, the CNOT gate has a direct transversal implementation. We now show that the CNOT gate can be realized using at most weight-three Hamiltonians.

If we want to apply a CNOT gate between two qubits each of which is, say, in the first row and column of its block, we can use as a starting Hamiltonian $-Z_{1,1}^t \otimes Z_{1,2}^t$, where the superscript t signifies that these are operators in the target block. We can then apply the CNOT gate as described in the previous section. After the operation, however, this gauge-group element will transform to $-Z_{1,1}^t \otimes Z_{1,1}^c \otimes Z_{1,2}^t$. If we now want to implement a CNOT gate between the qubits with index $\{1, 2\}$ using as a starting Hamiltonian the operator $-Z_{1,1}^t \otimes Z_{1,1}^c \otimes Z_{1,2}^t$ according to the same procedure, we will have to use a four-qubit Hamiltonian. Of course, at this point we can use the starting Hamiltonian $-Z_{1,2}^t \otimes Z_{1,3}^t$, but if we had also applied a CNOT between the qubits labeled $\{1,3\}$, this operator would not be available – it would have transformed to $-Z_{1,2}^t \otimes Z_{1,3}^t \otimes Z_{1,3}^c$.

What we can do instead, is to use as a starting Hamiltonian the operator $-Z_{1,1}^t \otimes Z_{1,2}^t \otimes Z_{1,2}^c$, which is obtained from the gauge-group element $Z_{1,1}^t \otimes Z_{1,1}^c \otimes Z_{1,2}^t \otimes Z_{1,2}^c$ after the application of the CNOT between the qubits with index $\{1,1\}$. Since the CNOT gate is its own inverse, we can regard the factor $Z_{1,1}^t$ as $\widetilde{G}$ in Eq. (17.48) and use this starting Hamiltonian to apply the procedure backwards. Thus we can implement any transversal CNOT gate using at most weight-three Hamiltonians.

Since the encoded X, Y, and Z operations have a bit-wise implementation, we can always apply them according to the described procedure using Hamiltonians of weight two. For the Bacon–Shor code, the encoded Hadamard gate can be applied via bit-wise Hadamard transformations followed by a rotation of the grid by a 90 degree angle [AC07]. The encoded P gate can be implemented by using the encoded CNOT and an ancilla.

The preparation and measurement of the "cat" state can also be done using Hamiltonians of weight two. To prepare the "cat" state, we prepare first all qubits in the state $(|0\rangle + |1\rangle)/\sqrt{2}$, which can be done by measuring each of them in the $\{|0\rangle, |1\rangle\}$ basis (this ability is assumed for any type of computation) and applying the transformation $-Z \to -X$ or $Z \to -X$ depending on the outcome. To complete the preparation of the "cat" state, apply a two-qubit transformation between the first qubit and each of the other qubits ($j > 1$) via the transformation

$$-I_1 \otimes X_j \to -Z_1 \otimes Z_j. \tag{17.54}$$

Single-qubit transformations on qubits from the "cat" state can be applied according to the method described in the previous section using at most weight-two Hamiltonians.

To measure the parity of the state, we need to apply successively CNOT operations from two different qubits in the "cat" state to the same ancillary qubit initially prepared in the state $|0\rangle$. This can also be done according to the method described in Section 17.2.4 and requires Hamiltonians of weight two.

For universal computation with the Bacon–Shor code, we also need to be able to apply one encoded transformation outside of the Clifford group. As we mentioned earlier, in order to implement the Toffoli gate or the $\pi/8$ gate, it is sufficient to be able to implement a CNOT gate conditioned on a "cat" state. For the Bacon–Shor code, the CNOT gate has a transversal implementation, so the conditioned CNOT gate can be realized by a series of transversal Toffoli operations between the "cat" state and the two encoded states. We will now show that this gate can be implemented using at most three-qubit Hamiltonians.

Reference [NC00] provides a circuit for implementing the Toffoli gate as a sequence of one- and two-qubit gates. We will use the same circuit, except that we flip the control and target qubits in every CNOT gate using the identity

$$(W_1 W_2) C X_{1,2} (W_1 W_2) = C X_{2,1}, \tag{17.55}$$

where W_i denotes a Hadamard gate on the qubit labeled by i and $CX_{i,j}$ denotes a CNOT gate between qubits i and j with i being the control and j being the target. Let Toffoli$_{i,j,k}$ denote the Toffoli gate on qubits i, j, and k with i and j being the two control qubits and k being the target qubit, and let P_i and T_i denote the phase and $\pi/8$ gates on qubit i, respectively. Then the Toffoli gate on three qubits (the first one of which we will assume to belong to the "cat" state), can be written as

$$\text{Toffoli}_{1,2,3} = W_2 C X_{3,2} W_3 T_3^\dagger W_3 W_1 C X_{3,1} W_3 T_3 W_3 C X_{3,2} W_3 T_3^\dagger$$
$$\times W_3 C X_{3,1} W_3 T_3 W_3 W_2 T_2^\dagger W_2 C X_{2,1} W_2 T_2^\dagger W_2 C X_{2,1} W_2 P_2 W_1 T_1. \quad (17.56)$$

To show that each of the above gates can be implemented adiabatically using Hamiltonians of weight at most three, we will need an implementation of the CNOT gate that is suitable for the case when we have a stabilizer or gauge-group element of the form

$$\widehat{G} = X \otimes \widetilde{G}, \quad (17.57)$$

where the factor X acts on the target qubit and $\widetilde{G}$ acts trivially on the control qubit. By a similar argument to the one in Section 17.2.2.3, one can verify that in this case the CNOT gate can be implemented as follows: Apply the operation $P^\dagger$ on the control qubit (we describe how to do this for our particular case below) together with the transformation

$$-X \otimes \widetilde{G} \to -Z \otimes \widetilde{G} \to X \otimes \widetilde{G} \quad (17.58)$$

on the target qubit, followed by the transformation

$$I^c \otimes X \otimes \widetilde{G} \to -(|0\rangle\langle 0|^c \otimes Z + |1\rangle\langle 1|^c \otimes Y) \otimes \widetilde{G} \to -I^c \otimes X \otimes \widetilde{G}. \quad (17.59)$$

Since the second and the third qubits belong to blocks encoded with the Bacon–Shor code, there are weight-two elements of the initial gauge group of the form $Z \otimes Z$ covering all qubits. The stabilizer generators on the "cat" state are also of this type. Following the transformation of these operators according to the sequence of operations (17.56), one can see that before every CNOT gate in this sequence there is an element of the form (17.57) with $\widetilde{G} = Z$ that can be used to implement the CNOT gate as described, provided that we can implement the gate $P^\dagger$ on the control qubit. We also point out that all single-qubit operations on qubit 1 in this sequence can be implemented according to the procedure described in Section 17.2.2.1, since at every step we have a weight-two stabilizer element on that qubit with a suitable form. Therefore, all we need to show is how to implement the necessary single-qubit operations on qubits 2 and 3. Due to the complicated transformation of the gauge-group elements during the sequence of operations (17.56), we will introduce a method of applying a single-qubit operation with a starting Hamiltonian that acts trivially on the qubit. For implementing single-qubit operations on qubits 2 and 3 we will use as a starting Hamiltonian the operator

$$\widehat{H}(0) = -I_i \otimes X_1 \otimes \widetilde{Z}, \quad i = 2, 3, \quad (17.60)$$

where the first factor (I_i) acts on the qubit on which we want to apply the operation (2 or 3), and $X_1 \otimes \widetilde{Z}$ is the transformed (after the Hadamard gate R_1) stabilizer element of the "cat" state that acts nontrivially on qubit 1 (the factor $\widetilde{Z}$ acts on some other qubit in the "cat" state).

To implement a single-qubit gate on qubit 3 for example, we first apply the interpolation

$$-I_3 \otimes X_1 \otimes \widetilde{Z} \to -Z_3 \otimes Z_1 \otimes \widetilde{Z}. \quad (17.61)$$

This results in a two-qubit transformation $U_{1,3}$ on qubits 1 and 3 in each eigenspace. We do not have to calculate this transformation exactly since we will undo it later, but the fact that each eigenspace undergoes the same two-qubit transformation can be verified similarly to the CNOT gate we described in Section 17.2.2.3.

At this point, the Hamiltonian is of the form (17.10) with respect to qubit 3, and we can apply any single-qubit unitary gate V_3 according to the method described in Section 17.2.2.1. This transforms the Hamiltonian to $-V_3 Z_3 V_3^\dagger \otimes Z_1 \otimes \widetilde{Z}$. We can now "undo" the transformation $U_{1,3}$ by the interpolation

$$-V_3 Z_3 V_3^\dagger \otimes Z_1 \otimes \widetilde{Z} \to -I_3 \otimes X_1 \otimes \widetilde{Z}. \tag{17.62}$$

The latter transformation is the inverse of Eq. (17.61) up to the single-qubit unitary transformation V_3, i.e., it results in the transformation $V_3 U_{1,3}^\dagger V_3^\dagger$. Thus, the net result is

$$V_3 U_{1,3}^\dagger V_3^\dagger V_3 U_{1,3} = V_3, \tag{17.63}$$

which is the desired single-qubit unitary transformation on qubit 3. Note that during this transformation, a single-qubit error can propagate between qubits 1 and 3, but this is not a problem since we are implementing a transversal Toffoli operation and such an error would not result in more than one error per block of the code.

We see that with the Bacon–Shor code the above scheme for fault-tolerant HQC can be implemented with at most 3-local Hamiltonians. This is optimal for this approach, because there are no nontrivial codes with stabilizer or gauge-group elements of weight smaller than two covering all qubits.

17.2.7 Effects on the accuracy threshold

Since the method we described conforms completely to a given fault-tolerant scheme, it would not affect the error threshold per qubit per operation for that scheme. However, the allowed errors per qubit per operation include both errors due to imperfectly applied transformations and errors due to interaction with the environment. It turns out that certain features of the method we described have an effect on the allowed distribution of these errors within the accuracy threshold.

First, observe that when applying the Hamiltonian (17.11), we cannot at the same time apply operations on the other qubits on which the factor $\widetilde{G}$ acts nontrivially. Thus, some operations at the lowest level of concatenation that would otherwise be implemented simultaneously might have to be implemented serially. The effect of this is equivalent to slowing down the circuit by a constant factor. (Note that we could also vary the factor $\widetilde{G}$ simultaneously with $H(t)$, but in order to obtain the same precision as that we would achieve by a serial implementation, we would have to slow down the change of the Hamiltonian by the same factor.) The slowdown factor resulting from this loss of parallelism is usually small, since this problem occurs only at the lowest level of concatenation. It can be verified that for the Bacon–Shor code, we can apply operations on up to six out of the nine qubits in a block simultaneously. For example, when applying encoded single-qubit operations, we can address simultaneously any two qubits in a row or column by taking $\widetilde{G}$ to be a single-qubit operator Z or X on the third qubit in the same row or column. The Hamiltonians used to apply operations on the two qubits commute with each other at all times and do not interfere. A similar phenomenon holds for the implementation of the encoded CNOT gate or the operations involving the "cat" state. Thus, for the Bacon–Shor code we have a slowdown due to parallelism by a factor of 1.5.

A more significant slowdown results from the fact that the evolution is adiabatic. In order to obtain a rough estimate of the slowdown due specifically to the adiabatic requirement, we will compare the time T_h needed for the implementation of a gate according to our method with precision $1 - \delta$ to the time T_d needed for a dynamical realization of the same gate with the same strength of the Hamiltonian. We will consider a realization of the X gate via the unitary interpolation

$$\widehat{H}(t) = -V_X(\tau(t))ZV_X^\dagger(\tau(t)) \otimes \widetilde{G}, \quad V_X(\tau(t)) = \exp\left(i\tau(t)\frac{\pi}{2T_h}X\right), \qquad (17.64)$$

where $\tau(0) = 0$, $\tau(T_h) = T_h$. The energy gap of this Hamiltonian is constant. The optimal dynamical implementation of the same gate is via the Hamiltonian $-X$ for time $T_d = \frac{\pi}{2}$.

As we argued in Section 17.2.2.2, the accuracy with which the adiabatic approximation holds for the Hamiltonian (17.64) is the same as that for the Hamiltonian

$$H(t) = V_X(\tau(t))ZV_X^\dagger(\tau(t)). \qquad (17.65)$$

We now present estimates for two different choices of the function $\tau(t)$. The first one is

$$\tau(t) = t. \qquad (17.66)$$

In this case the Schrödinger equation can be easily solved in the instantaneous eigenbasis of the Hamiltonian (17.65). For the probability that the initial ground state remains a ground state at the end of the evolution, we obtain

$$p = \frac{1}{1+\varepsilon^2} + \frac{\varepsilon^2}{1+\varepsilon^2}\cos^2\left(\frac{\pi}{4\varepsilon}\sqrt{1+\varepsilon^2}\right) = 1 - \delta, \qquad (17.67)$$

where

$$\varepsilon = \frac{T_d}{T_h}. \qquad (17.68)$$

Expanding in powers of ε and averaging the square of the cosine whose period is much smaller than T_h, we obtain the condition

$$\varepsilon^2 \leq 2\delta. \qquad (17.69)$$

Assuming, for example, that $\delta \approx 10^{-4}$ (approximately the threshold for the nine-qubit Bacon–Shor code [AC07]), we obtain that the time of evolution for the adiabatic case must be about 70 times longer than that in the dynamical case.

It is known, however, that if $H(t)$ is smooth and its derivatives vanish at $t = 0$ and $t = T_h$, the adiabatic error decreases super-polynomially with T_h [HJ02]. To achieve this, we choose

$$\tau(t) = \frac{1}{a}\int_0^t dt' e^{-1/\sin(\pi t'/T_h)}, \quad a = \int_0^{T_h} dt' e^{-1/\sin(\pi t'/T_h)}. \qquad (17.70)$$

For this interpolation, by a numerical solution we obtain that when $T_h/T_d \approx 17$ the error is already of the order of 10^{-6}, which is well below the threshold values obtained for the Bacon–Shor codes [AC07]. This is a remarkable improvement in comparison to the previous interpolation, which shows that the smoothness of the Hamiltonian plays an important role in the performance of the scheme.

An additional slowdown in comparison to a perfect dynamical scheme may result from the fact that the constructions for some of the standard gates we presented involve long sequences of loops. With more efficient parameter paths, however, it should be possible to reduce this slowdown to a minimum.

In comparison to a dynamical implementation, the allowed rate of environmental noise for the holonomic case would decrease by a factor similar to the slowdown factor. In practice, however, dynamical gates are not perfect and the holonomic approach may be advantageous if it allows for a better precision.

We finally point out that an error in the factor $H(t)$ in the Hamiltonian (17.11) would result in an error on the first qubit according to Eq. (17.24). Such an error clearly has to be below the accuracy threshold. More dangerous errors, however, are also possible. For example, if the degeneracy of the Hamiltonian is broken, this can result in an unwanted dynamical transformation affecting all qubits on which the Hamiltonian acts nontrivially. Such multi-qubit errors have to be of higher order in the threshold, which imposes more severe restrictions.

17.3 Conclusion and outlook

In this chapter we saw that HQC can be made fault tolerant by combining it with the techniques for fault-tolerant quantum error correction on stabilizer codes. This means that HQC is, at least in principle, a scalable method of computation. However, further research is needed in order to bring the presented ideas closer to experimental realization.

We presented a scheme which uses Hamiltonians that are elements of the stabilizer or the gauge group of the code. We saw that with the Bacon–Shor code, this scheme can be implemented with two- and two-qubit Hamiltonians. Since the scheme conforms completely to a given dynamical fault-tolerant scheme and does not require the use of extra qubits, it has the same error threshold as the dynamical scheme on which it is based. However, due to the fact that adiabatic gates are slower than dynamical gates and that at the lowest concatenation level this scheme requires certain gates to be implemented serially or more slowly, the allowed error threshold for environmental noise is lower in comparison to a dynamical scheme. The factor by which this threshold decreases depends on how smooth the adiabatic interpolations are, but it seems to be at most $\sim 10^2$. Therefore, if the robustness provided by the geometric nature of the gates is sufficiently higher than that achievable by dynamical means, this approach could be advantageous in comparison to dynamical schemes. The main challenge in the implementation of this approach, however, is that it requires the engineering of 3-local Hamiltonians. An alternative scheme for fault-tolerant HQC that uses Hamiltonians independent of the code at the expense of additional qubits [O09] has been proven to allow reducing the locality of the Hamiltonian with perturbative gadget techniques, showing that 2-local Hamiltonians are universal for fault-tolerant HQC. The disadvantage of using the gadgets is that they decrease the gap of the Hamiltonian by a very large factor, which requires a significant slowdown of the computation and decreases the allowed rate of environmental noise.

Applying the strategies we have described to actual physical systems will undoubtedly require modifications in accordance with the available interactions in those systems. A possible way of avoiding the use of multi-local Hamiltonians without the use of perturbative gadgets could be to use higher-dimensional systems (e.g., qutrits) between which 2-local interactions are naturally available. It may be possible to encode qubits in subspaces or subsystems of these higher-dimensional systems and use fault-tolerant techniques designed for stabilizer codes based on qudits [G99]. Given that simple quantum error-correcting codes and two-qubit geometric transformations have been realized using NMR [CPM+98, JVE+00] and ion-trap [CLS+04, LDM+03] techniques, these systems seem particularly suitable for hybrid HQC–QEC implementations.

Finally, it is interesting to point out that the adiabatic regime in which the holonomic schemes operate is consistent with the Markovian model of decoherence. In [ALZ06] it was argued that the standard dynamical paradigm of fault tolerance is based on assumptions that are in conflict with the rigorous derivation of the Markovian limit. Although the threshold theorem has been extended to non-Markovian models [TB05, AGP06, AKP06], the Markovian assumption is an accurate approximation for a wide range of physical scenarios [C93] and allows for a much simpler description of the evolution in comparison to non-Markovian models (see Chapter 8). In [ALZ06] it was shown that the weak-coupling-limit derivation of the Markovian approximation is consistent with computational methods that employ slow transformations, such as adiabatic quantum computation [FGG+00] or HQC. A theory of fault tolerance for the adiabatic model of computation at present is not known, although steps in this direction have been undertaken [JFS06, L08a]. The hybrid HQC–QEC schemes presented here provide solutions for the case of HQC. However, we point out that it is an open problem whether the Markovian approximation makes sense for a fixed value of the adiabatic slowness parameter when the circuit increases in size. Giving a definitive answer to this question requires a rigorous analysis of the accumulation of non-Markovian errors due to deviation from perfect adiabaticity.

The techniques described in this chapter may prove useful in other areas. It is possible that some combination of transversal adiabatic transformations and active correction could provide a solution to the problem of fault tolerance in the adiabatic model of computation.

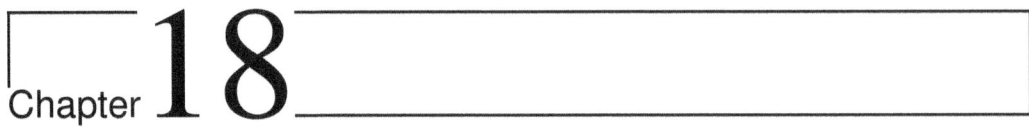

Fault-tolerant measurement-based quantum computing

Debbie Leung

18.1 Introduction

In the standard model of quantum computation [D95], some simple quantum state is prepared and evolved by a partially ordered set of *unitary* gates, and finally is measured in the computation basis. Up until the late 1990s, transformations that deviate from unitarity were commonly associated with irreversibility, and thus *incoherence* and harm induced in the quantum data of interest. However, this viewpoint was challenged by three notable results: (1) teleportation [BBC+93], (2) syndrome measurements [S95, S96e], and (3) the fault-tolerant Toffoli and $\pi/8$-gate constructions [S96b, BMP+99]. In these schemes, well-designed measurements project the quantum data of interest onto a subspace of the original ambient space for each outcome, up to some known unitary transformation that is either reversible (and harmless) or intended. Subsequently, researchers found many useful error-correction procedures and unitary gates that rely heavily on measurements.

Two "measurement models" of quantum computation were proposed in the early 2000s that make extreme use of measurements. In the cluster model [RB01], any quantum computation can be performed by applying a partially time-ordered set of single-qubit measurements on a certain entangled state. The state must be of appropriate size, but is otherwise independent of the computation. The related graph state model [RBB03] uses an initial state that depends on the computation itself. In the seemingly different teleportation-based model [N03, L01a, L04], any quantum computation can be implemented by applying one- or two-qubit *nondemolition* measurements on an arbitrary initial state, so as to "teleport" each gate in the simulated circuit. In all these models, the measurement bases are chosen according to earlier measurement outcomes. Recently, a unified derivation based on gate teleportation and connection to the circuit model was found [AL04, CLN05, JP05].

These measurement models of computation require unusual resources and have shallow circuits (at the expense of more memory space), making them useful analytic tools in the theory of

---

Quantum Error Correction, ed. Daniel A. Lidar and Todd A. Brun. Published by Cambridge University Press.
© Cambridge University Press 2013.

computation and communication complexity. Furthermore, the cluster and the graph state models offer potential practical advantages. For example, the idea has been adapted [N04b] to enhance the linear optics implementation of quantum computation [KLM01].

However, as in all implementations, noise and inaccuracies are inevitable. Also, even if it is possible to perform a set of operations almost perfectly, it may take fewer resources to perform the slightly imperfect versions followed by error correction. Thus, fault-tolerant techniques and tools of analysis can be useful. However, it is hard to determine how feasible these techniques are for measurement-based systems, or how they should be adapted. For example, adapting later measurements based on previous outcomes can propagate a single fault to many.

In this chapter, we summarize these models, and discuss why certain modifications in the cluster and the graph state models are necessary to achieve fault tolerance. Then, we explain how to make the modified model fault tolerant – simply by reducing the problem to that in the circuit model, and thus it can be solved by any solution for the latter. (This technique also applies to the teleportation-based model, though there is less practical interest.)

18.2 Common models for measurement-based quantum computation

Any quantum computation can be performed as a circuit in the standard model [D95], with copies of $|0\rangle, |1\rangle, |+\rangle, |-\rangle$ as the initial state, on which one- or two-qubit gates are applied to yield a state that is measured in the Z or the X bases (the measurements are denoted by M_Z or M_X). The gate set can be chosen to be X and Z rotations and CZ (to be defined in the next paragraph). We describe briefly how to effectively execute (or simulate) such a circuit (called the simulated circuit) in each measurement-based model.

We now define some gates. In the basis $\{|0\rangle, |1\rangle\}$, the matrix representations of the X and Z rotations are given by

$$X_\theta = e^{-i\theta X}, \quad Z_\theta = e^{-i\theta Z}, \tag{18.1}$$

where I, X, Y, Z represent the Pauli operators. The symbol W denotes the Hadamard matrix and we frequently use the fact

$$WXW = Z, \quad WZW = X. \tag{18.2}$$

We only use two-qubit gates within the Clifford group; for example, the *controlled-phase* and the *controlled-*NOT gates, denoted in the text as CZ and CX and in the circuit as

$$CZ: \quad \text{and} \quad CX: \tag{18.3}$$

The target of CX is taken to be the second qubit. In contrast, CZ is symmetric between the two qubits, as is evident in the notation of Eq. (18.3). In the basis $\{|00\rangle, |01\rangle, |10\rangle, |11\rangle\}$, the matrix representations of CZ and CX are given by

$$CZ = \begin{pmatrix} 1 & 0 & 0 & 0 \\ 0 & 1 & 0 & 0 \\ 0 & 0 & 1 & 0 \\ 0 & 0 & 0 & -1 \end{pmatrix}, \quad CX = \begin{pmatrix} 1 & 0 & 0 & 0 \\ 0 & 1 & 0 & 0 \\ 0 & 0 & 0 & 1 \\ 0 & 0 & 1 & 0 \end{pmatrix}. \tag{18.4}$$

We will repeatedly use the following identities involving CX and CZ:

$$(I \otimes W)\, CZ\, (I \otimes W) = CX\,, \tag{18.5}$$

$$CZ\, (X \otimes I)\, CZ = X \otimes Z\,, \tag{18.6}$$

$$CZ\, (Z \otimes I)\, CZ = Z \otimes I\,, \tag{18.7}$$

$$CX\, (X \otimes I)\, CX = X \otimes X\,, \tag{18.8}$$

$$CX\, (I \otimes X)\, CX = I \otimes X\,, \tag{18.9}$$

$$CX\, (Z \otimes I)\, CX = Z \otimes I\,, \tag{18.10}$$

$$CX\, (I \otimes Z)\, CX = Z \otimes Z\,. \tag{18.11}$$

18.2.1 Teleportation-based model

In the teleportation-based model of quantum computation [N03, L01a, L04], the allowed resource is nondemolition measurement, the simplest set of which consists of several one- or two-qubit measurements. Implicitly, quantum storage, classical computation, and the ability to condition the measurement bases on previous outcomes are needed.

To simulate the preparation of initial states such as $\{|0\rangle, |1\rangle\}$ or $\{|+\rangle, |-\rangle\}$, one simply applies the measurements M_Z or M_X. Even though one has no control over the measurement outcomes, the postmeasurement states differ only by a known Pauli operation. For example, if we want to prepare $|+\rangle$ but obtain $|-\rangle$ after M_X, we say that we have prepared $\sigma_z|+\rangle$. Similarly, throughout a computation in all measurement-based models, the actual state may differ from the intended one by an uncontrolled but known Pauli operation.

Final measurements in the simulated circuit M_Z, M_X are self-simulating. However, a known Pauli operation acting on the measured state affects the measurement outcome as follows. If the actual and the intended states differ by σ_z, applying M_X on the actual state yields an outcome opposite to that on the intended state. Similarly, a known σ_x flips the outcome of M_Z.

We now discuss how to perform unitary gates using measurements only, and how to deal with the known but uncontrolled Pauli operations.

The basic idea (informal discussion). We first discuss the simple but inefficient original idea. Following up on the previous example, if $\sigma_z|+\rangle$ is prepared in place of $|+\rangle$, and it goes through unitary transformations $U_1, U_2, \ldots$, etc, we can simulate $U_1 \sigma_z$ instead of U_1, again allowing the resulting state to deviate from the intended one by a known Pauli operation.

Unitary gates are simulated by gate teleportation [GC99]. State teleportation [BBC+93] transmits a qubit $|\psi\rangle$ via the following circuit:

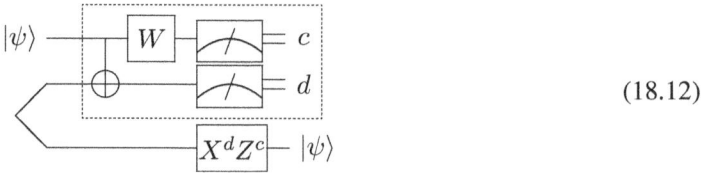
(18.12)

The two qubits connected in the left are prepared in the state $\frac{1}{\sqrt{2}}(|00\rangle + |11\rangle)$. Using the entanglement provided by this state, together with a measurement along the Bell basis (in the dashed box), classical communication of the measurement outcome c, d and a correction to Pauli operation $X^d Z^d$, the unknown state $|\psi\rangle$ is transmitted from the first qubit to the third one. Since the overall phase is preserved in this method, it can be used to transfer part of an entangled state.

Due to teleportation, a valid (albeit lengthy) way to apply a gate U to a state $|\psi\rangle$ is to prepend or append teleportation that has the Pauli correction omitted. For a single-qubit gate, the circuits are given by

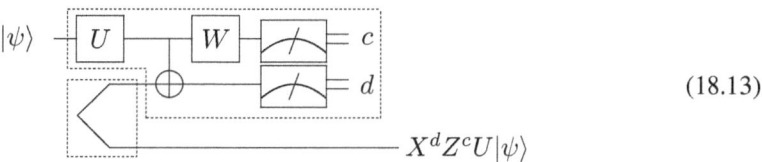

(18.13)

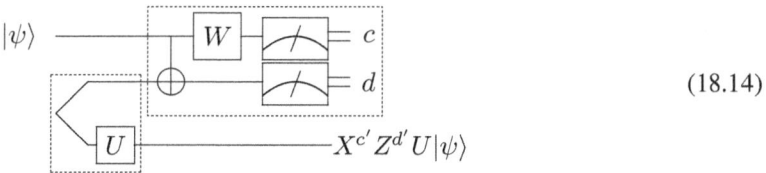

(18.14)

The second method is applicable to Clifford group gates only. By definition, a Clifford group gate congugates Pauli operators to Pauli operators. Thus for each c, d, one can find c', d' such that $U X^d Z^c U^\dagger = X^{c'} Z^{d'}$, and the output state $U X^d Z^c |\psi\rangle$ is simply $X^{c'} Z^{d'} U |\psi\rangle$.

When simulating a gate outside of the Clifford group, one has to adapt the simulated gate to undo the known Pauli operation that has acted on the actual state. For Clifford group gates, the known Pauli operation can simply be propagated as discussed in Eq. (18.14). This way, the intended gate is applied to the intended state despite the known Pauli operation on the actual state, again with a new known Pauli operation.

Any two-qubit gate in the Clifford group can be simulated similarly to Eq. (18.14), using a four-qubit state that can be prepared by a sequence of two-qubit measurements [L01a].

An efficient method. Gate teleportation is useful in that it allows a gate U to be applied to an unknown state via a measurement along a rotated Bell basis, as in Eq. (18.13), or via the preparation of a state $(I \otimes U)|\Phi_{00}\rangle$, as in Eq. (18.14). We now describe variants of teleportation that are not suitable for communication but will still give rise to useful gate teleportation circuits. These variants are called one-bit teleportation [ZLC00] and have circuits as follows:

Z-teleportation

(18.15)

X-teleportation

(18.16)

These circuits are analogous to teleportation in that they move a qubit from one register to another. The circuits are named after the Pauli corrections required to fully recover the input state. The circuits are easily verified.

Using one-bit teleportation, we obtain procedures analogous to the simulation circuits in Eqs. (18.13) and (18.14) [CLN05]. To simulate a single-qubit gate U acting on an intended state $|\psi\rangle$, when the actual state differs by $X^a Z^b$, a simulation circuit can consist of first applying $U' = U Z^b X^a$ before either form of one-bit teleportation. Since Z_b commutes with U if $U = Z_\theta$, we only undo X^a. But $Z_\theta X^a$ is not a Z rotation, so, instead we simulate $X^a Z_\theta X^a = Z_{(-1)^a \theta}$ (with an extra X^a acting on the output of the simulation). Thus, our simulation circuit takes the input state $X^a Z^b |\psi\rangle$ and applies $Z_{(-1)^a \theta}$, followed by Z-teleportation. When the measurement outcome is c, the output state is $Z^c Z_{(-1)^a \theta} X^a Z^b |\psi\rangle$. Using the identity $X^a Z_{(-1)^a \theta} X^a = Z_\theta$, the output state is $X^a Z^{b+c} Z_\theta |\psi\rangle$. This is summarized in the circuit

$$\begin{array}{c}\text{(circuit 18.17)}\end{array} \qquad (18.17)$$

where we have commuted CX and $Z_{(-1)^a \theta}$. Similarly, for the gate X_θ, consider a simulation circuit with an input state $X^a Z^b |\psi\rangle$, a gate $X_{(-1)^b \theta}$ applied to it, followed by X-teleportation. When the measurement outcome is d, the output state is $X^d X_{(-1)^b \theta} X^a Z^b |\psi\rangle = X^{a+d} Z^b X_\theta |\psi\rangle$. This is summarized in the circuit

$$\begin{array}{c}\text{(circuit 18.18)}\end{array} \qquad (18.18)$$

where we have commuted CX and $X_{(-1)^b \theta}$.

Finally, we consider a simulation circuit for CZ in which two circuits for X-teleportation (without correction) are applied to the two-qubit input state $X^{a_1} Z^{b_1} \otimes X^{a_2} Z^{b_2} |\psi\rangle$, followed by applying CZ:

$$\begin{array}{c}\text{(circuit 18.19)}\end{array} \qquad (18.19)$$

When the measurement outcomes of the two X-teleportation steps are d_1 and d_2, the output state of the circuit is $CZ(X^{a_1+d_1} Z^{b_1} \otimes X^{a_2+d_2} Z^{b_2})|\psi\rangle$. Due to Eqs. (18.6) and (18.7), the output state is equal to $(X^{a_1+d_1} Z^{b_1+a_2+d_2} \otimes X^{a_2+d_2} Z^{b_2+a_1+d_1}) CZ |\psi\rangle$. Thus in Eq. (18.19), $a'_1 = a_1 + d_1$, $b'_1 = b_1 + a_2 + d_2$, $a'_2 = a_2 + d_2$, and $b'_2 = b_2 + a_1 + d_1$.

We can derive useful simulation circuits from Eq. (18.19). Suppose we commute CZ to the left of the CXs, and reorder the qubits so that the second qubit from the top becomes the last:

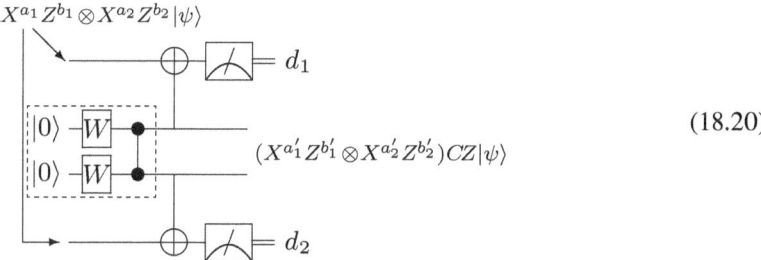
(18.20)

Furthermore, for the same input state, the following circuits produce the same outcomes and corresponding postmeasurement states:

$$\text{(circuit identity)} \quad (18.21)$$

Thus Eq. (18.20) implies the following:

$$\text{(circuit)} \quad (18.22)$$

where, according to Eq. (18.21), the output X errors in Eq. (18.22) are obtained by adding d_1, d_2 to a'_1, a'_2 defined in Eq. (18.20). The results are simply a_1, a_2. Finally, rewriting both CXs using Eq. (18.5) and noting that the state in the dashed box in Eq. (18.22) is stabilized by $W \otimes W$ gives a "remote CZ" construction:

$$\text{(circuit)} \quad (18.23)$$

Equations (18.17), (18.18), and (18.23) allow a universal set of gates to be performed in the teleportation-based model. They all involve two types of operations: (1) the preparation of some one- or two-qubit states, and (2) a two-qubit unitary gate that is immediately followed by measuring one of the two qubits. The first type of operation, state preparation, can be trivially performed by measuring along a basis including the state to be prepared. The second type of operations are effectively two-qubit measurements. The detail is out of scope of the current discussion and can be found in [CLN05]. Thus we obtain the result that two-qubit measurements are universal for quantum computation.

18.2.2 Graph state model

Given a graph G, the corresponding graph state is defined (and can be prepared) by assigning to each vertex a single qubit, initialized in the $|+\rangle$ state, and then applying a CZ to each pair of incident qubits for each edge in G.

A quantum computation in the graph state model starts with the preparation of a graph state, followed by single-qubit measurements [RBB03].

We now show that, to simulate any circuit in the graph state model, it suffices to employ Eqs. (18.17) and (18.18), together with appropriate direct applications of CZ [CLN05]. The simulated circuit can be decomposed as alternating steps of (1) Z and X rotations and (2) arbitrary CZs. To transcribe the simulated circuit to operations in the graph state model, every qubit that has a nontrivial transformation in step (1) will have both Eqs. (18.17) and (18.18) applied to it (possibly with zero angle of rotation in one of them). We demonstrate the graph state simulation on two qubits and two cycles of the alternating steps on an arbitrary input. In the first example, the single-qubit rotations on both qubits in both cycles are nontrivial.

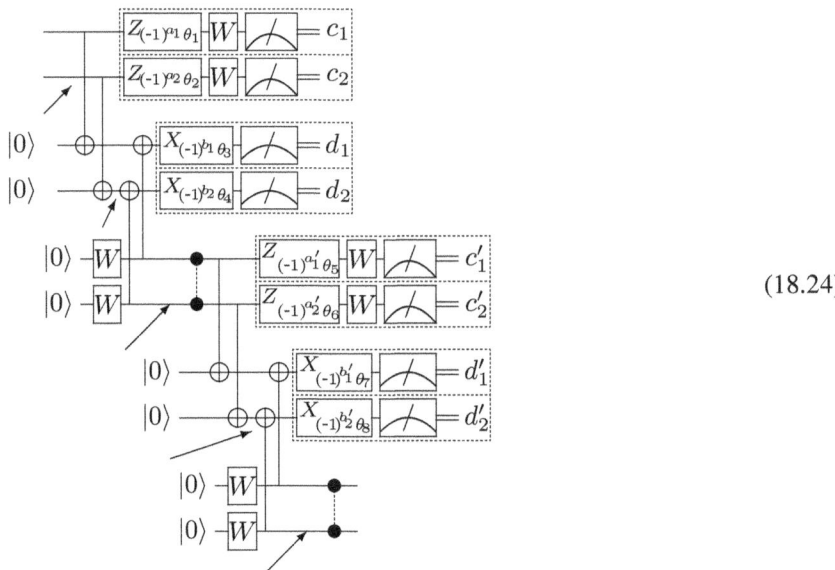

(18.24)

Each arrow in Eq. (18.24) indicates where the output of a certain teleportation step matches the input of the subsequent teleportation. The values of a'_i and b'_i can be read from Eqs. (18.17), (18.18), and (18.19). The CZ (in dashed line) is applied if and only if it is present in the simulated circuit. The circuit of Eq. (18.24) generalizes easily to n qubits with multiple possible CZ gates.

The simulation Eq. (18.24) can be simplified by (1) using the identity $CX = (I \otimes W)CZ(I \otimes W)$, (2) canceling out consecutive Hadamard gates (since $W^2 = I$), (3) rewriting

$W|0\rangle$ as $|+\rangle$, and (4) absorbing W before a single-qubit measurement as part of the measurement. We thus obtain a simpler simulation scheme:

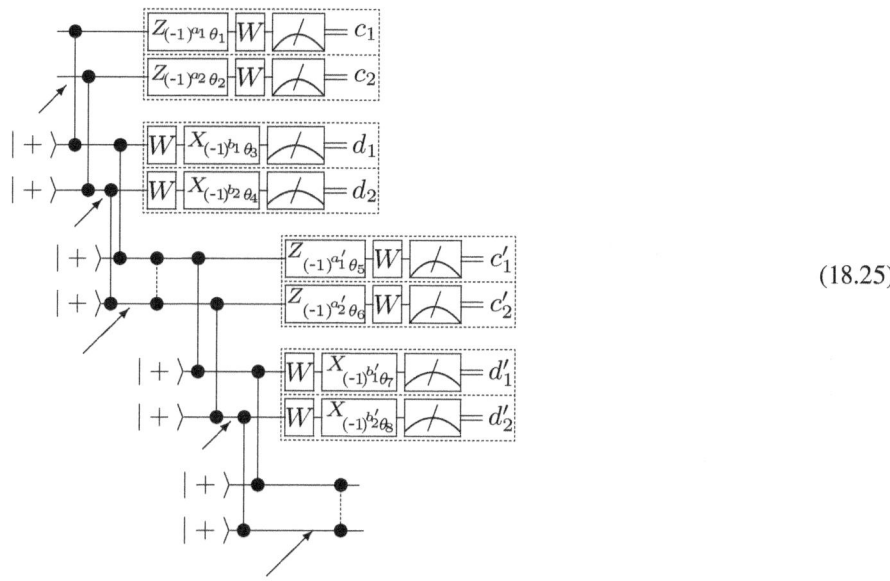

(18.25)

The operations in the dashed boxes are single-qubit measurements, and the rest of the circuit is the preparation of an initial graph state.

In the second example, the second qubit has a trivial rotation in the second cycle, and nothing is done on it. In the following diagram, the two solid circles denote the same space-time location in the circuit.

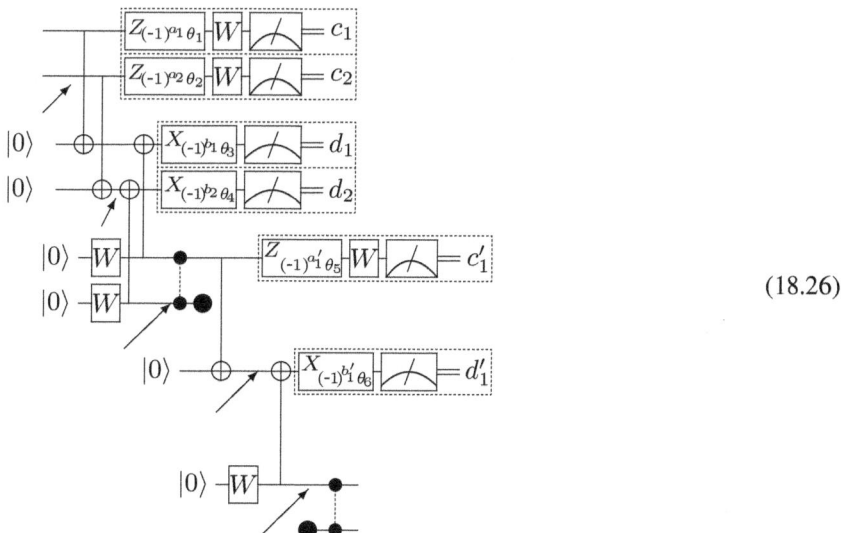

(18.26)

Simplifying the above circuit as in the first example gives:

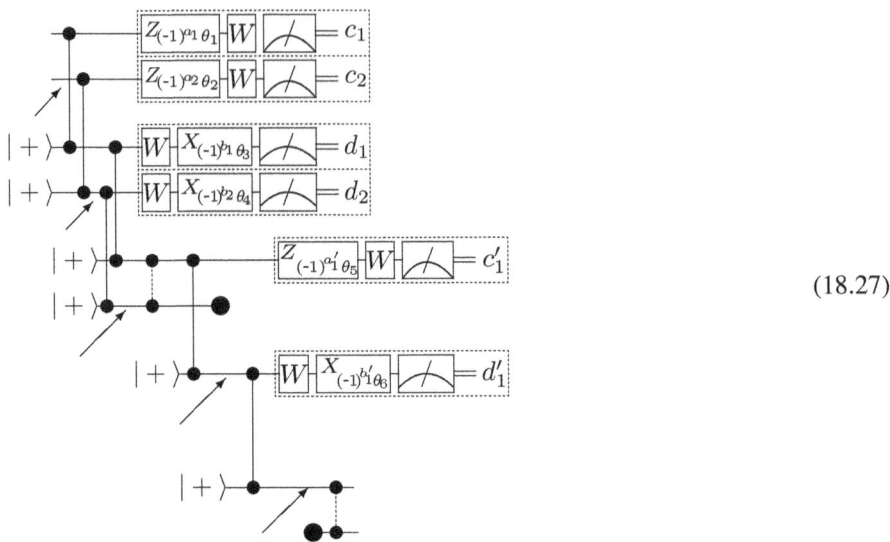 (18.27)

Once again, the circuit consists of single-qubit measurements in the dashboxes, and the rest of the circuit as the preparation of an initial graph state.

Note that the graph defining the graph state has a topology similar to the simulated circuit. When omitting the correction of the known Pauli operations, the rotation angles in the simulations are adapted accordingly. The CZs simply propagate the Pauli operations.

18.2.3 Cluster state model

The cluster state model can be viewed as a special case of the graph state model. Here, one starts with a cluster state, defined as the graph state corresponding to the square lattice [RB01]. This cluster state depends on the simulated circuit only in its dimensions. When the simulated circuit has n qubits and depth d, the cluster state has approximately $3n$ by $3d$ qubits.

A computation in the cluster state model is done by reduction to the graph state model – one simply "etches out" a desirable graph state from the cluster state. This is possible with several extra ingredients. First, one only uses nearest neighbor CZ gates in the simulated circuit. Second, even trivial single-qubit operations are simulated as a Z and an X rotation. The third ingredient is the deletion principle. A graph state can be transformed so that the underlying graph is converted to a subgraph with any vertex removed along with all incident edges. To do this, the qubit in that vertex is measured in the Z basis, projecting the $|+\rangle$ state to either $|0\rangle$ or $|1\rangle$. We can commute the measurement with the CZs applied to that qubit, and the CZs become Is or σ_z on the target qubits correspondingly. Thus, the "subgraph state" can be prepared up to extra known σ_z operations.

The last ingredient is the notion of an optional *CZ* [CLN05]. Consider the circuit Eq. (18.23) that simulates *CZ*:

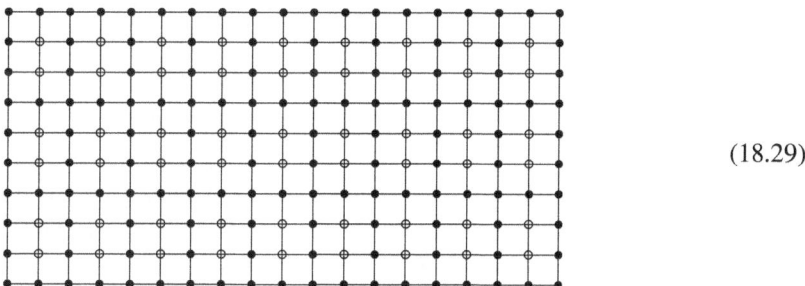
(18.28)

Note that we have explicitly labeled all the qubits. The circuit in Eq. (18.28) starts with a graph state, and applies the gate $Z^{d_2} \otimes Z^{d_1} CZ$ to qubits 1 and 4. On the other hand, using the deletion principle, if the W gates on qubits 2 and 3 are omitted, those qubits are measured in the Z basis, and are thus removed, disentangling qubits 1 and 4, and an identity gate is simulated instead. The optional *CZ* allows us to shrink three edges along a linear subgraph graph to one.

Starting from the cluster state

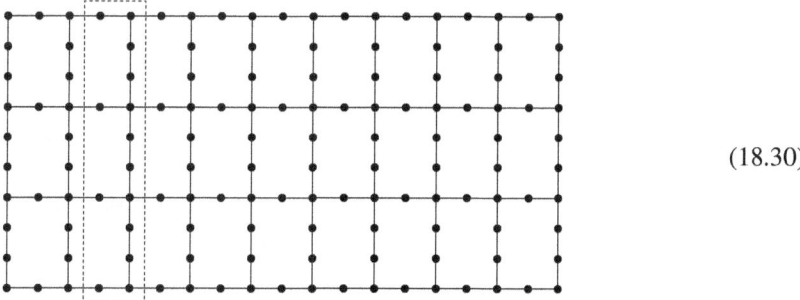
(18.29)

we can delete vertices corresponding to the empty circles to get another universal initial state of the form:

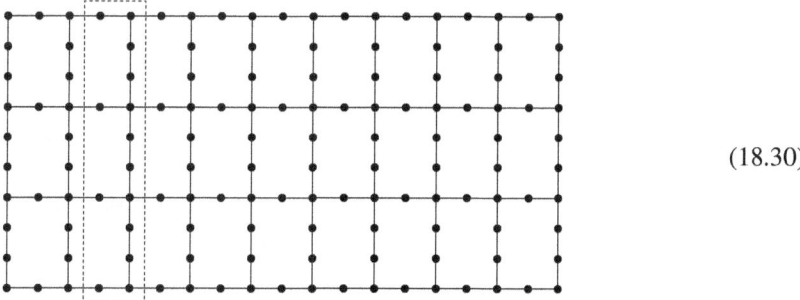
(18.30)

In the above state, a simulation for the optional *CZ* is built in wherever there may be a *CZ* in the simulated circuit. Together with the ability to simulate one-qubit gates between them (by measurement of the "horizontal" qubits), any circuit of a certain size can be simulated. This completes the derivation of the cluster state model.

18.3 Apparent issues concerning fault tolerance in measurement-based quantum computation

As mentioned earlier, noise and inaccuracies are inevitable in any model of computation. Most fault-tolerance results are derived for the standard circuit model, and there is little a-priori reason why they are applicable in the measurement-based models. Even more crucially, adapting measurements based on previous outcomes can potentially propagate a fault in one location to many, thus turning the local noise model to one in which fault tolerance may not be possible.

In the rest of this chapter, we discuss what interferes with fault tolerance in the measurement-based models, and what modifications are needed. Then, using the connection to the circuit model and better understanding of what constitutes errors in the measurement outcomes, we show that most fault-tolerance techniques in the circuit model apply to the modified measurement-based model [AL06]: If we simulate a circuit C in the modified measurement-based model with imperfect primitives, the simulated circuit has an effective noise model very similar to the physical noise model. Thus, if C is known to be fault tolerant in the circuit model, the correct computation result will be obtained if the noise strength is low enough, thus achieving fault-tolerant quantum computation in the measurement-based model itself. The proof will be given in Section 18.5.

18.4 Simulation of an operation

18.4.1 Simplified and informal notion

We first introduce the notation of operation simulation [AL06] informally. Suppose a quantum system with associated Hilbert space $\mathcal{H}$ undergoes an evolution described by a quantum operation $\mathcal{E}$, and consequently a subspace $\mathcal{H}_{\text{in}} \subset \mathcal{H}$ is transformed according to some other quantum operation $\mathcal{E}'$ and is moved to some other subspace $\mathcal{H}_{\text{out}} \subset \mathcal{H}$, we say that $\mathcal{E}$ simulates $\mathcal{E}'$ (with implicit underlying input/output Hilbert spaces). $\mathcal{E}$ is generally a circuit designed to simulate $\mathcal{E}'$. We call $\mathcal{E}$ the simulating circuit, $\mathcal{H}$ the ambient space, $\mathcal{E}'$ the simulated operation or circuit, and $\mathcal{H}_{\text{in,out}}$ the embedded input and output spaces. Whenever the distinction between $\mathcal{H}_{\text{in}}$ and $\mathcal{H}_{\text{out}}$ is unnecessary or undesirable, we use $\mathcal{H}'$ instead of both $\mathcal{H}_{\text{in}}$ and $\mathcal{H}_{\text{out}}$.

Note that an operation is simulated on $\mathcal{H}'$ independent of what state is supported on it. In some applications, it may be possible to have more flexibility in the design of $\mathcal{E}$ by demanding the simulation to be correct only on specific embedded states, but this line of discussion is beyond the scope of the present discussion, and is made unnecessary in most cases by a proper choice of $\mathcal{H}'$. In contrast, the simulation $\mathcal{E}$ acts on a state in $\mathcal{H}$ that is prepared in rather special ways (except the state supported on the embedded space $\mathcal{H}'$).

Note also that the simulated operation can be a state preparation, a measurement, or an operation that changes the dimension. Accordingly, the input or output space may be one-dimensional.

18.4.2 Examples: Teleportation, fault tolerance

Many well-known quantum information processing tasks can be phrased as simulations of operations.

A well-known example is teleportation [BBC+93] (see Eq. (18.12), and label the qubits 1–3 top-down). In this case, the simulated operation is the identity operation and the embedded input and output spaces $\mathcal{H}_{\text{in}}$ and $\mathcal{H}_{\text{out}}$ are both two-dimensional, and are naturally interpreted as qubits 1 and 3 in the eight-dimensional ambient space $\mathcal{H}$. The subspace orthogonal to $\mathcal{H}_{\text{in}}$ is prepared in the fixed state $|\Phi_2\rangle$ on qubits 2 and 3. $\mathcal{E}$ consists of a Bell measurement on qubits 1 and 2, whose outcome controls a Pauli operation on qubit 3.

Likewise, consider gate teleportation [GC99] Eq. (18.14), which simulates any Clifford group gate U. It is simulated by preparing the initial state $(I \otimes U)|\Phi_2\rangle$, performing the Bell measurement, and applying $U\sigma_k U^\dagger$ conditioned on the measurement outcome k. Once again the simulation is done with an associated change in the location of the embedded space. This simulation split the task of applying U to an unknown state to (1) preparing the fixed state $(I \otimes U)|\Phi_2\rangle$ and (2) applying $U\sigma_k U^\dagger$ to an unknown state instead. We have seen that Clifford group operations are much easier to perform in a fault-tolerant way, and adding any "C_3 gate" gives a universal set of gates. A C_3 gate U has the property that $U\sigma_k U^\dagger$ is in the Clifford group, and $(I \otimes U)|\Phi_2\rangle$ can be prepared by measuring similar operators. Thus this simulation provides a recipe for universal fault-tolerant operations using other accessible (Clifford) primitives.

In a different context, fault-tolerant quantum computation is also a form of simulation of operations. Here, the simulating operation $\mathcal{E}$ is the gadget consisting of noisy primitives. The code space supporting encoded quantum information is the embedded space, and the goal is to obtain a simulated noisy operation $\mathcal{E}'$ that has better noise parameters, such as lesser strength or weaker correlations.

As a final example, measurement-based quantum computation is built upon simulations that are basically gate teleportation, except the corrections conditioned on the measurement outcomes are suppressed. Instead, the information on the measurement outcomes is propagated and be used to control consequent measurement bases.

18.4.3 Composability

A simulation is rarely used as a stand-alone primitive. Rather, simulations for elementary operations are composed in various ways to simulate an arbitrarily complicated computation.

An example of recursive simulation is fault-tolerant quantum computation. We use 0-gadgets to build 1-gadgets, and use 1-gadgets to build 2-gadgets in a self-similar way, and so on, until the simulated operation is accurate enough. The bound of the noise threshold is based on the fact that, at each level, the simulated operation is self-similar to those one level lower, so that the analysis of a single level reduction can be applied recursively.

A different type of composition, sequential composition, is used whenever elements from an intended circuit are simulated one after another. Examples include the sequential application of the highest-level gadgets in a fault-tolerant quantum computation and the simulations in measurement-based quantum computation.

For now, we focus on sequential composition in measurement-based quantum computation. Later, we will turn to a different composition: that of a sufficient number of levels of fault-tolerant simulations, followed by a single step of graph state simulation.

Suppose the simulated circuit consists of elements $\mathcal{O}_1, \mathcal{O}_2, \ldots, \mathcal{O}_n$, and let $\mathcal{E}_{\mathcal{O}_i}$ be the stand-alone simulation for $\mathcal{O}_i$. We say that the simulation is composable if

$$\mathcal{E}_{\mathcal{O}_n} \cdots \mathcal{E}_{\mathcal{O}_2} \mathcal{E}_{\mathcal{O}_1} = \mathcal{E}_{\mathcal{O}_n \cdots \mathcal{O}_2 \mathcal{O}_1}. \tag{18.31}$$

In other words, the simulation of an arbitrary circuit can be done by composing the element-wise simulation.

For measurement-based quantum computation, the main difficulty in composing simulations is the omission of the Pauli correction, which can act as a known error that propagates through the rest of the circuit adversarially. A simple solution is to redefine simulation so that composability is automatic. In other words, one should employ a composable definition (which is backed by an actual set of element-wise simulations circuits for a universal set of simulated operations).

18.4.4 Definition for composable MBQC simulation

For measurement-based quantum computation, the main difficulty in composing simulations is handling the uncorrected but known Pauli error. A simple way to obtain a composable definition of simulation is to explicitly include the information on the uncorrected but known Pauli "error" as part of the input of the simulation, and ensure that the simulation propagates this known Pauli error properly. In particular, the input and output should be similar in structure, so that the output in a simulation is a valid input for the next simulation. Thus we extend the informal notion of a simulation to act on a "classical-quantum" input.

Let $\{|e\rangle\}$ denote the computational basis on two-qubits. Since it will be clear from the context, we will use the shorthands e, e_{in}, and e_{out} to denote both the two-bit classical information and also the corresponding pure state density matrices $|e\rangle\langle e|$, $|e_{\text{in}}\rangle\langle e_{\text{in}}|$, and $|e_{\text{out}}\rangle\langle e_{\text{out}}|$. Each value of e specifies a Pauli operation σ_e, and let $\mathscr{P}_e(\rho) = \sigma_e \rho \sigma_e^\dagger$. Likewise, $\mathscr{P}_{e_1,e_2}$ denotes a two-qubit Pauli operation.

Let $\mathscr{E}'$ be a single-qubit quantum operation. We say that $\mathscr{E}$ simulates $\mathscr{E}'$ in the measurement-based model if

$$\forall e_{\text{in}} \forall \rho \; \exists e_{\text{out}} \text{ s.t.} \quad \mathscr{E}(e_{\text{in}} \otimes \mathscr{P}_{e_{\text{in}}}(\rho)) = e_{\text{out}} \otimes \mathscr{P}_{e_{\text{out}}}(\mathscr{E}'(\rho)), \tag{18.32}$$

where ρ is an arbitrary qubit density matrix. Note that a fixed ancillary state α used in the simulation $\mathscr{E}$ is part of the description of $\mathscr{E}$ and not part of the input. $\mathscr{E}$ may produce some ancillary state β that is discarded afterwards (and again that is part of how $\mathscr{E}$ is implemented and not part of the output). Simulations for two-qubit operations are defined similarly.

The definition for simulation given by Eq. (18.32) is sequentially composable – it implies Eq. (18.31). Then, any measurement-based model in which a universal set of circuit operations can be simulated according to Eq. (18.32) is universal.

18.5 Fault tolerance in graph state model

18.5.1 A hybrid composition

A plausible approach to fault tolerance in measurement-based quantum computation is that, given any circuit $\mathscr{C}$ to be simulated (for example, an intended quantum algorithm to be executed ideally), we first write down a circuit $\mathscr{C}_{\text{FT}}$ simulating $\mathscr{C}$ that is fault tolerant, according to Chapter 5. Then, we simulate $\mathscr{C}_{\text{FT}}$ by $\mathscr{C}_{\text{MB}}$ which simulates operations in $\mathscr{C}_{\text{FT}}$ element by element, according to the composable definition. In other words, we concatenate a fault-tolerant simulation with a measurement-based simulation.

If the measurement-based simulation is performed using noisy primitives, and a quantum operation $\mathscr{E}_{\mathrm{MB}}$ is being performed instead of the ideal $\mathscr{C}_{\mathrm{MB}}$, is it simulating a noisy version $\mathscr{E}_{\mathrm{FT}}$ of $\mathscr{C}_{\mathrm{FT}}$? If so, what is the "simulated noise" defined by their difference?

18.5.2 Precise computation model

We consider the graph state model, with one modification. The simulating circuit $\mathscr{C}_{\mathrm{MB}}$ is the composition of many individual simulations for the circuit elements in $\mathscr{C}_{\mathrm{FT}}$ and each operation (bringing in $|+\rangle$ state, applying CZ, or measuring a single qubit) belongs to a specific circuit element in $\mathscr{C}_{\mathrm{FT}}$. We modify the graph state model, in that the operations in $\mathscr{C}_{\mathrm{MB}}$ follow the order of $\mathscr{C}_{\mathrm{FT}}$ (rather than having the entire graph state prepared upfront) [ND05].

The above modification is needed because, without the ability to bring in freshly prepared states throughout the computation, it is provably impossible to perform reliable quantum computation beyond logarithmic depth [AB.-OI+96].[1] Thus, in the original cluster and graph state models *that admit no further interaction*, it is impossible to bring the existing computing system to be in contact with fresh states, and one cannot compute with higher depth.

Our goal is to analyze the simulated noise for the composable simulations of a universal set of circuit elements. For concreteness, and to obtain an actual noise threshold in terms of the one derived for the fault-tolerant simulation $\mathscr{C}_{\mathrm{FT}}$, we specifically assume that $\mathscr{C}_{\mathrm{FT}}$ consists of the preparations of $|0\rangle, |+\rangle$, the measurements M_X, M_Z, and the universal set of gates W, CZ, and Z_θ for $\theta = \pi/4, \pi/8$. We can also include the Pauli operations and the SWAP gate.

Our analysis can be easily adapted to other universal sets of circuit elements used in $\mathscr{C}_{\mathrm{FT}}$. Recall that the threshold noise strength derived in Chapter 5 for a given noise model is with respect to the set of elementary operations used in $\mathscr{C}_{\mathrm{FT}}$. We will see that the graph state threshold differs from that in the standard model by a multiplicative factor given by the number of possible fault locations, maximized over the composable simulations for all the circuit elements in $\mathscr{C}_{\mathrm{FT}}$. For example, $\mathscr{C}_{\mathrm{FT}}$ may contain CX instead of CZ, or the universal gate set considered earlier in this chapter – that consisting of X_θ, Z_θ, and CZ.

18.5.3 Statement of the solution: concatenation of FT simulation with graph state simulation is fault tolerant

Suppose there is a fault-tolerant construction in the circuit model involving elementary gates W, CZ, Z_θ for $\theta = \pi/4, \pi/8$, the Pauli operations and the SWAP gate, the preparations of $|0\rangle, |+\rangle$, and the measurements M_X, M_Z. Further suppose that the threshold error probability is $\geq p_0$ under the independent stochastic noise model, and the threshold noise strength is $\geq \eta_0$ under the local non-Markovian noise model. Then, the corresponding thresholds are $\geq p_0/6$ and $\geq \eta_0/6$ in the modified graph state model. Switching CZ to CX in the fault-tolerant construction changes the scaling factor from 6 to 7.

---

[1] Without fresh states, the entropy of the entire computing system will be strictly increasing, and beyond a certain point, reliable computation is impossible.

18.5.4 Proof

- **Composable simulation circuits in the graph state model**

  Recall that $\mathscr{E}$ simulates $\mathscr{E}'$ in the measurement-based model if

  $$\forall e_{\text{in}} \forall \rho \exists e_{\text{out}} \text{ s.t.} \quad \mathscr{E}(e_{\text{in}} \otimes \mathscr{P}_{e_{\text{in}}}(\rho)) = e_{\text{out}} \otimes \mathscr{P}_{e_{\text{out}}}(\mathscr{E}'(\rho)). \tag{18.33}$$

  We now list the simulation circuits for our stated universal set of operations, given a supply of $|+\rangle$ states and CZ on demand. That the simulation circuits indeed satisfy the above requirement can be easily checked, and will be left as an exercise to the readers. In all cases, the dashed box defines the simulation $\mathscr{E}$ (only registers going in and out of the dashed box are inputs and outputs of $\mathscr{E}$). Also, we write $U(\rho)$ in place of $U\rho U^\dagger$ to avoid cluttering the circuit diagrams.

  (i) Preparation of $|+\rangle$, i.e., $\mathscr{E}' = |+\rangle\langle+|$, is given as a primitive. Thus, $\mathscr{E} = |00\rangle\langle00| \otimes \mathscr{E}'$ where the classical register is initialized to $|00\rangle$ because the Pauli correction is trivial. (In models in which the preparation is only up to a known σ_z operation on the output, $\mathscr{E} = p_0|00\rangle\langle00| \otimes \mathscr{E}' + p_1|01\rangle\langle01| \otimes \sigma_z\mathscr{E}'\sigma_z$.) Note that there is neither quantum nor classical input to a state preparation operation.

  $$\begin{array}{c} \text{[circuit diagram]} \end{array} \tag{18.34}$$

  (ii) Preparation of $|0\rangle$, i.e., $\mathscr{E}' = |0\rangle\langle0|$. $\mathscr{E}$ is given by

  $$\begin{array}{c} \text{[circuit diagram]} \end{array} \tag{18.35}$$

  Here, $\mathscr{E} = \frac{1}{2}|00\rangle\langle00| \otimes \mathscr{E}' + \frac{1}{2}|10\rangle\langle10| \otimes \sigma_x\mathscr{E}'\sigma_x$, where the second classical output is flipped if the measurement outcome is 1 (i.e., the $|-\rangle$ state is obtained).

  (iii) Measurement along the basis $|0\rangle, |1\rangle$, i.e., $\mathscr{E}'(\rho) = M_Z(\rho) = \langle0|\rho|0\rangle\,|0\rangle\langle0| + \langle1|\rho|1\rangle\,|1\rangle\langle1|$. The circuit $\mathscr{E}$ is given by

  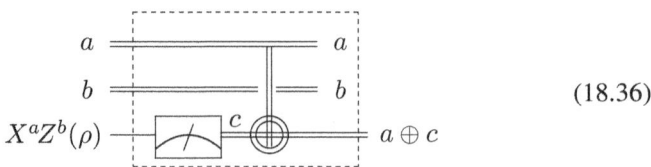

  $$\tag{18.36}$$

(iv) Measurement along the basis $|+\rangle, |-\rangle$, i.e., $\mathscr{E}'(\rho) = M_X(\rho) = \langle +|\rho|+\rangle\,|+\rangle\langle +| + \langle -|\rho|-\rangle\,|-\rangle\langle -|$:

$$\text{(18.37)}$$

(v) $\mathscr{E}'$ being the Hadamard gate W:

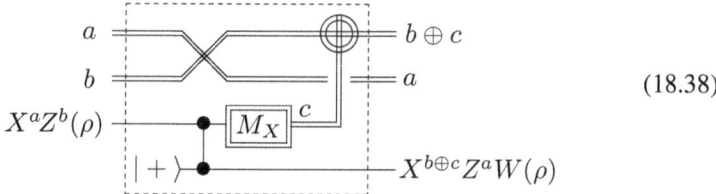

$$\text{(18.38)}$$

(vi) $\mathscr{E}'$ being a Z rotation Z_θ:

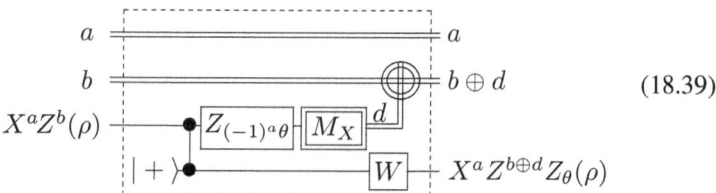

$$\text{(18.39)}$$

where an ideal W is used above, and when simulated by Eq. (18.38), we need to compose Eqs. (18.39) and (18.38) together:

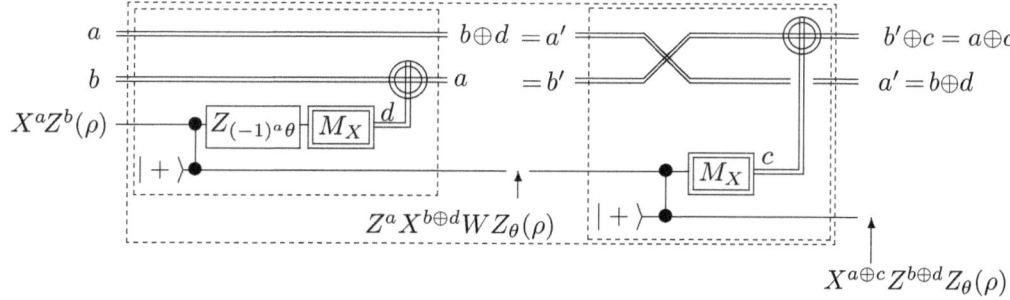

(vii) A CZ:

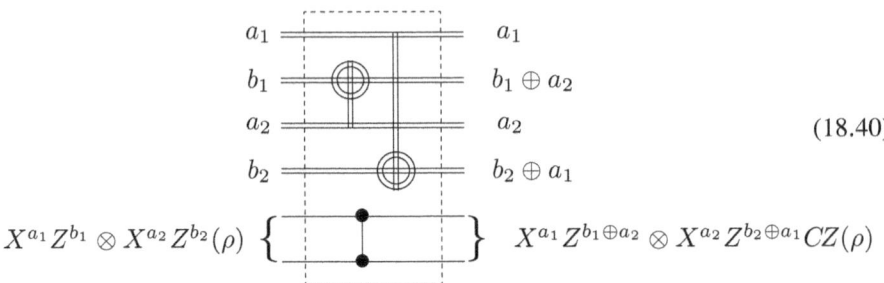
(18.40)

(viii) To simulate a Pauli operation $X^{a'}Z^{b'}$, we simply flipped the classical registers

$$\begin{array}{c} a \\ b \\ X^a Z^b(\rho) \end{array} \quad \begin{array}{c} \oplus \\ \oplus \\ \end{array} \quad \begin{array}{c} a \oplus a' \\ b \oplus b' \\ X^a Z^b(X^{a'} Z^{b'}(\rho)) \end{array} \qquad (18.41)$$

up to an overall sign conditioned on the classical registers.

(ix) The SWAP gate can be simulated by relabeling both quantum and classical registers.

- **Why classical registers can be assumed correct**

Consider Eq. (18.41) on how we can simulate the Pauli operation $X^{a'}Z^{b'}$. A known change in the classical register corresponds to a known Pauli operation on the qubit, and vice versa. They are equivalent and in fact indistinguishable. The same equivalence holds for unknown changes. Thus, any error on the classical register can instead be interpreted as an unknown Pauli error on the quantum system, and the former is *always error free*. In particular, using the classical measurement outcomes to control the future measurement basis does not spread errors from the simulation of one circuit element to another. Thus, the faults accumulating within a simulation are *localized* and cause errors of that simulation only; we call the combined error on the simulation the simulated fault.

- **How the physical noise model is transformed to one in the graph state simulation**

By the localization of error, one can define the simulated noise model based on the physical one. We say that the physical noise model is *transformed* to a simulated one. Moreover, we will see that this transformation preserves the noise model. For example, stochastic noise stays stochastic, and likewise for the local noise model.

In any noise model and without loss of generality, each noisy state preparation or noisy gate can be expressed as the ideal operation followed by a *noise operation*. Each noisy measurement is modeled as the ideal one preceeded by a noise operation. Hence, noise operations come between successive ideal operations. A noise operation is a system–environment coupling, and it can always be described by some unitary evolution acting jointly on the system and environment:

$$U_{\text{fault}} = I \otimes A_0 + \sum_i P_i \otimes A_i. \tag{18.42}$$

Here P_i ranges over all nontrivial Pauli operators indexed by i acting on the output system of the preceding ideal operation and each A_i is an arbitrary operator acting on the environment, subject to the condition that U_{fault} is unitary. We call the second term in Eq. (18.42) the *fault operator* or simply the fault. Noise operations given by Eq. (18.42) alternate with elementary operations throughout the entire computation. The joint operation on the system–environment can be expanded to a sum over all possible combinations of having a fault or not in each location. A *fault path* (see Chapter 5) is a term in this expansion.

For each simulation in the graph state computation, each term in the expansion of Eq. (18.42) consists of a Pauli operator acting on the simulation qubits that can be commuted to the end of the simulation (since each simulation is realized by a sequence of *CZ* and single-qubit measurements). The Pauli operators from different noise operations within a simulation can be combined, and can act on either classical or quantum output, but the net effect can always be interpreted as an error in the quantum part.

We can now prove an accuracy threshold for independent stochastic noise in the graph state model. By independence, each noise operation described by Eq. (18.42) acts on a different environmental register, and stochasticity means that the two terms in Eq. (18.42) are orthogonal, as if the environment keeps a record of all the fault locations. Thus, the fault paths *do not interfere* with one another, and each has a normalization representing its probability. Certain fault paths are "good" and they give the ideal computation results. Other fault paths are "bad." The probability of bad fault paths can be suppressed if the probability of a fault in each elementary physical operation is below the accuracy threshold. This forms the basis of the threshold theorem in the circuit model.

Consider the noisy graph state simulation $\mathscr{C}_{\text{MB}}$ of a *fault-tolerant circuit* $\mathscr{C}_{\text{FT}}$. The final output state is a sum over e_{out} for each qubit. In the absence of faults, each summand has an ideally evolved quantum state due to the correctness of the graph state simulation. In the noisy simulation $\mathscr{C}_{\text{MB}}$, elementary operations are interspersed with faults. Since each fault path in $\mathscr{C}_{\text{MB}}$ is mapped to a unique fault path in the simulated $\mathscr{C}_{\text{FT}}$ due to error localization, good fault paths in $\mathscr{C}_{\text{MB}}$ can be defined as those resulting in good fault paths in $\mathscr{C}_{\text{FT}}$. All other fault paths in the simulation $\mathscr{C}_{\text{MB}}$ are defined as bad. We now show that the sum of probabilities of all the bad fault paths in $\mathscr{C}_{\text{MB}}$ is highly suppressed if each elementary operation is accurate enough.

With independent stochastic noise, the fault paths in $\mathscr{C}_{\text{MB}}$ *do not interfere* with one another, and each has a normalization representing its probability. Each bad fault path has probability suppressed below a certain accuracy threshold just as in the circuit model, because a simulated fault appears after some simulated operation only if there is at least one fault in its simulation. Furthermore, the probability of a simulated fault is at most the sum of the fault probabilities of all the elementary steps in the simulation. Inspecting our composable simulation circuits, at most six elementary operations (*CZ*, measurements, preparation of $|+\rangle$) are used (in the $\pi/8$ gate simulation). Therefore, the probability of any simulated fault is at most $p_{\text{sim}} \leq 6p$ where p is the fault probability of each elementary operation. More specifically, if p_0 is the threshold value of the fault-tolerant architecture used in the circuit model and if

$p \leq p_0/6$ in the simulation, then $p_{\text{sim}} \leq p_0$ and the final measurement outcome will provide the correct computation results with the desired accuracy. This holds for all possible e_{out}, thereby establishing a threshold lower bound of $p_0/6$ for the graph state model.

In the above, we assumed that the fault-tolerant circuit $\mathscr{C}_{\text{FT}}$ used the same gate set used in our simulation circuits Eqs (18.34)–(18.40). In general, it may have a different underlying set of elementary gates. If so, they can be composed from our simulation circuits. For example, CX is often used instead of CZ as the elementary interaction. CX can be simulated by composing two simulations of W [Eq. (18.38)] with one of CZ [Eq. (18.40)], using seven elementary operations. Thus, $p \leq p_0/7$. In many cases of interest this lower bound is pessimistic. For example, in many fault-tolerant designs based on self-dual CSS codes, CZ can be used instead of CX.

In the local non-Markovian noise model, each noise operation still has the form given by Eq. (18.42). However, different noise operations can act on the same environmental register, thus making the noise non-Markovian. Also, different fault paths may not give rise to orthogonal states. A fault no longer corresponds to an "event" with a probability. Instead, one imposes that the *noise strength* of the fault operator at each location is bounded, i.e., $\|\sum_i P_i \otimes A_i\|_{\sup} \leq \eta$.

Once again, a threshold theorem can be obtained by reduction to proofs derived for the circuit model. These proofs employ a coherent description of the quantum computation [TB05, AL06]. This is an isometry V acting on an input state $|\psi_{\text{in}}\rangle$ consisting of copies of $|+\rangle$ and $|0\rangle$. Likewise, we consider a *purification* of $\mathscr{C}_{\text{MB}}$, replacing each measurement by a coherent operation acting on extra ancillary qubits. We have modeled noisy measurements as ideal measurements preceeded by noise operations. Thus, purifying the measurements does not affect the analysis. Likewise, the $2n$-bit classical registers can be replaced by a $2n$-qubit register, which is used to control the adaptive operations. The update of this register is now done coherently by controlling gates from the extra ancillary qubits, and also by doing the classical processing reversibly. We emphasize that this alternative coherent description is purely mathematical. The purification of $\mathscr{C}_{\text{MB}}$ is simply an isometry V_{sim} acting on two inputs $|e_{\text{in}}\rangle$ and $P_{e_{\text{in}}}|\psi_{\text{in}}\rangle$. (Ancillary qubits are incorporated in the description of the isometry.) Suppose $\{|\phi_i\rangle\}$ is the orthonormal basis on which measurements are to be performed, $\{|i\rangle\}$ is the computation basis with i labeling the possible measurement outcomes carried by the extra ancillas we have introduced, c_i is the *amplitude* of the ith term, and $|e_{\text{out}}\rangle$ is a $2n$-qubit state that depends on e_{in} and i. By the correctness of the composable simulation, $\forall e_{\text{in}}$, the output of the purified simulation is given by

$$V_{\text{sim}}(|e_{\text{in}}\rangle \otimes P_{e_{\text{in}}}|\psi_{\text{in}}\rangle) = \sum_i c_i |e_{\text{out}}\rangle \otimes |i\rangle \otimes |\phi_i\rangle \otimes P_{e_{\text{out}}} V|\psi_{\text{in}}\rangle. \tag{18.43}$$

Now, to the above ideal purified simulation, we insert a unitary noise operation given by Eq. (18.42) at *every* location. The noisy output is a linear superposition or a sum over states, each summand is labeled by a list of measurement outcomes (including measurements of the graph-state simulation and the simulated measurements in $\mathscr{C}_{\text{FT}}$) and a specific set of fault operators that evolve the state. Fault paths can again be "good" or "bad," defined as in our discussion for independent stochastic noise. For each good fault path, the state generates the correct statistics independently of the measurement outcome $|i\rangle$. This is because the classical registers can always be taken as being correct, and good fault paths in the simulation are mapped to good fault paths in the simulated circuit, on which the threshold theorem in the circuit model applies. Hence, by linearity, the coherent sum of these terms produces the ideal computation result.

It remains to bound the *sup norm* of the bad fault paths of the graph state simulation, which can combine coherently. Following the threshold theorem in the circuit model for local non-Markovian noise, it suffices to bound the sup norm of the "bad" part of a given simulation. But this sup norm is simply bounded by $\eta_{\text{sim}} \leq 6\eta$, where η is a bound on the sup norm of the fault operator acting on each location in the simulation (by the triangle inequality of the sum norm). Thus $\eta \leq \eta_0/6$ is the threshold condition for the graph state model if η_0 is the established threshold strength for the circuit model. Once again, if CX is used in place of CZ in $\mathscr{C}_{\text{FT}}$, $\eta \leq \eta_0/7$.

18.6 History and other approaches

The cluster state model of quantum computation was proposed in 2001 by Raussendorf and Briegel [RB01] in what was called a "one-way quantum computer." Its theory was fully explained in [RBB02]. The graph state model of quantum computation was explicitly discussed in [RBB03].

Independently of [RB01], Nielsen [N03] proposed a computation model that uses only projective measurements on up to four qubits, quantum memory, and preparation of the $|0\rangle$ state. Various simplifications were found by the author soon afterwards [L01a, L04]. These include the sufficiency of two-qubit projective measurements, and using the idea of [RB01] to adapt measurement bases to perform the operations with a deterministic and smaller number of steps. We use the name "teleportation-based model" to refer to the culmulative results in this area.

The cluster state model and graph state model of quantum computation were derived in the stabilizer formalism, and appear to be very different from the circuit model of quantum computation. In contrast, the teleportation-based model is just the circuit model with each gate performed by gate teleportation [GC99, ZLC00]. However, by 2004, Aliferis and Leung [AL06] showed that the cluster state model can be understood as a sequence of teleportations, and Childs, Nielsen, and Leung [CLN05] found that using one-bit teleportation, both models can be derived with much fewer resources; a similar result was obtained independently by Jorrand and Perdrix [JP05].

Thus, measurement-based quantum computation has a circuit interpretation that allows us to apply known results in the standard model. Meanwhile, its noncircuit description inspires various other results such as topological quantum computation, trade-off between circuit depth and space, and some simpler proofs in interactive proof systems.

The discussion in this chapter is primarily based on the circuit interpretation in [CLN05], so that known results for fault tolerance in the standard model can be directly applied.

The issue of fault tolerance in the cluster state and graph state models was first addressed in Raussendorf's Ph.D. thesis, in which a slightly restricted noise model was considered. Subsequently, Nielsen and Dawson [ND05] clarified the needs of the hybrid model (in which the CZ are applied just before the relevant qubits are measured) and employed the framework of Terhal and Burkard [TB05] for non-Markovian noise to handle the possible propagation of error in the classical register. The derivation given in this chapter is primarily based on a result by Aliferis and Leung [AL06]. Using the circuit description of the graph state model and considering composability, they found that the classical registers can be assumed noiseless and the threshold analysis of Aliferis, Gottesman, and Preskill [AGP06] can be applied. This provides an actual fault-tolerant scheme, a much shorter proof for fault tolerance (given [AGP06]), and also a higher error threshold. Subsequently, Aliferis provided another even shorter proof of fault tolerance for the same scheme in his Ph.D. thesis using level reduction [A07].

Part VI

Topological methods

Chapter 19

Topological codes

Héctor Bombín

19.1 Introduction

What a good code is depends on the particular constraints of the problem at hand. In this chapter we address a constraint that is relevant to many physical settings: locality. In particular, we are interested in situations where *geometrical locality* is relevant. This typically means that the physical qubits composing the code are placed in some lattice and only interactions between nearby qubits are possible. In this case, it is desirable that syndrome extraction also be local, so that fault tolerance can possibly be achieved. Topological codes offer a natural solution to locality constraints, as they have stabilizer generators with local support.

In topological codes information is stored in *global* degrees of freedom, so larger lattices provide larger code distances. The nature of these global degrees of freedom is illustrated in Fig. 19.1, where several closed curves in a torus are compared. Consider curves a and b. They look the same if examined locally, as in the region marked with dotted lines. However, curve a is the *boundary* of a region, whereas curve b is not. In order to decide whether a curve is a boundary or not we need global information about it. This is, as we will see, a core idea in topological codes.

This chapter only attempts to provide an introduction to the subject. In particular, we will only deal with two-dimensional codes and leave out the condensed-matter perspective.

19.2 Local codes

Since the emphasis of this chapter is on locality, we start by giving a formal definition of what it means for a code to be local. Intuitively, a local stabilizer code is one in which all stabilizer generators act only on a few nearby qubits. To formalize this idea, we could talk about n-local codes, those in which the support of each stabilizer generator is limited to n qubits. This is a notion that can be applied to individual codes. But if we want to talk about local codes, without further adjectives, we have to consider instead sets or *families* of codes.

Quantum Error Correction, ed. Daniel A. Lidar and Todd A. Brun. Published by Cambridge University Press.
© Cambridge University Press 2013.

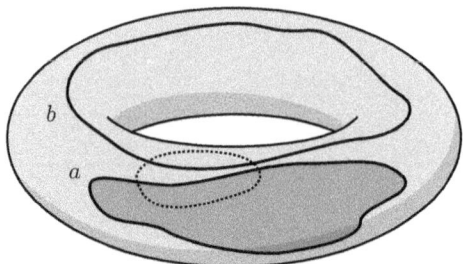

Fig. 19.1. Closed curves in a torus can be the boundary of a region, like curve a. However, curve b is not the boundary of a region. The difference cannot be detected by looking only at a local region such as the one marked with a dotted line.

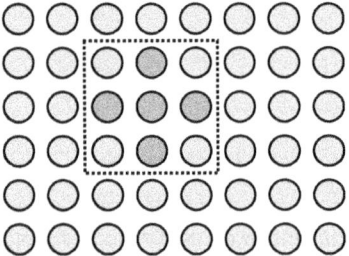

Fig. 19.2. In 2D local codes qubits are arranged in a 2D array. The support of any stabilizer generators must be contained in a box of a fixed size, here a 3×3 box. Circles represent qubits and darker ones form the support of a generator.

A family of stabilizer codes is *local* if we can choose the stabilizer generators so that:

(i) the number of qubits in the support of the stabilizer generators is bounded,
(ii) the number of stabilizers with support containing any given qubit is bounded, and
(iii) the family contains codes of arbitrary large distance.

This notion of locality can be put in terms of graph connectivity, without any further structure. But in practice we are often interested in locality from a geometrical point of view. A family of codes is *local in D dimensions* when, instead of (i) and (ii) above,

(i) qubits are placed in a D-dimensional array, and
(ii) the support of any stabilizer generator is contained in a hypercube of bounded size.

This is illustrated in Fig. 19.2. Notice that a code that is local in a geometric sense is also local in the more general sense above.

Locality does not come without a price. For example, in any family of two-dimensional (2D) local codes the distance is $d = O(\sqrt{n})$, with n the total number of qubits. We do not prove this here, but see [BT09]. This behavior of the code distance might appear undesirable, but indeed it is not harmful because the code distance alone does not dictate the error-correcting capability of a family of topological codes. The key for fault tolerance is statistics: an error that cannot be corrected but is unlikely to occur is not important. As we will see, topological codes can correct *most* errors of weight $O(n)$, which is reflected in the existence of an error threshold. For noise

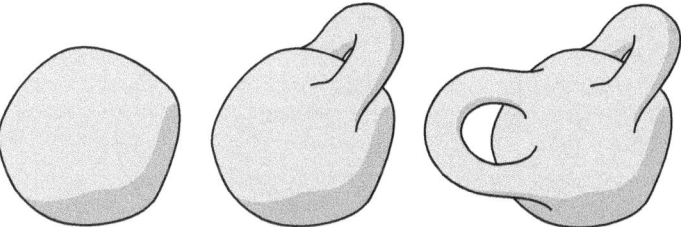

Fig. 19.3. From the topological perspective, orientable closed surfaces are classified by the genus g or number of handles. These are the three with lowest genus: the sphere, the sphere with a handle or torus, and the sphere with two handles or 2-torus.

below the threshold, error correction can be achieved with asymptotically perfect accuracy in the limit of large lattices.

19.3 Surface homology

In order to understand topological codes, it is convenient to have some background in algebraic topology. This section provides an elementary introduction to the topic, which will be sufficient for our purpose. For further reading, see for example [H02].

19.3.1 Topology of closed surfaces

Topology deals with spatial properties that are preserved under continuous deformations. It is sometimes called "rubber sheet geometry," because we are allowed to stretch or compress our objects of study, but not to tear them or glue their parts. The coffee mug and the donut are well-known examples that look the same to a topologist since, up to deformations, they are both nothing but a solid sphere with a hole.

In the topological codes that we will study, qubits are placed on 2D lattices. Such lattices will be embedded in surfaces, and it turns out that the topology of the surfaces is what matters to us. Since we only have a finite number of qubits at hand, there is no point in considering open surfaces, like the plane. Thus, we restrict ourselves to closed surfaces, like the sphere. For simplicity, we will focus on orientable closed surfaces here. These are the closed surfaces that have an inside and an outside, that is, those that are the boundaries of everyday solid objects. Moreover, we only consider connected surfaces, those in which we can move from any point to another without jumps.

From a topological perspective, the classification of connected orientable closed surfaces is pretty simple. First, we have the sphere. If we add a handle to a sphere, we obtain the torus (the surface of a donut). We can then add a second handle (2-torus), or a third (3-torus), and so on, see Fig. 19.3. This infinite process allows us to build all orientable closed surfaces, which are thus classified by the number of handles, known as their *genus*. A sphere has genus 0 and a g-torus has genus g. As we will see, the genus of a surface will dictate the number of encoded qubits in a topological code.

There is an interesting relationship between the number of elements of a lattice embedded in a surface and its genus. We will denote the number of vertices, edges and faces of the lattice

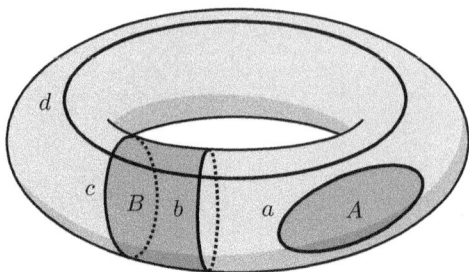

Fig. 19.4. Several closed curves in a torus. Curve a is the boundary of region A, so it is homologically trivial. Curves b and c form the boundary of region B, and thus are homologically equivalent. Curve d is homologically nontrivial and not equivalent to c.

as V, E, and F, respectively. The Euler characteristic is then defined as the quantity

$$\chi := V - E + F. \tag{19.1}$$

This is an important quantity because it only depends on the topology of the surface, not on the particular lattice. In particular, for closed orientable surfaces we have

$$\chi = 2(1 - g). \tag{19.2}$$

19.3.2 Homology of curves

Suppose that you are given two objects and asked whether they are topologically equivalent. If you can find a way to continuously transform one into the other, you will have shown that they are equivalent. But, if they are not equivalent, how can we show it? Topological *invariants* offer an aswer to this question. A topological invariant is a number or other kind of mathematical structure that is constructed from the objects of interest and depends only on their topology. If two objects return different values for any topological invariant, they cannot be topologically equivalent. Notice that we have already encountered an example of topological invariant, namely, the Euler characteristic.

The topological invariant that is involved in the construction of the most basic topological codes is the *first homology group* of a surface. Actually, the elements of this group label a basis for the encoded states, as we will see. So, what is homology about?

Consider the closed curves on a torus of Fig. 19.4. Curve a is the *boundary* of region A, whereas curves b, c, d are not the boundaries of any region (individually). We say that a is homologically trivial, and b, c, d homologically nontrivial. Moreover, b and c together enclose the region B, so they are said to be homologically equivalent. On the other hand, c and d together do not form the boundary of any region and are therefore not equivalent. We have thus a classification of the closed curves on a surface into different *homology classes*. We can further give to these classes the structure of an Abelian group, with the identity element corresponding to the trivial homology class of curves. This requires adopting a more formal approach.

We will work now with a particular lattice embedded in the surface. As is customary, we rename vertices as 0-cells, edges as 1-cells and faces as 2-cells. First, we want to form an Abelian group C_i out of the set of i-cells, for $i = 0, 1, 2$. Consider for example 1-cells. If we label the

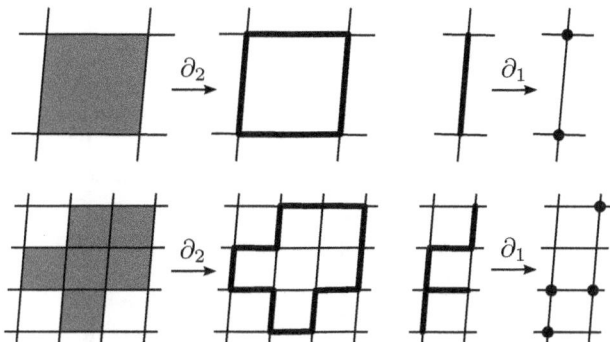

Fig. 19.5. The action of boundary operators. ∂_2 maps a set of faces to the set of edges that form its boundary. ∂_1 maps a set of edges to to the set of vertices where an odd number of these edges meet.

edges as $\{e_i\}_1^E$ we can represent any set of edges E' as a formal sum

$$c = \sum_i c_i e_i, \qquad c_i = \begin{cases} 0 & e_i \notin E' \\ 1 & e_i \in E' \end{cases}. \tag{19.3}$$

Such formal sums are called 1-chains. They can be added together to obtain another 1-chain taking into account the rule $e_i + e_i = 0$. This gives rise to an Abelian group structure on the 1-chains C_1, with the zero element corresponding to the empty set. We can represent 1-chains as binary vectors of length E, and indeed $C_1 \simeq \mathbf{Z}_2^E$, where $\mathbf{Z}_2$ is F_2 considered just as an additive group. A nice aspect of the notation (19.3) is that any edge e automatically denotes the 1-chain corresponding to the set $\{e\}$. Similarly, we can form the group of 0-chains $C_0 \simeq \mathbf{Z}_2^V$ and the group of 2-chains $C_2 \simeq \mathbf{Z}_2^F$.

Our next step is to introduce a family of group homomorphisms ∂, called boundary operators. There are actually two that are relevant to our discussion,

$$\partial_2 : C_2 \longrightarrow C_1, \qquad \partial_1 : C_1 \longrightarrow C_0, \tag{19.4}$$

but they are commonly referred as ∂. As its name suggests, ∂ takes objects to their boundaries, as illustrated in Fig. 19.5. Namely, if a face f has as boundary the set of edges $\{e'_1, \ldots, e'_k\}$, then $\partial_2 f = e'_1 + \cdots + e'_k$. What happens when we apply ∂_2 to a region, that is, to a set of faces $F' = \{f'_1, \ldots, f'_l\}$? As a 2-chain, the region has the expression $r = f'_1 + \cdots + f'_l$, so $\partial_2 r = \partial f'_1 + \cdots + \partial f'_l$ because ∂ respects the Abelian group structure. Since $e_i + e_i = 0$, the edges that are shared by neighboring faces in F' cancel and we have $\partial_2 r = e''_1 + \cdots + e''_m$, where $\{e''_1, \ldots, e''_m\}$ is the set of edges that form the boundary of the region. The definition of ∂_1 is analogous: if the edge e has as endpoints the vertices v'_1, v'_2, then $\partial_1 e = v'_1 + v'_2$. For sets of edges the boundary is composed of those vertices at which an odd number of these edges meet.

Now we define two subgroups of 1-chains $Z_1, B_1 \subset C_1$. Z_1 is the subgroup of 1-chains z that have no boundary, that is, $\partial_1 z = 0$, the kernel of ∂_1. The elements of Z_1, called cycles, are collections of closed curves. B_1 is the image of ∂_2, that is, the subgroup of 1-chains b that are a boundary of a 2-chain, $b = \partial_2 c$ for some $c \in C_2$. The crucial observation is that all boundaries are also cycles, namely $B_1 \subset Z_1$. In other words,

$$\partial^2 := \partial_1 \circ \partial_2 = 0. \tag{19.5}$$

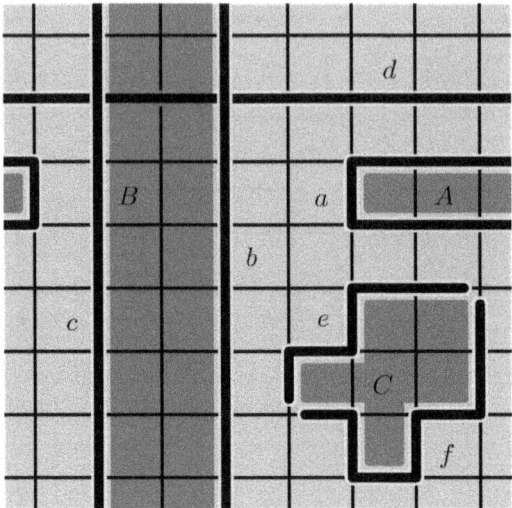

Fig. 19.6. Several 1-chains in a periodic 8×8 square lattice. $\bar{a} = 0$ because $a = \partial A$. $\bar{b} = \bar{c}$ because $b + c = \partial B$. $\bar{c} \neq \bar{d} \neq 0$ because neither $c + d = \partial D$, $c = \partial D$ nor $c = \partial D$ for any region D. $\bar{e} = \bar{f}$ because $e + f = \partial C$.

We can then define the first homology group H_1 as the quotient

$$H_1 := Z_1/B_1. \tag{19.6}$$

Let us recall the quotient group construction for Abelian groups. The elements of H_1 are cosets of the form $\bar{z} := \{z + b \mid b \in B_1\}$ for some $z \in Z_1$, the addition is that inherited from Z_1, namely $\bar{z} + \bar{z}' = \overline{z + z'}$, and the zero element is the coset $\bar{0} = B_1$. Technically, (19.6) defines $H_1(S; \mathbf{Z}_2)$, the first homology group of the surface S over $\mathbf{Z}_2$.

Algebraic topology teaches us that the group H_1 only depends, up to isomorphisms, on the topology of the surface. Indeed,

$$H_1 \simeq \mathbf{Z}_2^{2g}. \tag{19.7}$$

For a sphere, $g = 0$ and thus the first homology group is trivial: all cycles are boundaries, $B_1 = Z_1$. Notice how taking the quotient has erased all the information about the particular lattice used; this is the magic of topological invariants.

As an example, consider the square lattice embedded in a torus in Fig. 19.6. Here we are representing the torus in a convenient and conventional way, as a square where opposite edges are identified. The connected 1-chains in the figure are subject to the same homological equivalences as in Fig. 19.4. In the notation we have just developed, we have $\bar{a} = 0$, $\bar{b} = \bar{c}$ and $\bar{b} \neq \bar{d} \neq 0$. For a torus $H_1 \simeq \mathbf{Z}_2 \times \mathbf{Z}_2$, so the homology group has two generators. Here we can take, for example, $\bar{b}$ and $\bar{d}$ as the generators, and the elements of H_1 are $0, \bar{b}, \bar{d}$ and $\bar{b} + \bar{d}$.

The notion of equivalence up to homology is also useful for 1-chains that are not cycles. That is, we can consider the quotient C_1/B_1, and use the notation $\bar{c} := \{c + b \mid b \in B_1\}$, where

$c \in C_1$, to denote its elements. The open 1-chains e and f in Fig. 19.6 are homologous: they enclose the region C.

We have already encountered a quotient construction before, when studying stabilizer codes in Chapter 2, where logical operators are recovered from the quotient $N(S)/S$. As we will see in the next section, this quotient construction and the one in (19.6) can be fruitfully related.

19.4 Surface codes

Surface codes are the most basic examples of topological codes. In this section we will motivate their construction and study their main features.

19.4.1 Definition

The idea behind surface codes is to encode information in "homological degrees of freedom." To this end, our first step is to fix a lattice embedded in a given closed surface. We attach a qubit to each edge, so each element of the computational basis can be interpreted as a 1-chain $c \in C_1$ in the most obvious manner:

$$|c\rangle := \bigotimes_i |c_i\rangle, \qquad c \in C_1. \tag{19.8}$$

Here i runs over the physical qubits or, equivalently, the edges of the lattice $\{e_i\}$. We will find it convenient to label products of X and Z Pauli operators with 1-chains, too:

$$X_c := \bigotimes_i X_i^{c_i}, \qquad Z_c := \bigotimes_i Z_i^{c_i}, \qquad c \in C_1. \tag{19.9}$$

Notice that these labelings are indeed group homomorphisms from C_1 to G_n, because

$$X_c X_{c'} = X_{c+c'}, \qquad Z_c Z_{c'} = Z_{c+c'}. \tag{19.10}$$

The definition of the surface code is quite natural. It has a basis with elements (here and anywhere else we ignore normalization)

$$|\bar{z}\rangle := \sum_{b \in B_1} |z + b\rangle, \qquad \bar{z} \in H_1, \tag{19.11}$$

which are sums of all the cycles that form a given homology class. Clearly $\langle \bar{z} | \bar{z}' \rangle = 0$ for $\bar{z} \neq \bar{z}'$. Then from (19.7) we have $|H_1| = 2^{2g}$ and the number of encoded qubits is $k = 2g$.

To get a first flavor of the power of surface codes, we will study the effect of bit-flip errors. Notice that $X_c|b\rangle = |b + c\rangle$ and thus $X_c|\bar{z}\rangle = |\bar{z} + \bar{c}\rangle$. Let $z, z' \in Z_1$. If $\partial c \neq 0$ then $\partial(z + c) \neq 0$ and $\bar{z} + \bar{c} \neq \bar{z}'$. This implies $\langle \bar{z}' | X_c | \bar{z} \rangle = 0$, so X_c can map the code to itself only if c is a cycle. But errors X_b with b a boundary do nothing, as $X_b|\bar{z}\rangle = |\bar{z} + \bar{b}\rangle = |\bar{z}\rangle$. Therefore, only bit-flip errors X_z with $z \in Z_1$ are undetectable. It follows that the distance of a surface code for bit-flip errors is the length of the shortest nontrivial cycle. A similar analysis can be done for Z errors, but we will postpone the discussion until we can use the language of stabilizer operators. We anticipate that the distance for Z errors is given by the shortest nontrivial cycle in the *dual* lattice.

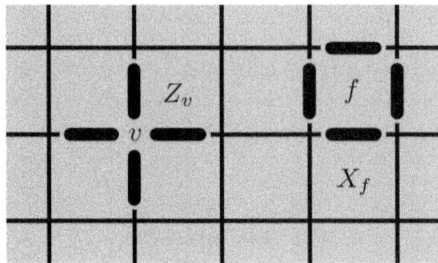

Fig. 19.7. The support of vertex and face stabilizer generators.

19.4.2 Stabilizer group

Given a vertex v and a face f, consider the face ("plaquette") and vertex ("star") Pauli operators

$$X_f := \prod_{e \in \partial_2 f} X_e, \qquad Z_v := \prod_{e | v \in \partial_1 e} Z_e, \qquad (19.12)$$

where ∂f and ∂e are understood as sets. These operators are depicted in Fig. 19.7, where we can see that a vertex operator has support on the links that meet at a vertex and a face operator has support on the edges that enclose the face. Using the notation (19.9), we could have written $X_f := X_{\partial f}$ in (19.12), but we wanted to remark on the symmetry between vertex and face operators.

Vertex and face operators commute with each other. We claim that they generate the stabilizer of surface codes, as defined in (19.11). To check this explicitly, consider the states

$$|\tilde{c}\rangle := \prod_f \frac{1 + X_f}{2} \prod_v \frac{1 + Z_v}{2} |c\rangle, \qquad c \in C_1. \qquad (19.13)$$

These states span the code defined by the stabilizers (19.12), because $|\tilde{c}\rangle$ is the projection of $|c\rangle$ onto the code subspace. Notice that $Z_v |c\rangle = -|c\rangle$ if $v \in \partial c$, and $Z_v |c\rangle = |c\rangle$ otherwise. Therefore, $|\tilde{c}\rangle = 0$ if $\partial c \neq 0$. We next notice that the first product in (19.13) can be expanded as a sum over subsets of faces $\{f_i\}$:

$$\prod_f (1 + X_f) = \sum_{\{f_i\}} \prod_i X_{\partial_2 f_i} = \sum_{\{f_i\}} X_{\partial_2 (\sum_i f_i)} = \sum_{c_2 \in C_2} X_{\partial_2 c_2}, \qquad (19.14)$$

where we have used (19.10). Since ∂_2 is a group homomorphism, we can replace the sum over 2-chains by a sum over 1-chains in the image of ∂_2, up to a factor. Therefore, for $z \in Z_1$ we have

$$|\tilde{z}\rangle := \prod_f \frac{(1 + X_f)}{2} |z\rangle \propto \sum_{b \in B_1} X_b |z\rangle = |\overline{z}\rangle. \qquad (19.15)$$

We get the same code subspace from the stabilizers (19.12) and the span of the basis (19.11).

Notice the different role played by vertex and face operators. Face operators are related to ∂_2, and they stabilize the subspace with basis $|\overline{c}\rangle := \sum_{b \in B_1} |c + b\rangle$, $c \in C_1$. That is, face operators enforce that states should be a uniform superposition of states on the same homology class. Vertex operators are related to ∂_1, and they stabilize the subspace with basis $|z\rangle$, $z \in Z_1$. That is, vertex operators enforce that states should have no boundary.

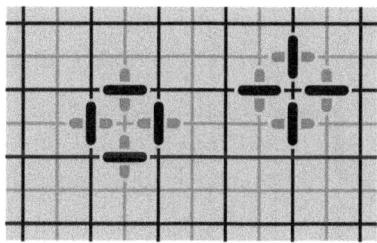

Fig. 19.8. A lattice and its dual. Under duality, vertex and faces are interchanged, and so are vertex and face operators.

The stabilizer generators (19.12) are not all independent. As one can easily check, they are subject to the following two conditions, and no more:

$$\prod_f X_f = 1, \qquad \prod_v Z_v = 1. \tag{19.16}$$

Therefore, there are $V + F - 2$ independent generators. It follows from the theory of stabilizer codes that the number of encoded qubits is

$$k = E - (V + F - 2) = 2 - \chi = 2g, \tag{19.17}$$

which of course agrees with the value given after (19.11).

19.4.3 Dual lattice

Given a lattice embedded in a surface, we can construct its dual lattice. This is illustrated in Fig. 19.8. The idea is that the faces of the original lattice get mapped to vertices in the dual lattice, edges to dual edges, and vertices to dual faces. We will use a hat * to denote dual vertices f^*, dual edges e^*, and dual faces v^*, in terms of their related faces f, edges e, and vertices v of the original lattice, respectively. Similarly, we have dual boundary operators

$$\partial_1^* : C_0^* \longrightarrow C_1^*, \qquad \partial_2^* : C_1^* \longrightarrow C_2^* \tag{19.18}$$

acting on dual chains c^*. *To simplify notation, for generic 1-chains we will consider c^* and c to be unrelated objects.* But, for single edges, e^* denotes the dual of e, so e^* and e refer to the same physical qubit. The boundary operators ∂^* produce the groups of dual cycles Z_1^* and dual boundaries B_1^*, and thus a homology group

$$H_1^* = \frac{Z_1^*}{B_1^*} \simeq H_1. \tag{19.19}$$

Comparing the action of ∂ and ∂^* on their respective lattices, we observe that

$$e^* \in \partial_1^* v^* \iff v \in \partial_1 e, \qquad f^* \in \partial_2^* e^* \iff e \in \partial_2 f. \tag{19.20}$$

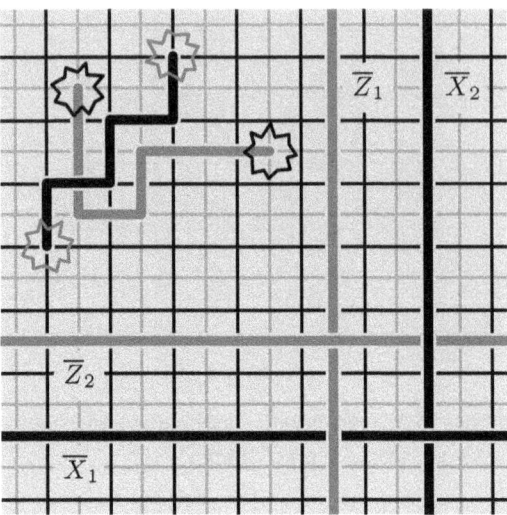

Fig. 19.9. String operators in a toric code. X-type (Z-type) string operators belong to the direct (dual) lattice and are displayed in a darker (softer) tone. The two open string operators anticommute with the vertex or face operators at their endpoints, marked with stars. Logical operators take the form of nontrivial closed string operators. The labeling agrees with the fact that crossing string operators of different types anticommute. This is a $[[128, 2, 8]]$ code.

Now, consider the effect of applying a transversal Hadamard gate $W^{\otimes E}$ across all qubits in a surface code. The code is mapped to a new subspace, described by the stabilizers

$$\begin{aligned} W^{\otimes E} X_f W^{\otimes E} &= \prod_{e \in \partial_2 f} Z_e = \prod_{e^* | f^* \in \partial_2^* e^*} Z_e =: Z_{f^*}, \\ W^{\otimes E} Z_v W^{\otimes E} &= \prod_{e | v \in \partial_1 e} X_e = \prod_{e^* \in \partial_1^* v^*} X_e =: X_{v^*}. \end{aligned} \qquad (19.21)$$

This is nothing but the surface code defined on the dual lattice! The moral is that we can deal with phase-flip errors as we already did with X errors, but working in the dual lattice. As a result, we have that the distance of a surface code is the length of the shortest nontrivial cycle in the original or the dual lattice.

As an example, consider periodic square lattices of size $d \times d$ embedded in a torus, like the one in Fig. 19.9. These lattices produce what was the first example of surface codes, the family of "toric codes." The dual of the $d \times d$ square lattice is again a $d \times d$ square lattice, and thus the distance is d because nontrivial cycles have to wind around the torus, as shown in the figure. It follows that these $[[2d^2, 2, d]]$ codes form a family of 2D local codes.

19.4.4 Logical operators

In the previous section we have learned that while it is useful to relate bit-flip errors X_c to 1-chains c, phase-flip errors should be related to 1-chains c^* in the dual lattice. Thus, we will use the notation $Z_{c^*} := \prod_i Z_{e_i}$, where $c^* = \sum_i e_i^*$ for some edge subset $\{e_i\}$. Any Pauli operator

A can be written as

$$A = i^\alpha X_c Z_{c^*}, \qquad (c, c^*) \in C_1 \times C_1^*, \quad \alpha \in Z_4. \tag{19.22}$$

Therefore, any Pauli operator can be visualized as a pair of chains (c, c^*) or, to give it a physical flavor, as a collection of strings. These strings are of two kinds, as they can live in the direct lattice (the X part) or the dual lattice (the Z part). Strings can be open, if they have endpoints, or closed, if they form a loop. An important property of closed string operators is that a direct and a dual string anticommute if and only if they cross an odd number of times. Notice that the oddness of the number of crossings is preserved up to homology. This has to be the case because, for $(b, b^*) \in B_1 \times B_1^*$ and $(z, z^*) \in Z_1 \times Z_1^*$, X_z and Z_{z^*} commute if and only if X_{z+b} and $Z_{z^*+b^*}$ commute.

The relationship between $N(S)/S$ and Z_1/B_1 will now become clear. First, it is easy to check that

$$[X_c, Z_v] = 0 \iff v \notin \partial_1 c, \qquad [Z_{c^*}, X_{f^*}] = 0 \iff f^* \notin \partial_2^* c^*, \tag{19.23}$$

as illustrated in Fig. 19.9. Consider any Pauli operator A as in (19.22). It follows from (19.23) that $A \in N(S)$ if and only if $(c, c^*) \in Z_1 \times Z_1^*$. What about the elements of the stabilizer S? An arbitrary stabilizer element will have the form $B = \prod_i X_{f_i} \prod_j Z_{v_j}$ for some subset of faces $\{f_i\}$ and some subset of vertices $\{v_j\}$. We can apply the same trick as in (19.14), obtaining $B = X_{\partial_2 c_2} Z_{\partial_1^* c_0^*}$, where $c_2 = \sum_i f_i$ and $c_0^* = \sum_i v_i^*$. Therefore, we see that A belongs to S if and only if $\alpha = 0$ and $(c, c^*) \in B_1 \times B_1^*$.

In summary, we have just seen that the elements of the normalizer $N(S)$ are labeled, up to a phase, by a cycle and a dual cycle, whereas the elements of the stabilizer are labeled by a boundary and a dual boundary. The parallelism is now apparent: we can label the elements of $N(S)/S$, up to a phase, with an element of $H_1 \times H_1^*$. The cosets indeed take the form

$$\{ i^\alpha X_{z+b} Z_{z^*+b^*} \mid (b, b^*) \in B_1 \times B_1^*, \alpha \in \mathbf{Z}_4 \}, \qquad (\overline{z}, \overline{z^*}) \in H_1 \times H_1^*. \tag{19.24}$$

Setting $S' := \langle iI \rangle S$, so that S' is the center of $N(S)$, we have

$$\frac{N(S)}{S'} \simeq H_1 \times H_1^* \simeq H_1^2. \tag{19.25}$$

Recall from the theory of stabilizer codes that $N(S)/S$ describes logical Pauli operators. In surface codes, we may therefore choose as generators of logical Pauli operators a set of closed string operators, as in the toric code of Fig. 19.9. Notice how the crossings and the required commutation relations match.

19.4.5 Boundaries

From a practical perspective, the 2D locality of surface codes can be very useful. However, the fact that we need nontrivial topologies complicates things from a geometrical perspective: it might be difficult to place qubits in a toroidal geometry. Fortunately, this obstacle can be overcome by introducing boundaries, as we will see. Boundaries can create nontrivial topologies even in a plane, which makes it possible to construct planar families of surface codes. Since a homological description of boundaries requires discussing the concept of relative homology, we will take an alternative route based on string operators.

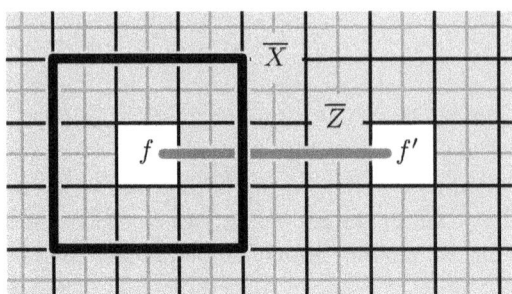

Fig. 19.10. When we remove two face stabilizer generators, a new encoded qubit appears. A logical operator takes the form of an open dual string connecting the missing faces and a direct closed string enclosing one of the holes.

In order to motivate the construction, we start with the following simple example. Suppose that, in a surface code, we remove the stabilizer generators corresponding to two separate faces f and f'. We know that this must increase the number of encoded qubits by one because, taking (19.16) into account, there is one generator less. Which are the string operators for this encoded qubit? The answer can be found in Fig. 19.10: a dual string operator with endpoints in the faces belongs now to $N(S)$. And a direct string that encloses f no longer belongs to S, since it is the product of either all the face operators "inside" it, which include X_f, or all the face operators "outside" it, which include $X_{f'}$. These two strings cross once and thus provide the $\overline{X}$ and $\overline{Z}$ operators for the new encoded qubit.

What is the distance of the new code? As we separate the two faces, the distance for phase-flip errors will grow accordingly. However, bit-flips do not behave like that, because a string that encloses f can be very small. Indeed, the smallest possible such string operator is X_f itself. Thus, we have failed at introducing an entirely global degree of freedom. Fortunately, the solution is at hand. If the perimeter of the faces f and f' is large, the distance for phase-flip errors will be large too.

The lesson is that we can introduce a nontrivial topology in the lattice by "erasing big faces." If we start with a sphere and remove $h + 1$ faces we will end up with a disc with h holes. The resulting surface code encodes h qubits. Naturally, we can do the same constructions in the dual lattice: removing $r + 1$ vertices will introduce r new encoded qubits. Because of their appearance, boundaries in the dual lattice are sometimes called "rough," whereas boundaries in the direct lattice are called "soft." Figure 19.11 shows an example of a geometry with dual boundaries.

It is possible to describe boundaries in terms of how they modify the notion of closed and boundary strings. For example, direct (soft) boundaries give rise to the following properties:

(i) dual string operators that have their two endpoints on a direct boundary belong to $N(S)$; and
(ii) dual string operators that enclose a region that only contains direct boundaries belong to S.

The second property implies that two dual string operators that together enclose such a region are equivalent up to stabilizers. Exchanging "direct" and "dual" we recover the defining properties of dual (rough) boundaries. All this is illustrated in Fig. 19.11.

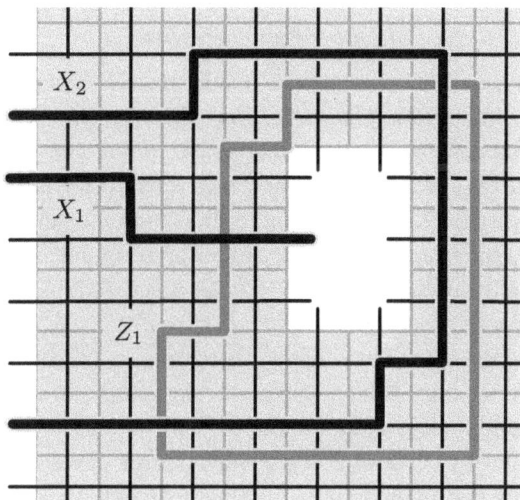

Fig. 19.11. A piece of a surface code with two dual or "rough" boundaries, one of them in the form of a hole. The direct string operator X_1 belongs to $N(S)$ because its endpoints lie on the dual boundaries, but does not belong to S as it does not enclose any region. The dual string operator Z_1 is closed, so $Z_1 \in N(S)$. It encloses a region, but this contains a piece of dual boundary and thus $Z_1 \notin S$. Indeed, $\{X_1, Z_1\} = 0$ because the strings cross. As for the direct string operator X_2, it encloses a region that only contains dual boundaries, so $X_2 \in S$.

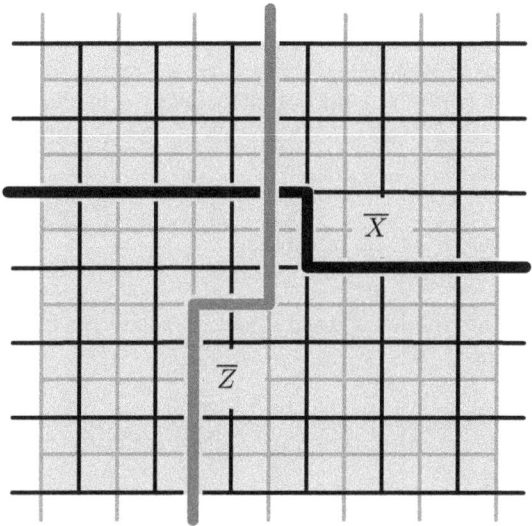

Fig. 19.12. A planar toric code. The top and bottom boundaries are direct, the left and right dual. Nontrivial direct (dual) strings connect left and right (top and bottom) boundaries, so that this is a [[85,1,7]] code.

In Fig. 19.12 we consider a somewhat more complicated geometry: four boundaries are combined to produce a planar toric code encoding a single qubit. Although this lattice can be obtained by removing two faces and a vertex in a sphere, there is no need to think this way: we only have to apply the rules above to understand the code in terms of string operators. Planar toric codes form a family of local $[[2d(d-1)+1, 1, d]]$ codes.

19.4.6 Error correction

Our next goal is the analysis of the correction of Pauli errors E in a surface code. A Pauli error E can be written, up to unimportant phases, as $E = X_c Z_{c^*}$. The first step in error correction is the measurement of stabilizer generators, in this case vertex and face operators. The resulting syndrome is dictated by (19.23): Vertex and face operators with eigenvalue -1 form the boundaries of c and c^*, respectively. According to the syndrome, any error $E' = X_d Z_{d^*}$ with $(\partial d, \partial^* d^*) = (\partial c, \partial^* c^*)$ may have happened. After choosing and applying such an E', error correction will be successful if and only if $E'E \in S$. Since $E'E \propto X_{c'+d'} Z_{c^*+d^*}$, it follows that error correction succeeds if and only if $(c' + d', c^* + d^*) \in B_1 \times B_1^*$. That is, when errors and corrections belong to the same homology class, $(\bar{c}, \bar{c^*}) = (\bar{d}, \bar{d^*})$. It is in this sense that in surface codes error correction must be done only up to homology, an advantage that has its origin in the fact that these are highly degenerate codes.

What is the best strategy to choose E'? Assume an error model in which Pauli errors $E = X_c Z_{c^*}$ follow a probability distribution $\{p_{c,c^*}\}$. Rather than individual error probabilities, we are interested in the probability for the whole set of errors with the same effect in the code, namely

$$\Pr(\bar{c}, \bar{c^*}) := \sum_{b \in B_1} \sum_{b^* \in B_1^*} p_{c+b, c^*+b^*}. \tag{19.26}$$

The probability to obtain a given syndrome $(\partial c, \partial^* c^*)$ is

$$\Pr(\partial c, \partial^* c^*) := \sum_{z^* \in H_1} \sum_{\overline{z^*} \in H_1^*} \Pr(\bar{c} + \bar{z}, \bar{c^*} + \overline{z^*}). \tag{19.27}$$

Given a particular syndrome $(\partial c, \partial^* c^*)$, the optimal strategy is to choose an E' from the class of errors $(\bar{d}, \bar{d^*})$ with highest conditional probability among those with $(\partial d, \partial^* d^*) = (\partial c, \partial^* c^*)$. The success probability is then

$$p_{\max}(\partial c, \partial^* c^*) := \max_{(\bar{z}, \overline{z^*}) \in H_1 \times H_1^*} \frac{\Pr(\bar{c} + \bar{z}, \bar{c^*} + \overline{z^*})}{\Pr(\partial c, \partial^* c^*)}. \tag{19.28}$$

The overall success probability for this optimal strategy is recovered by weighting each syndrome with its probability, obtaining

$$p_{\mathrm{succ}} := \sum_{\partial c, \partial^* c^*} \Pr(\partial c, \partial^* c^*) p_{\max}(\partial c, \partial^* c^*) = \sum_{\partial c, \partial^* c^*} \max_{(\bar{z}, \overline{z^*}) \in H_1 \times H_1^*} \Pr(\bar{c} + \bar{z}, \bar{c^*} + \overline{z^*}). \tag{19.29}$$

A very remarkable property of surface codes is that, in the limit of large lattices, $p_{\mathrm{succ}} \to 1$ when the noise is below a critical threshold. This will be the subject of next section.

In practice, computing which class has the maximal probability for a given syndrome might be costly, but there is an alternative approach. Notice that we can treat X and Z errors separately, ignoring any possible correlations. Then the problem reduces to choosing a 1-chain among those with a given boundary. When errors on physical qubits are independent, a good guess is to choose the chain c with minimal weight, a problem that can be solved on polynomial time in the size of the lattice with the so-called perfect matching algorithm [DKL+02]. Since it is only an approximation, this technique provides a suboptimal critical threshold.

19.4.7 Error threshold: random bond Ising model

There exists a useful connection between the error-correction threshold of surface codes and a phase transition in a 2D random bond Ising model. This connection appears when we separate, as above, the correction of bit-flip and phase-flip errors, ignoring any correlations. To fix ideas, we will study bit-flip errors, but phase-flip errors have an analogous treatment. We also fix the geometry to that of toric codes, of variable size. We assume an error model where bit-flip errors occur independently at each qubit with probability p. Since bit-flip errors are represented by 1-chains c, as above, we have a probability distribution $\{p_c\}$ with

$$p_c := (1-p)^{E-|c|} p^{|c|}. \tag{19.30}$$

Here $|c|$ denotes the number of edges of c and E the total number of edges.

19.4.7.1 Random bond Ising model Our first step is to define a family of classical Hamiltonian spin models. Classical spins are $s_i = \pm 1$ variables, with i a label, and we attach one of them to each face. Alternatively, spins live at the vertices of the dual lattice, so that they are connected by edges of the dual lattice. As it is customary, we denote neighboring pairs of spins as $\langle ij \rangle$. We are interested in the family of classical Hamiltonians of the form

$$H_\tau(s) := -\sum_{\langle ij \rangle} \tau_{ij}\, s_i s_j, \tag{19.31}$$

where the $\tau_{ij} = \pm 1$ are parameters of the Hamiltonian that define the ferromagnetic ($\tau_{ij} = 1$) or antiferromagnetic ($\tau_{ij} = -1$) nature of the interactions. We have thus a family of Ising models with arbitrary interaction signs. These models exhibit a $\mathbf{Z}_2$ global symmetry, since flipping all spins at once does not change the energy. The equilibrium statistics of the system are described by the partition functions, which are

$$Z(\beta, \tau) = \sum_s e^{-\beta H_\tau(s)}. \tag{19.32}$$

Here $\beta = 1/T$ is the inverse temperature.

Notice that the $\tau = \{\tau_{ij}\}$ are 1-chains in disguise: since edges are labeled by pairs $\langle ij \rangle$, we can label the coefficients of 1-forms with such pairs, $c = \{c_{ij}\}$ and then set $(\tau_c)_{ij} = (-1)^{c_{ij}}$. Similarly, we can attach to each $b \in B_1$ a spin configuration s_b by choosing a 2-chain d with $\partial d = b$ and setting $(s_b)_i := (-1)^{d_i}$. Notice that here we are labeling 2-chain coefficients with spin labels, which is fine because spins live at faces of the direct lattice. Also, let us define a product on spin configurations: $s'' = s' \cdot s$ stands for $s''_i = s'_i s_i$. We can now express in a simple way a crucial property of the model, illustrated in Fig. 19.13:

$$H_{\tau_{c+b}}(s) = H_{\tau_c}(s \cdot s_b). \tag{19.33}$$

Thus, homologically equivalent interaction configurations give equivalent systems, up to a transformation on the spin variables. On the other hand, if we change the sign of the interactions along a nontrivial loop, as the one of Fig. 19.13, this cannot be absorbed by a change of variables. For example, in a ferromagnetic system changing interactions along such a loop creates frustration: the new ground states gain an energy proportional to the length of the minimal loop with the

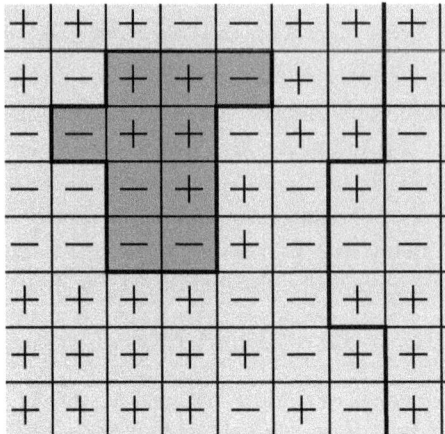

Fig. 19.13. The Ising model for a toric code. Classical ±1 spins live at plaquettes. Neighboring spins can interact ferro- or antiferromagnetically. If we switch all spins in the shaded region and at the same time also switch the sign of the interactions along its boundary, the energy is unchanged. But if we switch the sign along a nontrivial cycle like the one depicted by the bold line, this cannot be absorbed through spin switching.

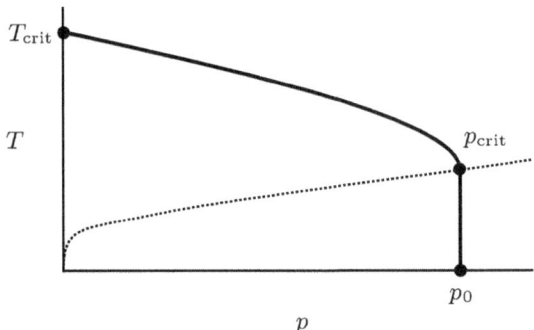

Fig. 19.14. Phase diagram of the random bond Ising model; p is the probability of antiferromagnetic bonds and T the temperature. The curve separates the low p, low T ordered phase from the disordered phase. The Nishimori line is shown as a dotted line. The critical probability p_{crit} along the Nishimori line gives the error threshold for error correction.

same homology. This suggests introducing the notion of domain wall free energy for a given $\bar{z} \in H_1, \bar{z} \neq 0$:

$$\Delta_{\bar{z}}(\beta_p, \tau_c) := F(\beta_p, \tau_{c+z}) - F(\beta_p, \tau_c), \tag{19.34}$$

where $F(\beta, \tau) = -T \log Z(\beta, \tau)$ is the free energy of a given interaction configuration τ.

Rather than individual systems with Hamiltonian H_τ, we are interested in the random bond Ising model, a statistical model obtained by making the parameter τ a quenched random variable. That is, τ is random but not subject to thermal fluctuations. We choose a probability distribution $\{p_\tau\}$ such that each τ_{ij} has an independent probability p of being antiferromagnetic. This will allow us later to connect with error correction, since $p_{\tau_c} = p_c$.

In thermal equilibrium the model has two parameters, the temperature T and the probability p. For $p = 0$, where we recover the standard Ising model, the system exhibits an order–disorder phase transition at a critical temperature T_{crit}. For low temperatures the system is ordered, as it spontaneously breaks the global $\mathbf{Z}_2$ symmetry, and for high temperatures it is disordered. Order can also be destroyed at $T = 0$ by increasing the disorder till we reach a critical value $p = p_0$. More generally, we can distinguish two regions in the pT plane, an ordered one at low T and p, and a disordered one, as sketched in Fig. 19.14. For the connection with error correction we will only be interested in the Nishimori line [N81], defined by

$$e^{-2\beta} = \frac{p}{1-p}. \tag{19.35}$$

(See also Section 11.2.2.) As we will see, the critical probability for error correction to be possible is given by the critical probability p_{crit} along the Nishimori line, see Fig. 19.14. A witness of the ordering is the domain wall free energy:

$$[\Delta_{\bar{z}}(\beta,\tau)]_p := \sum_{\tau} p_{\tau} \Delta_{\bar{z}}(\beta,\tau), \tag{19.36}$$

suitably averaged over quenched variables. This quantity diverges with the system size in the ordered phase and attains some finite limit in the disordered one.

19.4.7.2 Mapping and error threshold In order to express the homology class probabilities $\Pr(\bar{c})$ in terms of the partition function (19.32), we first observe that

$$p_c = (2\cosh\beta_p)^{-E} e^{-\beta_p H_{\tau_c}(s_0)}, \tag{19.37}$$

with $\beta_p := \log((1-p)/p)/2$ the inverse temperature in the Nishimori line (19.35) for a given p. Using (19.33) we get the desired result:

$$\Pr(\bar{c}) = 2^{-1}(2\cosh\beta_p)^{-E} Z(\beta_p, \tau_c), \tag{19.38}$$

where the extra factor of 2 comes from the constraint (19.16) or, equivalently, the global symmetry of the Ising model under spin inversion. We have then

$$\frac{\Pr(\bar{c}+\bar{z})}{\Pr(\bar{c})} = e^{-\beta\Delta_{\bar{z}}(\beta_p,p)}, \tag{19.39}$$

which shows that the homology class $\bar{c}$ is much more probable than the other candidates when the domain wall energy is high. For the average success probability p_{succ} in (19.29) to be close to one, those syndromes that are most probable must be such that one class dominates, which implies a big average domain wall energy (19.36). Indeed, it can be shown that if $p_{\text{succ}} \to 1$ then $[\Delta_z(\beta,\tau)]_p$ diverges [B10b], which establishes the connection between the error-correcting threshold and the critical probability along the Nishimori line, which is

$$p_{\text{crit}} \simeq 0.11. \tag{19.40}$$

19.5 Color codes

Surface codes are not very rich in terms of the gates that they allow to implement through transversal operations. Since they are CSS codes, they allow the transversal implementation of

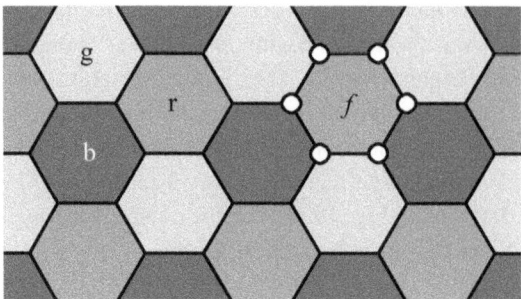

Fig. 19.15. Color codes are defined on 3-valent lattices with 3-colorable faces, like a honeycomb. We label the faces as red, green, and blue (r, g, b), as is customary, and distinguish them with three tones of gray. Qubits are placed on vertices. There are two generators of the stabilizer per face, X_f and Z_f, with the support shown.

CNOT gates. And, of course, we can implement $\overline{X}$ and $\overline{Z}$ operators transversally. But that is all there is to it.

To go beyond these gates we need to consider a different class of topological codes: color codes. As we will see, this class of codes includes a family of planar codes that allow the transversal implementation of the whole Clifford group of gates.

19.5.1 Lattice and stabilizer group

Surface codes can be built out of any lattice embedded in a closed manifold. In the case of color codes, we need to consider a particular kind of lattices: those that are 3-valent and have 3-colorable faces. That is, the lattice must be such that

(i) three edges meet at each vertex and
(ii) it is possible to assign one of three labels to each face in such a way that neighboring faces have different labels.

It is customary to choose as labels the colors red, green, and blue: r, g, b. The most basic example of such a lattice is the honeycomb lattice, which can be embedded in a torus, see Fig. 19.15. Notice that in a color code lattice we can also attach a color to edges: red edges are those that do not form part of red faces, and similarly for green and blue edges.

In order to construct a color code from such a lattice the first step is to attach a qubit to each vertex. Next we need the stabilizer generators, of which there are two per face f:

$$X_f := \prod_{v \in f} X_v, \qquad Z_f := \prod_{v \in f} Z_v, \qquad (19.41)$$

where X_v, Z_v are the X, Z Pauli operators on the qubit at the vertex v and $v \in f$ is a symbolic notation to denote that v is part of the face f. Notice that, as surface codes, color codes are CSS codes with local generators. The "plaquette" operators (19.41) are shown in Fig. 19.15.

As in (19.16), the stabilizer generators (19.41) are not independent. They are subject to four constraints. In order to write them, let us denote by F_r, F_g, and F_b the sets of red, green, and

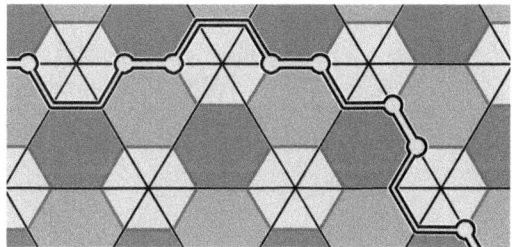

Fig. 19.16. Any shrunk lattice of a honeycomb lattice is triangular. Here we show the green shrunk lattice (light gray) and a string γ. The qubits in the support of X_γ^g and Z_γ^g are marked with circles along the string. They come in pairs, with each pair related to an edge of the shrunk lattice.

blue faces, respectively. It is not difficult to check that the constraints are

$$\prod_{f \in F_r} X_f = \prod_{f \in F_g} X_f = \prod_{f \in F_b} X_f,$$
$$\prod_{f \in F_r} Z_f = \prod_{f \in F_g} Z_f = \prod_{f \in F_b} Z_f. \tag{19.42}$$

Thus, the number of independent stabilizer generators is

$$g = 2(|F_r| + |F_g| + |F_b|) - 4. \tag{19.43}$$

This allows us to immediately compute the number of encoded qubits. Namely, since there are $2E$ physical qubits,

$$k = n - g = 2(E - |F_r| - |F_g| - |F_b| + 2). \tag{19.44}$$

However, to express k in terms of topological invariants we need a geometrical construction, which is our next topic.

19.5.2 Shrunk lattices

Out of a color code lattice we want to build three other "shrunk" lattices, labeled by the color of the faces that are actually shrunk. Let us focus on the red shrunk lattice; the green and blue are analogous. The vertex of the new lattice correspond to red faces, which are in this sense shrunk to points. Edges come from those edges that connect red faces, that is, red edges. As for faces, there is one for each green and blue face of the original lattice. The construction is demonstrated in Fig. 19.16.

Now, let V^r, E^r, and F^r denote the number of vertices, edges and faces of the red shrunk lattice. Using (19.1, 19.2, 19.44) the number of encoded qubits is found to be

$$k = 2(E^r - V^r - F^r + 2) = 2(2 - \chi) = 4g. \tag{19.45}$$

That is, encoded qubits have a topological origin! Notice that the number of encoded qubits doubles that of surface codes. The origin of this doubling will become clear when we explore string operators.

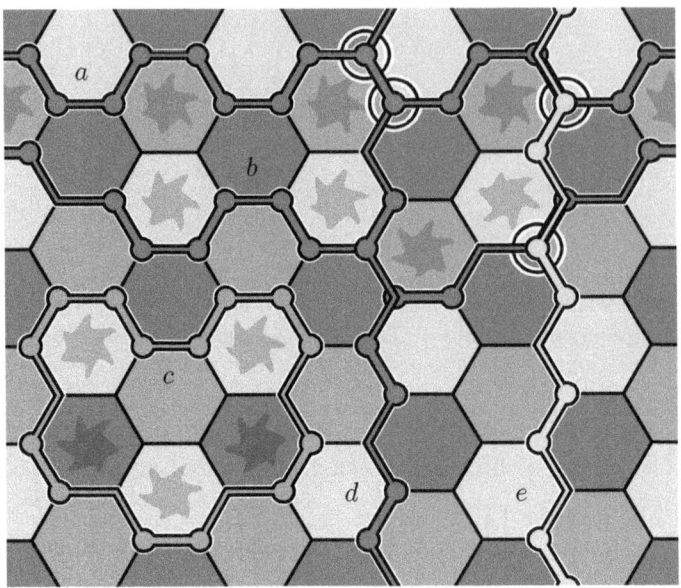

Fig. 19.17. A honeycomb color code in a torus, with five closed strings on display. For each string the support of the string operators for a given color are shown: blue for a, b, d, red for c and green for e. Those qubits at which two string operators share support are marked with large circles. Strings a and b are homologous and thus, for $\sigma = X, Z$, we have $\sigma_a^b = A\sigma_b^b$ with $A \in S$ the product of the face operators σ_f marked between a and b. String c is homologically trivial and thus $\sigma_c^r \in S$: it is obtained as the product of face operators σ_f marked in the region enclosed by c. Strings d and e are homologous, but due to the different colors $\sigma_d^b \sigma_e^g \notin S$. Strings a and d cross, but because the color is the same we get $[X_a^b, Z_d^b] = 0$, and similarly for b and d. Strings a and e cross, and due to the different coloring we have $\{X_a^b, Z_e^g\} = 0$, and similarly for strings b and e.

19.5.3 String operators

Given a loop or closed string γ in a color code lattice, we can construct out of it six different string operators: X_γ^c and Z_γ^c, with $c = $ r, g, b a color. They take the form

$$X_\gamma^c := \prod_{v \in V_c^\gamma} X_v, \qquad Z_\gamma^c := \prod_{v \in V_c^\gamma} Z_v, \qquad (19.46)$$

where the set V_c^γ contains those vertices that belong to a c-colored edge of γ. Indeed, the support of a string operator is best understood in terms of the shrunk lattice, as explained in Fig. 19.17.

We next list several properties of string operators that are easy to check. String operators obtained from closed strings belong to the normalizer $N(S)$. For a given closed string γ there are only four independent string operators because

$$X_\gamma^r X_\gamma^g X_\gamma^b = 1, \qquad Z_\gamma^r Z_\gamma^g Z_\gamma^b = 1. \qquad (19.47)$$

In this sense, there are only two independent colors, so that it suffices to consider, say, red and green strings. If two strings γ and γ' cross an even number of times their operators commute. If they cross an odd number of times, we have $[X_\gamma^c, Z_{\gamma'}^c] = 0$ and $\{X_\gamma^c, Z_{\gamma'}^{c'}\} = 0$ for $c \neq c'$, see Fig. 19.17.

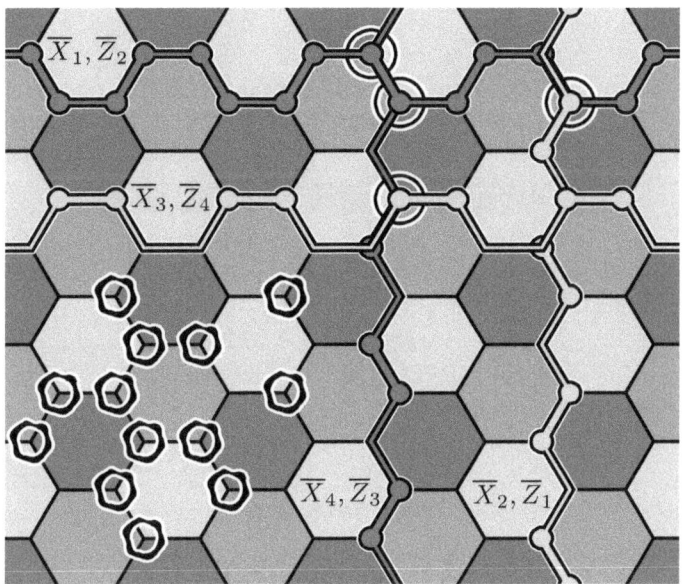

Fig. 19.18. A [[96, 4, 8]] color code in a torus. Logical operators $\overline X_i, \overline Z_i$ take the form of four blue string operators and four green string operators. The marked qubits form the support of a local operator. If such an operator belongs to $N(S)$, then it belongs to S because it clearly commutes with logical operators.

A question without such an immediate answer is: when does a string operator belong to the stabilizer and when are two string operators equivalent as logical operators in $N(S)/S$? As in surface codes, the answer lies in homology: the trick is to think in terms of the shrunk lattice. For example, consider X-type red string operators. We can regard any γ as a loop in the red shrunk lattice. If γ and γ' are homologically equivalent, they enclose a region in the shrunk lattice. This region corresponds to a set of green and blue faces in the original lattice. As one can easily check, we have then $X_\gamma^r = s X_{\gamma'}^r$ with $s \in S$ the product of the corresponding X-type face operators. It follows that boundary strings produce elements of S and that homologically equivalent strings produce, for each type of operator, equivalent operators up to stabilizers. All this is illustrated in Fig. 19.17.

We are now ready to choose logical operators for a given surface. Indeed, the task is almost the same as in surface codes, but now we have to take color into account. In particular, strings of two colors are needed, which is at the origin of the doubling of encoded qubits with respect to surface codes. As an example, we show a choice of logical operators for a torus in Fig.19.18. We adopt the convention that $\overline X_i$ and $\overline Z_i$ logical operators are obtained respectively from X-type and Z-type string operators.

The fact that logical operators can be chosen to be string operators has an important consequence that is not specific of color codes but common to 2D topological codes. If a region R is such that we can choose a set of logical operators with support out of it, any operator with support in R that belongs to $N(S)$ must belong to S. Since strings can be deformed, this is in particular true for any "local" region, such as the one in Fig. 19.18. In the case of regular lattices

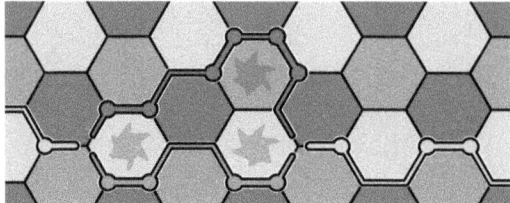

Fig. 19.19. A string-net operator. It can be transformed into a green string operator by taking the product with the face operators from the marked faces.

and due to the locality of stabilizer generators, this implies that the code distance will grow with the lattice. Thus, color codes are indeed local codes.

19.5.4 String-nets

We could be tempted to believe that the distance in a color code is given by the smallest weight among string operators of nontrivial homology. This holds in all the examples that we will present, but in general it is not true. The reason is that we can combine strings to form nets, resulting in smaller weights.

In order to understand what these string-nets are, take any green string and consider deforming part of it not by taking the product with the blue and red face operators in an adjacent region, as we should, but with the red and green face operators, as if it were a blue string. The result, as Fig. 19.19 shows, is not a string any more, but a net of three strings. The deformation has created a piece of blue string, leaving a red string where the piece of green string to be deformed was. The moral is that strings can have branching points and form string-nets. At each branching point three strings, one of each color, must meet. Although string-nets are not necessary to construct logical operators in closed surfaces, we will see how they can become essential in the presence of boundaries.

19.5.5 Boundaries

As in surface codes, in color codes we can obtain boundaries by erasing "big" faces from the lattice. There are thus three kinds of boundaries, one per color. To make the idea clear, we consider blue boundaries. These are obtained by erasing a blue face, so that blue strings can have endpoints at the boundaries, but not green or red ones. The properties of boundaries in color codes are analogous to those in Section 19.4.5:

(i) blue string operators that have their two endpoints on a blue boundary belong to $N(S)$;
(ii) blue string operators that enclose a region that only contains blue boundaries belong to S.

It is worth noting that these three kinds of boundaries are not exhaustive. For example, it is possible to have boundaries where only X-type string operators can have endpoints. But we will not need this more general cases.

From the constraints (19.42) we can infer the number of encoded qubits in geometries with boundaries: For each new face that we erase, we add two qubits, unless we have only erased a

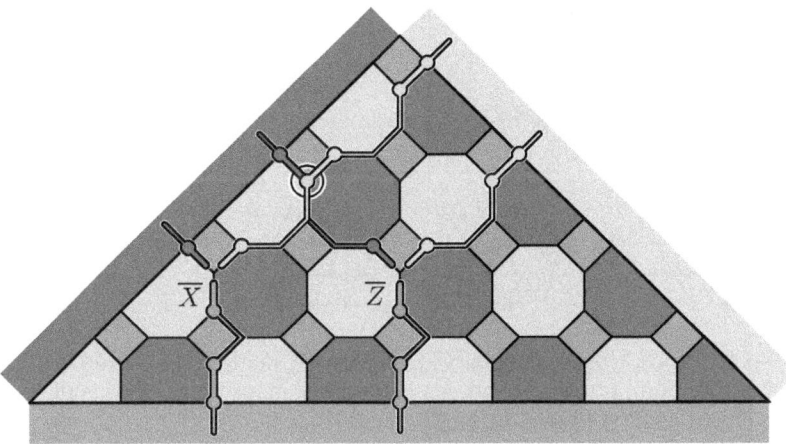

Fig. 19.20. A [[73, 1, 9]] triangular color code, based on a 4-8-8 lattice. The bottom boundary is red, the right one green and the left one blue. There is only one class of nontrivial string-nets, call it μ. That X_μ and Z_μ anticommute can be understood topologically by considering the string-net and a deformation of it, as in this figure, and noting that they cross once at a point where they have different colors.

single face of a different color previously. In this latter case we add no qubits. For example, if in a sphere we remove one face of each color, the resulting color code encodes two qubits. We will next discuss a closely related geometry that encodes a single qubit.

19.5.6 Transversal Clifford group

We can finally introduce the family of color codes that allows the transversal implementation of the whole Clifford group of gates: triangular codes. From a topological perspective, these are planar codes with the geometry of a triangle in which each side is a boundary of a different color, as depicted in Fig. 19.20. How many qubits are encoded with such a topology? A bit of experimentation can convince one that there is only one nontrivial class of string-nets, shown in the figure. There is something very special about this string-net. Denote it by μ. Then we have $\{X_\mu, Z_\mu\} = 0$, so that the encoded Pauli operators $\overline{X} = X_\mu$ and $\overline{Z} = Z_\mu$ can be chosen to have the same geometry!

Notice that any color code is invariant under the action of a transversal Hadamard gate $\widehat{W} := W^{\otimes V}$, where V is the number of vertices/qubits, because $\widehat{W} X_f \widehat{W} = Z_f$ and $\widehat{W} Z_f \widehat{W} = X_f$. For the triangular geometry we have in addition $\widehat{W} \overline{X} \widehat{W} = \overline{Z}$ and $\widehat{W} \overline{Z} \widehat{W} = \overline{X}$, simply because geometrically $\overline{X}$ and $\overline{Z}$ are the same.

Since CNOT gates are automatically transversal in a CSS code, all we need is to find a way to implement the phase gate P transversally. The obvious guess is to use $\widehat{P} := P^{\otimes V}$ but, does it leave the code invariant? In general no, because $\widehat{P} Z_f \widehat{P} = Z_f$ but $\widehat{P} X_f \widehat{P} = (-1)^{t/2} X_f Z_f$, with t the number of vertices of the face f. All we have to do then is to find lattices where all faces have a number of edges that is a multiple of four. This is indeed possible using the so-called 4-8-8 lattice, as in Fig. 19.20. As for the effect of $\widehat{P}$, it gives either an encoded P or $-P$, because $\overline{X}$ always has support on an odd number of qubits (otherwise, it could not have the same

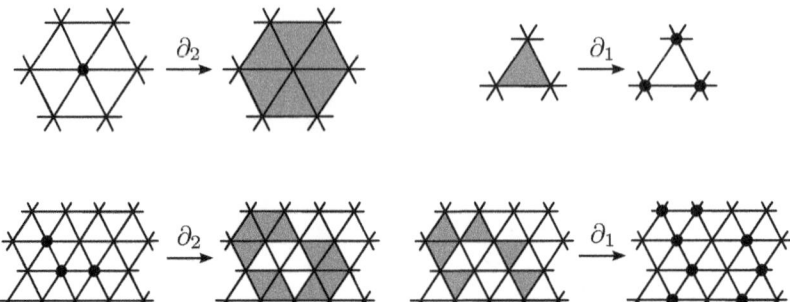

Fig. 19.21. The action of "color" boundary operators: ∂_2 maps a set of vertices to the set of triangles to which an odd number of vertices belong; ∂_1 maps a set of triangles to the set of vertices where an odd number of these triangles meet.

support as $\overline{Z}$). As a result, we have obtained a family of 2D local codes that allow the transversal implementation of any Clifford gate.

19.5.7 Homology

In the case of surface codes, we started by giving a homological definition and from that we obtained a description in terms of a stabilizer group. For color codes the definition has been in terms of the stabilizer, but we can now obtain from it a homological description by undoing our steps for surface codes.

Before we do this, it is useful to change our picture of color codes by switching to the dual lattice. The dual of a color code lattice has triangular faces and three colorable vertices. For example, the honeycomb lattice has as its dual the triangular lattice of Fig. 19.21. Notice that in the dual picture qubits are attached to triangles and stabilizer generators to vertices.

In our search for the "color homology," we start observing that, since qubits are attached to triangles, 1-chains should be composed of triangles. We have thus a group $C_1^\triangle$ with elements that are formal sums of triangles. As for 0-chains and 2-chains, they must be composed of the geometrical objects attached to Z and X stabilizer generators, respectively. For color codes they are the same: we take the elements of $C_0^\triangle = C_2^\triangle$ to be formal sums of vertices.

The next step is to build the boundary morphisms ∂. This is dictated by the geometry of stabilizer generators. The morphisms $\partial_2^\triangle : C_2^\triangle \longrightarrow C_1^\triangle$ and $\partial_1^\triangle : C_1^\triangle \longrightarrow C_0^\triangle$ are dual to each other:

$$\partial_2^\triangle v = \sum_{f | v \in f} f, \qquad \partial_1^\triangle f = \sum_{v \in f} v, \qquad (19.48)$$

where v stands for a vertex and f for a triangular face. The action of the boundary operators is illustrated in Fig. 19.21. Boundary morphisms give rise to a group of cycles $Z_1^\triangle$ and a group of boundaries $B_1^\triangle$, with $B_1^\triangle \subset Z_1^\triangle$. In a closed surface the resulting homology group must be

$$H_1^\triangle := \frac{Z_1^\triangle}{B_1^\triangle} \simeq H_1 \times H_1, \qquad (19.49)$$

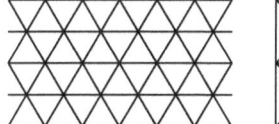 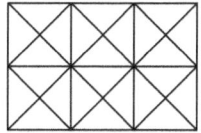

Fig. 19.22. Triangular lattice (left) and Union Jack lattice (right). These are duals of the honeycomb and 4-8-8 lattices, respectively.

because we know that there are two independent "homology structures," one per independent color.

Once we have a homological language for color codes, we can immediately apply all the results on error correction of Section 19.4.6. Color codes also attain in the limit of large lattices $p_{\text{succ}} \to 1$ when the noise is below a critical threshold. The main difference with error correction in surface codes is algorithmic, at least under the simplifying approach that led to length minimization. Here, this approach leads to the problem of finding a triangle chain of minimum weight for a given boundary, which cannot be solved using the perfect matching algorithm.

19.5.8 Error threshold: random three-body Ising model

As in the case of surface codes, the error-correction threshold of color codes is connected to a phase transition in a 2D statistical model: a random three-body Ising model. Given the similarities of the mappings, we will only discuss those aspects that are different with respect to surface codes, and the assumptions are the same. We consider two geometries, the honeycomb lattice and the 4-8-8 lattice that allows the transversal implementation of P. Or rather, the duals of these lattices, the triangular lattice and the so-called Union Jack lattice, shown in Fig. 19.22.

A question that comes to mind immediately is: do the properties have a cost in terms of the error threshold? Surprisingly, the answer turns out to be negative.

19.5.8.1 Random three-body Ising model This time, classical spins are attached to the vertices of the dual lattice, so that we can talk of red, green, and blue spins. Each triangular face can be given as a triad of vertices $\langle ijk \rangle$. The Hamiltonians take the form

$$H_\tau(s) := - \sum_{\langle ijk \rangle} \tau_{ijk}\, s_i s_j s_k, \qquad (19.50)$$

where the $\tau_{ijk} = \pm 1$ are again parameters of the Hamiltonian that determine the sign of the three-body interactions. Instead of the $\mathbf{Z}_2$ symmetry of the two-body Ising models, these models exhibit a $\mathbf{Z}_2 \times \mathbf{Z}_2$ global symmetry: for any color of choice, flipping all the spins of the two other colors leaves the energy unchanged. Equation (19.33) still holds, with the obvious definitions for τ_c and s_b, $c \in C_1^\triangle$, $b \in B_1^\triangle$. The rest of the details of the model are analogous to those for toric codes, including the phase diagram.

19.5.8.2 Mapping and error threshold The mapping works essentially as in toric codes. There is only a slight difference, that the factor due to global symmetry is now 4:

$$\Pr(\bar{c}) = 4^{-1}(2\cosh\beta_p)^{-F} Z(\beta_p, \tau_c), \qquad (19.51)$$

where F is the number of triangles. As for the critical probability, for both lattices we have

$$p_{\mathrm{crit}} \simeq 0.11, \qquad (19.52)$$

the same as for toric codes!

19.6 Conclusions

Topological codes are naturally local, which makes them appealing for practical implementations with locality constraints. We have described two classes of topological codes, surface codes and color codes. The main difference between them is that color codes allow the transversal implementation of more gates, even of all the gates in the Clifford group. Although topological codes were initially described in closed surfaces, it is possible to construct planar versions by introducing carefully designed boundaries.

We have emphasized the role of homology in topological codes, which offers a unified picture of surface and color codes. The first homology group is obtained as a quotient Z_1/B_1, and the very essence of topological codes is the identification of this quotient with $N(S)/S$, the quotient that gives logical operators in stabilizer codes. That is, stabilizer elements correspond to boundary loops, normalizer elements to closed loops, and logical operators to elements of the homology group.

From the point of view of error correction, topological codes exhibit two remarkable facts. One is that error correction must be done only up to homology, due to the high degeneracy of the codes. The other is the existence of an error threshold: for noise levels below this threshold, in the limit of large systems, error correction is asymptotically perfect. For error models with uncorrelated bit-flip and phase-flip errors, those for which the critical threshold is well understood, the critical error probability is $p_{\mathrm{crit}} \simeq 0.11$.

19.7 History and further reading

Topological codes started their history with the introduction by Kitaev of the toric code [K97b, K03a]. It was soon realized that boundaries allow one to build planar codes [BK98, FM01]. The basic reference on surface codes is [DKL+02]. This work introduced, among other things, the concept of topological quantum memory: in the limit of large surface codes, there exists an error threshold below which quantum information can be preserved arbitrarily well. It also discusses higher-dimensional toric codes and shows the connection between accuracy thresholds in error correction and phase transitions in statistical models, a subject developed in [WHP03, OAI+04, TN04] and other works. Another concept introduced in [DKL+02] is that of code deformation: the lattice defining the code can be progressively transformed locally, for example to encode information by "growing" the lattice as a crystal. It was later realized that this can be used to initialize, measure, and perform gates on encoded qubits [RHG07, BM.-D09], something closely related to the fault-tolerant one-way quantum computing scheme of [RHG07]. Examples of how research on toric codes continues more than a decade after their introduction are a study on loss errors [SBD09] and a new algorithm for error correction based on renormalization [D.-CP10].

Two-dimensional color codes were introduced in [BM.-D06] and soon a generalization to 3D followed [BM.-D07b]. The advantage of 3D color codes is that they allow the transversal implementation of the CNOT gate, the $\pi/8$ phase gate and X, Z measurements: a universal set of operations for quantum computing. The statistical models related to error correction in 2D color codes have been studied in [KBM.-D09, KBA+10].

Surface codes and color codes are not the end of the story. Other interesting topological codes might still be awaiting their discovery. Among recent developments we find topological subsystem codes [B10b] and Majorana fermion codes [BTL10]. On the other hand, bounds on the code distances in terms of the system sizes have been developed for large classes of codes [BPT10, B11c, BH11]. Moreover, all 2D topological stabilizer codes amount to copies of the toric code up to local transformations [BD.-CP11] and all 2D topological subsystem stabilizer codes have been classified [B11b]. New ways to exploit already known codes are also valuable. The "twists" introduced in [B10a] exemplify this, as they offer a new tool for constructing planar topological codes with enhanced code deformation capabilities. Indeed, in a suitable class of 2D topological subsystem codes all Clifford operations can be carried out using code deformation [B11a].

Topological codes give rise naturally to condensed matter systems in which the ground state corresponds to the encoded subspace [K03a]. These *topologically ordered* systems are stable against perturbations at $T = 0$: local modifications of the Hamiltonian, if not too big, do not affect the physics [BHM10]. The effect of thermal noise depends on the spatial dimension of the system: in two dimensions topological order is destroyed [AFH09], but in four dimensions it can be resilient to noise [AHH+10], giving rise to a self-correcting quantum memory. See Chapter 26 for more details. In six dimensions it is even possible to put together such *self-correcting* capabilities and the nice properties of 3D color codes [BCH+09] to obtain a self-correcting quantum computer. A class of 3D topological codes with unusual properties and that could be self-correcting has been introduced in [H11].

The excitations exhibited by 2D topologically ordered systems are gapped and localized. These quasiparticles, called *anyons*, have unusual statistics, neither bosonic nor fermionic. Indeed, they give rise to nonlocal, topological degrees of freedom that can be manipulated by moving the anyons around. This offers a new way to perform quantum computations: topological quantum computation [K03a, FKL+03]. A good introduction to this topic is [P04]. In higher dimensions excitations take the form of extended objects called branyons in [BM.-D07a].

Chapter 20

Fault-tolerant topological cluster state quantum computing

Austin Fowler and Kovid Goyal

20.1 Introduction

Cluster states were introduced in Chapter 18, along with some of the approaches to achieving fault tolerance in measurement-based quantum computing. In this chapter, we describe an extremely promising fault-tolerant cluster state quantum computing scheme [RH07, RHG07] with a threshold error rate of 7.46×10^{-3}, low overhead arbitrarily long-range logical gates and novel adjustable strength error correction capable of correcting general errors through the correction of Z errors only. Detailed proposed implementations of this scheme exist for ion traps [SJ09] and single photons with cavity mediated interactions [DGI+07].

The discussion is organized as follows. In Section 20.2, we describe the topological cluster state, which is a specific three-dimensional (3D) cluster state, and give a brief overview of what topological cluster state quantum computing involves. Section 20.3 describes logical qubits in more detail and how to initialize them to $|0_L\rangle$ and $|+_L\rangle$ and measure them in the Z_L and X_L bases. State injection, the non-fault-tolerant construction of arbitrary logical states, is covered in Section 20.4. Logical gates, namely the logical identity gate and the logical CNOT gate, are discussed in Section 20.5 along with their byproduct operators. Section 20.6 describes the error-correction procedure. In Section 20.7, we calculate an estimate for the threshold of this scheme. Section 20.8 presents an analysis of the overhead as a function of both the circuit size and the error rate.

20.2 Topological cluster states

A topological cluster state is a specific 3D cluster state. Figure 20.1a shows a single cell of a topological cluster state. This cell is tiled in 3D. Figure 20.1b shows a simplified picture of an

---

Quantum Error Correction, ed. Daniel A. Lidar and Todd A. Brun. Published by Cambridge University Press.
© Cambridge University Press 2013.

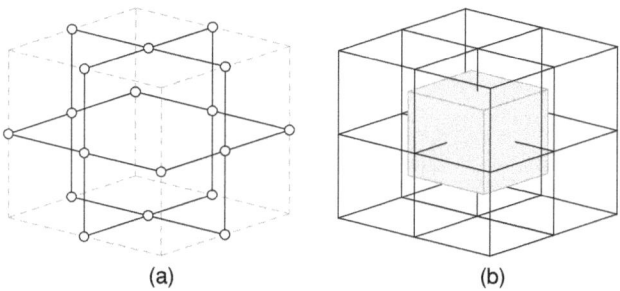

Fig. 20.1. (a) A 3D 18-qubit cluster state. This cell is tiled in 3D to form the topological cluster state. (b) A cube of eight cells, which we will call primal cells, each of the form shown in part (a) (qubits suppressed for clarity). Location of a dual cell (shaded) relative to its surrounding primal cells. A dual cell also contains exactly the arrangement of qubits shown in part (a). Reprinted with permission from [FG09].

eight-cell topological cluster state. The wireframe cubes as well as the central shaded region have exactly the same form as Fig. 20.1a. A topological cluster state can thus be thought of as of two interlocking cubic lattices. We arbitrarily label one of these lattices the "primal" lattice and the other the "dual" lattice. The boundaries of the lattice are also labeled primal or dual according to whether they consist of primal or dual cell faces. If we call the eight wireframe cells of Fig. 20.1b primal cells, then the lattice has only primal boundaries.

A few more definitions are required to discuss the effect of errors. All cluster states have stabilizers [G97a] of the form $X_i \otimes_{q_j \in \text{nghb}(q_i)} Z_j$ where $\text{nghb}(q_i)$ denotes the set of qubits linked to q_i, namely the neighborhood of qubit q_i. If we consider the product of the six stabilizers centered on the six face qubits of Fig. 20.1a, we find that all Z operators cancel, leaving us with the tensor product of X on each face qubit. This implies that if we measure each of these qubits in the X basis, we will obtain six bits of information with even parity (sum mod 2 = 0). A string of bits with odd parity (sum mod 2 = 1) tells us that one or more errors have occurred locally. This is how errors are detected. Error correction will be discussed in Section 20.6. For the moment we simply claim that erroneous measurement results can be corrected arbitrarily well given a sufficiently large topological cluster state and sufficiently low physical error rates. Errors are also defined to be primal or dual according to whether they occur on primal or dual face qubits.

Logical qubits are associated with pairs of "defects" – regions of qubits measured in the Z basis. A defect must have a boundary of a single type. There are thus two types of defects and logical qubits – primal and dual. Referring to Fig. 20.2, for both primal and dual logical qubits the initial U-shape or pair of individual beginnings corresponds to initialization, the middle section, braided with other defects of the opposite type, corresponds to computation, and the final U-shape or pair of individual endings corresponds to read-out. Full details will be given in later sections. Logical operators X_L and Z_L correspond to a ring or chain of single-qubit Z operators encircling a single defect or connecting the pair of defects. Examples of such rings and chains are shown in Fig. 20.2. An important point that will become clearer in Section 20.3 is that these rings and chains must be periodically chosen throughout the computation – they cannot be defined in a consistent manner as arbitrary rings and chains since different rings and chains are not equivalent. Furthermore, the logical operators cannot be defined at all during braiding. We will always choose primal Z_L to be a chain connecting two primal defects and dual Z_L to be a

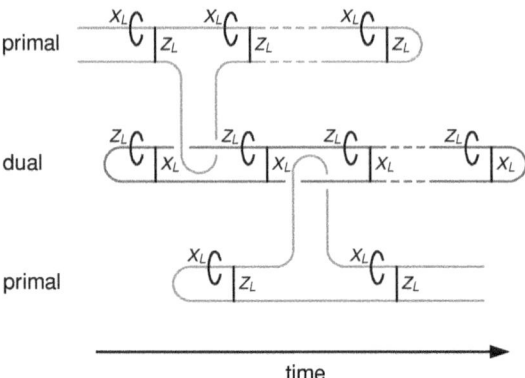

Fig. 20.2. The definitions of primal and dual Z_L and X_L – all rings or chains of Z operators. Note that the logical operators are undefined during braiding. Reprinted with permission from [FG09].

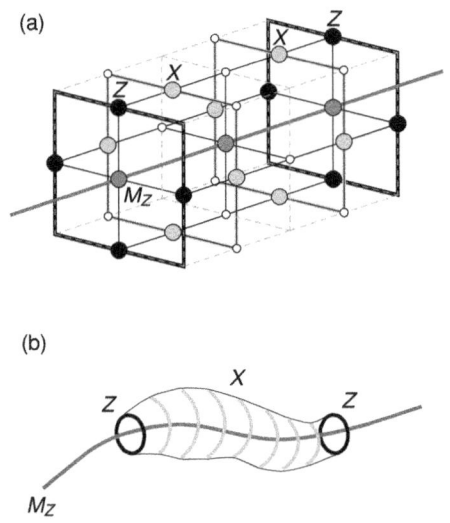

Fig. 20.3. Correlation surfaces beginning and ending with rings of Z operators. (a) Black qubits are associated with Z operators, light gray qubits with X operators. The collection of black and light gray qubits and their associated operators is a cluster state stabilizer. Middle gray highlighting indicates qubits measured in the Z basis, forming a defect. (b) Schematic representation. The surface of X operators can be arbitrarily deformed whereas we keep the initial and final rings of Z operators fixed. Reprinted with permission from [FG09].

ring encircling a single dual defect. The definitions of primal and dual X_L can be inferred from Fig. 20.2.

Computation makes use of "correlation surfaces" – large cluster state stabilizers connecting logical operators. For example, two rings of Z operators encircling the same defect can be connected with a tube of X operators such that a cluster state stabilizer is formed as shown in Fig. 20.3. Similarly, two chains of Z operators connecting two defects can be connected with a

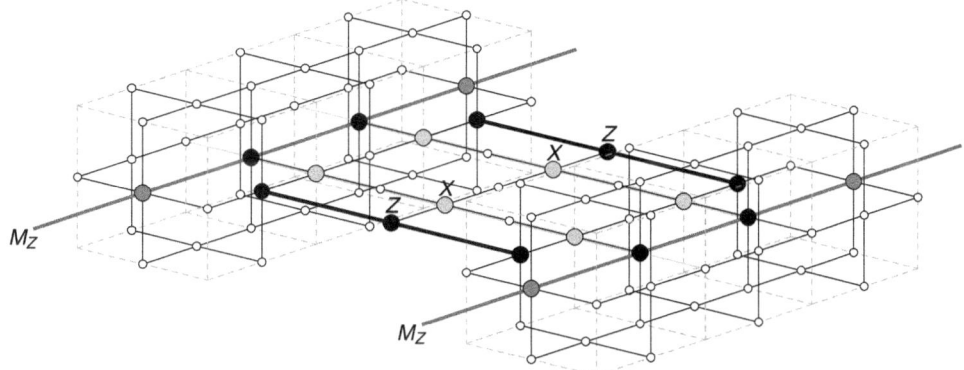

Fig. 20.4. A correlation surface beginning and ending with chains of Z operators. Black qubits are associated with Z operators, light gray qubits with X operators. The collection of black and light gray qubits and their associated operators is a cluster state stabilizer. Middle gray highlighting indicates qubits measured in the Z basis, forming defects. Reprinted with permission from [FG09].

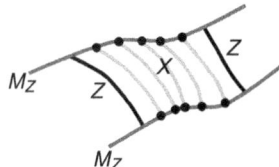

Fig. 20.5. Schematic representation of Fig. 20.4. The surface of X operators can be arbitrarily deformed provided the Z operators inside the defect remain in the defect. The initial and final chains of Z operators are kept fixed. Reprinted with permission from [FG09].

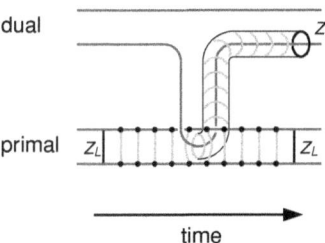

Fig. 20.6. A more complicated arrangement of defects and a correlation surface consistent with the arrangement. The connection of Z_L with $Z_L Z_L$ is suggestive of a CNOT gate, described in detail in Section 20.5. Reprinted with permission from [FG09].

surface of X operators bordered by Z operators as shown in Figs. 20.4 and 20.5. More complicated defect geometries lead to more complicated correlation surfaces. Figure 20.6 shows two logical qubits braided in such a way that Z_L on the lower logical qubit connects to $Z_L Z_L$ on both logical qubits – one of the four mappings associated with logical CNOT. Section 20.5 gives full details of the logical identity and logical CNOT gates.

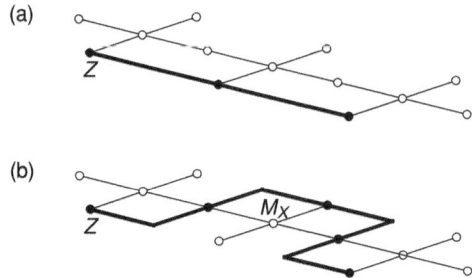

Fig. 20.7. Two nonequivalent chain operators. The second chain operator will have an eigenvalue $(-1)^{M_X}$ times the eigenvalue of the first chain operator. Reprinted with permission from [FG09].

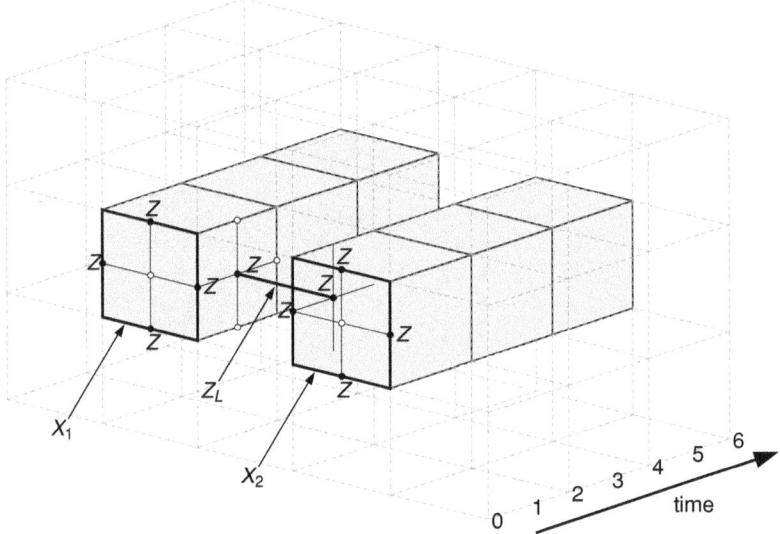

Fig. 20.8. A primal qubit consisting of two primal defects with Z_L indicated and X_L equal to X_1 or X_2. Reprinted with permission from [FG09].

20.3 Logical initialization and measurement

We now move on to the details of topological cluster state quantum computing, focusing on the initialization and measurement of logical qubits in this section. We wish to be able to initialize logical qubits to $|0_L\rangle$ and $|+_L\rangle$, the +1 eigenstates of Z_L and X_L. Take note that deforming a logical operator does not, in general, give an equivalent logical operator. For example, Fig. 20.7 shows two different chain operators. If the lattice is in the +1 eigenstate of the first chain, it will be in the $(-1)^{M_X}$ eigenstate of the second chain, where M_X is the result of the indicated X basis measurement. This issue can only be avoided by periodically choosing, by hand, specific rings and chains to represent primal and dual Z_L and X_L. The correlation surfaces connecting these logical operators can, however, take any shape consistent with the defects in the lattice.

To permit concrete discussion, we choose one dimension of the topological cluster state to be "simulated time" and arrange the defects of logical qubits not currently being braided to be parallel and in the direction of simulated time, as shown in Fig. 20.8. Note that we define a single

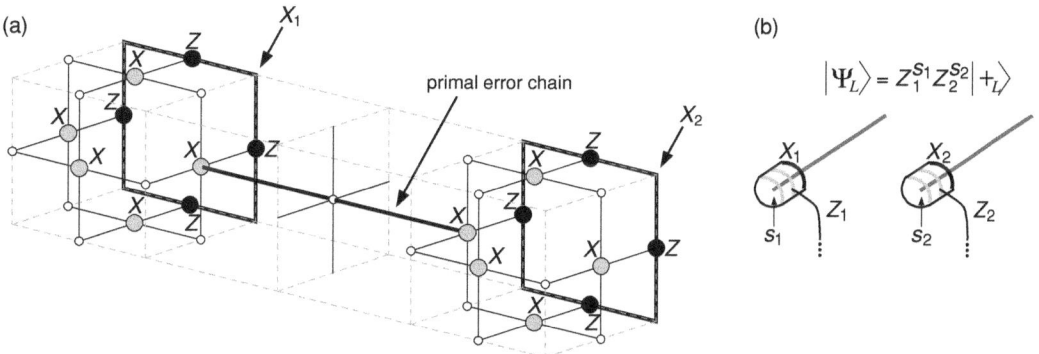

Fig. 20.9. Initializing a primal qubit to the $|+_L\rangle$ state. (a) After the indicated X measurements, the two defects are left in known eigenstates of their associated X_L operators, X_1 and X_2, which are both rings of single-qubit Z operators. (b) Schematic representation. Reprinted with permission from [FG09].

time step to correspond to the measurement of a single layer of the cluster state. We define a primal qubit to be in the $|+_L\rangle$ state if in a single even time step it is in the simultaneous $+1$ eigenstate of each of the two operators consisting of a ring of single-qubit Z operators encircling and on the boundary of each defect. Similarly, the simultaneous -1 eigenstate of these two boundary operators is defined to be $|-_L\rangle$.

There is some redundancy in the way we have defined $|+_L\rangle$ and $|-_L\rangle$. It would have been sufficient to focus on a single ring of Z operators around a single defect. Indeed, applying both of these Z rings simultaneously is the logical identity operation – X_L is just one of these rings, although it does not matter which ring. For later convenience, when we do not wish to specify which ring, we will use the notation X_L. When we need to discuss exactly which operator is being applied, we will write X_1 or X_2.

Primal qubits can be initialized to $|+_L\rangle$ up to byproduct operators via a measurement pattern similar to that shown in Fig. 20.9a. Measuring the indicated qubits in the X basis leaves the defects in either the $+1$ or -1 eigenstate of X_1 and X_2 depending on the parity of the associated X measurements. If we denote the parity of the X measurements associated with X_1 by s_1 as shown in Fig. 20.9b, the state of the logical qubit after initialization will be $Z_1^{s_1} Z_2^{s_2} |+_L\rangle$, with $Z_L = Z_1 Z_2$ and $\{X_1, Z_1\} = \{X_2, Z_2\} = 0$. The operators Z_1 and Z_2, while not physical unless at least one additional primal boundary is present in the system, are useful for keeping track of byproduct operators affecting a single defect. If an additional primal boundary is present, these operators can be represented by chains of Z starting on each defect and ending on this additional boundary.

Note that in the absence of errors all surfaces of X measurements bounded by either X_1 or X_2 will have the same parity, as the X stabilizer associated with the six faces of a single cell can be used to arbitrarily deform a surface without changing its parity. This implies that the initialization procedure is well-defined and fault tolerant when used in conjunction with the error correction described in Section 20.6.

Primal qubits can also be initialized to $|0_L\rangle$ up to byproduct operators via a measurement pattern similar to that shown in Fig. 20.10a. We choose Z_L to be any specific chain of Z in an odd time slice connecting two sections of defect. The parity s of the X and Z measurements in time slices earlier than the chosen logical operator determines the byproduct operator X_L^s.

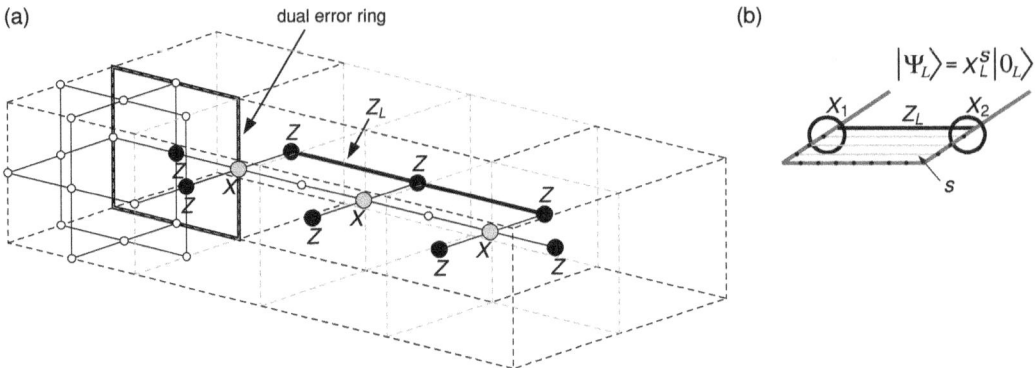

Fig. 20.10. Initializing a primal qubit to the $|0_L\rangle$ state. (a) After the indicated Z and X measurements, the U-shaped defect is left in a known eigenstate of the Z_L operator, which is a specific chosen chain of single-qubit Z operators. (b) Schematic representation. Reprinted with permission from [FG09].

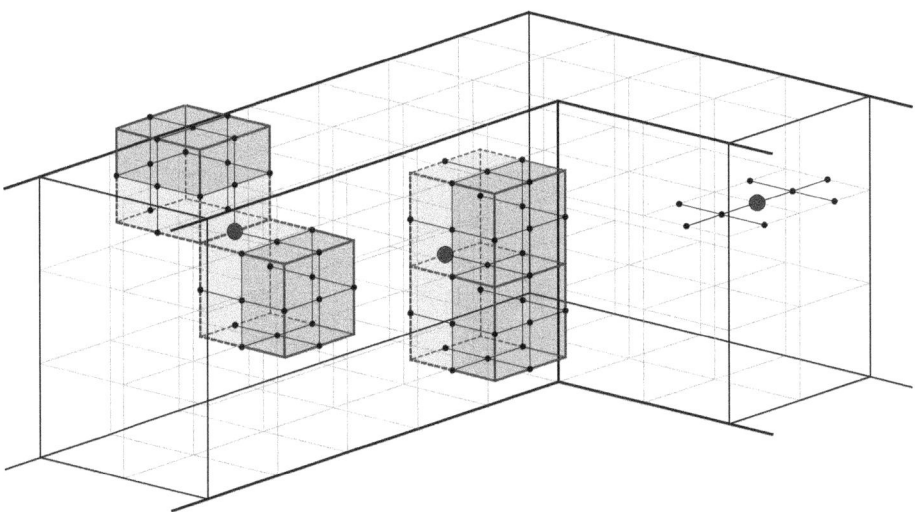

Fig. 20.11. Three examples of X or M_Z errors on qubits measured in the Z basis inside a defect. The leftmost examples are errors on the first layer of qubits inside the defect, which can be both detected and corrected by determining the parity of five-sided dual cells touching the error. The rightmost example is sufficiently deep inside the defect that no nontrivial stabilizers intersect it and therefore the error can be ignored. Reprinted with permission from [FG09].

As drawn, Fig. 20.10 is not fault tolerant. The defect is too narrow to provide any information about errors on the internal qubits measured in the Z basis. Figure 20.11 shows a larger defect and examples of odd parity five-sided dual cells resulting from errors on qubits inside the defect. For operations to be fault tolerant, defects must have minimum cross-section 2×2 cells.

In addition to demonstrating that primal-qubit initialization to $|0_L\rangle$ can be made fault tolerant, Fig. 20.11 shows how appropriate error information is extracted on the surface of a primal defect to permit dual error correction to continue and vice versa. Furthermore, note that

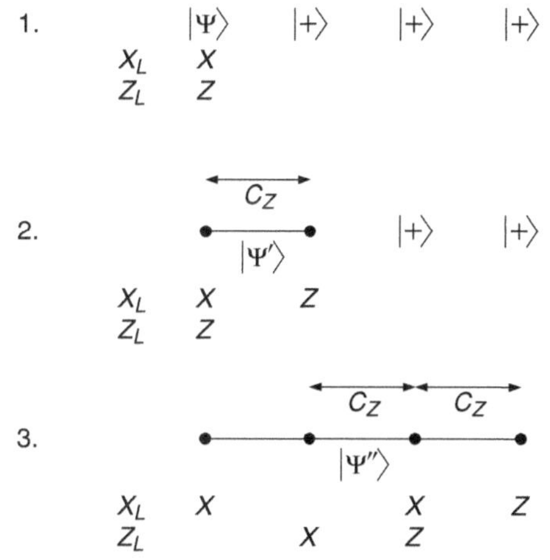

Fig. 20.12. Injecting an arbitrary state into a four-qubit cluster state. Information about the original state can be obtained from the parity of multiple single-qubit measurements. Reprinted with permission from [FG09].

Z measurements deeper inside the defect than the outermost layer are not used in any part of the computation or error-correction procedure and as such their results can be discarded.

Dual-qubit initialization, expressed in terms of dual cells, looks absolutely identical to primal-qubit initialization. The only difference lies in the interpretation of what the initialization procedures mean. A dual measurement pattern of the form shown in Fig. 20.9 initializes the dual qubit to $|0_L\rangle$. Similarly, a dual measurement pattern of the form shown in Fig. 20.10 initializes the dual qubit to $|+_L\rangle$. The definitions of all X and Z logical and byproduct operators are also reversed.

With logical qubits and logical operators defined, we can now discuss logical errors. In Fig. 20.9, any chain of primal errors, which can be thought of as Z errors on the underlying qubits before measurement or X basis measurement errors, that connects the two defects is undetectable and changes the state of the logical qubit from $|+_L\rangle$ to $|-_L\rangle$. To make this unlikely, defects must be kept well separated. In Fig. 20.10, any ring of dual Z and M_X errors encircling one of the defects is undetectable and changes the state of the logical qubit from $|0_L\rangle$ to $|1_L\rangle$. To make this unlikely, defects must have a sufficiently large perimeter. The situation is similar for dual qubits, with the meaning of the two types of logical errors interchanged.

Now that we have initialization, logical measurement follows in a straightforward manner. Figures 20.9 and 20.10, reversed in time, can be used to measure the logical operators of the qubit. The parity of the measurement results determines the sign of the eigenvalue of the logical operator.

20.4 State injection

We have discussed logical qubit initialization to states $|+_L\rangle$ and $|0_L\rangle$, measurement in the X_L and Z_L bases, and logical errors. We now turn our attention to state injection, specifically the preparation of logical states $\alpha|0_L\rangle + \beta|1_L\rangle$.

Consider Fig. 20.12. The first part of the figure shows a single qubit in an arbitrary state. The logical operators X_L and Z_L correspond to single-qubit X and Z respectively. The second part

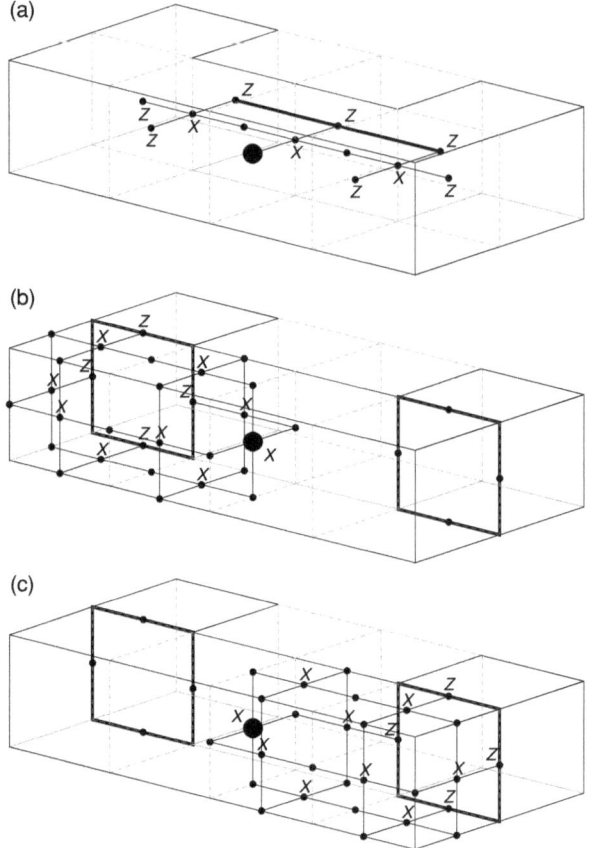

Fig. 20.13. Full form of (a) Z_L, (b) X_1, and (c) X_2 after state injection. Reprinted with permission from [FG09].

shows the effect of applying a single C_Z gate. A two-qubit entangled state is created; however, the parity of the two single-qubit measurements XZ gives the same information as the single-qubit measurement X before the C_Z gate – XZ is our new X_L operator. The C_Z gate transforms $+1$ eigenstates of X into $+1$ eigenstates of XZ. The third part of the figure includes a further two qubits. The essential idea is that cluster state stabilizers centered on the second, third, and fourth qubits can be used to extend the logical operators so they involve more qubits.

Consider Fig. 20.13. The enlarged qubit is the initial location of the arbitrary state. The three parts of the figure show Z_L and $X_L = X_1 = X_2$ after state injection. The parity of the results of measuring the indicated qubits in the indicated bases gives the same information as single-qubit measurements on the initial state. Note that the forms of Z_L, X_1, X_2 differ from those shown in Fig. 20.8. This is acceptable as all qubits associated with operators in black are measured in the same basis during computation, implying application of the operator has no effect. Nevertheless, the full form of the operators is important as only from the full form can it be seen that the logical operators anticommute. Furthermore, measuring the operators in black introduces logical byproduct operators. Let λ_Z, λ_1, and λ_2 denote the parities of the measurements indicated in black in the three parts of Fig. 20.13. After measurement we will be left with the state

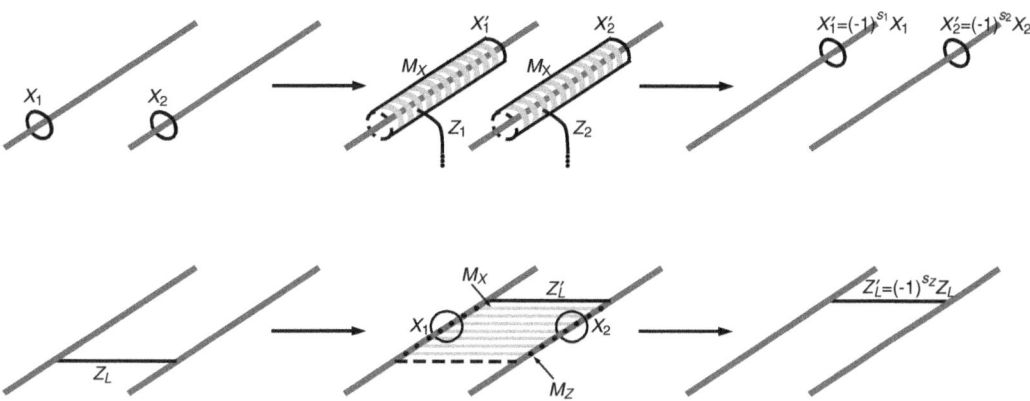

Fig. 20.14. The logical identity gate. Black lines and dots indicate Z operators, light gray lines indicate X operators. Measuring all qubits in the middle panels in the indicated bases results in a mapping between logical operators with byproduct operators dependent on the parity of the measurement results. Reprinted with permission from [FG09].

$X_L^{\lambda_z} Z_1^{\lambda_1} Z_2^{\lambda_2} |\psi_L\rangle$. Note that given the close proximity of the two defects around the enlarged qubit, Fig. 20.13 is not fault tolerant.

If the enlarged qubit is prepared in an arbitrary state $\alpha|0\rangle + \beta|1\rangle$ before being entangled with its neighboring qubits, an arbitrary logical state $\alpha|0_L\rangle + \beta|1_L\rangle$ can be obtained. In practice, it is likely that the cluster state will be prepared first, implying that the enlarged qubit can only be rotated in the Z basis, as such rotations commute with the controlled-Z operators used to construct the cluster state. Rotation before measurement could be replaced with measurement in a rotated basis. Either way, this would limit the class of injectable states to $(|0\rangle + e^{i\theta}|1\rangle)/\sqrt{2}$.

20.5 Logical gates

Only two logical gates, the identity gate and the CNOT gate, are required to complete the universal set of gates. These logical gates can be understood by examining their action on logical operators. For example, an ideal logical identity gate will have the property $X_L \mapsto X_L$, $Z_L \mapsto Z_L$.

Consider Fig. 20.14. The upper row shows X_1, X_2 followed by these same operators multiplied by a tubular cluster state stabilizer. Measuring the indicated qubits in the X basis results in new logical operators $X'_1 = (-1)^{s_1} X_1$, $X'_2 = (-1)^{s_2} X_2$. Similarly, the lower row relates input and output Z_L via $Z'_L = (-1)^{s_z} Z_L$. These logical operator mappings correspond to a logical identity gate with byproduct operators that map logical states according to $|\psi'_L\rangle = (I_L) Z_1^{s_1} Z_2^{s_2} X_L^{s_z} |\psi_L\rangle$.

Logical CNOT operates in a similar manner. An ideal CNOT maps control and target operators according to $X_c \mapsto X_c X_t$, $X_t \mapsto X_t$, $Z_c \mapsto Z_c$, $Z_t \mapsto Z_c Z_t$. Consider Fig. 20.15. The dual qubit is the control and the primal qubit is the target. From the figure it can be seen that $X_d \mapsto (-1)^{\lambda_{Xd}} X_d X_{1p}$, $X_{1p} \mapsto (-1)^{\lambda_{X1p}} X_{1p}$, $X_{2p} \mapsto (-1)^{\lambda_{X2p}} X_{2p}$, $Z_{1d} \mapsto (-1)^{\lambda_{Z1d}} Z_{1d}$, $Z_{2d} \mapsto (-1)^{\lambda_{Z2p}} Z_{2d}$, $Z_p \mapsto (-1)^{\lambda_{Zp}} Z_{2d} Z_p$. This corresponds to logical CNOT with byproduct operators mapping logical states according to

$$|\psi'_L\rangle = C_X Z_d^{\lambda_{Xd}} Z_{1p}^{\lambda_{X1p}} Z_{2p}^{\lambda_{X2p}} X_{1d}^{\lambda_{Z1d}} X_{2d}^{\lambda_{Z2d}} X_p^{\lambda_{Zp}} |\psi_L\rangle. \tag{20.1}$$

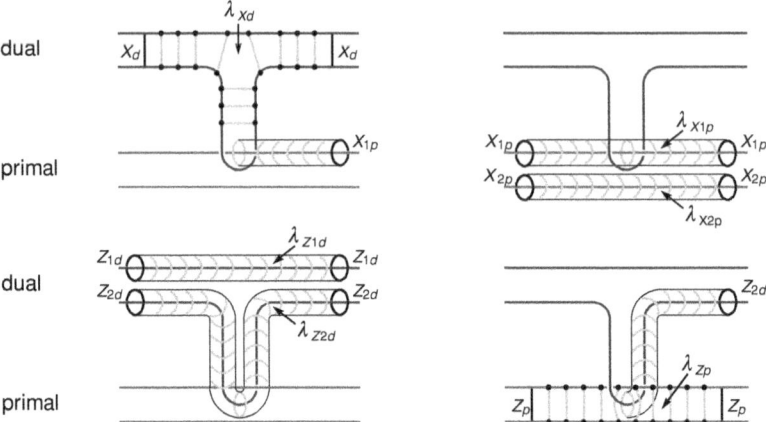

Fig. 20.15. Cluster state stabilizers consistent with the indicated braiding of defects and connecting the indicated logical operators in a manner corresponding to logical CNOT with byproduct operators. Reprinted with permission from [FG09].

We do not yet quite have what we need – a logical CNOT between two primal qubits. Consider Fig. 20.16a [RHG07]. This shows how an additional primal and dual ancilla qubit can be used to simulate logical CNOT between two primal qubits. Essentially, the first CNOT and associated measurement converts the control primal qubit into a dual qubit, the second CNOT performs the necessary logical operation and the third CNOT converts the dual qubit back into a primal qubit. Figure 20.16b shows a braiding of defects equivalent to Fig. 20.16a and a simplified braiding is shown in Fig. 20.16c [RHG07].

20.6 Topological cluster state error correction

Topological cluster state error correction is conceptually simple. As discussed in Section 20.2, measuring the six face qubits of a given cell in the X basis should yield six bits of information with even parity. Odd parity cells indicate the presence of errors. If we have a pair of cells with odd parity, we can connect the cells with a path running from face qubit to face qubit, then bit-flip the measurement results associated with the path. This will ensure every cell in the lattice has even parity once more. If we have many cells with odd parity, we can use an efficient classical algorithm, namely the minimum-weight matching algorithm [CR99], to pair up the cells using paths with minimum total length. Applying bit-flips to the measurement results along these paths again results in every cell having even parity.

There are, however, many important issues left unanswered by the above paragraph. Errors can occur in chains. A lattice with 64 cells and a number of errors is shown in Fig. 20.17. Only cells at the endpoints of error chains have odd parity (indicated by thick lines). No information about the path of the error chain is provided.

The boundaries of the lattice also require special consideration. Figure 20.17 shows a primal lattice of primal cells with primal boundaries containing primal errors. If an endpoint of a chain of at least two primal errors is located on a primal boundary, the boundary cell containing this endpoint will still have even parity. Primal error chains that begin and end on primal boundaries are

Fault-tolerant topological cluster state quantum computing 493

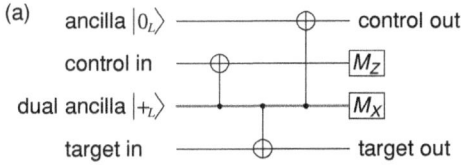

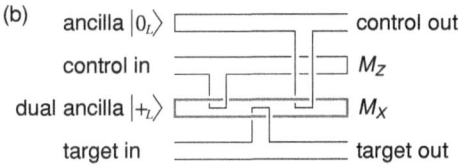

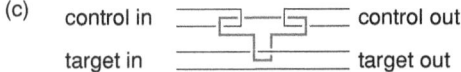

Fig. 20.16. (a) Circuit comprising of logical gates described in the text that simulates logical CNOT between two primal qubits. (b) Equivalent braiding of defects. (c) Equivalent simplified braiding of defects. Reprinted with permission from [FG09].

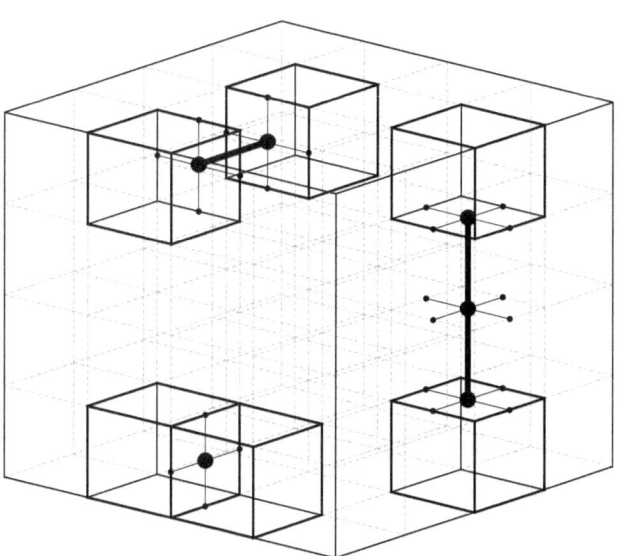

Fig. 20.17. A cluster state comprising 64 primal cells of the form shown in Fig. 20.1(a). Three different Z or M_X error chains of length 1, 2, and 3 are indicated by thick lines and enlarged qubits. Cells with odd parity are indicated with thick bounding lines. Most of the qubits in the cluster state have not been drawn for clarity. Reprinted with permission from [FG09].

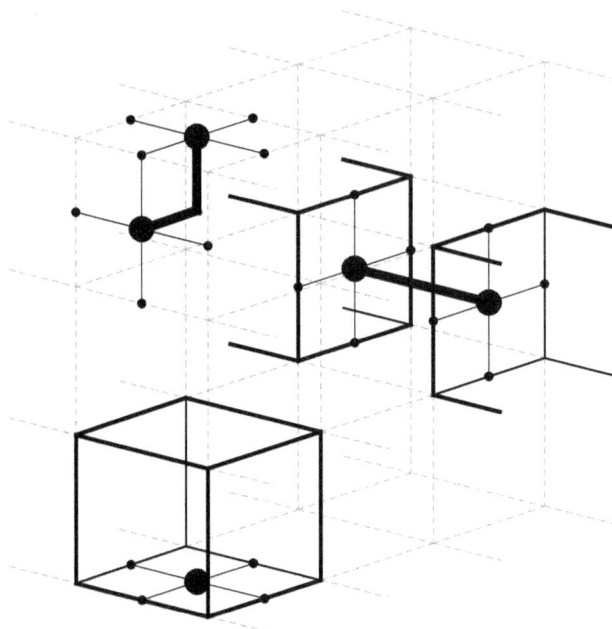

Fig. 20.18. A cluster state with both primal boundaries and dual boundaries, which consist of primal cells cut in half. Examples of the observable parity effects of primal error chains connected to the two types of boundaries are included with odd parity cells indicated by thick bounding lines. Reprinted with permission from [FG09].

thus undetectable and have the potential to cause logical errors as discussed in Section 20.3. Figure 20.18 contains examples of primal error chains connected to primal boundaries. Figure 20.18 also contains dual boundaries – lattice boundaries that pass through the centers of primal cells. A primal error chain connected to a dual boundary is always detectable as it changes the parity of the boundary cell containing the endpoint.

Dual cells are used in an identical manner to primal cells, meaning they also detect the presence of Z or M_X errors on their face qubits. Figure 20.19 shows a dual error chain starting and ending on dual boundaries. In an analogous manner to primal error chains, the parity of the dual boundary cells containing the chain endpoints remains unchanged.

Primal and dual error correction occur independently of one another. It may seem strange that both appear to only focus on Z and M_X errors. An X error that occurs just before an M_X measurement has no effect on the measurement result or the underlying cluster state after the measurement. An X error that occurs during the preparation of the cluster state, as shown in Fig. 20.20, is equivalent to one or more Z errors on the neighboring qubits as well as an X error just before measurement. As before, we can ignore the X error, and the error-correction scheme deals with Z errors.

We are now in a position to describe how correction proceeds. Note that only the classical measurement results will be corrected, not any remaining unmeasured qubits. Without loss of generality, let us focus on primal errors. The procedure for correcting dual errors is analogous. Suppose we have a connected lattice of (primal) cells with both primal and dual boundaries. Identify one dimension of the lattice as simulated time. Suppose we measure all qubits in the

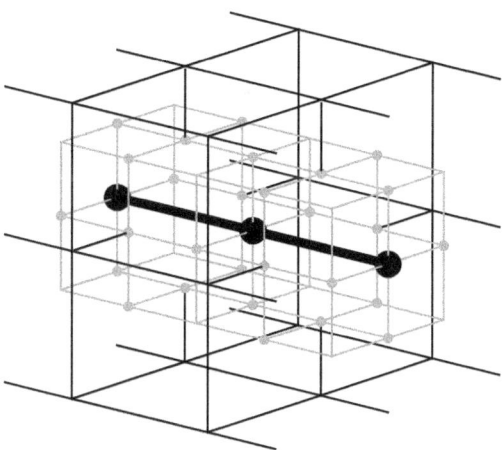

Fig. 20.19. An undetectable dual error chain connecting two dual boundaries. Reprinted with permission from [FG09].

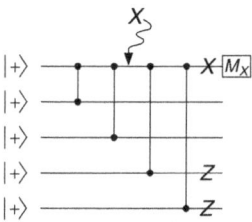

Fig. 20.20. Quantum circuit showing how X errors occurring at any point during the preparation of the cluster state are equivalent to potentially multiple Z errors and an X error just before measurement in the X basis, which can be ignored. Reprinted with permission from [FG09].

lattice up to some given simulated time t in the X basis and classically determine which cells in the measured region have odd parity. We need an algorithm to match odd parity cells with each other and with primal boundaries such that there is a high probability the matching corresponds to the errors that caused the odd parity cells. We have already mentioned that the algorithm we will use is called the minimum-weight matching algorithm [CR99].

The minimum-weight matching algorithm takes a weighted graph with an even number of vertices and produces a spanning list of disjoint edges with the property that no other list has lower total weight. The cells with odd parity become half the vertices we will feed into the algorithm. For every vertex in this list we add a vertex corresponding to the nearest point on the nearest primal boundary. We make an almost complete graph of these vertices according to the following rules: all boundary vertices are connected to all other boundary vertices with edge weight zero, odd parity cell vertices are connected to all odd parity cell vertices with edge weight equal to the sum of the absolute value of the differences of their three coordinates measured in cells, and odd parity cell vertices are connected to their nearest boundary vertex with edge weight equal to the number of cells that need to be passed through to reach the boundary plus one. When this graph is processed by the minimum-weight matching algorithm, the resulting edge list is highly likely to enable correction of the odd parity cells in a manner that does not introduce logical errors. The classical measurement results along an arbitrary path connecting the relevant pairs of cells are bit-flipped, resulting in all measured cells having even parity. Note that in a large computation such corrective bit-flips would only be applied between pairs of vertices such that at least one vertex of the pair is located at a time earlier than some $t - t_c$, where t_c depends on the size of the computation. This is to ensure that odd parity cells close to t have a chance to be matched with appropriate partner cells, which may not yet have been measured. Improved topological quantum error correction classical processing algorithms with optimal complexity were developed in [FWH12a, FWH12b, FWM+12].

20.7 Threshold

The analysis in this section and the next applies to a slight modification of the scheme described above. In this modification, the third dimension of the lattice is identified as time, and the computation is performed step-by-step on a two-dimensional physical lattice. This is to facilitate experimental implementation in a two-dimensional geometry. Furthermore, all qubits on the surface of a defect that are not face qubits are measured in the Z basis. It does not qualitatively change the threshold and overhead requirements of the scheme. For a full description of the modified scheme, see [RHG07].

Error model. In order to arrive at an estimate of the threshold, we must first specify an error model. We assume the following:

(i) Erroneous operations are modeled by perfect operations preceded/followed by a partially depolarizing single- or two-qubit error channel:

$$T_1 = (1-p_1)[I] + \frac{p_1}{3}([X] + [Y] + [Z]),$$

$$T_2 = (1-p_2)[I] + \frac{p_2}{15}([X_a X_b] + \cdots + [Z_a Z_b]).$$

(ii) The error sources are
- faulty preparation of individual qubit states $|+\rangle$, with error probability p_P,
- noisy Hadamard gates with error probability p_1 (Hadamard gates are used in the modified scheme of [RHG07] in place of $\Lambda(Z)$ gates in the time-like direction),
- noisy $\Lambda(Z)$ gates with error probability p_2, and
- noisy measurement with error probability p_M.

Note that in the modified scheme there is no storage error as no qubit is idle between preparation and measurement (see Fig. 8a of [RHG07]).

(iii) Classical processing is instantaneous.

When calculating a threshold, we assume all error sources to be equally strong, $p_1 = p_2 = p_M = p_P := p$, so that the noise strength is described by a single parameter p.

Error correction. When estimating the threshold, two key facts about error correction in the cluster state must be noted:

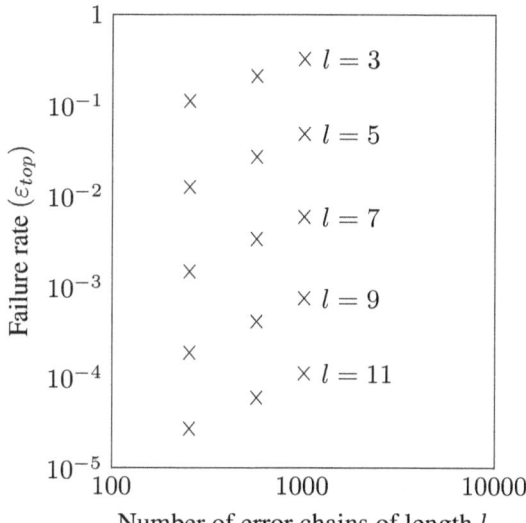

Fig. 20.21. Exponential decay of the failure rate ε of the topological error correction as a function of the length l of the shortest nontrivial error chain at $p = \frac{1}{3}p_{th}$. The failure rate also increases linearly with the number of error chains, as predicted by Eq. (20.3). Reprinted from [RHG06]. Copyright 2006, with permission from Elsevier.

(i) The error correction in the bulk region of the cluster state is topological. It can be mapped to the *random plaquette* $\mathbb{Z}_2$-*gauge model* (RPGM) in three dimensions [DKL+02]. If there are nontrivial error chains of finite smallest length l then, below the error threshold, the probability of error ε_{top} is

$$\varepsilon_{top} \sim \exp(-\kappa(p)l). \qquad (20.2)$$

$\kappa(p)$ is some function of the error rate p that depends on the details of the error model. For a simple model of purely local, independent errors on each qubit of the lattice, we can calculate $\kappa(p)$. Consider a defect of circumference u and length l. An error chain winding around the defect has weight at least $u + 4 \approx u$, and there are N such minimum weight chains. Thus, the probability $\varepsilon_{top}(u, N)$ for a logical error due to these types of chains is, to lowest contributing order,

$$\varepsilon_{top}(u, l) = N \frac{u!}{\frac{u}{2}!^2} q^{\frac{u}{2}} \approx N \exp\left(\frac{\ln 4q}{2} u\right) \frac{1}{\sqrt{\frac{\pi}{2}u}}. \qquad (20.3)$$

Where q is the local error probability. Here, $\kappa(p) = \frac{\ln 4q}{2}$. For the nonlocal error model described above, we first confirm numerically that ε_{top} is indeed suppressed exponentially in the lattice size (see Fig. 20.21). Then, we can estimate $\kappa(p)$ numerically [see Eq. (20.16)].

(ii) Topological error correction breaks down near the singular qubits. This results in an effective error on the singular qubits that needs to be taken care of by an additional correction method, namely magic state distillation. This effective error is *local* because the singular qubits are well separated from one another [RHG06].

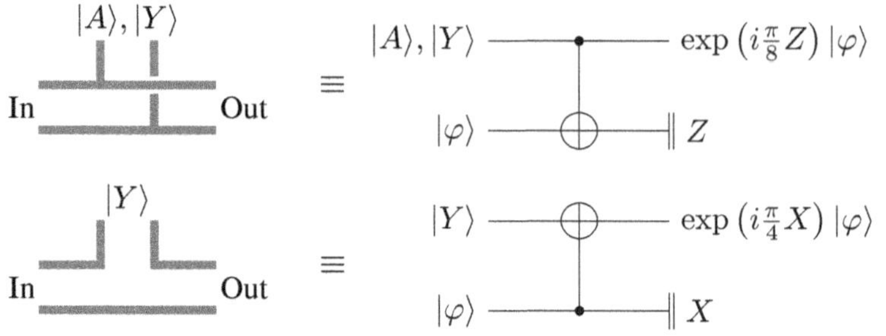

Fig. 20.22. Constructions to perform the non-Clifford gates using distilled ancillas $|A\rangle, |Y\rangle$. Reprinted from [RHG06]. Copyright 2006, with permission from Elsevier.

The fault-tolerance threshold associated with the RPGM is about 3.2×10^{-2} [OAI+04, TN04], for a strictly local error model with one source. The threshold estimates given in this section are based on the *minimum-weight-chain matching algorithm* [E65] for error correction. This algorithm yields a slightly smaller threshold of 2.9% [WHP03] but is computationally efficient.

We now consider the singular qubits (S-qubits). They are used to create the noisy ancilla states $\rho^A \approx |A\rangle\langle A|, \rho^Y \approx |Y\rangle\langle Y|$ encoded via the construction displayed in Fig. 20.22. $|A\rangle$, $|Y\rangle$ states are the eigenstates of the $X+Y$ and Y operators, respectively. See Section 20.4 for a discussion of the creation of ancilla states. Due to the effective error on the singular qubits, these ancilla states before distillation carry an error $\varepsilon_0^A := 1 - \langle A|\rho^A|A\rangle$, $\varepsilon_0^Y := 1 - \langle Y|\rho^Y|Y\rangle$ given by

$$\epsilon_0^A = \epsilon_0^Y = 6p. \tag{20.4}$$

Threshold. There are two types of threshold within the cluster, namely the topological one in the bulk of the lattice, far away from the singular qubits, and thresholds from $|A\rangle$ and $|Y\rangle$ state distillation. An estimate p_{th}^V for the topological threshold is found in numerical simulation of finite-size lattices to be

$$p_{th}^V = 7.46 \times 10^{-3}. \tag{20.5}$$

The results of the numerical simulation are shown in Fig. 20.23.

The recursion relations for state distillation, in the limit of negligible topological error, are to lowest contributing order $\varepsilon_{l+1}^A = 35(\varepsilon_l^A)^3$ (see [BK05]) and $\varepsilon_{l+1}^Y = 7(\varepsilon_l^Y)^3$. The circuits for state distillation are shown in Fig. 20.26. The corresponding distillation thresholds expressed in terms of the physical error rate p are

$$p_{th}^A = \frac{1}{6\sqrt{35}} \approx 2.8 \times 10^{-2}, \quad p_{th}^Y = \frac{1}{6\sqrt{7}} \approx 6.3 \times 10^{-2}.$$

The topological threshold is much smaller than the distillation threshold, and therefore the former sets the overall threshold for fault-tolerant quantum computation.

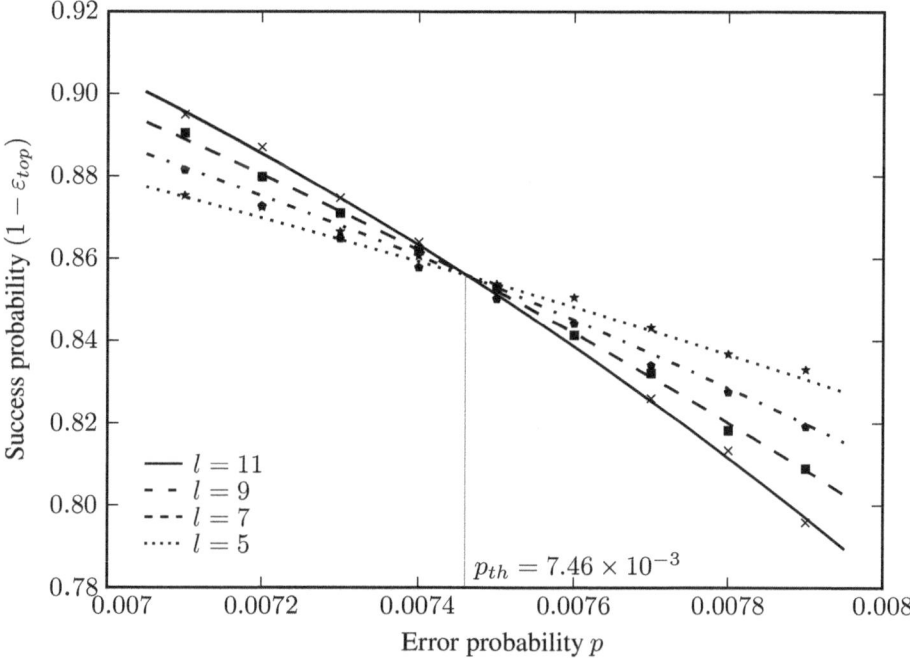

Fig. 20.23. Numerical simulation for the topological threshold. The curves are best fits taking into account finite-size effects of the lattice size l. Beyond the smallest lattices, these finite-size effects quickly vanish, and the curves intersect in a single point to a very good degree of accuracy. The value of p at the intersection gives the threshold. Reprinted from [RHG06]. Copyright 2006, with permission from Elsevier.

20.8 Overhead

Finally, we would like to analyze the resource requirements for this scheme. For the analysis, the metric we use is the *operational cost per gate*, O_3. It is defined as the number of physical operations that need to be performed per logical gate in a circuit of size N_G (i.e., a circuit that has N gates of type G).

Logical operations are realized by twisting defects around each other. If there were no noise, it would be sufficient to have line-like defects. Then, the elementary cells, Fig. 20.1, would be the building blocks from which gates are constructed. However, in the presence of noise, there will be short error cycles wrapping around the cells, leading to logical errors with high probability. In order to achieve fault tolerance, we have to rescale the elementary cell to a cube of $\lambda \times \lambda \times \lambda$. The cross-section of the defect with the plane perpendicular to the time direction becomes a square of area $d \times d$ (see Fig. 20.24). Remember that as per Eq. (20.2) errors are suppressed exponentially in the length of the shortest chain.

The gate length L_G is the total length of defect within a gate in units of the rescaled cell. The gate volume V_G is the number of rescaled cells the gate G occupies. Each rescaled cell consists of λ^3 elementary cells and each elementary cell is built with 12 operations. The number of two-qubit operations per elementary cell for the scheme described in this chapter is $(6 * 4)/2 = 12$.

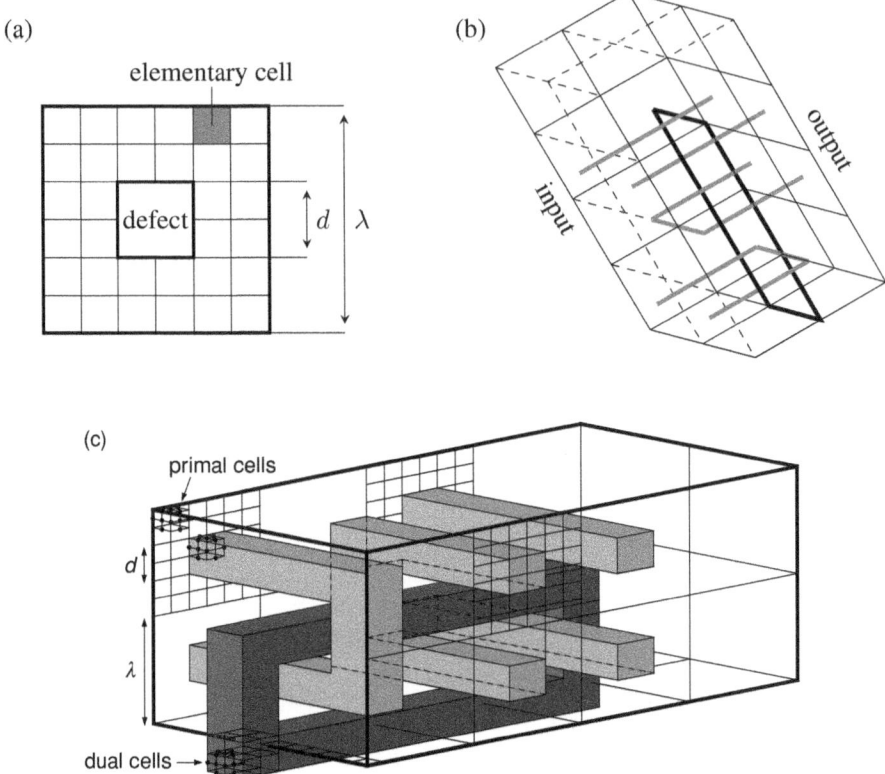

Fig. 20.24. (a) The logical cell. It is rescaled from the elementary cell of the lattice by a factor of λ in each direction. The defects have cross-section $d \times d$. The elementary cell can be either a primal or a dual cell (see Fig. 20.1), corresponding to the type of defect being considered. (b) Tileable schematic of the CNOT gate with size $V = 16$ and $L = 12 + 8 = 20$. (c) Example of a fault-tolerant CNOT gate with $d = 2$ and $\lambda = 6$. Note that all cells shown are primal cells except for the dual defect, which is made up of dual cells. Reprinted with permission from [FG09].

For the 2D modification of the scheme, this number is 8, so we will use 12 as an upper bound. Let $\varepsilon_{top}(G, \lambda, d)$ be the probability of failure of a topologically protected gate, such as the CNOT gate. It is a function of the layout (i.e., defect structure) of the gate, defect thickness d and scale factor λ. The operational overhead $O_3(G)$ is then

$$O_3(G) = \min_{\lambda, d} 12\lambda^3 V_G \exp\left(\varepsilon_{top}(G, \lambda, d) N_G\right). \tag{20.6}$$

The exponential factor comes from the expected number of repetitions for a circuit composed of N_G gates G. For a given N_G, the overhead should be minimized by choosing $\lambda(N_G)$ and $d(N_G)$.

20.8.1 Clifford gates

The overhead analysis is simplest for Clifford gates (i.e., gates that do not involve singular qubits and thus do not need distillation). To perform the minimization in Eq. (20.6), we need the gate

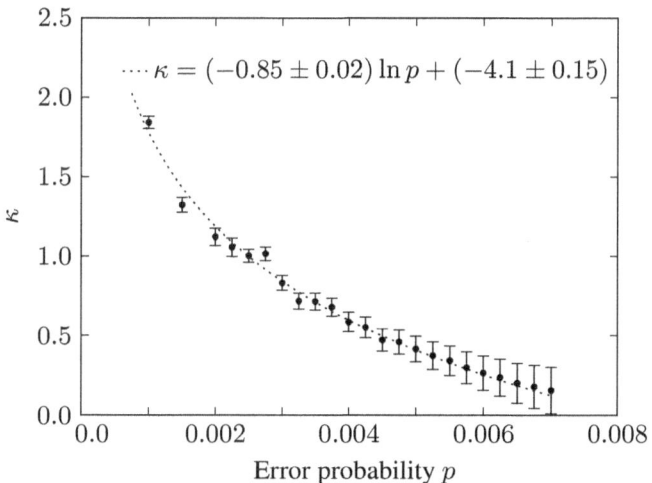

Fig. 20.25. The dependence of κ on the error rate p below the threshold. The curve is the best fit to the data points that come from numerical simulation on finite-size lattices. Reprinted from [RHG06]. Copyright 2006, with permission from Elsevier.

error ε_{top} as a function of G, λ, d. The error chains leading to gate failure can be either cycles wrapping around defects or chains that start and end on a defects (relative cycles). The probability of gate failure is exponential in the length of the shortest such chain and proportional to the number of such chains.

For a given defect thickness d, the minimal cycle length is $4(d+1)$ and the number of such cycles is λL_G; where L_G is the gate length. The minimal length of a relative cycle leading to an error is $\lambda - d$. It stretches between two neighboring defect segments one logical cell apart. The number of such relative cycles is at most $2L_G\lambda(d+1)$. There are shorter relative error cycles near junctions, but they are equivalent to the identity operation (see Eq. (25) of [RHG07]).

Thus, by Eq. (20.2), the gate failure rate is

$$\varepsilon_{top}(G, \lambda, d) = \lambda L_G \left(\exp\left(-4\kappa(d+1)\right) + 2(d+1)\exp(-\kappa(\lambda - d)) \right). \quad (20.7)$$

In Eq. (20.3), $\kappa = \frac{\ln 4q}{2}$. However, for nonlocal error models, κ must be estimated numerically, by simulation on finite-size lattices (see Fig. 20.25). In addition, Eq. (20.3) predicts a polynomial correction $l^{-1/2}$ to ε_{top}, for local noise. The numerical simulation finds a polynomial correction $l^{-0.85}$ for the nonlocal error model proposed above. We neglect the polynomial correction in our estimate of the overhead in Eq. (20.7). This is safe as including it would only reduce the overhead. We can now use this expression to perform the minimization in Eq. (20.6). For example, the overhead for the CNOT gate is plotted in Fig. 20.27.

20.8.2 Non-Clifford gates

Non-Clifford gates require the use of singular qubits to prepare the ancilla states needed to perform the gates via gate teleportation. Remember that this preparation in not fault tolerant. So we must use magic state distillation to purify the ancilla. The gates used in the distillation procedure are all Clifford gates and thus are topologically protected as above. However, the distillation

Table 20.1. *Performance of magic state distillation*

State at level l	Required states at level $l-1$	Performance
$\|A\rangle$	$15\|A\rangle$	$p_l^A = 1 - 15\varepsilon_{l-1}^A - \varepsilon_{top}(L_A, \lambda_{l-1}, d_{l-1})$ $\varepsilon_l^A = 35(\varepsilon_{l-1}^A)^3 + \varepsilon_{top}(L_A, \lambda_{l-1}, d_{l-1})$
$\|Y\rangle$	$7\|Y\rangle$	$p_l^Y = 1 - 7\varepsilon_{l-1}^Y - \varepsilon_{top}(L_Y, \lambda_{l-1}, d_{l-1})$ $\varepsilon_l^Y = 7(\varepsilon_{l-1}^Y)^3 + \varepsilon_{top}(L_Y, \lambda_{l-1}, d_{l-1})$

procedure is *concatenated*. At each level of concatenation, the optimal λ and d that minimize the overhead in Eq. (20.6) are different. Thus, the minimization is over the larger set of parameters, $\Lambda = \{\{\lambda_0, d_0, \lambda_1, d_1, \ldots, \lambda_{l_{max}}, d_{l_{max}}\}, l_{max}\}$.

In addition, there are now two types of error. First, the previously discussed error from nontrivial error cycles far away from the singular qubits. Second, the error associated with the singular qubits themselves, in whose neighborhood topological error correction breaks down, due to the presence of short error chains. The singular qubits are used for the distillation of $|A\rangle$ and $|Y\rangle$ states, the eigenstates of the $\frac{X+Y}{2}$ and Y operators, respectively.

Distillation of $|A\rangle$ and $|Y\rangle$ states is performed using the circuits shown in Fig. 20.26, of volumes V_A, V_Y and length L_A, L_Y (see Table 20.2). Let $p_l^{A,Y}$ be the probability of success for the distillation at level l of $|A\rangle, |Y\rangle$. On success, there will be a residual error at level l denoted by $\varepsilon_l^{A,Y}$ for $|A\rangle, |Y\rangle$, respectively. The performance of the distillation scheme is shown in Table 20.1. The expressions for success probability and residual errors hold to leading order in the contributing error probabilities $\varepsilon_l^A, \varepsilon_l^Y$. Further, a gate error cannot simultaneously lead to termination of the circuit and to a residual distillation error. Thus, we overestimate both error probabilities by adding the full weight $\varepsilon_{top}(L, \lambda_{l-1}, d_{l-1})$ to them.

The operational overheads for state distillation at level l, $O_{3,l}^A$ and $O_{3,l}^Y$, are given by the recursion relations

$$O_{3,l}^A = \frac{1}{p_l^A}\left(15 O_{3,l-1}^A + 12\lambda_{l-1}^3 V_A\right),$$

$$O_{3,l}^Y = \frac{1}{p_l^Y}\left(7 O_{3,l-1}^Y + 12\lambda_{l-1}^3 V_Y\right). \tag{20.8}$$

The recursion relations for $p_l^{A,Y}, \varepsilon_l^{A,Y}$ are given in Table 20.1. The initial conditions are $O_{3,0}^A = O_{3,0}^Y = 12$ and Eq. (20.4).

The distillation outputs states $|A\rangle$ and $|Y\rangle$ at level l_{max}. One such state $|A\rangle$ and, on average, 1/2 state $|Y\rangle$ is used to implement the $\exp(i\frac{\pi}{8}Z)$ gate via the construct displayed in Fig. 20.22, of volume $V_{1,z}$ and length $L_{1,z}$. Its overhead is

$$O_3^{\pi/8} = \left(O_{3,l_{max}}^A + \frac{1}{2}O_{3,l_{max}}^Y + 18\lambda_{l_{max}}^3 V_{1,z}\right)$$
$$\times \exp\left((\varepsilon_{l_{max}}^A + \varepsilon_{l_{max}}^Y + \varepsilon_{top}(L_{1,z}, \lambda_{l_{max}} d_{l_{max}}))N_G\right). \tag{20.9}$$

Table 20.2. *The gates sizes for various gates and subcircuits*

Gate	Volume	Length
CNOT Fig. 20.24b	$V_2 = 16$	$L_2 = 20$
$\exp(i\frac{\pi}{8}Z)$ Fig. 20.22	$V_{1,z} = 2$	$L_{1,z} = 3$
$\exp(i\frac{\pi}{4}X)$ Fig. 20.22	$V_{1,x} = 4$	$L_{1,x} = 4$
$|A\rangle$ distillation circuit	$V_A = 168$	$L_A = 266$
$|Y\rangle$ distillation circuit	$V_Y = 60$	$L_Y = 94$

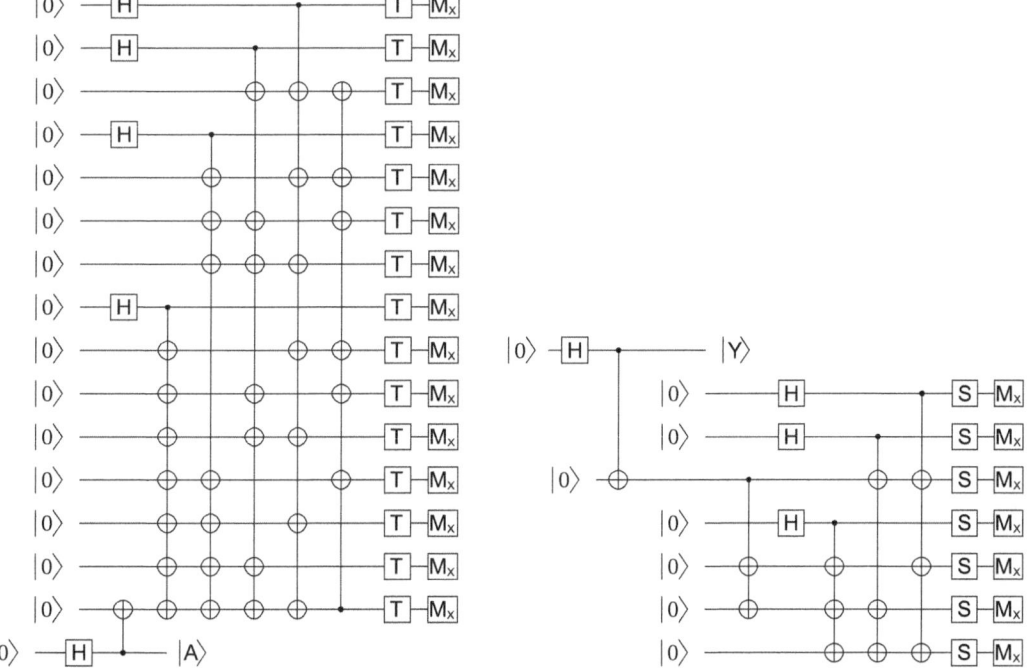

Fig. 20.26. Distillation circuits to produce purified $|A\rangle$ and $|Y\rangle$ states, respectively. These circuits use only operations that are topologically protected. Reprinted with permission from [FG09].

Similarly, the overhead for the $\exp(i\frac{\pi}{4}X)$ gate is

$$O_3^{\pi/4} = \left(O_{3,l_{max}}^Y + 12\lambda_{l_{max}}^3 V_{1,x}\right) \times \exp\left((\varepsilon_{l_{max}}^Y + \varepsilon_{top}(L_{1,x}, \lambda_{l_{max}} d_{l_{max}}))N_G\right). \quad (20.10)$$

The operational overhead must be minimized over the parameter set Λ. This has been done numerically[BLN95], and the result is shown in Fig. 20.27.

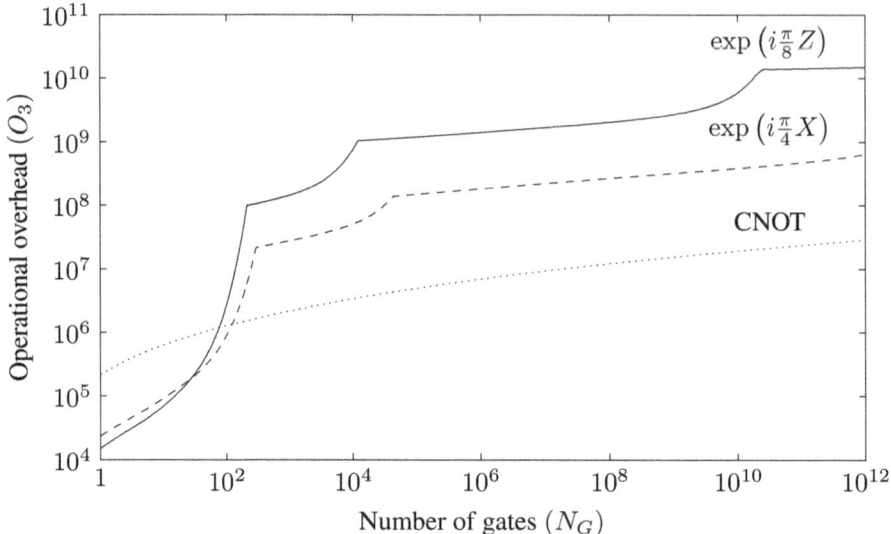

Fig. 20.27. Operational overhead as a function of the circuit size at $p = \frac{p_{th}}{3}$. The kinks in the curves correspond to increasing levels of concatenation l. Reprinted from [RHG06]. Copyright 2006, with permission from Elsevier.

20.8.3 Overhead scaling

The overhead for a given gate depends on two main parameters; the circuit size N_G and the underlying error probability p. In this section, we examine the behavior of the overhead with respect to each of these parameters.

20.8.3.1 The large N_G limit For Clifford gates, first notice that the residual topological error ε_{top} in Eq. (20.7) is minimized when both exponentials fall off equally fast, i.e., $d_{min} = \lambda_{min}/5$ for large d, λ. Further, the overhead O_3 in Eq. (20.6) is minimized near

$$\varepsilon(\lambda(N_G)) = 1/N_G. \tag{20.11}$$

Then, $\lambda_{min} \sim \ln N_G/\kappa$, and

$$O_3 \sim \frac{\ln^3 N_G}{\kappa^3}. \tag{20.12}$$

For non-Clifford gates, we first compare the two contributions to ε_l^A (see Table 20.1), $35(\varepsilon_{l-1}^A)^3$ and ε_{top}. If ε_{top} is much larger than $35(\varepsilon_{l-1}^A)^3$, it inhibits the convergence of ancilla distillation, forcing the use of additional distillation rounds, which are the most expensive component. On the other hand, if ε_{top} becomes much smaller than $35(\varepsilon_{l-1}^A)^3$, it blows up the size of the logical cell. Therefore, for optimal operational resources, both contributions should be comparable. Then, in the large size limit, $\ln \varepsilon_l^A = 3 \ln \varepsilon_{l-1}^A$, $\lambda_l = 3\lambda_{l-1}$, $d_l = 3d_{l-1}$. Further, the success probabilities $p_l^{A,Y}$ for ancilla distillation quickly approach unity with increasing distillation level l. Therefore, in the large size limit, for the point of minimal operational resources,

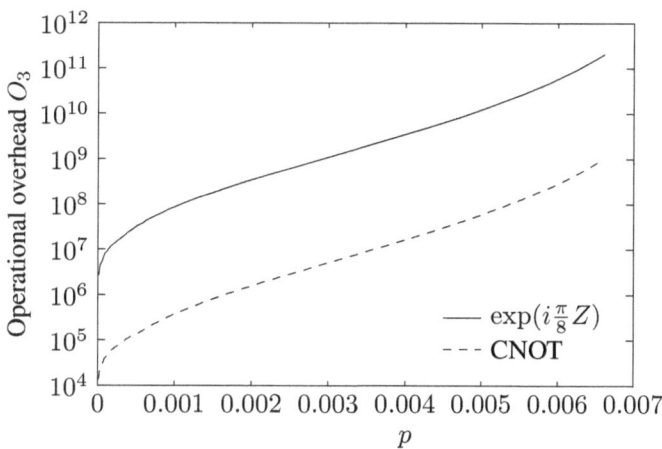

Fig. 20.28. The behavior of the operational overhead below threshold. The overhead decreases by four orders of magnitude at $\frac{p_{th}}{10}$. Reprinted from [RHG06]. Copyright 2006, with permission from Elsevier.

the recursion relations Eq. (20.10) can be replaced by

$$\begin{pmatrix} O_3^A \\ O_3^Y \\ \lambda^3 \\ d \\ \ln \varepsilon^A \\ \ln \varepsilon^Y \end{pmatrix}_l = \begin{pmatrix} 15 & 0 & 12V_A & & & \\ 0 & 7 & 12V_Y & & & \\ 0 & 0 & 27 & & & \\ & & & 3 & & \\ & & & & 3 & \\ & & & & & 3 \end{pmatrix} \begin{pmatrix} O_3^A \\ O_3^Y \\ \lambda^3 \\ d \\ \ln \varepsilon^A \\ \ln \varepsilon^Y \end{pmatrix}_{l-1}. \quad (20.13)$$

Thus, $O_{3,l}^{A,Y} \sim 27^l$, $\ln \varepsilon_l^{A,Y} \sim 3^l$. Then, with $\varepsilon \sim 1/N_G$ [Eq. (20.11)],

$$O_3^{A,Y} \sim \ln^3 N_G. \quad (20.14)$$

Note that the distillation operations, for the case of perfect Clifford gates, are associated with the more favorable scaling exponents $\log_3 15 \approx 2.46$ and $\log_3 7 \approx 1.77$, respectively. However, in our case the topological error protection of Clifford gates must keep step with the rapidly decreasing error of state distillation, by adjusting the scale factor λ. This leads to a scaling exponent of 3 for the Clifford operational resources [cf. Eq. (20.12)], which dominates the resource scaling of the entire state distillation procedure, yielding Eq. (20.14).

Thus, we have poly-logarithmic scaling of the overhead with circuit size

$$O_3 \sim \ln^3 N_G. \quad (20.15)$$

20.8.3.2 Behavior below threshold In the previous discussion, we saw that the large N_G scaling of the overhead is very good. However, the coefficients are quite large and this leads to very high actual overheads near the threshold. For example, see Fig. 20.27 for overhead numbers at an error rate of $p_{th}/3$.

Since the threshold Eq. (20.5) of the scheme is very high, it may make sense to run the computation at an error rate significantly below the threshold, thereby saving on overheads. In order to study the behavior of the overhead below the threshold, we need to find the behavior of

κ as a function of p by numerical simulation on finite-size lattices (See Fig. 20.25). The best fit for κ is

$$\kappa = (-0.85 \pm 0.02) \ln p + (-4.1 \pm 0.15). \qquad (20.16)$$

Using this result, we can estimate the behavior of the overhead below the threshold, as shown in Fig. 20.28. The overhead is reduced by four order of magnitude at $p_{th}/10$.

Part VII

Applications and implementations

Chapter 21

Experimental quantum error correction

Dave Bacon

Quantum error correction (QEC) is all for naught if it cannot be implemented experimentally in the laboratory. Further, even if it can be implemented in small laboratory experiments, these experiments need not lead to a technology that is scalable to large quantum computers (where large is defined as "big enough to contemporaneously outperform the best classical computer on some problem"). In this chapter we will survey some of the experiments that have been performed to implement QEC. These experiments are all a long way from demonstrating viable QEC, but demonstrate the proof-of-principle methods that will need to be implemented in the future if quantum computation is to be made viable. This chapter deals with experimental implementations whose goal is to implement QEC, and not experiments using passive or open-loop methods, which are covered in Chapter 22.

21.1 Experiments in liquid-state NMR

The first experiments that attempted to perform QEC were performed using room-temperature liquid-state NMR. Liquid-state NMR [CFH97, GC97] is a testbed for quantum computing ideas in which one uses the internal states of coupled nuclear spins from a molecule as the qubits in a quantum computer. In its original and experimentally implemented form, liquid-state NMR is not generally thought to be scalable: the signal used to read out the result of the quantum computation decays exponentially [W97] in the number of qubits in the system and the mixed states produced in these experiments can be shown to possess no quantum entanglement. That said, we must here say that it is not *thought* to be scalable, simply because known no-go proofs, for example those involving entanglement [BCJ+99], are not known to be loophole free and even the exponential decay of the signal can be overcome by suitably clever (but experimentally challenging) methods [SV99]. Given this consensus status of ultimate unscalability, however,

---

Quantum Error Correction, ed. Daniel A. Lidar and Todd A. Brun. Published by Cambridge University Press.
© Cambridge University Press 2013.

liquid-state NMR is a premiere testbed for many of the methods and techniques that will be necessary to build a quantum computer.

The first experiment to implement QEC on an NMR quantum computer was the experiment of Cory *et al.* [CPM+98]. For a good overview of NMR quantum computing the reader is referred to [LKC+02]. Here we briefly describe the model before turning to the experiment of [CPM+98]. A liquid-state NMR quantum computer experiment consists of an NMR spectrometer into which a sample of a liquid has been inserted, which contains approximately 10^{19} dissolved copies of the molecule to serve as the quantum computer. The spectrometer applies a strong (on the order of 10 T) magnetic field to the sample. Under these conditions, in the weak-coupling limit, the Hamiltonian for the internal evolution of the nuclei for a single molecule is given by

$$H = \sum_i \frac{\nu_i}{2} \sigma_z^{(i)} + \sum_{ij} J_{ij} \sigma_z^{(i)} \otimes \sigma_z^{(j)}, \qquad (21.1)$$

where ν_i and J_{ij} are suitable coupling constants and the sums run over the qubits (nuclei) used in the system. The molecules are assumed to be in thermal equilibrium and therefore are described by the density matrix $\rho = \exp(-\beta H)/\text{Tr}[\exp(-\beta H)]$. Because of the high temperature in a liquid-state NMR experiment (in comparison to the energies associated with the above Hamiltonian) this thermal state is very close to maximally mixed. However, there does exist some small deviation from the fully mixed state and, because one has a huge number of molecules in the system, one can see the effects of the differences in population in such a setup.

To perform quantum computation in such a setting one proceeds as follows. First, one must take the thermal equilibrium state of the system and distill out what is known as a "pseudo-pure" state. Pseudo-pure states are mixed states of the form $(1-\epsilon)\frac{I}{d} + \epsilon|\psi\rangle\langle\psi|$ where d is the dimension of the system. Such states are forced upon NMR quantum computation because the state of the system is a high-temperature thermal equilibrium state. However, while ϵ may be very small, on the order of 10^{-5}, such states behave under unitary evolution as pure states – albeit pure states with a huge amount of noise arising from the I component of the density matrix. Preparation of psuedo-pure states can proceed according to a number of different procedures: by the use of ancilla qubits [GC97] to distill out a pseudo-pure state, by suitable spatial averaging obtained via the use of gradient magnetic fields [CFH97], or by temporal averaging [KCL98].

Having prepared a suitable pseudo-pure state, one can then perform a quantum computation on liquid-state NMR by the application of weak but resonant oscillating magnetic fields in the xy plane. Important considerations here are that the Hamiltonian in Eq. (21.1) is always on, but by using suitable refocusing techniques, one can effectively turn off both one- and two-qubit interactions. Another complication is that, depending on whether one has qubits that are well separated in frequency, one must use either "soft" pulses, which are slow (similar frequencies), or "hard" pulses, which are fast (well-separated frequencies.) Finally, after the computation has been performed, a measurement of the spins is performed. This is done by detecting the magnetic field produced by the free induction decay of the magnetization of the nuclei in the applied magnetic field.

Given the above basics of NMR experiments, we can now describe the first NMR quantum error-correcting experiment. In fact there were actually two experiments in [CPM+98]. Both of these experiments used as the quantum error-correcting code the three-qubit phase-flip code. In this code, $|0_L\rangle = |+++\rangle$ and $|1_L\rangle = |---\rangle$, where $|\pm\rangle = \frac{1}{\sqrt{2}}(|0\rangle \pm |1\rangle)$ and it is a code that can correct a single Z error. The basic circuitry for QEC is given in Fig. 21.1.

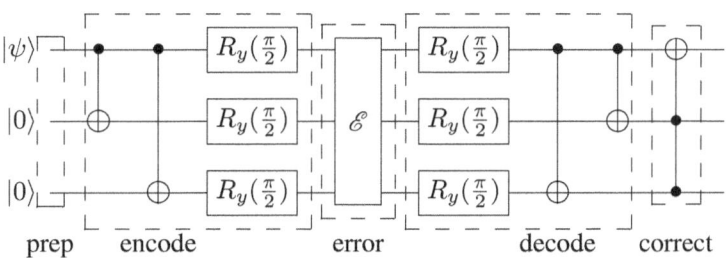

Fig. 21.1. Schematic of the circuits executed in the three-qubit phase-correcting code. Here $R_y(\theta)$ is the gate that rotates by angle θ about the y-axis.

The first of these experiments, call it experiment I, used the molecule alanine and for qubits, the three ^{13}C labeled carbons. To understand what this experiment did, it is useful to recall the product operator formalism of NMR. In this formalism one expresses the density matrix as a sum of the identity component and a remaining deviation density matrix. The identity component of this expression is the same for any state and thus is dropped, and one only deals with the deviation density matrix. Further, this deviation matrix is assumed to have a strength that is normalized out. Therefore, the state of the system is described not by a density matrix, but by a deviation density matrix. Thus, for instance, the density matrix $\frac{1}{2}(I + \gamma Z)$ ($\gamma \ll 1$) has a deviation density matrix given by Z. In experiment I, [CPM+98] considered how the three-qubit phase code would operate for a system that was initially prepared with deviation density matrix for the three labeled carbons of $Z_1 \frac{1}{2}(I_2 + Z_2)\frac{1}{2}(I_3 + Z_3)$. In particular, they considered how to prepare this state, then encode it, have errors occur on the system, and then decode the state. In this experiment, they did not carry out the correction step using the hardware, but instead did this off-line to simulate the effect of the Toffoli gate. Further, the experiment differed in three major ways from the standard procedures envisioned in the circuit of Fig. 21.1, which we now describe.

The first difference is that instead of preparing the state with a deviation matrix of $Z_1 \frac{1}{2}(I_2 + Z_2)\frac{1}{2}(I_3 + Z_3)$, they ran four experiments in which they prepared states with deviation matrices equal to one of the four terms in this deviation matrix. In other words, they ran experiments with deviation matrices of $Z_1 I_2 I_3$, $Z_1 I_2 Z_3$, $Z_1 Z_2 I_3$, and $Z_1 Z_2 Z_3$. They then added these experiments' results up to get the same result as if they had started with the proper deviation matrix.

The second difference in experiment I is that they used engineered noise. The noise was engineered using gradient magnetic fields applied along the quantization access of the NMR magnet. In particular, if one applies a gradient magnetic field along the z direction to an NMR sample, then molecules at different z locations in the sample will experience different magnetic fields and therefore precess about the z-axis at slightly different rates. If one then reverses the field one can refocus the magnetization. If the molecules have not moved during this process then no net phase will be picked up by the molecules. However, if the molecules change location during this time, a change in phase develops that is proportional to the displacement of the molecule. Because the molecules are in a liquid state they will diffuse around and thus, under these circumstances, such an effect occurs and a random phase decoherence is applied to the molecules. In experiment I with alanine, engineered noise was used to demonstrate proof-of-principle functioning of the quantum error-correcting circuit.

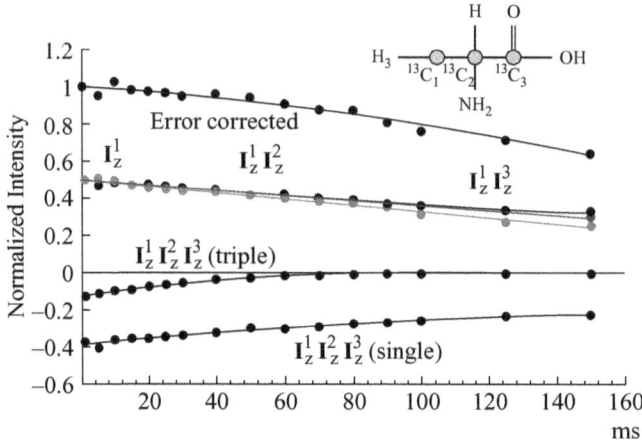

Fig. 21.2. The results of experiment I in [CPM+98] with the molecule alanine shown in the insert. The results for the four separate deviation density matrices are shown (here I_z^i represents Z acting on the ith qubit) and the three-qubit term is separated out into the triple and single quantum coherences. The error-corrected signal is then reconstructed by summation and the resulting evolution under the error correction is shown. That QEC is working as desired is witnessed by the fact that the reconstructed signal has a nearly flat initial slope. Reprinted with permission from [CPM+98]. Copyright 1998, American Physical Society.

The third difference in experiment I from what one normally expects in QEC was that the correct step was not performed. In this experiment, the results of the Toffoli gate performed at the end of the circuit were instead simply simulated by proper addition of the signals in the experiment.

The results of experiment I are presented in Fig. 21.2. The important thing to note in this plot is the top set of data points. These represent, effectively, the result of performing QEC on the state. So what is the indication that error correction is working as desired? Well, if error correction were not performed on this system, then one would expect the signal for the initially prepared state to decay exponentially as $\exp(-t/\tau)$. Error correction, for small enough t, however, should change this into a decay that is quadratic in t/τ: $1 - c\left(\frac{t}{\tau}\right)^2 + O(t^3)$. Thus, the indication in this figure that QEC is occurring is that the slope of the decay of the error-corrected signal is close to zero.

The second experiment performed in [CPM+98] enacted QEC but with some of the parts missing from experiment I alleviated. In particular, this experiment was designed to demonstrate the entire procedure of QEC, including the last error-correcting step. Experiment II was performed with the molecule trichloroethylene (TCE), and the qubits were the proton (hydrogen) and two labeled carbons. The following two procedures were performed. First, the system was initialized into one of four possible inputs with deviation density matrices of $P_1 \frac{1}{2}(I + Z_2)\frac{1}{2}(I + Z_3)$ where $P_1 \in \{I_1, X_1, Y_1, Z_1\}$. These possible inputs have operators that span the space of inputs to the quantum error-correcting circuit for the first qubit: thus they given an indication of how an arbitrary input will behave in the circuit. Then, given a particular preparation, a sequence of pulses was used to enact the encoding circuit. This was followed by allowing the system either to decohere under natural processes, or to induce Z errors on one

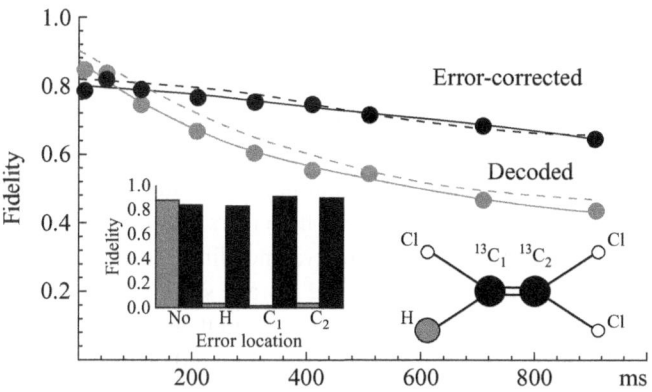

Fig. 21.3. The results of the experiment in [CPM+98] on TCE (shown in insert.) The result is the entanglement fidelity versus time. Entanglement fidelity is $\frac{1}{4}(1 + f_x + f_y + f_z)$, where f_α is the relative polarization along the α-axis. Entanglement fidelity is a good indication of how information is preserved in a quantum evolution. Above, the entanglement fidelity for the natural evolution is demonstrated, with the entire encoding, natural error, decoding, and optionally the error correction applied. We see that for short time scales the error-correcting scheme damages the fidelity, but for longer times the error correction does restore the quantum information. Finally, for the case when engineered noise is used, a bar graph represents the results when an error is applied to the specific location or to no location. Reprinted with permission from [CPM+98]. Copyright 1998, American Physical Society.

of the qubits. Then decoding was performed. Finally, either error correction was performed (the Toffoli gate in Fig. 21.1) or it was not performed and the amount of signal along the appropriate P_1 related direction was measured. By performing everything but the error correction, this experiment then allows for the demonstration of how the actual error-correction procedure degrades the signal. The results of this experiment are shown in Fig. 21.3.

The experiments in [CPM+98] represent the first example of an attempt to perform QEC in a system. They demonstrated some of basic pieces needed to perform QEC: a preparation step, errors, and and an error-correcting step. That said, there are certain deficiencies that should be kept in mind when thinking about this experiment. The first is that the NMR system used was in a state very close to the maximally mixed state. Such systems are not thought to be able to achieve quantum computational speedup. Second, the procedure used above consisted of encoding and decoding the quantum information. While this is good for proof-of-principle experiments, in a real quantum computation one should never decode or leave unencoded the relevant QEC. Instead, one should only use explicitly fault-tolerant *preparation* procedures. This experiment does not do this. Finally, one can note, for example, that the fidelity used in the experiments is normalized arbitrarily, and thus the experiment says nothing about the actual preservation of the coherence of the quantum state in comparison to an unencoded version of the experiment.

After the initial quantum error-correcting experiments of [CPM+98] (and a follow up in [SCS+00]), a series of other experiments were performed in NMR to perform proof-of-principle quantum error-correcting methods. Here we briefly review a few of the more important such experiments.

One drawback, as we have mentioned, to the first NMR quantum error-correcting experiment was that the experiment did not compare the results of error correction to what would happen without error correction. In order to overcome this, Leung *et al.* [LVZ+99] performed an experiment with a quantum error-*detecting* code in liquid-state NMR and compared the results of performing error detection and not performing any error detection. In particular, the authors used a two-qubit phase-detecting code: they get around not actually performing error correction by using this to post-select out the states where the errors had been detected. In these experiments, the authors performed an experiment where encoding, storing, and decoding were done and compared with an experiment where only the storage stage was performed. All of this was done in the presence of naturally occurring decoherence in their molecule (^{13}C labeled formate). The authors showed that while the encoding and decoding did change the effective error rate to second order, the result of the encoding and decoding degraded the signal below what would be achieved without the encoding and decoding, at least for short time scales.

Another important experiment in liquid-state NMR was the implementation of the five-qubit error-correcting code [KLM+01]. This result was important not so much for the demonstration of this important quantum error-correcting code, but for the introduction of the notion of using quantum error-correcting codes as benchmarks. A benchmark is used to compare different devices executing the same task. In [KLM+01] the authors used entanglement fidelity as the basic measurement for benchmarks involving the five-qubit error-correcting code. Importantly, however, the authors pointed out that the benchmark should be measured with respect to how well it achieves particular goals, and within the context of QEC, the error model used. For example, when one requires only that entanglement is preserved during the QEC procedure, the authors' NMR experiment achieved this level of fidelity. (Note, however, this demonstrates only that entanglement, if it were created, would be preserved. Since liquid NMR states are not entangled, an actual experiment performing such entanglement preservation would not work because of the difficulty of producing the initial entanglement.) However, for goals such as having the procedure result in an improvement of entanglement fidelity for a simple depolarization channel with and without encoding, the experiment in [KLM+01] does not achieve break-even.

Finally, methods combining QEC and decoherence-free subspace methods were demonstrated in [BVF+05]. In particular, an experiment was considered in which two qubits undergo collective decoherence, and thus need to be encoded into a DFS. This encoded qubit was then used with two independently decohering qubits in a three-qubit quantum error-correcting code. This was the first demonstration of concatenation of two quantum codes.

21.2 Ion trap quantum error correction

Liquid-state NMR provides a suitable testbed for exploring QEC but there are considerable challenges to scaling such systems up to a large quantum computer. A physical implementation that is not thought to suffer as severely is the instantiation of quantum computers, in (segmented) ion traps. In ion trap quantum computers, the internal levels of an ion serve as the states of a qubit, single-qubit gates and measurement are done using laser pulses and state-dependent resonance fluorescence, and two-qubit interactions are achieved via laser excitation of state-dependent spatial modes of the trapped ion crystal. Ion traps are distinguished as an implementation since nearly all of the components necessary to achieve scalable quantum computing have been built and tested in lab (with high fidelity), but only recently have integrated approaches that can achieve QEC become experimentally feasible. In Chiaverini *et al.* [CLS+04], the NIST ion

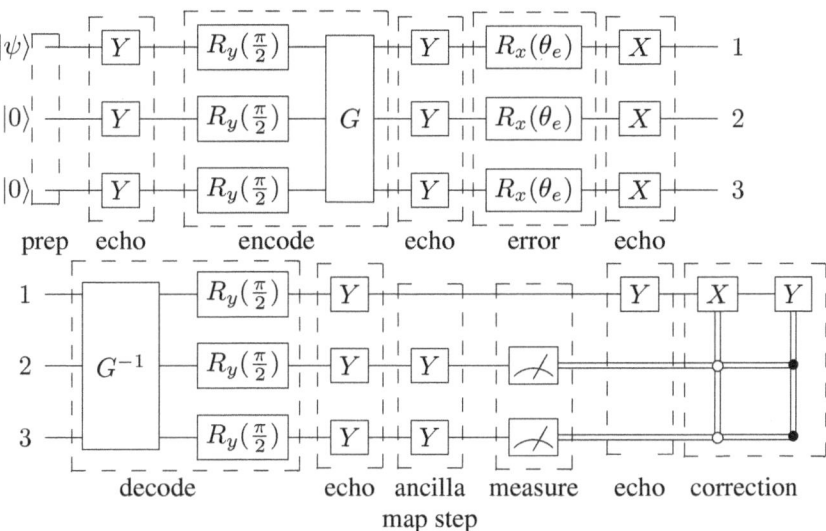

Fig. 21.4. Schematic of the circuits executed in the three-qubit phase-correcting code. Here $R_\alpha(\theta)$ is the gate that rotates by angle θ about the α-axis. The gate G is a gate that acts as $G|s_1, s_2, s_3\rangle = (-1)^{e(s_1,s_2,s_3)}|s_1, s_2, s_3\rangle$, where $e(s_1, s_2, s_3) = 1$ if $s_1 = s_2 = s_3$ and $e(s_1, s_2, s_3) = 0$ otherwise.

trap quantum computing group used the basic building blocks that they had developed over the prior decade and produced the first realization of QEC in an ion trap quantum computer.

The experiment of [CLS+04] used three ^{9}Be$^+$ ions in a multi-zone linear radiofrequency Paul trap [RB.-KD+02]. The trap allowed for different trapping regions where different components of the protocol could be achieved. In the protocol carried out in this experiment, for example, the ions were trapped in one of five regions, two of which were locations where gates could be applied to the ions, one of which was where the ions could be separated from each other, and the other regions were used as storage areas for ions not currently being acted upon. Movement of the ions between the regions and separation of groups of ions were achieved by changing the potentials applied to segmented electrodes of the trap.

The qubits used in this experiment were the electronic ground state hyperfine levels $|F = 1, m_F = -1\rangle$ and $|F = 1, m_F = -2\rangle$. Single-qubit gates were realized by two-photon Raman transitions. Measurements were performed using state-dependent resonance fluorescence (i.e., ions in one of the above two states can be made to fluoresce, while if the system is in the other state it does not fluoresce). The necessary multi-qubit gates needed for the encoding and decoding were achieved by coupling to the vibrational axial modes of the ion. In particular, the only multi-qubit gate necessary was achieved by a three-qubit geometric gate [LDM+03].

The code implemented in the experiment was the three-qubit stabilizer code with stabilizer generators $Z_1Z_2X_3$ and $Z_1X_2Z_3$. This code can correct single bit-flip errors (as was done in this experiment) but can also be adapted to other combinations of errors (such as a bit-flip on the first spin but phase-flips on the second and third qubits). Notably, this code cannot be thought of as simply a superposition of the classical repetition code. The circuit carried out in the experiment is given in Fig. 21.4.

A few notable tricks were used in the experiment, which bear noting. First, preparation was done by appying a single laser to all three ions (after they had been prepared in the $|000\rangle$ state) by spacing the qubits out in the trap and using the laser profile to address the first qubit differently than the other two qubits. Another method used to increase the fidelity of the circuit was that spin echoes were used to counteract dephasing caused by fluctuations in the local magnetic field between experiments. Finally, the ancillas were rotated such that the most common error for the experiment was in the state most easily detectable by the state-dependent resonance fluorescence.

Errors in the experiment were engineered. In particular, the error process consisted of rotating every qubit about the Pauli X-axis by an angle θ_e. Of course, the code cannot correct such arbitrary errors, but for small θ_e the errors have an amplitude that acts like independent errors. For perfect execution of the QEC, the fidelity of the initial state with the final state, if the initial state is $|\psi\rangle = \alpha|0\rangle + \beta|1\rangle$, is given by

$$F = 1 - |\alpha|^2|\beta|^2(2 - 3\cos\theta_e + \cos^3\theta_e), \tag{21.2}$$

which for small θ_e scales as

$$F = 1 - \frac{3}{4}|\alpha|^2|\beta|^2\theta_e^4 + O(\theta_e^6). \tag{21.3}$$

The sign of QEC is that the infidelity scales quadratically in θ_e^2 and not linearly in θ_e^2, which is what would occur if no error correction was performed.

In order to separate out the effects arising from the imperfection of the implementation of the gate elements from the effect of QEC, two experiments were run. In the first, everything in the circuit in Fig. 21.4 was executed, and in the second, all but the error-correcting components of the circuit were executed. In Fig. 21.5 a typical experimental result is shown giving the infidelity as a function of the error angle squared. The effects of error correction are clearly seen in this figure. Notice that the infidelities diagrammed here are the actual infidelities and are not hiding signal degradations as is done in NMR experiments. Thus, we see that while error correction does as it is supposed to there is a large error simply for carrying out the encoding and decoding operations along (the intercept with the y-axis is nonzero). Not shown are data that showed that, for large angles $\theta_e > 1$ radian, the actual fidelity is higher than a bare rate acting only on an unencoded qubit (as opposed to the data in Fig. 21.5, where encoding and decoding are done for the uncorrected data). Thus, for this particular error model, a *real* improvement over not doing error correction was achieved.

In conclusion, the experiment of [CLS+04] demonstrated the basic workings of ion trap QEC. Notably, the ancillas used in the experiment were "reset," i.e., measured and then could be used again. Another notable component of this experiment is that the figures of merit are for the actual pure state operation of the circuit, i.e., there is nothing hidden in a small signal averaged over a large number of quantum computers (as in liquid-state NMR). For a particular error type (large θ_e errors), an increase in fidelity over applying the error to the unencoded qubit was actually achieved. Finally, ion trap quantum computers are currently a leading physical implementation, and thus this experiment provides a prototype for future, larger, higher fidelity, fault-tolerant quantum error-correcting ion trap quantum computers.

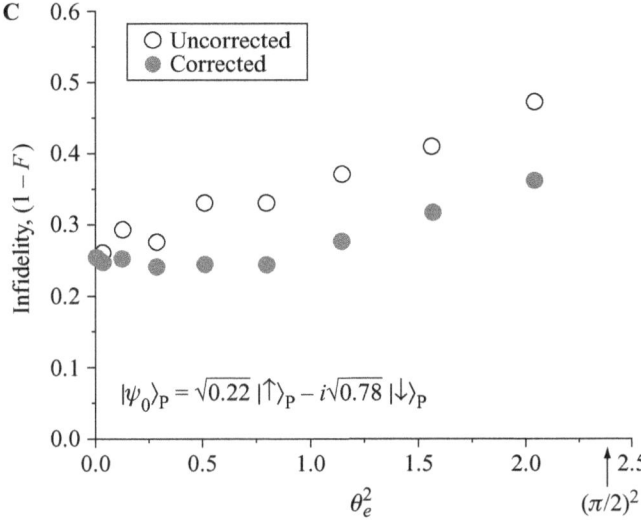

Fig. 21.5. Plot of the infidelity $1 - F$ versus the error angle squared for the state shown in the inset from the experiment of [CLS+04]. Reprinted with permission from Macmillan Publishers Ltd from [CLS+04].

21.3 Experiments using linear optics quantum computation

A final setting in which the methods of QEC have been implemented is that of linear optics quantum computing [KLM01]. The use of linear optical elements by themselves cannot perform universal quantum computation because they do not implement a large enough unitary group, but with the addition of elements such as the ability to create and detect a single photons it is possible to build a measurement-based scheme that implements universal quantum computation. This scheme is promising because it uses light for storage of quantum information, and thus has long-lived quantum memories, but still faces considerable challenges in the creation and detection of single photons, coupling of these photons into and out of components of the device, and large-scale integration. Current experiments in linear optics quantum computation, because they are not large enough and because they use elements that are not of high enough fidelity, do not reach a regime in which the scheme of [KLM01] is scalable, but technological improvements are possible that could make this a viable scheme for large-scale quantum computing. That said, however, one route to performing experiments *now* demonstrating the basic ideas of linear optics quantum computation is to use postselection to filter out places where the machine has behaved in a manner that carries out the desired computation. It is within this setting that experiments in QEC have been performed.

The first two experiments to implement QEC were performed in [PJF05] and [OPW+05]. The code used in the experiment is one that is normally not thought of as necessarily a quantum error-correcting code, but arises naturally in the linear optics scheme. Suppose that one encodes a qubit as $\alpha|0_L\rangle + \beta|1_L\rangle$, where $|0_L\rangle = \frac{1}{\sqrt{2}}(|00\rangle + |11\rangle)$ and $|1_L\rangle = \frac{1}{\sqrt{2}}(|01\rangle + |10\rangle)$. Next, suppose that a measurement is made on one of the qubits in the computational basis and further that *we know the outcome of the measurement*. Such measurement errors occur in linear optics algorithms, for instance, when one is attempting to teleport a quantum gate [KLM01]. This heralded outcome error can then easily be corrected (assuming one knows also *which* qubit was measured). If the measurement outcome corresponds to $|0\rangle$ then the qubit not measured is still in a proper superposition, i.e., $\alpha|0\rangle + \beta|1\rangle$. If the measurement outcome corresponds to $|1\rangle$ then the qubit not measured has a bit-flip applied to it, which can be corrected. It is in this latter sense that this is an error-correcting code. The experiments in [PJF05] and [OPW+05] proceed along roughly the same path, using photon polarization as the qubit. The errors were both engineered into the protocol and not naturally occurring. In [OPW+05] the error-correcting step itself (a possible bit-flip) is not performed, while in [PJF05] a feed-forward step with the possible correction is performed. Both experiments use coincidence counting to postselect out the portion of their experiment that performs encoding into the above basis. In other words, the results are recorded only when the encoding is deemed to have succeeded. Such postselected demonstrations are one of the current drawbacks of implementations of linear optics quantum computers, but, in much the same way as in NMR, allow for proof-of-principle experiments (and unlike in NMR, with suitable measurements and preparations, there are scalable protocols).

A more complicated example of an experiment implementing QEC in linear optics was the implementation of a qubit loss-correcting code in [LGZ+08]. Again, using postselection, the authors were able to encode into a four-qubit code using photon polarization as the qubits. They then tested the loss code by detecting a photon without detecting its polarization. Then by measuring the polarization of two of the other photons they could obtain the original qubit, up to a correction depending on the two measurement outcomes. The authors were able to demonstrate that the fidelity of the corrected state with the desired initial state was, in the worst case, $F = 0.745 \pm 0.15$. Finally, [LGZ+08] showed how to implement a five-qubit cluster-state loss code by a slight modification of the four-qubit loss code. One final, and the largest to date, experiment in QEC using linear optics was the experiment of [GFR+09]. In this experiment an eight-qubit topological code was used. Again polarizations were used as the qubit basis and postselection prepared the initial code states.

21.4 The future of experimental quantum error correction

Surely the experimental implementation of QEC is one of the great goals in the experimental implementation of quantum computers. We have seen that in three different physical implementations the basic ideas of QEC have been demonstrated. However, none of the experiments give results that function in a regime where both the natural source of error is present and an increase in actual fidelity is achieved. Clearly, the next goal for experimental QEC is to go beyond proof of principle and into proof of usefulness. Experimentally, a perhaps worthy next major step is the demonstration of basic fault-tolerant protocols for stabilizer codes in a setting where error correction improves over not encoding (for natural decoherence and control errors). To do this one should move away from experiments that encode the quantum information from bare qubits and instead work with fault-tolerant preparation methods: a suggestion for stabilizer codes is to focus on implementing the available Clifford group gates and Pauli eigenstates preparations in addition to QEC. This looks like a daunting task, but increases in gate fidelity certainly merit optimism for some physical implementations of quantum computers.

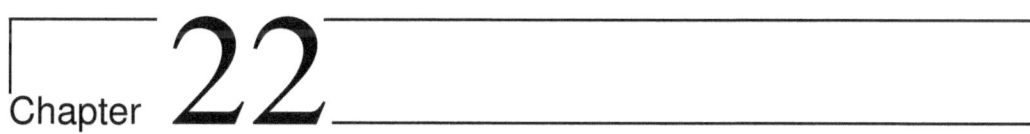

Chapter 22

Experimental dynamical decoupling

Lorenza Viola

22.1 Introduction and overview

As mentioned in Chapter 4, dynamical decoupling (DD) techniques have developed decades of tradition in the context of NMR spectroscopy, where they continue to play a pervasive role throughout all aspects of imaging and coherent control of multi-spin dynamics in the liquid and solid state. A dedicated experimental characterization and exploitation of DD as a tool for high-fidelity dynamical control and decoherence suppression in different quantum information processing (QIP) platforms has been undertaken only in more recent years, largely enabled by impressive laboratory progress in fast and ultrafast coherent control capabilities over the past decade. Without attempting to provide a comprehensive in-depth account, it is our goal in this chapter to highlight some key experimental advances to illustrate the significance of DD strategies for near-term practical quantum error control in QIP; and justify why, quoting from a recent experimental work [MTA+06], they "will likely form a quintessential element in real quantum computers."

The large majority of dynamical error-suppression implementations in QIP have aimed thus far to validate DD as a tool for enhanced *quantum memory* in different qubit devices in the presence of various decoherence mechanisms. Likewise, with the exception of a recent work reporting the suppression of collisional decoherence using a "continuous spin echo" close in spirit to Eulerian DD [SAD10], the majority of experiments have employed (sufficiently) "hard" pulses, such that any evolution other than that due to the control field could be taken as negligible and the bang-bang (BB) limit formalized in Chapter 4 invoked as an adequate approximation. In this context, implemented sequences have involved variants of periodic Hahn-echo or Carr–Purcell (CP) protocols [H50, CP54, GBC90], achieving "low-level" DD (in the terminology of Chapter 4), as well as single- and two-axis protocols aiming at "high-level" DD, notably concatenated DD (CDD) and Uhrig DD (UDD), discussed in Chapter 14. Recent years have witnessed a true explosion in the number of experiments aimed at quantitatively characterizing and

---

Quantum Error Correction, ed. Daniel A. Lidar and Todd A. Brun. Published by Cambridge University Press.
© Cambridge University Press 2013.

comparing the performance of different control sequences in different noise environments, resulting in dramatic advances and effectively taking experimental DD to a higher level of maturity and sophistication. In what follows, we will focus on illustrating some of the basic experimental achievements in different qubit platforms, and provide at the end of the chapter a brief summary of the most recent experiments along with an outlook on the remaining challenges for future implementation.

22.2 Single-axis decoupling

22.2.1 Optical polarization qubits

A simple proof-of-principle demonstration of BB decoherence control in a single-photon optical system was performed by Berglund and Kwiat in 2000 [B01]. In this case, the qubit computational basis states are identified with two mutually orthonormal linear polarization states of a single photon $\{|\chi_j\rangle\}$, $j = 1, 2$, which may for instance coincide with the $\{|H\rangle, |V\rangle\}$ basis provided by states with polarization parallel or perpendicular to the axis of a linear optical cavity. Assuming that an optical laser field with intensity reduced to about one photon per pulse and a Gaussian amplitude spectrum $A(\omega)$ is injected into the cavity, the photon input state may be written as

$$|\Psi\rangle_{\text{in}} = \int d\omega\, A(\omega)|\omega\rangle \otimes |\psi\rangle_{\text{in}}, \quad |\psi\rangle_{\text{in}} = \sum_{j=1,2} c_j |\chi_j\rangle, \qquad (22.1)$$

with the spectral amplitude profile normalized so that $\int d\omega |A(\omega)|^2 = 1$. While in the above state the polarization ($|\psi\rangle_{\text{in}}$) and frequency ($|\omega\rangle$) degrees of freedom are not entangled, any mechanism that introduces a coupling between them during propagation, followed by a frequency-insensitive detection of the polarization state, effectively transforms the pure input state $\rho_{\text{in}} = |\psi\rangle_{\text{in}}\langle\psi|$ into an output state ρ_{out}, which is mixed. That is, unobserved frequency modes may be regarded as a decohering environment for the system S. In particular, a purely dephasing environment as examined in Section 4.2 corresponds to the action of a joint unitary operator $\mathbf{U}(x)$, which preserves the polarization eigenstates,

$$\mathbf{U}(x)|\omega\rangle \otimes |\chi_j\rangle = e^{i\phi_j(x,\omega)}|\omega\rangle \otimes |\chi_j\rangle, \quad j = 1, 2, \qquad (22.2)$$

where $|\Psi\rangle_{\text{in}} = |\omega\rangle \times |\psi\rangle_{\text{in}}$ and the spatially dependent phase factor $\phi_j(x,\omega)$ reflects the details of the physical coupling mechanism. Thus,

$$\rho_{\text{out}}(x) = \int d\omega\, \langle\omega|\, \mathbf{U}(x)|\Psi\rangle_{\text{in}}\langle\Psi|\mathbf{U}^\dagger(x)\,|\omega\rangle, \qquad (22.3)$$

and the qubit coherence element after the photon has traveled a distance x (that is, after a time $t = x/c$) may be expressed as

$$\rho_{12}(x) = \rho_{12}(0) Z_0(x), \quad Z_0(x) = \int d\omega\, |A(\omega)|^2 e^{i[\phi_1(x,\omega) - \phi_2(x,\omega)]},$$

with $\rho_{12}(0) = c_1 c_2^*$ (note the formal analogy to Eq. (4.5) in Section 4.2). If $|A(\omega)|^2$ is peaked at a central frequency $\omega_0 = 2\pi c/\lambda_0$ with a finite width $\delta\omega$, phase coherence between polarization states will be lost once x exceeds the typical *coherence length* $L_c \sim c/\delta\omega \sim \lambda_0^2/\delta\lambda$. The resulting coherence loss can be counteracted by a BB control scheme that periodically interchanges the polarization eigenstates on a time scale sufficiently fast compared to the relevant correlation time $\tau_c = L_c/c \sim 1/\delta\omega$. That is, let each BB exchange operation be described by

$$\mathbf{R}|\chi_1\rangle = |\chi_2\rangle, \quad \mathbf{R}|\chi_2\rangle = \pm|\chi_1\rangle.$$

Then, in the notation of the first example in Section 4.4.3, a pair of such rotations clearly corresponds to a periodic DD (PDD) cycle based on the group $\mathscr{G} = \mathbb{Z}_2$.

In the experiment, different optical realizations of $\mathbf{U}$ and $\mathbf{R}$ were explored. Basically, the control element $\mathbf{R}$ may be realized either through a suitably oriented $\lambda/4$ or $\lambda/2$ wave-plate or through an optically active $90°$ quartz rotator, depending on whether prior knowledge of the coupling eigenstates $|\chi_j\rangle$ is included in the setting specification. Likewise, a stepwise implementation of the decoherence operator $\mathbf{U}$ may be obtained in terms of either single or multiple phase errors in the same basis. In the simplest case, a controllable ω-dependent phase shift in the fixed H/V basis was achieved by simply unbalancing, by a controlled amount x, the two arms of a Michelson polarization interferometer. In more elaborate settings, additional optical path differences were incorporated by including one or more birefringent quartz crystals with proper thickness and optical-axis alignment. In all cases, it is important to note that, due to the linear cavity geometry, each pass through the apparatus involves both forward and backward propagation, exemplifying a *two-way* communication channel; see also Section 22.3.2.

Representative experimental results for a generic linear-polarization basis $\{|\chi_j\rangle\}$ as well as for a time-dependent coupling basis (corresponding to a *non-stationary* open quantum system) are reported in Fig. 22.1. In the first case (top), a cavity with the photon passing through two birefringent quartz crystals of different thickness at each DD cycle is used, with $\delta\lambda = 1.5$ nm leading to an optical coherence length $L_c \sim 300\,\mu$m. In the second case (bottom), the optical axis of the first crystal is rotated with respect to that of the second crystal, which is aligned along the H cavity axis. At the output, the qubit density matrix ρ_{out} is tomographically reconstructed by determining the degree of polarization, $P = \text{Tr}(\rho_{\text{out}}^2)$ (that is, the state purity). The extent of decoherence suppression under BB control is evident from the significantly higher purity level maintained for the whole photon storage time in the cavity.

22.2.2 Liquid-state NMR spin qubits

As already remarked, it would be impossible to do justice to all the early NMR experiments that, in one way or the other, have incorporated decoupling (through either pulsed or continuous-wave irradiation) to enhance resolution during signal acquisition and/or to precisely implement desired effective Hamiltonians – notably, to achieve a so-called "time-suspension" of the qubit spins by (approximately) realizing a net identity propagator. While we refer the reader to the extensive NMR literature for a thorough account (see e.g., [H76, S90, EBW94, SK87], and references therein), a survey of liquid-state NMR QIP techniques as directly relevant to the present discussion is provided in [CLK+00].

In liquid-state NMR, the qubits are spin-$1/2$ nuclei in a molecule, each of the (order of 10^{18}) molecules in the sample acting as an independent information processor. Since a strong static magnetic field is applied to provide the dominant terms of the internal spin Hamiltonian, a preferred quantization axis is naturally defined by the (z) field direction. As a result, in the so-called *weak-coupling limit* (that is, when the differences between Zeeman interaction energies are

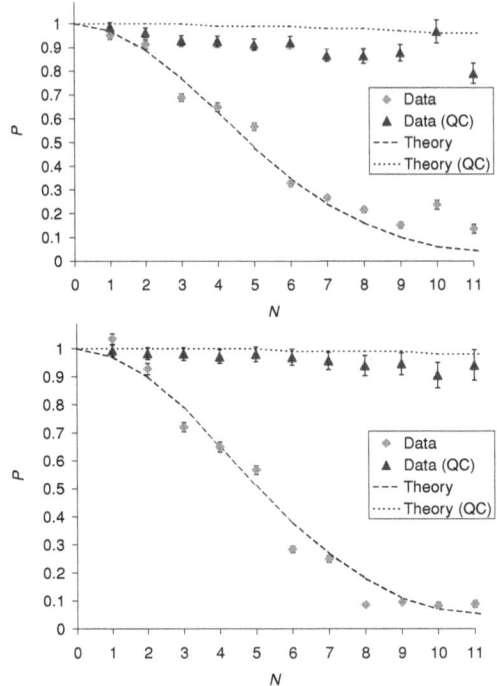

Fig. 22.1. Experimental and theoretical curves showing the qubit polarization (or purity) P vs. the number of passes N in a linear-cavity setup. "QC" refers to the cases where the BB control procedure was implemented. A linear cavity including two birefringent quartz crystals of thickness 0.850 and 1.046 mm was used, the exchange element **R** being implemented through an optically active quartz rotator. Top: Both birefringent crystals make an angle of 10° with the H longitudinal cavity axis, corresponding to a fixed system–environment coupling basis $|\chi_j\rangle$ offset by $-10°$ with respect to the horizontal. Bottom: The 1.0 mm crystal is oriented at 10° to the horizontal, while the 0.8 crystal is placed at 0°, effectively resulting in a system–environment coupling basis which changes with N. In both cases, the input state $|\psi\rangle_{\text{in}}$ is taken at 35° relative to the cavity axis. Reprinted with permission from [B01].

much larger than the coupling between spins) the internal Hamiltonian of a molecule's nuclear spins is well approximated by

$$H_{\text{int}} = \frac{1}{2}\sum_i \nu_i \sigma_z^{(i)} + \frac{\pi}{2}\sum_{i\neq j} J_{ij}\vec{\sigma}_i \cdot \vec{\sigma}_j \approx \frac{1}{2}\sum_i \nu_i \sigma_z^{(i)} + \frac{\pi}{2}\sum_{i\neq j} J_{ij}\sigma_z^{(i)} \otimes \sigma_z^{(j)}, \qquad (22.4)$$

where the summation is over the nuclear spins, and ν_i, J_{ij} are suitable resonance frequencies and scalar couplings, respectively. That is, off-diagonal (so-called "flip-flop" terms) in the Heisenberg coupling may be treated as a perturbation to the dominant Ising term. Likewise, with respect to a suitable product operator basis, constructed using $\sigma_z, \sigma_\pm$ operators of individual spins with respect to the quantization axis, the dominant modes of decoherence are associated with phase-damping processes that only attenuate the off-diagonal elements of the spin density operator, without energy exchange. These two factors – the simplicity of both the internal Hamiltonian and the dominant relaxation mechanisms – make it possible for *single-axis* sequences (based on either

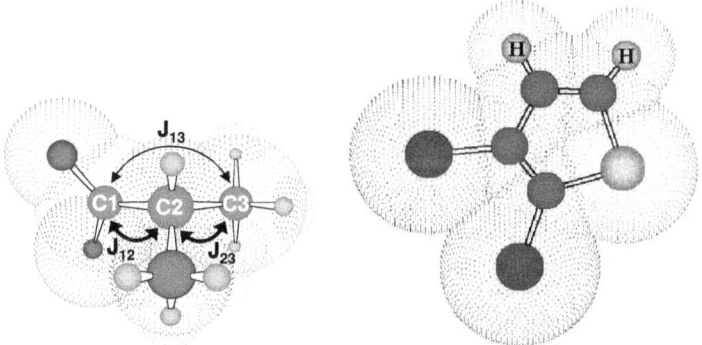

Fig. 22.2. Left: Molecular structure of alanine. The ^{13}C carbon nuclei (C1, C2, C3) represent the system, whereas the ^{1}H proton bound to C2 and each of the three methyl groups bound to C3 provide act as environment. Right: Molecular structure of dibromothiophene, with the two proton qubits employed in the experiment. Reprinted with permission of AAAS (left) and IOP (right) from [VFP+01, FVH+02].

x or y pulses, as mentioned in Section 4.4.3[1]) to be adequate and routinely used as building blocks for DD in liquid-state NMR QIP [JK99]. More interestingly from a QEC perspective, the latter is thus far the only experimental testbed where some steps have been taken toward *concatenating* different active or passive error-control strategies. While a DFS-QEC code was reported in [BVF+05], we specifically focus here on two experiments involving DD.

In the first experiment, carried out in 2002 by Boulant *et al.* [BPF+02], the usefulness of using DD in conjunction with standard QEC in the presence of multiple dephasing mechanisms was explored. The physical system consisted of an ensemble of alanine molecules in D$_2$O solution at room temperature (see Fig. 22.2), with the three ^{13}C labeled spin-1/2 nuclei being used as a three-qubit open quantum system, and the remaining protons (either directly coupled or coupled through a methyl group) being included in the local spin environment. While in normal spectroscopic applications protons would be decoupled by a strong continuous-wave field, in this setting pulses were intentionally applied to the carbon spins only, the coupling between protons and carbons being exploited to simulate an *engineered non-Markovian* dephasing process of the form

$$H_{SB}^{\text{deph}} = \sum_{i=1}^{3} \sum_{P} g_{i,P}\, \sigma_z^{(C_i)} \otimes \sigma_z^{(P)}, \quad g_{i,P} \in \mathbb{R}, \tag{22.5}$$

where P stands for either the proton- or the methyl-group nuclear spin. Provided that times short compared to the spin-lattice relaxation time are assumed, $T_1^{\min} \sim 1.5$ s, any internal dynamics due to the direct environment–spin couplings may be neglected – which allows the bath to be regarded as "quasi-static" and leaves the system–bath coupling strength $g_{i,P}$ as determining the relevant correlation time of the interaction. Stochastic phase errors, which may (to a good approximation) be taken as Markovian, are additionally induced by the natural T_2-relaxation, whose typical time scales for this molecule are $T_2^{\min} \sim 420$ ms. The implemented QEC code

---

[1] In NMR, external control is typically provided by amplitude- and phase-modulated radiofrequency magnetic fields, and qubits are realized in a frame that rotates about z at the carrier frequency. Thus, different rotation axes (say, x, $-x$, or y) are achieved by suitably adjusting the phase of the signal (to $0°$, $180°$, or $90°$, respectively), whereas different rotation angles are achieved by properly tuning the pulse amplitude and duration, see e.g., [H76, L01b].

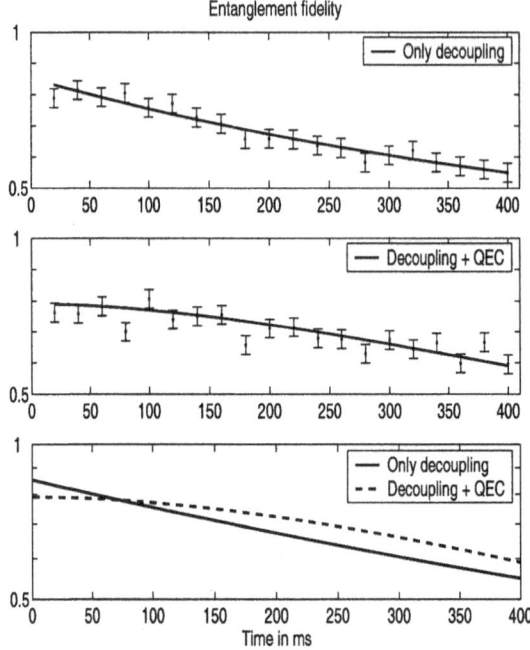

Fig. 22.3. Entanglement fidelity F_e as a function of evolution time in the presence of DD, with and without QEC. For DD only (top panel), the solid line was obtaining by fitting an exponential function, whereas for DD + QEC (middle panel) the short-time data points (up to 100 ms) were fitted to a function of the form $F_e = \gamma + (3e^{\alpha t} - e^{-3\alpha t})/2$, with γ, α fitting parameters. Reprinted with kind permission of Springer Science and Business Media from [BPF+02].

is the usual three-bit code for random independent phase errors, with one of the carbon spins initially carrying the data qubit $|\psi\rangle$ to be protected, and the two remaining ancilla carbons being initialized in the (pseudo-)pure state $|00\rangle$. The entanglement fidelity of the implemented "no-op" as a function of storage time was inferred after decoding and error correction, by contrasting the case where the above $\sigma_z^{(C)} \otimes \sigma_z^{(P)}$ coupling error had been implemented with one where it was refocused using suitable sequences of π-pulses DD.[2] Note that while QEC alone would correct for errors due to the system–bath coupling (22.5), repetition rates shorter than 1 ms would be needed to compensate for fidelity loss given the relatively strong coupling, $|g_{i,P}| \sim 100$ Hz. Experimental results are shown in Fig. 22.3. They indicate the benefits of using DD to suppress strong slowly correlated dephasing, while QEC takes care of the remaining weak but fast-correlated components of the error process, overall leading to a more effective protection scheme.

In the second experiment, reported in 2002 by Fortunato *et al.* [FVH+02], DD techniques have been used together with passive error protection, by providing a concrete realization of

---

[2] While nonselective "hard" BB pulses suffice in principle to refocus the coupling (22.5), spin-selective "soft" pulses were required in the experiment to both "extract" the contribution (22.5) from the total internal Hamiltonian via recoupling, and to implement the QEC network. So-called *strongly modulating pulses* were used to achieve sufficiently fast spin rotations by retaining the required degree of selectivity; see [FPB+02].

encoded DD as proposed in [LW02, V02] and by enabling the first demonstration of universal control of a DFS logical qubit. In this case, the system consisted of an ensemble of dibromothiophene in a solution of $CDCl_3$ at room temperature, the two proton spins now being used as physical qubits, subject to a predominantly collective dephasing process with typical time scale $T_2 \sim 3.5$ s ($T_1 \sim 7$ s for this molecule). As discussed in Chapter 3, the relevant DFS is the two-dimensional subspace

$$\mathscr{H}_L = \mathrm{span}_\mathbb{C}\{|0_L\rangle, |1_L\rangle\} = \mathrm{span}_\mathbb{C}\{|01\rangle, |10\rangle\},$$

that is, the "zero-quantum subspace" in NMR terminology, and a choice of DFS observables suitable for the implementation turns out to be given by $\sigma_z^L = -\sigma_z^{(2)}$, $\sigma_x^L = (\sigma_x^{(1)} \otimes \sigma_x^{(2)} + \sigma_y^{(1)} \otimes \sigma_y^{(2)})/2$.

By retaining the complete form of the Heisenberg coupling between the two proton spins, the internal Hamiltonian reads

$$H_\mathrm{int} = \frac{1}{2} \sum_{i=1,2} \nu_i \sigma_z^{(i)} + \frac{\pi}{2} J \left(\sigma_x^{(1)} \otimes \sigma_x^{(2)} + \sigma_y^{(1)} \otimes \sigma_y^{(2)} + \sigma_z^{(1)} \otimes \sigma_z^{(2)} \right),$$

which, using the actual molecule parameters, yields

$$H_\mathrm{int} =_L \pi(\Delta\nu \sigma_z^L + J \sigma_x^L) = -137.5\pi \sigma_z^L + 5.7\pi \sigma_x^L, \tag{22.6}$$

where Hz units are understood, and $=_L$ denotes restriction to the DFS $\mathscr{H}_L$ (note the large separation between the two time scales, consistent with weak-coupling conditions). Thus, the natural evolution implements a nontrivial logical operation within $\mathscr{H}_L$. The challenge is to achieve universality by supplementing this evolution with available control while, ideally, always remaining within the DFS.

The idea of encoded DD is simply to think of DD directly at the *logical* level, that is, to apply encoded DD pulses that refocus unwanted logical evolution as do the unencoded pulses on the corresponding physical evolutions. For the Hamiltonian in (22.6), applying an encoded CP sequence consisting of BB π_x^L-pulses would in principle remove the dominating σ_z^L-term, allowing logical x-rotations to be achieved. In practice, a complication arises due to the fact that the unitary propagator that describes an ideal BB π_x^L-pulse, $U_x^L = e^{-i\pi \sigma_x^L/2} = e^{-i\pi \sigma_x^{(1)} \otimes \sigma_x^{(2)}/2}$, is *not* realizable through the encoded Hamiltonian $\sigma_x^L/2$, but in NMR can only be *simulated* using physically available radio-frequency (RF) control Hamiltonians of the form

$$U_x^{1,2} = e^{-i\Omega\tau(\sigma_x^{(1)} + \sigma_x^{(2)})} = e^{-i\Omega\tau\sigma_x^{(1)}} e^{-i\Omega\tau\sigma_x^{(2)}} = -\sigma_x^{(1)} \sigma_x^{(2)} = U_x^L,$$

if $\Omega\tau = \pi/2$. While this is not harmful in the BB limit where $\tau \to 0$, real-life pulses with finite duration will necessarily cause departure from the DFS *during* control, hence exposure to phase errors. In the experiment, this issue was circumvented by both optimizing the duration of each π^L-pulse relative to the inter-pulse delay (62.4 μs vs. 630 μs, respectively), and by incorporating robustness against phase evolution into pulse design (with each π^L-pulse realized through a six-period composite-pulse sequence [FKL80, L01b]). In this way, entanglement fidelities higher than 90% were reported for logical rotations in the presence of engineered phase noise significantly stronger than that due to natural relaxation; see Fig. 22.4.

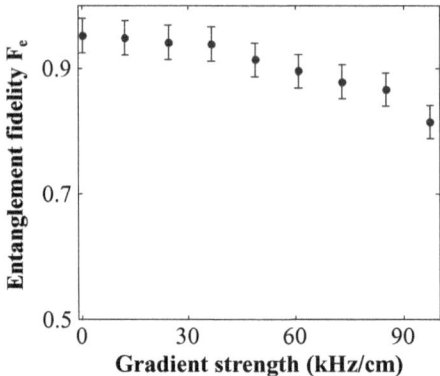

Fig. 22.4. Entanglement fidelity for an encoded $\pi/2$ rotation realized via the Euler-angle construction $e^{-i\pi\sigma_y^L/4} = e^{-i\pi\sigma_z^L/4}e^{-i\pi\sigma_x^L/4}e^{i\pi\sigma_z^L/4}$. Variable-strength phase noise was induced by applying a temporally modulated magnetic field gradient, with a correlation time $\tau_c \sim 50.6\,\mu\text{s}$ much shorter than the control cycle time ($\sim 700\,\mu\text{s}$). Reprinted with permission from [FVH+02].

22.2.3 Solid-state and hybrid spin qubits

22.2.3.1 DD as a diagnostic tool Solid-state devices tend to be characterized by complex environments whose decoherence mechanisms are typically not fully understood. In such situations, the application of simple DD sequences – most notably, "spin-echo spectroscopy" – may often serve as a valuable tool for *open-system identification*, allowing physical details of the environment to be inferred from the output response under varying conditions. Before turning to experimental demonstrations of DD as a proper error control method in the solid state, a few illustrative implementations are worth highlighting to exemplify the versatility of simple DD protocols for diagnostics (see also [YW11, AS11, dRD+11] for very recent contributions along these lines).

Spin-echo experiments involving the application of a single π-pulse separated by a variable delay time t_d from each of the two $\pi/2$-pulses involved in a standard Ramsey interference experiment [R50] were reported in 2002 for a superconducting charge qubit by Nakamura *et al.* [NPY+02], and in 2003 for a flux qubit by Chiorescu *et al.* [CNH+03]. In the first case, the qubit consisted of a Cooper-pair-box device, the desired rotations between charge states $|0\rangle, |1\rangle$ (corresponding to zero or one excess Cooper pair on the superconducting island) being accomplished by the application of a gate-voltage potential $V_g(t)$, hence fast modulation of the energy difference between the two charge states. The observed decay of the "charge-echo" signal as a function of the evolution time $2t_d$ is significantly slower than the free induction decay (FID) observed in the absence of the refocusing π-pulse, indicating that dephasing for the system is predominantly contributed by low-frequency energy-level fluctuations due to $1/f$ background charge noise.[3] For the flux qubit of [CNH+03], computational eigenstates $|0\rangle, |1\rangle$ correspond to states carrying oppositely circulating persistent currents, and control is achieved through resonant

---

[3] So-called $1/f$ noise processes are ubiquitous in nature, and are defined by a noise power spectrum of the form $J(|\omega|) = A/|\omega|$, $A > 0$, in a broad frequency range characterized by suitable spectral cutoffs; see e.g., [FI06, FI08] for recent theoretical studies in Josephson junction qubits.

microwave excitation. In this case, the fact that coherence is found to only marginally improve under spin echo, as compared to FID, has provided indication of the presence of high-frequency noise sources, well beyond 10 MHz.

In a similar vein, spin-echo spectroscopy has allowed us to experimentally probe coherence properties of a single electron spin associated to a nitrogen-vacancy center in a high-purity diamond sample [CDT+06]. Manipulation of the electronic spin state is possible via electron spin resonance (ESR) techniques [SJ01]. While the FID signal is found to decay very rapidly, $T_2^* \sim 1.7$ μs, decay of a typical Hahn-echo signal is almost an order of magnitude slower, indicating strongly non-Markovian behavior of the surrounding mesoscopic nuclear-spin environment. Additional short- and long-time structure of the spin-echo signal, as well as its dependence upon the strength of the external magnetic field, has enabled detailed quantitative information about the hyperfine interaction coupling electron and nuclear spins to be extracted.

22.2.3.2 DD for enhanced quantum storage A remarkably successful implementation of BB dynamic decoherence control was reported by Fraval *et al.* in 2005 [FSL05] in the context of QIP implementations using optically active rare-earth ions in crystalline hosts, where quantum information is stored in the centers' long-lived ground-state hyperfine transitions. The specific system under consideration consisted of a crystal of Y_2SiO_5 doped with praseodymium ions Pr^{3+}, held at liquid helium temperature. Praseodymium has a nuclear spin $I = 5/2$ and, as a result of a quadrupole interaction, its electronic ground state splits into three doubly degenerate levels corresponding to magnetic quantum number $m_I = \pm 5/2, \pm 3/2, \pm 1/2$, with splittings of the order of 10 MHz and final degeneracies being lifted by an applied Zeeman magnetic field. In the experiment, initialization into a pure-state ensemble was ensured by an optical/RF repump scheme, which forced all Pr ions into the $m_I = -1/2$ sub-level. Coherence was stored in the $m_I = -1/2 \leftrightarrow m_I = +3/2$ transition, whose coherence lifetime can be as long as $T_2^{\text{opt}} = 0.86$ s if the magnetic field is properly aligned and within 0.5 G of an optimal field value B^{opt}. Residual decoherence was primarily due to dephasing perturbations induced by effective magnetic fields that slowly fluctuate over time scales between 10 and 100 ms, as a result of host cross-relaxation. Controlled excitation and re-phasing of ground-state spin coherences was accomplished by a so-called "Raman heterodyne" technique, which relies on resonant optical Raman pulses and had been earlier experimentally demonstrated on Pr^{3+}:Y_2SiO_5 in [HSK+98]. In [FSL05], BB DD was implemented through sequences of π-pulses along a fixed axis but with alternating 0° and 180° phases in order to minimize systematic pulse-error accumulation.[4] Delay times τ_c between consecutive pulses varied from 20 ms to 0.5 ms, resulting in up to 1000 BB cycles. Some illustrative results are reported in Fig. 22.5. Under DD, the coherence time is found to increase dramatically, approaching a saturation value longer than 30 s.

In 2005, a successful implementation of a BB spin-echo sequence for suppressing hyperfine-induced dephasing in a double GaAs quantum dot was also reported by Petta *et al.* [PJT+05]. In

---

[4] Phase alternation would neither be required nor have any effect in the ideal BB limit of δ-pulses. Similarly to the idea of reversing the phase of alternating pairs of pulses mentioned in Section 4.6.3, however, reversing the phase of alternating pulses reduces the effect of finite-width errors, although in this case this may hold for specific initial states only; see [H76], pp. 74–76.

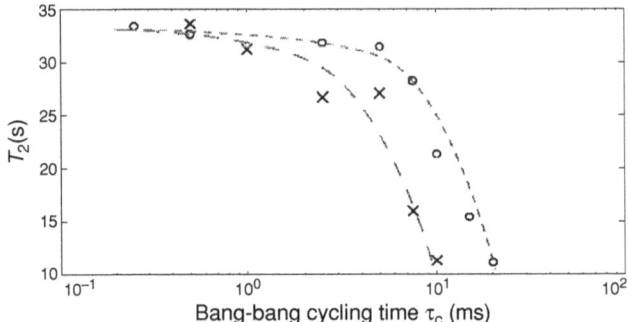

Fig. 22.5. Dependence of the decoherence time scale T_2 on the DD pulse interval for two applied magnetic field configurations. Circles: Optimal field alignment, corresponding to $T_2^{\text{opt}} = 0.86$ s in the absence of control. Crosses: Magnetic field misaligned, resulting in a free coherence time of $T_2 = 100$ ms. Reprinted with permission from [FSL05]. Copyright 2005, American Physical Society.

this case, at large applied bias fields ($B \gg B_{\text{nucl}} \sim 5$ mT), a logical qubit is encoded in the two-electron spin manifold by letting $|0_L\rangle = S$ and $|1_L\rangle = T_0$, with S and T_0 denoting the spatially separated singlet configuration where each electron resides at a different dot, and the $m_s = 0$ sub-level of the triplet subspace, respectively. The Hamiltonian on the S-T_0 subspace $\mathcal{H}_L$ has the form

$$H_L = J(\epsilon)(\sigma_z^L + I) + \Delta B_{\text{nucl}}^z \sigma_x^L = J(\epsilon)|S\rangle\langle S| + \Delta B_{\text{nucl}}^z |T_0\rangle\langle T_0|, \qquad (22.7)$$

where the energy difference $J(\epsilon)$ is determined by the voltage detuning between the right and left dot, and ΔB_{nucl}^z is the difference in random hyperfine fields along z, responsible for unwanted spin dephasing. This dephasing can be largely undone by the application of a DD sequence that periodically exchanges the singlet-triplet states, that is, effects an encoded π_z^L-pulse in the language used in the previous section. In the experiment, this was accomplished through fast electrical control of the exchange interaction, resulting in rapid modulation of the detuning parameter ϵ and extending the singlet coherence time by a factor of more than 100.

The last two experiments on single-axis DD we wish to survey have both been reported by Morton and co-workers [MTA+06, MTB+08], and share the important feature of exploiting a *hybrid* QIP platform provided by coupled electron–nuclear spin degrees of freedom. In the first experiment [MTA+06], a two-qubit system is supported by four levels out of the 12-level spin manifold in the endohedral fullerene molecule N@C_{60}, which consists of a single nitrogen atom in the $^4S_{3/2}$ electronic state incarcerated by a C_{60} fullerene cage. The electron qubit is formed from the levels with magnetic quantum number $M_S = +3/2, -3/2$, whereas the nuclear qubit corresponds to $M_I = 0, +1$. As the latter has excellent natural environmental isolation, a strong engineered coupling is introduced for demonstration purposes by applying a resonant RF field to drive nuclear Rabi oscillations, and this coupling is then successfully removed by DD on the electron. In practice, the required BB π-phase shifts are realized by resonantly driving a transition between one of the basis states (say, $|01\rangle$) through an auxiliary level ($|11\rangle$), so that $|01\rangle$ acquires

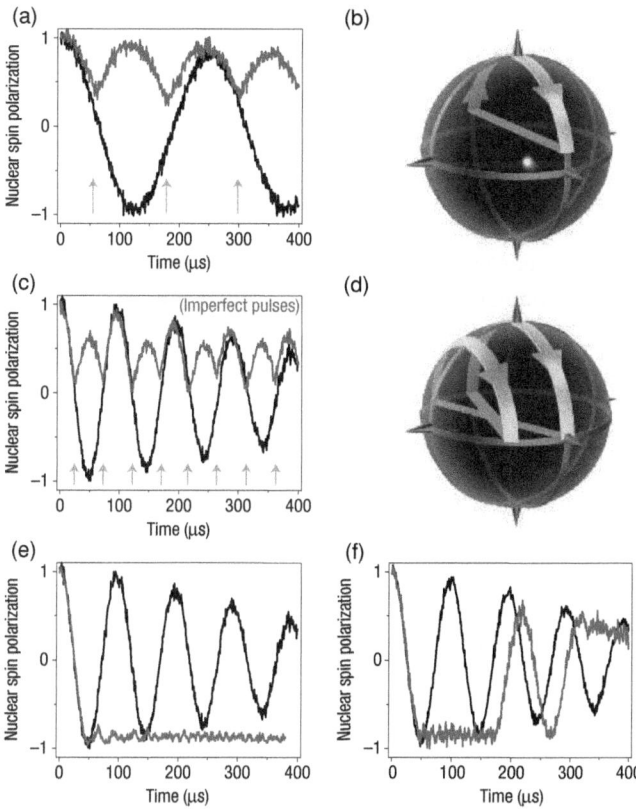

Fig. 22.6. Unperturbed vs. BB-controlled nuclear Rabi oscillations. (a), (b) Pulses are applied on the electron spin at regular intervals marked by arrows, reversing the evolution of the nuclear-spin qubit. (c), (d) If the applied phase shift is less than π, an odd–even behavior is observed, corresponding to different paths on the Bloch sphere. (e), (f) By controlling the repetition rate of the microwave pulses, the nuclear-spin qubit can be locked in a desired state and released at any point. Reprinted with permission from Macmillan Publishers Ltd from [MTA+06].

a -1 phase relative to the remaining basis state, $|00\rangle$.[5] Experimental results for Rabi oscillations in the absence and in the presence of DD pulses are shown in Fig. 22.6, successfully demonstrating the desired control.

In the second above-mentioned experiment [MTB+08], a CP DD sequence was exploited to suppress the effect of unwanted slowly fluctuating nuclear fields in a ^{31}P-doped ^{28}Si-enriched silicon single crystal. In the absence of control, an electron-spin coherence could be stored for up to 65 ms at temperatures below ~ 10 K, limited by the nuclear decoherence time T_{2n}. By applying DD pulses at a 1 kHz repetition rate on the nuclear spin during the storage period, it was possible to extend T_{2n} to 1.75 s at 5.5 K, thereby correspondingly increasing the lifetime of

---

[5] In a way, this setting may be thought of as the dual to the qubit-dephasing case discussed in Section 4.2, since now the unwanted interaction may be viewed as a purely off-diagonal "amplitude-damping" coupling along σ_x, which is removed by phase-flips along σ_z, that is, $\mathscr{G} = \mathbb{Z}_2$ is realized as $\{I, Z\}$.

the quantum memory element. Up to 1000 refocusing pulses were involved in a single CP DD protocol, the effect of systematic control imperfections being significantly mitigated by the use of suitable composite pulses [L01b].

22.2.4 Trapped ion qubits

A series of beautiful experiments aimed at testing the viability and usefulness of DD in scalable ion-trap device technology were carried out over the past few years by Biercuk *et al.* [BUV+09a, BUV+09b, UBB09]. In this case, the target qubit is a ground-state electron-spin-flip transition of ^{9}Be$^+$ ions in the $2s\,^2S_{1/2}$ manifold, corresponding to nuclear and total spin quantum numbers $m_I = 3/2$ and $m_J = \pm 1/2$, respectively. An ordered array of $\sim$1000 ions in a Penning ion trap operated at 4.5 T was formed in the experiment, with single-qubit operations being achieved by driving the $\sim$124 GHz transition via a quasi-optical microwave system.

Classical phase randomization was considered as the dominant source of errors. If the relevant qubit Hamiltonian is written as

$$H = \frac{\hbar}{2}\big[\Omega + \beta(t)\big]\sigma_z, \tag{22.8}$$

where Ω and $\beta(t)$ denote the qubit splitting and a classical Gaussian random variable, respectively, the buildup of such a random phase effectively results in decay of the ensemble-averaged coherence signal:

$$W(t) \equiv \mathbb{E}\{|\langle\sigma_y(t)\rangle|\} = e^{-\chi(t)}, \tag{22.9}$$

in a frame rotating with Ω. As shown in [U08] (see also Chapter 14), the coherence evolution in the presence of any DD sequence consisting of n BB π_x pulses may still formally be expressed as in Eq. (22.9), upon introducing an appropriate "filter function" $F_n(\omega\tau)$ that encapsulates the effect of the sequence in the frequency domain [CLN+08, BDU11, HKV+11], that is,

$$\chi(t) \equiv \chi_n(t) = \frac{2}{\pi}\int_0^\infty d\omega\, \frac{S_\beta(\omega)}{\omega^2}F_n(\omega t), \tag{22.10}$$

where $S_\beta(\omega)$ denotes the noise power spectrum and t is the total duration of the applied sequence.[6] In the experiment, natural phase noise due to ambient magnetic field fluctuations was measured directly, yielding a power spectrum $S_\beta^{\text{amb}} \propto 1/\omega^4$, whereas engineered noise with a tunable power spectrum $S_\beta^{\text{eng}}(\omega)$ was synthesized by effectively imposing a noisy modulation of the microwave frequency. High-fidelity π_x rotations of finite length were implemented ($\tau_\pi \sim 185\,\mu$s), but with the applied noise switched off during each pulse, effectively simulating the BB regime.

The performance of standard multi-pulse periodic CP DD was experimentally characterized and contrasted to the performance of UDD for different noise environments: recall that pulse timings for UDD obey Eq. (14.33). Representative results for sequences consisting of up to $n = 6$ pulses are depicted in Fig. 22.7. In each case, phase errors are manifested as nonzero fluorescence detection, a normalized count value at 0.5 thus implying complete phase randomization. UDD is found to perform similarly to the CP sequence (CPMG in the data, from

---

[6] Recall the related discussion in Section 14.5.2, where the definition of the filter function given in Eq. (14.58) differs by a factor of $\omega^2/4$.

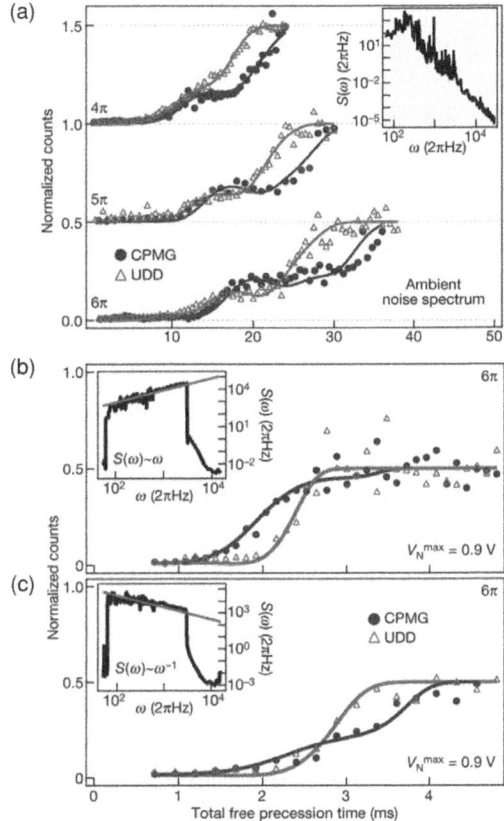

Fig. 22.7. Pulse sequence performance in the presence of different pure-dephasing noise spectra. (a) Decoherence signal as a function of the total free precession (storage) time in the presence of ambient phase noise. The experimentally measured spectrum is given in the inset, $S_\beta^{\mathrm{amb}} \propto 1/\omega^4$. CPMG DD and UDD are compared in each case for a fixed number of π_x-pulses, $n = 4, 5, 6$, top to bottom. (b) CPMG vs. UDD performance for Ohmic spectrum with a sharp spectral cutoff, $S_\beta^{\mathrm{eng}} \propto \omega$ and $\omega_c = 2\pi \times 500$ Hz. UDD clearly outperforms PDD in this regime. (c) CPMG vs. UDD performance for sub-Ohmic spectrum with a sharp spectral cutoff, $S_\beta^{\mathrm{eng}} \propto 1/\omega$ and ω_c as in (b). Overall, CPMG performs similarly to UDD in this regime. In both (b) and (c), the number of pulses $n = 6$, and V_N parametrizes the applied noise strength. Reprinted with permission from Macmillan Publishers Ltd from [BUV+09a].

"Carr–Purcell–Meiboom–Gill" [HW68]) in the presence of the ambient noise spectrum, which has a "soft" high-frequency cutoff. In contrast, the UDD sequence significantly outperforms PDD for noise spectra with a "sharp" spectral cutoff, the observed improvement being especially pronounced for the Ohmic case [U07] and increasing as the intensity of the applied noise is increased. Additional improvement was also experimentally demonstrated by tailoring UDD design to a given noise spectrum and storage time (so-called "locally optimized DD," LODD) [BUV+09b], as well as by further obtaining UDD-like sequences applicable to arbitrary pure-dephasing environments (so-called "universal" LODD) [UBB09].

22.3 Two-axis decoupling

22.3.1 Solid-state NMR spin qubits

In comparison to liquid-state implementations, solid-state NMR architectures can offer, in principle, key advantages in terms of scalability – including the possibility to reach nearly pure initial states and refreshing qubits, as well as much longer decoherence times. These advantages come, however, at the expense of a significantly higher complexity in the underlying internal-spin Hamiltonian and, correspondingly, the achievable level of control and DD. The main challenge originates from the fact that a (homonuclear) solid-state system is always in a *strong-coupling regime*, whereby the time scales of single-qubit addressing and qubit–qubit coupling are comparable, unlike in Eq. (22.4). While a detailed discussion of solid-state NMR qubits is beyond our scope, a brief mention is nevertheless in order since these systems naturally lend themselves to supporting one of the most important and widely used two-axis DD sequences in NMR, the so-called Waugh–Huber–Haeberlen (WAHUHA, or WHH-4) sequence [WHH68, H76].

In the notation of Section 4.4, each WAHUHA cycle consists of a time-symmetric sequence of four pulses,

$$\mathscr{P}_{\text{WHH-4}} = \{\tau_1, P_x, \tau_2, P_{-y}, \tau_3, P_y, \tau_4, P_{-x}, \tau_5\}, \quad \tau_k = \Delta t_k/T_c, \qquad (22.11)$$

where the relative pulse separations $\tau_1 = \tau_2 = \tau_4 = \tau_5 = 1/6$, $\tau_3 = 1/3$ and, in the ideal BB limit, each of the pulses effects an instantaneous $\pi/2$ rotation about the corresponding axis, $P_a = \exp(-i\pi\sigma_a/4)$.[7] Although, as mentioned in Section 4.4.2, WAHUHA averaging does not directly follow from a DD group, direct calculation using Eq. (4.23) shows that a nonselective application of the sequence removes (to the lowest order) an arbitrary *dipolar Hamiltonian* of the form

$$H^{\text{dip}} = \sum_{i,j} H_{ij}^{\text{dip}} = \sum_{i,j} d_{ij}\left(3\sigma_z^{(i)} \otimes \sigma_z^{(j)} - \vec{\sigma}_i \cdot \vec{\sigma}_j\right), \quad d_{ij} = \alpha\frac{1 - 3\cos^2\theta_{ij}}{r_{ij}^3}. \qquad (22.12)$$

In the above equation, the dipolar coupling strength d_{ij} between spins i,j is expressed in terms of a constant α proportional to the nuclear species gyromagnetic ratio, the length r_{ij} of the internuclear vector connecting the two spins, and the angle θ_{ij} between this vector and the external quantizing field (z) direction. In practice, the WAHUHA cycle in (22.11) is used as the starting point for designing more elaborate dipolar decoupling sequences, such as the so-called MREV-8 [H76] and Cory-48 [CMG90], which achieve higher-order DD and/or ensure robustness against relevant experimental pulse imperfections.

It is worth noting that in systems whose internal Hamiltonian includes both Zeeman (or "chemical shift") terms as in Eq. (22.4) and dipolar interactions as in Eq. (22.12), the former are not averaged out under a WAHUHA sequence although their intensity is effectively reduced by an appropriate "scaling factor" [H76]. In situations where the chemical shift terms are unimportant or undesired, and, conversely, retaining dipolar couplings may be of spectroscopic interest, two-axis DD based on sequences of π pulses (such as XY-4, XY-8 discussed in Section 4.4.3) becomes relevant, since bilinear terms of the form $\sigma_a^{(i)} \otimes \sigma_a^{(j)}$, $a = x, y, z$, are preserved under collective (nonselective) π pulses. For instance, applications of such "MOdified phase Cycled CArr–Purcell" sequences (MOCCA for short) have been discussed by Gullion and co-workers in the context of

---

[7] The distinction between clockwise or counterclockwise rotations about a given axis, $\pm x$ or $\pm y$, is irrelevant for ideal δ-pulses; however, it is essential to retain time-symmetry for realistic finite-width pulses; see [H76].

determining homonuclear coupling constants or heteronuclear dipole couplings in magic-angle spinning solids, as well as by Furrer and co-workers for coherence transfer in biological NMR [FKM+04].

As a concrete illustrative example within QIP, a three-qubit solid-state NMR quantum information processor was implemented in [BMR+06], using a single crystal of malonic acid with a diluted fraction of ^{13}C-labeled molecules, whose nuclear spins are used as qubits. In this case, typical strengths of both chemical shifts and *intra*molecular dipolar couplings in the ^{13}C internal Hamiltonian are in the range of a few kHz. Additional *inter*molecular ^{13}C–^{13}C dipolar fields introduce unwanted couplings responsible for decoherence in the diluted ensemble. Experimental results on the performance of a combined dipolar-plus-Hahn DD sequence are included in [BMR+06], demonstrating a 50-fold increase in the observed ensemble coherence time, and suggesting that further improvement could be achieved using more powerful DD sequences.

22.3.2 Optical polarization qubits revisited

Polarization qubits as described in Eq. (22.1) have been revisited in an experiment by Damodarakurup *et al.* [DLG+09], aimed at demonstrating the full power of BB control for suppressing *generic* single-qubit decoherence via Pauli-group DD (recall Section 4.4.3). As a further important difference from [B01], a ring as opposed to a linear cavity geometry was employed in this case, forcing the photon to always travel in the forward direction along a *one-way* channel. Let $|V\rangle$ ($|H\rangle$) denote the linear polarization state orthogonal (parallel) to the plane of a triangular ring cavity, where a pulsed laser field with $\lambda_0 \sim 800\,\mu\text{m}$, $\delta\lambda \sim 15$ nm is injected at a repetition rate of 100 kHz through a spherical mirror. Decoherence in the H/V basis is introduced by two cavity plane mirrors symmetrically oriented at 45° along the short arm of the cavity. If $Z = |H\rangle\langle H| - |V\rangle\langle V|$, the action of each plane mirror is described by a frequency-dependent propagator $\mathbf{U}_Z[\phi(\omega)] = Ze^{-i\phi(\omega)Z/2}$, for an appropriate relative phase $\phi(\omega) = \phi_H(\omega) - \phi_V(\omega)$ that depends on the mirrors' polarization-dispersion properties.[8] In analogy to Eqs. (22.2) and (22.3), the photon output polarization state after n round-trips through the cavity may then be expressed in the form

$$\rho_{\text{out}}^{(n)} = \int d\omega \, \langle\omega|\, \mathbf{U}_n[\phi(\omega)]|\Psi\rangle_{\text{in}}\langle\Psi|\mathbf{U}_n^\dagger[\phi(\omega)]\,|\omega\rangle,$$

where $|\Psi\rangle_{\text{in}} = |\omega\rangle \otimes |\psi\rangle_{\text{in}}$ and on each pass two identical contributions from the plane mirrors arise,

$$\mathbf{U}_n[\phi(\omega)] = \Big[\mathbf{U}_Z[\phi(\omega)]\, \mathbf{U}_Z[\phi(\omega)]\Big]^n = e^{-in\phi(\omega)Z}. \qquad (22.13)$$

Decoherence in an *arbitrary basis* can be engineered by inserting a so-called Soleil–Babinet (S-B) element in front of each of the 45°-oriented plane mirrors. Since the S-B acts on the polarization state as a rotation about the x-axis by an unknown angle θ, the combination of the S-B with the plane mirror implements decoherence in an unknown basis, modifying Eq. (22.13) to

$$\mathbf{U}_n[\phi(\omega)] = \Big[\mathbf{N}[\phi(\omega),\theta]\, \mathbf{N}[\phi(\omega),\theta]\Big]^n, \qquad (22.14)$$

---

[8] In the experiment, the spherical mirror additionally introduces a rotation $\mathbf{U}_Z \simeq Z$ in the photon free evolution. Since in practice this was compensated for by an auxiliary Z wave-plate, we only consider the effect of the plane mirrors in our simplified discussion.

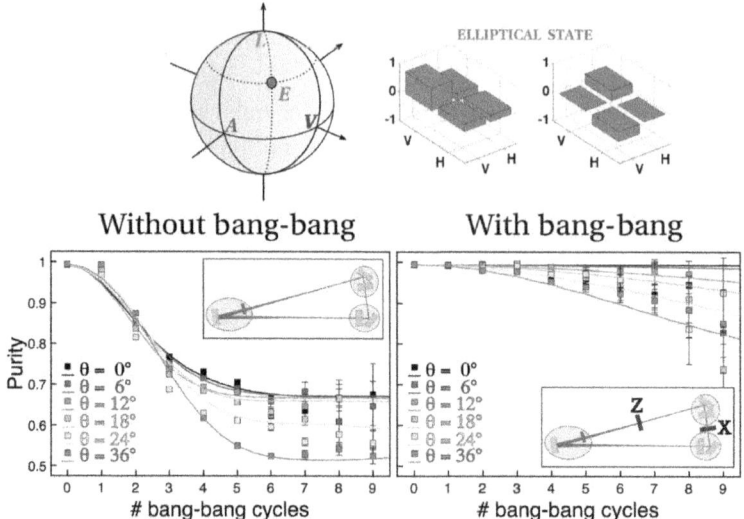

Fig. 22.8. Stroboscopic evolution of the output purity starting from an elliptical polarization state (labeled E in the Bloch sphere), in the presence (right) and absence (left) of BB control. Different curves correspond to a different orientation θ of the S-B, hence to a different unknown basis for decoherence entering Eq. (22.15). Worsening performance with increasing θ for a generic initial state like the one employed may be attributed to the fact that decoherence is only removed to first order in the variance of the relative phase difference ϕ, higher-order terms in the expansion growing with increasing θ. Reprinted with permission from [DLG+09]. Copyright 2009, American Physical Society.

for an appropriate "noise operator" of the form $\mathbf{N}[\phi(\omega),\theta] = [Ze^{-i\phi(\omega)Z/2}][e^{-i\theta X/2}]$, with $X = |H\rangle\langle V| + |V\rangle\langle H|$.

Bang-bang polarization control is implemented by performing two active operations during each universal PDD cycle: a Z $\lambda/2$ wave-plate is inserted in the long upper arm of the cavity, before the spherical mirror, so that a net Z operation is effected; and a phase delay equal to $\lambda/2$ is added to the S-B in the short arm, so that a net X operation is effected. The evolution under repeated DD cycles thus takes the form

$$\mathbf{U}_n^{\text{BB}}[\phi(\omega)] = \Big[Z\mathbf{N}[\phi(\omega),\theta]X\mathbf{N}[\phi(\omega),\theta]\Big]^n. \tag{22.15}$$

Experimental results are shown in Fig. 22.8, where a generic input elliptical polarization state $|\psi_{\text{in}}\rangle = E$ is used, and different decoherence bases are probed. In all cases, the output purity with BB DD is substantially higher throughout the whole storage time than the corresponding value without control. Possibly in conjunction with higher-order schemes, this demonstrates the potential for DD as a viable tool for significantly improving quantum communication in realistic settings.

22.4 Recent experimental progress and outlook

As mentioned in the introduction, the last couple of years have witnessed an impressive acceleration in experimental studies of DD across different qubit technologies.

Following the work by Biercuk and co-workers on trapped ions described earlier in this chapter, optimized DD schemes have been, in particular, carefully scrutinized in a variety of control settings. The first experimental realization of UDD in the solid-state was achieved by Du *et al.* for radical electron spins in irradiated malonic acid single crystals [DRZ+09], in which the electron spin coherence time was prolonged from 0.04 μs to about 30 μs by a seven-pulse UDD sequence. The experimental data from different samples under various conditions demonstrated that UDD performs better than CPMG in this solid-state system. For ^{13}C spin qubits in a ^{1}H purely dephasing spin bath whose high-frequency cutoff was not reached by the DD sequences, CPMG was instead found to outperform UDD [AAS11], as expected from theoretical predictions [CLN+08, HVD10]. Outside QIP, the advantages of nonperiodic sequences and UDD in particular have also been investigated by Jenista *et al.* [JSB+09] in the context of enhancing contrast in magnetic resonance imaging of tissue. In a GaAs double quantum dot, Bluhm *et al.* have extended coherence times of individual electron pairs to more than 200 μs [BFN+11] by using a 16-pulse CPMG sequence. Barthel *et al.* also demonstrated long coherence times in this system using CP, UDD, and CDD sequences [BMM+10].

Coherence preservation of nitrogen-vacancy centers in diamond by DD was also actively investigated by a number of groups. Ryan *et al.* demonstrated over two orders of magnitude increase in the room-temperature coherence time [RHC10]. Using 136 DD pulses, de Lange *et al.* observed a coherence time enhancement of more than 25 times compared to that obtained with a simple spin echo sequence [dWR+10]. By using DD, a coherence time T_2 of up to 2.44 ms was achieved by Naydenov *et al.* [NDH+11].

In the above-mentioned experiments on nitrogen-vacancy centers [RHC10, dWR+10] and ^{13}C spin qubits in a ^{1}H spin bath [AAP+10], the performance of PDD based on universal XY pulse sequences [GBC90] was investigated. References [WLF+10, PSL11] reported the first experimental demonstration of CDD against general noise on a qubit. In a comparative study reported in [AAP+10], CDD provided the best overall performance among PDD, UDD, and CPMG when the initial state is generic. The latter investigation also confirmed the theoretical predictions from [KL07, LYS07, NLP11, HVD10] that there is an *optimal* concatenation level in the presence of realistic pulse imperfections and a finite minimum pulse interval. The role of pulse imperfections has also been carefully scrutinized in a series of electron spin resonance experiments on phosphorus donor spins in silicon [TWZ+11, WZT+12], showing how the accumulation of systematic pulse errors may result in strongly biased, state-dependent performance of DD in realistic settings. Interestingly, these experiments have provided experimental confirmation of the possibility to engineer long-lasting "pointer states" by suitably manipulating the system bath coupling via DD-like quantum protocols [KDV11]. In recent experiments, also performed on phosphorus donors in silicon, Wang *et al.* experimentally increased the lifetime of bi-partite pseudo-entanglement from 0.4 μs in the absence of DD control to 30 μs under DD [WRF+11].

At the time of writing, experimental demonstrations of efficient DD schemes for arbitrary single- and many-qubit decoherence (QDD and NUDD, in particular) have yet to be performed. Beyond the use of DD for enhanced quantum memory, a main outstanding challenge is to experimentally validate dynamical quantum error suppression techniques for enhanced quantum information processing, in particular DD-protected [NLP11] and dynamically corrected gates [KV09b, KLV10]. It is interesting to note that periodic time-dependent modulation similar in spirit to DD has in fact been invoked to improve the fidelity of an entangling two-qubit gate mediated by laser light in a recent experiment on trapped ions [HCD+11], effectively realizing a simple dynamically corrected gate in the presence of classical frequency noise [HKV+11]. We anticipate significant experimental effort and progress in extending the benefits of DD to reducing decoherence errors in quantum-computing architectures at the physical layer.

Chapter 23

Architectures

Jacob Taylor

23.1 The principles of fault tolerance

In this chapter we will consider some of the practical difficulties in building a large-scale quantum computing device. This discussion relies heavily upon the prior chapters that introduced fault-tolerant quantum computation. Assuming that fault tolerance is possible allows us to focus on the physical realization of these ideas with a few specific examples.

Before going into detail, we review the governing ideas behind any fault-tolerant architecture. These are the necessary components that we analyze in this chapter: good quantum memory, high-fidelity quantum operations, long-range quantum gates, and highly parallel operation, such that error correction in different sections of the device can be accomplished at the same time. While a wide variety of potential implementations may be possible, the goal of a fault-tolerant architecture is not only to be scalable, i.e., to be able to run an arbitrarily large computation with at most polynomial overhead [S95, S96e], but also to be as efficient as possible. Efficiency in this context means using the fewest physical resources (quantum bits, time, control operations) necessary to accomplish the desired computation [S03b].

From this perspective, the recipe for fault tolerance is well established. First, we need to identify what quantum operations are available for the various quantum bits at our disposal. Developing error models for these operations forms the bulk of this chapter. Important questions about operations include noise, implementation time, and bandwidth (how many such operations may be performed in parallel).

Once these operations are established, a fault-tolerant computation scheme using standard quantum error correction (QEC) can be followed. In this sense, a logical bit comprises many physical bits and the associated error-correction apparatus. A set of replacement rules is then defined, in which the desired operation (e.g., wait, perform a two-qubit gate, move the qubit to a different location) is replaced by the encoded operation on the physical bits followed by an error-correction stage, as illustrated in Fig. 23.1.

Quantum Error Correction, ed. Daniel A. Lidar and Todd A. Brun. Published by Cambridge University Press.
© Cambridge University Press 2013.

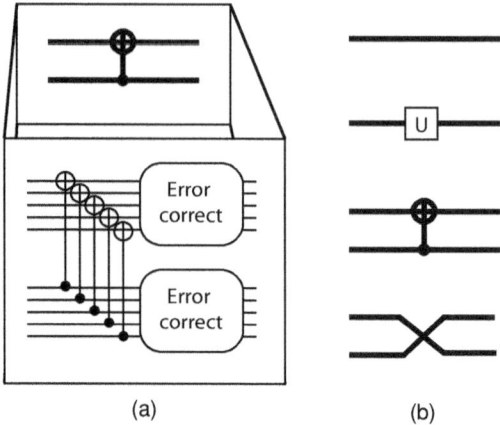

Fig. 23.1. (a) An example "replacement rule" for a two-qubit CNOT gate. The logical operation is replaced by the underlying physical operation (pairwise CNOT) and a following error detection and correction circuit. (b) An example set of operations for replacement. From top to bottom: wait (memory); one-bit gate; two-bit gate; move/swap two qubits.

We assume use of a CSS code with k logical bits encoded in n physical bits correcting t possible errors, denoted by $[[n, k, 2t+1]]$. Restricting the discussion this way has the advantage that all encoded two (logical) qubit gates can be applied transversally between the component physical qubits, which makes the analysis of transport overhead simple. Generally speaking, these codes also require $\mathcal{O}(n)$ ancillary bits per error-correction block in order to detect and correct errors.

Let us consider an encoded operation, with the logical operation applied, followed by a round of error detection and correction. In such a circuit, with N operations, we can estimate the number of potential fault paths [AGP06] with $t+1$ or more errors (which lead to an error in the encoded bit) as $C \approx \binom{N}{t+1}$ (see also Chapter 5). The error rate for the encoded system is then $\epsilon_1 \sim C\epsilon_0^{t+1}$, where ϵ_0 is the error probability for one operation at the physical level, and we have assumed each operation has an equal error probability.

By using the encoded bits to in turn encode a higher level single bit, we concatenate the code and further reduce the error. Repeating k times yields $\epsilon_k \sim C^{1+(t+1)+\cdots+(t+1)^{k-1}}\epsilon_0^{(t+1)^k} = \frac{1}{C^{1/t}}(C^{1/t}\epsilon_0)^{(t+1)^k}$. Following this naive argument, we can estimate that the threshold for fault tolerance occurs when $\epsilon_0 \lesssim \frac{1}{C^{1/t}}$; then, every level of concatenation reduces the error rate of the next level.

If we set a target error rate for the computational operations at the highest level, reducing C by reducing the number of error locations leads to fewer concatenation levels for achieving the target error rate. On the other hand, one could conceive of an architecture where the number of operations increases with concatenation level, such that C is a growing function of k. Then no real threshold exists: at some point the system becomes too large, and the growth of C overwhelms the error-reduction properties of any code. Such a system would not be scalable in a strict sense. For example, if quantum memory is imperfect, and if operations cannot be performed in parallel beyond some fixed number, C will start to grow for high levels of encoding. Thus, a system that does not have a high degree of parallelism is not scalable.

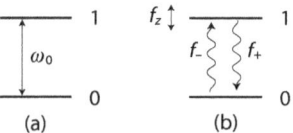

Fig. 23.2. (a) A two-level system, such as a spin, atomic energy levels, etc., separated by an energy gap with frequency ω_0. The two states, labeled 0 and 1, represent the qubit states. (b) Effects of coupling to the classical noise model. f_z shifts the resonance frequency ω_0, while $f_\pm$ flips the bit of the qubit; these latter terms change the energy of the qubit, and thus are due only to environmental couplings at the frequency ω_0.

The focus of the rest of this chapter will be on developing some intuition for the physical implementation of the necessary components for a real system: good quantum memory, high-fidelity single-qubit gates, and long-range two-qubit gates, either through transport or mediated by entanglement.

23.2 Memory

23.2.1 Practical quantum memory

Quantum bits do not appear in isolation. For example, the requirements of experimental control and measurement of the bits requires that they have some coupling to the outside world, and thus to some environment. Other, intrinsic couplings lead to additional noise. A commonly used error model assumes that qubits are effected by a depolarizing channel when not in use – single bits revert to a completely mixed state over a time $1/\gamma$. However, real-world systems can have substantially more complex behavior, which has a direct impact on the systems-level choices made for any computing architecture. We now develop a relatively simple model that captures the essential physical details of quantum memory, and illustrates the difficulties that may be overcome.

For many quantum bit implementations, there is a preferred physical axis (denoted z in what follows). For nuclear spins in NMR, the axis is defined by the eigenstates of the Zeeman interaction with an external magnetic field, with an energy difference ω_0 between the two levels (Fig. 23.2). We set $\hbar = 1$. With atoms, it could be the intrinsic difference between two levels set by the quantization of the electronic orbitals. Regardless of the system of choice, all implementations currently in use take advantage of such a preferred axis, both to allow for good quantum control, and to reduce the effects of the environment. In particular, environmental couplings that are parallel with this quantization axis, commonly associated with dephasing, behave differently than those perpendicular to the axis. The latter leads to bit-flip errors, due to coupling to the high-frequency components of the environmental degrees of freedom [GZ04].

As a model for this coupling, we start with an intrinsic Hamiltonian for the quantum bit $H_0 = \omega_0 \sigma_z/2$, where the z-axis is set by the intrinsic (physical) axis for the quantum bit. The bath couples to this quantum bit via an interaction $V = \vec{f}(t)\cdot\vec{\sigma}/2$, where $\vec{f}(t)$ is a (classical) noise vector and $\vec{\sigma}$ are the Pauli matrices. The assumption of classical noise is appropriate in the high-temperature limit, when quantum corrections to this model may be neglected. This simplistic model is useful as an illustrative tool – it does not seek to replicate the underlying physics of the bath, but rather to indicate how the intrinsic axis has a direct effect on the coherence properties of the system.

In what follows, we will further assume the noise vector to be Gaussian noise. Thus, $f_\mu(t) = \int F_\mu(\omega) e^{i\omega t} d\omega$, where each frequency component $F_\mu(\omega)$ is a an independent, Gaussian variable with zero mean. In particular, $\langle F_\mu(\omega) F_\nu(\omega')\rangle = \delta(\omega + \omega')\delta_{\mu\nu} S(\omega)$. This implies that the correlation function for the f_μ is given by $\langle f_\mu(t+\tau) f_\nu(t)\rangle = \delta_{\mu\nu} \int_{-\infty}^{\infty} S_\mu(\omega) e^{i\omega\tau} d\omega$.[1]

Moving to the interaction picture, the coupling separates into fast rotating terms (evolving at the angular velocity ω_0) and slow rotating terms. In particular,

$$V_I = f_z(t)\sigma_z/2 + \sigma_+ e^{i\omega_0 t} f_-(t) + \text{h.c.}, \tag{23.1}$$

where $f_\pm = f_x \pm i f_y$. For weak environmental couplings, the terms $f_\pm$ only contribute if they are varying on a time scale similar to $2\pi/\omega_0$, i.e., only the $F_\pm(\omega_0)$ component of the noise matters. This high-frequency component can cause spin flips, consistent with giving or receiving an energy ω_0 from the environment. Neglecting other frequency contributions to these spin-flip terms is an example of the rotating wave approximation, where energy nonconserving terms are eliminated. However, the term proportional to $f_z(t)$, which only produces a phase shift, has contributions across all frequencies. This difference in integrated power for transverse (spin-flip) terms and longitudinal (phase-error) terms is the basis for the general observation that phase errors tend to dominate over bit-flip errors in quantum memory.

We estimate the error from $f_z(t)$ by looking at the Lobschmidt echo: how much the actual evolution deviates from the desired unitary (in this case, the identity). Consider the propagator from $t = 0$ to T:

$$U = \hat{\mathcal{T}} \exp\left(-i \int_0^T f_z(t) dt \sigma_z/2\right). \tag{23.2}$$

Under the assumption of a classical noise term, the Hamiltonian commutes with itself during this time, and $U = \exp(-i\phi\sigma_z/2)$, with $\phi = \int_0^T f(t) dt$. For example, an initial state $|+\rangle = (|0\rangle + |1\rangle)/\sqrt{2}$ has a final fidelity $F = |\langle +|U(T)|+\rangle|^2 = \frac{1}{2}(1 + \langle\cos(\phi)\rangle)$, where the $\langle\,\rangle$ indicate averaging over the classical noise. For Gaussian noise, $\langle\exp(i\phi)\rangle = \exp(-\langle\phi^2\rangle/2)$. This yields

$$F = 1/2 + 1/2 \exp\left[-\int S(\omega) \frac{\sin^2(\omega T/2)}{(\omega/2)^2} d\omega\right]. \tag{23.3}$$

When the noise is low frequency, below $\omega_{\text{cutoff}} \ll 1/T$, we can expand the integrand about small ω and find a Gaussian-shaped decay with a rate given by $1/T_2^* = \sqrt{\int S(\omega) d\omega}$. The more general result for the Lobschmidt echo is given by $F = \text{Tr}[U]/2 = \langle\cos(\phi)\rangle$.

When there is a $1/\omega^2$ or greater tail in the spectral function $S(\omega)$, less generic results hold. For example, for a Lorentzian spectrum of noise, where the correlation function of f_z decays exponentially, an exponential decay of the fidelity can be found. We consider $S(\omega) = \frac{\Gamma/\pi}{(T_2^*)^2(\omega^2+\Gamma^2)}$. Then

$$\int_{-\infty}^{\infty} \frac{\Gamma/\pi}{(T_2^*)^2(\omega^2+\Gamma^2)} \frac{\sin^2(T\omega/2)}{(\omega/2)^2} d\omega = \frac{T\Gamma - (1 - e^{-T\Gamma})}{(T_2^*\Gamma)^2/2}. \tag{23.4}$$

---

[1] One can develop this model from a more microscopic perspective by considering $\vec{f}$ to correspond to the interaction picture for the environmental degrees of freedom.

When the correlation time is long compared to T ($T\Gamma \ll 1$) this behaves the same as the flat spectrum above, yielding initially quadratic decay at a rate $1/T_2^*$. However, for short-correlation time baths ($\Gamma T \gg 1$), it instead is dominated by the term proportional to T, giving decay as

$$\exp\left(-\frac{2T}{T_2^*(T_2^*\Gamma)}\right). \tag{23.5}$$

That is, the effective decay becomes exponential but the effective decay time is increased from the "bare" T_2^* by a factor $T_2^*\Gamma \gg 1$. This corresponds to so-called motional averaging: the bath changes too rapidly for phase to efficiently accumulate.

23.2.2 Improving quantum memory

In practice, several approaches can help mitigate these effects. One approach, covered in detail in the prior chapters, is to encode the quantum information in several quantum bits in such a way that the information is completely decoupled from the $f_z(t)$ degrees of freedom [LCW98]. This works well if the noise is correlated in both time and space, or if sufficiently fast quantum control is available to make the qubits sample all noise sources within its correlation time [WL02]. This approach may be used in conjunction with more standard quantum error-correction codes, and proves to be scalable [ZYZ+04].

Another approach is dynamical decoupling, such as the techniques investigated in Chapters 4, 14, and 15. This works to mitigate the effects of the lower-frequency terms in the bath. As a simple example, we consider the spin-echo sequence. Let us assume that for the quantum memory, the dominant error arises from dephasing terms ($\gamma \to 0$). If the memory storage time is T, at $T/2$ we apply a σ_x (bit-flip) to the quantum bit. We undo this bit-flip at T. Defining $\phi_1 = \int_0^T f(t)dt$ and $\phi_2 = \int_{T/2}^T f(t)dt$ and inserting the σ_x operation at $T/2$ gives the total propagator

$$U(T) = \sigma_x U' \sigma_x U = \exp[i(\phi_2 - \phi_1)\sigma_z/2]. \tag{23.6}$$

The corresponding fidelity becomes

$$F_{DD} = 1/2 + 1/2 \exp\left[-\int S(\omega)\frac{\sin^4(\omega T/4)}{(\omega/4)^2}\right]. \tag{23.7}$$

For low-frequency noise as defined above, the integral is approximated by $-T^4 \int S(\omega)\omega^2/16 d\omega \lesssim -T^4 \omega_{\text{cutoff}}^2 \int S(\omega)d\omega/16$, i.e., a rate going as the geometric mean of the original (not decoupled) dephasing rate and the high-frequency cutoff of the dephasing part of the environment.

To see the role high-frequency noise terms play, we can solve this same problem for the Lorentzian noise spectrum. A lengthy analysis finds the integrand of Eq. (23.7) to be

$$\frac{2}{(T_2^*\Gamma)^2}\left[(e^{-T\Gamma} - T\Gamma - 1) + 4(1 - e^{-T\Gamma/2})\right]. \tag{23.8}$$

For times $T \ll 1/\Gamma$, this function goes as $\Gamma T^3/6(T_2^*)^2$; decay is faster than exponential, but decreasing the decorrelation rate improves the spin echo. However, for $T \gg 1/\Gamma$, a different picture emerges: the function goes as $2T/\Gamma^2 T_2^*$. For this fast decorrelation rate regime, the spin echo is comparable to the lifetime of the uncorrected result from before. Thus, the success of

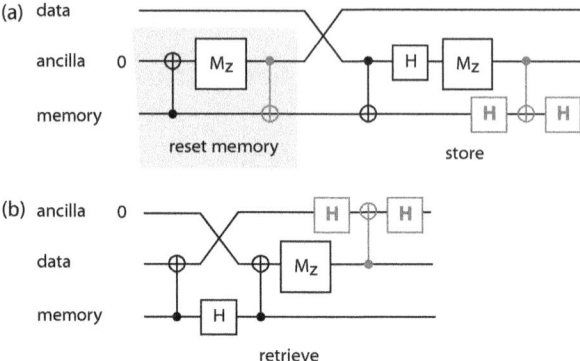

Fig. 23.3. (top) Storage of a data qubit into a "memory" qubit, with minimal number of operations on the "memory." Grayed out boxes indicate operations within the Clifford group that may be stored classically and performed after the data is retrieved. (bottom) Retrieval of data from the memory, again using a minimal number of operations.

dynamical decoupling is predicated on low-frequency noise being the dominant contribution to dephasing.

23.2.3 Hybrid architectures

One major side effect of dynamical decoupling is the introduction of noise from the bit-flip operations used. More advanced decoupling sequences use substantially more such control operations, making them susceptible to other errors, i.e., bit-flips. From an architecture perspective, a different approach may perform better.

We now investigate a so-called hybrid approach, in which one system, with very weak coupling to both the environment and to experimental controls, is used as a quantum memory, while another system (which may be easy to implement a universal set of gates upon and be straightforward to transport and measure) provides a temporary means for robust quantum control. A canonical example of this type of system is a coupled electron spin-nuclear spin system, such as nitrogen vacancy centers in diamond [DCJ+07].

To model this type of system, we can consider a Hamiltonian $H = \omega_0 \sigma_z/2 + \omega_n \tau_z/2 + g\sigma_z \tau_z/4$. We assume that we are only allowed to do Hadamard operations on the memory bit (with Pauli matrices $\vec{\tau}$), while full quantum control is possible for the processor bit (with Pauli matrices $\vec{\sigma}$). In general, the intrinsic frequency of the memory bit $\omega_n \ll \omega_0$, due to the reduced coupling of the memory bit to any perturbations. Storage is achieved using a standard quantum teleportation circuit, using controlled phase gates (the natural physical gate of the above Hamiltonian). Retrieval requires an additional ancillary bit (Fig. 23.3), and both rely on measurement. From a physical perspective the measurement provides a natural means of bridging the large energy difference between the two electron spin states and the corresponding nuclear spin states.

In general, storage and retrieval require a significant number of operations, increasing the potential errors induced by this procedure. However, the benefit arises from the substantially reduced coupling to the environment. For example, nuclear spins have an order 10^3 weaker magneton than electron spins, which corresponds to a 10^3 reduction of the environmental couplings f_z. This suggests a similar improvement in the potential coherence time of the system, consistent with experimental observations of nuclear spin coherence times from one to hundreds of seconds, compared to electron spin coherence times of tens of microseconds to tens of milliseconds.

To illustrate this potential advantage, consider how magnetic noise couples to an electron spin and a nuclear spin degree of freedom. The reduced gyromagnetic ratio leads to a reduction of the amplitude of $f_z(t)$ by around 10^3. For noise with a Lorentzian spectrum, this increases T_2^* without changing Γ; thus, we can move from the regime of $T\Gamma \ll 1$ to $T\Gamma \gg 1$ and get a reduction of noise due to motional averaging. Other improvements can arise for such hybrid memories due to the (neglected) high-frequency components $F_\pm(\omega_0)$, as the nuclear spin samples a different frequency than the electron spin.

Given the common asymmetry of quantum bits, and the impact it has upon their memory properties, we can conclude that any computing architecture will integrate some method for mitigating phase noise in memory. This can occur at the "software" level, by using an asymmetric quantum coding scheme [AP08]. It can occur at the hardware level, by building in dynamical decoupling or decoherence-free subspaces, or by using hybrid approaches. Finally, one may attempt to engineer it at the physical level by choice of qubit or by using quantum degrees of freedom that are decoupled from the environment due to their topological nature – see Part VI of this book, and also the review by Nayak and co-authors for more details [NSS+08].

23.3 Building gates

In the prior section we considered quantum bits in isolation. We now expand our analysis to include the necessary external controls for quantum gates. A naive approach is to use a constant or pulsed interaction. However, we will focus on the more robust approach of modulating the envelope of a fast oscillating control, in analogy to spin resonance techniques. Reducing this concept to the case of controlling a single quantum bit, and including how it couples to the environment of the qubit and other noise, is the focus of this section.

23.3.1 Oscillating control

We consider a slight expansion of the Hamiltonian from the prior section, in which we explicitly add to $f_x(t)$ an externally controlled component: $2\Omega(t)\cos(\omega t + \phi)$ where the variation of Ω is slow when compared to ω_0, ω. Working in a frame rotating with $\omega \sim \omega_0$, we find

$$V_I' = (\delta + f_z(t))\frac{\sigma_z}{2} + \Omega e^{i\phi}\sigma_+ + f_-(t)e^{i\omega t}\sigma_+ + \text{h.c.}, \qquad (23.9)$$

where we define the detuning $\delta \equiv \omega_0 - \omega$.

Written in this form, we see that control of Ω and its phase, ϕ, allows for rotations of the qubit about an axis in the xy plane with an angle ϕ from the x-axis. This is accomplished by setting $\delta = 0$ and turning Ω on and then off slowly over a time T such that the desired rotation angle $\theta = \int_0^T \Omega(t)dt$ is achieved.

As a side remark, we note that there is no intrinsic magnitude constraint on the size of Ω; even small Ω is sufficient to drive the system to the desired state. This is in direct contrast to a pulsed technique, in which the perturbation applied to the system does *not* vary periodically near the frequency ω_0. Such a broadband pulse has a strict power requirement for good quantum control: $\Omega \gtrsim \omega_0$, which may be hard to satisfy in practice. For most scenarios we rely instead upon modulation of a sinusoidal carrier.

23.3.2 Gate errors

Consider now the effects both of the intrinsic environmental noise (given by $f_z(t)$) and of classical errors in the control of Ω and of the carrier frequency, ω. We see that errors in Ω change the angle rotated, θ, and lead to under/over rotation. Errors in ω, or ϕ, have the same behavior as the intrinsic environmental term, $f_z(t)$. We focus first on errors in Ω. At a physical level, when the oscillating field is a laser or microwave source, errors in Ω arise due to power fluctuations of the source. For systems driven by a solid-state oscillator, these fluctuations typically exhibit a $1/\omega$ noise spectrum [W88], and thus have substantial power at low frequencies, corresponding to correlated under/over rotation errors.

Specifically, consider a rotation $U = \exp(i\theta\sigma_x/2)$ with $\theta = \int_0^\tau \Omega(t)dt$. We further write $\Omega(t) = \Omega_0 + g(t)$ and $\theta_0 = \tau\Omega_0$. The term $g(t)$ is classical noise with a $1/f$ spectrum $S_{1/f}(\omega)$. First, we evaluate the fidelity of U compared to the desired rotation using the so-called Lobschmidt echo:

$$F = \mathrm{Tr}[e^{-i\theta_0\sigma_x/2}e^{i\theta\sigma_x/2}]/2$$
$$= 1/2 + \langle e^{i\int_0^\tau g(t)dt}\rangle/2$$
$$= 1/2 + 1/2\exp\left[-\int_{-\infty}^\infty S_{1/f}(\omega)\frac{\sin^2(\omega\tau/2)}{(\omega/2)^2}\right]. \quad (23.10)$$

This is the same mathematical result as for the case of phase noise in a quantum memory. Now, however, the infidelity corresponds to a bit-flip error.

The interplay of phase noise and the rotations caused by Ω are somewhat more complex, as the two terms do not commute. For low-frequency noise we can take advantage of the long correlation time of the noise to find a useful approximate solution [TL06]. Then we set $\tau\bar{\delta} = \int_{n\tau}^{(n+1)\tau} \delta(t) + f_z(t)dt$ and similarly for $\tau\bar{\Omega}$. The unitary generated becomes $U = e^{i\tau(\bar{\delta}\sigma_z + \bar{\Omega}\sigma_x)/2}$, where the total angle rotated is $\Theta = \tau\sqrt{\bar{\delta}^2 + \bar{\Omega}^2}$. The fidelity is

$$F = \mathrm{Tr}[(\cos\theta_0\hat{1} - \sin\theta_0\sigma_x)(\cos\Theta\hat{1} + \sin\Theta\frac{\tau}{\Theta}(\bar{\delta}\sigma_z + \bar{\Omega}\sigma_x))]/2$$
$$= \cos\theta_0\cos\Theta + \sin\theta_0\sin\Theta\frac{\tau\bar{\Omega}}{\Theta}. \quad (23.11)$$

Writing $\Theta = \theta_0 + \epsilon$ and assuming $\bar{\delta} \ll \bar{\Omega}$, $\bar{\Omega} = \Omega_0$ allows for the simplification: $\epsilon \approx \tau\frac{\bar{\delta}^2}{2\bar{\Omega}} \ll 1$ and $\tau\bar{\Omega}/\Theta \approx 1 - \epsilon$. Then,

$$F \approx \cos\theta_0(\cos\theta_0\cos\epsilon - \sin\theta_0\sin\epsilon) + \sin\theta_0(\cos\theta_0\sin\epsilon + \cos\epsilon\sin\theta_0)(1-\epsilon)$$
$$= 1 - \sin^2\theta_0\epsilon + O(\epsilon^2). \quad (23.12)$$

Several remarks are in order. First, the average error (for arbitrary angle θ_0 is given by $\sim \langle\bar{\delta}^2\rangle/\Omega_0^2$, i.e., goes as the mean square value of the phase noise, and is suppressed by large Ω. This is the effect of "power broadening" – with sufficient driving force, the location of the resonance becomes less pronounced. Second, for rotations that are integer multiples of π, the first-order error disappears entirely, and only second-order errors remain. From a practical perspective, this means that the π-pulses used commonly in spin-echo decoupling sequences are particularly insensitive to phase noise.

We may be concerned about the correlations these errors have in time; this, after all, violates the premise that errors are uncorrelated in time for the same quantum bit. However, such correlations can, in fact, dramatically improve the situation, much as the correlations of phase noise allow for spin echo to successfully remove the effects of the unknown field. To this end we can use so-called composite pulse sequences, in which we use a series of rotations with the same control electronics in a time much faster than the decorrelation time of the noise. The rotations are designed such that an under/over rotation in one is compensated by the *same* under/over rotation in a subsequent element of the pulse. Details of this can be found in the review article by Vandersypen and Chuang [VC04].

The primary conclusions of this section indicate that any computing architecture benefits from high-frequency control with modulation, which needs to be added without letting in additional environmental couplings. In scenarios where one axis of qubit control, such as the phase of the modulation, is extremely robust, composite pulse sequences can be used to improve the performance of these operations. However, when the quantum control cannot take advantage of the benefits of modulating the field, it becomes difficult to evaluate approaches to appropriately removing the noise while simultaneously allowing for sufficiently broadband power to be applied to the bit for creating a complete set of gates. Finally, the effects of static inhomogeneities, e.g., that each qubit have its own resonance frequency, may make building a large-scale device difficult, as a common "clock" carrier can no longer be used for all bits.

23.4 Entangling operations and transport

So far we have outlined physical methods for understanding quantum memory and operations on single quantum bits. We now extend these ideas to pairs of quantum bits, and focus particularly on approaches for producing two-qubit gates at long distances. This is of importance for quantum error correction, in which non-local, many bit measurements (determination of the stabilizers of the code) require quantum gates between distant bits. Thus, the transport errors and transport time will have a significant impact on the scaling properties of any fault-tolerant computation.

23.4.1 Exchange operations

One approach to transporting quantum information is to implement a coherent operation, such as an exchange gate between adjacent qubits. The can be accomplished directly for two qubits with a controlled interaction $H_{\text{ex}} = \frac{J(t)}{4} \vec{\sigma}_1 \cdot \vec{\sigma}_2$. For a gate of length T, $\theta = \int_0^T J(t) dt$ is the rotation angle between the state $|01\rangle$ and $|10\rangle$, leading to a SWAP with fidelity $\sin^2(\theta/2)$. Errors from phase noise on individual quantum bits is suppressed by the exchange energy during the operation, while the gap protection provided by the intrinsic energy splitting of individual qubits is preserved as $[H_0, H_{\text{ex}}] = 0$. However, this approach is sensitive to rotation and timing errors, for which there is no easy detection or correction beyond full error correction. Furthermore, these effects only grow with distance.

Still, SWAP-type operations can be scalable [STD05]. In particular, consider the nearest neighbor SWAP between logical qubits. For an $[[n, k, 2t + 1]]$ code, we noted before that the average number of physical bits per block is $\mathcal{O}(n)$. For a D-dimensional array of quantum bits, grouping bits within a block together, the average distance to travel for an encoded nearest neighbor gate is $\mathcal{O}(n^{1/D})$, requiring $\mathcal{O}(n^{1+1/D})$ transport operations to SWAP all of the bits in one block with all of the bits in the other. In addition, the standard error-correction circuitry must be applied. With zero transport cost, this takes a minimum of $\mathcal{O}(n)$ operations (assuming efficient preparation of ancillas). This leads again to $\mathcal{O}(n^{1+1/D})$ operations (compared to the original $\mathcal{O}(n)$) when the cost of transporting the bits via SWAP gates is included.

Specifically, the cost at a given level of encoding does not increase when going to the next level of encoding, thus satisfying the scalability requirement. However, from an efficiency perspective, the overhead of SWAP-type operations is high: on average, every two-qubit operation (one error location) is replaced with $\mathcal{O}(n^{1/D})$ operations. In addition, each associated memory location is increased to many locations. This is an increase of the total number of error locations, N, by a factor $n^{1/D}$, which for the distance 3 ($t = 1$) code corresponds to a reduction of the threshold by at least $n^{2/D}$.

23.4.2 Shuttling quantum bits

For some scenarios, quantum bits have another handle, e.g., quantum information is stored in spin while charge is independent, as first suggested for ions [KMW02] and later for electrons [TED+05]. Moving bits by controlling this additional, decoupled handle, can lead to lower error in these scenarios. Immediate benefits include the insensitivity of the quantum bit degree of freedom to these operations and the potential for easy detection and correction of transport errors. Parallelism is straightforward to achieve, with quantum bits moving in sequence. On the other hand, this approach still takes a time that grows linearly in the distance. And, while keeping the quantum bit in motion may be straightforward, loading and unloading the quantum bit from the transport channel can be nontrivial.

In particular, classical charge detection would allow for the immediate determination of certain errors in transport, e.g., missing or lost bits. Other errors, such as heating of the orbital degrees of freedom due to nonadiabatic effects, can also be corrected using techniques such as sympathetic cooling. However, additional errors can still arise, particularly errors due to coupling between the quantum and spin degrees of freedom.

As an example of this approach we consider, in addition to a one-dimensional confining potential for the quantum bits, a periodic potential created and controlled, e.g., by external metallic gates. We assume three interlinked gate arrays, each producing a sinusoidal potential with period a, $2\pi/3$ out of phase with the other two. When the three arrays are modulated by $2\pi/3$ out of phase with each other at an angular frequency ν, the resulting total potential is $V(x,t) = V_0 \cos(2\pi/a(x - vt))$, a traveling wave with velocity $v = 2\pi\nu a$, as shown in Fig. 23.4.

We now estimate two errors in the transport: a nonarrival error where the bit does not make it to the destination, and errors due to a spin-orbit coupling, $p\vec{\alpha} \cdot \vec{\sigma}$. We start with $H = p^2/2m + V(x,t) + p\vec{\alpha} \cdot \vec{\sigma}$, then apply two unitary transformations. The first is a time-dependent displacement, $U_1 = \exp(ipvt)$. This removes the time dependence from V, but adds a finite background momentum term pv. The second transform is a momentum boost, $U_2 = \exp(ixvm)$, which removes the background momentum term, leading to a net rotation

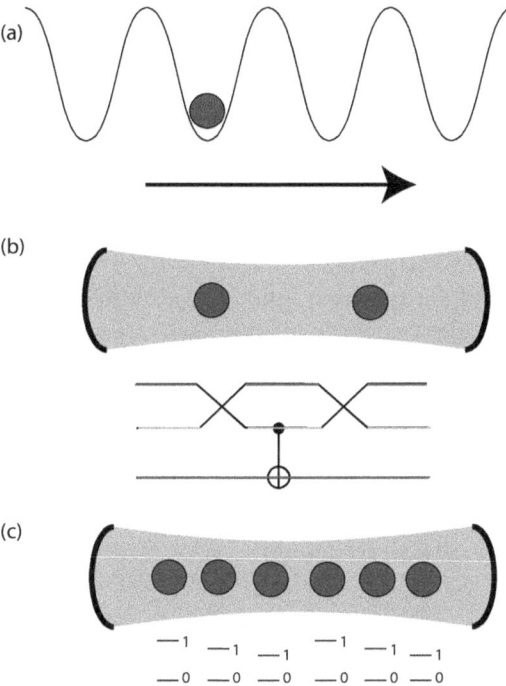

Fig. 23.4. (a) The traveling wave potential used in the text, with a qubit (circle) in transport. Many qubits, in adjacent wells, can be used to efficiently move large numbers of bits in parallel. (b) A high finesse cavity mode couples two quantum bits. A quantum gate between the two bits can be accomplished by swapping one bit into the cavity, then performing a controlled phase or NOT gate on the other bit, followed by a swap of the cavity photon state back into the control bit. (c) Schematic for multiple, simultaneous gates between three pairs of quantum bits using a single cavity mode. Each pair is tuned to a different frequency, and no pair frequencies are in resonance with the cavity frequency. Cavity photons are only excited virtually, leading to minimal interference between different pairs and reduced reliance on the quality factor of the cavity.

of the quantum bit. The final, transformed Hamiltonian is

$$H = p^2/2m + V_0 \cos(2\pi x/a) + \vec{\alpha} \cdot \vec{\sigma}(p - mv) + \text{const.} \tag{23.13}$$

This describes a nonmoving periodic potential, which for large V_0 leads to the well-known Hubbard model.

In this limit, each minimum corresponds to a localized excitation indexed by the period of the potential. Tunneling between sites occurs at a rate $J \ll V_0$, and the corresponding Hamiltonian becomes $H = -J \sum_i |i\rangle\langle i+1|$. Transforming to a Fourier basis gives $H = -J \sum_q \cos(qa)|q\rangle\langle q|$. Initial preparation of a state at the site 0 yields an overlap with that same state later in time going as $|1/N \sum_q e^{-2iJ\cos(qa)t}|^2$, indicating an error for $Jt \ll 1$ going as $(Jt)^2$. Like memory errors from low-frequency noise, this is a nonexponential error as a function of distance traveled (vt). Together, this indicates that the number of error locations at the lowest level is effectively reduced for small to intermediate distances, thus improving the efficiency of the architecture.

Spin-orbit coupling has a static contribution, due to the finite velocity of translation, but this is a known unitary rotation and can easily be corrected. More pernicious is the coupling via p to orbitally excited states with finite momentum, particularly when the qubit energy scale ω is comparable to the tunneling between sites J. Then a spin-flip combined with an orbital excitation through p can occur while conserving energy, leading to simultaneous location and spin errors. These terms can be neglected when the qubit energy is larger than J and smaller than the Bloch band energy ω. In this situation there are no states of the orbital wave function that allow for a spin-flip while still conserving energy.

One additional benefit of this approach is the intrinsic filtering it provides to local perturbations, such as position-dependent magnetic fields. Consider a field $f(x,t)$. Assuming a contact-type interaction of this field with the quantum bit, we find that the field the bit feels is given by $f_z(t) = \int |w(x)|^2 f(x,t) dx$, where $w(x)$ is the Wannier function for a single site of the periodic portion. For simplicity, we take $|w(x)|^2 \approx \exp(-x^2/(2\sigma^2))/\sqrt{2\pi\sigma^2}$. We also assume the correlation function to be $\langle f(x+\xi, t+\tau) f(x,t)\rangle = G(\xi) F(\tau)$. This yields, for example,

$$\langle f_z(t+\tau) f(t)\rangle = F(\tau) \int \frac{e^{-\frac{\xi^2}{4\sigma^2}}}{\sqrt{4\pi\sigma^2}} G(\xi) d\xi. \qquad (23.14)$$

This indicates that only the short correlation length component of $G(\xi)$ enters. Transforming this scenario into the co-moving, boosted frame corresponds to a change $F(\tau)G(\xi) \to F(\tau)G(\xi - v\tau)$. If $G(\xi)$ has a characteristic correlation length ξ_0, we find that the co-moving $f_z(t)$ has a modified correlation function; when the distance $v\tau \gg \xi_0$, the time correlations are destroyed due to the distance moved. For example, for $G(\xi) = \exp(-|x|/\xi)$, we find $F(\tau) \to F(\tau) e^{-|v\tau|/\xi_0}$. This imposes a natural low-frequency cutoff on the noise at a frequency $|v|/\xi_0$. Position decorrelation becomes time decorrelation as the quantum bit moves. For sufficiently fast motion and short-range correlated noise, this leads to an overall reduction of phase noise from low-frequency terms.

Further improvements can be obtained by using a so-called decoherence-free subspace in the co-moving frame, such as two-qubit states $|10\rangle$ and $|01\rangle$: the phase accumulated by the first is opposite to the phase accumulated by the second qubit, leading to a protected subspace for the pair. More details on this type of approach are available elsewhere [TDZ+05].

23.4.3 Cavity QED approaches

So far we have concentrated on approaches where the time and error rates for performing a long-distance quantum gate scale polynomially with the distance. A different approach is to couple quantum bits with good memory properties to so-called "flying" quantum bits. As a prototype system to explore this idea, we consider quantum bits coupled to photons in a cavity [BHW+04, IAB+99]. This type of system has been experimentally investigated both with atoms in cavities and with superconducting quantum bits coupled to a strip-line microwave resonator. In many respects, linear chains of ions use a similar formalism, where the cavity mode is a high-Q vibrational mode of the ionic crystal.

We start with the illustrative example of two quantum bits coupled to a cavity mode with bosonic creation operator a. For simplicity, we use the total angular momentum vector of the two bits, $\vec{J} = (\vec{\sigma}^{(1)} + \vec{\sigma}^{(2)})/2$ in what follows. The Jaynes–Cummings Hamiltonian, which describes the interaction between two-level systems and a bosonic field, is

$$H_{JC} = \omega_0 J_z + \omega a^\dagger a + g J_- a^\dagger + \text{h.c.}, \qquad (23.15)$$

where ω_0 are the qubits' frequencies, ω is the cavity frequency, and g is the vacuum Rabi coupling between the cavity and qubit system.

One approach to a two-qubit gate when g can be varied in time is to transfer the state of one qubit into the cavity mode, perform a gate between the cavity photon and the other qubit, then transfer the state back to the first qubit. In this approach, g would depend on the qubit index. This has the disadvantage of not allowing more than one operation at a time in the cavity. Clearly, as the number of bits coupled to the cavity increases, this cannot be scalable, as simultaneous error correction in different logical blocks cannot be accomplished in a fixed time. One simple solution is to have some qubits coupled to multiple cavities; then, the system is in principle scalable as long as the long time overhead for simultaneous operations within a logical block does not induce measurement error.

However, another approach is to use the cavity in an off-resonant mode. Then, multiple gates may be performed at once through the same cavity. It is this latter picture that we now focus on. To evaluate the performance of this system for multiple quantum bits undergoing simultaneous gates, we consider the following toy scenario. We start by pairing qubits, where the goal is to perform a gate within each pair. We assume each pair of qubits to undergo a gate is indexed by an integer $n \in [n_{min}, n_{max}]$ and both qubits in the pair have an intrinsic frequency $\omega_0 + \delta n$. The cavity is tuned to the frequency ω_0 and prepared in a vacuum state (no photons in the cavity). The interaction between the cavity and the atoms is given by the Jaynes–Cummings Hamiltonian in the interaction picture:

$$V_{JC} = \sum_n g_n e^{i\delta nt} J_+^n \hat{a} + \text{h.c.}, \qquad (23.16)$$

where $\vec{J}^n$ is the angular momentum vector associated with the pair indexed by n.

Assuming that the time scale for a single pair interaction $1/g_n$ is slow compared to $1/\delta$, we can use a Magnus expansion (section 1.2.2.3) to evaluate the unitary evolution induced by this periodic Hamiltonian. The Magnus expansion for a time-dependent Hamiltonian with a period τ is given by

$$U(\tau) = \exp[-i\tau(H_0 + H_1 + H_2 + \cdots)],$$
$$H_0 = \frac{1}{\tau} \int_0^\tau V(t) dt,$$
$$H_1 = \frac{1}{2i\tau} \int_0^\tau \int_0^t [V(t), V(t')] dt' dt,$$

with higher-order terms derived by comparing the expansion of the exponent with the time-ordered unitary $U(\tau) = \hat{\mathcal{T}} \exp(-i \int_0^\tau V(t) dt)$.

For our case, $V_{JC}(\tau) = V_{JC}$ for $\tau = 2\pi/\delta$. Then,

$$H_0 = 0, \qquad (23.17)$$
$$H_1 = \sum_n \frac{2g_n^2}{n\delta} (J_+^n J_-^n + 2a^\dagger a J_z^n). \qquad (23.18)$$

For the cavity in an initial vacuum state, the last term does not contribute to the dynamics. Additional coupling between pairs is absent: the different time dependence of each pair from the others prevents cross-talk between different pairs, while a given pair has an induced interaction going as $2g_n^2/(n\delta)$. When we set $g_n = G\sqrt{n}$, a constant interaction can be achieved for all pairs, allowing for all operations to occur for the same interaction time $\sim \pi\delta/G^2$.

Two sources of error become apparent. First, the virtual elimination of the photon from the problem is not exact. Higher-order corrections will make it probable that the state with at least one photon is occupied. For cavities with a finite decay rate κ, this will lead to dephasing of the quantum bits. From a perturbative expansion, we can estimate the probability of having at least one photon excited to be $\sim \sum_n g_n^2/(n\delta)^2 = G^2/\delta^2 \sum_n 1/n$. The sum, running from n_{min} to n_{max}, is bounded for a finite number of pairs, and grows only logarithmically with the number of pairs involved: $\sim \log[\frac{n_{max}}{n_{min}-1}]$. We still require $n_{min} \gg 1$ for the logarithmic approximation to be valid.

The logarithmic growth of the cavity photonic number with increasing number of qubit pairs suggests that many quantum bits can use the same cavity as a simultaneous, high-bandwidth bus. We can estimate the error if the photon decays at a rate κ by setting the time for the gate to $\pi\delta/G^2$, and find $\epsilon_{cav} \sim \pi\frac{\kappa}{\delta}\log[\frac{n_{max}}{n_{min}-1}]$. This constrains $\delta \gg \kappa$; for simplicity, we set $\delta = \kappa/\eta$, where η is an estimate of the error probability during the gate.

A second constraint arises from the requirement that $g_n \ll \delta$, i.e., that all pair–pair couplings are negligible. This implies $\kappa/\eta = G\sqrt{n_{\max}}/\eta'$, where $\eta' \sim \eta \ll 1$ is a small parameter estimating the cross-coupling effects between different bits. For qubits with an intrinsic decay rate γ (i.e., the memory error rate), we require $\pi\gamma\kappa/G^2\eta = \pi\gamma\sqrt{n_{max}}/G\eta' \ll 1$. Thus we need G sufficiently large: $G \gg \gamma\sqrt{n_{max}}/\eta'$.

Taken together, and setting all error probabilities to be η, we have $g_{n_{max}} \sim \gamma n_{max}/\eta$ and $\delta \sim \kappa/\eta$. In practice, the first constraints may be difficult to achieve; $g_{n_{max}}$ is limited by the maximum coupling to the cavity that a qubit can achieve $= \sqrt{\alpha}\omega$, where $\alpha \approx 1/137$ is the fine structure constant. We are ignoring the constraint that $n_{max}\delta$ (the largest detuning used) is still in a physical regime, i.e., not too low.

From the perspective of building a quantum computing device, we have illustrated that even nearest neighbor operations are, in principle, sufficient for scalable computation. However, using transport, which does not necessarily incur any qubit errors beyond memory error, proves to be beneficial for systems where the qubit degree of freedom is connected to another handle, such as charge, while being large decoupled. Another approach is to use photons in high-Q cavities, where it may be possible to do many, simultaneous gates through the same intermediary. However, this approach is limited by the decay rate of photons in the cavity and the maximum coupling one can achieve between a qubit and the cavity. As such, it eventually must be supplemented either by using multiple cavities or by some other method.

23.5 Quantum networking

As a final approach to the complete architecture, we consider is using a nondeterministic method for entanglement generation followed by teleportation-based gates, as outlined by Gottesman and co-workers [GC99]. This idea bypasses much of the systems-level difficulties encountered in building up physical structures, such as the periodic potentials for moving bits or the high-Q cavities for coupling them deterministically. However, it incurs an overhead, both in the

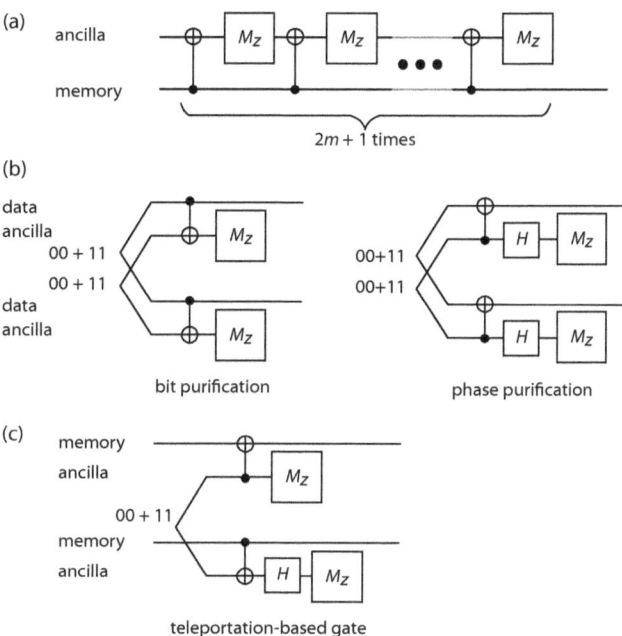

Fig. 23.5. (a) A fault-tolerant measurement circuit when local operations (i.e., CNOT) are good but measurement is faulty. Majority outcomes of the $2m+1$ measurements give the correct result, suppressing measurement errors ϵ to order ϵ^{m+1}. (b) Phase- and bit-purification circuits for improving an initial, low-fidelity pair into a high-fidelity pair. These can be interpreted as faulty measurements of the ZZ and XX stabilizers. (c) Teleportation-based CNOT circuit, with the correction to the Pauli frame not shown.

necessity of purifying the entanglement before use, and in the time to complete a nonlocal two-qubit gate.

The essence of the idea is to use one qubit to connect one small register of qubits to another such register. This qubit is, in turn, coupled to a memory qubit in the register to perform a teleportation-based gate. Additional qubits within the register can be used to purify the entanglement before it is used [DB03]. More specifically, consider a small quantum register, with one qubit coupled to a lossy channel (i.e., an atom coupled to an optical fiber). The one qubit connected to the channel can be entangled with another such qubit some distance away through an entanglement generation scheme, of which many varieties exist [CTS+05]. Such schemes generate entanglement with some finite probability p, but the entanglement event is heralded, i.e., a classical detector indicates the success of the protocol. Within the register, four other quantum bits are necessary. One is the quantum memory. The other three are used for fault-tolerant measurement, and a two-stage entanglement purification protocol.

Robust measurement is a simple bit-verification procedure, necessary when direct measurement of qubits that are not the one coupled to the lossy channel is difficult. This is outlined in Fig. 23.5; in essence, several local, high-fidelity operations are used to improve the fidelity of the measurement process. This leads to an exponential suppression of the errors in measurement with a linear overhead in local errors. Details of this, and other components, can be found in [JTS+07].

The next step of the procedure is purification. Say we wish to improve the fidelity of a Bell pair between two distant registers, such as $|00\rangle + |11\rangle$. This state is the $+1$ eigenstate of the stabilizers $ZZ = \sigma_z^1 \sigma_z^2$ and $XX = \sigma_x^1 \sigma_x^2$. If we make another, low-fidelity Bell pair, we can make a Shor-style measurement of the stabilizer ZZ (for phase error purification), which teleports bit errors onto the pair that we wish to measure [NC00], as shown in Fig. 23.5. However, if we first purify the low-fidelity Bell pair against bit errors (using another such pair), this improved pair can be fault tolerant with respect to bit errors.

In this way we can improve the entanglement between distant registers with an overhead in time (it takes many entangled pair generation events to create one high-fidelity pair) and in the number of local operations. However, such an approach dramatically simplifies the difficulty of building a quantum computing device: the engineering problem becomes one of mass producing few-qubit registers with extremely high fidelity operation, while the interconnection between these registers may be relatively low fidelity. This idea, first recognized by Oskin and collaborators [OCC02], may prove a powerful simplifying step in producing large-scale quantum computing devices.

Chapter 24

Error correction in quantum communication

Mark Wilde

24.1 Introduction

Many of the chapters in this book have developed the theory of quantum error correction (QEC) for the specific purpose of building a fault-tolerant, reliable quantum computer. Specifically, Chapter 5 shows how the theory of QEC is a core component in the theory of fault-tolerant quantum computation. Fault-tolerant quantum computation has received much attention from the theoretical quantum information community because it is a key requirement for a scalable quantum computer. The most celebrated tasks that one could perform with a fault-tolerant quantum computer are factoring a prime number [S94b, S97], searching a database quickly [G96b, G97b], or simulating a quantum system efficiently [F82, L96].

One can exploit quantum phenomena to enhance communication as well. The field of *quantum communication* encompasses any aspect of quantum theory that is exploited for communicative purposes. In this chapter, we discuss four main topics involving QEC in quantum communication:

(i) entanglement distillation,
(ii) a security proof of quantum key expansion[1],
(iii) continuous-variable systems, and
(iv) implementation of QEC for the purpose of quantum communication.

We first comment briefly on each of the above topics. *Entanglement distillation* converts a set of noisy ebits to a smaller set of noiseless ones. It is useful to have a procedure for entanglement distillation because noiseless entanglement is the core resource in several quantum communication protocols. We show how the techniques from QEC apply to entanglement distillation. A *security proof* for quantum key expansion is a mathematical proof that shows that a

---

[1] Quantum key expansion is similar to quantum key distribution. In quantum key expansion, the two parties share a secret key before the protocol begins. The quantum key expansion protocol expands this key so that the resulting shared secret key is larger than the original shared secret key.

Quantum Error Correction, ed. Daniel A. Lidar and Todd A. Brun. Published by Cambridge University Press.
© Cambridge University Press 2013.

quantum key expansion protocol is secure against an arbitrary attack by an eavesdropper. It is perhaps surprising at first glance that QEC plays a role in a security proof, but the methods that we highlight later on should make this connection clear. A *continuous-variable* system is one that has continuous degrees of freedom, as opposed to the discrete degrees of freedom of a qubit. We highlight some QEC routines for continuous-variable systems because these systems may be useful for the communication of quantum information. Our final topic is the *implementation* of QEC. Several experimental groups have implemented the routines. We highlight the progress for optical systems, because these systems are the likely candidates for quantum communication.

24.2 Entanglement distillation

The goal of entanglement distillation is similar to the goal of QEC. In QEC, we use n physical qubits to encode k information qubits. We can recover the quantum information in the k qubits provided that the channel is not too noisy. The *rate* of such an $[[n,k]]$ QEC code is k/n. In entanglement distillation, we begin with n physical ebits where we know that a noisy channel has probabilistically corrupted half of each ebit. An entanglement distillation protocol is a procedure to extract k noiseless ebits from the original n noisy ebits. The *yield* of an $[[n,k]]$ entanglement distillation protocol is k/n.

Entanglement distillation is a useful protocol on its own, but we also have another purpose for introducing it. It turns out that entanglement distillation is relevant for proving the security of a quantum key expansion protocol. We get to this application of entanglement distillation in the next section.

In the current section, we outline several methods for performing entanglement distillation. We first outline a simple example of an entanglement distillation protocol that uses the three-qubit bit-flip code to distill entanglement established with a noisy bit-flip channel. We generalize this example to show how to convert an arbitrary $[[n,k]]$ stabilizer code into an $[[n,k]]$ entanglement distillation protocol. It is then straightforward to show how to convert an arbitrary $[[n,k;c]]$ entanglement-assisted code into an $[[n,k;c]]$ *entanglement breeding protocol*. The final technical development of this section is a discussion of the detailed operation of a CSS entanglement breeding protocol. We wrap up with some historical notes about entanglement distillation.

24.2.1 Three-qubit bit-flip code example

Recall that the logical codewords for the three-qubit bit-flip code are

$$|0_L\rangle \equiv |000\rangle, \quad |1_L\rangle \equiv |111\rangle.$$

Its stabilizer operators are

$$Z \otimes Z \otimes I, \quad I \otimes Z \otimes Z,$$

and its logical $\bar{X}$ and $\bar{Z}$ operators are respectively

$$\bar{X} \equiv X \otimes X \otimes X, \quad \bar{Z} \equiv Z \otimes Z \otimes Z.$$

The set of correctable errors are $X \otimes I \otimes I$, $I \otimes X \otimes I$, and $I \otimes I \otimes X$.

The entanglement distillation protocol consists of several steps (illustrated in Fig. 24.1):

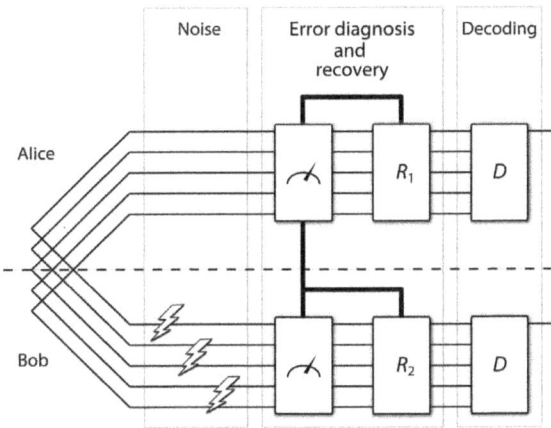

Fig. 24.1. The essential steps in an entanglement distillation protocol. Alice first prepares Bell states and sends half of each state over a noisy channel. Alice and Bob both perform measurements. Alice sends the results of her measurements to Bob. Bob corrects the error from the channel. They both restore their state to a set of logical ebits. In the above figure, we write Alice's restoration as a conditional unitary R_1 and we write Bob's correction and restoration as a conditional unitary R_2. They finish by decoding to extract some number of noiseless ebits.

Preparation. Alice locally prepares three Bell states $|\Phi^+\rangle^{\otimes 3}$ in her laboratory. She sends half of each Bell state over a noisy bit-flip channel and Bob receives the three transmitted qubits. Suppose first that no noise acts on the state. The state that they share is then

$$|\Phi^+\rangle^{A_1 B_1} |\Phi^+\rangle^{A_2 B_2} |\Phi^+\rangle^{A_3 B_3},$$

where Alice and Bob possess the respective systems labeled A_i and B_i for $i = 1, 2, 3$. In what follows, we abbreviate Alice's and Bob's respective systems as $A^3 \equiv A_1 A_2 A_3$ and $B^3 \equiv B_1 B_2 B_3$. We can rewrite their shared state as follows:

$$|\Phi\rangle^{A^3 B^3} \equiv \frac{1}{\sqrt{8}} \sum_{i=0}^{7} |i\rangle^{A^3} |i\rangle^{B^3},$$

where i is the decimal representation of the binary numbers $0, \ldots, 7$. Let us introduce the notation for an "encoded" ebit in terms of the logical codewords as follows:

$$|\Phi_L^+\rangle^{A^3 B^3} \equiv \frac{|0_L\rangle^{A^3} |0_L\rangle^{B^3} + |1_L\rangle^{A^3} |1_L\rangle^{B^3}}{\sqrt{2}}.$$

The "ebit stabilizer operators" for the above logical ebit are as follows:

$$\bar{X}^{A^3} \otimes \bar{X}^{B^3}, \quad \bar{Z}^{A^3} \otimes \bar{Z}^{B^3}.$$

We can rewrite Alice and Bob's state $|\Phi\rangle^{A^3 B^3}$ in terms of encoded ebits and the error operators of the three-qubit code:

$$\begin{aligned} |\Phi\rangle^{A^3 B^3} &= |\Phi_L^+\rangle^{A^3 B^3} + X^{A_1} X^{B_1} |\Phi_L^+\rangle^{A^3 B^3} \\ &\quad + X^{A_2} X^{B_2} |\Phi_L^+\rangle^{A^3 B^3} + X^{A_3} X^{B_3} |\Phi_L^+\rangle^{A^3 B^3}. \end{aligned}$$

Noise. Now we consider the possibility of noise from the channel. We assume that the noisy channel is a bit-flip channel whose error probability is p. The probability of no error is $(1-p)^3$, the probability of a single-qubit error is $3p(1-p)^2$, the probability of a double-qubit error is $3p^2(1-p)$, and the probability of a triple-qubit error is p^3. We further assume that p is small so that the probability of double- and triple-qubit errors is negligible. Thus, the errors that we are concerned with are as follows:

$$I^{B^3},\quad X^{B_1},\quad X^{B_2},\quad X^{B_3}.$$

Suppose the noisy channel applies one of the above errors to Bob's qubits. Let E^{B_i} denote the random error that the noisy channel applies on Bob's qubits. The state is now as follows:

$$(I^{A^3}\otimes E^{B_i})|\Phi\rangle^{A^3 B^3}.$$

Measurement. Alice performs a measurement of the stabilizer operators $Z^{A_1}Z^{A_2}$ and $Z^{A_2}Z^{A_3}$. This measurement randomly collapses the state $|\Phi\rangle^{A^3 B^3}$ into one of the following four states: $E^{B_i}|\Phi_L^+\rangle^{A^3 B^3}$, $X^{A_1}E^{B_i}X^{B_1}|\Phi_L^+\rangle^{A^3 B^3}$, $X^{A_2}E^{B_i}X^{B_2}|\Phi_L^+\rangle^{A^3 B^3}$, $X^{A_3}E^{B_i}X^{B_3}|\Phi_L^+\rangle^{A^3 B^3}$. Alice records the results of measurements as two bits $a_1 a_2$. These bits identify to which of the above states the state has collapsed and therefore identify which operator in the set $\{I^{B^3}, X^{B_1}, X^{B_2}, X^{B_3}\}$ acts on Bob's side, but they do not identify the error E^{B_i}. Bob then measures the same stabilizer operators and records the results as two bits $b_1 b_2$. These bits identify which of the operators in the set $\{E^{B_i}, E^{B_i}X^{B_1}, E^{B_i}X^{B_2}, E^{B_i}X^{B_3}\}$ act on his side, but they do not identify the error E^{B_i}.

Classical communication. Notice that it is possible for Bob to identify the error E^{B_i} if he has Alice's bits $a_1 a_2$. So, Alice sends him her bits $a_1 a_2$.

Correction. Bob computes the syndrome bits $(a_1\oplus b_1)(a_2\oplus b_2)$. Bob can identify the error E^{B_i} based on these syndrome bits and he then corrects the error E^{B_i}.

Restoration. Alice and Bob are left with one of the states $|\Phi_L^+\rangle^{A^3 B^3}$, $X^{A_1}X^{B_1}|\Phi_L^+\rangle^{A^3 B^3}$, $X^{A_2}X^{B_2}|\Phi_L^+\rangle^{A^3 B^3}$, $X^{A_3}X^{B_3}|\Phi_L^+\rangle^{A^3 B^3}$, identified by the bits $a_1 a_2$. They know which of the above states they possess because they both now have bits $a_1 a_2$. Therefore, they restore the state to the standard encoded ebit state $|\Phi_L^+\rangle^{A^3 B^3}$.

Decoding. They both perform the decoding circuit for the three-qubit code and are left with the state $|\Phi^+\rangle^{A_1 B_1}|0000\rangle^{A_2 A_3 B_2 B_3}$. Thus, they distill one noiseless ebit from the original noisy three ebits and the yield of this entanglement distillation protocol is $1/3$.

24.2.2 Stabilizer entanglement distillation

We generalize the above example to perform an $[[n, k]]$ entanglement distillation protocol with an arbitrary $[[n, k]]$ stabilizer code. We assume that the stabilizer code corrects the errors in an error set $\mathscr{E}\subset G^n$. The protocol has steps similar to the above example:

Preparation. The sender Alice first prepares n Bell states $|\Phi^+\rangle^{\otimes n}$ locally. She sends the second qubit of each Bell state over a noisy quantum channel to the receiver Bob. Without noise, the state shared between Alice and Bob is $(|\Phi^+\rangle^{AB})^{\otimes n}$. Let $|\Phi\rangle^{A^n B^n}$ be the state $(|\Phi^+\rangle^{AB})^{\otimes n}$

rearranged so that all of Alice's qubits are on the "left" and all of Bob's qubits are on the "right":

$$|\Phi\rangle^{A^n B^n} \equiv \frac{1}{\sqrt{2^n}} \sum_{i=0}^{2^n-1} |i\rangle^{A^n} |i\rangle^{B^n},$$

where i is the decimal representation of all the binary numbers between 0 and $2^n - 1$.

Noise. The noisy channel applies a Pauli error E^{B^n} from the error set $\mathscr{E}$ to the n qubits sent over the channel. The sender and receiver then share n noisy ebits in the state

$$(I^{A^n} \otimes E^{B^n})|\Phi\rangle^{A^n B^n},$$

where the identity I^{A^n} acts on Alice's qubits and $E^{B^n} \in \mathscr{E}$ acts on Bob's qubits.

Measurement. Suppose the stabilizer S for an $[[n, k, d]]$ quantum error-correcting code has generators $g_1, \ldots, g_{n-k}$. The distillation procedure begins with Alice measuring the $n - k$ generators in S and obtaining a bit vector a from these measurements. Let $\{\Pi_i\}$ be the set of 2^{n-k} projectors that project onto the 2^{n-k} orthogonal joint eigenspaces of the generators in S. The measurement projects $|\Phi\rangle^{A^n B^n}$ randomly onto one of the i subspaces. Each Π_i commutes with the noisy operator E^{B^n} on Bob's side so that

$$\Pi_i^{A^n} E^{B^n} |\Phi\rangle^{A^n B^n} = E^{B^n} \Pi_i^{A^n} |\Phi\rangle^{A^n B^n}. \tag{24.1}$$

The following important identity[2] holds for an arbitrary matrix M and the maximally entangled state $|\Phi\rangle^{A^n B^n}$:

$$(M^{A^n} \otimes I^{B^n})|\Phi\rangle^{A^n B^n} = (I^{A^n} \otimes (M^{B^n})^T)|\Phi\rangle^{A^n B^n}. \tag{24.2}$$

Then the state in (24.1) is equal to the following state:

$$(I^{A^n} \otimes E^{B^n})(\Pi_i^{A^n} \otimes I^{B^n})|\Phi\rangle^{A^n B^n}$$

$$= (I^{A^n} \otimes E^{B^n})(\Pi_i^{A^n} \otimes I^{B^n})(\Pi_i^{A^n} \otimes I^{B^n})|\Phi\rangle^{A^n B^n} \tag{24.3}$$

$$= (I^{A^n} \otimes E^{B^n})(\Pi_i^{A^n} \otimes I^{B^n})(I^{A^n} \otimes (\Pi_i^{B^n})^T)|\Phi\rangle^{A^n B^n} \tag{24.4}$$

$$= (I^{A^n} \otimes E^{B^n})(\Pi_i^{A^n} \otimes (\Pi_i^{B^n})^T)|\Phi\rangle^{A^n B^n}, \tag{24.5}$$

where the first equality follows because $\Pi_i^{A^n}$ is a projector and the second equality follows by applying the identity in (24.2). Therefore, each of Alice's projectors $\Pi_i^{A^n}$ projects Bob's qubits

---

[2] We give a simple proof of the identity for a maximally entangled qudit state $|\Phi^+\rangle^{AB} = \frac{1}{\sqrt{d}} \sum_{i'} |i'\rangle^A |i'\rangle^B$. Consider applying the operator $M^A \otimes I^B$ to $|\Phi^+\rangle^{AB}$:

$$(M^A \otimes I^B)|\Phi^+\rangle^{AB} = \left(\sum_{i,j} m_{ij} |i\rangle \langle j|^A \otimes I^B\right) \frac{1}{\sqrt{d}} \sum_{i'} |i'\rangle^A |i'\rangle^B$$

$$= \frac{1}{\sqrt{d}} \sum_{i,j,i'} m_{ij} |i\rangle^A \langle j|i'\rangle \otimes |i'\rangle^B = \frac{1}{\sqrt{d}} \sum_{i,i'} m_{ii'} |i\rangle^A \otimes |i'\rangle^B.$$

Consider applying the operator $I^A \otimes (M^T)^B$ to $|\Phi^+\rangle^{AB}$:

$$(I^A \otimes (M^T)^B)|\Phi^+\rangle^{AB} = \left(I^A \otimes \sum_{i,j} m_{ji} |i\rangle \langle j|^B\right) \frac{1}{\sqrt{d}} \sum_{i'} |i'\rangle^A |i'\rangle^B$$

$$= \frac{1}{\sqrt{d}} \sum_{i,j,i'} m_{ji} |i'\rangle^A \otimes |i\rangle^B \langle j|i'\rangle = \frac{1}{\sqrt{d}} \sum_{i,i'} m_{i'i} |i'\rangle^A \otimes |i\rangle^B.$$

Finally, inspection of the two resulting states reveals that they are equal.

onto a subspace $\left(\Pi_i^{B^n}\right)^T$ corresponding to Alice's projected subspace $\Pi_i^{A^n}$. Alice's bit vector a identifies into which subspace the measurement projected the state. Bob measures the generators in S and obtains a bit vector b containing his measurement results. The bit vector b identifies to which subspace the error E^{B^n} takes the state $\Pi_i^{A^n}|\Phi\rangle^{A^n B^n}$, but it cannot alone identify what the error E^{B^n} is. Notice that the bit-wise XOR of the bit vector a and the bit vector b does identify the error E^{B^n}.

Classical communication. Alice sends her bit vector a to Bob.

Correction. Bob combines his bit vector b with Alice's bit vector a to determine a syndrome $e = a \oplus b$ for the error E^{B^n}. He performs a correction operation on his qubits to reverse the error.

Restoration. Both Alice and Bob restore their qubits to the simultaneous +1-eigenspace of the following ebit stabilizer operators:

$$\bar{X}_1^A \bar{X}_1^B, \; \bar{Z}_1^A \bar{Z}_1^B, \; \ldots, \; \bar{X}_k^A \bar{X}_k^B, \; \bar{Z}_k^A \bar{Z}_k^B, \tag{24.6}$$

where the operators

$$\bar{X}_1, \; \bar{Z}_1, \; \ldots, \; \bar{X}_k, \; \bar{Z}_k,$$

are the logical operators for the stabilizer code S, and the labels A and B indicate whether the operators act on Alice's or Bob's respective qubits. Alice and Bob now possess k logical ebits – the operators in (24.6) stabilize a set of k logical ebits.

Decoding. Alice and Bob both perform the decoding unitary corresponding to stabilizer S to convert their k logical ebits into k physical ebits.

24.2.3 The breeding protocol

We now show how to produce an $[[n, k; c]]$ entanglement-assisted entanglement distillation protocol from an $[[n, k; c]]$ entanglement-assisted stabilizer code. The notation for the entanglement-assisted entanglement distillation protocol means that Alice and Bob share c noiseless ebits before the protocol begins. The protocol produces $k > c$ noiseless ebits by exploiting the initial c noiseless ebits. For this reason, an entanglement-assisted entanglement distillation protocol is called a *breeding protocol for entanglement distillation*. In this chapter, we simply call it the *breeding protocol*.

The steps in the breeding protocol are essentially the same as those in the stabilizer entanglement distillation protocol in the previous section, with a few notable differences that we discuss briefly. The "Preparation" and "Noise" steps are the same, with the crucial exception that Alice and Bob possess c noiseless ebits in addition to their n noisy ebits. After the "Noise" step, the state of the noisy and noiseless ebits is

$$\left(I^{A^{n+c}} \otimes \left(E^{B^n} \otimes I^{B^c}\right)\right) |\Phi\rangle^{A^{n+c} B^{n+c}}, \tag{24.7}$$

where $I^{A^{n+c}}$ is the $2^{n+c} \times 2^{n+c}$ identity matrix acting on Alice's qubits and the noisy Pauli operator $\left(E^{B^n} \otimes I^{B^c}\right)$ affects Bob's first n qubits only. Thus the last c ebits are noiseless, and Alice and Bob have to correct for errors on the first n ebits only. In the "Measurement" step, Alice and Bob measure the generators in an $[[n, k; c]]$ entanglement-assisted stabilizer code. Each generator spans over $n + c$ qubits where the last c qubits are noiseless. The remaining steps proceed in a similar fashion.

The breeding protocol produces $k + c$ ebits if it arises from an $[[n, k; c]]$ entanglement-assisted stabilizer code (note that it distills k ebits from the code in addition to the c initial ebits with which the code began). But it consumes c initial noiseless ebits as a catalyst. Therefore, the net yield of the breeding protocol is k/n.

24.2.4 The CSS breeding protocol

Suppose now that we would like to use an $[n, k_1, d]$ classical linear code C_1 and an $[n, k_2, d]$ classical linear code C_2 for entanglement distillation. The next subsection details how to construct an $[[n, k_1 + k_2 - n + c, d; c]]$ CSS entanglement-assisted code. We then show how to formulate a *CSS breeding protocol* for this case.

We go into detail in the subsection because the notation is useful later on in providing a rigorous security proof for quantum key expansion. We express everything with symplectic binary vectors and matrices. We first establish some notation for CSS entanglement-assisted codes and then give the CSS breeding protocol.

24.2.4.1 The structure of a CSS entanglement-assisted code We now show how to produce an $[[n, k_1 + k_2 - n + c, d; c]]$ entanglement-assisted quantum code from an $[n, k_1, d]$ classical linear code C_1 and an $[n, k_2, d]$ classical linear code C_2. Suppose C_1 has an $n \times (n - k_1)$-dimensional check matrix H_1 and C_2 has an $n \times (n - k_2)$-dimensional check matrix H_2. Suppose that we have performed the Gram–Schmidt symplectic orthogonalization procedure from Chapter 7 on check matrices H_1 and H_2. We can then define a quantum check matrix H as follows:

$$H \equiv \begin{bmatrix} 0 & | & H_1 \\ H_2 & | & 0 \end{bmatrix}.$$

In general, the above quantum check matrix does not yet form a valid quantum code because its rows may not all be orthogonal with respect to the symplectic product. The matrix $H_1 H_2^T$ contains the symplectic orthogonality relations of the quantum check matrix H. The matrix $H_1 H_2^T$ has the following *standard* symplectic orthogonality relations after performing the Gram–Schmidt symplectic orthogonalization procedure:

$$H_1 H_2^T = \begin{bmatrix} \overbrace{I}^{c} & \overbrace{0}^{n-k_2-c} \\ 0 & 0 \end{bmatrix} \begin{matrix} \} \, c \\ \} \, n - k_1 - c \end{matrix}.$$

Let us divide each of the matrices H_i where $i = 1, 2$ into submatrices $H_{i,a}$ and $H_{i,b}$ as follows:

$$H_i = \begin{bmatrix} \overbrace{H_{i,a}}^{n} \\ H_{i,b} \end{bmatrix} \begin{matrix} \} \, c \\ \} \, n - k_i - c \end{matrix}.$$

We can then write the quantum check matrix H equivalently as follows:

$$H = \begin{bmatrix} 0 & | & H_{1,a} \\ 0 & | & H_{1,b} \\ H_{2,a} & | & 0 \\ H_{2,b} & | & 0 \end{bmatrix}.$$

A simple way to make the quantum check matrix H be orthogonal with respect to the symplectic product is to augment it as follows to form a new matrix H':

$$H' \equiv \begin{bmatrix} \overbrace{0}^{n} & \overbrace{0}^{c} & \overbrace{H_{1,a}}^{n} & \overbrace{I}^{c} \\ 0 & 0 & H_{1,b} & 0 \\ H_{2,a} & I & 0 & 0 \\ H_{2,b} & 0 & 0 & 0 \end{bmatrix}.$$

The n entries on the left in both submatrices act on Alice's qubits and the c entries on the right in both submatrices act on Bob's qubits. The augmented matrix H' has the property that it is orthogonal with respect to the symplectic product and thus forms a valid quantum code. The rows in $H_{1,a}$ and $H_{1,b}$ correspond to halves of c ebits that Alice possesses and the rows in $H_{1,b}$ and $H_{2,b}$ correspond to $n - k_1 - c$ and $n - k_2 - c$ respective ancilla qubits that Alice uses for encoding. We know that the following formula should hold for any entanglement-assisted code:

channel qubits = # information qubits + # ancilla qubits + # ebits.

Thus, the code has room for the following number of information qubits:

$$n - (n - k_1 - c) - (n - k_2 - c) - c = k_1 + k_2 - n + c,$$

because the number of channel qubits is n, the number of ancilla qubits is $(n - k_1 - c) + (n - k_2 - c)$, and the number of ebits is c.

The augmentation procedure produces an $[[n, k_1 + k_2 - n + c, d; c]]$ entanglement-assisted quantum code. We introduce the following shorthand:

$$H'_1 \equiv \begin{bmatrix} H_{1,a} & I \\ H_{1,b} & 0 \end{bmatrix}, \quad H'_2 \equiv \begin{bmatrix} H_{2,a} & I \\ H_{2,b} & 0 \end{bmatrix},$$

so that we can write the augmented check matrix H' as follows:

$$H' = \begin{bmatrix} 0 & H'_1 \\ H'_2 & 0 \end{bmatrix}.$$

The above entanglement-assisted code encodes $k_1 + k_2 - n + c$ information qubits. Therefore, the code has $2(k_1 + k_2 - n + c)$ logical operators, each corresponding to the logical X and Z operator for an information qubit. We can write the logical operators for this entanglement-assisted code as follows:

$$\begin{bmatrix} \overbrace{0}^{n+c} & \overbrace{E_1}^{n+c} \\ E_2 & 0 \end{bmatrix} \begin{matrix} \} \, k_1 + k_2 - n + c \\ \} \, k_1 + k_2 - n + c \end{matrix}.$$

The matrix E_1 represents the logical Z operators and the matrix E_2 represents the logical X operators. The logical operators act only on the first n qubits. Therefore, the entries in the last c columns of both E_1 and E_2 are equal to zero.

It is also possible to find a $(k_2 + c) \times (n + c)$-dimensional matrix F_1 whose rows are linearly independent of the rows in both H'_1 and E_1. In addition, the operators corresponding to the entries in F_1 act only on Alice's n qubits so that the last c columns of F_1 are zero. Similarly, it is possible to find a $(k_1 + c) \times (n + c)$-dimensional matrix F_2 whose rows are linearly independent of the rows in both H'_2 and E_2 and whose last c columns are zero. These matrices F_1 and F_2 in a sense correspond to a basis for the error operators that affect the encoded qubits.

We can arrange all of our matrices into a full-rank matrix of dimension $2(n+c) \times 2(n+c)$:

$$\begin{bmatrix} 0 & H_1' \\ 0 & E_1 \\ 0 & F_1 \\ H_2' & 0 \\ E_2 & 0 \\ F_2 & 0 \end{bmatrix}.$$

The following symplectic orthogonality relations hold for the above matrices:

$$\begin{array}{lll} H_1' H_2'^T = 0 & H_1' E_2^T = 0 & H_1' F_2^T = I \\ E_1 H_2'^T = 0 & E_1 E_2^T = I & E_1 F_2^T = 0 \\ F_1 H_2'^T = I & F_1 E_2^T = 0 & F_1 F_2^T = K \end{array} \tag{24.8}$$

where K is some arbitrary binary matrix (we do not care so much for the symplectic orthogonality relations of the matrices F_1 and F_2 because they correspond to "error operators"). The above structure of a CSS entanglement-assisted code is important for our development of the CSS breeding protocol below and the development in Section 24.3 of a security proof for quantum key expansion.

24.2.4.2 Steps of the CSS breeding protocol The CSS breeding protocol consists of several steps:

Preparation. Alice and Bob initially share c noiseless ebits in the state $|\Phi^+\rangle^{\otimes c}$. Alice prepares n Bell states locally and sends half of each state to Bob. For now, let us suppose that they then share $n+c$ noiseless ebits (we consider the effects of noise later). The stabilizer matrix for their shared entanglement is as follows:

$$\begin{bmatrix} 0 & 0 & I^A & I^B \\ I^A & I^B & 0 & 0 \end{bmatrix},$$

where each identity matrix has dimension $(n+c) \times (n+c)$ and the labels A and B refer to Alice and Bob's qubits respectively. We can use row operations to obtain an equivalent stabilizer for the ebits:

$$\begin{bmatrix} 0 & 0 & H_1'^A & H_1'^B \\ 0 & 0 & E_1^A & E_1^B \\ 0 & 0 & F_1^A & F_1^B \\ H_2'^A & H_2'^B & 0 & 0 \\ E_2^A & E_2^B & 0 & 0 \\ F_2^A & F_2^B & 0 & 0 \end{bmatrix},$$

where the above matrices for a CSS entanglement-assisted code are defined in Section 24.2.4.1. The above two representations of the stabilizer are equivalent because the following matrices are both full rank:

$$\begin{bmatrix} H_1' \\ E_1 \\ F_1 \end{bmatrix}, \quad \begin{bmatrix} H_2' \\ E_2 \\ F_2 \end{bmatrix}.$$

Noise. Suppose now that noise does affect Bob's halves of the ebits that Alice sent over the channel. Alice and Bob then possess the following ensemble:

$$\left\{p_i, |\Phi\rangle_i^{A^{n+c}B^{n+c}}\right\}. \tag{24.9}$$

In the above, p_i is the probability that the state is $|\Phi\rangle_i^{A^{n+c}B^{n+c}}$,

$$|\Phi\rangle_i^{A^{n+c}B^{n+c}} \equiv (I^{A^{n+c}} \otimes (E_i^{B^n} \otimes I^{B^c}))|\Phi\rangle^{A^{n+c}B^{n+c}}, \tag{24.10}$$

and $E_i^{B^n} \in G^n$ is a Pauli error acting on Bob's side that results from the noisy channel. For simplicity, we assume that this error is in the correctable set of the entanglement-assisted code. Alice and Bob correct for a particular error set in order to distill noiseless ebits.

Measurement. Alice measures the operators corresponding to the rows of matrices $H_1'^A$ and $H_2'^A$. These measurements randomly collapse the noisy $n+c$ ebits into one of $2^{2n-k_1-k_2}$ orthogonal subspaces. Alice's measurement yields $2n - k_1 - k_2$ bits, and uniquely identifies in which subspace the logical ebits are. Let a denote the bit vector of her measurement results. Bob then measures the operators corresponding to matrices $H_1'^B$ and $H_2'^B$, and obtains a bit vector b with his measurement results. Bob's bit vector b matches Alice's bit vector a if there are no errors in the channel, and differs from Alice's if there are errors. The syndrome vector $e \equiv a \oplus b$ determines which errors have occurred on Bob's halves of the ebits.

Classical communication. Alice sends Bob her bit vector a.

Correction. Bob computes the syndrome e. Based on this syndrome, he performs an error recovery operation that corrects both the bit errors and the phase errors that the channel introduces.

Restoration. Their ebits are now in a subspace corresponding to the bit vector a. They each perform local operations based on the bit vector a to rotate their ebits to the standard logical ebits. The stabilizer matrix for the resulting logical ebits is as follows:

$$\begin{bmatrix} 0 & 0 & E_1^A & E_1^B \\ E_2^A & E_2^B & 0 & 0 \end{bmatrix}. \tag{24.11}$$

Decoding. Alice and Bob then each perform the decoding circuit that transforms the logical ebits to physical ebits so that the following matrix stabilizes them:

$$\begin{bmatrix} \overbrace{0^A}^{m} & \overbrace{0^A}^{n+c-m} & \overbrace{0^B}^{m} & \overbrace{0^B}^{n+c-m} & \overbrace{I^A}^{m} & \overbrace{0^A}^{n+c-m} & \overbrace{I^B}^{m} & \overbrace{0^B}^{n+c-m} \\ I^A & 0^A & I^B & 0^B & 0^A & 0^A & 0^B & 0^B \end{bmatrix} \begin{matrix} \} m \\ \} m \end{matrix}. \tag{24.12}$$

Thus they obtain $m = k_1 + k_2 - n + c$ ebits.

24.3 Quantum key expansion

In this section, we introduce the notion of a quantum key expansion protocol. A quantum key expansion protocol assumes that Alice and Bob begin with some amount of a secret key and their aim is to produce a larger key or expand the size of the original secret key.

Our main goal in this section is to provide a proof of the security of quantum key expansion by appealing to the CSS breeding protocol. This method of proof remains the most intuitive method, in our opinion, to a quantum information theorist, and highlights the connection between quantum error correction, entanglement distillation, the security of an entanglement-based quantum key expansion protocol, and the security of a "prepare-and-measure" quantum key expansion protocol. It may seem like a roundabout way to arrive at a security proof, but it is a beautiful path, and should provide the reader with a unified understanding of many of the elements of quantum information theory.

We structure this section as follows. We first review the prepare-and-measure BB84 protocol for quantum key distribution and discuss classical postprocessing procedures that improve the security of the protocol. Section 24.3.2 then discusses an entanglement-based version of this protocol. We finally discuss the use of the CSS breeding protocol to prove the security of the BB84 quantum key expansion protocol. The last section wraps up with some historical notes on quantum key expansion and security proofs.

24.3.1 The BB84 protocol

The first and perhaps most well-known example of exploiting uniquely quantum phenomena in communication is the Bennett–Brassard 1984 (BB84) protocol for creating a shared secret key between two parties.[3] The intuitive, physical arguments for the security of the BB84 protocol rest on fundamentally "quantum" principles: the Heisenberg uncertainty principle and the no-cloning theorem. These physical arguments are actually not sufficient to prove the security of the protocol against an arbitrary attack by an eavesdropper. But one can prove that it is possible to distill a secret key when the BB84 protocol has the aid of classical postprocessing.

The essence of the BB84 protocol is the use of nonorthogonal quantum states. Specifically, the protocol employs the following four nonorthogonal states:

$$|0\rangle, \quad |1\rangle, \quad |+\rangle \equiv \frac{|0\rangle + |1\rangle}{\sqrt{2}}, \quad |-\rangle \equiv \frac{|0\rangle - |1\rangle}{\sqrt{2}},$$

where the states $|0\rangle$ and $|1\rangle$ are the respective $+1$ and -1 eigenstates of the Pauli Z operator and the states $|+\rangle$ and $|-\rangle$ are the respective $+1$ and -1 eigenstates of the Pauli X operator. The protocol uses the above four states randomly in creating a secret key.

The first intuitive physical argument for security of the BB84 protocol is the no-cloning theorem. Suppose that Alice produces one of the above states at random and sends it over a quantum channel to Bob. Suppose further that an eavesdropper Eve is trying to figure out which state Alice produced. The no-cloning theorem forbids Eve copying the state in general if she has no knowledge of the state that Alice produced. This no-cloning property holds because the above states are nonorthogonal. The other physical argument for security is the Heisenberg uncertainty principle. If Eve tries to measure the random state in either the X or Z basis, then she inevitably disturbs the state some of the time because the states are eigenstates of noncommuting observables. If Alice produces a large number of random states, Alice and Bob can determine if Eve is performing measurements because of the resulting disruption to some of the states. Alice and Bob then terminate the protocol if they determine that Eve is listening in.

The BB84 protocol assumes that a public authenticated classical channel connects Alice to Bob in addition to the quantum channel. The meaning of "authenticated" is that Eve can listen in on the communication but cannot tamper with the transmission of the classical channel. It

---
[3] Wiesner introduced the notion of "quantum money" that first highlighted the connection between quantum mechanics and secrecy [W83]. Bennett and Brassard later exploited this connection to form the [BB84] protocol.

actually requires some secret key to authenticate a channel. In this sense, a quantum key distribution protocol is really a *quantum key expansion* protocol because the key that quantum key distribution produces is larger than the one consumed for authentication of the classical channel (but keep in mind that we also use the term quantum key expansion to refer to a protocol that expands any amount of secret key that is left over after the authentication process consumes some secret key). It is possible to authenticate the classical channel using, e.g., a Wegman–Carter authentication scheme; for details, see [WC79, WC81].

Here are the basic steps of the BB84 protocol:

Preparation. Alice generates two random bit strings r and s each having length $(4+\delta)n$, where δ is some small positive number and n is large. She prepares the following state for transmission over the quantum channel:

$$|\psi\rangle \equiv \bigotimes_{i=1}^{(4+\delta)n} |\psi_{r_i s_i}\rangle, \qquad (24.13)$$

where

$$|\psi_{00}\rangle \equiv |0\rangle, \quad |\psi_{10}\rangle \equiv |1\rangle, \quad |\psi_{01}\rangle \equiv |+\rangle, \quad |\psi_{11}\rangle \equiv |-\rangle.$$

The bits r are the random bits intended to become part of the key and the bits s are the "basis" bits that encode r in either the Z or X basis.

Noise. Alice sends the state $|\psi\rangle$ over the quantum channel to Bob. We assume that Eve interacts with the transmission in some way before the state reaches Bob. Also, the quantum channel may corrupt the state with noise. We model both actions as a noisy quantum operation $\mathcal{E}$.

Measurement. Bob receives the state $\mathcal{E}(|\psi\rangle\langle\psi|)$, generates a random basis bit string s' of length $(4+\delta)n$, measures each received qubit in the X or Z basis according to the basis bit string s', and stores his measurement results in a bit string r' of length $(4+\delta)n$.

Classical communication. Alice and Bob exchange their respective basis bit strings s and s' over the authenticated classical channel.

Sifting. They discard their respective bits in r and r' where s and s' disagree; i.e., Alice and Bob respectively discard all bits r_i and r'_i where $s_i \oplus s'_i = 1$. This step of the protocol is called the *sifting* step. According to the law of large numbers, there should be around $2n$ bits left after sifting, and we call these bits the *raw key*. (It is possible to choose δ large enough so that the probability of not having $2n$ bits left is exponentially small.) If there are not $2n$ bits left, they abort the protocol.

Channel estimation. Alice chooses a random subset consisting of n bits, called *channel estimation bits*, from the raw key and notifies Bob which ones she chose. They use this random subset to perform *channel estimation*, i.e., to determine how much noise the channel introduces and how much information Eve may have gained about the raw key. After channel estimation, n bits remain for use in producing a shared secret key. Let x and x' denote Alice and Bob's respective remaining n bits. Alice and Bob perform classical postprocessing on x and x'. If more than an acceptable number of errors have occurred, Alice and Bob abort the protocol.

Correction. *Information reconciliation* corrects the errors on x and x'. It is simply an error-correction protocol that Alice and Bob employ to correct for bit errors. Alice and Bob can use any classical error-correcting code.

Privacy amplification. This procedure produces m bits from the remaining n bits where $m < n$. The aim of the privacy amplification protocol is to decouple Eve from the picture by making her knowledge independent of the m secret key bits.

The protocol that we outlined above is actually a quantum key distribution protocol. A simple modification turns it into a quantum key expansion protocol where Alice and Bob expand a shared secret key. In key expansion, Alice and Bob include their shared secret key in the information reconciliation and privacy amplifications steps. They then produce a secret key that is larger than the original key, so that the net key gain is positive.

Notice that the names of some steps in the above protocol are similar to the names of steps in the entanglement distillation protocols outlined in the previous section. This similarity is not a coincidence. There are strong ties between quantum key expansion and the breeding protocol. In fact, it is possible to prove the security of a quantum key expansion protocol by appealing to the ability of the CSS breeding protocol to produce high-fidelity ebits. We make these connections explicit in the next section.

One might think at this point that the above procedure should be able to produce a secret key, but it is not clear yet that it does. We have left out many portions that require more detail to establish a firm security proof. Some questions remain. Is it guaranteed that the n channel estimation bits give a good indication of the noise and Eve's information about the remaining n bits? How do Alice and Bob perform privacy amplification? What is a good security criterion? We answer these questions below and establish a rigorous security proof in the upcoming sections.

The channel estimation step assumes that Alice and Bob have generated a large amount of raw key (n is large). This step allows Alice and Bob to put a high-probability upper bound on the number of errors that the channel introduces. Specifically, we can show that the probability of obtaining less than $\epsilon_1 n$ errors on the channel estimation bits and more than $(\epsilon_1 + \epsilon_2) n$ errors on the raw key is less than $\exp\left\{-O\left(\epsilon_2^2 n\right)\right\}$ for large n and for any $\epsilon_1, \epsilon_2 > 0$. The method for showing this result involves large deviation techniques.

There are several methods for performing privacy amplification. One method is for Alice and Bob to use a two-universal hash function. A two-universal hash function is a random function from a domain $\mathscr{X}$ to a range $\mathscr{Y}$, i.e., it is a random variable whose realizations are in the set of functions with domain $\mathscr{X}$ and range $\mathscr{Y}$. It has the following "two-universal" property for any distinct $x, x' \in \mathscr{X}$:

$$\Pr\{f(x) = f(x')\} \leq |\mathscr{Y}|,$$

where the probability is with respect to the random function f. Alice and Bob employ privacy amplification with a two-universal hash function in the following way. Suppose that Alice would like to share a private message k consisting of m bits with Bob, starting from a message x consisting of n bits that are known to Bob and partially known to Eve, where $n > m$. Let x_E denote Eve's bit string that is partially correlated with x. Alice begins by randomly selecting a realization of the random function f. She applies the realization of f to x and publicly announces which function she chose. Bob can then determine the private message k because he knows f and x. Eve, however, has only partial information on x. Even if she knows the hash function f, there is only a small probability that applying the hash function to x_E gives the secret message k. In fact, it is possible for Alice and Bob to make Eve's information about their key be exponentially small in $n - m$.

The criterion for a quantum key expansion protocol to be ϵ-secure is as follows:

$$\|\rho^{KE} - \pi^K \otimes \rho^E\| \leq \epsilon \tag{24.14}$$

where ρ^{KE} represents a state shared between the key system K and Eve's system E, π^K is the maximally mixed state on the key system K, and ρ^E is some state on Eve's system E. The criterion states that the shared state ρ^{KE} should be close to a state for which the key is uniformly distributed and for which Eve's knowledge in ρ^E is completely independent of the key. This criterion is known as the *composable* definition of security. A key that satisfies the above criterion is composable because the generated key is secure in any application, whether directly after generation of the key or at some later point. This security holds because any subsequent action of an attacker is either appending an additional quantum system, performing a CPTP map, or both. The key is still ϵ-secure because the trace-norm distance does not increase under these operations.

24.3.2 Entanglement-based quantum key expansion

Suppose that Alice and Bob share an ebit $|\Phi^+\rangle^{AB}$ where

$$|\Phi^+\rangle^{AB} \equiv \frac{|00\rangle^{AB} + |11\rangle^{AB}}{\sqrt{2}}.$$

The state is pure and this purity implies that no third party such as an eavesdropper Eve has any correlation with it; i.e., whatever state the third party may have is completely independent of the ebit that Alice and Bob share. We can write the global state of Alice, Bob, and a third party Eve as follows:

$$\left(\Phi^+\right)^{AB} \otimes \rho^E,$$

where ρ^E is some density operator for Eve that is independent of the shared ebit. Suppose now that Alice and Bob each measure this state in the Z basis. They either both obtain the outcome "0" with probability $1/2$ or they both obtain the outcome "1" with probability $1/2$. After their measurement, the global state is as follows:

$$\frac{1}{2}\left(|0\rangle\langle 0|^A \otimes |0\rangle\langle 0|^B + |1\rangle\langle 1|^A \otimes |1\rangle\langle 1|^B\right) \otimes \rho^E.$$

This procedure is a simple way for Alice and Bob to generate a secret key. The resulting key automatically satisfies the security criterion in (24.14) with $\epsilon = 0$ because the original state is pure and thus decoupled from any third party. Also, the resulting key has a uniform distribution. Eve's best strategy then is simply to guess randomly at the value of the key.

The fact that we can generate a secret key from entanglement immediately leads to a protocol for quantum key distribution called the *entanglement-based protocol*. The main requirement of this protocol is that Alice and Bob possess entangled states that are close to being pure, so that Eve shares little to no correlation with their entanglement. The steps of this entanglement-based protocol are similar to the steps for the BB84 protocol, with the difference that Alice and Bob operate on entangled states. We examine an idealized version of this protocol in more detail in the next section.

24.3.3 The breeding protocol and quantum key expansion

The first thing we notice is that we can convert the CSS breeding protocol in Section 24.2.4 into a secret key distillation protocol. Let t be the number of errors that the CSS breeding protocol corrects. We assume for now that the noise introduces no more than t errors into Alice's transmitted qubits. The result of the CSS breeding protocol is to produce m noiseless ebits in the state $|\Phi^+\rangle^{\otimes m}$. The state $|\Phi^+\rangle^{\otimes m}$ is pure and thus is decoupled from Eve or any other third party, as argued in the previous section. Alice can measure the Z operator on her m qubits and Bob can measure the Z operator on his m qubits. The result is a common secret key k of m bits.

The above protocol for secret key distillation is secure against attacks of the eavesdropper that introduce no more than t errors. But the practical problem with it is that it requires both Alice and Bob to possess a fault-tolerant quantum computer. It also requires them to possess pure shared entanglement before the protocol begins. These requirements are difficult to achieve in practice and it would be better to simplify the physical requirements for such a protocol.

We now show how to simplify the CSS breeding protocol so that it requires Alice and Bob to possess a shared secret key and to perform single-qubit operations instead of respectively possessing shared entanglement and fault-tolerant quantum computers. First, notice that there is no need to perform the decoding circuit in the "Decoding" step where Alice and Bob decode and then measure the operators in (24.12). They can just directly measure the operators in (24.11) to obtain the secret key k. We can thus eliminate the "Decoding" step. Notice also that correction of phase errors does not change the value of the secret key k because correction of phase errors commutes with the measurement of the operators corresponding to the rows of E_1^A and E_1^B in (24.11). Thus, it is not necessary to correct phase errors in the "Correction" and "Restoration" steps. As a result, it is not necessary for Alice and Bob to measure the operators $H_2'^A$ and $H_2'^B$ that give the syndromes for the phase errors. Also, in the CSS breeding protocol, Bob measures $H_1'^B$, obtains a syndrome, performs bit error correction based on this syndrome, and measures the operator E_1^B to obtain k. He gets the same result if he measures $H_1'^B$, obtains a syndrome, measures E_1^B, and corrects the measurement result to obtain k.

Thus, we can modify the last few steps of the CSS breeding protocol as follows to give the following modified secret key distillation protocol:

Measurement. Alice measures the operators corresponding to the rows of $H_1'^A$. Let a denote the bit vector of her measurement results. Bob then measures the operators corresponding to the rows of $H_1'^B$ and obtains a bit vector b with his measurement results.

Classical communication. Alice sends Bob her bit vector a.

Correction. Bob computes the syndrome e. Based on this syndrome, he performs an error recovery operation that corrects the bit errors.

Restoration. Their ebits are now in a subspace corresponding to the bit vector a (without concern for the effect of the phase errors). They each perform local operations based on the syndrome a to rotate their ebits to the +1-eigenspace of the following operators:

$$\begin{bmatrix} 0 & 0 \mid E_1^A & E_1^B \end{bmatrix}. \qquad (24.15)$$

Secret key generation. Alice measures the operators E_1^A, obtains the secret key k, Bob measures the operators E_1^B, and obtains the bit vector k'. Bob uses Alice's vector a and his vector b to correct the bit vector k' so that it is equal to the secret key k.

We have simplified the secret key distillation protocol, but it still requires that Alice and Bob possess entanglement before it begins and it still requires them both to have a fault-tolerant quantum computer. We now simplify the secret key distillation protocol even more. Notice that Alice can perform her measurements of $H_1'^A$ and E_1^A before the "Noise" step because these measurements commute with Bob's measurements in the "Measurement" step and with the noise in the "Noise" step. She can also measure the operators F_1^A before the "Noise" step as well without affecting the protocol. All of these measurements are actually equivalent to Alice just measuring the Z operator on her individual qubits because the three matrices $H_1'^A$, E_1^A, and F_1^A form a full-rank matrix. Let u denote the string that results from her individual measurements. The following relations hold for the bit vectors a and k:

$$a = H_1'^A u, \qquad k = E_1^A u.$$

Similarly, Bob can also measure the operators in F_1^B at the end of the protocol without changing the results because this measurement commutes with the measurements of $H_1'^B$ and E_1^B. These measurements are equivalent to Bob just measuring the Z operator on his individual qubits because the three matrices $H_1'^B$, E_1^B, and F_1^B form a full-rank matrix. Let v denote the string that results from his individual measurements. The following relations hold for Bob's bit vectors b and k':

$$b = H_1'^B v, \qquad k' = E_1^B v.$$

In the above reduction, we showed that Alice can measure her qubits as soon as she has them. Bob has the last c qubits at the beginning of the protocol at the same time that Alice does. The last c qubits for both Alice and Bob are halves of pure, noiseless ebits. So each of them performing individual measurements on their last c qubits is mathematically equivalent to them possessing a uniformly distributed shared secret key κ of c bits. Also, Alice measuring half of each ebit in the Z basis and sending the other half through the channel is mathematically equivalent to preparing one of $|0\rangle$ or $|1\rangle$ at random and sending it through the channel. Thus we can reduce the above secret key distillation protocol to an $[n, k_1 + k_2 - n, d; c]$ quantum key expansion protocol that uses only single-qubit operations and requires Alice and Bob to share a secret key before communication begins. We assume again for now that Eve introduces no more than t errors in the transmitted qubits. The steps are as follows:

Preparation. Alice and Bob share a secret key κ of c bits. Alice generates a random string x of n bits. The concatenation of these two bit strings x and κ form Alice's bit string u:

$$u \equiv \begin{bmatrix} x \\ \kappa \end{bmatrix}. \tag{24.16}$$

Noise. Alice prepares the quantum state $|x\rangle$ encoded in the Z basis and sends it over the noisy channel.

Measurement. Bob receives the noisy state and performs measurements of the Z operator on each qubit. He obtains a bit string x'. The concatenation of this bit string x' and the secret key κ form Bob's bit string v:

$$v \equiv \begin{bmatrix} x' \\ \kappa \end{bmatrix}. \tag{24.17}$$

Alice computes $a = H_1' u$ and Bob computes $b = H_1' v$.

Classical communication. Alice sends her bit string a to Bob.

Correction. Bob combines his bit string b with Alice's bit string a to determine a syndrome for the channel errors.

Privacy amplification. Alice computes the secret key $k = E_1 u$ and Bob computes $k' = E_1 v$. Bob uses the syndrome to correct his bit string k' so that it is equal to the secret key k.

With the above quantum key expansion protocol, we have dramatically simplified the practical requirements for expanding a secret key, while having the nice property that the protocol is formally equivalent in a strict mathematical sense to the CSS breeding protocol. The quantum key expansion protocol requires Alice and Bob to possess a secret key and act with single-qubit operations. These requirements are much simpler to realize in practice than shared noiseless entanglement and fault-tolerant quantum computing. Also, if the CSS breeding protocol is secure against attacks from an eavesdropper, then the quantum key expansion protocol is also secure because of their formal mathematical equivalence.

Note that we have made the unrealistic assumption that Eve introduces no more than t errors on the channel. We can modify the quantum key expansion protocol to include a channel estimation step. Using channel estimation, Alice and Bob can determine with high probability the actual number of errors that Eve introduces, and then choose classical codes that correct this number of errors.

We finally give the steps of a secure quantum key expansion protocol:

Preparation. Alice generates two random strings r and s, each consisting of $(4 + \delta) n$ bits. She prepares the state $|\psi\rangle$ in (24.13) for transmission over the quantum channel.

Noise. Alice sends $|\psi\rangle$ to Bob over the quantum channel.

Measurement. Bob receives the noisy state and measures each qubit randomly in the X or Z basis.

Classical communication. Alice uses the public authenticated classical channel to announce her basis bit string s.

Sifting. Bob discards any results where he measured in a basis different from the one that Alice prepared in. With high probability, there are at least $2n$ bits left after discarding. Alice randomly chooses a subset consisting of n bits to be her string x and the corresponding bits belonging to Bob form his bit string x'. Alice and Bob use the remaining bits for channel estimation.

Channel estimation. Alice publicly announces her channel estimation bits and they perform channel estimation. If they estimate that the channel introduces more than t errors, Alice and Bob abort the protocol.

Correction. Alice computes $a = H_1' u$ where u is defined as in (24.16). Bob computes $b = H_1' v$ where v is defined as in (24.17).

Classical communication. Alice sends her bit string a to Bob.

Privacy amplification. Alice computes the secret key $k = E_1 u$ and Bob computes $k' = E_1 v$. Bob combines his bit string b with Alice's bit string a to correct his bit string k' so that it is equal to k.

Notice that Alice and Bob only exploit their shared secret key in the "Correction" and "Privacy amplification" steps. Therefore, the secret key is uncompromised and still available for use if the quantum key expansion protocol aborts in any of the previous steps.

In the secure protocol, Alice randomly chooses and announces the channel estimation bits after she has transmitted all of her qubits over the channel and Bob has received all of her qubits.

Alice merely performs a random permutation on her bits and announces the random permutation in the first "Classical communication" step. This random permutation symmetrizes the action of Eve's attacks on the qubits and effectively reduces her actions and the channel noise to be independent and identically distributed (IID). Before Bob receives all of the qubits, Eve has time to do any interaction that she would like, but after he receives, she cannot do anything else. So, by the time channel estimation comes around, Eve is out of the picture.

It is possible to use codes that do not necessarily correct a fixed number t of errors but instead perform well on a channel with IID noise. In fact, the symmetrization of the noise reduces it to be identical to two binary symmetric channels producing X and Z errors at some rate q.

The advantage of making the connection with entanglement-assisted CSS codes through the CSS breeding protocol is that Alice and Bob can choose any two classical codes with respective check matrices H_1 and H_2 for use in producing a secure quantum key expansion protocol. In particular, this result implies that they can use modern coding techniques that are efficiently decodable in polynomial time. These codes perform well on a binary symmetric channel with error rate q and can essentially achieve the Shannon capacity $1 - H(q)$ where $H(q)$ is the binary entropy function. If Alice and Bob estimate a rate q for both the Pauli X and Z noise, then the two classical codes that Alice and Bob choose for their quantum key expansion protocol should perform well on a binary symmetric channel with error parameter slightly greater than q. The key rate that they attain with this technique is approximately

$$\frac{m-c}{n} \approx 2(1 - H(q)) - 1 = 1 - 2H(q).$$

The above key rate is positive whenever the error rate q is less than the noise threshold of 0.11.

24.4 Continuous-variable quantum error correction

Many of the chapters in this book discuss methods for protecting quantum information in the form of qubits. These techniques fall under the category of *discrete-variable quantum information processing* because the basis states for a qubit form a discrete basis. These methods are useful, but it turns out that there is another way that we can encode quantum information.

Continuous-variable quantum information is an alternative to discrete-variable quantum information. It is called thus because a continuous parameter spans the basis states for a continuous quantum variable. An example of a continuous quantum variable is the position of a particle. We can describe a particle's position quantum-mechanically with a wave function $\psi(x)$ and write it in Dirac notation as follows:

$$\int dx \ \psi(x) |x\rangle,$$

where the states $|x\rangle$ are position eigenstates. The direct quantum-optical analogy of position and momentum of a particle is the respective position quadrature and momentum quadrature of an optical mode. The typical implementation for continuous-variable quantum information protocols is with quantum optics because it is somewhat straightforward to manipulate the position quadrature and momentum quadrature of an optical mode. Thus, in this chapter, we concern ourselves with the application in quantum optics.

Continuous-variable systems may be useful for quantum communication because their most prominent physical implementation is with optics. One of the current advantages of continuous-variable quantum information is that the corresponding experiments are less difficult to perform than discrete-variable optical ones. These experiments do not require single-photon sources and detectors and usually require linear optical devices only – offline squeezers, passive optical devices, feedforward control, conditional modulation, and homodyne measurements. An offline squeezer is a device that prepares a standard squeezed state for use in an optical circuit and an online squeezer is a nonlinear optical device used in an optical circuit.

Error correction is necessary for a continuous-variable quantum device to operate properly. In this section, we briefly review some of the mathematics behind continuous-variable QEC. The form of the codes is analogous to that of stabilizer codes for discrete variables. These continuous-variable error correction schemes should prove useful as a testbed for theoretical ideas even if the final form of a quantum computer is not a continuous-variable optical device.

24.4.1 Symplectic algebra for continuous variables

We first review some mathematical preliminaries. Chapter 2 of this book demonstrates that the Pauli group is important for the theory of discrete-variable QEC. The continuous-variable analog of the Pauli group is the Heisenberg–Weyl group, and thus, it is not surprising that many of its properties are important in the theory of continuous-variable stabilizer codes.

We first relate the n-mode phase-free Heisenberg–Weyl group $([\mathscr{W}^n], *)$ to the additive group $(\mathbb{R}^{2n}, +)$. Let $X(x)$ be a single-mode position translation by x and let $Z(p)$ be a single-mode momentum kick by p where

$$\begin{aligned} X(x) &\equiv \exp\{-i\pi x \hat{p}\}, \\ Z(p) &\equiv \exp\{i\pi p\, \hat{x}\}, \end{aligned} \qquad (24.18)$$

and $\hat{x}$ and $\hat{p}$ are the position-quadrature and momentum-quadrature operators respectively. The canonical commutation relations are $[\hat{x}, \hat{p}] = i$. Let $\mathscr{W}$ denote the single-mode Heisenberg–Weyl group:

$$\mathscr{W} \equiv \{X(x)Z(p) \mid x, p \in \mathbb{R}\}. \qquad (24.19)$$

Let $\mathscr{W}^n$ be the set of all n-mode operators of the form $\mathbf{A} \equiv A_1 \otimes \cdots \otimes A_n$ where $A_j \in \mathscr{W}\ \ \forall j \in \{1, \ldots, n\}$. Define the equivalence class

$$[\mathbf{A}] \equiv \{\beta \mathbf{A} \mid \beta \in \mathbb{C}, |\beta| = 1\} \qquad (24.20)$$

with representative operator having $\beta = 1$. The above equivalence class is useful because global phases are not relevant in the formulation of our codes. The group operation $*$ for the above equivalence class is as follows:

$$\begin{aligned}[] [\mathbf{A}] * [\mathbf{B}] &\equiv [A_1] * [B_1] \otimes \cdots \otimes [A_n] * [B_n] \\ &= [A_1 B_1] \otimes \cdots \otimes [A_n B_n] = [\mathbf{AB}]. \end{aligned} \qquad (24.21)$$

The equivalence class $[\mathscr{W}^n] = \{[\mathbf{A}] : \mathbf{A} \in \mathscr{W}^n\}$ forms a commutative group $([\mathscr{W}^n], *)$. We name $([\mathscr{W}^n], *)$ the *phase-free Heisenberg–Weyl group*.

Consider the $2n$-dimensional real vector space $\mathbb{R}^{2n}$. It forms the commutative group $(\mathbb{R}^{2n}, +)$ with operation $+$ defined as vector addition. We employ the notation $\mathbf{u} = (\mathbf{p}|\mathbf{x})$,

$\mathbf{v} = (\mathbf{p}'|\mathbf{x}')$ to represent any vectors $\mathbf{u}, \mathbf{v} \in \mathbb{R}^{2n}$ respectively. Each vector $\mathbf{p}$ and $\mathbf{x}$ has elements $(p_1, \ldots, p_n)$ and $(x_1, \ldots, x_n)$ respectively with similar representations for $\mathbf{p}'$ and $\mathbf{x}'$. The *symplectic product* $\odot$ of $\mathbf{u}$ and $\mathbf{v}$ is

$$\mathbf{u} \odot \mathbf{v} \equiv \mathbf{p} \cdot \mathbf{x}' - \mathbf{x} \cdot \mathbf{p}' = \sum_{i=1}^{n} p_i x_i' - x_i p_i', \qquad (24.22)$$

where $\cdot$ is the standard inner product. Define a map $\mathbf{D} : \mathbb{R}^{2n} \to \mathscr{W}^n$ as follows:

$$\mathbf{D}(\mathbf{u}) \equiv \exp\left\{ i\sqrt{\pi} \sum_{i=1}^{n} (p_i \hat{x}_i - x_i \hat{p}_i) \right\}. \qquad (24.23)$$

The operator resulting from the above map $\mathbf{D}$ is the same as the displacement operator from quantum optics. Let

$$\begin{aligned} \mathbf{X}(\mathbf{x}) &\equiv X(x_1) \otimes \cdots \otimes X(x_n), \\ \mathbf{Z}(\mathbf{p}) &\equiv Z(p_1) \otimes \cdots \otimes Z(p_n), \end{aligned} \qquad (24.24)$$

so that $\mathbf{D}(\mathbf{p}|\mathbf{x})$ and $\mathbf{Z}(\mathbf{p})\mathbf{X}(\mathbf{x})$ belong to the same equivalence class:

$$[\mathbf{D}(\mathbf{p}|\mathbf{x})] = [\mathbf{Z}(\mathbf{p})\mathbf{X}(\mathbf{x})]. \qquad (24.25)$$

The map $[\mathbf{D}] : \mathbb{R}^{2n} \to [\mathscr{W}^n]$ is an isomorphism

$$[\mathbf{D}(\mathbf{u} + \mathbf{v})] = [\mathbf{D}(\mathbf{u})][\mathbf{D}(\mathbf{v})], \qquad (24.26)$$

where $\mathbf{u}, \mathbf{v} \in \mathbb{R}^{2n}$. We use the Baker–Campbell–Hausdorff theorem $e^A e^B = e^B e^A e^{[A,B]}$ and the symplectic product to capture the commutation relations of any operators $\mathbf{D}(\mathbf{u})$ and $\mathbf{D}(\mathbf{v})$:

$$\mathbf{D}(\mathbf{u})\mathbf{D}(\mathbf{v}) = \exp\{i\pi(\mathbf{u} \odot \mathbf{v})\}\mathbf{D}(\mathbf{v})\mathbf{D}(\mathbf{u}). \qquad (24.27)$$

The operators $\mathbf{D}(\mathbf{u})$ and $\mathbf{D}(\mathbf{v})$ commute if $\mathbf{u} \odot \mathbf{v} = 2n$ and anticommute if $\mathbf{u} \odot \mathbf{v} = 2n + 1$ for any $n \in \mathbb{Z}$. The set of canonical operators $\hat{x}_i, \hat{p}_i$ for all $i \in \{1, \ldots, n\}$ have the canonical commutation relations

$$[\hat{x}_i, \hat{x}_j] = 0, \quad [\hat{p}_i, \hat{p}_j] = 0, \quad [\hat{x}_i, \hat{p}_j] = i\delta_{ij}.$$

Let $\mathscr{T}^n$ be the set of all linear combinations of the canonical operators:

$$\mathscr{T}^n \equiv \left\{ \sum_{i=1}^{n} \alpha_i \hat{x}_i + \beta_i \hat{p}_i : \forall i, \ \alpha_i, \beta_i \in \mathbb{R} \right\}. \qquad (24.28)$$

Define the map $\mathbf{M} : \mathbb{R}^{2n} \to \mathscr{T}^n$ as

$$\mathbf{M}(\mathbf{u}) \equiv \mathbf{u} \cdot \hat{\mathbf{R}}^n, \qquad (24.29)$$

where $\mathbf{u} = (\mathbf{p}|\mathbf{x}) \in \mathbb{R}^{2n}$,

$$\hat{\mathbf{R}}^n = \begin{bmatrix} \hat{x}_1 & \cdots & \hat{x}_n & | & \hat{p}_1 & \cdots & \hat{p}_n \end{bmatrix}^T, \qquad (24.30)$$

and $\cdot$ is the inner product. We can now write $\mathscr{T}^n \equiv \{\mathbf{M}(\mathbf{u}) : \mathbf{u} \in \mathbb{R}^{2n}\}$. The symplectic product gives the commutation relations of elements of $\mathscr{T}^n$:

$$[\mathbf{M}(\mathbf{u}), \mathbf{M}(\mathbf{v})] = (\mathbf{u} \odot \mathbf{v}) i. \qquad (24.31)$$

The definitions given below provide terminology used in the construction of our continuous-variable stabilizer codes.

A subspace V of a space W is *symplectic* if there is no $\mathbf{v} \in V$ such that
$$\forall\, \mathbf{u} \in V : \mathbf{u} \odot \mathbf{v} = 0.$$

A subspace V of a space W is *isotropic* if
$$\forall\, \mathbf{u} \in W, \mathbf{v} \in V : \mathbf{u} \odot \mathbf{v} = 0.$$

The *symplectic dual* $V^\perp$ of a subspace V is
$$V^\perp \equiv \{\mathbf{w} : \mathbf{w} \odot \mathbf{u} = 0,\ \forall\, \mathbf{u} \in V\}.$$

A *symplectic matrix* $\boldsymbol{\Upsilon} : \mathbb{R}^{2n} \to \mathbb{R}^{2n}$ preserves the symplectic product:
$$\boldsymbol{\Upsilon}\mathbf{u} \odot \boldsymbol{\Upsilon}\mathbf{v} = \mathbf{u} \odot \mathbf{v} \qquad \forall\, \mathbf{u}, \mathbf{v} \in \mathbb{R}^{2n}. \tag{24.32}$$

It satisfies the condition $\boldsymbol{\Upsilon}^T \mathbf{J} \boldsymbol{\Upsilon} = \mathbf{J}$ where
$$\mathbf{J} = \begin{bmatrix} \mathbf{0}_{n \times n} & \mathbf{I}_{n \times n} \\ -\mathbf{I}_{n \times n} & \mathbf{0}_{n \times n} \end{bmatrix}. \tag{24.33}$$

24.4.2 Stabilizer formalism for continuous variables

We begin our development of continuous-variable stabilizer coding by introducing a canonical code. We then show how to think about the canonical code in terms of a parity check matrix. The last part of this development shows how to relate an arbitrary continuous-variable stabilizer code to the canonical continuous-variable stabilizer code. We then present two simple examples of a continuous-variable stabilizer code.

24.4.2.1 Canonical code for continuous-variable stabilizer codes Suppose Alice wishes to protect a k-mode quantum state $|\psi\rangle$:
$$|\psi\rangle = \int \cdots \int dx_1 \cdots dx_k\, \psi(x_1, \ldots, x_k)\, |x_1\rangle \cdots |x_k\rangle. \tag{24.34}$$

Alice possesses $n - k$ ancilla modes initialized to infinitely squeezed zero-position eigenstates[4] of the position observables $\hat{x}_{k+1}, \ldots, \hat{x}_n$: $|\mathbf{0}\rangle = |0\rangle^{\otimes n-k}$. She encodes the state $|\psi\rangle$ with the canonical isometric encoder as follows:
$$U_0 : |\psi\rangle\langle\psi| \to |\psi\rangle\langle\psi| \otimes |\mathbf{0}\rangle\langle\mathbf{0}|. \tag{24.35}$$

The canonical encoder merely appends the $n - k$ ancilla modes to the k information modes.

Continuous-variable errors are equivalent to translations in position and kicks in momentum [B98c, GKP01]. The canonical code corrects the error set
$$S_0 = \{(\alpha(\mathbf{a}), \mathbf{b} \mid \beta(\mathbf{a}), \mathbf{a}) : \mathbf{b}, \mathbf{a} \in \mathbb{R}^{n-k}\}, \tag{24.36}$$

---

[4] These position eigenstates play a similar role as ancilla qubits, initialized in the qubit state $|0\rangle$, in discrete-variable quantum information processing. Though, note that the position eigenstate $|0\rangle$ corresponds to a vacuum state infinitely squeezed with respect to the position quadrature.

for any known functions $\alpha, \beta : \mathbb{R}^{n-k} \to \mathbb{R}^k$. Alice and Bob choose the functions α and β in order to correct a particular error set. Consider an arbitrary error $\mathbf{D}(\mathbf{u})$ where

$$\mathbf{u} = (\alpha(\mathbf{a}), \mathbf{b} \mid \beta(\mathbf{a}), \mathbf{a}). \tag{24.37}$$

Suppose an error $\mathbf{D}(\mathbf{u})$ occurs. State $|\psi\rangle\langle\psi| \otimes |0\rangle\langle 0|$ becomes as follows (up to a global phase):

$$\mathbf{Z}(\alpha)\mathbf{X}(\beta)|\psi\rangle\langle\psi|\mathbf{X}(-\beta)\mathbf{Z}(-\alpha) \otimes |\mathbf{a}\rangle\langle\mathbf{a}|, \tag{24.38}$$

where $|\mathbf{a}\rangle = \mathbf{X}(\mathbf{a})|\mathbf{0}\rangle$. Bob measures the position observables of the ancillas $|\mathbf{a}\rangle$. He obtains a reduced error syndrome $\mathbf{r} = \mathbf{a}$. The reduced error syndrome specifies the error up to an irrelevant value of $\mathbf{b}$ in (24.37). The $\mathbf{b}$ errors are irrelevant because the ancilla modes absorb these errors. (The ancillas are eigenstates of these error operators, and hence are unaffected by them.) Bob reverses the error $\mathbf{D}(\mathbf{u})$ by applying the map $\mathbf{D}(-\mathbf{u}')$ where

$$\mathbf{u}' = (\alpha(\mathbf{a}), \mathbf{0} \mid \beta(\mathbf{a}), \mathbf{a}). \tag{24.39}$$

The canonical code is a simple example of a continuous-variable stabilizer code, but it illustrates all of the principles that are at work in the operation of a continuous-variable stabilizer code.

24.4.2.2 Parity check matrix for the canonical code We now illustrate how the canonical code operates in the Heisenberg picture by using a parity check matrix. A parity check matrix F_0 characterizes the operators that Bob measures:

$$F_0 = \begin{bmatrix} 0 & I \mid 0 & 0 \end{bmatrix}. \tag{24.40}$$

The first zero in the left and right submatrices represents k columns of zeros and corresponds to the k information modes. The identity matrix in the left submatrix and rightmost zero in the right submatrix correspond to the $n-k$ ancilla modes. The map $\mathbf{M}$ in (24.29) determines the observables that Bob measures to learn about the errors. Each row $\mathbf{f}$ of F_0 corresponds to an element of the set

$$\mathcal{M}_0 \equiv \{\mathbf{M}(\mathbf{f}) : \mathbf{f} \text{ is a row of } F_0\}. \tag{24.41}$$

Therefore, the $n-k$ rows of F_0 correspond to the $n-k$ position observables. Matrix F_0 thus gives another way of describing the measurements performed in the canonical code. Bob measures the observables in $\mathcal{M}_0$ to learn about the error without disturbing the encoded state.

The canonical code can correct an error set $\mathcal{E}_0$ that consists of all pairs of errors obeying the following condition: $\forall \mathbf{D}(\mathbf{e}), \mathbf{D}(\mathbf{e}') \in \mathcal{E}_0$ with $\mathbf{e} \neq \mathbf{e}'$ either

$$\mathbf{e} - \mathbf{e}' \notin \text{rowspace}(F_0)^\perp, \quad \text{or} \quad \mathbf{e} - \mathbf{e}' \in \text{rowspace}(F_0), \tag{24.42}$$

where $\perp$ denotes the symplectic dual.

24.4.2.3 General continuous-variable stabilizer codes We describe the relation between the canonical continuous-variable stabilizer code and an arbitrary one. Alice can perform the encoding of an arbitrary code with a unitary U. This unitary U preserves operators in the phase-free Heisenberg–Weyl group under conjugation. The encoding unitary is analogous to a Clifford unitary in discrete-variable quantum information processing. An equivalent representation of U is with a symplectic matrix Υ that operates on the real vectors that result from the inverse maps $\mathbf{D}^{-1}$ and $\mathbf{M}^{-1}$. We state this result formally with the following theorem.

Theorem 24.1. *There exists a unitary operator U_Υ corresponding to a symplectic matrix Υ so that the following two conditions hold $\forall\, \mathbf{u} \in R^{2n}$:*

$$[\mathbf{D}(\Upsilon \mathbf{u})] = [U_\Upsilon \mathbf{D}(\mathbf{u})\, U_\Upsilon^{-1}], \\ \mathbf{M}(\Upsilon \mathbf{u}) = U_\Upsilon \mathbf{M}(\mathbf{u})\, U_\Upsilon^{-1}. \tag{24.43}$$

Theorem 24.1 is a consequence of the Stone–von Neumann theorem [EP03].

In the Heisenberg picture, the symplectic matrix Υ is a $(2n \times 2n)$-dimensional matrix that takes the canonical parity check matrix F_0 to a general check matrix F. The symplectic matrix Υ then performs the following transformation:

$$F_0 \Upsilon^T = F. \tag{24.44}$$

The parity check matrix F for a general code has the following form:

$$F = [F_Z | F_X]. \tag{24.45}$$

Bob measures the observables in the set

$$\mathscr{M} \equiv \{\mathbf{M}(\mathbf{f}) : \mathbf{f} \text{ is a row of } F\}, \tag{24.46}$$

to diagnose and correct for errors [with $\mathbf{M}$ defined in (24.29)].

A general code can correct an error set $\mathscr{E}$ that consists of all pairs of errors obeying the following condition: $\forall\, \mathbf{D}(\mathbf{e}), \mathbf{D}(\mathbf{e})' \in \mathscr{E}$ with $\mathbf{e} \neq \mathbf{e}'$ either

$$\mathbf{e} - \mathbf{e}' \notin \text{rowspace}(F)^{\perp}, \quad \text{or} \quad \mathbf{e} - \mathbf{e}' \in \text{rowspace}(F). \tag{24.47}$$

24.4.2.4 Examples of a continuous-variable stabilizer code   We briefly give two examples of a continuous-variable stabilizer code. The first example is analogous to the three-qubit bit-flip code and the second is analogous to the nine-qubit Shor code.

Our first example is a three-mode continuous-variable stabilizer code. Alice begins with one information mode in the state $|\psi\rangle = \int dx\, \psi(x)\,|x\rangle$ and two ideally infinitely squeezed ancilla modes in the state $|0\rangle^{\otimes 2}$. She appends the two ancilla modes to her information mode so that the state is as follows:

$$\int dx\, \psi(x)\,|x\rangle\,|0\rangle\,|0\rangle.$$

The above state corresponds to that for the canonical code. The operators that stabilize the above state are the position operators for the second and third modes: $\hat{x}_2$ and $\hat{x}_3$. We readily form a parity check matrix for this canonical code by using the inverse map $\mathbf{M}^{-1}$ in (24.29):

$$\begin{bmatrix} 0 & 1 & 0 & 0 & 0 & 0 \\ 0 & 0 & 1 & 0 & 0 & 0 \end{bmatrix}.$$

The above check matrix is similar to that of the discrete-variable three-qubit code, but keep in mind that the entries in the above matrix are real numbers.

Alice now needs to encode the above state so that it can protect against some errors. She performs two position-quadrature nondemolition interactions[5] from the first mode to the second and from the first mode to the third. A position-quadrature nondemolition interaction acts as follows on ideal position eigenstates:

$$|x\rangle |y\rangle \to |x\rangle |x+y\rangle.$$

It acts as follows on the mode operators in the Heisenberg picture:

$$\hat{x}_1 \to \hat{x}_1, \quad \hat{p}_1 \to \hat{p}_1 - g\hat{p}_2, \quad \hat{x}_2 \to \hat{x}_2 + g\hat{x}_1, \quad \hat{p}_2 \to \hat{p}_2,$$

where g is the strength of the interaction and an ideal interaction for implementing a controlled displacement has $g = 1$. The state becomes as follows after these encoding operations:

$$\int dx\, \psi(x) |x\rangle |x\rangle |x\rangle,$$

and the check matrix becomes as follows:

$$\begin{bmatrix} -1 & 1 & 0 & 0 & 0 & 0 \\ -1 & 0 & 1 & 0 & 0 & 0 \end{bmatrix},$$

so that the check matrix corresponds to the operators $\hat{x}_2 - \hat{x}_1$ and $\hat{x}_3 - \hat{x}_1$, where we use the map **M** in (24.29). Note that the inverse transformation acts on the measurement operators in the continuous-variable stabilizer because any stabilizer operator $\hat{s}$ evolves under a unitary transformation U according to $U\hat{s}U^\dagger$ as in Theorem 24.1, as opposed to the usual Heisenberg-picture evolution $U^\dagger \hat{s} U$. For this reason, $\hat{x}_2$ evolves to $\hat{x}_2 - \hat{x}_1$ under the first position-quadrature quantum nondemolition interaction and likewise for $\hat{x}_3$.

Suppose now that a "position" error occurs on the first mode and displaces its position quadrature by an amount y. The encoded state changes as follows:

$$\int dx\, \psi(x) |x+y\rangle |x\rangle |x\rangle.$$

Bob measures the operators $\hat{x}_2 - \hat{x}_1$ and $\hat{x}_3 - \hat{x}_1$ and obtains the respective results $-y$ and 0. This syndrome uniquely identifies the error to be a displacement of y on the first mode and Bob can reverse the error. Similar arguments apply to show that Bob can reverse any single-mode displacement error.

The above code has no ability to correct momentum "kick" errors. We can remedy this problem by concatenating the above three-mode code with a Fourier-transformed version of itself to produce a nine-mode continuous-variable (just as we can concatenate a three-qubit bit-flip code with a three-qubit phase-flip to produce the discrete-variable Shor code). The encoding circuit is similar to that for the discrete-variable Shor code with the exception that one uses position-quadrature nondemolition interactions instead of CNOT gates and one uses a single-mode Fourier transform operation (implementable with a phase shifter) instead of a Hadamard gate. We detail the encoding operations. We begin with one information mode and eight ancilla

---

[5] A position-quadrature nondemolition interaction is the continuous-variable analog of the CNOT gate.

modes. The check matrix for this unencoded state is as follows:

$$
\left[\begin{array}{ccccccccc|ccccccccc}
0 & 1 & 0 & 0 & 0 & 0 & 0 & 0 & 0 & 0 & 0 & 0 & 0 & 0 & 0 & 0 & 0 & 0 \\
0 & 0 & 1 & 0 & 0 & 0 & 0 & 0 & 0 & 0 & 0 & 0 & 0 & 0 & 0 & 0 & 0 & 0 \\
0 & 0 & 0 & 1 & 0 & 0 & 0 & 0 & 0 & 0 & 0 & 0 & 0 & 0 & 0 & 0 & 0 & 0 \\
0 & 0 & 0 & 0 & 1 & 0 & 0 & 0 & 0 & 0 & 0 & 0 & 0 & 0 & 0 & 0 & 0 & 0 \\
0 & 0 & 0 & 0 & 0 & 1 & 0 & 0 & 0 & 0 & 0 & 0 & 0 & 0 & 0 & 0 & 0 & 0 \\
0 & 0 & 0 & 0 & 0 & 0 & 1 & 0 & 0 & 0 & 0 & 0 & 0 & 0 & 0 & 0 & 0 & 0 \\
0 & 0 & 0 & 0 & 0 & 0 & 0 & 1 & 0 & 0 & 0 & 0 & 0 & 0 & 0 & 0 & 0 & 0 \\
0 & 0 & 0 & 0 & 0 & 0 & 0 & 0 & 1 & 0 & 0 & 0 & 0 & 0 & 0 & 0 & 0 & 0
\end{array}\right].
$$

We then encode modes one, four, and seven with the three-mode code and perform a Fourier transform on those modes. The Fourier transform is a single-mode operation that takes $\hat{x} \to \hat{p}$ and $\hat{p} \to -\hat{x}$ (but recall that we perform the inverse transformation on the stabilizer). The check matrix becomes as follows:

$$
\left[\begin{array}{ccccccccc|ccccccccc}
0 & 1 & 0 & 0 & 0 & 0 & 0 & 0 & 0 & 0 & 0 & 0 & 0 & 0 & 0 & 0 & 0 & 0 \\
0 & 0 & 1 & 0 & 0 & 0 & 0 & 0 & 0 & 0 & 0 & 0 & 0 & 0 & 0 & 0 & 0 & 0 \\
0 & 0 & 0 & 0 & 0 & 0 & 0 & 0 & 0 & 1 & 0 & 0 & -1 & 0 & 0 & 0 & 0 & 0 \\
0 & 0 & 0 & 0 & 1 & 0 & 0 & 0 & 0 & 0 & 0 & 0 & 0 & 0 & 0 & 0 & 0 & 0 \\
0 & 0 & 0 & 0 & 0 & 1 & 0 & 0 & 0 & 0 & 0 & 0 & 0 & 0 & 0 & 0 & 0 & 0 \\
0 & 0 & 0 & 0 & 0 & 0 & 0 & 0 & 0 & 1 & 0 & 0 & 0 & 0 & 0 & -1 & 0 & 0 \\
0 & 0 & 0 & 0 & 0 & 0 & 0 & 1 & 0 & 0 & 0 & 0 & 0 & 0 & 0 & 0 & 0 & 0 \\
0 & 0 & 0 & 0 & 0 & 0 & 0 & 0 & 1 & 0 & 0 & 0 & 0 & 0 & 0 & 0 & 0 & 0
\end{array}\right].
$$

We finally encode modes one, two, and three with the three-mode code, and do the same for the modes four, five, and six, and modes seven, eight, and nine. The check matrix for the resulting continuous-variable Shor code is as follows:

$$
\left[\begin{array}{ccccccccc|ccccccccc}
-1 & 1 & 0 & 0 & 0 & 0 & 0 & 0 & 0 & 0 & 0 & 0 & 0 & 0 & 0 & 0 & 0 & 0 \\
-1 & 0 & 1 & 0 & 0 & 0 & 0 & 0 & 0 & 0 & 0 & 0 & 0 & 0 & 0 & 0 & 0 & 0 \\
0 & 0 & 0 & 0 & 0 & 0 & 0 & 0 & 0 & 1 & 1 & 1 & -1 & -1 & -1 & 0 & 0 & 0 \\
0 & 0 & 0 & -1 & 1 & 0 & 0 & 0 & 0 & 0 & 0 & 0 & 0 & 0 & 0 & 0 & 0 & 0 \\
0 & 0 & 0 & -1 & 0 & 1 & 0 & 0 & 0 & 0 & 0 & 0 & 0 & 0 & 0 & 0 & 0 & 0 \\
0 & 0 & 0 & 0 & 0 & 0 & 0 & 0 & 0 & 1 & 1 & 1 & 0 & 0 & 0 & -1 & -1 & -1 \\
0 & 0 & 0 & 0 & 0 & 0 & -1 & 1 & 0 & 0 & 0 & 0 & 0 & 0 & 0 & 0 & 0 & 0 \\
0 & 0 & 0 & 0 & 0 & 0 & -1 & 0 & 1 & 0 & 0 & 0 & 0 & 0 & 0 & 0 & 0 & 0
\end{array}\right].
$$

Its ability to correct a single-mode displacement or "kick" error follows from the ability of the discrete-variable Shor code to correct an arbitrary single-qubit error.

24.4.3 Discussion

There is a beautiful correspondence between the formalism for continuous-variable QEC and that for discrete-variable error correction, but it may be more difficult in practice to implement robust continuous-variable codes. The major concern is that the errors inherent in the measurement of a continuous quantum variable may throw off the error-correction procedure (similar to the concern over encoding analog classical information). Continuous-variable codes perform well only if the actual errors are larger than the errors in measurement of a continuous variable. It is not clear that continuous-variable codes would ever be able to overcome these obstacles because it would take a large amount of squeezing and high-performance photodetectors to have robust operation of a continuous-variable code (though the discussion in Appendix F of [ATK+09] lends some credence to the argument that continuous-variable codes will give an improvement in fidelity when compared to no coding at all).

24.5 Implementations of quantum error correction for communication

The last section of this chapter concerns progress on the implementation of QEC. We keep this section brief and mainly just summarize the recent progress.

In an influential article, David DiVincenzo [D00] listed five requirements that are essential for quantum computation. In addition to his outline of the requirements for a reliable quantum computer, he listed two requirements that are essential for the realization of quantum communication. We quote these two requirements from his article:

(i) "The ability to interconvert stationary and flying qubits."
(ii) "The ability faithfully to transmit flying qubits between specified locations."

A "stationary qubit" refers to a physical implementation for a qubit that is better suited for staying in one place. Examples are quantum dots, ion traps, or superconducting qubits. A "flying qubit" is one that propagates easily. The clear implementation for flying qubits is the polarization of a photon or the spatial mode of a photon. Photons are an excellent choice for flying qubits because they do not easily interact with their environment [KMN+07], but if they do interact, they are lost.

We do not discuss DiVincenzo's first requirement for quantum communication here. It does not apply to quantum key distribution and it does not apply to the other two topics of this chapter. His second requirement applies because the theory of QEC plays a fundamental role in faithfully transmitting "flying qubits."

Photon loss is the major source of error in quantum communication. A simple way to model this loss is with the quantum erasure channel introduced by Grassl *et al.* [GBP97]. Several researchers have contributed to the theory of quantum error correction to protect against photon loss [GKL+03, WB07, BW06, SRZ05].

24.5.1 Experiments for quantum communication

Several experimental groups have contributed experiments that realize QEC with an optical implementation. Pittman, Jacobs, and Franson realized a simple two-qubit code that protects against Z-measurement errors [PJF05], and O'Brien *et al.* conducted a similar experiment [OPW+05]. Both of these experiments employ the following two-qubit logical codewords:

$$|0_L\rangle \equiv \frac{1}{\sqrt{2}}\left(|00\rangle + |11\rangle\right),$$
$$|1_L\rangle \equiv \frac{1}{\sqrt{2}}\left(|01\rangle + |10\rangle\right).$$

The error rejection protocol in [CZZ+06] highlights another simple proof-of-principle experiment for protecting quantum information with a two-mode code. A more recent experiment [LGZ+08] protects against photon loss by employing the quantum erasure channel code of Grassl *et al.* [GBP97]. These experiments are at an early stage and indicate that we really have a long way to go before we are transmitting the large amount of quantum data necessary to make a quantum Internet useful.

24.6 Closing remarks

We have discussed four topics relevant to QEC and quantum communication: entanglement distillation, a security proof for quantum key expansion, continuous-variable systems, and optical implementations of QEC. The theory that is closest to being practical is that for quantum key expansion because the requirements are not as demanding as those for entanglement distillation or QEC. Other topics within quantum communication that we do not discuss here include entanglement concentration [BBP+96a], quantum secret sharing [CGL99], distributed computation [CEH+99], game playing [BH01], and quantum communication complexity [CB97].

24.7 Historical notes

We wrap up with some historical notes on the developments in this chapter. Two landmark papers by the authors Bennett, Brassard, DiVincenzo, Popescu, Schumacher, Smolin, and Wootters introduced the notion of entanglement distillation [BBP+96b, BDS+96]. These papers established many of the fundamental notions, including how to use a QEC code for entanglement distillation. They introduced the breeding protocol by showing that one could use an initial set of noiseless ebits and apply parity checks. Luo and Devetak explicitly showed how to perform the breeding protocol in the stabilizer formalism using the notion of entanglement-assisted coding [LD07].

Dür and Briegel have written an expository article on entanglement distillation and its relation to QEC for the interested reader [DB07]. They also discuss multi-party entanglement distillation. Nielsen and Chuang discuss entanglement distillation with a stabilizer code in Exercise 12.34 of their text [NC00]. Both Gaitan [G08] and Bruß and Leuchs [BL07] discuss entanglement distillation in their respective texts.

Bennett and Brassard published the BB84 quantum key distribution protocol in a conference paper that went mostly unnoticed by the physics community [BB84]. The basis of their argument was the no-cloning theorem [WZ82] and the Heisenberg uncertainty principle. In 1991, Ekert independently published his entanglement-based quantum key distribution protocol and proved the security of it with Bell's inequalities [E91]. Bennett, Brassard, and Mermin later showed the sense in which these schemes are equivalent [BBM92]. Bennett later developed the B92 protocol for quantum key distribution using any two nonorthogonal states [B92]. The security proof for B92 is in [TKI03]. Scarani, Acín, Ribordy, and Gisin later developed the SARG04 protocol as a modification of the original BB84 protocol [SAR+04]. It is simply the BB84 protocol with the role of the encoded bits and the basis bits exchanged. With this simple modification, the SARG04 protocol is more robust in practice when experimentalists employ attenuated laser pulses instead of perfect single-photon sources. We point the reader to some recent interesting experiments in quantum key distribution [RHR+07, RPH+09].

Wegman and Carter have developed some of theory behind two-universal hash functions in the context of creating a secret key for the authentication of a classical channel [WC79, WC81]. This Wegman–Carter authentication scheme is useful for the authentication of the classical channel in the BB84 protocol.

We describe some of the history of security proofs for quantum key expansion that appeal to QEC. Lo and Chau provided a security proof by exploiting the connection between entanglement distillation and quantum key distribution [LC99]. The drawback of their proof is that it requires Alice and Bob to possess fault-tolerant quantum computers. Shor and Preskill simplified the Lo–Chau proof by showing how to reduce a CSS entanglement distillation protocol to the BB84 protocol with classical postprocessing [SP00]. This simplification requires Alice and Bob to use single-qubit operations and thus dramatically simplifies the physical requirements for quantum key expansion to those that the original BB84 protocol employs. The Shor–Preskill result is helpful, but in practice, it is difficult to find the large dual-containing codes that it requires. The Luo-Devetak result [LD07] connects entanglement-assisted QEC to quantum key expansion and develops the security proof that we present in this section. The Shor–Preskill result is a special case of the Luo–Devetak result. It is easier in practice to find large entanglement-assisted codes by producing them from arbitrary classical codes that do not necessarily have to satisfy the restrictive dual-containing condition. The noise threshold of 0.11 that Luo and Devetak obtained is the same threshold that Shor and Preskill obtained.

We mention that there do exist other methods to prove the security of quantum key distribution or quantum key expansion. The first method rests on an uncertainty principle argument [M96, M01, K06b, K06c, K07b] and the second method rests on pure information theoretic arguments [B.-O02, KGR05, RGK05, R05]. We do not discuss these methods here because they do not employ the tight connection to the theory of QEC.

The notion of a good security criterion developed quite a bit after the original BB84 proposal [RK05, B.-OHL+05]. Many researchers originally thought that it was acceptable to minimize an attacker's accessible information about the key (including Shor and Preskill's proof). But this definition does not provide a composable definition of security. Alice and Bob can make Eve's accessible information small at the end of their protocol, but it is possible for Eve to wait to make measurements until a later time when she may gain more information about the key. She can then adapt her measurements according to this new information and increase her accessible information beyond the bound that Alice and Bob thought they had imposed on it. We further note that any prepare-and-measure protocol whose security derives from an entanglement-based protocol automatically meets the universally composable definition of security [R05].

Scarani, Bechmann-Pasquinucci, Cerf, Dušek, and Lütkenhaus have written a review article on quantum key distribution [SB.-PC+09]. They discuss many of the more practical aspects of quantum key distribution that our security analysis does not take into account. One major assumption in our proof is that it relies on asymptotic results where the data length and the size of the resulting key are large. The more recent focus of quantum cryptotheorists is to bridge the gap between experiment and theory by determining how secure a finite-length key is [H07, SR08].

We make some comments regarding the connection of QEC to quantum privacy. Later on in the development of quantum information theory, theorists began to think about the connections between quantum information and privacy [CP02, SW98]. These early connections were somewhat rough and were later further developed [SN96, SW02b]. Eventually, Devetak proved the quantum capacity theorem by "coherifying" private classical codes. Devetak first showed how to construct private classical codes that transmit private classical information over a noisy quantum channel. Devetak's essential insight was to "coherify" these private classical codes into quantum codes and then to use the privacy condition in (24.14) to construct a decoder for a QEC code. The decoder decouples Eve (the environment in the case of a quantum code) from the picture and allows Alice to transmit coherent quantum information to Bob. The decoupling method then took off from there in different shapes and forms and has now become the "hammer for many nails" in quantum information theory [ADH+09, HHW+08, HOW05, HOW07, K07a, DY08, YD09, YBW08]. Interestingly, Luo has shown shown how to "coherify" a classical error-correction code and a privacy amplification protocol to construct quantum stabilizer codes that have excellent asymptotic performance (they attain the hashing bound on noisy Pauli channels) [L08b].

Several authors have suggested methods for error correction of continuous-variable quantum information [B98b, B98c, LS98, WKB07, NAC08]. Barnes suggested a stabilizer formalism for continuous-variable systems in [B04]. Some of these schemes [B98b, B98c, LS98, WKB07] are vulnerable to small displacement errors that occur in a continuous-variable quantum system [GKP01]. They operate well only when the squeezing of optical parametric oscillators is high and the homodyne detectors are high efficiency. The group of Furusawa has implemented the continuous-variable Shor code [ATK+09]. They used a scheme of Braunstein for performing the encoding that uses linear optics only and avoids the use of quantum nondemolition interactions [B98b].

Gottesman, Kitaev, and Preskill (GKP) devised a scheme for encoding a qubit in an optical mode [GKP01]. With this scheme, there are again many obstacles before it could become practical. Nevertheless, some authors have contributed later work to the GKP scheme [HP01, TM02, PMV+04].

One of the fundamental components of a continuous-variable encoder is the online squeezer. It is possible to simulate an online squeezer using a linear-optical circuit [FMA05]. The theoretical proposal [FMA05] and experimental implementation of an online squeezer [YHA+07] and a quantum nondemolition interaction [YMH+08] using linear optics are examples of recent progress in continuous-variable quantum information processing.

One way to determine an encoding circuit for a continuous-variable stabilizer code is to use the Bloch–Messiah decomposition [B05]. A Gaussian elimination-like procedure can also determine an encoding circuit that uses quantum nondemolition interactions and single-mode operations [WKB07].

The extension of the continuous-variable stabilizer formalism to entanglement-assisted codes [WKB07] and subsystem codes [WB08b] is fairly straightforward and again involves exploiting the correspondences between the Pauli group and the Heisenberg–Weyl group.

We briefly mention that there are interesting quantum Shannon-theoretic results for continuous-variable systems. These results are the classical capacity [GGL+04], the entanglement-assisted classical capacity [GLM+03], the quantum capacity [WP.-GG07], and the capacity of the bosonic wiretap channel for sending private information [GSE08].

The optical implementation of a quantum computer has focused mainly on the scheme of Knill, Laflamme, and Milburn [KLM01]. They incorporated error correction as a component in their theoretical proposal for a linear-optical quantum computer. See also the review article of Kok *et al.* for progress in linear-optical quantum computing [KMN+07]. The recent development of low-cost number-resolving photodetectors should give a boost to the implementation of a linear-optical quantum computer [KYS08].

Part VIII

Critical evaluation of fault tolerance

Chapter 25

Hamiltonian methods in quantum error correction and fault tolerance

Eduardo Novais, Eduardo R. Mucciolo, and Harold U. Baranger

25.1 Introduction

The existence of efficient quantum error correction (QEC), combined with the concept of the accuracy threshold [G98, KLZ98a], inspired confidence that reliable quantum computation is achievable in principle. However, a key point to understand is whether there are *physical* limitations to resilient quantum computation within this framework. In this chapter, we discuss one of the few situations that still poses some doubts [AHH+02, CSG.-B04, KF05, A06, ALZ06] about the effectiveness of QEC codes: critical environments.

The term "critical environment" originates from condensed matter physics, where it refers to physical systems in which quantum correlations decay as power laws. In such an environment, the Born–Markov approximation used to evaluate decoherence rates cannot be formally justified [W99]. For quantum computation, this fact translates into the appearance of errors that can depend on previous events in the computer history. The ultimate nightmare [AHH+02] is that this memory effect may eventually lead to error probabilities above the threshold value and therefore to the breakdown of resilient quantum computation. In this chapter, our goal is to find the minimal conditions for the existence of a finite threshold value; finding a particular value for the error threshold is a more detailed question which we leave for future work.

We start our study by formulating the dynamics of the computer and the environment through a Hamiltonian. This is the most natural way to discuss some elements crucial to the system dynamics such as spatial dimension, correlation function exponents, and coupling constants. The first step is to separate the total Hamiltonian into two distinct parts:

$$H = H_0(t) + V. \tag{25.1}$$

The first part, H_0, describes the "free" system, by which we mean the ideal evolution of the quantum computer and the dynamics of the environment (bath) while isolated from

---

Quantum Error Correction, ed. Daniel A. Lidar and Todd A. Brun. Published by Cambridge University Press.
© Cambridge University Press 2013.

each other:
$$H_0(t) = H_{\text{QC}}(t) + H_{\text{bath}}. \qquad (25.2)$$

Note that $H_0(t)$ has an explicit time dependence because of the computation being performed through $H_{\text{QC}}(t)$. We assume that the form of the operator H_0 is simple enough to allow an explicit evaluation of the corresponding quantum evolution operator. The second part of the Hamiltonian in (25.1) is the "interaction," V, which includes everything that precludes writing explicitly the complete evolution operator. In particular, V represents the coupling between the computer and the environment.

Since V destroys our ability to write down explicitly the quantum evolution of the entire system, we have to resort to an expansion of the evolution operator in powers of the interaction. The result is the Dyson series (Section 1.2.2.2),
$$U(t_c, 0) = T_+ \, e^{-i \int_0^{t_c} dt' \, V(t')}, \qquad (25.3)$$

where T_+ denotes the time-ordering operator, t_c is the duration of the computation, and $V(t)$ already incorporates the evolution generated by H_0 (the "interaction representation" – explicit expression given below). In this series, every insertion of V is a deviation of the computer evolution from the path that we envisaged in $H_0(t)$. Of course, some of these deviations are harmless since they are "good" paths, which do not change the result of the computation. Our problem is to evaluate the likelihood and influence of "bad" paths [TB05].

A reasonable idea is to consider the evolution of the system with at least one insertion of V [KLV00, TB05, AGP06, AKP06]:
$$\mathscr{E}(t) = U(t, 0) - 1 = -i \int_0^t dt' \, V(t') \, U(t', 0). \qquad (25.4)$$

It has been shown that using the operator norm one can derive an upper bound on the error probability of "bad" paths [KLV00, TB05, AGP06]. For Eq. (25.4), this approach implies that
$$\|\mathscr{E}(t)\|_\infty \leq \int_0^t dt' \, \|V(t')\|_\infty \leq \Lambda t, \qquad (25.5)$$

where Λ is the largest eigenvalue of V [KLV00, TB05, AGP06].

Without QEC, the bound Eq. (25.5) does not help much: the computational time t must be regarded as large, so that after a time $\propto 1/\Lambda$ the computation certainly fails. The situation changes when QEC is added: error correction introduces another time scale into the problem, namely, the periodicity Δ at which error correction operations are carried out repeatedly. In this case, the relevant time is not the total computation time but rather Δ [AGP06, AKP06]. The remaining issue is, then, the magnitude of Λ.

In many relevant physical situations, Λ can be extremely large, since it usually grows with the number of degrees of freedom of the environment [TB05]. An illustrative example is that of a single qubit, σ, interacting with N two-level systems, $\{\tau^{(j)}\}$, through
$$V = \lambda \sum_{j=1}^N \sigma_z \tau_z^{(j)}. \qquad (25.6)$$

This is a simplified version of the central spin problem [G76, KLG02, A.-HDD+06] that has been studied in the context of decoherence [CPZ05]. In this case, $\Lambda = \lambda N$, thus diverging with the number of degrees of freedom of the spin bath.

Physical interactions are ultimately mediated by gauge fields [W67, K89]. Hence, a natural and generic form for V is the bilinear minimal coupling model:

$$V = \sum_{\mathbf{x}} \sum_{\alpha=\{x,y,z\}} \frac{\lambda_\alpha}{2} f_\alpha(\mathbf{x}) \, \sigma_\alpha(\mathbf{x}), \qquad (25.7)$$

where $\sigma_{x,y,z}$ are Pauli matrices representing the qubit degrees of freedom at positions $\mathbf{x}$, and $\mathbf{f}$ is some (vector) function of the environmental variables (the bath), which will be represented by gauge fields. In this case as well, Λ diverges with the number of bath degrees of freedom, which here is the number of modes in the field determined by the short time and space cutoffs.

In order to deal with V while including the possible self-interaction and retardation effects that were neglected previously [AKP06], we study the stability of the Dyson series [NB06, NMB07, NMB08]. For this purpose, we assume that the environment is described by a free-field theory with the relevant two-point correlation function given by $\langle \Psi_{\text{env}} | f_\alpha(\mathbf{x}_1, t_1) f_\beta(\mathbf{x}_2, t_2) | \Psi_{\text{env}} \rangle$. By "free-field theory" we mean that the fluctuations are Gaussian so that Wick's theorem can be used to calculate the higher-order correlation functions [P98b, M00]. The correlations naturally decay in both space and time. For a critical environment, this decay is power-law (*not* exponential). For equal times, the spatial correlation decays as an integer power of $1/\Delta x^{2\delta}$ for large separations, while for a given location the temporal correlations decay as an integer power of $1/\Delta t^{2\delta/z}$. A general two-point correlation function depends on the form of the function $\mathbf{f}$, and we summarize its behavior by writing

$$\langle \Psi_{\text{env}} | f_\alpha(\mathbf{x}_1, t_1) f_\beta(\mathbf{x}_2, t_2) | \Psi_{\text{env}} \rangle \sim \mathscr{O}\left(\frac{1}{|\mathbf{x}_2 - \mathbf{x}_1|^{2\delta}}, \frac{1}{|t_2 - t_1|^{2\delta/z}} \right). \qquad (25.8)$$

These power-law decays hold at large separations and long times; for short times and distances, their singular nature must be cutoff or "regularized", an issue to which we return below. The parameters δ and z are usually called the scaling dimension and the dynamical exponent, respectively [S99a]. Notice that z sets the relation between the decay exponents of temporal and spatial correlations. To compare spatial and temporal correlations on the same footing, we introduce the model-dependent (dimensionful) parameter v such that

$$\xi \equiv (\text{v}\Delta)^{1/z} \qquad (25.9)$$

is the distance traveled by a disturbance in the bath in time Δ. For instance, for $z = 1$, v is the velocity of the environmental field, while for $z = 2$, it is the diffusion constant.

Another critical assumption we make is that the qubits are separated by a minimum distance so that an entire error-correction cycle can be performed before correlations between neighboring qubits develop. This assumption allows for the introduction of a well-defined error probability for a single qubit during a QEC cycle. We call this the *"hypercube" assumption*; it allows for a connection to the usual derivation of the threshold theorem.

The basic strategy that we will follow is, first, to use the Dyson series to find a reasonable way to evaluate the probability of an error in a single qubit. We will then determine the conditions necessary to reduce the problem to a stochastic error problem, i.e., find when correlations between errors do not fundamentally change the long-time dynamics of the quantum computer.

25.2 Microscopic Hamiltonian models

There are a variety of physical realizations of qubits where the interaction between the environmental degrees of freedom and the qubits can be cast in the form Eq. (25.7). First, the spin of an electron confined in a GaAs lateral quantum dot couples to nuclear spins through a hyperfine interaction, in which case the field $f_\alpha(\mathbf{x})$ represents a component of the local nuclear magnetization, also known as the Overhauser field [BLD99, A61]. Second, the minimal coupling model also appears in superconducting (Josephson) qubits where the field $f_\alpha(\mathbf{x})$ accounts for the coupling to electromagnetic fluctuations. These fluctuations typically arise from the Johnson–Nyquist noise in currents and voltages and can be described by a bath of harmonic oscillators, that is, by bosonic fields [MSS01]. Third, another common situation where Eq. (25.7) applies is in qubits based on charge motion (e.g., double-dot charge qubits or impurities embedded in a semiconductor matrix) [VMB05, HDW+04]. In this case the field $f_\alpha(\mathbf{x})$ accounts for the coupling to acoustic phonons and can also be represented by bosonic degrees of freedom. Finally, if the qubit is a localized magnetic moment embedded in a conducting medium, Eq. (25.7) can be used to represent the coupling between the spin of itinerant electrons and the local moment (the so-called Kondo problem [AL91, H93]).

For many qubit systems, particularly in solid-state contexts, the typical environment is bosonic. A ubiquitous interaction is

$$V = \sum_{\mathbf{x}} \sum_{\alpha=\{x,y,z\}} \sigma_\alpha(\mathbf{x}) \sum_{\mathbf{q}} \lambda_{\alpha,\mathbf{q}} \, e^{i\mathbf{q}\cdot\mathbf{x}} \left(a_\mathbf{q} + a^\dagger_{-\mathbf{q}} \right), \qquad (25.10)$$

where the bosonic field $a_\mathbf{q}$ is usually assumed to have free dynamics described by a quadratic Hamiltonian,

$$H_{\text{bath}} = \sum_{\mathbf{q}} \omega_\mathbf{q} \, a^\dagger_\mathbf{q} a_\mathbf{q}. \qquad (25.11)$$

(Generalizations where multiple bosonic baths couple to the qubits have been studied, see for instance [AL91, S00, NC. NB+05].) Equations (25.10) and (25.11) define the so-called spin-boson model [LCD+87]. This model has been intensively studied in the contexts of dissipative quantum mechanics and condensed matter physics (for detailed discussions see [LCD+87, W99]). The spin-boson model is representative of the kind of physical constraints faced by a wide variety of qubit implementations. It should thus be regarded as a paradigmatic model, and we use it below to provide technical details of our calculations. However, it is important to remark that our approach applies to more general interaction models, as defined by Eq. (25.7).

Before including QEC, we need to discuss the form of $V(t)$ in the interaction picture. This is the operator that enters into the Dyson series Eq. (25.3) and thus defines the dynamics of the system. As we argue below, this issue is related to the duration of the quantum gates used in the computation.

The "free" Hamiltonian H_0 in Eq. (25.2) is composed of two parts: the Hamiltonian that dictates the time evolution of the environment, H_{bath}, and the control Hamiltonian that implements quantum gates, $H_{\text{QC}}(t)$. These terms act on different Hilbert spaces, hence $[H_{\text{bath}}, H_{\text{QC}}] = 0$. In this case, it is straightforward to write an interaction picture that takes into account not only the environment but also the free evolution of the computer:

$$V(t) = \sum_{\mathbf{x}} \sum_{\alpha=\{x,y,z\}} \frac{\lambda_\alpha}{2} \left[e^{iH_{\text{bath}}t} f_\alpha(\mathbf{x}) e^{-iH_{\text{bath}}t} \right] W^\dagger(t) \sigma_\alpha(\mathbf{x}) W(t)$$

$$= \sum_{\mathbf{x}} \sum_{\alpha=\{x,y,z\}} \frac{\lambda_\alpha}{2} f_\alpha(\mathbf{x},t) G_\alpha(\mathbf{x},t), \qquad (25.12)$$

where $W(t) = Te^{-i\int_0^t dt' H_{\text{QC}}(t')}$, $f_\alpha(\mathbf{x},t) = e^{iH_{\text{bath}}t} f_\alpha(\mathbf{x}) e^{-iH_{\text{bath}}t}$, $G_\alpha(\mathbf{x},t) = W^\dagger(t)\sigma_\alpha \times (\mathbf{x})W(t)$. The operator G_α is a $SU(2^N)$ matrix that depends on the particular sequence of quantum gates that is being performed. In order to keep the discussion general, it is necessary to introduce some simplification. There are two possible paths:

(i) Suppose the quantum gates are performed much faster than the environment's shortest response time. In this limit, the action of a gate on the qubits is effectively instantaneous, and it is straightforward to see that the evolution between the gates is given by the microscopic form Eq. (25.7) despite the gate operations in Eq. (25.12):

$$V(t) = \sum_{\mathbf{x}} \sum_{\alpha=\{x,y,z\}} \frac{\lambda_\alpha}{2} f_\alpha(\mathbf{x},t) \sigma_\alpha(\mathbf{x}) . \qquad (25.13)$$

(ii) The other possibility is to derive an effective Hamiltonian that provides an upper estimate to the effect of errors. The point here is to realize that the information encoded in the qubits is exposed to different components of the environment depending on the particular gate being performed. For instance, single-qubit gates can be written as

$$G_{\alpha,1}(\mathbf{x},t) = \sum_{\beta=\{x,y,z\}} g_{\alpha\beta}(\mathbf{x},t) \sigma_\beta(\mathbf{x}) , \qquad (25.14)$$

where $g_{\alpha\beta}(\mathbf{x},t)$ are ordinary functions. Consequently, the "interaction" Hamiltonian can be written as

$$V_1(t) = \sum_{\mathbf{x}} \sum_{\alpha=\{x,y,z\}} \sum_{\beta=\{x,y,z\}} \frac{\lambda_\alpha}{2} f_\alpha(\mathbf{x},t) g_{\alpha\beta}(\mathbf{x},t) \sigma_\beta(\mathbf{x}), \qquad (25.15)$$

which tells us that all components of the qubits mix with all components of the environment. In order to define a suitable upper bound estimate, we can make all functions $g_{\alpha\beta}(\mathbf{x},t)$ constant and set them to unity. This obviously breaks the unitarity of the gates but has the virtue of simplicity. More accurate upper bounds could in principle be obtained by making use of the functional form of $g_{\alpha\beta}(\mathbf{x},t)$, but this would also make any calculation considerably more difficult and model dependent.

A similar argument can be made for two-qubit gates. In this case, the evolution operator for a two-qubit rotation can be written as

$$W(\mathbf{x}_1, \mathbf{x}_2, t) = \cos\left[\theta(\mathbf{x}_1, \mathbf{x}_2, t)\right] + i \sin\left[\theta(\mathbf{x}_1, \mathbf{x}_2, t)\right] \sigma_a(\mathbf{x}_1) \sigma_b(\mathbf{x}_2), \qquad (25.16)$$

where $\theta(\mathbf{x}_1, \mathbf{x}_2, t)$ is an ordinary function, $\mathbf{x}_1$ and $\mathbf{x}_2$ tag the positions of the qubits involved in the gate, and a and b are the components used for the gate. This implies that the interaction between the bath and the qubit at $\mathbf{x}_1$ (the target, say) will also depend on the state of the qubit at

$\mathbf{x}_2$ (the control),

$$G_{\alpha,2}(\mathbf{x}_1, t) = \sin[2\theta(\mathbf{x}_1, \mathbf{x}_2, t)] \sum_\gamma \epsilon_{a\alpha\gamma} \sigma_\gamma(\mathbf{x}_1) \sigma_b(\mathbf{x}_2)$$
$$+ \delta_{a,\alpha} \sigma_\alpha(\mathbf{x}_1) + \cos[2\theta(\mathbf{x}_1, \mathbf{x}_2, t)] (1 - \delta_{a,\alpha}) \sigma_\alpha(\mathbf{x}_1), \quad (25.17)$$

where $\epsilon_{\alpha\beta\gamma}$ is the usual antisymmetric tensor. Once again we can define a suitable upper bound by setting all the functions in the prefactors to unity, yielding

$$G_{\alpha,2}(\mathbf{x}_1, t) = \sum_\gamma \epsilon_{a\alpha\gamma} \sigma_\gamma(\mathbf{x}_1) \sigma_b(\mathbf{x}_2) + \sigma_\alpha(\mathbf{x}_1). \quad (25.18)$$

The corresponding interaction Hamiltonian is then

$$V_2(t) = \sum_{\mathbf{x}} \sum_{\alpha=\{x,y,z\}} \frac{\lambda_\alpha}{2} f_\alpha(\mathbf{x}, t) G_{\alpha,2}(\mathbf{x}, t), \quad (25.19)$$

where each pair of qubits for which there is a gate contributes a term to the sum.

While the second term on the right-hand-side of Eq. (25.18) is just a local noise, the first term is more worrisome since it leads to errors propagating between target and control qubits. Obviously, it is not unique to have the error occurring during the two-qubit gate. In fact, we can re-interpret Eq. (25.19) as an error on the qubit $\mathbf{x}$ before the gate is performed. This error is then propagated by a perfect two-qubit gate. Propagation of errors is in general unavoidable in a quantum circuit. Nevertheless, we can assume that it can be handled by fault-tolerant procedures. As a result, we conclude that an upper bound estimate to the action of gates in the microscopic model is accounted for by the interaction Hamiltonian

$$V_{\text{eff}}(t) = \sum_{\mathbf{x}} \sum_{\alpha=\{x,y,z\}} \frac{\lambda}{2} f_{\text{eff}}(\mathbf{x}, t) \sigma_\alpha(\mathbf{x}), \quad (25.20)$$

where

$$f_{\text{eff}}(\mathbf{x}, t) = \frac{1}{\lambda} \left[\sum_{\beta=\{x,y,z\}} \lambda_\beta f_\beta(\mathbf{x}, t) \right] \quad (25.21)$$

and $\lambda = \sqrt{\sum_{\beta=\{x,y,z\}} \lambda_\beta^2}$ is the new coupling parameter.

Because Eqs. (25.13) and (25.20) have the same functional form, we hereafter drop the subscript "eff."

25.3 Time evolution with quantum error correction

In this section we consider the case where the initial states of both the computer, ψ_0, and the environment, φ_0, are pure. Furthermore, since one of the basic assumptions of quantum computation is the ability to prepare the initial state of the computer, it is reasonable to suppose that the computer is not entangled with the environment at the beginning of the calculation. Hence, we assume that the initial state of the system is

$$|\Psi(t=0)\rangle = |\psi_0\rangle \otimes |\varphi_0\rangle. \quad (25.22)$$

By using a pure state for the environment, we are assuming that the temperature is zero, $T = 0$. However, generalization of our discussion to the case of mixed initial states is straightforward. One may think that considering a small but nonzero temperature would ameliorate the situation. In many situations, a bath temperature T simply introduces a thermal coherence length ζ_T for the bath modes. For distances smaller than ζ_T, the correlation functions have a power-law behavior that is potentially troublesome; however, for distances larger than ζ_T, the correlation functions decay exponentially. Hence, why not simply operate the computer at $T \neq 0$ and use the results of Aharonov and Ben-Or [AB.-O08]? The answer is that the same mechanism that is setting T for the environment is also affecting the qubits. Thus, this is likely to introduce an exponentially short coherence time for the qubits as well.

During a QEC cycle, the state Ψ will evolve according to the unitary operator $U(\Delta, 0)$, Eq. (25.3). At time Δ, the syndrome is extracted and the system wave function is projected, becoming

$$P_m U(\Delta, 0) |\Psi(0)\rangle, \tag{25.23}$$

where m labels a particular syndrome. The projection operators obey $\sum_m P_m = I$ and $P_m^2 = P_m$. (In the case of many logical qubits evolving together, m denotes the set of all the syndromes extracted at time Δ.) Finally, as required by QEC, an appropriate recovery operation R_m is performed:

$$|\Psi(\Delta)\rangle = R_m (\Delta + \delta_r, \Delta) \, P_m \, U(\Delta, 0) |\Psi(0)\rangle, \tag{25.24}$$

where δ_r is the duration of the recovery operation.

It is well known that QEC can also be performed without measuring the syndrome [NC00, TB05]. However, in that case a fresh supply of cold ancillas must be available at each QEC step. Thus, one should consider two possible scenarios: (i) If the ancillas are only briefly in contact with the computer and the bath, then this procedure is completely equivalent for our purposes to the use of a measurement. (ii) If, however, the ancillas cannot be separated from the computer and the bath, then their dynamics must be followed. Although the inclusion of such ancillas would not change our conclusion, it would introduce nonessential elements to the discussion (such as where and how they are stored). Therefore, we will consider only the more usual case of QEC using syndrome extraction.

The generalization of Eq. (25.24) to a sequence of QEC cycles is

$$\Upsilon_{\mathbf{w}} = v_{w_N}(N\Delta, (N-1)\Delta) \cdots v_{w_1}(\Delta, 0), \tag{25.25}$$

where $\mathbf{w}$ is the particular history of syndromes for all the qubits and

$$v_{w_j}(j\Delta, (j-1)\Delta) = R_{w_j}(j\Delta + \delta_r, j\Delta) P_{w_j} U(j\Delta, (j-1)\Delta) \tag{25.26}$$

is the QEC evolution for cycle j.

There are two useful quantities that can now be calculated. The first one is the probability to have a particular history of syndromes:

$$\Pr(\Upsilon_{\mathbf{w}}) = \{\langle \varphi_0 | \otimes \langle \psi_0 | \} \Upsilon_{\mathbf{w}}^{\dagger} \Upsilon_{\mathbf{w}} \{ |\psi_0\rangle \otimes |\varphi_0\rangle \}. \tag{25.27}$$

The second quantity is the residual decoherence, which can be read from the reduced density matrix. For a single QEC cycle, the reduced density matrix corresponding to states $\vec{r}$ and $\vec{s}$ in the

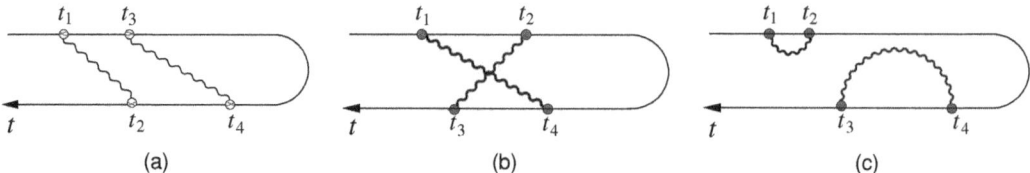

Fig. 25.1. Graphical representation of fourth-order terms in a "time-loop" expansion for either the probability of a given evolution or the reduced density matrix (spatial dimensions are suppressed for clarity). Points of interaction with the bath (circles) are connected by the propagation of environmental modes (wiggly lines). Adapted with permission from [NMB08]. Copyright 2008, American Physical Society.

computer Hilbert space is

$$\rho_{\vec{r},\vec{s}}^{w_1}(t=\Delta) = \frac{\text{tr}_\varepsilon \left[\langle \vec{r} | v_{w_1} | \Psi(0) \rangle \langle \Psi(0) | v_{w_1}^\dagger | \vec{s} \rangle \right]}{\langle \Psi(0) | v_{w_1}^\dagger v_{w_1} | \Psi(0) \rangle}$$
$$= \frac{\{\langle \varphi_0 | \otimes \langle \psi_0 | \} v_{w_1}^\dagger | \vec{s} \rangle \langle \vec{r} | v_{w_1} \{ | \psi_0 \rangle \otimes | \varphi_0 \rangle \}}{\{\langle \varphi_0 | \otimes \langle \psi_0 | \} v_{w_1}^\dagger v_{w_1} \{ | \psi_0 \rangle \otimes | \varphi_0 \rangle \}}, \qquad (25.28)$$

where tr_ε is the trace over the environment Hilbert space. Generalizing to a sequence of QEC cycles, we have

$$\rho_{\vec{r},\vec{s}}(\Upsilon_\mathbf{w}) = \frac{\{\langle \varphi_0 | \otimes \langle \psi_0 | \} \Upsilon_\mathbf{w}^\dagger | \vec{s} \rangle \langle \vec{r} | \Upsilon_\mathbf{w} \{ | \psi_0 \rangle \otimes | \varphi_0 \rangle \}}{\{\langle \varphi_0 | \otimes \langle \psi_0 | \} \Upsilon_\mathbf{w}^\dagger \Upsilon_\mathbf{w} \{ | \psi_0 \rangle \otimes | \varphi_0 \rangle \}} . \qquad (25.29)$$

The presence of the interaction V in the Hamiltonian Eq. (25.1) precludes explicitly writing the exact quantum evolution, except in special cases such as that considered in the example below (Section 25.6). Therefore, the best one can do with Eqs. (25.27) and (25.29) is to write them as a double series in V, one for $\Upsilon_\mathbf{w}$ and one for $\Upsilon_\mathbf{w}^\dagger$. These are usually represented graphically as a double contour in time (see Fig. 25.1) [K05a]. The upper leg represents the time-ordered series ($\Upsilon_\mathbf{w}$), while the lower leg represents the anti-time-ordered series ($\Upsilon_\mathbf{w}^\dagger$). In the out-of-equilibrium literature, this is called a diagram on a Keldysh contour [M00, K05a]. There are six interdependent Green functions in such a representation: the usual advanced and retarded functions for the time-ordered series, the advanced and retarded functions for the anti-time-ordered, and the greater ($<$) and lesser ($>$) functions, which involve contracting a term from the time-ordered series with one from the anti-time-ordered series. There are good reviews of the Keldysh formalism [K05a], but unfortunately the diagrammatic rules can be cumbersome. Hence, before we start a general discussion, it is instructive to consider a simple case. For clarity we will focus on the simplest quantity to calculate, $\text{Pr}(\Upsilon_\mathbf{w})$.

25.3.1 Qualitative discussion

Our goal for this section is to develop some intuitive understanding of how QEC works in a critical environment. Hence, we assume for the moment two simplifications: (i) for QEC periods where an error is diagnosed, we expand the evolution to lowest order in V; and (ii) for QEC periods where a "nonerror" is diagnosed, we approximate $U \approx I$. Neither of these assumptions is valid in general (and we will discard them in the next section), but they strip the initial discussion

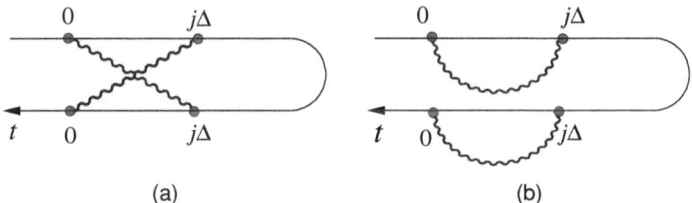

Fig. 25.2. Graphical representation of the two leading fourth-order corrections to the uncorrelated probability [see Eqs. (25.33) and (25.36)]. Adapted with permission from [NMB08]. Copyright 2008, American Physical Society.

of unimportant details. An immediate consequence of these assumptions is that the computer degrees of freedom do not appear explicitly in the time evolution v_m: since the QEC code corrects a single error perfectly, the recovery operation engineers that the computer's time evolution is simply I.

The simplest case is a single error and a single QEC step. Suppose that the syndrome m_1 tell us that a Z error occurred at qubit 1 in the first QEC cycle. Then, Eqs. (25.26) and (25.27) together with the two simplifying assumptions imply that the probability of this history is

$$\Pr\left(v_{m_1}(\Delta,0)\right) = \left(\frac{\lambda_z}{2}\right)^2 \int_0^\Delta dt_2 \int_0^\Delta dt_1 \, \langle f_z^\dagger(\mathbf{x}_1,t_2) f_z(\mathbf{x}_1,t_1) \rangle + \mathscr{O}\left(\lambda_z^4\right). \quad (25.30)$$

The use of power-law correlations, Eq. (25.8), in this integral is problemtic – the short-time behavior must be cut off. Since it is the *long* time behavior that is of interest, we assume that some appropriate "ultraviolet regularization" is made. Then, Eq. (25.30) is well defined, and we set $\epsilon \equiv \Pr\left(v_{m_1}(\Delta,0)\right)$.

The next case to consider is two Z errors at different periods. For instance, let us assume the history points to an error in the first period at qubit 1 and another error in period $j+1$ at qubit 3. Under the simplifying assumptions, the quantum evolution now reads

$$v_{m_3} v_{m_1} = \left(\frac{\lambda_z}{2}\right)^2 \int_{j\Delta}^{(j+1)\Delta} dt_3 \int_0^\Delta dt_1 \, f_z(\mathbf{x}_3,t_3) f_z(\mathbf{x}_1,t_1) + \mathscr{O}\left(\lambda_z^4\right), \quad (25.31)$$

which implies a probability

$$\Pr\left(v_{m_3} v_{m_1}\right) = \left(\frac{\lambda_z}{2}\right)^4 \int_{j\Delta}^{(j+1)\Delta} dt_4 \, dt_3 \int_0^\Delta dt_1 \, dt_2$$
$$\times \langle f_z^\dagger(\mathbf{x}_1,t_2) f_z^\dagger(\mathbf{x}_3,t_4) f_z(\mathbf{x}_3,t_3) f_z(\mathbf{x}_1,t_1) \rangle$$
$$+ \mathscr{O}\left(\lambda_z^6\right). \quad (25.32)$$

Further progress can be made by using the properties of Gaussian integrals. Using Wick's theorem, we can write the four-point correlation function occurring in Eq. (25.32) in terms of products of two-point correlation functions, yielding

$$\langle f_z^\dagger(\mathbf{x}_1,t_2) f_z^\dagger(\mathbf{x}_3,t_4) \, f_z(\mathbf{x}_3,t_3) f_z(\mathbf{x}_1,t_1) \rangle$$

$$= \langle \overline{f_z^\dagger(\mathbf{x}_1,t_2) f_z^\dagger(\mathbf{x}_3,t_4)} \, \overline{f_z(\mathbf{x}_3,t_3) f_z(\mathbf{x}_1,t_1)} \rangle$$

$$+ \langle \overline{f_z^\dagger(\mathbf{x}_1,t_2) \overline{f_z^\dagger(\mathbf{x}_3,t_4) f_z(\mathbf{x}_3,t_3)} f_z(\mathbf{x}_1,t_1)} \rangle$$

$$+ \langle \overline{f_z^\dagger(\mathbf{x}_1,t_2) f_z^\dagger(\mathbf{x}_3,t_4)} \overline{f_z(\mathbf{x}_3,t_3) f_z(\mathbf{x}_1,t_1)} \rangle \quad (25.33)$$

where the bars indicate the corresponding two-point functions. The first term is the simplest to understand: the domains of integration are disjoint yielding a result proportional to ϵ^2 – the probability of having two "uncorrelated" errors. The other terms are corrections to ϵ^2 due to "correlations" between errors and correspond to the situations in Figs. 25.2(a) and (b), respectively. It is expected that these terms will produce small corrections to the "uncorrelated" value. Hence, the strategy that we will follow is to derive a perturbative expansion for these corrections.

There are, of course, many terms to be calculated, but these leading terms already reveal a very important pattern. Since we know from the syndrome that a particular event (error or no-error) has happened, both branches of the Keldysh contour (the time-ordered and anti-time-ordered series) must have insertions of V in the same QEC period (Fig. 25.2).

This becomes critical when considering the probability of having two errors at any time in the full calculation. Then the probability Eq. (25.32) must be summed over all possible intervals. In the absence of QEC, there would be *four* sums as the time arguments of the operators coming from the expansion of $\Upsilon_\mathbf{w}$ are unrelated to those coming from $\Upsilon_\mathbf{w}^\dagger$ (Fig. 25.1). In contrast, when QEC is applied, there are only *two* sums because of the constraint imposed by the syndrome. This reduction of the number of sums corresponds to an enormous reduction in the number of cases (or diagrams) that contribute to any physical process. *In particular, the effect of correlations in the bath is much reduced when QEC is used.*

25.3.2 Quantitative discussion

To make our qualitative discussion quantitative, the key step is to coarse-grain. The idea is that if errors occur far away in space-time, the precise position of each one is not very relevant to the calculation. Hence, we wish to coarse-grain space-time using volumes of size $\Delta \times (v\Delta)^{D/z}$.

But coarse-graining introduces a potential conceptual problem. If two qubits are separated by a distance smaller than $(v\Delta)^{1/z}$, then we can define neither an independent probability for an event nor a unique "long-range" operator. To understand this problem, imagine for instance that two errors of the α-type were diagnosed in two physical qubits belonging to different logical qubits (let us again use physical qubits 1 and 3 for that purpose). If the two errors occur within the fundamental volume $\Delta \times (v\Delta)^{D/z}$, then the probability of this event, to lowest order in V, is

$$\Pr = \epsilon^2 + \left(\frac{\lambda_\alpha}{2}\right)^4 \int_0^\Delta dt_1\, dt_2\, dt_3\, dt_4$$

$$\times \left[\langle f_\alpha^\dagger(\mathbf{x_1},t_2) f_\alpha^\dagger(\mathbf{x_3},t_4) f_\alpha(\mathbf{x_3},t_3) f_\alpha(\mathbf{x_1},t_1)\rangle\right.$$

$$\left. + \langle f_\alpha^\dagger(\mathbf{x_1},t_2) f_\alpha^\dagger(\mathbf{x_3},t_4) f_\alpha(\mathbf{x_3},t_3) f_\alpha(\mathbf{x_1},t_1)\rangle\right]. \tag{25.34}$$

The last two terms are corrections to the probability ϵ of a qubit error that are conditional on events in the other qubits inside the fundamental volume. Similarly, if we try to coarse-grain space-time, we find a different long-range operator for each possible set of events.

To avoid dealing with such conditional probabilities, we assume the single most important simplifying hypothesis of our discussion. We assume hereafter that the qubits are separated by at least the distance ξ given in Eq. (25.9), the distance that a disturbance in the bath propagates in the time Δ. Thus, for all qubits separated by at least this minimum distance, the correlator of bath operators vanishes. For instance, in the previous example of two qubits, if $\mathbf{x}_1 \neq \mathbf{x}_3$ and $|t_1 - t_3| < \Delta$, we have $\langle f_\alpha(\mathbf{x}_3,t_3) f_\alpha(\mathbf{x}_1,t_1)\rangle \approx 0$. The hypercube assumption reduces Eq. (25.34) to simply $\Pr = \epsilon^2$, regardless of the spatial distance between qubits 1 and 3, since the errors occurred in the same QEC cycle so there is not enough time for correlations to propagate. In conclusion, the hypercube assumption allows us to assign an independent probability for an error to each qubit. Likewise, a unique long-range coarse-grained operator can be defined.

Perhaps the simplest method to carry out the coarse-graining is to use a technique from quantum field theory known as the operator product expansion (OPE) [P98b]. It allows us to rewrite $\Upsilon(\mathbf{w})$ keeping only the terms that are most divergent at *long* times. For instance, the OPE for the environmental operator in the first time interval is

$$f_z(\mathbf{x_1},t_1) \sim f_z(\mathbf{x_1},0) + \partial_t f_z(\mathbf{x_1},0)\, t_1 + \text{l.r.t.}, \tag{25.35}$$

where $\sim$ stands for "equal up to nonsingular terms" and the abbreviation l.r.t. means "less relevant terms." (Because the distance between qubits is greater than ξ, no spatial derivative terms appear.) Notice that we have replaced the dependence of the field in a small interval of time by its value at one point in that interval, neglecting all rapid variations of the field, which will be unimportant when viewed from far away. The OPE method provides a rigorous way of carrying this out for any field theory of the type considered here.

Returning to the example of the last subsection, Eq. (25.32), we can write the *leading* corrections to the uncorrelated probability,

$$\Pr(v_{m_3} v_{m_1}) \approx \epsilon^2$$
$$+ \frac{\lambda_z^4 \Delta^2}{2^4} \langle f_z^\dagger(\mathbf{x_1},0) f_z(\mathbf{x_3},j\Delta)\rangle \langle f_z^\dagger(\mathbf{x_3},j\Delta) f_z(\mathbf{x_1},0)\rangle$$
$$+ \frac{\lambda_z^4 \Delta^2}{2^4} \langle f_z^\dagger(\mathbf{x_1},0) f_z^\dagger(\mathbf{x_3},j\Delta)\rangle \langle f_z(\mathbf{x_3},j\Delta) f_z(\mathbf{x_1},0)\rangle. \tag{25.36}$$

The dependence on only two time arguments is now transparent. Technically, this means that the effective scaling dimension for the infrared component of the probability is doubled with respect to the naive expectation.

Until this point, our discussion was based on using an expansion to the evolution operator to lowest nontrivial order. However, in order to obtain a better quantitative result, it is important to try to improve the expansion by taking into account higher-order contributions. For this purpose, we start by writing the evolution operator at the end of a QEC cycle. Using the hypercube assumption, one finds that the evolution operator when an error α is diagnosed on qubit 1 in the first QEC cycle is given by

$$v_\alpha(\mathbf{x}_1, \lambda_\alpha) \approx -i\lambda_\alpha \int_0^\Delta dt \, f_\alpha(\mathbf{x}_1, t) - \frac{1}{2} \sum_{\beta,\gamma} |\epsilon_{\alpha\beta\gamma}| \lambda_\beta \lambda_\gamma \, \sigma_\alpha(\Delta)$$

$$\times T \int_0^\Delta dt_1 \, dt_2 \, f_\beta(\mathbf{x}_1, t_1) f_\gamma(\mathbf{x}_1, t_2) \sigma_\beta(t_1) \sigma_\gamma(t_2)$$

$$+ \frac{i}{6} \sum_\beta \lambda_\alpha \lambda_\beta^2 \, \sigma_\alpha(\Delta) \, T \int_0^\Delta dt_1 \, dt_2 \, dt_3$$

$$\times f_\alpha(\mathbf{x}_1, t_1) f_\beta(\mathbf{x}_1, t_2) f_\beta(\mathbf{x}_1, t_3) \sigma_\alpha(t_1) \sigma_\beta(t_2) \sigma_\beta(t_3)$$

$$+ \cdots, \tag{25.37}$$

where $\epsilon_{\alpha\beta\gamma}$ is the antisymmetric tensor. The second term accounts for two interactions that together produce the same syndrome as the first-order term. Note the appearance of operators in the computer's Hilbert space; in particular, the factor $\sigma_\alpha(\Delta)$ is the recovery operation in this case. Likewise, in the third term, three interactions in the qubit give the same syndrome. Unless the coupling λ_α is extremely small, the sum must be performed in order to have a reasonable estimate of ϵ.

At this point, one can use the renormalization group to systematically take into account higher-order terms by "dressing" the lower-order ones [S94a, GNT98], as has been done in [NMB07, NMB08]. This is certainly a compact and elegant way to proceed, but a conceptually simpler approach is also possible in many important cases by summing the infinite series of so-called "bubble" diagrams [M00, BF04]. In both cases, one ends up with an expression for the effective evolution operator where only the most significant lowest-order term appears, and its prefactor is renormalized. Thus, one is led to an effective truncation of the series in Eq. (25.37) and a substantial simplification of the problem, even though an infinite number of terms are being taken into account. This approach is sound as long as the bare qubit–bath coupling is sufficiently weak, which we assume to be the case.

From the qualitative discussion, we know that there are two very different frequency regimes: (i) a high-frequency domain (from Δ^{-1} to ω_Λ, which is the cutoff frequency of V) corresponding to summing up the most divergent diagrams inside a hypercube, and (ii) a low-frequency domain related to contractions (i.e., correlations) between hypercubes. Hence, it is convenient to separate these two domains at the bath operator level,

$$f_\alpha = f_\alpha^{(<)} + f_\alpha^{(>)}, \tag{25.38}$$

and integrate out of the problem the high-frequency components of the environment [S94a]. This integration produces two effects. First, the sum of "bubble" diagrams containing only high-frequency terms usually produces a geometric progression. Assuming, for the sake of simplicity, that the bath operators commute (bosonic commutation relations), the series adds to

$$\epsilon_\alpha \sim \frac{\lambda_\alpha^2 \, (\Omega\Delta)^{2(z-\delta)}}{1 + \lambda_\alpha^2 \, (\Omega\Delta)^{2(z-\delta)}}, \tag{25.39}$$

where Ω is a function of ω_Λ and other microscopic parameters. The second effect is to dress the low-frequency part and create an effective coupling constant:

$$\lambda_\alpha^* \sim \frac{\lambda_\alpha}{\sqrt{1 + \lambda^2 \, (\Omega\Delta)^{2(D+z-\delta)}}}. \tag{25.40}$$

Note that in obtaining these last two equations, no QEC is involved; indeed, because of the hypercube assumption, only a single qubit is involved for frequencies $> 1/\Delta$. Thus, these are essentially standard results in condensed matter many-body theory and can be found, for instance, in [M00] and [BF04].

These two contributions can be incorporated in a set of low-frequency operators, using the OPE, defined on the coarse-grained space-time grid,

$$v_\alpha^2(\mathbf{x}, \Delta, 0) \approx \epsilon_\alpha + (\lambda_\alpha^* \Delta)^2 : [f_\alpha^{(<)}(\mathbf{x},0)]^\dagger f_\alpha^{(<)}(\mathbf{x},0): \tag{25.41}$$

$$v_0^2(\mathbf{x}, \Delta, 0) \approx 1 - \sum_\alpha \epsilon_\alpha - \sum_\alpha (\lambda_\alpha^* \Delta)^2 : [f_\alpha^{(<)}(\mathbf{x},0)]^\dagger f_\alpha^{(<)}(\mathbf{x},0): \tag{25.42}$$

with : : denoting normal ordering with respect to the bath state φ_0. Here we have immediately combined the individual insertions into products, assuming that all the bath operators commute. This drastic assumption leads to consideration of probabilities rather than interfering amplitudes. In a more realistic situation, the commutator structure of the quantum bath will intervene and lead to other terms neglected in any error model [NP09] that focuses on probability alone. The simple spin-boson model is an example that can be worked out in detail. It is treated below in Section 25.6; see in particular the discussion in Section 25.6.3.

For further analysis, it is convenient to work with the slightly modified operators F_α defined by

$$F_\alpha(\mathbf{x},0) = \frac{1}{\epsilon_\alpha} (\lambda_\alpha^* \Delta)^2 : [f_\alpha^{(<)}(\mathbf{x},0)]^\dagger f_\alpha^{(<)}(\mathbf{x},0) : \tag{25.43}$$

$$F_0(\mathbf{x},0) = 1 - \frac{\sum_\alpha (\lambda_\alpha^* \Delta)^2}{1 - \sum_\alpha \epsilon_\alpha} : [f_\alpha^{(<)}(\mathbf{x},0)]^\dagger f_\alpha^{(<)}(\mathbf{x},0) : \tag{25.44}$$

so that

$$v_\alpha^2(\mathbf{x}, \Delta, 0) \approx \epsilon_\alpha \left[1 + F_\alpha(\mathbf{x},0) \right] \tag{25.45}$$

$$v_0^2(\mathbf{x}, \Delta, 0) \approx \left(1 - \sum_\alpha \epsilon_\alpha\right) F_0(\mathbf{x},0). \tag{25.46}$$

If ϵ_α and λ_α^* are small parameters, Eqs. (25.43)–(25.46) will give good approximations to the expectation values that we are trying to evaluate. They separate the local (stochastic) contribution to the probability of a particular syndrome from the long-range (correlated) part.

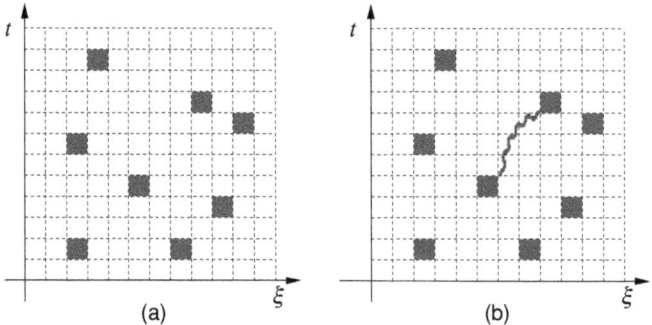

Fig. 25.3. Graphical representation of Eqs. (25.48) and (25.49). The wiggly line represents a pair contraction $\langle F_\alpha(\mathbf{x}_i, t_i) F_\alpha(\mathbf{x}_j, t_j) \rangle$.

25.4 The threshold theorem in a critical environment

Using the results from the previous section and the hypercube assumption, one can calculate the probability of a particular history of syndromes. In this section, we use such a calculation to address the issue of the resilience of quantum computation in a critical environment. The basic strategy is rather simple: If a correlated error model can be reasonably approximated by a stochastic error model, then we can make use of the traditional derivation of the threshold theorem [G98, AB.-O08, KLZ98a]. Thus, our goal is to find out when this approximation is possible.

We start by asking what is the probability of a computer with R qubits to have m errors of type α diagnosed after N QEC cycles. Based on Eqs. (25.45) and (25.46), this probability can be written as

$$\Pr_m^\alpha = p_m \int \frac{d\mathbf{x}_1}{(\mathrm{v}\Delta)^{D/z}} \cdots \frac{d\mathbf{x}_m}{(\mathrm{v}\Delta)^{D/z}} \int_0^{N\Delta} \frac{dt_1}{\Delta} \cdots \int_0^{t_{m-1}} \frac{dt_m}{\Delta}$$
$$\times \left\langle \left[\prod_\zeta F_0(\mathbf{x}_\zeta, t_\zeta) \right] [1 + F_\alpha(\mathbf{x}_1, t_1)] \cdots [1 + F_\alpha(\mathbf{x}_m, t_m)] \right\rangle, \qquad (25.47)$$

where the integrals incorporate sums over all possible space-time grid positions of m errors (i.e., the possible location of the corresponding hypercubes), ζ denotes the set of remaining hypercubes (free of errors), and $p_m = (1 - \sum_\alpha \epsilon_\alpha)^{RN-m} (\epsilon_\alpha)^m$ is the probability of a particular configuration in the uncorrelated error model.

We now expand and reorganize the expectation value shown in Eq. (25.47) in powers of $(\lambda^*, \epsilon_\alpha)$ and invoke Wick's theorem again. The first, leading term we find is just the stochastic contribution to the probability (see Fig. 25.3a), namely,

$$p_m \int \prod_{k=1}^m \frac{d\mathbf{x}_k}{(\mathrm{v}\Delta)^{D/z}} \frac{dt_k}{\Delta} = p_m \binom{NR}{m} \sim p_m (NR)^m, \qquad (25.48)$$

where the integrals account for the number of ways we can distribute m errors in NR hypercubes. The next, sub-leading terms are typically of the form (see Fig. 25.3b)

$$p_m \int \prod_{k=1}^{m} \frac{d\mathbf{x}_k}{(v\Delta)^{D/z}} \frac{dt_k}{\Delta} \langle F_\alpha(\mathbf{x}_i, t_i) F_\alpha(\mathbf{x}_j, t_j) \rangle. \tag{25.49}$$

Other higher-order terms can be systematically taken into account.

A perturbative expansion in λ_α^* is stable only if the term represented by (25.49) remains finite and smaller than the leading term (25.48). Thus, as the last step in our calculation, we invoke Wick's theorem again and use Eq. (25.8) to find that the two-point correlation function for F_α has the general form

$$\langle F_\alpha(\mathbf{x}_i, t_i) F_\alpha(\mathbf{x}_j, t_j) \rangle \sim \mathcal{O}\left(\frac{1}{|\mathbf{x}_i - \mathbf{x}_j|^{4\delta_\alpha}}, \frac{1}{|t_i - t_j|^{4\delta_\alpha/z}}\right). \tag{25.50}$$

Substituting this correlation function into (25.49) and counting powers, we see that a sufficient condition for the stability of the perturbative expansion is the dimensional inequality

$$D + z - 2\delta_\alpha < 0. \tag{25.51}$$

In other words, whenever (25.51) is satisfied, correlations between hypercubes are expected to produce small corrections to Eq. (25.48) as long as ϵ_α is sufficiently small. This means that as long as (25.51) holds, one can employ the traditional proof of resilience based on stochastic errors, which is applicable whenever the single-qubit error probability ϵ_α stays below a certain (model-dependent) threshold value.

The opposite situation, namely, $D + z - 2\delta_\alpha > 0$, is much less clear. In this case, the expansion is not stable and no conclusion can be draw. It is important to emphasize that the calculation we just did does not preclude QEC from still being effective. The instability is only telling us that a perturbative expansion in λ_α^* is not well defined and that the threshold theorem, as we stated, is not valid. It is conceivable that some different derivation of a condition for resilience still exists, but it would likely require a nonperturbative construction.

25.5 The threshold theorem and quantum phase transitions

It is possible to make an analogy between our discussion so far and the theory of quantum phase transitions [S99a]. In fact, the traditional threshold theorem can be thought as a quantum/classical phase transition: as argued in [A00b], the error probability ϵ plays to fault tolerance a role similar to the temperature in a physical system. This is how the analogy works:

(i) For $\epsilon < \epsilon_c$, fault-tolerant procedures allow the computer to maintain a large entanglement among its components and at the same time maintain weak entanglement to the environment. Hence, due to this large internal entanglement, the quantum computer departs from the classical computer model and cannot be efficiently simulated by a Turing machine.

(ii) For $\epsilon > \epsilon_c$, the computer components are weakly entangled and, therefore, can be efficiently simulated by a Turing machine. In other words, the computer density matrix no longer represents a pure state, but rather a statistical mixture, and the computer components are strongly entangled with the environment.

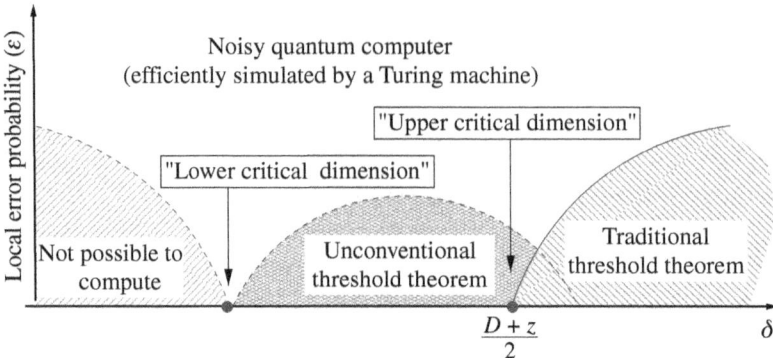

Fig. 25.4. Phase diagram of a quantum computer running QEC. The parameter δ denotes the scaling dimension of the environment operator in the system–environment interaction [see Eq. (25.8)]. Reprinted with permission from [NMB08]. Copyright 2008, American Physical Society.

In this sense, the threshold theorem defines a "phase transition" from a high-temperature phase, where qubits are "independent" from each other (case i), to a low-temperature phase (case ii) where quantum coherence and entanglement are possible. Our analysis of a critical environment adds another parameter to this interpretation: The scaling dimension δ. This quantum parameter defines another axis in the "phase diagram" of a quantum computer (see Fig. 25.4).

In the theory of quantum phase transitions, a dimensional criterion such as (25.51) defines what is usually referred to as the *upper critical dimension* of the model [S99a]. Above the upper critical dimension [when an inequality such as (25.51) is satisfied], the model is essentially free, perturbation theory works, and critical exponents follow their mean-field values. Below the upper critical dimension, there are two possibilities: (i) If the model is above its *lower critical dimension*, some infinite re-summation of diagrams is necessary, but the model still preserves its weak coupling character (although no longer with mean-field exponents); (ii) if, however, the model is below its lower critical dimension, it is not possible to determine its characteristics by re-summing of diagrams. Fluctuations are usually too strong and the system does not order.

In the case of a quantum computer, what matters as far resilience is concerned is how fast the bath correlation functions decay, namely, how large the scaling dimension 2δ of the F operators is in comparison to the effective dimension of the system, $D + z$. When (25.51) is satisfied, correlation functions decay sufficiently fast so as to allow for a just a minor, perturbative correction to the stochastic model. Thus, if we take the analogy with the theory of quantum phase transitions literally, we could argue that below the upper critical scaling dimension of the computer, i.e., when $\delta < (D + z)/2$, there are two possible scenarios. It is possible that for a range of scaling dimensions δ another proof of resilience can be achieved, albeit nonperturbative. However, there may also be a range of δ, below a lower critical value, where there is never resilience to errors. Although this analogy is very speculative and no proof of its validity is known, the similarity between the threshold theorem and quantum phase transitions is so striking and intuitive that the existence of these extra "phases" is rather tantalizing and likely worth exploring.

25.6 An example: the simplified spin-boson model

In the previous sections we presented a general approach to the problem of critical environments. Let us now demonstrate how a calculation is actually done through an example. For that purpose, some assumptions about the computer geometry and the Hamiltonians that govern its evolution have to be made. We will adopt a simplified version of the spin-boson model that has been extensively used by the quantum information community [U95, PSE96, DG98b, KF05, NB06, NP09]. But before we start to employ this model, let us define the layout of the computer. We will assume that the qubits are located at the vertices of a D-dimensional lattice of constant η, where $D = 1, 2,$ or 3. This lattice is immersed in a gapless bosonic environment that evolves according to the noninteracting Hamiltonian

$$H_{\mathrm{CL}} = \sum_{\mathbf{k}\neq 0} \omega_{\mathbf{k}} a_{\mathbf{k}}^{\dagger} a_{\mathbf{k}}, \qquad (25.52)$$

where $[a_{\mathbf{k}}, a_{\mathbf{k}'}^{\dagger}] = \delta_{\mathbf{k},\mathbf{k}'}$, with $\mathbf{k} = \frac{2\pi}{L}\mathbf{n}$. Here, L denotes the physical size of the environment – taken to have the same dimension as the computer – and $\mathbf{n}$ is a vector made of integers. We choose modes with a linear dispersion relation, $\omega_{\mathbf{k}} = v|\mathbf{k}|$, thus $z = 1$. For simplicity, we exclude zero modes. The coupling between the bosonic degrees of freedom and the qubits is given by

$$V = \frac{\lambda}{2} \sum_{j} \sum_{\mathbf{k}\neq 0} c|\mathbf{k}|^{s} \left(e^{i\mathbf{k}\cdot\mathbf{x}_j} a_{\mathbf{k}}^{\dagger} + e^{-i\mathbf{k}\cdot\mathbf{x}_j} a_{\mathbf{k}} \right) \sigma_z(\mathbf{x}_j), \qquad (25.53)$$

with c being a dimensionful and s a dimensionless constant. There are two roughly equivalent methods to regularize the high-frequency dependence of the bosonic component of V [GNT98]. It is possible to define a "hard" cutoff wave-vector Λ, or to consider a "smooth" cutoff by introducing in the interaction Hamiltonian the exponential damping $e^{-\frac{|\mathbf{k}|}{\Lambda}}$. All relations hereafter should be understood as regularized at high frequencies.

The interaction Hamiltonian (25.53) defines the pointer basis for the qubits as the direction z [U95, BP02, Z03]. Hence, it is natural to also define z as the computational basis. The reason is that the z direction corresponds to the natural classical basis that would emerge if the qubits were left by themselves to decohere. In order to obtain the complete spin-boson model, we should also include a finite tunneling splitting between the two states in the computational basis [LCD+87].

$$\tilde{V} = \sum_{j} h_j \sigma_x(\mathbf{x}_j). \qquad (25.54)$$

Such a coupling accounts for the case when the qubit is not perfect (even in the absence of a bath). In the standard spin-boson model, the coupling constants h_j are the smallest parameters in the problem [LCD+87]. Although small, they are fundamental to the dynamics and make the problem very hard to study analytically [GPW99]. Thus, we take a pragmatic approach. We assume that with good "engineering" of the physical system, one can set $h_j = 0$ for all j. Finally, we also need to define how precisely we perform the quantum gates. It is clear from our previous discussion [see Eq. (25.20)] that a minimal error model that accounts for "slow" gates should have the form

$$V_{\text{eff}} = \frac{\lambda}{2} \sum_j \sum_{\mathbf{k} \neq 0} c |\mathbf{k}|^s \left(e^{i\mathbf{k}\cdot\mathbf{x}_j} a_{\mathbf{k}}^\dagger + e^{-i\mathbf{k}\cdot\mathbf{x}_j} a_{\mathbf{k}} \right) \left[\sum_{\alpha=x,y,z} \sigma_\alpha(\mathbf{x}_j) \right], \quad (25.55)$$

instead of Eq. (25.53). However, in the spirit of keeping the discussion as simple as possible, we assume that gates can be done perfectly and in a time smaller that Λ^{-1}. Hence, we will use the microscopic Hamiltonian in Eq. (25.53) as our error model.

A disclaimer is now in order. If all these assumptions were true, we would not need to use QEC. A more cost-effective method would be to use dynamical decoupling (in the case where $h_j \neq 0$) [VKL99, VS98] or decoherence-free subspaces (in the case where $h_j = 0$) [LCW98]. However, we want to learn about QEC, and these assumptions lead to two very useful facts. First, they allow for a closed form of the quantum evolution operator to all orders in λ. Second, since this is a pure dephasing error model, we can use a simple three-qubit code to protect the quantum information [S96d, NC00]. Hence, we emphasize that this is more of a pedagogical introduction to the general approach than the solution of a practical problem. For a realistic problem, the use of perturbative methods as introduced in the previous sections is unavoidable.

25.6.1 Evolution operator

In order to make the notation compact, it is useful to define the bosonic field

$$\theta(\mathbf{x},t) = -i\frac{c}{v} \sum_{\mathbf{k} \neq 0} |\mathbf{k}|^{s-1} \left(e^{i\mathbf{k}\cdot\mathbf{x}+iv|\mathbf{k}|t} a_{\mathbf{k}}^\dagger - e^{-i\mathbf{k}\cdot\mathbf{x}-iv|\mathbf{k}|t} a_{\mathbf{k}} \right) \quad (25.56)$$

and the commutators

$$\Xi(\mathbf{x}-\mathbf{y},t-t') = [\theta(\mathbf{x},t),\theta(\mathbf{y},t')]$$
$$= -2i \left(\frac{c}{v}\right)^2 \sum_{\mathbf{k} \neq 0} |\mathbf{k}|^{2s-2} \sin\left[\mathbf{k}\cdot(\mathbf{x}-\mathbf{y}) + v|\mathbf{k}|(t-t')\right] \quad (25.57)$$

$$\Theta(\mathbf{x}-\mathbf{y},t-t') = [\theta(\mathbf{x},t),\partial_{t'}\theta(\mathbf{y},t')]$$
$$= 2i\frac{c^2}{v} \sum_{\mathbf{k} \neq 0} |\mathbf{k}|^{2s-1} \cos\left[\mathbf{k}\cdot(\mathbf{x}-\mathbf{y}) + v|\mathbf{k}|(t-t')\right] \quad (25.58)$$

$$\Phi(\mathbf{x}-\mathbf{y},t-t') = [\partial_t\theta(\mathbf{x},t),\partial_{t'}\theta(\mathbf{y},t')]$$
$$= -2ic^2 \sum_{\mathbf{k} \neq 0} |\mathbf{k}|^{2s} \sin\left[\mathbf{k}\cdot(\mathbf{x}-\mathbf{y}) + v|\mathbf{k}|(t-t')\right]. \quad (25.59)$$

(An interesting case is given by $s = 1 - D/2$; for $D = 1$, this corresponds to the so-called Ohmic case of the spin-boson model [LCD+87].) Using this notation, the interaction Hamiltonian can be written as

$$V(t) = \frac{\lambda}{2} \sum_j \partial_t \theta(\mathbf{x}_j,t') \sigma_z(\mathbf{x}_j). \quad (25.60)$$

We recall now the Dyson series Eq. (25.3),

$$U(t,0) = Te^{-i\int_0^t dt' V(t')}. \quad (25.61)$$

The model defined by Eq. (25.60) is attractive for analytical calculations because $[V(t), V(t')]$ is a c-number. Hence, it is straightforward to rewrite the Dyson series as a simple exponential (the Magnus expansion, see Section 1.2.2.3),

$$U(t,0) = e^{-\left(\frac{\lambda}{2}\right)^2 \sum_{j,l} \int_0^t dt' \int_0^{t'} dt'' \Phi(\mathbf{x}_j, \mathbf{x}_l, t', t'') \sigma_z(\mathbf{x}_j) \sigma_z(\mathbf{x}_l)}$$
$$\times e^{-i\frac{\lambda}{2} \sum_j \int_0^t dt' \partial_t \theta(\mathbf{x}_j, t') \sigma_z(\mathbf{x}_j)}, \quad (25.62)$$

where we used Eq. (25.59). The first exponential is sometimes called the "lambda shift" [DG98b] and in general introduces a rapidly oscillating phase between the qubits. Now we evoke our *hypercube assumption*. Hereafter, we will *assume* that the distance η between nearest neighbor qubits is much larger than $\xi \equiv v\Delta$, where Δ is the QEC cycle period. This assumption has a simple physical interpretation: A qubit will only enter the causality cone of another qubit in future QEC cycles. The immediate consequence is that Eqs. (25.57) and (25.59) simplify to

$$\Xi(\mathbf{x}_j - \mathbf{x}_l, t' - t'' < \Delta) \propto \delta_{j,l}, \quad (25.63)$$
$$\Phi(\mathbf{x}_j - \mathbf{x}_l, t' - t'' < \Delta) \propto \delta_{j,l}. \quad (25.64)$$

Therefore, the evolution operator just before the syndrome extraction is given by

$$U(\Delta, 0) = e^{-\left(\frac{\lambda}{2}\right)^2 \sum_j \int_0^\Delta dt' \int_0^{t'} dt'' \Phi(\mathbf{x}_j - \mathbf{x}_j, t' - t'')}$$
$$\times \prod_j e^{-i\frac{\lambda}{2} \int_0^\Delta dt' \partial_t \theta(\mathbf{x}_j, t') \sigma_z(\mathbf{x}_j)}. \quad (25.65)$$

We can then perform one of the time integrals and obtain the evolution operator [BP02]:

$$U(\Delta, 0) = e^{-\left(\frac{\lambda}{2}\right)^2 \sum_j \int_0^\Delta dt' \int_0^{t'} dt'' \Phi(\mathbf{x}_j - \mathbf{x}_j, t' - t'')}$$
$$\times \prod_j e^{-i\frac{\lambda}{2}[\theta(\mathbf{x}_j, \Delta) - \theta(\mathbf{x}_j, 0)] \sigma_z(\mathbf{x}_j)}. \quad (25.66)$$

For the QEC discussion below, it is also convenient to break down the last exponential:

$$U(\Delta, 0) = e^{-\left(\frac{\lambda}{2}\right)^2 \sum_j \int_0^\Delta dt' \int_0^{t'} dt'' \Phi(\mathbf{x}_j - \mathbf{x}_j, t' - t'')}$$
$$\times \prod_j \left\{ \cos\left\{\frac{\lambda}{2}[\theta(\mathbf{x}_j, \Delta) - \theta(\mathbf{x}_j, 0)]\right\} \right.$$
$$\left. - i \sin\left\{\frac{\lambda}{2}[\theta(\mathbf{x}_j, \Delta) - \theta(\mathbf{x}_j, 0)]\right\} \sigma_z(\mathbf{x}_j) \right\}. \quad (25.67)$$

In order to make the equations more compact, we introduce the notation

$$c(\lambda, \mathbf{x}_j, t) = \cos\left\{\frac{\lambda}{2}[\theta(\mathbf{x}_j, t + \Delta) - \theta(\mathbf{x}_j, t)]\right\}, \quad (25.68)$$

$$s(\lambda, \mathbf{x}_j, t) = \sin\left\{\frac{\lambda}{2}[\theta(\mathbf{x}_j, t + \Delta) - \theta(\mathbf{x}_j, t)]\right\}. \quad (25.69)$$

25.6.2 The computer evolution with a three-qubits QEC code

Having the evolution operator in a closed form, we can now discuss how the computer evolves in the presence of a QEC code. Since only phase-flip errors occur, we can use a three-qubit code

Fig. 25.5. Three-qubit QEC code [NC00]. The initial wave function, $|\psi_0\rangle \otimes [(|\uparrow\rangle + |\downarrow\rangle)/2] \otimes [(|\uparrow\rangle + |\downarrow\rangle)/2]$, is encoded by two CNOT gates, $R_{\text{CNOT}} = \sigma_i^- \sigma_i^+ \sigma_j^x + \sigma_i^+ \sigma_i^-$, into an entangled state $|\psi_{\text{encode}}\rangle = \alpha|\bar{\uparrow}\rangle + \beta|\bar{\downarrow}\rangle$, with $|\bar{\uparrow}\rangle = (|\uparrow\uparrow\uparrow\rangle + |\uparrow\downarrow\downarrow\rangle + |\downarrow\uparrow\downarrow\rangle + |\downarrow\downarrow\uparrow\rangle)/2$ and $|\bar{\downarrow}\rangle = (|\downarrow\downarrow\downarrow\rangle + |\downarrow\uparrow\uparrow\rangle + |\uparrow\downarrow\uparrow\rangle + |\uparrow\uparrow\downarrow\rangle)/2$. After some time, the information is decoded by a second pair of CNOT gates. An error in $|\psi\rangle$ is identified by measuring the value of σ_2^x and σ_3^x (rectangle). The QEC cycle ends with the correction of a possible phase-flip (arrow).

[S96d, NC00]. The code circuit is presented in Fig. 25.5. The relevant Pauli group for the three-qubit code is G_3. Within G_3 there is a subset of all possible errors that can be introduced by the evolution operator of a logical qubit:

$$U(\Delta, 0) = e^{-i\varphi'} \prod_{j=1}^{3} [c(\lambda, \mathbf{x}_j, 0) - i s(\lambda, \mathbf{x}_j, 0) \sigma_z(\mathbf{x}_j)], \qquad (25.70)$$

where we defined the phase

$$\varphi' = -i \left(\frac{\lambda}{2}\right)^2 \sum_{j=1}^{3} \int_0^{\Delta} dt' \int_0^{t'} dt'' \, \Phi(\mathbf{x}_j - \mathbf{x}_j, t' - t''). \qquad (25.71)$$

We call this set the "error set" [NB06], which, for this simple example, is

$$E = \{\sigma_1^z, \sigma_2^z, \sigma_3^z, \sigma_1^z \sigma_2^z, \sigma_1^z \sigma_3^z, \sigma_2^z \sigma_3^z, \sigma_1^z \sigma_2^z \sigma_3^z\}. \qquad (25.72)$$

The logical operation $\bar{Z} = \sigma_1^z \sigma_2^z \sigma_3^z$ is contained in the error set. This means this error keeps the code words in the logical Hilbert and therefore cannot be distinguished from a nonerror event by the code. The subgroup $E_0 = \{I, \bar{Z}\}$ defines a partition of E:

$$E = \left\{ \{I, \bar{Z}\}, \{\sigma_1^z, \sigma_2^z \sigma_3^z\}, \{\sigma_2^z, \sigma_1^z \sigma_3^z\}, \{\sigma_3^z, \sigma_1^z \sigma_2^z\} \right\}. \qquad (25.73)$$

Each left/right coset is associated with a given syndrome and can be used to write the different quantum evolutions at the end of a QEC cycle of time Δ. After the recovery operation is performed, the four possible evolutions of a logical qubit are given by the operators

$$v_0(\mathbf{x}_1, \mathbf{x}_2, \mathbf{x}_3, 0) = e^{-i\varphi'} \left[\prod_{j=1}^{3} c(\lambda, \mathbf{x}_j, 0) + i \prod_{j=1}^{3} s(\lambda, \mathbf{x}_j, 0) \bar{Z} \right], \qquad (25.74)$$

$$v_{l=1,2,3}(\mathbf{x}_1, \mathbf{x}_2, \mathbf{x}_3, 0) = -i e^{-i\varphi'} \left[s(\lambda, \mathbf{x}_l, 0) \prod_{j \neq l} c(\lambda, \mathbf{x}_j, 0) \right.$$

$$\left. - c(\lambda, \mathbf{x}_l, 0) \prod_{j \neq l} s(\lambda, \mathbf{x}_j, 0) \bar{Z} \right]. \qquad (25.75)$$

After the first QEC step, the evolution of the system is given by the operator

$$v_w(0) = \prod_m \mathrm{v}_{w(m)}(\mathbf{m}, 0), \qquad (25.76)$$

where $\mathbf{m}$ labels the positions of the physical qubits, $\{\mathbf{x}_{m_1}, \mathbf{x}_{m_2}, \mathbf{x}_{m_3}\}$, that encode the logical qubit m and $w(m) = \{0,1,2,3\}$ is the corresponding syndrome. Note that the hypercube assumption renders the particular order of the qubits in the product irrelevant.

Following the notation of the previous sections, we define $\mathbf{w}$ to be the history of syndromes after N QEC steps and the "time-ordered" evolution operator is given by Eq. (25.25).

25.6.3 Calculating the probability of a history of syndromes

The simplest quantity to calculate is the probability of a history of syndromes [see Eq. (25.27)]. In fact, such a calculation is sufficient to illustrate all the main issues that appear when environments are critical. For clarity, let us consider the case when, after N QEC cycles, all syndromes returned a "no error" result. The probability of this "flawless" evolution is given by

$$\Pr(0) = \left\langle \left\{ \prod_{i=1}^{N} \left[\prod_m \mathrm{v}_0^\dagger(\mathbf{m}, t_i) \right] \right\} \left\{ \prod_{j=N}^{1} \left[\prod_m \mathrm{v}_0(\mathbf{m}, t_j) \right] \right\} \right\rangle. \qquad (25.77)$$

The "i"-product is "anti-time-ordered," or in other words, the earliest QEC cycle is the utmost term to the left while the latest is the utmost to the right. Conversely, the "j"-product is "time-ordered." The simplest way to proceed is to follow the perturbative prescription explained in the previous sections and evoke Wick's theorem (see, for instance, [NMB08] for an explanation of this approach in the context of the spin-boson model) and the OPE. However, in our case, we can gain a better insight by pushing the exact calculation a little further.

The main strategy is to rewrite Eq. (25.77) as a single time-ordered product. Therefore, we need to carry forward each term in the left brackets of Eq. (25.77) until it reaches its counterpart in the right brackets. There is nothing to be done with the Nth QEC cycle, since corresponding terms in the left and right brackets are already in contact. Furthermore, due to the "hypercube assumption," we can reorder the product and write

$$\left[\prod_m \mathrm{v}_0^\dagger(\mathbf{m}, t_N) \right] \left[\prod_m \mathrm{v}_0(\mathbf{m}, t_N) \right] = \prod_m \mathrm{v}_0^\dagger(\mathbf{m}, t_N) \mathrm{v}_0(\mathbf{m}, t_N)$$

$$= \prod_m \left[\prod_{j=1}^{3} \mathrm{c}^2(\lambda, \mathbf{x}_{m_j}, t_N) + \prod_{j=1}^{3} \mathrm{s}^2(\lambda, \mathbf{x}_{m_j}, t_N) \right],$$

$$= \prod_m \left[\frac{1 + \sum_k \sum_{j \neq k} \mathrm{c}(2\lambda, \mathbf{x}_{m_j}, t_N) \mathrm{c}(2\lambda, \mathbf{x}_{m_k}, t_N)}{4} \right].$$

$$(25.78)$$

The next term that we must consider is $\left[\prod_m \mathrm{v}_0^\dagger(\mathbf{m}, t_{N-1}) \right]$. It should be moved across the operators at the Nth QEC cycle. Using the results from the appendix, Eqs. (25.101)–(25.104), we see that every commutation generates new operators. The algebra is simple but tedious. Here we present a few terms in detail. Suppose we would like to commute a single operator at the

$(N-1)$th QEC cycle with a term in the Nth cycle:

$$c(\lambda, \mathbf{x}_{n_1}, t_{N-1})\left[1 + \sum_k \sum_{j \neq i} c\left(2\lambda, \mathbf{x}_{m_j}, t_N\right) c\left(2\lambda, \mathbf{x}_{m_k}, t_N\right)\right]$$

$$= \left\{\sum_{j \neq k} \cos\left[\frac{\lambda^2}{4}\tilde{\Xi}\left(\mathbf{x}_{n_1}-\mathbf{x}_{m_j}, t_{N-1}-t_N\right)\right] \cos\left[\frac{\lambda^2}{4}\tilde{\Xi}\left(\mathbf{x}_{n_1}-\mathbf{x}_{m_k}, t_{N-1}-t_N\right)\right]\right.$$

$$\left. \times c\left(2\lambda, \mathbf{x}_{m_j}, t_N\right) c\left(2\lambda, \mathbf{x}_{m_k}, t_N\right) + 1\right\} c\left(\lambda, \mathbf{x}_{n_1}, t_{N-1}\right)$$

$$+ i \sum_{j \neq k} \cos\left[\frac{\lambda^2}{4}\tilde{\Xi}\left(\mathbf{x}_{n_1}-\mathbf{x}_{m_k}, t_{N-1}-t_N\right)\right] \sin\left[\frac{\lambda^2}{4}\tilde{\Xi}\left(\mathbf{x}_{n_1}-\mathbf{x}_{m_j}, t_{N-1}-t_N\right)\right]$$

$$\times c\left(2\lambda, \mathbf{x}_{m_k}, t_N\right) s\left(2\lambda, \mathbf{x}_{m_j}, t_N\right) s\left(\lambda, \mathbf{x}_{n_1}, t_{N-1}\right)$$

$$+ i \sum_{j \neq k} \cos\left[\frac{\lambda^2}{4}\tilde{\Xi}\left(\mathbf{x}_{n_1}-\mathbf{x}_{m_j}, t_{N-1}-t_N\right)\right] \sin\left[\frac{\lambda^2}{4}\tilde{\Xi}\left(\mathbf{x}_{n_1}-\mathbf{x}_{m_k}, t_{N-1}-t_N\right)\right]$$

$$\times c\left(2\lambda, \mathbf{x}_{m_j}, t_N\right) s\left(2\lambda, \mathbf{x}_{m_k}, t_N\right) s\left(\lambda, \mathbf{x}_{n_1}, t_{N-1}\right)$$

$$- \sum_{j \neq k} \sin\left[\frac{\lambda^2}{4}\tilde{\Xi}\left(\mathbf{x}_{n_1}-\mathbf{x}_{m_j}, t_{N-1}-t_N\right)\right] \sin\left[\frac{\lambda^2}{4}\tilde{\Xi}\left(\mathbf{x}_{n_1}-\mathbf{x}_{m_k}, t_{N-1}-t_N\right)\right]$$

$$\times s\left(2\lambda, \mathbf{x}_{m_j}, t_N\right) s\left(2\lambda, \mathbf{x}_{m_k}, t_N\right) c\left(\lambda, \mathbf{x}_{n_1}, t_{N-1}\right). \tag{25.79}$$

It is instructive to count powers of λ for each term of Eq. (25.79). The last two terms on the right-hand side correspond to higher-order events in λ and originated from the time-reordering of the Keldysh contour. They are somewhat unsettling since they seem to indicate "errors" in places that we did not expect them to occur. For instance, the term $c\left(\lambda, \mathbf{x}_{n_1}, t_{N-1}\right)$ on the left-hand side came from a syndrome of "no error," thus suggesting a series of the form $1, \lambda^2, \lambda^4, \ldots$ at position $(\mathbf{x}_{n_1}, t_{N-1})$. However, the second term of the right-hand side indicates a series of the form $\lambda, \lambda^3, \lambda^5, \ldots$ at positions $(\mathbf{x}_{n_1}, t_{N-1})$ and $(\mathbf{x}_{m_j}, t_N)$ because of the sine factors. It is tempting to interpret this as propagation of an error from one position to the other, but this is not what is going on. The missing piece is that $\sin\left[\frac{\lambda^2}{4}\tilde{\Xi}\left(\mathbf{x}_{n_1}-\mathbf{x}_{m_k}, t_{N-1}-t_N\right)\right]$ is actually an interaction term from $(\mathbf{x}_{n_1}, t_{N-1})$ to $(\mathbf{x}_{m_k}, t_N)$. It is equivalent to the "lambda shift" term discussed within a QEC cycle, but now these terms represent interactions between different QEC cycles. *Therefore, when reordering the Keldysh contour, an extremely large set of terms is uncovered.*

The expectation value for the probability can now be written as

$$\Pr(0) = \left\langle \prod_{i=N}^{1} \prod_m \left[v_0^\dagger(\mathbf{m}, t_i) v_0(\mathbf{m}, t_i)\right]\right\rangle + \text{"lambda shift" terms}. \tag{25.80}$$

The next step in the calculation is to normal order each term inside the square brackets of Eq. (25.80). Using Eqs. (25.74) and (25.96), we rewrite the probability as

$$\Pr(0) = \left\langle \prod_{i=N}^{1} \prod_m \left[\frac{1 + \sum_{j \neq k} e^{-2\epsilon} : c\left(2\lambda, \mathbf{x}_{m_j}, t_i\right) :: c\left(2\lambda, \mathbf{x}_{m_k}, t_i\right) :}{4}\right]\right\rangle$$
$$+ \text{"lambda shift" terms}. \tag{25.81}$$

To understand what the first term of Eq. (25.81) means, we need to expand it in powers of λ. A very convenient way to do that is to use the OPE. A good description of the OPE is given in the book by Polchinski [P98b]. The OPE is to a quantum field theory what a Taylor series is to calculus. Using Eqs. (25.106) and (25.107), we finally rewrite the probability in lowest order in λ as

$$\Pr(0) = \left\langle \prod_{i=N}^{1} \prod_{m} \left\{ 1 - \frac{3}{2}\epsilon - \lambda^2 \sum_{j=1}^{3} :\left[\partial_t \theta\left(\mathbf{x}_{m_j}, t_i\right)\right]^2 : \Delta^2 + \text{l.r.t.} \right\} \right\rangle$$
$$+ \text{"lambda shift" terms.} \tag{25.82}$$

We see that $p_0 = 1 - \frac{3}{2}\epsilon$ is the single-qubit error probability usually considered in the "threshold analysis." This term is ubiquitous in all analyses where probabilities, rather than wave-function amplitudes, are evolved in time. It indicates that in the limit of an infinite number of qubits or an infinite computational time, an infinite number of interference terms is disregarded. Although they all have higher powers of λ^2, there is absolutely no reason to assume that they are not relevant. The usual claim is that all the "lambda shift" terms and contractions between $:\left[\partial_t \theta\left(\mathbf{x}_{m_j}, t\right)\right]^2:$ in Eq. (25.82) will add incoherently and cancel each other. Although this is possible, an explicit calculation with a specific model ($D = 1$ and $s = 1/2$) shows that this is not in general true. The issue at hand is now quite clear: How can one justify the dismissal of an infinite number of terms that could in principle contribute to the final probability? The same is also true for the calculation of the residual decoherence, thus casting doubts on the entire "threshold" argument.

To address this issue we must go back to Eq. (25.82) and try to find out the meaning of l.r.t. The basic correlation function in this problem can be written as

$$\langle :\partial_t \theta(\mathbf{x}, t): :\partial_t \theta(\mathbf{y}, t'): \rangle \sim \frac{1}{|\mathbf{r}|^{2\delta}}, \tag{25.83}$$

where $\mathbf{r} = (\mathbf{x} - \mathbf{y}, v(t - t'))$ is the "space-time" distance (remember that we chose $z = 1$), and δ is a function of D and s. Thus, the first correction to p_0 due to correlations has the form

$$(\lambda\Delta)^4 \left\langle :[\partial_t \theta(\mathbf{x}, t)]^2: :[\partial_t \theta(\mathbf{y}, t')]^2: \right\rangle \sim \frac{(\lambda\Delta)^4}{|\mathbf{r}|^{4\delta}}. \tag{25.84}$$

Now let us consider all possible positions in the space-time grid $(v\Delta) \times \eta^D$ where we could place the factors $:\partial_t \theta(\mathbf{x}, t):$ and $:\partial_t \theta(\mathbf{y}, t'):$. As is clear from Eq. (25.82), we must sum over all these positions in order to find the contributions of order $(\lambda\Delta)^4$ to the correction of p_0, namely,

$$(\lambda\Delta)^4 \sum_{\mathbf{x},\mathbf{y},t,t'} \left\langle [:\partial_t \theta(\mathbf{x}, t):]^2 [:\partial_t \theta(\mathbf{y}, t'):]^2 \right\rangle \sim \lambda^4 \Delta^2 \int d^D\mathbf{x} d^D\mathbf{y}\, dt\, dt' \frac{1}{|\mathbf{r}|^{4\delta}}. \tag{25.85}$$

This integral converges in the limit of an infinite number of qubits and an infinite computational time only if

$$D + 1 - 2\delta < 0. \tag{25.86}$$

When this inequality is satisfied, it also ensures that higher-order terms in the OPE will also converge. For instance, a term such as

$$(\lambda\Delta)^8 \sum_{\mathbf{x},\mathbf{y},t,t'} \left\langle [:\partial_t \theta(\mathbf{x}, t):]^4 [:\partial_t \theta(\mathbf{y}, t'):]^4 \right\rangle \sim \lambda^8 \Delta^6 \int d^D\mathbf{x} d^D\mathbf{y}\, dt\, dt' \frac{1}{|\mathbf{r}|^{8\delta}} \tag{25.87}$$

will certainly converge. We also note that when $\mathbf{r} \to 0$, all the terms of the OPE expansion diverge in the same way (essentially as $\Delta^{1-\delta}$ to some power). Thus, in this sense, they are "less relevant."

We are now left to understand how the "lambda shift" terms compare to the others that we just discussed. We start by going back to Eq. (25.79) and picking one of the "lambda shift" terms. For instance, a typical "lambda shift term" would have the form

$$\langle c(\lambda, \mathbf{x}, t) c(2, \mathbf{z}, t') c(2\lambda, \mathbf{y}, t') c(\lambda, \mathbf{x}, t) \rangle$$
$$\sim \ldots + \cos\left[\frac{\lambda^2}{4}\tilde{\Xi}(\mathbf{x} - \mathbf{z}, t - t')\right] \sin\left[\frac{\lambda^2}{4}\tilde{\Xi}(\mathbf{x} - \mathbf{y}, t' - t)\right]$$
$$\times \langle c(2\lambda, \mathbf{z}, t') s(2\lambda, \mathbf{y}, t') s(2\lambda, \mathbf{x}, t) \rangle + \cdots, \quad (25.88)$$

where the qubits at positions $\mathbf{y}$ and $\mathbf{z}$ belong to the *same* logical qubit. Thus, when we would eventually like to sum over all possible qubits, we only have *two* independent position variables in the coarse grain grid. Now, as we did before, we normal order,

$$\mathscr{G} = \cos\left[\frac{\lambda^2}{4}\tilde{\Xi}(\mathbf{x} - \mathbf{z}, t - t')\right] \sin\left[\frac{\lambda^2}{4}\tilde{\Xi}(\mathbf{x} - \mathbf{y}, t' - t)\right]$$
$$\times e^{-3\epsilon} \langle : c(2\lambda, \mathbf{z}, t') :: s(2\lambda, \mathbf{y}, t') :: s(2\lambda, \mathbf{x}, t) : \rangle, \quad (25.89)$$

and OPE each term,

$$\mathscr{G} = \cos\left[\frac{\lambda^2}{4}\tilde{\Xi}(\mathbf{x} - \mathbf{z}, t - t')\right] \sin\left[\frac{\lambda^2}{4}\tilde{\Xi}(\mathbf{x} - \mathbf{y}, t' - t)\right]$$
$$\times e^{-3\epsilon} \langle : \cos[\lambda \partial_t \theta(\mathbf{z}, t') \Delta] :: \sin[\lambda \partial_t \theta(\mathbf{y}, t') \Delta] :$$
$$\times : \sin[\lambda \partial_t \theta(\mathbf{x}, t) \Delta] : \rangle. \quad (25.90)$$

Expanding again in power of λ, we obtain

$$\mathscr{G} = \frac{\lambda^4 \Delta^2}{4} \tilde{\Xi}(\mathbf{x} - \mathbf{y}, t' - t) \langle : \partial_t \theta(\mathbf{y}, t') :: \partial_t \theta(\mathbf{x}, t) : \rangle + \cdots. \quad (25.91)$$

It is straightforward to see that for $t' - t \gg \Delta$, $\tilde{\Xi}$ behaves at most as $1/|\mathbf{r}|^{2\delta}$ [see Eq. (25.99) for an example and Eq. (25.100) for the general case]. Hence,

$$\mathscr{G} = \frac{\lambda^6 \Delta^4}{4} \frac{1}{|\mathbf{r}|^{4\delta}} + \cdots, \quad (25.92)$$

and the convergence criterion is still valid. This example is quite instructive to show us the meaning of these "lambda shift" terms. They correspond to higher-order terms in λ that are dressing Eq. (25.85).

There is one last issue to be discussed. When we reorder the Keldysh contour, there is another archetypical term within our label "lambda shift." It involves three physical qubits in three different logical qubits. Since the algebra is very cumbersome, we present here only the lowest-order terms after normal ordering and OPE. They have the form

$$\sim \int d^D\mathbf{x}_{1,2,3} \, dt_{1,2,3} \, \langle : \partial_t \theta(\mathbf{x}_1, t_1) :: \partial_t \theta(\mathbf{x}_2, t_2) : \rangle$$
$$\times \langle : \partial_t \theta(\mathbf{x}_1, t_1) :: \partial_t \theta(\mathbf{x}_3, t_3) : \rangle \langle : \partial_t \theta(\mathbf{x}_2, t_2) :: \partial_t \theta(\mathbf{x}_3, t_3) : \rangle$$
$$\sim \int d^D\mathbf{x}_{1,2,3} \, dt_{1,2,3} \frac{1}{|\mathbf{r}_{12}|^{2\delta} |\mathbf{r}_{13}|^{2\delta} |\mathbf{r}_{23}|^{2\delta}}. \quad (25.93)$$

Once again, these integrals will converge for infinite computational time and infinite number of qubits as long as $D + 1 - 2\delta < 0$. Furthermore, they are less divergent for short distances in "space-time" than any of the previous cases.

In conclusion, as long as the dimensional criterion is satisfied, the perturbative expansion on λ is well behaved and systematic corrections to the usual perturbative statement of the "threshold theorem" can be calculated. Nevertheless, even in this extremely simple problem, it was not possible to give a specific value for the "threshold value," but only to prove that there is one if the dimensional criterion is satisfied.

25.7 Conclusions

It is believed that some sort of QEC will have to be implemented at every logical level [Z03] for quantum computation to become reality. This fact is one of the cornerstones of quantum computation theory. However, it has been argued that the proof that resilient quantum computation relies on a set of unphysical assumptions [AHH+02, AHH+04, ALZ06], namely: (i) "fast" measurements, (ii) "fast" gates, and (iii) stochastic (uncorrelated) errors. Our perspective is that all these problems do not fundamentally limit resilience. They are legitimate concerns, but have now been addressed in the literature. First, in [DA07] DiVincenzo and Aliferis demonstrated that resilient circuits can be constructed with slow measurements. Second, we showed here that, under some reasonable assumptions, it is possible to treat "slow" gates. In this process, we developed a theoretical framework that connects microscopic Hamiltonians with realistic error models in correlated environments. Finally, our results show that a large class of critical environments are already properly treated within the QEC framework.

Some useful results

Normal ordering: The basic object used in the example is the "vertex" $e^{i\alpha\theta(\mathbf{x},t)}$, where α is a real number. The normal ordering with respect to the bosonic vacuum can be done by using the basic properties of the Gaussian integral,

$$\left\langle e^{i\alpha\theta(\mathbf{x},t)} \right\rangle = \prod_{\mathbf{k}\neq 0} \left\langle e^{-i\alpha\frac{c}{v}|\mathbf{k}|^{s-1}\left(e^{i\mathbf{k}\mathbf{x}+iv|\mathbf{k}|t}a_{\mathbf{k}}^{\dagger} - e^{-i\mathbf{k}\mathbf{x}-iv|\mathbf{k}|t}a_{\mathbf{k}}\right)} \right\rangle,$$

$$= e^{-\frac{1}{2}\left(\alpha\frac{c}{v}\right)^2 \sum_{\mathbf{k}\neq 0}|\mathbf{k}|^{2s-2}}. \tag{25.94}$$

Hence, by definition of normal ordering,

$$e^{i\alpha\theta(\mathbf{x},t)} = e^{-\left(\alpha\frac{c}{v}\right)^2 \sum_{\mathbf{k}\neq 0}|\mathbf{k}|^{2s-2}} : e^{i\alpha\theta(\mathbf{x},t)} :, \tag{25.95}$$

where $\left\langle : e^{i\alpha\theta(\mathbf{x},t)} : \right\rangle = 1$.

Another useful normal ordering is

$$e^{i\alpha\theta(\mathbf{x},t)+i\beta\theta(\mathbf{y},t')} = e^{-\epsilon} : e^{i\alpha\theta(\mathbf{x},t)+i\beta\theta(\mathbf{y},t')} :, \tag{25.96}$$

where

$$\epsilon = \frac{1}{2}\left(\frac{c}{v}\right)^2 \sum_{\mathbf{k}\neq 0} |\mathbf{k}|^{2s-2} \left\{\alpha^2 + \beta^2 + 2\alpha\beta \cos\left[\mathbf{k}\left(\mathbf{x}-\mathbf{y}\right) + iv\left|\mathbf{k}\right|\left(t-t'\right)\right]\right\}.$$

Commutation of vertices

$$e^{i\alpha\theta(\mathbf{x},t)}e^{i\beta\theta(\mathbf{y},t')} = e^{i\beta\theta(\mathbf{y},t')}e^{i\alpha\theta(\mathbf{x},t)}e^{-\alpha\beta\Xi(\mathbf{x},\mathbf{y},t,t')},$$

$$e^{i\alpha[\theta(\mathbf{x},t+\Delta)-\theta(\mathbf{x},t)]}e^{i\beta[\theta(\mathbf{y},t'+\Delta)-\theta(\mathbf{y},t')]} = e^{i\beta[\theta(\mathbf{y},t'+\Delta)-\theta(\mathbf{y},t')]}$$
$$\times e^{i\alpha[\theta(\mathbf{x},t+\Delta)-\theta(\mathbf{x},t)]}e^{i\alpha\beta\tilde{\Xi}},$$

with

$$\begin{aligned}\tilde{\Xi}(\mathbf{x}-\mathbf{y},t-t') &= i\Xi(\mathbf{x},\mathbf{y},t+\Delta,t'+\Delta)+i\Xi(\mathbf{x},\mathbf{y},t,t')\\
&\quad -i\Xi(\mathbf{x},\mathbf{y},t+\Delta,t')-i\Xi(\mathbf{x},\mathbf{y},t,t'+\Delta)\\
&= 4\left(\frac{c}{v}\right)^2\sum_{\mathbf{k}\neq 0}|\mathbf{k}|^{2s-2}\cos[\mathbf{k}(\mathbf{x}-\mathbf{y})]\sin[v|\mathbf{k}|(t-t')]\\
&\quad \times\{1-\cos[v|\mathbf{k}|\Delta]\}\end{aligned}\qquad(25.97)$$

Some special cases are:
- if $t-t'=0$,

$$\tilde{\Xi}(\mathbf{x}-\mathbf{y},0)=0;\qquad(25.98)$$

- for $D=1$ and $|t-t'|\gg\Delta$,

$$\tilde{\Xi}(x-y,t-t')\approx\left(\frac{c\Delta}{v}\right)^2\frac{L}{\pi}\sin\left((2s+1)\frac{\pi}{2}\right)\Gamma(2s+1)$$
$$\times\left\{\frac{\text{sgn}(v(t-t')+|x-y|)}{|v(t-t'))+|x-y||^{2s+1}}+\frac{\text{sgn}(v(t-t')-|x-y|)}{|v(t-t')-|x-y||^{2s+1}}\right\},\qquad(25.99)$$

where $\Gamma(z)$ is the Euler Gamma function and L is the length of the system.

In general, we can write

$$\tilde{\Xi}(x-y,t-t')\approx\langle\partial_t\theta(x,t)\partial_t\theta(y,t')\rangle\Delta^2\,.\qquad(25.100)$$

Commutation of "sin" and "cos": Another important point is the commutation of these two terms,

$$c(\lambda,\mathbf{x},t)c(\lambda,\mathbf{y},t') = \cos\left[\frac{\lambda^2}{4}\tilde{\Xi}(\mathbf{x}-\mathbf{y},t-t')\right]c(\lambda,\mathbf{y},t')c(\lambda,\mathbf{x},t)$$
$$+i\sin\left[\frac{\lambda^2}{4}\tilde{\Xi}(\mathbf{x}-\mathbf{y},t-t')\right]s(\lambda,\mathbf{y},t')s(\lambda,\mathbf{x},t)\qquad(25.101)$$

$$s(\lambda, \mathbf{x}, t) s(\lambda, \mathbf{y}, t') = \cos\left[\frac{\lambda^2}{4}\tilde{\Xi}(\mathbf{x} - \mathbf{y}, t - t')\right] s(\lambda, \mathbf{y}, t') s(\lambda, \mathbf{x}, t)$$
$$+ i \sin\left[\frac{\lambda^2}{4}\tilde{\Xi}(\mathbf{x} - \mathbf{y}, t - t')\right] c(\lambda, \mathbf{y}, t') c(\lambda, \mathbf{x}, t) \quad (25.102)$$

$$s(\lambda, \mathbf{x}, t) c(\lambda, \mathbf{y}, t') = \cos\left[\frac{\lambda^2}{4}\tilde{\Xi}(\mathbf{x} - \mathbf{y}, t - t')\right] c(\lambda, \mathbf{y}, t') s(\lambda, \mathbf{x}, t)$$
$$+ i \sin\left[\frac{\lambda^2}{4}\tilde{\Xi}(\mathbf{x} - \mathbf{y}, t - t')\right] s(\lambda, \mathbf{x}, t) c(\lambda, \mathbf{x}, t) \quad (25.103)$$

$$c(\lambda, \mathbf{x}, t) s(\lambda, \mathbf{y}, t') = \cos\left[\frac{\lambda^2}{4}\tilde{\Xi}(\mathbf{x} - \mathbf{y}, t - t')\right] s(\lambda, \mathbf{y}, t') c(\lambda, \mathbf{x}, t)$$
$$+ i \sin\left[\frac{\lambda^2}{4}\tilde{\Xi}(\mathbf{x} - \mathbf{y}, t - t')\right] c(\lambda, \mathbf{y}, t') s(\lambda, \mathbf{x}, t) \quad (25.104)$$

OPEs

$$\lim_{t \to 0} \theta(\mathbf{x}, t) - \theta(\mathbf{x}, 0) = -i\frac{c}{v} \sum_{\mathbf{k} \neq 0} |\mathbf{k}|^{s-1} \left[e^{i\mathbf{k}\mathbf{x}} \left(e^{iv|\mathbf{k}|t} - 1 \right) a_{\mathbf{k}}^{\dagger} \right.$$
$$\left. - e^{-i\mathbf{k}} \left(e^{-iv|\mathbf{k}|t} - 1 \right) a_{\mathbf{k}} \right]$$
$$= \sum_{n=1}^{\infty} \partial_t \theta(\mathbf{x}, 0) \frac{t^n}{n!}, \quad (25.105)$$

$$\lim_{t \to 0} : e^{i\lambda[\theta(\mathbf{x},t) - \theta(\mathbf{x},0)]} := \; : e^{i\lambda[\partial_t \theta(\mathbf{x},0) t + \text{l.r.t.}]} : \quad (25.106)$$

$$\lim_{t \to 0} : \cos\left[\lambda\left[\theta(\mathbf{x}, t) - \theta(\mathbf{x}, 0)\right]\right] := 1 - \frac{\lambda^2}{2} : [\partial_t \theta(\mathbf{x}, 0)]^2 : t^2 + \text{l.r.t.} \quad (25.107)$$

Chapter 26

Critique of fault-tolerant quantum information processing

Robert Alicki

26.1 Introduction

The idea of fault-tolerant quantum computation presents a bold challenge to the rather well established principles called the strong Church–Turing thesis and the Bohr correspondence principle. One of the variations of the strong Church–Turing thesis (SCTT), which is not due to Church or Turing, but rather gradually developed in the field of computational complexity theory, is the following [BV97]:

Any "reasonable" model of computation can be efficiently simulated on a probabilistic Turing machine.

This can be expressed even less formally but more practically as:

No computer can be more efficient than a digital one equipped with a random number generator.

Here, a computer A is "more efficient" than a computer B if A can solve, in polynomial time, a problem that cannot be solved in a polynomial time by computer B.

On the other hand the Bohr correspondence principle (BCP) demands that:

Classical physics and quantum physics give the same answer when the systems become large.

A more rigorous form of the BCP is the following:

For large systems the experimental data are consistent with classical probabilistic models.

Here, by a *classical probabilistic model* we mean a model in which all observables are described by real functions, and all states by probability distributions on a certain set ("phase-space").

The conditions under which quantum and classical physics agree are referred to as the correspondence limit, or the classical limit. Bohr provided a rough prescription for the correspondence limit: it occurs when the quantum numbers describing the system are large, meaning some quantum numbers of the system are excited to a very large value. Note that the number of particles that form a physical system should also be considered a quantum number.

Quantum Error Correction, ed. Daniel A. Lidar and Todd A. Brun. Published by CAMBRIDGE UNIVERSITY PRESS.
© Cambridge University Press 2013.

One should not expect that the principles above can be proved in the sense of mathematical theorems. They are rather similar (and perhaps related) to the second law of thermodynamics, which is supported by numerous theoretical models of very different levels of generality analyzed within more or less rigorous frameworks, and a large body of experimental data.

First of all, one should notice that, in practice, *the BCP implies the SCTT*. Indeed, in classical physics one has, essentially, two models of computers: digital and analog. The latter could in principle be more powerful because they use continuous variables described by real numbers. However, the finite accuracy of state preparation and measurement combined with the chaotic behavior of generic classical dynamics implies that analog computers cannot out-perform digital ones. The only known resource of Nature that remains is "quantumness," i.e., superpositions and entanglement of quantum states. However, all complexity notions are asymptotic, valid in the limit of large input and consequently large computer. Therefore, if the BCP is universally valid, a large quantum computer is equivalent to a classical analog machine.

An often used heuristic argument for the robustness of quantum information processing relies on the linearity of the Schrödinger equation, in contrast to the nonlinearity of the evolution equation for a generic classical system. This is, however, a misunderstanding as both theories should be compared on the level of dynamical equations for states or observables, which are linear in both the quantum and classical cases. Moreover, there exists a large body of evidence that on a logarithmic (with respect to Hilbert space dimension) time scale, quantum dynamics follows its corresponding semiclassical limit with the same sensitivity to external perturbations [ZP94, AF01].

In the next sections we discuss two challenges to the BCP related to quantum information processing: threshold theorems for fault-tolerant quantum computation (FTQC), and the idea of quantum memory based on topological degrees of freedom. In contrast to often used phenomenological approaches, we scrutinize the basic assumptions using first-principle Hamiltonian models. In the case of FTQC, we argue that the basic assumptions of the phenomenological models disagree with the fundamental features of the Hamiltonian approach. Concerning topological quantum information processing, although some interesting phenomena are observed, e.g., for four-dimensional Kitaev models, it seems unlikely that all the desired properties of a quantum memory are achievable.

26.2 Fault-tolerant quantum computation

The most important results of the theory of quantum error correction (QEC) and FTQC are the threshold theorems (compare Chapter 3). Here we reproduce a less formal presentation of this theory following the review article [KLA+02].

Theorem 26.1. Assume the requirements for scalable quantum information processing (see below). If the error per gate is less than a threshold, then it is possible to efficiently quantum compute arbitrarily accurately.

The assumptions of the above theorem are the following.
(i) Scalability: The systems must be able to support any number of qubits.
(ii) State preparation and measurement: One must be able to prepare any qubit in a standard initial state with probability $1 - \epsilon$ and measure any qubit in the logical basis with accuracy $1 - \epsilon$ at the end of computation, where ϵ is a small number.

(iii) Quantum control: One must be able to implement a universal set of unitary gates acting on a small number of qubits (typically one and two). A certain amount of parallelism in gate application is also required.

(iv) Errors: The error probability per gate must be below a threshold and satisfy certain independence and locality properties.

All these conditions present formidable technological challenges to experimentalists and engineers. From the point of view of fundamental physical principles, the last condition concerning the properties of noise seems to be the most important and also the most questionable one. In the following, commonly used phenomenological models of noise will be compared with first-principles ones.

26.2.1 Phenomenological vs. Hamiltonian models of FTQC

We begin with the phenomenological model of quantum computation as presented, for example, in [AB.-O08]. One assumes that the quantum computer (QC) consists of $N = N_r + N_a$ qubits where N_r belongs to the *register* and N_a form an ancillary system used in the error-correction procedures. The difference between register and ancillas becomes important only at the end of the computation when the information is extracted only from register qubits. One now treats the whole computer as an open system with the Hilbert space $\mathcal{H}_N$. Gates are maps acting on the density matrices ρ as $\mathcal{U}\rho = U\rho U^\dagger$ where U is a unitary matrix from the universal set of gates (say, one- or two-qubit gates). One divides the computation time into time steps and $l(k)$ denotes a *location*, i.e., space-time coordinates of the qubits participating in the same gate (including trivial gates) at time step k. The influence of the noise is described by a map $\mathcal{E}_k$ acting between time steps $k-1$ and k. The assumptions concerning the *error maps* $\mathcal{E}_k$ that allow one to prove threshold theorems are the following:

(A1) $\mathcal{E}_k$ is linear.

(A2) The error map can be always written as $\mathcal{E}_k = I + \sum_L \Phi_L$ where Φ_L is a linear map acting only on qubits from a subset L containing $|L|$ qubits. There exists a constant η, called the *error per gate*, and an overall constant C such that $\|\Phi_L\| \leq C\eta^{|L|}$, where the norm is an appropriate one for superoperators, such as the diamond norm [Chapter 1, Eq. (1.94)].

In the literature it is usually assumed that the error maps are completely positive. This is not necessary, as in the proofs only linearity and the norm estimates are used, and it is even not desirable, as shown below.

We now compare the phenomenological model above with the standard description in terms of the reduced dynamics of an open quantum system [AL07, BP02]. One can treat the whole computer as an open system with the Hilbert space $\mathcal{H}_N$ weakly interacting with a bath described by the Hilbert space $\mathcal{H}_B$. The dynamics of the total system is governed by the Hamiltonian

$$H(t) = H'_Q(t) + H_B + \lambda H_{int}, \tag{26.1}$$

where $H'_Q(t)$ describes a *bare* time-dependent control over the QC, H_B is the bath Hamiltonian, and λ is the coupling constant of the QC–bath interaction Hamiltonian H_{int}. The bare Hamiltonian differs from the *physical Hamiltonian* $H_Q(t)$ by the presence of (generally frequency cutoff dependent) counterterms, which compensate for Hamiltonian corrections due to the interaction with an environment (i.e., terms compensating for the Lamb and Stark shifts). In the following we assume that the physical Hamiltonian can be perfectly designed and implemented. In all formulas below we use a renormalized picture, the evolution is always governed by the physical (renormalized) Hamiltonian, and all Hamiltonian corrections are removed.

A standard assumption is that the initial state is a product state $\rho(0) \otimes \rho_B$, and that $[H_B, \rho_B] = 0$, which is consistent with the weak coupling regime. This leads to the following expression for the reduced dynamics of the QC:

$$\rho(t) = \mathscr{E}(t)\rho(0) = \mathrm{Tr}_B\left(U(t)\rho(0) \otimes \rho_B U^\dagger(t)\right), \tag{26.2}$$

where

$$\frac{dU(t)}{dt} = -iH(t)U(t), \qquad U(0) = I. \tag{26.3}$$

Provided $\mathscr{E}(t)^{-1}$ exists, which is the generic case, one can always treat $\rho(t)$ as the solution of the following *time-convolutionless master equation* [BP02]:

$$\frac{d}{dt}\rho(t) = \mathscr{L}(t)\rho(t), \qquad \mathscr{L}(t) = \left(\frac{d}{dt}\mathscr{E}(t)\right)\mathscr{E}(t)^{-1}. \tag{26.4}$$

The following notation for unitary and nonunitary *superpropagators*, defined in terms of ordered exponentials, will be used:

$$\mathscr{U}(t_2, t_1) = T_+ \exp\left\{-i \int_{t_1}^{t_2} \mathscr{H}_Q(t)dt\right\}, \qquad \mathscr{H}_Q(t)\rho \equiv [H_Q(t), \rho], \tag{26.5}$$

$$\mathscr{E}(t_2, t_1) = T_+ \exp\left\{\int_{t_1}^{t_2} \mathscr{L}(t)dt\right\}. \tag{26.6}$$

It is usually assumed that the QC works according to a clock with time step τ, such that for the total time of computation $t = K\tau$ we can represent the dynamical map $\mathscr{E}(K\tau)$ as a product of unitary and nonunitary maps:

$$\mathscr{E}(K\tau) = \mathscr{E}_K \mathscr{U}_K \cdots \mathscr{E}_2 \mathscr{U}_2 \mathscr{E}_1 \mathscr{U}_1, \tag{26.7}$$

where

$$\mathscr{U}_k = \mathscr{U}(t_k, t_{k-1}), \qquad \mathscr{E}_k = \mathscr{E}(t_k, t_{k-1})\mathscr{U}_k^{-1}. \tag{26.8}$$

The implementation of the quantum algorithm in terms of gates is performed in such a way that for a given computation step k the unitary superoperator $\mathscr{U}_k$ can be decomposed into a product of commuting (say one- or two-qubit) superoperators corresponding to disjoint *locations* $l(k)$, which involve all qubits of the QC:

$$\mathscr{U}_k = \prod_{l(k)} \mathscr{U}_{l(k)}, \qquad \mathscr{U}_{l(k)} = T_+ \exp\left\{-i \int_{t_{k-1}}^{t_k} \mathscr{H}_{l(k)}(t)\, dt\right\}, \tag{26.9}$$

where $H_{l(k)}(t)$ is a one- or two-qubit Hamiltonian implementing the gate.

We discuss now the properties of error maps defined by Eqs. (26.6) and (26.8). Although the total evolution map $\mathscr{E}(t)$ is completely positive, the error maps $\mathscr{E}_k$ need not be, since the joint system–bath state may be nonclassically correlated at the time instants t_k. They are obviously linear, i.e., satisfy A1, but in general there is no reason to assume the validity of the condition A2.

This follows from the fact that the structure of the noise map depends not only on the interaction Hamiltonian, which usually has a local structure, but also on the system Hamiltonian $H_Q(t)$. Assuming that the interaction Hamiltonian is given by a single-qubit coupling to the bath:

$$H_{int} = \sum_{\mu=x,y,z} \sum_{j=1}^{N} \sigma_j^\mu \otimes B_j^\mu \equiv \sum_\alpha \sigma^\alpha \otimes B_\alpha, \qquad (26.10)$$

we can derive the approximate form of the error map $\mathscr{E}_k$ in the lowest order (Born) approximation. We begin with the definition of reduced dynamics in the interaction picture with respect to the free dynamics $\mathscr{U}_0(t)$ generated by the Hamiltonian $H_0(t) = H_Q(t) + H_B$:

$$\mathscr{E}^{int}(t)\rho = \mathrm{Tr}_B\big((\mathscr{U}_0(t))^{-1}\mathscr{U}(t)\rho \otimes \rho_B\big) = T_+ \exp\left\{\int_0^t \mathscr{L}^{int}(s)\,ds\right\}. \qquad (26.11)$$

Applying van Kampen's cumulant expansion technique [BP02], one can write $\mathscr{L}^{int}(t) = \sum_n \lambda^n \mathscr{L}_n^{int}(t)$ and explicitly compute the leading (Born) term as

$$\mathscr{L}_2^{int}(t)\rho = -\lambda^2 \int_0^t ds\, \mathrm{Tr}_B[H_{int}(t),[H_{int}(s),\rho \otimes \rho_B]]. \qquad (26.12)$$

Comparing the definitions (26.7), (26.8), and (26.11), one can show that the error map can be expressed as

$$\mathscr{E}_k = \mathscr{U}(t_k,t_{k-1})\, T_+ \exp\left\{\int_{t_{k-1}}^{t_k} \mathscr{L}^{int}(t)\,dt\right\} \mathscr{U}^{-1}(t_k,t_{k-1}). \qquad (26.13)$$

Combining (26.10), (26.11), (26.12), and (26.13), one obtains the structure of the leading terms of the error map:

$$\mathscr{E}_k = I + \lambda^2 \sum_{\alpha\beta} \int_{t_{k-1}}^{t_k} dt \int_0^t ds\, F_{\alpha\beta}(t-s)\sigma^\alpha \cdot \sigma^\beta(t-s)$$
$$+ \text{similar terms} + \mathscr{O}(\lambda^4), \qquad (26.14)$$

with the correlation function

$$F_{\alpha\beta}(t) = \mathrm{Tr}_B\big(\rho_B B_\alpha(t) B_\beta\big). \qquad (26.15)$$

Note that only in the case of Dirac-delta correlations of the reservoir, i.e., $F_{\alpha\beta}(t-s) = \delta_{\alpha\beta}(t-s)$, does the single-qubit coupling to the bath produce single-qubit error maps to leading order. Otherwise, $\sigma^\beta(t-s)$ contains multi-qubit contributions to the noise "coming from the past" [AHH+02]. The Hamiltonian $H_Q(t)$ should allow coupling between all qubits of the QC, for otherwise the system could be decomposed into completely independent components. Therefore, one can expect that the number of qubits that may contribute to the noise operator $\sigma^\beta(\tau)$ grows roughly linearly with τ. According to Eq. (26.14), the weight of terms that contain n-qubit operators with $n \sim \tau$ is proportional to $|F_{\alpha\beta}(\tau)|$. Hence, only if $|F_{\alpha\beta}(\tau)| \sim e^{-a\tau} \simeq e^{-bn}$ can the condition A2 be satisfied. Introducing the *spectral density matrix of the bath*, $R_{\alpha\beta}(\omega) = (1/2\pi)\int F_{\alpha\beta}(t)e^{-i\omega t}dt > 0$ (Fourier transform of the correlation function), one can use the Kubo–Martin–Schwinger (KMS) condition satisfied by all heat baths (we reintroduce $\hbar$ to indicate the quantum character of the KMS condition):

$$R_{\alpha\beta}(-\omega) = e^{-\hbar\omega/kT} R_{\beta\alpha}(\omega). \qquad (26.16)$$

Consider a single diagonal element $R_{\alpha\alpha}(\omega)$ corresponding to $F_{\alpha\alpha}(t)$. Replace the actual model of a heat bath by a model with a simplified spectral density, still satisfying the KMS condition:

$$R_{\alpha\alpha}(\omega) = R \text{ if } \omega \geq 0 \text{ , or } e^{\hbar\omega/kT} R \text{ if } \omega < 0. \tag{26.17}$$

The asymptotic behavior of the autocorrelation function is then given by

$$|F_{\alpha\alpha}(t)| \simeq \frac{\hbar R}{kT} \frac{1}{t^2}, \tag{26.18}$$

and, obviously, is not exponentially decaying. A similar $\sim 1/t^2$ tail is observed for a generic heat bath, because it is related to the jump $\hbar/kT$ of the first derivative $R'_{\alpha\alpha}(\omega)$ at $\omega = 0$, which is a consequence of the KMS condition. This property of the thermal quantum autocorrelation function is often attributed to the *thermal memory* time $\hbar/kT$. The presence of this nonexponential tail leads to the substantial contribution of many-qubit errors. These considerations illustrate the challenges faced in applying versions of the threshold theorem that rely on the "locality assumption" A2.

26.2.2 Generalized threshold theorems and generic environments

There exist attempts to prove threshold theorems under weaker assumptions, starting with Hamiltonian models and avoiding the unrealistic assumption A2 [TB05, NMB08]. Before discussing these models, one should stress the fundamental assumption that should be imposed on the dynamics of the bath and the interaction Hamiltonian in order to make the problem of FTQC nontrivial, and which was briefly introduced in the previous section. In the phenomenological approach, it is assumed that the error per gate is fixed and cannot be scaled with the size of the computer. The corresponding assumption in the Hamiltonian approach with the interaction Hamiltonian of the form (26.10) is that the spectral density matrix for the reservoir

$$R_{\alpha\beta}(\omega) = \frac{1}{2\pi} \int_{-\infty}^{\infty} e^{-i\omega t} \text{Tr}(\rho_B B_\alpha(t) B_\beta) dt \tag{26.19}$$

satisfies two conditions (for $\omega \in [0, \Omega]$):
(R1)

$$\mathscr{R}(\omega) - R(\omega)I \geq 0, \tag{26.20}$$

where the matrix $\mathscr{R}(\omega)$ has matrix elements $R_{\alpha\beta}(\omega)$;
(R2)

$$R(\omega) \geq \omega\eta, \quad \eta > 0. \tag{26.21}$$

The first condition means that, for a fixed ω, the eigenvalues of $\mathscr{R}(\omega)$ are bounded from below by $R(\omega)$. On the other hand, those eigenvalues describe the dissipation rates for the degrees of freedom of the open system that oscillate with the frequency ω. This relation will be better understood in the Markovian limit (see the next section). A simple argument can involve the Fermi Golden Rule, which associates the transition probabilities, and hence dissipation (decoherence) rates, to the effective density of the bath's excitations at the energy $E = \hbar\omega$, strictly related to the spectral density. Therefore, the condition R1 eliminates the situations where a certain system's degrees of freedom do not dissipate (or decohere) at all. A particular example is decoherence-free subspaces [LCW98] generated by a collective coupling to a single bath of the form

$$H_{int} = \sum_{\mu=x,y,z} \left(\sum_{j=1}^{N} \sigma_j^\mu \right) \otimes B_\mu. \tag{26.22}$$

Indeed, the collective coupling (26.22) produces a correlation matrix of the tensor form $R_{\mu j,\nu k}(\omega) = R_{\mu\nu}(\omega)$ which has all but three eigenvalues equal to zero.

The condition R1 means that the relaxation rates of the original model are larger then the relaxation rates of the simplified one with diagonal correlation matrix $R(\omega)\delta_{\alpha\beta}$. The latter corresponds to a model with identical "private baths" coupled to each σ^α. Such a simplified model shows at most slower relaxation to the equilibrium state, and is very useful if we want to estimate the slowest relaxation time of the system from above.

The condition R2 corresponds to the "fixed error η for a single gate" assumption in the phenomenological approach. To show this, one can use the Margolus–Levitin theorem [ML98]: *A quantum system of energy E needs at least a time $\simeq \frac{\hbar}{E} = \frac{1}{\omega}$ to go from one state to an orthogonal state.* On the other hand, as argued above, $R(\omega)$ can be seen as the lower bound for the relaxation rate of the modes with frequency ω (i.e., corresponding to the energy difference $E = \hbar\omega$). As a consequence, a typical gate needs a time of the order of $\tau \simeq \frac{\hbar}{E} = \frac{1}{\omega}$, and hence the error due to the interaction with a bath during the gate's execution is roughly bounded from below by $\tau R(\omega) = R(\omega)/\omega$.

Note that a commonly used anzatz for the spectral density, of the form

$$R(\omega) = C\omega^d e^{-\omega/\Omega}, \quad \text{with} \quad d \geq 1, \tag{26.23}$$

does not satisfy the condition R2. For such a coupling one can effectively eliminate errors by appropriately scaling down the energy used to implement gates [AHH+02]. On the other hand, such a spectral density is typical for linear coupling to bosonic heat baths (photons, phonons, etc.). Unfortunately, there always exist other mechanisms, such as elastic scattering, that lead to a finite dephasing rate characterized by the value $R(0) > 0$.

We now discuss briefly two Hamiltonian models of FTQC. In the non-Markovian model of Terhal and Burkard [TB05], the assumption of "small norm" of the interaction Hamiltonian implies that the operator norm (largest eigenvalue) of each of the bath operators B in the interaction Hamiltonian (26.10) satisfies

$$\|B\|_\infty \tau \leq \epsilon, \tag{26.24}$$

where $\epsilon \ll 1$ is a dimensionless small constant characterizing the decoherence strength, and where τ is the execution time of a logic gate. Adding the definition of the spectral density and the condition R2, together with $\omega \in [0, \Omega]$, one obtains a sequence of inequalities:

$$\frac{1}{2}\eta\Omega^2 < \int_{-\infty}^{\infty} R(\omega)d\omega = \text{Tr}(\rho_B B^2) \leq \|B\|_\infty^2 \leq \epsilon^2 \tau^{-2}. \tag{26.25}$$

This implies a cutoff-dependent bound on the gate time τ. As the physics should not depend strongly on the particular value of the cutoff frequency Ω, the small norm assumption is difficult to defend.

The second example of the Hamiltonian modeling is presented in [NMB08] (see Chapter 25), where the ideas of renormalization group techniques are applied. The arguments are not fully rigorous, certain simplifying hypotheses concerning the computer-bath interaction are used without proofs. An example of such a condition is the *"hypercube" assumption*, which is difficult to justify, as the correlations between neighboring qubits are due to the same interactions that are needed to couple the qubits in the process of error correction. There are a number of other delicate points that one must be sure to treat carefully. For example, the authors of [TB05] and [NMB08] use the basic ingredients of the theory of fault tolerance, including, e.g., the constant supply of *fresh qubits*, and the assumption that error propagation is handled by the quantum code. However, these ideas need a rigorous first-principles background. Not all FTQC analyses fully conform to these requirements.

26.3 Fault tolerance and quantum memory

The previous discussion demonstrates that a rigorous proof of the validity of quantum fault tolerance, based on a first-principles Hamiltonian analysis, is far from being complete. The main problem is related to the new time scale introduced by the external control, which is not well separated from the other time scales of the problem such as thermal memory time and the inverse of the cutoff frequency for the bath, or relaxation time scales for the computer. Therefore, standard approximation techniques, such as the Markovian or adiabatic limits, cannot be directly applied [ALZ06]. Moreover, the problem of fault tolerance belongs to the category of *subtle problems*, in the sense that even reasonable but not rigorously controlled approximations can produce completely false results (compare mean-field approximation in the theory of phase transitions). As a consequence, it is prudent first to try to solve rigorously a simpler problem: the existence (or perhaps nonexistence) of a stable quantum memory. Any quantum computer could be used as a quantum memory, and the preservation of an arbitrary state of an encoded qubit for a long time can be treated as the simplest quantum algorithm.

One can consider two cases of quantum memory: a dynamical one, based on the standard model of a quantum computer with unitary gates, ancillas, etc., and a self-correcting one, with a properly designed time-independent Hamiltonian that protects certain degrees of freedom. Actually, these two cases should be equivalent from both the physical and mathematical points of view. The first type of memory is described by time-periodic Hamiltonians, which correspond to state recovery cycles by error-correcting procedures. However, periodic Hamiltonians are mathematically very similar to time-independent ones [CFK+87], and in the theory of open systems there exists a construction of Markovian dynamics for periodic Hamiltonians that is very similar to the derivation of the Davies generators used for self-correcting model Hamiltonians [ALZ06]. Therefore, in the following, only self-correcting models of quantum memory will be considered.

26.3.1 Definition of quantum memory

A many-body quantum system consisting of N elementary subsystems (e.g., qubits), and hence described by an algebra of bounded operators $\mathscr{A}_N$ and a Hamiltonian H_N, provides a model of a *scalable quantum memory* if there exists at least one pair of Hermitian operators $X, Z \in \mathscr{A}_N$ corresponding to an encoded robust qubit satisfying the following conditions:

(M1) They generate a qubit algebra, i.e., $X^2 = Z^2 = 1$, $XZ + ZX = 0$.

(M2) They are physically implementable, i.e., one can construct perturbed Hamiltonians of the form

$$H_N(t) = H_N + f_x(t)X + f_y(t)Y + f_z(t)Z, \tag{26.26}$$

where $\{f_i\}$ represent external classical fields that allow control over the qubit, and $Y = iZX$.

(M3) They are stable with respect to thermal noise. This can be described in terms of autocorrelation function decay

$$|\langle X(t) X \rangle_{eq}| \sim e^{-\gamma_N t} \tag{26.27}$$

and similarly for Z. The decay rate should satisfy

$$\lim_{N \to \infty} \gamma_N = 0, \tag{26.28}$$

preferably exponentially fast (*exponentially stable memory*).

The average in (26.27) is taken with respect to the thermal equilibrium state of the total system consisting of our candidate for the memory and a heat bath. The Heisenberg evolution of $X(t)$, etc., is also governed by the total Hamiltonian of the system and bath in the weak coupling regime. Note that the decay of autocorrelation functions (26.27) implies a similar decay of the averaged state's fidelity, which is a common measure characterizing the quality of quantum information stored in a noisy environment [AF09, AHH+10].

26.3.2 Markovian model of self-correcting quantum memory

Following [AL07], consider the scheme presented in Section 1.2.1 but with a constant bare Hamiltonian $H_Q{'}$ and a system–bath interaction Hamiltonian of the form

$$H_{int} = \lambda \sum_\alpha S_\alpha \otimes B_\alpha, \tag{26.29}$$

with an explicitly small coupling constant λ, and S_α denoting, for example, $\sigma_j^{x,y,z}$ from Eq. (26.10). Denote by $\{\omega\}$ the set of eigenfrequencies of the renormalized, physical Hamiltonian H_Q, and let $S_\alpha(\omega)$ be the discrete Fourier components of S_α in the interaction picture, i.e.,

$$S_\alpha(t) = \exp(iH_Q t) S_\alpha \exp(-iH_Q t) = \sum_{\{\omega\}} S_\alpha(\omega) \exp(i\omega t). \tag{26.30}$$

According to the nontriviality condition (26.20), it suffices to consider models with *private baths*, i.e., independent, identical heat baths for each degree of freedom corresponding to S_α. A sequence of approximations, discussed for example in [ALZ06], leads to the following Markovian master equation of the Lindblad–Gorini–Kossakowski–Sudarshan [L76, GKS76] type, derived rigorously in terms of the van Hove *weak coupling limit* by Davies [D74]:

$$\frac{d\rho}{dt} = -i[H_Q, \rho] + \mathscr{L}\rho, \tag{26.31}$$

$$\mathscr{L}\rho \equiv \frac{1}{2}\lambda^2 \sum_\alpha \sum_{\{\omega\}} R(\omega)\Big([S_\alpha(\omega), \rho S_\alpha(\omega)^\dagger] + [S_\alpha(\omega)\rho, S_\alpha(\omega)^\dagger]\Big), \tag{26.32}$$

with the spectral density satisfying the KMS condition $R(-\omega) = e^{-\omega/kT} R(\omega)$.

It is convenient to use the Heisenberg picture version of the evolution Eq. (26.32)

$$\frac{dA}{dt} = i\mathcal{H}A + \mathcal{L}^*A, \quad \mathcal{H}A \equiv [H_Q, A], \tag{26.33}$$

$$\mathcal{L}^*A \equiv \frac{1}{2}\lambda^2 \sum_\alpha \sum_{\{\omega\}} R(\omega)\big(S_\alpha(\omega)^\dagger[A, S_\alpha(\omega)] + [S_\alpha(\omega)^\dagger, A]S_\alpha(\omega)\big). \tag{26.34}$$

The sum $\mathcal{G} = i\mathcal{H} + \mathcal{L}^*$ generates a semi-group of completely positive, identity-preserving transformations on the algebra of observables. However, due to its specific form, it enjoys a number of important additional properties [AL07]:

(D1) The canonical Gibbs state is stationary:

$$\operatorname{Tr}\!\left(\rho_\beta\, e^{t\mathcal{G}}(X)\right) = \operatorname{Tr}(\rho_\beta\, X) \tag{26.35}$$

with

$$\rho_\beta = \frac{e^{-\beta H_Q}}{\operatorname{Tr}\!\left(e^{-\beta H_Q}\right)}. \tag{26.36}$$

(D2) The semi-group is relaxing: any initial state ρ evolves to ρ_β:

$$\lim_{t\to\infty} \operatorname{Tr}\!\left(\rho\, e^{t\mathcal{G}}(X)\right) = \operatorname{Tr}(\rho_\beta\, X). \tag{26.37}$$

(D3) $\mathcal{L}^*$ satisfies the detailed balance condition, often called reversibility:

$$\mathcal{H}\mathcal{L}^* = \mathcal{L}^*\mathcal{H} \tag{26.38}$$

and

$$\operatorname{Tr}\!\left(\rho_\beta\, Y^\dagger\, \mathcal{L}^*(X)\right) = \operatorname{Tr}\!\left(\rho_\beta\, (\mathcal{L}^*(Y))^\dagger\, X\right). \tag{26.39}$$

Equation (26.39) expresses the self-adjointness of $\mathcal{L}^*$ with respect to the Liouville scalar product

$$\langle X, Y \rangle_\beta := \operatorname{Tr}\rho_\beta\, X^\dagger Y. \tag{26.40}$$

(D4) The dissipative part $\mathcal{L}^*$ of the generator is negative definite.

Due to D2 any initial state of a system will eventually relax to equilibrium. Information can be encoded by perturbing the equilibrium state of the system and, in order to retrieve this information, one must single out observables that detect the perturbation of the state. To encode a single qubit one needs metastable observables X, Y, Z satisfying conditions M1–M3 of the previous section. It is natural to search for such observables among the constants of motion for the Hamiltonian that have zero expectation values in the Gibbs state:

$$\langle R \rangle_\beta = 0. \tag{26.41}$$

Hence, for the Markovian model above, the following estimation holds ($R = X, Y, Z$):

$$\langle R, e^{t\mathcal{G}} R \rangle_\beta = \langle R, e^{t\mathcal{L}^*} R \rangle_\beta \geq \exp\{t\langle R, \mathcal{L}^* R \rangle_\beta\}. \tag{26.42}$$

One proves it easily, by decomposing R into normalized eigenvectors of $\mathcal{L}^*$, and using convexity of the function e^{-x}.

It follows from Eq. (26.42) that the necessary condition for the existence of such (exponentially) metastable observables is exponentially fast vanishing of the matrix elements

$\langle R, \mathscr{L}^* R \rangle_\beta$. In particular, one can expect the following scaling:

$$|\langle R, \mathscr{L}^* R \rangle_\beta| \sim e^{-cN^p}, \qquad (26.43)$$

with constants $c, p > 0$ independent of N. As R is orthogonal to I (the nondegenerate eigenvector of $\mathscr{L}^*$ with eigenvalue 0), and $\|R\|_\beta = 1$, the matrix element $\langle R, -\mathscr{L}^* R \rangle_\beta$ is bounded from below by a spectral gap for the self-adjoint operator $-\mathscr{L}^*$ (i.e., its lowest eigenvalue is different from 0). It shows that the stability analysis of encoded qubits relies on the investigation of the spectrum of $-\mathscr{L}^*$ in the neighborhood of zero.

Finally, one should also remember the condition M2, which requires that the encoded qubit observables must be efficiently implementable.

26.3.3 Kitaev models

The family of Kitaev models in $D = 2, 3, 4$ dimensions [K03a, DKL+02] consists of spin-1/2 models on a D-dimensional lattice with a toric topology, and with a Hamiltonian exhibiting the special structure:

$$H = -\sum_s X_s - \sum_c Z_c. \qquad (26.44)$$

Here, $X_s = \otimes_{j \in s} \sigma_j^x$, $Z_c = \otimes_{j \in c} \sigma_j^z$ are products of Pauli matrices belonging to certain finite sets on the lattice: "stars" and "cubes." They are chosen in such a way that all X_s, Z_c commute and form an Abelian subalgebra $\mathscr{A}_{ab}$ in the total algebra of $2^N \times 2^N$ matrices. The commutant of $\mathscr{A}_{ab}$, denoted by $\mathscr{C}$, is noncommutative, and provides a natural candidate for the subalgebra containing encoded qubit observables. Indeed, for all $D = 2, 3, 4$ one can define *bare qubit observables* $X^\mu, Z^\mu \in \mathscr{C}$, where $\mu = 2, 3, 4$ correspond to D independent encoded qubits. They are products of the corresponding Pauli matrices over topologically nontrivial loops (surfaces). The choice of loops is, of course, not unique.

To discuss the question of stability with respect to thermal noise one can use the Markovian models with Davies generators described in the previous section. In this context, Kitaev models are particularly simple [AFH07]. The commutation of all Hamiltonian terms implies a strict locality of the model (absence of wave propagation), and implies that the Fourier components in (26.30) are local and correspond to only a few Bohr frequencies, independent of the size of the system. This makes the analysis of spectral properties of the Davies generator feasible. Despite this simplification, the proofs of the results are too involved to be reproduced here; we refer the reader to [AFH07, AFH09] for details.

For the two-dimensional Kitaev model it is enough to take the terms containing σ^x, σ^z in the interaction Hamiltonian (26.10). Then the form of the Markovian master equation in the Heisenberg picture is the following [AFH07]:

$$\frac{dA}{dt} = i[H, A]$$

$$+ \frac{1}{2} \sum_{j=1}^N \left\{ \left(a_j^\dagger [A, a_j] + [a_j^\dagger, A] a_j + e^{-2\beta} a_j [A, a_j^\dagger] + e^{-2\beta} [a_j, A] a_j^\dagger \right) - [a_j^0, [a_j^0, A]] \right\}$$

$$+ \frac{1}{2} \sum_{j=1}^N \left\{ \left(b_j^\dagger [A, b_j] + [b_j^\dagger, A] b_j + e^{-2\beta} b_j [A, b_j^\dagger] + e^{-2\beta} [b_j, A] b_j^\dagger \right) - [b_j^0, [b_j^0, X]] \right\}.$$

$$(26.45)$$

We do not define the operators a_j, a_j^0, b_j, b_j^0 here, but rather give their physical interpretation. The operator a_j ($a_j^\dagger$) annihilates (creates) a pair of excitations (or anyons) attached to the site j, and corresponding to the part of the Hamiltonian $-\sum Z_c$ in (26.44) (type-Z anyons), while a_j^0 generates diffusion of anyons of the same type. Similarly, the operators $b_j, b_j^\dagger, b_j^0$ correspond to the type-X anyons. From the form of the Hamiltonian, it follows that the two-dimensional Kitaev model is equivalent to a gas of noninteracting particles (anyons of two types), which are created/annihilated in pairs, and diffuse. Hence, heuristically, no mechanism of macroscopic free-energy barrier between different phases is present that could be used to protect even classical information. Mathematically, it was proved that the dissipative part of the Davies generator possesses a spectral gap independent of the size N, and therefore no metastable observables exist in this system [AFH09]. The main tool used in the proof is the fact that, for a positive operator K acting on the Hilbert space, the inequality $K^2 \geq cK$, $c > 0$ implies that the spectral gap of K is bounded from below by the number c. Another useful property of the Davies generator is that it is a sum of many negatively defined terms. Hence, skipping some of them can simplify estimations without increasing the spectral gap.

The four-dimensional Kitaev model is much more interesting. Instead of noninteracting particles, a picture similar to droplets in the two-dimensional Ising model appears [DKL+02]. The excitations of the system are represented by closed loops, with energy proportional to the loops' lengths. This provides the sought-after mechanism of a macroscopic energy barrier separating topologically nonequivalent spin configurations. The three-dimensional Kitaev model provides this mechanism for one type of excitations only. Therefore, only the encoded "bit" is protected, but not the "phase." The structure of the evolution equation is always similar to (26.45), with the operators $a_j^\dagger, b_j^\dagger$ creating excitations of two types, and a_j^0, b_j^0 changing the shape of excitations but not their energy. It seems necessary to use the full interaction Hamiltonian (26.10), which leads to additional processes of energy transfer between the two types of anyons.

In the paper [AHH+10] it is proved that, for the four-dimensional model, there exist exponentially metastable *dressed* qubit observables $\tilde{X}^\mu, \tilde{Z}^\mu \in \mathscr{C}$ with $\mu = 1, 2, 3, 4$, related to the bare ones by the formulas

$$\tilde{X}^\mu = X^\mu F_x^\mu, \quad \tilde{Z}^\mu = Z^\mu F_z^\mu, \tag{26.46}$$

where F_z^μ, F_z^μ are Hermitian elements of the algebra $\mathscr{A}_{ab}$ with eigenvalues ± 1. On the other hand, bare qubit observables are highly unstable, with relaxation times $\sim \sqrt{N}$. The metastability of (26.46) is proved using the Peierls argument applied to classical "submodels" of the four-dimensional Kitaev model generated by either $-\sum_s X_s$ or $-\sum_c Z_c$. The main mathematical tool is the following inequality [AHH+10]:

$$-\langle A, \mathscr{L}^* A \rangle_\beta \leq 2 \max_{\{\omega\}}\{R(\omega)\} \sum_\alpha \langle [S_\alpha, A], [S_\alpha, A] \rangle_\beta, \tag{26.47}$$

valid for any Davies generator (26.32) and any A in the eigenspace of $[H, \cdot]$. The advantage of this formula is the absence of Fourier components $S_\alpha(\omega)$, replaced now by much simpler S_α.

The metastable observable (say $\tilde{X}^\mu$) is constructed by the following operational procedure, which determines its outcomes:

(i) Perform a measurement of all observables σ_j^x.
(ii) Compute the value of the bare observable X^μ (multiply previous outcomes for spins belonging to the "surface" that defines X^μ).

(iii) Perform a certain classical algorithm (polynomial in N) that allows one to compute from the σ_j^x-measurement data the value ± 1 of "correction," i.e., the eigenvalue of F_x^μ.
(iv) Multiply the bare value by the correction to get the outcome of $\tilde{X}^\mu$.

Although the observables defined by the operational procedure above satisfy M1 and M3, the condition M2 appears problematic. It is hard to imagine any efficient construction of the corresponding operators that could be used to design the control Hamiltonians (26.26). The measurement of individual spins that is necessary for the extraction of the $\tilde{X}^\mu$'s outcome is destructive and not repeatable, at least for the model with the full interaction Hamiltonian (26.10). Therefore, at present the four-dimensional Kitaev model seems problematic even as a classical memory.

An attempt to solve the problems for a quantum memory is presented in [BCH+09], where six-dimensional topological color codes are discussed. Those systems admit both the stable encoding of qubits based on a similar mechanism to the four-dimensional Kitaev model, and local transversal unitary gates. However, the gate Hamiltonians do not commute with the protecting Hamiltonian, and therefore the effective map acting on the encoded qubits is dissipative. To avoid this phenomenon, the authors proposed switching off part of the protective Hamiltonian, but the consequences of this procedure were not discussed rigorously.

26.4 Concluding remarks

As shown in this chapter, the generic nonexponential decay of thermal autocorrelation functions and the four-dimensional Kitaev model illustrate the serious difficulties associated with the idea of FTQC. For the quantum computer Hamiltonian model, the lower the temperature, the more correlated the noise ("thermal memory"). For the Kitaev model, the better the protection of the encoded qubit observables, the more difficult is their control and accessibility. It is quite plausible that these difficulties are fundamental and not technical, and that a kind of "Heisenberg relation" is at work. This could be related to the fact that the same physical interactions used to control a system provide the coupling of that system to an environment.

Another observation valid for topological memories is that the mechanism of information protection is essentially the same for the classical and quantum cases: the existence of free-energy barriers separating metastable states. It seems that to perform a gate one has to overcome such a barrier, which involves a suitable amount of work that is then dissipated into the environment. Therefore, it is plausible to expect that there exists a fundamental conflict between stability of the encoded information and reversibility (in the sense of Hamiltonian, nondissipative dynamics) of its processing. It does not harm classical information processing, as it can be and is, in all practical implementations, performed by strongly dissipative dynamical maps; but quantum information based on quantum coherence needs unitary (Hamiltonian) gates.

We are still far from a quantitative understanding of these relations and associated bounds, and further studies are necessary to clarify these questions. It is quite possible that the ultimate bounds on the efficiency of quantum information processing will be provided by phenomenological thermodynamics, in particular by its second law [A09].

References

[ADH+09] A. Abeyesinghe, I. Devetak, P. Hayden, and A. Winter. 2009. The mother of all protocols: Restructuring quantum information's family tree. *Proc. R. Soc. London Ser. A*, **465**, 2537.

[A61] A. Abragam. 1961. *The Principles of Nuclear Magnetism*. International Series of Monographs on Physics. London: Oxford University Press.

[AL91] I. Affleck and A. W. W. Ludwig. 1991. Critical theory of overscreened Kondo fixed points. *Nucl. Phys. B*, **360**, 641.

[A00a] G. Agarwal. 2000. Control of decoherence and relaxation by frequency modulation of a heat bath. *Phys. Rev. A*, **61**, 013809.

[A10] G. S. Agarwal. 2010. Saving entanglement via a nonuniform sequence of π pulses. *Phys. Scr.*, **82**, 038103.

[AC08] V. Aggarwal and A. R. Calderbank. 2008. Boolean functions, projection operators and quantum error correcting codes. *IEEE Trans. Inf. Theory*, **54**, 1700.

[A00b] D. Aharonov. 2000. Quantum to classical phase transition in noisy quantum computers. *Phys. Rev. A*, **6206**, 062311.

[AB.-O08] D. Aharonov and M. Ben-Or. 2008. Fault-tolerant quantum computation with constant error rate. *SIAM J. Comput.*, **38**, 1207.

[AB.-OI+96] D. Aharonov, M. Ben-Or, R. Impagliazzo, and N. Nisan. 1996. Limitations of noisy reversible computation. eprint arXiv:quant-ph/9611028.

[AKP06] D. Aharonov, A. Kitaev, and J. Preskill. 2006. Fault-tolerant quantum computation with long-range correlated noise. *Phys. Rev. Lett.*, **96**, 050504.

[AV89] Y. Aharonov and L. Vaidman. 1989. Aharonov and Vaidman reply. *Phys. Rev. Lett.*, **62**, 2327.

[AV90] Y. Aharonov and L. Vaidman. 1990. Properties of a quantum system during the time interval between two measurements. *Phys. Rev. A*, **41**, 11.

[AAV88] Y. Aharonov, D. Z. Albert, and L. Vaidman. 1988. How the result of a measurement of a component of the spin of a spin-1/2 particle can turn out to be 100. *Phys. Rev. Lett.*, **60**, 1351.

[ADL02] C. Ahn, A. C. Doherty, and A. J. Landahl. 2002. Continuous quantum error correction via quantum feedback control. *Phys. Rev. A*, **65**, 042301.

[AWM03] C. Ahn, H. W. Wiseman, and G. J. Milburn. 2003. Quantum error correction for continuously detected errors. *Phys. Rev. A*, **67**, 052310.

[AM00] S. Aji and R. McEliece. 2000. The generalized distributive law. *IEEE Trans. Inf. Theory*, **46**, 325.

[AAS11] A. Ajoy, G. A. Álvarez, and D. Suter. 2011. Optimal pulse spacing for dynamical decoupling in the presence of a purely dephasing spin bath. *Phys. Rev. A*, **83**, 032303.

[A.-HDD+06] K. A. Al-Hassanieh, V. V. Dobrovitski, E. Dagotto, and B. N. Harmon. 2006. Numerical modeling of the central spin problem using the spin-coherent-state P representation. *Phys. Rev. Lett.*, **97**, 037204.

[A88] R. Alicki. 1988. Limited thermalization for the Markov mean-field model of N atoms in thermal field. *Physica A*, **150**, 455.

[A06] R. Alicki. 2006. Quantum error correction fails for Hamiltonian models. *Fluct. Noise Lett.*, **6**, C23.

[A09] R. Alicki. 2009. Quantum memory as a perpetuum mobile of the second kind. eprint arXiv:0901.0811.

[AF01] R. Alicki and M. Fannes. 2001. *Quantum Dynamical Systems*. Oxford: Oxford University Press.

[AF09] R. Alicki and M. Fannes. 2009. Decay of fidelity in terms of correlation functions. *Phys. Rev. A*, **79**, 012316.

[AL07] R. Alicki and K. Lendi. 2007. *Quantum Dynamical Semigroups and Applications* 2nd edn., Lecture Notes in Physics, Volume 717. Berlin: Springer.

[AHH+02] R. Alicki, M. Horodecki, P. Horodecki, and R. Horodecki. 2002. Dynamical description of quantum computing: Generic nonlocality of quantum noise. *Phys. Rev. A*, **65**, 062101.

[AHH+04] R. Alicki, M. L. Horodecki, P. L. Horodecki, and R. Horodecki. 2004. Thermodynamics of quantum information systems. Hamiltonian description. *Open Syst. Inf. Dyn.*, **11**, 205.

[ALZ06] R. Alicki, D. A. Lidar, and P. Zanardi. 2006. Internal consistency of fault-tolerant quantum error correction in light of rigorous derivations of the quantum Markovian limit. *Phys. Rev. A*, **73**, 052311.

[AFH07] R. Alicki, M. Fannes, and M. Horodecki. 2007. A statistical mechanics view on Kitaev's proposal for quantum memories. *J. Phys. A*, **40**, 6451.

[AFH09] R. Alicki, M. Fannes, and M. Horodecki. 2009. On thermalization in Kitaev's 2D model. *J. Phys. A*, **42**, 065303.

[AHH+10] R. Alicki, M. Horodecki, P. Horodecki, and R. Horodecki. 2010. On thermal stability of topological qubit in Kitaev's 4D model. *Open Syst. Inf. Dyn.*, **17**, 1.

[A07] P. Aliferis. 2007. *Level Reduction and the Quantum Threshold Theorem*. Ph.D. thesis, California Institute of Technology. eprint arXiv:quant-ph/0703230.

[AC07] P. Aliferis and A. W. Cross. 2007. Subsystem fault tolerance with the Bacon–Shor code. *Phys. Rev. Lett.*, **98**, 220502.

[AL04] P. Aliferis and D. W. Leung. 2004. Computation by measurements: A unifying picture. *Phys. Rev. A*, **70**, 062314.

[AL06] P. Aliferis and D. W. Leung. 2006. Simple proof of fault tolerance in the graph-state model. *Phys. Rev. A*, **73**, 032308.

[AP08] P. Aliferis and J. Preskill. 2008. Fault-tolerant quantum computation against biased noise. *Phys. Rev. A*, **78**, 052331.

[AP09] P. Aliferis and J. Preskill. 2009. The Fibonacci scheme for fault-tolerant quantum computation. *Phys. Rev. A*, **79**, 012332.

[AT07] P. Aliferis and B. M. Terhal. 2007. Fault-tolerant quantum computation for local leakage faults. *Quant. Inf. Comput.*, **7**, 139.

[AGP06] P. Aliferis, D. Gottesman, and J. Preskill. 2006. Quantum accuracy threshold for concatenated distance-3 codes. *Quant. Inf. Comput.*, **6**, 97.

[AGP08] P. Aliferis, D. Gottesman, and J. Preskill. 2008. Accuracy threshold for postselected quantum computation. *Quant. Inf. Comput.*, **8**, 181.

[AS11] G. A. Alvarez and D. Suter. 2011. Dynamical decoupling noise spectroscopy. eprint arXiv:1106.3463.

[AAP+10] G. A. Álvarez, A. Ajoy, X. Peng, and D. Suter. 2010. Performance comparison of dynamical decoupling sequences for a qubit in a rapidly fluctuating spin bath. *Phys. Rev. A*, **82**, 042306.

[A08] S. Aly. 2008. A class of quantum LDPC codes constructed from finite geometries. *IEEE Global Telecommunications Conference*, p. 1.

[A54] P. W. Anderson. 1954. A mathematical model for the narrowing of spectral lines by exchange or motion. *J. Phys. Soc. Jpn.*, **9**, 316.

[ADD+07] K. S. Andrews, D. Divsalar, S. Dolinar, J. Hamkins, C. R. Jones, and F. Pollara. 2007. The development of turbo and LDPC codes for deep space applications. *Proc. IEEE, Special Issue on "Technical Advances in Deep Space Communications and Tracking"*, **95**, 2142.

[ATK+09] T. Aoki, G. Takahashi, T. Kajiya, J. Yoshikawa, S. L. Braunstein, P. van Loock, and A. Furusawa. 2009. Quantum error correction beyond qubits. *Nature Phys.*, **5**, 541.

[AL70] H. Araki and E. H. Lieb. 1970. Entropy inequalities. *Commun. Math. Phys.*, **18**, 160.

[AK01] A. Ashikhmin and E. Knill. 2001. Nonbinary quantum stabilizer codes. *IEEE Trans. Inf. Theory*, **47**, 3065.

[AD02] K. Audenaert and B. De Moor. 2002. Optimizing completely positive maps using semidefinite programming. *Phys. Rev. A*, **65**, 030302(R).

[B06] D. Bacon. 2006. Operator quantum error-correcting subsystems for self-correcting quantum memories. *Phys. Rev. A*, **73**, 12340.

[BC06] D. Bacon and A. Casaccino. 2006. Quantum error correcting subsystem codes from two classical linear codes. *44th Annual Alerton Conferences.* eprint arXiv:quant-ph/0610088.

[BL+99] D. Bacon, D. A. Lidar, and K. Whaley. 1999. Robustness of decoherence-free subspaces for quantum computation. *Phys. Rev. A*, **60**, 1944.

[BKL+00] D. Bacon, J. Kempe, D. A. Lidar, and K. B. Whaley. 2000. Universal fault-tolerant computation on decoherence-free subspaces. *Phys. Rev. Lett.*, **85**, 1758.

[BBW01] D. Bacon, K. Brown, and K. Whaley. 2001. Coherence-preserving quantum bits. *Phys. Rev. Lett.*, **87**, 247902.

[BLW83] R. D. Baker, vJ. H. Lint, and R. M. Wilson. 1983. On the Preparata and Goethals codes. *IEEE Trans. Inf. Theory*, **29**, 342.

[BW06] K. Banaszek and W. Wasilewski. 2006. Linear-optics manipulations of photon-loss codes. *Proceedings of NATO Advanced Research Workshop "Quantum Communication and Security"*. Institute of Theoretical Physics and Astrophysics, Gdansk, Poland.

[B82] F. Barahona. 1982. On the computational complexity of Ising spin glass models. *J. Phys. A*, **15**, 3241.

[BBD+97] A. Barenco, A. Berthiaume, D. Deutsch, A. Eckert, R. Jozsa, and C. Macchiavello. 1997. Stabilization of quantum computations by symmetrization. *SIAM J. Comput.*, **26**, 1541.

[B04] R. Barnes. 2004. Stabilizer codes for continuous-variable quantum error correction. arXiv:quant-ph/0405064.

[BK02] H. Barnum and E. Knill. 2002. Reversing quantum dynamics with near-optimal quantum and classical fidelity. *J. Math. Phys.*, **43**, 2097.

[BKN00] H. Barnum, E. Knill, and M. A. Nielsen. 2000. On quantum fidelities and channel capacities. *IEEE Trans. Inf. Theory*, **46**, 1317.

[BMM+10] C. Barthel, J. Medford, C. M. Marcus, M. P. Hanson, and A. C. Gossard. 2010. Interlaced dynamical decoupling and coherent operation of a singlet-triplet qubit. *Phys. Rev. Lett.*, **105**, 266808.

[BMR+06] J. Baugh, O. Moussa, C. A. Ryan, R. Laflamme, C. Ramanathan, T. F. Havel, and D. G. Cory. 2006. A solid-state NMR three-qubit homonuclear system for quantum information processing: Control and characterization. *Phys. Rev. A*, **73**, 022305.

[B.-O02] M. Ben-Or. 2002. Security of BB84 QKD Protocol. MSRI: www.msri.org/publications/ln/msri/2002/quantumintro/ben-or/2/.

[B.-OHL+05] M. Ben-Or, M. Horodecki, D. W. Leung, D. Mayers, and J. Oppenheim. 2005. The universal composable security of quantum key distribution. *Theory of Cryptography: Second Theory of Cryptography Conference*. Lecture Notes in Computer Science, Vol. 3378, p. 387. Berlin: Springer Verlag.

[BH01] S. C. Benjamin and P. M. Hayden. 2001. Multiplayer quantum games. *Phys. Rev. A*, **64**, 030301.

[B92] C. H. Bennett. 1992. Quantum cryptography using any two nonorthogonal states. *Phys. Rev. Lett.*, **68**, 3121.

[BB84] C. H. Bennett and G. Brassard. 1984. Quantum cryptography: Public key distribution and coin tossing. *Proceedings of IEEE International Conference on Computers Systems and Signal Processing*, p. 175.

[BBM92] C. H. Bennett, G. Brassard, and N. D. Mermin. 1992. Quantum cryptography without Bell's theorem. *Phys. Rev. Lett.*, **68**, 557.

[BBC+93] C. H. Bennett, G. Brassard, C. Crépeau, R. Jozsa, A. Peres, and W. K. Wootters. 1993. Teleporting an unknown quantum state via dual classical and Einstein–Podolsky–Rosen channels. *Phys. Rev. Lett.*, **70**, 1895.

[BBP+96a] C. H. Bennett, H. J. Bernstein, S. Popescu, and B. Schumacher. 1996. Concentrating partial entanglement by local operations. *Phys. Rev. A*, **53**, 2046.

[BDS+96] C. H. Bennett, D. P. DiVincenzo, J. A. Smolin, and W. K. Wootters. 1996. Mixed state entanglement and quantum error correction. *Phys. Rev. A*, **54**, 3824.

[BBP+96b] C. H. Bennett, G. Brassard, S. Popescu, B. Schumacher, J. A. Smolin, and W. K. Wootters. 1996. Purification of noisy entanglement and faithful teleportation via noisy channels. *Phys. Rev. Lett.*, **76**, 722.

[BCL+02] C. H. Bennett, J. I. Cirac, M. S. Leifer, D. W. Leung, N. Linden, S. Popescu, and G. Vidal. 2002. Optimal simulation of two-qubit Hamiltonians using general local operations. *Phys. Rev. A*, **66**, 012305.

[BKK07a] C. Beny, A. Kempf, and D. W. Kribs. 2007. Generalization of quantum error correction via the Heisenberg picture. *Phys. Rev. Lett.*, **98**, 100502.

[BKK07b] C. Beny, A. Kempf, and D. W. Kribs. 2007. Quantum error correction of observables. *Phys. Rev. A*, **76**, 042303.

[BKK09] C. Beny, A. Kempf, and D. W. Kribs. 2009. Quantum error correction on infinite-dimensional Hilbert spaces. *J. Math. Phys.*, **50**, 062108.

[B01] A. J. Berglund. 2001. Quantum coherence and control in one- and two-photon optical systems. eprint arXiv:quant-ph/0010001.

[BMT78] E. R. Berlekamp, R. J. McEliece, and H. C. A. Van Tilborg. 1978. On the inherent intractability of certain coding problems. *IEEE Trans. Inf. Theory*, **24**, 384.

[BV97] E. Bernstein and U. Vazirani. 1997. Quantum complexity theory. *SIAM J. Comput.*, **26**, 11.

[BGT93] C. Berrou, A. Glavieux, and P. Thitimajshima. 1993. Near Shannon limit error-correcting coding and decoding. *ICC'93*, p. 1064.

[B84] M. Berry. 1984. Quantal phase factors accompanying adiabatic changes. *Proc. R. Soc. London Ser. A*, **392**, 45.

[BJL99] T. Beth, D. Jungnickel, and H. Lenz. 1999. *Design Theory*. Encyclopedia of Mathematics and Its Applications, Vol. I. Cambridge: Cambridge University Press.

[B97] R. Bhatia. 1997. *Matrix Analysis*. Graduate Texts in Mathematics. New York: Springer-Verlag.

[BUV+09a] M. J. Biercuk, H. Uys, A. P. VanDevender, N. Shiga, W. M. Itano, and J. J. Bollinger. 2009. Optimized dynamical decoupling in a model quantum memory. *Nature*, **458**, 996.

[BUV+09b] M. J. Biercuk, H. Uys, A. P. VanDevender, N. Shiga, W. M. Itano, and J. J. Bollinger. 2009. Experimental Uhrig dynamical decoupling using trapped ions. *Phys. Rev. A*, **79**, 062324.

[BDU11] M. J. Biercuk, A. C. Doherty, and H. Uys. 2011. Dynamical decoupling as a filter design problem. *J. Phys. B*, **44**, 154002.

[BIZ+00] E. Biolatti, R. Iotti, P. Zanardi, and F. Rossi. 2000. Quantum information processing with semiconductor macroatoms. *Phys. Rev. Lett.*, **85**, 5647.

[BHW+04] A. Blais, R.-S. Huang, A. Wallraff, S. M. Girvin, and R. J. Schoelkopf. 2004. Cavity quantum electrodynamics for superconducting electrical circuits: An architecture for quantum computation. *Phys. Rev. A*, **69**, 062320.

[BCO+09] S. Blanes, F. Casas, J. Oteo, and J. Ros. 2009. The Magnus expansion and some of its applications. *Phys. Rep.*, **470**, 151.

[BFN+11] H. Bluhm, S. Foletti, I. Neder, M. Rudner, D. Mahalu, V. Umansky, and A. Yacoby. 2011. Dephasing time of GaAs electron-spin qubits coupled to a nuclear bath exceeding 200 μs. *Nature Phys.*, **7**, 109.

[B.-KNP+08] R. Blume-Kohout, H. K. Ng, D. Poulin, and L. Viola. 2008. Characterizing the structure of preserved information in quantum processes. *Phys. Rev. Lett.*, **100**, 030501.

[B98a] B. Bollobás. 1998. *Modern Graph Theory*. New York: Springer-Verlag.

[B10a] H. Bombin. 2010. Topological order with a twist: Ising anyons from an Abelian model. *Phys. Rev. Lett.*, **105**, 030403.

[B10b] H. Bombin. 2010. Topological subsystem codes. *Phys. Rev. A*, **81**, 032301.

[B11a] H. Bombin. 2011. Clifford gates by code deformation. *New J. Phys.*, **13**, 043005.

[B11b] H. Bombin. 2011. Structure of 2D topological stabilizer codes. eprint arXiv:1107.2707.

[BM.-D06] H. Bombin and M. Martin-Delgado. 2006. Topological quantum distillation. *Phys. Rev. Lett.*, **97**, 180501.

[BM.-D07a] H. Bombin and M. Martin-Delgado. 2007. Exact topological quantum order in $D=3$ and beyond: Branyons and brane-net condensates. *Phys. Rev. B*, **75**, 75103.

[BM.-D07b] H. Bombin and M. Martin-Delgado. 2007. Topological computation without braiding. *Phys. Rev. Lett.*, **98**, 160502.

[BM.-D09] H. Bombin and M. Martin-Delgado. 2009. Quantum measurements and gates by code deformation. *J. Phys. A*, **42**, 095302.

[BCH+09] H. Bombin, R. Chhajlany, M. Horodecki, and M. Martin-Delgado. 2009. Self-correcting quantum computers. eprint arXiv:0907.5228.

[BD.-CP11] H. Bombin, G. Duclos-Cianci, and D. Poulin. 2011. Universal topological phase of 2D stabilizer codes. eprint arXiv:1103.4606.

[BCP97] W. Bosma, J. J. Cannon, and C. Playoust. 1997. The MAGMA algebra system I: The user language. *J. Symb. Comput.*, **24**, 235.

[BPF+02] N. Boulant, M. A. Pravia, E. M. Fortunato, T. F. Havel, and D. G. Cory. 2002. Experimental concatenation of quantum error correction with decoupling. *Quant. Inf. Proc.*, **1**, 135.

[BVF+05] N. Boulant, L. Viola, E. M. Fortunato, and D. G. Cory. 2005. Experimental implementation of a concatenated quantum error-correcting code. *Phys. Rev. Lett.*, **94**, 130501.

[B02a] G. Bowen. 2002. Entanglement required in achieving entanglement-assisted channel capacities. *Phys. Rev. A*, **66**, 052313.

[BV04a] S. Boyd and L. Vandenberghe. 2004. *Convex Optimization*. Cambridge: Cambridge University Press.
[BMP+99] P. O. Boykin, T. Mor, M. Pulver, V. Roychowdhury, and F. Vatan. 1999. On universal and fault-tolerant quantum computing. *Proceedings 40th FOCS*. Society Press, p. 486.
[BV04b] F. G. S. L. Brandao and R. O. Vianna. 2004. Separable multipartite mixed states: Operational asymptotically necessary and sufficient conditions. *Phys. Rev. Lett.*, **93**, 220503.
[B98b] S. L. Braunstein. 1998. Error correction for continuous quantum variables. *Phys. Rev. Lett.*, **80**, 4084.
[B98c] S. L. Braunstein. 1998. Quantum error correction for communication with linear optics. *Nature*, **394**, 47.
[B05] S. L. Braunstein. 2005. Squeezing as an irreducible resource. *Phys. Rev. A*, **71**, 055801.
[BCJ+99] S. L. Braunstein, C. M. Caves, R. Jozsa, N. Linden, S. Popescu, and R. Schack. 1999. Separability of very noisy mixed states and implications for NMR quantum computing. *Phys. Rev. Lett.*, **83**, 1054.
[B11c] S. Bravyi. 2011. Subsystem codes with spatially local generators. *Phys. Rev. A*, **83**, 012320.
[BH11] S. Bravyi and J. Haah. 2011. On the energy landscape of 3D spin Hamiltonians with topological order. eprint arXiv:1105.4159.
[BK98] S. Bravyi and A. Kitaev. 1998. Quantum codes on a lattice with boundary. eprint arXiv:quant-ph/9811052.
[BK05] S. Bravyi and A. Kitaev. 2005. Universal quantum computation with ideal Clifford gates and noisy ancillas. *Phys. Rev. A*, **71**, 022316.
[BT09] S. Bravyi and B. Terhal. 2009. A no-go theorem for a two-dimensional self-correcting quantum memory based on stabilizer codes. *New J. Phys.*, **11**, 043029.
[BFG06] S. Bravyi, D. Fattal, and D. Gottesman. 2006. GHZ extraction yield for multipartite stabilizer states. *J. Math. Phys.*, **47**, 062106.
[BHM10] S. Bravyi, M. Hastings, and S. Michalakis. 2010. Topological quantum order: Stability under local perturbations. *J. Math. Phys.*, **51**, 093512.
[BPT10] S. Bravyi, D. Poulin, and B. Terhal. 2010. Tradeoffs for reliable quantum information storage in 2D systems. *Phys. Rev. Lett.*, **104**, 50503.
[BTL10] S. Bravyi, B. M. Terhal, and B. Leemhuis. 2010. Majorana fermion codes. *New J. Phys.*, **12**, 083039.
[BDN+04] M. J. Bremner, J. L. Dodd, M. A. Nielsen, and D. Bacon. 2004. Fungible dynamics: There are only two types of entangling multiple-qubit interactions. *Phys. Rev. A*, **69**, 012313.
[BBN05] M. J. Bremner, D. Bacon, and M. A. Nielsen. 2005. Simulating Hamiltonian dynamics using many-qubit Hamiltonians and local unitary control. *Phys. Rev. A*, **71**, 052312.
[BP02] H.-P. Breuer and F. Petruccione. 2002. *The Theory of Open Quantum Systems*. Oxford: Oxford University Press.
[BBP04] H.-P. Breuer, D. Burgarth, and F. Petruccione. 2004. Non-Markovian dynamics in a spin star system: Exact solution and approximation techniques. *Phys. Rev. B*, **70**, 045323.
[BE93] H.-J. Briegel and B.-G. Englert. 1993. Quantum optical master equations: The use of damping bases. *Phys. Rev. A*, **47**, 3311.
[B02b] T. A. Brun. 2002. A simple model of quantum trajectories. *Am. J. Phys.*, **70**, 719.
[BDH06a] T. A. Brun, I. Devetak, and M.-H. Hsieh. 2006. Catalytic quantum error correction. eprint arXiv:quant-ph/0608027.
[BDH06b] T. A. Brun, I. Devetak, and M.-H. Hsieh. 2006. Correcting quantum errors with entanglement. *Science*, **314**, 436.

[BL07] D. Bruss and G. Leuchs (eds.). 2007. *Lectures on Quantum Information*. Berlin: Wiley-VCH.
[BF04] H. Bruus and K. Flensberg. 2004. *Many-Body Quantum Theory in Condensed Matter Physics*. Oxford: Oxford University Press.
[BLD99] G. Burkard, D. Loss, and D. P. DiVincenzo. 1999. Coupled quantum dots as quantum gates. *Phys. Rev. B*, **59**, 2070.
[BL02a] M. S. Byrd and D. A. Lidar. 2002. Comprehensive encoding and decoupling solution to problems of decoherence and design in solid-state quantum computing. *Phys. Rev. Lett.*, **89**, 047901.
[BL02b] M. Byrd and D. Lidar. 2002. Bang-bang operations from a geometric perspective. *Quant. Inf. Proc.*, **1**, 19.
[BLN95] R. H. Byrd, P. Lu, and J. Nocedal. 1995. A limited memory algorithm for bound constrained optimization. *SIAM J. Sci. Statist. Comput.*, **16**, 1190.
[CS96] A. R. Calderbank and P. W. Shor. 1996. Good quantum error-correcting codes exist. *Phys. Rev. A*, **54**, 1098.
[CRS+97] A. R. Calderbank, E. M. Rains, P. W. Shor, and N. J. A. Sloane. 1997. Quantum error correction and orthogonal geometry. *Phys. Rev. Lett.*, **78**, 405.
[CRS+98] A. R. Calderbank, E. M. Rains, P. W. Shor, and N. J. A. Sloane. 1998. Quantum error correction via codes over GF(4). *IEEE Trans. Inf. Theory*, **44**, 1369.
[COT07] T. Camara, H. Ollivier, and J.-P. Tillich. 2007. A class of quantum LDPC codes: Construction and performances under iterative decoding. *ISIT*, p. 811.
[C93] H. J. Carmichael. 1993. *An Open System Approach to Quantum Optics*. Berlin: Springer.
[CP54] H. Y. Carr and E. M. Purcell. 1954. Effects of diffusion on free precession in nuclear magnetic resonance experiments. *Phys. Rev.*, **94**, 630.
[C99] C. M. Caves. 1999. Quantum error correction and reversible operations. *J. Superconductivity*, **12**, 707.
[CLG08] B. A. Chase, A. J. Landahl, and J. M. Geremia. 2008. Efficient feedback controllers for continuous-time quantum error correction. *Phys. Rev. A.*, **77**, 032304.
[CL10] K. Chen and R.-B. Liu. 2010. Dynamical decoupling for a qubit in telegraphlike noises. *Phys. Rev. A*, **82**, 052324.
[C07] P. Chen. 2007. Dynamical decoupling induced renormalization of the non-Markovian dynamics. *Phys. Rev. A*, **75**, 062301.
[CZC08] X. Chen, B. Zeng, and I. L. Chuang. 2008. Nonbinary codeword-stabilized quantum codes. *Phys. Rev. A*, **78**, 062315.
[CZZ+06] Y.-A. Chen, A.-N. Zhang, Z. Zhao, X.-Q. Zhou, and J.-W. Pan. 2006. Experimental quantum error rejection for quantum communication. *Phys. Rev. Lett.*, **96**, 220504.
[CLS+04] J. Chiaverini, D. Leibfried, T. Schaetz, M. D. Barrett, R. B. Blakestad, J. Britton, W. M. Itano, J. D. Jost, E. Knill, C. Langer, R. Ozeri, and D. J. Wineland. 2004. Realization of quantum error correction. *Nature*, **432**, 602.
[CTS+05] L. Childress, J. M. Taylor, A. S. Sorensen, and M. D. Lukin. 2005. Fault-tolerant quantum repeaters with minimal physical resources and implementations based on single-photon emitters. *Phys. Rev. A*, **72**, 052330.
[CDT+06] L. Childress, M. G. Dutt, J. Taylor, A. Zibrov, F. Jelezko, J. Wrachtrup, P. R. Hemmer, and M. D. Lukin. 2006. Coherent dynamics of coupled electron and nuclear spin qubits in diamond. *Science*, **314**, 281.
[CLV04] A. M. Childs, D. W. Leung, and G. Vidal. 2004. Reversible simulation of bipartite product Hamiltonians. *IEEE Trans. Inf. Theory*, **50**, 1189.
[CLN05] A. M. Childs, D. W. Leung, and M. A. Nielsen. 2005. Unified derivations of measurement-based schemes for quantum computation. *Phys. Rev. A*, **71**, 032318.
[CLP07] A. M. Childs, A. J. Landahl, and P. A. Parrilo. 2007. Quantum algorithms for the ordered search problem via semidefinite programming. *Phys. Rev. A*, **75**, 032335.

[CNH+03] I. Chiorescu, Y. Nakamura, C. Harmans, and J. Mooij. 2003. Coherent quantum dynamics of a superconducting flux qubit. *Science*, **299**, 1869.

[C74] M.-D. Choi. 1974. A Schwarz inequality for positive linear maps on C*-algebras. *Illinois J. Math.*, **18**, 565.

[C75] M.-D. Choi. 1975. Completely positive linear maps on complex matrices. *Lin. Alg. Appl.*, **10**, 285.

[CK06] M.-D. Choi and D. W. Kribs. 2006. Method to find quantum noiseless subsystems. *Phys. Rev. Lett.*, **96**, 050501.

[CJK09] M.-D. Choi, N. Johnston, and D. W. Kribs. 2009. The multiplicative domain in quantum error correction. *J. Phys. A*, **42**, 245303.

[CK10] D. Chruścinki and A. Kossakowski. 2010. Non-Markovian quantum dynamics: Local versus nonlocal. *Phys. Rev. Lett.*, **104**, 070406.

[CCS+09] I. Chuang, A. Cross, G. Smith, J. Smolin, and B. Zeng. 2009. Codeword stabilized quantum codes: Algorithms and structure. *J. Math. Phys.*, **50**, 042109.

[CEH+99] J. I. Cirac, A. K. Ekert, S. F. Huelga, and C. Macchiavello. 1999. Distributed quantum computation over noisy channels. *Phys. Rev. A*, **59**, 4249.

[CBK10] J. Clausen, G. Bensky, and G. Kurizki. 2010. Bath-optimized minimal-energy protection of quantum operations from decoherence. *Phys. Rev. Lett.*, **104**, 040401.

[CSG.-B04] J. P. Clemens, S. Siddiqui, and J. Gea-Banacloche. 2004. Quantum error correction against correlated noise. *Phys. Rev. A*, **70**, 069902.

[CB97] R. Cleve and H. Buhrman. 1997. Substituting quantum entanglement for communication. *Phys. Rev. A*, **56**, 1201.

[CG97] R. Cleve and D. Gottesman. 1997. Efficient computations of encodings for quantum error correction. *Phys. Rev. A*, **56**, 76.

[CGL99] R. Cleve, D. Gottesman, and H.-K. Lo. 1999. How to share a quantum secret. *Phys. Rev. Lett.*, **83**, 648.

[CEL99] G. Cohen, S. Encheva, and S. Litsyn. 1999. On binary constructions of quantum codes. *IEEE Trans. Inf. Theory*, **45**, 2495.

[CD96] C. J. Colbourn and J. H. Dinitz (eds.). 1996. *The CRC Handbook of Combinatorial Designs*. Boca Raton: CRC Press.

[CP02] D. Collins and S. Popescu. 2002. Classical analog of entanglement. *Phys. Rev. A*, **65**, 032321.

[CR99] W. Cook and A. Rohe. 1999. Computing minimum-weight perfect matchings. *INFORMS J. Comput.*, **11**, 138.

[CMG90] D. G. Cory, J. B. Miller, and A. N. Garroway. 1990. Time-suspension multiple-pulse sequences: Applications to solid-state imaging. *J. Magn. Reson.*, **90**, 205.

[CFH97] D. G. Cory, A. F. Fahmy, and T. F. Havel. 1997. Ensemble quantum computing by NMR spectroscopy. *Proc. Natl. Acad. Sci. USA*, **94**, 1634.

[CPM+98] D. G. Cory, M. D. Price, W. Maas, E. Knill, R. Laflamme, W. H. Zurek, T. F. Havel, and S. S. Somaroo. 1998. Experimental quantum error correction. *Phys. Rev. Lett.*, **81**, 2152.

[CLK+00] D. G. Cory, R. Laflamme, E. Knill, L. Viola, T. F. Havel, N. Boulant, G. Boutis, E. M. Fortunato, S. Lloyd, R. Martinez, C. Negrevergne, Y. Sharf, G. Teklemariam, Y. S. Weinstein, and W. H. Zurek. 2000. NMR based quantum information processing: Achievements and prospects. *Fortschr. Phys.*, **48**, 875.

[CT91] T. M. Cover and J. A. Thomas. 1991. *Elements of Information Theory*. New York: Wiley.

[CSS+09] A. Cross, G. Smith, J. A. Smolin, and B. Zeng. 2009. Codewords stabilized quantum codes. *IEEE Trans. Inf. Theory*, **55**, 433.

[CPZ05] F. M. Cucchietti, J. P. Paz, and W. H. Zurek. 2005. Decoherence from spin environments. *Phys. Rev. A*, **72**, 052113.

[CFK+87] H. Cycon, R. Froese, W. Kirsch, and B. Simon. 1987. *Schrodinger Operators with Applications to Quantum Mechanics and Global Geometry*. Berlin: Springer.

[CLN+08] L. Cywiński, R. M. Lutchyn, C. P. Nave, and S. Das Sarma. 2008. How to enhance dephasing time in superconducting qubits. *Phys. Rev. B*, **77**, 174509.

[DVB10] B. Dakić, V. Vedral, and Č. Brukner. 2010. Necessary and sufficient condition for nonzero quantum discord. *Phys. Rev. Lett.*, **105**, 190502.

[D07] D. D'Alessandro. 2007. *Introduction to Quantum Control and Dynamics*. Boca Raton: CRC Press.

[DCM92] J. Dalibard, Y. Castin, and K. Mølmer. 1992. Wave-function approach to dissipative processes in quantum optics. *Phys. Rev. Lett.*, **68**, 580.

[DLG+09] S. Damodarakurup, M. Lucamarini, G. D. Giuseppe, D. Vitali, and P. Tombesi. 2009. Experimental inhibition of decoherence on flying qubits via bang-bang control. *Phys. Rev. Lett.*, **103**, 040502.

[DL01] G. M. D'Ariano and P. Lo Presti. 2001. Optimal nonuniversally covariant cloning. *Phys. Rev. A*, **64**, 042308.

[DKS+07] G. D'Ariano, D. Kretschmann, D. Schlingemann, and R. Werner. 2007. Reexamination of quantum bit commitment: The possible and the impossible. *Phys. Rev. A*, **76**, 032328.

[D96] K. R. Davidson. 1996. C^*-*Algebras by Example*. Providence: American Mathematical Society.

[D74] E. Davies. 1974. Markovian Master Equations. *Commun. Math. Phys.*, **39**, 91.

[dWR+10] G. de Lange, Z. H. Wang, D. Ristè, V. V. Dobrovitski, and R. Hanson. 2010. Universal dynamical decoupling of a single solid-state spin from a spin bath. *Science*, **330**, 60.

[dRD+11] G. de Lange, D. Ristè, V. V. Dobrovitski, and R. Hanson. 2011. Single-spin magnetometry with multipulse sensing sequences. *Phys. Rev. Lett.*, **106**, 080802.

[d67] J. de Pillis. 1967. Linear transformations which preserve Hermitian and positive semidefinite operations. *Pacific J. Math.*, **23**, 129.

[D86] W. de Launey. 1986. A survey of generalized Hadamard matrices and difference matrices $D(k, \lambda; G)$ with large k. *Utilitas Math.*, **30**, 5.

[DKL+02] E. Dennis, A. Kitaev, A. Landahl, and J. Preskill. 2002. Topological quantum memory. *J. Math. Phys.*, **43**, 4452.

[D05] I. Devetak. 2005. The private classical capacity and quantum capacity of a quantum channel. *IEEE Trans. Inf. Theory*, **51**, 44.

[DY08] I. Devetak and J. Yard. 2008. Exact cost of redistributing multipartite quantum states. *Phys. Rev. Lett.*, **100**, 230501.

[DHW04] I. Devetak, A. W. Harrow, and A. Winter. 2004. A family of quantum protocols. *Phys. Rev. Lett.*, **93**, 230504.

[DHW08] I. Devetak, A. W. Harrow, and A. Winter. 2008. A resource framework for quantum Shannon theory. *IEEE Trans. Inf. Theory*, **54**, 4587.

[DGI+07] S. J. Devitt, A. D. Greentree, R. Ionicioiu, J. L. O'Brien, W. J. Munro, and L. C. L. Hollenberg. 2007. The photonic module: An on-demand resource for photonic entanglement. *Phys. Rev. A*, **76**, 052312.

[DGR06] D. Dhar, L. K. Grover, and S. M. Roy. 2006. Preserving quantum states using inverting pulses: A super-Zeno effect. *Phys. Rev. Lett.*, **96**, 100405.

[D54] R. Dicke. 1954. Coherence in spontaneous radiation processes. *Phys. Rev.*, **93**, 99.

[D95] D. P. DiVincenzo. 1995. Quantum computation. *Science*, **270**, 255.

[D00] D. P. DiVincenzo. 2000. The physical implementation of quantum computation. *Fortschr. Phys.*, **48**, 771.

[DA07] D. P. DiVincenzo and P. Aliferis. 2007. Effective fault-tolerant quantum computation with slow measurements. *Phys. Rev. Lett.*, **98**, 020501.

[DNB+02] J. L. Dodd, M. A. Nielsen, M. J. Bremner, and R. T. Thew. 2002. Universal quantum computation and simulation using any entangling Hamiltonian and local unitaries. *Phys. Rev. A*, **65**, 040301.

[DPS05] A. C. Doherty, P. A. Parrilo, and F. M. Spedalieri. 2005. Detecting multipartite entanglement. *Phys. Rev. A*, **71**, 03233.

[DHR02] M. J. Donald, M. Horodecki, and O. Rudolph. 2002. The uniqueness theorem for entanglement measures. *J. Math. Phys.*, **43**, 4252.

[DRZ+09] J. Du, X. Rong, N. Zhao, Y. Wang, J. Yang, and R.-B. Liu. 2009. Preserving electron spin coherence in solids by optimal dynamical decoupling. *Nature*, **461**, 1265.

[DG97] L. M. Duan and G. C. Guo. 1997. Preserving coherence in quantum computation by pairing quantum bits. *Phys. Rev. Lett.*, **79**, 1953.

[DG98a] L.-M. Duan and G.-C. Guo. 1998. Optimal quantum codes for preventing collective amplitude damping. *Phys. Rev. A*, **58**, 3491.

[DG98b] L. M. Duan and G. C. Guo. 1998. Reducing decoherence in quantum-computer memory with all quantum bits coupling to the same environment. *Phys. Rev. A*, **57**, 737.

[DG99] L.-M. Duan and G.-C. Guo. 1999. Suppressing environmental noise in quantum computation through pulse control. *Phys. Lett. A*, **261**, 139.

[DCZ01] L.-M. Duan, J. Cirac, and P. Zoller. 2001. Geometric manipulation of trapped ions for quantum computation. *Science*, **292**, 1695.

[D.-CP10] G. Duclos-Cianci and D. Poulin. 2010. Fast decoders for topological quantum codes. *Phys. Rev. Lett.*, **104**, 050504.

[DB03] W. Dür and H.-J. Briegel. 2003. Entanglement purification for quantum computation. *Phys. Rev. Lett.*, **90**, 067901.

[DB07] W. Dür and H. J. Briegel. 2007. Entanglement purification and quantum error correction. *Rep. Prog. Phys.*, **70**, 1381.

[DVC+01] W. Dür, G. Vidal, J. I. Cirac, N. Linden, and S. Popescu. 2001. Entanglement capabilities of non-local Hamiltonians. *Phys. Rev. Lett.*, **87**, 137901.

[DMO01] M. Durdevich, H. E. Makaruk, and R. Owczarek. 2001. Generalized noiseless quantum codes utilizing quantum enveloping algebras. *J. Phys. A*, **34**, 1423.

[DCJ+07] M. V. G. Dutt, L. Childress, L. Jiang, E. Togan, J. Maze, F. Jelezko, A. S. Zibrov, P. R. Hemmer, and M. D. Lukin. 2007. Quantum register based on individual electronic and nuclear spin qubits in diamond. *Science*, **316**, 1312.

[E65] J. Edmonds. 1965. Paths, trees, and flowers. *Can. J. Math.*, **17**, 449.

[EP03] J. Eisert and M. B. Plenio. 2003. Introduction to the basics of entanglement theory in continuous-variable systems. *Int. J. Quant. Inf.*, **1**, 479.

[E91] A. K. Ekert. 1991. Quantum cryptography based on Bell's theorem. *Phys. Rev. Lett.*, **67**, 661.

[EM96] A. Ekert and C. Macchiavello. 1996. Error correction in quantum communication. *Phys. Rev. Lett.*, **77**, 2585.

[E03a] Y. C. Eldar. 2003. Mixed-quantum-state detection with inconclusive results. *Phys. Rev. A*, **67**, 042309.

[E03b] Y. C. Eldar. 2003. A semidefinite programming approach to optimal unambiguous discrimination of quantum states. *IEEE Trans. Inf. Theory*, **49**, 446.

[EMV03] Y. Eldar, A. Megretski, and G. Verghese. 2003. Designing optimal quantum detectors via semidefinite programming. *IEEE Trans. Inf. Theory*, **49**, 1007.

[ESH04] Y. C. Eldar, M. Stojnic, and B. Hassibi. 2004. Optimal quantum detectors for unambiguous detection of mixed states. *Phys. Rev. A*, **69**, 062318.

[EBW94] R. Ernst, G. Bodenhausen, and A. Wokaun. 1994. *Principles of Nuclear Magnetic Resonance in One and Two Dimensions*. Oxford: Oxford University Press.

[FLP04] P. Facchi, D. A. Lidar, and S. Pascazio. 2004. Unification of dynamical decoupling and the quantum zeno effect. *Phys. Rev. A*, **69**, 032314.

[FI06] L. Faoro and L. B. Ioffe. 2006. Quantum two level systems and Kondo-like traps as possible sources of decoherence in superconducting qubits. *Phys. Rev. Lett.*, **96**, 047001.

[FI08] L. Faoro and L. B. Ioffe. 2008. Microscopic origin of low-frequency flux noise in Josephson circuits. *Phys. Rev. Lett.*, **100**, 227005.

[FV04] L. Faoro and L. Viola. 2004. Dynamical suppression of $1/f$ noise processes in qubit systems. *Phys. Rev. Lett.*, **92**, 117905.

[F01] D. Farenick. 2001. *Algebras of Linear Transformations*. New York: Springer-Verlag.

[FGG+00] E. Farhi, J. Goldstone, S. Gutmann, and M. Sipser. 2000. Quantum computation by adiabatic evolution. eprint arXiv:quant-ph/0001106.

[FCY+04] D. Fattal, T. S. Cubitt, Y. Yamamoto, S. Bravyi, and I. L. Chuang. 2004. Entanglement in the stabilizer formalism. eprint arXiv:quant-ph/0406168.

[FHB01] M. Fazel, H. Hindi, and S. Boyd. 2001. A rank minimization heuristic with application to minimum order system approximation. *American Control Conference, 2001*, **6**, 4734.

[FHB03] M. Fazel, H. Hindi, and S. Boyd. 2003. Log-det heuristic for matrix rank minimization with applications to Hankel and Euclidean distance matrices. *American Control Conference, 2003*, **3**, 2156.

[FHB04] M. Fazel, H. Hindi, and S. Boyd. 2004. Rank minimization and applications in system theory. *American Control Conference, 2004*, **4**, 3273.

[F82] R. P. Feynman. 1982. Simulating physics with computers. *Int. J. Theor. Phys.*, **21**, 467.

[FMA05] R. Filip, P. Marek, and U. L. Andersen. 2005. Measurement-induced continuous-variable quantum interactions. *Phys. Rev. A*, **71**, 042308.

[F00] S. D. Filippo. 2000. Quantum computation using decoherence-free states of the physical operator algebra. *Phys. Rev. A*, **62**, 052307.

[FSW07] A. S. Fletcher, P. W. Shor, and M. Z. Win. 2007. Optimum quantum error recovery using semidefinite programming. *Phys. Rev. A*, **75**, 012338.

[FSW08] A. S. Fletcher, P. W. Shor, and M. Z. Win. 2008. Structured near-optimal channel-adapted quantum error recovery. *Phys. Rev. A*, **77**, 012320.

[FC07] G. Forney and D. Costello. 2007. Channel coding: The road to channel capacity. *Proc. IEEE*, **95**, 1150.

[FG05] G. D. Forney and S. Guha. 2005. Simple rate-1/3 convolutional and tail-biting quantum error-correcting codes. *Proceedings of the IEEE International Symposium on Information Theory*, p. 1028.

[FGG07] G. D. Forney, M. Grassl, and S. Guha. 2007. Convolutional and tail-biting quantum error-correcting codes. *IEEE Trans. Inf. Theory*, **53**, 865.

[FVH+02] E. M. Fortunato, L. Viola, J. Hodges, G. Teklemariam, and D. G. Cory. 2002. Implementation of universal control on a decoherence-free qubit. *New J. Phys.*, **4**, 5.1.

[FPB+02] E. Fortunato, M. Pravia, N. Boulant, T. Havel, and D. Cory. 2002. Design of strongly modulating pulses to implement precise effective Hamiltonians for quantum information processing. *J. Chem. Phys.*, **116**, 7599.

[FG09] A. G. Fowler and K. Goyal. 2009. Topological cluster state quantum computing. *Quant. Inf. Comput.*, **9**, 721.

[FWM+12] A. G. Fowler, A. C. Whiteside, A. L. McInnes, and A. Rabbani. 2012. Topological code Autotune. eprint arXiv:1202.6111.

[FWH12a] A. G. Fowler, A. C. Whiteside, and L. C. L. Hollenberg. 2012. Towards practical classical processing for the surface code. *Phys. Rev. Lett.*, **108**, 180501.

[FWH12b] A. G. Fowler, A. C. Whiteside, and L. C. L. Hollenberg. 2012. Towards practical classical processing for the surface code: timing analysis. *Phys. Rev. A*, **86**, 042313.

[FSL05] E. Fraval, M. J. Sellars, and J. J. Longdell. 2005. Dynamic decoherence control of a solid-state nuclear quadrupole qubit. *Phys. Rev. Lett.*, **95**, 030506.

[FM01] M. Freedman and D. Meyer. 2001. Projective plane and planar quantum codes. *Found. Comp. Math.*, **1**, 325.

[FKL+03] M. Freedman, A. Kitaev, M. Larsen, and Z. Wang. 2003. Topological quantum computation. *Bull. Amer. Math. Soc.*, **40**, 31.

[FKL80] R. Freeman, S. P. Kempsell, and M. H. Levitt. 1980. Radio-frequency pulse sequences which compensate their own imperfections. *J. Magn. Reson*, **38**, 453.

[FKM+04] J. Furrer, F. Kramer, J. P. Marino, S. J. Glaser, and B. Luy. 2004. Homonuclear Hartmann-Hahn transfer with reduced relaxation losses by use of the MOCCA-XY16 multiple pulse sequence. *J. Magn. Reson*, **166**, 39.

[G83] P. Gács. 1983. Reliable computation with cellular automata. *Proceedings 15th Annual ACM Symposium on Theory of Computing*. New York: ACM Press, p. 32.

[G01a] P. Gács. 2001. Reliable cellular automata with self-organization. *J. Stat. Phys.*, **103**, 45.

[G05] P. Gács. 2005. Reliable computation. Online at Gács' website at Boston University. www.cs.bu.edu/~gacs/.

[G08] F. Gaitan. 2008. *Quantum Error Correction and Fault Tolerant Quantum Computing*. Boca Raton: CRC Press.

[G63] R. G. Gallager. 1963. *Low Density Parity Check Codes*. Cambridge, MA: MIT Press.

[GFR+09] W. Gao, A. G. Fowler, R. Raussendorf, X. Yao, H. Lu, P. Xu, C. Lu, C. Peng, Y. Deng, Z. Chen, and J. Pan. 2009. Experimental demonstration of topological error correction. eprint arXiv:0905.1542.

[GZ04] C. W. Gardiner and P. Zoller. 2004. *Quantum Noise: A Handbook of Markovian and Non-Markovian Quantum Stochastic Methods with Applications to Quantum Optics*. 3rd edn. Berlin, Heidelberg: Springer-Verlag.

[GJ79] M. R. Garey and D. S. Johnson. 1979. *Computers and Intractability: A Guide to the Theory of NP-Completeness*. New York: W. H. Freeman & Co.

[G76] M. Gaudin. 1976. Diagonalisation d'une classe d'hamiltoniens de spin. *J. Physique*, **37**, 1087.

[GC97] N. A. Gershenfeld and I. L. Chuang. 1997. Bulk spin-resonance quantum computation. *Science*, **275**, 350.

[G.-S02] S. Gheorghiu-Svirschevski. 2002. Suppression of decoherence in quantum registers by entanglement with a nonequilibrium environment. *Phys. Rev. A*, **66**, 032101.

[GKL+03] R. M. Gingrich, P. Kok, H. Lee, F. Vatan, and J. P. Dowling. 2003. All linear optical quantum memory based on quantum error correction. *Phys. Rev. Lett.*, **91**, 217901.

[GLM+03] V. Giovannetti, S. Lloyd, L. Maccone, and P. W. Shor. 2003. Broadband channel capacities. *Phys. Rev. A*, **68**, 062323.

[GGL+04] V. Giovannetti, S. Guha, S. Lloyd, L. Maccone, J. H. Shapiro, and H. P. Yuen. 2004. Classical capacity of the lossy bosonic channel: The exact solution. *Phys. Rev. Lett.*, **92**, 027902.

[G74] J.-M. Goethals. 1974. Two families of nonlinear binary codes. *Electron. Lett.*, **10**, 471.

[GNT98] A. O. Gogolin, A. A. Nersesyan, and A. M. Tsvelik. 1998. *Bosonization and Strongly Correlated Systems*. Cambridge: Cambridge University Press.

[GKL08] G. Gordon, G. Kurizki, and D. Lidar. 2008. Optimal dynamical decoherence control of a qubit. *Phys. Rev. Lett.*, **101**, 010403.

[GKS76] V. Gorini, A. Kossakowski, and E. Sudarshan. 1976. Completely positive dynamical semigroups of N-level systems. *J. Math. Phys.*, **17**, 821.

[G96a] D. Gottesman. 1996. Class of quantum error-correcting codes saturating the quantum Hamming bound. *Phys. Rev. A*, **54**, 1862.

[G97a] D. Gottesman. 1997. *Stabilizer Codes and Quantum Error Correction*. Ph.D. thesis, California Institute of Technology. eprint arXiv:quant-ph/9705052.

[G98] D. Gottesman. 1998. Theory of fault-tolerant quantum computation. *Phys. Rev. A*, **57**, 127.

[G99] D. Gottesman. 1999. Fault-tolerant quantum computation with higher-dimensional systems. *Chaos Solitons Fractals*, **10**, 1749.

[G00] D. Gottesman. 2000. Fault-tolerant quantum computation with local gates. *J. Mod. Optics*, **47**, 333.

[G09] D. Gottesman. 2009. An introduction to quantum error correction and fault-tolerant quantum computation. eprint arXiv:0904.2557.

[GC99] D. Gottesman and I. Chuang. 1999. Demonstrating the viability of universal quantum computation using teleportation and single-qubit operations. *Nature*, **402**, 390.

[GKP01] D. Gottesman, A. Kitaev, and J. Preskill. 2001. Encoding a qubit in an oscillator. *Phys. Rev. A*, **64**, 012310.

[GB97] M. Grassl and T. Beth. 1997. A note on non-additive quantum codes. eprint arXiv:quant-ph/9703016.

[GB99] M. Grassl and T. Beth. 1999. Quantum BCH codes. *Proceedings Xth International Symposium on Theoretical Electrical Engineering, Magdeburg*, p. 207.

[GR06a] M. Grassl and M. Rötteler. 2006. Noncatastrophic encoders and encoder inverses for quantum convolutional codes. *Proceedings of the IEEE International Symposium on Information Theory*, p. 1109.

[GR06b] M. Grassl and M. Rotteler. 2006. Quantum convolutional codes: Encoders and structural properties. *Proceedings of the Forty-Fourth Annual Allerton Conference*, p. 510.

[GR07] M. Grassl and M. Rotteler. 2007. Constructions of quantum convolutional codes. *Proceedings of the IEEE International Symposium on Information Theory*, p. 816.

[GR08a] M. Grassl and M. Rötteler. 2008. Non-additive quantum codes from Goethals and Preparata codes. *Proceedings of the IEEE Information Theory Workshop (ITW 08)*, p. 396.

[GR08b] M. Grassl and M. Rötteler. 2008. Quantum Goethals-Preparata codes. *Proceedings of the IEEE International Symposium on Information Theory*, p. 300.

[GBP97] M. Grassl, T. Beth, and T. Pellizzari. 1997. Codes for the quantum erasure channel. *Phys. Rev. A*, **56**, 33.

[GKR02] M. Grassl, A. Klappenecker, and M. Rötteler. 2002. Graphs, quadratic forms, and quantum codes. *Proceedings of the IEEE International Symposium on Information Theory*, p. 45.

[GRB03] M. Grassl, M. Rötteler, and T. Beth. 2003. Efficient quantum circuits for non-qubit quantum error-correcting codes. *Int. J. Found. Comput. Sci.*, **14**, 757.

[GBR04] M. Grassl, T. Beth, and M. Rötteler. 2004. On optimal quantum codes. *Int. J. Quant. Inf.*, **2**, 757.

[GSS+09] M. Grassl, P. Shor, G. Smith, J. Smolin, and B. Zeng. 2009. Generalized concatenated quantum codes. *Phys. Rev. A*, **79**, 050306(R).

[G01b] L. F. Gray. 2001. A reader's guide to Gacs's "positive rates" paper. Online at Gray's website at the University of Minnesota. www.math.umn.edu/~gray/.

[GPW99] M. Grifoni, E. Paladino, and U. Weiss. 1999. Dissipation, decoherence and preparation effects in the spin-boson system. *Eur. Phys. J. B*, **10**, 719.

[G96b] L. K. Grover. 1996. A fast quantum mechanical algorithm for database search. *Proceedings of the 28th Annual ACM Symposium on the Theory of Computing (STOC)*, p. 212.

[G97b] L. K. Grover. 1997. Quantum mechanics helps in searching for a needle in a haystack. *Phys. Rev. Lett.*, **79**, 325.

[GSE08] S. Guha, J. H. Shapiro, and B. I. Erkmen. 2008. Capacity of the bosonic wiretap channel and the entropy photon-number inequality. *Proceedings of the 2008 International Symposium on Information Theory*, p. 91.

[GBC90] T. Gullion, D. B. Baker, and M. S. Conradi. 1990. New, compensated Carr-Purcell sequences. *J. Magn. Reson.*, **89**, 479.

[H11] J. Haah. 2011. Local stabilizer codes in three dimensions without string logical operators. *Phys. Rev. A*, **83**, 042330.

[H76] U. Haeberlen. 1976. *High Resolution NMR in Solids: Selective Averaging*. New York: Academic Press.

[HW68] U. Haeberlen and J. S. Waugh. 1968. Coherent averaging effect in magnetic resonance. *Phys. Rev.*, **175**, 453.

[HJ02] G. A. Hagedorn and A. Joye. 2002. Elementary exponential error estimates for the adiabatic approximation. *J. Math. Anal. Appl.*, **267**, 235.

[HI07] M. Hagiwara and H. Imai. 2007. Quantum quasi-cyclic LDPC codes. *Proceedings of the IEEE International Symposium on Information Theory*, p. 806.

[H50] E. L. Hahn. 1950. Spin echoes. *Phys. Rev.*, **80**, 580.

[HSK+98] B. Ham, M. Shahriar, M. Kim, and P. Hemmer. 1998. Spin coherence excitation and rephasing with optically shelved atoms. *Phys. Rev. B*, **58**, R11825.

[HKC+94] A. R. Hammons, Jr., P. V. Kumar, A. R. Calderbank, N. J. A. Sloane, and P. Solé. 1994. The $\mathbb{Z}_4$-Linearity of Kerdock, Preparata, Goethals, and related codes. *IEEE Trans. Inf. Theory*, **40**, 301.

[HP01] J. Harrington and J. Preskill. 2001. Achievable rates for the Gaussian quantum channel. *Phys. Rev. A*, **64**, 062301.

[H02] A. Hatcher. 2002. *Algebraic Topology*. Cambridge: Cambridge University Press.

[H03] T. F. Havel. 2003. Robust procedures for converting among Lindblad, Kraus and matrix representations of quantum dynamical semigroups. *J. Math. Phys.*, **44**, 534.

[H07] M. Hayashi. 2007. Upper bounds of eavesdropper's performances in finite-length code with the decoy method. *Phys. Rev. A*, **76**, 012329.

[HHW+08] P. Hayden, M. Horodecki, A. Winter, and J. Yard. 2008. A decoupling approach to the quantum capacity. *Open Syst. Inf. Dynam.*, **15**, 7.

[HSW08] P. Hayden, P. W. Shor, and A. Winter. 2008. Random quantum codes from Gaussian ensembles and an uncertainty relation. *Open Syst. Inf. Dynam.*, **15**, 71.

[HCD+11] D. Hayes, S. M. Clark, S. Debnath, D. Hucul, Q. Quraishi, and C. M. Monroe. 2011. Coherent error suppression in spin-dependent force quantum gates. eprint arXiv:1104.1347.

[HKV+11] D. Hayes, K. Khodjasteh, L. Viola, and M. J. Biercuk. 2011. Reducing sequencing complexity in quantum dynamical error suppression by Walsh modulation. eprint arXiv:1109.6002.

[HSS99] A. S. Hedayat, N. J. A. Sloane, and J. Stufken. 1999. *Orthogonal Arrays*. Springer Series in Statistics. New York: Springer.

[H90] F. B. Hergert. 1990. On the Delsarte-Goethals codes and their formal duals. *Discrete Math.*, **83**, 249.

[H93] A. C. Hewson. 1993. *The Kondo Problem*. Cambridge: Cambridge University Press.

[H73] R. Hill. 1973. Linear transformations which preserve Hermitian matrices. *Linear Algebr. Appl.*, **6**, 257.

[HVD10] T. E. Hodgson, L. Viola, and I. D'Amico. 2010. Towards optimized suppression of dephasing in systems subject to pulse timing constraints. *Phys. Rev. A*, **81**, 062321.

[HDW+04] L. C. L. Hollenberg, A. Dzurak, C. Wellard, A. Hamilton, D. Reilly, G. Milburn, and R. Clark. 2004. Charge-based quantum computing using single donors in semiconductors. *Phys. Rev. B*, **69**, 113301.

[HOW05] M. Horodecki, J. Oppenheim, and A. Winter. 2005. Partial quantum information. *Nature*, **436**, 673.

[HOW07] M. Horodecki, J. Oppenheim, and A. Winter. 2007. Quantum state merging and negative information. *Commun. Math. Phys.*, **269**, 107.

[HLW08] M. Horodecki, S. Lloyd, and A. Winter. 2008. Quantum coding theorem from privacy and distinguishability. *Open Syst. Inf. Dynam.*, **15**, 47.

[HHH+09] R. Horodecki, P. Horodecki, M. Horodecki, and K. Horodecki. 2009. Quantum entanglement. *Rev. Mod. Phys.*, **81**, 865.

[HH.-KW10] M. Houshmand, S. Hosseini-Khayat, and M. Wilde. 2010. Minimal memory requirements for pearl-necklace encoders of quantum convolutional codes. *IEEE Trans. Comput.*, **PP**, 1.

[HH.-KW11] M. Houshmand, S. Hosseini-Khayat, and M. M. Wilde. 2011. Minimal-memory, non-catastrophic, polynomial-depth quantum convolutional encoders. eprint arXiv:1105.0649.

[HW10] M.-H. Hsieh and M. M. Wilde. 2010. Trading classical communication, quantum communication, and entanglement in quantum Shannon theory. *IEEE Trans. Inf. Theory*, **56**, 4705.

[HDB07] M.-H. Hsieh, I. Devetak, and T. A. Brun. 2007. General entanglement-assisted quantum error-correcting codes. *Phys. Rev. A*, **76**, 062313.

[HBD09] M.-H. Hsieh, T. A. Brun, and I. Devetak. 2009. Quantum quasi-cyclic low-density parity-check codes. *Phys. Rev. A*, **79**, 032340.

[HYH11] M.-H. Hsieh, W.-T. Yen, and L.-Y. Hsu. 2011. High performance entanglement-assisted quantum LDPC codes need little entanglement. *IEEE Trans. Inf. Theory*, **57**, 1761.

[HTZ+08] D. Hu, W. Tang, M. Zhao, Q. Chen, S. Yu, and C. Hiap. 2008. Graphical nonbinary quantum error-correcting codes. *Phys. Rev. A*, **78**, 012306.

[HP03] W. C. Huffman and V. Pless. 2003. *Fundamentals of Error-Correcting Codes*. Cambridge: Cambridge University Press.

[IAB+99] A. Imamoglu, D. D. Awschalom, G. Burkard, D. P. DiVincenzo, D. Loss, M. Sherwin, and A. Small. 1999. Quantum information processing using quantum dot spins and cavity QED. *Phys. Rev. Lett.*, **83**, 4204.

[I02] A. Iserles. 2002. Expansions that grow on trees. *Not. AMS*, **49**, 430.

[J04] K. Jacobs. 2004. Optimal feedback control for the rapid preparation of a single qubit. *Proc. SPIE*, **5468**, 355.

[JSB+09] E. R. Jenista, A. M. Stokes, R. T. Branca, and W. Warren. 2009. Optimized, unequal pulse spacing in multiple echo sequences improves refocusing in magnetic resonance. *J. Chem. Phys.*, **131**, 240510.

[JŘF02] M. Ježek, J. Řeháček, and J. Fiurášek. 2002. Finding optimal strategies for minimum-error quantum-state discrimination. *Phys. Rev. A*, **65**, 060301.

[JI11] L. Jiang and A. Imambekov. 2011. Universal dynamical decoupling of multiqubit states from environment. *Phys. Rev. A*, **84**, 060302.

[JTS+07] L. Jiang, J. M. Taylor, A. Sørensen, and M. D. Lukin. 2007. Distributed quantum computation based on small quantum registers. *Phys. Rev. A*, **76**, 062323.

[JZ99] R. Johannesson and K. S. Zigangirov. 1999. *Fundamentals of Convolutional Coding*. New York: Wiley-IEEE Press.

[JK99] J. A. Jones and E. Knill. 1999. Efficient refocusing of one-spin and two-spin interactions for NMR quantum computation. *J. Magn. Reson.*, **141**, 322.

[JVE+00] J. A. Jones, V. Vedral, A. Ekert, and G. Castagnoli. 2000. Geometric quantum computation using nuclear magnetic resonance. *Nature*, **403**, 869.

[JFS06] S. Jordan, E. Farhi, and P. Shor. 2006. Error-correcting codes for adiabatic quantum computation. *Phys. Rev. A*, **74**, 052322.

[JSS04] T. Jordan, A. Shaji, and E. Sudarshan. 2004. Dynamics of initially entangled open quantum systems. *Phys. Rev. A*, **70**, 052110.

[JP05] P. Jorrand and S. Perdrix. 2005. Unifying quantum computation with projective measurements only and one-way quantum computation. *Proceedings SPIE, Quantum Informatics (QI'04)*, Vol. 5833, p. 44.

[KU98] N. Kahale and R. Urbanke. 1998. On the minimum distance of parallel and serially concatenated codes. *Proceedings IEEE International Symposium on Information Theory (ISIT'98)*, p. 31.

[K05a] A. Kamenev. 2005. Many-body theory of non-equilibrium systems. In H. Bouchiat, Y. Gefen, S. Guron, G. Montambaux, and J. Dalibard (eds.), *Nanophysics: Coherence and Transport*, Les Houches Summer School Proceedings, Vol. 81, p. 177. Amsterdam: Elsevier.

[KYS08] B. E. Kardynal, Z. L. Yuan, and A. J. Shields. 2008. An avalanche-photodiode-based photon-number-resolving detector. *Nature Photonics*, **2**, 425.

[KBM.-D09] H. Katzgraber, H. Bombin, and M. Martin-Delgado. 2009. Error threshold for color codes and random 3-body ising models. *Phys. Rev. Lett.*, **103**, 090501.

[KBA+10] H. Katzgraber, H. Bombin, R. Andrist, and M. Martin-Delgado. 2010. Topological color codes on Union Jack lattices: A stable implementation of the whole Clifford group. *Phys. Rev. A*, **81**, 012319.

[KBL+01] J. Kempe, D. Bacon, D. A. Lidar, and K. B. Whaley. 2001. Theory of decoherence-free, fault-tolerant, universal quantum computation. *Phys. Rev. A*, **63**, 042307.

[KBS+09] J. Kerckhoff, L. Bouten, A. Silberfarb, and H. Mabuchi. 2009. Physical model of continuous two-qubit parity measurement in a cavity-QED network. *Phys. Rev. A*, **79**, 024305.

[K09] O. Kern. 2009. *Randomized Dynamical Decoupling Strategies and Improved One-Way Key Rates for Quantum Cryptography*. Ph.D. thesis, Technische Universität Darmstadt. eprint arXiv:0906.2927.

[KA05] O. Kern and G. Alber. 2005. Controlling quantum systems by embedded dynamical decoupling schemes. *Phys. Rev. Lett.*, **95**, 250501.

[KA06] O. Kern and G. Alber. 2006. Selective recoupling and stochastic dynamical decoupling. *Phys. Rev. A*, **73**, 062302.

[KAS05] O. Kern, G. Alber, and D. L. Shepelyansky. 2005. Quantum error correction of coherent errors by randomization. *Eur. Phys. J. D*, **32**, 153.

[KKK+06] A. Ketkar, A. Klappenecker, S. Kumar, and P. Sarvepalli. 2006. Nonbinary stabilizer codes over finite fields. *IEEE Trans. Inf. Theory*, **52**, 4892.

[KLG02] A. V. Khaetskii, D. Loss, and L. Glazman. 2002. Electron spin decoherence in quantum dots due to interaction with nuclei. *Phys. Rev. Lett.*, **88**, 186802.

[KL05] K. Khodjasteh and D. A. Lidar. 2005. Fault-tolerant quantum dynamical decoupling. *Phys. Rev. Lett.*, **95**, 180501.

[KL07] K. Khodjasteh and D. A. Lidar. 2007. Performance of deterministic dynamical decoupling schemes: Concatenated and periodic pulse sequences. *Phys. Rev. A*, **75**, 062310.

[KL08] K. Khodjasteh and D. A. Lidar. 2008. Rigorous bounds on the performance of a hybrid Dynamical decoupling–quantum computing scheme. *Phys. Rev. A*, **78**, 012355.

[KV09a] K. Khodjasteh and L. Viola. 2009. Dynamical quantum error correction of unitary operations with bounded controls. *Phys. Rev. A*, **80**, 032314.

[KV09b] K. Khodjasteh and L. Viola. 2009. Dynamically error-corrected gates for universal quantum computation. *Phys. Rev. Lett.*, **101**, 080501.

[KLV10] K. Khodjasteh, D. A. Lidar, and L. Viola. 2010. Arbitrarily accurate dynamical control in open quantum systems. *Phys. Rev. Lett.*, **104**, 090501.
[KEV11] K. Khodjasteh, T. Erdélyi, and L. Viola. 2011. Limits on preserving quantum coherence using multipulse control. *Phys. Rev. A*, **83**, 020305.
[KDV11] K. Khodjasteh, V. V. Dobrovitski, and L. Viola. 2011. Pointer states via engineered dissipation. *Phys. Rev. A*, **84**, 022336.
[KMW02] D. Kielpinski, C. Monroe, and D. Wineland. 2002. Architecture for a large-scale ion-trap quantum computer. *Nature*, **417**, 709.
[KK05] G. Kimura and A. Kossakowski. 2005. A note on positive maps and classification of states. *Open Syst. Inf. Dyn.*, **12**, 207.
[K97a] A. Kitaev. 1997. Quantum computation: Algorithms and error correction. *Russ. Math. Surv.*, **52**, 1191.
[K97b] A. Kitaev. 1997. Quantum error correction with imperfect gates. In O. Hirota, A. S. Holevo, and C. M. Caves (eds.), *Proceeding of the Third International Conference on Quantum Communication and Measurement*. New York: Plenum, p. 181.
[K03a] A. Kitaev. 2003. Fault-tolerant quantum computation by anyons. *Ann. Phys.*, **303**, 2.
[KS07] A. Klappenecker and P. K. Sarvepalli. 2007. On subsystem codes beating the quantum Hamming or Singleton bound. *Proc. R. Soc. A*, **463**, 2887.
[KS08] A. Klappenecker and P. Sarvepalli. 2008. Clifford code construction of operator quantum error-correcting codes. *IEEE Trans. Inf. Theory*, **54**, 5760.
[K89] H. Kleinert. 1989. *Gauge Fields in Condensed Matter*. Singapore; Teaneck, NJ: World Scientific.
[K07a] R. Klesse. 2007. Approximate quantum error correction, random codes, and quantum channel capacity. *Phys. Rev. A*, **75**, 062315.
[K08] R. Klesse. 2008. A random coding based proof for the quantum coding theorem. *Open Syst. Inf. Dyn.*, **15**, 21.
[KF05] R. Klesse and S. Frank. 2005. Quantum error correction in spatially correlated quantum noise. *Phys. Rev. Lett.*, **95**, 230503.
[K05b] E. Knill. 2005. Quantum computing with realistically noisy devices. *Nature*, **434**, 39.
[K06a] E. Knill. 2006. Protected realizations of quantum information. *Phys. Rev. A*, **74**, 042301.
[KL97] E. Knill and R. Laflamme. 1997. A theory of quantum error-correcting codes. *Phys. Rev. A*, **55**, 900.
[KCL98] E. Knill, I. Chuang, and R. Laflamme. 1998. Effective pure states for bulk quantum computation. *Phys. Rev. A*, **57**, 3348.
[KLZ98a] E. Knill, R. Laflamme, and W. H. Zurek. 1998. Resilient quantum computation. *Science*, **279**, 342.
[KLZ98b] E. Knill, R. Laflamme, and W. Zurek. 1998. Resilient quantum computation: error models and thresholds. *Proc. R. Soc. A*, **454**, 365.
[KLV00] E. Knill, R. Laflamme, and L. Viola. 2000. Theory of quantum error correction for general noise. *Phys. Rev. Lett.*, **84**, 2525.
[KLM01] E. Knill, R. Laflamme, and G. J. Milburn. 2001. A scheme for efficient quantum computation with linear optics. *Nature*, **409**, 46.
[KLM+01] E. H. Knill, R. Laflamme, R. Martinez, and C. Negrevergne. 2001. Benchmarking quantum computers: The five-qubit error correcting code. *Phys. Rev. Lett.*, **86**, 5811.
[KLA+02] E. Knill, R. Laflamme, A. Ashikhmin, H. Barnum, L. Viola, and W. Zurek. 2002. Introduction to quantum error correction. *LA Science*, **27**, 188.
[K06b] M. Koashi. 2006. Efficient quantum key distribution with practical sources and detectors. eprint arXiv:quant-ph/0609180.

[K06c] M. Koashi. 2006. Unconditional security of quantum key distribution and the uncertainty principle. *J. Phys: Conf. Ser.*, **36**, 98.

[K07b] M. Koashi. 2007. Complementarity, distillable secret key, and distillable entanglement. eprint arXiv:0704.3661.

[KK01] A. G. Kofman and G. Kurizki. 2001. Universal dynamical control of quantum mechanical decay: modulation of the coupling to the continuum. *Phys. Rev. Lett.*, **87**, 270405.

[KK04] A. G. Kofman and G. Kurizki. 2004. Unified theory of dynamically suppressed qubit decoherence in thermal baths. *Phys. Rev. Lett.*, **93**, 130406.

[KMN+07] P. Kok, W. J. Munro, K. Nemoto, T. C. Ralph, J. P. Dowling, and G. J. Milburn. 2007. Linear optical quantum computing with photonic qubits. *Rev. Mod. Phys.*, **79**, 135.

[KL09] R. Kosut and D. Lidar. 2009. Quantum error correction via convex optimization. *Quant. Inf. Proc.*, **8**, 443.

[KSL08] R. L. Kosut, A. Shabani, and D. A. Lidar. 2008. Robust quantum error correction via convex optimization. *Phys. Rev. Lett.*, **100**, 020502.

[KGR05] B. Kraus, N. Gisin, and R. Renner. 2005. Lower and upper bounds on the secret-key rate for quantum key distribution protocols using one-way classical communication. *Phys. Rev. Lett.*, **95**, 080501.

[K83] K. Kraus. 1983. *States, Effects and Operations: Fundamental Notions of Quantum Theory.* Berlin: Academic.

[KHB08] I. Kremsky, M.-H. Hsieh, and T. A. Brun. 2008. Classical enhancement of quantum error-correcting codes. *Phys. Rev. A*, **78**, 012341.

[K03b] D. W. Kribs. 2003. Quantum channels, wavelets, dilations and representations of O_n. *Proc. Edinburgh Math. Soc.*, **46**, 421.

[KS06] D. W. Kribs and R. W. Spekkens. 2006. Quantum error correcting subsystems are unitarily recoverable subsystems. *Phys. Rev. A*, **74**, 042329.

[KLP05] D. Kribs, R. Laflamme, and D. Poulin. 2005. Unified and generalized approach to quantum error correction. *Phys. Rev. Lett.*, **94**, 180501.

[KLP+06] D. W. Kribs, R. Laflamme, D. Poulin, and M. Lesosky. 2006. Operator quantum error correction. *Quant. Inf. Comput.*, **6**, 382.

[KOR+07] H. Krovi, O. Oreshkov, M. Ryazanov, and D. A. Lidar. 2007. Non-Markovian dynamics of a qubit coupled to an Ising spin bath. *Phys. Rev. A*, **76**, 052117.

[K54] R. Kubo. 1954. Note on the stochastic theory of resonance absorption. *J. Phys. Soc. Jpn.*, **9**, 935.

[KL11] W.-J. Kuo and D. A. Lidar. 2011. Quadratic dynamical decoupling: Universality proof and error analysis. *Phys. Rev. A*, **84**, 042329.

[KQP+12] W.-J. Kuo, G. Quiroz, G. A. Paz-Silva, and D. A. Lidar. 2012. Universality proof and analysis of generalized nested Uhrig dynamical decoupling. *J. Math. Phys.*, **53**, 122207.

[KBA+00] P. Kwiat, A. Berglund, J. Altepeter, and A. White. 2000. Experimental verification of decoherence-free subspaces. *Science*, **290**, 498.

[LJL+10] T. D. Ladd, F. Jelezko, R. Laflamme, Y. Nakamura, C. Monroe, and J. L. O'Brien. 2010. Quantum computers. *Nature*, **464**, 45.

[LMP+96] R. Laflamme, C. Miquel, J. P. Paz, and W. H. Zurek. 1996. Perfect quantum error correcting code. *Phys. Rev. Lett.*, **77**, 198.

[LKC+02] R. Laflamme, E. Knill, D. G. Cory, E. M. Fortunato, T. Havel, C. Miquel, R. Martinez, C. Negrevergne, G. Ortiz, M. A. Pravia, Y. Sharf, S. Sinha, R. Somma, and L. Viola. 2002. Introduction to NMR quantum information processing. *Los Alamos Science*, 226.

[LB10] C.-Y. Lai and T. A. Brun. 2010. Entanglement increases the error-correcting ability of quantum error-correcting codes. eprint arXiv:1008.2598.

[LBW10]   C.-Y. Lai, T. A. Brun, and M. M. Wilde. 2010. Dualities and identities for entanglement-assisted quantum codes. eprint arXiv:1010.5506.

[LWD08]   B. Lee, W. M. Witzel, and S. Das Sarma. 2008. Universal pulse sequence to minimize spin dephasing in the central spin decoherence problem. *Phys. Rev. Lett.*, **100**, 160505.

[L89]   A. J. Leggett. 1989. Comment on "How the result of a measurement of a component of the spin of a spin-1/2 particle can turn out to be 100". *Phys. Rev. Lett.*, **62**, 2325.

[LCD+87]   A. J. Leggett, S. Chakravarty, A. T. Dorsey, M. P. A. Fisher, A. Garg, and W. Zwerger. 1987. Dynamics of the dissipative two-state system. *Rev. Mod. Phys.*, **59**, 1.

[LDM+03]   D. Leibfried, B. DeMarco, V. Meyer, D. Lucas, M. Barrett, J. Britton, W. M. Itano, B. Jelenković, C. Langer, T. Rosenband, and D. J. Wineland. 2003. Experimental demonstration of a robust, high-fidelity geometric two ion-qubit phase gate. *Nature*, **422**, 412.

[LP08]   M. Leifer and D. Poulin. 2008. Quantum graphical models and belief propagation. *Ann. Phys.*, **323**, 1899.

[L02]   D. Leung. 2002. Simulation and reversal of n-qubit Hamiltonians using Hadamard matrices. *J. Mod. Opt.*, **49**, 1199.

[L01a]   D. W. Leung. 2001. Two-qubit projective measurements are universal for quantum computation. eprint eprint arXiv:quant-ph/0111122.

[L04]   D. W. Leung. 2004. Quantum computation by measurements. *Int. J. Quant. Inf.*, **2**, 33.

[LNC+97]   D. W. Leung, M. A. Nielsen, I. L. Chuang, and Y. Yamamoto. 1997. Approximate quantum error correction can lead to better codes. *Phys. Rev. A*, **56**, 2567.

[LVZ+99]   D. Leung, L. Vandersypen, X. Zhou, M. Sherwood, C. Yannoni, M. Kubinec, and I. Chuang. 1999. Experimental realization of a two-bit phase damping quantum code. *Phys. Rev. A*, **60**, 1924.

[LCY+00]   D. W. Leung, I. L. Chuang, Y. Yamaguchi, and Y. Yamamoto. 2000. Efficient implementation of coupled logic gates for quantum computing using Hadamard matrices. *Phys. Rev. A*, **61**, 042310.

[L83]   M. Levitt. 1983. Broadband decoupling in high-resolution NMR spectroscopy. *Adv. Magn. Reson.*, **11**, 47.

[L01b]   M. H. Levitt. 2001. *Spin Dynamics: Basics of Nuclear Magnetic Resonance.* Chichester: John Wiley & Sons.

[LDP10]   Y. Li, I. Dumer, and L. P. Pryadko. 2010. Clustered error correction of codeword-stabilized quantum codes. *Phys. Rev. Lett.*, **104**, 190501.

[LDG+10]   Y. Li, I. Dumer, M. Grassl, and L. P. Pryadko. 2010. Structured error recovery for codeword-stabilized quantum codes. *Phys. Rev. A*, **81**, 052337.

[L08a]   D. A. Lidar. 2008. Towards fault tolerant adiabatic quantum computation. *Phys. Rev. Lett.*, **100**, 160506.

[LW02]   D. A. Lidar and L.-A. Wu. 2002. Reducing constraints on quantum computer design by encoded selective recoupling. *Phys. Rev. Lett.*, **88**, 017905.

[LCW98]   D. A. Lidar, I. L. Chuang, and K. B. Whaley. 1998. Decoherence-free subspaces for quantum computation. *Phys. Rev. Lett.*, **81**, 2594.

[LBW99]   D. A. Lidar, D. Bacon, and K. B. Whaley. 1999. Concatenating decoherence-free subspaces with quantum error correcting codes. *Phys. Rev. Lett.*, **82**, 4556.

[LBW01]   D. Lidar, Z. Bihary, and K. Whaley. 2001. From completely positive maps to the quantum Markovian semigroup master equation. *Chem. Phys.*, **268**, 35.

[LBK+01a]   D. A. Lidar, D. Bacon, J. Kempe, and K. B. Whaley. 2001. Decoherence-free subspaces for multiple-qubit errors. I. Characterization. *Phys. Rev. A*, **63**, 022306.

[LBK+01b] D. A. Lidar, D. Bacon, J. Kempe, and K. B. Whaley. 2001. Decoherence-free subspaces for multiple-qubit errors. II. Universal, fault-tolerant quantum computation. *Phys. Rev. A*, **63**, 022307.

[LZK08] D. A. Lidar, P. Zanardi, and K. Khodjasteh. 2008. Distance bounds on quantum dynamics. *Phys. Rev. A*, **78**, 012308.

[LR73] E. H. Lieb and M. B. Ruskai. 1973. A fundamental property of quantum-mechanical entropy. *Phys. Rev. Lett.*, **30**, 434.

[L76] G. Lindblad. 1976. On the generators of quantum dynamical semigroups. *Commun. Math. Phys.*, **48**, 119.

[LS08] S. Ling and P. Solé. 2008. Nonadditive quantum codes from $\mathbb{Z}_4$-codes. http://hal.archives-ouvertes.fr/hal-00338309/fr/.

[LYS07] R.-B. Liu, W. Yao, and L. J. Sham. 2007. Control of electron spin decoherence caused by electron-nuclear spin dynamics in a quantum dot. *New J. Phys.*, **9**, 226.

[L96] S. Lloyd. 1996. Universal quantum simulators. *Science*, **273**, 1073.

[L97] S. Lloyd. 1997. Capacity of the noisy quantum channel. *Phys. Rev. A*, **55**, 1613.

[LS98] S. Lloyd and J.-J. E. Slotine. 1998. Analog quantum error correction. *Phys. Rev. Lett.*, **80**, 4088.

[LC99] H.-K. Lo and H. F. Chau. 1999. Unconditional security of quantum key distribution over arbitrarily long distances. *Science*, **283**, 2050.

[LW06] S. Loepp and W. Wootters. 2006. *Protecting Information: From Classical Error Correction to Quantum Cryptography*. New York: Cambridge University Press.

[LYG+08] S. Y. Looi, L. Yu, V. Gheorghiu, and R. B. Griffiths. 2008. Quantum-error-correcting codes using qudit graph states. *Phys. Rev. A*, **78**, 042303.

[LGZ+08] C.-Y. Lu, W.-B. Gao, J. Zhang, X.-Q. Zhou, T. Yang, and J.-W. Pan. 2008. Experimental quantum coding against qubit loss error. *Proc. Nat. USA Acad. Sci.*, **105**, 11050.

[L08b] Z. Luo. 2008. Quantum error correcting codes based on privacy amplification. eprint arXiv:0808.1392.

[LD07] Z. Luo and I. Devetak. 2007. Efficiently implementable codes for quantum key expansion. *Phys. Rev. A*, **75**, 010303.

[M09] H. Mabuchi. 2009. Continuous quantum error correction as classical hybrid control. *New J. Phys.*, **11**, 105044.

[MMM04] D. J. C. MacKay, G. Mitchison, and P. L. McFadden. 2004. Sparse graph codes for quantum error-correction. *IEEE Trans. Inf. Theory*, **50**, 2315.

[MS77] F. J. MacWilliams and N. J. A. Sloane. 1977. *The Theory of Error-Correcting Codes*. Amsterdam: North-Holland.

[MK06] P. K. Madhu and N. D. Kurur. 2006. Fer expansion for effective propagators and Hamiltonians in NMR. *Chem. Phys. Lett.*, **418**, 235.

[MKS04] W. Magnus, A. Karrass, and D. Solitar. 2004. *Combinatorial Group Theory*. New York: Dover.

[M00] G. D. Mahan. 2000. *Many-Particle Physics*, 3rd edn. New York: Kluwer Academic/Plenum Publishers.

[MSS01] Y. Makhlin, G. Schön, and A. Shnirman. 2001. Quantum-state engineering with Josephson-junction devices. *Rev. Mod. Phys.*, **73**, 357.

[ML98] N. Margolus and L. Levitin. 1998. The maximum speed of dynamical evolution. *Physica D*, **120**, 188.

[M82] M. M. Maricq. 1982. Application of average Hamiltonian theory to the NMR of solids. *Phys. Rev. B*, **25**, 6622.

[M96] D. Mayers. 1996. Quantum key distribution and string oblivious transfer in noisy channels. *Lecture Notes In Computer Science; Vol. 1109, Proceedings of the 16th Annual International Cryptology Conference on Advances in Cryptology*. London: Springer-Verlag, p. 343.

[M01] D. Mayers. 2001. Unconditional security in quantum cryptography. *J. ACM*, **48**, 351.
[MG58] S. Meiboom and D. Gill. 1958. Modified spin-echo method for measuring nuclear relaxation times. *Rev. Sci. Instrum.*, **29**, 688.
[M07] N. Mermin. 2007. *Quantum Computer Science: An Introduction*. New York: Cambridge University Press.
[M65] A. Messiah. 1965. *Quantum Mechanics, Vol. II*. Amsterdam: North-Holland.
[MM08] M. Mézard and A. Montanari. 2008. *Information, Physics and Computation*. Oxford: Oxford University Press.
[MS97] B. Misra and E. Sudarshan. 1997. The Zeno's paradox in quantum theory. *J. Math. Phys.*, **18**, 756.
[MS00] T. K. Moon and W. C. Stirling. 2000. *Mathematical Methods and Algorithms for Signal Processing*. Upper Saddle River: Prentice Hall.
[MTA+06] J. Morton, A. Tyryshkin, A. Ardavan, S. Benjamin, K. Porfyrakis, S. Lyon, and G. Briggs. 2006. Bang-bang control of fullerene qubits using ultra-fast phase gates. *Nature Phys.*, **2**, 40.
[MTB+08] J. Morton, A. Tyryshkin, R. Brown, S. Shankar, B. Lovett, A. Ardavan, T. Schenkel, E. Haller, J. Ager, and S. A. Lyon. 2008. Solid state quantum memory using the 31P nuclear spin. *Nature*, **455**, 1085.
[MSS+10a] M. Mukhtar, W. T. Soh, T. B. Saw, and J. Gong. 2010. Protecting unknown two-qubit entangled states by nesting Uhrig's dynamical decoupling sequences. *Phys. Rev. A*, **82**, 052338.
[MSS+10b] M. Mukhtar, T. B. Saw, W. T. Soh, and J. Gong. 2010. Universal dynamical decoupling: two-qubit states and beyond. *Phys. Rev. A*, **81**, 012331.
[N90] M. Nakahara. 1990. *Geometry, Topology and Physics*. Bristol: Institute of Physics Publishing.
[N58] S. Nakajima. 1958. On quantum theory of transport phenomena: Steady diffusion. *Prog. Theor. Phys.*, **20**, 948.
[NPY+02] Y. Nakamura, Y. A. Pashkin, T. Yamamoto, and J. S. Tsai. 2002. Charge echo in a Cooper-pair box. *Phys. Rev. Lett.*, **88**, 047901.
[NNP96] H. Nakazato, M. Namiki, and S. Pascazio. 1996. Temporal behavior of quantum mechanical systems. *Int. J. Mod. Phys. B*, **10**, 247.
[NSS+08] C. Nayak, S. H. Simon, A. Stern, M. Freedman, and S. Das Sarma. 2008. Non-Abelian anyons and topological quantum computation. *Rev. Mod. Phys.*, **80**, 1083.
[NDH+11] B. Naydenov, F. Dolde, L. T. Hall, C. Shin, H. Fedder, L. C. L. Hollenberg, F. Jelezko, and J. Wrachtrup. 2011. Dynamical decoupling of a single-electron spin at room temperature. *Phys. Rev. B*, **83**, 081201(R).
[N04a] R. E. Neapolitan. 2004. *Learning Bayesian Networks*. Pearson Prentice Hall.
[N55] J. von Neumann. 1955. Probabilistic logics and the synthesis of reliable organisms from unreliable components. In C. E. Shannon and J. McCarthy (eds.), *Automata Studies*. Princeton, NJ: Princeton University Press, p. 43.
[N66] J. von Neumann. 1966. *Theory of Self-Reproducing Automata*. Champaign, IL: University of Illinois Press.
[NP09] H.-K. Ng and J. Preskill. 2009. Fault-tolerant quantum computation versus Gaussian noise. *Phys. Rev. A*, **79**, 032318.
[NLP11] H. K. Ng, D. A. Lidar, and J. Preskill. 2011. Combining dynamical decoupling with fault-tolerant quantum computation. *Phys. Rev. A*, **84**, 012305.
[N03] M. A. Nielsen. 2003. Universal quantum computation using only projective measurement, quantum memory, and preparation of the 0 state. *Phys. Lett. A*, **308**, 96.
[N04b] M. A. Nielsen. 2004. Optical quantum computation using cluster states. *Phys. Rev. Lett.*, **93**.

[NC00] M. A. Nielsen and I. L. Chuang. 2000. *Quantum Computation and Quantum Information*. Cambridge: Cambridge University Press.

[ND05] M. A. Nielsen and C. M. Dawson. 2005. Fault-tolerant quantum computation with cluster states. *Phys. Rev. A*, **71**, 042323.

[NP07] M. A. Nielsen and D. Poulin. 2007. Algebraic and information-theoretic conditions for operator quantum error correction. *Phys. Rev. A*, **75**, 064304.

[NBD+02] M. A. Nielsen, M. J. Bremner, J. L. Dodd, A. M. Childs, and C. M. Dawson. 2002. Universal simulation of Hamiltonian dynamics for quantum systems with finite-dimensional state spaces. *Phys. Rev. A*, **66**, 022317.

[NAC08] J. Niset, U. L. Andersen, and N. J. Cerf. 2008. Experimentally feasible quantum erasure-correcting code for continuous variables. *Phys. Rev. Lett.*, **101**, 130503.

[N81] H. Nishimori. 1981. Internal energy, specific heat and correlation function of the bond-random Ising model. *Prog. Theor. Phys.*, **66**, 1169.

[N01] H. Nishimori. 2001. *Statistical Physics of Spin Glasses and Information Processing: An Introduction*. Oxford: Clarendon Press.

[NÖ03] S. Niskanen and P. R. J. Östergård. 2003. *Cliquer User's Guide, Version 1.0*. Communications Laboratory, Helsinki University of Technology, Espoo, Finland. Technical Report T48.

[NB06] E. Novais and H. U. Baranger. 2006. Decoherence by correlated noise and quantum error correction. *Phys. Rev. Lett.*, **97**, 040501.

[NC. NB+05] E. Novais, A. H. Castro Neto, L. Borda, I. Affleck, and G. Zarand. 2005. Frustration of decoherence in open quantum systems. *Phys. Rev. B*, **72**, 014417.

[NMB07] E. Novais, E. R. Mucciolo, and H. U. Baranger. 2007. Resilient quantum computation in correlated environments: A quantum phase transition perspective. *Phys. Rev. Lett.*, **98**, 040501.

[NMB08] E. Novais, E. Mucciolo, and H. U. Baranger. 2008. Hamiltonian formulation of quantum error correction and correlated noise: Effects of syndrome extraction in the long-time limit. *Phys. Rev. A*, **78**, 012314.

[OPW+05] J. L. O'Brien, G. J. Pryde, A. G. White, and T. C. Ralph. 2005. High-fidelity Z-measurement error encoding of optical qubits. *Phys. Rev. A*, **71**, 060303(R).

[OAI+04] T. Ohno, G. Arakawa, I. Ichinose, and T. Matsui. 2004. Phase structure of the random-plaquette Z2 gauge model: accuracy threshold for a toric quantum memory. *Nucl. Phys. B*, **697**, 462.

[OT05] H. Ollivier and J.-P. Tillich. 2005. Interleaved serial concatenation of quantum convolutional codes: Gate implementation and iterative error estimation algorithm. *Proceedings of the 26th Symposium on Information Theory in the Benelux*, p. 149.

[OZ02] H. Ollivier and W. Zurek. 2002. Quantum discord: A measure of the quantumness of correlations. *Phys. Rev. Lett.*, **88**, 017901.

[OT03] H. Ollivier and J.-P. Tillich. 2003. Description of a quantum convolutional code. *Phys. Rev. Lett.*, **91**, 177902.

[OT04] H. Ollivier and J.-P. Tillich. 2004. Quantum convolutional codes: Fundamentals. eprint arXiv:quant-ph/0401134.

[OT06] H. Ollivier and J.-P. Tillich. 2006. Trellises for stabilizer codes: Definition and uses. *Phys. Rev. A*, **74**, 032304.

[O08] O. Oreshkov. 2008. *Topics in Quantum Information and the Theory of Open Quantum Systems*. Ph.D. thesis, University of Southern California. eprint arXiv:0812:4682.

[O09] O. Oreshkov. 2009. Holonomic quantum computation in subsystems. *Phys. Rev. Lett.*, **103**, 090502.

[OB05] O. Oreshkov and T. A. Brun. 2005. Weak measurements are universal. *Phys. Rev. Lett.*, **95**, 110409.

[OB07] O. Oreshkov and T. A. Brun. 2007. Continuous quantum error correction for non-Markovian decoherence. *Phys. Rev. A*, **76**, 022318.

[OBL08] O. Oreshkov, T. A. Brun, and D. A. Lidar. 2008. Fault-tolerant holonomic quantum computation. *Phys. Rev. Lett.*, **102**, 070502.

[OLB08] O. Oreshkov, D. A. Lidar, and T. A. Brun. 2008. Operator quantum error correction for continuous dynamics. *Phys. Rev. A*, **78**, 022333.

[OBL09] O. Oreshkov, T. A. Brun, and D. A. Lidar. 2009. Scheme for fault-tolerant holonomic computation on stabilizer codes. *Phys. Rev. A*, **80**, 022325.

[OCC02] M. Oskin, F. T. Chong, and I. L. Chuang. 2002. A practical architecture for reliable quantum computers. *IEEE Computing*, **18**, 79.

[PZ01] J. Pachos and P. Zanardi. 2001. Quantum holonomies for quantum computing. *Int. J. Mod. Phys. B*, **15**, 1257.

[PSE96] M. Palma, K.-A. Suominen, and A. K. Ekert. 1996. Quantum computers and dissipation. *Proc. R. Soc. London A*, **452**, 567.

[PSS+06] D. Parodi, M. Sassetti, P. Solinas, P. Zanardi, and N. Zanghí. 2006. Fidelity optimization for holonomic quantum gates in dissipative environments. *Phys. Rev. A*, **73**, 052304.

[PU08] S. Pasini and G. S. Uhrig. 2008. Generalization of short coherent control pulses: Extension to arbitrary rotations. *J. Phys. A*, **41**, 312005.

[PU10a] S. Pasini and G. S. Uhrig. 2010. Optimized dynamical decoupling for power-law noise spectra. *Phys. Rev. A*, **81**, 012309.

[PU10b] S. Pasini and G. S. Uhrig. 2010. Optimized dynamical decoupling for time-dependent Hamiltonians. *J. Phys. A*, **43**, 132001.

[PFK+08] S. Pasini, T. Fischer, P. Karbach, and G. S. Uhrig. 2008. Optimization of short coherent control pulses. *Phys. Rev. A*, **77**, 032315.

[PZ98] J. P. Paz and W. H. Zurek. 1998. Continuous error correction. *Proc. R. Soc. London A*, **454**, 355.

[PSL11] X. Peng, D. Suter, and D. A. Lidar. 2011. High fidelity quantum memory via dynamical decoupling: Theory and experiment. *J. Phys. B*, **44**, 154003.

[P89] A. Peres. 1989. Quantum measurements with postselection. *Phys. Rev. Lett.*, **62**, 2326.

[P98a] A. Peres. 1998. *Quantum Theory: Concepts and Methods*. Dordrecht: Kluwer.

[PJT+05] J. Petta, A. Johnson, J. Taylor, E. Laird, A. Yacoby, M. D. Lukin, C. Marcus, M. Hanson, and A. C. Gossard. 2005. Coherent manipulation of coupled electron spins in semiconductor quantum dots. *Science*, **309**, 2180.

[PMV+04] S. Pirandola, S. Mancini, D. Vitali, and P. Tombesi. 2004. Constructing finite-dimensional codes with optical continuous variables. *Europhys. Lett.*, **68**, 323.

[PJF05] T. B. Pittman, B. C. Jacobs, and J. D. Franson. 2005. Demonstration of quantum error correction using linear optics. *Phys. Rev. A*, **71**, 052332.

[P98b] J. G. Polchinski. 1998. *String Theory*. Vol. 1. Cambridge: Cambridge University Press.

[P01] M. S. Postol. 2001. A proposed quantum low density parity check code. eprint quant-ph/0108131.

[P05] D. Poulin. 2005. Stabilizer formalism for operator quantum error correction. *Phys. Rev. Lett.*, **95**, 230504.

[P06] D. Poulin. 2006. Optimal and efficient decoding of concatenated quantum block codes. *Phys. Rev. A*, **74**, 052333.

[PTO09] D. Poulin, J.-P. Tillich, and H. Ollivier. 2009. Quantum serial turbo-codes. *IEEE Trans. Inf. Theory*, **55**, 2776.

[P68] F. P. Preparata. 1968. A class of optimum nonlinear double-error-correcting codes. *Inf. Control*, **13**, 378.

[P96] J. Preskill. 1996. Fault-tolerant quantum computation. In H.-K. Lo, S. Popescu, and T. P. Spiller (eds.), *Introduction to Quantum Computation and Information*: Singapore: World Scientific.

[P04] J. Preskill. 2004. Lecture notes on topological quantum computation. www.theory.caltech.edu/ preskill/ph219/topological.pdf.

[QRZ02] B. Qiao, H. Ruda, and M. Zhan. 2002. Two-qubit quantum computing in a projected subspace. *Phys. Rev. A*, **65**, 042325.

[QWJ+97] T. Quang, M. Woldeyohannes, S. John, and G. S. Agarwal. 1997. Coherent control of spontaneous emission near a photonic band edge: A single-atom optical memory device. *Phys. Rev. Lett.*, **79**, 5238.

[QL11] G. Quiroz and D. A. Lidar. 2011. Quadratic dynamical decoupling with nonuniform error suppression. *Phys. Rev. A*, **84**, 042328.

[R99] E. Rains. 1999. Nonbinary quantum codes. *IEEE Trans. Inf. Theory*, **45**, 1827.

[R01] E. M. Rains. 2001. A semidefinite program for distillable entanglement. *IEEE Trans. Inf. Theory*, **47**, 2921.

[RHS+97] E. M. Rains, R. H. Hardin, P. W. Shor, and N. J. A. Sloane. 1997. Nonadditive quantum code. *Phys. Rev. Lett.*, **79**, 953.

[R50] N. F. Ramsey. 1950. A molecular beam resonance method with separated oscillating fields. *Phys. Rev.*, **78**, 695.

[RB01] R. Raussendorf and H. J. Briegel. 2001. A one-way quantum computer. *Phys. Rev. Lett.*, **86**, 5188.

[RH07] R. Raussendorf and J. Harrington. 2007. Fault-tolerant quantum computation with high threshold in two dimensions. *Phys. Rev. Lett.*, **98**, 190504.

[RBB02] R. Raussendorf, D. E. Browne, and H. J. Briegel. 2002. The one-way quantum computer: A non-network model of quantum computation. *J. Mod. Opt.*, **49**, 1299.

[RBB03] R. Raussendorf, D. E. Browne, and H. J. Briegel. 2003. Measurement-based quantum computation with cluster states. *Phys. Rev. A*, **68**, 022312.

[RHG06] R. Raussendorf, J. Harrington, and K. Goyal. 2006. A fault-tolerant one-way quantum computer. *Ann. Phys.*, **321**, 2242.

[RHG07] R. Raussendorf, J. Harrington, and K. Goyal. 2007. Topological fault-tolerance in cluster state quantum computation. *New J. Phys.*, **9**, 199.

[R06] B. W. Reichardt. 2006. *Error-Detection-Based Quantum Fault Tolerance Against Discrete Pauli Noise*. Ph.D. thesis, University of California, Berkeley. eprint arXiv:quant-ph/0612004.

[RW05] M. Reimpell and R. F. Werner. 2005. Iterative optimization of error correcting codes. *Phys. Rev. Lett.*, **94**, 080501.

[RWA06] M. Reimpell, R. F. Werner, and K. Audenaert. 2006. Comment on "Optimum quantum error recovery using semidefinite programming". eprint arXiv/quant-ph/0606059.

[R05] R. Renner. 2005. *Security of Quantum Key Distribution*. Ph.D. thesis, ETH Zurich. eprint arXiv:quant-ph/0512258.

[RK05] R. Renner and R. Konig. 2005. Universally composable privacy amplification against quantum adversaries. In *Theory of Cryptography: Second Theory of Cryptography Conference*. Lecture Notes in Computer Science, Vol. 3378. Berlin: Springer Verlag, p. 407.

[RGK05] R. Renner, N. Gisin, and B. Kraus. 2005. Information-theoretic security proof for quantum-key-distribution protocols. *Phys. Rev. A*, **72**, 012332.

[R.-RMK+08] C. Rodríguez-Rosario, K. Modi, A.-M. Kuah, E. Sudarshan, and A. Shaji. 2008. Completely positive maps and classical correlations. *J. Phys. A*, **41**, 205301.

[RW06] M. Roetteler and P. Wocjan. 2006. Equivalence of decoupling schemes and orthogonal arrays. *IEEE Trans. Inf. Theory*, **52**, 4171.

[RHR+07] D. Rosenberg, J. W. Harrington, P. R. Rice, P. A. Hiskett, C. G. Peterson, R. J. Hughes, A. E. Lita, S. W. Nam, and J. E. Nordholt. 2007. Long-distance decoy-state quantum key distribution in optical fiber. *Phys. Rev. Lett.*, **98**, 010503.

[RPH+09] D. Rosenberg, C. G. Peterson, J. W. Harrington, P. R. Rice, N. Dallmann, K. T. Tyagi, K. P. McCabe, S. Nam, B. Baek, R. H. Hadfield, R. J. Hughes, and J. E. Nordholt. 2009. Practical long-distance quantum key distribution system using decoy levels. *New J. Phys.*, **11**, 045009.

[R08] M. Rötteler. 2008. Dynamical decoupling schemes derived from Hamilton cycles. *J. Math. Phys.*, **49**, 042106.

[RB.-KD+02] M. Rowe, A. Ben-Kish, B. DeMarco, D. Leibfried, V. Meyer, J. Beall, J. Britton, J. Hughes, W. Itano, B. Jelenkovic, C. Langer, T. Rosenband, and D. Wineland. 2002. Transport of quantum states and separation of ions in a dual RF ion trap. *Quant. Inf. Comput.*, **2**, 257.

[RHC10] C. A. Ryan, J. S. Hodges, and D. G. Cory. 2010. Robust decoupling techniques to extend quantum coherence in diamond. *Phys. Rev. Lett.*, **105**, 200402.

[S99a] S. Sachdev. 1999. *Quantum Phase Transitions*. Cambridge: Cambridge University Press.

[SAD10] Y. Sagi, I. Almog, and N. Davidson. 2010. Observation of collisional narrowing in an ensemble of cold atoms. *Quantum Electronics and Laser Science Conference*. Optical Society of America, p. QFE1.

[SV05] L. Santos and L. Viola. 2005. Dynamical control of qubit coherence: Random versus deterministic schemes. *Phys. Rev. A*, **72**, 062303.

[SV06] L. Santos and L. Viola. 2006. Enhanced convergence and robust performance of randomized dynamical decoupling. *Phys. Rev. Lett.*, **97**, 150501.

[SV08] L. Santos and L. Viola. 2008. Advantages of randomization in coherent quantum dynamical control. *New J. Phys.*, **10**, 083009.

[SM05] M. Sarovar and G. J. Milburn. 2005. Continuous quantum error correction by cooling. *Phys. Rev. A.*, **72**, 012306.

[SAJ+04] M. Sarovar, C. Ahn, K. Jacobs, and G. J. Milburn. 2004. A practical scheme for error control using feedback. *Phys. Rev. A.*, **69**, 052324.

[SR08] V. Scarani and R. Renner. 2008. Quantum cryptography with finite resources: Unconditional security bound for discrete-variable protocols with one-way postprocessing. *Phys. Rev. Lett.*, **100**, 200501.

[SAR+04] V. Scarani, A. Acín, G. Ribordy, and N. Gisin. 2004. Quantum cryptography protocols robust against photon number splitting attacks for weak laser pulse implementations. *Phys. Rev. Lett.*, **92**, 057901.

[SB.-PC+09] V. Scarani, H. Bechmann-Pasquinucci, N. J. Cerf, M. Dusek, N. Lutkenhaus, and M. Peev. 2009. The security of practical quantum key distribution. *Rev. Mod. Phys.*, **81**, 1301.

[S02a] D. Schlingemann. 2002. Stabilizer codes can be realized as graph codes. *Quantum Inf. Comput.*, **2**, 307.

[SW02a] D. Schlingemann and R. F. Werner. 2002. Quantum error-correcting codes associated with graphs. *Phys. Rev. A*, **65**, 012308.

[SV99] L. J. Schulman and U. V. Vazirani. 1999. Molecular scale heat engines and scalable quantum computation. *STOC '99: Proceedings of the Thirty-First Annual ACM Symposium on Theory of Computing*. New York: ACM, p. 322.

[S96a] B. Schumacher. 1996. Sending entanglement through noisy quantum channels. *Phys. Rev. A*, **54**, 2615.

[SN96] B. Schumacher and M. A. Nielsen. 1996. Quantum data processing and error correction. *Phys. Rev. A*, **54**, 2629.

[SW98] B. Schumacher and M. D. Westmoreland. 1998. Quantum privacy and quantum coherence. *Phys. Rev. Lett.*, **80**, 5695.

[SW02b] B. Schumacher and M. D. Westmoreland. 2002. Approximate quantum error correction. *Quant. Inf. Proc.*, **1**, 5.

[SJ01] A. Schweiger and G. Jeschke. 2001. *Principles of Pulse Electron Paramagnetic Resonance*. Oxford: Oxford University Press.

[S00] A. M. Sengupta. 2000. Spin in a fluctuating field: The Bose(+Fermi) Kondo models. *Phys. Rev. B*, **61**, 4041.
[SL05] A. Shabani and D. A. Lidar. 2005. Completely positive post-Markovian master equation via a measurement approach. *Phys. Rev. A*, **71**, 020101(R).
[SL09a] A. Shabani and D. A. Lidar. 2009. Maps for general open quantum systems and a theory of linear quantum error correction. *Phys. Rev. A*, **80**, 012309.
[SL09b] A. Shabani and D. A. Lidar. 2009. Vanishing quantum discord is necessary and sufficient for completely positive maps. *Phys. Rev. Lett.*, **102**, 100402.
[SK87] A. Shaka and J. Keeler. 1987. Broadband spin decoupling in isotropic liquids. *Progr. NMR Spectrosc.*, **19**, 47.
[S94a] R. Shankar. 1994. Renormalization-group approach to interacting fermions. *Rev. Mod. Phys.*, **66**, 129.
[S40] C. E. Shannon. 1940. *A Symbolic Analysis of Relay and Switching Circuits*. M.Phil. thesis, Massachusetts Institute of Technology.
[S48] C. E. Shannon. 1948. A mathematical theory of communication. *Bell Syst. Tech. J.*, **27**, 379.
[SW89] A. Shapere and F. Wilczek (eds). 1989. *Geometric Phases in Physics*. Singapore: World Scientific.
[SB02] M. Shapiro and P. Brumer. 2002. S-matrix approach to the construction of decoherence-free subspaces. *Phys. Rev. A*, **66**, 052308.
[SCS+00] Y. Sharif, D. G. Cory, S. S. Somaroo, T. F. Havel, E. Knill, R. Laflamme, and W. Zurek. 2000. A study of quantum error correction by geometric algebra and liquid-state NMR spectroscopy. *Mol. Phys.*, **98**, 1347.
[SA80] F. Shibata and T. Arimitsu. 1980. Expansion formulas in nonequilibrium statistical mechanics. *J. Phys. Soc. Jpn.*, **49**, 891.
[STH77] F. Shibata, Y. Takahashi, and N. Hashitsume. 1977. A generalized stochastic Liouville equation: Non-Markovian versus memoryless master equations. *J. Stat. Phys.*, **17**, 171.
[SL04] K. Shiokawa and D. A. Lidar. 2004. Dynamical decoupling using slow pulses: Efficient suppression of $1/f$ noise. *Phys. Rev. A*, **69**, 030302.
[S94b] P. W. Shor. 1994. Algorithms for quantum computation: Discrete logarithms and factoring. *Proceedings of the 35th Annual Symposium on Foundations of Computer Science*. Los Alamitos, CA: IEEE Computer Society Press, p. 124.
[S95] P. W. Shor. 1995. Scheme for reducing decoherence in quantum memory. *Phys. Rev. A*, **52**, R2493.
[S96b] P. Shor. 1996. Fault-tolerant quantum computation. *Proceedings of the 37th Annual Symposium on Foundations of Computer Science*. Los Alamitos, CA: IEEE Computer Society Press, p. 56.
[S97] P. W. Shor. 1997. Polynomial-time algorithms for prime factorization and discrete logarithms on a quantum computer. *SIAM J. Sci. Comput.*, **26**, 1484.
[SP00] P. W. Shor and J. Preskill. 2000. Simple proof of security of the BB84 quantum key distribution protocol. *Phys. Rev. Lett.*, **85**, 441.
[S02b] P. Shor. 2002. The quantum channel capacity and coherent information. Lecture Notes, MSRI Workshop on Quantum Computation, www.msri.org/publications/ln/msri/2002/quantumcrypto/shor/1/.
[S03a] P. W. Shor. 2003. Capacities of quantum channels and how to find them. *Math. Program., Ser. B*, **97**, 311.
[SRZ05] M. Silva, M. Rotteler, and C. Zalka. 2005. Thresholds for linear optics quantum computing with photon loss at the detectors. *Phys. Rev. A*, **72**, 032307.
[SMK+08] M. Silva, E. Magesan, D. W. Kribs, and J. Emerson. 2008. Scalable protocol for identification of correctable codes. *Phys. Rev. A*, **78**, 012347.
[S90] C. P. Slichter. 1990. *Principles of Magnetic Resonance*. 3rd edn. Berlin: Springer.

[SZZ+03a] P. Solinas, P. Zanardi, N. Zanghì, and F. Rossi. 2003. Holonomic quantum gates: A semiconductor-based implementation. *Phys. Rev. A*, **67**, 062315.

[SZZ+03b] P. Solinas, P. Zanardi, N. Zanghi, and F. Rossi. 2003. Semiconductor-based geometrical quantum gates. *Phys. Rev. B*, **67**, 121307.

[SZZ04] P. Solinas, P. Zanardi, and N. Zangh. 2004. Robustness of non-Abelian holonomic quantum gates against parametric noise. *Phys. Rev. A*, **70**, 042316.

[SBD09] T. Stace, S. Barrett, and A. Doherty. 2009. Thresholds for topological codes in the presence of loss. *Phys. Rev. Lett.*, **102**, 200501.

[S96c] A. M. Steane. 1996. Multiple-particle interference and quantum error correction. *Proc. R. Soc. London A*, **452**, 2551.

[S96d] A. M. Steane. 1996. Simple quantum error correcting codes. *Phys. Rev. A*, **54**, 4741.

[S96e] A. M. Steane. 1996. Error correcting codes in quantum theory. *Phys. Rev. Lett.*, **77**, 793.

[S99b] A. M. Steane. 1999. Enlargement of Calderbank–Shor–Steane quantum codes. *IEEE Trans. Inf. Theory*, **45**, 2492.

[S99c] A. M. Steane. 1999. Quantum Reed–Muller codes. *IEEE Trans. Inf. Theory*, **45**, 1701.

[S03b] A. M. Steane. 2003. Overhead and noise threshold of fault-tolerant quantum error correction. *Phys. Rev. A*, **68**, 042322.

[S55] W. F. Stinespring. 1955. Positive functions on C*-algebras. *Proc. Amer. Math. Soc.*, **6**, 211.

[S03c] D. Stinson. 2003. *Combinatorial Designs*. Berlin: Springer.

[SJ09] R. Stock and D. F. V. James. 2009. Scalable, high-speed measurement-based quantum computer using trapped ions. *Phys. Rev. Lett.*, **102**, 170501.

[SM01] M. Stollsteimer and G. Mahler. 2001. Suppression of arbitrary internal coupling in a quantum register. *Phys. Rev. A*, **64**, 052301.

[STD05] K. M. Svore, B. M. Terhal, and D. P. DiVincenzo. 2005. Local fault-tolerant quantum computation. *Phys. Rev. A*, **72**, 022317.

[TKL10] S. Taghavi, R. Kosut, and D. Lidar. 2010. Channel-optimized quantum error correction. *IEEE Trans. Inf. Theory*, **56**, 1461.

[TN04] K. Takeda and H. Nishimori. 2004. Self-dual random-plaquette gauge model and the quantum toric code. *Nucl. Phys. B*, **686**, 377.

[TKI03] K. Tamaki, M. Koashi, and N. Imoto. 2003. Unconditionally secure key distribution based on two nonorthogonal states. *Phys. Rev. Lett.*, **90**, 167904.

[TL06] J. M. Taylor and M. D. Lukin. 2006. Dephasing of quantum bits by a quasi-static mesoscopic environment. *Quant. Inf. Proc.*, **5**, 503.

[TED+05] J. M. Taylor, H.-A. Engel, W. Dür, A. Yacoby, C. M. Marcus, P. Zoller, and M. D. Lukin. 2005. Fault-tolerant architecture for quantum computation using elecrically controlled semicondcutor spins. *Nature Phys.*, **1**, 177.

[TDZ+05] J. M. Taylor, W. Dür, P. Zoller, A. Yacoby, C. M. Marcus, and M. D. Lukin. 2005. Solid-state circuit for spin entanglement generation and purification. *Phys. Rev. Lett.*, **94**, 236803.

[TB05] B. M. Terhal and G. Burkard. 2005. Fault-tolerant quantum computation for local non-Markovian noise. *Phys. Rev. A*, **71**, 012336.

[TV06] F. Ticozzi and L. Viola. 2006. Single-bit feedback and quantum-dynamical decoupling. *Phys. Rev. A*, **74**, 052328.

[TM02] B. C. Travaglione and G. J. Milburn. 2002. Preparing encoded states in an oscillator. *Phys. Rev. A*, **66**, 052322.

[TWZ+11] A. M. Tyryshkin, Z.-H. Wang, W. Zhang, E. E. Haller, J. W. Ager, V. V. Dobrovitski, and S. A. Lyon. 2011. Dynamical decoupling in the presence of realistic pulse errors. eprint arXiv:1011.1903.

[T03] J. Tyson. 2003. Operator-Schmidt decompositions and the Fourier transform, with applications to the operator-Schmidt numbers of unitaries. *J. Phys. A*, **36**, 10101.

[UA02] C. Uchiyama and M. Aihara. 2002. Multipulse control of decoherence. *Phys. Rev. A*, **66**, 032313.

[U07] G. Uhrig. 2007. Keeping a quantum bit alive by optimized π-pulse sequences. *Phys. Rev. Lett.*, **98**, 100504. Erratum: 2011. *Phys. Rev. Lett.*, **106**, 129901.

[U08] G. S. Uhrig. 2008. Exact results on dynamical decoupling by π-pulses in quantum information processes. *New J. Phys.*, **10**, 083024.

[U09] G. S. Uhrig. 2009. Concatenated control sequences based on optimized dynamic decoupling. *Phys. Rev. Lett.*, **102**, 120502.

[UL10] G. S. Uhrig and D. A. Lidar. 2010. Rigorous bounds for optimal dynamical decoupling. *Phys. Rev. A*, **82**, 012301.

[UP10] G. S. Uhrig and S. Pasini. 2010. Efficient coherent control by sequences of pulses of finite duration. *New J. Phys.*, **12**, 045001.

[U95] W. G. Unruh. 1995. Maintaining coherence in quantum computers. *Phys. Rev. A*, **51**, 992.

[UBB09] H. Uys, M. J. Biercuk, and J. J. Bollinger. 2009. Optimized noise filtration through dynamical decoupling. *Phys. Rev. Lett.*, **103**, 040501.

[VGW96] L. Vaidman, L. Goldenberg, and S. Wiesner. 1996. Error prevention scheme with four particles. *Phys. Rev. A*, **54**, R1745.

[VB96] L. Vandenberghe and S. Boyd. 1996. Semidefinite programming. *SIAM Rev.*, **38**, 49.

[VC04] L. M. Vandersypen and I. L. Chuang. 2004. NMR techniques for quantum control and computation. *Rev. Mod. Phys.*, **76**, 1037.

[V02] L. Viola. 2002. Quantum control via encoded dynamical decoupling. *Phys. Rev. A*, **66**, 012307.

[V04] L. Viola. 2004. Advances in decoherence control. *J. Mod. Opt.*, **51**, 2357.

[V06] L. Viola. 2006. Randomized control of open quantum systems. *Proceedings 44th IEEE Conference on Decision and Control*, p. 1794.

[VK03] L. Viola and E. Knill. 2003. Robust dynamical decoupling with bounded controls. *Phys. Rev. Lett.*, **90**, 037901.

[VK05] L. Viola and E. Knill. 2005. Random decoupling schemes for quantum dynamical control and error suppression. *Phys. Rev. Lett.*, **94**, 060502.

[VS98] L. Viola and S. Lloyd. 1998. Dynamical suppression of decoherence in two-state quantum systems. *Phys. Rev. A*, **58**, 2733.

[VS06] L. Viola and L. F. Santos. 2006. Randomized dynamical decoupling techniques for coherent quantum control. *J. Mod. Opt.*, **53**, 2559.

[VKL99] L. Viola, E. Knill, and S. Lloyd. 1999. Dynamical decoupling of open quantum systems. *Phys. Rev. Lett.*, **82**, 2417.

[VLK99] L. Viola, S. Lloyd, and E. Knill. 1999. Universal control of decoupled quantum systems. *Phys. Rev. Lett.*, **83**, 4888.

[VKL00] L. Viola, E. Knill, and S. Lloyd. 2000. Dynamical generation of noiseless quantum subsystems. *Phys. Rev. Lett.*, **85**, 3520.

[VFP+01] L. Viola, E. Fortunato, M. Pravia, E. Knill, R. Laflamme, and D. G. Cory. 2001. Experimental realization of noiseless subsystems for quantum information processing. *Science*, **293**, 5537.

[VT99] D. Vitali and P. Tombesi. 1999. Using parity kicks for decoherence control. *Phys. Rev. A*, **59**, 4178.

[V67] A. J. Viterbi. 1967. Error bounds for convolutional codes and an asymptotically optimum decoding algorithm. *IEEE Trans. Inf. Theory*, **13**, 260.

[VMB05] S. Vorojtsov, E. R. Mucciolo, and H. U. Baranger. 2005. Phonon decoherence of a double quantum dot charge qubit. *Phys. Rev. B*, **71**, 205322.

[WAS+03] Z. D. Walton, A. F. Abouraddy, A. V. Sergienko, B. E. A. Saleh, and M. C. Teich. 2003. Decoherence-free subspaces in quantum key distribution. *Phys. Rev. Lett.*, **91**, 087901.

[WHP03] C. Wang, J. Harrington, and J. Preskill. 2003. Confinement-Higgs transition in a disordered gauge theory and the accuracy threshold for quantum memory. *Ann. Phys.*, **303**, 31.

[WRF+11] Y. Wang, X. Rong, P. Feng, W. Xu, B. Chong, J.-H. Su, J. Gong, and J. Du. 2011. Preservation of bipartite pseudoentanglement in solids using dynamical decoupling. *Phys. Rev. Lett.*, **106**, 040501.

[WZT+12] Z.-H. Wang, W. Zhang, A. M. Tyryshkin, S. A. Lyon, J. W. Ager, E. E. Haller, and V. V. Dobrovitski. 2012. Effect of pulse error accumulation on dynamical decoupling of the electron spins of phosphorus donors in silicon. *Phys. Rev. B*, **85**, 085206.

[WL11a] Z.-Y. Wang and R.-B. Liu. 2011. Extending quantum control of time-independent systems to time-dependent systems. *Phys. Rev. A*, **83**, 062313.

[WL11b] Z.-Y. Wang and R.-B. Liu. 2011. Protection of quantum systems by nested dynamical decoupling. *Phys. Rev. A*, **83**, 022306.

[W97] W. S. Warren. 1997. The usefulness of NMR quantum computing. *Science*, **277**, 1688.

[WB07] W. Wasilewski and K. Banaszek. 2007. Protecting an optical qubit against photon loss. *Phys. Rev. A*, **75**, 042316.

[W04] J. Watrous. 2004. www.cs.uwaterloo.ca/watrous/lecture-notes/701/all.pdf.

[WHH68] J. S. Waugh, L. M. Huber, and U. Haeberlen. 1968. Approach to high-resolution NMR in solids. *Phys. Rev. Lett.*, **20**, 180.

[WC79] M. N. Wegman and J. L. Carter. 1979. Universal classes of hash functions. *J. Comput. Syst. Sci.*, **18**, 143.

[WC81] M. N. Wegman and J. L. Carter. 1981. New hash functions and their use in authentication and set equality. *J. Comput. Syst. Sci.*, **22**, 265.

[W67] S. Weinberg. 1967. A model of leptons. *Phys. Rev. Lett.*, **19**, 1264.

[W99] S. Weiss. 1999. *Quantum Dissipative Systems*. 2nd edn. Singapore: World Scientific.

[W88] M. B. Weissman. 1988. $1/f$ noise and other slow, nonexponential kinetics in condensed matter. *Rev. Mod. Phys.*, **60**, 537.

[W89] R. Werner. 1989. Quantum states with Einstein–Podolsky–Rosen correlations admitting a hidden-variable model. *Phys. Rev. A*, **40**, 4277.

[WLF+10] J. R. West, D. A. Lidar, B. H. Fong, M. F. Gyure, X. Peng, and D. Suter. 2010. Quantum gates via concatenated dynamical decoupling: Theory and experiment. eprint arXiv:0911.2398.

[WFL10] J. R. West, B. H. Fong, and D. A. Lidar. 2010. Near-optimal dynamical decoupling of a qubit. *Phys. Rev. Lett.*, **104**, 130501.

[W83] S. Wiesner. 1983. Conjugate coding. *SIGACT News*, **15**, 78.

[WZ84] F. Wilczek and A. Zee. 1984. Appearance of gauge structure in simple dynamical systems. *Phys. Rev. Lett.*, **52**, 2111.

[W09a] M. M. Wilde. 2009. Logical operators of quantum codes. *Phys. Rev. A*, **79**, 062322.

[W09b] M. M. Wilde. 2009. Quantum-shift-register circuits. *Phys. Rev. A*, **79**, 062325.

[WB08a] M. M. Wilde and T. A. Brun. 2008. Optimal entanglement formulae for entanglement-assisted quantum coding. *Phys. Rev. A*, **77**, 064302.

[WB08b] M. M. Wilde and T. A. Brun. 2008. Protecting quantum information with entanglement and noisy optical modes. *Quant. Inf. Proc.*, **8**, 401.

[WB08c] M. M. Wilde and T. A. Brun. 2008. Unified quantum convolutional coding. *Proceedings of the IEEE International Symposium on Information Theory*, p. 359.

[WB09] M. M. Wilde and T. A. Brun. 2009. Extra shared entanglement reduces memory demand in quantum convolutional coding. *Phys. Rev. A*, **79**, 032313.

[WB10a] M. M. Wilde and T. A. Brun. 2010. Entanglement-assisted quantum convolutional coding. *Phys. Rev. A*, **81**, 042333.

[WB10b] M. M. Wilde and T. A. Brun. 2010. Quantum convolutional coding with shared entanglement: General structure. *Quant. Inf. Proc.*, **9**, 509.

[WH10] M. M. Wilde and M.-H. Hsieh. 2010. Entanglement boosts quantum turbo codes. eprint arXiv:1010.1256.

[WKB07] M. M. Wilde, H. Krovi, and T. A. Brun. 2007. Entanglement-assisted quantum error correction with linear optics. *Phys. Rev. A*, **76**, 052308.

[WHH.-K11] M. M. Wilde, M. Houshmand, and S. Hosseini-Khayat. 2011. Examples of minimal-memory, non-catastrophic quantum convolutional encoders. *Proceedings of the 2011 International Symposium on Information Theory*, p. 376.

[WR06] H. M. Wiseman and J. F. Ralph. 2006. Reconsidering rapid qubit purification by feedback. *New J. Phys.*, **8**, 90.

[WD07] W. M. Witzel and S. Das Sarma. 2007. Concatenated dynamical decoupling in a solid-state spin bath. *Phys. Rev. B*, **76**, 241303(R).

[W06] P. Wocjan. 2006. Efficient decoupling schemes with bounded controls based on Eulerian arrays. *Phys. Rev. A*, **73**, 062317.

[WJB02] P. Wocjan, D. Janzing, and T. Beth. 2002. Simulating arbitrary pair-interactions by a given Hamiltonian: Graph-theoretical bounds on the time complexity. *Quant. Inf. Comput.*, **2**, 117.

[WRJ+02a] P. Wocjan, M. Rötteler, D. Janzing, and T. Beth. 2002. Simulating Hamiltonians in quantum networks: efficient schemes and complexity bounds. *Phys. Rev. A*, **65**, 042309.

[WRJ+02b] P. Wocjan, M. Roetteler, D. Janzing, and T. Beth. 2002. Universal simulation of Hamiltonians using a finite set of control operations. *Quant. Inf. Comput.*, **2**, 133.

[WP.-GG07] M. M. Wolf, D. Pérez-García, and G. Giedke. 2007. Quantum capacities of bosonic channels. *Phys. Rev. Lett.*, **98**, 130501.

[WZ82] W. K. Wootters and W. H. Zurek. 1982. A single quantum cannot be cloned. *Nature*, **299**, 802.

[WL02] L.-A. Wu and D. A. Lidar. 2002. Creating decoherence-free subspaces using strong and fast pulses. *Phys. Rev. Lett.*, **88**, 207902.

[WZL05] L.-A. Wu, P. Zanardi, and D. A. Lidar. 2005. Holonomic quantum computation in decoherence-free subspaces. *Phys. Rev. Lett.*, **95**, 130501.

[YF07] N. Yamamoto and M. Fazel. 2007. Computational approach to quantum encoder design for purity optimization. *Phys. Rev. A*, **76**, 012327.

[YHT05] N. Yamamoto, S. Hara, and K. Tsumura. 2005. Supoptimal quantum-error-correcting procedure based on semidefinite programming. *Phys. Rev. A*, **71**, 022322.

[YG.-B01] C.-P. Yang and J. Gea-Banacloche. 2001. Three-qubit quantum error-correction scheme for collective decoherence. *Phys. Rev. A*, **63**, 022311.

[YL08] W. Yang and R.-B. Liu. 2008. Universality of Uhrig dynamical decoupling for suppressing qubit pure dephasing and relaxation. *Phys. Rev. Lett.*, **101**, 180403.

[YWL11] W. Yang, Z.-Y. Wang, and R.-B. Liu. 2011. Preserving qubit coherence by dynamical decoupling. *Front. Phys.*, **6**, 2.

[YLS07] W. Yao, R.-B. Liu, and L. J. Sham. 2007. Restoring coherence lost to a slow interacting mesoscopic spin bath. *Phys. Rev. Lett.*, **98**, 077602.

[YD09] J. Yard and I. Devetak. 2009. Optimal quantum source coding with quantum side information at the encoder and decoder. *IEEE Trans. Inf. Theory*, **55**, 5339.

[YBW08] M.-Y. Ye, Y.-K. Bai, and Z. D. Wang. 2008. Quantum state redistribution based on a generalized decoupling. *Phys. Rev. A*, **78**, 030302.

[Y01] J. S. Yedidia. 2001. An idiosyncratic journey beyond mean field theory. In *Advanced Mean Field Methods: Theory and Practice*. Cambridge, MA: MIT Press, p. 21.

[YHA+07] J.-I. Yoshikawa, T. Hayashi, T. Akiyama, N. Takei, A. Huck, U. L. Andersen, and A. Furusawa. 2007. Demonstration of deterministic and high fidelity squeezing of quantum information. *Phys. Rev. A*, **76**, 060301.

[YMH+08] J.-I. Yoshikawa, Y. Miwa, A. Huck, U. L. Andersen, P. van Loock, and A. Furusawa. 2008. Demonstration of a quantum nondemolition sum gate. *Phys. Rev. Lett.*, **101**, 250501.

[YW11] K. C. Young and B. K. Whaley. 2011. Qubits as spectrometers of dephasing noise. eprint arXiv:1102.5115.

[YCO07] S. Yu, Q. Chen, and C. H. Oh. 2007. Graphical quantum error-correcting codes. Preprint arXiv:0709.1780v1 [quant-ph].

[YCL+09] S. Yu, Q. Chen, C. H. Lai, and C. H. Oh. 2009. Nonadditive quantum error-correcting code. *Phys. Rev. Lett.*, **101**, 090501.

[Z97] P. Zanardi. 1997. Dissipative dynamics in a quantum register. *Phys. Rev. A*, **56**, 4445.

[Z98] P. Zanardi. 1998. Dissipation and decoherence in a quantum register. *Phys. Rev. A*, **57**, 3276.

[Z99] P. Zanardi. 1999. Symmetrizing evolutions. *Phys. Lett. A*, **258**, 77.

[Z00] P. Zanardi. 2000. Stabilizing quantum information. *Phys. Rev. A*, **63**, 012301.

[Z01] P. Zanardi. 2001. Virtual quantum subsystems. *Phys. Rev. Lett.*, **87**, 077901.

[ZL03] P. Zanardi and S. Lloyd. 2003. Topological protection and quantum noiseless subsystems. *Phys. Rev. Lett.*, **90**, 067902.

[ZR97a] P. Zanardi and M. Rasetti. 1997. Error avoiding quantum codes. *Mod. Phys. Lett. B*, **11**, 1085.

[ZR97b] P. Zanardi and M. Rasetti. 1997. Noiseless quantum codes. *Phys. Rev. Lett.*, **79**, 3306.

[ZR99a] P. Zanardi and M. Rasetti. 1999. Holonomic quantum computation. *Phys. Lett. A*, **264**, 94.

[ZR98] P. Zanardi and F. Rossi. 1998. Quantum information in semiconductors: Noiseless encoding in a quantum-dot array. *Phys. Rev. Lett.*, **81**, 4752.

[ZR99b] P. Zanardi and F. Rossi. 1999. Subdecoherent information encoding in a quantum-dot array. *Phys. Rev. B*, **59**, 8170.

[ZLL04] P. Zanardi, D. Lidar, and S. Lloyd. 2004. Quantum tensor product structures are observable-induced. *Phys. Rev. Lett.*, **92**, 060402.

[ZDS+07] W. Zhang, V. V. Dobrovitski, L. F. Santos, L. Viola, and B. N. Harmon. 2007. Dynamical control of electron spin coherence in a quantum dot: A theoretical study. *Phys. Rev. B*, **75**, 201302.

[ZKD+08] W. Zhang, N. P. Konstantinidis, V. V. Dobrovitski, B. N. Harmon, L. F. Santos, and L. Viola. 2008. Long-time electron spin storage via dynamical suppression of hyperfine-induced decoherence in a quantum dot. *Phys. Rev. B*, **77**, 125336.

[ZHH+11] N. Zhao, J.-L. Hu, S.-W. Ho, J. T. K. Wan, and R.-B. Liu. 2011. Atomic-scale magnetometry of distant nuclear spin clusters via nitrogen-vacancy spin in diamond. *Nature Nanotech.*, **6**, 242.

[ZLC00] X. Zhou, D. W. Leung, and I. L. Chuang. 2000. Methodology for quantum logic gate constructions. *Phys. Rev. A*, **62**, 052316.

[ZYZ+04] Z.-W. Zhou, B. Yu, X. Zhou, M. J. Feldman, and G.-C. Guo. 2004. Scalable fault-tolerant quantum computation in decoherence-free subspaces. *Phys. Rev. Lett.*, **93**, 010501.

[Z84] W. H. Zurek. 1984. Reversibility and stability of information processing systems. *Phys. Rev. Lett.*, **53**, 391.

[Z03] W. H. Zurek. 2003. Decoherence, einselection, and the quantum origins of the classical. *Rev. Mod. Phys.*, **75**, 715.
[ZP94] W. Zurek and J. P. Paz. 1994. Decoherence, chaos, and the second law. *Phys. Rev. Lett.*, **72**, 2508.
[Z60] R. Zwanzig. 1960. Ensemble method in the theory of irreversibility. *J. Chem. Phys.*, **33**, 1338.

Index

$((5,6,2))$ nonadditive code, 261
$((9,12,3))$ nonadditive code, 261, 275
$((10,24,3))$ optimal code, 261, 268
$((n,K,d))$ code, 261
$((n,K,d))_q$ code, 312
$((n,K,d))_q$ stabilizer code, 314
$[n,k,d]$ classical code, 63, 64, 196
$[n,k,d]_q$ classical code, 177, 315, 318, 324, 325, 389, 393
$[n,k,t]$ classical code, 63
$[n,k]$ random classical code, 280, 281
$[n,k]_q$ classical MDS code, 326
$[n,k]_q$ classical cyclic code, 317
$[[n,k,d]]$ code, 59, 169, 263, 279, 334
$[[n,k,d]]_q$ code, 312, 325, 326
$[[n,k,r,d]]$ subsystem code, 170
$[[n,k,r,d]]_q$ subsystem code, 178
$[[n,1,r,d]]$ operator stabilizer code, 205
$[[n,k]]$ random stabilizer code, 198, 281
$[4,2,3]$ classical quaternary code, 195, 196
$[4,2]$ classical quaternary code, 195
$[7,4,3]$ classical Hamming code, 65
$[[4,0,2]]$ code, 196
$[[4,1,2]]$ code, 328, 332, 333
$[[4,1,3;1]]$ code, 189, 196
$[[4,1;1]]$ code, 188, 195
$[[5,1,3]]$ code, 71, 189, 328, 332, 333, 343, 345
$[[5,1,3]]_2$ code, 326
$[[6,0,4]]_2$ code, 326
$[[7,0,3]]$ code, 267
$[[7,1,3]]$ code, 65, 71, 73, 267, 343
$[[9,0,3]]$ graph state, 275
$[[9,1,3]]$ code, 61, 70, 275, 343, 345
$[[9,1,4,3]]_2$ Bacon–Shor code, 178
$[[10,0,4]]$ graph state, 268
$[[10,1,3]]$ code, 268
$[[12,2,3]]$ code, 298
$[[128,2,8]]$ toric code, 464
$[[73,1,9]]$ triangular color code, 477
$[[85,1,7]]$ planar toric code, 467
$[[96,4,8]]$ color code, 475

accessible information, 580
accuracy threshold, 449, 480
additive
    classical error-correcting codes, 261
    quantum error-correcting codes, 241, 263, 270
Aharonov–Bohm effect, 397
alanine, 511, 512, 523
amplitude, **6**
amplitude damping, 135
ancilla, 27, 165, 182, 183, 188, 192, 210, 226, 516
ancillary qubits, 159, 182, 184, 412, 450, 516, 591, 614
angular momentum, 95, 549
angular momentum algebra, 84
annihilation operator, 83, 85
anticommutation, 186
anticommutator, 28
anyon, 397, 481, 623
approximate quantum error correction, 328
Araki–Lieb inequality, 15
atom trap, 6
augmented check matrix, 560
average Hamiltonian theory, 105, 113, 115, 379

B92 protocol, 579
ball, 63
    billiard, 283
    tennis, 283
Bayes' rule, 15, 284–286
BB84 protocol, **563**, 563, 564, 566, 579, 580
BCH (Bose–Chaudhuri–Hocquenghem) code, 318
belief propagation, 283, 302
Bell basis, 434
Bell states, 12
Bernoulli numbers, 9
Berry phase, 397
billiard ball, 283
binary quantum BCH code, 318
binary symmetric channel, 280, 286, 570
bit-flip, 47, 54, 195, 541
bit-flip channel, 48
bit-flip code, 59
Bloch space, 8

657

Index

Bloch sphere, 6, 7, 208, 529
  mixed states in, 7
  pure states on, 7
Bloch vector, 7, 8
Bohr correspondence principle, 612
Bohr frequency, 622
Boolean function, 276
bootstrapping, 198
Born approximation, 616
bound
  BCH, 323
  classical
    Gilbert–Varshamov, 280
    Hamming, 280
    Singleton, 320
  dual, 345
  iterated, 345
  quantum
    Gilbert–Varshamov, 281
    Hamming, 60, 77, 178
    Singleton, 177, 178
  sphere-packing, 323
boundary, 455
boundary operators, 459
bounded control, 351, 374
bounded linear operators, 5, 329
bounded-strength dynamical decoupling, 120
bra, **5**
bra-ket notation, **5**
branyon, 481
Bratteli diagram, 95
breeding protocol, 558, 579
  CSS, 561, 565, 567, 569

Cauchy sequence, 4
cavity, 85, 547
Cayley graph, 120, 122, 391, 393
CDD (concatenated dynamical decoupling), **355**, 355–358, 361, 362, 364–366, 374, 375, 535
cell
  dual, 494
  primal, 494
cellular automata, 159
centralizer, 69, 116, 177, 184
channel
  amplitude damping, 328, 333, 342
  authenticated classical, 563
  binary symmetric, 280, 570
  depolarizing, 176, 281, 295, 514, 539
  quantum, 44, 53, 164, 556
channel estimation, 565, 569
Choi matrix, 331, 339
classical DFS, 78
classical information over noisy quantum channel, 581
Clebsch–Gordan coefficients, 84
Clifford
  group, **74**, 75, 76, 242, 417, 423, 433, 443, 518
    transversal, 477
  operation, 244, 421, 423, 424, 481, 505
  subsystem code, 178
clique, 268
closed quantum system, 112
CNOT (controlled-NOT) gate, **10**

code
  additive
    over $GF(4)$, 262
  algebraic, 282
  as eigenspace of commuting operators, 193, 262
  Bacon–Shor, 172, 178, 425, 428, 429
  BCH (Bose–Chaudhuri–Hocquenghem), 318
    narrow-sense, 319
  bit-flip, 173, 217
  canonical, 191, 194, 573, 574
  classical
    BCH, 318
    cyclic code, 317
    dual of, 63
    error-correcting, 47
    generator matrix of, 63
    Goethals, 271
    sparse, 176
  Clifford subsystem, 178
  color, 472, 474, 476, 481, 624
  concatenation, 172, 173, 279, 504
  convolutional, 182, 200, 231, 232, 234–236, 240, 244, 250, 253, 255, 260, 297
  CRSS (Calderbank–Rains–Shor–Steane), 195, 196
  CSS (Calderbank–Shor–Steane), 43, 65
  CSS entanglement-assisted, 559, 561, 570
  degeneracy, 175, 480
  dual, 63, 234, 263, 320, 389
  entanglement-assisted quantum error-correcting (EAQECC), 182
  error-correcting, 47
  Goethals, 269
  Gray, 389
  Hamming, 65, 387
  LDPC (low-density parity check), 176, 199, 304
  linear, 61, 199, 269, 280, 315, 389
  local, 456, 478
  parameters, 299, 325
  planar, 472
  Preparata, 269
  primitive, 319
  private classical, 581
  probabilistic, 282
  purity, 178
  quantum
    BCH, 318
    error-correcting, 232, 261, 301, 319
      concatenated, 181
      degenerate, 72, 89, 163
      distance, 59, 175, 196, 261, 272
      error probability, 182
      nondegenerate, 59, 89, 195
    Goethals–Preparata, 274
    MDS, 325
  quaternary, 195, 197
  rate, 296
  Reed–Muller, 269, 270, 272, 273
  Reed–Solomon, 319
    narrow-sense, 319
  single-qubit, 225
  space, 65–67, 69, 71, 73, 129, 130, 143–145, 170, 176, 184, 187, 188, 195, 205, 206, 215, 217–220, 222, 226, 277, 289, 443
  sparse, 279, 297, 298, 300, 304

stabilizer, 76, 261
subsystem, 163, 165, 169, 177, 178, 481, 581
surface, 159, 461, 465, 468, 479
three-qubit bit-flip, 207, 554, 576
topological, 455–458, 461, 472, 475, 480, 481, 518
toric, 464, 467, 480
triangular, 477
turbo, 200, 292, 293, 295, 297, 305
    classical, 279, 293
    quantum, 292, 293, 295, 305, 306
coherence-preserving states, 103
coherent information, 280
collective
    decoherence, 83–86, 95, 96, 102, 103, 119, 406, 407, 514
    dephasing, 79, 81, 84, 85, 103, 406, 525
    dissipation, 103
    spin operators, 84
communication complexity, 433
commutant, 94
commutator, 8, 9, 184, 313, 597
complete positivity, 32
completely mixed state, 7
completely positive (CP) map, 17, 33, 43, 180, 208
completely positive trace-preserving (CPTP) map, 3, **17**, 34, 191, 201, 202, 208, 328, 330
completeness, 16
completeness relation, 24
complex conjugate, 5
complex optimization, 327, 329, 334
computational basis, 48, 74, 401
computational complexity, 348, 612
computing with linear optics, 433
concatenation
    experimental of DFS and QECC, 514
    for fault tolerance of holonomic QC, 428
    of a convolutional code, 295
    of bit-flip and phase-flip codes, 70
    of classical codes, 173
    of convolutional code, 292
    of DD sequences, 366
    of DFS and QECC, 102, 104
    of strings, 568
conditional entropy, 15
conditional probability, 15, 285, 303
connection, 400
    $u(1)$-valued, 408
    $u(3)$-valued, 403
    adiabatic, 413
    form, 408, 409
    irreducible, 400
    non-Abelian gauge, 407
continuous-variable system, 554, 571, 581
control cycle, 119, 380, 526
control propagator, 115, 120, 380, 392
control sequence, 520
Cooper pair, 526
correctable errors, 59, 166, 193, 203, 211, 277, 554
correlation function, 107, 141, 360, 540, 585, 616
    two-point, 587, 599
correlations
    Dirac-delta, 616
coset, 64, 265, 315, 460

CP (Carr–Purcell) sequence, 105–108, 110, 114, 115, 118, 123, 356, 377, 530
CP (completely positive) map, **17**, 33, 43, 180, 208
CPMG (Carr–Purcell–Meiboom–Gill) sequence, 373, 530
CPTP (completely positive trace-preserving) map, 3, 34, 191, 201, 202, 208, 328, 330
creation operator, 83
critical environment, 585
cryptography, 81
CSS (Calderbank–Shor–Steane), **43**
CSW (codeword stabilized codes), 261, **266**
cumulant expansion, 616
curvature, 400
cycle, 114, 115, 120, 121, 390, 392, 438, 458–460, 465, 477, 483, 495, 501
cyclic group, 310, 313
cyclic property of trace, 332

dark state, 401, 403
Davies generator, 623
DD (dynamical decoupling), 4, 36, 78, 104, 105, 351, 375, 376, 541–543
    bang-bang, 377
    combinatorial, 376
    concatenated (CDD), 355–358, 361, 362, 364–366, 374, 375, 535
    concatenated Uhrig (CUDD), 365
    embedded (EMD), 372
    experimental, 519
    naive random (NRD), 372
    nested Uhrig (NUDD), 367
    periodic (PDD), 113, 114, 118, 119, 353, 372, 535
    quadratic (QDD), 366
    random path (RPD), 372
    randomized (RDD), 355, 372
    symmetric (SDD), 118, 373
    symmetric random path (SRPD), 373
    Uhrig (UDD), 357, 360, 361, 363, 365, 366, 368, 373, 530
decoder
    bit-wise, 285
    maximum-likelihood, 287
    soft, 286
decoding, 47, 48, 145, 170, 176, 190, 231, 237, 257, 260, 279, 282, 284, 285, 287, 289, 293, 303, 513
    in exponential time, 282
    in polynomial time, 282
decoherence, 3, 42, 76, 104, 108, 158, 202, 213, 224, 327, 331, 360, 380, 522, 533, 585, 618
    control, 106, 204, 520
    estimates of, 117
decoherence problem, 46
decoupling scheme
    bang-bang, 379, 380
    bounded strength, 377
    Eulerian, 391
    periodic, 377
    regular, 380
degeneracy, 82, 289, 398
density matrix, 5
dephasing errors, 43
depolarizing channel, 281
depolarizing noise, 44
deviation density matrix, 511

DFS (decoherence-free subspace), 4, **78**, 78–92, 94, 96–99, 101–104, 106, 127, 163–165, 179, 180, 346, 406, 412, 514, 523, 525, 543, 602
   for atoms in a cavity, 85
   for collective decoherence, 102
   for collective dephasing, 81
   for stabilizer errors, 88
   NMR experiment, 514
diamond norm, 132, 614
dibromothiophene, 523, 525
difference matrix, 384, 391
dimension of a vector space, 61
dipolar coupling, 532
Dirac notation, **5**
discord, quantum, 12, 15, 16, 18, 26, 339
discretization of errors, 42
distance
   Kolmogorov, 23
   trace-norm, 23
distillation
   entanglement, 182, 553, 554, 579
   entanglement-assisted entanglement, 558
   stabilizer entanglement, 556
divisor, 322
drift Hamiltonian, 111, 377
dual
   bound, 345
   distance, 323, 389
   feasible point, 345
   lattice, 464, 466
   optimization, 336
dual of a classical linear code, **63**
dynamical exponent, 587
dynamics in the bang-bang limit, 108
Dyson series, 8–10, 23, 586–588, 602

EAQECC (entanglement-assisted quantum error-correcting codes), **182–184**, 186–191, 194–196, 198–200
ebit, 182, 183, 186, 188, 191, 198, 254–256, 553, 558, 562, 567
   logical, 555, 558, 562
   physical, 554, 558, 562
effective Hamiltonian, 9, 82, 116
EigQER (eigen quantum error recovery), 341, 343
electron spin, 373, 535, 542
encoded basis, 204, 217
encoding operation, 182, 188, 250, 330, 346
energy eigenstate, 136
ENIAC, 126
ensemble, 34, 169, 283, 329, 405, 523
entanglement, **12**, 13, 25, 35, 39, 76
   distillable, 334
   entropy, 13
   fidelity, 112
   generation, 550
   measure, 14
   pseudo, 374, 535
   relative to choice of observables, 13
   shared, 182, 198, 561
entanglement distillation protocol yield, 556
entanglement-assisted convolutional codes, 253
entanglement-assisted quantum error correction, 189

entropy, 13
   binary, 570
   of entanglement, 14, 38
   Shannon, 280
   von Neumann, 13, 14, 22
environment, 578, 585, 588, 596, 605
EPR pair, 182, 198
error
   classical, 62, 544
   map, 616
   model, 33, 41, 45, 220, 294, 480, 497, 539, 602
      changes under concatenation, 176
   propagation in fault-tolerant quantum circuits, 619
   rate, 225, 287, 299, 403, 482, 505, 514
   rate reduction, 225
   syndromes, 237, 276
   threshold, 177, 428, 470, 480
error-avoiding codes, 103, 164
Euler
   characteristic, 458
   function, 610
Eulerian
   circuit, 121
   cycle, 121, 124, 392, 393
   graph, 121
exchange gate, 545
exchange interaction, 102
expectation value, **19**
exponential resources, 291
extended gadget, 146–154
   bad, 150
   good, 150
extraspecial p-group, 310

factoring, 553
fault path, 132, 134, 137, 147, 154, 449
   bad, 152, 449, 450
   expansion, 133, 150
   good, 152, 449, 450
   sum over, 134
   trivial, 152
fault-tolerant design, 450
fault-tolerant operation, 443
   Toffoli gate, 417
Fermi Golden Rule, 617
Feynman path integral, 397
fiber bundle, 408
Fibonacci scheme, 159
fidelity
   average entanglement, 331, 336
   entanglement, 329, 331, 333, 343, 513, 514, 524, 526
   maximum output, 24
   minimum, 329, 336
   quantum, 327
   Uhlman, 23
filter function, 358, 530
finite field, 177, 271, 307, 309, 388
first homology group, 460, 480
Fourier expansion, 363, 364
Fourier transform, 107, 359, 360, 381, 576, 577, 616
   discrete, 381
free-field theory, 587
Frobenius norm, **22**
fullerene molecule, 528

Gacs
  and noisy cellular automata, 159
  and reliable classical computation, 159
gadget
  bad, 148
  extended, 146, 152
  fault-tolerant, 127, 147
  for measuring an observable, 145
  for pure state preparation, 144
  good, 146
  measurement extended, 147
  noisy, 146
  preparation, 147
  unitary-gate, 147
gadgets
  and organs, 143
  and rectangles, 143
  and robustness to noise, 146
gauge field, 407, 587
gauge qubits, 206
Gaussian
  noise, 159, 540
  random variable, 42, 530
generalized measurement, 24, 33, 35, 129, 144
generator matrix, 62, 65, 233, 244, 266, 284, 389
  of classical linear code, **63**
genus, 457
$GF(4)$, 195, 262
Gibbs
  distribution, 287
  state, 621
Gilbert–Varshamov bound, 280
global phase, 6, 359, 420, 571
Goethals code, 269, 271
Goethals–Preparata code, 272
good coding scheme, 282
Gram–Schmidt procedure, 185
Gram–Schmidt symplectic orthogonalization procedure, 559
graph
  Cayley, 120, 122, 391, 393
  Eulerian, 121
  Tanner, 199, 297–299, 302–305
Grassmann manifold, 410
Green functions, 592
ground state, 39, 287
group
  Abelian, 174, 187, 262, 300
  centralizer, 116
  Clifford, 75, 76, 242, 417, 423, 433, 443, 518
  finite, 120, 309
  generators, 67, 121
  homomorphism, 459
  index, 381
  non-Abelian, 66, 90, 188
  normalizer, 265
  quotient, 69, 310, 460
  stabilizer, 263, 289, 311
  subgroup, 74, 262
group theory, 120, 313

Hadamard matrix, 382, 433
Hadamard transform, 65
Hamiltonian, **8**
  bare, 614
  decoupling, 390
  dynamics, 18
  engineering, 106
  internal molecular, 522
  interpolating, 419
  physical, 614
  system, 377
Hamming
  bound, **60**
  code, 62
  distance, 388
  weight, 177, 315
hard cutoff, 358
hash function, 580
hashing bound, 581
Heisenberg coupling, 522
Heisenberg uncertainty principle, 242, 563
Hermitian map, 18
Hermitian operator, **5**
Hilbert space, **4**
  quiet corner of, 78, 90
  separable, 13
Hilbert–Schmidt norm, **22**
Hilbert–Schmidt inner product, 330, 380
history of syndromes, 591, 605
holonomy, **399–401**, 414–416, 425
  group, 400
  inverse, 400
  map, 399
homology, 458, 460–462, 480
HQC (holonomic quantum computation), **398**
Hubbard model, 547
hypercube, 456, 595, 598, 619
hyperfine
  field, 528
  interaction, 527, 588
  level, 130
  splitting, 130
  state, 6, 515
hyperfine-induced dephasing, 527

independent generators, 185
inequality
  Araki–Lieb, 15
  Lieb–Ruskai, 15
  linear matrix, 332, 346
information reconciliation, 565
information theory, 76
initialization-free
  decoherence-free subspace, 90
  noiseless subsystem, 98
inner product space, 4, 22
interference, 46, 108, 127, 136, 547, 607
interleaver, 292–297
ion trap quantum computer, 514
irreducible representations (irreps), 82
Ising model, 286, 287, 469, 471, 623
  and decoding, 287
  for toric code, 470
  ground state, 287
  phase diagram, 470
  random three-body, 479
  random bond, 469

isometry, 143, 169, 316, 317, 340, 450
isotropic subgroup, 185, 196
iterated bound, 345
iterative optimization, 334
itinerant electrons, 588

Jamiolkowski isomorphism, 330, 332, 346
Jaynes–Cummings Hamiltonian, 85, 103, 548, 549
Johnson–Nyquist noise, 588
Josephson junction, 526, 588

Keldysh
   contour, 592, 594, 606, 608
   formalism, 592
ket, **5**, 292
Kitaev model
   four-dimensional, 613, 623
   three-dimensional, 623
   two-dimensional, 622
Kondo problem, 588
Kraus operator, **16**
   block-diagonal representation, 86
   sum representation, 16, 17, 28–30, 58, 338
Kronecker product, 12

Lamb shift, 615
   Hamiltonian, 30
lateral quantum dot, 588
lattice
   D-dimensional, 601
   for Bacon–Shor code, 172
   for Kitaev models, 622
   triangular, 479
   Union Jack, 479
law of large numbers, 564
leakage error, 91, 92, 102, 141, 403
leakage space, 141, 142
level reduction, 154
Lie algebra, 82–84, 95, 400
Lieb–Ruskai inequality, 15
lift of a curve, 408
limited thermalization, 102
Lindblad generator, 32, 202, 218
Lindblad–Gorini–Kossakowski–Sudarshan master
    equation, 620
Lindbladian, 202, 203
LMI (linear matrix inequality), 332, 346
local leakage noise, 141
local Markovian noise, 134
locality, 117, 130, 132, 139, 142, 227, 377, 386, 387, 430, 455, 456, 465, 476, 480, 614, 622
LODD (locally optimized dynamical decoupling), 531
log det heuristic, 347
logarithmic time scale, 613
logic gates, 181, 397, 618

magnetic resonance imaging, 535
magnetization, 287
Magnus expansion, **9**, 9, 10, 113, 114, 119, 549, 603
malonic acid, 373
map, **17**
Markovian dynamics, 32, 619
   defining feature, 27
master equation, 26
   Born–Markov, 405
   discrete, 29
   Heisenberg picture, 622
   Lindblad, 28, 30, 32, 41, 202, 620
   Markovian bit-flip, 207
   Nakajima–Zwanzig, 220
   post-Markovian, 32
   quantum, 26
   stochastic, 41
   time-convolutionless, 220, 615
   time-local, 31
matrix algebra, 95
maximally entangled, **14**
maximum-likelihood decoding, 285
MDS (maximum distance deparable) code, 325
measurement
   continuous, 203, 213
   current, 214
   fault-tolerant, 417
   projective, 20, 33, 132, 340, 451
   weak, 203, 208, 216
memory kernel, 32
metastable observable, 621
methyl group, 523
metric, 112, 316
metric space, **4**
minimum distance decoder, 283
mirror
   plane, 533
   spherical, 533, 534
mixed state, **5**, 6, 19, 136, 329, 539
   rank, 7
MOOS (mutually orthogonal operation set), **353**
mutual information, 15, 167, 280
   classical, 15
   quantum, 15

Nakajima–Zwanzig
   master equation, 220
   projection operator technique, 26
narrow-sense
   BCH code, 319
   Reed–Solomon code, 319
nice error basis, 381, 391
Nishimori
   line, 470, 471
   temperature, 286, 287, 303
NMR (nuclear magnetic resonance), 105, 356, 430, 510
   experimental quantum error correction, 514
no-cloning theorem, 47, 51, 563
noise, 33, 45, 126, 130, 132, 135, 138, 155, 206, 287, 361, 433, 540, 568, 614
   classical, 46, 373, 539
   collective, 164
   depolarizing, 176
   fully coherent, 138
   Gaussian, 159, 540
   Johnson–Nyquist, 588
   non-Markovian, 138
   power spectrum, 110, 526, 530
   spectrum, 358, 361, 544
noiseless subsystem, 88, 93–95, 97, 98, 101, 179
noisy quantum channel, 198
non-CP maps, 222

non-Markovian dynamics, 27, 204
norm, 112, 133, 370, 614
　diamond, 25, 132, 614
　Euclidean, 22
　Frobenius, 22
　Hilbert–Schmidt, 22
　operator, 22
　trace, 22, 43, 137
　unitarily invariant, 22
normal ordering, 597, 609
normalizer, 69, 74, 75, 264, 265, 272
NP-complete problem, 281, 286
NP-hard problem, 268, 283, 286, 347
NS (noiseless subsystem), **78**, 78, 95, 96, 99, 101, 102, 104, 163–165, 179, 180, 406, 407
　initialization-free, 98
nuclear spin, 130, 355, 522, 529, 539, 543, 588
NUDD (nested Uhrig dynamical decoupling), **367**

observable, **19**
　expectation value of, 19, 215
　metastable, 623
OPE (operator product expansion), 595, 597, 605–608
operation element, **16**
operator
　boundary, 463, 478
　density, 40, 205
　ebit stabilizer, 558
　error, 43, 57, 207, 377, 386, 555
　logical, 70, 75, 170, 172, 174, 258, 264, 288, 464, 475, 476, 484, 490, 558
　Pauli, 10, 59, 68, 69, 124, 186, 195, 244, 262, 303, 383, 449, 472
　projection, 16, 98, 340
　string, 464–466, 474, 475
　time-ordering, 9, 352, 359
　unitary, 136
operator algebras, 93, 163
operator basis, 352
operator norm, **22**
order of decoupling, 357
orthogonal
　arrays, 377, 386, 388, 390
　complement, 86, 205
　decomposition of the identity, 19
orthogonality constraint, 341
outer product, **6**
Overhauser field, 588

pair-interaction Hamiltonian, 377, 387
parameter drift, 113
parity check, 196, 297
parity check matrix, 63, 193, 574
parity kicks, 110
partial matrix element, 17, 100
partial trace, **13**, 330
path ordering, 399
Pauli group, 66, 67, 69, 74, 75, 169, 262, 288, 417, 571, 604
Peierls argument, 623
perfect matching algorithm, 468
permutation symmetry, 79, 83, 103, 164

phase
　coherence, 520
　decoherence, 511
　flip, 54, 174, 515
　gate, 187, 243, 422
　randomization, 530
phonon
　acoustic, 588
　bath, 83
　induced decoherence, 405
　wavelength, 84
photodetector, 39, 578
photon polarization, 164
Planck constant, 8
Plotkin sum, 271
Poisson process, 32
polarization, 401, 518, 533
polynomial time
　decoding, 282, 570
　solution, 468, 612
population relaxation, 358, 366
positive operator, **17**, 346
positive semidefinite operator, **5**, 331
postselection, 17, 159, 518
POVM (positive operator valued measure), 20, 21, 23, 24, 27, 38
power spectrum, 107, 110, 526, 530
Preparata code, 269
preparation gadget, 147
prime number, 177
primitive code, 319
principal bundle, 408, 410
privacy amplification, 565, 569
probability leakage, 17
projection superoperator, 116
projective representation, 116, 381
projective syndrome measurement, 340
projector, 16, 19, 133, 290, 414, 419, 557
pulse shaping, 368
pure dephasing, 117, 355, 358, 366, 602
pure state, **5**, 6, 87, 336, 510
purification, 175, 280, 450
　entanglement, 183, 551
purity, 20, 216, 323, 346, 347, 521, 534

QDD (quadratic dynamical decoupling), **366**, 367
QECC (quantum error-correcting code), **43**
quantum check matrix, 189, 190, 240, 256–257, 258, 559
quantum circuit, 128, 138, 231, 495
quantum codeword, 397
quantum communication, 181, 199, 231, 254, 553, 571, 578, 579
quantum computer architecture, 536
quantum convolutional coding, 232
quantum discord, 12, **15**, 16, 18, 26, 339
quantum dots, 398, 578
quantum error correction, 43, 46, 48, 56, 65, 76, 77, 90, 126, 144, 160, 169, 199, 327, 509, 510, 512, 514, 516, 517, 537, 553, 554, 578, 579, 585, 596, 609, 613
　continuous-time, 202, 206
　continuous-variable, 577
　operator, 163, 413

quantum gate, 11, 76, 131, 335, 403, 545, 588
   $\pi/8$, 417, 423, 480
   CNOT, 11, 242, 417, 480, 500
   controlled-phase, 243
   Hadamard, 53, 208, 259, 420, 464
   identity, 482
   NOT, 10, 547
   SWAP, 11
quantum information theory, 17, 183, 563
quantum jumps, 37
quantum key
   distribution, 563, 564, 566, 580
   entanglement-based, 566
   expansion, 553, 554, 562, 564, 565, 569, 570, 580
quantum memory, 103, 352, 530, 613, 619
quantum money, 563
quantum mutual information, 18
quantum operation, 42, 89, 166, 203, 291, 328, 442, 537, 564
quantum phase transition, 599, 600
quantum process tomography, 4
quantum register, 119
quantum serial-turbo coding, 232
quantum state, **5**
quantum trajectory, 34
qubit, 6, 21, 83, 130, 158, 188, 238, 404, 511, 601
   charge, 526, 588
   dark state, 401
   electron, 528
   flux, 526
   harmonic oscillator representation, 140
   in fullerene, 528
   Josephson junction, 526, 588
   memory, 543
   nuclear, 528
   proton, 523
   spin, 521
   trapped ion, 530
qudit, 379–381, 383, 386, 387, 389, 392, 393, 430
   maximally entangled, 557
quiet corner of Hilbert space, 78, 90
qutrit, 430

Rabi
   coupling, 401, 549
   frequency, 403, 405
   oscillation, 528, 529
Ramsey interference, 526
random coding
   quantum, 280
   Shannon's, 279
rank, **5**
raw key, 564, 565
RDD (randomized dynamical decoupling), 355, **372**
recovery superoperator, **57**
reduced density matrix, **13**, 14, 131, 220, 221, 359, 405, 591, 592
Reed–Muller code, 269, 270, 272, 273
Reed–Solomon code, 319, 320
refocusing, 106, 107, 127, 510, 526, 530
relative phase, 533, 534
renormalization group, 596, 619
representation theorem for noiseless subsystems, 93
ribbon, 295

SARG04 protocol, 579
scalar
   coupling, 522
   Liouville product, 621
   operator acting as, 82, 166
   product structure, 408
   state transforming as, 84
   system–bath interaction, 84
scaling dimension, 587, 600
Schmidt decomposition, 14, 344
Schrödinger equation, **8**
Schur product, 383
SDP (semidefinite program), 332–334, 336, 338, 340, 342, 344, 345, 347
secret key, 553, 562–568
   distillation, 567, 568
   generation, 567
   uncompromised, 569
security
   analysis, 580
   criterion, 566, 580
   of quantum key expansion, 562
security proof, 553, 554, 559, 561, 563, 565, 579, 580
   Lo–Chau, 580
   Luo–Devetak, 580
   Shor–Preskill, 580
self-correcting quantum memory, 620
separable
   Hilbert space, 5, 13, 408
   state, **12**, 15, 16
seven bridges of Königsberg, 121
Shannon
   capacity, 280, 570
   entropy, 14
shift register, 234, 235, 244
   classical, 231, 235
   quantum, 245, 247, 249–251, 253
shifted symplectic product, 239, 240, 254
   matrix, 240, 241
Shor code, 61, 70, 275, 343, 345
   continuous-variable, 581
sifting, 564
sign matrix, 383
single photons, 482, 517, 520
single-qubit operation, 206, 212, 405, 427, 440, 568
Singleton bound, 320, 325, 326
singular values, 14, 24, 338, 341
soft high-frequency cutoff, 358, 531
soft pulse, 510
spectral decomposition, **7**, 17, 415
spectral density, 108, 359, 405, 617
spin-boson model, 83, 103, 357, 358, 588, 601
spin-glass, 286
spin-orbit coupling, 405, 546, 548
spontaneous emission, 40
stabilizer code, **65**, 311
   centralizer, 69
   circuit for encoding into, 74
   cyclic code, 271
   degenerate errors, 284
   distance, 261
   entanglement-assisted, 253
   normalizer, 262
   quantum convolutional, 231

union, **263**, 265
  via privacy amplification, 581
stabilizer formalism, 76, 169, 182, 199, 262, 579, 581
stabilizer generator, 68, 71, 72, 169, 172–175, 196, 208, 217, 288, 290, 297, 301, 456, 466, 478
stabilizer group, **67–69**, 75, 76, 189, 190, 263, 265, 289, 300, 311, 312, 478
  normalizer of, 75
standard form
  for codeword stabilized codes, 266
  for group generators in EAQECC, 186
  for semidefinite program, 334
  for stabilizer code generators, 194
Stark shift, 615
state space, 110, 309, 407
state vector, 399, 409, 410
statistical model, 470
Steane code, 54, 65, 71, 75, 76, 181, 267, 270, 343
Stiefel manifold, 410
Stinespring representation, 25, 167
stochastic
  differential equation, 203, 214
  equation, 41
  error model, 32
  errors, 4, 32, 33, 37, 38, 41, 587, 598, 599
  jump, 33
  master equation, 41
  Schrödinger equation, 39–41
  variable, 37
stochasticity, 449
string-net, 476, 477
strong Church–Turing thesis, 612
submultiplicativity of unitarily invariant norms, 22, 23
subsystem principle, 226
superconducting qubits, 130
superdense coding, 183, 191
super-Ohmic, 405
superoperator, **5**, 53, 55, 56, 58, 61, 121, 131, 133–135, 165
  for amplitude damping, 135
  for Eulerian dynamical decoupling, 123, 124
  projection, 116, 201
  recovery, 57
  unital, 115
  unitary, 615
superoperators
  commuting, 615
  composition of, 131, 132
  distance between, 24
  extensions of, 25
  limitations of, 136
  norm for, 25
superposition, 39, 64, 65, 127, 341, 450, 462, 515, 518, 613
superpropagator, 615
surface
  closed, 457, 478
  code, 159, 461, 464, 465, 468, 475, 478
  connected orientable closed, 457
  correlation, 484, 486
  first homology group, 458, 460
  genus, 457
  of a defect, 496
  of operators, 485

open, 457
orientable closed, 457
topology, 457
SVD (singular value decomposition), **14**, 338, 341
SWAP gate, 11, 545
symmetrization
  in dynamical decoupling, 116, 122, 123, 352, 355, 393
  of noise channels, 570
symmetry and conservation laws, 78
symmetry-breaking perturbations, 101
symplectic
  inner product, 189, 262
  partners, 186
  product, 252, 560, 572
  representation, 189, 196, 244
  subgroup, 185
syndrome, 63, 72, 144, 148, 171, 175, 176, 205, 211, 261, 277, 284, 288, 290, 304, 341, 468, 567, 604
  extraction, 455, 603
  qubit, 148, 205, 206, 211, 227
  reduced, 191, 193, 194
syndrome measurement, 73, 78, 129, 144–146, 148, 163, 178, 289, 290, 340, 341, 432
system–bath coupling
  diagonal, 355
  off-diagonal, 355
system–bath Hamiltonian, 94, 101, 351

Tanner graph, 199, 297–299, 302–305
TCL (time-convolutionless), 220
teleportation, 183, 435, 436, 438, 442, 451, 542
  for implementing quantum error correction, 160
  gate, 160, 432, 434, 435, 443, 451, 501, 550, 551
  model for quantum computation, 432–434, 437, 451
  one-bit, 435, 436, 451
  state, 434
tennis ball, 283
tensor product, **12**
theorem
  Baker–Campbell–Hausdorff, 572
  fixed point, 180
  Margolus–Levitin, 618
  Neumark's, 38
  no-cloning, 47, 563, 579
  quantum capacity, 581
  Stinespring, 167
  Stokes', 409
  Stone–von Neumann, 575
  threshold, 155, 157, 431, 449–451, 598–600, 609, 613, 614, 617
  Wick's, 141, 587, 593, 598, 599, 605
thermal coherence length, 591
thermal equilibrium, 107, 471
threshold condition, 451
time-homogeneous, 27, 28
time-local
  generator, 26
  master equation, 26, 27
time-ordering operator, **9**
topological code, 455–458, 461, 472, 475, 480, 481, 518
topology, 304, 399, 457, 460
toric topology, 622
torus, 455–458, 460, 464, 472, 474, 475

trace
  alternating form, 316
  form, 316
trace inner product, 380
trace norm, **22**
trace-norm distance, 23
  between pure states, 137
transversal
  gate, 181, 481, 538
    P, 477, 479
    $\pi/8$, 481
    adiabatic, 423
    Clifford, 421, 472
    CNOT, 416, 417, 425, 472, 481
    Hadamard, 464, 477
    Toffoli, 428
  operation, 227, 412, 416, 417, 424, 471
transversality, 479, 481
turbo
  code, 279, 292, 293, 295, 297, 305, 306
  decoder, 294, 295
Turing machine, 138, 599
two-universal hash function, 565

UDD (Uhrig dynamical decoupling), 357, 362
unification of methods for overcoming quantum errors, 78
union stabilizer code, **263**, 264, 265, 268–270, 274–277
unit vector, **8**
unital channel, 164, 180
unitarily invariant norm, **22**, 113, 140
unitary
  error basis, 380
  operator, **8**
  freedom, 164
universal
  dynamical control protocol, 369
  dynamical decoupling, 118, 354
  electric computer, 126
  fault-tolerant quantum computation, 423
  holonomic control, 403
  initial state, 441
  recovery operation, 166
  set of quantum gates, 128, 416, 437, 443, 444, 491, 542, 614
  set of quantum operations, 481
  set of weak operations, 227

universal quantum computation
  on a DFS, 104, 406, 525
    experimental, 525
  on a stabilizer code, 76
  using gadgets, 159
  via linear optics, 517
universality
  computational, 399
  encoded, 400, 407
  in holonomic quantum computation, 400
  of 2-local Hamiltonians, 430
  of DD for time-dependent Hamiltonians, 369
  of NUDD, 368
  of QDD, 367
  of two-qubit measurements, 438
  of UDD, 358

van Hove limit, 620
vanishing discord, 15
virtual degrees of freedom, 94
von Neumann
  and classical fault tolerance, 74, 142
  and noisy cellular automata, 159
von Neumann entropy, **13**

wavefunction, 5
weak coupling limit, 620
weak coupling regime, 615
weak measurement, 20, 22, 31
weight enumerator, 282
Werner state, 15
Wiener increment, 214

Young
  diagram, 97
  tableaux, 97

Zeeman
  interaction, 521, 539
  magnetic field, 527
  splitting, 105
  terms, 377, 532
Zeno
  effect, 4, 36, 110, 219, 220, 357
  regime, 204, 219, 220, 225, 226
  time, 225